D1405973

MRI of the Musculoskeletal System

A Teaching File

Second Edition

MRI of the Musculoskeletal System

A Teaching File

Second Edition

Editors

Andrew L. Deutsch, M.D.
Tower Imaging
Los Angeles, California
Associate Clinical Professor of Radiology
University of California, San Diego
La Jolla, California

Jerrold H. Mink, M.D., F.A.C.R.
Tower Imaging
Los Angeles, California
Associate Clinical Professor
Department of Radiology
University of California, San Francisco
San Francisco, California

Acquisitions Editor: James D. Ryan
Developmental Editor: Brian Brown
Manufacturing Manager: Dennis Teston
Production Manager: Larry Bernstein
Production Editor: Jodi Borgenicht
Cover Designer: Diana Andrews
Indexer: Leon Kremzner
Compositor: Tapsco, Inc.
Printer: Courier-Westford

Printed in the United States of America

9 8 7 6 5 4 3 2 1

Library of Congress Cataloging-in-Publication Data

MRI of the musculoskeletal system: a teaching file/editors, Andrew L. Deutsch, Jerrold H. Mink.—2nd ed.
p. cm.
Includes bibliographical references and index.
ISBN 0-397-51672-X
1. Musculoskeletal system—Magnetic resonance imaging—Case studies. I. Deutsch, Andrew L. II. Mink, Jerrold H.
[DNLM: 1. Musculoskeletal System—radiography. 2. Magnetic Resonance Imaging. 3. Musculoskeletal System—anatomy & histology.
WE 141 M939 1996]
RC925.7.M753 1996
616.7′07548—dc20
DNLM/DLC
for Library of Congress 96-16185
CIP

“A baby is God’s opinion that life should go on.”

Carl Sandberg

To Peter, Phillip, and Dana; their mom, Jeanne;
and their grandmother’s, Adelaide and Anne.

ALD

To my babies, Justin and Samantha, and their mom, Barbara.

JHM

Contents

Contributing Authors ... ix

Preface to the First Edition ... xiii

Preface ... xv

1. The Shoulder ... 1
Mahvash Rafii, Cornelia N. Golimbu, and Hossein Firooznia

2. The Elbow ... 77
Russell C. Fritz

3. The Wrist ... 149
Lawrence M. White, Mark E. Schweitzer, Patrick T. Liu, Andrew L. Deutsch, and Michael E. Timins

4. The Hip ... 197
Mini N. Pathria, Andrew L. Deutsch, and Dennis M. Wilcox

5. The Spine ... 273
Andrew L. Deutsch, J. Randy Jinkins, Carl V. Bundschuh, Farid F. Shafaie, Jerrold H. Mink, Michael I. Rothman, and Greg H. Zoarski

6. The Knee ... 347
Andrew L. Deutsch, Jerrold H. Mink, Arthur A. DeSmet, Michael J. Tuite, David W. Stoller, Charles G. Peterfy, and Michael P. Recht

7. The Foot and Ankle ... 461
Andrew L. Deutsch, Mitchell A. Klein, and Jerrold H. Mink

8. Muscle ... 539
James L. Fleckenstein and Andrew L. Deutsch

9. Marrow ... 585
Bruce A. Porter and Andrew L. Deutsch

10. Soft Tissue Tumors ... 613
Mark J. Kransdorf and Mark D. Murphey

11. Bone Tumors ... 655
Johan L. Bloem, Henk-Jan van der Woude, and Pancras C. W. Hogendoorn

Subject Index ... 721

Contributing Authors

Johan L. Bloem, M.D.
Professor and Chairman of Radiology
Departments of Radiology and Nuclear Medicine
University Hospital Leiden
Rijnsburgerweg 10
2333 AA Leiden
The Netherlands

Carl V. Bundschuh, M.D.
Associate Professor
Department of Radiology
Eastern Virginia Medical School
Director of Neuroradiology and MRI
Sentara Hospitals and Medical Center Radiologists
Halifax Building, Suite 330
6161 Kempsville Circle
Norfolk, Virginia 23502

Arthur A. DeSmet, M.D.
Professor
Department of Radiology
University of Wisconsin
600 Highland Avenue
Madison, Wisconsin 53792

Andrew L. Deutsch, M.D.
Tower Musculoskeletal Imaging Center
444 San Vicente Boulevard
Los Angeles, California 90048
Associate Clinical Professor of Radiology
University of California, San Diego
La Jolla, California 92093

Hossein Firooznia, M.D.
Professor
Department of Radiology
New York University Medical Center
560 First Avenue
New York, New York 10016

James L. Fleckenstein, M.D.
Associate Professor
Department of Radiology
University of Texas Southwestern Medical Center
5323 Harry Hines Boulevard
Dallas, Texas 75235-8896

Russell C. Fritz, M.D.
Clinical Assistant Professor of Radiology
University of California, San Francisco
National Orthopaedic Imaging Associates
1260 South Eliseo Drive
Greenbrae, California 94904

Cornelia N. Golimbu, M.D.
Professor of Radiology
New York University Medical Center
560 First Avenue
New York, New York 10016

Pancras C. W. Hogendoorn, M.D.
Attending Pathologist
Department of Pathology
University Hospital Leiden
Rijnsburgerweg 10, Building 1
2333 AA Leiden
The Netherlands

J. Randy Jinkins, M.D., F.A.C.R.
Associate Professor of Radiology
Director of Neuroradiology
Department of Radiology
University of Texas Health Science Center at San Antonio
7703 Floyd Curl Drive
San Antonio, Texas 78284-7800

Mitchell A. Klein, M.D.
Assistant Professor
Department of Diagnostic Radiology
Medical College of Wisconsin
Senior Attending Radiologist
Froedtert Memorial Lutheran Hospital
Zablocki Veterans Administration Medical Center
Milwaukee Center for Diagnostic Imaging
9200 West Wisconsin Avenue
Milwaukee, Wisconsin 53226

Mark J. Kransdorf, M.D.
Clinical Associate Professor
Department of Radiology
University of Virginia
Charlottesville, Virginia 22908
Department of Radiology
Saint Mary's Hospital
5801 Bremo Road
Richmond, Virginia 23226

Patrick T. Liu, M.D.
Instructor of Radiology
Mayo Medical School
Department of Diagnostic Radiology
Mayo Clinic Scottsdale
13400 East Shea Boulevard
Scottsdale, Arizona 85259

Jerrold H. Mink, M.D., F.A.C.R.
Tower Musculoskeletal Imaging Center
444 San Vicente Boulevard
Los Angeles, California 90048
Associate Clinical Professor
Department of Radiology
University of California, San Francisco
San Francisco, California 94143

Mark D. Murphey, M.D.
Chief, Musculoskeletal Radiology
Department of Radiologic Pathology
Armed Forces Institute of Pathology
6825 16th Street N.W.
Washington, D.C. 20306

Mini N. Pathria, M.D.
Associate Professor
Department of Radiology
University of California San Diego Medical Center
200 West Arbor
San Diego, California 92103

Charles G. Peterfy, M.D., Ph.D.
Assistant Professor
Department of Radiology
University of California, San Francisco
505 Parnassus Avenue
San Francisco, California 94143-0628

Bruce A. Porter, M.D.
Medical Director
First Hill Diagnostic Imaging
Clinical Associate Professor
Department of Radiology
University of Washington
1001 Boylston Avenue
Seattle, Washington 98104

Mahvash Rafii, M.D.
Professor of Clinical Radiology
Department of Diagnostic Radiology
New York University School of Medicine
New York University Medical Center
Tisch Hospital/Bellevue Hospital
560 First Avenue
New York, New York 10016

Michael P. Recht, M.D.
Staff Radiologist
Department of Radiology
The Cleveland Clinic Foundation
9500 Euclid Avenue
Cleveland, Ohio 44195

Michael I. Rothman, M.D.
Assistant Professor of Radiology, Neurosurgery, and Otolaryngology
Medical Director, Anna Gudelsky MRI Facility
University of Maryland Medical Systems
22 South Greene Street
Baltimore, Maryland 21201

Mark E. Schweitzer, M.D.
Associate Professor
Department of Radiology
Thomas Jefferson University Hospital
1096 Main Building
132 South 10th Boulevard
Philadelphia, Pennsylvania 19107

Farid F. Shafaie, M.D.
Neuroradiology Fellow
Mallinckrodt Institute of Radiology
Washington University Medical Center
510 South Kingshighway Boulevard
St. Louis, Missouri 63110

David W. Stoller, M.D.
Director, California Advanced Imaging
Director, Marin Radiology and National Orthopedic Imaging Associates
Assistant Clinical Professor of Radiology
University of California at San Francisco
San Francisco, California 94143

Michael E. Timins, M.D.
Assistant Professor of Radiology
Department of Diagnostic Radiology
Medical College of Wisconsin
8701 Watertown Plank Road
Milwaukee, Wisconsin 53226

Michael J. Tuite, M.D.
Assistant Professor
Department of Radiology
University of Wisconsin
600 Highland Avenue
Madison, Wisconsin 53792

Lawrence M. White, M.D.
Assistant Professor
Department of Radiology
University of Toronto
Mount Sinai Hospital
600 University Avenue
Toronto, Ontario M5G 1X5
Canada

Dennis M. Wilcox, M.D.
Clinical Assistant Professor
Department of Radiology
University of Missouri
Kansas City, Missouri 64111

Henk-Jan van der Woude, M.D., Ph.D.
Radiologist-in-Training
Departments of Diagnostic Radiology and Nuclear Medicine
Leiden University Hospital
Rijnsburgerweg 10
2333 AA Leiden
The Netherlands

Greg H. Zoarski, M.D.
Department of Radiology
University of Maryland Medical Systems
22 South Greene Street
Baltimore, Maryland 21201

Preface to the First Edition

The inspiration for this book grew out of our perceived need for a comprehensive text on musculoskeletal MR. There is rather scant authoritative material available on the "how to read 'em" aspects of musculoskeletal MR. The large major references on MR devote surprisingly little space to this topic. There are, of course, subspecialty texts on knee and spine MR, but the other major joints are ignored. This lack of consolidated information is most surprising when one realizes that musculoskeletal imaging is by far the fastest growing application of MR. When we began to perform MR in 1985, virtually all of the 10 examinations per day were of the head (7–8) and spine (1–2). By 1988, our volume of neuroimaging had increased to a total of 12 examinations per day, but we were performing 9 bone and joint (nonspine) studies. At the time of this writing, musculoskeletal MR accounts for 12–14 studies, which is approximately 35% of our daily work load. Our experience is not unique; interaction with our colleagues at conferences suggests that this rapid rate of growth can be expected in radiological practices with available scanner time and a radiologist interested in and conversant with osteoradiology.

We hope to benefit as much from this text as will the reader; it has given us the chance to spend time in the library, consolidate information, organize our thoughts, and learn more about the fascinating topic of musculoskeletal MR.

This book is organized as a case study, but by no means is this a text that asks a question, answers it, and then gets on to the next case. Rather, this book is more closely akin to the ACR syllabi. A question is asked, based on the findings in one or more MR images; there is a detailed description of the findings in the following paragraph. In general, the case centers primarily on one clinical entity. Just below the answer is a comprehensive analysis of the entity under discussion, a differential diagnosis and follow-up imaging studies (CT, MR, arthrograms, bone scans) that are pertinent to the case. The discussions are not limited to the *imaging* aspects of the case; we believe that it is essential for the skeletal radiologist to understand what it is that the clinician needs to know.

This text is designed primarily for the practicing radiologist who desires a reference text that he can keep next to the view box. But the case study format also allows one to digest a small kernel of knowledge in the evening, without having to devote a "chapters worth" of time to learn something of value.

Orthopedic surgeons will also find a great value in this book. MR is now a daily part of orthopedic practice, and it is rare to find an orthopedic surgeon who does not order or review an MR on a daily basis. Surgeons first became familiar with MR in regards to the detection of femoral head osteonecrosis and in assessment of spinal disorders. The rapidly expanding spectrum of musculoskeletal applications makes a working knowledge of MR essential to the orthopedist.

The major articulations—hip, knee, shoulder, and spine—are covered as individual chapters. Smaller joints like the foot and ankle and the elbow and wrist are combined together. Three special chapters are included in the text. The first covers the difficult topic of physics; most authors feel obliged to include a detailed analysis of the physical basis of MR. Being practical, practicing radiologists, we abhored the idea of repeating the concepts of T1 and T2, gradient imaging, and flow. We, instead, chose to instruct radiologists, like ourselves, on the important factors that govern the quality of the images that one produces and to accomplish this in the most painless manner (for us, physics can be painful) possible. We therefore utilized the case study style that characterizes the book.

We have established the chapter on tumors without regard for the body part in which the lesion occurred so as to better consolidate one's thinking about this complex issue. The other special chapter covers dynamic joint imaging, a topic that requires a new presentation of technical considerations. While most of the cases presented in the dynamic chapter do cover the knee and TMJ, it is our belief that the ability of MR to image physiologic motion will be a primary use of this modality.

Preface

Our goal in undertaking a new edition of this text was to produce a pragmatic and highly readable teaching tool directed primarily at the practicing radiologist actively involved in the interpretation of musculoskeletal MR examinations. Our objective in this effort, as in the First Edition, was to produce a work that could directly communicate and teach in a forum less rigid and constrained than standard textbooks. In contrast to many Second Editions, the text has not been updated but rather entirely rewritten, with all new cases and contributors. Areas that received only minor attention in the First Edition (e.g., elbow) now represent major sections and reflect the progress made in musculoskeletal MR over the past five years.

This edition covers nearly the entire gamut of the field, and in many areas does so in considerable depth. Attempts were made by the editors to avoid the repetition that can occur in multi-authored works and, in particular, by nature of the organization of the work by anatomic site. Where overlap unavoidably occurs (e.g., discussion of tendon disorders at different sites), the approach to the material is directed at considerations related to the specific area, and the treatment of the subject matter is often presented differently reflecting the author's individual experience and perspective. This text is not an introductory work and presumes a working knowledge of the field. Attempts were made to emphasize the practical and to specifically address areas of interpretive difficulty. For example, the invariably difficult task of postoperative evaluation (e.g., postoperative carpal tunnel release, postoperative spine, and post-therapy tumor evaluation) was addressed in several of the individual sections. In addition, specific attempts were made to address recently introduced techniques including new pulse sequences, dynamic perfusion imaging for tumor evaluation, and MR arthrography.

The text is cross-referenced to allow the interested reader to readily explore different perspectives on related themes. While in many sections cases build upon each other, the book is designed to allow random access to cases and each case can be studied independently. In contrast to the First Edition, each case is individually referenced. The editors' objective in each case was to provide the reader with a reasonably comprehensive treatment of the subject matter chosen and one that could be assimilated in the approximate time course required to enjoy a cup of coffee (or in Los Angeles, more properly a double-decaf nonfat latte). We hope that we have succeeded.

Andrew L. Deutsch, M.D.
Jerrold H. Mink, M.D., F.A.C.R.

MRI of the Musculoskeletal System

A Teaching File

Second Edition

CHAPTER 1

The Shoulder

Mahvash Rafii, Hossein Firooznia, and Cornelia Golimbu

Magnetic resonance (MR) imaging of the shoulder joint has been a topic of great interest and considerable controversy. Unlike the case for other joints, such as the knee, technical advances including off-center zoom technique, oblique plane imaging capability, and the development of dedicated surface coils were needed for adequate display of the complex anatomy of the shoulder joint (1–5). Availability of high-field MR imaging systems and improved software further enhanced the image quality and the display of abnormalities of this joint.

The rotator cuff, by virtue of its peculiar anatomy and its susceptibility to degeneration and tears, is the predominant site of shoulder pain and dysfunction. Through the work of many investigators, the MR imaging criteria for the diagnosis of rotator cuff tears were soon established, and the accuracy of this modality was suggested to surpass that of conventional arthrography (5–19). Furthermore, these studies evaluated various other aspects of impingement syndrome and rotator cuff tendinopathies before the development of cuff tears. Histopathologic studies, although scant, added insight to the underlying nature of MR imaging findings; other studies described variations of rotator cuff signal intensity due to anatomic peculiarities, technical factors, or perhaps asymptomatic tendon degeneration (20–26).

Imaging of shoulder instability has been of considerable interest to orthopedic surgeons who deal with this often disabling condition in young individuals. Conventional MR imaging has been found by most accounts to be reliable for evaluation of glenohumeral instability and for detection of various lesions of the glenoid labrum, although the consensus is not nearly as strong as in rotator cuff imaging (14,27–31). Developmental variations of the capsulolabral complex of the glenohumeral joint and some normal morphologic and signal intensity variations of the labrum are the underlying factors (20,23,32–37). One of the great advantages of MR imaging is its noninvasive nature. Lack of distension of the joint capsule, without effusion being present in the joint, may conceal subtle lesions of the capsulolabral complex. MR arthrography has therefore been introduced to improve imaging accuracy for labral tears as well as for improved delineation and differentiation of partial synovial surface and small full-thickness rotator cuff tears (38–42).

IMAGING TECHNIQUES

Utilization of a dedicated surface coil is mandatory for shoulder imaging (5,12). Imaging is performed with the patient supine and the arm in neutral rotation; the palm is placed against the thigh and wrapped to the trunk for immobilization (43). Axial imaging in both internal and external rotation of the arm has been advocated for improved visualization of capsulolabral lesions (41). Recently, additional imaging with the arm fully abducted (palm placed under the head) has been used in MR arthrography for visualization of partial tears of the posterosuperior aspect of the rotator cuff (44).

Imaging is performed in the axial, oblique coronal, and oblique sagittal planes. First a series of coronal scout images is obtained. Axial images are then obtained from the acromion process to below the glenoid margin using a coronal scout image as localizer. We routinely use a T1-weighted or intermediate spin-echo sequence in conjunction with a gradient-echo sequence for imaging in this plane. When shoulder instability and labrum tears are suspected (and routinely in younger patients), we utilize a double-echo T2-weighted sequence (27).

The oblique coronal images are obtained in an orientation parallel to the main axis of the supraspinatus; they span the joint from the tip of the coracoid process anteriorly to the scapular spine posteriorly. The axial image of the supraspinatus is used as a localizer. A T2-weighted sequence in this plane is essential for accurate evaluation and differentiation of the rotator cuff lesions. An additional T1-weighted sequence is helpful for improved visualization of the subacromial peribursal fat, but this sequence is not essential, and we do not routinely perform it (15).

Oblique sagittal images are obtained perpendicular to the long axis of the supraspinatus or perpendicular to oblique coronal images. This plane is roughly parallel to the surface of the glenoid fossa. These images display the sectional anatomy of the coracoacromial arch and the rotator cuff in the sagittal plane of the shoulder. A double-echo T2-weighted sequence is preferred for imaging in this plane. A combination of T1-weighted, proton-density, and T2-weighted or double-echo T2-weighted sequences is most often used for shoulder imaging.

T1-weighted images provide excellent osseous and soft tissue anatomic detail in the least imaging time. They are specifically useful for depiction of peri-articular fat planes and for assessing the bone marrow. Imaging parameters include repetition time (TR) of 500 to 800 msec, echo time (TE) of 12 to 30 msec, and two to four excitations.

Proton-density images display the highest signal-to-noise ratio, improving depiction of anatomic detail and increasing sensitivity in detection of intrinsic signal alterations of the rotator cuff and caspulolabral structures. T2-weighted images characteristically have a low signal-to-noise ratio owing to longer GD (8,200 msec); these images, however, by virtue of higher signal intensity of fluid collections and soft tissue edema, are essential for characterization of signal alterations within and around the rotator cuff and capsulolabral complex. The long TR (2,000 msec) in proton-density and T2-weighted sequences prolongs imaging time and may increase the likelihood of involuntary patient motion and degradation of images. Other imaging parameters include 3- to 4-mm slice thickness, 128 × 256 to 196 × 256 imaging matrix, and a field of view of 14 to 16 cm.

Several newer pulse sequences have been introduced for reducing imaging time and improving image quality (45). Fast spin-echo sequences reduce the imaging time. The major differences in tissue contrast with these sequence include increased signal intensity of fat on T2-weighted images and reduced sensitivity to susceptibility effects related to blood breakdown products (45). Other characteristics of fast spin-echo sequences include image blurring with the use of short TE or long echo trains and reduced image quality with the use of a low-resolution imaging matrix (45). Because signal characteristics in fast spin-echo sequences may reduce the conspicuousness of some lesions, we have not routinely utilized this sequence for evaluation of the rotator cuff, particularly in the coronal plane.

Gradient-echo sequences also reduce imaging time. The image contrast with these sequences is considerably different from that of spin-echo sequences; generally, signal intensity from muscle is higher and signal from bone marrow lower (susceptibility artifact). The use of gradient-echo sequences for rotator cuff imaging has been reported (46). The data are insufficient to compare the accuracy of these sequences with that of spin-echo sequences, however. Visualization of the glenoid labrum is enhanced on gradient-echo sequences owing to a high signal intensity of the articular cartilage, which results in an arthrographic effect (37). It is our impression that the generally higher signal intensity of soft tissue elements on gradient-echo images compared with spin-echo images may result in false-positive readings. Fat suppression sequences have been used in shoulder imaging to improve visualization of bone marrow abnormalities and rotator cuff lesions (45–49).

CASE 1

History: A 39-year-old man was referred for MR imaging of his nondominant left shoulder for recurrent discomfort and limited range of motion. On examination the impingement test was positive. What are the findings in Fig. 1A and B?

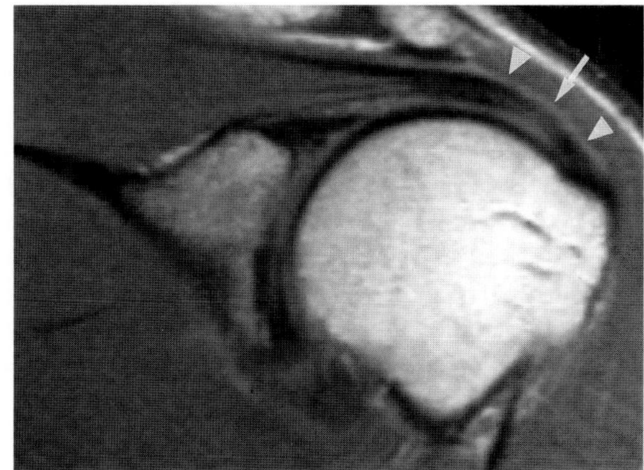

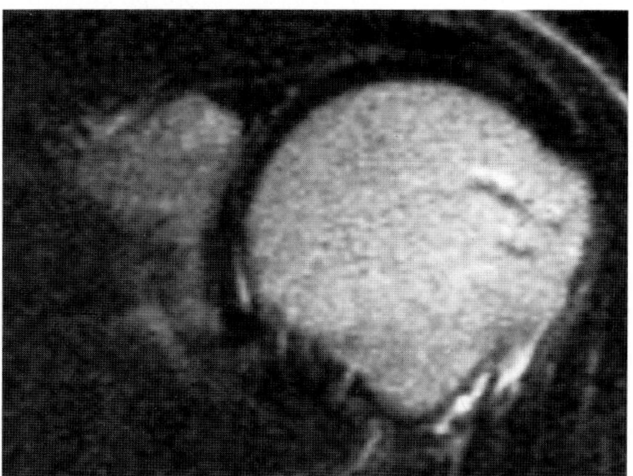

FIG. 1.

Findings: In Fig. 1A, a proton-density oblique coronal image (TR/TE 2,000/30 msec) of the left shoulder, demonstrates increased signal intensity of the central zone of the supraspinatus tendon (arrow). The peribursal fat is preserved (arrowheads). The tendon morphology is normal in this region. In Fig. 1B, a T2-weighted image (TR/TE 2,000/80 msec) in the same plane shows no abnormal signal in the central zone with overall signal void appearance of the entire tendon.

Diagnosis: Normal signal variation versus focal cuff degeneration.

Discussion: The homogeneously hypointense appearance of the normal rotator cuff is an uncommon occurrence (Fig. 1C). On T1-weighted or proton-density images, focal or linear regions of increased signal intensity of the supraspinatus tendon are often observed in asymptomatic individuals. These are attributed to anatomic characteristics of this tendon, technical factors, or subclinical tendon degeneration (1–7).

Anatomic configuration of the musculotendinous cuff varies from childhood to old age. The normal cuff in children and young adults is more muscular. With aging there is a progressive transformation of muscle to tendon (8). Peripheral extension of muscle fibers interposed between tendinous slips of the supraspinatus (or tendons of supraspinatus and infraspinatus) results in a linear or trapezoidal region of increased signal intensity on T1-weighted or proton-density oblique coronal images at the middle to posterior aspect of the joint (Fig. 1D,F) (9). This normal variation is distinguished from normal muscle by its isointense appearance on all pulse sequences.

Other anatomic characteristics of the rotator cuff may also contribute to variations in its signal intensity. As demonstrated in some anatomy texts and stressed by Neer, the tendons of the supraspinatus and infraspinatus may overlap just before their greater tuberosity insertion (10). The connective tissue at the junction of the supraspinatus and infraspinatus tendons, or between tendinous slips of the supraspinatus and adjacent muscle fibers, may also appear as linear or oblique regions of increased signal intensity (5,9,11). Overlap between the infraspinatus and supraspinatus tendons, which is observed in some individuals with neutral rotation of the arm, is accentuated with internal rotation (3). In this position, the supraspinatus tendon is visualized superior and lateral to the infraspinatus tendon and may result in apparent discontinuity of the tendon (Fig. 1F) (3).

Another common variation is a focal region of increased (intermediate) signal intensity within the critical zone of the supraspinatus tendon (4–7,9). This well-defined round or oval region of intermediate signal involves the anterior component of the tendon approximately 1 cm from its greater tuberosity insertion (4–7). It exhibits signal intensity isointense with skeletal muscle; the signal intensity is highest on proton-density fat suppression and gradient-echo sequences and is low or absent on T2-weighted sequences (2,5). This variation in supraspinatus signal intensity is not associated with morphologic alterations of the tendon, such as thinning or outline irregularity.

Several explanations have been proposed as possible etiologies of increased signal intensity of the critical zone of the supraspinatus tendon. The critical zone is a region of relative hypovascularity. It is also the area most susceptible to impingement by the coracoacromial arch and is the predominant site for the development of cuff pathology, including cuff tears (10). It is therefore postulated that the critical zone is prone to the development of tendon degeneration. Focal regions of altered signal intensity similar to one demonstrated in our patient in fact may represent early or subclinical manifestation of tendon degeneration (9,11–13). Histologic studies of cadaver tendons have revealed fibrillary degeneration or scarring of the tendon, although the specimens studied were mostly those of elderly individuals (14). The aberrant vascularity of the critical zone and poor hydration have also been suggested as possible etiologic factors (6,9).

The focal increased signal intensity of the critical zone has also been related to the magic angle theory. It has been demonstrated that poorly hydrated tendons exhibit higher T1 relaxation values in oblique orientation within the magnetic field (15). Intermediate signal intensity of tendons in the wrist and ankle has been observed in tendons oriented at 45° to 55° angles in relation to the constant magnetic induction field (16). The anterior component of the supraspinatus tendon deviates anteriorly near its insertion on the greater tuberosity, and therefore the magic angle theory may contribute to increased signal intensity of the critical zone of the supraspinatus tendon (4,6,7,9,17). The high prevalence of this focal intermediate signal intensity in asymptomatic individuals has been cited as strong evidence for its nonpathologic nature (4,6). Without histologic proof, however, the exact nature of this finding remains obscure and debatable.

It is therefore essential that characteristics of signal intensity and morphologic appearance of the rotator cuff tendon be carefully evaluated on both short and long TR/TE sequences. It is important that the tendon remain homogeneously low signal on T2-weighted images. Higher signal intensities on long TR sequences are specifically associated with pathologic changes such as tendinitis or tendon tear (12,13). Morphologic alterations such as thinning, irregularity, or expansion of the tendon also correlate with pathologic processes (12,13). In our patient presented here, the intermediate signal intensity of the central zone was present on several more posterior oblique coronal images, and in addition several images demonstrated thinning of the tendon and erosion of its synovial surface (Fig. 1G). Despite low signal on T2-weighted images, the morphologic changes indicate the presence of tendon degeneration and early superficial tear of the tendon. These findings correlate well with this patient's clinical presentation.

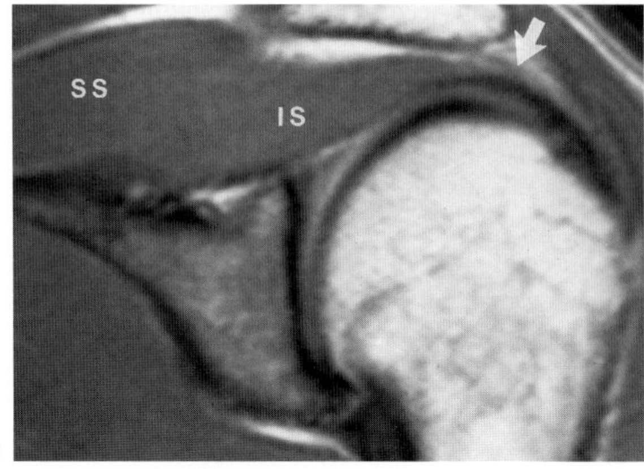

FIG. 1. C: Normal rotator cuff. Proton-density (TR/TE 2,100/30 msec) oblique coronal image of left shoulder at mid-posterior glenohumeral joint. At this level both supraspinatus (SS) and infraspinatus (IS) muscles are visualized. The homogeneously low signal tendon of the IS, with its musculotendinous junction at the center of the humeral head (*arrow*), is visualized.

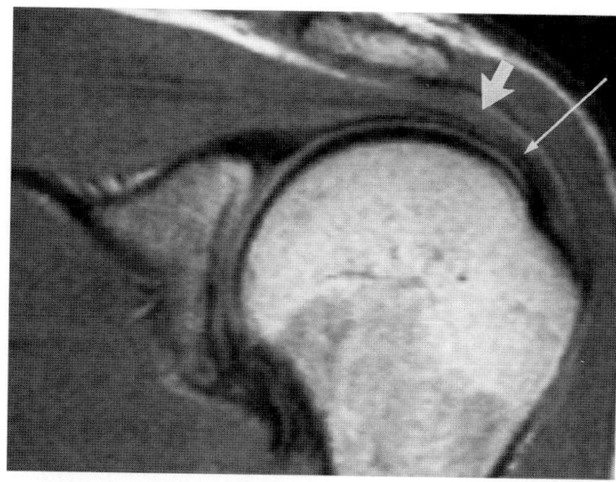

D

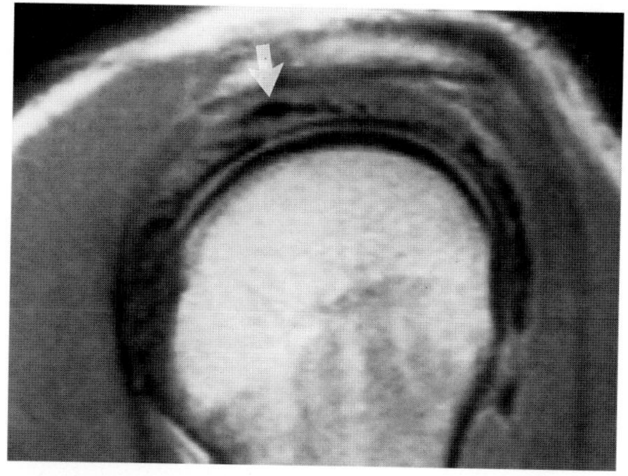

E

FIG. 1. D,E: Normal rotator cuff. **D:** Proton-density (TR/TE 2,100/30 msec) oblique coronal image, left shoulder. Region of intermediate signal at bursal surface is isointense with surrounding normal muscles (*thick arrow*). The linear intermediate signal of substance of the tendon (*thin arrow*) is either interposed muscle fibers or connective tissue between tendinous slips. **E:** Proton-density (TR/TE 1,800/29 msec) oblique sagittal image demonstrates heterogeneity of normal rotator cuff. The supraspinatus at this level consists of the prominent anterior component (*arrow*) as well as several smaller tendinous slips and muscle fibers.

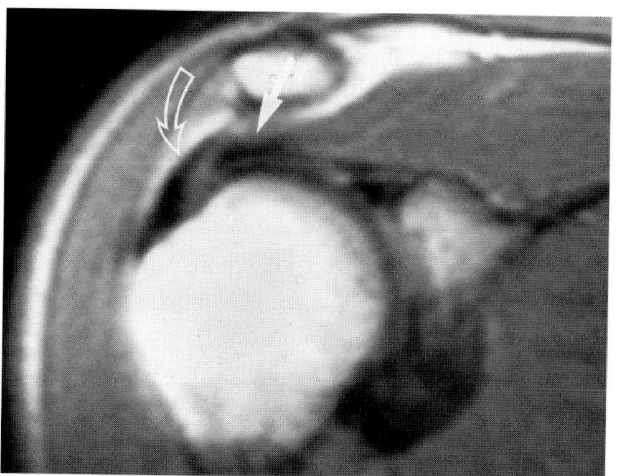

F

FIG. 1. F: Normal rotator cuff. Proton-density oblique coronal image of right shoulder in a 29-year-old asymptomatic individual. At this level the infraspinatus tendon (*open arrow*) is visualized superior and lateral to the supraspinatus tendon (*solid arrow*). The overlap of two tendons demonstrates intermediate signal.

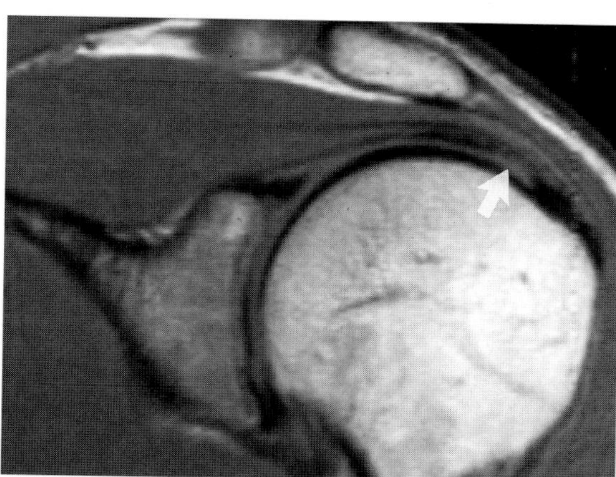

G

FIG. 1. G: Same patient as in A and B. Oblique coronal image posterior to A and at mid-glenohumeral level. There is slight erosion of the synovial surface in the central zone (*arrow*). This is consistent with superficial tendon degeneration.

CASE 2

History: A 25-year-old woman presented with shoulder pain and soreness. She is a recreational volleyball player and a professional photographer. She often carries and handles heavy photographic equipment. What are the findings in Fig. 2A and 2B? How specific are they?

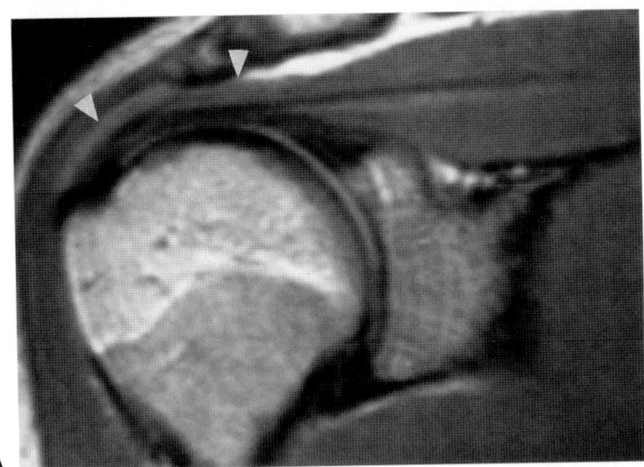

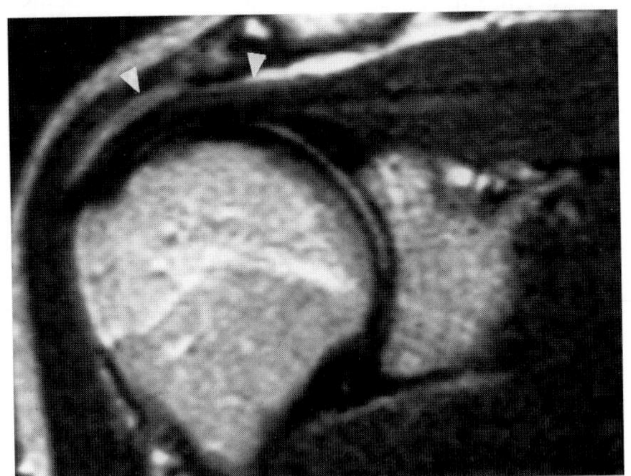

FIG. 2.

Findings: Fig. 2A, an oblique coronal proton-density (TR/TE 2,100/30 msec) image of the right shoulder at the mid-glenohumeral joint level demonstrates slight increased signal intensity of the supraspinatus tendon. The tendon morphology is within normal limits. The subacromial–subdeltoid peribursal fat is replaced by intermediate signal tissue (arrowheads). Fig. 2B is a T2-weighted (TR/TE 2,100/80 msec) image obtained in the same plane as Fig. 2A. There is no increased signal intensity of the supraspinatus tendon. There is slight heterogeneously increased signal intensity of the peribursal fat compared with Fig. 2A.

Diagnosis: Supraspinatus rotator cuff tendinitis and subacromial–subdeltoid bursitis.

Discussion: Pain associated with disorders of the rotator cuff is the most common cause of shoulder disability in all age groups, including young athletes (1–6). Tendinitis is the earliest stage in a continuing pathologic process that leads to gradual degeneration and eventual tear of the rotator cuff. As observed by Codman, (1) the great majority of cuff lesions, including cuff tears, involve the distal area of the supraspinatus tendon just proximal to the greater tuberosity insertion, referred to as the critical zone. This region is further identified as an area of hypovascularity or watershed, promoting cuff degeneration as a result of nonhealing of traumatic lesions (7,8).

Rotator cuff tendinitis is primarily a manifestation of intrinsic overload and repeated microinjury of the supraspinatus tendon. Neer proposed mechanical impingement by the coracoacromial arch as the primary mechanism of injury to the supraspinatus tendon (9,10). Mechanical impingement of the rotator cuff may also occur as a result of a decrease in the supraspinatus outlet in glenohumeral instability, particularly in young athletes (11,12). Others have proposed primary degeneration of the tendon as the main etiology of cuff disorders (1,13,14). The role of intrinsic overload of the tendon (overuse syndrome), as in repeated athletic or job-related overhead activities, has been emphasized in young individuals (10,11).

Compounding the prevalent studies, Fu et al. (15) divide the etiologic factors in rotator cuff tendinitis into two groups: extrinsic (including primary and secondary mechanical impingement) and intrinsic (primary degeneration). Overuse of the shoulder is considered the trigger for the initiation of the sequence of events resulting in rotator cuff injury and tendinitis (15).

Neer (10) classified impingement syndrome into three clinical stages. Stage I is observed in individuals younger than age 25, in whom symptoms respond to conservative therapy. In stage II, which is observed in individuals 25 to 40 years of age, symptoms are recurrent and fail to respond to conservative treatment. Stage III is manifested by progressive disability in individuals older than 40 years. The pathologic changes are suspected, but not histologically verified, to be hemorrhage and edema in stage I, bursal thickening and fibrosis as well as tendinitis in stage II, and rotator cuff tear in stage III (10). Nirschl (12) attributes rotator cuff tendinitis to repetitive microtrauma of the supraspinatus tendon during overhead sports activities that results in mechanical microruptures. A specific histologic appearance referred to as angiofibroblastic hyperplasia then ensues, which this author believes is consistent with an avascular element in rotator cuff tendinitis (12).

Various pathologic changes of the rotator cuff are diagnosed on MR images based on alterations in morphology and signal intensity of the involved tendons (16,17). The normal rotator cuff is ordinarily void of signal or hypointense with some normal variations observed (see Case 1). Tendinitis or tendon degeneration on MR images is recognized when the signal void appearance of the tendon is replaced by mild to moderate increased signal intensity on T1-weighted or proton-density images with no or minimal increased signal intensity on T2-weighted images. The morphology of the tendon remains normal (Fig. 2C,D).

Increased signal intensity of the rotator cuff is an indication of chemical or histologic alterations within the tendon and may reflect a variety of tendinous disorders, including degenerative or posttraumatic reparative processes (18). Histologic studies of supraspinatus tendons with increased signal intensity on MR imaging have demonstrated manifestations of tendon degeneration including disorganization of the tendinous fibers and regions of fatty and myxoid degeneration (18–20). Inflammatory changes of unspecified nature have been described in biopsy specimens, although the presence of true inflammatory cells has not been documented in these cases (21). Other investigators, indicating a lack of histologic evidence to support an inflammatory process, have concluded that *tendinitis* is a misnomer and that *tendinosis* is a more appropriate term (12,20). Increased signal intensity of the rotator cuff without morphologic abnormality therefore is a nonspecific finding, and it is not possible to differentiate edema, hemorrhage, interstitial tear, or degenerative change based on signal intensity alone. This concept is further supported by MR examinations of asymptomatic shoulders, which have revealed tendons with increased signal intensity on T1-weighted and proton-density images, probably reflecting the degenerative process of aging or anatomic variations (22).

In our experience, several MR imaging findings are more specifically indicative of tendinitis than asymptomatic tendon degeneration or normal variations in signal intensity. These include more diffuse and/or heterogeneous increased signal intensity on T1-weighted or proton-density images, tendinous enlargement, and associated abnormality of the peribursal fat with or without fluid in the bursa (Fig. 2E,F). Tendinous enlargement and increased signal intensity probably reflect edema and cellular proliferation in the initial stage of tendon injury (11,19). Codman (1) observed that bursal inflammation or thickening is frequently associated with rotator cuff tendinitis. Bursal thickening appears as loss of fat signal on proton-density and T2-weighted images. Mild increased signal intensity may be observed on T2-weighting, indicative of granulation tissue formation (Fig. 2A,B). Small amounts of bursal fluid may also be observed in tendinitis (Fig. 2G,H).

Tendon degeneration (or mild tendinitis) may manifest as increased signal intensity of the tendon without associated tendinous enlargement or abnormality of the peribursal fat. In more advanced stages of impingement syndrome, tendon degeneration may imply loss of tendon substance and is diagnosed more specifically when, in addition to increased signal intensity on T1-weighted or proton-density images, abnormal morphology of the tendon (attenuation or surface irregularity) is also observed. These morphologic alterations of tendon outline are also regarded as partial tears of the synovial or bursal surfaces (17).

Heterogeneous increased signal intensity on T2-weighted images (less than fluid signal) with intact tendon surfaces is occasionally observed and probably represents a more advanced stage of tendon injury (Fig. 2C,D). This pattern of increased signal intensity of the cuff could be referred to as *interstitial tear.* We have used this term in distinction from *partial intrasubstance tear,* which indicates visualization of a focal defect.

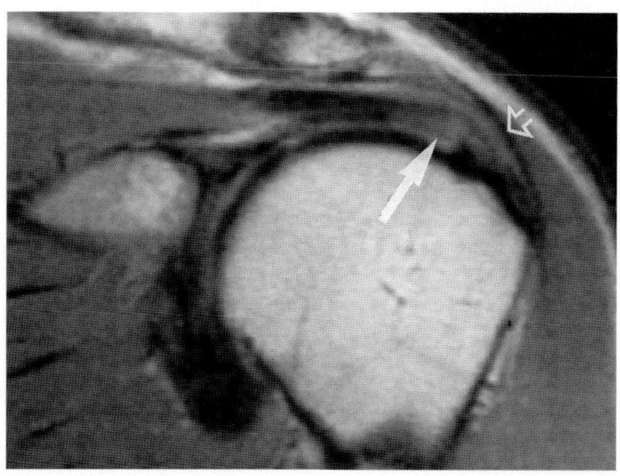

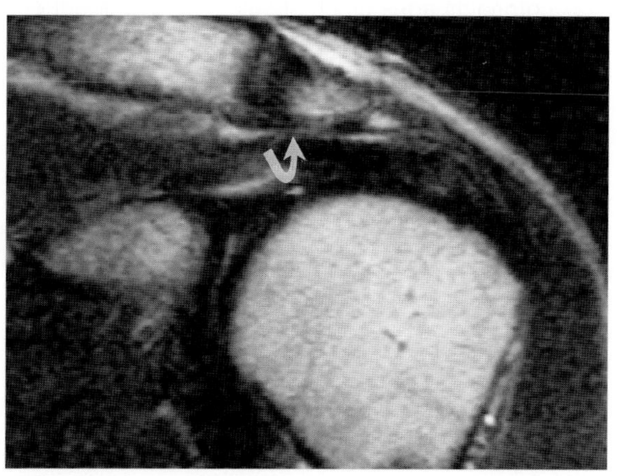

FIG. 2. C,D: Tendinitis. **C:** Proton-density (TR/TE 2,100/30 msec) oblique coronal image demonstrates moderate increased signal intensity of distal supraspinatus tendon (*solid arrow*). The tendon morphology is normal. The subdeltoid peribursal fat signal is diminished (*open arrow*). **D:** T2-weighted (TR/TE 2,100/80 msec) demonstrates minimal increased signal intensity compared with C. Note soft tissue proliferation of acromial undersurface (*curved arrow*).

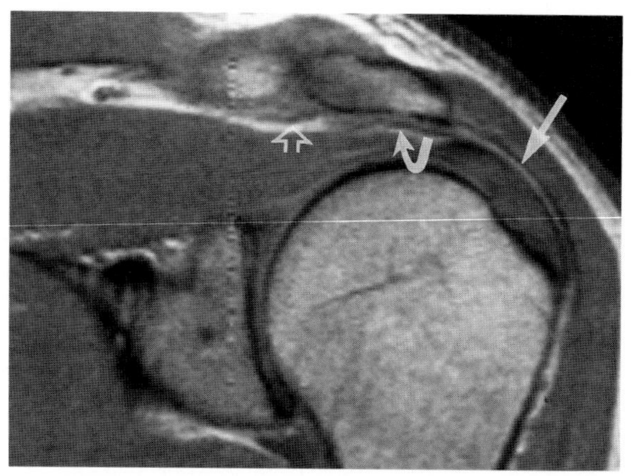
E

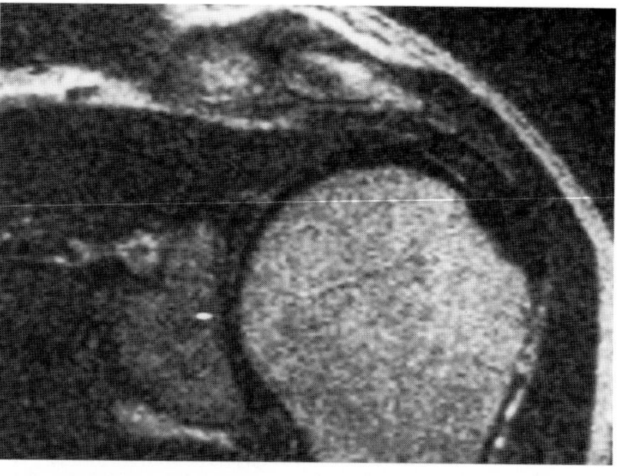
F

FIG. 2. E,F: Tendinitis. Proton-density (**E**) and T2-weighted (**F**) (TR/TE 2,100/30/80 msec) oblique coronal images of left shoulder. The supraspinatus tendon shows diffuse increased signal intensity on proton-density image (E, *straight arrow*) with no increased signal on T2 weighting (F). The tendon is moderately enlarged. The synovial and bursal surfaces are normal. Note soft tissue proliferation of the acromial undersurface (E, *open arrow*) and expansion of acromioclavicular joint capsule (E, *open arrow*) with mild impingement of subacromial space. The peribursal fat is normal.

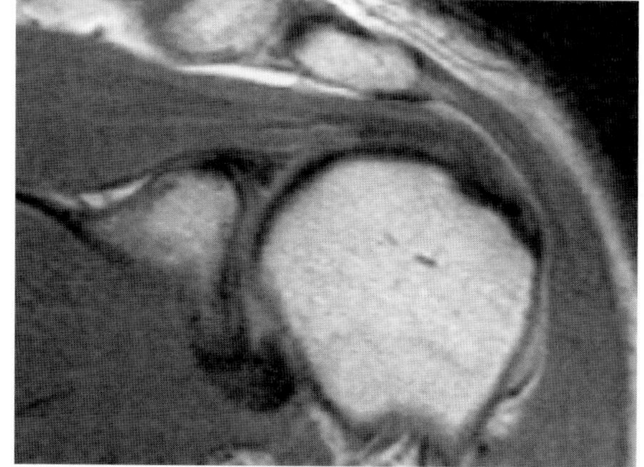
G

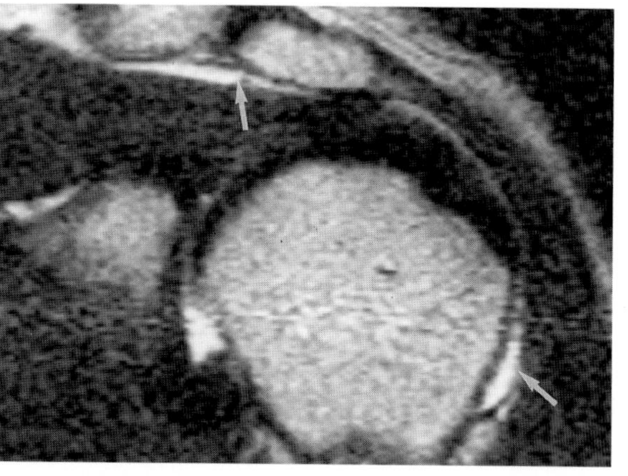
H

FIG. 2. G,H: Tendinitis. Proton-density (**G**) and T2-weighted (**H**) (TR/TE 2,100/80/30 msec) oblique coronal images of left shoulder. There is generalized increased signal intensity of the supraspinatus tendon on proton-density image (G). A small amount of fluid is present in the subacromial-subdeltoid bursa (*arrows* in H).

CASE 3

History: A 45-year-old radiologist presented with recurrent right shoulder pain for the past 2 years. The latest episode followed a week of rather heavy review of large stacks of radiographic examinations. Pain is persistent at night. What is your diagnosis? What is the relevant osseous abnormality?

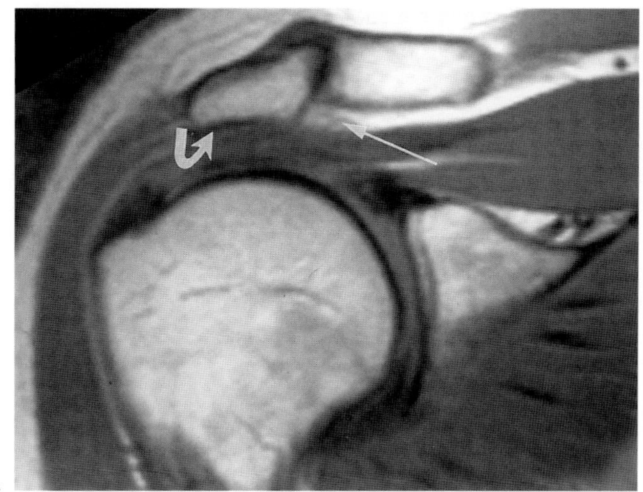

FIG. 3.

Findings: Fig. 3A, a relative proton-density oblique coronal image (TR/TE 1,070/25 msec) of the right shoulder, demonstrates heterogeneous intermediate signal intensity of the supraspinatus tendon, which is also thickened. The subdeltoid peribursal fat is preserved, but the bursal surface of the tendon appears adherent to the undersurface of the acromion (curved arrow). Medial to the acromion process, the peribursal fat is obliterated. Note the inferior position of the acromion in relation to the clavicle (straight arrow). The irregular medium-signal tissue at the tip of the arrow is probably thickened bursal tissue.

Diagnosis: Supraspinatus cuff tendinitis and low-lying acromion.

Discussion: Several developmental and acquired variations of the acromion process and acromioclavicular joint have been identified in association with impingement syndrome. The rotator cuff is surrounded by the partly osseous, partly ligamentous coracoacromial arch. When the arm is in anatomic position, the supraspinatus tendon is mostly peripheral to the acromion (1). When the arm is elevated, the central zone of the supraspinatus tendon first glides under the coracoacromial ligament and then passes beneath the acromion and the acromioclavicular joint. The movement of the supraspinatus tendon under the coracoacromial arch is cushioned by the subacromial–subdeltoid bursa. Neer (1) proposed mechanical impingement of the rotator cuff as a result of developmental and acquired changes of the anterior acromion and the acromioclavicular joint as the primary etiology in the development of the rotator cuff disorders. Based on his observations, Neer devised the surgical procedure (i.e., anterior acromioplasty) for treatment of patients with recurrent symptoms of impingement. Further promoting the concept of anterior impingement by an abnormal acromion, Bigliani et al. (2) identified three configurations of acromion based on visualization on outlet view of the shoulder. These configurations, which reflect the undersurface of the acromion process, include type I (flat), type II (curved), and type III (hooked). The hooked acromion showed higher prevalence in patients with rotator cuff tears (2).

The concept of a hooked acromion, although distinguished from secondary spurs by Bigliani et al., has not been corroborated by a number of subsequent studies. In contrast to Bigliani et al., (2), Edelson and Taitz (3), in a study of scapular specimens, determined that the concept of hooked acromion, although readily translated into a schematic drawing, is difficult to measure in practice and that secondary degenerative spurs of the anterior acromion may be mislabeled as such.

They concluded that developmental hooks rarely occur and represent an unusually large preacromial epiphysis (Fig. 3B). Spurs of the anterior acromion represent traction spurs formed within the substance of the coracoacromion ligament as a result of tensile forces transmitted through the ligament (Fig. 3C).

The slope of the acromion, in addition to its size, was indicated by Neer (1) as a factor in producing impingement, although it was not further clarified. Aoki et al. (4) in a study of scapular specimens, found that acromions with spurs had a flatter (decreased) scapular angle compared with those without spurs and concluded that degenerative spurs are the result of impingement in the former group.

Edelson and Taitz found, in addition to the slope of the acromion, a correlation between the length of the acromion and the height of the coracoacromial arch and the development of degenerative changes of the undersurface of the acromion process (3). Decreased slope (more horizontal orientation) of the acromion was found to have the greatest degree of association with spur formations. The slope of the acromion in turn was directly related to the angle of the base of the scapular spine with the blade of the scapula. In this study, the size of the acromion process correlated positively with formation of osteophytes. The height of the coracoacromial arch was also determined and found to be dependent on the slope of the acromion and length and slope of the base of the coracoid process. A lower height of the arch was associated with a greater degree of acromion osteophyte formation.

MR imaging allows a three-dimensional depiction of the acromion and coracoacromial arch. Osteophytes of the anterior acromion are depicted either as mature bone containing bone marrow signal intensity (Fig. 3C) or as bands of low-signal thickening at the site of insertion of the coracoacromial ligament or occasionally covering the entire undersurface of the acromion (Fig. 3D). A higher incidence of low-lying acromions as identified on oblique coronal images has been reported in patients with impingement syndrome (Fig. 3A) (5). Anterior downsloping of the acromion, as identified on oblique sagittal images, is identified as resulting in diminution of the anterior subacromial space and possible impingement of the rotator cuff (Fig. 3E). This variation has been related to decreased slope of the acromion (5). Inferior sloping of the peripheral margin of the anterior acromion, as identified on oblique coronal images, has also been implicated in impingement syndrome (Fig. 3F,G). The downsloping of the peripheral margin of the acromion and the low-lying acromion have not been clearly correlated with the acromial shape as identified on oblique sagittal images, and their position in the scheme of classification of acromial shapes remains uncertain. The low-lying acromion, however, is probably a consequence of a flat (decreased) acromion slope rather than a hooked acromion.

The prevalence of a hooked acromion as depicted on oblique sagittal images has accordingly been reported to be high in several MR imaging studies in relation to impingement syndrome and rotator cuff tears (6–11). Nevertheless, the high incidence of hooked acromion in patients with impingement syndrome and rotator cuff tear is suggested to reflect at least partially the development of enthesopathy of the undersurface of the acromion process (10). We agree with Edelson and Taitz (3), who considered the true hook-shaped acromion an uncommon anatomic variation Ozaki et al., (12) in a study of a group of patients with partial- and full-thickness rotator cuff tears, demonstrated that acromial spurs were only observed in patients with partial bursal surface and full-thickness tears of the rotator cuff (12). Those with synovial surface tears did not demonstrate alterations of the acromion. They concluded that rotator cuff tears are degenerative in nature and secondarily result in alterations of the acromion. In a study of a group of patients with unilateral cuff tears, Zuckerman et al. (13) demonstrated similar anatomy of the coracoacromial arch bilaterally and concluded that other predisposing factors are important in producing cuff pathology (Fig. 3H,I).

Several recent studies have outlined difficulties in determining acromial shape based on the radiographic outlet view or by MR imaging. Haygood et al. (14) found MR imaging to be more diagnostic than the radiographic outlet view for demonstration of acromial shape, although interobserver agreement for MR examinations was poor. MR images often demonstrate variable shapes of the acromion in the same individuals on various oblique sagittal sections. Peh et al. (15) similarly noted the varied shape of the acromion on oblique sagittal sections but also pointed out that acromial shape on radiographic studies is sensitive to minor changes in radiographic technique. These investigators also found poor correlation of acromial shape as determined on MR images and outlet view, and no correlation between acromial shape as seen on oblique sagittal MR images, and the presence of supraspinatus tendon abnormality in their patients. They concluded that MR imaging should not be a substitute for the outlet view for evaluation of acromial shape.

Another developmental variation of the acromion process, os acromiale, has been associated with rotator cuff tears (3,16,17). Up to three ossification centers in the acromion appear at about age 15 to 18 and fuse at about 22 to 25 years (17). Failure of union of these ossification centers forms the os acromiale, which, based on the number of ossification centers and the site of nonunion, may be divided into seven types (17). The unstable segment of the acromion may therefore result in impingement of the rotator cuff and eventual tear due to movement by the strong function of the deltoid muscle inferiorly. Motion at the site of a nonunited epiphysis usually results in formation of hypertrophic spurs that independently erode the surface of the rotator cuff. MR imaging can identify this developmental variation and assist in identifying the nonunited form and possible osteophytic changes at its undersurface (Fig. 3J,K) (5). The variation os acromiale of the anterior aspect of the acromion is the common form observed in practice, and most appear to be stable. The os acromiale, including the entire acromion, may occasionally be observed (Fig. 3L).

In conclusion, anatomic variations and secondary changes of the acromion and the coracoacromial arch are implicated in the development of impingement syndrome and rotator cuff tendinopathies. These issues are controversial, however, MR imaging displays in detail the anatomic relation of various structures in the subacromial space, although only in the static state. The role of MR imaging is therefore to outline regions of subacromial space narrowing with specification as to the nature of the abnormality (i.e., acromion shape, size, or secondary hypertrophic process) and the degree of superior migration of the humeral head, if present.

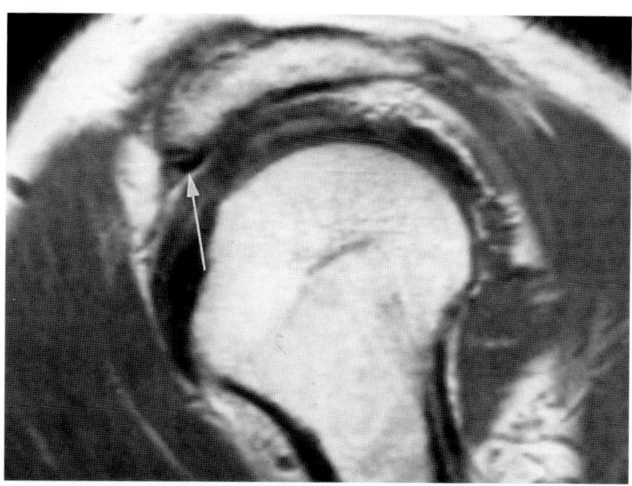

FIG. 3. B: Proton-density (TR/TE 1,800/29 msec) oblique sagittal images, left shoulder. The anterior acromion is large and hook shaped. There is a low signal spur at the insertion of the coracoacromial ligament (*arrow*). Impingement of the subacromial space and supraspinatus is present in this region.

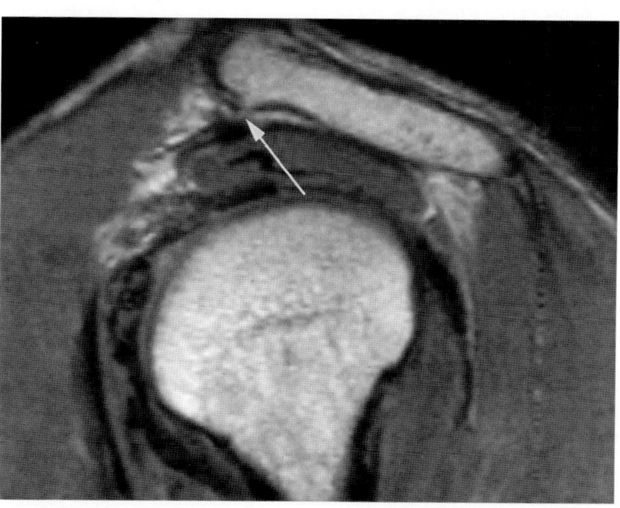

FIG. 3. C: Acromial traction spur. Proton-density (TR/TE 1,800/29 msec) oblique sagittal image shows a small traction spur of the anterior acromion process (*arrow*). Note bone marrow signal within the spur.

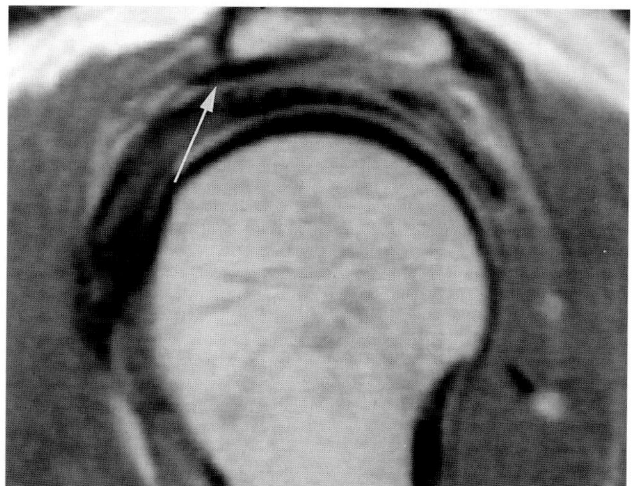

FIG. 3. D: Acromial spur. Proton-density (TR/TE 1,800/29 msec) oblique sagittal image, left shoulder. There is a low signal spur at the insertion site of the coracoacromial ligament (*arrow*).

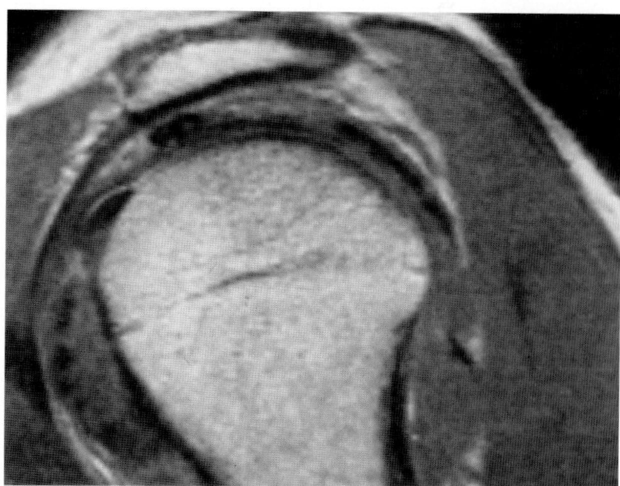

FIG. 3. E: Anterior down-sloping acromion. Proton-density (1,800/29 msec) oblique sagittal image, left shoulder. There is downward orientation of the acromion, resulting in narrowing of the anterior subacromial space. The coracoacromial ligament is rather prominent.

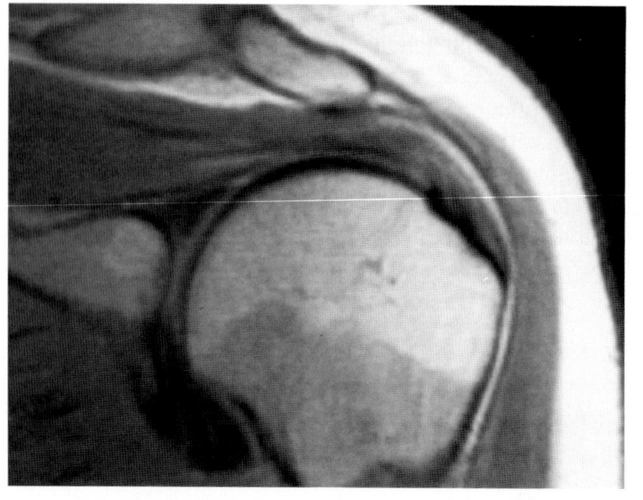

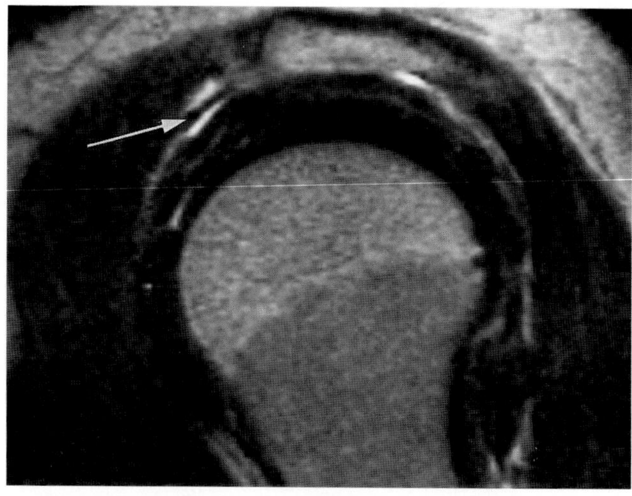

FIG. 3. F: Proton-density (TR/TE 2,100/30 msec) oblique coronal image, left shoulder. Acromial process shows a downward peripheral orientation, impinging on the supraspinatus tendon which shows increased signal and width due to tendinitis. **G:** T2-weighted (TR/TE 2,100/80 msec) oblique sagittal image shows the acromion to be flat in the anteroposterior orientation. Note thickening of coracoacromial ligament surrounded by inflammatory reaction (*arrow*).

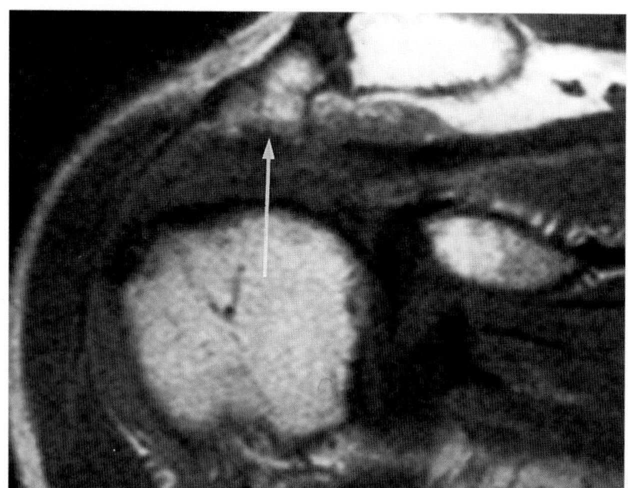

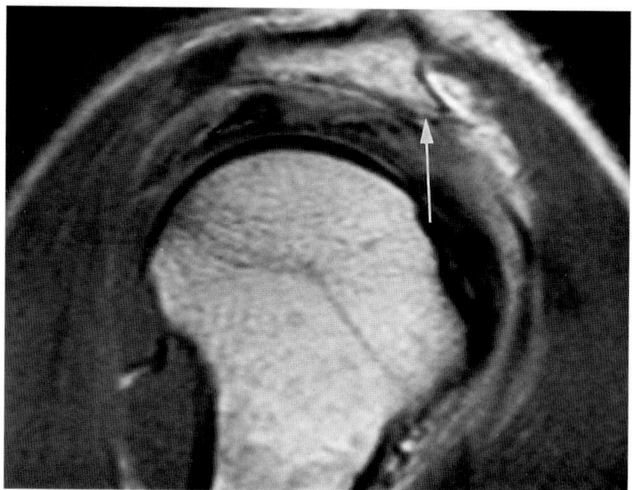

FIG. 3. H,I: Hooked acromion secondary to large acromial spur. Proton-density (TR/TE 2,100/29 msec) oblique coronal (**H**) and oblique sagittal (**I**) images, right shoulder. The large spur of the anterior margin of the acromion (*arrows*) has the appearance of a hook on oblique sagittal image. Note extensive abnormality in signal and morphology of the supraspinatus tendon and SA-SD bursa on oblique coronal image. A cuff tear was identified on T2-weighted images (not shown).

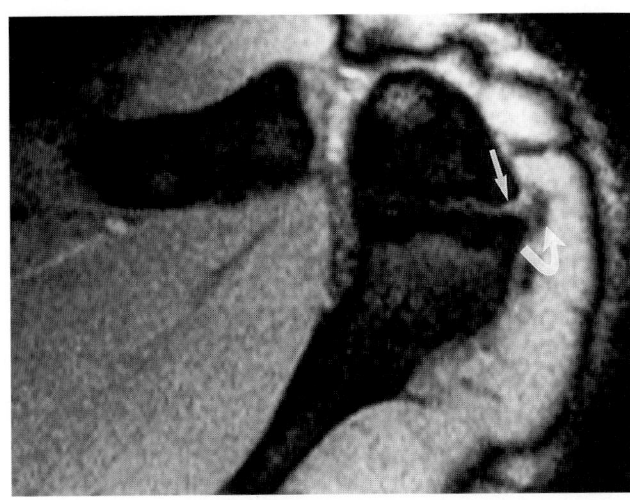

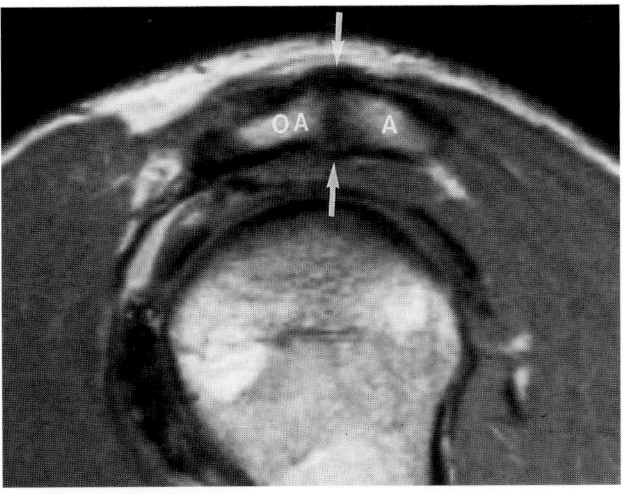

FIG. 3. J,K: Os acromiale. **J:** Gradient-echo (TR/TE 540/15/30 msec) axial image, left shoulder. The secondary ossification center of the anterior acromion is suspected to be partly nonunited, as suggested by slight widening and high signal intensity at its peripheral aspect (*straight arrow*). Low signal area adjacent to this region is bony proliferation (*curved arrow*). **K:** Proton-density (TR/TE 1,800/29 msec) oblique sagittal image shows os acromiale (OA) with irregular margin consistent with a fibrous union with posterior acromion (A). Note hypertrophic changes both superiorly and inferi-orly (*arrows*), with the latter impinging on the subacromial space.

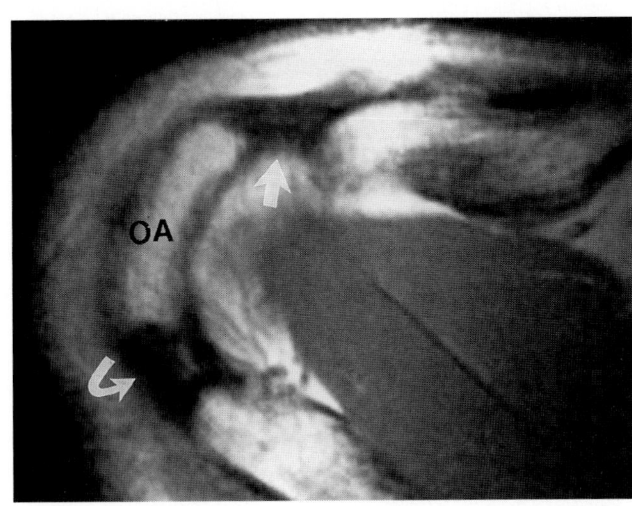

FIG. 3. L: Os acromiale. Proton-density (TR/TE 1,600/35 msec) axial image of the right shoulder demonstrates a large os acromiale (OA). This secondary ossification center shows a wide, probably fibrous union with the scapular spine (*short arrow*). The AC joint is also unusually wide (*curved arrow*). The acromial process had a peripheral down-sloping appearance on oblique coronal images (not shown).

CASE 4

History: This 67-year-old man had been experiencing pain and weakness of the right shoulder for the past 6 months after a fall. Pain was more severe on external rotation in the abducted position of the arm. A recreational golfer, he had been unable to play. On physical examination, Jobe's test for pain was positive.

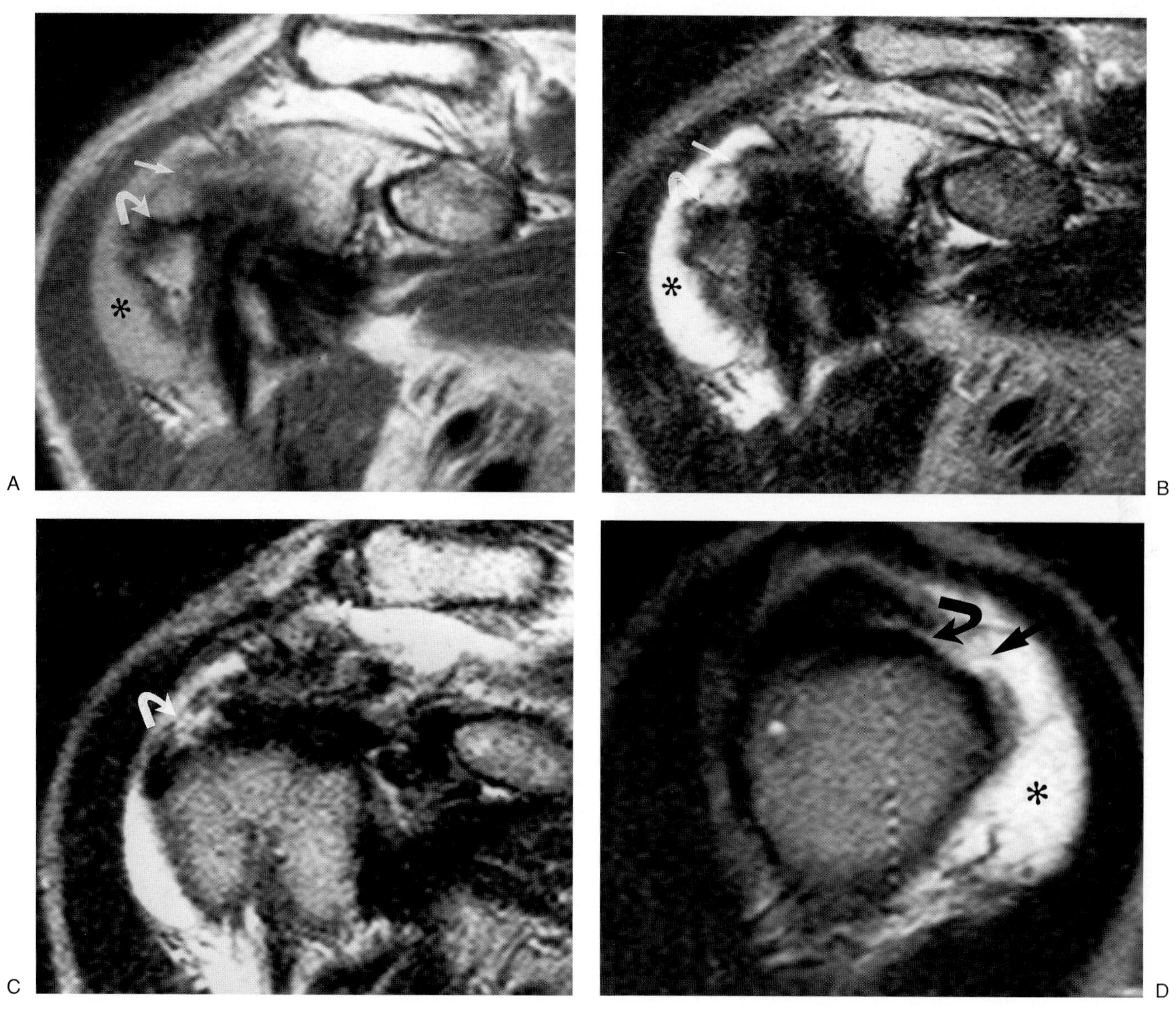

FIG. 4.

Findings: Proton-density (Fig. 4A) and T2-weighted (Fig. 4B) oblique coronal images (TR/TE 2,100/30/80 msec) of the right shoulder, anterior aspect, show detachment of the anterior margin of the supraspinatus (straight arrows) from the greater tuberosity (curved arrows), as indicated by intermediate signal intensity on the proton-density image and intense signal on the T2-weighted image. Note the characteristic horizontal appearance of the greater tuberosity adjacent to the bicipital groove on these images (curved arrows). The subdeltoid bursa (*) is distended with fluid and communicates with the cuff tear. Fig. 4C, a T2-weighted (TR/TE 2,100/80 msec) oblique coronal image immediately posterior to Fig. 4B, demonstrates the posterior extent of the cuff tear (arrow). Fig. 4D, a T2-weighted (TR/TE 1,800/80 msec) oblique sagittal image, shows the small full-thickness tear of the ante-

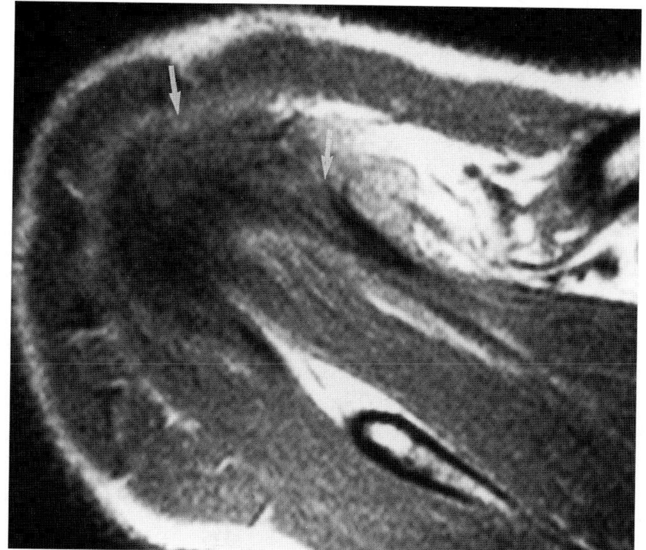

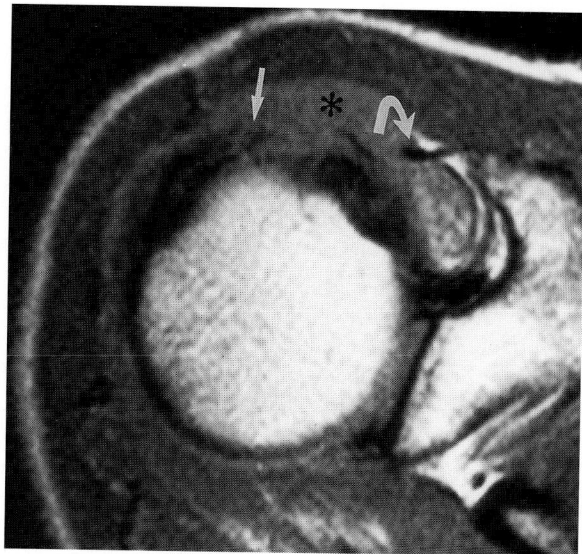

FIG. 4.

rior supraspinatus tendon (straight arrow) and posterior extension of a partial component of the tear (curved arrow). Note the marked expansion of the anterior component of the subdeltoid bursa (*). Fig. 4E,F shows T1-weighted (TR/TE 900/29 msec) axial images of the superior aspect of the shoulder demonstrating the torn anterior component of the supraspinatus tendon (straight arrows). There is apparent interruption of the coracohumeral ligament, indicating injury to the rotator interval capsule (curved arrow in Fig. 4F). Note the distended subdeltoid bursa (*).

Diagnosis: Small full-thickness tear of the supraspinatus tendon and the rotator interval capsule. Bursitis with distension of the subdeltoid bursa.

Discussion: Except for a small percentage of cuff tears that involve normal tendons and could be attributed to acute trauma, the majority are the result of tendon degeneration even though a traumatic event, however trivial, may precede tendon injury (1–4). A significant predilection exists for involvement of the central zone of the supraspinatus tendon. Increasing frequency of rotator cuff tears with advancing age and concentration of tears within the central zone–anterior component of the supraspinatus tendon correlate with hypovascularity of this region and concepts of overuse and chronic impingement of the cuff by the coracoacromial arch (1,4–7). Most often a tear originates at the greater tuberosity insertion of the supraspinatus, with the tendon retracting from the anatomic neck (sulcus) of the humerus. These insertion tears evolve from partial tears at the synovial surface into full-thickness detachments (Fig. 4G,H) (1,5). Cuff tears usually extend posteriorly and may eventually involve the infraspinatus tendon. Anteriorly the rotator interval capsule may be involved.

By definition, a full-thickness tear of the rotator cuff is a defect that bridges the entire substance of the tendon in the cephalocaudad direction, establishing communication between the glenohumeral joint and the subacromial–subdeltoid bursa (Fig. 4I,J). Full-thickness cuff tears are divided according to their size; those less than 1 cm are considered small tears (Fig. 4D) (3). A more recent classification takes into consideration the size as well as the location of the tear for surgical planning and prognostic purposes (8).

MR imaging corroborates the long-established anatomic distribution and morphology of rotator cuff tears (9,10). Characteristically, the anterior component of the supraspinatus is detached (Fig. 4A–F) from the greater tuberosity, or there is a defect within the central zone with a stump remaining attached to the greater tuberosity (Fig. 4I,J). These defects are visualized on one or more than one oblique coronal image. On oblique coronal images, the most anterior edge of the greater tuberosity is visualized immediately lateral and posterior to the bicipital groove as a horizontal cortical margin resembling a shelf (Fig. 4A,B). The attachment of normal tendon to bone is continuous and void of signal. Detachment of the tendon is manifested by intermediate signal on proton-density images and high signal intensity on T2-weighted images (Fig. 4A,B,D,G,H). The anteroposterior dimension of the detached tendon is best visualized on oblique sagittal images (Fig. 4D). Axial images also usually demonstrate the torn tendon. The attachment of the cuff on the greater tuberosity on axial images appears as a low-signal semicircle surrounding the periphery of the humeral head. In case of an insertion tear and cuff detachment, a defect is visualized along the anterior boundaries of the greater tuberosity (Fig. 4E,F).

The characteristic and most specific MR imaging sign of a cuff tear is visualization of a defect or interruption of the low-signal tendon depicted as intermediate signal on T1-weighted and proton-density images and high or intense signal on T2-weighted images (9–15). This MR appearance is considered the primary criterion for cuff tears and is the finding most often seen in full-thickness tears. Consistently high accuracies are reported using these criteria in the diagnosis of full-thickness rotator cuff tears (14,15). Depending on the degree of inflammatory reaction at the time of imaging, fluid may be present within the glenohumeral joint and within the subacromial–subdeltoid bursa, although fluid may be seen in one compartment only. Subacromial–subdeltoid peribursal fat is frequently disrupted when a cuff tear is present. Although it had been postulated that the bursal fluid generally originates from the glenohumeral joint and is an indication of a full-thickness cuff tear, not infrequently bursal fluid is present without a joint effusion. Also, bursal fluid may be present with intact tendons, indicating that it may originate within an inflamed bursa (Fig. 4A–F). There is a tendency for bursal fluid to accumulate within the anterior component of the subacromial–subdeltoid bursa. Fluid may also expand medially under the acromioclavicular joint or laterally over the greater tuberosity. Bursal thickening due to granulation tissue or scar formation, which is observed in chronic impingement syndrome and in association with tendinitis, may also be observed with rotator cuff tears. On MR imaging, the peribursal fat signal is replaced by intermediate- or low-signal tissue on all pulse sequences. It should be noted that the peribursal fat signal is a variable finding in normal individuals and cannot be considered a highly specific finding in association with rotator cuff disorders. In our judgment, however, if the peribursal fat is thickened and appears as low signal on all sequences, it is an accurate manifestation of chronic bursitis.

Although the presence of a full-thickness rotator cuff tear is considered an indication for surgical intervention to repair the defect, small tears may respond well to conservative treatment. These measures include various methods of physical therapy and subacromial injection of steroids (one or two injections only) along with oral nonsteroidal antiinflammatory medications. Our patient (Fig. 4A–F) in fact responded well to a regimen of conservative treatment and experienced decreasing pain. Weakness of the shoulder also dramatically improved.

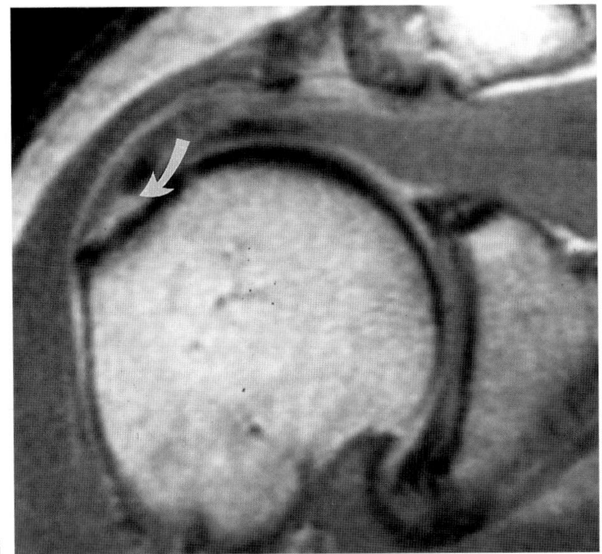

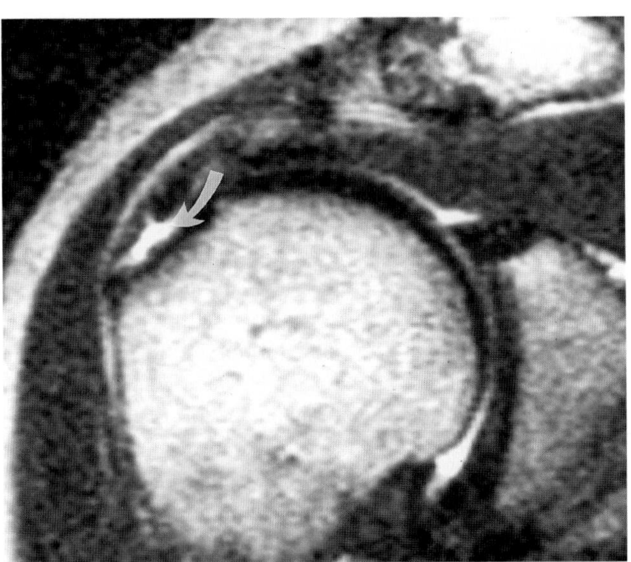

FIG. 4. G,H: Partial (near full-thickness) insertion tear. Proton-density and T2-weighted (TR/TE 2,100/30/80 msec) oblique coronal images of the right shoulder. There is detachment of the synovial aspect of the supraspinatus tendon from the greater tuberosity. Only a thin layer at the bursal surface is intact. Note AC joint arthritic process and small acromial spur.

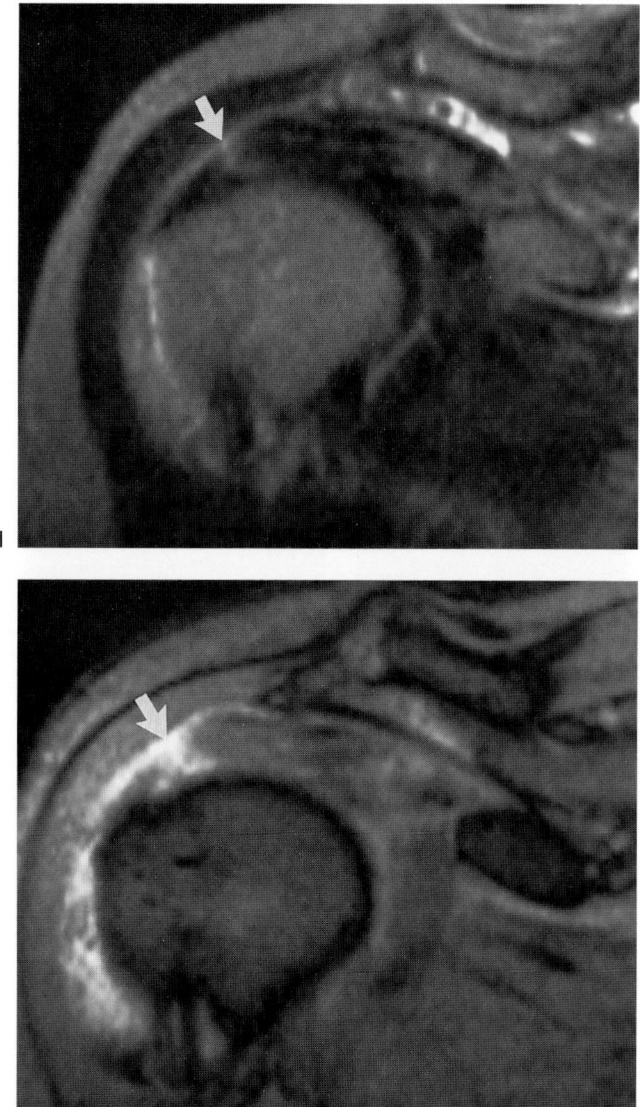

FIG. 4. I,J: Small full-thickness rotator cuff tear. T2-weighted (TR/TE 2,000/80 msec) spin-echo and gradient-echo (TR/TE 563/117/40 msec) oblique coronal images, right shoulder, demonstrate a small full-thickness tear involving the central zone of the supraspinatus tendon (*arrows*). Note overall higher signal intensity than the soft tissues and an apparently longer appearance of the cuff tear on gradient-echo image.

CASE 5

History: A 45-year-old male recreational athlete had a 4-month history of shoulder pain and weakness. Pain was getting increasingly worse and was incapacitating. On physical examination, he had painful restriction of motion. What are the MR imaging findings, and what is your diagnosis?

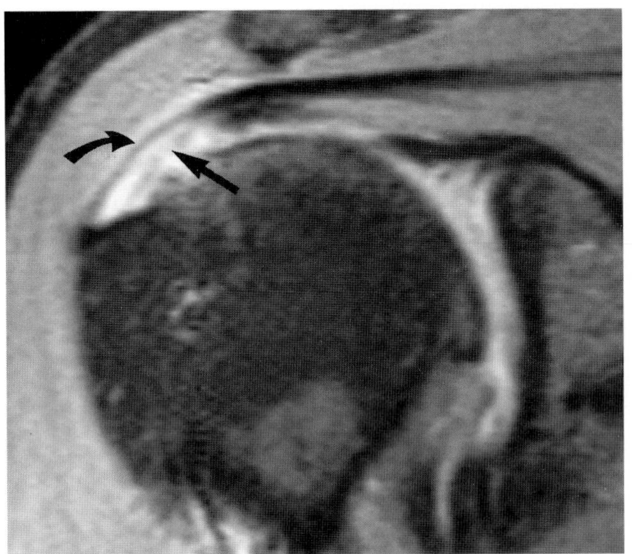

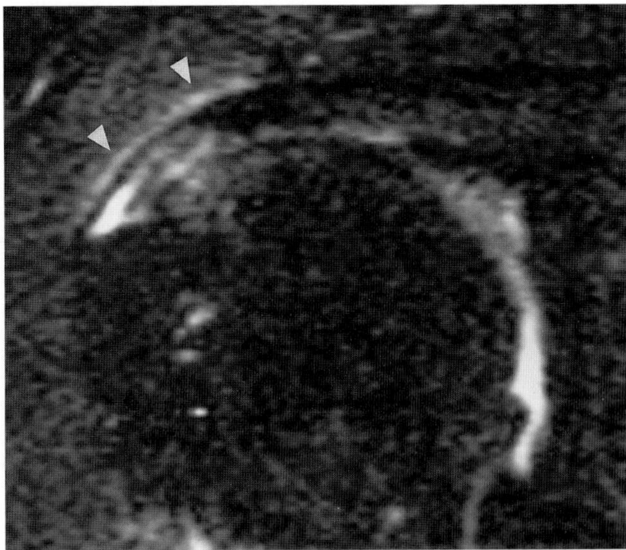

FIG. 5.

Findings: Fig. 5A, a proton-density (TR/TE 2,000/20 msec), spin-echo, fat-saturated, oblique coronal image of the right shoulder, shows increased signal intensity of the distal segment of the anterior component of the supraspinatus tendon (straight arrow). Abnormal signal involves the synovial surface and the underlying substance, including the greater tuberosity site of insertion of the tendon. The bursal surface is intact (curved arrow). Note a small, low-signal spur of the inferior acromion. Fig. 5B, a T2-weighted (TR/TE 2,000/80 msec), spin-echo, fat-saturated image at the same level, demonstrates further increase in signal intensity of the region, confirming a partial tear of the synovial surface and the tuberosity insertion site and extending within the tendinous substance. Tendon morphology is abnormal with depression and irregularity of the synovial surface. Note the increased signal intensity of the subdeltoid bursa (arrowheads), an indication of mild inflammatory reaction.

Diagnosis: Large partial-thickness tear of the supraspinatus tendon.

Discussion: Codman (1) first observed partial-thickness tears of the supraspinatus tendon and described the progressive tearing of the tendon fibers at the synovial surface with gradual retraction of the tendon from the greater tuberosity site of insertion. He also observed and described as partial intrasubstance tears clefts within the cuff tendons in necropsy material. Partial tears may also independently involve the bursal surface of the cuff (1). Partial tears are infrequently reported in clinical series, reflecting the difficulty in visualizing some partial-thickness tears at surgery (2,3). Most partial-thickness tears represent one stage in evolution of pathologic changes of the rotator cuff in impingement syndrome. According to Neer's (4) classification, partial tears are observed late in the second stage of impingement. Tears of the synovial surface are the result of repeated tensile microtrauma. The bursal surface tears result from more direct impact by the coracoacromial arch.

Arthrographic diagnosis of partial-thickness tears is limited to visualization of synovial surface tears. MR imaging is therefore considered the imaging modality capable of visualizing all partial-thickness tears (5–7). MR imaging has consistently been less accurate in the diagnosis of partial thickness tears, however, compared with its accuracy in the diagnosis of full-thickness tears (7–12). The discrepancy is multifactoral; the nature of partial-thickness tears, the varying criteria used for diagnosis by MR imaging and at surgery, and finally the difficulty in visualizing some partial- and even small full-thickness tears at surgery are to be considered.

The most specific criterion used for the diagnosis of cuff tears is the high or intense signal on long-TR, long-

TE images depicting the cuff defect (5–10). Iannotti et al. (9) introduced a grading system in which tendon morphology, tendon signal intensity, and abnormality of the peribursal fat signal were used for evaluation of rotator cuff lesions. They concluded that some partial tears may not be differentiated from full-thickness tears or tendinitis. Combining alterations in morphology and signal intensity of the rotator cuff for diagnosis of cuff disorders, we concluded that less than 50% of partial tears demonstrate high or intense signal on T2-weighted images (7). The remainder, mostly shallow tears of the synovial or bursal surfaces, demonstrate low to intermediate signal on T2-weighted images. These tears may be regarded as focal regions of tendon degeneration both at surgery and on MR imaging. The central zone is a common site for these partial tears (Fig. 5C,D). Evaluation of tendon morphology (outline irregularity or thinning) is important in detecting partial cuff tears at bursal or synovial surfaces (Fig. 5C–F).

An abrupt change in signal intensity is also suggestive of a partial tear. Recent MR imaging studies show improved diagnostic accuracy in the diagnosis of partial tears with the use of fat suppression imaging sequences (11,12).

Intrasubstance partial tears are also most accurately diagnosed when a linear area of increased signal intensity is present within the tendinous substance (Fig. 5E,F) The presence of the fluid signal within intrasubstance tears, however, may in fact indicate that a communication exists with either the synovial or the bursal surface of the tendon. Arthroscopy may also fail in visualizing intrasubstance partial tears or small full-thickness tears. The presence of a partial tear usually does not influence the treatment protocol for a painful shoulder. When conservative treatment fails, usually an arthroscopic decompression of the subacromial space is performed. The presence of a full-thickness tear, however, is by most accounts treated by cuff repair. Therefore, failure of MR imaging to visualize small full-thickness tears will not adversely affect proper treatment, whereas overestimation of partial tears as full-thickness tears may subject the patient to unnecessary exploration of the cuff.

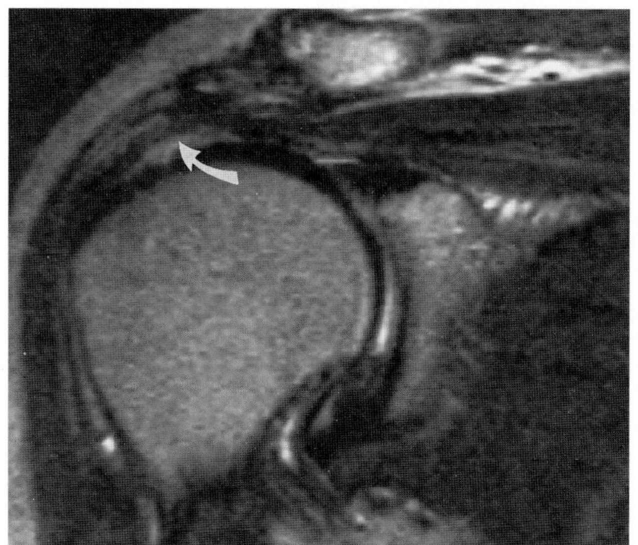

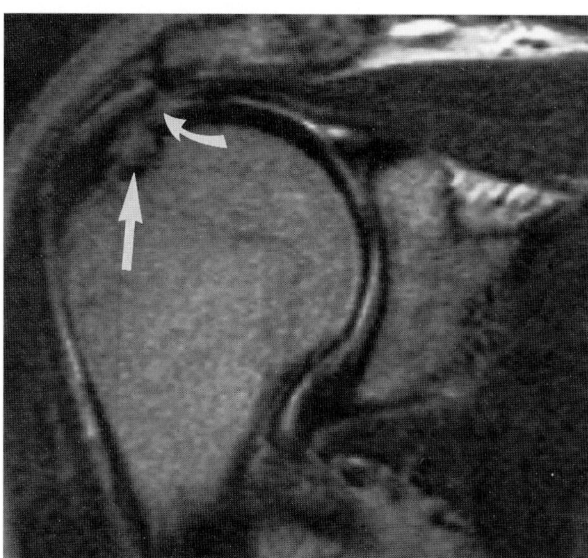

FIG. 5. C,D: Partial tear/degeneration, supraspinatus tendon, synovial surface. T2-weighted (TR/TE 2,100/80 msec) oblique coronal images of the right shoulder. There is irregularity and loss of tendinous substance at the synovial surface of the central zone of the supraspinatus tendon (*curved arrow*). This is manifested by intermediate signal alteration outlining the morphologic change, which is more pronounced in D. There is mild superior migration of the humeral head, further diminishing the subacromial space, which is already compromised by hypertrophic change of the acromion and the acromioclavicular joint. Note cystic erosion of the anatomic neck in D, most likely due to synovial hyperplasia (*straight arrow* in D).

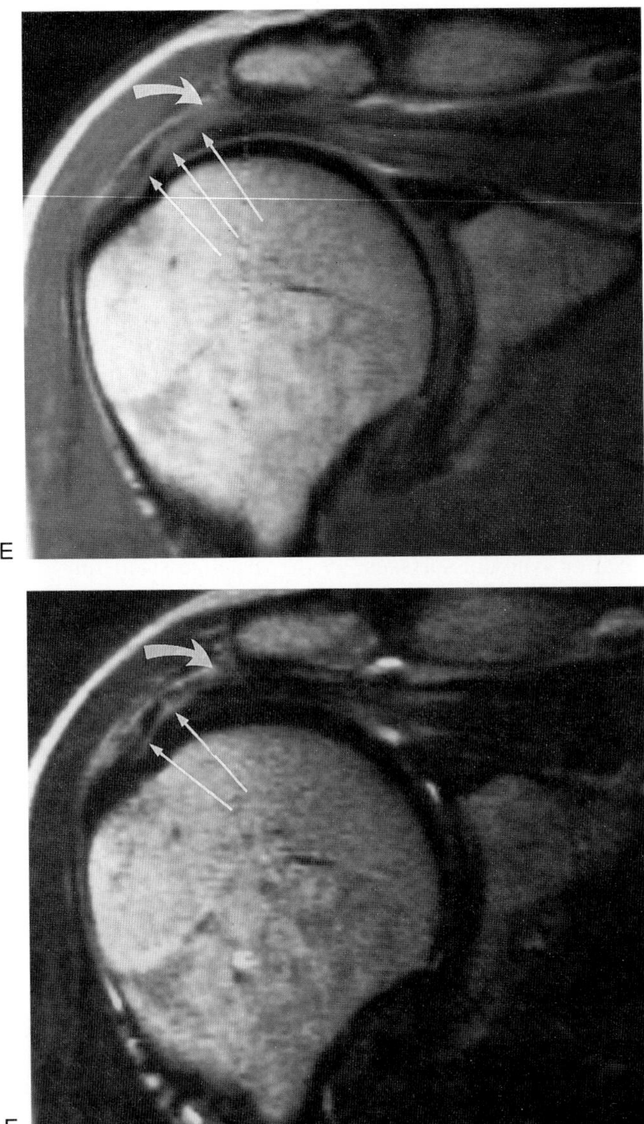

FIG. 5. E,F: Partial intrasubstance-bursal surface tear. Proton-density (**E**) and T2-weighted (**F**) (TR/TE 2,100/80/30 msec) image of right shoulder. A linear tear extends obliquely through the substance of the supraspinatus tendon (*straight arrows*), focally interrupting the bursal surface (*curved arrows*). Note low signal thickening of the acromial undersurface and slight superior migration of the humeral head.

CASE 6

History: A 51-year-old man was injured 4 months ago while playing basketball. He heard a ''pop'' and felt severe pain. During the following weeks there was some improvement in pain, which was worse at night. Sharp pain accompanied certain movements, particularly internal rotation and forward flexion. Movement of the arm above horizontal was also painful. Supraspinatus atrophy was observed on physical examination.

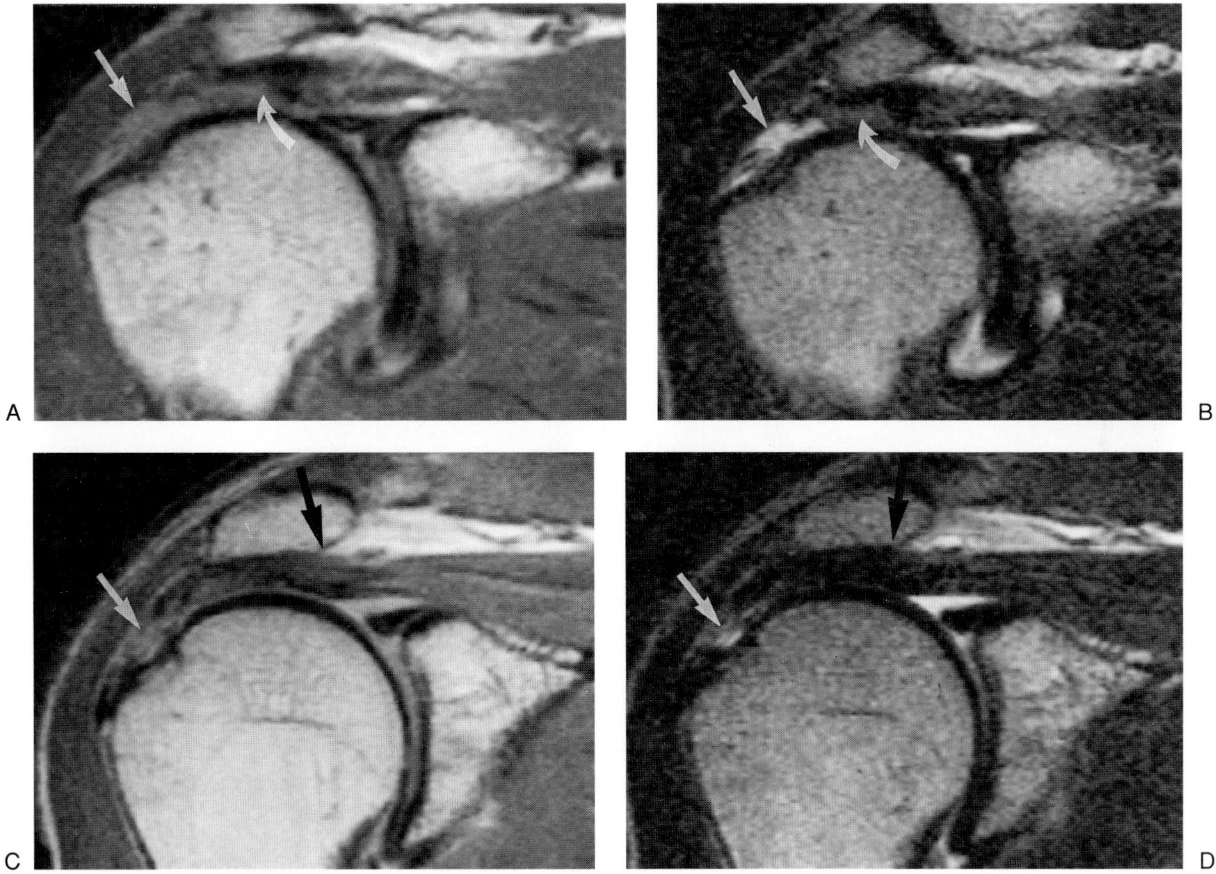

FIG. 6.

Findings: Fig. 6A,B shows proton-density and T2-weighted (TR/TE 2,100/30/80 msec) oblique coronal images, respectively, of the right shoulder at the anterior aspect of the glenohumeral joint. They demonstrate a cuff defect manifested by intermediate signal on the proton-density image and intense signal on the T2-weighted image (straight arrows). This full-thickness cuff defect measures approximately 2 cm in lateral-medial dimension. The tendon edge is irregular (Fig. 6B). The inferior tendon bundle–capsule is more prominently retracted and is replaced by medium-signal tissue (curved arrows).

Fig. 6C,D, proton density and T2-weighted (TR/TE 2,110/30/80 msec) oblique coronal images at the mid-glenohumeral joint, show the full-thickness cuff tear extending posteriorly, although its medial-lateral dimension is reduced. At this level, few tendon fibers remain intact (white arrows). The musculotendinous junction is retracted (black arrow). The subacromial bursal surface is thickened and appears adherent to the acromial undersurface. Also note mild atrophy of the supraspinatus muscle on all images.

Diagnosis: Medium full-thickness supraspinatus cuff tear.

Discussion: Rotator cuff tears characteristically involve the anterior aspect of the supraspinatus tendon. The cuff defect may extend posteriorly and involve the upper aspect of the infraspinatus tendon. Cuff tears may also extend medially. The medial extension of a cuff tear is often more pronounced anteriorly, as shown in this patient. The outcome is a triangular defect with its apex located medially (1). The presence of a full-thickness rotator cuff tear and the size of the cuff defect are important factors in treatment planning. The consensus appears to favor repair of a torn cuff along with an anterior acromioplasty (2–4). This patient did not improve with conservative measures and underwent arthroscopic decompression and rotator cuff repair.

Accurate diagnosis of a full-thickness rotator cuff tear and accurate assessment of the size of the defect not only are important in surgical planning but also have prognostic value. MR imaging has consistently shown high accuracies in detection of full-thickness rotator cuff tears (5–12). The dimensions of the cuff defect are also by most accounts accurately determined by MR imaging (10,12). Occasionally, however, a discrepancy is encountered and is usually due to underestimation of the extent of degeneration of the torn margin of the cuff on MR images (Fig. 6E,F) (13). A thickened or intact floor of the subacromial–subdeltoid bursa with or without residual degenerated tendon is occasionally observed and may conceal a full-thickness tear (Fig. 6G). An abrupt demarcation of the relatively normal remaining tendon from the abnormal region is often observed. Skinner (14) describes perforation of the floor of the subacromial–subdeltoid bursa establishing communication with the glenohumeral joint as the last stage in the development of a full-thickness rotator cuff tear. Retained bursal floor or degenerated tendon and granulation tissue may contribute to reported false-negative diagnosis of full-thickness cuff tears on arthrography (10). A cuff tear may conceivably appear larger at surgery because of the necessity for distention of the joint and traction on the arm during the procedure (13). Careful analysis of the cuff on MR imaging should suggest the correct diagnosis. A detailed description of the torn cuff defect, including an assessment of the remaining tendon, is helpful. A thin layer of the bursal surface and a severely degenerated layer of the tendon with an abrupt demarcation from the remaining normal tendon are usually functionally deficient and are readily removed at the time of bursectomy.

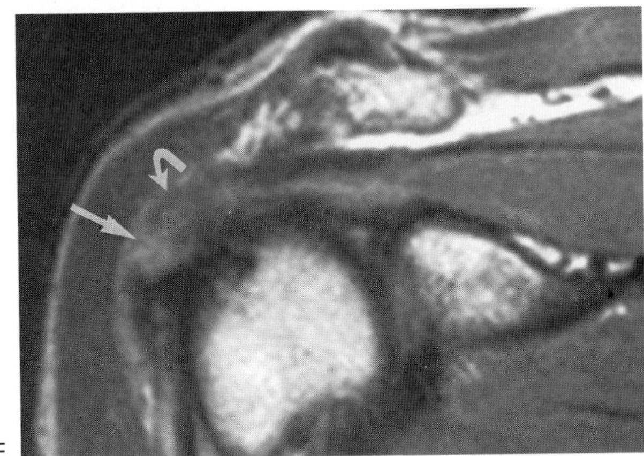

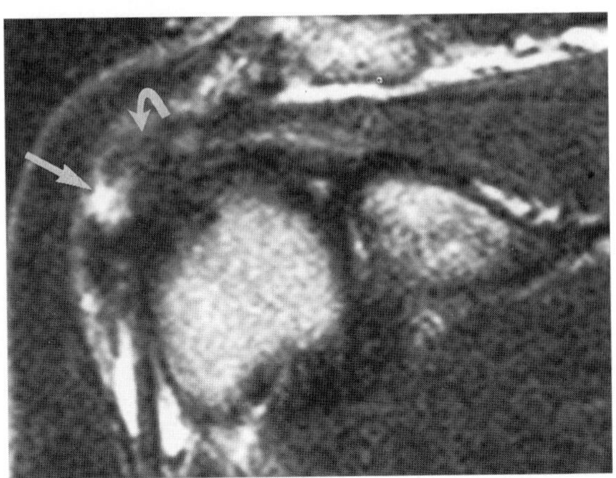

FIG. 6. E,F: Proton-density (**E**) and T2-weighted (**F**) (TR/TE 2,100/30/80 msec) oblique coronal images demonstrate small full-thickness detachment of the anterior supraspinatus tendon from the greater tuberosity (*arrows*). The tendon proximal to the cuff defect shows increased signal intensity and is wavy in contour (*curved arrow* in E) but appears intact at the bursal surface (*curved arrows*). The peribursal fat signal is diminished on both images. At surgery the bursa was found to be markedly thickened and scarred. Debridement of the bursal floor was carried out, which then revealed a 2-cm full-thickness cuff defect.

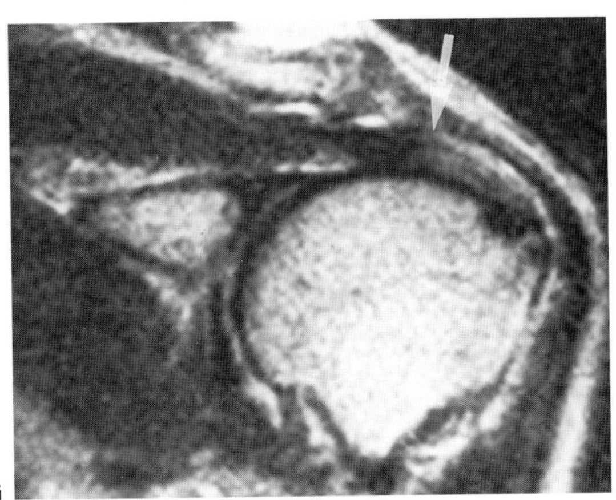

FIG. 6. G: T2-weighted (TR/TE 2,100/80 msec) oblique coronal image of the left shoulder. There is abrupt demarcation of the normal proximal supraspinatus tendon and the distal two-thirds, which shows marked attenuation, increased signal intensity, and irregular outline. At surgery the bursal surface appeared intact, but upon probing it disintegrated, and a large full-thickness tear became apparent. (From Rafii M. MR imaging of the shoulder. In: Beltran J, ed. Current review of MRI. Hong Kong: Paramount Printing Group; 1995:203–214, with permission.)

CASE 7

History: A 48-year-old man presented with gradual onset of pain and swelling of the right acromioclavicular joint. Discomfort originally started because of an injury playing squash. He had been an avid squash player for many years. Pain was worse when he used the arm in abducted position. Radiographic examination revealed a slight irregularity of the acromioclavicular joint articular surfaces.

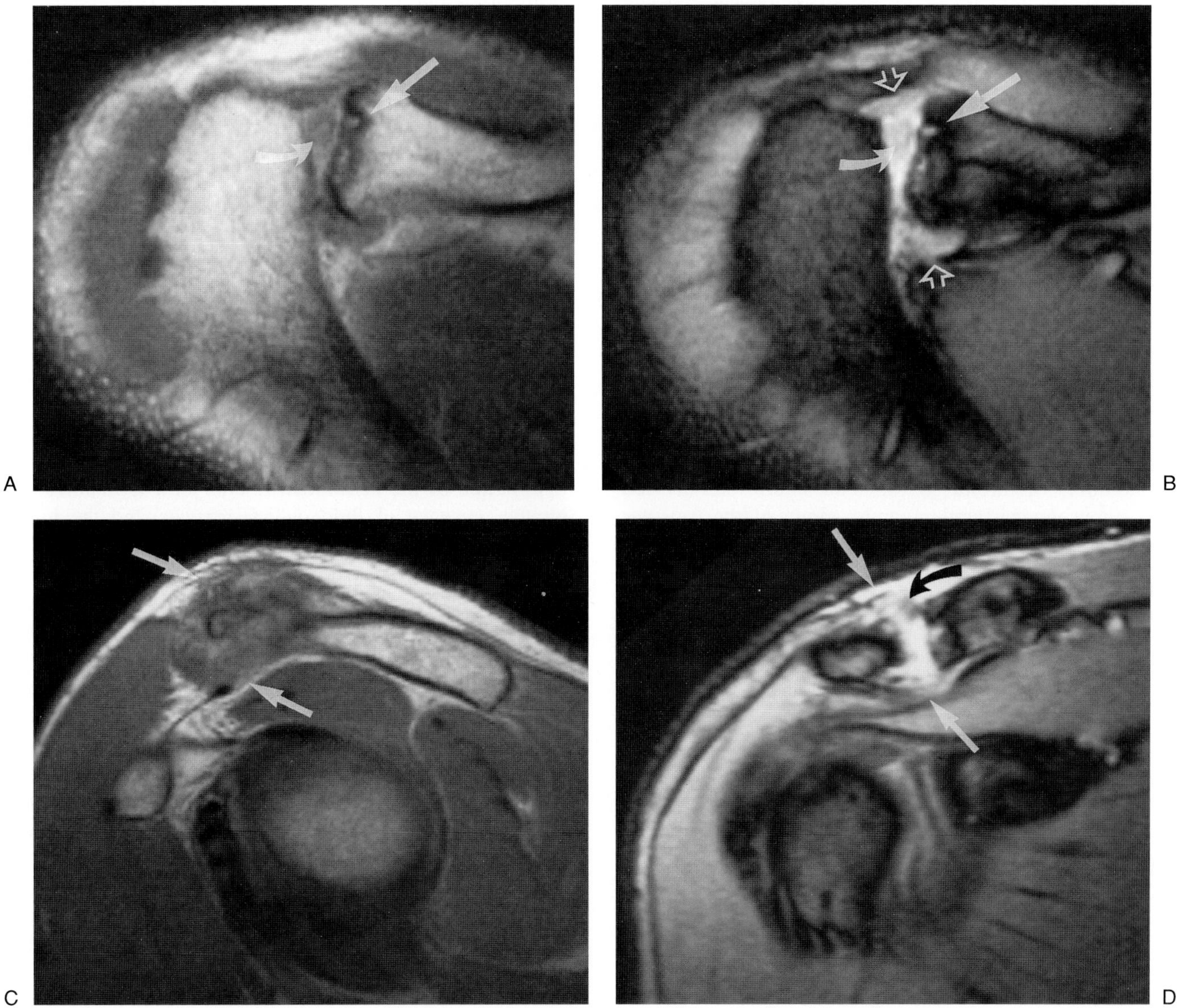

FIG. 7.

Findings: Fig. 7A, a proton-density (TR/TE 1,600/30 msec) axial image, and Fig. 7B, a gradient-echo (TR/TE 940/18/30 msec) axial image, demonstrate irregularity and subcortical cystic changes of the clavicular articular aspect of the acromioclavicular joint (straight arrows). The degenerated articular disk is visualized on both images (curved arrows). Joint swelling and capsular expansion are well seen (hollow arrows in B). Fig. 7C, a proton-density (TR/TE 3,200/17 msec) oblique sagittal fast spin-echo image, and Fig. 7D, gradient-echo (TR/TE 705/18/30 msec) oblique coronal image, demonstrate expansion and irregularity of the capsular ligaments and impingement of the subacromial space (straight arrows). The degenerated articular disk (curved arrow in Fig. 7D) is visualized. Note irregularity and erosions of the articular surfaces on both images.

Diagnosis: Posttraumatic degenerative arthritis of the acromioclavicular joint.

Discussion: MR imaging in this patient demonstrates in detail the osteoarticular abnormalities that were clinically and radiographically suspected. Perhaps the more significant role of MR imaging in this patient is evaluation of the subacromial space and the rotator cuff for detection of possible impingement of these structures by the arthritic acromioclavicular joint.

The acromioclavicular articulation is a relatively flat, diarthrodial joint that can be variably oriented from vertical to horizontal (1). This joint is only slightly movable, with a range of motion of 20° (2,3). The articular surfaces are covered by fibrocartilage and are separated by a fibrocartilaginous disk, which is incomplete or more pronounced superiorly (4,5). Variations in the developmental anatomy of the fibrocartilaginous meniscus have correlated with degenerative disease of the acromioclavicular joint (1). The synovium-lined articular capsule is supported by the superior and inferior acromioclavicular ligaments. The superior ligament is reinforced by fibers from the trapezius and deltoid muscles. The inferior capsule is reinforced by fibers from the coracoacromial ligament (4).

The acromioclavicular joint is frequently subjected to acute injury or chronic stress. Acute injuries are divided into three grades: grade I, mild sprain of the capsular ligaments; grade II, tear of the capsular ligaments that results in slight instability, and grade III, associated with destruction of the coracoclavicular ligament, rendering the joint totally unstable. Acute injuries of the acromioclavicular joint and surrounding soft tissues are seldom evaluated by MR imaging. Although grade I and II injuries often heal, a small percentage may remain symptomatic because of the development of arthritic changes (6). Osteolysis of the distal end of the clavicle either may follow acute injury or may be seen in individuals with repeated stress injury to the shoulders. Weightlifting is a common etiology (7). Various degrees of bone resorption or spur formation are observed. The acromial surface is usually not affected.

Degenerative disease of the acromioclavicular joint is frequently observed in middle-age and elderly individuals and may accompany osteoarthritis of other joints. Soft tissue and osseous proliferative changes are often present. Symptomatic degenerative disorders of the acromioclavicular joints are usually responsive to conservative treatment, although intractable pain may require surgical intervention.

Soft tissue or osseous proliferative changes of the acromioclavicular joints may compromise the subacromial space and contribute to the development of impingement syndrome (Fig. 7E,F) (8). Inferiorly pointing osteophytes of the acromioclavicular joint are frequently observed in patients with rotator cuff tears (9,10). Acromioclavicular osteophytes correlate specifically with the subtype of superior impingement and rotator cuff degeneration. These patients generally require decompressive surgery (10).

Although many of osseous changes of the acromioclavicular joint described above are radiographically evident, MR imaging is valuable in depicting the extent of these changes in greater detail. MR imaging confirms correlation of these osseous changes with impingement syndrome and rotator cuff pathology (11). More important, the soft tissue abnormalities (i.e., disruption of the capsular ligaments, degeneration of the articular disk and articular fibrocartilage, or proliferative process of the synovium and capsular ligaments) are visualized. Inflammation of the acromioclavicular joint is manifested by the presence of fluid, which may expand the joint capsule (Fig. 7B–D), and by medium-signal proliferative changes of the capsular elements in chronic stage (5). With progressive tear and degeneration of the rotator cuff and superior migration of the humeral head, the joint osteophytes and acromial undersurface spurs are gradually eroded, conforming to the contour of the humeral head (Fig. 7G,H). Furthermore, a communication may develop between the glenohumeral joint and the acromioclavicular joint. Synovial cysts of the acromioclavicular joint have occasionally been observed resulting from flow and accumulation of the synovial fluid from the glenohumeral joint. Ganglion cysts are also rarely observed (Fig. 7I,J) (12).

A review of MR images of asymptomatic acromioclavicular joints demonstrated that small amounts of fluid may normally be present.

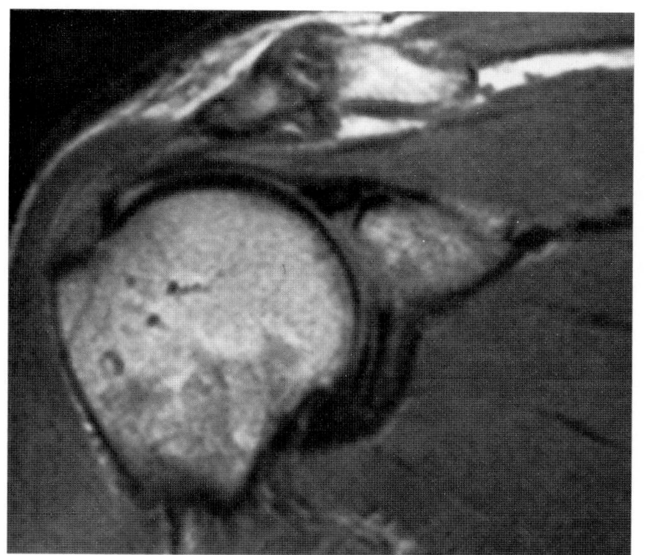

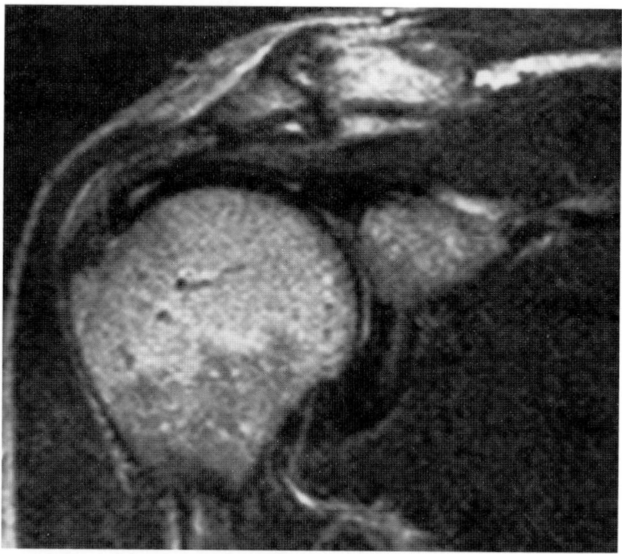

FIG. 7. E,F: Proton-density and T2-weighted (TR/TE 2,100/30/80 msec) oblique coronal images of the right shoulder demonstrate a prominent inferiorly pointing osteophyte of the acromioclavicular joint with impingement of the supraspinatus musculotendinous junction. The supraspinatus tendon demonstrates increased signal intensity, consistent with tendinitis. The bursal surface of the tendon is ill defined due to irregular loss of peribursal fat signal, an indication of chronic bursitis and tendinitis, and superficial tendon degeneration.

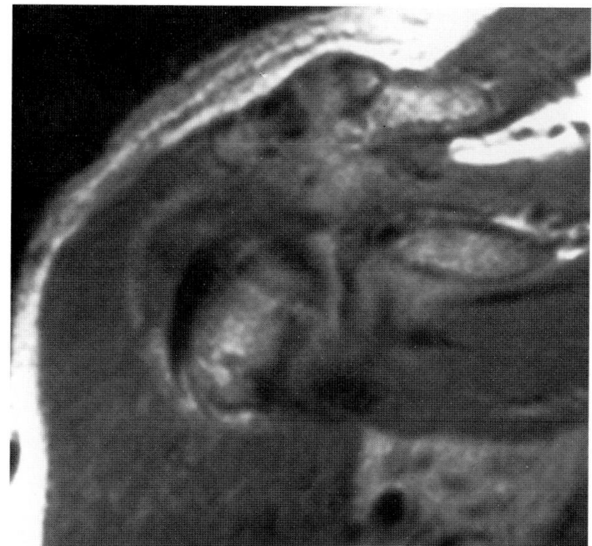

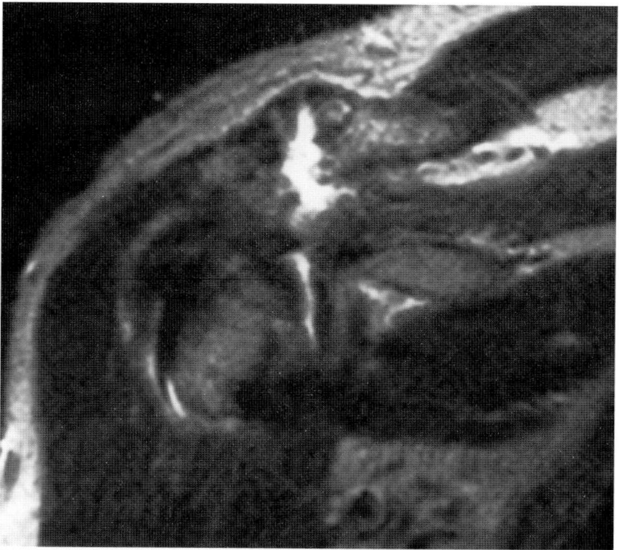

FIG. 7. G,H: Proton-density (**G**) and T2-weighted (**H**) (TR/TE 2,100/30/80 msec) oblique coronal image, right shoulder. There is a medium-sized full-thickness tear of the supraspinatus muscle near its musculotendinous junction communicating with the acromioclavicular joint, due to destruction of the inferior capsular ligament.

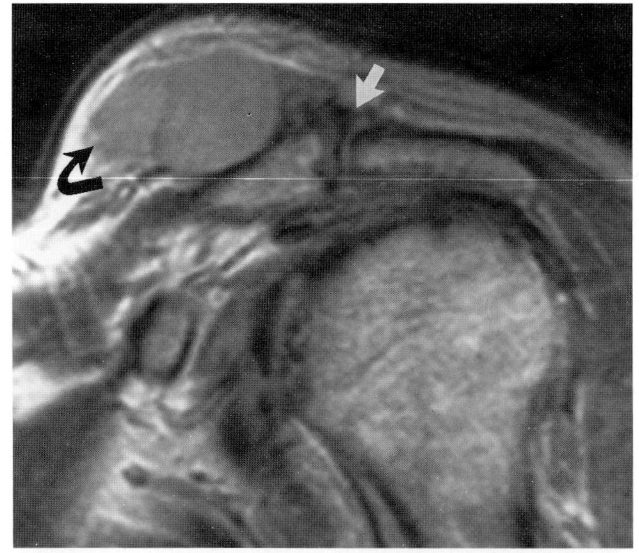

I

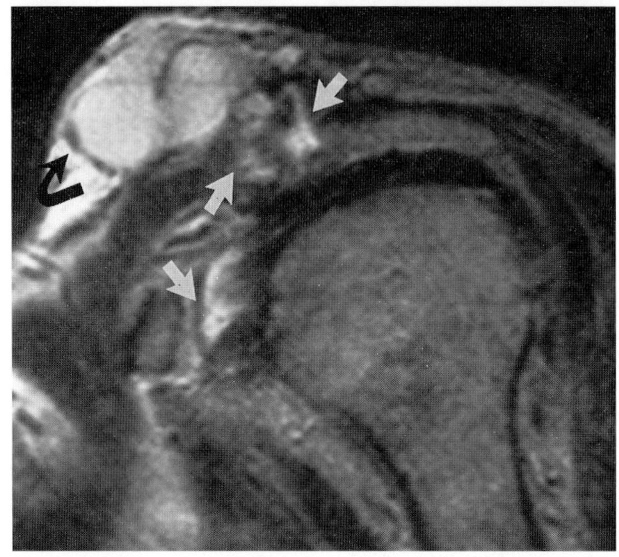

J

FIG. 7. I,J: Para-articular synovial cyst of the acromioclavicular joint. **I:** Oblique sagittal proton-density (TR/TE 1,800/30 msec) image. A multiloculated cyst with signal characteristics suggestive of fluid is visualized above and anterior to the distal clavicle (*black arrow*) and the AC joint (*white arrow*). There is apparent soft tissue continuity between the cyst wall and the AC joint. **J:** Oblique sagittal, T2-weighted image lateral to A (at the level of the acromion process). The high-intensity fluid-filled cyst (*black arrow*) as well as other smaller collections anterior to the acromion (and AC joint) and within the glenohumeral joint (*white arrows*) is visualized. The humeral head is moderately arthritic. The origin of the cyst from the glenohumeral joint and extending through the AC joint was confirmed at surgery. (From Rafii M. Shoulder. In: Firooznia H, et al., eds. *MRI and CT of the musculoskeletal system.* St. Louis: Mosby Yearbook Publishers; 1992, with permission.)

CASE 8

History: A 47-year-old amateur competitive basketball player was injured during a game as a result of an awkward throw of the ball. He felt a "snap" in the shoulder. In spite of pain and some weakness of the arm, he continued playing the tournament while taking antiinflammatory drugs. These symptoms persisted, and he consulted his physician.

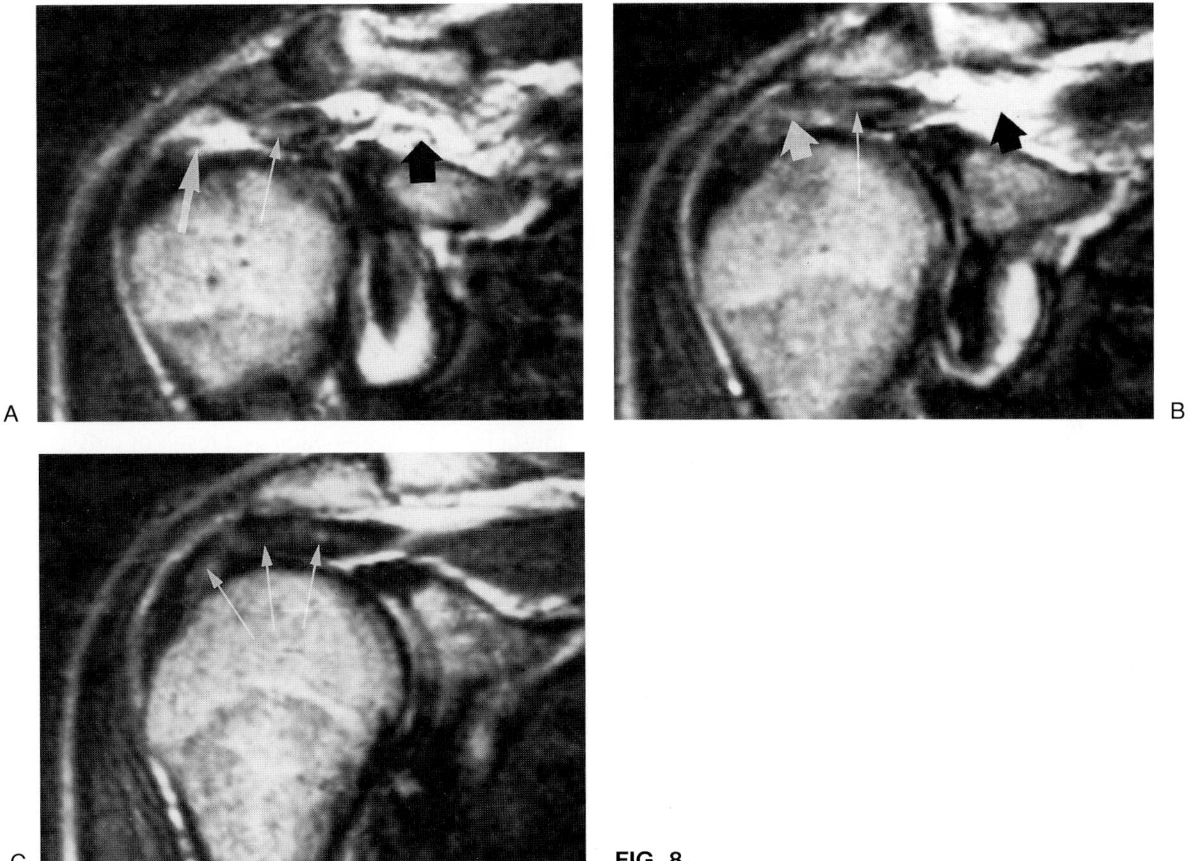

FIG. 8.

Findings: Fig. 8A, a T2-weighted (TR/TE 2,100/80 msec) oblique coronal image of the right shoulder, demonstrates discontinuity of the anterior supraspinatus tendon (central zone) manifested by intense signal indicative of a full-thickness tear. The longitudinal dimension of the tear is approximately 1 cm. An attenuated distal stump is visualized (thick arrow). The proximal margin of the defect is thickened and shows a horizontal intrasubstance band of increased signal intensity (thin arrow). There is fluid within the subacromial and subdeltoid bursae and the glenohumeral joint. Note the apparent discontinuity of the musculotendinous junction (black arrow). Fig. 8B is the same sequence and immediately posterior to Fig. 8A. The posterior margin of the cuff defect (short arrow) shows marked attenuation and irregularity at the synovial surface. There is a persistent band of high signal within the remaining tendon (thin arrow) and the apparent discontinuity of the musculotendinous junction (black arrow). Fig. 8C is the same sequence, posterior to Fig. 8A. At this level the cuff appears continuous, but there is abnormal heterogeneous signal intensity within the substance of the tendon and along the synovial surface (arrows). At this level the musculotendinous junction is normally visualized. The supraspinatus muscle is moderately atrophied.

Diagnosis: Subacute full-thickness tear of the central zone of the supraspinatus, degeneration of the torn margin, and intrasubstance partial tear of the remaining tendon.

Discussion: This case is a characteristic full-thickness rotator cuff tear that involves the central zone of the supraspinatus tendon. Although a traumatic incident clearly indicates the onset of tear, the patient's age, the location of the defect, the history of athletic participation, and finally the trivial nature of the injury all conform to the scenario of failure of an already degenerative cuff. The cuff defect, however, is of recent onset, which along with continuing physical activity on the patient's part resulted in dramatic changes observed within the remaining tendon and the bursal cavity. The intrasubstance high-signal band of the remaining tendon is an indication of an intrasubstance partial tear proximal and posterior to the cuff defect. Overall thickening of the torn tendon in this patient and similar cases is due to cellular proliferation and attempted healing response observed in early stages of a cuff tear. Furthermore, surrounding a cuff tear there may be characteristic signal changes indicative of a preexisting tendinitis (2). The severity of synovial and bursal inflammatory response in patients with rotator cuff tear is variable with time and perhaps the degree of physical activity.

In time, gradual attenuation of the margin of the torn tendon occurs as a result of impingement by the elevated humeral head (Fig. 8D). In chronic tears, the quality of the remaining tendon may vary from significant degeneration to normal hypointense tendon. Frequently, the remaining tendon demonstrates various degrees of degeneration or partial tear (Fig. 8E).

The size and extent of the cuff defect and the nature of the remaining tendon are valuable in surgical planning for treatment of a cuff tear and are all accurately depicted by MR imaging (2–4). The presence of a full-thickness rotator cuff tear is considered an indication for surgical repair (5). Some patients may respond well to conservative treatment, however, or simply arthroscopic decompression may alleviate symptoms of impingement (6). This approach, however, was reported to be adequate in less than 50% of patients in one group (7). Patients described in this chapter were all treated without a cuff repair. The patient in Case 1 underwent arthroscopic subacromial decompression and vigorous rehabilitation of the cuff. In less than 1 year the patient was pain free and had gained complete range of motion. MR imaging at this time demonstrated marked improvement of all abnormalities seen on the original study (Fig. 8F,G). This anecdotal case certainly supports the philosophy of treating patients with small full-thickness cuff tears without a cuff repair and, if needed, with acromioplasty alone. Rotator cuff tears are frequently discovered in cadaver studies. In addition, various pathologic changes of the rotator cuff, including partial- and full-thickness tears, are visualized by MR imaging in asymptomatic individuals (8). In our experience, however, upon questioning these patients may remember past episodes of shoulder pain and disability.

Figure 8A,B demonstrates an apparent discontinuity of the supraspinatus musculotendinous junction. This finding, on oblique coronal images depicting the anterior margin of the supraspinatus, is due to an indentation of the musculotendinous junction as a result of cuff atrophy and may be mistaken for a tear (9,10). Indentation of the anterior margin of the supraspinatus is usually well visualized on axial and oblique sagittal images and is further characterized by signal intensity of fibrofatty tissue. In the images shown here, however, there is evidence of edema in this region due to seepage of bursal fluid into the surrounding tissues.

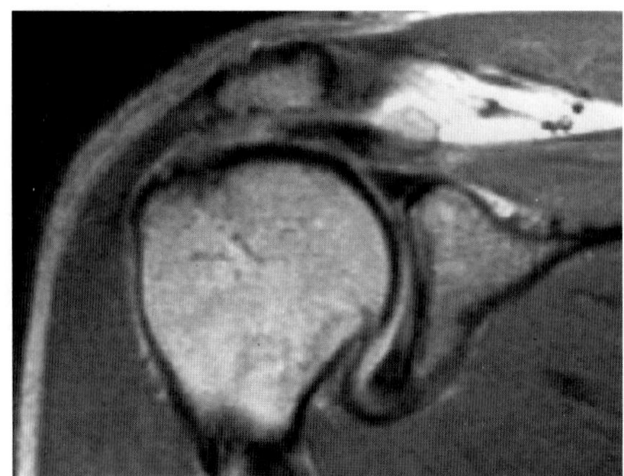

FIG. 8. D: Proton-density (TR/TE 2,100/30 msec) oblique coronal image of the right shoulder of a 48-year-old recreational tennis player with several months of pain and weakness of the arm. There is a moderate-sized full-thickness tear of the supraspinatus tendon. The remaining tendon is wedged between the high-riding humeral head and acromion and shows an attenuated and irregular margin. The supraspinatus muscle shows moderate atrophy.

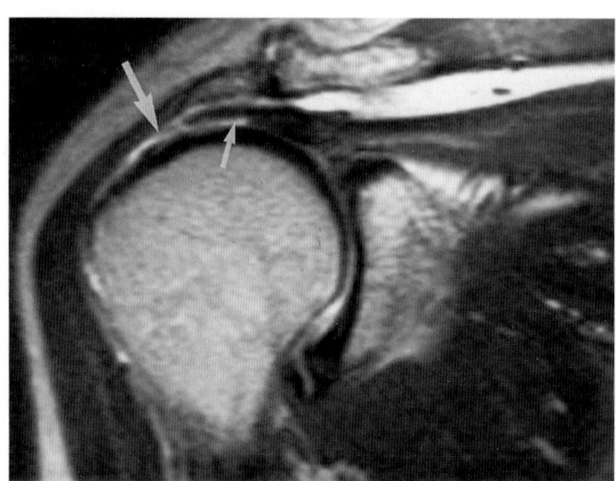

FIG. 8. E: T2-weighted (TR/TE 2,000/90 msec) oblique coronal image of the right shoulder in an elderly woman with long-standing shoulder pain and disability. There is medium-sized full-thickness tear of the supraspinatus tendon (*large arrow*) and a partial tear of the synovial surface of the remaining tendon (*small arrow*).

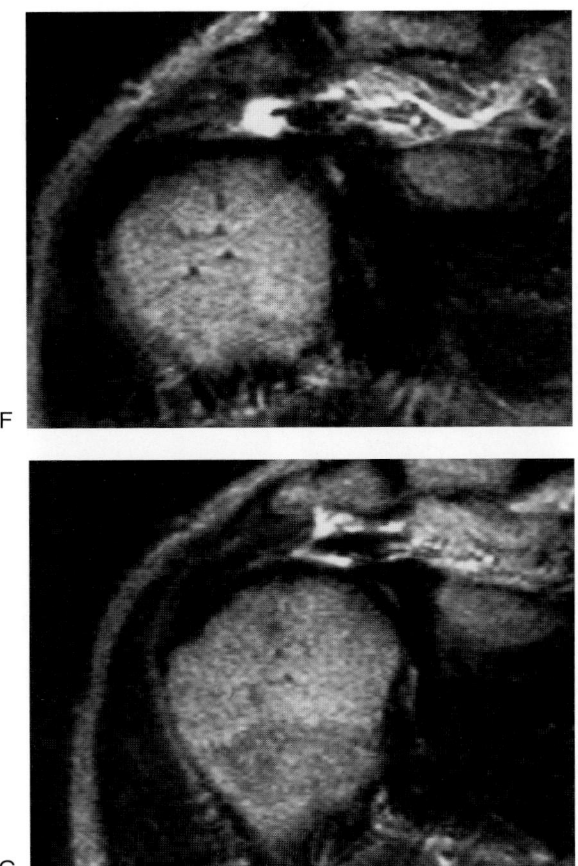

FIG. 8. F,G: Same patient as in Fig. 8A–C. T2-weighted (TR/TE 2,100/80 msec) oblique coronal images at the anterior aspect of the right shoulder, approximately the same level as in Fig. 8A,B. One year following arthroscopic acromioplasty and rehabilitation of the shoulder, the patient is asymptomatic with full range of motion. These images demonstrate complete resolution of inflammatory reaction of the glenohumeral joint and bursal cavities. The remaining supraspinatus tendon has reverted to normal width, and the longitudinal intrasubstance tear is no longer visualized. The overall size of the full-thickness tear is somewhat smaller, mainly due to formation of scar tissue.

CASE 9

History: This 37-year-old woman presented with a history of recurrent pain of both shoulders, worse on the right side. She had been on the varsity volleyball team in high school. On physical examination, impingement sign was positive. Also, a moderate degree of glenohumeral joint laxity was observed.

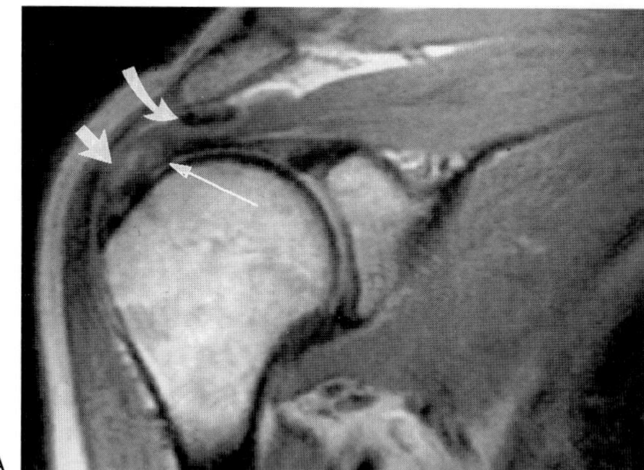
A

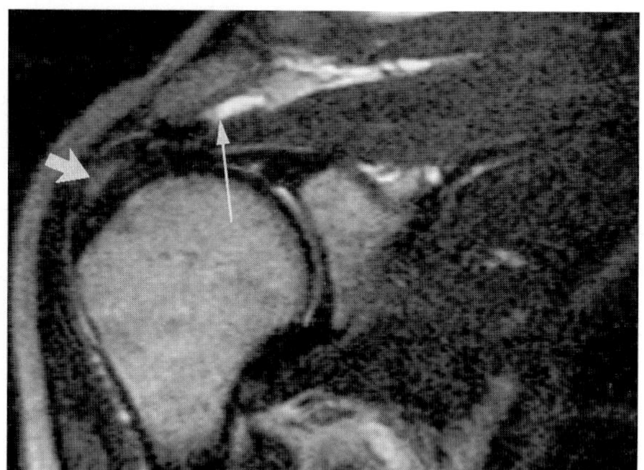
B

FIG. 9.

Findings: Fig. 9A is a proton-density (TR/TE 2,100/30 msec) oblique coronal image of the right shoulder demonstrating an irregular area of increased signal intensity of the midsubstance and bursal surface involving the distal aspect of the supraspinatus tendon, which is also enlarged (short arrow). There is a downsloping acromion with a prominent spur impinging on the supraspinatus tendon (curved arrow). The peribursal fat signal is decreased. The synovial surface of the tendon is well defined (long arrow). Fig. 9B, a T2-weighted (TR/TE 2,100/80 msec) image at the same level, shows persistent intermediate signal and distortion of the bursal surface with morphology indicative of a moderate-size partial cuff tear (short arrow). The subdeltoid peribursal fat is diminished. There is a small amount of fluid in the subacromial component of the bursa medial to the acromial spur (thin arrow). Axial images (not shown) did not reveal capsulolabral pathology.

Diagnosis: Moderate-size partial-thickness tear of the bursal surface of the supraspinatus tendon. Rotator cuff impingement with downsloping acromion and large acromial spur. Clinical glenohumeral stability.

Discussion: This patient presented with clinicalmanifestations and MR imaging findings of late second-stage impingement syndrome. At arthroscopy, the synovial surface was intact. Arthroscopic subacromial decompression and debridement of the bursal surface tear were performed. After an initial postoperative period of improvement, there was recurrence of symptoms after an injury to the shoulder. Repeat arthroscopy demonstrated progression of cuff pathology to a full-thickness tear. At this time, examination under anesthesia revealed rather marked global instability of the glenohumeral joint. A capsular shift procedure was performed, and the cuff defect was repaired.

Glenohumeral instability, either traumatic or due to congenital laxity of the capsular complex, is a known underlying etiology in impingement syndrome and appears to have been the original factor in producing cuff pathology in this patient (1,2). Downsloping of the acromion process has also been recognized as a contributing factor in evolution of impingement syndrome. Acromial spurs, however, are likely to represent a secondary manifestation of bursal surface or full-thickness tears and an additional cause of supraspinatus outlet obstruction. Ozaki et al., (3), in a study of 200 cadaver shoulders, observed various osseous changes of the acromial undersurface only when a partial-thickness tear of the bursal surface or a full-thickness tear of the rotator cuff was present. Synovial surface tears in their observations were not associated with acromial pathology.

Synovial surface partial tears most frequently develop in the form of gradual separation of deep tendon fibers from the greater tuberosity (4). Tensile forces may also result in tearing of the synovial surface of the tendon in the central zone (Fig. 9C–F). This patient had, in common with the patient in Case 1, manifestations of glenohumeral laxity on physical examination, which was confirmed by examination under general anesthesia. Both patients had normal capsulolabral complexes on conventional MR imaging and on arthroscopy, consistent with atraumatic or congenital glenohumeral instability. Clinical manifestations of impingement and partial rotator cuff tear may also accompany traumatic instabilities. Without clinically observed flagrant instability, the true etiology of cuff pathology may be overlooked.

Another patient, a 16-year-old high school football player, presented with dull pain and weakness during overhead movements of the arm. His impingement test was positive. No definite glenohumeral instability was detected on physical examination. A period of conservative treatment and physical therapy failed to relieve his symptoms. MR imaging demonstrated, in addition to the cuff pathology (Fig. 9G), a partial Bankart lesion. Subsequently, examination under anesthesia confirmed glenohumeral instability in this patient. The clinical presentation in these patients is often confusing because symptoms of pain, weakness, and dysfunction due to instability, secondary impingement of the cuff, and overuse of the shoulder may overlap. MR imaging is shown to be useful in evaluation of these patients because of its ability to image the rotator cuff and the capsulo-labral complex simultaneously (5).

Anterior instability is also suggested to contribute to the development of posterosuperior glenoid impingement in the throwing athlete (see Fig. 5A). Abnormalities of the undersurface of the cuff, the posterosuperior labrum, and the posterosuperior aspect of the humeral head are observed as a result of repeated impingement of the cuff and the humeral head against the posterosuperior glenoid.

Cases presented in this chapter reflect the wide morphologic and anatomic variations of partial tears as depicted by MR imaging. There is unanimity among investigators about the lesser accuracy of MR imaging in diagnosis of partial tears compared with its consistently greater accuracy in diagnosing full-thickness tears (6–11). The discrepancy is multifactoral. The most specific criterion, a region of high signal intensity on T2-weighted images in a cuff defect, in our experience is observed only in approximately 50% of partial tears (8). Morphologic abnormality of the tendon in the form of contour irregularity or contour defect observed on MR images is helpful in recognition of synovial or bursal surface tears, as demonstrated in Cases 1 and 2. The use of fat suppression sequences has been suggested to improve MR accuracy in the diagnosis of partial tears (10,12).

The treatment of partial tears is controversial (13). It has been suggested that the detection of partial tears by MR imaging is insignificant in treatment planning for these patients (7). Most patients in the late second stage of impingement syndrome respond to acromioplasty when conservative methods of treatment fail. A large partial tear, however, involving most of the width of the tendon, may require debridement or even a cuff repair because of the increased probability of progression to a full-thickness tear, even after a subacromial decompression procedure has been performed. The MR imaging diagnosis of these large or deep partial tears is less problematic when both signal intensity and morphologic alterations are used as criteria in diagnosis (see Fig. 8A,B), although differentiation from a small full-thickness tear is not always possible (7,8,11).

In addition to fat suppression sequences, MR arthrography has been advocated for improved visualization of synovial surface partial-thickness and small full-thickness tears (14–16). This technique fails to improve visualization of intrasubstance or bursal surface tears, however. MR arthrography with an abducted position of the arm is shown to increase sensitivity in detection of partial tears of the rotator cuff and other intraarticular abnormalities in patients with posterosuperior glenoid impingement (17).

C

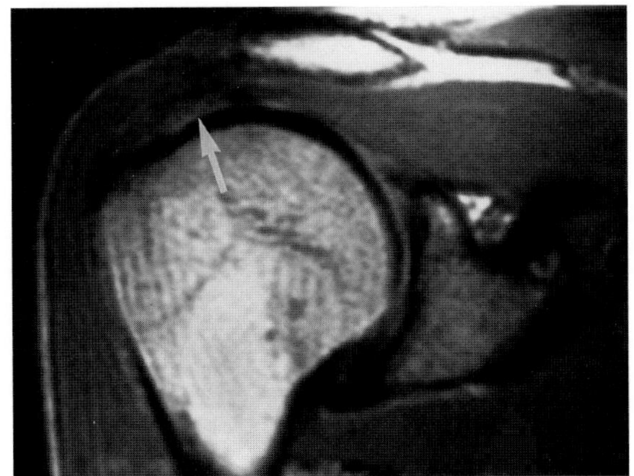

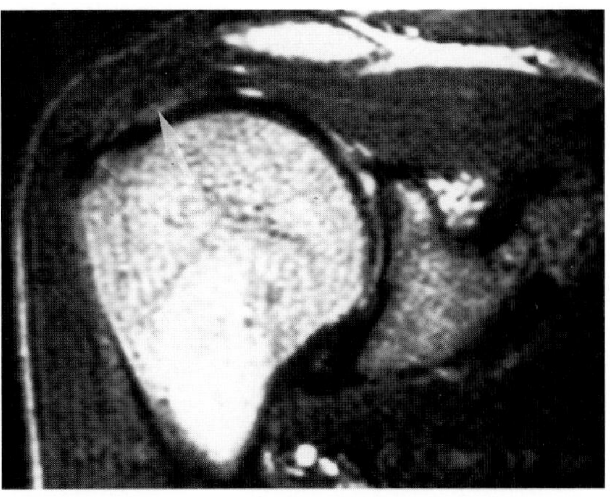

D

FIG. 9. C–F: Partial rotator cuff tear. Proton-density (**C**) and T2-weighted (**D**) (TR/TE 2,100/30/80 msec) oblique coronal images, right shoulder. There is heterogeneous increased signal intensity of the distal supraspinatus muscle, which is more pronounced at the synovial surface and shows further increase on T2-weighted images indicative of partial synovial surface tear. The peribursal fat is totally obliterated on both images, indicative of bursal scarring.

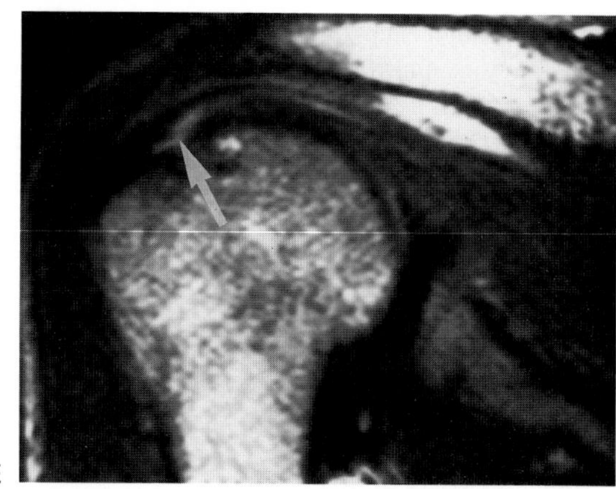

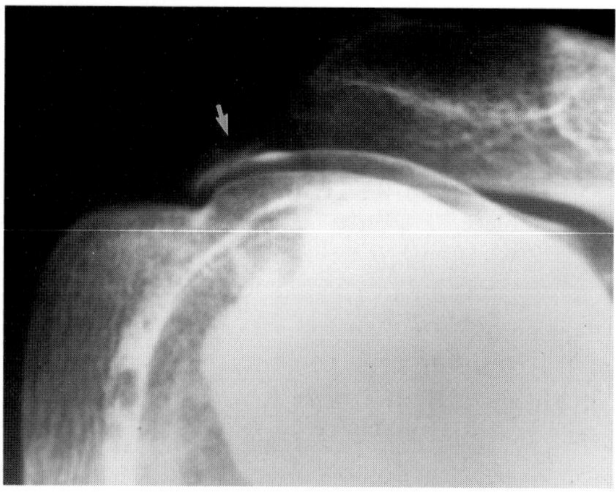

FIG. 9. E: Same sequence and posterior to **D.** Partial tear of the synovial surface and focal detachment from the greater tuberosity are manifested by a bright linear zone of increased signal intensity (*arrow*). **F:** Conventional arthrogram shows linear tears at the synovial surface of the supraspinatus muscle (*arrow*) corresponding to the abnormality on MR images. However, MR images accurately display extensive cuff pathology with diffuse abnormal signal and bursal thickening. Also note slight superior migration of the humeral head (C – E) indicative of a weakened tendon. At surgery, extensive debridement of the cuff necessitated cuff repair. Flagrant instability found under general anesthesia was treated with a capsular shift.

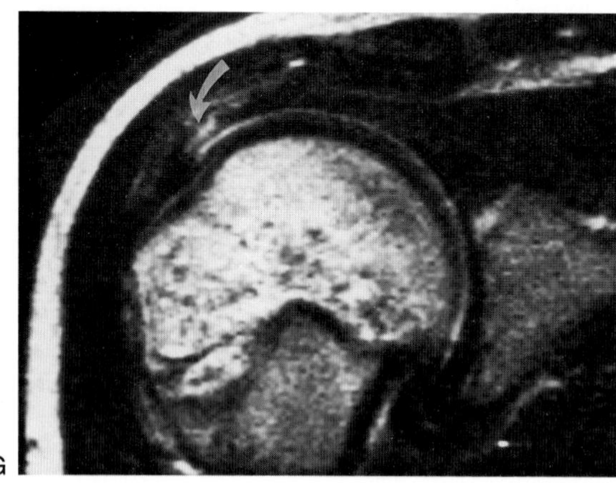

FIG. 9. G: Partial intrasubstance cuff tear. T2-weighted (TR/TE 2,000/90 msec) oblique coronal image, right shoulder. Tear is manifested by an irregular band of increased signal intensity with the tendon (*arrow*).

CASE 10

History: A 48-year-old male recreational tennis player presented with increasing pain and weakness of the right shoulder. He had a long history of recurrent shoulder soreness. The current episode began 6 months ago during a game, when he felt a "snap" in the shoulder when hitting the ball hard. He had been unresponsive to physical therapy. On physical examination he had limited abduction and elevation of the arm.

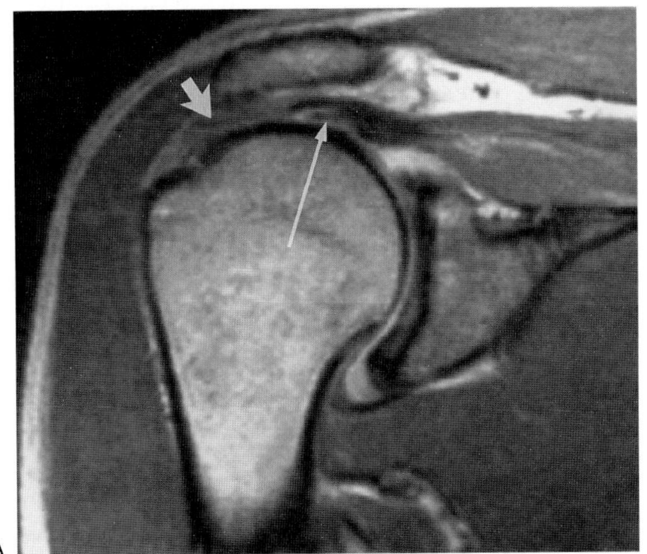
A

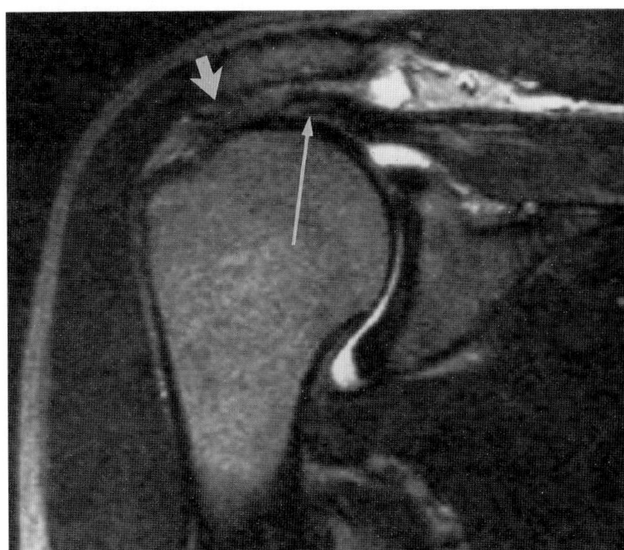
B

FIG. 10.

Findings: Fig. 10A,B shows proton-density and T2-weighted (TR/TE 2,100/30/80 msec) oblique coronal images, respectively, of the right shoulder at the midglenohumeral joint level. There is a medium-size defect of the supraspinatus tendon, manifested by medium signal intensity (short arrow in Fig. 10A) on the proton-density image and relatively low signal intensity on the T2-weighted image (short arrow Fig. 10B). The torn margin of the tendon is irregular and attenuated on both images (long arrows). The musculotendinous junction is retracted to the level of the glenohumeral joint. The peribursal fat signal is not identified on either image and is isointense with the cuff defect. There is, however, a small fluid collection in the medial extension of the subacromial bursa. There is a small joint effusion and mild supraspinatus atrophy.

Diagnosis: Moderate-size full-thickness tear of the supraspinatus tendon with retracted musculotendinous junction and distention of the subcoracoid bursa.

Discussion: The most specific, accurate, and common criterion in diagnosis of a cuff tear is visualization of a defect characterized by intermediate signal intensity on proton-density images and high or intense signal on T2-weighted images (1–5). Although these findings are present in the majority of full-thickness tears, at times there is difficulty in distinguishing small full-thickness tears from partial tears or even tendinitis. The so-called low-signal pattern of cuff tears is due to obliteration of all or part of the cuff defect by scar and granulation tissue or degenerated remnants of the torn tendon (4). Occasionally, medium or large cuff defects may also be obliterated by scar or granulation tissue (Fig. 10A,B). Previous surgical procedures may predispose to excessive scar formation. Alterations in tendon morphology, along with lesser degrees of increased signal within the cuff defect, help in recognition of some of these cuff tears. Also, a number of secondary signs are recognized that may be helpful in the diagnosis and characterization of rotator cuff tears.

Alterations in the peribursal fat signal and, more specifically, fluid in the subacromial–subdeltoid bursa observed on T2-weighted images were used as secondary criteria in the diagnosis of rotator cuff tears (6) (Fig. 10A,B). Bursal fluid, however, was subsequently recognized in association with tendinitis and, in small amounts, even in asymptomatic individuals (4,5,7). Obliteration of the peribursal fat signal on short TR/TE and long TR/TE images due to scar is also often observed in chronic cuff tears as well as tendinitis (6). Although variations in the peribursal fat signal have also been reported in asymptomatic individuals (7), a thickened and low-signal bursa may be distinguished from smoothly diminished peribursal fat (see Case 1). Furthermore, variations of the peribursal fat are often observed in muscular and athletic individuals with small amounts of subcutaneous fat.

Retraction of the supraspinatus–musculotendinous junction is another secondary sign observed with full-thickness rotator cuff tears. On oblique coronal images, the supraspinatus–musculotendinous junction is normally visualized above the center of the humeral head or roughly at the peripheral margin of the acromion process (5,8,9). Variations in positioning of the musculotendinous junction of up to 15° lateral or medial to the center of the humeral head are observed in normal tendons (8). These reported normal variations on MR imaging of the supraspinatus–musculotendinous junction are consistent with the observations of Skinner, who described the rotator cuff as more muscular in young individuals, with transformation of muscle to tendon at the supraspinatus–musculotendinous junction from childhood to adult life (9). Retraction of the musculotendinous junction is observed in full-thickness tears with increasing frequency and severity relative to the size of cuff tear (1,6). Retraction is usually either not present or difficult to ascertain in partial and small tears but is always present in complete ruptures involving the entire anteroposterior dimension of the supraspinatus tendon (4). Retraction of the musculotendinous junction as a result of acute tears may give rise to a globular appearance of supraspinatus muscle (6).

Atrophy of the rotator muscles in general, and of the supraspinatus and infraspinatus in particular, is frequently observed in association with cuff tears as well as cuff degeneration. Atrophy is manifested by the appearance of fatty streaks within the muscle bundles, best seen on T1-weighted images, as well as by generalized loss of muscle mass. The severity of these changes is dependent on the extent and chronicity of the cuff lesions.

Farley et al. (5) reviewed the primary and secondary signs of rotator cuff tears in a group of patients with rotator cuff tears. The primary sign (i.e., tendon defect with fluid signal) was found to be the most accurate criterion; bursal fluid was the most sensitive, though least specific, finding. Supraspinatus muscle atrophy was the most specific finding. According to this review, supraspinatus muscle atrophy and retraction of the musculotendinous junction had a statistically significant association with a tear of 2 cm or larger. It is therefore clear that the secondary signs have limited value in the diagnosis and differentiation of small full-thickness and partial cuff tears. Quantification of muscle atrophy and degree of retraction of the musculotendinous junction, however, along with other variables such as the size of the cuff defect and the nature of the remaining tendon, are valuable findings afforded by MR imaging in the process of treatment, particularly in surgical planning for cuff repair (3). Consistently favorable operative results are achieved with smaller tears and lesser degrees of muscle atrophy and retraction (10).

C

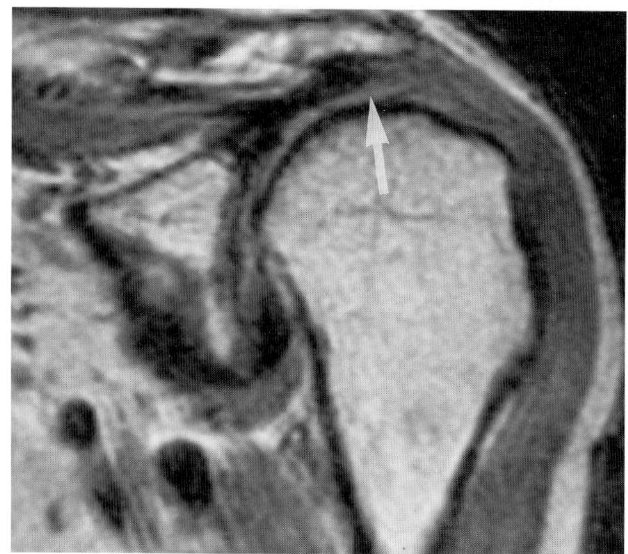

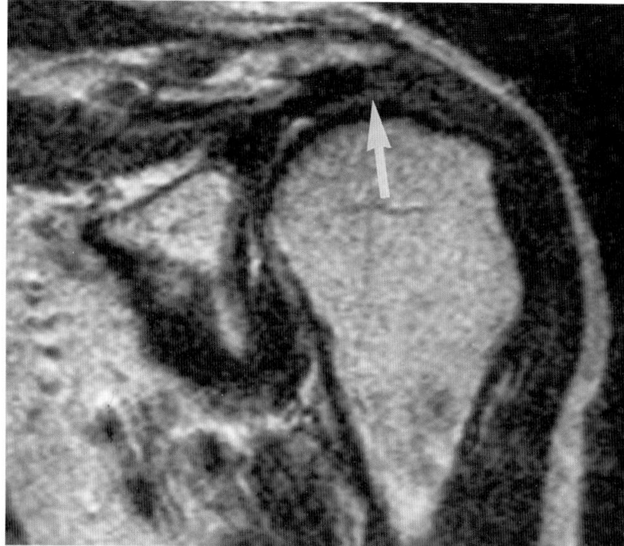

D

FIG. 10. C,D: Proton-density (**F**) and T2-weighted (**G**) (TR/TE 2,100/30/80 msec) oblique coronal images of the left shoulder at mid-glenohumeral level in a patient with prior rotator cuff repair with recurrence of symptoms after an injury. The supraspinatus tendon terminates at the level of center of the humeral head and beneath the acromion process (*arrows*). The cuff defect is replaced by medium signal intensity on proton density and relatively low signal tissue on T2-weighted image. The musculotendinous junction is retracted to the level of the glenoid margin. The appearance is that of a large cuff tear obliterated by scar tissue, which was proved at surgery. Note deformity of the humeral head and greater tuberosity, the result of previous fracture.

CASE 11

History: This 70-year-old woman was first evaluated for right shoulder pain in 1979 and had a history of previous injury. An arthrogram had revealed a rotator cuff tear, although the size of the cuff defect was not reported. She refused surgery and was treated conservatively with good results for a period of 10 years, when weakness of the arm and pain became progressively worse. On physical examination she had painful passive range of motion in abduction and elevation.

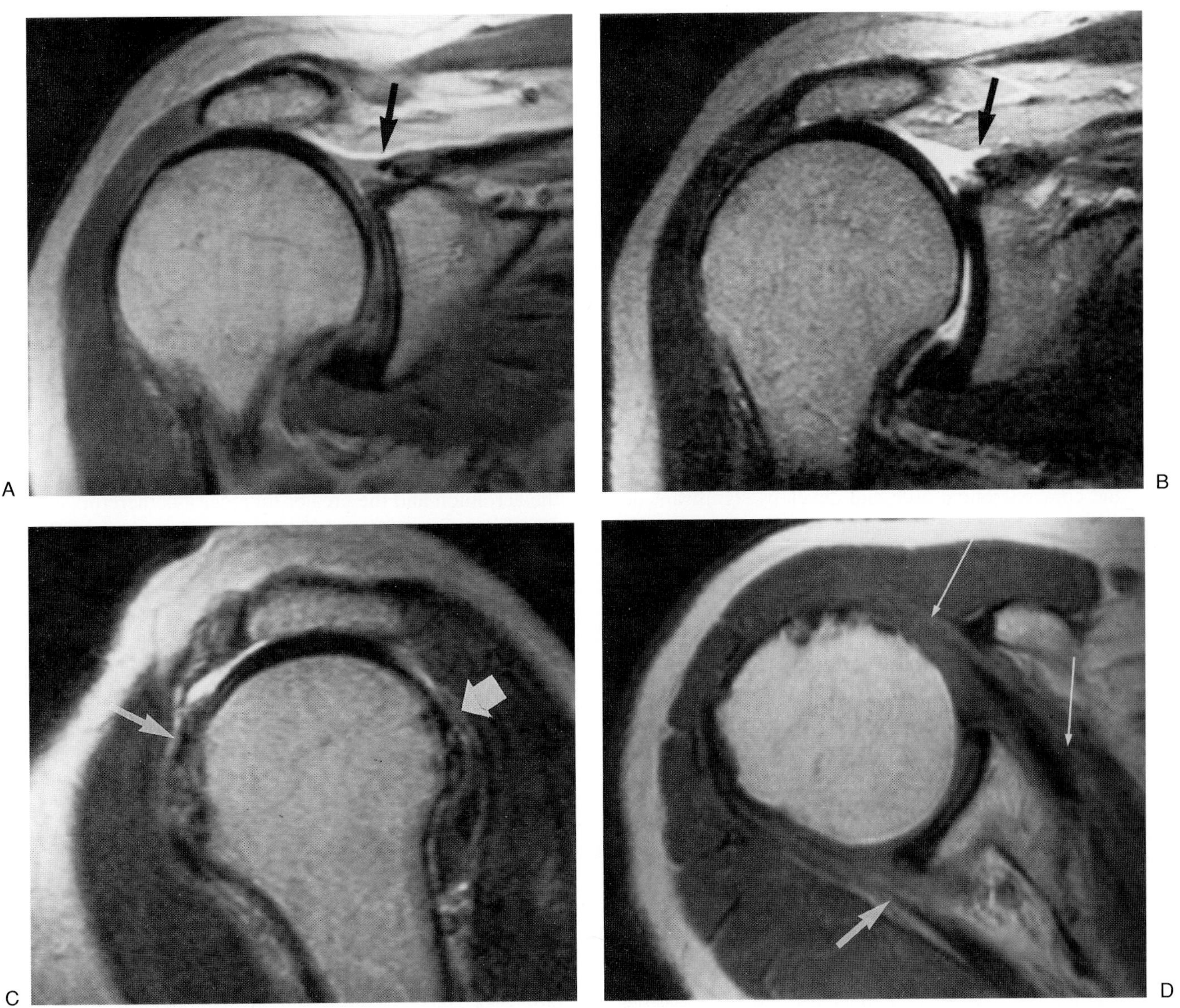

FIG. 11.

Findings: Fig. 11A,B shows proton-density and T2-weighted (TR/TE 2,100/30/80 msec) oblique coronal images, respectively, of the right shoulder demonstrating a massive tear of the supraspinatus tendon. The moderately degenerated remnant of the tendon and musculotendinous junction are retracted medially (arrow). There is moderate atrophy of the supraspinatus muscle. Also, there is superior migration of the humeral head under the acromion and significant flattening of the greater tuberosity. Note the lack of hypertrophic changes of the acromion. Fig. 11C is a T2-weighted (TR/TE 1,800/80 msec) oblique sagittal image medial to the lesser tuberosity. The humeral head is almost entirely uncovered. There is a residual band of the anterior component of the supraspinatus tendon, which has slipped inferiorly in continuity with the rotator interval (thin arrow). The rotator interval is thickened. A tear of the infraspinatus tendon is demonstrated on this image (thick arrow). The long head of the biceps tendon, visualized on other images (not shown) was degenerated and partially torn. Fig. 11D, a T1-weighted (TR/TE 1,000/29 msec) axial image, shows degeneration and tear of the upper aspect of the subscapularis tendon and medial retraction of its musculotendinous junction (thin arrows). Tear and retraction of the infraspinatus tendon are also identified (thick arrow). Note the irregularity and hypertrophic changes in the vicinity of the lesser tuberosity and the bicipital groove.

Diagnosis: Massive tear of the supraspinatus tendon and the adjacent components of the infraspinatus and subscapularis tendons.

Discussion: This patient's history reflects the progressive nature of rotator cuff tears over a period of 15 years. Although the size of the (full-thickness) tear was not originally demonstrated by the arthrogram, it is likely that it involved only a portion of the supraspinatus tendon. This is suggested by the favorable response to conservative therapy. Supraspinatus tears, however, tend to propagate medially toward the musculotendinous junction and posteriorly to involve the upper aspect of the infraspinatus tendon (1). The superior aspect of the infraspinatus and teres minor may be involved also. An associated tear of the subscapularis tendon is uncommon (1). At this stage, the intracapsular segment of the long head of the biceps tendon is increasingly subjected to repeated impingement by the coracoacromial arch and develops significant degeneration and partial or complete tears. Occasionally, however, the long head of the biceps may remain intact and become hypertrophied.

As the cuff defect expands, the functional balance of the glenohumeral joint is increasingly diminished. The supraspinatus is the primary elevator of the arm (up to 90°) and also helps in holding the humeral head against the glenoid fossa. Tears of the supraspinatus tendon result in weakness or complete loss of range of motion in abduction and elevation. Furthermore, muscle imbalance results in gradual elevation of the humeral head by the normal deltoid as rotator cuff function is diminished. Tear and degeneration of the cuff also remove the barrier against upward displacement of the humeral head. At this stage, a secondary mechanical impingement develops as a result of approximation of the humeral head and the greater tuberosity to the undersurface of the acromion through the cuff defect. This painful impingement results in osseous changes observed in advanced cuff tears. There is gradual erosion of the undersurface of the acromion and preexisting osteophytes, resulting in a concave surface conforming to the contour of the humeral head. The thickened bursal tissue also gradually atrophies, although this is not a rule, and in some cases massive scar formation may be present around the torn tendon. The greater tuberosity of the humeral head is also gradually flattened as a result of impingement against the acromial undersurface (Fig. 11A,B). After superior migration of the humeral head in large or massive tears, the remaining anterior and posterior components of the cuff slip inferiorly below the equator of the humeral head. With the development of fibrosis, the head may be fixed in superior subluxation. This phenomenon may prevent reduction of the head at surgery (2). The sagittal images in our patient clearly portray this phenomenon and demonstrate the abnormal alignment of the remaining cuff tendon relative to the humeral head as well as thickening and fibrotic changes of the rotator interval region (Fig. 11C). At this time the patient underwent cuff repair due to persistent and severe of pain. At surgery, the defect was found to be primarily a longitudinal split. The displaced anterior band was identified and the defect was repaired. However, as is often the case with massive and chronic cuff tears, the repair was unsuccessful. A recurrence of the tear was diagnosed by repeat MR imaging several months after surgery when the patient's symptoms recurred during the post-operative physical therapy. This case underscores the importance of early repair of cuff tears before the secondary changes takes place (2–6).

A number of radiographic studies have evaluated the acromiohumeral distance and its relation to rotator cuff tears. A decrease to less than 7 to 5 mm, according to various reports, is considered diagnostic for the presence of a rotator cuff tear. A radiographic study demonstrating superior migration of the head with atrophic changes of the acromion and the greater tuberosity is therefore characteristic of massive and chronic supraspinatus tears. The value of MR imaging at this stage of the disease therefore could be questioned. Nevertheless, in a persistently symptomatic patient, a decision has to be made as to whether surgical intervention would be helpful in alleviating pain or in improving the function of the arm. The interpretation of MR images therefore should take into account not only the presence and the size of the cuff defect but also the degree of retraction and atrophy of the supraspinatus, the condition of the other rotator cuff tendons, and the status of the humeral head and glenoid articular surfaces. Degenerative disease of the glenohumeral joint adversely affects surgical treatment of cuff tears.

As indicated above, abnormal alignment of the glenohumeral joint and superior migration of the humeral head are common findings in advanced and chronic cuff tears. With the loss of the stabilizing function of the rotator cuff, gross instability of the joint may develop and result in a progressive deterioration of the articular surfaces (7). According to Neer et al. (7), cuff tear arthropathy is due to a combination of factors, including inactivity and disuse of the shoulder, leaking of the synovial fluid, and instability of the humeral head. These events result in both nutritional and mechanical factors that cause atrophy of the glenohumeral articular cartilage and osteoporosis of the subchondral bone of the humeral head (7). Incongruity of the glenohumeral joint articular surfaces along with gross instability as a result of massive cuff tear also result in progressive deterioration and marked erosive changes of the articular surfaces (Fig. 11E,F). The humeral head will gradually collapse, and erosion of the glenoid may extend deeply to involve the coracoid process. MR imaging at this stage of the disease is again helpful for evaluation of inflammatory changes of the glenohumeral joint and its recesses as well as for detection of the extent of cuff pathology and a detailed evaluation of osseous changes.

Treatment of patients with cuff arthropathy is extremely difficult. Neer et al. (7) have reported resurfacing arthroplasty with rotator cuff reconstruction with relatively good but limited results, MR imaging is helpful at this stage of the disease in presurgical evaluation of the extent of cuff pathology and the severity of inflammatory reaction and osseous deterioration as well as for differential diagnosis of other arthritides involving the glenohumeral joint that may resemble cuff tear arthropathy.

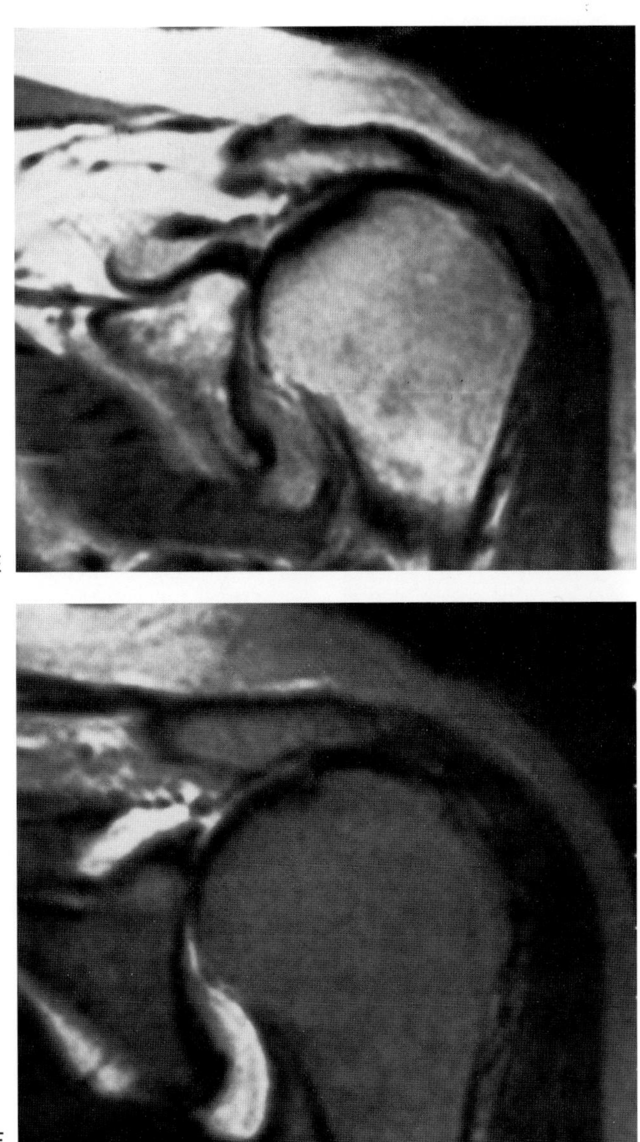

FIG. 11. E,F: Cuff arthroplasty. Proton-density and T2-weighted (TR/TE 2,100/80/30 msec) oblique coronal images demonstrate massive tear with severe retraction and atrophy of the supraspinatus tendon and muscle. There is superior migration of the humeral head. Erosions of the acromion undersurface and humeral head are also visualized.

CASE 12

History: This 45-year-old woman had acute onset of severe right shoulder pain and refused to move her arm. She had had several episodes of pain and discomfort in the past but did not require treatment. A radiographic examination was performed and followed by MR imaging.

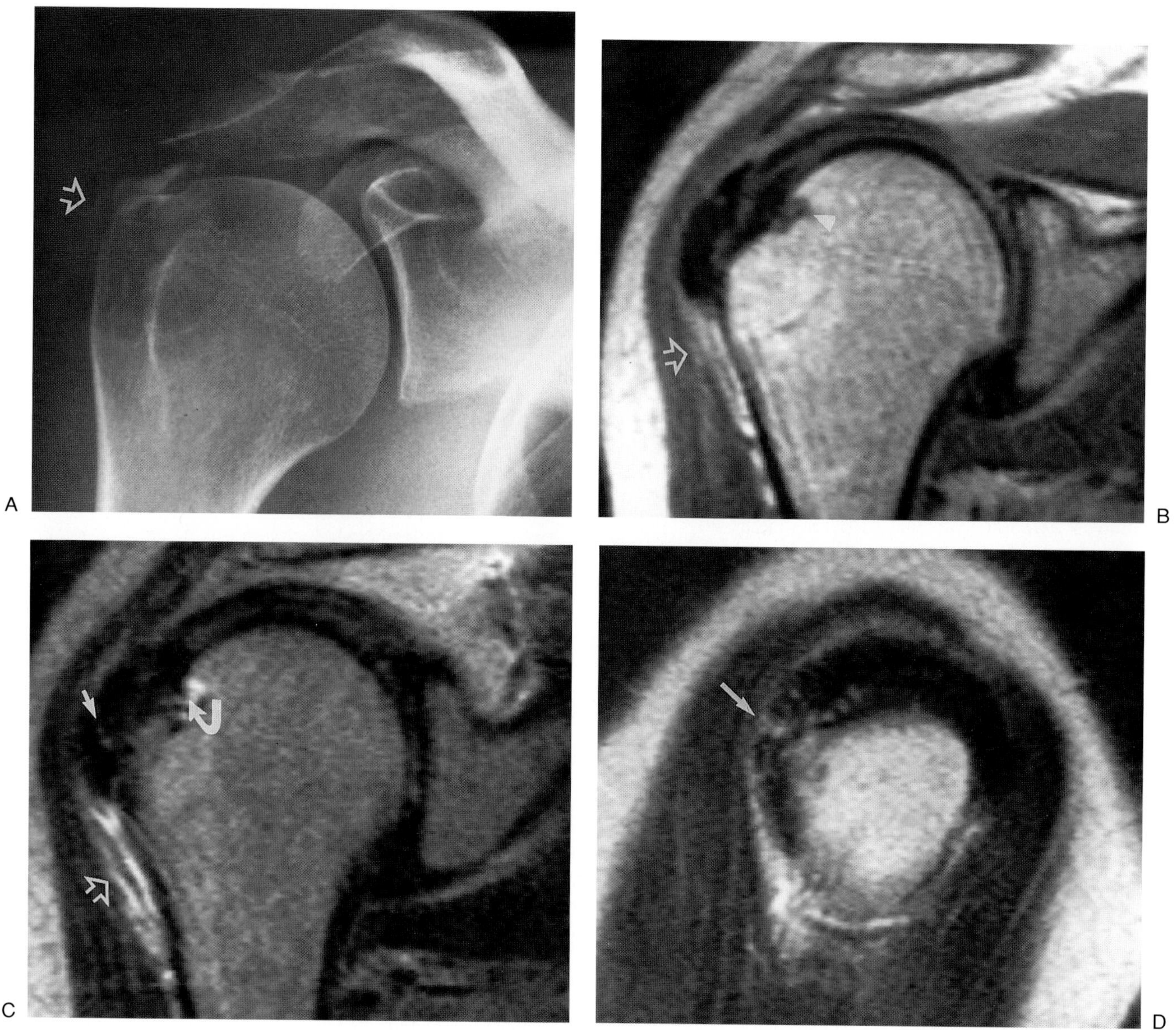

FIG. 12.

Findings: Fig. 12A is a posterior oblique projection of the right shoulder. One large and several small calcific densities are visualized in the subacromial space above the greater tuberosity. An ill-defined calcific density is visualized peripheral to the greater tuberosity (arrow). There is a subcortical cyst in the anatomic neck. Fig. 12B is a proton-density (TR/TE 2,100/29 msec) oblique coronal image of the right shoulder. The right shoulder at the mid-posterior glenohumeral joint level demonstrates a slightly heterogeneous low-signal mass above and slightly over the greater tuberosity, replacing the normal tendon. There is obliteration and elevation of the peribursal fat in the subdeltoid region. Distally, the bursa is expanded (hollow arrow). A subcortical cyst is also present in the anatomic neck (arrowhead). Fig. 12C, a T2-weighted (TR/TE 2,100/80 msec) oblique coronal image posterior to Fig. 12B, demonstrates focal areas of increased signal within the low-signal mass (solid arrow) and distention of the subdeltoid bursa with inflammatory reaction (hollow arrow). The subcortical cyst of the anatomic neck shows bright signal with a focal low-signal structure protruding into it (curved arrow). Fig. 12D, a T2-weighted (TR/TE 1,800/80 msec) oblique sagittal image at the level of the greater tuberosity, demonstrates focal tear of the posterosuperior cuff (straight arrow) and other punctate areas of detachment over the greater tuberosity. Inflammatory reaction within the subdeltoid bursa is demonstrated on this image as well.

Diagnosis: Calcific tendinitis of the posterosuperior rotator cuff with subdeltoid bursal inflammation. Small tear of the cuff and subcortical cyst of the greater tuberosity. Probable protrusion of the calcific deposit within the cyst.

Discussion: Calcific tendinitis of the rotator cuff is a common disorder of uncertain etiology (1). A degenerative etiology as a result of "wear and tear" and hypoxia of the rotator cuff was originally suggested by Codman (2). This is not universally accepted by other investigators, who have cited the varied age group of patients with calcific tendinitis and the natural history of the disorder, which is different from that of other degenerative disorders of the rotator cuff (1,3). Calcific tendinitis has a higher incidence in women, and most cases are reported in the third and fourth decades of life (1,3). In these large series, no cases were found after age 71. No overlap is found with rotator cuff tears, although partial tears of the bursal surface may remain after calcific deposits disappear (2). Nevertheless, calcific tendinitis of the rotator cuff has distributions similar to those of other cuff tendinopathies (i.e., they are most common within the critical zone of the supraspinatus, less frequent in the infraspinatus and teres minor, and rare in the subscapularis).

The evolutionary nature of calcific tendinitis (i.e., deposition of hydroxyapatite crystals within the tendon and resorption later in life) recognized by Codman (2) prompted Uthoff and Kiriti (1) to advocate *calcifying tendinitis* as the more appropriate term for this disorder. They disagreed with the notion of a degenerative etiology and described an active fibrocartilaginous transformation of the tendon, where calcium crystals are then deposited (1). There is no clear explanation as to what triggers this process, however. After a period of inactivity, spontaneous resorption of calcific deposits occurs; at this stage the deposits have a creamy consistency and are surrounded by vascular channels and macrophages (1). This stage is associated with increased pressure within the tendon. Release of calcific deposits and resorption by the bursal surface are also likely to occur (1). After resorption of the calcific focus, granulation tissue is formed, and the defect often heals.

Clinically, patients are asymptomatic or have minimal symptoms during the formative phase (1,2). Large calcific deposits may induce symptoms of secondary impingement syndrome. The chronic stage is followed by an acute stage during the resorptive phase. At this stage, there is elevation of the bursa as the calcific deposit enlarges (4). The symptoms are primarily due to increased intratendinous pressure or, occasionally, bursal inflammation (1).

Conventional radiographic examination readily demonstrates the presence, the location, and even the stage of evolution of calcific tendinitis. In the chronic stage, calcific deposits within the tendon are dense and well defined; in the acute or resorptive stage, they appear cloudy and ill defined or fleecy and may also appear in the bursa (Fig. 12A). MR imaging does not have a primary role in evaluation of calcific tendinitis. Nevertheless, MR imaging is performed for evaluation of the rotator cuff, particularly when symptoms become refractory and surgery is contemplated. On MR imaging, nodular or lobulated deposits of hydroxyapatite crystals are low in signal intensity on all pulse sequences (5). Heterogeneity of signal within the calcified mass and inflammatory reaction within the adjacent bursa are expected findings in the acute phase (Fig. 12B–D). In this stage, low-signal calcific foci may also be visualized within the inflamed bursa (6). A subcortical cyst of the sulcus of the humeral head, resulting from erosion by calcific deposits, may be observed (Fig. 12C) (1). Coexistence of calcific deposits with a cuff tear is uncommon; when present, the deposits are usually small or involve only the bursal surface (Fig. 12D). Heterogeneity of signal intensity of calcific tendinitis on MR imaging is variable. Occasionally marked heterogeneity is present, and calcific deposits appear as linear or multiple foci of low signal within the tendon, forming a cluster (Fig. 12E,F).

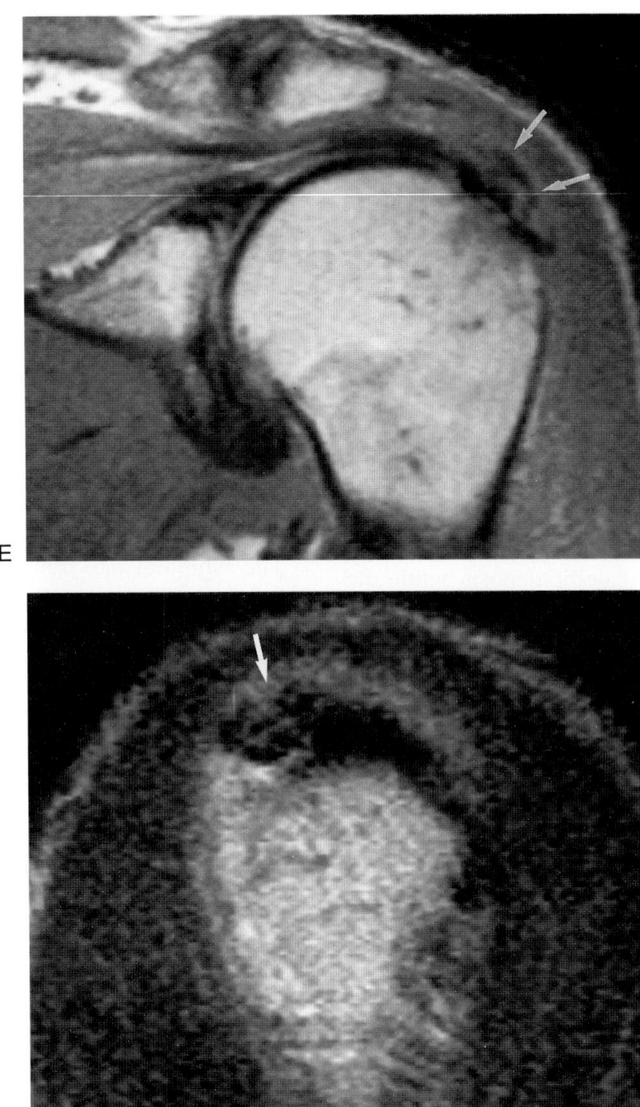

FIG. 12. E,F: E: Proton-density (2,100/30 msec) oblique coronal image of left shoulder. There is heterogeneous low signal appearance involving the bursal aspect of the supraspinatus tendon (*arrows*). **F:** T2-weighted (TR/TE 2,100/80 msec) oblique sagittal image at the level of the greater tuberosity demonstrates a cluster of low signal foci adjacent to the greater tuberosity (*arrow*). There is no inflammatory reaction within the bursa. The peribursal fat signal, as noted on the oblique coronal image, is diminished. This is suggestive of a chronic type of bursitis. This patient with calcific tendinitis had presented with acute symptoms.

CASE 13

History: This 60-year-old woman presented with a long history of ''problems'' with her right shoulder. Her major complaints included painful and restricted range of motion. Pain was more severe at night and at times was localized in the anterior aspect of the shoulder and extended down the arm. External rotation of the arm was also more painful.

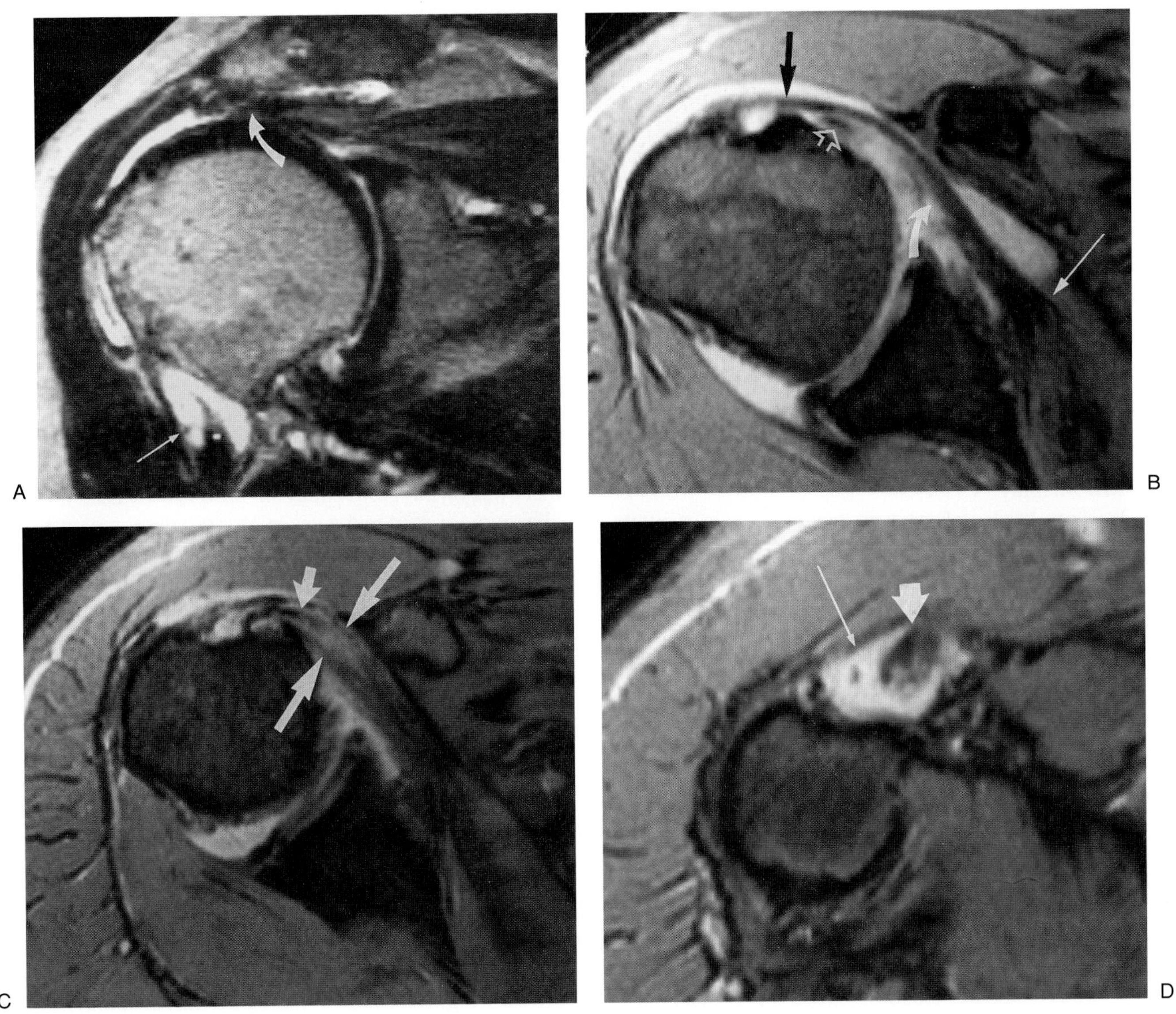

FIG. 13.

Findings: Fig. 13A is a T2-weighted (TR/TE 2,000/90 msec) oblique coronal image of the right shoulder. There is a large supraspinatus tear with the retracted margin visualized below the acromion (curved arrow). Note the acromial spur, bursal fluid, and mild superior migration of the humeral head. Fluid is also present within the bicipital tendon sheath (straight arrow). Fig. 13B is a gradient-echo (TR/TE 896/15/30 msec) axial image of the superior aspect of the glenohumeral joint. The subscapularis-musculotendinous junction is retracted medially (thin arrow), and its elongated tendon shows graduated tearing of the inner fibers (curved arrow). The thin superficial remnant of the tendon is detached from the lesser tuberosity but remains in continuity with the transverse humeral ligament, which is also separated from the lesser tuberosity (black arrow). The long head of the biceps tendon is dislocated medially (hollow arrow) under the intact layer of the subscapularis tendon. Fig. 13C, an axial image below Fig. 13B, shows the ill-defined biceps tendon with regions of increased signal (long arrow) embedded within the substance of the subscapularis tendon with intact superficial and deep fibers (short arrows) at the medial margin of the lesser tuberosity. In Fig. 13D, distal to the bicipital groove the biceps tendon is enlarged and shows heterogeneous increased signal intensity representing substance tear and tendinitis (thick arrow). The tendon sheath is distended with fluid (thin arrow).

Diagnosis: Large full-thickness supraspinatus and partial-thickness subscapularis rotator cuff tears. Medial dislocation of the long head of the biceps tendon with tendon degeneration, partial tear, and tenosynovitis.

Discussion: Medial dislocation of the long head of the biceps tendon is an uncommon disorder that is usually associated with degenerative or traumatic tears of the anterior aspect of the rotator cuff (1–3). The biceps tendon may dislocate over the subscapularis tendon (Fig. 13E,F) or it may slip under a partially–torn and detached subscapularis tendon (Fig. 13A–C), a condition that may go unrecognized at surgery. Displacement of the long head of the biceps tendon is readily depicted by MR imaging (4–7).

The long head of the biceps tendon, by virtue of its anatomic characteristics, is predisposed to a variety of painful disorders. This tendon originates from the supraglenoid tubercle of the scapula. The intracapsular segment of the tendon extends in front of the humeral head beneath the rotator interval capsule (the coracohumeral ligament) and exits the joint through an opening in the capsule above the bicipital groove. In this region, the tendon is reinforced by the coracohumeral ligament and slips from the subscapularis and supraspinatus muscles (8). Below this region, the tendon is secured within the bicipital groove by the transverse humeral ligament and farther inferiorly by the tendinous expansion of the pectoralis major (the falciform ligament). A synovial sheath covers the long head of the biceps tendon from its origin and forms the synovial sleeve of the tendon. The long head of the biceps tendon is a humeral head depressor and, along with rotator cuff tendons, plays a role in stabilizing the humeral head against the glenoid. Anatomic studies demonstrate that the coracohumeral ligament is the major stabilizer of the biceps tendon within the groove (2). The normal tendon of the long head of the biceps is well visualized on MR images. The intracapsular segment is best seen on oblique sagittal images as a signal-void, ovoid structure. The intertubercular segment is usually round and is best seen on axial images (6).

Disorders of the long head of the biceps are among the common causes of pain and tenderness in the anterior aspect of the shoulder and frequently coexist with disorders of the rotator cuff tendons. The clinical presentation may therefore be confusing. Biceps lesions may be divided into two major categories: biceps tendinitis and instability (9). The most common etiology of long head biceps tendon disorders is impingement by the coracoacromial arch and the acromioclavicular joints (9–12). In addition, diminished vascularity of the biceps tendon with a critical zone similar to that of the supraspinatus tendon as a predisposing factor to degeneration and tear has been described (13).

Instability of the long head of the biceps is most commonly the result of degenerative tear of the anterior aspect of the cuff including the coracohumeral ligament. Acute traumatic dislocation of the biceps has also been reported (Fig. 13E,F) (14). Depending on the nature and extent of the subscapularis tendon tear, the biceps may be displaced beneath the torn tendon proximally while appearing ventral to the subscapularis, where it remains intact. The dislocated biceps tendon may demonstrate manifestations of degeneration and partial tear (see Fig. 1B–D) or may show no significant abnormalities (Fig. 13E,F). In longstanding intraarticular dislocations, the biceps tendon may appear enlarged or pear-shaped (Fig. 13G,H).

A subluxing biceps tendon due to intermittent displacement out of the groove may be observed clinically with elicitation of pain and a snapping sensation on elevation and rotation of the arm. This condition may be particularly disturbing for the throwing athlete (15). Proliferative changes of the groove and spur formation may be observed with, and are considered predisposing factors in, a subluxed tendon (Fig. 13I,J).

Bicipital tendinitis is most commonly secondary in nature, notably due to impingement syndrome or less frequently to arthritic disorders of the glenohumeral joint, such as osteoarthritis and rheumatoid arthritis. Primary bicipital tendinitis is an uncommon condition (11). There are no specific studies addressing MR imaging findings in bicipital tendinitis (16). Increased signal intensity of morphologically normal tendon on T1-weighted or proton-density images may represent tendinitis or tendon degeneration (7,16). In tendon degeneration, increased signal intensity is not expected on T2-weighted images. Increased signal intensity of the tendon above the bicipital groove has also been attributed to the magic angle phenomenon (17). Not infrequently, enlargement of the biceps tendon with increased signal intensity is observed (Fig. 13K,L). This MR imaging finding, particularly with an intact rotator cuff, is probably a specific appearance of bicipital tendinitis (6,18,19).

With progressive degeneration and tear of the rotator cuff, the long head of the biceps is further subjected to forces of impingement. Gradual fraying and eventual tear of the intracapsular segment of the tendon may then develop. With degenerative tears, however, the biceps is often reattached to the floor of the groove, and distal retraction will not occur (Fig. 13M,N). On the other hand, in many cases of large or massive cuff tears the intracapsular segment of the biceps may enlarge and flatten against the humeral head or it may actually become hypertrophied (Fig. 13O,P).

Acute traumatic tear of the long head of the biceps tendon is an uncommon event; it is usually observed in a young individual and involves the musculotendinous junction. On MR imaging the torn or retracted tendon is visualized near the musculotendinous junction (16).

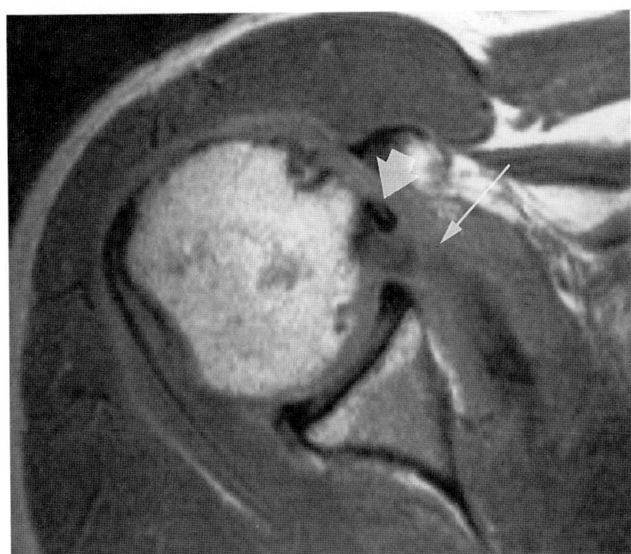

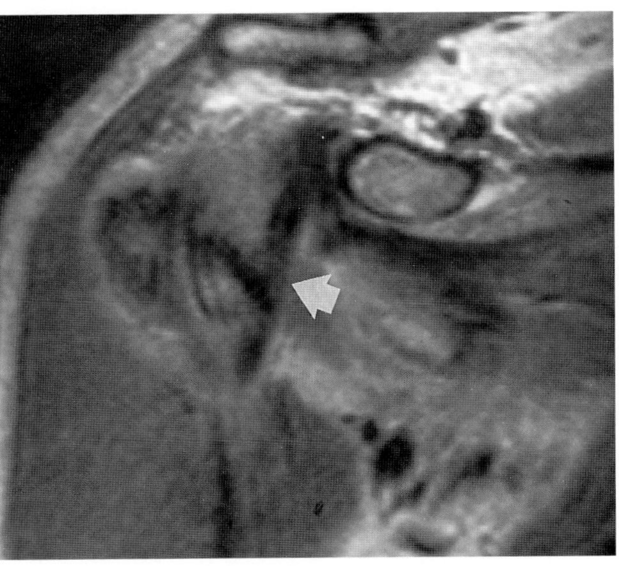

FIG. 13. E,F: Dislocated biceps tendon in a 48-year-old construction worker after acute trauma. **E:** Proton-density (TR/TE 1,500/29 msec) axial image. There is displacement of the long head of the biceps tendon medial to the lesser tuberosity and ventral to the subscapular tendon (*short arrow*). The subscapular tendon is torn (*long arrow*). **F:** Proton-density (TR/TE 2,100/30 msec) oblique coronal image. The dislocated biceps tendon is visualized medial to the humeral head (*arrow*). The supraspinatus tendon is also torn.

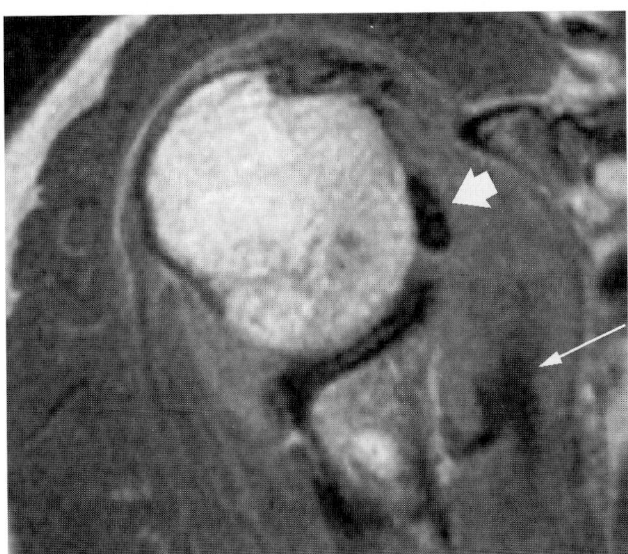

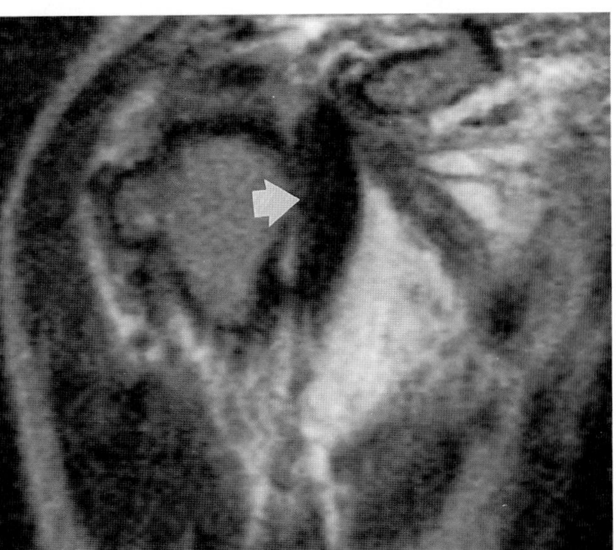

FIG. 13. G,H: Dislocated biceps tendon. **G:** Relative proton-density (TR/TE 1,100/29 msec) axial image. There is a tear with marked retraction of the subscapularis tendon (*long arrow*). The dislocated biceps is visualized within the glenohumeral joint and is hypertrophied (*short arrow*). **H:** T2-weighted (TR/TE 2,100/80 msec) oblique coronal image showing the dislocated biceps within the joint (*arrow*).

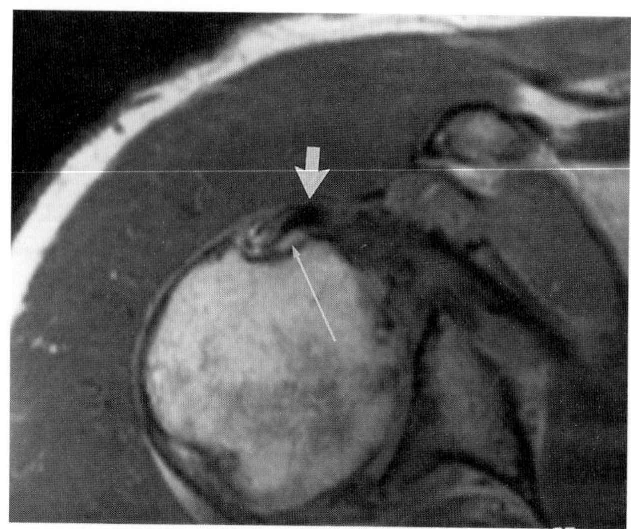
I

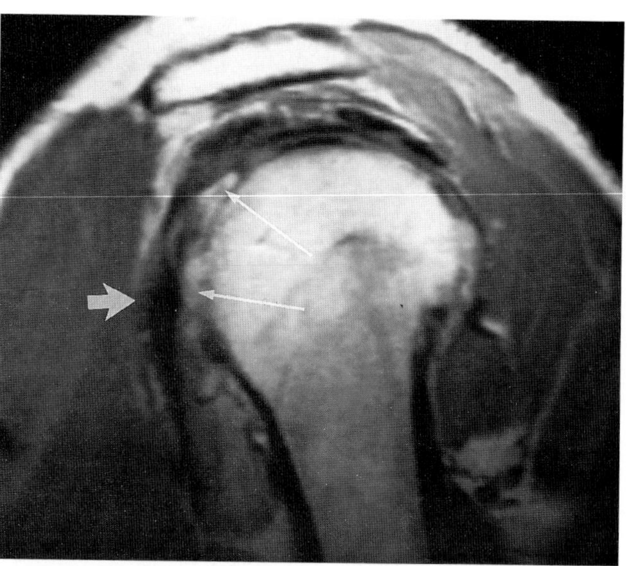
J

FIG. 13. I,J: Subluxed biceps tendon in a patient with degenerative arthritis. Proton-density axial (**I**) and oblique sagittal (**J**) images demonstrate the biceps tendon to be partially subluxed over the lesser tuberosity (*short arrows*). Hypertrophic bone formation with bone marrow signal intensity covers the floor of the groove and elevates the biceps tendon (*long arrows*).

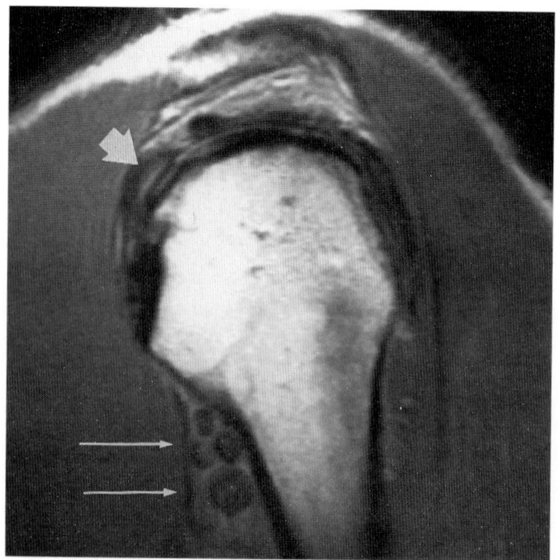
K

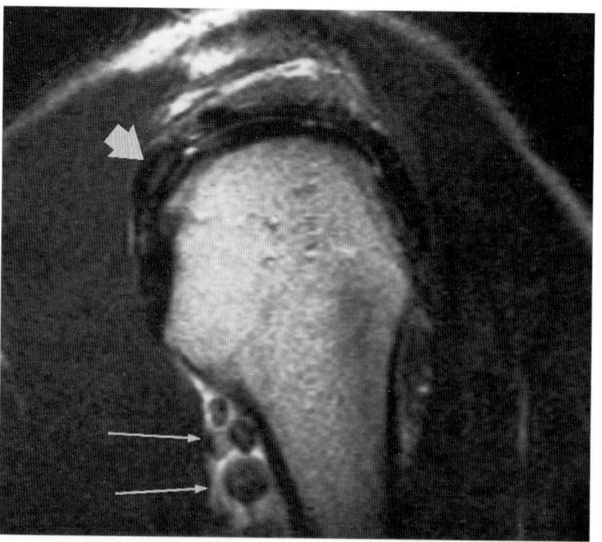
L

FIG. 13. K,L: Bicipital tendinitis. Elderly patient with prior history of dislocation and degenerative disease. Proton-density (**K**) with T2-weighted (**L**) (1,800/30/80 msec) oblique sagittal images. The intracapsular segment of the biceps is enlarged and shows increased signal intensity on proton density and to a lesser extent on T2-weighted images (*short arrows*). There are multiple loose bodies within the distended bicipital sheath (*long arrows*). The supraspinatus tendon is intact.

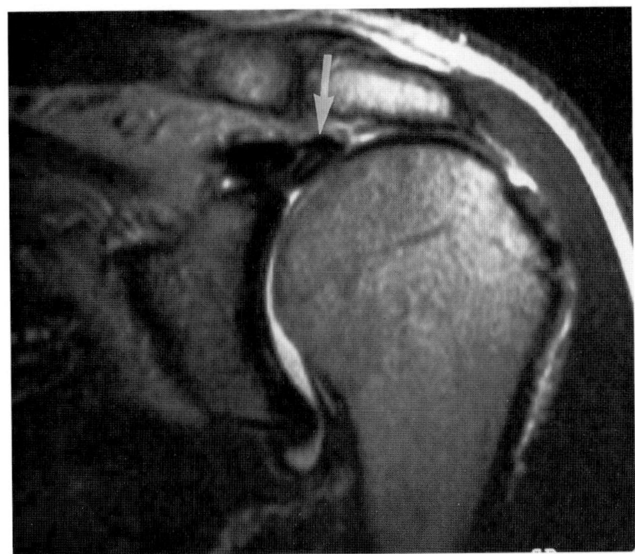
M

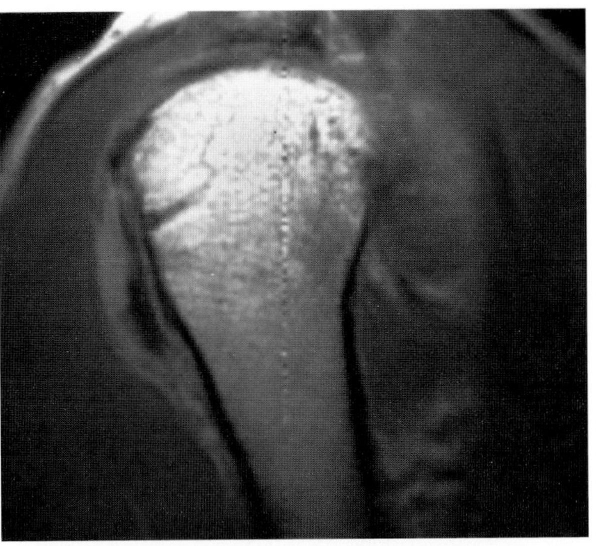
N

FIG. 13. M,N: Intracapsular tear of biceps tendon. **M:** T2-weighted (2,000/90 msec) oblique coronal image in a patient with massive cuff tear. The retracted biceps tendon is visualized adjacent to the margin of the cuff defect (*arrow*). **N:** Proton-density (1,800/29 msec) oblique sagittal image shows the graduated appearance of the biceps within the groove (*arrow*). The tendon is probably anchored within the groove.

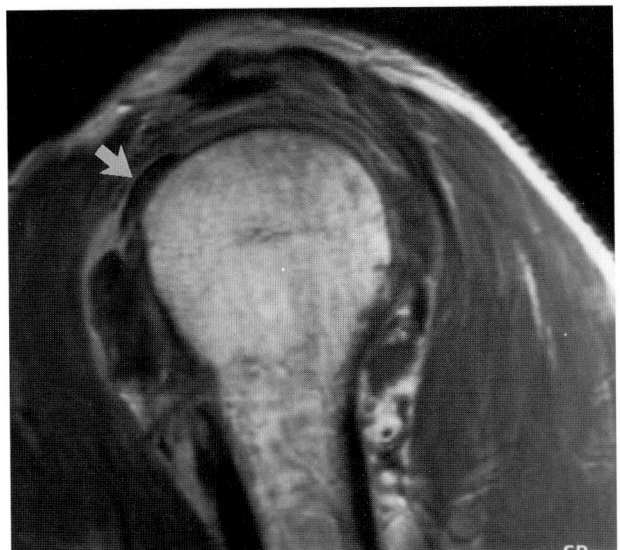
O

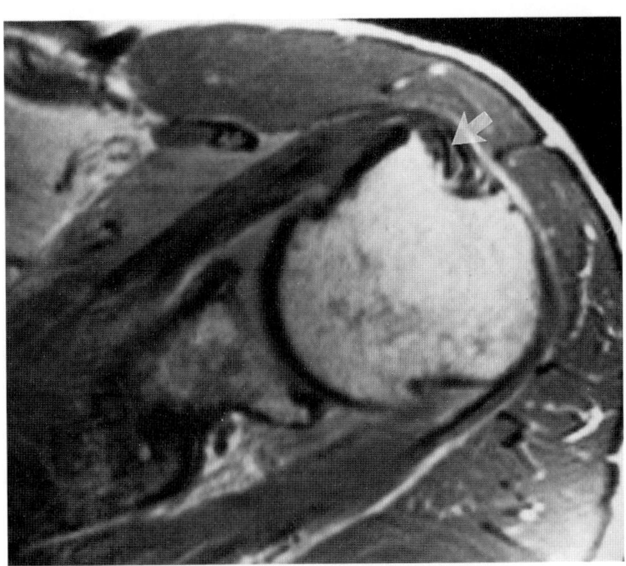
P

FIG. 13. O,P: Flattened biceps tendon. **O:** Proton-density (200/30 msec) oblique sagittal image in a patient with massive cuff tear. The intracapsular segment of the biceps tendon is widened and flattened against the humeral head (*arrow*). **P:** Proton-density (1,500/29 msec) axial image shows the flat biceps tendon within the groove (*arrow*).

CASE 14

History: This 75-year-old woman presented with pain and persistent weakness of her shoulder 6 months after conventional acromioplasty and rotator cuff repair for massive tear of the supraspinatus tendon. Her original symptoms temporarily improved after surgery, only to return and gradually worsen as she began rehabilitation and using the arm for light housework. On physical examination there was pain on internal rotation and elevation, and arc of motion was significantly reduced.

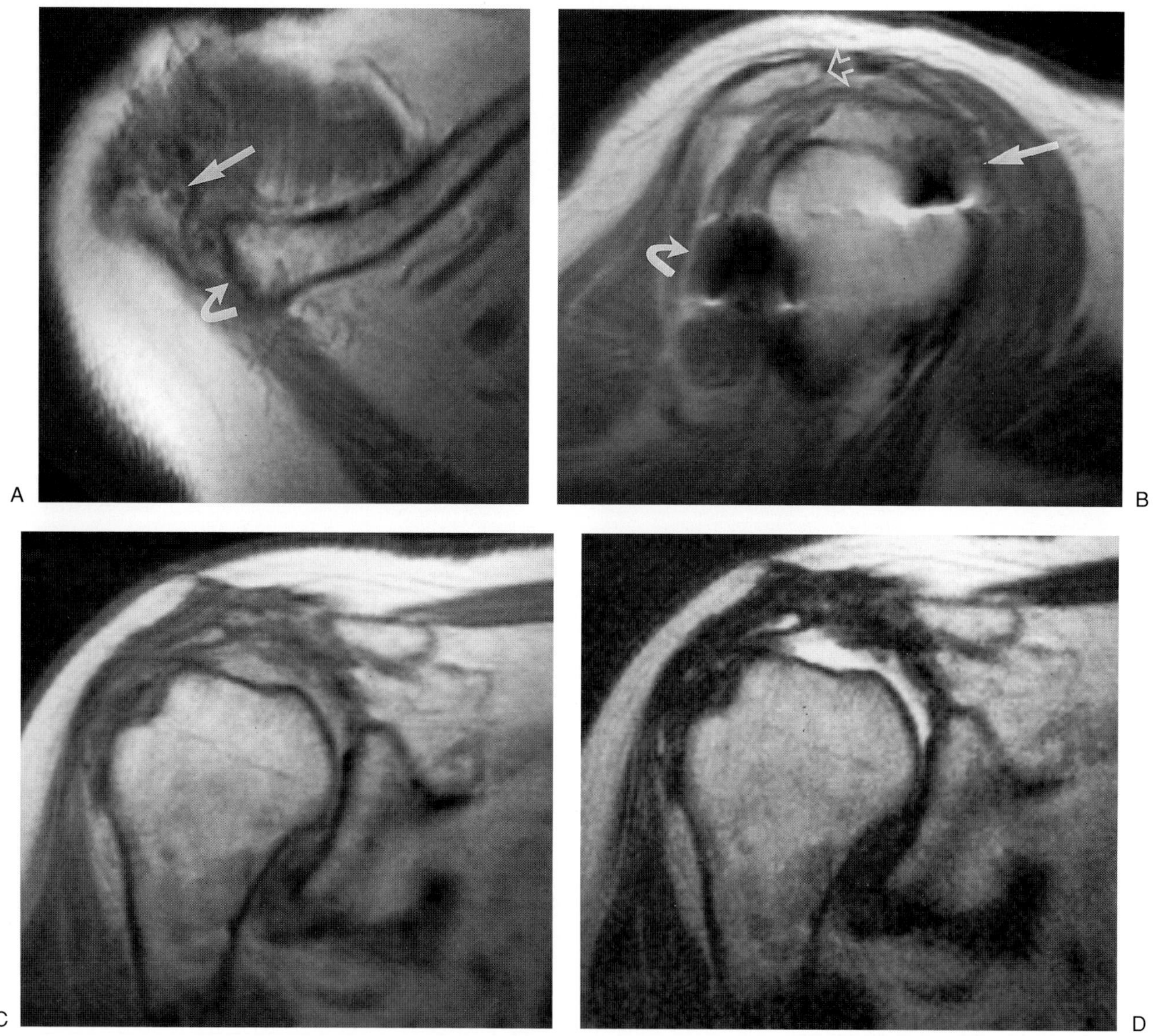

FIG. 14.

Findings: Fig. 14A,B shows proton-density (TR/TE 1,500/15 msec) axial images of the right shoulder. At the level of the acromioclavicular joint (Fig. 14A), a large segment of the anterior acromion process and the acromioclavicular joint has been resected. The remaining posterior segment of the acromion is visualized (straight arrow). The distal end of the clavicle is displaced posteriorly against the scapular spine and remaining acromion (curved arrow). At the level of Fig. 14B (superior glenoid), the irregular tendon of the supraspinatus (straight arrow) is displaced posterior to the humeral head. The distal segment of the tendon is replaced by amorphous low-signal tissue, probably fibrosis and scar (curved arrow). The anterior metallic artifact indicates the site of suture placed in the humeral head at the time of repair; note the impingement of the head against the acromion process.

Fig. 14C is a proton-density (TR/TE 3,200/17 msec) fast spin-echo oblique sagittal image. In addition to the metallic suture artifact within the anterior aspect of the humeral head (straight arrow), another similar artifact is visualized posteroinferiorly peripheral to the humeral head (curved arrow). This has the appearance of a dislodged suture. The top of the humeral head is flat. The posterosuperior and superior cuff appear markedly distorted at this level. Note the foreshortening of the acromion (open arrow). Fig. 14D is a T2-weighted (TR/TE 3,300/119 msec) fast spin-echo oblique coronal image. The humeral head shows a defect at the superior aspect. There are no recognizable cuff elements. The humeral head is instead covered by low-signal, irregular tissue, most likely a mixture of scar, fibrosis, and degenerated tendon. Note the retraction of the deltoid muscle.

Diagnosis: Failed rotator cuff repair and synovitis.

Discussion: The outcome of a rotator cuff repair is based on many factors, including the size and chronicity of the cuff defect, the nature of the remaining tendon, and the patient's age (1). Acute traumatic tears without a prolonged history of cuff problems and without significant weakness and superior migration of the humeral head respond most favorably. Long-standing tears with severe loss of strength and superior migration of the humeral head often fail the surgical repair, although some relief of pain may be achieved. Failure of cuff repair may result from numerous causes, including inadequate decompression, excessive removal of acromial bone and loss of arm lever, failure at the repair site, or development of a new tear at a different site (1,2). A number of complications may also contribute to the failure of acromioplasty and cuff repair. These complications include hematoma, infection, subacromial subdeltoid adhesions, deltoid detachment, and denervation of the deltoid and cuff (1–3).

The patient presented above displayed a number of factors contributing to a poor surgical outcome. Preoperative MR imaging had shown a massive tear and atrophy of the supraspinatus and superior migration of the humeral head. Excessive removal of bone from the acromion and acromioclavicular joint resulted in loss of the anterosuperior arch and posterior displacement of the end of the clavicle (Fig. 14A). At surgery, the degenerative remnants of cuff were mobilized and advanced, and sutures into a trough were taken at the periphery of the humeral head using three Tevdek sutures to help anchor the cuff. Although it is desirable to advance the cuff adequately to achieve an anatomic repair, this is often not possible when massive tears exist (1). In these instances, the cuff margin is anchored into a trough made at the periphery or the center of the humeral head, as appears to have been the case in this patient (Fig. 14D). Only the anterior suture remains in place, however. The metallic artifact shown posteriorly indicates a dislodged and displaced suture, leaving a defect at the superior aspect of the head (Fig 14C,D). No functional cuff elements cover the humeral head, and retraction of the deltoid, another postoperative complication, has further contributed to this patient's disability after surgery.

MR imaging of the postoperative shoulder is likely to be a difficult task. Knowledge of the type of surgical procedure performed and familiarity with expected anatomic changes relevant to each procedure are extremely helpful in accurate interpretation of MR imaging of the shoulder after surgery.

Conventional acromioplasty for treatment of impingement syndrome is an open procedure (4). Through a vertebral split in the lateral deltoid, the subdeltoid region is exposed. Decompression of the subacromial–subdeltoid space is carried out by debridement of the inflamed, thickened bursal tissue and resection of the anterior margin of the acromion and any hypertrophic spur or soft tissue that may be present at the undersurface of the acromion and the acromioclavicular joint. An acromioclavicular joint resection may also be performed. This procedure is now routinely performed arthroscopically. Rotator cuff repair is ordinarily an open procedure, performed concurrently with acromioplasty. Occasionally small tears are repaired through a "mini-incision" after arthroscopic subacromial decompression.

A number of MR imaging findings are expected after arthroscopic subacromial decompression (5–8). These include loss of the normal peribursal fat stripe, morphologic changes of the acromion due to resection of bone, interruption of the coracoacromial ligament, and decreased signal intensity of the acromial bone marrow due to fibrosis or sclerosis. Tiny metallic particles released from the surgical instruments used for bone debridement often result in small low-signal or bright artifacts. Excessive granulation tissue and scar formation after surgery may replace the resected bursae and the peribursal fat (Fig. 14E). Increased signal intensity of the cuff (without a cuff defect) is commonly observed and may represent postsurgical edema or preexisting tendinopathy (8). The adequacy of surgical decompression may be assessed by evaluating the acromion process and the subacromial space on both coronal and sagittal images. Inadequate removal of bone or bone spurs and excessive scar are important findings on MR images and potential causes of persistence or recurrence of impingement symptoms. Our limited experience with several symptomatic individuals examined after surgery also indicates that abnormal signal intensity of the rotator cuff is usually irreversible.

After a rotator cuff repair, irregularity or increased signal intensity may be present at the site of tendon-to-tendon repair and may resemble a partial tear (8). Tendon-to-bone repair may be accomplished using sutures anchored in bone (Fig. 14F) or placed within a trough at the supralateral humeral head (Fig. 14G) (5,7). Intermediate signal intensity of the repaired tendon on proton-density and T2-weighted images may be attributable to granulation tissue or disorganized collagen bundles at the repair site (8). MR imaging is therefore not reliable for the diagnosis of tendinopathies or partial tears after cuff repairs. Recurrent or new full-thickness tears in a postoperative shoulder are most accurately diagnosed when a full-thickness defect with high signal intensity on T2-weighted images is present. Retraction of the musculotendinous junction is the only one of the secondary criteria that can be used reliably in MR diagnosis of retears. The data reported so far indicate that full-thickness recurrent or new tears in a postoperative shoulder are diagnosed with sufficiently high accuracy (90%) in the diagnosis of a retear (5). The reported data, however, are scant and mostly include patients with large defects (5). The difficulty arises when a cuff defect is small or excessive scar is present, which may fill the tendinous gap or blend with the margins of the repaired tendon (Fig. 14G,H). Small cuff defects, however, may not be incompatible with good surgical results. An arthrographic study of a group of patients after successful cuff repair detected leakage of contrast material through the cuff in many such patients and concluded that a repaired cuff need not be watertight to be functional (9).

Complications related to the surgical procedure, such as synovitis, hematoma, or muscle denervation, have rarely been reported (6,8). Postoperative synovitis, either infectious or reactive, may present with large effusion with or without a synovial proliferative process (Fig. 14I,J). Nerve injury becomes apparent with atrophy (delayed finding) or diffuse increased signal intensity of denervated muscle(s) on T2-weighted images (early finding) (8).

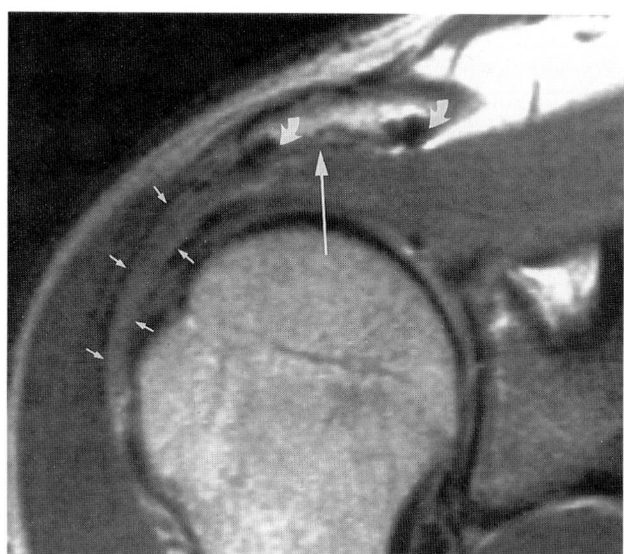

FIG. 14. E: Postoperative arthroscopic subacromial decompression. Proton-density (2,100/30 msec) oblique coronal image, right shoulder. There is irregularity and focal concavity of the acromion undersurface (*long arrow*) due to bone debridement. The subdeltoid space is expanded and replaced with low signal intensity tissue (*small arrowheads*), which remains low signal on T2-weighted images (not shown), indicative of fibrosis and scar tissue formation. Micrometallic artifacts are visualized (*curved arrows*). (From Rafii M. Shoulder. In: Firooznia H, et al., eds. *MRI and CT of the musculoskeletal system.* St. Louis: Mosby Yearbook Publishers; 1992:464–549, with permission.)

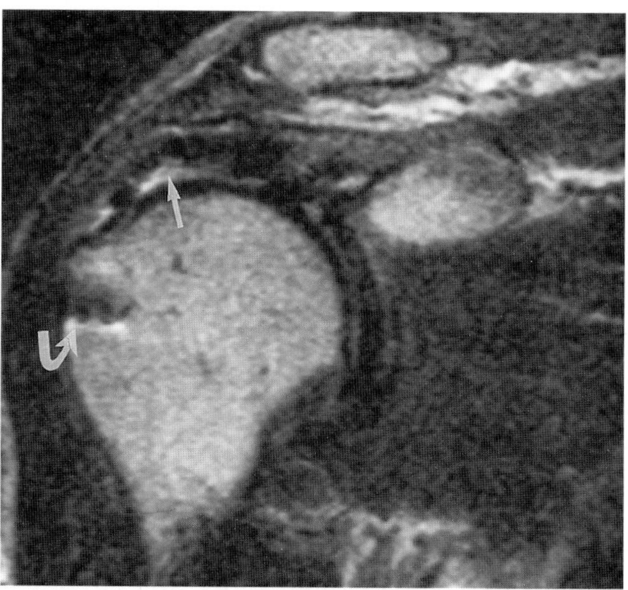

FIG. 14. F: Postoperative cuff repair (same as in Fig. 6A–D). T2-weighted (2,100/80 msec) irregularity and focal increased signal intensity of the tendon is present (*straight arrow*). The metallic artifact in the greater tuberosity is the site where the suture is anchored in bone (*curved arrow*). (From Rafii M. MR imaging of the shoulder joint. In: Beltran J, (ed). *Current review of MRI.* HongKong: Paramount Printing Group; 1995:203–213, with permission.)

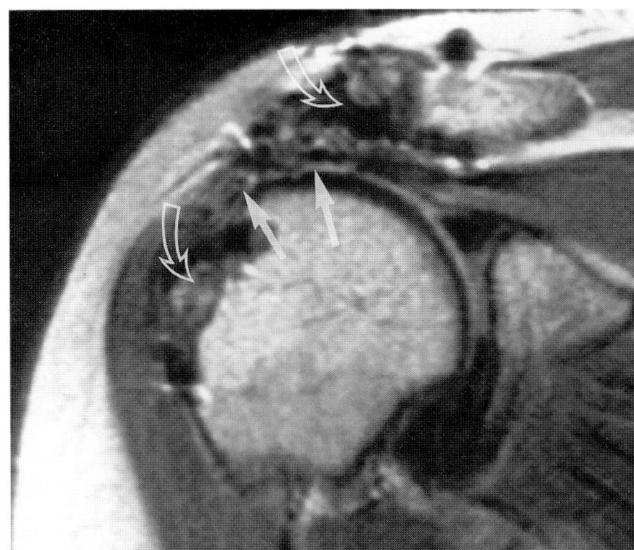
G

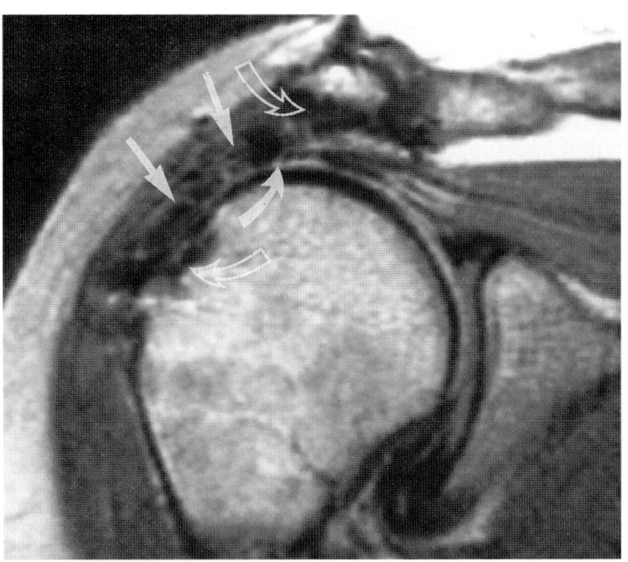
H

FIG. 14. G,H: Postoperative acromioplasty and rotator cuff repair, right shoulder. MR examination was performed due to injury and pain several months following an uneventful surgical procedure. **G,H:** Oblique coronal, proton-density-weighted (2,100/30 msec) images at anterior aspect of the glenohumeral joint. The site of the repaired supraspinatus tendon is manifested by marked inhomogeneous low signal intensity (*straight arrows*). The peribursal fat signal is absent. Some bright signal, (*curved solid arrow* in **G**) suggests interruption of the tendon; however, a tendon defect is not visualized. A presumed linear tear in this region may be a residual lesion or, conceivably, a new tear. Note foreshortened acromion due to surgical resection and trough of the greater tuberosity, the site of tendon reattachment (*curved open arrows*). Micrometallic artifacts are also noted. Patient improved with conservative treatment. (From Rafii M. Shoulder. In: Firooznia H, et al., eds. *MRI and CT of the Musculoskeletal System.* St. Louis: Mosby Yearbook Publishers; 1992:464–549, with permission.)

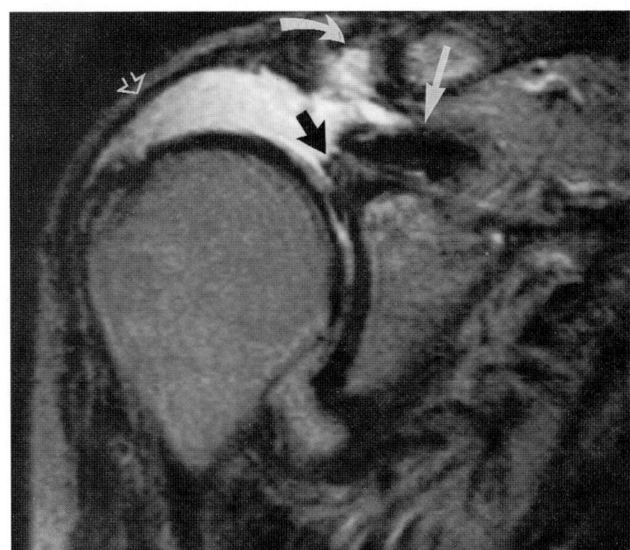
I

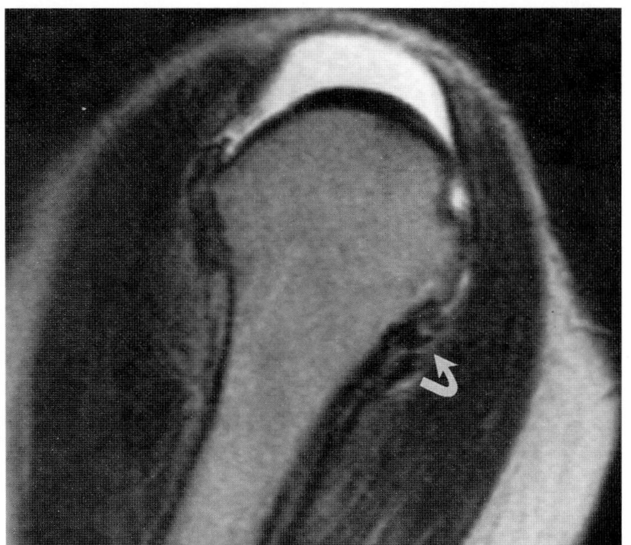
J

FIG. 14. I,J: I: T2-weighted (TR/TE 2,100/80 msec) oblique coronal image of right shoulder. The cuff is retorn and markedly retracted (*straight arrow*). The torn stump of the intracapsular segment of the biceps is also visualized (*black arrow*). There is expansion of the subdeltoid region due to large joint effusion. Note changes secondary to acromioplasty and AC joint resection (*curved arrow*). The deltoid is retracted and its insertion is reduced to a low-signal fibrous band (*hollow arrow*). **J:** T2-weighted (TR/TE 1,800/80 msec) oblique sagittal image at the level of the lesser tuberosity bicipital groove. Absence of the superior rotator cuff and intracapsular segment of the biceps tendon is noted. Tenodesis of the biceps was performed at the time of cuff repair (*arrow*).

CASE 15

History: This 19-year-old college hockey player had experienced several episodes of direct injury to his left shoulder joint, including one episode of acromioclavicular joint separation. He complained of pain localized to the anterior aspect of the shoulder. On physical examination there was no frank instability. ''Clicking'' and ''popping'' were felt anteriorly, and strength was markedly decreased. Impingement signs were positive. A conventional MR examination was followed by MR arthrography.

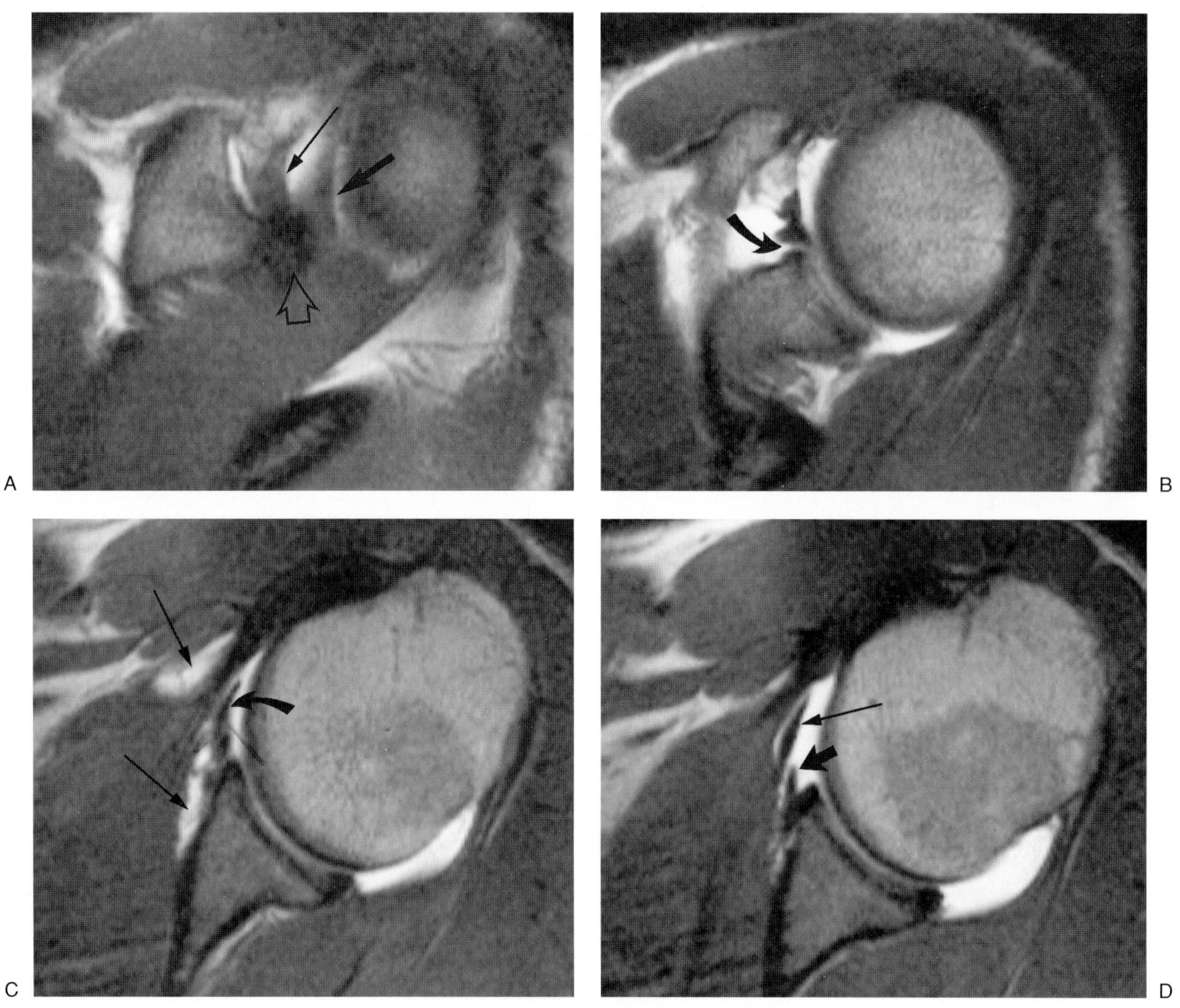

FIG. 15.

Findings: Fig. 15A–D shows T2-weighted (TR/TE 800/29 msec) MR arthrograms of the left shoulder in the axial plane. At the level of the superior glenoid (Fig. 15A), the superior glenohumeral ligament (thin arrow) and the long head of the biceps tendon (thick arrow) are visualized converging toward their attachment on the superior glenoid labrum (hollow arrow). In Fig. 15B, taken below the level of Fig. 15A, the combination of the superior glenohumeral ligament and the anterosuperior labrum is slightly detached from the glenoid margin (arrow). Fig. 15C, below the tip of the coracoid process and at the superior margin of the subscapularis tendon, shows the middle glenohumeral ligament (curved arrow) directly attached to the anterior labrum. Normal extensions of the subscapular bursa are visualized (straight arrows). In Fig. 15D (midglenoid level), the middle glenohumeral ligament is flat and partly attached to the capsule (long arrow). The attachment of the inferior glenohumeral ligament to the peripheral aspect of the glenoid labrum is also visualized (short arrow). The anterior and posterior labrum and the capsular attachments on the glenoid are normal. (Fig. 15C from Rafii M, Firooznia H. Technical aspects: magnetic resonance imaging with normal anatomy. In: Sartoris J, ed. *Principles of shoulder imaging.* New York: McGraw-Hill; 1995:45–64, with permission. Fig. 15D from Turtle AH, et al. Diagnostic imaging of the shoulder. In: Nicholas JA, Hershman BH, eds. *Upper extremity in sports medicine.* New York: Mosby; 1990:85–148, with permission.)

Diagnosis: Nonpathologic anterosuperior labral separation (sublabral hole). Prominent middle glenohumeral ligament with direct labral insertion.

Discussion: This case demonstrates the complexity of symptoms and physical findings in the athletic shoulder. Pain and physical findings attributable to the acromioclavicular joint and the rotator cuff were present in this patient. In addition, clicking and popping suggested a labrum tears, although glenohumeral instability was not observed on physical examination. MR imaging was performed in this patient to evaluate the rotator cuff and the capsulolabral complex, specifically to search for a labrum tear as the etiology of pain and clicking. Although no frank instability was observed clinically, a subtle form of instability as the underlying etiology of impingement was a consideration. Conventional MR imaging excluded mechanical impingement or a rotator cuff tear; it raised the possibility of a superior labrum anterior and posterior (SLAP) lesion. MR arthrography was performed to assess the superior labrum and confirm the integrity of the anteroinferior capsulolabral complex.

The capsular anatomy of the glenohumeral joint has been studied by numerous anatomic, CT arthrographic, MR imaging, and MR arthrographic investigations (1–16). MR arthrography has been advocated for improved visualization of the capsulolabral complex (12,13,15,16). MR arthrography is performed with a diluted paramagnetic contrast material or normal saline (14,15,17). Gadopentetate dimeglumine (Magnevist), in a concentration of 469.01 mg/mL, has been used in variable dilutions ranging from 1 ml gadopentetate dimeglumine in 100 to 250 ml normal saline. Using the standard arthrographic technique and under fluoroscopic guidance, a 20-gauge spinal needle is placed in the shoulder joint at the junction of the middle and the distal third of the anterior glenohumeral joint. The intra-articular position of the needle tip is confirmed by injection of 0.5 ml iodinated contrast agent. Contrast material (12 to 18 mL) is injected into the joint, with caution used to avoid overdistending the joint capsule or introducing air bubbles (8). The intra-articular use of paramagnetic contrast has not been approved by the US Food and Drug Administration, so that it is necessary to obtain institutional approval and informed consent to perform this procedure. No adverse effect on the synovial lining and articular cartilage as a result of intra-articular administation of such contrast agents has been reported (13). MR imaging is then performed in axial, oblique coronal, and oblique sagittal planes utilizing a T1-weighted sequence. Additional oblique coronal plane images are also obtained using a T1-weighted fat-suppressed sequence for differentiation of the bright peribursal fat from extravasated paramagnetic contrast agent. Additional axial plane imaging also using a T1-weighted fat-suppressed sequence has been advocated (16).

The capsular mechanism of the shoulder joint consists of the rotator cuff, a redundant joint capsule and the capsular ligaments, the synovial recesses, the glenoid labrum, and the scapular periosteum (1). The attachment of the long head of the biceps tendon at the supraglenoid tubercle (conjoined insertion with the superior labrum) and the long head of the triceps inferiorly divide the capsular mechanism into the anterior and posterior components (Fig. 15E,F). Anteriorly, the joint capsule is reinforced by three capsular ligaments: the superior, middle, and inferior glenohumeral ligaments (Fig.15E,F) (1,2). These ligaments are variably thickened fibrous bands of the capsule, and their primary function is to limit external rotation of the humeral head (1). There are normal variations in the development of the glenohumeral ligaments; the middle glenohumeral ligament in particular may be ill defined or absent, and its glenoid attachment may vary from directly on the labrum to the scapular neck (Fig. 15B,C) (3,4). The superior and middle glenohumeral ligaments frequently have a direct labral or near-labral attachment. Palmer et al., (16) in a recent study, found conjoined labral insertion of the middle and superior glenohumeral ligaments in 40 of 41 patients examined by MR arthrography (Fig. 15A). The inferior glenohumeral ligament is less variable. Functionally, it is the most important of the glenohumeral ligaments in providing joint stability. It is a broad ligament and consists of the anterior and posterior bands and the axillary pouch. The more prominent anterior band is usually contiguous with the anterior inferior labrum, and the site of attachment may appear as a notch on axial images (Fig. 15D).

The normal glenoid labrum may vary in size, morphology, and signal intensity. The normal labrum is signal void. The anterior labrum is often triangular and larger than the posterior labrum, which is usually rounded (2,6,14,19). Reported variations of the labrum most often involve the anterior labrum. The labrum may be round, flat, notched, or cleaved, or may demonstrate a cleft (Fig. 15G–J) (20–22). The morphologic variations of the labrum are frequently observed at the midglenoid level or anterosuperiorly. Most likely these reflect or are associated with developmental variations of the capsular ligaments (3,7,8). The inferior glenoid labrum is often less well defined as it becomes confluent with the anterior band of the inferior glenohumeral ligament (Fig. 15J) (8).

The glenoid labrum is normally attached to the glenoid margin; its peripheral surface is contiguous with the glenoid periosteum, while its base extends over and is loosely adherent to the glenoid articular cartilage (5). The signal intensity of the articular cartilage under the labrum is a potential pitfall and may mimic detachment of the labrum (Fig. 15H) (19,20). Detachment of the articular aspect of the superior glenoid labrum often occurs due to aging. A sulcus or recess has also been observed between the glenoid articular cartilage and the labrum at arthroscopy (22,23). In a cadaveric study, Cooper et al. (5) concluded "that there may be an actual defect in the labral attachment to the anterosuperior glenoid rim and this should not be considered evidence of disease or instability." This observation has also been made on imaging studies (15,24).

Williams et al. (25) in 200 arthroscopies identified sublabral holes in 12% and a cord-like middle glenohumeral ligament in 9% of his patients. In three patients (1.5%) they observed a cord-like middle glenohumeral ligament originating from the superior labrum without any labral tissue existing between this attachment and the mid-glenoid notch (Fig. 15K,L). This so-called Buford complex is considered a normal variation and is significant in that it may be mistaken for a sublabral hole or a pathologic labral detachment (26).

In conclusion, the finding of a focal anterosuperior labrum separation, without an associated detachment of the superior or anteroinferior labrum as shown in Fig. 15A–D, was considered nonpathologic. The patient was conservatively treated and gradual experienced diminution of pain and return of strength to his arm.

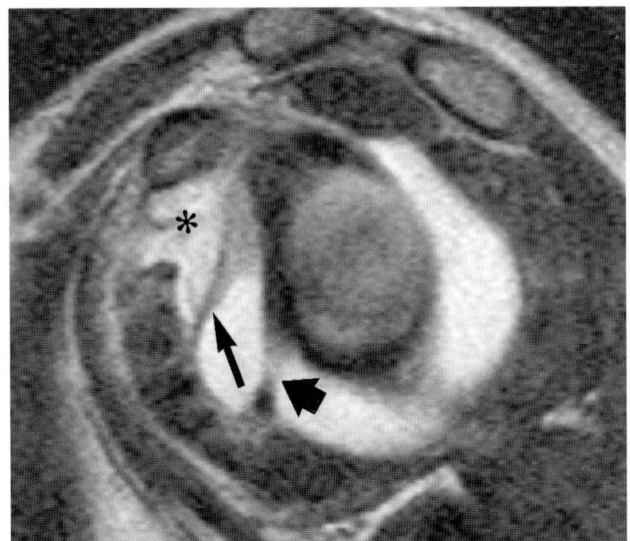

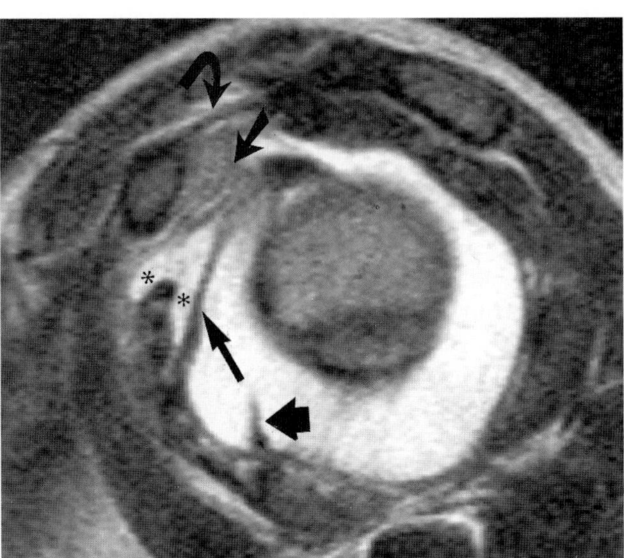

FIG. 15. E,F: T2-weighted (TR/TE 1,800/80 msec) oblique sagittal images, right shoulder distend with effusion. The superior (*slanted arrow*), middle (*thin arrow*), and anterior band of the inferior glenohumeral ligaments are visualized. The subscapularis bursa (*asterisk*) and the coracohumeral ligament (*curved arrow* in F) are also visualized. (From Rafii M, Firooznia H. Technical aspects: magnetic resonance imaging with normal anatomy. In: Sartoris DJ, ed. *Principles of shoulder imaging.* New York: McGraw-Hill; 1995:45–64, with permission.)

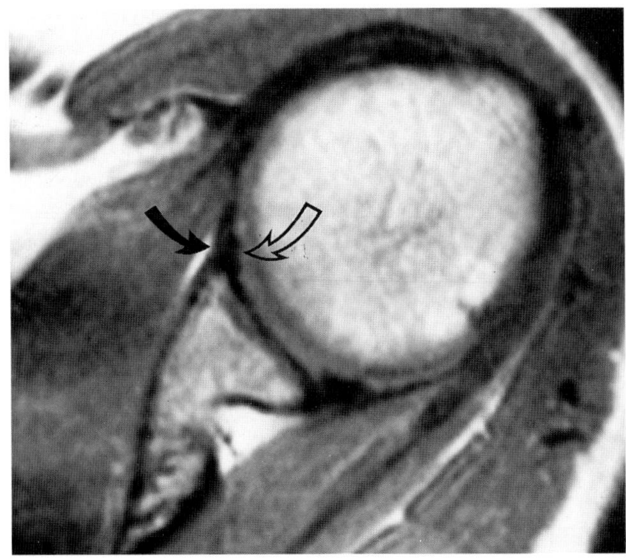

FIG. 15. G–J: Normal variations of the glenoid labrum. **G:** T1-weighted axial image, midglenoid. The middle glenohumeral ligament has a near labral attachment (*solid arrow*). The labrum is notched (*open arrow*).

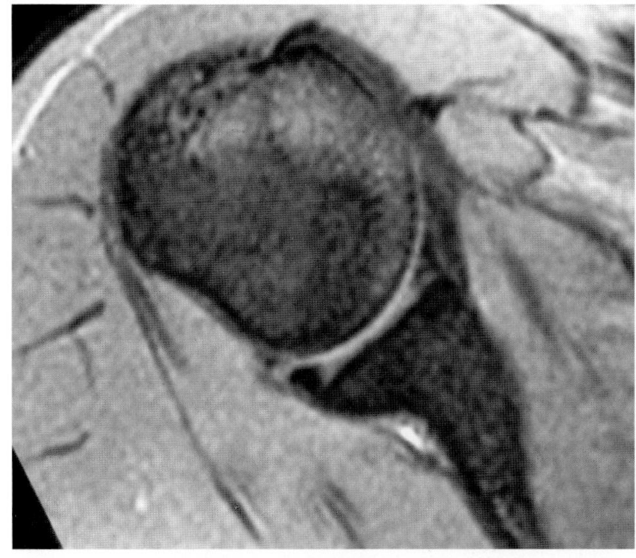

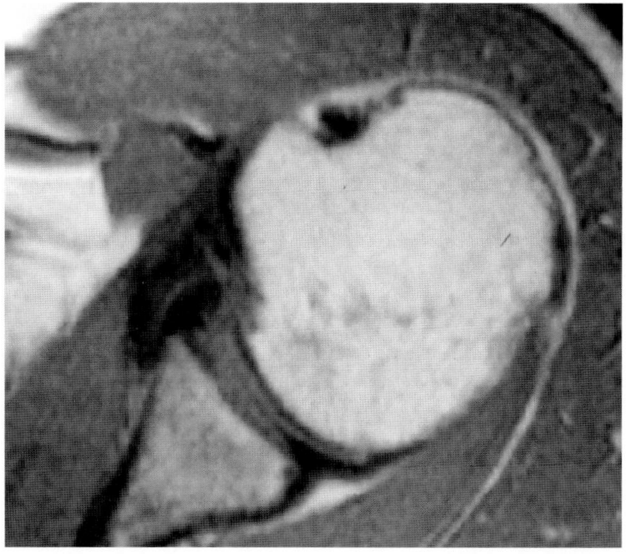

FIG. 15. H: Axial gradient-echo image. Undercutting of the glenoid articular cartilage and notched appearance of the labrum are more pronounced on gradient-echo sequence. **I:** Axial T1-weighted image. The anterior labrum is larger than the posterior labrum. It is also rounded and notched.

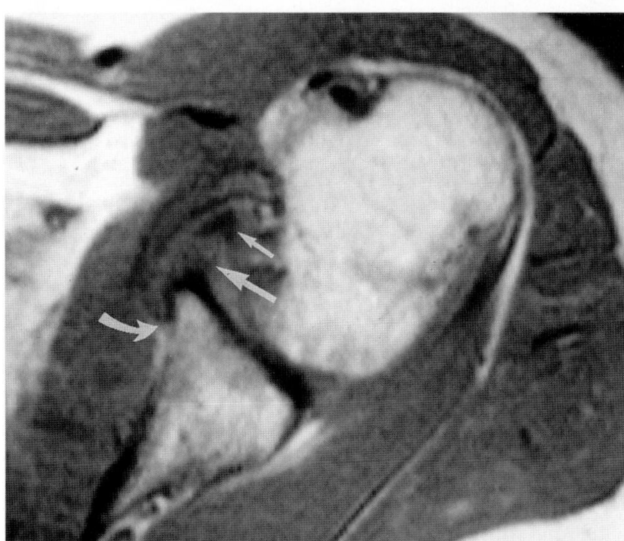

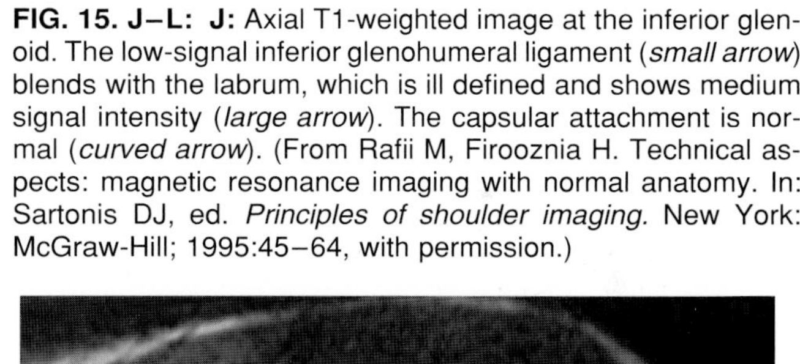

FIG. 15. J–L: J: Axial T1-weighted image at the inferior glenoid. The low-signal inferior glenohumeral ligament (*small arrow*) blends with the labrum, which is ill defined and shows medium signal intensity (*large arrow*). The capsular attachment is normal (*curved arrow*). (From Rafii M, Firooznia H. Technical aspects: magnetic resonance imaging with normal anatomy. In: Sartonis DJ, ed. *Principles of shoulder imaging.* New York: McGraw-Hill; 1995:45–64, with permission.)

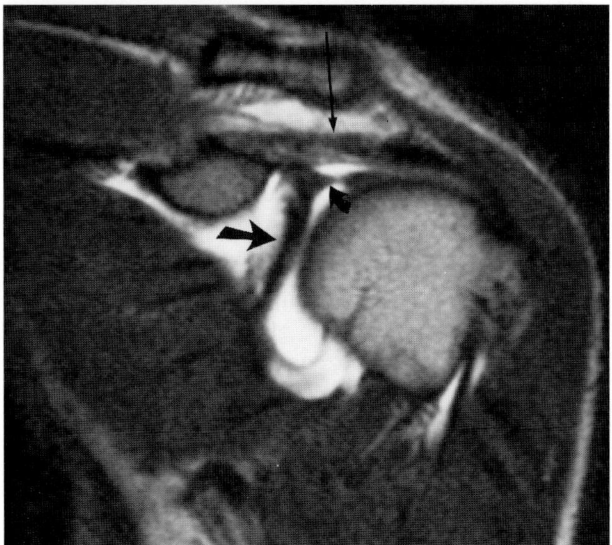

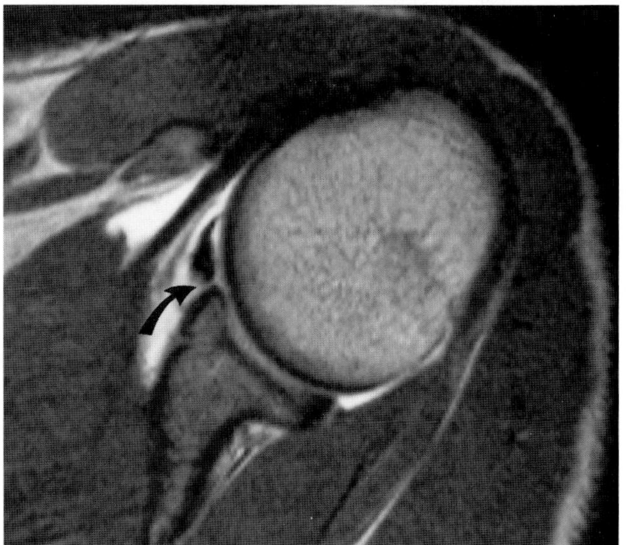

FIG. 15. K,L: Prominent middle glenohumeral ligament (the Buford complex). MR arthrogram. **K:** T1-weighted oblique coronal image. The middle glenohumeral ligament is prominent and shows insertion on the superior glenoid (*slanted arrow*). The biceps tendon (*curved arrow*) and the coracohumeral ligament (*thin arrow*) are also visualized. **L:** Axial image at the superior aspect of the joint. The prominent middle glenohumeral ligament appears to be confluent with the anterior labrum and separated from the glenoid margin at this level (sublabral hole) (*arrow*).

CASE 16

History: This 24-year-old man has a history of recurrent anterior dislocation for several years. The original injury was a fall off a ladder on his outstretched hand. The latest incident occurred when he attempted to reach up to a shelf and dilocated his shoulder. MR imaging was performed presurgically to determine the extent of soft tissue and osseous abnormalities.

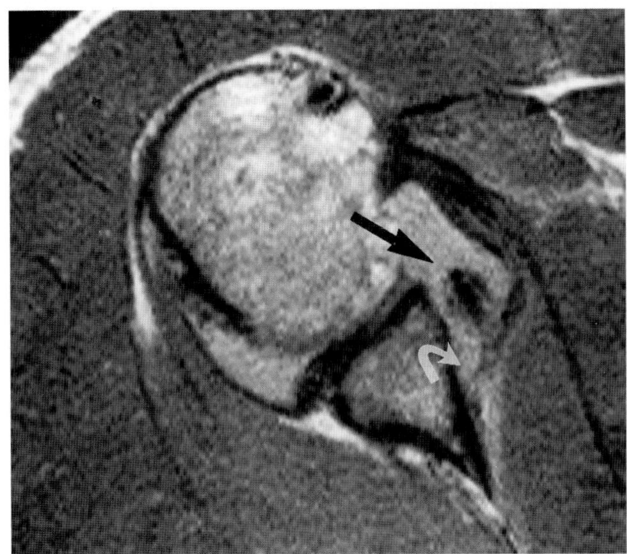

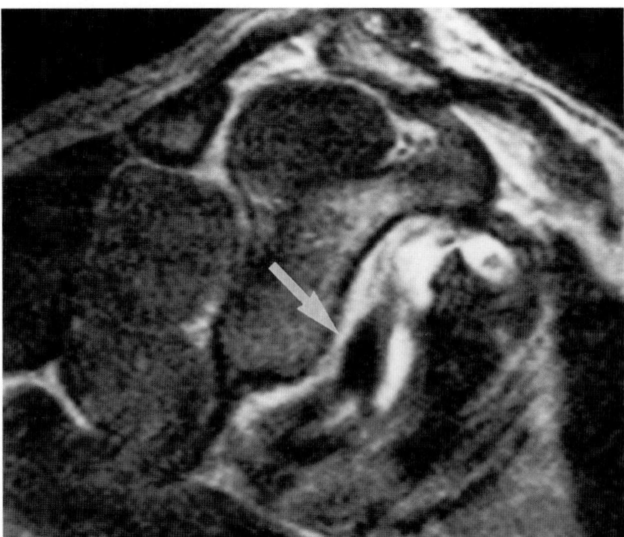

FIG. 16.

Fig. 16A shows a proton-density (TR/TE 1,100/29 msec) axial image at the inferior glenoid level. There is complete detachment of the anterior capsulolabral complex (black arrow). The capsule is stripped to the scapular neck (curved arrow). Figure 16B shows a T2-weighted TR/TE 1,800/80 msec) oblique sagittal image at the glenoid level. This image shows detachment of the entire anterior and anteroinferior glenoid labrum along with all three glenohumeral ligaments (arrow).

Diagnosis: Anterior glenohumeral instability. Complete Bankart lesion.

Discussion: Glenohumeral instability is a recurrence of dislocation or a subluxation (or a combination of both) primarily seen in young individuals after an initial episode of dislocation or repeated episodes of less severe injury (1–5). Developmental predisposition to instability such as glenoid hypoplasia may exist (1,2,5–7). Instabilities may be undirectional in two or three directions including inferiorly.

Matsen et al. (8) divide instabilities into two categories. The TUBS group includes patients with traumatic unidirectional instability who have a Bankart lesion and require surgery for treatment. The AMBRII group includes patients with no history of trauma (atraumatic type with multidirection and bilateral instability) who are primarily treated with rehabilitation or if need be by an inferior capsular shift procedure. The TUBS group (traumatic anterior glenohumeral instability) consititutes over 95% of all instabilities observed in clinical practice.

Traumatic glenohumeral instability is the consequence of failure of soft tissue structures, which stabilize this inherently unstable joint (9–13). Referred to as ''the capsular mechanism,'' these structures include the rotator cuff, the fibrous joint capsule and the capsular ligaments, the glenoid labrum, and the scapular periosteum (12,13). Anteriorly, the fibrous capsule and in particular the inferior glenohumeral ligament are considered the most important elements in preventing outward displacement of the humeral head (14). The glenoid labrum increases the articulating surface and depth of the glenoid fossa, but its contribution to stability of the joint has been debated (12,13).

Traumatic anterior glenohumeral dislocation or subluxation usually occurs when there is an applied force to the arm with the humeral head in abduction and external rotation or when a force is applied directly to the back of the shoulder. Both mechanisms result in tensile failure of the anterior capsular mechanism (9,11,13). This failure is in the form of capsular rupture in the elderly (with or without subscapularis tendon tear), a lesion that does not predispose to instability. In young individuals the anterior capsulolabral complex, always including the inferior glenohumeral ligament, is avulsed from the glenoid. The labrum may be avulsed along with the capsule or it is secondarily torn by the dislocating humeral head (Fig. 16C,D). This so-called Bankart lesion is usually not a reversible injury and results in recurrent instability in over 80% of cases (15,16). The typical patient with anterior glenohumeral instability is therefore young and at surgery will reveal a Bankart lesion or a lesser form of anterior capsulolabral injury. The predominant and most specific site of a labrum tear or detachment of the capsulolabral complex in anterior instability is the anteroinferior glenoid quadrant below the glenoid notch. However, the labrum tear may extend proximally along the entire anterior glenoid margin to the biceps insertion (Fig. 16B). Tear of the rotator interval capsule may also be observed. It is important to recognize that isolated tears of the labrum limited to the anterosuperior glenoid margin, (above the glenoid notch) are not affiliates of anterior glenohumeral instability (see Case 15).

The MR imaging appearance of the detached capsulolabral complex is dependent on the severity of the initial injury and the duration of instability (17–19). The torn labrum may be irregular in outline or in chronic stages may be degenerated or entirely absent (Fig. 16E,F) (18). A bare or eroded glenoid margin is often observed in chronic cases (Fig. 16A). Stripping of the capsule from the scapular cortex is often well visualized on MR images, especially when joint effusion is present (Fig. 16A–D). In acute injury disruption of the capsule and extracapsular leakage of fluid or hemorrhage is visualized (Fig. 16C,D). In chronic cases the capsule is ill defined and irregular (Fig. 16A). Also it is likely that the stripped capsule becomes loosely adherent to the scapular cortex by adhesions or scar. At surgery these ''healed capsules'' are readily elevated off the scapular cortex. Differentiation from a normal variant of type III capsule is based on loss of smooth reflection of the capsule over the scapular cortex and frequent association with a labrum tear (19). Detachment of the capsule without a torn labrum is a rare occurrence (Fig. 16G,H).

Expansion and redundancy of the capsule are also best visualized when joint effusion is present (Fig. 16E,F). Variable degrees of joint effusion are observed in more than 50% of our patient population undergoing MR imaging. Redundancy of the capsule in a dry joint may be manifested by the wavy contour of the capsule, which is secondary to a recoiled, stretched glenohumeral ligament (Fig. 16I).

Osseous injuries may also be observed following anterior dislocation or subluxation as a result of impaction of the posterosuperior aspect of the humeral head against the anterior or anteroinferior glenoid margin. A superolateral defect of the humeral head, i.e., the Hill-Sachs lesion, is frequently observed (Fig. 16J) (20,21). A fracture of the anterior glenoid margin may additionally be observed (Fig. 16I). Another finding relevant to capsular detachment is ectopic bone formation at the site of anterior capsular detachment due to scapular periosteal injury (4). Ectopic bone formation is readily observed on axillary views of the shoulder, and, like Hill-Sachs lesion, is a pathognomonic sign of anterior glenohumeral instability on conventional radiography. On MR imaging, ectopic bone formation is low signal on all pulse sequences and is further recognized by its characteristic location at the peripheral margin of the glenoid (Fig. 16K) (22).

In a patient with recurrent dislocation or subluxation, the objective of MR imaging is to define the type and extent of soft tissue injury for surgical planning. Currently after acute injury surgical intervention for reattachment of a detached capsulolabral complex is recommended, to prevent recurrence. Imaging after the first episode of injury is therefore indented to determine the presence of a surgically correctable lesion, particularly in young individuals (16,23).

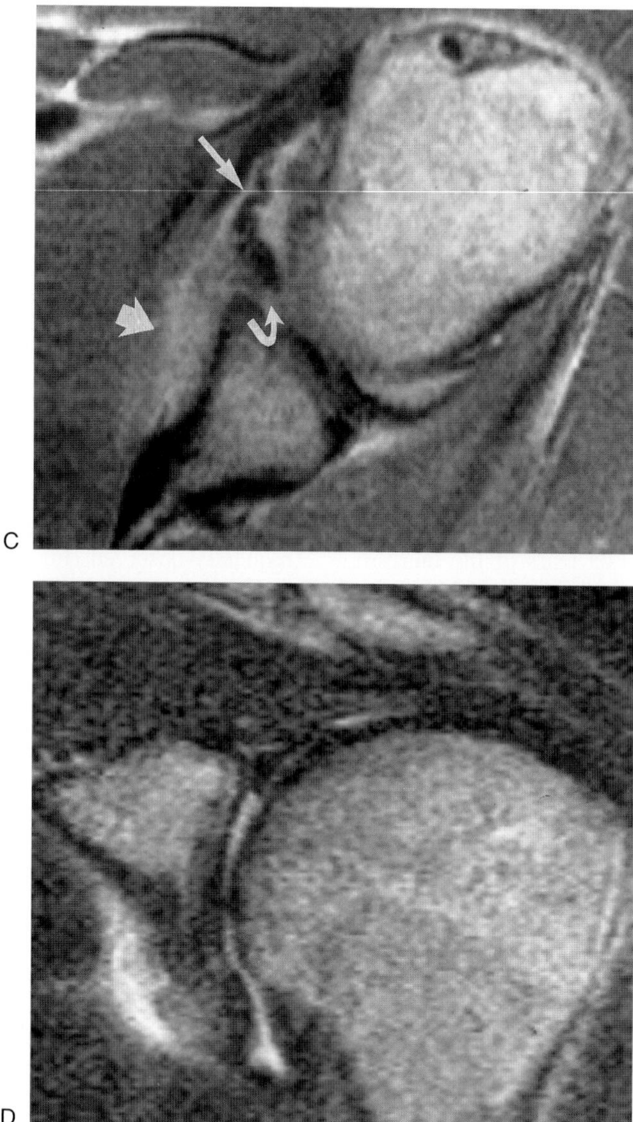

FIG. 16. C,D: Anterior instability in a 25-year-old professional hockey player following the initial episode of dislocation. **C:** Proton-density (TR/TE 1,600/30 msec) axial image, inferior aspect of glenohumeral joint. The joint capsule is disrupted; edema (and hemorrhage) extends to the scapular neck, beneath the subscapularis (*thick arrows*). The inferior glenohumeral ligament is markedly redundant (*thin arrow*). The ligament–labrum complex is avulsed from the glenoid, but not displaced (*curved arrow*). **D:** T2-weighted (TR/TE 2,100/80 msec) oblique coronal image shows the edema and hemorrhage as a result of anteroinferior capsular tear. (C, from Turtle A, et al. Diagnostic imaging of the shoulder. In: Nicholas JA, Hershman EB, eds. *Upper extremity in sports medicine.* St. Louis: CV Mosby; 1995:85–148, with permission.)

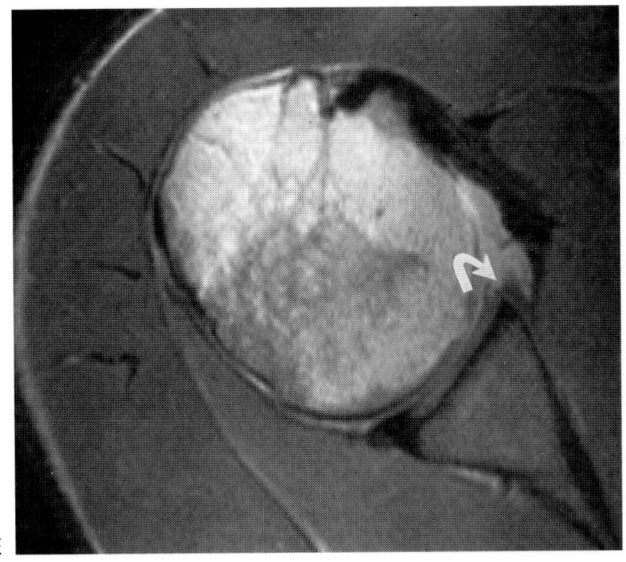

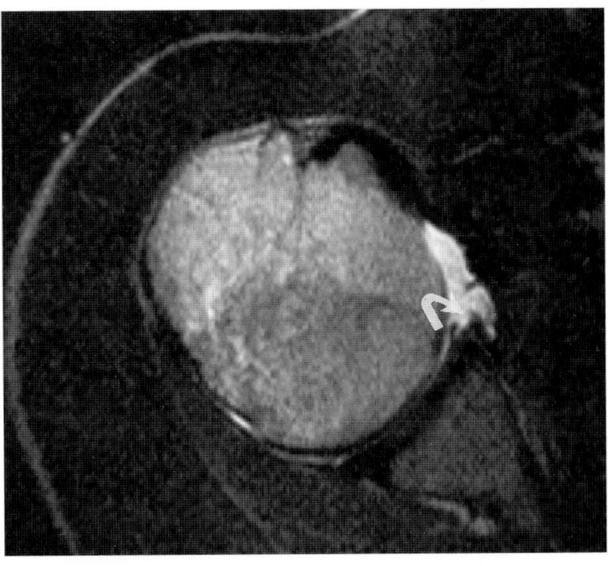

Fig. 16. E,F: Anterior glenohumeral instability. Proton-density and T2-weighted (TR/TE 2,100/30/80 msec) images at distal glenoid level. The anteroinferior labrum is torn and degenerated (*arrow*). There is no stripping of the joint capsule, although it appears lax and wavy, in contour.

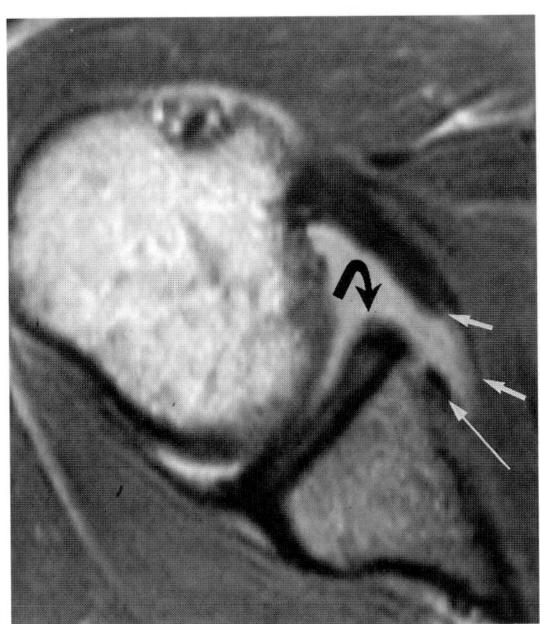

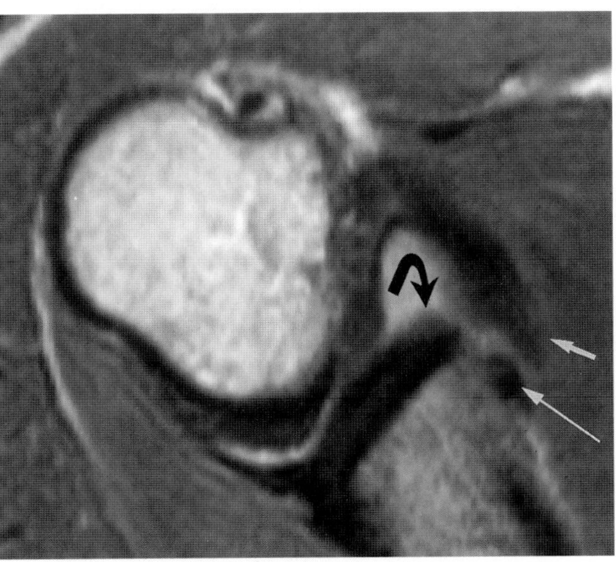

FIG. 16. G,H: Anterior glenohumeral instability (recurrent subluxation). Proton-density (TR/TE 2,100/30 msec) axial images. There is detachment of the joint capsule-inferior glenohumeral ligament from scapular cortex (*curved arrow*). Focal low-signal elevation over the scapular cortex is ectopic calcification (*long arrows*).

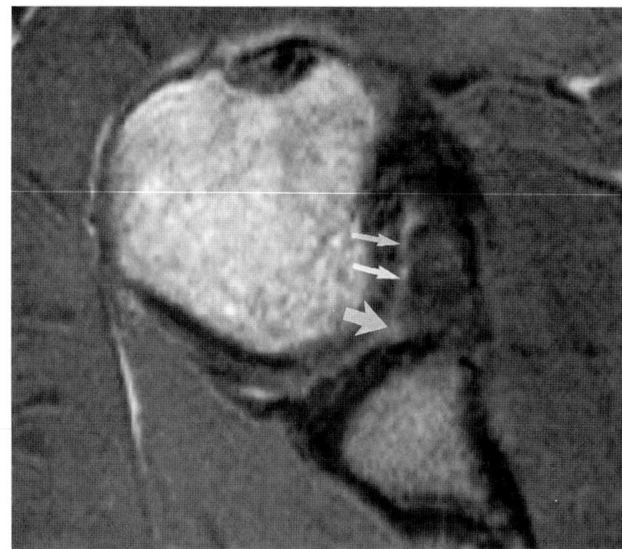

I

FIG. 16. I: Anterior instability and recurrent subluxations in a 24-year-old professional hockey player. Proton-density (TR/TE 2,100/30 msec) axial images at inferior aspect of the glenohumeral joint. There is deformity of the anteroinferior glenoid margin due to an osseous Bankart lesion, which is at least partly healing, leaving a defect at the articular surface (*thick arrow*). The labrum is not identified apart from the capsule. The capsule is thick. The wavy contour represents the redundant inferior glenohumeral ligament (*thin arrows*). (From Rafii M, Firooznia H, Golimbu C, Weinreb J. *MRI Clin North Am* 1993;1:87–104, with permission.)

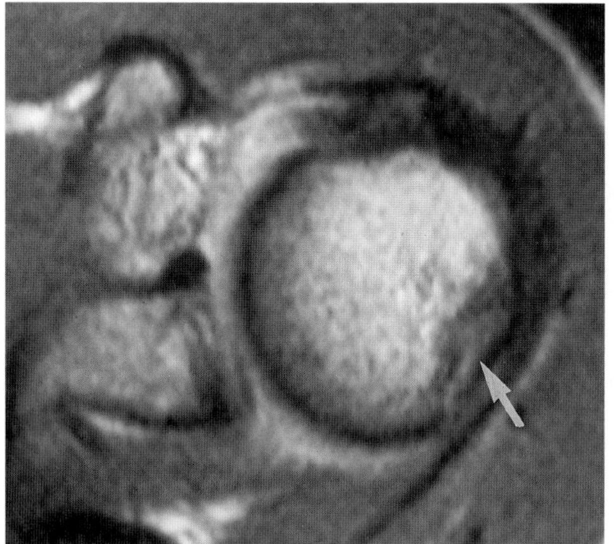

J

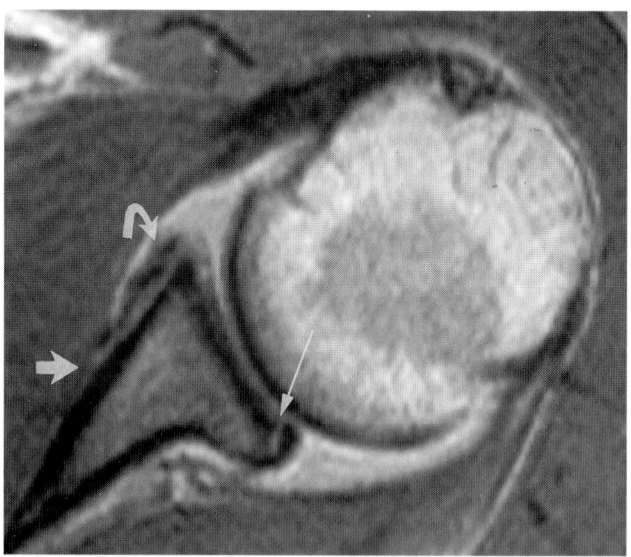

K

FIG. 16. J,K: Anterior glenohumeral instability. Proton-density (TR/TE 2,100/30 msec) axial images. **J:** A Hill-Sachs lesion is present (*arrow*). **K:** Stripping of the anterior capsule (*short arrow*) and ectopic bone formation (*curved arrow*) are present. There is also a tear of the posterior glenoid labrum (*long arrow*).

CASE 17

History: A 25-year-old man first sustained injury that to his left shoulder at age 15 while playing ice hockey. He felt that his shoulder popped out, although it did not dislocate. Since then he has intermittently experienced severe pain in the anterior aspect of the joint with a feeling of instability when doing overhead movements. On physical examination the apprehension test was positive and as a result the glenohumeral joint could not be thoroughly examined for presence of instability.

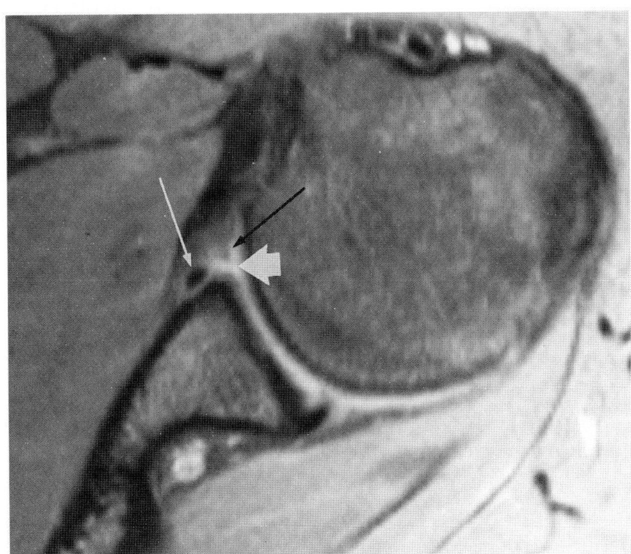

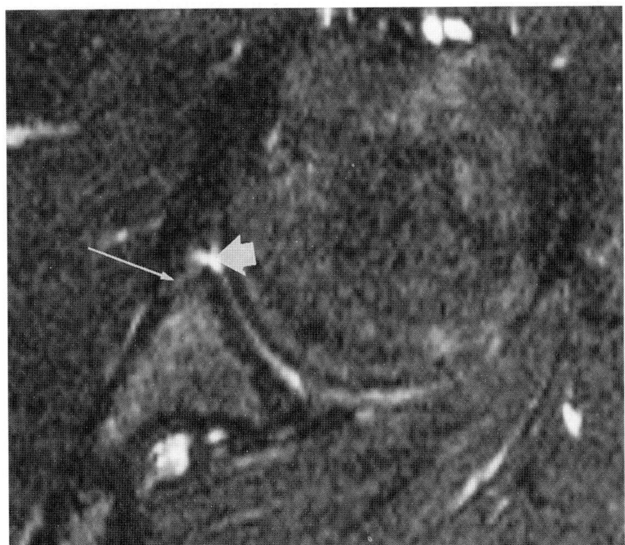

FIG. 17.

Findings: Fig. 17A,B gives proton-density and T2-weighted (TR/TE 2,000/15/90 msec) fat-suppressed axial images at the distal aspect of the glenoid. There is detachment of the anterior capsulolabral complex (short arrows). The low signal intensity focus seen on proton density image (long arrow in Fig. 17A) is a small ectopic calcification. The capsular insertion otherwise shows heterogeneous intermediate signal on the T2-weighted image (long arrow in Fig. 17B). The detached labrum is small, with increased signal intensity (black arrow in Fig. 17A).

Diagnosis: Anterior glenohumeral instability (recurrent subluxation).

Discussion: Recurrent subluxation of the glenohumeral joint was first described by Blazina and Saltzman (1) in a group of athletes. Subsequently transient subluxation of the glenohumeral joint was recognized as a cause of anterior shoulder pain or ''dead arm syndrome'' without awareness of instability on the patient's part. The mechanism of injury was described by some as excessive ''throwing'' or serving in tennis (2). The observed soft tissue pathology (detachment of the anterior capsulolabral complex) in these patients is similar to that found in recurrent dislocation, although the frequency is variably reported and may be decreased in comparison with recurrent dislocation (2,3). A milder form of glenohumeral instability (microinstability) has also been described, primarily in athletes who participate in throwing activities (4). Even though a typical Bankart lesion may not be observed in these patients, capsular laxity or tear and degeneration of the glenoid labrum are often present due to excessive translation of the humeral head over the glenoid margin (4). Progression to frank instability is likely. The role of MR imaging in patients with milder forms of instability is to confirm the presence of instability and, more importantly, to determine whether or not a detached capsulolabral complex is present for surgical planning.

Although the function of the glenoid labrum and its significance in glenohumeral stability or instability has been a subject of controversy, its high association as a component of the Bankart lesion with glenohumeral instabilities has been appreciated (3,5–7). Visualization of labrum tears by tomographic imaging modalities has proved highly rewarding in the diagnosis of glenohumeral instabilities (8–12). Conventional MR imaging is also, by most accounts, highly accurate in delineation of the capsulolabral lesions, with results comparable to those obtained with CT-arthrography (13–21). However, low sensitivity in the diagnosis of labrum tears by conventional MR imaging has been reported by some authors (22). MR-arthrography has been recommended for improved visualization of labrum tears and abnormalities of the joint capsule (23,24). We concur with the opinion that conventional MR imaging, if performed with a high-resolution technique using appropriate imaging sequences, is highly accurate in depicting abnormalities of the capsulolabral complex. The invasive and more costly MR-arthrography could be selectively performed, even as a sequel to conventional MR imaging (25,26).

Lesions of the glenoid labrum on MR imaging are manifested by morphologic and signal alterations. Most often a detached labrum is identified on T2-weighted images manifested by a band or zone of high signal intensity separating the labrum from the glenoid margin (Fig. 17B). This is due to fluid trapped within the separated region and is even frequently observed in a dry joint. Fat-saturated sequences are more sensitive in depiction of small lesions; conventional T2-weighted images obtained at the same time as Fig. 17A,B failed to show the abnormal zone of high signal in this patient.

Other morphologic abnormalities of the labrum may be observed such as outline irregularity and attenuation, indicating a chronically torn and degenerated labrum secondary to recurrent instability (see Fig. 16E,F). Presence of a low-signal ovoid structure adjacent to the anterosuperior margin of the glenoid in the subcoracoid region has been described in some patients with a detached labrum. Termed the *GLOM sign,* it probably represents a detached and redundant middle glenohumeral ligament. Occasionally a detached labrum may fail to demonstrate high signal intensity on T2-weighted images either due to total lack of joint effusion or obliteration of tear by granulation tissue. These lesions are recognized by their abnormal anatomic relation to the glenoid margin, which may in turn show a fracture or erosion (Fig. 17C,D). This appearance of a detached labrum should be distinguished from normal undercutting of the glenoid hyaline articular cartilage by its width, which is greater than that of articular cartilage and, more importantly, by extension of high signal zone over the glenoid margin and beneath the scapular aspect of the labrum (26). The normally thin and uniform articular cartilage terminates at the tip of the glenoid margin (see Case 15). A labrum tear may also be distinguished by irregularity of its margins (27).

Increased signal intensity of the glenoid labrum may be of degenerative or traumatic origin. Labrum degeneration due to aging is characterized by the development of focal or more conspicuous regions of increased signal intensity within the labral substance. Ordinarily labrum degeneration due to aging is a generalized process involving all aspects of the glenoid labrum, while traumatic lesions are more focal.

MR imaging is highly accurate in the diagnosis of Hill-Sachs lesions (27). These lesions are best seen on axial images at the level of the tip of the coracoid process (see Fig. 16J). They are also well visualized on coronal and sagittal images (Fig. 17E,F). This multiplanar capability allows for visualization of subtle and at time cartilaginous Hill-Sachs lesions as an aid in the diagnosis of subtle glenohumeral instabilities (25).

In conclusion, labrum detachment in anterior glenohumeral instability is a manifestation of a more extensive injury of the inferior glenohumeral ligament–capsule–labrum complex and is characteristically and most commonly, but not exclusively, found at the anterior–inferior glenoid margin. It cannot be overstated that accurate diagnosis of the capsulolabral lesions requires the use of a high-resolution imaging technique including appropriate imaging sequences. The use of a T2-weighted sequence is mandatory, and fat saturation is likely to add to the sensitivity of MR imaging in this aspect.

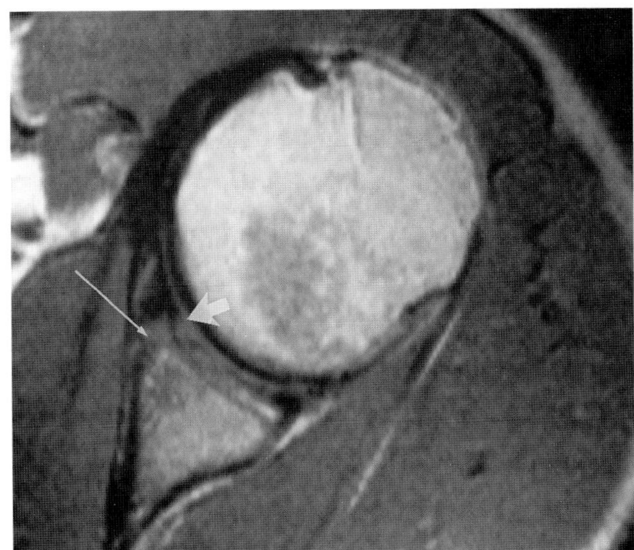
C

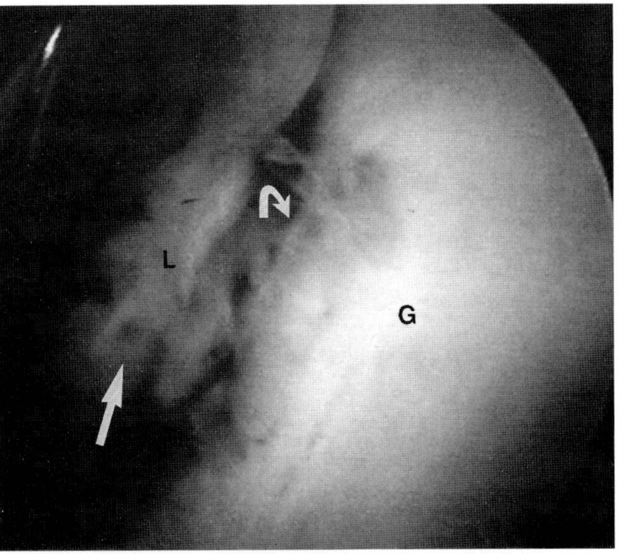

D

FIG. 17. C,D: Bankart lesion, in patient with recurrent dislocation. **C:** Proton-density (TR/TE 1,500/30 msec) axial image at distal third of glenoid. There is detachment of the anteroinferior labrum manifested by intermediate signal intensity (*short arrow*). This region showed only minimal increased signal intensity on T2-weighted images (not shown). The glenoid margin is rounded and deficient (*long arrow*). **D:** Arthroscopic view of anteroinferior glenoid margin. The irregularly torn and detached labrum is visualized (*straight arrow*). Note irregularity of the glenoid margin (*curved arrow*). L., labrum; G., glenoid. (From Turtle A, et al. Diagnostic imaging of the shoulder. In: Nicholas JA, and Hershman EB (eds). *Upper extremity in sports medicine.* St. Louis: CV Mosby; 1995:85–148, with permission.)

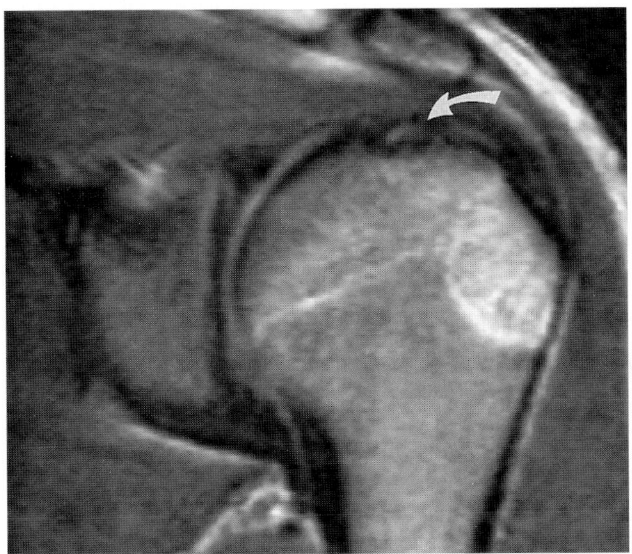
E

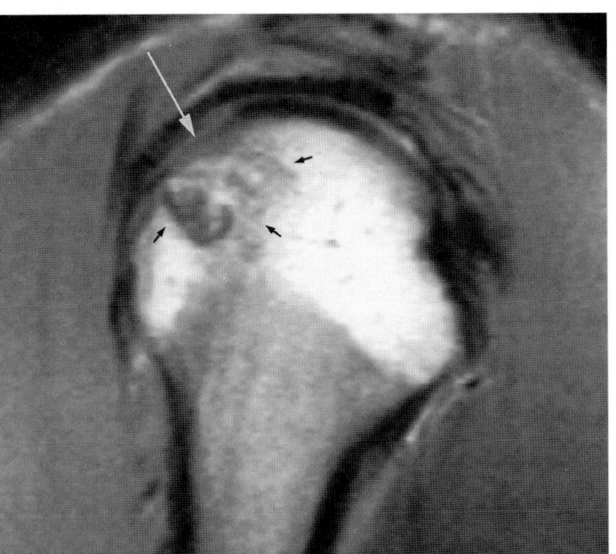
F

FIG. 17. E,F: Hill-Sachs lesions, two cases. **E:** Proton-density oblique coronal image, left shoulder. A superficial Hill-Sachs lesion is present (*arrow*). **F:** Proton density oblique sagittal image. In this patient the Hill-Sachs lesion is manifested by a slightly flattened articular surface (*long arrow*) and cystic resorption of the subarticular bone (*small arrows*). (From Rafii M. Shoulder. In: Firooznia H, Golimbu C, Rafii M, eds. *MRI and CT of the musculoskeletal system.* St. Louis: CV Mosby; 1990:465–549, with permission.)

CASE 18

History: A 16-year-old high school football player sustained an injury to his right shoulder while being tackled. He felt his shoulder temporally "went out" but was unsure of the direction of displacement. On physical examination his shoulder was generally sore, and a thorough examination could not be done. The overall impression from history and limited physical examination was that of self-reduced anterior dislocation. MR imaging was performed to confirm that impression and to determine the extent of soft tissue injury and whether or not it necessitated a surgical repair.

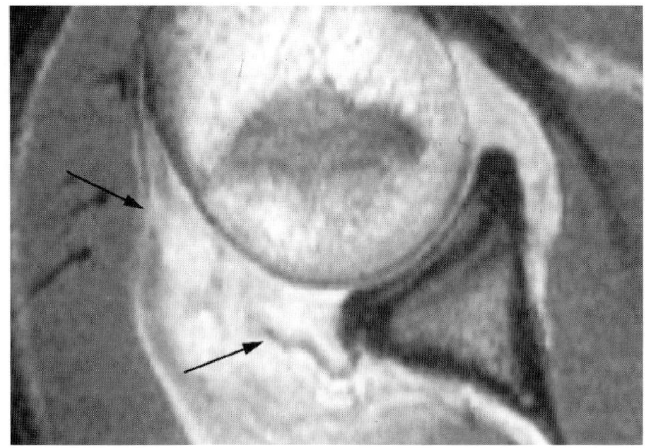

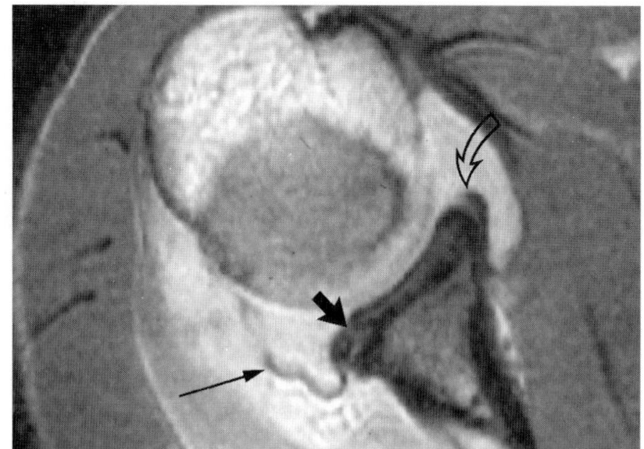

FIG. 18.

Findings: Figure 18A,B gives proton-density (TR/TE 2,100/30 msec) axial images at the mid- and distal glenoid levels. There is disruption of the posterior cuff and capsule (thin arrows), which are also separated by hemarthrosis. The posterior labrum is torn (thick arrow in Fig. 18B). The glenoid margin beneath the torn labrum is slightly impacted. There is a large hemarthrosis expanding the anterior capsule and extending into the posterior soft tissues. The anterior labrum is intact. The inferior glenohumeral ligament (open arrow in Fig. 18B) is visualized at its insertion to the anterior labrum. (From Rafii M, Firooznia H, Golimbu C, Weinreb J. Magnetic resonance imaging of glenohumeral instability. *MRI Clin North Am* 1993;1:87–104, with permission.)

Diagnosis: Acute posterior subluxation (or dislocation, self reduced).

Discussion: This case demonstrates the difficulty in assessing the shoulder clinically after acute injury. Although the mechanism of injury often provides helpful clues for diagnosis, in a contact sport such as football the mode of injury may not always be clear. Also, examination of the shoulder for instability in a well-built, muscular athlete is difficult and may be compounded by guarding due to pain. In these cases conventional MR imaging is most suited as a diagnostic tool for assessment of the site and the extent of soft tissue injury.

Posterior shoulder dislocation is uncommon, with an incidence of 2% and 3.8% observed in two large series (2,3). Posterior instability in the form of a recurrent subluxation is increasingly recognized as the cause of pain and shoulder disability, particularly among young athletes (4–7). The mechanism of injury is either a lone incident of trauma or repetitive applied athletic force (swimming, throwing, punching) of the abducting, flexing, and internally rotating humerus. Pre-existing laxity of the capsule may play a role as a factor predisposing to instability, or it may constitute the main etiologic factor (6).

Posterior glenohumeral instability, although uncommon, frequently presents the clinician with a diagnostic and therapeutic dilemma. Posterior instabilities are also less likely to be corrected by surgery (8). Pathologic findings at surgery or by diagnostic imaging modalities are often minimal; a tear or shredding of the labrum may be the only observation made (Fig. 18C). A detached capsulolabral complex, similar to a Bankart lesion, is an infrequent finding (Fig. 18D–F). Capsular tear and disruption of the posterior cuff may occur with more severe injuries and may result in formation of a subcapsular synovial recess or cyst (1,9).

Although the major forms of capsulolabral pathology in posterior instability are readily identified by MR imaging, the accuracy of MR imaging in the diagnosis of posterior labrum tears compared with other segments of the labrum is reduced (10). This may be related to the subtlety of most posterior labral lesions as well as the proximity of the capsule to the peripheral margin of the labrum. Not infrequently posterior labrum tears are also present in patients with frank anterior instability (or SLAP lesions) due to the traction force applied to the posterior capsulolabral complex at the time of anterior dislocation (10,11). These lesions do not necessarily represent a component of multidirectional instability, although they may result in excessive posterior laxity (see Fig. 16K). Treatment in these cases is primarily focused on correcting the anterior lesions (8,10).

This concept is also relevant in the treatment of multidirectional instability when the instability is more prominent in one direction. The role of MR imaging in these cases is to identify the predominant direction of instability. In unidirectional or multidirectional instability due to congenital laxity of the capsule, frequently the capsulolabral complex is intact, or only degeneration or fraying of the labrum is observed. Laxity of the capsule is more accurately diagnosed by using an arthrographic (MR or CT) technique. The role of MR imaging in these instances is to exclude positively the presence of a Bankart lesion and therefore direct the treatment to a capsular shift procedure (12).

Posterior dislocation may result in an impaction fracture of the anteromedial aspect of the humeral head (Fig. 18G,H). The "reverse Hill-Sachs" or McLaughlin fracture is usually vertically oriented (3). In recurrent posterior subluxation articular cartilage erosion and subarticular sclerosis of the anterior aspect of the humeral head may be observed. Following an acute event, bone contusion (either posterior or anterior subluxation or dislocation) may also be observed and in subtle cases may be useful in documenting the direction of instability. Various changes of the posterior glenoid margin including erosion, sclerosis, or ectopic bone formation may occasionally be present (7).

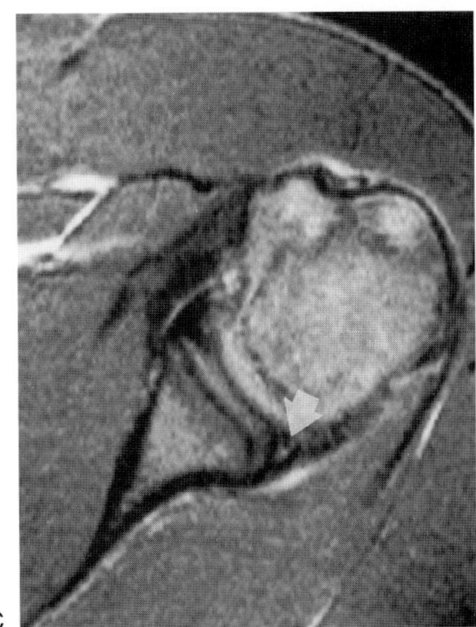

FIG. 18. C: Posterior instability. Proton-density (TR/TE 2,100/30 msec) axial image, left shoulder demonstrates a tear of the posterior labrum (*arrow*). This tear extended along the entire posterior glenoid margin (not shown).

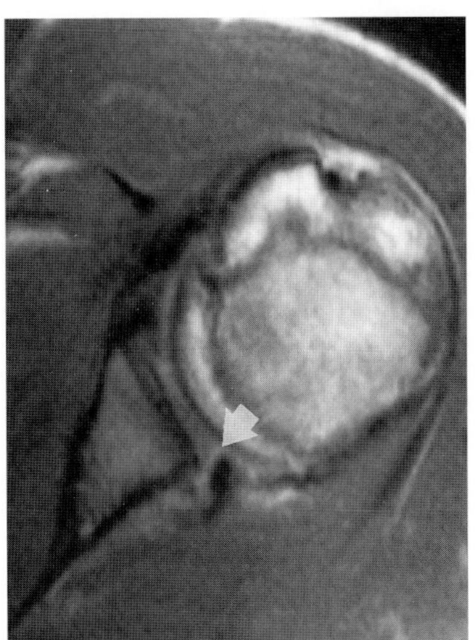

Fig. 18. D–F: Posterior instability, "reverse Bankart" lesion.

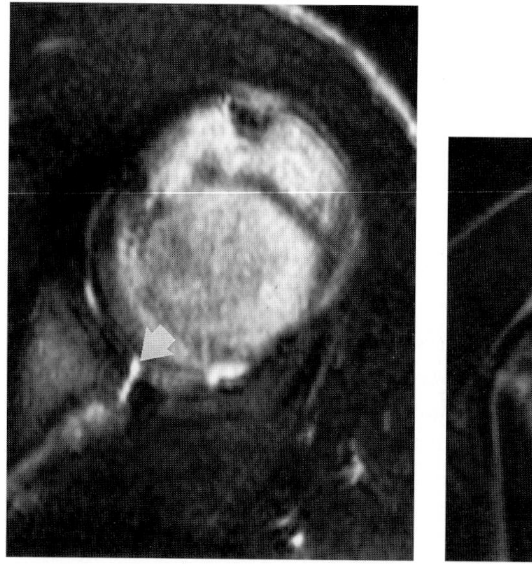
E

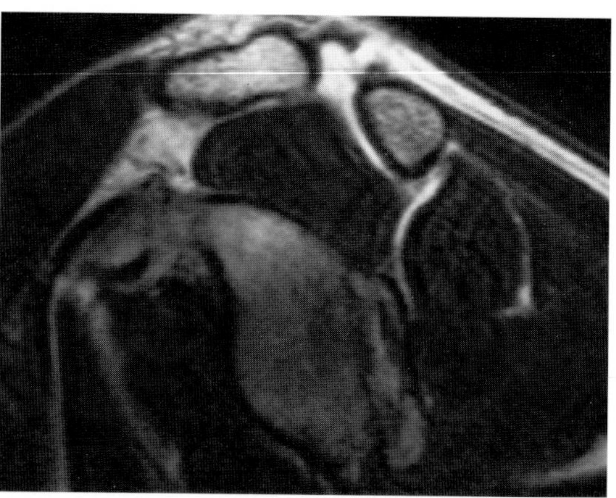
F

Fig. 18. D,E: Proton-density and T2-weighted (2,100/30/80 msec) axial images at the mid-glenohumeral joint, left shoulder. There is detachment of the posterior capsulolabral complex (*arrow*). **F:** Proton-density (1,800/29 msec) oblique sagittal image. The posterior capsulolateral detachment extends along the entire posterior glenoid margin.

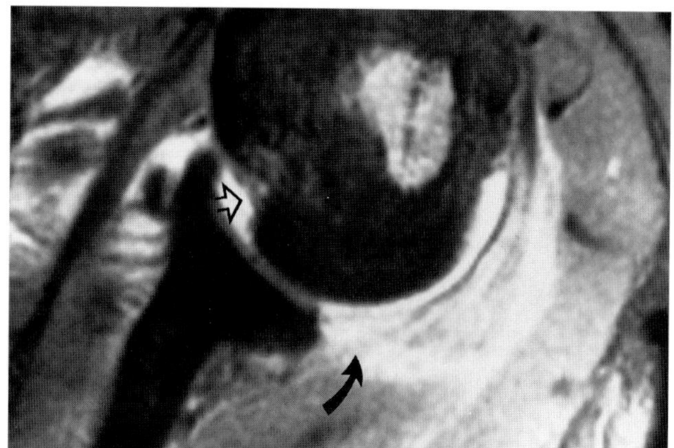
G

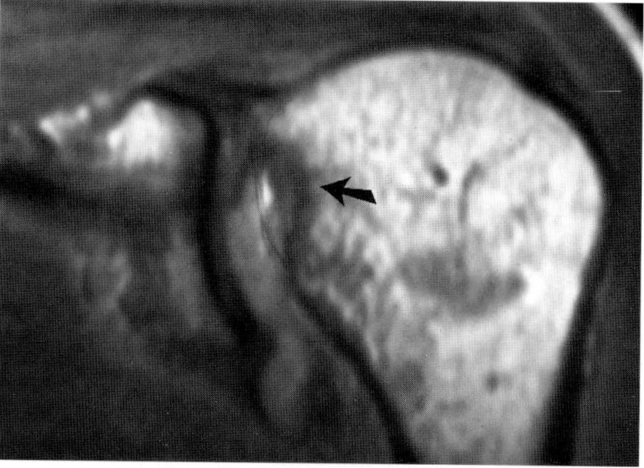
H

Fig. 18. G,H: Reversed Hill-Sachs lesion following posterior dislocation **G:** MPGR (450/15/20 msec) axial image at the mid-glenoid, left shoulder. There is a compression fracture of the anterior humeral head (*open arrow*). The posterior labrum is avulsed, and the capsule is disrupted (*curved arrow*) **H:** Proton-density (2,200/15 msec) oblique coronal image. The extent of the vertically oriented defect is visualized (*arrow*). The medullary lesion of the humeral head is presumed to represent a benign enchondroma. (Courtesy of Dr. Edward Lubat.) (From Rafii M, Firooznia H, Golimbu C, Weinreb J: Magnetic resonance imaging of glenohumeral instability. *MRI Clin North Am* 1993;1:87–104, with permission.)

CASE 19

History: A 35-year-old male injured his right shoulder while swimming in shallow waters in the ocean. He was lifted by a wave and landed on his extended arm on the bottom of the ocean. On physical examination he had tenderness over anterolateral and posterior aspect of the shoulder. Impingement signs were positive. He was treated with conservative measures for 2 months; he had persistent pain and tenderness over the greater tuberosity region. MR imaging was performed to assess the rotator cuff.

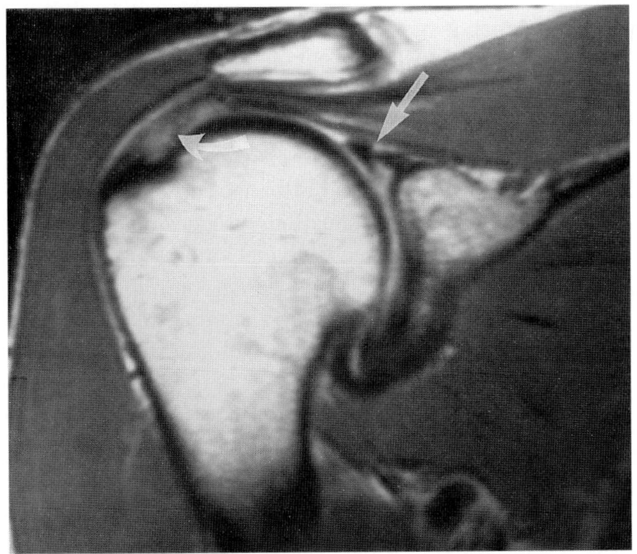
A

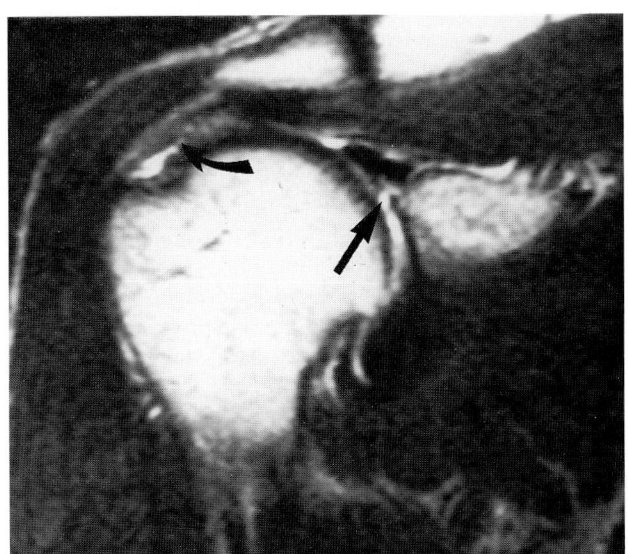
B

FIG. 19.

Findings: Figure 19A,B, proton-density and T2-weighted (TR/TE 2,000/15/90 msec) oblique coronal images of right shoulder. There is an oblique tear through the substance of the superior labrum (straight arrows). The torn segment, which includes attachment of the long head of the biceps tendon, is centrally displaced. There is also a large partial tear of the synovial surface and tuberosity insertion of the supraspinatus tendon (curved arrows). Other images showed labrum tear from the anterosuperior to posterosuperior aspect of the glenoid.

Diagnosis: Complete superior labrum–biceps anchor tear (SLAP lesion). Large partial supraspinatus rotator cuff tear.

Discussion: At surgery the presence of a complete tear of the superior labrum–biceps anchor was confirmed. Both the unstable labrum segment and the rotator cuff tear were repaired.

Isolated tear or detachment of the superior glenoid labrum was first described by Andrews et al. (1) in baseball pitchers and was suggested to be the result of forceful contraction of the long head of the biceps tendon in the follow-through phase of the throwing motion. More recently Synder et al. (2) described a distinct pattern of traumatic lesions of the superior labrum in young individuals. These SLAP injuries involve the superior labrum–biceps anchor with extension posterosuperiorly as well as anterosuperiorly to or about the glenoid notch. SLAP lesions can result from two basic mechanisms of injury. Most common is a compression force due to a fall on an outstretched arm with the shoulder in abduction and slight forward flexion. The second mechanism is a traction force as a result of either sudden pull on the arm or throwing or overhead sports motions (2–4). In some patients, however, the onset of symptoms is insidious and a specific injury is not found (2).

Based on arthroscopic visualization in 27 patients, Synder et al. (2) classified SLAP lesions into four types. In type I the superior labrum is frayed with no tear or detachment from the glenoid, and the biceps anchor is intact. In type II the superior labrum is frayed and detached from the glenoid along with the biceps anchor. A type III lesion is a bucket-handle tear of the superior labrum. The central portion of the torn labrum is unstable. The peripheral segment, which includes the biceps anchor, remains attached to the glenoid in this type of injury. The type IV lesion is a bucket-handle tear of the labrum that extends into the biceps tendon, forming an unstable flap. In summary the biceps tendon and anchor are intact in type I and III, the biceps anchor is disrupted in type II, and the tendon is partially torn in type IV lesions (5). Types II and IV constitute unstable lesions.

MR imaging affords improved visualization of the superior glenoid labrum in the oblique coronal plane and has therefore enhanced depiction of superior labrum–biceps anchor lesions. However, the accuracy of MR imaging in the diagnosis and classification of SLAP lesions has not been definitively determined (3,4,4A). In a retrospective study of nine patients, eight of whom had undergone MR arthrography, Hodler et al. (4) concluded that early traumatic lesions of the superior labrum such as fraying cannot be accurately detected with MR imaging. Based on this study, complete detachments may be correctly diagnosed if the patient's age and history are taken into consideration, although differentiation between partial or complete detachments may not be possible. Both lesions require surgical intervention (2,4). In another prospective study in ten patients, Cartland et al. (3) concluded that MR images correlated well with surgical findings and were able to classify SLAP lesions fairly accurately. MR-arthrography was not considered helpful in visualizing these lesions (3). However, a larger and prospective study is required to determine the MR imaging criteria for SLAP lesions and their correlation with surgical findings.

Fraying of the superior labrum is manifested on MR images as an irregular and ill-defined outline of the labrum. Globular regions of increased signal intensity interposed between the labrum and the glenoid margin were reported by Cartland et al. (3) as type II SLAP lesions. At arthroscopy, we have seen that a globular region of intermediate signal within the labral substance represents labral fraying with a flap tear (Fig. 19C) (6).

Detachment of the superior labrum–biceps anchor is readily diagnosed when there is fluid entrapped within the detached region or a wide zone of increased signal intensity is present (Fig. 19D). Partial or nondisplaced complete tears of the labral substance may appear as linear intermediate signal intensity interrupting the articular surface of the labrum or extending from the articular surface to the capsular surface of the labrum respectively (Fig. 19E). Interruption of both labral surfaces is an indication of a bucket-handle tear or type III lesion (Fig. 19E). The type IV lesion is characterized by visualization of a labrum tear that also extends longitudinally into the biceps tendon (Fig. 19F,G). Generalized increased signal intensity of the superior labrum–biceps anchor has also been reported in type IV SLAP lesions (3). Although SLAP lesions are best visualized on oblique coronal images, correlative abnormalities have also been reported on axial and oblique sagittal images (4A).

For proper evaluation of the superior glenoid labrum lesions, anatomic features and normal variations of this labral segment must be considered. On MR images the normal intermediate signal intensity glenoid hyaline articular cartilage is visualized extending beneath the superior labrum over the glenoid margin (see Case 1). A recess may also be normally present beneath the articular aspect of the superior glenoid labrum (7). This may mimic a partial detachment (3). Age-related degenerative changes of the glenoid labrum should also be recognized and differentiated from traumatic lesion. As shown by DePalma (8), a degenerative process of the glenoid labrum is usually more prominent superiorly and includes partial detachment from the glenoid articular surface, fissuring of the labral substance, attenuation, or occasionally enlargement of the labrum. These changes may appear on MR images as irregularity of the labral outline as well as increased signal intensity of the labrum substance (6,9). Correlation with patient's age as well as history are therefore important in differentiating age-related changes of the labrum from traumatic lesion and for more accurate diagnosis of SLAP lesions.

Although commonly an isolated finding, other pathologic changes may accompany injuries of the superior glenoid labrum. The most important of these include partial or full-thickness rotator cuff tears (Fig. 19A,B) and anterior glenohumeral instability (Fig. 19F,G) (2,4). Ganglion cysts are also observed in association with SLAP lesions, probably reflecting their common traumatic origin (Fig. 19H,I) (6). Large ganglion cysts may result in suprascapular nerve entrapment and may further complicate the clinical presentation in these patients (10,11). One expects to observe atrophy of the supraspinatus and infraspinatus muscles in long-standing cases. No other MR imaging findings relevant to spinati muscles have been reported, and the presence of suprascapular nerve entrapment must be further confirmed by electromyographic studies. Large ganglion cysts in the suprascapular fossa may also result in secondary impingement syndrome by decreasing the subacromial space (Fig. 19I). Ganglion or labral cysts are also observed at other sites in the vicinity of the glenohumeral joint; association with labrum tear in most, and glenohumeral instability in a significant number of patients has been reported (6,12).

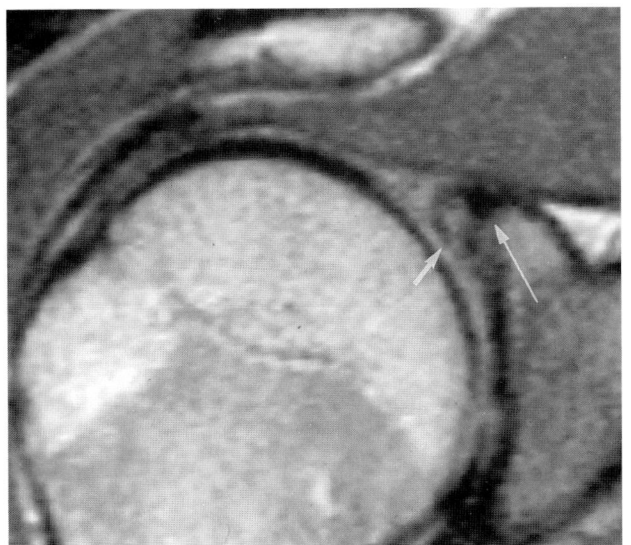

FIG. 19. C: SLAP lesion in a 29-year-old recreational athlete with insidious onset of shoulder pain. Proton-density (TR/TE 2,100/30 msec) oblique coronal image. There is a globular intermediate signal within the articular aspect of the superior labrum, which is also superficially irregular (*short arrow*). Fraying of this segment of the labrum and a flap tear were observed at surgery. The peripheral aspect of the labrum including the biceps anchor was intact and attached to the glenoid, as is seen in this image (*long arrow*).

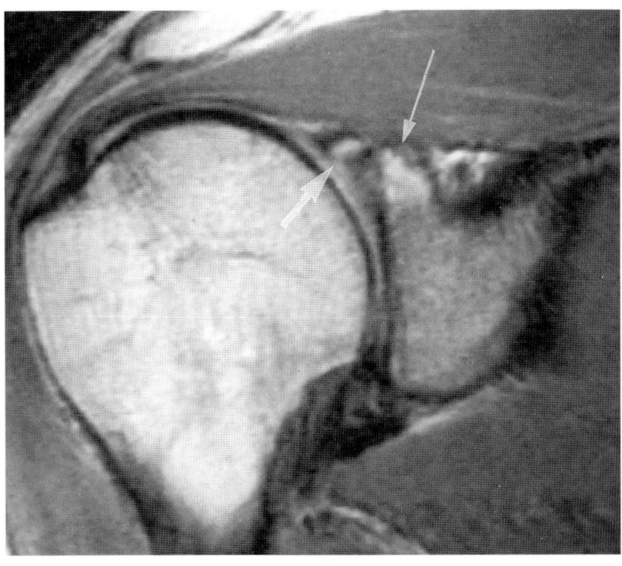

FIG. 19. D: SLAP lesion, with complete detachment in a 42-year-old recreational tennis player. Proton-density (TR/TE 2,100/30 msec) oblique coronal image shows detachment and displacement of the superior labrum from the glenoid margin (*short arrow*). This is consistent with a type II (unstable) lesion. Note irregularity of the bare superior glenoid margin and alteration in subcortical bone marrow signal, indicating chronicity of injury (*long arrow*). (From Rafii M. Magnetic resonance imaging of the shoulder joint. In: Beltran H, ed. *Current review of MRI.* Hong Kong: Paramount Printing Group; 1995:203–214, with permission.)

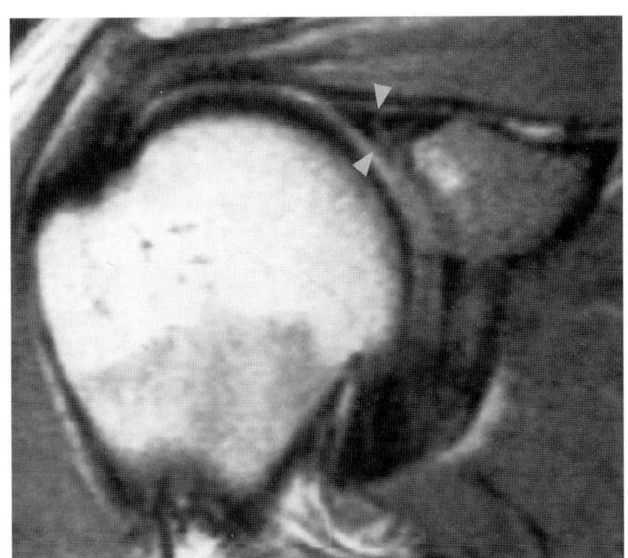

FIG. 19. E: SLAP lesion, bucket-handle tear in a 27-year-old recreational athlete. Proton-density (TR/TE 2,100/ 30 msec) oblique coronal image. Tear of the superior labrum is shown as an irregular line of intermediate signal interrupting both the capsular and the articular surface of the superior labrum (*arrowheads*). A ganglion cyst was present posterosuperiorly (not shown). (From Rafii M. Shoulder. In: Firooznia H, Golimbu C, Rafii M, eds. *MRI and CT of the musculoskeletal system.* St. Louis: CV Mosby; 1990:465–549, with permission.)

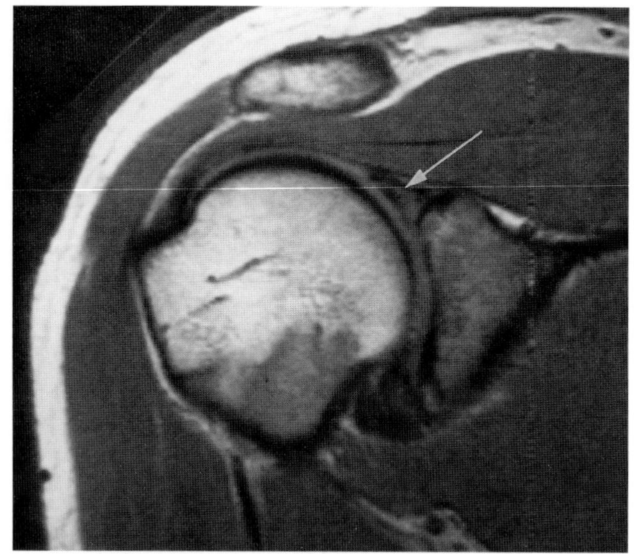
F

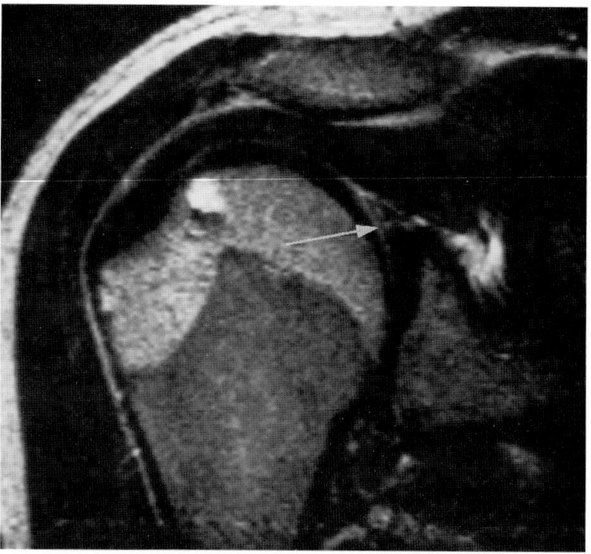
G

FIG. 19. F,G: SLAP lesion, type IV in a 16-year-old varsity baseball player (pitcher) with insidious onset of shoulder pain. **F:** T2-weighted (TR/TE 2,000/90 msec) obllique coronal image, posterior to F. At this level complete detachment of the superior-posterior labrum is present (*arrow*). Note the subcortical cyst of the posterosuperior aspect of the humeral head, an indication of posterosuperior impingement in this young athlete. Proton-density (TR/TE 2,000/15 msec) oblique cornal image demonstrates a superior labrum biceps anchor tear (*arrow*).

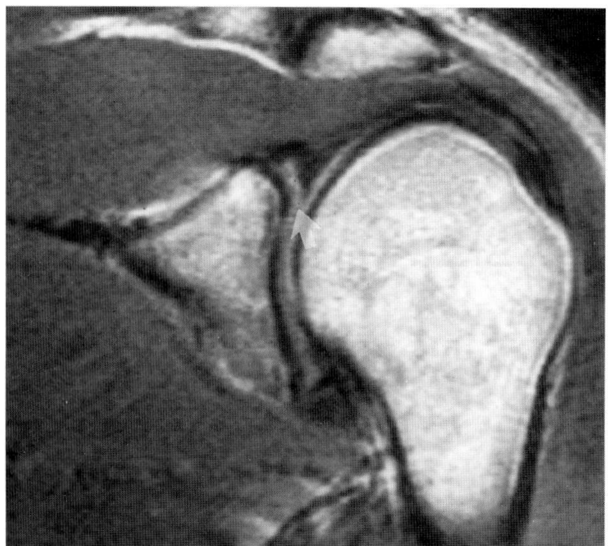
H

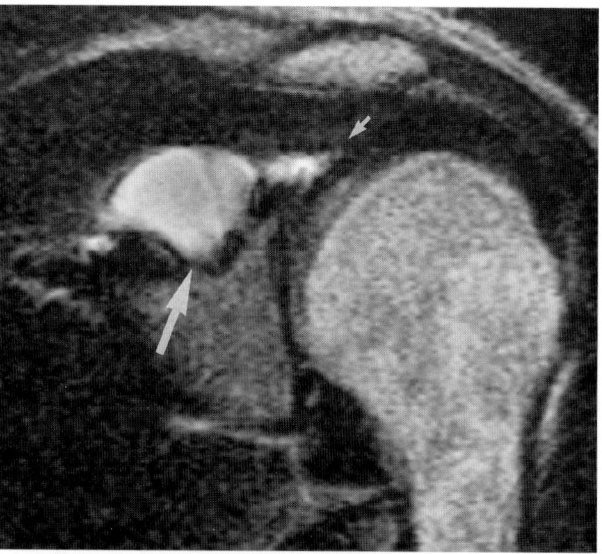
I

FIG. 19. H,I: SLAP lesion and ganglion cyst in a 26-year-old man with shoulder pain. **H:** Proton-density (TR/TE 2,100/30 msec) oblique coronal image demonstrates detachment of the superior glenoid labrum (*arrow*). This lesion was present on other images anteriorly and appeared as high signal on T2-weighted images (not shown). **I:** T2-weighted (TR/TE 2,100/80 msec) oblique coronal image posterior to H and at the level of the spinoglenoid notch. There is a well-demarcated fluid-filled cyst at the posterosuperior aspect of the glenohumeral joint. This ganglion cyst extends laterally between the joint capsule and the supraspinatus muscle (*small arrow*). The major component of the cyst extends medially into the spinoglenoid notch where the suprascapular neurovascular bundle is not clearly visualized (*long arrow*). It also results in elevation of the supraspinatus muscle. This patient did not exhibit clinical manifestations of suprascapular nerve palsy.

CASE 20

History: A 2-month-old male infant presented with irritability and soft tissue swelling in the region of the scapula.

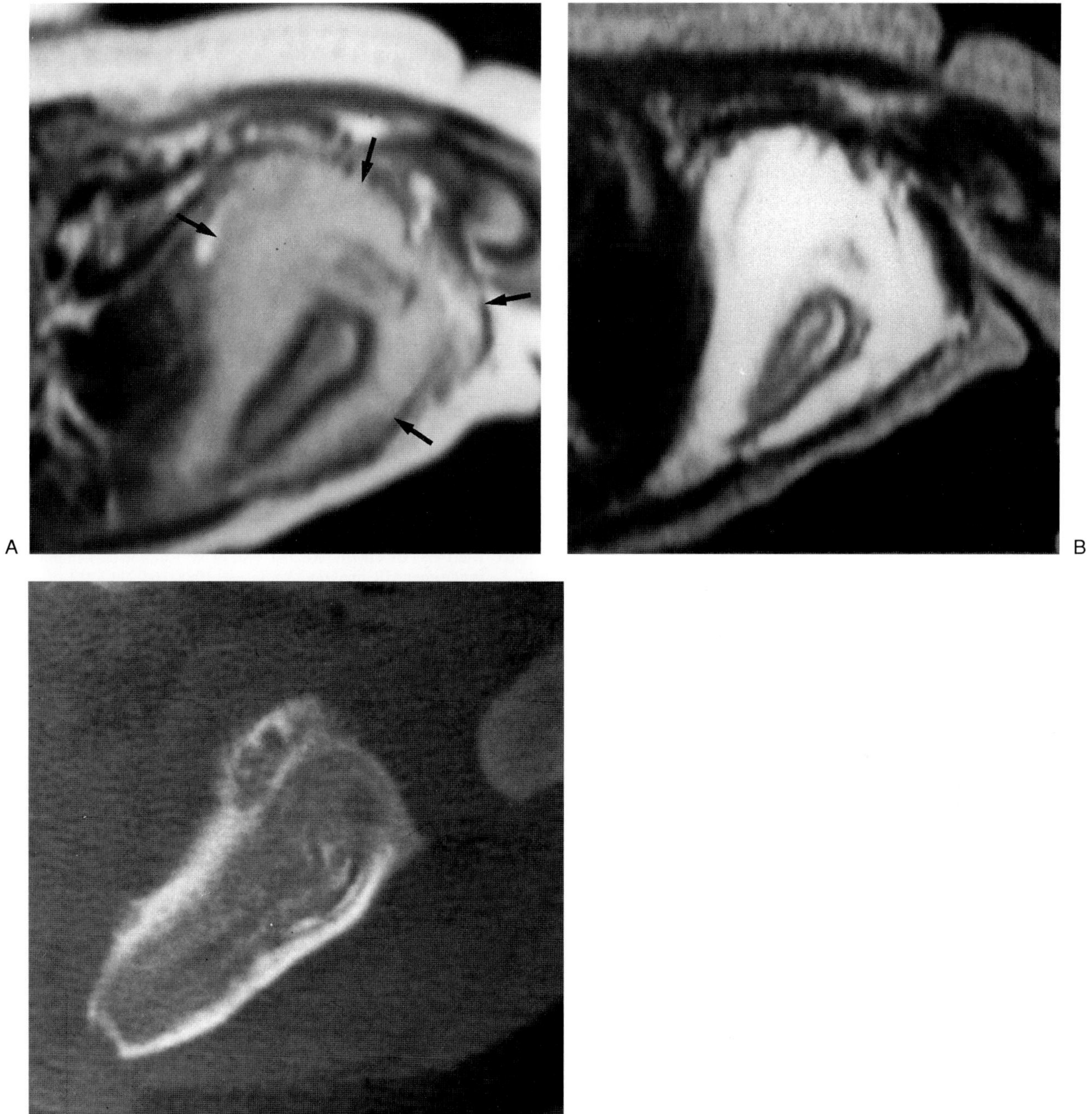

FIG. 20.

Findings: Axial proton-density image (Fig. 20A) (TR/TE 2,000/20 msec) demonstrates a thick ''rim'' of intermediate signal intensity entirely surrounding the scapula (arrows). Axial T2-weighted spin-echo image demonstrates that the signal encasing the scapula markedly increases in intensity (Fig. 20B). Axial CT demonstrates a thick cloak of new bone encasing the scapula (Fig. 20C). (Courtesy of Lionel Young, MD, Loma Linda, California.)

Diagnosis: Infantile cortical hyperostosis (Caffey's disease).

Discussion: First reported by Caffey and Silverman (1) in 1945, this uncommon condition commences in infancy and is characterized clinically by fever (often abrupt onset), hyperirritability, and soft tissue swelling. Physical examination may reveal hard, indurated, and tender soft tissue masses, reflecting soft tissue extension of the exuberant periosteal reaction that is one of the hallmarks of this condition. The soft tissue changes may antedate radiographically evident findings. The clinical course is extremely variable. Clinical and radiographic features, as in the present case, may resolve over a period of months to years but on occasion, active and recurring disease may be encountered that can contribute to delay in musculoskeletal development and crippling deformities.

The mandible, clavicles, and ribs are most commonly involved (the scapula is the fourth most commonly involved bone), and involvement is typically but not invariably polyostotic. The scapula is the most common site of monostotic involvement, and changes at this site may be particularly exuberant and associated with neurologic deficit and diaphragmatic elevation. Hyperostosis of the cortex is the most characteristic radiographic feature of the disease. New bone formation begins in the soft tissue swelling directly contiguous to the original cortex of the involved bone. Progressive deposition of new bone may reach dramatic proportion, often doubling or tripling the normal bone width. The epiphyseal ossification centers are generally spared. Endosteal new bone may impinge on the medullary canal. With "healing" of the condition, radiographic findings may entirely resolve or residual changes (including diaphyseal expansion, longitudinal overgrowth, bowing deformities, and osseous bridging) may be evident.

Pathologically, acute inflammatory changes are demonstrated in the periosteal membrane characterized by edema and cellular and polymorphonuclear leukocyte infiltration. The altered periosteum blends with the adjacent fasciae, tendons, and muscles. Beneath the periosteum, coarsened trabeculae become evident. Despite intense investigation, the etiology of infantile cortical hyperostosis remains unknown. An infectious (possibly viral) basis has been suggested and is supported by many of the clinical and pathologic features of the disease. An allergic response to altered collagen has also been suggested as a possible cause of the condition.

No prior experience with the MR findings in this condition have been reported. The rind of soft tissue involvement encasing the markedly expanded although not destroyed underlying bone precisely mirrors the known pathologic changes. In the differential diagnosis of the MR findings the possibility of infantile Ewing's tumor, neuroblastoma, and infection could be entertained. Additionally, both rickets and scurvy can result in periostitis and hyperostosis in the infant, but the absence of epiphyseal or metaphyseal changes and the clinical course of infantile hyperostosis should allow ready distinction of these entities. The manifestations of hypervitaminosis A, another diagnostic consideration, become apparent toward the end of the first year of life; mandibular involvement and facial swelling is uncommon, and metatarsal predilection is encountered. Cortical hyperostosis and periostitis, similar to that seen in infantile cortical hyperostosis, is also encountered in infants receiving long-term administration of prostaglandin E_1 for ductus-dependent cyanotic congenital heart disease. In this regard, increased levels of prostaglandin E_1 have been described in patients with Caffey's disease and may provide an etiologic basis for the condition.

REFERENCES

Introduction

1. Huber DJ, Sauter RS, Muller E, et al. MR imaging of the normal shoulder. *Radiology* 1986;158:405–408.
2. Middleton WD, Kneeland JB, Carrera GF. High resolution MR imaging of the normal rotator cuff. *AJR* 1987;148:559–564.
3. Kieft GJ, Bloem JL, Oberman WR. Normal shoulder: MR imaging. 1986;159:741–745.
4. Seeger LL, Ruszkowski JT, Bassett LW, et al. MR imaging of the normal shoulder: anatomic correlation. *AJR* 1987;148:83–91.
5. Kneeland JB, Carrera GF, Middleton WD, et al. Rotator cuff tears: preliminary application of high-resonators. *Radiology* 1986;160: 695–699.
6. Kneeland BJ, Middleton WD, Carrera GF, et al. MR imaging of the shoulder: diagnosis of rotator cuff tears. *AJR* 1987;149:333–337.
7. Zlatkin MB, Reicher MA, Kellerhouse LE, et al. The painful shoulder: MR imaging of the glenohumeral joint. *J Comput Assist Tomogr* 1988;12:995.
8. Seeger LL, Cold RH, Bassett LW, Ellman H. Shoulder impingement syndrome. MR findings in 53 shoulders. *AJR* 1988;150:343–347.
9. Evancho AM, Stiles RG, Fajman WA, et al. MR imaging diagnosis of rotator cuff tears. *AJR* 1988;151:751–754.
10. Kieft GJ, Bloem JL, Rozing PM, et al. Rotator cuff impingement syndrome: MR imaging. *Radiology* 1988;166:211–214.
11. Kieft GJ, Sartoris DJ, Bloem JL, et al. Magnetic resonance imaging of the glenohumeral joint disease. *Skeletal Radiol* 1987;26:285–290.
12. Zlatkin MB, Reicher MA, Kellerhouse LE, et al. The painful shoulder: MR imaging of the glenohumderal joint. *J Comput Assist Tomogr* 1988;12:995.
13. Zlatkin MB, Dalinka MK, Kressel H. Magnetic resonance imaging of the shoulder. *Magn Reson Q* 1989;5:3.
14. Zlatkin MB, Ionnatti JP, Robert MC, et al. Rotator cuff tears: diagnostic performance of MR imaging. *Radiology* 1989;172:223.
15. Iannotti JP, Zlatkin MB, Esterhai JL, et al. Magnetic resonance imaging of the shoulder. Sensitivity, specificity and predictive value. *J Bone Surg* 1991;73A:17–29.
16. Rafii M, Firooznia H, Sherman O, et al. Rotator cuff lesions: signal pattern at MR imaging. *Radiology* 1990;177:817.
17. Farley TE, Neumann CH, Steinbach LS, et al. Full-thickness tears of the rotator cuff of the shoulder: diagnosis with MR imaging. *AJR* 1992;158:347–351.
18. Burk DL, Karasick DK, Kurtz AB, et al. Rotator cuff tears: Prospective comparison of MR imaging with arthrography, sonography, and surgery. *AJR* 1989;153:87–92.
19. Habibian A, Staufer A, Resnick D, et al. Comparison of Conventional and completed arthrotomography with MR imaging of shoulder. *J Comput Axial Tomogr* 1989;13:968–975.
20. Kaplan PA, Bryans KC, Davik JP, et al. MR imaging of the normal shoulder: variants and pitfalls. *Radiology* 1992;184:519.
21. Neumann CH, Holt RG, Steinbach LS, et al. MR imaging of the shoulder: appearance of supraspinatus tendon in asymptomatic volunteers. *AJR* 1992;158:1281.
22. Vahlensieck M, Pollack M, Lang P, et al. Two segments of supraspinatus muscle: cause of high signal intensity at MR imaging? *Radiology* 1993;186:449.
23. Liou JT, Wilson AJ, Totty WG, Brown JJ. The normal shoulder: common variations that stimulate pathologic conditions at MR imaging. *Radiology* 1993;186:435.
24. Erickson SJ, Cox IH, Hyde JS, et al. Effect of tendon orientation on MR imaging signal intensity: a manifestation of the "magic angle" phenomenon. *Radiology* 1991;181:389.
25. Chandnani V, HO C, Gerharter J, et al. MR findings in asymptomatic shoulders: a blind analysis using symptomatic shoulders as controls. *Clin Imaging* 1992;16:25.

26. Kjellin I, Ho CP, Cervilla V, et al. Alterations in the supraspinatus tendon at MR imaging: correlation with histopathologic findings in cadavers. *Radiology* 1991;181:837–841.
27. Seeger LL, Gold RH, Bassett LW: Shoulder instability: evaluation with MR imaging. *Radiology* 1988;168:695–697.
28. Kieft GJ, Bloem JL, Rozing PM, et al. MR imaging of recurrent anterior dislocation of the shoulder: comparison with CT arthrography. *AJR* 1988;150:1083–1087.
29. Legan JM, Burkhard TK, Goff WBII, et al. Tears of the glenoid labrum: MR imaging of 88 arthroscopically confirmed cases. *Radiology* 1991;179:241.
30. Garneau RA, Renfrew DL, Moore TE, et al. Glenoid labrum: evaluation with MR imaging. *Radiology* 1991;179:519–522.
31. Cartland JP, Crues JVI, Stauffer A, et al. MR imaging in evaluation of SLAP injuries of shoulder: findings in 10 patients. *Am J Radiol* 1992;159:787.
32. Moseley HF, Overgaard B. The anterior capsular mechanism in recurrent anterior dislocation of shoulder. Morphological and clinical studies with special reference to the glenoid labrum and the glenohumeral ligaments. *J Bone Joint Surg* 1962;44B:913.
33. Depalma AF. Regional variational and surgical anatomy. In: DePalma AF, ed. *Surgery of the shoulder.* Philadelphia: JB Lippincott, 1983.
34. Zlatkin MB, Bjorkengren AG, Gylys-Morin V, et al. Cross-sectional imaging of the capsular mechanism of the glenohumeral joint. *AJR* 1988;150:151.
35. McNiesh LM, Callaghan JJ. CT arthrography of the shoulder: variations of the glenoid labrum. *AJR* 1987;149:963.
36. Neumann CH, Petersen SA, Jahnke AH. MR imaging of the labral-capsular complex: normal variations. *AJR* 1991;157:1015.
37. McCauley TR, Pope CF, Jokl P. Normal and abnormal glenoid labrum: assessment with multiplanar gradient-echo MR imaging. *Radiology* 1992;183:35.
38. Flanningan B, Kursunoglu-Brahme S, Synder S, et al. MR arthrography of the shoulder: comparison with conventional MR imaging. *AJR* 1990;155:829.
39. Palmer WE, Brown JH, Rosenthal DI. Rotator cuff: evaluation with fat-suppressed MR arthrography. *Radiology* 1993;188:683.
40. Hodler J, Kursunoglu-Brahme S, Flanningan B, et al. Injuries of the superior portion of the glenoid labrum involving the insertion of the biceps tendon. *AJR* 1992;159:565–568.
41. Palmer WE, Caslowitz PL. Anterior shoulder instability. Diagnostic criteria determined from prospective analysis of 121 MR arthrograms. *Radiology* 1995;197:819.
42. Chandnani VP, Gagliardi JA, Murnane TG, Bradley YC, DeBerardino TA, Spaeth J, Hansen MF. Glenohumeral ligaments and shoulder capsular mechanism: evaluation with MR arthrography. *Radiology* 1995;196:27.
43. Rafii M. Shoulder. In: Firooznia et al., eds. *MRI and CT of musculoskeletal system.* St. Louis: Mosby-Year Book; 1992:645.
44. Tirman PFJ, Applegate GR, Flannigan BD, et al. Magnetic resonance arthrography of the shoulder. *MRI Clin North Am* 1993;1:125.
45. Mirowitz SA. Imaging techniques, normal variations and diagnostic pitfalls in shoulder magnetic resonance imaging. *MRI Clin North Am* 1992;1:19.
46. Resendes M, Helms CA, Eddy R, et al. Double-echo MPGR imaging of the rotator cuff. *J Comput Assist Tomogr* 1991;15:1077.
47. Mirowitz SA, Shady K, Reinus WR. Diagnostic performance of fat suppression MR imaging for detection of rotator cuff pathology. *Radiology* 1991;181:247.
48. Traughber PD, Goodwin TE. Shoulder MRI: arthroscopic correlation with emphasis on partial tears. *J Comput Assist Tomogr* 1992;16:129.
49. Mirowitz SA. Normal rotator cuff: MR imaging with conventional and fat-suppression techniques. *Radiology* 1991;180:735.

Case 1

1. Tsai JC, Zlatkin MB. Magnetic resonance imaging of the shoulder. *Radiol Clin North Am* 1990;28:279.
2. Mirowitz SA. Normal rotator cuff: MR imaging with conventional and fat-suppression techniques. *Radiology* 1991;180:735–740.
3. Davis SJ, Teresi LM, Bradley WG, et al. Effect of arm rotation on MR imaging of the rotator cuff. *Radiology* 1991;181:265–268.
4. Kaplan PA, Bryans KC, Davick JP, et al. MR imaging of the normal shoulder: variants and pitfalls. *Radiology* 1992;184:519–524.
5. Neumann CH, Holt RG, Steinbach LS, et al. MR imaging of the shoulder: appearance of the supraspinatus tendon in asymptomatic volunteers. *AJR* 1992;158:1281–1287.
6. Vahlensieck M, Pollack M, Lang P, et al. Two segments of supraspinatus muscle: cause of high signal intensity at MR imaging? *Radiology* 1993;186:449–454.
7. Liou JT, Wilson AJ, Totty WG, Brown JJ. The normal shoulder: common variations that simulate pathologic conditions at MR imaging. *Radiology* 1993;186:435–441.
8. Skinner HA. Anatomical considerations relative to rupture of the supraspinatus tendon. *J Bone Joint Surg* 1937;19:137–151.
9. Mirowitz SA. Imaging techniques, normal variations and diagnostic pitfalls in shoulder magnetic resonance imaging. *MRI Clin North Am* 1993;1:19–36.
10. Neer CS. Anterior acromioplasty for chronic impingement syndrome in the shoulder: a preliminary report. *J Bone Joint Surg [Am]* 1972;54:41–50.
11. Rafii M. Shoulder. In: Firooznia H, et al., eds. *MRI and CT of the musculoskeletal system.* St. Louis: Mosby Yearbook Publisher; 1992:465.
12. Rafii M, Firooznia H, Sherman O, et al. Rotator cuff lesions: signal pattern at MR imaging. *Radiology* 1990;177:817–823.
13. Zlatkin MB, Ionnatti JP, Roberts MC, et al. Rotator cuff tears: diagnostic performance of MR imaging. *Radiology* 1989;172:223–229.
14. Kejellin I, Ho CP, Cervilla V, et al. Alterations in the supraspinatus tendon at MR imaging: correlation with histopathologic findings in cadavers. *Radiology* 1991;181:837–841.
15. Fullerton GD, Cameron IL, Ord VA. Orientations of tendons in magnetic field and its effect on T2 relaxation times. *Radiology* 1985;155:433–435.
16. Erickson SJ, Cox IH, Hyde JS, et al. Effect of tendon orientation on MR imaging signal intensity: a manifestation of the "magic angle" phenomenon. *Radiology* 1991;181:389–392.
17. Timins, ME, et al. Increased signal intensity of the normal supraspinatus tendon on MR imaging: Diagnostic pitfall caused by the magic angel effect. *AJR* 1995;164:109–114.

Case 2

1. Codman EA. *The shoulder. Rupture of the supraspinatus tendon and other lesions in or about the subacromial space.* Boston: Thomas Todd; 1934.
2. Bland JH, Merrit JA, Boushey DR. The painful shoulder. *Semin Arthritis Rheum* 1977;7:21–47.
3. Rogers LF, Hendrix RW. The painful shoulder. *Radiol Clin North Am* 1988;26:1359–1371.
4. Moseley HF. Athletic injuries to the shoulder region. *Am J Surg* 1959;98:401–422.
5. Jobe FW, Jobe CM. Painful athletic injuries of the shoulder. *Clin Orthop* 1983;173:124.
6. Leach RE, Schepsis AA. Shoulder pain. *Clin Sports Med* 1983;2:123–135.
7. Rothman RH, Parke WW. The vascular anatomy of the rotator cuff. *Clin Orthop* 1965;41:176–186.
8. Rathbun JB, Macnab I. The microvascular pattern of the rotator cuff. *J Bone Surg [Br]* 1970;52:540–553.
9. Neer CS. Anterior acromioplasty for chronic impingement syndrome in the shoulder: a preliminary report. *J Bone Joint Surg [Am]* 1972;54:41–50.
10. Neer CS II. Impingement lesions. *Clin Orthop* 1983;173:70–77.
11. Jobe FW, Kvitne RS. Shoulder pain in the overhand or throwing athlete: the relationship of anterior instability and rotator cuff impingement. *Orthop Rev* 1989;18:963.
12. Nirschl RP. Rotator cuff tendinitis: basic concepts of pathoetiology. *Instr Course Lect* 1989;38:439–445.
13. Uhthoff HK, Loehr J, Sarkar K. The pathogenesis of rotator cuff tears. In: Takagishi N, ed. *The shoulder.* Tokyo: Tokyo Professional Postgraduate Services; 1987.
14. Ozaki J, Fujimato S, Nakagawa Y, et al. Tears of the rotator cuff of the shoulder associated with pathological changes in the acromion: a study in cadavera. *J Bone Joint Surg [Am]* 1988;70A:1224.

15. Fu FH, Horner CD, Klein AH. Shoulder impingement syndrome. A critical review. *Clin Orthop* 1991;269:162–173.
16. Seeger LL, Gold RH, Bassett LW, Ellman H. Shoulder impingement syndrome. MR findings in 53 shoulders. *AJR* 1988;150:343–347.
17. Zlatkin MB, Reicher MA, Kellerhouse LE, et al. The painful shoulder: MR imaging of the glenohumeral joint. *J Comput Assist Tomogr* 1988;12:995–1001.
18. Rafii M, Firooznia H, Sherman O, et al. Rotator cuff lesions: signal patterns at MR imaging. *Radiology* 1990;177:817–823.
19. Zaslav K, Rafii M, Sherman O, et al. Magnetic resonance imaging of rotator cuff. Anatomic and histologic correlations of alteration in signal intensity. Presented at the meeting of American Society of Sports Medicine, New Orleans, February 9, 1990.
20. Kjellin I, Ho CP, Cervilla V, et al. Alterations in the supraspinatus tendon at MR imaging: correlation with histopathologic findings in cadavers. *Radiology* 1991;181:837–841.
21. Kieft GJ, Bloem JL, Rozing PM, et al. Rotator cuff impingement syndrome: MR imaging. *Radiology* 1988;166:211–214.
22. Chandnani V, Ho C, Gerharter J. MR findings in asymptomatic shoulders: a Blins analysis using symptomatic shoulders as controls. *Clin Imaging* 1992;16:25–30.

Case 3

1. Neer CS II. Anterior acromioplasty for the chronic impingement syndrome in the shoulder. A preliminary report. *J Bone Surg [AM]* 1972;54A:41–50.
2. Bigliani LU, Morrisson BS, April AW. The morphology of the acromion and its relationship to anterior cuff tears. *Orthop Trans* 1986;10:216.
3. Edelson JG, Taitz C. Anatomy of the coracoacromial arch. Relation to degeneration of the acromion. *J Bone Joint Surg [Br]* 1992;74B: 589–591.
4. Aoki M, Ishii S, Usui M. The slope of the acromion and rotator cuff impingement. *Orthop Trans* 1986;10:228.
5. Tyson LL, Crues JV III. Pathogenesis of rotator cuff disorders: magnetic resonance imaging characteristics. *MRI Clin North Am* 1993;1:37–46.
6. Seeger LL, Gold RH, Bassett LW, et al. Shoulder impingement syndrome. MR findings in 53 shoulders. *AJR* 1988;150:343–347.
7. Zlatkin MB, Iannotti JT, Roberts MC, et al. Rotator cuff tears: diagnostic performance of MR imaging. *Radiology* 1989;172:223–229.
8. Zlatkin MB, Reicher MA, Kellerhouse LE, et al. The painful shoulder: MR imaging of the glenohumeral joint. *J Comput Assist Tomogr* 1988;12:995–1001.
9. Newhouse KE, El-Khoury GY, Nepola JV. Morphology of the acromion: comparison of asymptomatic persons and those with shoulder impingement syndrome. In: *Proceedings of the Radiological Society of North America.* Chicago: Radiological Society of North America; 1992:240.
10. Epstein RE, Schweitzer ME, Frieman AG, et al. Hooked acromion: prevalence on MR images of painful shoulders. *Radiology* 1993; 187:479–481.
11. Farley TE, Neumann CH, Steinbach LS, et al. The coracoacromial arch: MR evaluation and correlation with rotator cuff pathology. *Skeletal Radiol* 1994;23:641–645.
12. Ozaki J, Fujimoto S, Nakagaway, Masuhara K, Tamai S. Tears of the rotator cuff of the shoulder associated with pathological changes in the acromion: studies in cadaver. *J Bone Joint Surg [Am]* 1988; 70:1224–1230.
13. Zuckerman EA, Kummer FG, Cuomo F, Simon J, Rosenblum S, Cotz N. The influence of coracoacromial arch anatomy on rotator cuff tears. *J Shoulder Elbow Surg* 1992;1:4–12.
14. Haygood TM, Langlotz CP, Kneeland JB et al. Categorization of acromial shape: interobserver variability with MR imaging and conventional radiography. *AJR* 1994;162:1377–1382.
15. Peh WCG, Farmer THR, Totty WG. Acromial arch shape: assessment with MR imaging. *Radiology* 1995;195:501–505.
16. Mudge MK, Wood VE, Frykman GK. Rotator cuff tears associated with os acromiale. *J Bone Joint Surg [Am]* 1984;66:427–429.
17. Park JG, Lee JK, Phelps CT. Os acromiale associated with rotator cuff impingement: MR imaging of the shoulder. *Radiology* 1994; 193:255–257.

Case 4

1. Codman EA. *The shoulder. Rupture of the supraspinatus tendon and other lesions in or about the subacromial space.* Boston: Thomas Todd: 1934.
2. Cofield RH. Tears of the rotator cuff. *Instr Course Lect* 1981;30: 258–273.
3. DePalma AF. *Surgery of the shoulder.* 3rd ed. Philadelphia: JB Lippincott; 1983;51–63, 100–173, 211–241.
4. Neer CS. Anterior acromioplasty for chronic impingement syndrome in the shoulder: a preliminary report. *J Bone Joint Surg [Am]* 1972;54:41–50.
5. Brewer BJ. Aging of the rotator cuff. *Am J Sports Med* 1979;7: 102–110.
6. Rathbun JB, MacNab I. The microvascular pattern of the rotator cuff. *J Bone Surg* 1970;52B:540–553.
7. Cotton RE, Rideout DF. Tears of the humeral rotator cuff. A radiological and pathological necropsy survey. *J Bone Joint Surg [Br]* 1964;46B:314–328.
8. Gschwend N, Ivosevic-Radovanovic D, Patte D. Rotator cuff tear—relationship between clinical and anatomopathological findings. *Arch Orthop Trauma Surg* 1988;107:7–15.
9. Seegar LL, Gold RH, Bassett LW, Ellman H. Shoulder impingement syndrome. MR findings in 53 shoulders. *AJR* 1988;15:343.
10. Evancho AM, Stiles RG, Fajman WA, et al. MR imaging diagnosis of rotator cuff tears. *AJR* 1988;151:751.
11. Kieft GJ, Bloem JL, Rozing PM, et al. Rotator cuff impingement syndrome: MR imaging. *Radiology* 1988;166:211.
12. Zlatkin MB, Dalinka MK, Kressel H. Magnetic resonance imaging of the shoulder. *Magn Reson Q* 1989;5:3–22.
14. Zlatkin MB, Ionnatti JP, Roberts MC, et al. Rotator cuff tears: diagnostic performance of MR imaging. *Radiology* 1989;172:223–229.
15. Rafii M, Firooznia H, Sherman O, et al. Rotator cuff lesions: signal patterns at MR imaging. *Radiology* 1990;177:817–823.

Case 5

1. Codman EA. *The shoulder. Rupture of the supraspinatus tendon and other lesions in or about the subacromial space.* Boston: Thomas Todd 1934.
2. Fukuda H, Mikasa M, Yamanaka K. Incomplete thickness rotator cuff tears diagnosed by subacromial bursography. *Clin Orthop* 1987;223:51–58.
3. Tamai K, Ogawa K. Intratendinous tear of the supraspinatus tendon exhibiting winging of the scapula. *Clin Orthop* 1985;194:159–163.
4. Neer CS. Anterior acromioplasty for chronic impingement syndrome in the shoulder: a preliminary report. *J Bone Joint Surg [Am]* 1972;54:41–50.
5. Seeger LL, Gold RH, Bassett LW, Ellman H. Shoulder impingement syndrome. MR findings in 53 shoulders. *AJR* 1988;150:343–347.
6. Zlatkin MB, Dalinka MK, Kressel H. Magnetic resonance imaging of the shoulder. *Magn Reson* 1989;5:3–22.
7. Rafii M, Firooznia H, Sherman O, et al. Rotator cuff lesions: signal patterns at MR imaging. *Radiology* 1990;177:817–823.
8. Nelson MC, Leather GP, Nirschl RP, Pettrone FA, Freedman MT. Evaluation of painful shoulder. A prospective comparison of magnetic resonance imaging, computerized tomographic arthrography, ultrasonography and operative findings. *J Bone Joint Surg [Am]* 1991;73:707–716.
9. Iannotti JP, Zlatkin MB, Esterhai JL, et al. Magnetic resonance imaging of the shoulder. Sensitivity, specificity and predictive value. *J Bone Joint Surg [Am]* 1991;73A:17–29.
10. Evancho AM, Stiles Rg, Fajman WA, et al. MR imaging diagnosis of rotator cuff tears. *AJR* 1988;151:751–754.
11. Quinn SF, Sheley RC, Demlow TA. Rotator cuff tendon tears: evaluation with fat-suppressed MR imaging with arthroscopic correlation in 100 patients. *Radiology* 1995;195:497–501.
12. Reinus WR, Shady KL, Mirowitz SA, Totty WG. MR diagnosis of rotator cuff tears of the shoulder: value of using T2-weighted fat-saturated images. *AJR* 1995;164:1451–1455.
13. Zaslav K, Rafii M, Sherman O, et al. Magnetic resonance imaging of rotator cuff. Anatomic and histologic correlations of alteration

in signal intensity. Presented at the meeting of the American Society of Sports Medicine, New Orleans, February 9, 1990.
14. Hawkins RJ, Kennedy JC. Impingement syndrome in athletes. *Am J Sports Med* 1980;8:151–158.

Case 6

1. DePalma AF. *Surgery of the shoulder.* 3rd ed. Philadelphia: JB Lippincott; 1983:51–63, 100–173, 211–241.
2. Neer CS. Anterior acromioplasty for chronic impingement syndrome in the shoulder: a preliminary report. *J Bone Joint Surg [Am]* 1972;54:41–50.
3. Neviaser RJ. Tears of the rotator cuff. *Orthop Clin* 1980;11:295–306.
4. Post M, Silver R, Singh M. Rotator cuff tear. Diagnosis and treatment. *Clin Orthop* 1983;173:78–91.
5. Kneeland BJ, Middleton WD, Carrera GF, et al. MR imaging of the shoulder: diagnosis of rotator cuff tears. *AJR* 1987;149:333–337.
6. Zlatkin MB, Reicher MA, Kellerhouse LE, et al. The painful shoulder: MR imaging of the glenohumeral joint. *J Comput Assist Tomogr* 1988;12:995–1001.
7. Seeger LL, Gold RH, Bassett LW, Ellman H. Shoulder impingement syndrome. MR findings in 53 shoulders. *AJR* 1988;150:343–347.
8. Evancho AM, Stiles RG, Fajman WA, et al. MR imaging diagnosis of rotator cuff tears. *AJR* 1988;151:751–754.
9. Kieft GJ, Bloem JL, Rozing PM, et al. Rotator cuff impingement syndrome: MR imaging. *Radiology* 1988;166:211–214.
10. Zlatkin MB, Ionnatti JP, Roberts MC, et al. Rotator cuff tears: diagnostic performance of MR imaging. *Radiology* 1989;172:223–229.
11. Burk DL, Karasick DK, Kurtz AB, et al. Rotator cuff tears: prospective comparison of MR imaging with arthrography, sonography and surgery. *AJR* 1989;153:87–92.
12. Rafii M, Firooznia H, Sherman O, et al. Rotator cuff lesions: signal patterns at MR imaging. *Radiology* 1990;177:817–823.
13. Rafii M. Shoulder. In: Firooznia et al., eds. *MRI and CT of musculoskeletal system.* St. Louis: Mosby Yearbook Medical Publishers; 1992:465.
14. Skinner HA. Anatomical considerations relative to rupture of the supraspinatus tendon. *J Bone Joint Surg* 1937;19:137–151.

Case 7

1. DePalma AF. *Surgery of the shoulder.* 3rd ed. Philadelphia: JB Lippincott; 1983.
2. Codman EA. *The shoulder. Rupture of the supraspinatus tendon and other lesions in or about the subacromial space.* Boston: Thomas Todd; 1934.
3. Inman VT, Saunders JB, Abbott LC. Observations on the function of the shoulder joint. *J Bone Joint Surg* 1944;26:130.
4. Salter EG, Nasca RJ, Shelley BS. Anatomical observations on the acromioclavicular joint and supporting ligaments. *MJ Sports Med* 1987;15:199–206.
5. Rafii M. Shoulder. In: Firooznia H, et al., eds. *MRI and CT of the musculoskeletal system.* St. Louis: Mosby Yearbook Publishers; 1992.
6. Cox JS. The faith of the acromioclavicular joint in athletic injuries. *MJ Sports Med* 1981;9:50–53.
6a. Bergfeld JA, Andrish JT, Clancy WG. Evaluation of the acromioclavicular joint following first and second degree sprains. *MJ Sports Med* 1978;6:153–159.
7. Cahill BR. Osteolysis of the distal part of the clavicle in male athletes. *J Bone Joint Surg [Am]* 1982;64A:1053–1058.
8. Neer II CS. Anterior acromioplasty for the chronic impingement syndrome: a preliminary report. *J Bone Joint Surg [Am]* 1972;54A:41.
9. Petersson CJ, Gentz CF. Significance of distally pointing acromio clavicular osteophytes in ruptures of the supraspinatus tendon. *Acta Orthop Can* 1983;54:490–491.
10. Kessel L, Watson M. The painful arc syndrome. *J Bone Joint Surg [Br]* 1977;59B:166–172.
11. Iannotti JP, Zlatkin MB, Esterhai JL, et al. Magnetic resonance imaging of the shoulder. *J Bone Joint Surg [Am]* 1991;73A:17.
12. Craig EV. The acromioclavicular joint cyst: an unusual presentation of rotator cuff tear. *Clin Orthop* 1989;202:189–192.

Case 8

1. DePalma AF. *Surgery of the shoulder.* 3rd ed. Philadelphia: JB Lippincott; 1983.
2. Rafii M. Shoulder. In: Firooznia H, et al. eds. *MRI and CT of musculoskeletal system.* St. Louis: CV Mosby; 1990:465–549.
3. Zlatkin MB, Iannotti JP, Roberts MC, et al. Rotator cuff tears: diagnostic performance of MR imaging. *Radiology* 1989;172:223–229.
4. Iannotti JP, Zlatkin MB, Esterhai JL, et al. Magnetic resonance imaging of the shoulder: sensitivity, specificity and predictive value. *J Bone Joint Surg [Am]* 1991;73A:17–29.
5. Neviaser RJ. Tears of the rotator cuff. *Orthop Clin* 1980;11:295–306.
6. Rafii M, Minkoff J, Destefano V. Diagnostic imaging of the shoulder. In: Nicholas JA, Hershman EB, Posner MA, eds. *The upper extremity in sports medicine.* St. Louis: CV Mosby; 1990:91–158.
7. Cofield RH. Current concepts review. Rotator cuff disease of the shoulder. *J Bone Joint Surg [Am]* 1985;67A:974–969.
8. Sher JS, Uribe JW, Posada A, et al. Abnormal findings on magnetic resonance images of asymptomatic shoulders. *J Bone Joint Surg [Am]* 1995;77A:10–14.
9. Tsai JC, Zlatkin MB. Magnetic resonance imaging of the shoulder. *Radiol Clin North Am* 1990;28:279.
10. Rafii M, Firooznia H. Technical aspects: magnetic resonance imaging with normal anatomy. In: Sartoris DJ, ed. *Principals of shoulder imaging.* New York: McGraw-Hill; 1995:45–64.

Case 9

1. Fu FH, Horner CD, Klein AH. Shoulder impingement syndrome. A critical review. *Clin Orthop* 1991;269:162–173.
2. Jobe FW, Jobe CM. Painful athletic injuries of the shoulder. *Clin Orthop* 1983;173:117–124.
3. Ozaki J, Fujimoto S, Nakagawa Y, et al. Tears of the rotator cuff of the shoulder associated with pathological changes in the acromion: a study in cadavera. *J Bone Joint Surg [Am]* 1988;70A:1224.
4. Codman EA. *The shoulder. Rupture of the supraspinatus tendon and other lesions in or about the subacromial space.* Boston: Thomas Todd; 1934.
5. Iannotti JP, Zlatkin MB, Esterhai JL, et al. Magnetic resonance imaging of the shoulder. Sensitivity, specificity and predictive value. *J Bone Joint Surg [Am]* 1991;73A:17–29.
6. Jobe CM. Evidence for a superior glenoid impingement upon the rotator cuff. *J Shoulder Elbow Surg* 1993;2(Pt2):S19.
7. Evancho AM, Stiles RG, Fajman WA, et al. MR imaging diagnosis of rotator cuff tears. *AJR* 1988;151:751–754.
8. Zlatkin MB, Ionnatti JP, Roberts MC, et al. Rotator cuff tears: diagnostic performance of MR imaging. *Radiology* 1989;172:223–229.
9. Rafii M, Firooznia H, Sherman O, et al. Rotator cuff lesions: signal patterns of MR imaging. *Radiology* 1990;177:817–823.
10. Tuite MJ, Yandow DR, DeSmet AA, et al. Diagnosis of partial and complete rotator cuff tears using combined gradient echo and spin echo imaging. *Skeletal Radiol* 1994;23:541–546.
11. Traughber PD, Goodwin TE. Shoulder MRI: arthroscopic correlation with emphasis on partial tears. *J Comput Assist Tomogr* 1992;16:129–133.
12. Quinn SF, Sheley RC, Demlow TA. Rotator cuff tendon tears: evaluation with fat-suppressed MR imaging with arthroscopic correlation in 100 patients. *Radiology* 1995;195:497–501.
13. Reinus WR, Shady KL, Mirowitz SA, Totty WG. MR diagnosis of rotator cuff tears of the shoulder. Value of using T2-weighted fat-saturated images. *AJR* 1955;164:1451–1455.
14. Matsen FA III, Arntz CT. Rotator cuff tendon failure. In: Rockwood CA Jr, Matsen FA III, eds. *The shoulder.* Philadelphia: WB Saunders; 1990:637–677.
15. Flannigan B, Kursunoglu-Brahme S, Snyder S, et al. MR-arthrography of the shoulder: comparison with conventional MR imaging. *AJR* 1990;155:829–832.
16. Hodler J, Kursunoglu-Brahme S, Flannigan B, et al. Injuries of the superior portion of the glenoid labrum involving the insertion of the biceps tendon: MR imaging findings in nine cases. *AJR* 1992;159:565–568.

17. Palmer WE, Brown JH, Rosenthal DI. Rotator cuff: evaluation of fat-suppressed MR arthrography. *Radiology* 1993;188:683–687.
18. Tirman PFJ, Bost FW, Garvin GJ, et al. Posterosuperior glenoid impingement of the shoulder: findings at MR imaging and MR-arthrography with arthroscopic correlation. *Radiology* 1994;193: 431–436.

Case 10

1. Seegar LL, Gold RH, Bassett LW, Ellman H. Shoulder impingement syndrome. MR findings in 53 shoulders. *AJR* 1988;150:343–347.
2. Zlatkin MB, Iannotti JP, Roberts MC, et al. Rotator cuff tears: diagnostic performance of MR imaging. *Radiology* 1989;172:223–229.
3. Iannotti JP, Zlatkin MB, Esterhai JL, et al. Magnetic resonance imaging of the shoulder: sensitivity and predicted value. *J Bone Joint Surg [Am]* 1991;73A:17–29.
4. Rafii M, Firooznia H, Sherman O, et al. Rotator cuff lesions: signal patterns at MR imaging. *Radiology* 1990;177:817–823.
5. Farley PE, Neumann CH, Steinbach LS, et al. Full-thickness tears of the rotator cuff of the shoulder: diagnosis with MR imaging. *AJR* 1992;158:347–351.
6. Zlatkin MB, Reicher MA, Kellerhouse LE, et al. The painful shoulder: MR imaging of the glenohumeral joint. *J Comput Assist Tomogr* 1988;12:995–1001.
7. Mirowitz SA. Normal rotator cuff: MR imaging with conventional and fat suppression techniques. *Radiology* 1991;180:735–740.
8. Neumann CH, Holt RG, Steinbach LS, et al. MR imaging of the shoulder: appearance of the supraspinatus tendon in asymptomatic volunteers. *AJR* 1992;158:1281.
9. Skinner HA. Anatomical considerations relative to rupture of the supraspinatus tendon. *J Bone Joint Surg* 1937;19:137–151.
10. Matsen FA III, Arntz CT. Rotator cuff tendon failure. In: Rockwood CA Jr, Matsen FA III, eds. *The shoulder.* Philadelphia: WB Saunders, 1990;637–677.

Case 11

1. DePalma AF. *Surgery of the shoulder.* 3rd ed. Philadelphia: JB Lippincott; 1983:51–63, 100–173, 211–241.
2. Matsen III FA, Arntz CT. Rotator cuff tendon failure. In: Rockwood CA, Matsen FA, eds. The shoulder. Philadelphia: WB Saunders; 1990:647–677.
3. Post M, Silva R, Singh M. Rotator cuff tears. *Clin Orthop* 1983; 173:78–91.
4. Fu FH, Harner CD, Klein AH. Shoulder impingement syndrome. A critical review. *Clin Orthop* 1991;269:162–173.
5. Ellman H, Hanker G, Bayer M. Repair of the rotator cuff. *J Bone Joint Surg [Am]* 1986;68A:1136–1144.
6. Neviassen RJ. Tears of the rotator cuff. *MRI Clin North Am* 1980; 11:295–306.
7. Neer CS, Craig EV, Fukuda H. Cuff tear arthropathy. *J Bone Joint Surg [Am]* 1983;65A:1232–1244.

Case 12

1. Uhthoff HK, Sarkar K. Calcifying tendinitis. In: Rockwood CR, Matsen FA, eds. *The shoulder.* Philadelphia: WB Saunders; 1990: 774–790.
2. Codman FA. *The shoulder. Rupture of supraspinatus tendon and other lesions in or about the subacromial space.* Boston: Thomas Todd; 1934.
3. DePalma AF, Kruper JS. Long term study of shoulder joints afflicted with and treated for calcific tendinitis. *Clin Orthop* 1961;20:61–72.
4. Hays CW, Conway WF. Calcium hydroxyapatite deposition disease. *Radiographics* 1990;10:1032–1048.
5. Burke DL, Karasick D, Mitchell DG, et al. MR imaging of the shoulder: correlation with plain radiography. *AJR* 1991;157:1023–1027.
6. Stoller DW, Fritz R. Magnetic resonance imaging of impingement and rotator cuff tears. *MRI Clin North Am* 1993;1:47–63.

Case 13

1. Meyer AW. Spontaneous dislocation of the long head of biceps brachii. *Arch Surg* 1926;13:109–119.
2. Slatis P, Aalto K. Medial dislocation of the tendon of the long head of the biceps brachii. *Acta Orthop Scand* 1979;50:73–77.
3. Petersson CJ. Spontaneous medial dislocation of the tendon of long biceps brachii. *Clin Orthop* 1986;211:224–227.
4. Chan TW, Dalinka MK, Kneeland JB, et al. Biceps tendon dislocation: evaluation with MR imaging. *Radiology* 1991;179:649–652.
5. Cervilla A, Schweitzer ME, Ho C, et al. Medial dislocation of the biceps brachii tendon: appearance at MR imaging. *Radiology* 1991; 180:523–526.
6. Rafii M. Shoulder. In: Firooznia H, et al., eds. *MRI and CT of the musculoskeletal system.* St. Louis: Mosby Yearbook Publishers; 1992:465–549.
7. Leersum MV, Schweitzer ME. Magnetic resonance imaging of the biceps complex. *MRI Clin North Am* 1993;1:77–86.
8. Petersilge CA, Witte DH, Sewell BO, et al. Normal regional anatomy of the shoulder. *MRI Clin North Am* 1993;1:1–18.
9. Burkhead Jr WZ. The biceps tendon. In: Rockwood CA, Matsen III FA, eds. *The shoulder.* Philadelphia: WB Saunders; 1990:791–836.
10. Neer CS II. Impingement lesions. *Clin Orthop* 1983;173:70–77.
11. Post M, Benca P. Primary tendinitis of the long head of the biceps. *Clin Orthop* 1989;246:117–125.
12. Dines D, Warren RF, Inglis AE. Surgical treatment of lesions of the long head of the biceps. *Clin Orthop* 1982;164:165–171.

12a.Neviaser RJ. Lesions of the biceps and tendinitis of the shoulder. *Orthop Clin* 1980;11:343–348.

13. Rathbun JB, Macnab I. The microvascular pattern of the rotator cuff. *J Bone Surg [Br]* 1970;52:540–553.
14. Collier SG, Wynn-Jones CH. Displacement of the biceps with subscapularis avulsion. *J Bone Joint Surg [Br]* 1990;72B:145.
15. O'Donoghue DH. Subluxating biceps tendon in athlete. *Clin Orthop* 1982;164:26–29.
16. Zlatkin MB. *MRI of the shoulder.* New York: Raven Press; 1991.
17. Erickson SJ, Fitzgerald SW, Quinn SF, et al. Long bicipital tendon of the shoulder on MR imaging. *AJR* 1992;158:1091–1096.
18. Rafii M. MR imaging of the shoulder. In: Beltran J, ed. *Current review of MRI.* Hong Kong: Paramount Printing Group; 1995:203–213.
19. Tuckerman GA. Abnormalities of the long head of the biceps tendon of the shoulder: MR imaging findings. *AJR* 1994;163:1183–1188.

Case 14

1. Masten Fa III, Arntz CT. Rotator cuff tendon failure. In: Rockwood CA JR, Matsen FA III, eds. *The shoulder.* Philadelphia: WB Saunders; 1990:637–677.
2. Ogilvie-Harris DJ, Wiley AM, Sattarian J. Failed acromioplasty for impingement syndrome. *J Bone Joint Surg [Br]* 1990;72B:1070–1072.
3. Post M. Complications of rotator cuff surgery. *Clin Orthop* 1990; 254:97–104.
4. Neer CS. Anterior acromioplasty for chronic impingement syndrome in the shoulder: a preliminary report. *J Bone Joint Surg [Am]* 1972;54:41–50.
5. Owen RS, Iannotti JP, Kneeland JB, et al. Shoulder after surgery: MR imaging with surgical validation. *Radiology* 1993;186:443–447.
6. Zlatkin MB. MRI of the shoulder. New York: Raven Press; 1991: 90–92.
7. Raffi M. Shoulder. In Firooznia H, et al. eds. *MRI and CT of the musculoskeletal system.* St. Louis: Mosby Yearbook Publishers; 1992:464–549.
8. Haygood TM, Oxner KG, Kneeland JB, Danlink MK. Magnetic resonance imaging of the postoperative shoulder. *MRI Clin North Am* 1993;1:143–155.
9. Calvert PT, Packer NP, Stoker DJ, et al. Arthrography of the shoulder after operative repair of the torn rotator cuff. *J Bone Joint Surg [Am]* 1986;68:147–150.

Case 15

1. Moseley HF, Overgaard B. The anterior capsular mechanism in recurrent anterior dislocation of the shoulder. Morphological and

clinical studies with special reference to the glenoid labrum and the glenohumeral ligaments. *J Bone Joint Surg* 1962;44B:913.
2. Warwick R, Williams PL. *Gray's anatomy.* 36th ed. Philadelphia: WB Saunders; 1989:456.
3. DePalma AF. Regional variational surgical anatomy. In: DePalma AF, ed. *Surgery of the shoulder.* 3rd ed. Philadelphia: JB Lippincott; 1983.
4. Turkel SJ, Panio MW, Marshal JL, Girgis FG. Stabilizing mechanisms preventing anterior dislocation of the glenohumeral joint. *J Bone Joint Surg [Am]* 1981;63A:1208.
5. Cooper DE, Armoczky SP, O'Brien SJ, et al. Anatomy, histology, and vascularity of the glenoid labrum. *J Bone Joint Surg [Am]* 1992;74A:46.
6. Raffi M, Firooznia H, Golimbu C, et al. CT arthrography of capsular structures of the shoulder. *AJR* 1986;146:361.
7. Rafii M. Shoulder. In: Firooznia et al., eds. *MRI and CT of musculoskeletal system.* St. Louis: Mosby-Year Book; 1992:465.
8. Rafii M, Firooznia H. Technical aspects: magnetic resonance imaging with normal anatomy. In: Sartoris DJ, ed. *Principles of shoulder imaging.* New York: McGraw-Hill; 1995:43–64.
9. Huber DJ, Sauter RS, Muller E, et al. MR imaging of the normal shoulder. *Radiology* 1986;158:405.
10. Kieft GJ, Bloem JL, Oberman WR. Normal shoulder: MR imaging. *Radiology* 1986;159:741.
11. Seeger LL, Ruszkowski JT, Bassett LW, et al. MR imaging of the normal shoulder: anatomic correlation. *AJR* 1987;148:83–91.
12. Coumas JM, Waite RJ, Goss TP, et al. CT and MR evaluation of the labral capsular ligamentous complex of the shoulder. *AJR* 1992; 158:591–597.
13. Flannigan B, Kursunoglu-Brahme S, Synder S, et al. MR arthrography of the shoulder: comparison with conventional MR imaging. *AJR* 1990;155:829.
14. Petersilge CA, Witte DH, Sewell BO. Normal regional anatomy of the shoulder. *MRI Clin North Am* 1993;1:1.
15. Tirman PFJ, Applegate GR, Flannigan BD, et al. Magnetic resonance arthrography of the shoulder. *MRI Clin North Am* 1993;1:125.
16. Palmer WE, Brown JH, Rosenthal DI. Labral ligamentous complex of the shoulder: evaluation with MR arthrography. *Radiology* 1994; 190:645–651.
17. Bieze J. Saline injection enhances shoulder defect diagnosis. *Diagn Imaging* 1992;14:13.
18. Mirowitz SA. Imaging techniques, normal variations and diagnostic pitfalls in shoulder magnetic resonance imaging. *MRI Clin North Am* 1993;1:19.
19. Kaplan PA, Bryans KC, Davick JP, et al. MR imaging of the normal shoulder: variants and pitfalls. *Radiology* 1992;184:519–1992.
20. McCauley TR, Pope CF, Jokl P. Normal and abnormal glenoid labrum: assessment with multiplanar gradient-echo MR imaging. *Radiology* 1992;183:35.
21. Neumann CH, Petersen SA, Jahnke AH. MR imaging of the labral-capsular complex: normal variations. *AJR* 1991;157:1015.
22. Johnson LL. *Arthroscopic surgery: principles and practice.* 3rd ed. St. Louis: CV Mosby; 1986.
23. Detrisac DA, Johnson LL. *Arthroscopic shoulder anatomy: pathologic and surgical implications.* Thorofare, NJ: SLACK, 1987.
24. McNiesh LM, Callaghan JJ. CT arthrography of the shoulder: variations of the glenoid labrum. *AJR* 1987;149:963.
25. Williams MM, Snyder SJ, Buford DJ. The Buford Complex: the cord like middle glenohumeral ligament and absent anterosuperior labrum complex: a normal anatomic capsulolabral variant. *Arthroscopy* 1994;10:241–247.
26. Tirman PFJ, Feller JF, Palmar WE, et al. The Buford Complex—A variation of normal shoulder anatomy: MR arthrographic imaging features. *AJR* 1996;166:869–873.

Case 16

1. Cofield RH. Current concepts review. Rotator cuff disease of the shoulder. *J Bone Joint Surg [Am]* 1985;67A:974–979.
2. DePalma AF. *Surgery of the shoulder.* 3rd ed. Philadelphia: JP Lippincott, 1983:51–63, 100–173, 211–241.
3. Protzman RR. Anterior instability of shoulder. *J Bone Joint Surg [Am]* 1980;62A:909–918.
4. Rowe CR, Zarins B. Recurrent transient subluxation of the shoulder. *J Bone Joint Surg [Am]* 1981;63A:863–871.
5. Skyhar MR, Warren RF, Altchek DW. Instability of the shoulder. In: Nicholas JA, Hershman EB, eds. *The upper extremity in sports medicine.* St. Louis: CV Mosby; 1990:181–212.
6. Fronek J, Warren R, Bowen M. Posterior subluxation of the glenohumeral joint. *J Bone Joint Surg [Am]* 1989;71A:205–216.
7. Uhthoff HK, Piscopo M. Anterior capsular redundancy of the shoulder: congenital or traumatic. *J Bone Joint Surg [Br]* 1985;65B: 363–366.
8. Matsen FA, Harryman DT, Sidles JA. Mechanics of glenohumeral instability. *Clin Sports Med* 1990;10:783–788.
9. Bankart ASB. The pathology and treatment of dislocation of the shoulder joint. *Br J Surg* 1938;26:23–29.
10. Bost FC, Inman VT. The pathological changes in recurrent dislocation of the shoulder. A report of Bankart's operative procedure. *J Bone Joint Surg [Am]* 1942;15:595–613.
11. Du Toit GT, Roux D. Recurrent dislocation of the shoulder. A twenty-four year study of the Johannesburg stapling operating. *J Bone Joint Surg [Am]* 1956;38A:1–12.
12. Moseley HF, Overgaard B. The anterior capsular mechanism in recurrent anterior dislocation of shoulder. Morphological and clinical studies with special reference to the glenoid labrum and the glenohumeral ligaments. *J Bone Joint Surg [Br]* 1962;44B:913–927.
13. Townley CO. The capsular mechanism in recurrent dislocation of the shoulder. *J Bone Joint Surg [Am]* 1950;32A:370–380.
14. Turkel SJ, Panio MW, Marshal JL, et al. Stabilizing mechanism preventing anterior dislocation of the glenohumeral joint. *J Bone Joint Surg [Am]* 1981;63A:1208–1217.
15. Reeves B. Experiments on the tensile strength of the anterior capsular structures of the shoulder in man. *J Bone Joint Surg [Br]* 1968; 50B:858–865.
16. Baker CL, Uribe JW, Whitman C II. Arthroscopic evaluation of acute initial anterior shoulder dislocation. *Am J Sports Med* 1990; 18:25–28.
17. Seeger LL, Gold RH, Bassett LW, et al. Shoulder instability: evaluation with MR imaging. *Radiology* 1988;168:695–695.
18. Habibian A, Staufer A, Resnick D, et al. Comparison of conventional and computed arthrotomography with MR imaging of shoulder. *J Comput Axial Tomogr* 1989;13:968–975.
19. Rafii M, Firooznia H, Golimbu C, Weinreb J. *MRI Clin North Am* 1993;1:87–104.
20. Danzig LA, Greenway G, Resnick D. The Hill-Sachs lesion: an experimental study. *Am J Sports Med* 1980;8:328–332.
21. Warren RF. Instability of shoulder in throwing sports. *Instr Course Lect* 1985;34:337–348.
22. Kieft GJ, Bloom JL, Rozing PM, et al. MR imaging of recurrent anterior dislocation of the shoulder: comparison with CT-arthrography. *AJR* 1988;150:1083–1087.
23. Minkoff J, Cavaliere G. Glenohumeral instabilities and the role of magnetic resonance imaging. *MRI Clin North Am* 1993;1:105–123.
24. Gross M, Seeger LL, Smith J, et al. Magnetic resonance imaging of the glenoid labrum. *Am J Sports Med* 1990;18:229–234.

Case 17

1. Blazina E, Satzman JS. Recurrent anterior subluxation of the shoulder in athletes. A distinct entity. *J Bone Joint Surg [Am]* 1969;5A: 1037–1038.
2. Rowe CR, Zarins B. Recurrent transient subluxation of the shoulder. *J Bone Joint Surg [Am]* 1981;63A:863–871.
3. Rowe CR, Patel D. The Bankart procedure. A long-term end-result study. *J Bone Joint Surg [Am]* 1978;60A:1–16.
4. Skyhar MR, Warren RF, Altchek DW. Instability of the shoulder. In: Nicholas JA, Hershman EB, eds. *The upper extremity in sports medicine.* St. Louis: CV Mosby; 1990:181–212.
5. Bost FC, Inman VT. The pathological changes in recurrent dislocation of the shoulder. A report of Bankart's operative procedure. *J Bone Joint Surg [Am]* 1942;15:595–613.
6. Townley CO. The capsular mechanism in recurrent dislocation of the shoulder. *J Bone Joint Surg [Am]* 1950;32A:370–380.
7. Mizuno K, Hirohata K. Diagnosis of recurrent traumatic anterior subluxation of the shoulder. *Clin Orthop* 1983;179:160–167.

8. El-Khoury GY, Albright JP, Abu-Yousef MM, et al. Arthrotomography of the glenoid labrum. *Radiology* 1979;131:333–337.
9. Deutch AL, Resnick D, Mink JH, et al. Computed and conventional arthrotomography of the glenohumeral joint: normal anatomy and clinical experience. *Radiology* 1984;153:603–609.
10. Rafii M, Firooznia H, Golimbu C, et al. CT-arthrography of capsular structures of the shoulder. *AJR* 1986;146:361–367.
11. Rafii M, Firooznia H, Bonamo JJ, et al. Athlete shoulder injuries: CT arthrographic findings. *Radiology* 1987;62:599–564.
12. Rafii M, Minkoff J, Bonamo J, et al. CT arthrography of shoulder instabilities in athletes. *Am J Sports Med* 1988;16:352–361.
13. Seeger LL, Gold RH, Bassett LW, et al. Shoulder instability: evaluation with MR imaging. *Radiology* 1988;168:695–697.
14. Kieft GJ, Bloom JL, Rozing PM, et al. MR imaging of recurrent anterior dislocation of the shoulder: comparison with CT-arthrography. *AJR* 1988;150:1083–1087.
15. Habibian A, Staufer A, Resnick D, et al. Comparison of conventional and computed arthrotomography with MR imaging of shoulder. *J Comput Axial Tomogr* 1989;13:968–975.
16. Legan JM, Burkhard TK, Goff II WB, et al. Tears of the glenoid labrum: MR imaging of 88 arthroscopically confirmed cases. *Radiology* 1991;179:241–246.
17. Gross M, Seeger LL, Smith J, et al. Magnetic resonance imaging of the glenoid labrum. *Am J Sports Med* 1990;18:229–234.
18. Rafii M, Firooznia M, Sherman O, et al. High resolution glenohumeral instability lesions. *Radiology* 1991;181P:154 (abst).
19. Zlatkin MB, Reicher MA, Kellerhouse LE, et al. The painful shoulder: MR imaging of the glenohumeral joint. *J Comput Assist Tomogr* 1988;12:955–1001.
20. Iannotti JP, Zlatkin MB, Esterhai JL, et al. Magnetic resonance imaging of the shoulder. Sensitivity, specificity and predictive value. *J Bone Joint Surg [Am]* 1991;73A:17–29.
21. Chadnani VP, Yeager TD, DeBernardino T, et al. Glenoid labrum tears: prospective evaluation with MR imaging, MR arthrography and CT arthrography. *AJR* 1993;161:1229–1235.
22. Garneau RA, Renfrew DL, Moore TE, et al. Glenoid labrum: evaluation with MR imaging. *Radiology* 1991;179:519–522.
23. Flannigan B, Kursunoglu-Brahme S, Snyder S, et al. MR arthrography of the shoulder: comparison with conventional MR imaging. *AJR* 1990;155:829–832.
24. Palmer WE, Brown JH, Rosenthal DI. Labral ligamentous complex of the shoulder: evaluation with MR arthrography. *Radiology* 1994;190:645–651.
25. Rafii M. Shoulder. In: Firooznia H, Golimbu C, Rafii M, eds. *MRI and CT of the musculoskeletal system.* St. Louis: CV Mosby; 1990: 465–549.
26. Rafii M, Firooznia H, Golimbu C. Magnetic resonance imaging of the glenohumeral instability. *MRI Clin North Am* 1993;1:87–104.
27. Workman TL, Burkhard TK, Resnick D, et al. Hill-Sachs lesion: comparison of detection with MR imaging, radiography and arthroscopy. *Radiology* 1992;185:847–852.

Case 18

1. Rafii M, Firooznia H, Golimbu C, Weinreb J. Magnetic resonance imaging of glenohumeral instability. *MRI Clin North Am* 1993;1: 87–104.
2. Rowe CR. Prognosis in dislocations of the shoulder. *J Bone Joint Surg [Am]* 1956;38A:957–977.
3. McLaughlin HL. Posterior dislocation of the shoulder. *J Bone Joint Surg [Am]* 1952;34A:584–590.
4. O'Donoghue DH. *Treatment of injuries to athletes.* 4th ed. Philadelphia: WB Saunders; 1988:118–220.
5. Hawkins RJ, Koppert G, Johnston G. Recurrent posterior instability (subluxation) of the shoulder. *J Bone Joint Surg [Am]* 1984;66A: 169–174.
6. Fronek J, Warren R, Bowen M. Posterior subluxation of the glenohumeral joint. *J Bone Joint Surg [Am]* 1989;71A:205–216.
7. Warren RF. Instability of shoulder in throwing sports. *Instr Course Lect* 1985;34:337–348.
8. Minkoff J, Cavaliere G. Glenohumeral instabilities and the role of magnetic resonance imaging techniques: the orthopedics surgeon's perspective. *MRI Clin North Am* 1993;1:105–123.
9. Rafii M, Firooznia H, Bonamo JJ, et al. Athlete shoulder injuries: CT arthrographic findings. *Radiology* 1987;162:559–564.
10. Legan JM, Burkhard TK, Goff II WB, et al. Tears of the glenoid labrum: MR imaging of 88 arthroscopically confirmed cases. *Radiology* 1991;179:241–246.
11. Skyhar MR, Warren RF, Altchek DW. Instability of the shoulder. In: Nicholas JA, Hershman EB, eds. *The upper extremity in sports medicine.* St. Louis: CV Mosby; 1990:181–121.
12. Neer CS II, Foster CF. Inferior capsular shift for involuntary inferior and multidirectional instability of the shoulder. A preliminary report. *J Bone Joint Surg* 1980;62:897.

Case 19

1. Andrews JR, Carson WG, McLeod WD. Glenoid labrum tears related to the long head of the biceps. *Am J Sports Med* 1985;13: 237–241.
2. Snyder SJ, Karzel RP, Del Pizzo WD, et al. SLAP lesion of the shoulder. *Arthroscopy* 1990;6:274–279.
3. Cartland JP, Crues III, JV, Stauffer A, et al. MR imaging in the evaluation of SLAP injuries of the shoulder: findings in 10 patients. *AJR* 1992;159:787–792.
4. Hodler J, Kursunoglu-Brahme S, Flannigan B, et al. Injuries of the superior portion of the glenoid labrum involving the insertion of the biceps tendon: MR imaging findings in nine cases. *AJR* 1992; 159:565–568.
5. Tirman PFJ, Applegate B, Flannigan BD, et al. Magnetic resonance arthrography of the shoulder. *MRI Clin North Am* 1993;1:125–142.
6. Rafii M, Firooznia H, Golimbu C, Weinreb J. Magnetic resonance imaging of shoulder instability. *MRI Clin North Am* 1993;1:87–104.
7. Detrisac DA, Johnson LL. Arthroscopic shoulder anatomy: pathologic and surgical implications. Thorofare, NJ: SLACK; 1987.
8. DePalma AF. *Surgery of the shoulder.* 3rd ed. Philadelphia: JB Lippincott; 1983.
9. Rafii M. Shoulder. In: Firooznia H, Golimbu C, Rafii M, eds. *MRI and CT of the musculoskeletal system.* St. Louis: CV Mosby; 1990: 465–549.
10. Fritz RC, Helms CA, Steinbach LS. Suprascapular nerve entrapment: evaluation with MR imaging. *Radiology* 1992;182:437–444.
11. Deutsch AL, Mink JH. Magnetic resonance imaging of miscellaneous disorders of the shoulder. *MRI Clin North Am* 1993;1:171–183.

Case 20

1. Caffey J, Silverman WA. Infantile cortical hyperostoses. Preliminary report on a new syndrome. *AJR* 1945;54:1–6.
2. Resnick D. Enostosis, hyperostosis, and periostitis. In: Resnick D, ed. *Diagnosis of bone and joint disorders.* 3rd ed. WB Saunders; Philadelphia: 1995:4436–4442.
3. Caffey J. *Pediatric x-ray diagnosis.* 7th ed. Chicago: Year Book Medical Publishers; 1978:1430–1436.
4. Holtzman D. Infantile cortical hyperostosis of the scapula presenting as an ipsilateral Erb's palsy. *J Pediatr* 1972;81:785–787.
5. Padfield E, Hicken P. Cortical hyperostosis in infants: a radiological study of sixteen patients. *Br J Radiol* 1970;43:231–234.
6. Ueda K, Saito A, Nakano H, et al. Cortical hyperostosis following long-term administration of prostaglandin E1 in infants with cyanotic congenital heart disease. *J Pediatr* 1980;97:834–838.

CHAPTER 2

The Elbow

Russell C. Fritz

Magnetic resonance (MR) imaging provides clinically useful information in assessing the elbow joint. Superior depiction of muscles, ligaments, and tendons as well as the ability to visualize nerves, bone marrow, and hyaline cartilage directly are advantages of MR imaging relative to conventional imaging techniques. These features may help to establish the cause of elbow pain by accurately depicting the presence and extent of bone and soft tissue pathology. Ongoing improvements in surface coil design and newer pulse sequences have resulted in higher quality MR images of the elbow that can be obtained more rapidly. Recent clinical experience has shown the utility of MR imaging in detecting and characterizing disorders of the elbow in a noninvasive fashion (1–8).

IMAGING TECHNIQUES

The elbow is typically scanned with the patient in a supine position and the arm at the side. A surface coil is essential for obtaining high-quality images. Depending on the size of the patient and the size of the surface coil relative to the bore of the magnet, it may be necessary to scan the patient in a prone position with the arm extended overhead. In general, the prone position is less well tolerated and results in a greater number of motion-degraded studies. The elbow should be scanned in a comfortable position to avoid motion artifact. The elbow is typically extended and the wrist placed in a neutral position. Patients who cannot extend the elbow are more difficult to position and require more time and skill to obtain optimal images. Taping a vitamin E capsule or other marker to the skin at the site of tenderness or at the site of a palpable mass is useful to ensure that the area of interest has been included in the study, especially when there is no pathology identified on the images.

Excellent images may be obtained with both mid-field and high-field MR systems. Proton-density and T2-weighted images are typically obtained in the axial and sagittal planes using spin-echo or fast spin-echo techniques. T1-weighted and short T1 inversion recovery (STIR) sequences are usually obtained in the coronal plane. Although the STIR sequence has relatively poor signal-to-noise ratio because of the suppression of signal from fat, pathology is often more conspicuous due to the effects of additive T1 and T2 contrast.

The axial images, in general, should extend from the distal humeral metaphysis to the radial tuberosity. The common flexor and extensor origins from the medial and lateral humeral epicondyles and the biceps insertion on the radial tuberosity are routinely imaged with this coverage. This coverage is usually obtained with 3- or 4-mm-thick slices using a long repetition time (TR) sequence. The coronal images are angled parallel to a line through the humeral epicondyles on the axial images. The sagittal images are angled perpendicular to a line through the humeral epicondyles on the axial images.

The field of view on the axial images should be as small as the signal of the surface coil and the size of the patient's elbow allows. The field of view selected on the coronal and sagittal sequences is usually larger than the field of view on the axial images, to include more of the anatomy about the elbow. This guideline is especially true when imaging a ruptured biceps tendon that may retract to the normal superior margin of coverage. The slice thickness, interslice gap, and TR may be increased on the axial sequences, just as the field of view is increased on the coronal and sagittal sequences as long as the surface coil provides adequate signal to image the entire length of the pathology.

Additional sequences may be added or substituted depending on the clinical problem to be solved. T2*-weighted gradient-echo sequences provide useful supplemental information for identifying loose bodies within the elbow. Gradient-echo volume sequences allow acquisition of multiple very thin axial images that may subsequently be reformatted in any plane using a computer workstation. These volume sequences have not been routinely useful in our experience because of their relatively poor soft tissue contrast compared with the spin-echo technique. Fast spin-echo volume sequences will soon become available and may be more useful than the currently available gradient-echo volume technique. In general, gradient-echo sequences are to be avoided after elbow surgery because magnetic susceptibility artifacts associated with micrometallic debris may obscure the images and also may be mistaken for loose bodies. Furthermore, the degree of artifact surrounding orthopedic hardware is most prominent on gradient-echo sequences due to the lack of a 180° refocusing pulse and is least prominent on fast spin-echo sequences due to the presence of multiple 180° pulses. Fast spin-echo and fast STIR sequences may be substituted for conventional T2-weighted spin-echo and conventional STIR sequences if available; these newer sequences allow greater flexibility in imaging the elbow while continuing to provide information comparable with that of the conventional spin-echo and conventional STIR sequences. The speed of fast spin-echo sequences may be used to obtain higher resolution T2-weighted images in the same amount of time as the conventional spin-echo sequences or may simply be used to increase the speed of the examination. The ability to shorten the examination with fast spin-echo has been useful when scanning claustrophobic patients or when scanning patients who become uncomfortable in the prone position with the arm overhead.

Fat suppression may be added to various pulse sequences to improve visualization of the hyaline articular cartilage. Avoidance of chemical shift artifact at the interface of cortical bone and fat-containing marrow permits a more accurate depiction of the overlying hyaline cartilage. T1-weighted images with fat suppression are useful whenever gadolinium is administered, either intravenously or directly into the elbow joint. Intravenous gadolinium may provide additional information in the assessment of neoplastic or inflammatory processes about the elbow. Articular injection of saline or dilute gadolinium may be useful in patients without a joint effusion to detect loose bodies, to determine if the capsule is disrupted, or to determine if an osteochondral fracture fragment is stable.

ELBOW ANATOMY

A thorough understanding of the anatomy and function of the elbow is essential for interpretation of the MR images. The anatomic structures of the elbow are depicted reliably with MR imaging. Knowledge of the relative functional significance of these structures allows assessment of the clinically important anatomy. Focusing on the relevant anatomic structures leads to more meaningful interpretation of the images and facilitates clinical problem solving.

The elbow is composed of three articulations contained within a common joint cavity. The radial head rotates within the radial notch of the ulna allowing supination and pronation distally. The radial head is surrounded by the annular ligament, which is best seen on the axial images. Disruption of the annular ligament results in proximal radioulnar joint instability. The radius articulates with the capitellum, and the ulna articulates with the trochlea in a hinge fashion. The anterior and posterior portions of the joint capsule are relatively thin, whereas the medial and lateral portions are thickened to form the collateral ligaments. The medial collateral ligament complex consists of anterior and posterior bundles as well as an oblique band also known as the transverse ligament. The functionally important anterior bundle of the medial collateral ligament extends from the medial epicondyle to the medial aspect of the coronoid process. The anterior bundle is well seen on coronal images. The anterior bundle provides the primary constraint to valgus stress and is commonly damaged in throwing athletes (9–11).

The lateral collateral ligament complex is more variable and less well understood than the medial collateral ligament (12). The radial collateral ligament proper arises from the lateral epicondyle anteriorly and blends with the fibers of the annular ligament, which surrounds the radial head. The annular ligament is the primary stabilizer of the proximal radioulnar joint and is evaluated best on axial images. A more posterior bundle, known as the lateral ulnar collateral ligament or the ulnar part of the lateral collateral ligament, arises from the lateral epicondyle and extends along the posterior aspect of the radius to insert on the supinator crest of the ulna. The lateral ulnar collateral ligament acts as a sling or guy wire that provides the primary ligamentous constraint to varus stress (13–16). Disruption of the lateral ulnar collateral ligament also results in the recently recognized pivot-shift phenomenon and posterolateral rotatory instability of the elbow (13,17). Both the radial collateral ligament proper and the lateral ulnar collateral ligament are well seen progressing from anterior to posterior on the coronal images and should be considered separately because of the difference in functional significance of these structures.

The muscles of the elbow are divided into anterior, posterior, medial, and lateral compartments. The anterior compartment contains the biceps and brachialis muscles, which are best evaluated on sagittal and axial images. The brachialis extends along the anterior joint capsule and inserts on the ulnar tuberosity. The biceps lies superficial to the brachialis and inserts on the radial tuberosity. The posterior compartment contains the triceps and anconeus muscles, which are best evaluated on sagittal and axial images. The triceps inserts on the proximal aspect of the olecranon. The anconeus arises from the posterior aspect of the lateral epicondyle and inserts more distally on the olecranon. The anconeus provides dynamic support to the lateral collateral ligament in resisting varus stress. The medial and lateral compartment muscles are best seen on coronal and axial images. The medial compartment structures include the pronator teres and the flexors of the hand and wrist, which arise from the medial epicondyle as the common flexor tendon. The common flexor tendon provides dynamic support to the medial collateral ligament in resisting valgus stress. The lateral compartment structures include the supinator, the brachioradialis, and the extensors of the hand and wrist that arise from the lateral epicondyle as the common extensor tendon.

The ulnar, median, and radial nerves are subject to entrapment in the elbow region. These nerves are normally surrounded by fat and are best seen on the axial images.

CASE 1

History: This 38-year-old man was lifting weights when he experienced acute pain and felt a pop in his right elbow. This was shortly followed by antecubital swelling and ecchymosis. He delayed seeking medical attention until 4 weeks after the injury. What is the diagnosis? Does this require surgery?

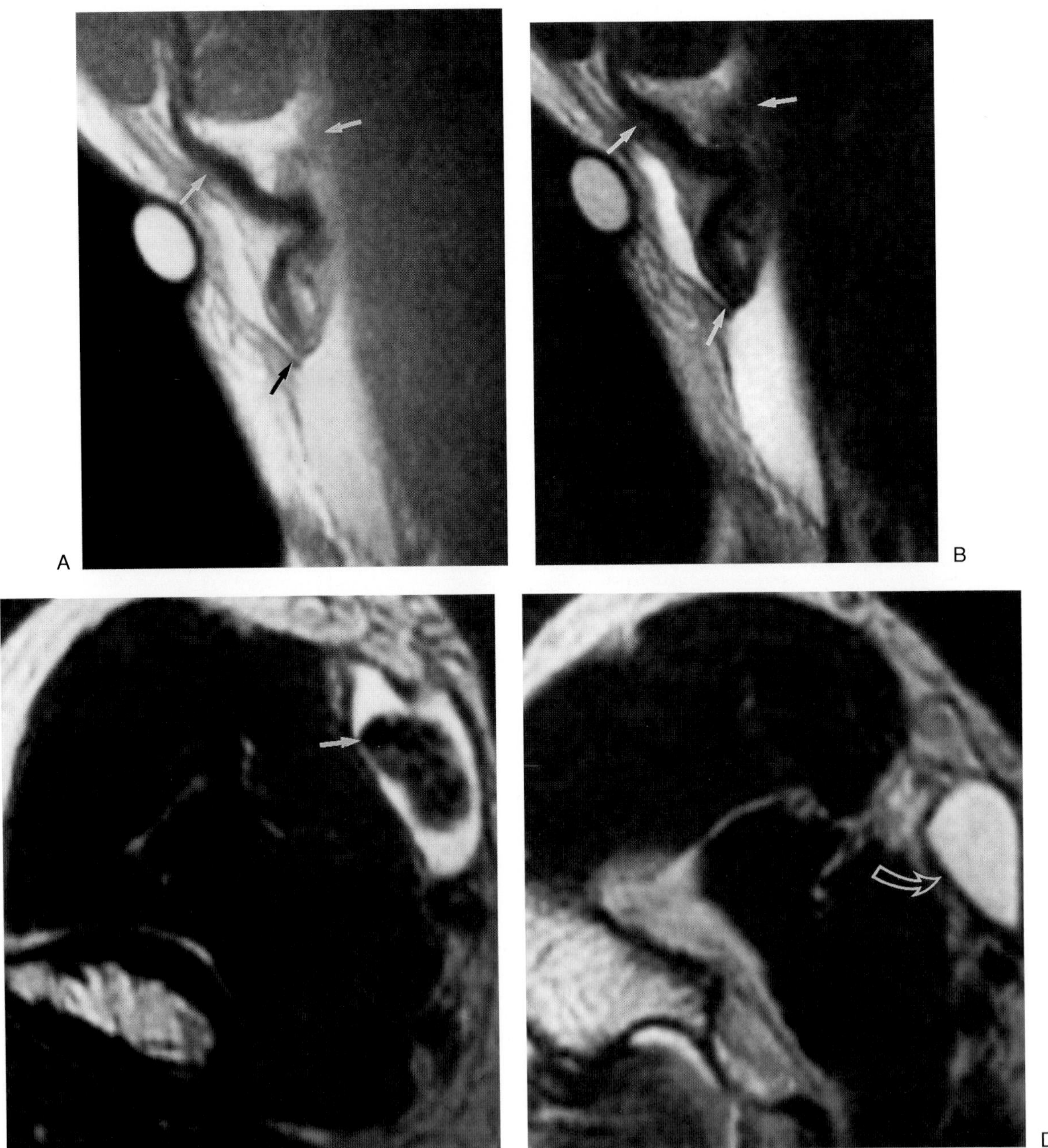

FIG. 1.

Findings: Proton density (Fig. 1A) and T2-weighted (Fig. 1B) sagittal images as well as a T2-weighted axial image (Fig. 1C) reveal a proximally retracted distal biceps tendon (arrows) with surrounding fluid. The tendon is thickened and folded back on itself. Fluid is seen (curved arrow) instead of the tendon at the level of the humeral epicondyles (Fig. 1D).

Diagnosis: Subacute rupture of the distal biceps tendon

Discussion: Rupture of the distal biceps tendon was once thought to be an unusual injury. Recently, however, it has become more commonly diagnosed and reported (1–8). The vast majority of distal biceps ruptures occur in men, with the injury involving the dominant arm in 80% (9). The average age at rupture is 55 years, with bodybuilders and weightlifters usually presenting at a younger age (2). Anabolic steroid abuse has been implicated in some of these younger patients (10). Complete rupture of the tendon from its insertion on the radial tuberosity is most commonly observed. Complete tears of the distal biceps are thought to be much more common than partial tears (3,11). MR imaging is useful in evaluating these injuries because degenerative tendinosis, partial tears, and complete ruptures may be distinguished (4–7,12,13).

The biceps brachii is a long fusiform muscle that has two heads proximally and one tendon distally. The long head arises from the supraglenoid tubercle and the superior labrum at the shoulder joint, whereas the short head arises from the coracoid process in a conjoined fashion with the coracobrachialis. The two heads join to form a common muscle belly that ends in a flattened, horizontal distal tendon at the elbow. The distal biceps tendon averages approximately 7 cm in length and rotates laterally about 90° before inserting on the radial tuberosity (Fig. 1E) (10,14). The tendon is coronally oriented proximally and sagittally oriented distally at the radial tuberosity. The bicipital radial bursa separates the distal tendon from the anterior aspect of the radial tuberosity just proximal to the tendon insertion (15). The distal tendon also has a flattened aponeurotic attachment known as the lacertus fibrosus that extends from the myotendinous junction to the medial deep fascia of the forearm. The lacertus fibrosus covers the median nerve and brachial artery that lie medial to the distal biceps tendon. The primary function of the biceps brachii is flexion of the elbow and supination of the forearm. Flexion of the elbow is assisted by the brachialis, which lies just posterior to the biceps. Supination of the forearm is assisted by the supinator muscle.

Rupture of the tendon of the long head of the biceps commonly occurs, either avulsing from the superior labrum or tearing within the bicipital groove of the proximal humerus. Rupture of the short head of the biceps is rare. Rupture of the biceps muscle belly or the myotendinous junction as well as midsubstance rupture of the distal tendon are considered rare injuries that may occur secondary to direct trauma (16,17).

The mechanism of distal biceps tendon injury usually involves eccentric contraction of the biceps against resistance, typically in weightlifters or manual laborers who are attempting to lift a heavy object. The bicipital aponeurosis (lacertus fibrosus) is typically damaged to varying degrees at the time of biceps tendon rupture, but may remain intact (Fig. 1F,G). In some cases the distal biceps tendon may rupture in stages, first with avulsion of the tendon followed by secondary tearing of the lacertus fibrosus, which allows proximal retraction of the biceps (Fig. 1H–J) (3,11). Clinical diagnosis can be difficult when the lacertus fibrosus remains intact, since retraction of the muscle is minimal (9). Flexion power at the elbow may be preserved if the lacertus fibrosus remains intact; however, supination of the forearm is usually weakened due to the biceps tendon detachment from the radial tuberosity.

Distal biceps tendinosis is common and has been shown to precede spontaneous tendon rupture (18). Tendinosis of the distal biceps is probably a multifactorial process that involves repetitive mechanical impingement of a poorly vascularized distal segment of the tendon (Fig. 1K). Irregularity of the radial tuberosity and chronic inflammation of the adjacent radial bicipital bursa may also be contributory (19,20). A zone of relatively poor blood supply exists within the distal biceps tendon approximately 10 mm from its insertion on the radial tuberosity (14). In addition, this hypovascular zone may be impinged between the radius and the ulna during pronation. The space between the radius and ulna progressively narrows by 50% during pronation, with average measurements of approximately 8 mm in supination, 6 mm in neutral position, and 4 mm in pronation recorded in asymptomatic volunteers with computed tomography (CT) and MR imaging (14). Repetitive impingement during pronation coupled with an intrinsically poor blood supply of the distal biceps tendon may result in a failed healing response and degenerative tendinosis. Enlargement of the degenerated tendon as well as irregularity and hypertrophy of the radial tuberosity may lead to inflammation of the adjacent bursa. Each of these factors may contribute to worsening impingement between the radial tuberosity and the ulna, leading to further degeneration of the distal biceps tendon (Fig. 1L,M). Ultimately this process may result in complete tendon rupture or, less commonly, partial tendon rupture or bursitis.

The distal biceps tendon is covered by an extrasynovial paratenon and is separated from the radial tuberosity by the bicipital radial bursa. Inflammation of this cubital bursa may accompany tendinosis and tearing of the distal biceps (Fig. 1N,O). Enlargement of the bicipital radial bursa may occasionally present as a nonspecific antecubital fossa mass as large as 5 cm in diameter (20,21). Intravenous administration of gadolinium may aid in recognition of this enlarged bursa on MR imaging and may allow differentiation of this benign entity from a solid neoplasm (21). Cubital bursitis, tendinosis, and partial tendon rupture may coexist to differing degrees and may be impossible to distinguish clinically (3,20). Cubital bursitis and partial tendon rupture may both cause irritation of the adjacent median nerve, further complicating the clinical findings (20,22).

The T2-weighted axial images are most useful for determining the degree of tendon tearing. The axial images should extend from the musculotendinous junction to the insertion of the tendon on the radial tuberosity. The axial images also are useful for evaluating the lacertus fibrosus (Fig. 1H,J). The status of the lacertus on MR imaging is usually not a critical issue because this structure is typically not repaired during the surgical repair of a ruptured biceps tendon. However, surgical repair of a symptomatic lacertus fibrosus rupture has been reported along with repair of a partial tear of the biceps tendon (11).

MR imaging provides useful information regarding the degree of tearing, the size of the gap, and the location of the tear for preoperative planning (Fig. 1A–D). The tendon typically tears from its attachment on the radial tuberosity as a result of attempted elbow flexion against resistance (23). Other injuries that may occur via the same mechanism include avulsion and strain of the brachialis as well as disruption of the annular ligament with anterior dislocation of the radial head. These rare injuries also may be identified with MR imaging (4).

Rupture of the distal biceps tendon is generally treated with prompt surgical repair and reattachment to the radial tuberosity to restore flexion and supination strength. Surgical repair may be complicated by radial nerve injury or heterotopic bone formation resulting in radioulnar synostosis (24). Early diagnosis of biceps tendon rupture is important, since surgical outcome is improved in patients treated during the first several weeks after injury (8). After several months, the tendon retracts into the substance of the biceps muscle, making retrieval and reattachment more complicated. MR imaging may be useful in such cases to confirm the clinical diagnosis and plan reconstructive surgery. Reattachment of a chronically retracted biceps to the radial tuberosity has a significantly higher risk of radial nerve injury (9). In chronic ruptures, therefore, the retracted distal tendon may be attached to the brachialis muscle or the ulnar tuberosity to avoid injury to the radial nerve. These delayed repairs restore some flexion strength but do not improve supination weakness.

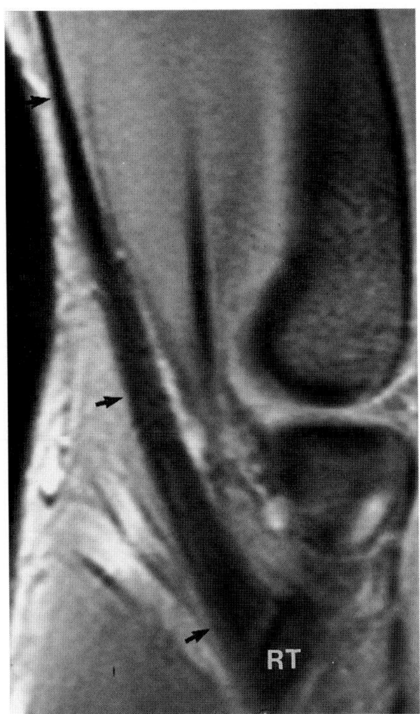

FIG. 1. E: Normal distal biceps tendon. The biceps tendon (*arrows*) undergoes a 90° rotation as it extends along a 7-cm length from the myotendinous junction to the radial tuberosity (RT) as seen on this gradient-echo T2*-weighted sagittal image.

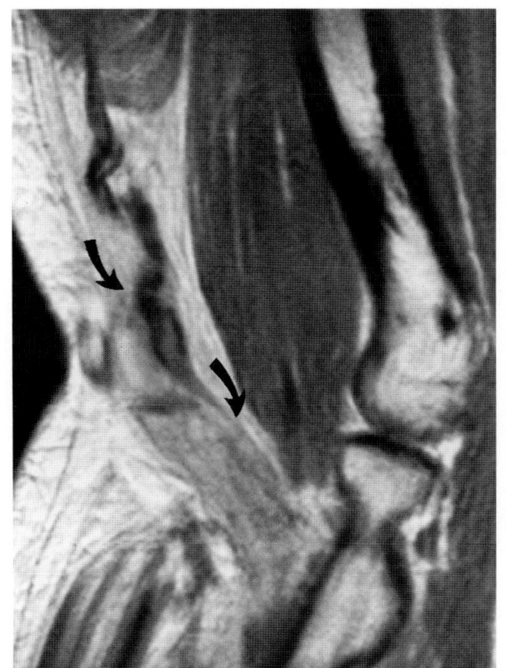
F

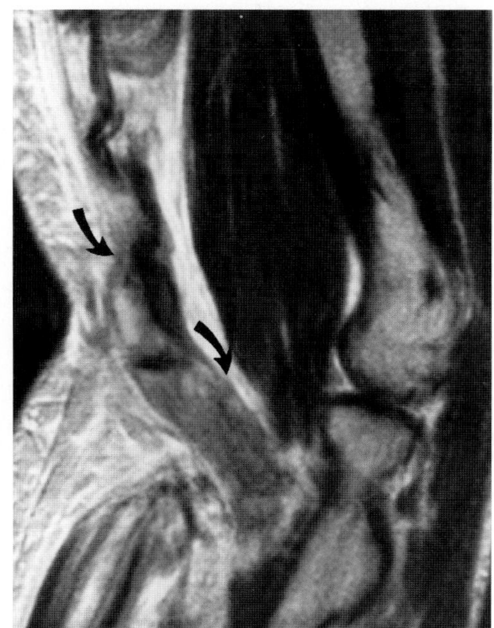
G

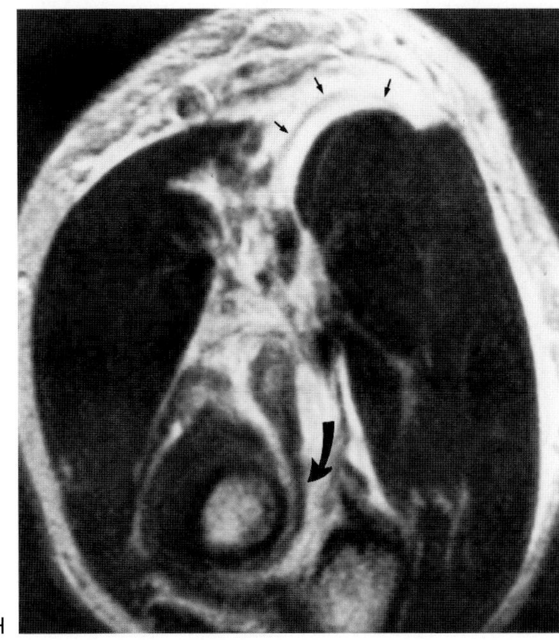
H

FIG. 1. F–H: High-grade partial tear of the distal biceps tendon. Proton density **(F)** and T2-weighted **(G)** sagittal images reveal a lax and redundant-appearing biceps tendon (*curved arrows*) with surrounding edema. A T2-weighted **(H)** axial image reveals a thin strand of the biceps tendon (*curved arrow*) that remains attached to the radial tuberosity. There is increased signal delineating rupture of the lacertus fibrosus (*small black arrows*).

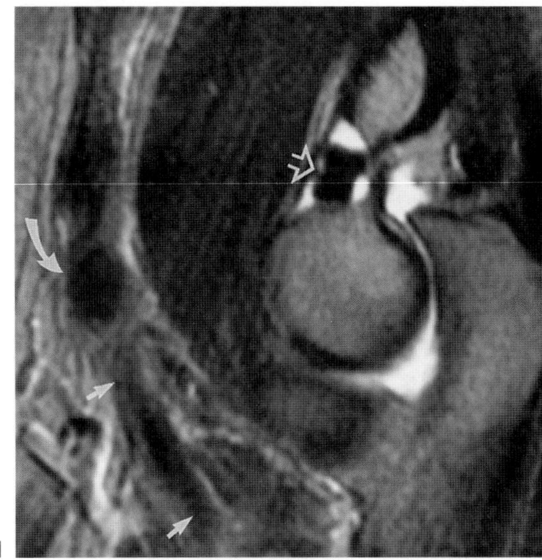

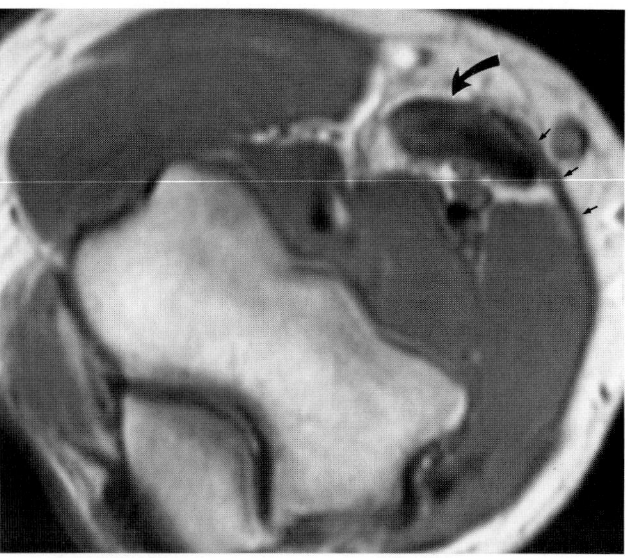

FIG. 1. I,J: Old partial biceps tendon tear in a 42-year-old man with painful limitation of elbow flexion. A T2-weighted **(I)** sagittal image reveals an unsuspected loose body (*open arrow*) in the anterior aspect of the elbow joint, which explains the loss of full flexion. The proximal biceps tendon is thickened (*curved arrow*), whereas the distal portion of the tendon is abnormally thin (*small white arrows*). A proton density **(J)** axial image reveals increased signal and thickening of the biceps tendon (*curved black arrow*) as well as thickening of the lacertus fibrosus (*small black arrows*) consistent with scarring from a prior partial tear.

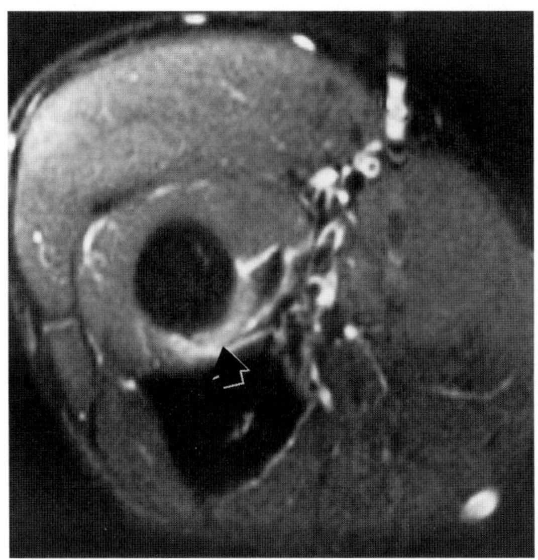

FIG. 1. K: Distal biceps tendinosis. There is prominent increased signal within distal fibers of the biceps tendon (*arrow*) on this STIR axial image consistent with severe degenerative tendinosis. A discrete partial tear or rupture is not identified.

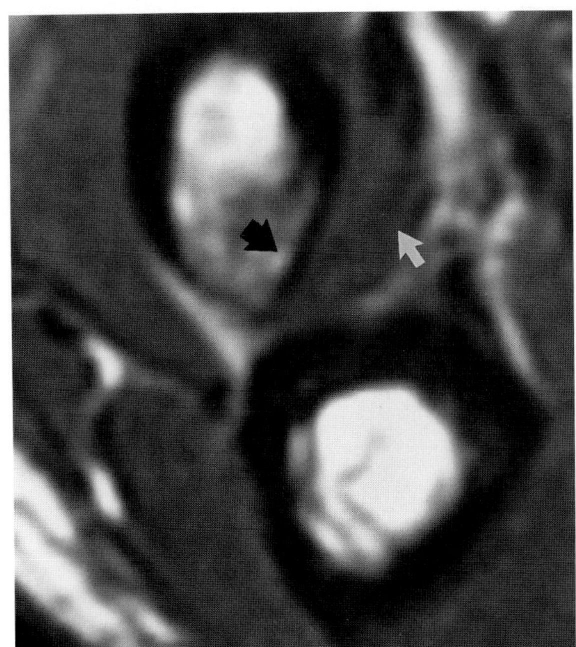

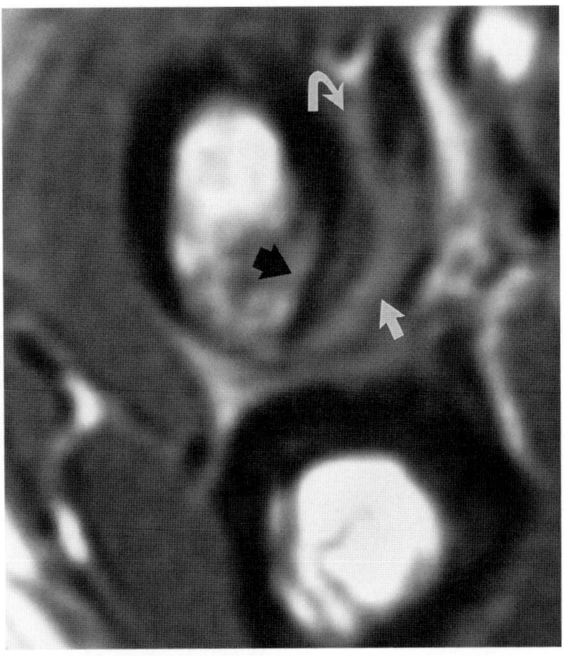

FIG. 1. L,M: Distal biceps impingement syndrome. T1-weighted axial images before **(L)** and after **(M)** intravenous administration of gadolinium reveal prominence of the radial tuberosity (*black arrows*) on images performed with the arm not fully pronated. Further pronation causes impingement of the biceps tendon between the radial tuberosity and the ulna. Abnormal signal is noted in the biceps tendon, which enhances with gadolinium, consistent with tendinosis (*straight white arrow*). The synovium within the bicipital radial bursa also enhances with contrast (*curved arrow*).

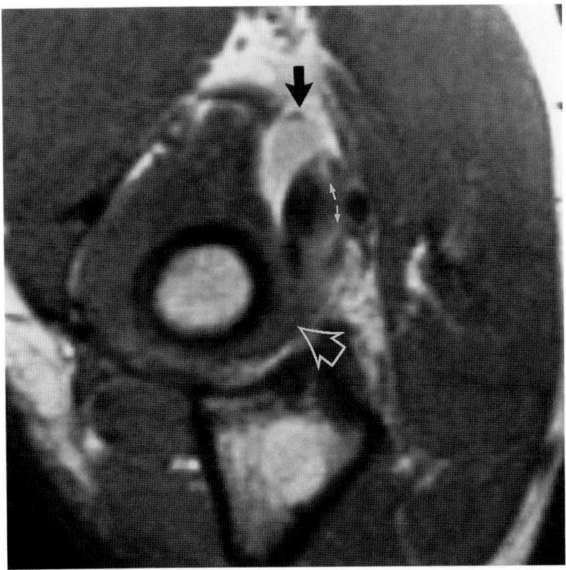

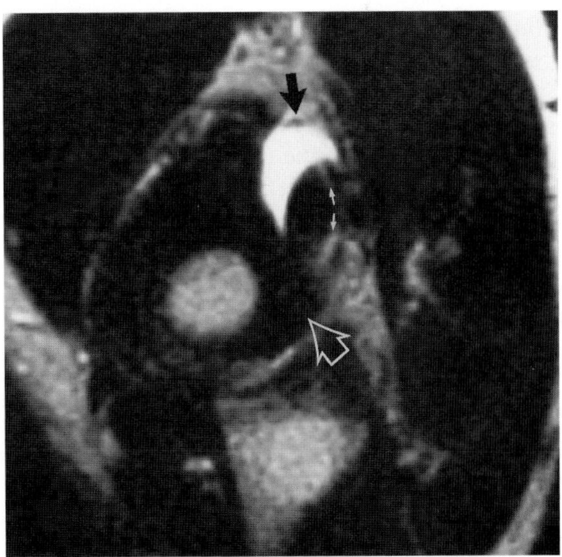

FIG. 1. N,O: Bursitis, tendinosis, and mild intrasubstance partial tearing of the distal biceps tendon. Proton density **(N)** and T2-weighted **(O)** axial images reveal prominent distention of the bicipital radial bursa (*black arrows*). The bursa separates the biceps tendon from the radial tuberosity further distally. Moderate increased signal in the biceps tendon (*open arrows*) is noted, which normalizes in signal on the T2-weighted image, consistent with degenerative tendinosis. A small longitudinal split (*small white arrows*) is seen in the medial aspect of the thickened tendon.

CASE 2

History: A 40-year-old man experienced sudden posterior elbow pain while shooting a basketball. He was unable to continue playing. His past medical history was remarkable for olecranon bursitis, which had been refractory to repeated drainage and local steroid injections. He had recovered uneventfully from surgical excision of the olecranon bursa 4 months before. What is the current diagnosis? Will this require further surgery? What is the significance of this patient's prior elbow problems?

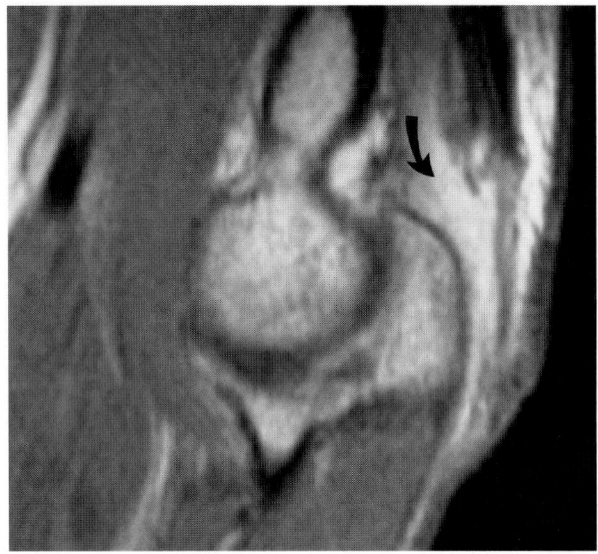

FIG. 2.

Findings: A proton-density sagittal image (Fig. 2A) reveals complete detachment (arrow) and retraction of the triceps tendon from the olecranon.

Diagnosis: Acute triceps tendon rupture.

Discussion: Rupture of the distal triceps tendon has been considered one of the least common tendon injuries, with approximately 60 cases reported in the literature to date. Partial tears may occur but have generally been considered less common than complete ruptures of the triceps tendon (1–3). Triceps tendon injury has recently been reported with increasing frequency and has not been uncommon in our MR imaging practice (4–9).

The triceps muscle lies in the posterior compartment of the arm and is composed of three heads. The long head arises from the infraglenoid tubercle of the scapula, the lateral head arises from the lateral and posterior aspect of the humerus, and the medial head arises further distally from the medial and posterior aspect of the humerus. The triceps is innervated by the radial nerve, which passes between the lateral and medial heads in a bony groove along the posterior aspect of the humerus. The distal triceps tendon begins in the middle of the muscle and is initially composed of a small superficial layer and a more substantial deep layer that combine to form the tendon that inserts on the posterior superior surface of the olecranon. The triceps tendon typically tears at or adjacent to its insertion on the olecranon (Fig. 2A). The triceps tendon usually retracts with a small fleck of bone imbedded in it. This small avulsion fracture may be detected radiographically in approximately 80% of reported cases (2,3–10). Partial tears usually involve the central third of the tendon adjacent to the olecranon (Fig. 2B) (11,12). Rupture of the muscle belly or rupture at the myotendinous junction of the triceps may also occur but is relatively rare (13).

The mechanism of triceps tendon rupture usually involves forced flexion of the elbow against the resistance of a contracting triceps muscle. Such eccentric contraction of the triceps most commonly occurs during a fall on an outstretched arm that results in avulsion of the tendon at its bony attachment to the olecranon.

Rupture of the triceps tendon may also occur secondary to a direct blow (Fig. 2C) (2,3,12,14). The mechanisms of injury in sports-related triceps ruptures include a fall on an outstretched arm, a direct blow to the tendon, a decelerating counterforce during active extension, or some combination of these factors. Recent case reports have indicated that the triceps tendon may also rupture during sustained extreme concentric contraction while weightlifting (4,10,15). Both anabolic steroid abuse and local steroid injections have been implicated in rupture of the triceps tendon (4,8,10,15). The detrimental effects of either systemic anabolic steroids (16) or local corticosteroid injections (17) on the strength of tendons has been documented. Athletes who use anabolic steroids are at risk of tendon ruptures, since their excessive muscle strength is exerted on tendons that have become stiffer and absorb less energy prior to failure (1,16).

The triceps tendon may rupture secondary to minor trauma if there is pre-existing degenerative tendinosis (18). Spontaneous rupture of the triceps has been reported in association with chronic renal failure and secondary hyperparathyroidism, Marfan's syndrome, and osteogenesis imperfecta as well as in patients treated with oral corticosteroids for systemic lupus erythematosus or rheumatoid arthritis (19,20).

Olecranon bursitis may mimic or accompany triceps tendon tears. It has also been suggested that olecranon bursitis may predispose to triceps rupture (21). The presence of underlying triceps tendinosis may explain the association between olecranon bursitis and triceps tendon rupture. An additional factor may be the frequent use of local steroid injections in the treatment of olecranon bursitis. We have seen several cases of triceps rupture in patients who had been treated for olecranon bursitis. Moreover, we have seen both complete and partial tears of the triceps tendon in patients who previously underwent surgery for olecranon bursitis (Fig. 2A,B).

Clinical diagnosis of triceps tendon injury may be difficult due to pain and swelling, which limit the physical examination. The clinical assessment centers on the loss of extension power that is present with complete rupture. Unlike the biceps, there are no other muscles that substantially assist in extending the elbow. Indeed, a missed rupture of the triceps may result in severe functional impairment (Fig. 2D,E) (20). Immediate surgery is considered the treatment of choice for complete rupture, whereas partial rupture may not necessarily require surgery. Conservatively treated partial tears must be followed closely, however, to ensure that a complete disruption and retraction of the tendon does not develop. Cases that present late with a large gap in the triceps require reconstruction, which has less reliable results compared with simple repair at the time of injury (14,20).

In patients who fall on an outstretched arm there may be other associated injuries in addition to rupture of the triceps. A group of 16 patients has been reported with concomitant radial head fractures and triceps tendon ruptures (22). Posterior compartment syndrome has been reported with more proximal triceps muscle injury (23). An interesting association of triceps tendon rupture and ulnar neuritis has been reported in two cases. In both cases, delayed surgery resulted in scar formation about the ulnar nerve (12,15).

The consequences of overloading the extensor mechanism of the elbow depend largely on the age of the patient and the presence of pre-existing tendon degeneration. Most often, the tendon ruptures at the site of degenerative tendinosis. In skeletally immature individuals, separation of the olecranon growth plate may occur and may require internal fixation. The growth plate usually fuses at about age 14 in girls and age 16 in boys in an anterior to posterior direction. Closure of the olecranon physis may be delayed in pitchers secondary to overuse and the repetitive traction of the triceps tendon (24). This painful persistence of the olecranon growth plate in the throwing elbow usually responds to rest but may require internal fixation (25). Acute overload of the extensor mechanism in an adolescent with a partially closed olecranon growth plate may result in a Salter-Harris type II fracture that may be radiographically subtle. MR imaging may be useful in this setting to evaluate the extensor mechanism and detect occult injury to the growth plate.

Injuries of the triceps tendon and muscle are well seen with MR imaging (6,7,26,27). The normal triceps tendon often appears lax and redundant when the elbow is imaged in full extension or mild hyperextension (Fig. 2F). This appearance resolves when the elbow is imaged in mild degrees of flexion and should not be mistaken for pathology. Degenerative tendinosis is characterized by thickening and signal alteration of the distal tendon fibers (Fig. 2G,H). Acute rupture is well seen on T2-weighted or STIR images due to surrounding fluid (Fig. 2I,J). Partial tears are much less common than complete rupture and are more difficult to diagnose clinically (11). MR imaging can distinguish between complete tears that require surgery and partial tears that may do well with protection and rehabilitation. MR imaging can also help to delineate the degree of tendon retraction and muscular atrophy that is present when rupture of the triceps has been missed and a more extensive reconstruction of the defect is required (Fig. 2D,E).

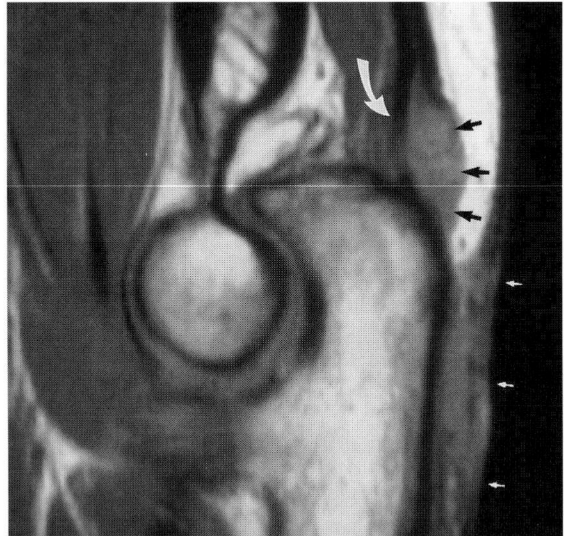

FIG. 2. B: Partial triceps tendon rupture in a college football player in whom the olecranon bursa was previously removed. A T1-weighted sagittal image reveals a gap (*black arrows*) separating the superficial fibers of the triceps tendon from the olecranon. The deep layer of the tendon (*curved arrow*) remains intact. Infiltration of the fat is noted at the site of a prior olecranon bursectomy (*small arrows*).

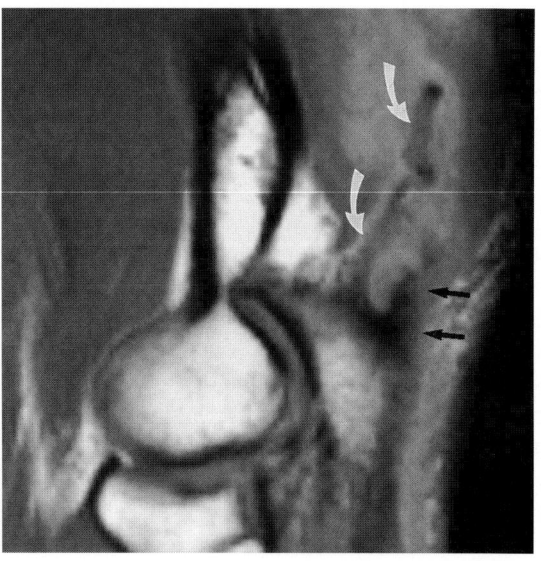

FIG. 2. C: Subacute triceps tendon rupture in a professional football player who suffered a direct blow to the tendon. A T1-weighted sagittal image reveals a shredded distal triceps tendon imaged 6 weeks after the original injury. The bulk of the tendon has retracted proximally. A small stump of the superficial tendon fibers (*black arrows*) remains attached to the olecranon. The deep layer of the triceps tendon (*curved arrows*) is also torn.

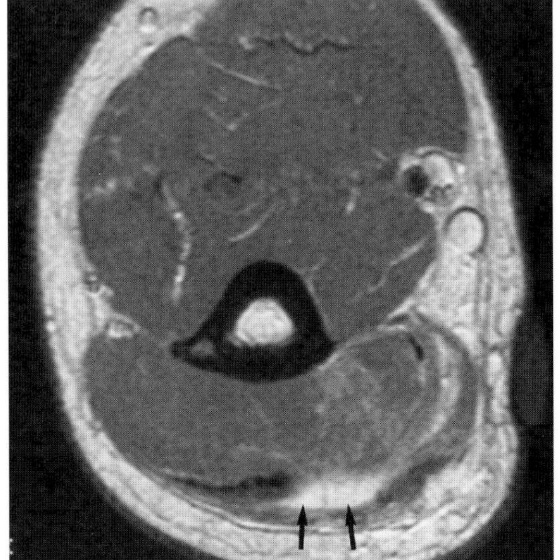

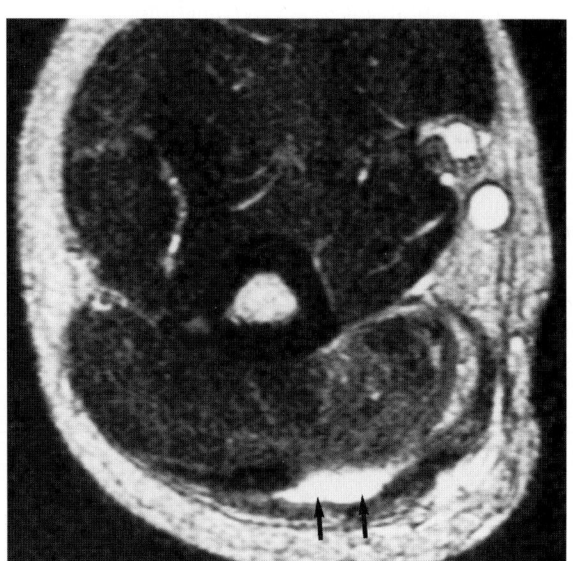

FIG. 2. D,E: Chronic, retracted tear of the triceps tendon in a professional football player. Proton-density **(D)** and T2-weighted **(E)** axial images through the mid-portion of the upper arm reveal a fluid-filled defect within the central third of the triceps (*arrows*). This portion of the triceps tendon had retracted even further proximally, resulting in a large gap, atrophy, and functional disability.

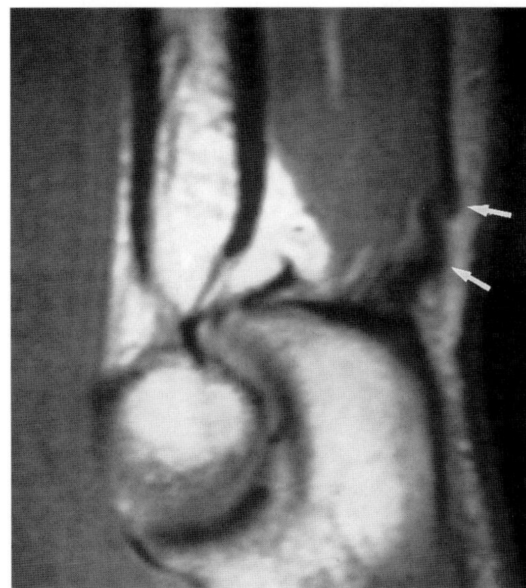

F

FIG. 2. F: Normal triceps tendon imaged in mild hyperextension. A T1-weighted sagittal image reveals apparent laxity. This redundant appearance of the tendon is normal and resolves with flexion.

FIG. 2. G,H: Degenerative tendinosis in a 57-year-old man. There is prominent thickening and increased signal intensity involving the triceps tendon on these proton-density **(G)** and T2-weighted **(H)** sagittal images. A small focus of heterotopic ossification (*arrows*) is seen within the degenerated tendon. An os patella cubiti is a sesamoid bone of the triceps tendon that may have a similar appearance.

FIG. 2. I,J: Acute triceps tendon avulsion in a professional football player. T1-weighted **(I)** and STIR **(J)** sagittal images reveal a small fluid-filled gap (*arrows*) that separates the distal triceps tendon from the olecranon. This patient did well with surgical repair 2 days after the injury.

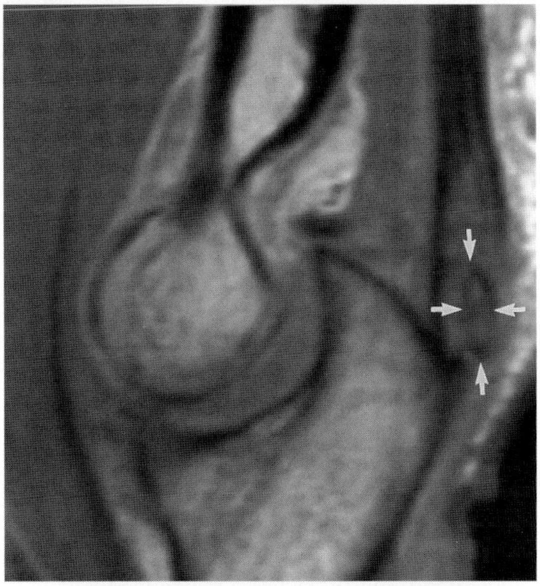

G

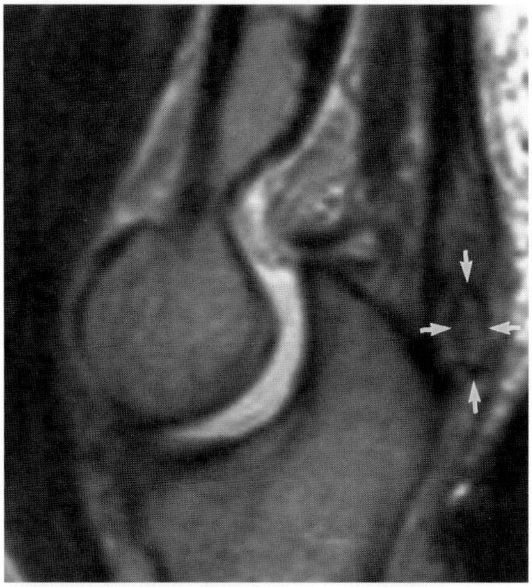

H

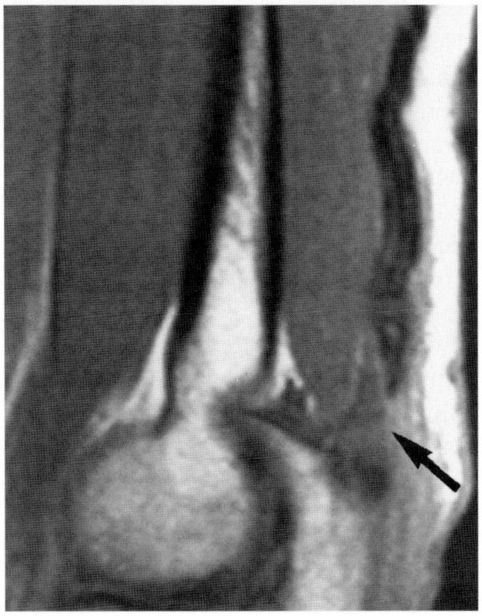

I

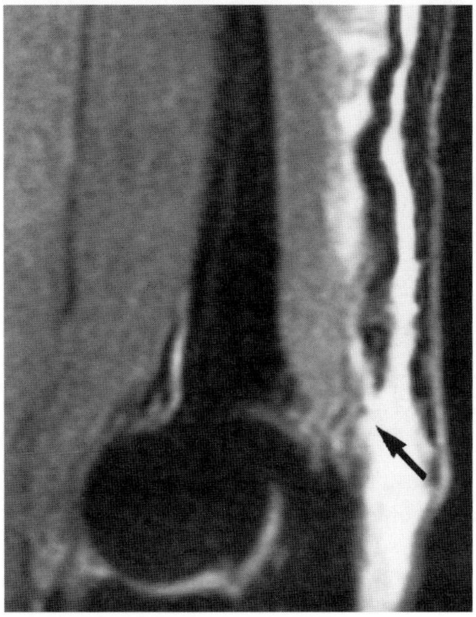

J

CASE 3

History: This 32-year-old tennis player complained of persistent pain while serving. She had suffered a direct blow to the olecranon during a fall 3 weeks previously. Radiographs of the elbow (not shown) were normal. What is the diagnosis? What other conditions could cause this appearance?

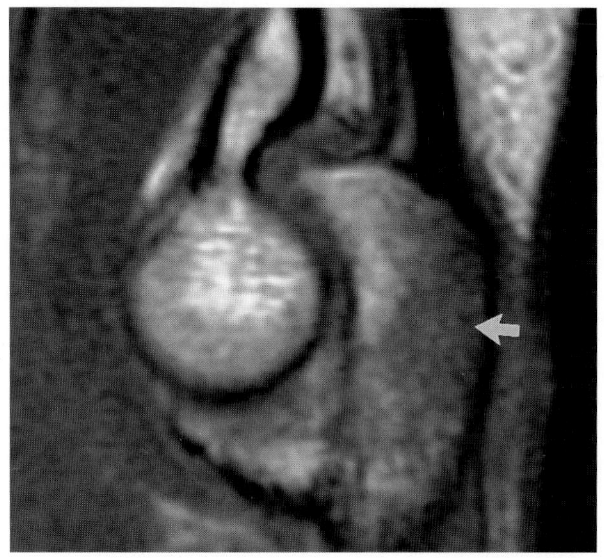

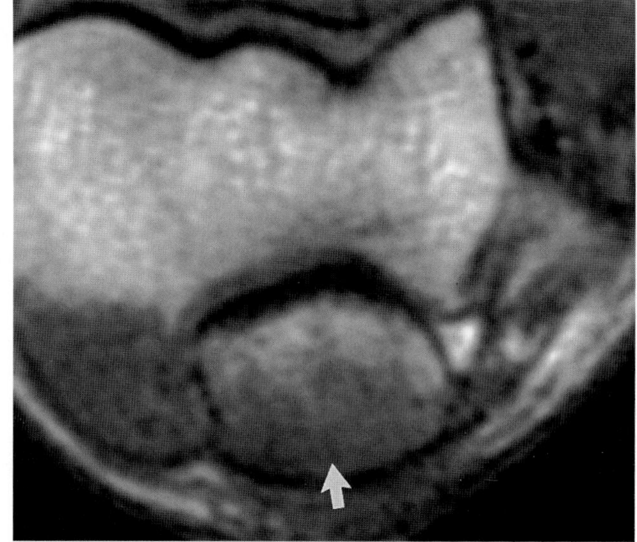

FIG. 3.

Findings: T1-weighted sagittal (Fig. 3A) and axial (Fig. 3B) images reveal abnormal decreased signal (arrows) throughout the posterior aspect of the olecranon as well as infiltration of the overlying subcutaneous fat at the site of the olecranon bursa.

Diagnosis: Olecranon bone contusion.

Discussion: In this case, MR imaging was helpful in diagnosing this radiographically occult bone injury that accounted for the symptoms of posterior elbow pain during full extension. MR imaging was also helpful in excluding other injuries about the elbow. This information enabled proper therapy and rehabilitation to be instituted in a timely manner. Further more invasive tests and expense were avoided. The patient was asymptomatic 3 weeks later.

This case illustrates the typical changes seen in the medullary space with significant traumas that have been termed bone contusion, bone bruise, trabecular microfracture, or bony trabecular injury (1–6). The exact nature of the abnormal signal remains speculative, but it is probably secondary to trabecular disruption (microfracture) of the cancellous bone with hemorrhage and edema extending into the medullary space. Typical findings are poorly marginated loss of signal intensity relative to fatty marrow on T1-weighted images that is localized to the site on impaction injury. This area of decreased signal in T1-weighted images increases in signal intensity in T2-weighted images. Bone contusions are most conspicuous in STIR and other fat-suppressed sequences (6). These injuries are also visualized on T1-weighted images if predominately fatty marrow provides background contrast. Bone contusions, as well as other types of marrow pathology, may be inconspicuous on proton density and T2-weighted images because marrow edema and marrow fat may be of similar signal intensity. Artifactual loss of signal intensity due to magnetic susceptibility differences among trabecular bone, edema, and fat also results in unreliable marrow contrast on gradient-echo T2*-weighted sequences (7). Bone contusions may enhance with gadolinium; in this setting, enhancement is most conspicuous when fat-suppressed T1-weighted images are performed. In the limited number of patients with bone contusions that have been followed in our experience and in the literature, symptoms have resolved within 3 months, and the MR appearance has returned to normal between 6 weeks and 3 months after the injury (1,2). Similarly, bone contusions have been identified about the knee in 71% of 98 patients with clinically diagnosed anterior cruciate ligament injuries on T1- and T2-weighted images, whereas no contusions were seen on scans done longer than 6 weeks after injury (4). Bone contusions may represent regions of bone theoretically at risk for the subsequent development of insufficiency fractures or osteochondral sequelae if these regions are not adequately protected during trabecular healing (1,3). In the lower extremity, some authors have advocated a delay before resuming full weight bearing and resumption of sports activities, both for the alleviation of pain and to lessen the risk of potential progression of these regions of trabecular disruption into complete fractures (1,8).

The differential diagnosis of poorly defined marrow edema includes small cortical avulsions that may be radiographically subtle or occult (Fig. 3C). Avulsion of the joint capsule, ligaments, or tendons may result in marrow edema, which is the most conspicuous finding on MR imaging (9,10). Careful inspection of the cortex on MR imaging as well as review of the plain films or additional radiographic views may be necessary to recognize these avulsion fractures.

Other processes may result in a nonspecific appearance of bone marrow edema on MR imaging. Differential considerations include transient bone marrow edema syndrome, early osteonecrosis, osteomyelitis, edema associated with primary or metastatic tumors, and edema associated with osteochondral defects. Transient bone marrow edema syndrome is more common in the lower extremity, especially the hip, but may occasionally be seen in the bones about the elbow. Marrow edema may be seen adjacent to osteochondritis dessicans and Panner's disease in the capitellum. Osteomyelitis may also result in an appearance of bone marrow edema on MR imaging (11) (also see Case 7–19). Identification of bony destruction (Fig. 3D,E), an abcess, a sequestrum, or a sinus tract may enable recognition of this process. ^{67}G scans or ^{111}In-labeled white blood cell scans may add specificity when osteomyelitis is suspected. Bone marrow edema may be a conspicuous finding of many neoplasms, especially infiltrative neoplasms and chondroblastoma (12). Bone marrow edema may be the sole finding of an osteoid osteoma on MR imaging; this may result in an erroneous diagnosis unless the plain films are reviewed (13–15). Osteoid osteoma may involve the elbow and may present a confusing appearance of marrow edema on MR imaging (14,16,17) (see Case 4–2). CT may be necessary to detect the small nidus of an osteoid osteoma. Reactive marrow edema is often seen adjacent to foci of articular cartilage loss associated with various arthropathies about the elbow.

Plain films have typically been performed before MR imaging and should be reviewed for evidence of a fracture, bony destruction, sclerosis, osteochondral defects, or other findings that may explain the cause of bone marrow edema on MR imaging. Bone scans are usually abnormal when STIR images reveal marrow edema; however, bone scans may be useful when unexplained marrow edema is identified on MR imaging if a multifocal process such as metastatic disease is suspected.

In cases of repetitive trauma and overuse, a stress reaction or stress response can also result in a nonspecific pattern of marrow edema. This process may be followed with limited STIR scans to evaluate various treatment regimens (Fig. 3F–I). A stress reaction is differentiated from a stress fracture on MR imaging by the presence of a fracture line. A stress reaction is differentiated from a bone contusion by a history of overuse rather than acute trauma. As in all of the differential considerations described above, the clinical history is extremely important in identifying the cause of marrow edema and the correct diagnosis on MR imaging. (see also Case 4-4).

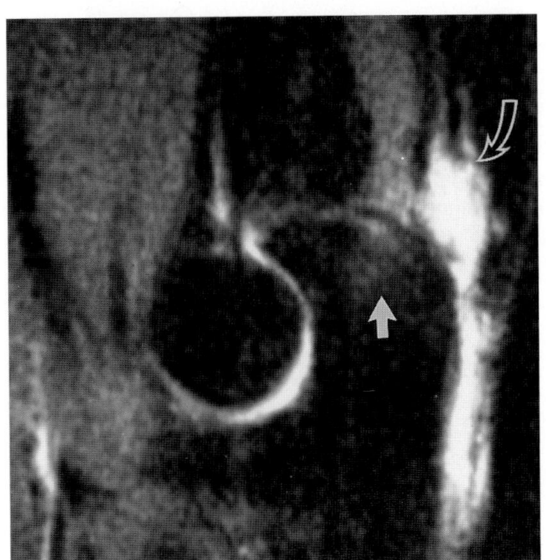

FIG. 3. C: Partial triceps tendon avulsion. A STIR sagittal image reveals a partial rupture of the superficial tendon fibers (*curved arrow*) as well as a small area of increased signal (*solid arrow*) within the superior aspect of the olecranon. This signal alteration within the adjacent bone marrow is probably due to a small periosteal or cortical avulsion.

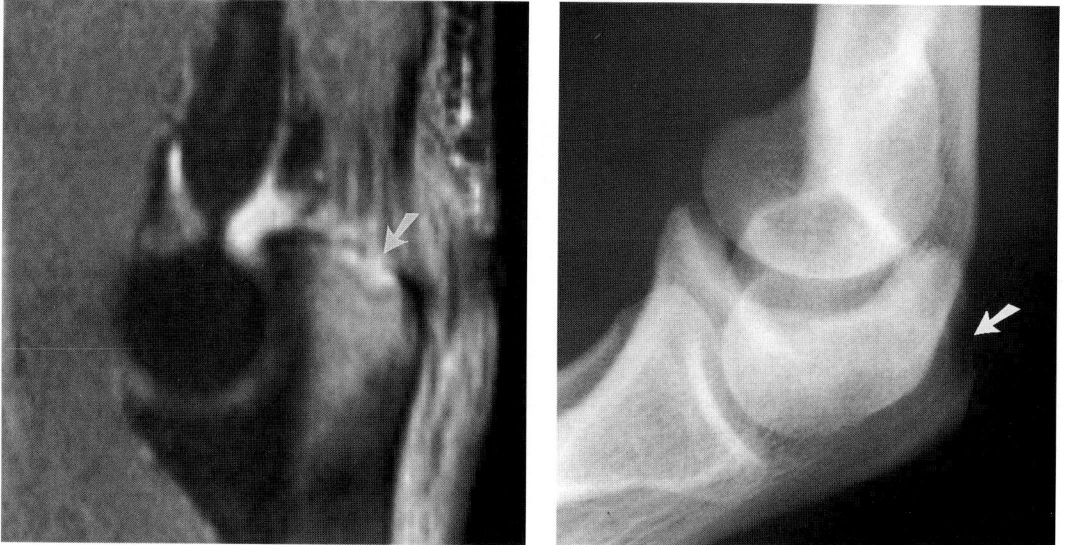

FIG. 3. D,E: Osteomyelitis of the olecranon secondary to cellulitis and septic bursitis. A STIR sagittal image **(D)** reveals increased signal throughout the olecranon with a cortical defect noted superiorly (*arrow*). There is increased signal and thickening of the posterior soft tissues compatible with cellulitis. Bony destruction (*arrow*) was subsequently visualized on a lateral plain film **(E)**.

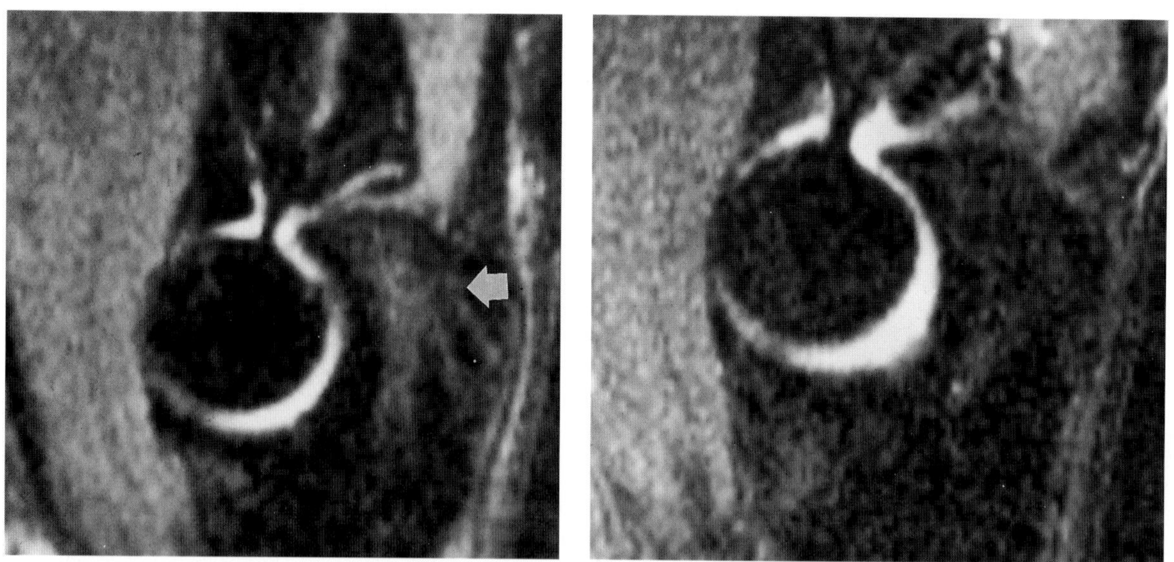

FIG. 3. F–I: Olecranon stress reaction secondary to posterior impingement and overuse in a professional baseball pitcher. A STIR sagittal image **(F)** reveals increased signal throughout the olecranon. A bone scan was also positive at this time. The abnormal signal has resolved on this STIR sagittal image **(G)** after 6 weeks of rest. Spurring of the anterior olecranon is evident (*open arrow*). With the resumption of pitching, the patient's symptoms returned, and arthroscopic posterior decompression was then performed.

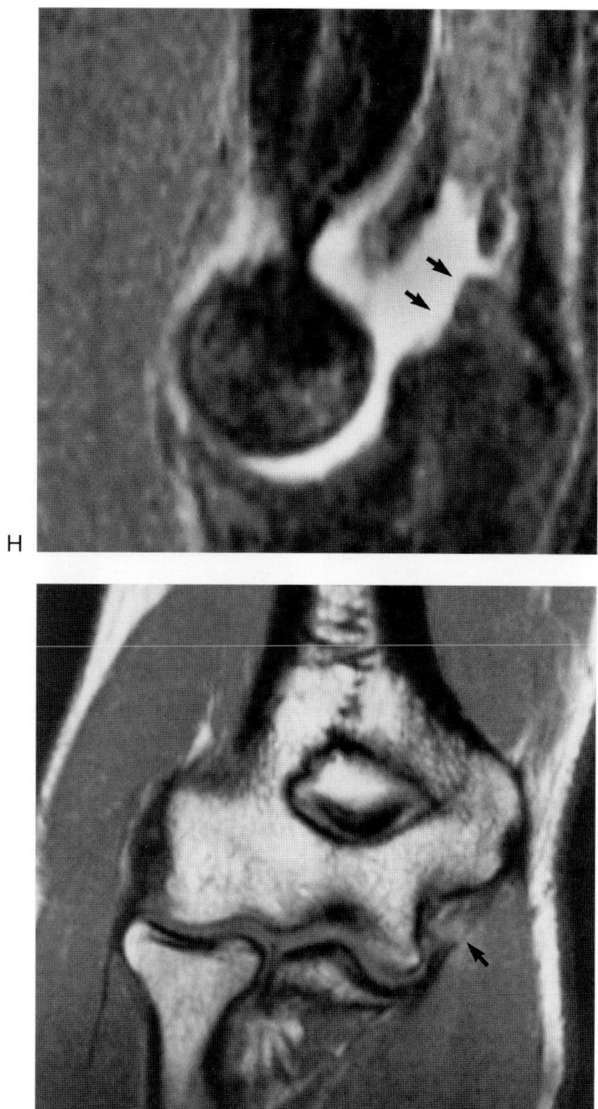

FIG. 3. H: A postoperative STIR sagittal image shows the site of bony resection (*arrows*). The patient was able to pitch without pain for several months; however, he ruptured his medial collateral ligament while pitching the following season. Midsubstance rupture of the anterior bundle of the medial collateral ligament (*arrow*) is seen on this T1-weighted coronal image **(I)**.

CASE 4

History: This 40-year-old man presented with a persistently painful elbow 3 weeks after falling on his outstretched arm while skiing. Plain films (not shown) showed no evidence of a joint effusion or fracture.

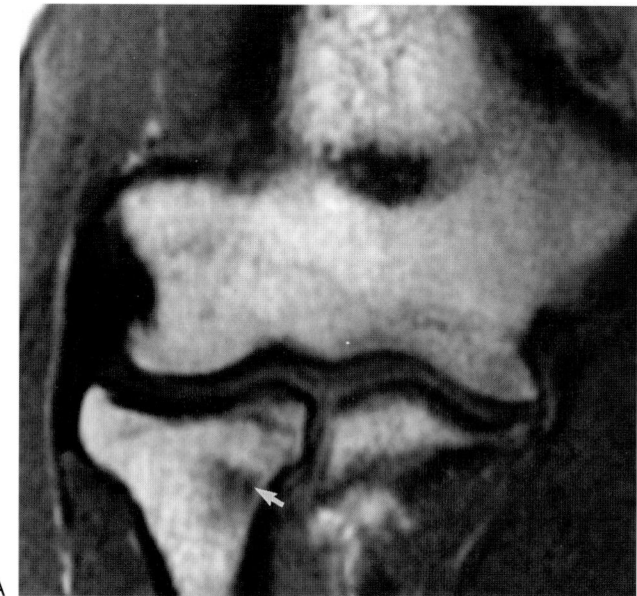

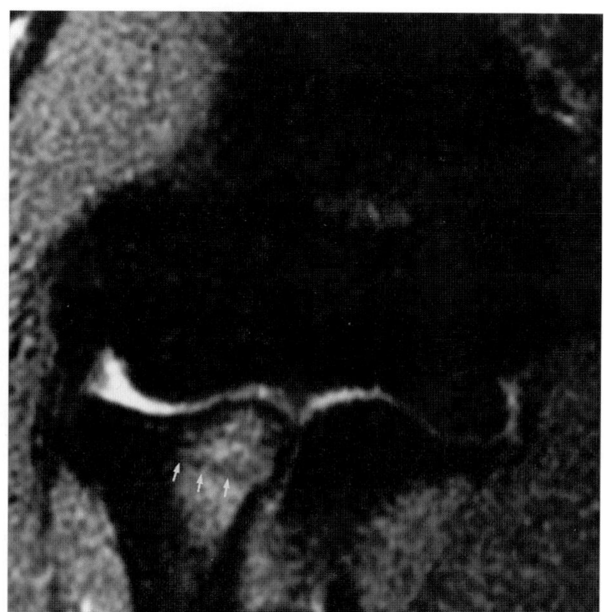

FIG. 4.

Findings: A T1-weighted coronal image (Fig. 4A) reveals a small focus of decreased signal (arrow) in the medial aspect of the radial neck. A STIR coronal image (Fig. 4B) reveals more prominent increased signal surrounding a linear focus of decreased signal (small arrows) in the medial aspect of the radial neck.

Diagnosis: Trabecular fracture of the radial neck.

Discussion: In this case, the T1-weighted image (Fig. 4A) shows a subtle area of abnormal signal whereas the STIR image (Fig. 4B) shows a trabecular fracture line with surrounding edema that is quite conspicuous. This abnormality is easily recognized on the STIR image but could have missed on standard spin-echo or gradient-echo images, as discussed previously in Case 3.

This is another case of radiographically occult bone injury that is presented to emphasize the usefulness of STIR sequences in musculoskeletal trauma cases. I have found fat-suppressed sequences in general, and STIR or fast STIR sequences in particular, quite useful in routine MR imaging of the elbow. The fast-spin-echo STIR technique can be substituted for the conventional STIR sequence without significant penalty according to preliminary experience in the literature (1). I have found fast STIR sequences helpful on several 1.0-T and 1.5-T systems and recommend their routine use when available.

STIR and fast STIR sequences are especially useful for identifying bone marrow and soft tissue edema during MR imaging of the elbow. Subarticular marrow edema on STIR images may direct attention to the adjacent articular cartilage on other pulse sequences, allowing recognition of full-thickness chondral defects and osteochondral fractures. Similarly, recognition of soft tissue edema may direct attention to adjacent ligament and tendons, allowing recognition and characterization of these injuries on the other pulse sequences.

Fat-suppressed sequences sacrifice signal-to-noise ratios for contrast-to-noise ratios and conspicuity (2). Although the fat-suppressed images are often the least pretty, they are frequently the most diagnostic and important images that I obtain. These images are not without problems, however. One should be aware of artifacts related to incomplete fat suppression throughout the image. Heterogeneous fat suppression may be confusing if it is not recognized as a technical pitfall and is instead mistaken for pathology (Fig. 4C,D) (see introductory paragraphs). Furthermore, the ability to identify abnormalities confidently is lost in that portion of the image with incomplete fat suppression. In general, I have had greater difficulty in obtaining homogeneous fat suppression consistently with the spectroscopic chemical-shift methods compared with the STIR and fast STIR techniques. This is especially true when imaging the elbow with the patient supine and the arm at the side. In this position, the elbow lies in the periphery of the magnet where the magnetic field is less uniform and suboptimal fat suppression may occur.

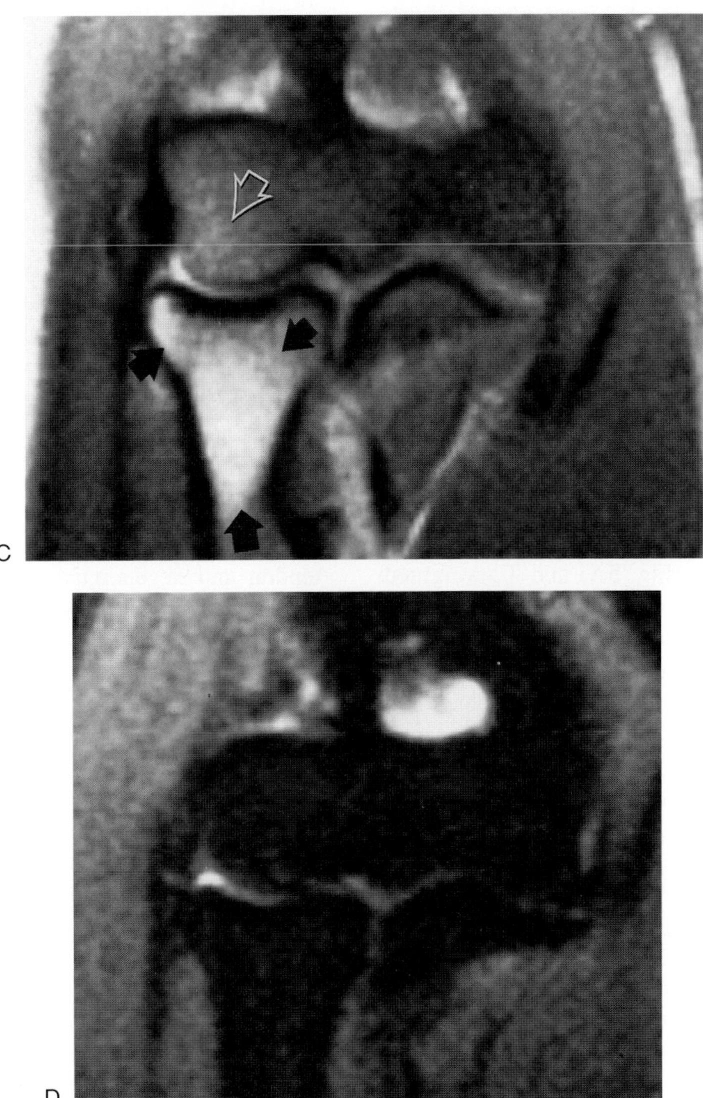

FIG. 4. C,D: Technically inadequate fat suppression with the chemical-shift technique resulting in potential misdiagnosis. A fat-suppressed fast-spin-echo proton-density (TR/TE 4,000/40 msec) coronal image **(C)** reveals increased signal throughout the proximal radius (*solid black arrows*). Less prominent signal alteration is seen in the lateral aspect of the capitellum (*open arrow*). The signal within the lateral subcutaneous fat was also not suppressed, providing a clue to the artifactual nature of the increased signal in the radius and capitellum. A fast STIR coronal image **(D)** with uniform fat suppression throughout the image reveals no abnormalities.

CASE 5

History: This 35-year-old man presented with a painful, gradually enlarging soft tissue mass in the posterior aspect of the elbow. He recalled no specific trauma. What is the diagnosis? Can you tell if this mass is cystic or solid on MR imaging?

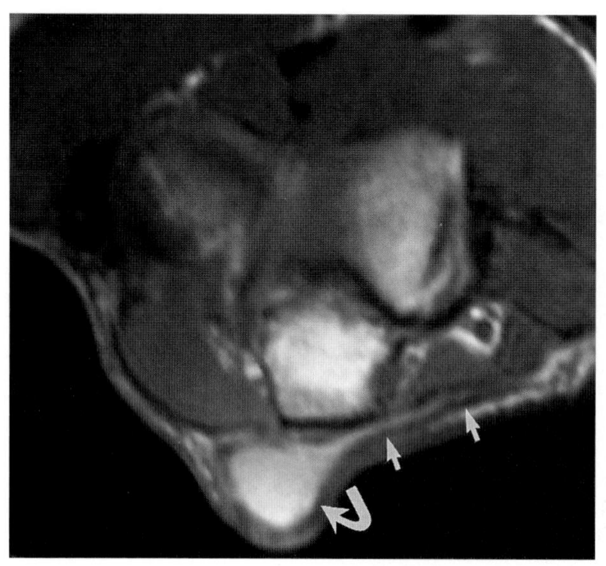

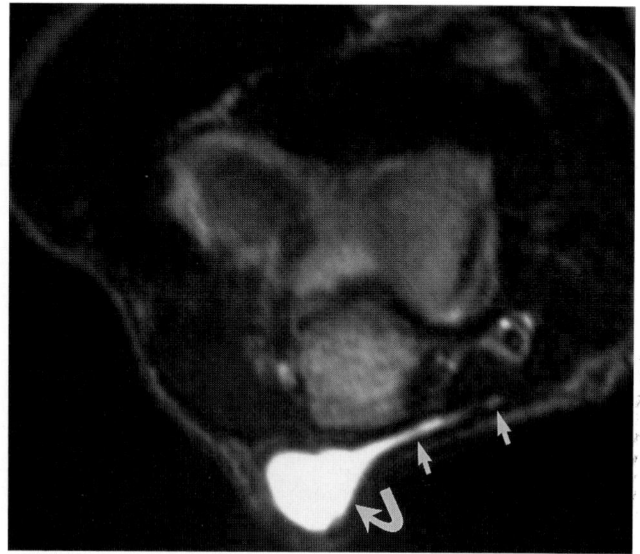

FIG. 5.

Findings: Proton density (Fig. 5A) and T2-weighted (Fig. 5B) axial images reveal a well-marginated area of homogeneous increased signal (curved arrows) as well as a thin pedicle that extends further medially (small arrows), consistent with fluid. The fluid lies within the olecranon bursa along the posterior margin of the olecranon and triceps tendon.

Diagnosis: Acute olecranon bursitis, secondary to gout.

Discussion: In this case, fluid was subsequently aspirated that was remarkable for monosodium urate crystals. The aspirate revealed no evidence of infection. The diagnosis is therefore acute olecranon bursitis secondary to gout. Other MR images (not shown) revealed no evidence of joint effusion, synovitis, osteomyelitis, or abnormalities of the triceps tendon.

The bursae of the human body are generally divided into deep and superficial types. Although most of the deeper bursae are present at birth, the olecranon bursa, like most other subcutaneous bursae, does not form until childhood, in response to movement and function (1). The subcutaneous olecranon bursa is the most common superficial site of bursitis in the body (2). Deeper bursae have also been described in the olecranon region. An intratendinous bursa, in the substance of the distal triceps tendon, may possibly be involved in tears of the triceps tendon and degenerative tendinosis (2). An association has been also been suggested between subcutaneous olecranon bursitis and triceps tendon rupture, as was discussed in Case 2 (3). A subtendinous bursa has been described between the triceps tendon and the posterior joint capsule; however, pathology of this deep bursa is rarely identified.

Olecranon bursitis is most commonly due to trauma, either acute or repetitive (4,5). This condition has been called miner's elbow or student's elbow. Traumatic olecranon bursitis is a common football injury that is usually associated with artificial turf (6). Other sports that commonly result in olecranon bursitis include ice hockey and wrestling.

Olecranon bursitis may be secondary to systemic diseases such as rheumatoid arthritis, gout, hydroxyapatite deposition, and calcium pyrophosphate deposition. Olecranon bursitis is also commonly seen in patients undergoing hemodialysis. In patients with rheumatoid arthritis, the bursa may communicate with the joint, may rupture, or may dissect into the forearm similar to a popliteal cyst about the knee (2).

About 20% of patients presenting with acute bursitis have infection, most commonly due to *Staphylococcus aureus* (7,8). Trauma usually precedes septic olecranon bursitis. In addition, steroid injections have preceded infection in about 10% of cases (8,9). The source of infection may be hematogenous or direct spread from abrasions and cellulitis. Osteomyelitis may develop in the underlying olecranon (Fig. 5C); this is thought to be uncommon, however (2). MR imaging may be useful when osteomyelitis is suspected (also see Case 3, Figs. 3D,E).

In chronic olecranon bursitis, there is usually chronic synovitis and fibrosis. Nodules of granulation tissue are often present (7). The synovitis and fibrosis may result in a complex appearance on MR imaging that may be difficult to distinguish from a solid mass. Alternatively, a solid neoplasm may occasionally be mistaken for olecranon bursitis (Fig. 5D–G). There is often spurring of the adjacent olecranon in chronic bursitis (4,5). These spurs as well as the tip of the olecranon are usually resected in patients who have failed conservative treatment and require removal of the bursa.

MR imaging of olecranon bursitis may have a range of appearances due to the various conditions that may affect the bursa. Acute and chronic hemorrhage as well as acute and chronic synovitis may result in a complex appearance of the olecranon bursa. Infiltration of the subcutaneous fat and cellulitis may accompany septic bursitis. Osteomyelitis and abnormalities of the adjacent triceps tendon may also be seen on MR imaging.

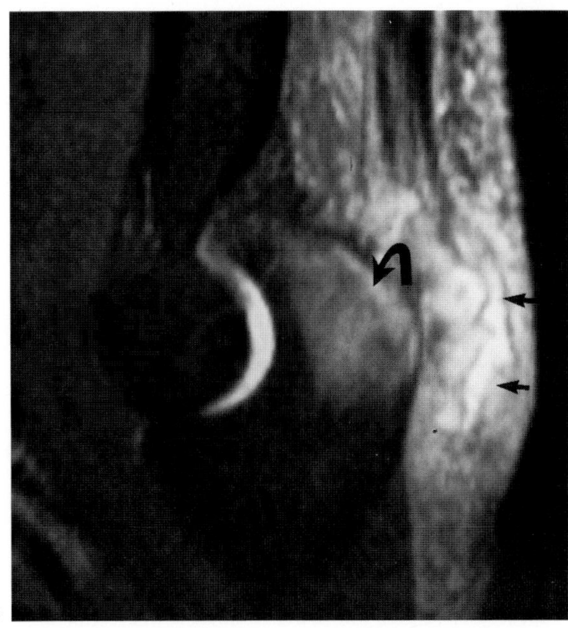

FIG. 5. C: Cellulitis, septic bursitis, and osteomyelitis. A STIR sagittal image reveals infiltration of the subcutaneous fat, distension of the olecranon bursa (*straight arrows*), and increased signal within the olecranon (*curved arrow*) compatible with osteomyeliti

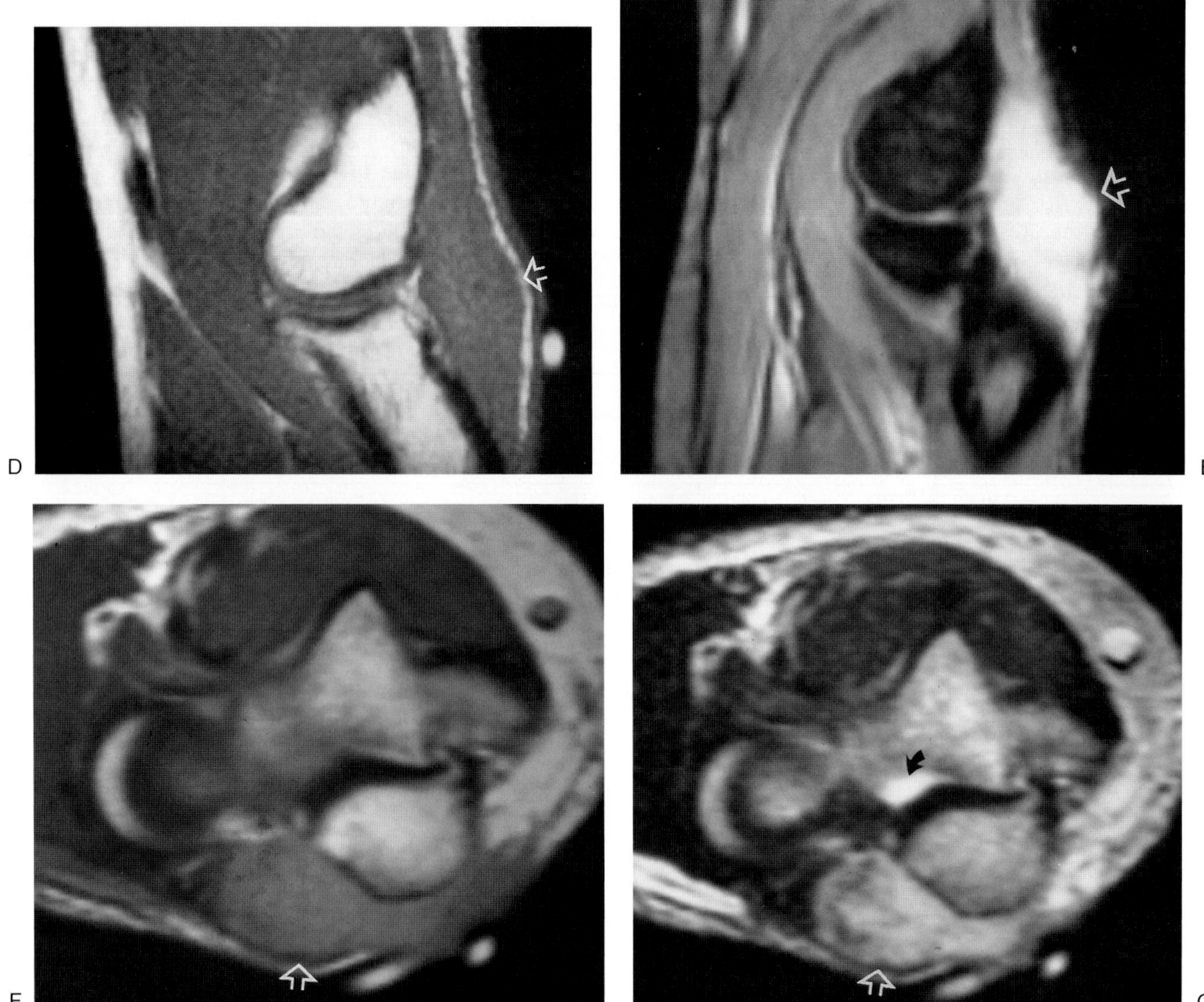

FIG. 5. D–G: Clear cell sarcoma initially misdiagnosed as chronic olecranon bursitis on MR imaging. T1-weighted **(D)** and gradient-echo T2*-weighted **(E)** sagittal images reveal a seemingly homogeneous mass (*open arrows*) of increased signal that was mistaken for fluid. Proton-density **(F)** and T2-weighted **(G)** axial images reveal a mass (*open arrows*) of heterogeneous increased signal that is less intense than joint fluid (*curved black arrow*).

CASE 6

History: This 55-year-old man presented to the emergency room with elbow pain after a fall. No fractures were found, but an apparent lytic lesion was seen on an anteroposterior (AP) radiograph. What is this lesion? Does this need biopsy or further evaluation?

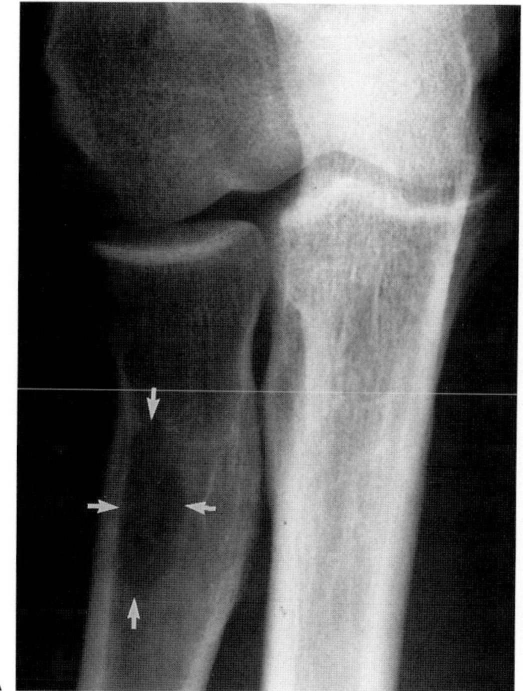

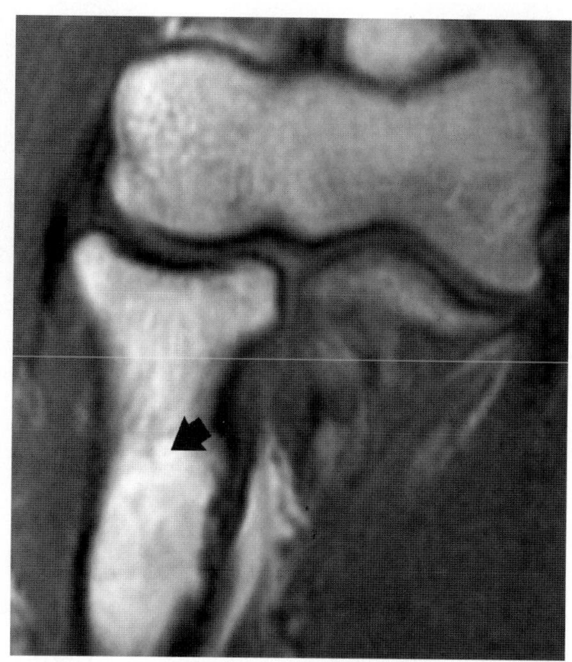

FIG. 6.

Findings: An AP view of the elbow (Fig. 6A) reveals a lucency (arrows) within the proximal radius. No abnormalities were seen on the lateral view (not shown). A T1-weighted coronal image (Fig. 6B) reveals normal marrow in this location (black arrow). No lesion is present.

Diagnosis: Normal radiolucency secondary to the radial tuberosity (radiographic pitfall).

Discussion: MR imaging was probably not necessary in this case, since this radiographic pitfall of the proximal radius has a fairly characteristic appearance (1). The radiographic appearance is thought to be secondary to the projection of the radial tuberosity through the radial shaft. This radiolucency may be seen in various radiographic projections depending on the degree of pronation and the relative position of the radial tuberosity to the x-ray beam. This appearance is occasionally thought to represent an area of bony destruction such as a metastatic lesion. Again, in this case, MR imaging was probably not necessary but was useful to ensure normality and exclude a bony lesion.

CASE 7

History: This 33-year-old woman presented with a several-month history of elbow pain and tenderness. Radiographs (not shown) revealed displacement of the anterior and posterior fat pads, suggesting a joint effusion. No fluid was obtained with attempted aspiration. What is the diagnosis?

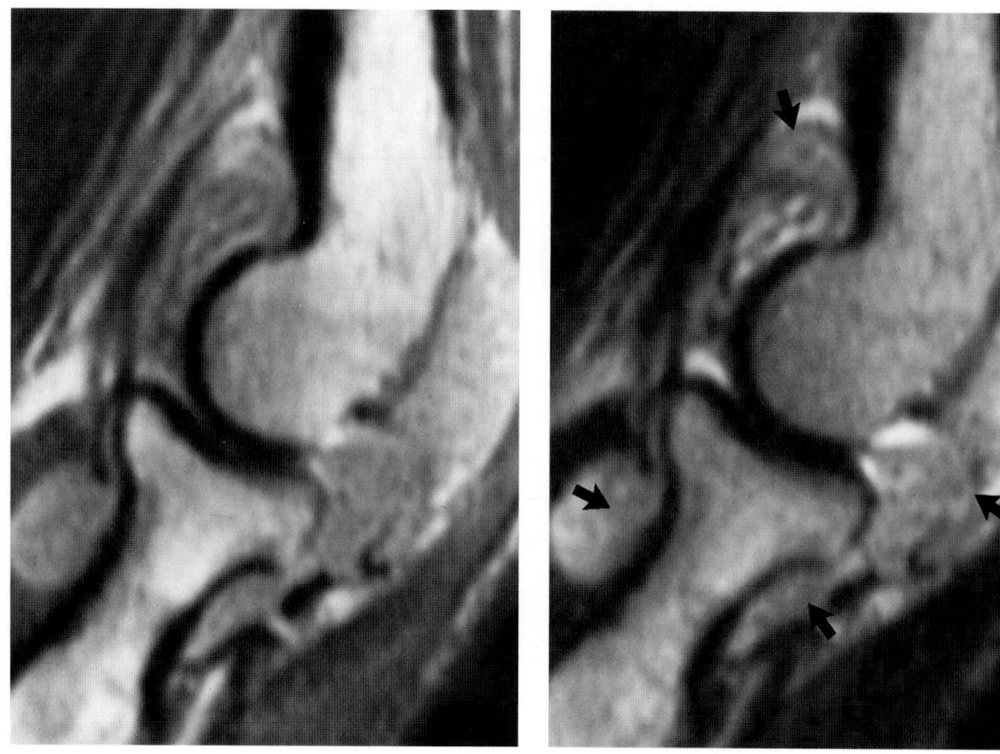

FIG. 7.

Findings: Proton-density (Fig. 7A) and T2-weighted (Fig. 7B) sagittal images reveal a minimal amount of joint fluid. There is prominent distension of the elbow joint secondary to intermediate signal intensity tissue that represents thickened synovium and pannus (arrows).

Diagnosis: Rheumatoid arthritis of the elbow.

Discussion: The appearance of thickened synovium throughout the elbow in this case explains the displacement of the intracapsular/extrasynovial fat pads on the lateral radiograph. Displacement of the fat pads is a reliable radiographic indicator of joint space distension. Joint space distension is most commonly due to a joint effusion, especially in the setting of trauma. On further workup, this patient was found to have rheumatoid arthritis. Thickened synovium is often recognizable on conventional MR imaging (Fig. 7C,D). In such cases, the synovium is greater in signal intensity relative to fluid on T1-weighted images and decreased in signal intensity relative to fluid on T2-weighted images (1). Gadolinium-enhanced images may be necessary to differentiate fluid from acute synovitis in certain cases (2–4). Actively inflamed synovium will enhance after intravenous administration of gadolinium (Fig. 7E–G) (see Case 3–6). Synovitis can also be identified with MR arthrography using either saline or dilute gadolinium (Fig. 7H). MR imaging is especially useful for providing objective evidence of synovitis in the early stages of rheumatoid arthritis so that the diagnosis can be established and treatment can be instituted in a timely fashion (5).

Rheumatoid arthritis usually involves the elbow joint within the first 5 years of onset (6). Of patients with rheumatoid arthritis, 20% to 50% eventually have involvement of the elbow joint. Patients usually present with painful distension of the joint capsule. If the synovitis is uncontrolled, there is erosion of the hyaline cartilage on the joint surfaces. Progressive destruction results in joint space narrowing and instability. The synovitis may herniate into the periarticular soft tissues, resulting in compression of the peripheral nerves about the elbow (6). Rheumatoid synovitis may also produce large intraosseous synovial cysts (Fig. 7E–G). These features of rheumatoid synovitis may be well seen with MR imaging. Chronic pannus and fibrosis may occasionally result in low signal intensity synovitis on MR imaging; however, this appearance is more characteristic of the recurrent intra-articular hemorrhage that is commonly seen in hemophilia (7–9) and pigmented villonodular synovitis (Fig. 7I) (10,11). This appearance may also be seen in hemodialysis-related amyloid arthropathy (12). This arthropathy is a recently recognized complication of long-term hemodialysis. It is caused by the deposition of a unique form of amyloid derived from circulating $\beta 2$-microglobulin that may result in low-signal-intensity nodular synovial lesions similar to the appearance of the hemosiderin-laden synovium seen in pigmented villonodular synovitis and hemophilia.

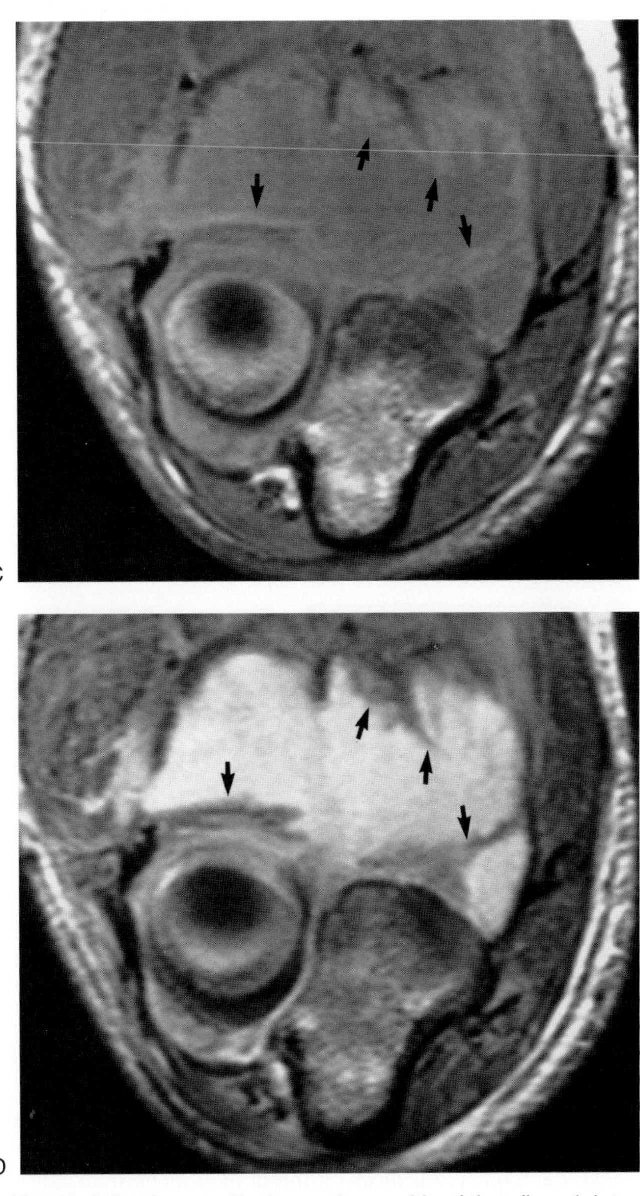

FIG. 7. C,D: Large effusion and synovitis of the elbow joint. Proton-density **(C)** and T2-weighted **(D)** axial images reveal thickened synovium (*arrows*) compatible with synovitis. (Courtesy of John Feller, MD.)

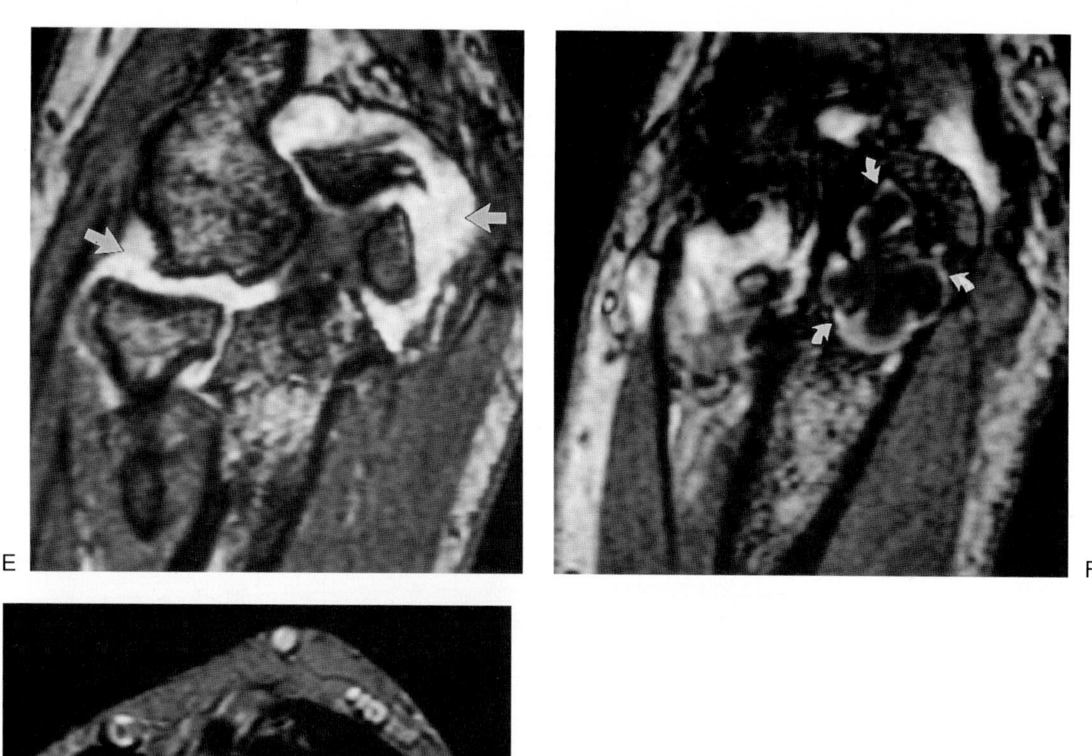

FIG. 7. E–G: Enhancing rheumatoid synovitis and a large intraosseous synovial cyst of the olecranon. T1-weighted gradient-echo coronal images **(E,F)** reveal generalized increased signal throughout the elbow joint secondary to enhancing pannus (*straight arrows*). There is rim enhancement (*curved arrows*) of a synovial cyst within the olecranon. A T2-weighted axial image **(G)** shows the lobular septated synovial cyst within the olecranon (*open arrow*). (Courtesy of Peter Munk, MD.)

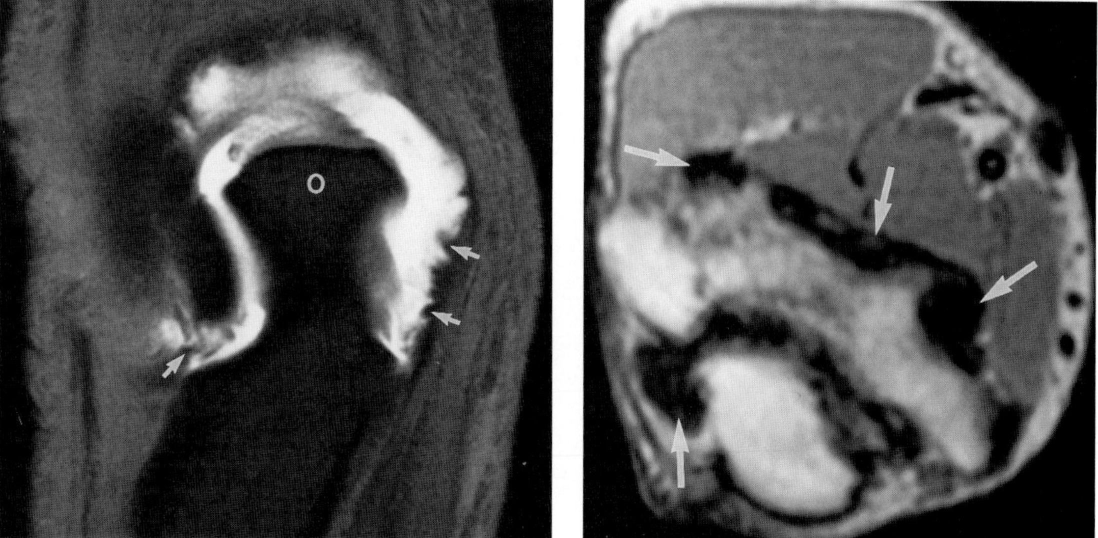

FIG. 7. H: Mild synovitis demonstated with MR arthrography using intra-articular dilute gadolinium. A fat-suppressed T1-weighted coronal image at the level of the olecranon (O) reveals mild corregation and synovial thickening (*arrows*) in the posterior compartment compatible with synovitis. **I:** Pigmented villonodular synovitis. A proton-density axial image shows irregular low-signal-intensity synovial thickening (*arrows*) consistent with pigmented villonodular synovitis. (Courtesy of Barbara Weissman, MD)

CASE 8

History: This 35-year-old man presented with painless limitation of motion at the elbow. Radiographs (not shown) revealed displacement of the fat pads, suggesting a large joint effusion. No fluid was obtained with attempted aspiration. CT (not shown) revealed a mass of approximately water density without evidence of calcification or ossification. What is the diagnosis? How is this condition treated?

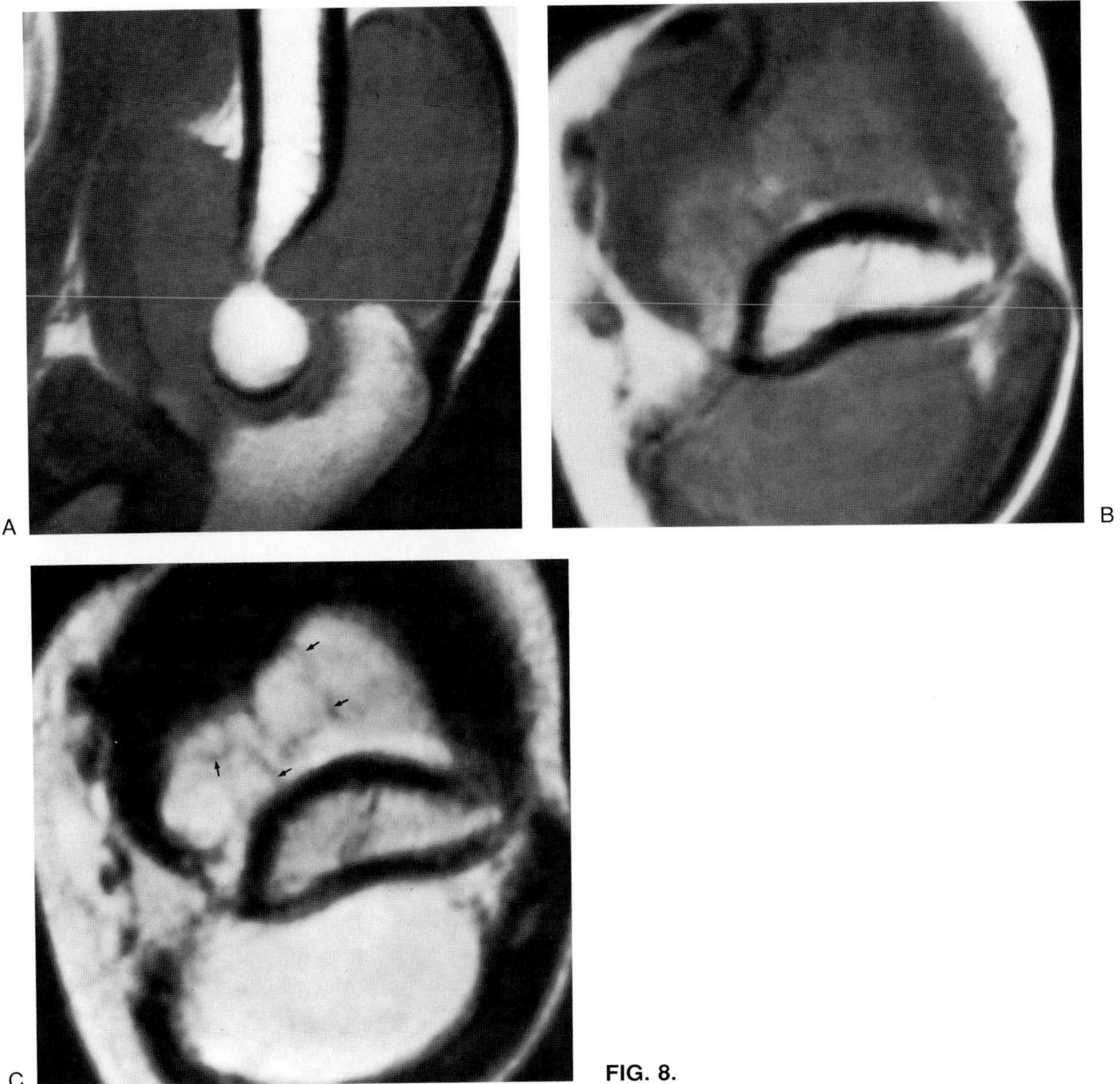

FIG. 8.

Findings: A T1-weighted sagittal image (Fig. 8A) as well as proton-density (Fig. 8B) and T2-weighted (Fig. 8C) axial images reveal prominent distension of the elbow joint secondary to material that at first glance looks like fluid. Inhomogeneity of the material within the joint as well as thin septations (small arrows) are seen on the T2-weighted axial image, however, providing a clue to the diagnosis.

Diagnosis: Synovial chondromatosis of the elbow.

Discussion: Multiple tightly packed nodules of cartilage were removed from the elbow joint at surgery. Active proliferation of cartilage was present within the synovium at histology. These findings are diagnostic of synovial chondromatosis. A complete synovectomy was performed to prevent recurrence of this condition.

Synovial chondromatosis is an uncommon disorder characterized by metaplasia of the subsynovial soft tissues that results in cartilage formation within the synovium (1). This benign proliferative process may involve any joint but predominates in the knee, hip, or elbow. It is almost invariably a monarticular process that occasionally arises within a bursa or tendon sheath. It may be sharply localized, multifocal, or diffusely present within the synovium of a particular joint (Fig. 8A–F).

The average age of patients is about 40 years, and the condition occurs more commonly in males than females (2:1) (1). The disease process is typically predictable and self-limiting. In the elbow, there is usually a several-year history of pain and swelling with progressive limitation of motion. Gross distension of the joint space may occasionally cause entrapment of various nerves about the elbow (2,3). Malignant degeneration has been rarely reported (1).

This process has been called synovial chondromatosis or synovial osteochondromatosis depending on the presence of enchondral bone formation within the multiple cartilaginous nodules. Others have exclusively used the term synovial osteochondromatosis to encompass each of the progressive stages of this condition. The terms idiopathic synovial osteochondromatosis and primary synovial osteochondromatosis have been used to differentiate this condition from the more common secondary causes of loose bodies such as osteochondral fractures and osteoarthritis with degenerative fragmentation of the articular surface (4–6).

Three progressive phases of idiopathic or primary synovial osteochondromatosis have been identified (7). In the initial phase there is active intrasynovial disease without loose body formation. In the transition phase both active intrasynovial proliferation and multiple loose bodies, which may or may not be ossified, are present. In the final phase, the process may apparently become quiescent with multiple osteochondral loose bodies and no active intrasynovial disease. The latter stages of primary synovial osteochondromatosis frequently result in destruction of the articular surfaces and secondary osteoarthritis (1).

Patients with disease in the initial or transition stages of primary synovial osteochondromatosis are usually treated with complete synovectomy to prevent recurrence. Focal recurrence after surgery is not uncommon, however, as nests of synovium may be left behind (1). Patients with disease in the final phase may not require synovectomy and may simply be treated with removal of the multiple osteochondral loose bodies (7). Patients with loose bodies secondary to degeneration or trauma also do not require synovectomy, since the nidus for loose body formation is unrelated to metaplasia within the synovium.

Radiographically, there is usually calcification or ossification of the chondromatosis within the elbow that allows recognition of this condition. Widening of the joint space, bony erosions, and displacement of the intraarticular fat pads may also be seen on plain films. Secondary degenerative changes are frequently visible in the latter stages of the disease. In as many as one-third of reported cases no calcification or ossification is present, making radiographic diagnosis difficult (1). CT reveals an intraarticular mass of approximately water density in such cases, contributing to the diagnostic difficulty and the erroneous impression of a large effusion. Arthrography or MR imaging is useful in identifying the mass of uncalcified chondromatosis in these cases.

The MR imaging appearance of idiopathic synovial osteochondromatosis reflects the variable gross pathologic appearance of this condition (8). The most difficult cases to recognize are those that do not have visible calcification or ossification. The closely packed nodules of cartilage are bright on T2-weighted images and may mimic fluid on MR imaging, especially if the images are not properly windowed. Thin septations of decreased signal and somewhat decreased, heterogenous signal intensity of the chondromatosis allow for differentiation from a simple effusion (Fig. 8A–C). Foci of signal void are present on all pulse sequences when there is calcification within the osteochondromatosis (8). These foci of signal void are most prominent on gradient-echo T2*-weighted images. Multiple bony loose bodies with a low-signal-intensity cortical rim and central marrow fat are visible when there is ossification of the synovial chondromatosis in the more advanced stages of this condition (Fig. 8D–F). Inhomogeneous enhancement of synovial osteochondromatosis may be seen after intravenous gadolinium administration (8).

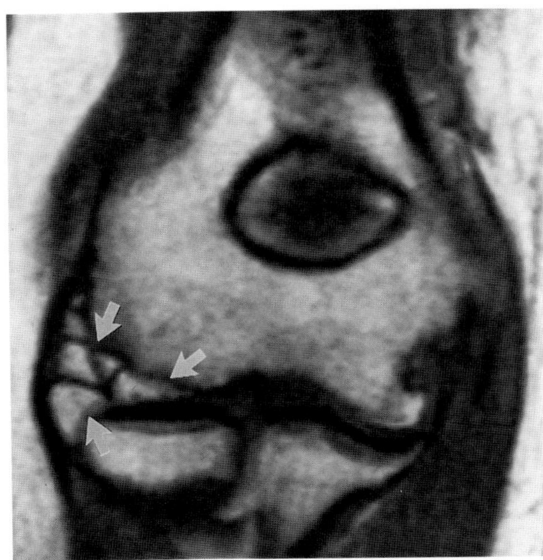

FIG. 8. D: Localized primary synovial osteochondromatosis. A T1-weighted coronal image reveals multiple faceted bony loose bodies in the lateral compartment of the elbow (*arrows*).

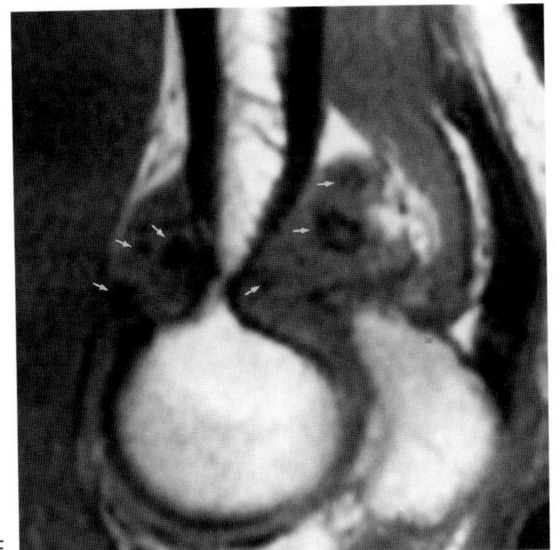

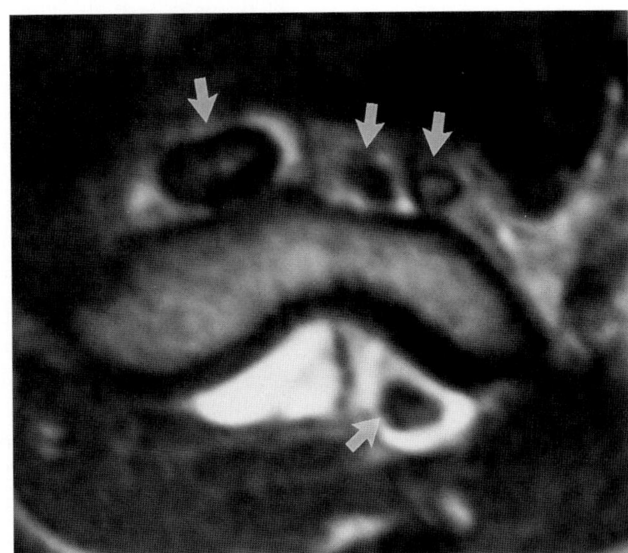

FIG. 8. E,F: Diffuse primary synovial osteochondromatosis. T1-weighted sagittal **(E)** and T2-weighted axial **(F)** images reveal multiple osteochondral loose bodies (*arrows*) throughout the elbow joint. Thirty loose bodies were arthroscopically removed in this case.

CASE 9

History: This 18-year-old man presented with posterior elbow pain and gradual loss of extension. Radiographs (not shown) revealed a large ossicle in the posterior region of the elbow. What is the diagnosis? How is this condition treated?

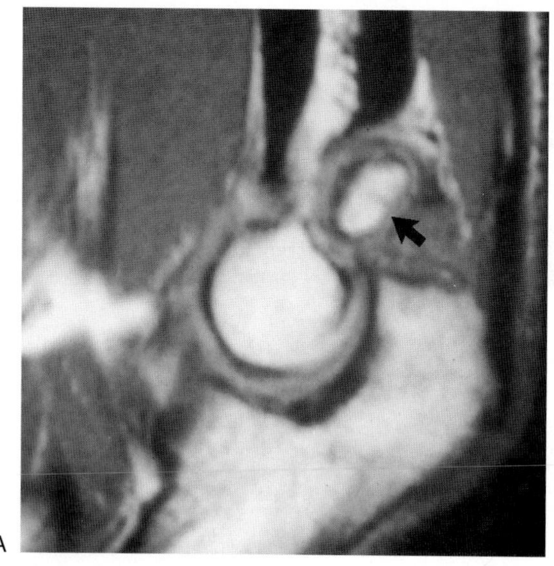

A

FIG. 9.

Findings: A T1-weighted sagittal image (Fig. 9A) reveals a large ossicle (arrow) within the posterior compartment of the elbow joint.

Diagnosis: Symptomatic os supratrochleare dorsale.

Discussion: The os supratrochleare dorsale is an accessory ossicle that lies within the olecranon fossa of the humerus (1–4). Although it may be asymptomatic and discovered as an incidental finding on radiographs or MR imaging, it may be associated with pain and progressive loss of elbow extension, as in this case. The os supratrochleare dorsale may be subjected to trauma and impaction during forced elbow extension or hyperextension. Although it almost always involves the dominant arm and is more common in males, it has been generally considered an accessory bone that is congenital or developmental rather than post-traumatic.

The precise origin of this ossicle is controversial because of its intra-articular location (4,5). It is thought to arise from a separate ossification center that is partially within the olecranon fossa. In asymptomatic cases, the os supratrochleare dorsale tilts to allow the olecranon to enter the olecranon fossa during full extension (5). In symptomatic cases, there is progressive enlargement of the ossicle through synovial nutrition that results in loss of elbow extension. Deepening and remodeling of the olecranon fossa may occur as the ossicle enlarges (Fig. 9A–C). Fragmentation and sclerosis of the ossicle may occur as a result of forced extension or hyperextension of the elbow. In such cases, the appearance of the olecranon fossa on MR imaging may be useful to differentiate post-traumatic osteochondral loose bodies from a fragmented os supratrochleare dorsale (Fig. 9D). Differention of these entities is not clinically significant, however, since the treatment is the same whether or not the loose body was caused by trauma or was an os supratrochleare dorsale that was subjected to trauma. If an os supratrochleare dorsale is painful owing to a direct impact, the symptoms should resolve with conservative treatment. If persistent pain, progressive loss of extension, locking, or catching are present, the ossicle can be removed arthroscopically or through a limited arthrotomy (3).

MR imaging can be useful to confirm the intra-articular location of a symptomatic os supratrochleare dorsale prior to arthroscopic removal. Posterior and superior displacement of the posterior intra-articular fat pad of the elbow and joint fluid along the margins of the ossicle on sagittal images confirm its location within the joint (Fig. 9A–C). The relatively large size of these ossicles at the time of presentation allows for accurate characterization with MR imaging. Small loose bodies, in general, are more difficult to identify accurately with MR imaging, since foci of synovitis, scarring, and small loose bodies appear as foci of signal void on T2-weighted images (Fig. 9B,C,E,F). Foci of synovitis may be mistaken for loose bodies on MR imaging (6). Routine radiographic correlation may be useful when attempting to identify loose bodies with MR imaging (Fig. 9E,F). CT is occasionally useful as a problem-solving examination when there is uncertainty regarding the presence of a small loose body on MR imaging. CT arthrography or MR arthrography may also be useful to differentiate periarticular ossification from intra-articular loose bodies by clearly delineating the joint space with contrast. In general, when there is a high clinical suspicion of loose bodies and the initial radiographs are negative or equivocal, conventional MR imaging is the next best study. MR arthrography is the next best study when there is little fluid in the elbow on conventional MR imaging and the results are negative or uncertain.

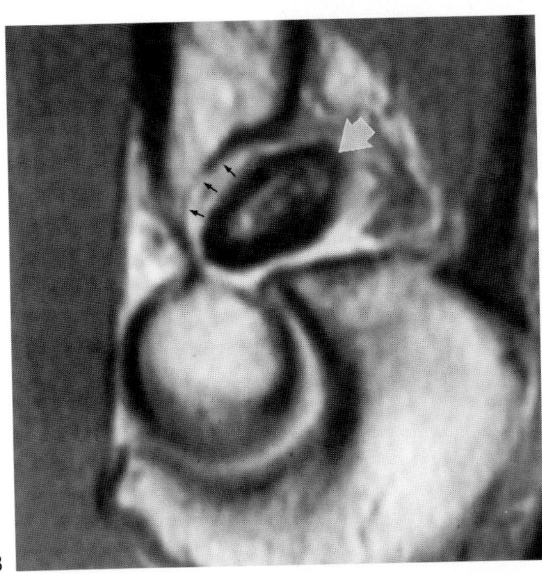
B

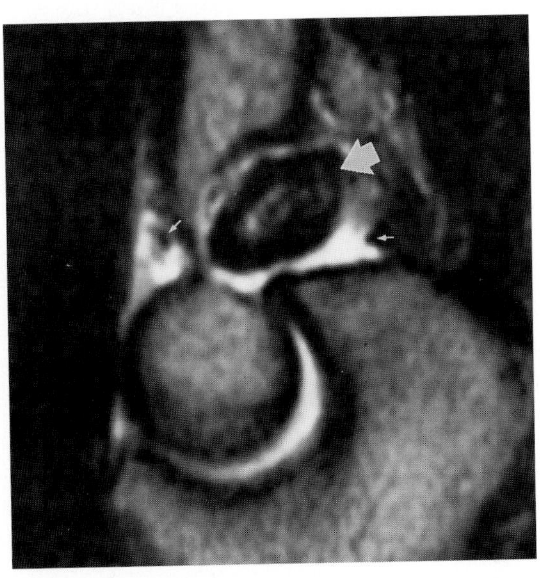
C

FIG. 9. B,C: Symptomatic os supratrochleare dorsale and synovitis. Proton-density **(B)** and T2-weighted axial **(C)** images reveal a large intraarticular ossicle (*large white arrows*). There is enlargement and remodeling of the olecranon fossa (*small black arrows*), probably on the basis of gradual enlargement of the adjacent os supratrochleare dorsale. Foci of synovial thickening (*small white arrows*) are noted that should not be confused for loose bodies.

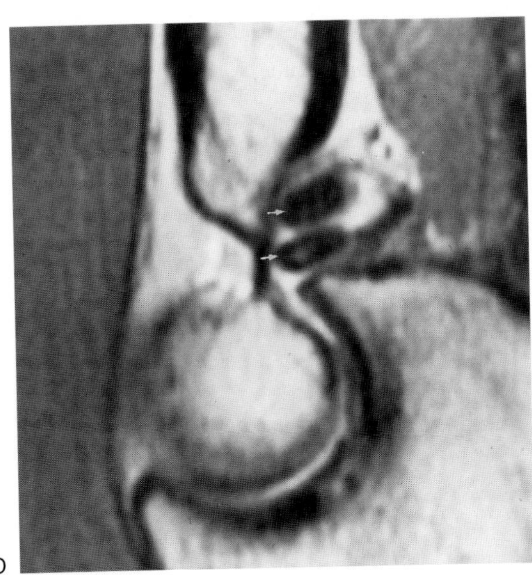
D

FIG. 9. D: Posterior compartment loose bodies. A proton-density sagittal image reveals two sclerotic loose bodies (*arrows*) that were treated with subsequent arthroscopic removal. There is no deepening and remodeling of the olecranon fossa in this case to suggest fragmentation of a previous os supratrochleare dorsale.

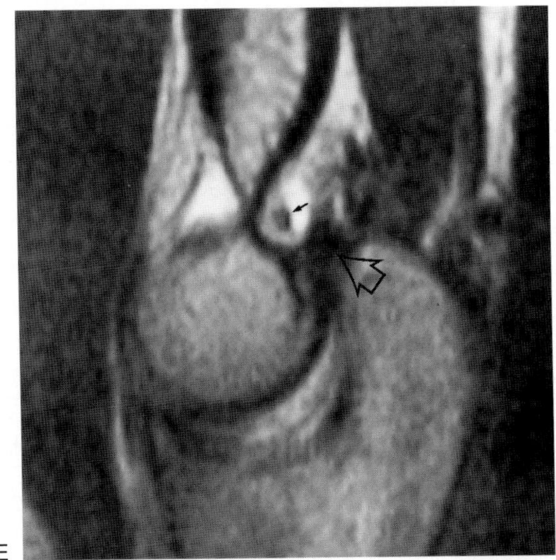

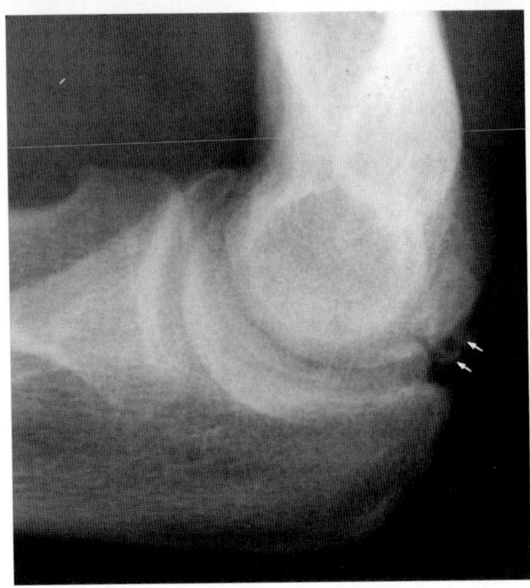

FIG. 9. E,F: Small recurrent loose bodies. This professional baseball pitcher developed intermittent locking and catching. He had previously undergone removal of loose bodies and olecranon spurs. A T2-weighted sagittal image **(E)** reveals a small focus of decreased signal (*small black arrow*) suspicious for a loose body. There is low signal scarring at the site of prior surgery (*open arrow*). A lateral radiograph **(F)** reveals several loose bodies (*small white arrows*) within the olecranon fossa.

CASE 10

History: This 44-year-old man presented with intermittent locking and catching of the elbow. Radiographs (not shown) revealed spurring and possible loose bodies. What are the MR findings? How is this condition treated?

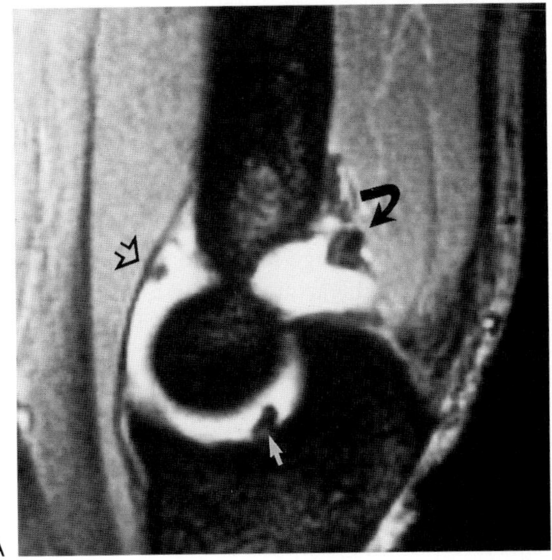

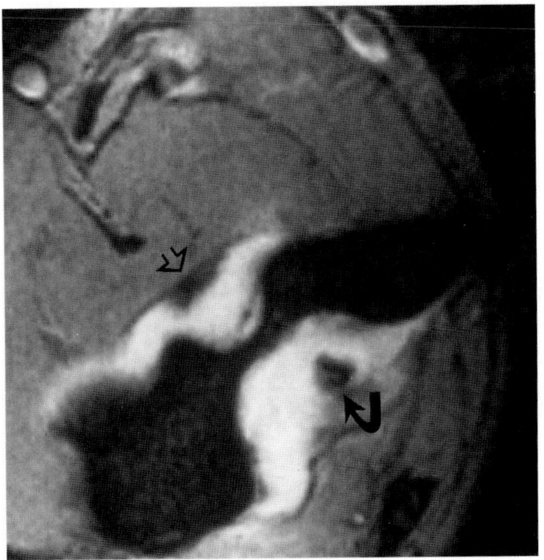

FIG. 10.

Findings: Gradient-echo T2*-weighted sagittal (Fig. 10A) and axial (Fig. 10B) images reveal foci of signal void in the anterior compartment (open black arrows) and the posterior compartment (curved black arrows). A focus of signal void is also noted in the central aspect of the trochlear notch (straight white arrow).

Diagnosis: Multiple osteochondral loose bodies.

Discussion: The elbow is the second most common site of loose bodies after the knee joint (1). Loose bodies are thought to arise from a small nidus of bone or cartilage within the joint. The nidus may result from fragmentation of the articular cartilage associated with osteoarthritis or from an osteochondral fracture (Fig. C–E). The small nidus may grow in a laminar fashion receiving nutrition from the synovial fluid (2,3). The growth process may continue as long as the loose body is exposed to synovial fluid. Loose bodies may attach to the synovium or may float freely within the joint space. A change in position of a loose body over sequential imaging studies indicates that it is freely mobile. Similarly, movement of a loose body with changes in position of the elbow joint on a particular imaging study also indicates that the loose body is not firmly attached to the synovium. In the case illustrated above, the loose body in the anterior compartment is apparently attached to the synovium, since it does not appear to respect gravity and sink posteriorly toward the humerus.

Large loose bodies are well seen with MR imaging, especially when an effusion is present (4). Small loose bodies may be more difficult to detect and differentiate from other foci of signal void on MR imaging such as thickened synovium (Fig. 10F,G). Air bubbles may also mimic loose bodies on MR imaging (5). Small air bubbles may arise naturally from the vacuum phenomenon or may be introduced iatrogenically during aspiration or injection of fluid. The vacuum phenomenon is unusual in the elbow joint, whereas small bubbles are commonly seen with MR arthrography.

Loose bodies may become quite large and may result in mechanical symptoms such as locking and limitation of motion. Most patients present with loss of motion, usually extension. Pain is variably present and usually occurs with a sensation of grating or locking (1). Symptomatic loose bodies are usually arthroscopically removed when detected, because they may lead to premature degenerative arthritis in addition to their effects on joint function (6).

The group of patients that benefits most from elbow arthroscopy is those with loose bodies (7,8). Accurate diagnosis of loose bodies is important prior to arthroscopy to avoid the unnecessary expense and potential complications of a surgical procedure. Plain films are routinely obtained prior to arthrography, but they may be unreliable. Indeed, radiographs did not demonstrate loose bodies in 7 of 23 patients treated with arthroscopic removal in a recent series (7). Osteophytes and periarticular ossification may be confused for intra-articular loose bodies on radiographs. Noncalcified chondral loose bodies cannot be visualized on CT or radiographs but can be identified on MR imaging. Calcified loose bodies are quite conspicuous on MR imaging, especially with gradient-echo T2*-weighted sequences. Calcified loose bodies may appear slightly larger than their actual size on gradient-echo T2*-weighted images as a result of magnetic suspectibility effects that are normally dampened by the 180° refocusing pulse on spin-echo images.

Loose bodies may lie anywhere within the elbow joint but are most commonly seen anteriorly (9). In throwing athletes, loose bodies are typically found in the posterior compartment as a result of the incongruity of the olecranon and the olecranon fossa that develops from chronic valgus stress (10,11). Loose bodies may also lodge in the midportion of the trochlear notch of the ulna (Fig. 10A–G). The predilection of loose bodies for this location may be explainable by the normal anatomy of the trochlear notch, also referred to as the greater sigmoid notch of the ulna. The mid-portion of the trochlear notch contains a variably sized bare area that is normally devoid of articular cartilage. A variably deep groove with a strip of synovium and fat extends transversely across the trochlear notch at the junction of the olecranon and the coronoid portions of the proximal ulna (Fig. 10H–J) (12). Loose bodies may lodge in this transverse groove or at the margins of this groove, where the ulna is normally somewhat constricted. This normal waist or constriction in the midportion of the trochlear notch should not be mistaken for an osteochondral defect (Fig. 10H–J). A thin transverse bony ridge may also extend across this portion of the trochlear notch, which lacks articular cartilage, and should not be mistaken for an intra-articular osteophyte or a loose body on a single sagittal MR image (13).

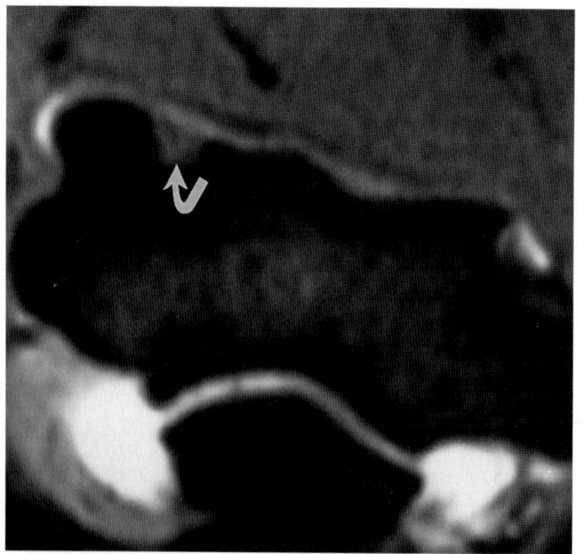
C

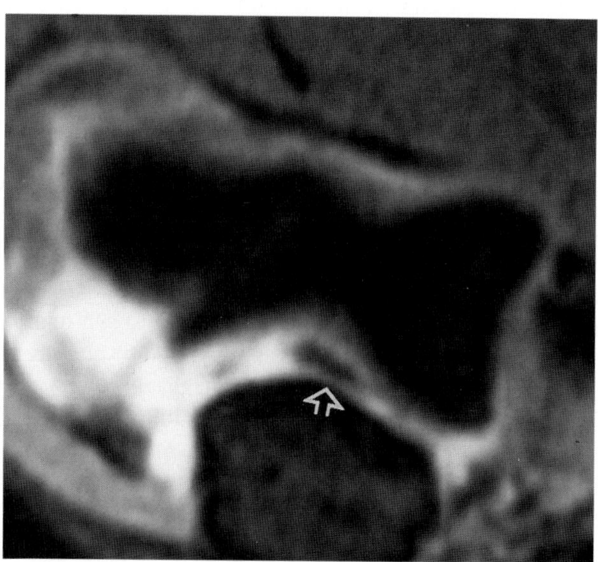
D

FIG. 10. C–E: Displaced osteochondral defect of the capitellum. Gradient-echo T2*-weighted axial **(C,D)** and sagittal **(E)** images reveal an osteochondral defect in the anterior aspect of the capitellum (*curved arrow*) and a loose body (*open arrows*) in a groove within the mid-portion of the trochlear notch of the ulna.

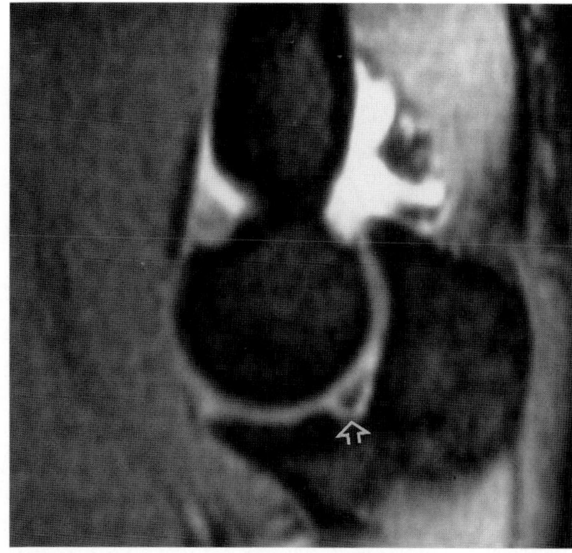

E

FIG. 10. C–E (*continued.*)

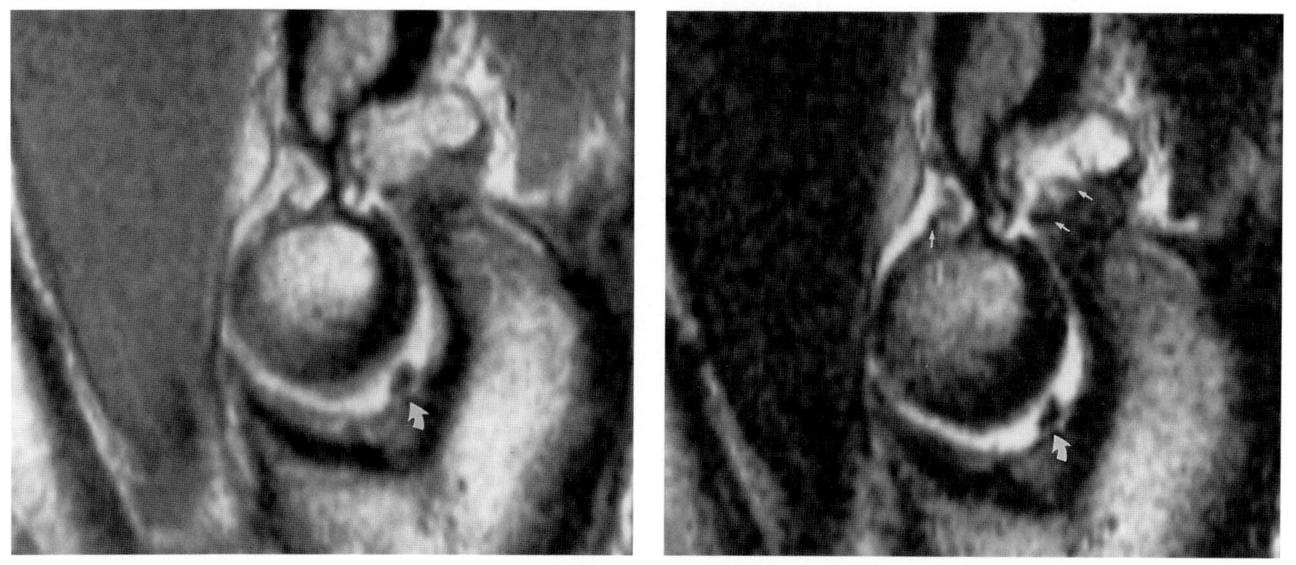

F G

FIG. 10. F,G: Loose body and synovitis. Proton-density **(F)** and T2-weighted sagittal **(G)** images reveal a small loose body in the central aspect of the trochlear notch of the ulna (*curved white arrows*). Curvilinear foci of synovial thickening (*small white arrows*) are noted that should not be mistaken for loose bodies.

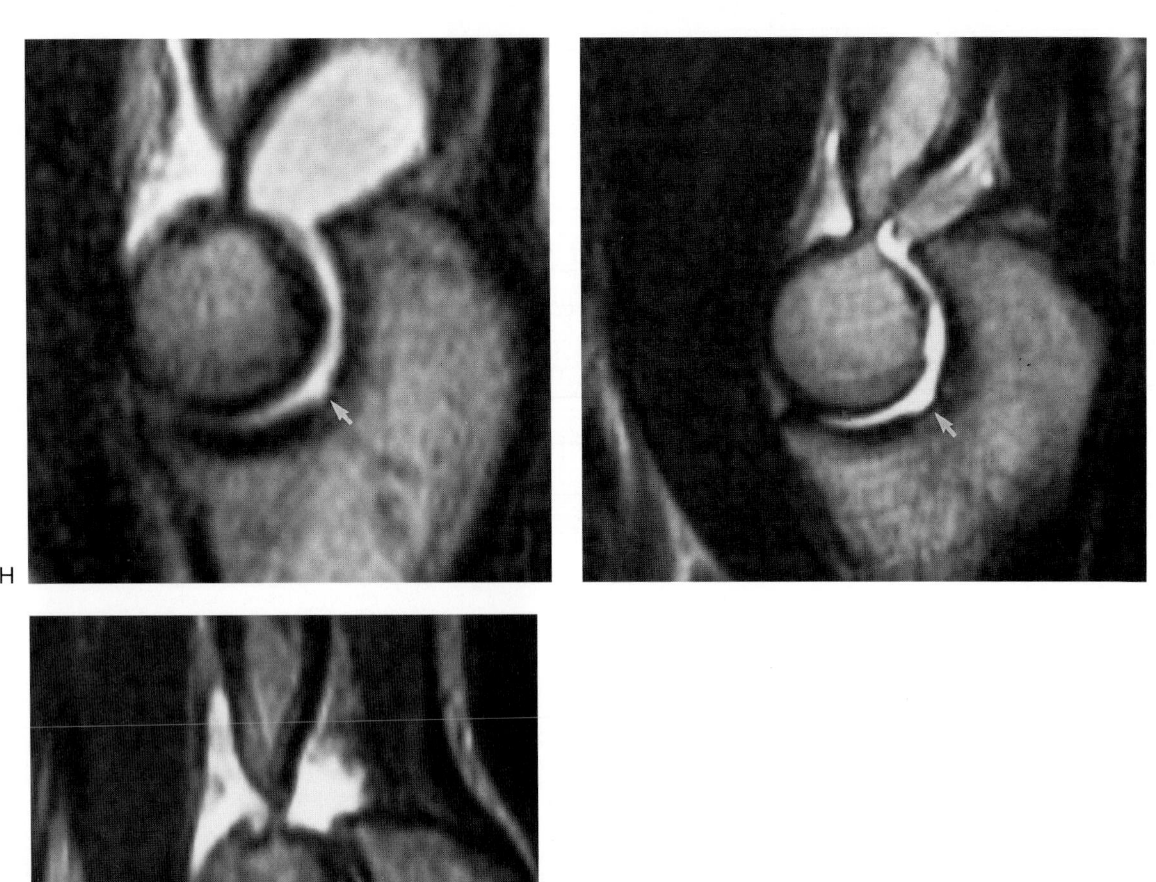

H

I

J

FIG. 10. H–J: Normal groove within the central aspect of the trochlear notch that should not be mistaken for an osteochondral defect of the ulna. A variably sized nonarticular groove is seen in the midportion of the trochlear notch (*arrows*) in three different patients. Loose bodies may lodge in this groove, which is normally devoid of articular cartilage.

CASE 11

History: This 16-year-old baseball pitcher presented with lateral elbow pain that worsened with throwing. What are the MR findings? How is this condition treated?

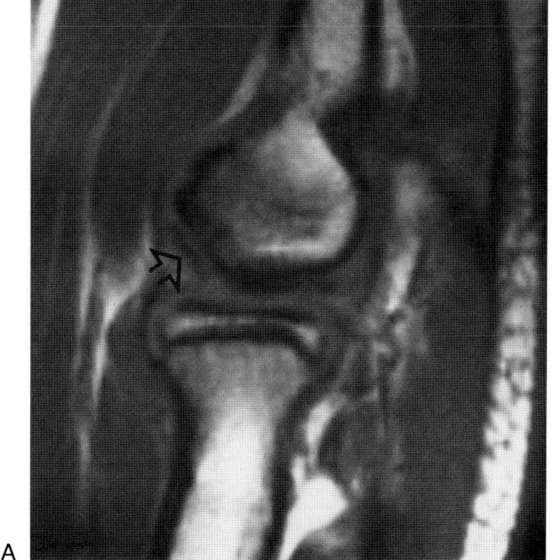

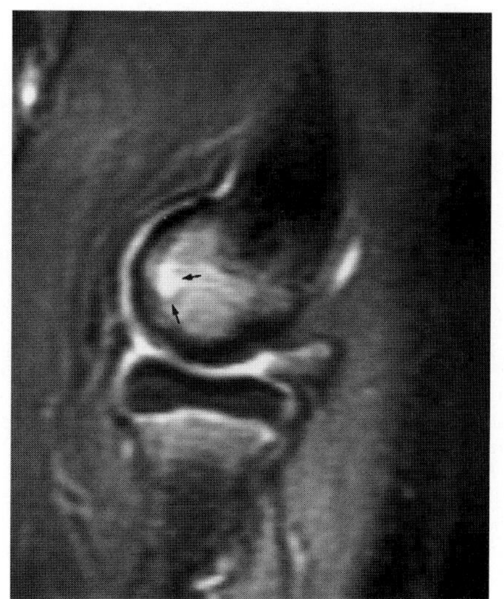

FIG. 11.

Findings: A T1-weighted sagittal image (Fig. 11A) reveals a linear focus of decreased signal (arrow) at the anterior surface of the capitellum. A STIR sagittal image (Fig. 11B) shows generalized increased signal surrounding a more well-defined cystic-appearing area (small arrows) in the anterior aspect of the capitellum.

Diagnosis: Osteochondritis dissecans of the capitellum.

Discussion: A nondisplaced osteochondral lesion of the anterior capitellum is present in this case. The underlying cystic-appearing lesion suggests instability. This patient subsequently developed locking and catching. A repeat scan demonstrated displacement of the lesion (Fig. 11C). The thin osteochondral fragment was then arthroscopically removed.

Chronic lateral impaction may lead to osteochondritis dissecans of the capitellum or radial head in adolescent pitchers or gymnasts (1–3). Repeated valgus stress and a relatively tenuous blood supply within the capitellum has been proposed to explain the frequent occurrence of osteochondritis dissecans in this location (4). Stable osteochondral lesions are usually treated with rest and splinting, whereas unstable lesions and loose bodies are usually excised (5–7). Abrasion chondroplasty or microfracture of the osteochondral defect to stimulate a healing response is generally performed rather than internal fixation or bone grafting unless a large acutely displaced defect is encounted (6–8). Osteochondritis dissecans of the capitellum ultimately leads to osteoarthritis in more than half of patients at long-term follow-up (9).

MR imaging can reliably detect and stage osteochondritis dissecans, but the accuracy of staging is improved by performing MR arthrography using dilute gadolinium (10). Unstable lesions are characterized by fluid or contrast encircling the osteochondral fragment on T2-weighted images. Loose in situ lesions may also be diagnosed by identifying a cyst-like lesion beneath the osteochondral fragment (11). These cyst-like lesions typically contain loose granulation tissue at surgery, explaining why they may enhance after intravenous administration of gadolinium (Fig. 11D,E). There is limited published experience with intravenous gadolinium-enhanced scans to evaluate osteonecrosis and osteochondritis dissecans in the elbow (12). When scans are delayed after intravenous gadolinium injection, enhancement of fluid within the joint predictably results (13). This technique may potentially provide a less invasive method of opacifying the joint fluid and staging osteochondral lesions.

Osteochondritis dissecans should be distinguished from osteochondrosis of the capitellum, known as Panner's disease. Age is an important discriminator, since osteochondritis dissecans typically is seen in the 13- to 16-year-old age range, whereas Panner's disease typically is seen in the 5- to 11-year-old range, before ossification of the capitellum is complete. Loose body formation and significant residual deformity of the capitellum are concerns in osteochondritis dissecans but are usually not seen in Panner's disease (14,15). Panner's disease is characterized by fragmentation and abnormally decreased signal intensity within the ossifying capitellar epiphysis on T1-weighted images, similar in appearance to Legg-Calve-Perthes disease in the hip. Panner's disease is believed to represent avascular necrosis of the capitellar ossification center that occurs secondary to trauma. Subsequent scans reveal normalization of these changes with little or no residual deformity of the capitellar articular surface. The articular surface typically remains intact and does not undergo fragmentation or loose body formation (Fig. 11F–H).

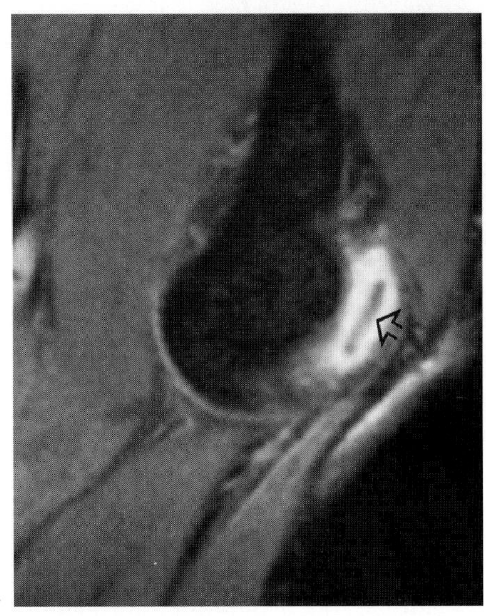

FIG. 11. C: Displaced osteochondral defect of the capitellum. A gradient-echo T2*-weighted sagittal **(C)** image reveals a loose body (*open arrow*) in the posteromedial aspect of the elbow joint.

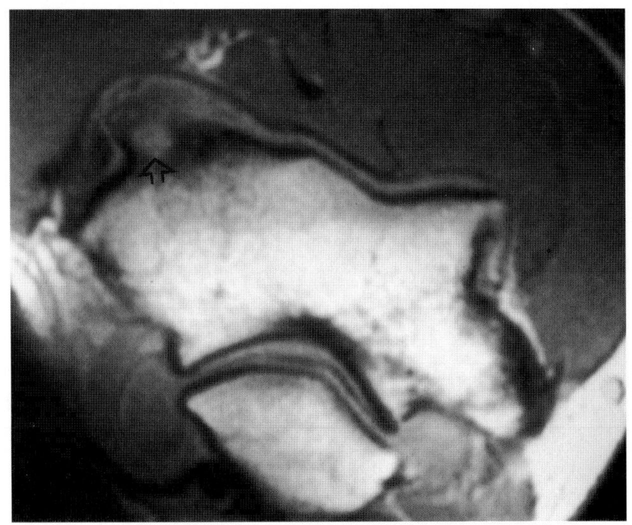

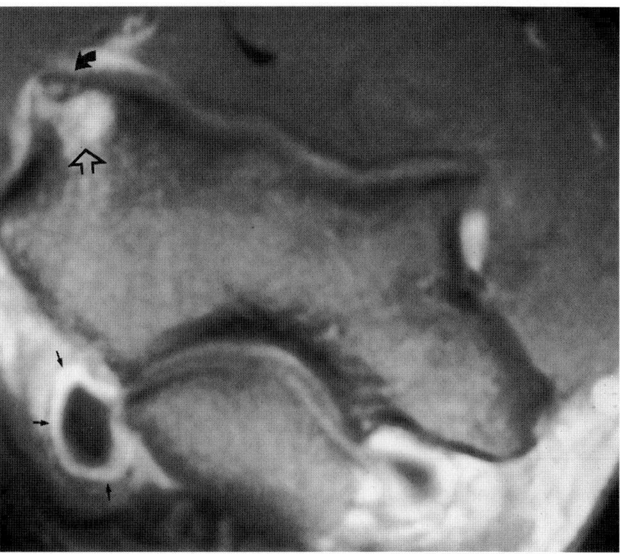

FIG. 11. D,E: Osteochondritis dissecans in a 14-year-old boy. A T1-weighted axial image **(D)** reveals an area of osteochondritis dissecans (*open arrow*) in the anterolateral aspect of the capitellum. A fat-suppressed T1-weighted axial image after intravenous gadolinium administration **(E)** reveals an enhancing area of granulation tissue (*open arrow*) beneath a thin osteochondral flap (*curved arrow*). Enhancement of thickened synovium (*small arrows*) compatible with synovitis is noted.

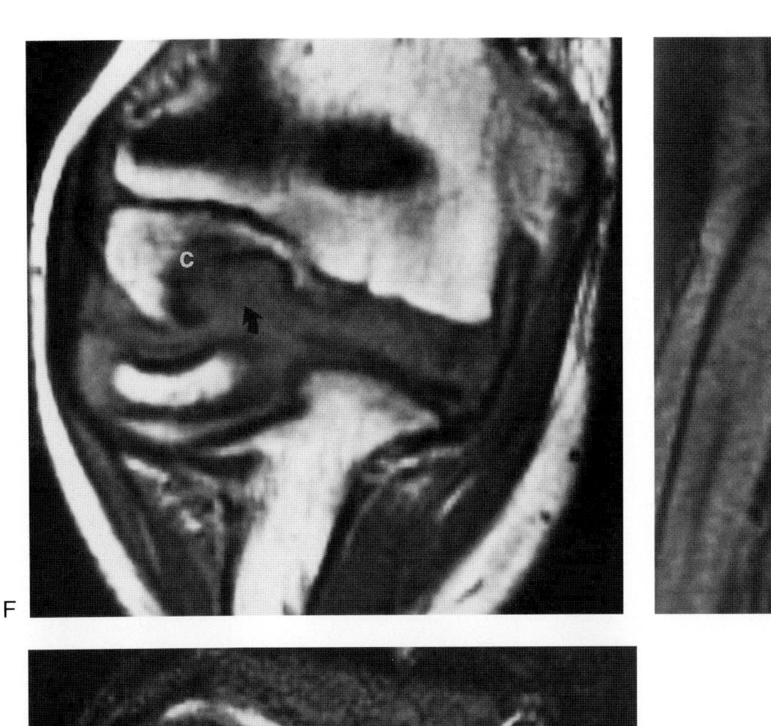

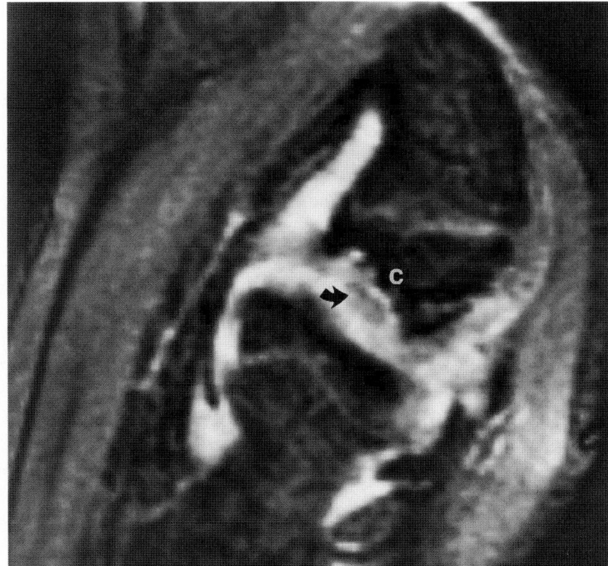

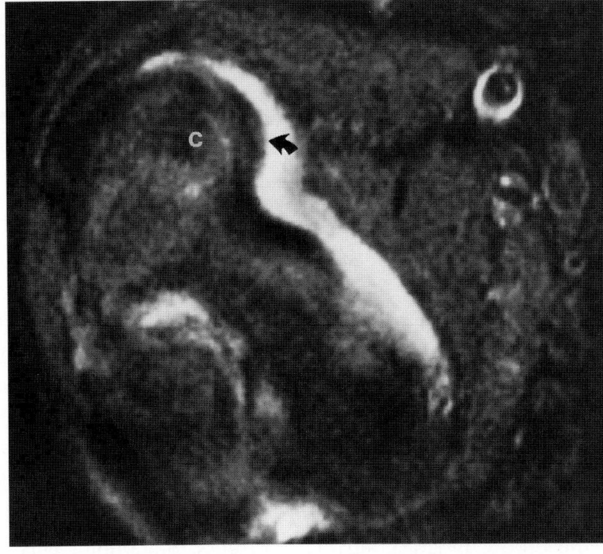

FIG. 11. F–H: Panner's disease in a 10-year-old girl. A T1-weighted axial image **(F)**)and a gradient-echo T2*-weighted sagittal **(G)** image show irregular ossification of the anteromedial aspect of the capitellum (*arrows*). A STIR axial image **(H)** shows a smooth articular surface, however, in this portion of the capitellum. There is insufficient contrast between the signal intensity of fluid and the unossified epiphyseal cartilage in F and G to determine if the surface of the cartilage is intact. C, capitellum.

CASE 12

History: This 25-year-old professional baseball player presented with lateral elbow pain that worsened with throwing. What is the anatomic explanation for this apparent osteochondral defect?

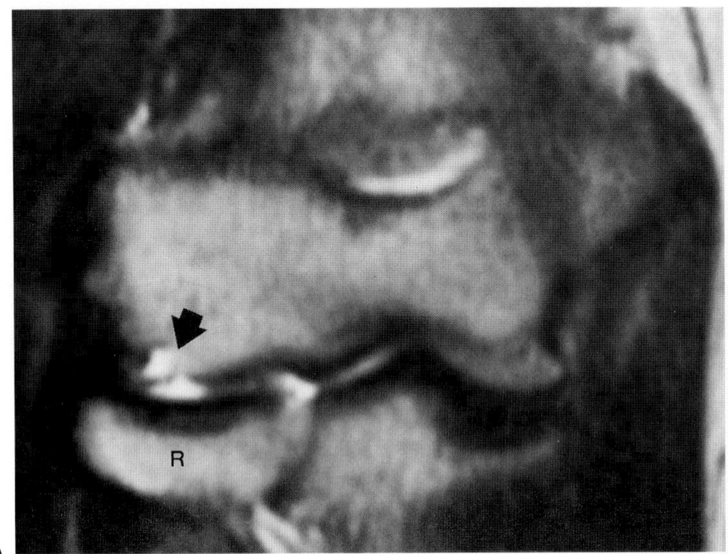

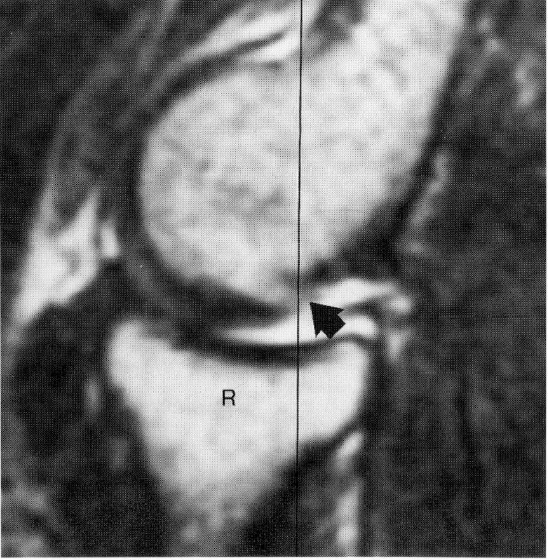

FIG. 12.

Findings: A T2-weighted coronal image (Fig. 12A) reveals an apparent discontinuity of the cortex at the inferolateral margin of the capitellum (arrow). A T2-weighted sagittal image (Fig. 12B) shows the abrupt posterior termination of the cortex and articular cartilage of the capitellum (arrow). A vertically orientated black line shows the location of the coronal image illustrated in Figure 12A (R, radius).

Diagnosis: Pseudodefect of the capitellum.

Discussion: The pseudodefect of the capitellum is an MR imaging pitfall that is related to the normal anatomy of the radiocapitellar articulation. Familiarity with the typical appearance and location of the pseudodefect is important to avoid an erroneous diagnosis of an osteochondral defect or impaction fracture of the capitellum.

The capitellum is an anteriorly directed prominence that arises from the lateral aspect of the distal humerus and resembles one-half of a sphere. The articular cartilage of the capitellum extends through an arc of approximately 180° from superior to inferior (1,2). Osteochondritis dissecans and osteochondral defects typically involve the anterior aspect of the capitellum, (3,4) whereas the pseudodefect of the capitellum occurs at the abrupt transition between the posterolateral margin of the capitellum and the adjacent nonarticular portion of the lateral humeral condyle (Fig. 12C). Fluid or contrast in the posterior aspect of the lateral compartment further highlights this abrupt transition between the normal overhanging margins of the capitellum and the rough nonarticular portion of the humerus, which simulates an osteochondral defect (Fig. 12D–G). This pseudodefect is also conspicuous because of the normal incomplete articulation between the capitellum and the radial head that occurs when the elbow is extended. The absence of contact between the posterolateral aspect of the radial head and the capitellum is a normal feature of the extended elbow joint; however, this further creates the illusion of an osteochondral defect, since the cartilage of the radial head opposes the rough nonarticular portion of the humerus.

The capitellum has a tapered appearance from anterior to posterior that accounts for the lack of contact with the posterolateral radius and the appearance of the pseudodefect (Fig. 12G). Sagittal images through the lateral aspect of the capitellum may have the appearance of a posteroinferior defect (Fig. 12B,E), whereas coronal images through the posteror aspect of the capitellum may have the appearance of a posteroinferior defect (Fig. 12A,D). A recent study of the capitellar pseudodefect found that the variable conspicuity of the pseudodefect depended on the presence of fluid and the angle of the sagittal and coronal images (5). The pseudodefect was present and more conspicuous when the coronal images were truly parallel to the plane of the humeral epicondyles (Fig. 12G), and the sagittal images were truly perpendicular to these "oblique" coronal images.

Another important pitfall of MR imaging is illustrated in Fig. 12D–G. Multiple foci of signal void with a thin margin of increased signal are seen secondary to injection of small air bubbles along with the dilute gadolinium contrast agent. Even with good arthrographic technique, it is not uncommon to inject several small air bubbles into the joint that may mimic loose bodies on MR imaging. These air bubbles can be recognized by a characteristic margin of high signal adjacent to the signal void that is due to a magnetic susceptibility artifact and will not be found along the margins of a real loose body (also see Case 4–16). A similar appearance of multiple foci of magnetic susceptibility artifact may also be seen at the site of micrometallic deposition associated with prior surgery. These foci of magnetic susceptibility artifact are most prominent on gradient-echo T2*-weighted images.

Another normal variant that has been described is the radiohumeral meniscus, which is occasionally prominent on MR imaging (Fig. 12H) (6). This low-signal-intensity structure has an appearance similar to the meniscal homologue in the wrist or the glenoid labrum in the shoulder. Chronic trauma and fibrosis of this meniscus-like invagination of the lateral capsule have been advanced as a possible cause of lateral elbow pain that may mimic symptoms of a loose body or lateral epicondylitis (6). The radiohumeral meniscus seen on MR imaging is probably the same structure that has been termed a symptomatic lateral synovial plica of the elbow at arthroscopy (7). The lateral synovial plica of the elbow, like the more familar symptomatic medial patella plica in the knee, may be encountered at arthroscopy as a thickened fold or fibrotic fringe of synovial tissue at the lateral margin of the radiocapitellar articulation (8). Relief of symptoms has been described in several cases after arthroscopic resection of a fibrotic lateral synovial plica of the elbow (7,9).

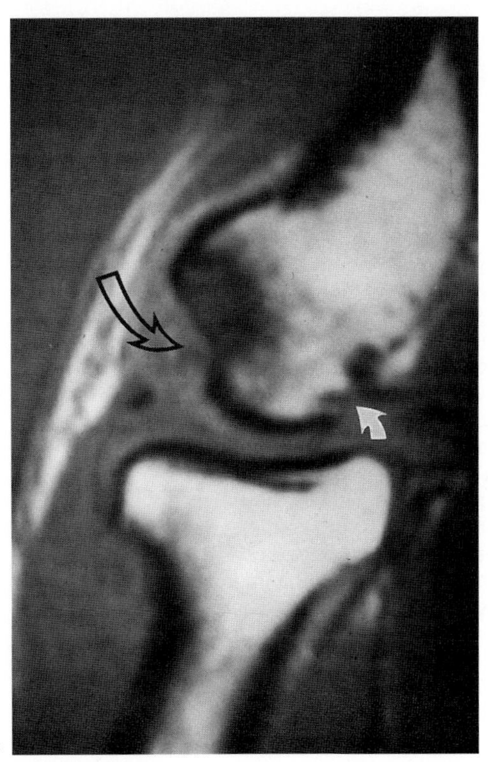

FIG. 12. C: Osteochondritis dissecans in a 14-year-old boy. A T1-weighted sagittal image **(C)** reveals a typical defect in the anterior aspect of the capitellum (*arrow*). The pseudodefect of the capitellum occurs posteriorly (*white arrow*) on sagittal images through the lateral aspect of the capitellum.

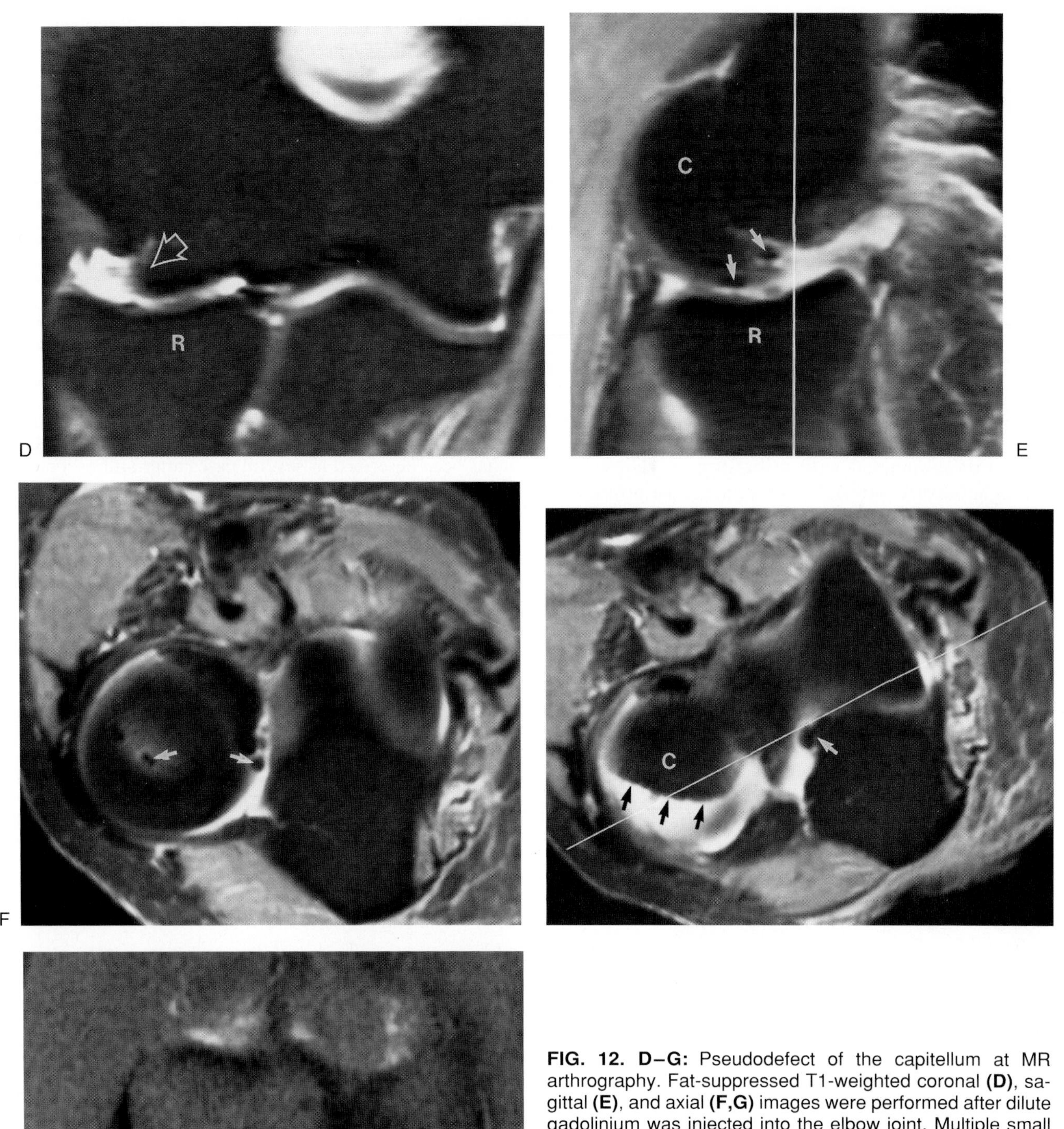

FIG. 12. D–G: Pseudodefect of the capitellum at MR arthrography. Fat-suppressed T1-weighted coronal **(D)**, sagittal **(E)**, and axial **(F,G)** images were performed after dilute gadolinium was injected into the elbow joint. Multiple small air bubbles are noted that should not be mistaken for loose bodies (*small white arrows*). The coronal image **(D)** depicts a typical pseudodefect of the capitellum (*open arrow*). The normal posterior tapering of the capitellum (*small black arrows*) is well seen in G and accounts for the pseudodefect at its inferolateral margin. A white line shows the location of the coronal section in D. C, capitellum; R, radius. **H:** Radiohumeral meniscus. A fat-suppressed T2-weighted coronal image **(H)** shows a prominent meniscus-like structure (*arrow*) that extends into the lateral margin of the radiohumeral joint.

CASE 13

History: This 18-year-old man presented with medial soft tissue swelling and pain. Erythema and tenderness were present on examination. Rupture of the common flexor tendon was suspected clinically because of a questionable history of trauma. What are the MR findings? What does the central area of fat (curved arrows; see Fig. 13A–D) within this mass signify?

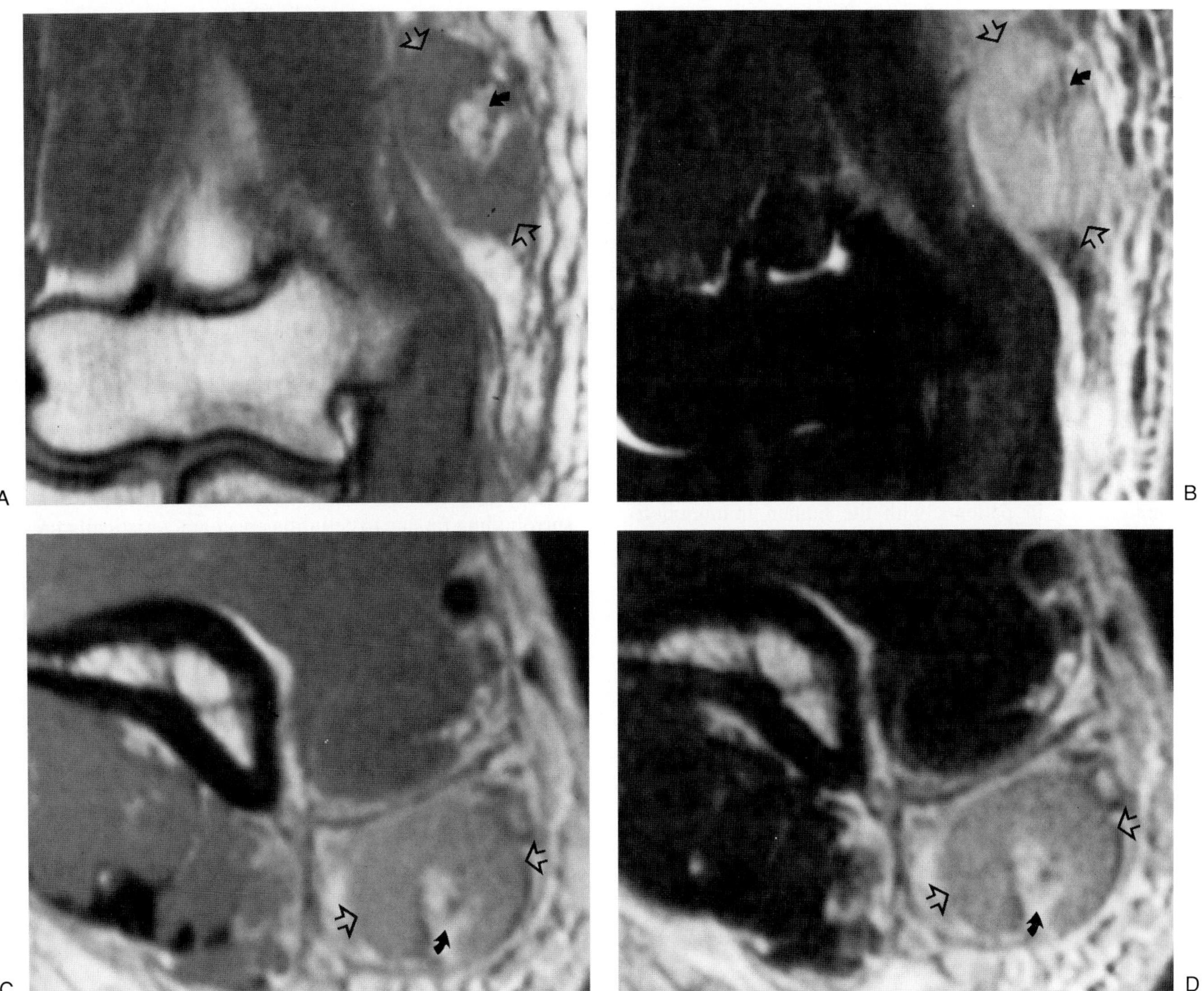

FIG. 13.

Findings: T1-weighted (Fig. 13A) and STIR (Fig. 13B) coronal images as well as proton-density (Fig. 13C) and T2-weighted (Fig. 13D) axial images reveal a 3-cm mass (open arrows) in the medial subcutaneous fat just superior to the medial epicondyle. A central area of fat (curved arrows) that represents the hilum of an enlarged epitrochlear lymph node is noted. There is lymphatic infiltration and marked swelling of the adjacent tissue compatible with cellulitis.

Diagnosis: Cat-scratch disease.

Discussion: After the MR findings were reviewed, further examination revealed several small puncture wounds in the patient's hand that were caused 3 weeks before by his cat. Antibiotic treatment was begun without improvement. Excision of the enlarged epitrochlear node and drainage of pus was performed several days later. Serologic studies and histology in this case were diagnostic for cat-scratch disease.

Cat-scratch disease is a bacterial infection that results

in regional adenopathy after inoculation from a scratch or puncture wound (1–4). The site of inoculation is most commonly in the hands and forearms, resulting in more proximal adenopathy about the elbow, axilla, and neck. Although a cat scratch is not always discovered, 93% of affected patients have a history of exposure to cats. An estimated 22,000 cases are diagnosed each year in the United States, resulting in more than 2,000 hospital admissions, with most cases occurring in children and adolescents (1,2,4).

Cat-scratch disease typically begins with a papule that appears 4 to 6 days after inoculation. This progresses to a pustule that is followed in 3 to 4 weeks by regional adenopathy. A single enlarged node occurs in most cases, with multiple nodes present in 24% of cases (1). The adenopathy typically resolves within 3 months but may persist for as long as 1 year (1,4). Involvement of multiple nodal sites suggests either multifocal inoculation or dissemination of the disease (1).

Disseminated infection is unusual in immunocompetent patients with cat-scratch disease, but it may result in splenic and hepatic granulomata, mesenteric adenitis, multifocal osteomyelitis, encephalitis, and meningitis (1–5). A systemic form of cat-scratch disease may occur in immunocompromised patients following organ transplantation or in patients with the human immunodeficiency virus infection. Multiple lytic bony lesions may develop along with cutaneous proliferative vascular lesions in immunocompromised patients with the acquired immunodeficiency syndrome (AIDS) (6). This systemic infection in patients with AIDS is secondary to the same organism that produces isolated regional adenopathy in immunocompetent patients with cat-scratch disease. The organism is a gram-negative bacillus that was originally named *Rochalimaea henselae* and has recently been renamed *Bartonella henselae* (7).

Histologic study of the affected lymph nodes in cat-scratch disease shows granulomas, stellate abscesses, and a nonspecific inflammatory cell infiltrate. The presence of each of these findings in the same specimen is highly suggestive of cat-scratch disease (4). Gram-negative organisms may be identified on tissue stains. Isolation of the organism responsible for cat-scratch disease has recently led to the development of an indirect fluorescent antibody test that has become widely available in the United States for serologic diagnosis (4). Treatment of uncomplicated cat-scratch disease is controversial, since some authors believe the disease is self-limited and antibiotics have not been shown to alter the course of the disease in immunocompetent patients (1,2,4,8).

Epitrochlear adenopathy from cat-scratch disease may be mistaken clinically for a hematoma or sarcoma about the elbow, leading to MR imaging for further characterization of the mass (Fig. 13E,F) (5,8). MR imaging typically reveals a nonspecific mass of low to intermediate signal intensity on T1-weighted images and intermediate to high signal intensity on T2-weighted sequences. There is enhancement of the mass and surrounding lymphedema with gadolinium (Fig. 13H). There may be central fluid that does not enhance with gadolinium if there is central necrosis and liquifaction (8). Suppuration of lymph nodes, as in the case presented above, occurs in about 15% of cases in the literature (4). Nodes typically range in size from 1 to 5 cm (1). Lymphedema and infiltration of the surrounding fat due to cellulitis are characteristically present (8). The mass is located at the site of epitrochlear lymph nodes, adjacent to the medial neurovascular bundle and just proximal to the elbow. There may be associated axillary adenopathy. Additional clues to diagnosis on MR imaging include a central fatty hilum indicating a lymph node (Fig. 13A–D) or a series of contiguous soft tissue masses indicating a chain of nodes (Fig. 13E–H). Clinical correlation and serologic studies will usually allow diagnosis of cat-scratch disease presenting as an epitrochlear mass on MR imaging. Recognition of the characteristic appearance of this condition may avert an unnecessary biopsy (Fig. 13G,H).

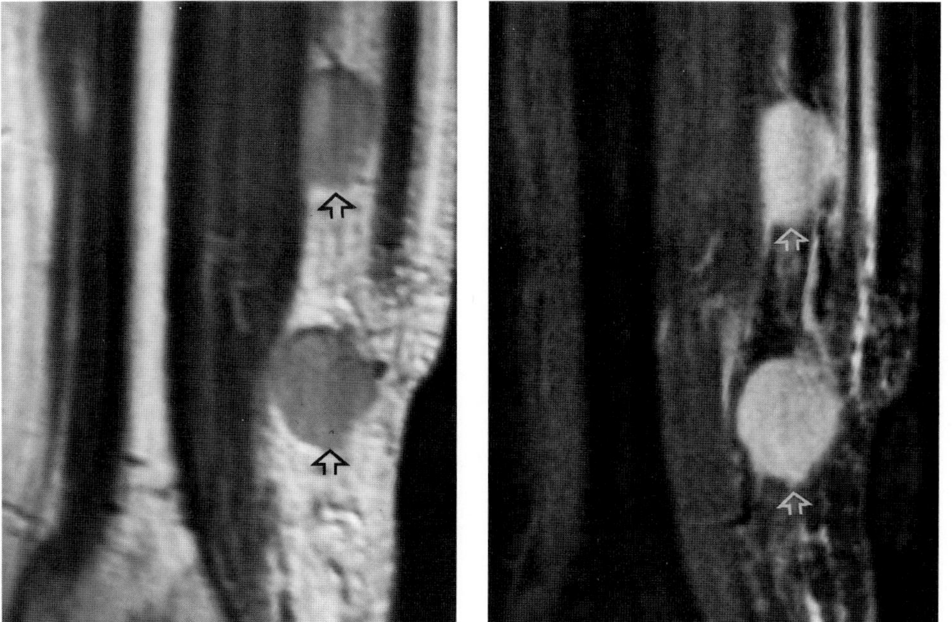

FIG. 13. E,F: Biopsy-proven cat-scratch disease in a 27-year-old woman suspected of having a soft tissue sarcoma. T1-weighted **(E)** and STIR **(F)** coronal images reveal two adjacent masses (*arrows*). Mild swelling and lymphatic infiltration is seen in the adjacent subcutaneous fat, suggesting an inflammatory process.

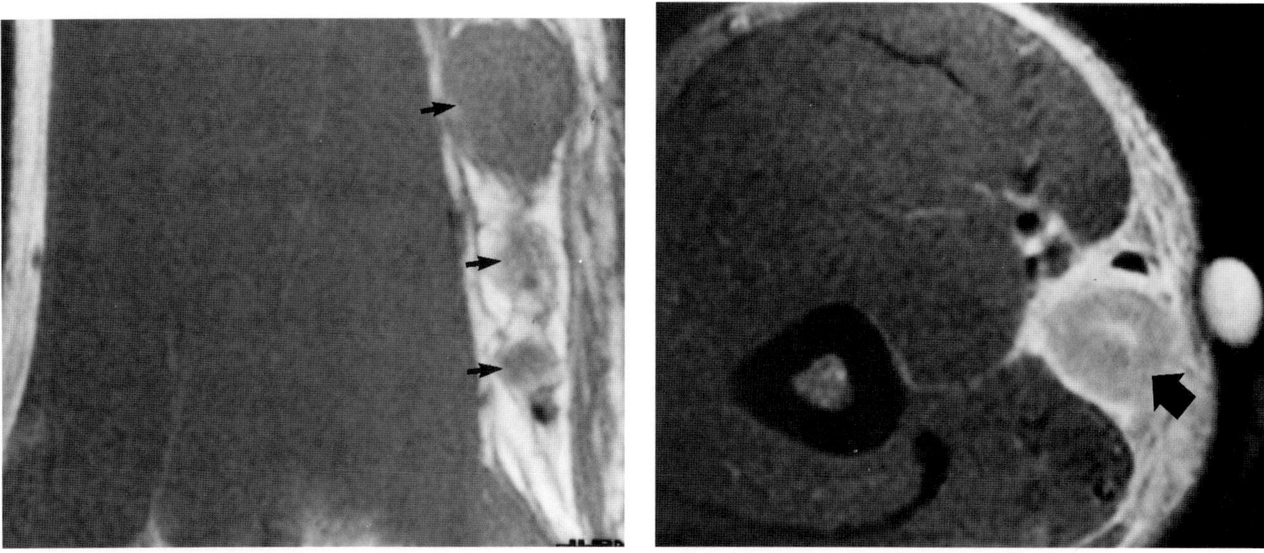

FIG. 13. G,H: Cat-scratch disease in a 13-year-old boy suspected of having a soft tissue sarcoma. A T1-weighted coronal image **(G)** reveals a chain of three epitrochlear lymph nodes (*arrows*) with swelling and lymphatic infiltration of the adjacent subcutaneous fat. A fat-suppressed T1-weighted axial image **(H)** obtained after intravenous gadolinium administration shows enhancement of the largest epitrochlear lymph node (*arrow*) and enhancement of the surrounding cellulitis. The adenopathy gradually resolved without treatment over the next 2 weeks.

CASE 14

History: This 5-year-old boy presented with medial elbow pain after falling with his arm outstretched in front of him. Radiographs (not shown) revealed partial ossification of the capitellum, radial head, and medial epicondyle with apparent distal displacement of the medial epicondyle compared with the uninjured elbow. What are the MR findings? Does this fracture involve an articular surface?

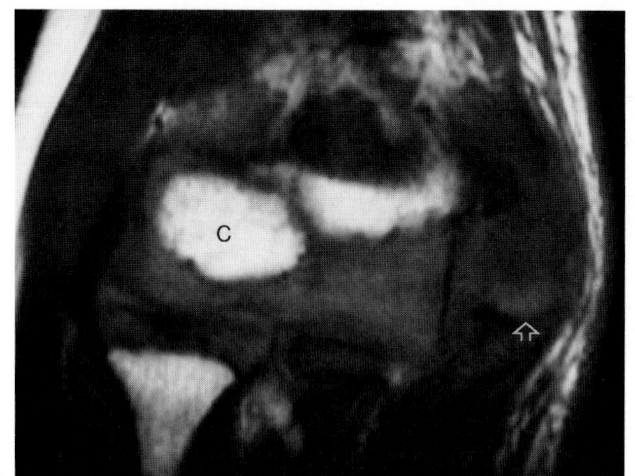

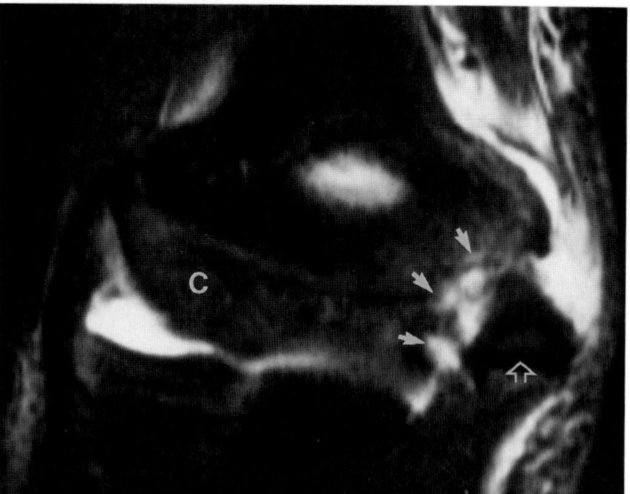

FIG. 14.

Findings: T1-weighted (Fig. 14A) and STIR (Fig. 14B) coronal images reveal distal avulsion of the medial epicondylar apophysis (open arrows) that is just beginning to ossify. The fracture (solid arrows) is well delineated on the STIR sequence (C, capitellum).

Diagnosis: Avulsion fracture of the medial epicondyle.

Discussion: The fracture in this case does not extend into the unossified cartilage of the trochlea to involve the articular surface. In other words, it is purely an extra-articular fracture of the medial epicondyle rather than a more complicated intra-articular fracture of the medial humeral condyle. Intra-articular fractures of the medial humeral condyle are unusual, but they can be confused with avulsion fractures of the medial epicondyle in younger children, in whom the trochlea is unossified (1). The presence of an effusion or significant displacement of the medial epicondyle are plain film clues to diagnosing a fracture of the medial humeral condyle (2). If this diagnosis is not made, and the injury is treated nonoperatively, a poor functional result can be anticipated (3). MR imaging can be used rather than arthrography to exclude involvement of the unossified trochlear cartilage, as in this case.

Fractures of the medial epicondyle usually occur between the ages of 9 and 15 secondary to a valgus stress, which produces traction and avulsion at the unfused apophyseal growth plate. The epicondyle is displaced inferiorly due to the pull of the attached flexor muscles. The medial epicondyle may become entrapped within the medial aspect of the joint if there is associated rupture of the medial collateral ligament and opening of the joint space at the time of valgus stress. Entrapment of the medial epicondyle is typically associated with posterior dislocation of the elbow and rupture of the collateral ligaments (1,4,5). This diagnosis can ordinarily be made on plain films by remembering the sequential appearance of the ossification centers about the elbow so that the entrapped medial epicondyle is not mistaken for a normal trochlear ossification center. MR imaging may be useful to evaluate this complication when the medial epicondyle has not yet ossified. The ulnar nerve is commonly displaced within the joint along with the medial epicondyle. MR imaging may evaluate the status of the ulnar nerve in such cases. Moreover, chronic tension stress injury and avulsion fractures in young baseball pitchers (Little League elbow) may also be evaluated with MR imaging (6).

Treatment of medial epicondyle fractures is controversial and depends on the degree of displacement as well as the functional requirements of the patient (1,7,8). Cases with significant displacement or young athletes who require a stable elbow for throwing are generally treated surgically. The function and range of motion of the elbow was uniformly good in a series of 56 children followed for 21 to 48 years after conservatively treated fractures of the medial epicondyle (7). However, the late development of ulnar neuropathy is a common complication of conservatively treated cases (Fig. 14C–F) (7).

MR imaging may identify or exclude the more common radial head fractures in adults and supracondylar fractures in children when radiographic evidence of a joint effusion is present and a fracture is not visualized (Fig. 14G). In children, supracondylar fractures that do not involve the physis are more common than all physeal injuries about the elbow combined (3,9). However, the elbow is a relatively common site of physeal injury, occurring most frequently after distal radial and distal tibial physeal fractures are considered (10). Fractures of the lateral humeral condyle are the most common specific type of physeal injury about the elbow (Fig. 14H). They occur in children between the ages of 2 and 14 years but are most common between 6 and 10 years of age. A fracture of the lateral humeral condyle is usually a longitudinal Salter-Harris type IV fracture that is both intra-articular and transphyseal (Fig. 14I) (3). The intra-articular fracture line is usually entirely through cartilage and is therefore not visible on radiographs. Injury to the physis and the unossified epiphyseal cartilage may be assessed with MR imaging in these cases (10–12). This information is important, since Salter-Harris type IV fractures of the lateral humeral condyle tend to be unstable and require surgical intervention, whereas Salter-Harris type II fractures can be treated successfully with closed reduction (Fig. 14H,I). Unrecognized Salter-Harris type IV fractures of the lateral humeral condyle are frequently complicated by malunion and nonunion that results in deformity, loss of motion, degenerative arthrosis, and tardy ulnar neuropathy (3).

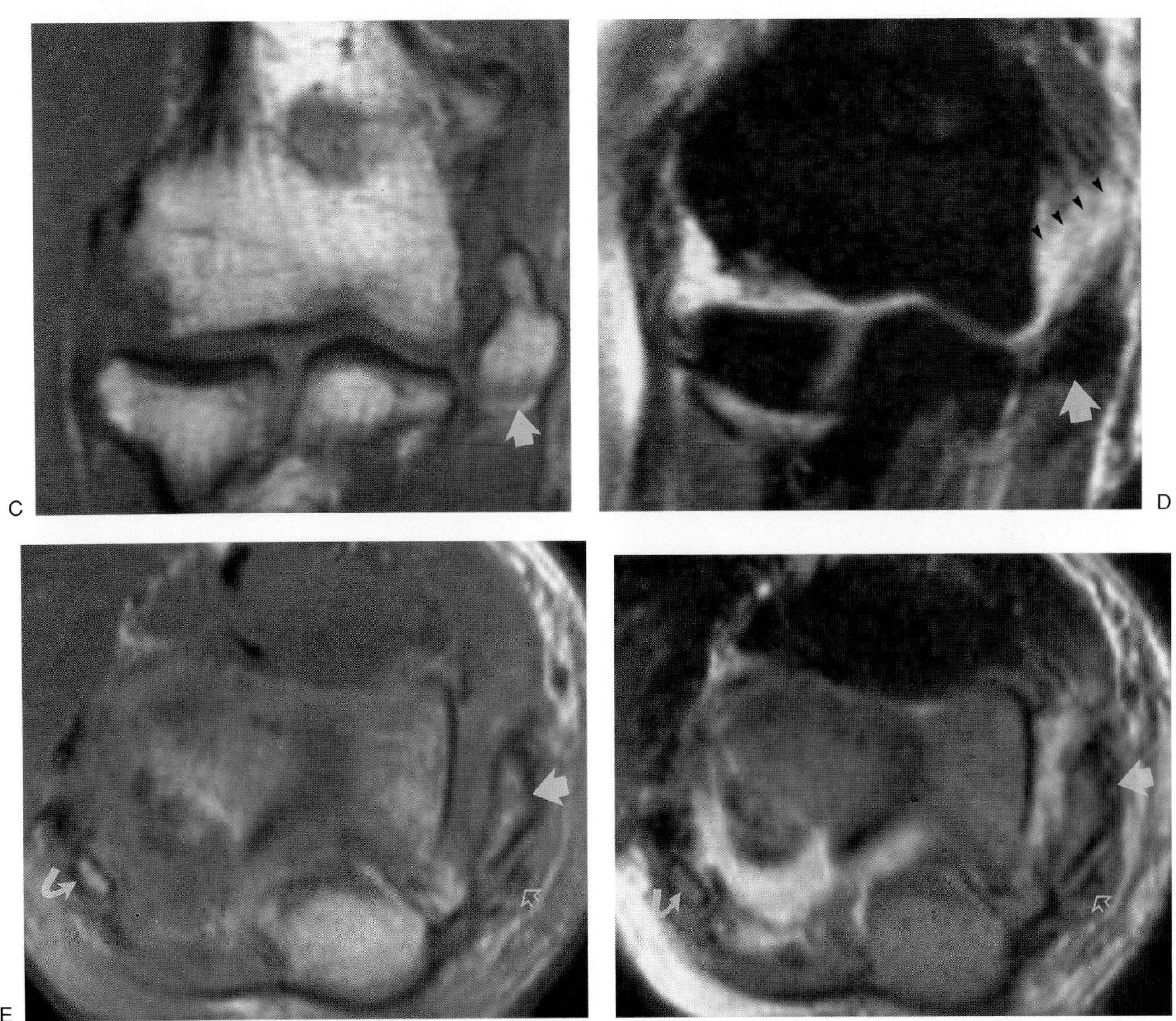

FIG. 14. C–F: Old united fracture of the medial epicondyle presenting as late ulnar neuritis. T1-weighted **(C)** and STIR **(D)** coronal images as well as proton-density **(E)** and T2-weighted **(F)** axial images reveal a large displaced fracture of the medial epicondyle (*large arrows*) with fluid delineating a chronic pseudarthrosis (*arrowheads*). Heterotopic ossification (*curved arrows*) is incidentally noted within the lateral ulnar collateral ligament as a result of prior dislocation and sprain injury. The relationship of the ulnar nerve (*open arrows*) to the adjacent fracture fragment is well seen on the axial images.

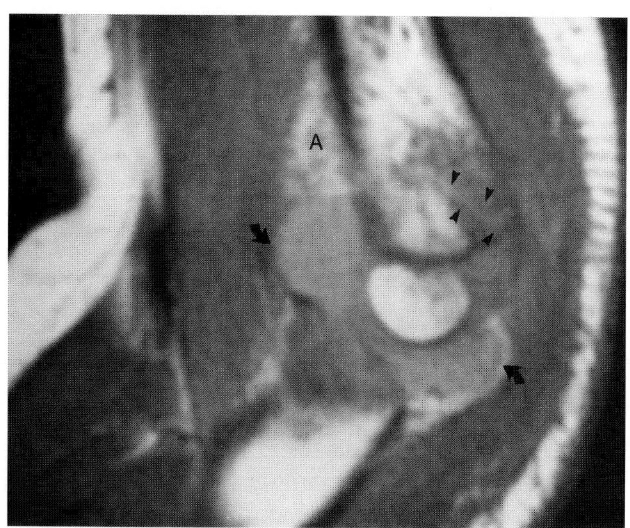

FIG. 14. G: Radiographically occult elbow fracture in an 8-year-old boy with a positive fat pad sign on radiograph. A T1-weighted sagittal image reveals a supracondylar fracture (*arrowheads*) with adjacent decreased signal. A large effusion is noted (*curved arrows*) elevating the anterior fat pad. A, anterior fat pad.

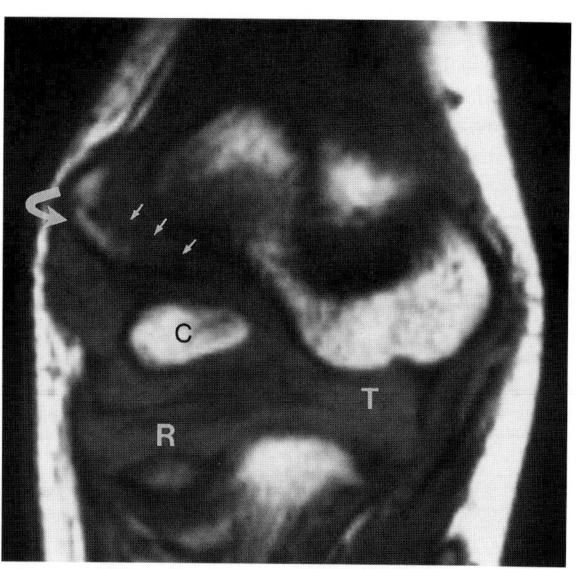

FIG. 14. H: Radiographically subtle elbow fracture. T1-weighted coronal image reveals a Salter-Harris type II fracture (*arrows*) with mild lateral displacement of the lateral humeral condyle (*curved arrow*). The fracture does not extend into the capitellum **(C)**. T2-weighted images (not shown) excluded extension of the fracture into the unossified trochlear epiphysis. This 5-year-old child did well with closed reduction. R, unossified radial head cartilage; T, unossified trochlear cartilage.

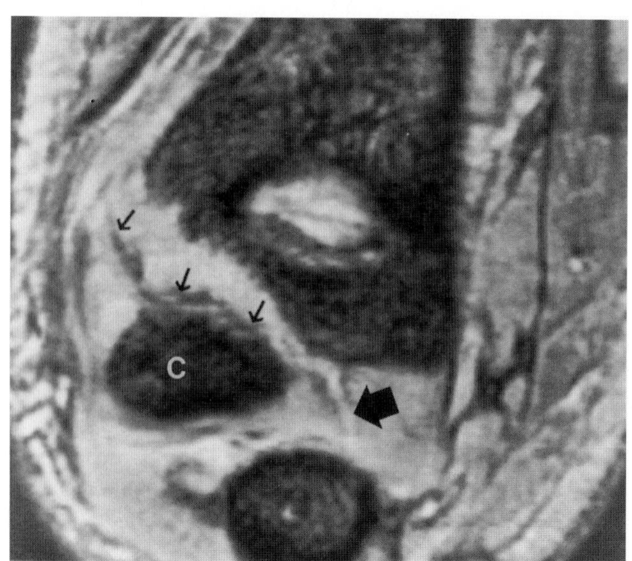

FIG. 14. I: Salter-Harris type IV fracture of the lateral humeral condyle. T2*-weighted gradient-echo coronal image of a partially flexed elbow reveals the thin metaphyseal fracture fragment (*small arrows*) as well as extension of the fracture through the unossified trochlear epiphysis (*large arrow*). These fractures usually require open reduction and internal fixation. C, capitellum. (Courtesy of Phoebe Kaplan, MD.)

CASE 15

History: This 42-year-old man presented with elbow pain that radiated into the ulnar aspect of the hand. Ulnar neuritis was suspected clinically. The ulnar nerve was difficult to palpate posterior to the medial epicondyle, suggesting a possible mass superficial to the ulnar nerve. Is there a mass present on MR imaging? What structures lie deep and superficial to the ulnar nerve in this case?

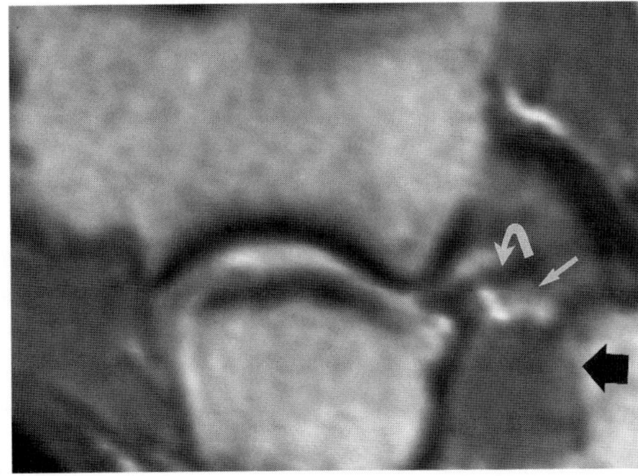
A

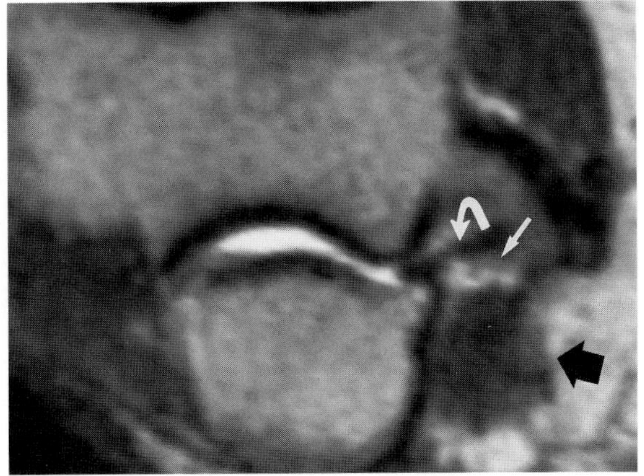
B

FIG. 15.

Findings: Proton-density (Fig. 15A) and T2-weighted (Fig. 15B) axial images reveal the ulnar nerve (straight arrows) along the posterior margin of the medial epicondyle. The posterior bundle of the medial collateral ligament (curved arrow) lies along the deep margin of the ulnar nerve and forms the floor of the cubital tunnel. An anomalous muscle, the anconeus epitrochlearis (black arrows), lies superficial to the ulnar nerve instead of the normal cubital tunnel retinaculum.

Diagnosis: Anconeus epitrochlearis muscle contributing to narrowing of the cubital tunnel and ulnar neuritis.

Discussion: The ulnar nerve is well seen on axial MR images as it passes through the cubital tunnel (1,2). The roof of the cubital tunnel is formed by the deep fibers of the flexor carpi ulnaris aponeurosis distally and the cubital tunnel retinaculum proximally (3). The flexor carpi ulnaris aponeurosis is a triangular, tendinous arch that extends between the humeral and ulnar heads of the flexor carpi ulnaris muscle and forms the roof of the cubital tunnel just distal to the medial epicondyle and the cubital tunnel retinaculum. The flexor carpi ulnaris aponeurosis has also been termed the arcuate ligament, although this term is sometimes confused with the cubital tunnel retinaculum. The flexor carpi ulnaris aponeurosis normally tenses as the medial collateral ligament relaxes and bulges superficially during elbow flexion (4). These changes result in decreased volume and increased pressure within the cubital tunnel during elbow flexion.

The cubital tunnel retinaculum is normally a thin fibrous structure that extends from the medial epicondyle to the olecranon (Fig. 15C). The cubital tunnel retinaculum has also been termed the epicondylo-olecranon ligament and the Osborne ligament or band (5,6). When the cubital tunnel retinaculum is thickened, it has been referred to as the Osborne lesion (7). Anatomic variations of the cubital tunnel retinaculum may contribute to ulnar neuropathy (3). These variations in the cubital tunnel retinaculum and the appearance of the ulnar nerve itself can be identified with MR imaging. The retinaculum may be thickened in 22% of the population, resulting in dynamic compression of the ulnar nerve during elbow flexion. In 11% of the population, the cubital tunnel retinaculum may be replaced by an anomalous muscle, the anconeus epitrochlearis, resulting in static compression of the ulnar nerve (Fig. 15A,B) (3). The cubital tunnel retinaculum may be absent in 10% of the population, allowing anterior dislocation of the nerve over the medial epicondyle during flexion with subsequent friction neuritis (Fig. 15D) (8).

The floor of the cubital tunnel is formed by the capsule of the elbow and the posterior and transverse portions of the medial collateral ligament. Thickening of the medial collateral ligament and medial bony spurring may undermine the floor of the cubital tunnel and result in ulnar neuropathy (Fig. 15E) (3,9,10). Heterotopic ossification in the medial collateral ligament, underlying loose bodies, tumors, ganglion cysts, scarring, or displaced fracture fragments also may result in ulnar nerve entrapment.

MR imaging signs of ulnar neuritis and entrapment

include displacement and flattening of the nerve adjacent to a mass, swelling and enlargement of the nerve proximal or distal to a mass, infiltration of the perineural fat, and increased signal intensity within the nerve on T2-weighted images (11). Peripheral nerves are normally intermediate in signal intensity on T2-weighted images. The ulnar nerve must be followed carefully to avoid mistaking it for enlargement of the adjacent veins. The posterior ulnar recurrent artery and the deep veins that accompany it are normally small structures that course with the ulnar nerve through the cubital tunnel. Enlargement of a deep vein may appear as a bright tubular structure on T2-weighted, gradient-echo, or STIR sequences and may mimic an edematous ulnar nerve (11).

Surgical procedures for ulnar nerve entrapment include medial epicondylectomy, decompression of the nerve, and translocation of the nerve (12,13). Translocation or transfer of the nerve may be subcutaneous, intramuscular, or submuscular (7). Low-signal-intensity scarring may be seen along the margins of the surgically translocated ulnar nerve; this finding also has been observed in patients with ulnar nerve subluxation and friction neuritis.

Entrapment of the median nerve and radial nerve also may be evaluated with MR imaging. Median nerve entrapment may be due to a variety of uncommon anatomic variations about the elbow, including the presence of a supracondyloid process with a ligament of Struthers, anomalous muscles, an accessory bicipital aponeurosis, and hypertrophy of the ulnar head of the pronator teres (12,14). These anatomic variants and pathologic mass lesions such as an enlarged radial bicipital bursa may entrap the median nerve and may be identified with MR imaging. Radial nerve entrapment may occur due to thickening of the arcade of Frohse along the proximal edge of the supinator muscle. Parosteal lipomas may arise from the proximal radius and entrap the radial nerve (Fig. 15F) (15). Ganglion cysts may arise from the anterior margin of the elbow joint and compress the radial nerve (16).

MR imaging may be complementary to electromyography and nerve conduction studies in cases of nerve entrapment about the elbow (2). In subacute denervation the affected muscles have prolongation of T1 and T2 relaxation times secondary to muscle fiber shrinkage and associated increases in extracellular water (17). Entrapment of a nerve about the elbow may therefore cause increased signal within the muscles innervated by that nerve on T2-weighted or STIR images. These changes may be followed to resolution or progressive atrophy and fatty infiltration (Fig. 15G) (18,19). Moreover, the site and cause of entrapment may be discovered with MR imaging by following the nerve implicated from the distribution of abnormal muscles on MR imaging (20).

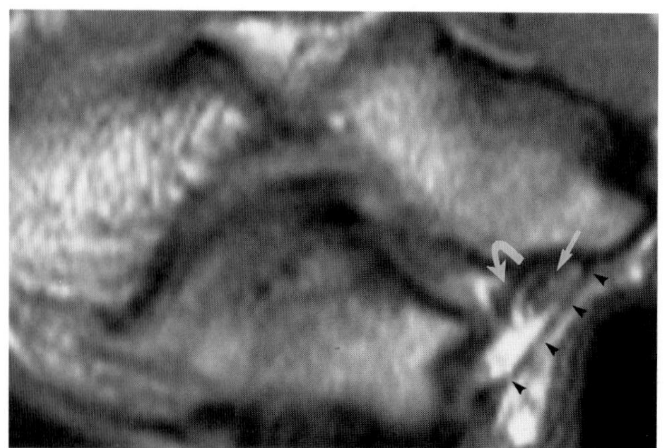
C

FIG. 15. C: Anatomy of the cubital tunnel. A proton-density axial image reveals the ulnar nerve (*white arrow*) deep to a normal, thin cubital tunnel retinaculum (*arrowheads*) and superficial to the posterior bundle of the medial collateral ligament (*curved arrow*).

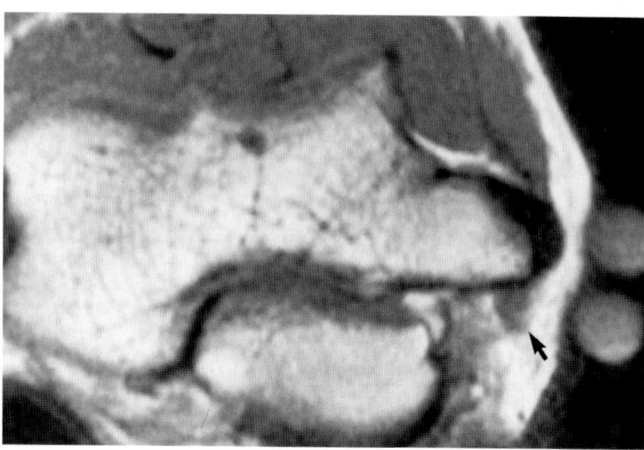
D

FIG. 15. D: Ulnar neuritis in a patient with questionable ulnar nerve subluxation on physical exam. A T1-weighted axial image reveals prominent enlargement and medial subluxation of the ulnar nerve (*arrow*). The overlying cubital tunnel retinaculum is developmentally absent, allowing anterior dislocation of the ulnar nerve during elbow flexion with subsequent friction neuritis. T2-weighted images (not shown) revealed increased signal intensity within the ulnar nerve.

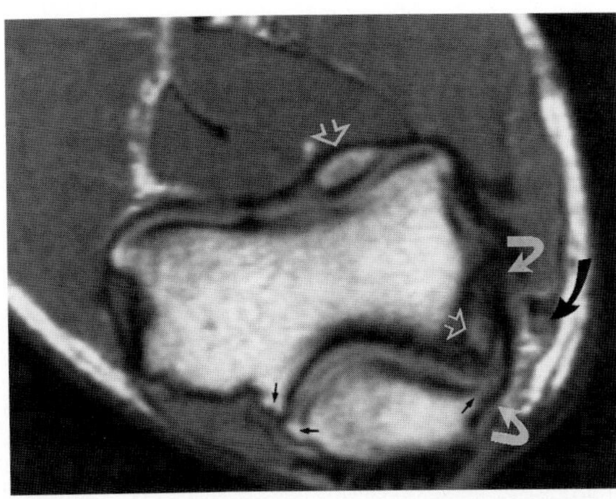

FIG. 15. E: Ulnar neuritis in a 27-year-old professional pitcher. This proton-density axial image shows thickening and medial bowing of the medial collateral ligament (*curved arrows*) as well as a medial loose body (*small open arrow*) and small spurs (*small arrows*) that undermine the floor of the cubital tunnel distal to the medial epicondyle. An anterior compartment loose body is incidentally noted (*large open arrow*) eroding the surface of the trochlea.

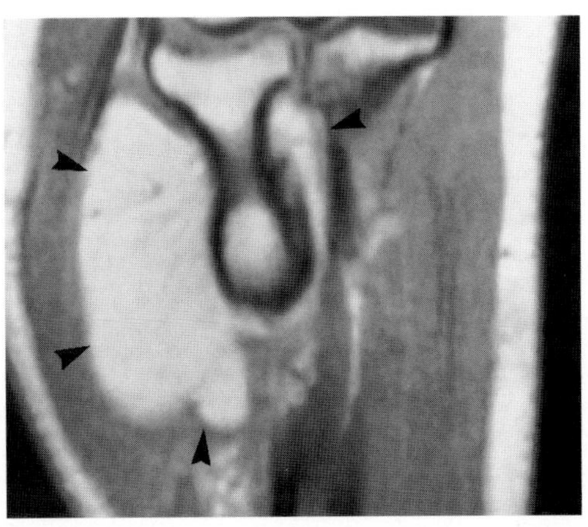

FIG. 15. F: Entrapment of the radial nerve secondary to a lipoma at the level of the elbow. T1-weighted coronal image delineates a mass with fat signal intensity (*arrowheads*) adjacent to the radius, the branches of the radial nerve, and the extensor musculature of the proximal forearm.

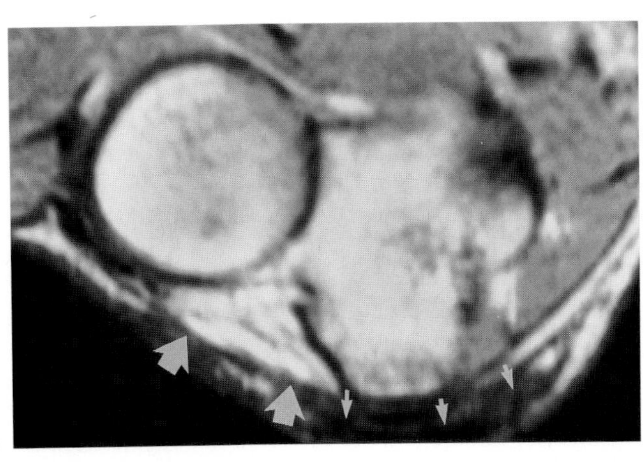

FIG. 15. G: Fatty infiltration and atrophy of the anconeus muscle secondary to chronic denervation. There is prominent fatty replacement of the anconeus muscle (*large arrows*) on this proton-density axial image. Low-signal-intensity scarring (*small arrows*) is noted at the site of a prior olecranon bursectomy that was complicated by damage to the innervation of the anconeus muscle.

CASE 16

History: This 25-year-old professional baseball player complained of acute medial elbow pain while throwing a pitch. He was unable to continue pitching and had to leave the game. What are the MR findings? How is this injury treated?

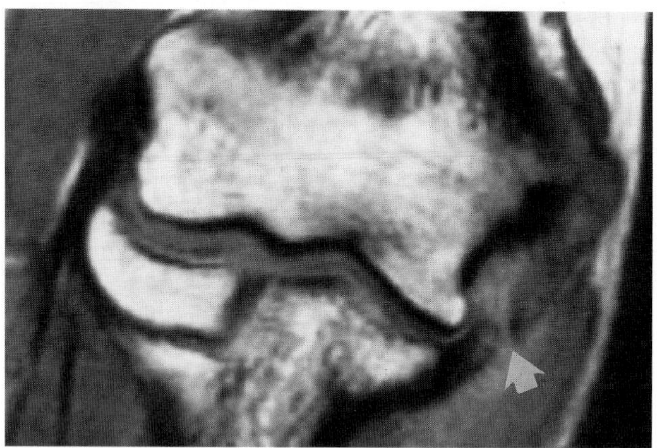

A

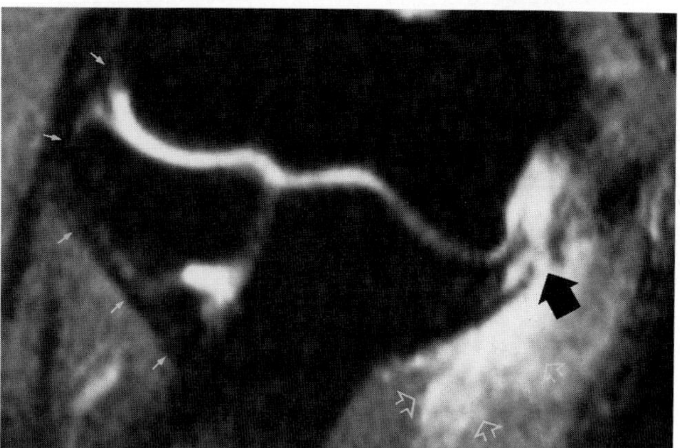

B

FIG. 16.

Findings: T1-weighted (Fig. 16A) and STIR (Fig. 16B) coronal images reveal disruption of the mid-portion of the anterior bundle of the medial collateral ligament (large arrows). A strain of the adjacent flexor digitorum superficialis muscle is also noted (open arrows). The normal lateral ulnar collateral ligament (small arrows) is well seen on the STIR coronal image.

Diagnosis: Midsubstance rupture of the anterior bundle of the medial collateral ligament.

Discussion: The medial collateral ligament injury in this case required reconstruction with a palmaris longus tendon allograft. This patient returned to professional baseball 10 months after the injury and has pitched successfully in the past 3 seasons.

Acute injury of the medial collateral ligament can be detected, localized, and graded with MR imaging. The medial collateral ligament complex consists of anterior and posterior bundles as well as an oblique band also known as the transverse ligament. The anterior bundle extends from the medial epicondyle to the medial aspect of the coronoid process and provides the primary constraint to valgus stress (1–3). The status of the functionally important anterior bundle of the medial collateral ligament complex may be determined by assessing the axial and coronal images.

Degeneration and tearing of the medial collateral ligament with or without concomitant injury of the common flexor tendon commonly occurs in throwing athletes. Injury of these medial stabilizing structures is due to chronic microtrauma from repetitive valgus stress during the acceleration phase of throwing (4–6). Complete rupture of the anterior bundle of the medial collateral ligament usually occurs as a sudden event.

Acute ruptures of the medial collateral ligament are well seen with standard MR imaging (Fig. 16A,B). Partial detachment of the deep undersurface fibers of the anterior bundle may also occur in pitchers with medial elbow pain and are more difficult to diagnose with standard MR imaging. These partial tears of the medial collateral ligament characteristically spare the superficial fibers of the anterior bundle and are therefore not visible from an open surgical approach unless the ligament is incised to inspect the torn capsular fibers (7,8). As a result, MR imaging is important to localize these partial tears, which are treated with repair or reconstruction. Detection of these undersurface partial tears is improved when intra-articular contrast is administered and CT or MR arthrography is performed (9). The capsular fibers of the anterior bundle of the medial collateral ligament normally insert on the medial margin of the coronoid process (Fig. 16C). Undersurface partial tears of the anterior bundle are characterized by distal extension of fluid or contrast along the medial margin of the coronoid (Fig. 16D–G) (9,10).

Midsubstance medial collateral ligament ruptures can be differentiated from proximal or distal avulsions (Fig. 16H). Midsubstance ruptures of the medial collateral ligament accounted for 87%, whereas distal and proximal avulsions were found in 10% and 3%, respectively, in a large series of surgically treated throwing athletes (1). Others have found a lesser percentage of midsubstance ruptures (11,12).

The fibers of the flexor digitorum superficialis muscle blend with the anterior bundle of the medial collateral ligament (8,13,14). A strain of the flexor digitorum superficialis muscle commonly is seen when the medial collateral ligament is injured (Fig. 16A,B). Ulnar traction spurs commonly are seen at the insertion of the medial collateral ligament on the coronoid process due to repetitive valgus stress, which occurs in 75% of professional baseball pitchers (15). Chronic degeneration of the medial collateral ligament is characterized by thickening of the ligament secondary to scarring often accompanied by foci of calcification or heterotopic bone (Fig. 16I) (1). The findings are similar to those seen after healing of medial collateral ligament sprains in the knee, in which the development of heterotopic ossification has been termed the Pellegrini-Stieda phenomenon.

Patients with symptomatic medial collateral ligament insufficiency usually are treated with reconstruction using a palmaris tendon graft. Graft failure is unusual but also may be evaluated with MR imaging (Fig. 16J). Lateral compartment bone contusions usually are seen in association with acute medial collateral ligament tears and may provide useful confirmation of recent lateral compartment impaction secondary to valgus instability (Fig. 16J).

A number of different conditions may occur secondary to the repeated valgus stress to the elbow that occurs with throwing. Medial tension overload typically produces extra-articular injury such as flexor/pronator strain, medial collateral ligament sprain, ulnar traction spurring, and ulnar neuropathy. Lateral compression overload typically produces intra-articular injury such as osteochondritis dissecans of the capitellum or radial head, degenerative arthritis, and loose body formation. MR imaging can be used for assessment of each of these related pathologic processes associated with repeated valgus stress (16,17). The additional information provided by MR imaging can be helpful in formulating a logical treatment plan, especially when surgery is being considered.

Rupture of the medial collateral ligament is also commonly encountered as a result of posterior dislocation of the elbow (18). After the shoulder, the elbow is the second most common joint to be dislocated (19). The mechanism of posterior elbow dislocation usually involves falling on an outstretched arm. There is typically rupture of the medial and lateral collateral ligaments as well as the anterior and posterior capsule during posterior elbow dislocation (20). Associated rupture of the common extensor tendon or the common flexor tendon may also occur. The extent of injury secondary to elbow dislocation is well delineated with MR imaging.

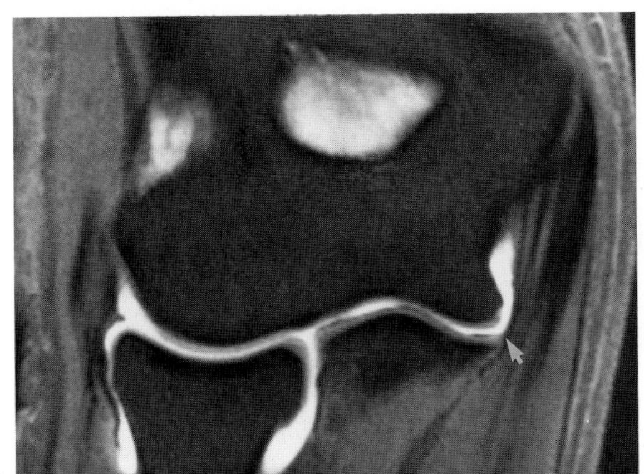

FIG. 16. C: Normal MR arthrography of the elbow. A fat-suppressed T1-weighted coronal image was performed after intra-articular injection of dilute gadolinium. There is no evidence of contrast extending distal to the normal attachment of the MCL at the medial margin of coronoid (*arrow*).

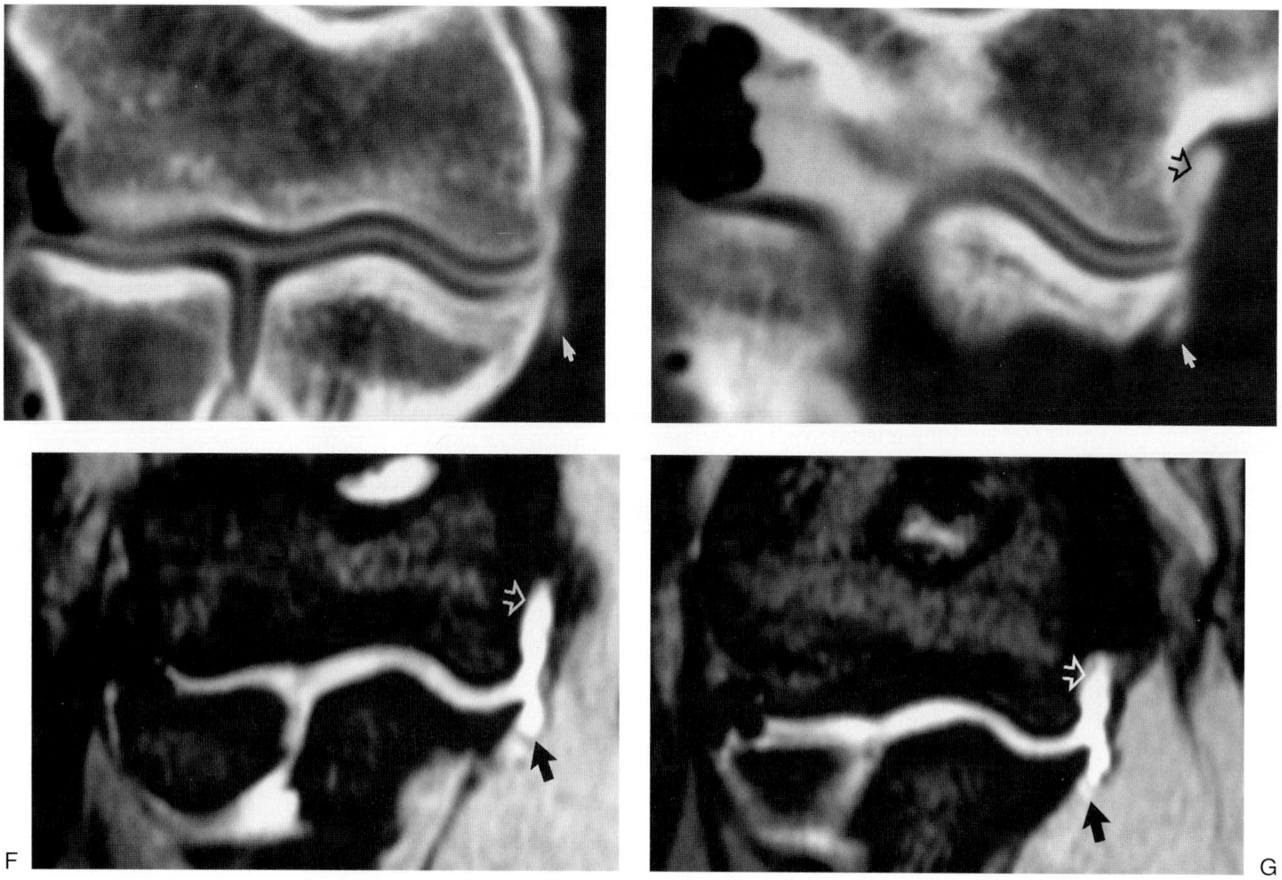

D E F G

FIG. 16. D–G: CT and MR-arthrography in a patient with nearly complete detachment of the anterior bundle of the MCL from the coronoid process. Coronal CT **(D,E)** and T2*-weighted gradient echo coronal MR **(F,G)** images were performed after intra-articular injection of radiographic contrast. There is leakage of contrast distal to the normal attachment of the anterior bundle of the MCL to the coronoid (*solid arrows*). The contrast remains contained within the intact superficial layer of the ligament and capsule. The proximal extension of contrast (*open arrow*) beneath the MCL is normal. (Courtesy of Martin Schwartz, MD.)

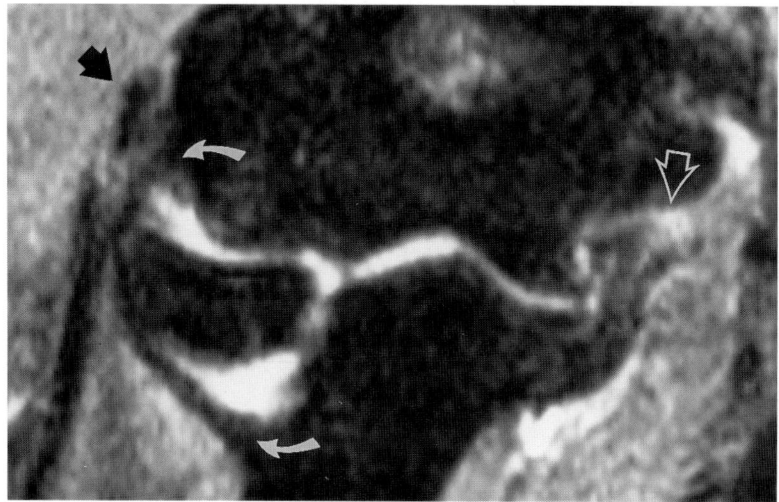

FIG. 16. H: A 26-year-old professional baseball player with medial elbow pain. A STIR coronal image reveals thickening of the anterior bundle of the MCL with increased signal delineating the site of proximal detachment (*open arrow*). A focus of calcification is incidentally seen in the common extensor tendon due to tendinosis and previous steroid injections (*black arrow*). The lateral ulnar collateral ligament is also well seen (*curved arrows*).

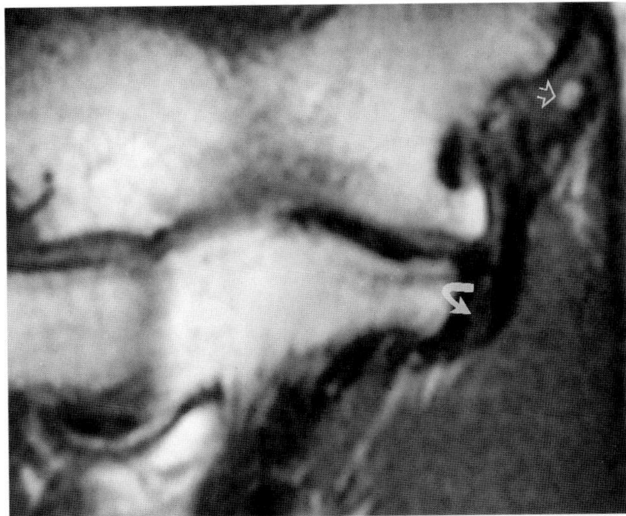
I

FIG. 16. I: Degeneration and partial tearing of the medial collateral ligament in a 36-year-old professional baseball player. A T1-weighted coronal image reveals partial detachment of the medial collateral ligament from the medial margin of the coronoid process (*curved arrow*) as well as thickening and irregularity of the proximal medial collateral ligament. Heterotopic ossification (*open arrow*) is noted within the degenerated medial collateral ligament just distal to the medial epicondyle.

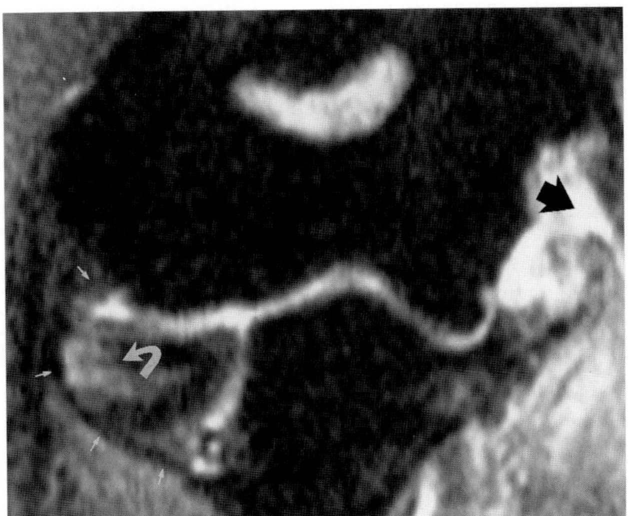
J

FIG. 16. J: A 33-year-old professional baseball player after medial collateral ligament reconstruction with acute graft rupture. A STIR coronal image reveals increased signal and poor definition of an medial collateral ligament graft (*black arrow*). A contusion of the radial head is noted (*curved arrow*) due to lateral impaction and valgus insufficiency. The normal lateral ulnar collateral ligament (*small arrows*) is well seen on this image.

CASE 17

History: This 36-year-old golfer presented with a several-year history of medial elbow pain that improved with local steroid injections. What structure is abnormal in this case? To what normal structure does the open arrow (Fig. 17A) point?

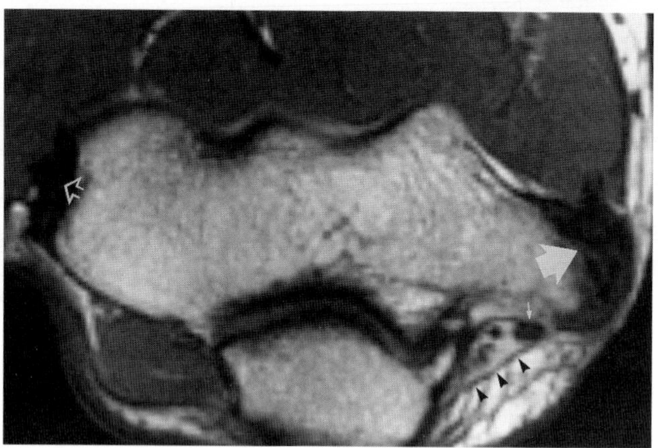

FIG. 17.

Findings: A T1-weighted axial image (Fig. 17A) reveals prominent thickening and mild increased signal within the common flexor tendon (large white arrow). The open arrow points to the normal common extensor tendon adjacent to the lateral epicondyle. The ulnar nerve (small white arrow) and the cubital tunnel retinaculum (arrowheads) are also normal.

Diagnosis: Medial epicondylitis (degenerative tendinosis of the common flexor tendon).

Discussion: Medial epicondylitis, also known as golfer's elbow, pitcher's elbow, or medial tennis elbow is less common than lateral epicondylitis (1,2). This condition is caused by degeneration of the common flexor tendon secondary to overload of the flexor/pronator muscle group that arises from the medial epicondyle (3–6). The spectrum of damage to the muscle–tendon unit that may be characterized with MR imaging includes muscle strain injury, tendon degeneration (tendinosis), and macroscopic tendon disruption.

MR imaging is useful for detecting and characterizing acute muscle injury as well as following its resolution (7–11). The STIR sequence is the most sensitive for detecting muscle pathology (Fig. 17B). The common flexor tendon and medial collateral ligament should be evaluated carefully for associated tearing when there is evidence of medial muscle strain injury on MR imaging. Alternatively, increased signal intensity on STIR and T2-weighted sequences may be seen after an intramuscular injection and may persist for as long as 1 month (12). Abnormal signal intensity within a muscle may simply be due to the effect of a therapeutic injection for epicondylitis rather than an indication of muscle strain. Steroid injections ideally should be done after MR imaging to avoid the confounding appearance of the injection on the structures about the elbow.

A normal muscle–tendon unit will tear at the myotendinous junction (Fig. 17C,D) (13). A much more common clinical entity, however, is failure of a muscle–tendon unit through an area of tendinosis (14). Tendon degeneration (tendinosis) is common about the elbow (1,6). Medial and lateral epicondylitis may occur concurrently secondary to flexor and extensor tendinosis (Fig. 17E,F). MR imaging can determine if there is tendinosis (secondary to degeneration, microscopic partial tearing, and repair) versus macroscopic partial tearing or complete rupture. This distinction is made by identifying fluid signal intensity delineating the presence or absence of tendon fibers on T2-weighted images (Fig. 17G–I). The appearance of medial and lateral epicondylitis about the elbow is similar to the appearance of other common degenerative tendinopathies that involve the attachment of tendons to bone. Similar criteria can be used to evaluate the common flexor and common extensor tendons in the elbow, the supraspinatus tendon in the shoulder, the patellar tendon in the knee, and the plantar fascia in the foot on MR imaging. In each of these conditions, there is degenerative tendinosis and a failed healing response that precedes rupture (14–17).

The coronal, sagittal, and axial sequences are all useful for assessing the degree of tendon injury (Fig. 17E–

I). MR imaging facilitates surgical planning by delineating and grading tears of the common flexor tendon as well as evaluating the underlying medial collateral ligament and adjacent ulnar nerve. Ulnar neuritis commonly accompanies common flexor tendinosis and may be difficult to identify clinically (Fig. 17E,F). Patients with concomitant ulnar neuropathy have a significantly poorer prognosis after surgery compared with patients who have isolated medial epicondylitis (18,19). Patients with coexisting ulnar neuritis and common flexor tendinosis (25% to 50% of patients who undergo surgery for medial epicondylitis) require transposition or decompression of the ulnar nerve in addition to debridement and repair of the abnormal flexor tendon (3,5,18,19). The increased preoperative diagnostic information may lessen the need for extensive surgical exploration in cases in which the medial collateral ligament is clearly intact on MR imaging (Fig. 17J,K). In addition, MR imaging may be useful for problem solving in patients who develop recurrent symptoms after surgery for medial or lateral epicondylitis.

In skeletally immature individuals, the flexor muscle–tendon unit may fail at the unfused apophysis of the medial epicondyle. Stress fracture, avulsion, or delayed closure of the medial epicondylar apophysis may occur in young baseball players secondary to overuse (Little League elbow) (20). MR imaging may detect these injuries before complete avulsion and displacement by revealing soft tissue or marrow edema about the medial epicondylar apophysis on the STIR images (21).

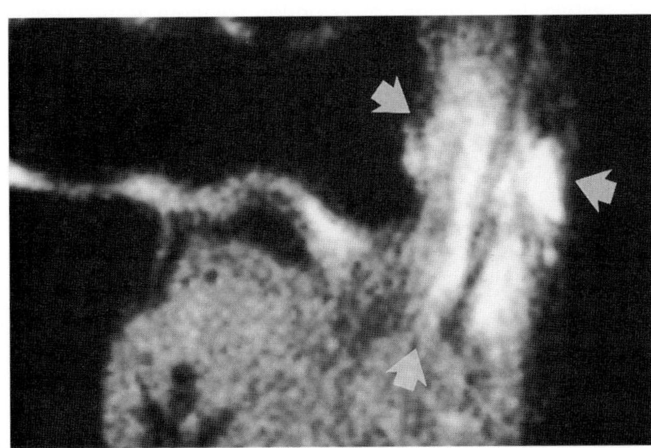

FIG. 17. B: Muscle strain in a baseball player. A STIR coronal image reveals increased signal intensity throughout the pronator teres and adjacent flexors (*arrows*) compatible with a muscle strain injury.

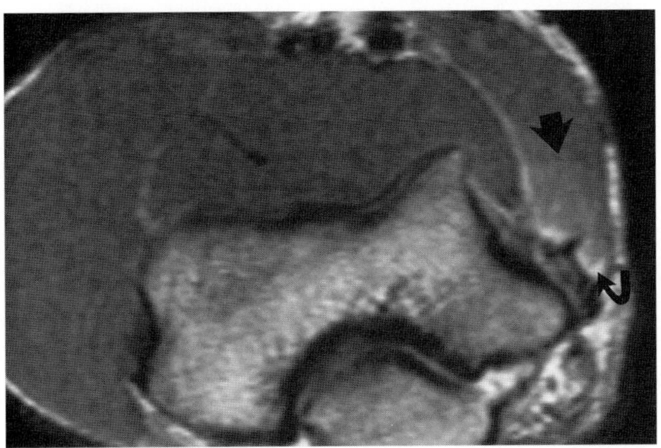

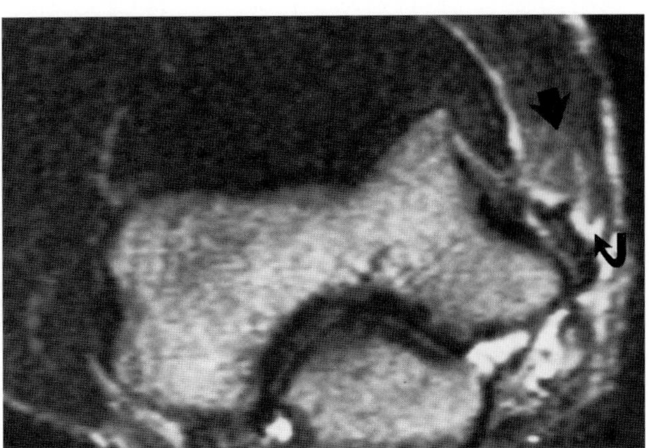

FIG. 17. C,D: Acute muscle strain injury secondary to resisted supination. Proton-density **(C)** and T2-weighted **(D)** axial images reveal increased signal within the pronator teres muscle (*large arrow*) as well as fluid delineating tearing of the muscle from the common flexor tendon (*curved arrow*).

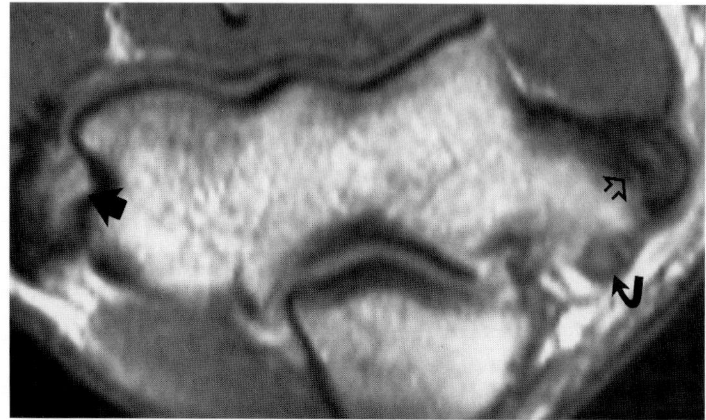

E

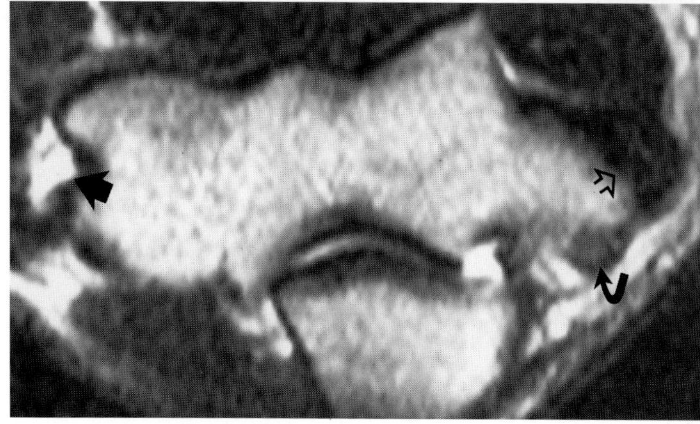

F

FIG. 17. E,F: Medial and lateral epicondylitis in a 48-year-old golfer. Proton-density **(E)** and T2-weighted **(F)** axial images reveal thickening and mild increased signal within the common flexor tendon adjacent to the medial epicondyle (*open arrows*) compatible with mild tendinosis. The ulnar nerve (*curved arrows*) is mildly enlarged, suggesting ulnar neuritis. There is fluid (*solid arrows*) delineating rupture of the common extensor tendon from the lateral epicondyle.

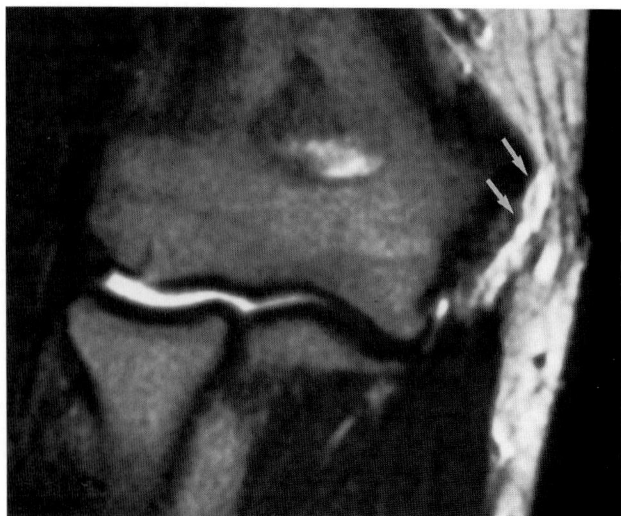

G

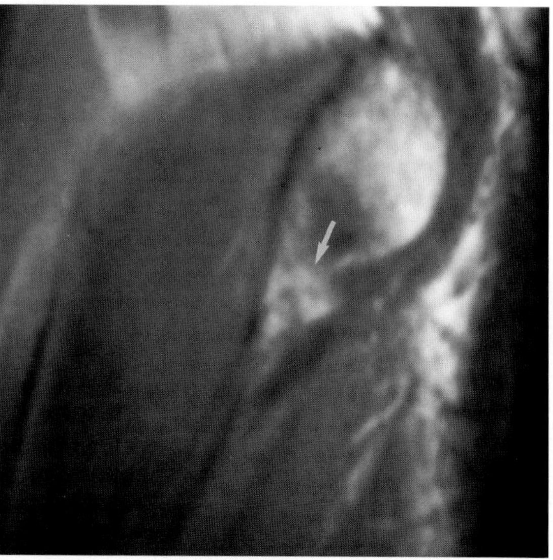

H

FIG. 17. G-I: Rupture of the common flexor tendon in a 35-year-old softball player. T2-weighted coronal **(G)** as well as proton-density **(H)** and T2-weighted sagittal images reveal fluid separating the fibers of the common flexor tendon from the medial epicondyle (*arrows*) pt, pronator teres muscle.

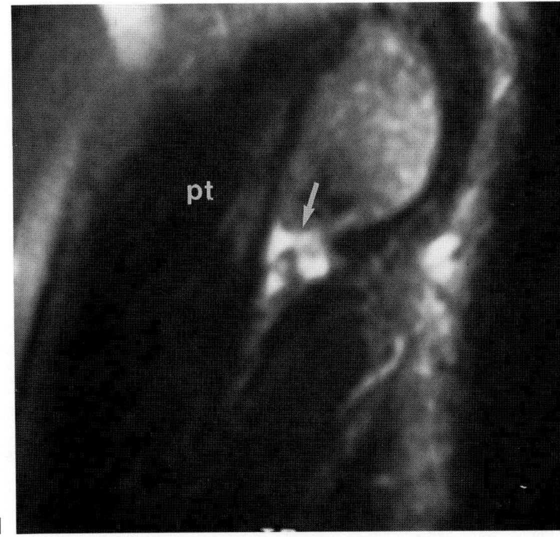

FIG. 17. G-I: (*continued.*)

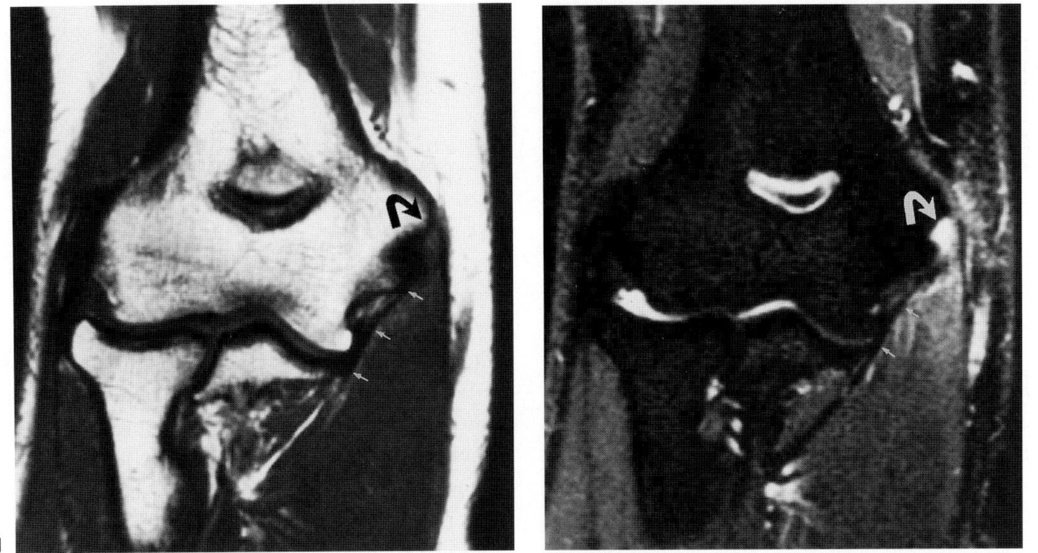

FIG. 17. J,K: Medial elbow pain in a 45-year-old tennis player. T1-weighted **(J)** and STIR **(K)** coronal images reveal complete rupture of the common flexor tendon from the medial epicondyle (*curved arrows*). The anterior bundle of the medial collateral ligament (*straight arrows*) appears to be intact.

CASE 18

History: This 35-year-old tennis player presented with a 3-year history of progressive lateral elbow pain that did not improve after a local steroid injection. On physical examination there was tenderness over the lateral epicondyle and lateral elbow pain with resisted extension of the wrist. What is the diagnosis? To what structure does the curved arrow (Fig. 18A,B) point?

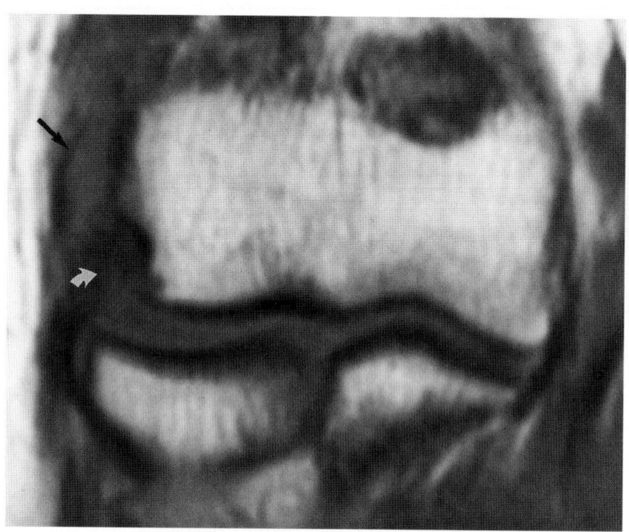
A

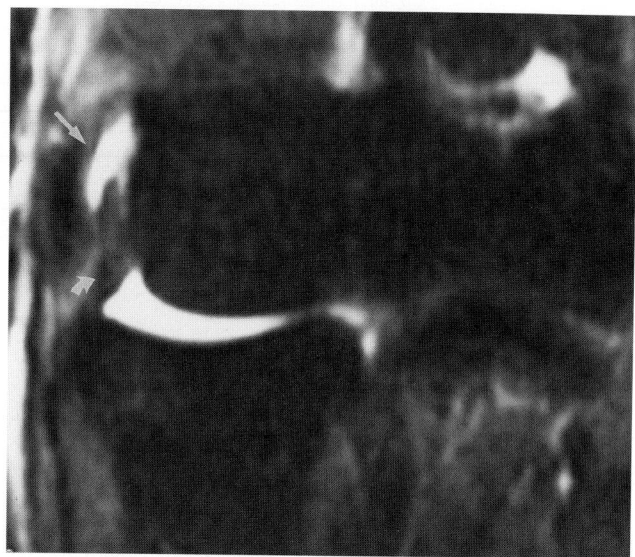
B

FIG. 18.

Findings: T1-weighted (Fig. 18A) and STIR (Fig. 18B) coronal images reveal fluid (straight arrows) that delineates tearing of the common extensor tendon from the lateral epicondyle. The underlying ulnar part of the lateral collateral ligament, the lateral ulnar collateral ligament, is mildly increased in signal (curved arrow) but appears intact.

Diagnosis: Lateral epicondylitis (tennis elbow) secondary to rupture of the common extensor tendon. Mild degeneration of the lateral ulnar collateral ligament.

Discussion: Lateral epicondylitis, also referred to as tennis elbow, is caused by degeneration and tearing of the common extensor tendon (1,2). This condition often occurs as a result of repetitive sports-related trauma to the tendon, although it is seen far more commonly in nonathletes (3–5). In the typical case, the degenerated extensor carpi radialis brevis tendon is partially avulsed from the lateral epicondyle (2). Scar tissue forms in response to this partial avulsion, which is then susceptible to further tearing with repeated trauma. Recent histologic studies have shown angiofibroblastic tendinosis with a lack of inflammation in the surgical specimens of patients with lateral epicondylitis; this finding suggests that the abnormal signal seen on MR images is secondary to tendon degeneration and repair rather than tendinitis (1,6–8). Local steroid injections commonly are used to treat lateral epicondylitis and may increase the risk of tendon rupture (9,10). Signal alteration in the region of a local steroid injection should not be confused for primary muscle pathology on MR imaging (Fig. 18C,D).

Overall, 4% to 10% of cases of lateral epicondylitis are resistant to conservative therapy (1,4); MR imaging is useful in assessing the degree of tendon damage in such cases. Tendon degeneration (tendinosis) is manifested by normal to increased tendon thickness with increased signal intensity on T1-weighted images that does not further increase in signal intensity on the T2-weighted images. Partial tears are characterized by thinning of the tendon that is outlined by adjacent fluid on the T2-weighted images. Tendinosis and tearing typically involve the extensor carpi radialis brevis portion of the common extensor tendon anteriorly (Fig. 18E). Complete tears may be diagnosed on MR imaging by identifying a fluid-filled gap separating the tendon from its adjacent bony attachment site (Fig. 18A,B).

At surgery for lateral epicondylitis, 97% of the tendons appear scarred and edematous secondary to tendinosis, and 35% have macroscopic tears (1,2). MR imaging is useful in identifying high-grade partial tears and complete tears that are unlikely to improve with rest and repeated steroid injections. In addition to determining the degree of tendon damage, MR imaging also provides a more global assessment of the elbow and therefore is able to detect additional pathologic conditions that may explain the lack of a therapeutic response. For example, unsuspected ruptures of the lateral collateral ligament complex may occur in association with tears of the common extensor tendon (Fig. 18F,G). Morrey (11) has recently reported a series of 13 patients who underwent reoperation for failed lateral epicondylitis surgery; stabilization procedures were required in 4 patients with either iatrogenic or unrecognized lateral ligament insufficiency. Iatrogenic tears of the lateral ulnar collateral ligament may occur secondary to an overaggressive release of the common extensor tendon (Fig. 18H,I) (12). MR imaging can reveal concurrent tears of the lateral ulnar collateral ligament and common extensor tendon in patients with lateral epicondylitis as well as isolated lateral ulnar collateral ligament tears in patients with posterolateral rotatory instability. Moreover, the lack of a significant abnormality involving the common extensor tendon on MR imaging may prompt consideration of an alternate diagnosis such as radial nerve entrapment, which may mimic or accompany lateral epicondylitis (13–15).

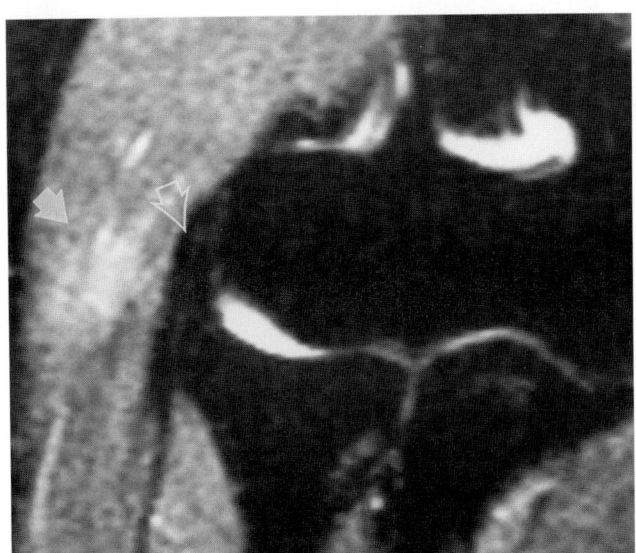

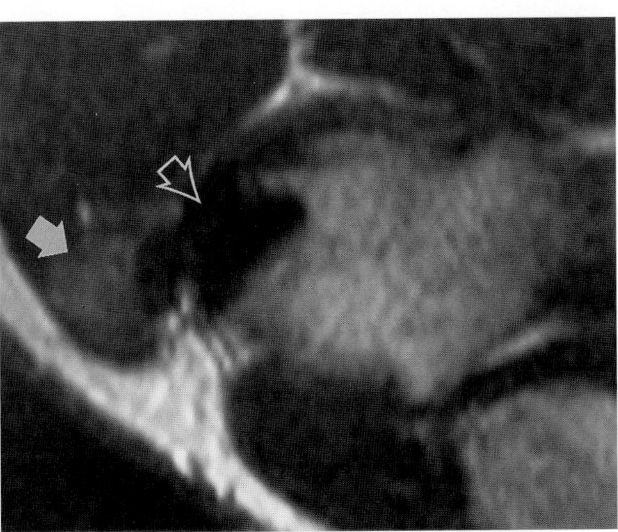

FIG. 18. C,D: Clinically suspected tennis elbow in a patient who did not respond to a local steroid injection. A STIR coronal image **(C)** and a T2-weighted axial image **(D)** reveal a completely normal common extensor tendon (*open arrows*) as well as increased signal within the adjacent extensor carpi radialis longus muscle (*solid arrows*) secondary to a recent steroid injection. Abnormal signal may persist for weeks after an injection and can be mistaken for primary muscle pathology on MR imaging.

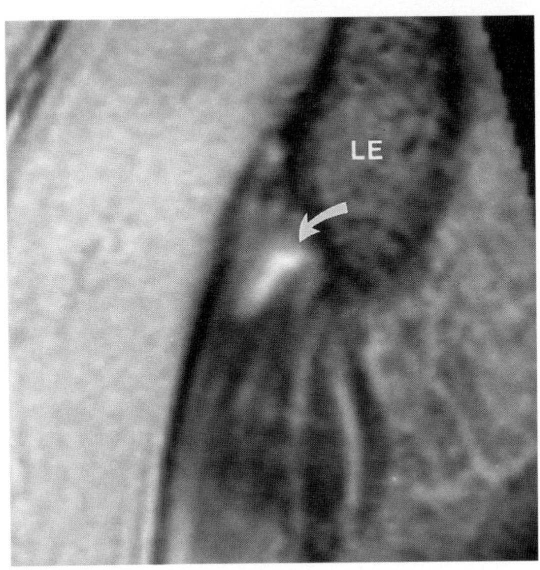

FIG. 18. E: Partial tear of the extensor carpi radialis brevis. A T2*-weighted gradient-echo sagittal image reveals a small tear (*arrow*) involving the anterior attachment of the common extensor tendon to the lateral epicondyle (LE).

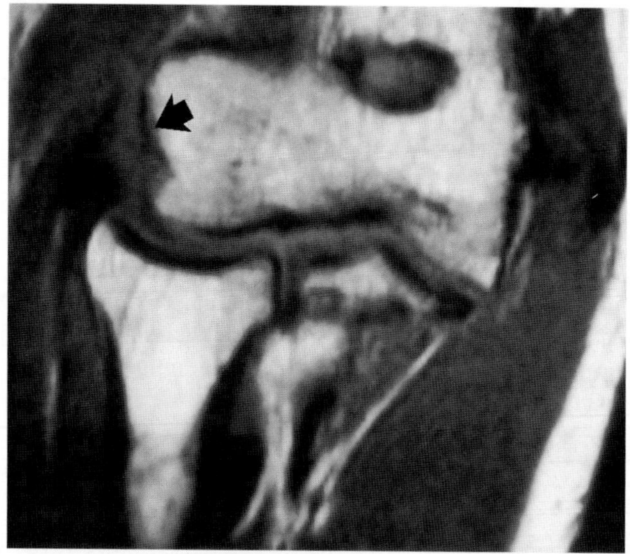
F

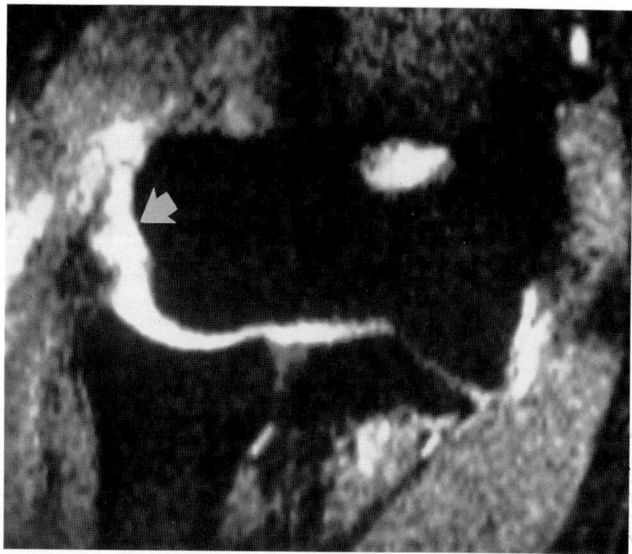
G

FIG. 18. F,G: Lateral epicondylitis in a 43-year-old golfer. T1-weighted **(F)** and STIR **(G)** coronal images reveal a complete tear of the common extensor tendon and underlying lateral collateral ligament complex (*arrows*) from the lateral epicondyle.

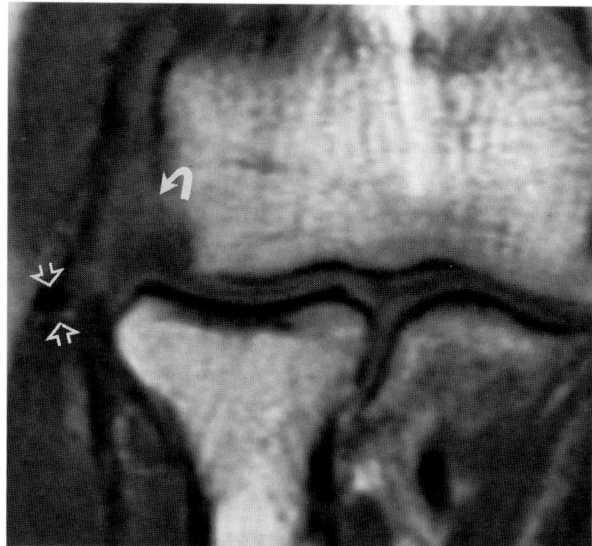
H

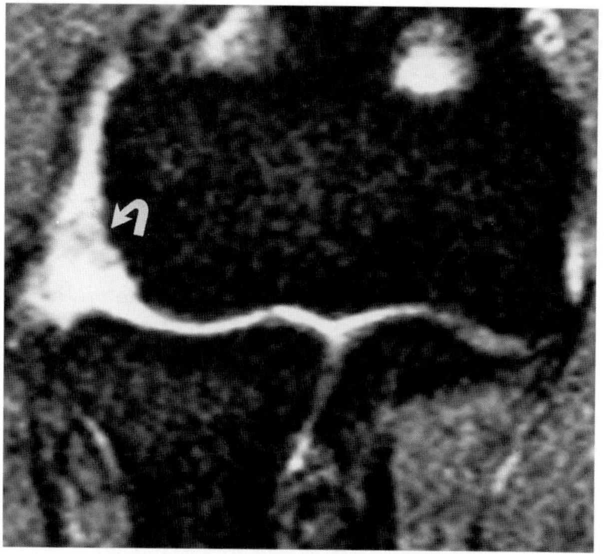
I

FIG. 18. H,I: Torn lateral ulnar collateral ligament (LUCL) in a patient who developed posterolateral rotatory instability after extensor tendon release. T1-weighted **(H)** and STIR **(I)** coronal images reveal complete absence of the common extensor tendon and LUCL adjacent to the lateral epicondyle (*curved white arrow*). Micrometallic artifact is noted from prior surgical release (*open arrows*).

CASE 19

History: This 42-year-old man presented with elbow pain after falling with his arm outstretched in front of him. What are the MR findings? What is the mechanism and significance of this injury?

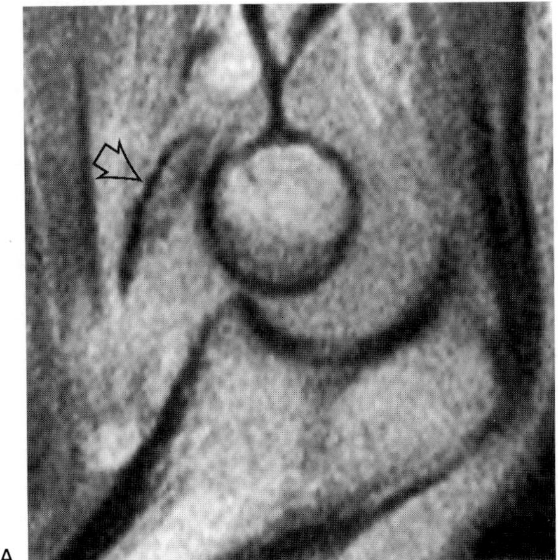

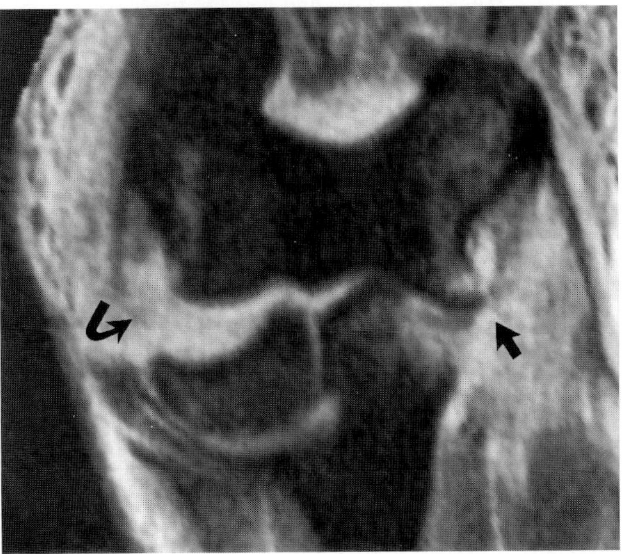

FIG. 19.

Findings: A proton-density sagittal image (Fig. 19A) reveals posterior subluxation of the ulna and a large fracture of the coronoid process that is superiorly displaced (open arrow). Increased signal is also seen within the brachialis muscle and the anterior capsule just anterior to the fracture fragment. A STIR coronal image (Fig. 19B) reveals disruption of the lateral collateral ligament proximally (curved arrow) and disruption of the anterior bundle of the medial collateral ligament distally (straight arrow).

Diagnosis: Coronoid process fracture and collateral ligament rupture secondary to posterior dislocation of the elbow.

Discussion: A fracture of the coranoid process is highly characteristic of a previous posterior dislocation or subluxation of the elbow (1). These fractures may be subtle on standard radiographs, especially when they are small or nondisplaced. Fractures of the coronoid process may predispose to recurrent posterior instability depending on the size of the fracture fragment and the presence of associated collateral ligament rupture (2). Coronoid process fractures occur as a result of direct shear injury by the trochlea during posterior dislocation or subluxation (3). These fractures are not hyperextension avulsion injuries, since the tip of the coronoid is an intra-articular structure that does not have a capsular attachment. The anterior capsule and the brachialis muscle insert further distally on the ulna. Anterior capsular injury and contusion or strain of the adjacent brachialis muscle (Fig. 19A) as well as medial and lateral collateral ligament injury (Fig. 19B) are commonly seen after posterior elbow dislocation (4–7). The common flexor and extensor tendons also commonly rupture with posterior dislocation (Fig. 19C,D). If the common flexor tendon is torn or the medial epicondyle is avulsed, the median nerve may rarely become entrapped within the elbow joint during posterior dislocation (8,9). Therefore, each of these structures should be carefully evaluated when a coronoid process fracture is identified on MR imaging.

The coronoid is an important structure for stability of the elbow. Fractures of the coronoid process have been classified by Regan and Morrey (2). Type I fractures are small shear fractures that do not destabilize the joint (Fig. 19E,F). They should be recognized, however, as an indicator of posterior elbow dislocation/subluxation injury that may be associated with significant soft tissue disruption. Type II fractures involve less than 50% of the coronoid. They should be fixed if the joint remains dislocated or subluxed. Type III fractures involve greater than 50% of the coronoid and have a poor prognosis. Type III fractures, as well as malunions and nonunions of the coronoid in patients with instability, also require fixation (2,3).

Approximately 10% of elbow dislocations result in fractures of the radial head; conversely, about 10% of patients with a radial head fracture have an elbow dislocation (Fig. 19G,H) (10). Displaced fractures of the radial head are best treated with internal fixation when there is ligamentous disruption and instability (2,3). Prosthetic replacement may be necessary to maintain stability when the radial head is comminuted and is not repairable.

Dislocation of the elbow is an unusual event; it is, however, the second most common major joint dislocation (after the shoulder) in adults and it is the most common dislocation in children younger than 10 years old (11). Children are predisposed to elbow dislocation due to the relative lack of congruity of the immature cartilaginous articulation compared with the constrained bony articulation of adults. Recurrent complete dislocation of the elbow is usual but is more common in children and adolescents (12,13). Many of the dislocations that occur in children go unrecognized because of spontaneous reduction with the only finding being a swollen tender elbow (14). MR imaging in such cases usually shows an effusion as well as a contusion of the brachialis muscle. Bone contusions may be seen at the posterior margin of the capitellum as well as the radial head and coronoid process.

The usual mechanism of dislocation involves a fall on the outstretched hand (15). A hyperextension force has been classically proposed to explain posterior dislocation of the elbow. More recent investigation by O'Driscoll and associates (16,17) has resulted in a clearer understanding of how the flexed elbow may subluxate posterolaterally and then dislocate. This mechanism involves hypersupination, valgus stress, and axial compressive loading of the elbow that may occur during a fall on the outstretched hand.

Elbow instability occurs as a spectrum from subluxation to dislocation that has been divided into three stages (16). Each of these stages is associated with progressive soft tissue injury that extends from lateral to medial. In stage 1, there is posterolateral subluxation of the ulna and radius relative to the humerus with disruption of the ulnar part of the lateral collateral ligament, which has been termed the lateral ulnar collateral ligament (LUCL). Rupture of the LUCL is considered the essential lesion of posterolateral rotatory instability. In stage 2, there is incomplete dislocation so that the coronoid appears to be perched on the trochlea. There is further disruption of the lateral ligamentous structures in stage 2 as well as tearing of the anterior and posterior joint capsule (Fig. 19I). In stage 3, there is complete posterior dislocation with progressive disruption of the medial collateral ligament (MCL) complex. There may be disruption of the posterior bundle of the MCL only (stage 3A), in which case the elbow is stable to valgus stress. Stage 3B is characterized by disruption of the anterior bundle of the MCL so that the elbow is unstable in all directions. Complete disruption of the MCL (stage 3B) is the most common clinical situation after complete dislocation. The common flexor and extensor tendons are often disrupted when there is complete posterior dislocation of the elbow (3,5,6).

The clinical assessment of elbow instability is difficult, since the physical examination is often compromised by guarding and pain. The pivot-shift test of the elbow has recently been described as a clinical test for posterolateral rotatory instability of the elbow due to insufficiency of the LUCL (17). The pivot-shift test of the elbow is analogous to the widely known pivot-shift test of the knee to determine the integrity of the anterior cruciate ligament. A supination/valgus moment is applied during flexion, causing the radius and ulna to subluxate posteriorly. Further flexion produces a palpable and visible clunk as the elbow reduces. This subluxation/reduction manuver creates apprehension, however, and it usually is not possible to perform it in the awake patient. Thus, clinical confirmation of recurrent instability of the elbow may require examination under anesthesia to elicit a pivot-shift manuver (3).

I have found MR imaging quite reliable in detecting rupture of the LUCL. This ligament usually tears proximally at the lateral margin of the capitellum and is best evaluated on coronal and axial images. The LUCL may tear as an isolated finding on MR imaging in patients with posterolateral rotatory instability (stage 1) or may be seen in association with rupture of the MCL in stage 3B (Fig. 19B). Disruption of the LUCL is also commonly seen in patients with severe tennis elbow who have tears of the common extensor tendon on MR imaging. Iatrogenic causes of LUCL disruption resulting in posterolateral rotatory instability include overaggressive extensor tendon release for lateral epicondylitis (common extensor tendinosis) and radial head excision for comminuted fractures of the radial head (18).

Recurrent instability of the elbow involves a common pathway of posterolateral rotatory subluxation due to insufficiency of the LUCL (3). Surgical correction is performed by reattaching the avulsed LUCL to the humerus or reconstructing it with a tendon graft placed isometrically through tunnels in the ulna and the humerus (19,20).

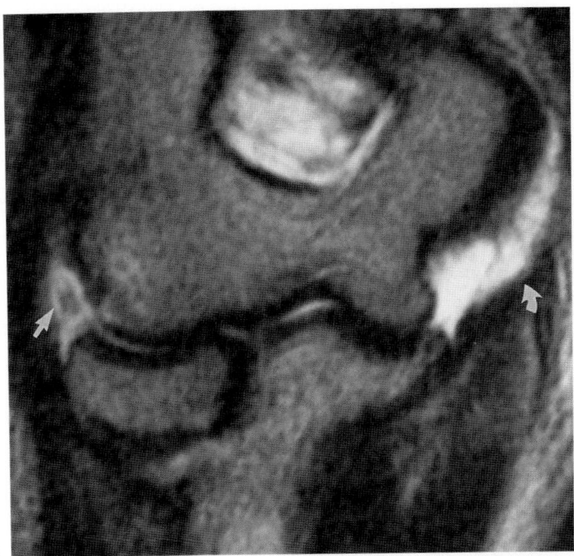

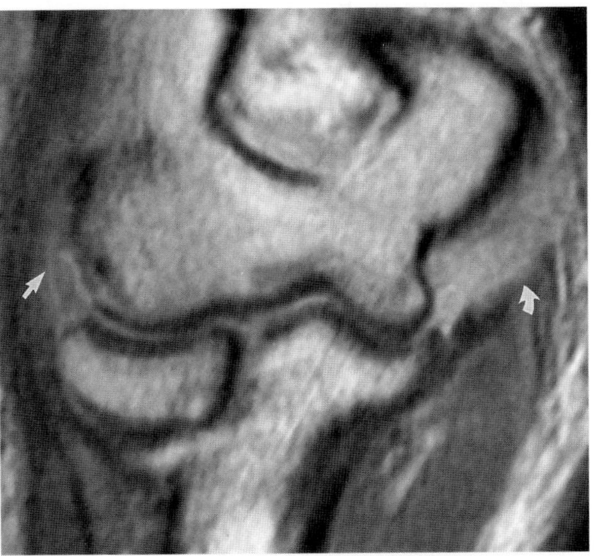

FIG. 19. C,D: A 45-year-old woman with a history of recurrent elbow dislocation. Proton-density **(C)** and T2-weighted **(D)** coronal images reveal complete avulsion of the medial collateral ligament and the common flexor tendon (*curved arrows*) from the medial epicondyle. Avulsion of the lateral collateral ligament (*straight arrows*) and partial tearing of the common extensor tendon are also seen.

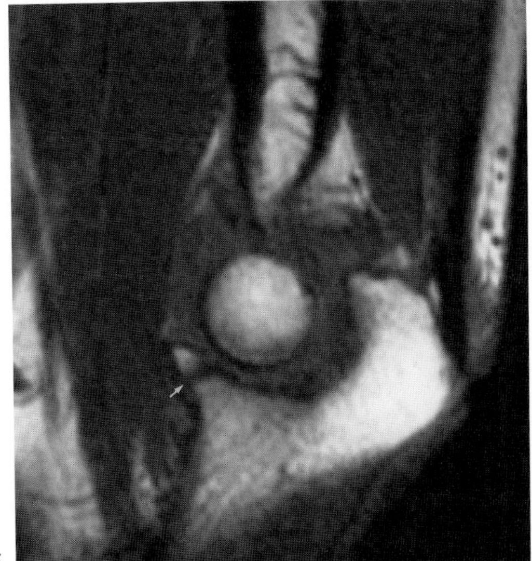

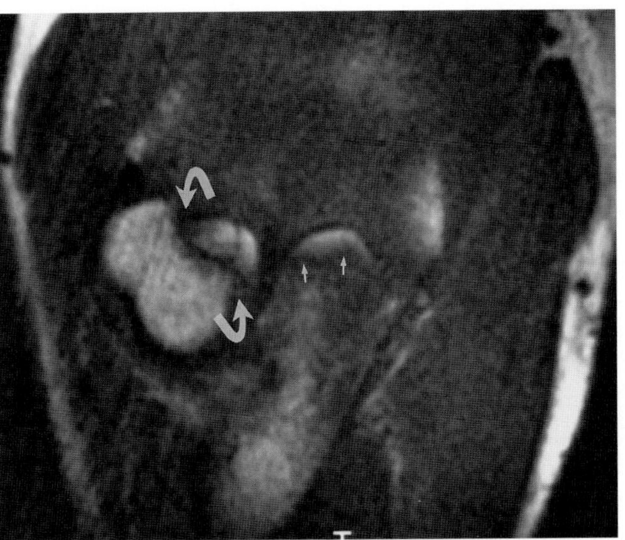

FIG. 19. E,F: Posterior dislocation/subluxation injury from a fall on the outstretched hand in a 30-year-old jogger. T1-weighted sagittal **(E)** and coronal **(F)** images reveal a nondisplaced fracture of the tip of the coronoid process (*small arrows*) and an impacted fracture of the radial head (*curved arrows*). The fracture of the radial head is caused by impaction of the capitellum, whereas the coronoid fracture is caused by the adjacent trochlea.

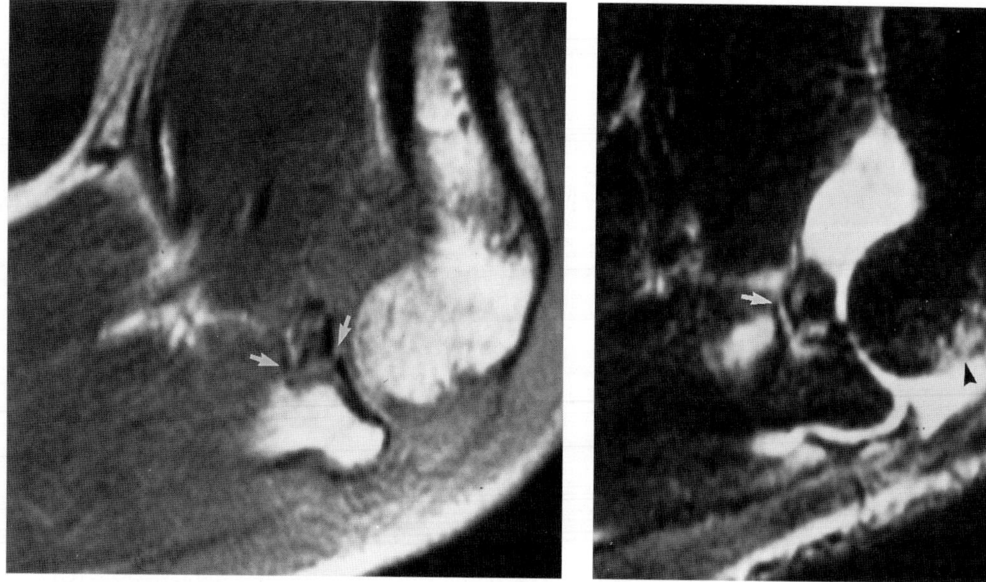

FIG. 19. G,H: Signs of posterior elbow dislocation injury on MR imaging. T1-weighted **(H)** and STIR **(I)** sagittal images reveal an anterior radial head fracture (*white arrows*) and a contusion (*black arrowhead*) at the posterior margin of the capitellum as a consequence of posterior dislocation/ relocation injury. (Courtesy of Mark Anderson, MD.)

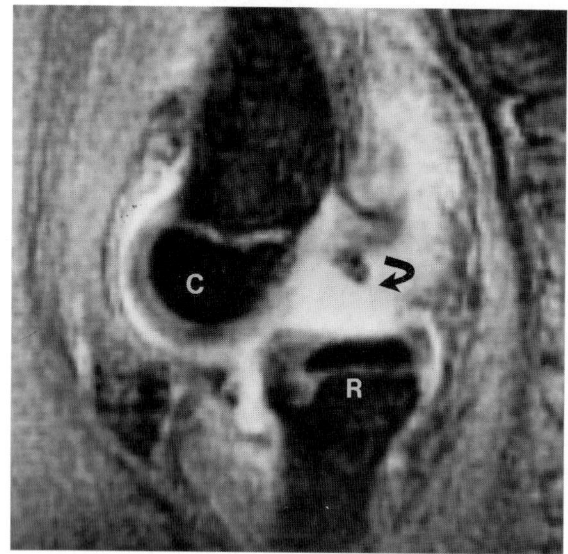

FIG. 19. I: A 13-year-old child with a history of instability and recurrent elbow dislocation. A T2*-weighted gradient-echo sagittal image reveals posterior subluxation of the radius (R) relative to the capitellum **(C)**. Rupture of the lateral ulnar collateral ligament (*arrow*) is seen at the posterolateral aspect of the joint capsule.

REFERENCES

Introduction

1. Patten RM. Overuse syndromes and injuries involving the elbow: MR imaging findings. *AJR* 1995;164:1205–1211.
2. Herzog RJ. Efficacy of magnetic resonance imaging of the elbow. *Med Sci Sports Exerc* 1994;26:1193–202.
3. Beltran J, Rosenberg ZS, Kawelblum M, Montes L, Bergman AG, Strongwater A. Pediatric elbow fractures: MRI evaluation. *Skeletal Radiol* 1994;23:277–281.
4. Potter HG, Hannafin JA, Morwessel RM, DiCarlo EF, O'Brien SJ, Altchek DW. Lateral epicondylitis: correlation of MR imaging, surgical, and histopathologic findings. *Radiology* 1995;196:43–46.
5. Schwartz ML, Al-Zahrani S, Morwessel RM, Andrews JR. Ulnar collateral ligament injury in the throwing athlete: evaluation with saline-enhanced MR arthrography. *Radiology* 1995;197:297–299.
6. Kingston S. Diagnostic imaging of the upper extremity. In: Jobe FW, ed. *Operative techniques in upper extremity sports injuries.* St. Louis: Mosby-Year Book; 1996:31–98.
7. Ho CP. Sports and occupational injuries of the elbow: MR imaging findings. *AJR* 1995;164:1465–1471.
8. Fritz RC. MR imaging of the elbow. *Semin Roentgenol* 1995;30: 241–264.
9. Conway JE, Jobe FW, Glousman RE, Pink M. Medial instability of the elbow in throwing athletes. Treatment by repair or reconstruction of the ulnar collateral ligament. *J Bone Joint Surg [Am]* 1992; 74:67–83.
10. Morrey BF. Applied anatomy and biomechanics of the elbow joint. *Instr Course Lect* 1986;35:59–68.
11. Morrey BF, An KN. Functional anatomy of the ligaments of the elbow. *Clin Orthop* 1985;201:84–90.
12. Jordan SE. Surgical anatomy of the elbow. In: Jobe FW, ed. *Operative techniques in upper extremity sports injuries.* St. Louis: Mosby-Year Book; 1996:402–410.
13. O'Driscoll SW, Bell DF, Morrey BF. Posterolateral rotatory instability of the elbow. *J Bone Joint Surg [Am]* 1991;73:440–446.
14. O'Driscoll SW, Morrey BF, Korinek S, An KN. Elbow subluxation and dislocation. A spectrum of instability. *Clin Orthop* 1992;280: 186–197.
15. Cohen MS, Hastings H. Rotatory stabilizers of the elbow: the lateral stabilizers. *J Shoulder Elbow Surg* 1995;4:S10.
16. Olsen BS, Vaesel MT. The lateral collateral ligament of the elbow joint. Anatomy and kinematics. *J Shoulder Elbow Surg* 1995;4: S21.
17. Nestor BJ, O'Driscoll SW, Morrey BF. Ligamentous reconstruction for posterolateral rotatory instability of the elbow. *J Bone Joint Surg [Am]* 1992;74:1235–1241.

Case 1

1. Morrey BF. Distal biceps tendon rupture. In: Morrey BF, ed. *The elbow.* New York: Raven Press; 1994:115–128.
2. D'Alessandro DF, Shields CL Jr, Tibone JE, Chandler RW. Repair of distal biceps tendon ruptures in athletes. *Am J Sports Med* 1993; 21:114–119.
3. Bourne MH, Morrey BF. Partial rupture of the distal biceps tendon. *Clin Orthop* 1991;271:143–148.
4. Fitzgerald SW, Curry DR, Erickson SJ, Quinn SF, Friedman H. Distal biceps tendon injury: MR imaging diagnosis. *Radiology* 1994;191:203–206.
5. Strong JA, Melamed JW, Martinez S, Burk DL, Harrelson JM, Spritzer CE. MRI of acute and chronic distal biceps tendon injuries. Presented at 94th annual meeting of the American Roentgen Ray Society, New Orleans, LA, 1994, pp 152–153.
6. Falchook FS, Zlatkin MB, Erbacher GE, Moulton JS, Bisset GS, Murphy BJ. Rupture of the distal biceps tendon: evaluation with MR imaging. *Radiology* 1994;190:659–663.
7. Mayer DP, Schmidt RG, Ruiz S. MRI diagnosis of biceps tendon rupture. *Comput Med Imaging Graph* 1992;16:345–347.
8. Agins HJ, Chess JL, Goekstra DV, Teitge RA. Rupture of the distal insertion of the biceps brachii tendon. *Clin Orthop* 1988;234:34–38.
9. Morrey BF. Tendon injuries about the elbow. In: Morrey BF, ed. *The elbow and its disorders.* 2nd ed. Philadelphia: WB Saunders; 1993:492–504.
10. Whiteside JA, Andrews JR. Tendinopathies of the elbow. *Sports Med Arthosc Rev* 1995;3:195–203.
11. Nielsen K. Partial rupture of the distal biceps brachii tendon. A case report. *Acta Orthop Scand* 1987;58:287–288.
12. Sotje G, Besch L. [Distal rupture of the biceps tendon—a magnetic resonance tomography follow-up]. *Aktuel Traumatol* 1993;23:105–107.
13. Murphy BJ. MR imaging of the elbow. *Radiology* 1992;184:525–529.
14. Seiler JG, Parker LM, Chamberland PDC, Sherbourne GM, Carpenter WA. The distal biceps tendon. Two potential mechanisms involved in its rupture: arterial supply and mechanical impingement. *J Shoulder Elbow Surg* 1995;4:149–156.
15. Morrey BF. Anatomy of the elbow joint. In: Morrey BF, ed. *The elbow and its disorders.* 2nd ed. Philadelphia: WB Saunders; 1993: 16–52.
16. Heckman JD, Levine MI. Traumatic closed transection of the biceps brachii in the military parachutist. *J Bone Joint Surg [Am]* 1978; 60:369—372.
17. Kelly JD, Elattrache NS. Muscle ruptures of the shoulder girdle. In: Jobe FW, ed. *Operative techniques in upper extremity sports injuries.* St. Louis: Mosby-Year Book; 1996:360–72.
18. Kannus P, Jozsa L. Histopathological changes preceding spontaneous rupture of a tendon. A controlled study of 891 patients. *J Bone Joint Surg [Am]* 1991;73:1507–1525.
19. Davis WM, Yassine Z. An etiological factor in tear of the distal tendon of the biceps brachii. *J Bone Joint Surg [Am]* 1956;37: 1365–1368.
20. Karanjia ND, Stiles PJ. Cubital bursitis. *J Bone Joint Surg [Br]* 1988;70:832–833.
21. Kosarek FJ, Hoffman CJ, Martinez S. Distal bicipital bursitis: MR imaging characteristics. *Radiology* 1995;197(P):398.
22. Foxworthy M, Kinninmonth AW. Median nerve compression in the proximal forearm as a complication of partial rupture of the distal biceps brachii tendon. *J Hand Surg [Br]* 1992;17:515–517.
23. Coonrad RW. Tendinopathies at the elbow. *Instr Course Lect* 1991; 40:25–42.
24. Failla JM, Amadio PC, Morrey BF, Beckenbaugh RD. Proximal radioulnar synostosis after repair of distal biceps brachii rupture by the two-incision technique. Report of four cases. *Clin Orthop* 1990: 133–136.

Case 2

1. D'Alessandro DF, Shields CL. Biceps rupture and triceps avulsion. In: Jobe FW, ed. *Operative techniques in upper extremity sports injuries.* St. Louis: Mosby-Year Book; 1996:506–517.
2. Tarnsey FF. Rupture and avulsion of the triceps. *Clin Orthop* 1972; 83:177–183.
3. Farrar ELI, Lippert FGI. Avulsion of the triceps tendon. *Clin Orthop* 1981;161:242–246.
4. Klemme WR, Petersen S. Avulsion of the triceps brachii with selective radial neuropathy. *Orthopedics* 1995;18:285–287.
5. Kessler KJ, Uribe JW, Vargas L. Triceps rupture: a new mechanism of injury. *Contemp Orthop* 1994;29:134–136.
6. Bos CF, Nelissen RG, Bloem JL. Incomplete rupture of the tendon of triceps brachii. A case report. *Int Orthop* 1994;18:273–275.
7. Tiger E, Mayer DP, Glazer R. Complete avulsion of the triceps tendon: MRI diagnosis. *Comput Med Imaging Graph* 1993;17:51–54.
8. Stannard JP, Bucknell AL. Rupture of the triceps tendon associated with steroid injections. *Am J Sports Med* 1993;21:482–485.
9. Viegas SF. Avulsion of the triceps tendon. *Orthop Rev* 1990;19: 533–536.
10. Bach BRJ, Warren RF, Wickiewicz TL. Triceps rupture, a case report and literature review. *Am J Sports Med* 1984;15:285–289.
11. Morrey BF, Regan WD. Tendinopathies about the elbow. In: DeLee JC, Drez D Jr, eds. *Orthopaedic sports medicine: principles and practice.* Philadelphia: WB Saunders; 1994:860–881.
12. Anderson KJ, LeCocq JF. Rupture of the triceps tendon. *J Bone Joint Surg [Am]* 1957;39:444–446.

13. Aso K, Torisu T. Muscle belly tear of the triceps. *Am J Sports Med* 1984;12:485–487.
14. Sherman OH, Snyder SJ, Fox JM. Triceps tendon avulsion in a professional body builder. A case report. *Am J Sports Med* 1984; 12:328–329.
15. Herrick RT, Herrick S. Ruptured triceps in a powerlifter presenting as cubital tunnel syndrome. A case report. *Am J Sports Med* 1987; 15:514–516.
16. Miles JW, Grana WA, Egle D, Min KW, Chitwood J. The effect of anabolic steroids on the biomechanical and histological properties of rat tendon. *J Bone Joint Surg [Am]* 1992;744:411–422.
17. Unverferth LJ, Olix ML. The effect of local steroid injections on tendon. *Am J Sports Med* 1973;1:31–37.
18. Kannus P, Jozsa L. Histopathological changes preceding spontaneous rupture of a tendon. A controlled study of 891 patients. *J Bone Joint Surg [Am]* 1991;73:1507–1525.
19. Kelly JD, Elattrache NS. Muscle ruptures of the shoulder girdle. In: Jobe FW, ed. *Operative techniques in upper extremity sports injuries.* St. Louis: Mosby-Year Book; 1996:360–372.
20. Morrey BF. Tendon injuries about the elbow. In: Morrey BF, ed. *The elbow and its disorders.* 2nd ed. Philadelphia: WB Saunders; 1993:492–504.
21. Clayton ML, Thirupathi RG. Rupture of the triceps tendon with olecranon bursitis. *Clin Orthop* 1984;184:183–185.
22. Levy M, Goldberg I, Meir I. Fracture of the head of the radius with a tear or avulsion of the triceps tendon. A new syndrome? *J Bone Joint Surg [Br]* 1982;64:70–72.
23. Brumback RJ. Compartment syndrome complicating avulsion of the origin of the triceps muscle. A case report. *J Bone Joint Surg [Am]* 1987;69:1445–1447.
24. Whiteside JA, Andrews JR. Tendinopathies of the elbow. *Sports Med Arthosc Rev* 1995;3:195–203.
25. Lowery WD, Kurzweil PR, Forman SK, Morrison DS. Persistence of the olecranon physis: a cause of ''Little League elbow''. *J Shoulder Elbow Surg* 1995;4:143–147.
26. Murphy BJ. MR imaging of the elbow. *Radiology* 1992;184:525–529.
27. Fritz RC, Brody GA. MR imaging of the wrist and elbow. *Clin Sports Med* 1995;14:315–352.

Case 3

1. Lynch TC, Crues JVI, Morgan FW, Sheehan WE, Harter LP, Ryu R. Bone abnormalities of the knee: prevalence and significance at MR imaging. *Radiology* 1989;171:761–766.
2. Mink JH, Deutsch AL. Occult cartilage and bone injuries of the knee: detection, classification, and assessment with MR imaging. *Radiology* 1989;170:823–829.
3. Vellet AD, Marks PH, Fowler PJ, Munro TG. Occult posttraumatic osteochondral lesions of the knee: prevalence, classification, and short-term sequelae evaluated with MR imaging. *Radiology* 1991; 178:271–276.
4. Graf BK, Cook DA, De Smet AA, Keene JS. Bone bruises on magnetic resonance imaging evaluation of anterior cruciate ligament injuries. *Am J Sports Med* 1993;21:220–223.
5. Berger PE, Ofstein RA, Jackson DW, Morrison DS, Silvino N, Amador R. MRI demonstration of radiographically occult fractures: what have we been missing? *Radiographics* 1989;9:407–436.
6. Kapelov SR, Teresi LM, Bradley WG, et al. Bone contusions of the knee: increased lesion detection with fast spin-echo MR imaging with spectroscopic fat saturation. *Radiology* 1993;189:901–904.
7. Sebag G, Moore S. Effect of trabecular bone on the appearance of marrow in gradient-echo imaging of the appendicular skeleton. *Radiology* 1990;174:855–859.
8. Anderson IF, Crichton KJ, Grattan-Smith T, Cooper RA, Brazier D. Osteochondral fractures of the dome of the talus. *J Bone Joint Surg [Am]* 1989;71:1143–1152.
9. Kovalovich AM, Schweitzer ME, Wapner K, Hecht P. Occult calcaneal avulsions: possible cause of pain in recalcitrant plantar fascitis. *Radiology* 1995;197(P):156.
10. Weber WN, Neumann CH, Barakos JA, Petersen SA, Steinbach LS, Genant HK. Lateral tibial rim (Segond) fractures: MR imaging characteristics. *Radiology* 1991;180:731–734.
11. Morrison WB, Schweitzer ME, Bock GW, et al. Diagnosis of osteomyelitis: utility of fat-suppressed contrast-enhanced MR imaging. *Radiology* 1993;189:251–257.
12. Weatherall PT, Maale GE, Mendelsohn DB, Sherry CS, Erdman WE, Pascoe HR. Chondroblastoma: classic and confusing appearance at MR imaging. *Radiology* 1994;190:467–474.
13. Goldman AB, Schneider R, Pavlov H. Osteoid osteomas of the femoral neck: report of four cases evaluated with isotopic bone scanning, CT, and MR imaging [see comments]. *Radiology* 1993; 186:227–232.
14. Woods ER, Martel W, Mandell SH, Crabbe JP. Reactive soft-tissue mass associated with osteoid osteoma: correlation of MR imaging features with pathologic findings. *Radiology* 1993;186:221–225.
15. Assoun J, Richardi G, Railhac JJ, et al. Osteoid osteoma: MR imaging versus CT. *Radiology* 1994;191:217–223.
16. Moser R Jr, Kransdorf MJ, Brower AC, et al. Osteoid osteoma of the elbow. A review of six cases. *Skeletal Radiol* 1990;19:181–186.
17. Otsuka NY, Hastings DE, Fornasier VL. Osteoid osteoma of the elbow: a report of six cases. *J Hand Surg [Am]* 1992;17:458–461.

Case 4

1. Weinberger E, Shaw DWW, White KS, et al. Nontraumatic pediatric musculoskeletal MR imaging: comparison of conventional and fast-spin-echo short inversion time inversion-recovery technique. *Radiology* 1995;194:721–726.
2. Kapelov SR, Teresi LM, Bradley WG, et al. Bone contusions of the knee: increased lesion detection with fast spin-echo MR imaging with spectroscopic fat saturation. *Radiology* 1993;189:901–904.

Case 5

1. Chen J, Alk D, Eventov I, Wientroub S. Development of the olecranon bursa: an anatomic cadaver study. *Acta Orthop Scand* 1987; 58:408–409.
2. Morrey BF. Bursitis. In: Morrey BF, ed. *The elbow and its disorders.* 2nd ed. Philadelphia: WB Saunders; 1993:872–880.
3. Clayton ML, Thirupathi RG. Rupture of the triceps tendon with olecranon bursitis. *Clin Orthop* 1984;184:183–185.
4. Saini M, Canoso JJ. Traumatic olecranon bursitis: radiologic observations. *Acta Radiol Diagn* 1982;23:255–258.
5. Canoso JJ. Idiopathic or traumatic olecranon bursitis. *Arthritis Rheum* 1977;20:1213–1216.
6. Larson RL, Osternig LR. Traumatic bursitis and artificial turf. *J Sports Med* 1974;2:183–188.
7. Singer KM, Butters KP. Olecranon bursitis. In: DeLee JC, Drez D Jr, eds. *Orthopaedic sports medicine: principles and practice.* Philadelphia: WB Saunders; 1994:890–895.
8. Soderquist B, Hedstrom SA. Predisposing factors, bacteriology, and antibiotic therapy in 35 cases of septic bursitis. *Scand J Infect Dis* 1986;18:305–311.
9. Weinstein PS, Canosos JJ, Wohlgethan JR. Long-term follow-up of corticosteroid injection for traumatic olecranon bursitis. *Ann Rheum Dis* 1984;43:44–46.

Case 6

1. Keats TE. *Atlas of normal roentgen variants that may simulate disease.* 5th ed. St. Louis: Mosby-Year Book; 1992.

Case 7

1. Singson RD, Zalduondo FM. Value of unenhanced spin-echo MR imaging in distinguishing between synovitis and effusion of the knee. *AJR* 1992;159:569–571.
2. Adam G, Dammer M, Bohndorf K, Christoph R, Fenke F, Gunther RW. Rheumatoid arthritis of the knee: value of gadopentetate dimeglumine-enhanced MR imaging. *AJR* 1991;156:125–129.
3. Bjorkengren AG, Geborek P, Rydholm U, Holtas S, Petterson H. MR imaging of the knee in acute rheumatoid arthritis: synovial uptake of gadolinium-DOTA. *AJR* 1990;155:329–332.
4. Kursunoglu-Brahme S, Riccio T, Weisman MH, et al. Rheumatoid

knee: role of gadopentetate-enhanced MR imaging. *Radiology* 1990; 176:831–835.
5. Sugimoto H, Takeda A, Masuyama J, Furuse M. Early-stage rheumatoid arthritis: diagnostic accuracy of MR imaging. *Radiology* 1996;198:185–192.
6. Inglis AE, Figgle MP. Rheumatoid arthritis. In: Morrey BF, ed. *The elbow and its disorders.* 2nd ed. Philadelphia: WB Saunders; 1993: 751–766.
7. Yulish BS, Lieberman JM, Strandjorf SE, Bryan PJ, Mulopulos GP, Modic MT. Hemophilic arthritis. Assessment with MR imaging. *Radiology* 1987;164:759–762.
8. Idy-Peretti I, Le Balc'h T, Yvart J, Bittoun J. MR imaging of hemophilic arthropathy of the knee: classification and evolution of the subchondral cysts. *Magn Reson Imaging* 1992;10:67–75.
9. Baunin C, Railhac JJ, Younes I, et al. MR imaging in hemophilic arthropathy. *Eur J Pediatr Surg* 1991;1:358–363.
10. Steinbach LS, Neumann CH, Mills CM, et al. MRI of the knee in diffuse pigmented villonodular synovitis. *Clin Imaging* 1989;13: 305–316.
11. Goldman AB, DiCarlo EF. Pigmented villonodular synovitis. Diagnosis and differential diagnosis. *Radiol Clin North Am* 1988;26: 1327–1347.
12. Cobby MJ, Adler RS, Swartz R, Martel W. Dialysis-related amyloid arthropathy: MR findings in four patients. *AJR* 1991;157:1023–1027.

Case 8

1. Pritchard DJ, Unni KK. Neoplasms of the elbow. In: Morrey BF, ed. *The elbow and its disorders.* 2nd ed. Philadelphia: WB Saunders; 1993:843–859.
2. Field JH. Posterior interosseous nerve palsy secondary to synovial chondromatosis of the elbow joint. *J Hand Surg [Am]* 1981;6:336–338.
3. Fahmy NR, Noble J. Ulnar nerve palsy as a complication of synovial osteochondromatosis of the elbow. *Hand* 1981;13:308–310.
4. Milgram JW, Rogers LF, Miller H. Osteochondral fractures: mechanism of injury and fate of fragments. *AJR* 1978;130:651–658.
5. Morrey BF. Loose bodies. In: Morrey BF, ed. *The elbow and its disorders.* 2nd ed. Philadelphia: WB Saunders; 1993:860–871.
6. Milgram JW. Secondary synovial osteochondromatosis. *Bull Hosp Joint Dis* 1979;40:38–54.
7. Milgram JW. Synovial osteochondromatosis: a histopathological study of thirty cases. *J Bone Joint Surg [Am]* 1977;59:792–801.
8. Kramer J, Recht M, Deely DM, et al. MR appearance of idiopathic synovial osteochondromatosis. *J Comput Assist Tomogr* 1993;17: 772–776.

Case 9

1. Wood VE, Campbell GS. The supratrochleare dorsale accessory ossicle in the elbow. *J Shoulder Elbow Surg* 1994;3:395–398.
2. Gudmundsen TE, Østensen H. Accessory ossicles in the elbow. *Acta Orthop Scand* 1987;58:130–132.
3. Morrey BF. Loose bodies. In: Morrey BF, ed. *The elbow and its disorders.* 2nd ed. Philadelphia; WB Saunders: 1993:860–871.
4. Lawson JP. Clinically significant radiologic anatomic variants of the skeleton. *AJR* 1994;163:249–255.
5. Obermann WR, Loose HWC. The os supratrochleare dorsale: a normal variant that may cause symptoms. *AJR* 1983;141:123–127.
6. Quinn SF, Haberman JJ, Fitzgerald SW, Traughber PD, Belkin RI, Murray WT. Evaluation of loose bodies in the elbow with MR imaging. *J Magn Reson Imaging* 1994;4:169–172.

Case 10

1. Morrey BF. Loose bodies. In: Morrey BF, ed. *The elbow and its disorders.* 2nd ed. Philadelphia: WB Saunders; 1993:860–871.
2. Milgram JW. The classification of loose bodies in human joints. *Clin Orthop* 1977;124:282–291.
3. Milgram JW. The development of loose bodies in human joints. *Clin Orthop* 1977;124:292–303.
4. Quinn SF, Haberman JJ, Fitzgerald SW, Traughber PD, Belkin RI, Murray WT. Evaluation of loose bodies in the elbow with MR imaging. *J Magn Reson Imaging* 1994;4:169–172.
5. Patten RM. Vacuum phenomenon: a potential pitfall in the interpretation of gradient-recalled-echo MR images of the shoulder. *AJR* 1994;162:1383–1386.
6. Ogilvie-Harris DJ, Schemitsch E. Arthroscopy of the elbow for removal of loose bodies. *Arthroscopy* 1993;9:5–8.
7. O'Driscoll SW, Morrey BF. Arthroscopy of the elbow. Diagnostic and therapeutic benefits and hazards. *J Bone Joint Surg [Am]* 1992; 74:84–94.
8. O'Driscoll SW. Elbow arthroscopy for loose bodies. *Orthopaedics* 1992;15:855–859.
9. Bell MS. Loose bodies in the elbow. *Br J Surg* 1975;62:921–924.
10. Andrews JR, Craven WM. Lesions of the posterior compartment of the elbow. *Clin Sports Med* 1991;10:632–657.
11. Andrews JR, Timmerman LA. Outcome of elbow surgery in professional baseball players. *Am J Sports Med* 1995;23:407–413.
12. Morrey BF. Anatomy of the elbow joint. In: Morrey BF, ed. *The elbow and its disorders.* 2nd ed. Philadelphia: WB Saunders; 1993: 16–52.
13. Rosenberg ZS, Beltran J, Cheung Y, Broker M. MR imaging of the elbow: normal variant and potential diagnostic pitfalls of the trochlear groove and cubital tunnel. *AJR* 1995;164:415–418.

Case 11

1. Milgram J, Rogers LF, Miller H. Osteochondral fractures: mechanism of injury and fate of fragments. *AJR* 1978;130:651–658.
2. Mitsunaga MM, Adashian DA, Bianco AJJ. Osteochondritis dissecans of the capitellum. *J Trauma* 1982;22:53–55.
3. Ruch DS, Poehling GG. Arthroscopic treatment of Panner's disease. *Clin Sports Med* 1991;10:629–636.
4. Singer KM, Roy SP. Osteochondrosis of the humeral capitellum. *Am J Sports Med* 1984;12:351–360.
5. Morrey BF. Osteochondritis dissecans. In: DeLee JC, Drez D Jr, ed. *Orthopaedic sports medicine: principles and practice.* Philadelphia: WB Saunders; 1994:908–912.
6. Baumgarten TE. Osteochondritis dissecans of the capitellum. *Sports Med Arthosc Rev* 1995;3:219–223.
7. McManama GB, Micheli LJ, Berry MV, Sohn RS. The surgical treatment of osteochondritis of the capitellum. *Am J Sports Med* 1985;13:11–21.
8. Martin SD, Baumgarten TE. Elbow arthroscopy in sports medicine. *Sports Med Arthosc Rev* 1995;3:187–194.
9. Bauer M, Jonsson K, Josefsson PO, Linden B. Osteochondritis dissecans of the elbow. A long-term follow-up study. *Clin Orthop* 1992;284:156–160.
10. Kramer J, Stiglbauer R, Engel A, Prayer L, Imhof H. MR contrast arthrography (MRA) in osteochondrosis dissecans. *J Comput Assist Tomogr* 1992;16:254–260.
11. DeSmet AA, Fisher DR, Burnstein MI, Graf B, Lange RH. Value of MR imaging in staging osteochondral lesions of the talus (osteochondritis dissecans). *AJR* 1990;154:555–558.
12. Peiss J, Adam G, Casser R, Urhahn R, Gunther RW. Gadopentetate-dimeglumine enhanced MR imaging of osteonecrosis and osteochondritis dissecans of the elbow: initial experience. *Skeletal Radiol* 1995;24:17–20.
13. Winalski CS, Aliabadi P, Wright RJ, Shortkroff S, Sledge CB, Weissman BN. Enhancement of joint fluid with intravenously administered gadopentetate dimeglumine: technique, rationale, and implications. *Radiology* 1993;187:179–185.
14. Vispo-Seara J, Loehr JF, Krauspe R, Gohlke F, Eulert J. Osteochondritis dissecans in children and adolescents. *J Shoulder Elbow Surg* 1995;4:S21.
15. Shaughnessy WJ, Bianco AJ. Osteochondritis dissecans. In: Morrey BF, ed. *The elbow and its disorders.* 2nd ed. Philadelphia: WB Saunders; 1993:282–287.

Case 12

1. Morrey BF. Anatomy of the elbow joint. In: Morrey BF, ed. *The elbow and its disorders.* 2nd ed. Philadelphia: WB Saunders; 1993: 16–52.

2. Morrey BF. Anatomy and kinematics of the elbow. *Instr Course Lect* 1991;40:11–16.
3. Tivnon MC, Anzel SH, Waugh TR. Surgical management of osteochondritis dissecans of the capitellum. *Am J Sports Med* 1976;4:121–128.
4. Gryzlo SM. Bony disorders: clinical assessment and treatment. In: Jobe FW, ed. *Operative techniques in upper extremity sports injuries.* St. Louis: Mosby-Year Book; 1996:496–505.
5. Rosenberg ZS, Beltran J, Cheung YY. Pseudodefect of the capitellum: potential MR imaging pitfall. *Radiology* 1994;191:821–823.
6. Rosenberg ZS, Beltran J, Shankman S, Cheung Y. MR imaging of the elbow: potential pitfalls and normal variants. *Radiology* 1993;189(P):310.
7. Clark RP. Symptomatic lateral synovial fringe (plica) of the elbow joint. *Arthroscopy* 1988;4:112–116.
8. Martin SD, Baumgarten TE. Elbow arthroscopy in sports medicine. *Sports Med Arthosc Rev* 1995;3:187–194.
9. Soffer SR, Andrews JR. Arthroscopic surgical procedures of the elbow: common cases. In: Andrews JR, Soffer SR, eds. *Elbow arthroscopy.* St. Louis: Mosby-Year Book; 1994:74.

Case 13

1. Margileth AM. Cat-scratch disease. *Adv Pediatr Infect Dis* 1993;8:1–21.
2. Shinall EA. Cat-scratch disease: a review of the literature. *Pediatr Dermatol* 1990;7:165.
3. Spires JR, Smith RJ. Cat-scratch disease. *Otolaryngol Head Neck Surg* 1986;94:622–627.
4. Chen SC, Gilbert GL. Cat-scratch disease: past and present. *J Paediatr Child Health* 1994;30:467–469.
5. Hopkins KL, Simoneaux SF, Patrick LE, Wyly JB, Dalton MJ, Snitzer JA. Imaging manifestations of cat-scratch disease. *AJR* 1996;166:435–438.
6. Baron AL, Steinbach LS, LeBoit PE, Mills CM, Gee JH, Berger TG. Osteolytic lesions and bacillary angiomatosis in HIV infection: radiologic differentiation from AIDS-related Kaposi sarcoma. *Radiology* 1990;177:77–81.
7. Bergmans AMC, Groothedde JW, Schellekens JFP, et al. Etiology of cat-scratch disease: comparison of polymerase chain-reaction detection of *Bartonella* (formerly *Rochalimaea*) and *Afipia felis* DNA with serology and skin tests. *J Infect Dis* 1995;171:916–923.
8. Dong PR, Seeger LL, Yao L, Panosian CB, Johnson BL, Eckardt JJ. Uncomplicated cat-scratch disease: findings at CT, MR imaging, and radiography. *Radiology* 1995;165:837–839.

Case 14

1. Wilkins KE. Fractures of the medial epicondyle in children. *Instr Course Lect* 1991;40:3–10.
2. Harrison RB, Keats TE, Frankel CJ, et al. Radiographic clues to fractures of the unossified medial humeral condyle in young children. *Skeletal Radiol* 1984;11:209–212.
3. Peterson HA. Physeal fractures of the elbow. In: Morrey BF, ed. *The elbow and its disorders.* 2nd ed. Philadelphia: WB Saunders; 1993:248–265.
4. Inoue G. Neglected intraarticular entrapment of the medial epicondyle after dislocation of the elbow. *J Shoulder Elbow Surg* 1994;3:320–322.
5. Bede WB, Lefebure AR, Rosman MA. Fractures of the medial humeral epicondyle in children. *Can J Surg* 1975;18:137–142.
6. Patten RM. Overuse syndromes and injuries involving the elbow: MR imaging findings. *AJR* 1995;164:1205–1211.
7. Josefsson PO, Danielsson LG. Epicondylar elbow fracture in children. *Acta Orthop Scand* 1986;57:313–315.
8. Hines RF, Herndon WA, Evans JP. Operative treatment of medial epicondyle fractures in children. *Clin Orthop* 1987;223:170–174.
9. Klassen RA. Supracondylar fractures of the elbow in children. In: Morrey BF, ed. *The elbow and its disorders.* 2nd ed. Philadelphia: WB Saunders; 1993:206–247.
10. Beltran J, Rosenberg ZS, Kawelblum M, Montes L, Bergman AG, Strongwater A. Pediatric elbow fractures: MRI evaluation. *Skeletal Radiol* 1994;23:277–281.
11. Jaramillo D, Hoffer FA. Cartilaginous epiphysis and growth plate: normal and abnormal MR imaging findings. *AJR* 1992;158:1105–1110.
12. Jaramillo D, Waters PM. Abnormalities of the pediatric elbow: evaluation with MR imaging. *Radiology* 1992;185(P):137.

Case 15

1. Wirth BA. High-resolution MR imaging of the ulnar nerve within the ulnar canal: normal and pathologic appearance of postoperative changes. *Radiology* 1992;185(P):115.
2. Rosenberg ZS, Beltran J, Cheung YY, Ro SY, Green SM, Lenzo SR. The elbow: MR features of nerve disorders. *Radiology* 1993;188:235–240.
3. O'Driscoll SW, Horii E, Carmichael SW, Morrey BF. The cubital tunnel and ulnar neuropathy. *J Bone Joint Surg [Br]* 1991;73:613–617.
4. Apfelberg DB, Larson SL. Dynamic anatomy of the ulnar nerve at the elbow. *Plast Reconstr Surg* 1973;51:76–81.
5. Pecina MM, Krmpotic-Nemanic J, Markiewitz AD. Tunnel syndromes in the upper extremity. In: Pecina MM, Krmpotic-Nemanic J, Markiewitz AD, eds. *Tunnel syndromes.* Boca Raton, FL: CRC Press; 1991:29–53.
6. Osborne GV. The surgical treatment of tardy ulnar neuritis. *J Bone Joint Surg [Br]* 1957;39:782.
7. Jobe FW, Fanton GS, El Attrache NS. Ulnar nerve injury. In: Morrey BF, ed. *The elbow and its disorders.* 2nd ed. Philadelphia: WB Saunders; 1993:560–565.
8. Morrey BF. Applied anatomy and biomechanics of the elbow joint. *Instr Course Lect* 1986;35:59–68.
9. McPherson SA, Meals RA. Cubital tunnel syndrome. *Orthop Clin North Am* 1992;23:111–123.
10. Kurosawa H, Nakashita K, Nakashita H, Sasaki S. Pathogenesis and treatment of cubital tunnel syndrome caused by osteoarthosis of the elbow joint. *J Shoulder Elbow Surg* 1995;4:30–34.
11. Rosenberg ZS, Beltran J, Cheung Y, Broker M. MR imaging of the elbow: normal variant and potential diagnostic pitfalls of the trochlear groove and cubital tunnel. *AJR* 1995;164:415–418.
12. Spinner M, Linscheid RL. Nerve entrapment syndromes. In: Morrey BF, ed. *The elbow and its disorders.* 2nd ed. Philadelphia: WB Saunders; 1993:813–832.
13. Spinner M. Nerve decompression. In: Morrey BF, ed. *The elbow.* New York: Raven Press; 1994:183–208.
14. Spinner RJ, Carmichael SW, Spinner M. Partial median nerve entrapment in the distal arm because of an accessory bicipital aponeurosis. *J Hand Surg* 1991;16A:236–244.
15. Murphey MD, Johnson DL, Bhatia PS, Neff JR, Rosenthal HG, Walker CW. Parosteal lipoma: MR imaging characteristics. *AJR* 1994;162:105–110.
16. Ogino T, Minami A, Kato H. Diagnosis of radial nerve palsy caused by ganglion with use of different imaging techniques. *J Hand Surg [Am]* 1991;16:230–235.
17. Polak JF, Jolesz FA, Adams DF. Magnetic resonance imaging examination of skeletal muscle prolongation of T1 and T2 subsequent to denervation. *Invest Radiol* 1988;23:365–369.
18. Fleckenstein JL, Watumull D, Conner KE, et al. Denervated human skeletal muscle: MR imaging evaluation. *Radiology* 1993;187:213–218.
19. Shabas D, Gerard G, Rossi D. Magnetic resonance imaging of denervated muscle. *Comput Radiol* 1987;11:9–13.
20. Uetani M, Hayash K, Matosunaga N, Imamura K, Ito N. Denervated skeletal muscle: MR imaging. *Radiology* 1993;189:511–515.

Case 16

1. Conway JE, Jobe FW, Glousman RE, Pink M. Medial instability of the elbow in throwing athletes. Treatment by repair or reconstruction of the ulnar collateral ligament. *J Bone Joint Surg [Am]* 1992;74:67–83.
2. Morrey BF. Applied anatomy and biomechanics of the elbow joint. *Instr Course Lect* 1986;35:59–68.

3. Morrey BF, An KN. Functional anatomy of the ligaments of the elbow. *Clin Orthop* 1985;201:84–90.
4. Kvitne RS, Jobe FW. Ligamentous and posterior compartment injuries. In: Jobe FW, ed. *Operative techniques in upper extremity sports injuries.* St. Louis: Mosby-Year Book; 1996:411–430.
5. Fleisig GS, Andrews JR, Dillman C, Escamilla RF. Kinetics of baseball pitching with implications about injury mechanisms. *Am J Sports Med* 1995;23:233–239.
6. Joyce ME, Jelsma RD, Andrews JR. Throwing injuries to the elbow. *Sports Med Arthosc Rev* 1995;3:224–236.
7. Timmerman LA, Andrews JR. Undersurface tear of the ulnar collateral ligament in baseball players. A newly recognized lesion. *Am J Sports Med* 1994;22:33–36.
8. Timmerman LA, Andrews JR. Histology and arthroscopic anatomy of the ulnar collateral ligament of the elbow. *Am J Sports Med* 1994;22:667–673.
9. Timmerman LA, Schwartz ML, Andrews JR. Preoperative evaluation of the ulnar collateral ligament by magnetic resonance imaging and computed tomography arthrography: evaluation in 25 baseball players with surgical confirmation. *Am J Sports Med* 1994;22:26–32.
10. Schwartz ML, Al-Zahrani S, Morwessel RM, Andrews JR. Ulnar collateral ligament injury in the throwing athlete: evaluation with saline-enhanced MR arthrography. *Radiology* 1995;197:297–299.
11. Bennett JB, Green MS, Tullos HS. Surgical management of chronic medial elbow instability. *Clin Orthop* 1992;278:62–68.
12. Sugimoto H, Hyodo K, Shinozaki T, Furuse M. Evaluation of 3D fourier transform imaging for assessing throw injuries of the elbow. *Radiology* 1994;193(P):413.
13. Morrey BF. Anatomy of the elbow joint. In: Morrey BF, ed. *The elbow and its disorders.* 2nd ed. Philadelphia: WB Saunders; 1993: 16–52.
14. Jordan SE. Surgical anatomy of the elbow. In: Jobe FW, ed. *Operative techniques in upper extremity sports injuries.* St. Louis: Mosby-Year Book; 1996:402–410.
15. Gore RM, Rogers LF, Bowerman J, Suker J, Compere CL. Osseous manifestations of elbow stress associated with sports activities. *AJR* 1980;134:971–977.
16. Murphy BJ. MR imaging of the elbow. *Radiology* 1992;184:525–529.
17. Fritz RC. MR imaging of the elbow. *Semin Roentgenol* 1995;30: 241–264.
18. Linscheid RL, O'Driscoll SW. Elbow dislocations. In: Morrey BF, ed. *The elbow and its disorders.* 2nd ed. Philadelphia: WB Saunders; 1993:441–452.
19. Josefsson PO, Nilsson BE. Incidence of elbow dislocation. *Acta Orthop Scand* 1986;57:537–538.
20. O'Driscoll SW, Morrey BF, Korinek S, An KN. Elbow subluxation and dislocation. A spectrum of instability. *Clin Orthop* 1992;280: 186–197.

Case 17

1. Nirschl RP. Elbow tendinosis/tennis elbow. *Clin Sports Med* 1992; 11:851–870.
2. Coonrad RW, Hooper WR. Tennis elbow, its course, natural history, conservative and surgical management. *J Bone Joint Surg [Am]* 1973;55:1177–1182.
3. Ollivierre CO, Nirschl RP, Pettrone FA. Resection and repair for medial tennis elbow. A prospective analysis. *Am J Sports Med* 1995; 23:214–221.
4. Nirschl RP. Muscle and tendon trauma: tennis elbow. In: Morrey BF, ed. *The elbow and its disorders.* 2nd ed. Philadelphia: WB Saunders; 1993:537–552.
5. Vangsness CT, Jobe FW. Surgical treatment of medial epicondylitis. *J Bone Joint Surg [Br]* 1992;73:409–411.
6. Coonrad RW. Tendinopathies at the elbow. *Instr Course Lect* 1991; 40:25–42.
7. DeSmet AA, Fisher DR, Heiner JP, Keene JS. Magnetic resonance imaging of muscle tears. *Skeletal Radiol* 1990;19:283–286.
8. Fleckenstein JL, Weatherall PT, Parkey RW, Payne JA, Peshock RM. Sports related muscle injuries: evaluation with MR imaging. *Radiology* 1989;172:793–798.
9. Fleckenstein JL, Shellock FG. Exertional muscle injuries: magnetic resonance imaging evaluation. *Top Magn Reson Imaging* 1991;3: 50–70.
10. Fleckenstein JL, Weatherall PT, Bertocci LA, et al. Locomotor system assessment by muscle magnetic resonance imaging. *Magn Reson Q* 1991;7:79–103.
11. Speer KP, Lohnes J, Garrett WE. Radiographic imaging of muscle strain injury. *Am J Sports Med* 1993;21:89–96.
12. Resendes M, Helms CA, Fritz RC, Genant HK. MR appearance of intramuscular injections. *AJR* 1992;158:1293–1294.
13. Garrett WEJ. Injuries to the muscle-tendon unit. *Instr Course Lect* 1988;37:275–282.
14. Kannus P, Jozsa L. Histopathological changes preceding spontaneous rupture of a tendon. A controlled study of 891 patients. *J Bone Joint Surg [Am]* 1991;73:1507–1525.
15. Doran A, Gresham GA, Rushton N, Watson C. Tennis elbow. A clinicopathologic study of 22 cases followed for 2 years. *Acta Orthop Scand* 1990;61:535–538.
16. Jozsa L, Kvist M, Balint BJ, et al. Alterations in dry mass content of collagen fibers in degenerative tendinopathy and tendon rupture. *Matrix* 1989;9:140–146.
17. Regan W, Wold LE, Coonrad R, Morrey BF. Microscopic histopathology of chronic refractory lateral epicondylitis. *Am J Sports Med* 1992;20:746–749.
18. Gabel G, Morrey BF. Operative treatment of medial epicondylitis. Influence of concomitant ulnar neuropathy at the elbow. *J Bone Joint Surg [Am]* 1995;77:1065–1069.
19. Kurvers H, Verhaar J. The results of operative treatment of medial epicondylitis. *J Bone Joint Surg [Am]* 1995;77:1374–1379.
20. Brogdon BG, Crow NE. Little Leaguer's elbow. *AJR* 1960;83:671–675.
21. Patten RM. Overuse syndromes and injuries involving the elbow: MR imaging findings. *AJR* 1995;164:1205–1211.

Case 18

1. Nirschl RP. Elbow tendinosis/tennis elbow. *Clin Sports Med* 1992; 11:851–870.
2. Nirschl RP, Pettrone FA. Tennis elbow: the surgical treatment of lateral epicondylitis. *J Bone Joint Surg [Am]* 1979;61:832–839.
3. Boyd HB, McLoed AC. Tennis elbow. *J Bone Joint Surg [Am]* 1973;55:1183–1187.
4. Coonrad RW, Hooper WR. Tennis elbow, its course, natural history, conservative and surgical management. *J Bone Joint Surg [Am]* 1973;55:1177–1182.
5. Coonrad RW. Tendinopathies at the elbow. *Instr Course Lect* 1991; 40:25–42.
6. Regan W, Wold LE, Coonrad R, Morrey BF. Microscopic histopathology of chronic refractory lateral epicondylitis. *Am J Sports Med* 1992;20:746–749.
7. Doran A, Gresham GA, Rushton N, Watson C. Tennis elbow. A clinicopathologic study of 22 cases followed for 2 years. *Acta Orthop Scand* 1990;61:535–538.
8. Potter HG, Hannafin JA, Morwessel RM, DiCarlo EF, O'Brien SJ, Altchek DW. Lateral epicondylitis: correlation of MR imaging, surgical, and histopathologic findings. *Radiology* 1995;196:43–46.
9. Halpern AA, Horowitz BG, Nagel DA. Tendon ruptures associated with corticosteroid therapy. *West J Med* 1977;127:378–382.
10. Unverferth LJ, Olix ML. The effect of local steroid injections on tendon. *Am J Sports Med* 1973;1:31–37.
11. Morrey BF. Reoperation for failed surgical treatment of refractory lateral epicondylitis. *J Shoulder Elbow Surg* 1992;1:47–55.
12. Morrey BF. Surgical failure of the tennis elbow. In: Morrey BF, ed. *The elbow and its disorders.* 2nd ed. Philadelphia: WB Saunders; 1993:553–559.
13. Wittenberg RH, Schaal S, Muhr G. Surgical treatment of persistent elbow epicondylitis. *Clin Orthop* 1992;278:73–80.
14. Verhaar J, Spaans F. Radial tunnel syndrome. *J Bone Joint Surg [Am]* 1991;73:539–544.
15. Werner CO. Lateral elbow pain and posterior interosseous nerve entrapment. *Acta Orthop Scand* 1979 [Suppl] 174:1–62.

Case 19

1. Regan WD, Morrey BF. Fractures of the coronoid process of the ulna. *J Bone Joint Surg [Am]* 1989;71:1348–1354.

2. Morrey BF. Current concepts in the treatment of fractures of the radial head, the olecranon, and the coronoid. *J Bone Joint Surg [Am]* 1995;77:316–327.
3. O'Driscoll SW. Elbow instability. *Hand Clin* 1994;10:405–416.
4. Josefsson PO, Johnell O, Wendleberg B. Ligamentous injuries in dislocations of the elbow joint. *Clin Orthop* 1987;221:221–225.
5. Josefsson PO, Gentz CF, Johnell O, et al. Surgical versus nonsurgical treatment of ligamentous injuries following dislocation of the elbow joint. *Clin Orthop* 1987;214:165–169.
6. Josefsson PO, Gentz CF, Johnell O, et al. Surgical versus nonsurgical treatment of ligamentous injuries following dislocation of the elbow joint. A prospective randomized study. *J Bone Joint Surg [Am]* 1987;69:605–608.
7. O'Driscoll SW, Morrey BF, Korinek S, An KN. Elbow subluxation and dislocation. A spectrum of instability. *Clin Orthop* 1992;280:186–197.
8. Limb D, Hodkinson SL, Brown RF. Median nerve palsy after posterolateral elbow dislocation. *J Bone Joint Surg [Br]* 1994;76:987–988.
9. Hallett J. Entrapment of the median nerve after dislocation of the elbow: a case report. *J Bone Joint Surg [Br]* 1981;63:408–12.
10. Morrey BF. Radial head fracture. In: Morrey BF, ed. *The elbow and its disorders.* 2nd ed. Philadelphia: WB Saunders; 1993:383–404.
11. Linscheid RL, O'Driscoll SW. Elbow dislocations. In: Morrey BF, ed. *The elbow and its disorders.* 2nd ed. Philadelphia: WB Saunders; 1993:441–452.
12. Osborne G, Cotterill P. Recurrent dislocation of the elbow. *J Bone Joint Surg [Br]* 1965;48:340–346.
13. Nevaiser JS, Wickstrom JK. Dislocation of the elbow: a retrospective study of 115 patients. *South Med J* 1977;70:172–173.
14. Letts M. Dislocations of the child's elbow. In: Morrey BF, ed. *The elbow and its disorders.* 2nd ed. Philadelphia: WB Saunders; 1993:288–315.
15. Timmerman LA, McBride DG. Elbow dislocations in sports. *Sports Med Arthosc Rev* 1995;3:210–218.
16. O'Driscoll SW. Classification and spectrum of elbow instability: recurrent instability. In: Morrey BF, ed. *The elbow and its disorders.* 2nd ed. Philadelphia: WB Saunders; 1993:453–463.
17. O'Driscoll SW, Bell DF, Morrey BF. Posterolateral rotatory instability of the elbow. *J Bone Joint Surg [Am]* 1991;73:440–446.
18. Morrey BF. Reoperation for failed surgical treatment of refractory lateral epicondylitis. *J Shoulder Elbow Surg* 1992;1:47–55.
19. O'Driscoll SW, Morrey BF. Surgical reconstruction of the lateral collateral ligament. In: Morrey BF, ed. *The elbow.* New York: Raven Press; 1994:169–182.
20. Nestor BJ, O'Driscoll SW, Morrey BF. Ligamentous reconstruction for posterolateral rotatory instability of the elbow. *J Bone Joint Surg [Am]* 1992;74:1235–1241.

CHAPTER 3

The Wrist

Lawrence M. White, Mark E. Schweitzer, Patrick T. Liu, Andrew L. Deutsch, and Michael E. Timins

Over the past decade, the subject of magnetic resonance (MR) imaging of the hand and wrist has developed from the experimental to the practical. Clinical acceptance of the technique has paralleled improvements in MR hardware and software that now allow the routine production of virtually life-sized images of the critical anatomic structures of this region. Presently MR is commonly applied toward the assessment of a wide spectrum of hand and wrist conditions including ligamentous injuries, osteonecrosis of the carpal bones, entrapment neuropathies including carpal tunnel syndrome, soft tissue tumors, and traumatic and inflammatory conditions of tendons and synovium.

Given the complex anatomy of the region, the production of high-quality images is critical to the success of MR. In no other area has the development of local receiver coils had greater impact than in MR visualization and assessment of the ligaments of the wrist. Greater demands are placed on the receiver coil when it is utilized to image small anatomic structures at the periphery of the magnet (as in most wrist examinations) than for larger structures at isocenter. While spectacular images have been reported with custom-designed receiver coils, excellent images can be produced with readily available general purpose small parts coils (1–3).

Imaging of the wrist can be accomplished in two ways; with the patient positioned prone and the wrist placed overhead at magnet ioscenter, or the patient positioned supine and the wrist placed at the patient's side. (1–3). While the signal-to-noise ratio is usually greater when the wrist is at isocenter, many patients cannot tolerate this position for prolonged periods. For this reason, we perform most examinations with the wrist placed in a local coil by the patient's side. The coil should be positioned with the lunate in its center (1). The dorsum of the wrist should be positioned parallel to the coronal plane of the magnet to facilitate assessment of the volar and dorsal capsular ligaments. Dorsiflexion of the wrist, which will produce oblique orientation of these ligaments, is to be avoided (1). With a circular or flexible wraparound coil, cylindric padding can be placed in the patient's hand to help avoid muscle twitching, and tight padding placed around the wrist limits involuntary motion.

When circular coils are utilized (e.g, paired temporomandibular joint coils), the palm of the hand should be placed flat against the padding and the coils taped to the hand (1).

Multiple pulse sequences are available for imaging the hand and wrist. Because of the small size of the carpal ligaments, 3-DFT (3-dimensional Fourier-transform) gradient-echo sequences are frequently utilized to produce high-resolution images in a reasonable scan time. Images with in-plane resolution of 0.3 $\times$ 0.3 mm are currently obtainable, and these can be viewed interactively in planes most useful for evaluating specific ligaments (1,2). These sequences are best utilized to assess the presence or absence of a specific structure as well as its morphology. Assessment of intrinsic signal is more problematic and less reliable (1). Because contrast between normal and abnormal ligaments is limited utilizing 3-DFT gradient-echo techniques, spin-echo and fast-inversion recovery techniques are also employed for comprehensive evaluation. Presently we routinely obtain T1-weighted images in the coronal plane for general marrow assessment, coronal 3-DFT images, and axial fast spin-echo proton-density weighted images (with fat suppression) to assess the carpal tunnel and evaluate for soft tissue masses including ganglion cysts. While we frequently obtain at least one sagittal image through the wrist, this plane is primarily used for assessment of the hand and fingers.

CASE 1

History: A 26-year-old professional football player sustained an abduction injury of the thumb. What is the nature of the injury depicted?

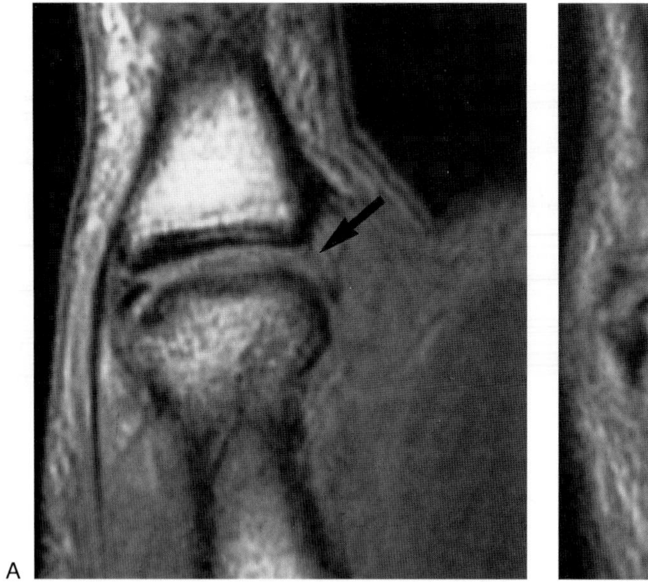
A

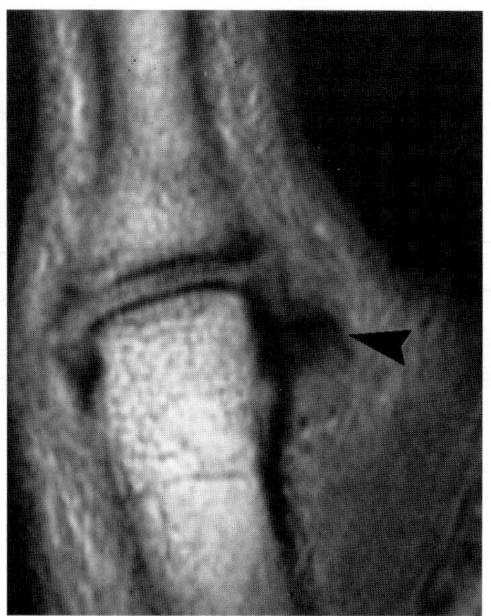
B

FIG. 1.

Findings: Oblique coronal proton-density image (TR/TE 2,000/20 msec) of the first metacarpophalangeal joint demonstrates the absence of the ulnar collateral ligament in its normally expected location (arrow) Fig. 1A. Contiguous oblique coronal section demonstrates the ulnar collateral ligament to be proximally retracted (arrowhead) and displaced external to the adductor aponeurosis (Fig. 1B).

Diagnosis: Gamekeeper's thumb complicated by the Stener lesion.

Discussion: The term *Gamekeeper's thumb* refers to the disruption of the ulnar collateral ligament (UCL) at the first metacarpal–phalangeal (MCP) joint and the resulting instability of the thumb. Originally described as an overuse injury in Scottish gamekeepers, the injury is now recognized as a more common complication of acute sports injuries than the repeated breaking of rabbits' necks (1). *Skier's thumb* may be a more appropriate name, because the most frequent cause of the injury is downhill skiing accidents (2). The usual mechanism is hyperabduction and hyperextension of the thumb of the skier who has fallen forward onto outstretched hands with ski poles still in his grasp. Other common sources of similarly violent injuries to the UCL include motor vehicle accidents and football.

Partial tears of the UCL usually heal with conservative therapy, such as immobilization. On physical examination a complete tear of the UCL can be distinguished from a partial tear by the finding of instability of the thumb at the MCP joint. Complete tears that are not properly treated will result in delayed instability. Anatomically, the UCL lies immediately adjacent to the articular capsule of the MCP joint, just deep to the aponeurosis of the adductor pollicis brevis muscle. In cases of complete rupture of the UCL in which the free ends of the torn ligament are widely separated, this aponeurosis may become interposed between the proximal and distal fragments of the UCL and interfere with proper healing.

This complication was originally described by Stener (3), and it necessitates surgical intervention. Long-term instability leads to premature osteoarthritis, an undesirable result. Neither clinical examination, stress radiography nor arthrography have been able to distinguish the presence of the Stener lesion reliably. Therefore the popular current belief among orthopedic surgeons is that all unstable thumbs should be treated surgically. As a result, the number of cases operated on and found not to have the Stener lesion is high, ranging from 48% to 85% (4).

Recently several authors have examined the effectiveness of MR imaging and ultrasound (US) in the diagnosis of UCL tears and Stener lesions. US was found to be useful when examining for the Stener lesion and for demonstrating avulsions of the UCL (5,6). US is a simple and widely available technique, but MR imaging is better for demonstrating associated injuries to cartilage, bone cortex, and marrow. In studies using anatomic correlation with cadaveric specimens and surgical correlation in symptomatic patients, MR has been shown to be accurate at depicting the normal and the torn UCL, as well as the presence or absence of Stener lesions (7–10).

The best plane for imaging the UCL is the coronal plane. It may be useful to utilize three perpendicular localizing scans to prescribe the scanning parameters adequately for imaging the thumb, because it lies in a plane that is oblique to the other digits of the hand. On MR imaging the normal UCL appears as a band-shaped structure adjacent to the medial or ulnar aspect of the MCP joint, of very low intensity on all sequences. Overlying the superficial aspect of the distal portion of the UCL is the aponeurosis of the adductor pollicis brevis. This aponeurosis also has very low signal on all sequences, is very thin, and inserts on the base of the proximal phalanx.

Incomplete tears of the UCL may display attenuation of the fibers, thickening of the ligament with intraligamentous edema that is high signal on T2-weighted images, joint effusion, periarticular soft tissue edema or hemorrhage. Complete UCL tears will initially have discontinuity of the ligament and possibly retraction of the proximal or distal ends. These tears may appear incomplete if low-signal scar tissue has filled in the gap between the torn ligament ends (10). However, if a complete tear has resulted in displacement of the proximal fragment, the Stener lesion may be present. The free end of this fragment may be found on the superficial aspect of the thin adductor aponeurosis (Fig. 1C,D). Occasionally this proximal fragment may be retracted and rolled up into a nodular, low-signal mass.

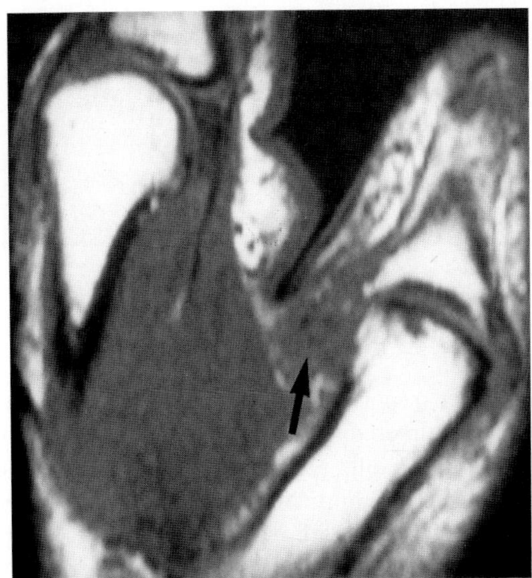

C

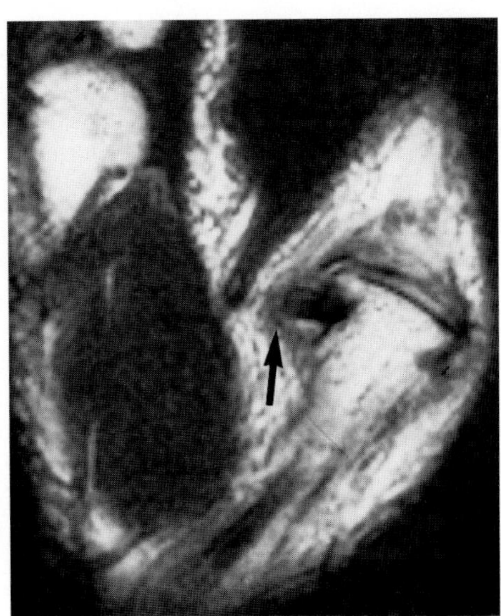

D

FIG. 1. C,D: Gamekeeper's thumb complicated by the Stener lesion. This 42-year-old man fell while skiing 1 week before presenting with the complaints of pain and swelling at the MCP joint of his thumb. On physical examination, the referring doctor found that the patient's thumb was "unstable." On the coronal T1-weighted image (TR/TE 400/16 msec) **(C)**, there is disruption and increased signal of the UCL (arrow). The proximal fragment of the ligament is displaced away from the joint, oriented perpendicular to its usual orientation, a finding best seen on the T2-weighted image (TR/TE 2,133/105 msec) **(D)**.

CASE 2

History: A 27-year-old man presented with numbness, tingling, and pain in the digits of his hand. What is your diagnosis?

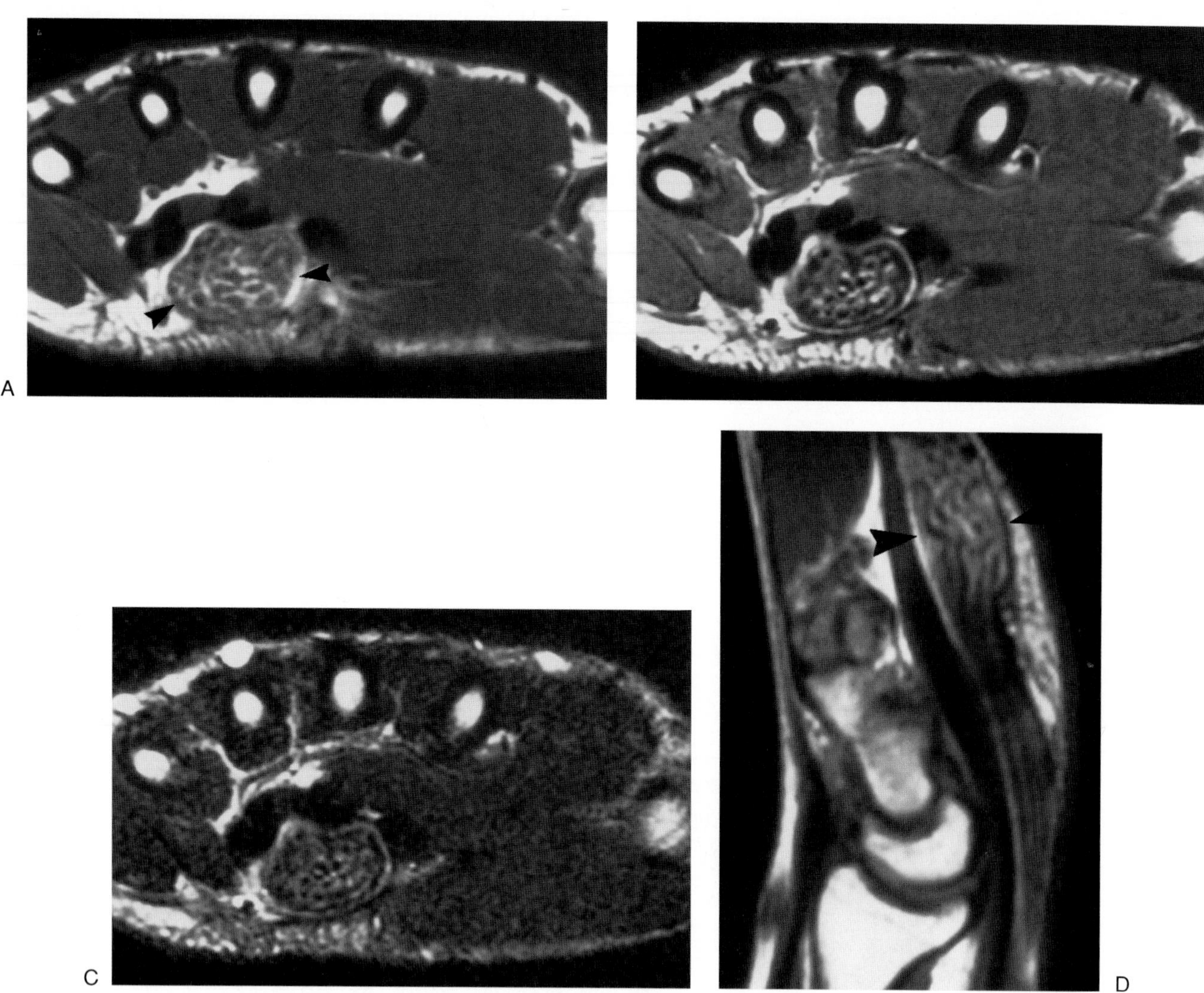

FIG. 2.

Findings: T1-weighted spin-echo (Fig. 2A) and proton-density (Fig. 2B) and T2-weighted spin-echo (Fig. 2C) axial images, as well as a T1-weighted spin-echo sagittal image (Fig. 2D), demonstrate a markedly enlarged median nerve (arrowheads) with heterogeneous signal intensity consisting of tubes of signal void (nerve fascicles and epi- and perineural fibrosis) and fat signal.

Diagnosis: Fibrolipomatous infiltration of the median nerve.

Discussion: Fibrolipomatous infiltration of the nerve (hamartoma) is a rare benign lesion that involves the median nerve in 80% of cases (1–6). Tumor involvement of the ulnar and radial nerves and dorsum of the foot has been reported (1,3,5,7). This entity is most commonly encountered in infants and less commonly in children and young adults (1,6). Early signs and symptoms are often absent or minimal. Later findings of nerve compression are evident, with pain, motor deficit, and paresthesia (1). With involvement of the median nerve, signs and symptoms of carpal tunnel syndrome may be evident. Macrodactyly of the involved body part may be present in nearly two-thirds of cases (1,3,5).

The tumor is seen as a rubbery, yellowish tan mass within the nerve sheath. Characteristic histologic findings include abundant perineural and epineural fibrosis and infiltration of mature fat cells in the interfascicular connective tissue (1). Atrophy of nerve fibers has been reported as a late finding. The tumor demonstrates a characteristic appearance on MR (1,2). The longitudinally oriented cylindric regions of signal void seen on all pulse sequences are thought to represent the nerve fascicles and accompanying epineural and perineural fibrosis. Separating these structures are areas of high signal on T1-weighted spin-echo images reflective of infiltrating mature fat cells in the interfascicular connective tissue.

The differential diagnosis of this entity includes intraneural lipoma, ganglion cyst, traumatic neuroma, and vascular malformations in which the signal voids on MR could potentially mimic the appearance of the signal voids in fibrolipomatous harmatoma (1). Treatment most frequently includes exploration, biopsy, and carpal tunnel release (1,3,4,6). Surgical excision of the primary lesion is controversial, with both satisfactory results and significant neurologic complications reported.

CASE 3

History: This 34-year-old man complained of numbness and tingling in his hand, especially at night. He had recently altered his weightlifting routine, adding extra wrist curl exercises. What abnormalities do you see on his MR images? What is the likely diagnosis?

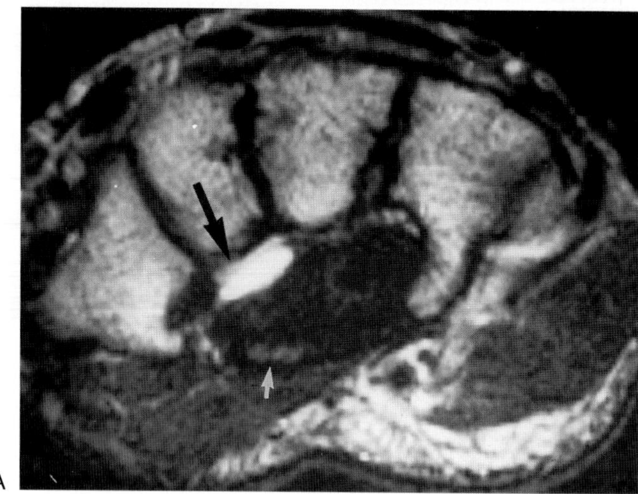

FIG. 3.

Findings: A T2-weighted spin-echo image demonstrates an ovoid, well-defined, high signal collection (large arrow) deep to the flexor tendons in the carpal tunnel (Fig. 3A). The flexor tendons are displaced and the median nerve has a flattened appearance (small arrow).

Diagnosis: Ganglion of the carpal tunnel with apparent compression of the median nerve in a patient with clinical findings of carpal tunnel syndrome.

Discussion: Carpal tunnel syndrome is typically encountered throughout a wide age range (30 to 70 year-old age group) and is more common in women than in men. The dominant hand is usually affected, but the syndrome may be bilateral in up to two-thirds of patients. It is frequently seen in occupational overuse injuries and is also the most common sports-related compression neuropathy.

Patients with carpal tunnel syndrome usually complain of paresthesias of the palmar aspects of the first through fourth digits, as well as weakness of the thenar muscles. The paresthesias are often exacerbated by motions that involve repeated flexion of the fingers or grasping. On physical examination the patients classically have positive Tinel and Phalen tests, which attempt to elicit the symptoms by percussion of the median nerve as it passes through the carpal tunnel and by forced flexion of the wrist to 90°. Thenar eminence atrophy with a weak grasp may be evident in chronic cases. Electromyographic studies usually show a delay of nerve conduction velocities localized to the wrist.

The carpal tunnel may be thought of as a fibro-osseous tunnel on the volar aspect of the wrist, through which travel the flexor tendons of the fingers from the forearm to the wrist. Four tendons of the flexor digitorum superficialis, four tendons of the flexor digitorum profundus, the tendon of the flexor pollicis longus, and the median nerve pass through the tunnel. Synovial tissue and two bursae separate the tendons and allow for their smooth gliding. Occasional anatomic variants, such as a persistent median artery or a proximal origin of a lumbrical muscle, may also be found by MR imaging in the carpal tunnel and should not be mistaken for masses (1). The confines of the carpal tunnel are stiff and relatively inflexible, giving it a nearly fixed diameter. Superficially, the tunnel is bordered by the stiff flexor retinaculum, also known as the transverse carpal ligament. The remaining three walls are bony, comprised dorsally by the carpal bones, medially by the hook of the hamate and the pisiform, and laterally by the tubercles of the scaphoid and the trapezium. Disorders that cause either an increase in the size of the contents of the tunnel or a decrease in the volume of the tunnel will result in compression of the median nerve, giving rise to the carpal tunnel syndrome.

Axial plane MR imaging, perpendicular to the orientation of the flexor tendons and median nerve, has been found to demonstrate the structures of the carpal tunnel best. The normal flexor retinaculum has very low signal intensity on all sequences and is slightly bowed, convex toward the volar aspect of the wrist. The flexor tendons also characteristically display uniformly dark signal. The median nerve usually appears round to ovoid in shape, with signal isointense to muscle on T1-weighted images, and isointense to slightly hyperintense on T2-weighted images. Its usual location is just deep to the flexor retinaculum, on the ulnar aspect of the flexor pollicis longus tendon. MR imaging has been found to be of use in demonstrating many of the various causes of median nerve compression. Most commonly, tenosynovitis of the flexor tendons is the cause of crowding of the structures in the carpal tunnel, resulting in median nerve compression. On axial images the tendons will appear to be widely separated by an excess of soft tissue, representing hypertrophied synovium. Synovium will have homogeneously intermediate signal intensity on T1- and T2-weighted images, similar to muscle. Occasionally, if the synovium is hypervascular or inflammatory, it may appear bright on T2-weighted sequences. If gadolinium is administered, synovium will enhance rapidly (2). Excess fluid in the tendon sheaths may also be seen, appearing as high signal on the T2-weighted images.

Flexor tenosynovitis is usually due to overuse, but inflammatory arthropathies, such as rheumatoid arthritis, gout, and pseudogout, can have a similar effect. Systemic conditions that may increase the volume of the contents in the carpal tunnel include amyloidosis, pregnancy, acromegaly, diabetes mellitus, hypothyroidism, and lupus. Crowding of the carpal tunnel contents and the median nerve may be also be due to focal masses such as ganglion cysts, lipomas, hemangiomas, and nerve-related tumors. These entities can be easily visualized on MR and may be clinically unsuspected. Entities that reduce the fixed volume of the tunnel may be more subtle, due mostly to shortening of the wrist along its long axis, as seen in volar segmental instability, scapholunate advanced collapse, and after distal radius fractures.

Indirect MR findings have been described in retrospective studies of patients with carpal tunnel syndrome and have been best demonstrated on axial images. They include increased palmar bowing of the flexor retinaculum, flattening of the median nerve at the level of the pisiform, swelling of the nerve at the level of the hamate hook, and high signal within the nerve on T2-weighted images. The use of median nerve swelling ratios, flattening ratios, and carpal tunnel content/volume ratios on MR has been described (3–6).

In a study analyzing data from a large cohort of patients referred for wrist MR imaging for reasons other than carpal tunnel syndrome, we have found these criteria to be insensitive and nonspecific. A significant percentage of patients with the clinical diagnosis of carpal tunnel syndrome had normal MR examinations, and a similarly large percentage of patients with unrelated symptoms displayed some of the findings said to be associated with carpal tunnel syndrome. In light of these findings, we feel that the radiologist's role when evaluating patients with a diagnosis of carpal tunnel syndrome should be to search for a possible underlying cause that may require an altered treatment regimen. If a patient has neither the symptoms nor diagnosis of carpal tunnel syndrome, MR findings of median nerve swelling or flexor tenosynovitis should not prompt the radiologist to suggest such a diagnosis.

Treatment of carpal tunnel syndrome that is not related to a mass lesion initially consists of splinting the wrist and administering nonsteroidal anti-inflammatory agents. Cases that are refractory to conservative therapy may be treated surgically by dividing the flexor retinaculum at its ulnar aspect. The MR images of the wrist after carpal tunnel surgery should demonstrate discontinuity of the retinaculum and palmar displacement of the flexor tendons and median nerve (Fig. 3B–E). If the flexor retinaculum appears continuous in the postoperative patient, the release may have been incomplete or, more likely, scar tissue may have bridged the separated ends of the retinaculum. Such an appearance is thought to be related to persistent carpal tunnel syndrome after release (Fig. 3F).

Most cases of carpal tunnel syndrome do not require MR imaging because they are relatively straightforward clinical diagnoses and treatment is conservative. MR imaging is useful in cases with atypical histories, refractory symptoms, and suspicion of a mass lesion. If surgery is being considered, then the MR image may be used to evaluate for the presence of significant flexor tenosynovitis, which, if present, may be treated with a synovectomy. MR may also be of value in assessment of patients with persistent symptoms following carpal tunnel release as well as in assessment of possible complications related to carpal tunnel surgery.

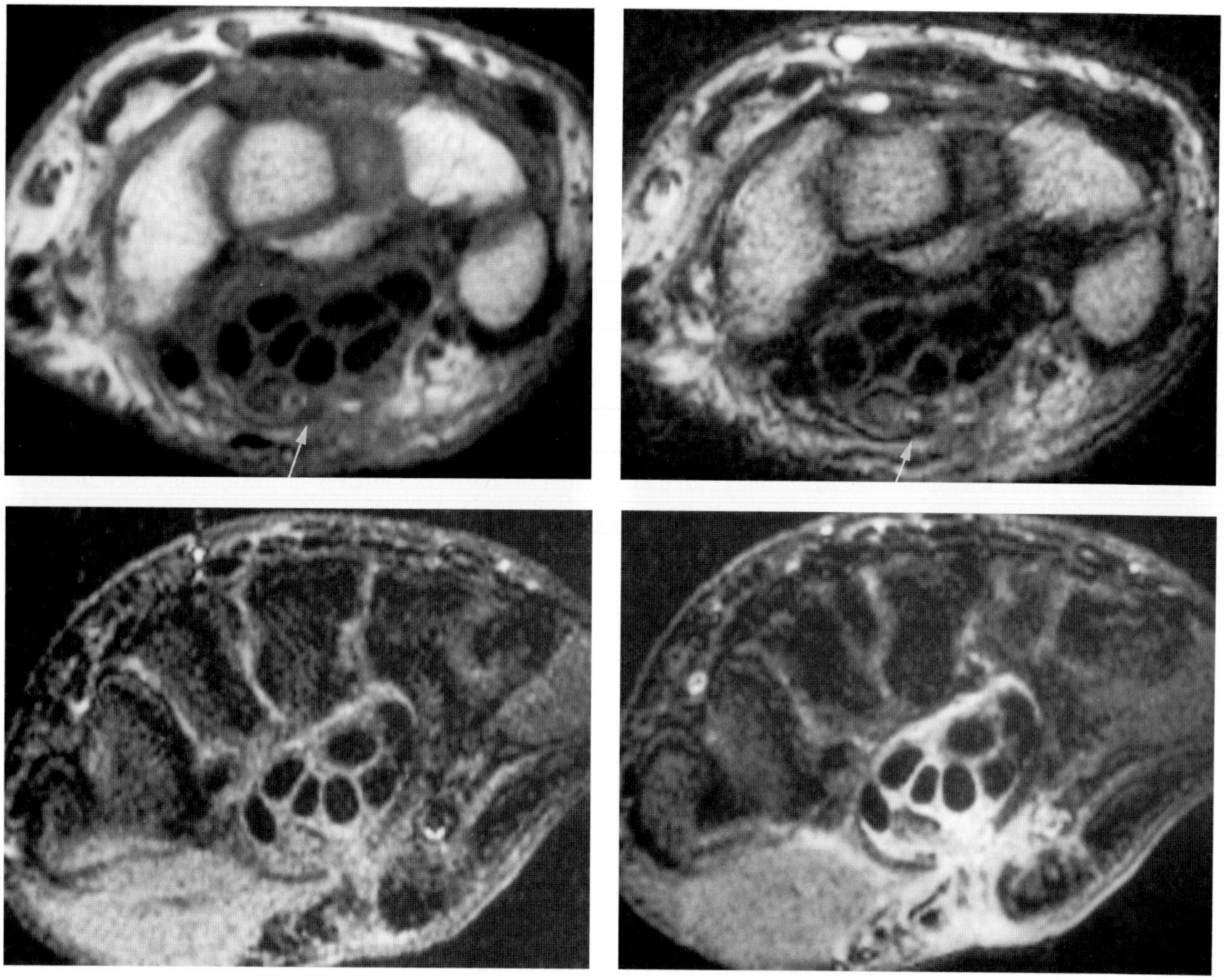

FIG. 3. B–E: T1-weighted (TR/TE 400/16 msec) image of this patient with persistent carpal tunnel syndrome 3 months after carpal tunnel release **(B)** demonstrates discontinuity of the flexor retinaculum (*arrow*) and palmar displacement of the flexor tendons and median nerve. Excess soft tissue is seen surrounding the flexor tendons **(C)**. Corresponding axial T2-weighted fast spin echo (TR/TE 3,883/92 msec) demonstrates a slight increase in intensity in the signal surrounding the flexor tendons. *Arrow* indicates discontinity of the flexor retinaculum. Axial fat-suppressed T1-weighted images (TR/TE 566/17 msec) before **(D)** and after **(E)** intravenous administration of gadopentetate dimeglumine demonstrates marked enhancement of the tissue surrounding the flexor tendons, consistent with synovitis.

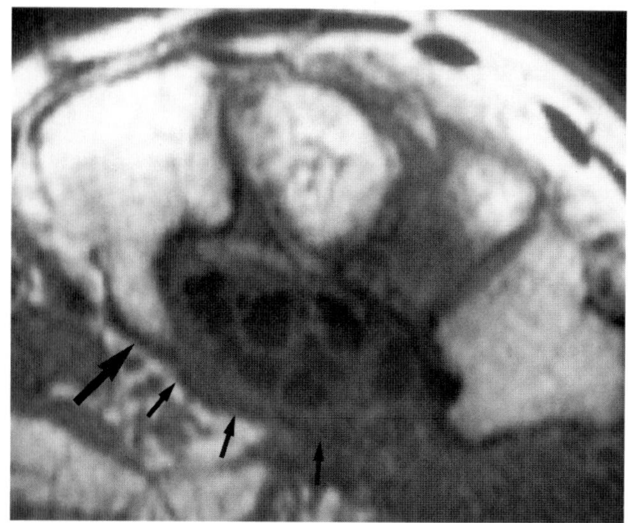

F

FIG. 3. F: The T1-weighted (TR/TE 450/15 msec) image of this patient with persistent carpal tunnel syndrome 6 months after carpal tunnel release demonstrates a normal appearance of the flexor retinaculum (*small arrows*). After release, the flexor retinaculum should have a gap at its ulnar apect, adjacent to its attachment to the hook of the hamate (*large arrow*). The palmar displacement of the flexor tendons and median nerve seen in Fig. 3B–E is not found in this patient, consistent with incomplete release or reattachment of the retinaculum from scarring.

CASE 4

History: This 36-year-old woman has had wrist pain and clicking for the past 6 months, ever since she fell while rollerblading. She neglected to seek medical attention initially, thinking she had only sprained her wrist. What does her MR image show?

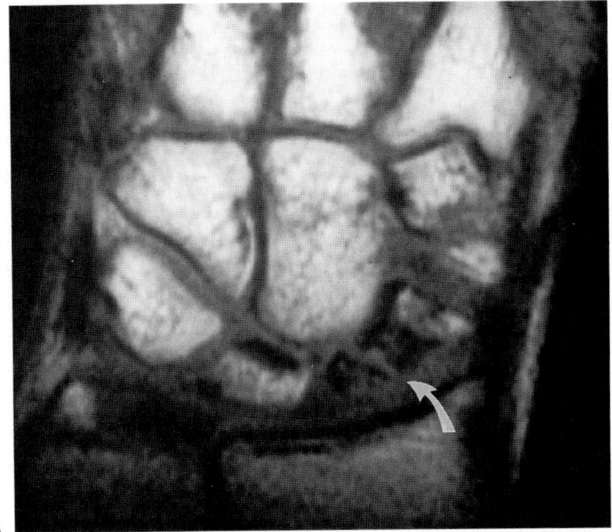
A

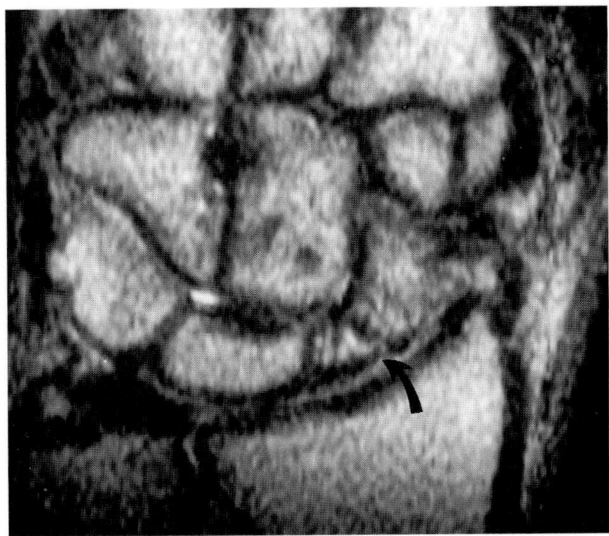
B

FIG. 4.

Findings: A fracture line (arrow) across the waist of the scaphoid, with incongruity of the fracture fragments, is seen on the T1-weighted image (TR/TE 500/16 msec) (Fig. 4A). The T2-weighted image (TR/TE 2,000/80 msec) demonstrates high signal intensity reflecting fluid filling the gap between the slightly separated fracture fragments (arrow) (Fig. 4B). Sclerosis can also be seen at the opposing faces of the fracture fragments.

Diagnosis: Nonunion of a prior scaphoid waist fracture with pseudarthrosis formation.

Discussion: Scaphoid fractures usually result from forced hyperextension of the wrist during an uncontrolled fall onto an outstretched hand. The sudden dorsiflexion force is transmitted from the hand to the wrist through the scaphoid, which is the only bony link between the proximal and distal rows of carpal bones. Since the proximal pole of the scaphoid is fixed to the radius by the strong radiocapitate and radioscaphoid ligaments, it tends to fracture on impaction with the dorsal lip of the radius. The degree of radial or ulnar deviation of the wrist at the time of injury is thought to determine the location of the fracture in the proximal, mid-, or distal portion of the scaphoid, with the waist being the most common site (1).

The great majority of scaphoid fractures heal with the appropriate treatment. Nonunion of scaphoid fractures occurs in the minority of cases, with large case series demonstrating a 95% average healing rate in uncomplicated fractures. Risk factors influencing the development of nonunion include inadequate fixation, uncorrected displacement of the fracture by more than 1 mm, angulation by more than 45°, delay of treatment by more than 4 weeks, coexisting carpal instability, avascular necrosis, and location of the fracture in the proximal pole (2,3). The poor healing of proximal pole fractures is thought to be due to the incomplete internal blood supply to the isolated proximal pole after fracture at the waist (4).

Most scaphoid fractures heal with immobilization in approximately 7 to 12 weeks. Delayed union, however, is not uncommon, and the diagnosis of nonunion should not be made until at least 6 months after the initial trauma. *Nonosseous union* has been suggested as a more appropriate term to describe this phase of fracture healing. Fracture healing may also be partial but not demonstrable on plain radiographs related to a state of fibrous union. This fibrous tissue may provide partial stability or even progress to osseous union with conservative therapy. If synovium appears in the fracture site and is bounded by a fibrous pseudocapsule, fluid is produced and a pseudarthrosis is said to have formed. These cases rarely lead to true osseous union without surgical intervention.

Scaphoid fracture nonunion has been classified into four types based on the presence or absence of displacement, carpal instability, and degenerative change on plain radiography (3). In type I nonunion there is less than 1 mm displacement and no degenerative change or instability. Type II nonunion has significant displacement or instability but no degenerative findings. (Instability is classified as a scapholunate angle greater than 70° or a radiolunate angle greater than 10°.) In type III nonunion degenerative changes are seen at the radioscaphoid articulation, with joint space narrowing, subchondral sclerosis, and osteophytosis of the radial styloid. Displacement or instability may or may not be present. Type IV nonunion is distinguished by the development of arthritis in additional carpal joints.

On MR imaging, a healing fracture will be surrounded by bright signal in the marrow on T2-weighted or short tau inversion recovery (STIR) images, representing some combination of marrow edema, hyperemia, hemorrhage, and trabecular injury. The fracture line, which initially appears dark on T1- and T2-weighted images, may persist for an indeterminate length of time. Periosteal reaction appears as a thin stripe of bright signal on T2-weighted images, paralleling the cortex. Completely healed fractures may have a fibrous scar or fat replacement of the marrow, or may even display high marrow signal on T2-weighted images at the fracture site for many years (5) (Fig. 4C,D). Nonunion should be suspected if there is no marrow continuity across a prior fracture site, or if the fracture fragments are distracted, sclerotic, or resorbed. In one study that used MR–surgical correlation, fibrous union was found to have low signal material in the fracture defect on both T1- and T2-weighted images. Fluid-type signal was found bridging the fracture site, with low intensity on T1 and high intensity on T2, in nine of ten cases of pseudarthrosis (6). It is to be emphasized, however, that granulation tissue may display similar signal characteristics, a finding contributing to difficulty in using this methodology to differentiate delayed union from pseudarthrosis (7).

Intraosseous cystic lesions at a site of prior fracture may be seen in nonunion but are a nonspecific finding since similar "pseudocysts" may be found in the early phases of normal fracture healing. These lesions may be filled with granulation or fibrous tissue and are thought to be related to post-traumatic resorption of bone adjacent to a fracture site. Post-traumatic cysts have been described in healing fractures as early as 6 weeks after injury, (8) (Fig. 4E–G). On MR imaging both cysts and pseudocysts will appear as low signal on T1- and as high signal on T2-weighted images (if they are fluid filled). The treatment of delayed scaphoid fractures (or nonunion) has proved difficult, and many techniques have been tested. Some surgeons have recommended prolonged cast immobilization with or without electrical stimulation for delayed or fibrous union. Surgical intervention may involve internal fixation, bone grafting, or removal of the nonunited proximal pole and replacement with a silicone or soft tissue spacer.

With the exception of the femoral head, the scaphoid is the most common bone in the human body to be affected by avascular necrosis (AVN) (9). The proximal pole of the scaphoid is the portion usually affected, and fracture is the most common cause. Since the proximal pole is completely covered with articular cartilage, its blood supply must originate distally. Cadaveric studies have found that branches of the radial artery, entering at the dorsal aspect of the scaphoid distal pole or waist, were responsible for the internal blood supply to 70% to 80% of the proximal pole of the scaphoid (10). Fractures at, or proximal to, the scaphoid waist may interrupt this internal vascular supply and may therefore result in osteonecrosis. Penetrating blood vessels entering at the proximal pole of the scaphoid were found in only a small percentage of wrists.

Approximately 30% of fractures at the waist of the scaphoid and nearly all fractures of the proximal 1/5th proceed eventually to AVN (11). The incidence of AVN after fractures is closely related to that of nonunion. While almost all cases of AVN are post-traumatic, an idiopathic form of scaphoid AVN has also been reported and given the name Preiser's disease (12). These cases, however, are extremely rare. AVN has also been associated with corticosteroid therapy, alcoholism, systemic lupus erythematosus, progressive systemic sclerosis, sickle cell disease, pancreatitis, and other systemic disorders.

The appearance of scaphoid AVN on MR imaging has been described as replacement of the normal fatty marrow signal with hypointense signal on both T1- and T2-weighted sequences. (Fig. 4H,I). While this is the most common pattern in AVN, the signal on the T2-weighted images can range from hypo- to hyperintense (13,14). This variability is thought to be dependent on the different degrees of edema, hemorrhage, fibrosis, cystic degeneration, and sclerosis present within the bone and is similar to the signal variability of AVN seen in the femoral head (15–17).

In late stages of AVN, the proximal pole appears collapsed, with an irregular cortical margin and possible fragmentation. The loss of articular congruity leads to osteoarthiritis of the radioscaphoid joint, with osteophyte formation and subchondral sclerosis. Cysts or pseudocysts may be seen in AVN; however, they may be simply the result of postfracture bone resorption or osteoarthritis.

The utility of MR imaging for diagnosing AVN of the carpal bones has been described in a small number of patients; these investigations have reported that the sensitivity of MR is equal to that of scintigraphy, but with higher specificity (9,13). The loss of the normal marrow signal on T1-weighted sequences was found to have the highest sensitivity, at 87.5% to 89%. Utilizing both the T1- and T2-weighted sequences, the specificity of MR imaging approached 100% in these small series. In the period since these early studies were performed, improvements in resolution and signal-to-noise ratio can reasonably be expected to have improved accuracy; however, no follow-up studies have been published.

Several investigators have examined the use of intravenous contrast enhancement in the evaluation of bone marrow perfusion (18,19). The use of contrast-enhanced MR has been studied (in the hip) with preliminary results of both experimental and clinical investigations demonstrating the ability of MR to detect differences in perfusion between femoral heads in which osteonecrosis subsequently developed and those that have remained viable (18,19). Patients in whom osteonecrosis did not develop demonstrated contrast enhancement in the affected femoral head that was similar to that of the femoral neck and shaft on the same side as well as to that of the uninvolved contralateral side. The utility of this technique, both in the hip as well as at other sites, remains to be validated in larger investigations.

Early scaphoid AVN may be reversible and can be treated conservatively (20). Surgical treatment of AVN involves insertion of a vascularized bone graft harvested from the distal radius. If degenerative changes are advanced, however, then the proximal pole can be resected or carpal arthrodesis may be performed. MR imaging can be effectively utilized for both occult carpal fracture detection and assessment of fracture complications including the development of pseudarthrosis and osteonecrosis.

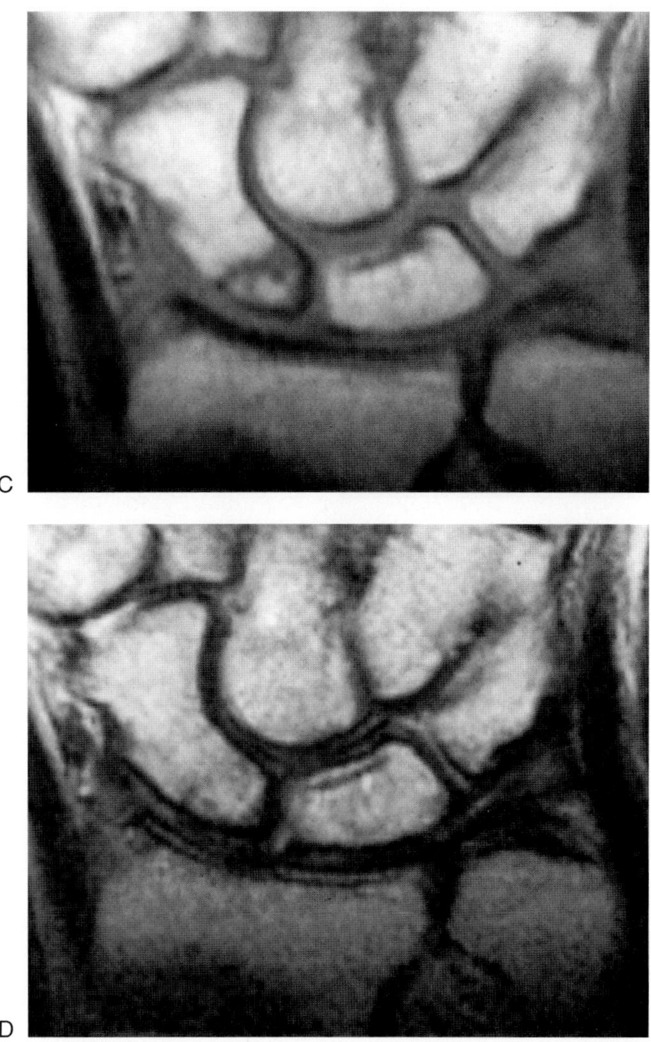

FIG. 4. C,D: Healed scaphoid fracture with complete osseous union. T1-weighted (TR/TE 500/14 msec) **(C)** and T2-weighted images (TR/TE 2,500/100 msec) **(D)** show an irregular, low signal line, probably representing a scar at the site of a fracture of the scaphoid waist.

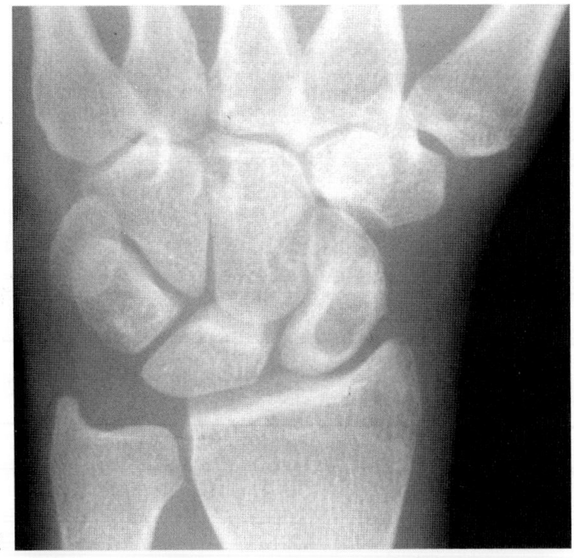

E

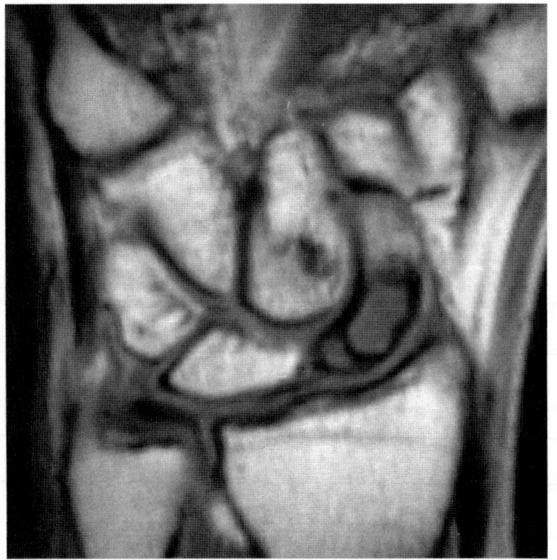

F

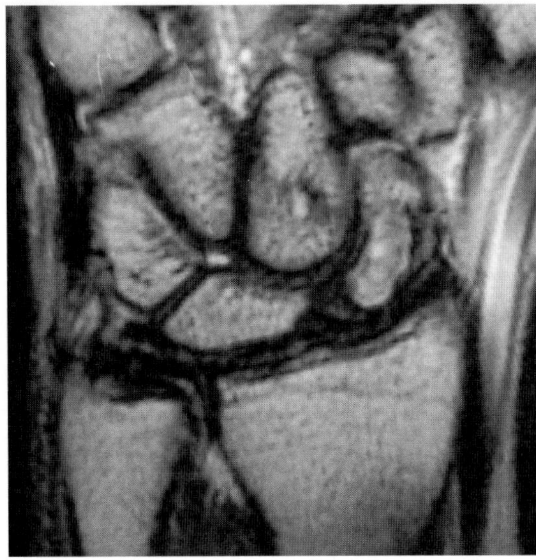

G

FIG. 4. E–G: Healed scaphoid 4 months after-fracture, with a pseudocyst. The radiograph **(E)** demonstrates a large, well-defined cyst in the mid-portion of the scaphoid surrounded by intact cortex. T1-weighted (TR/TE 500/14 msec) **(F)** and T2-weighted (TR/TE 2,500/100 msec) **(G)** images reveal that the cyst contains heterogeneous material that does not have simple fluid characteristics. At biopsy, this lesion was found to contain fibrous tissue.

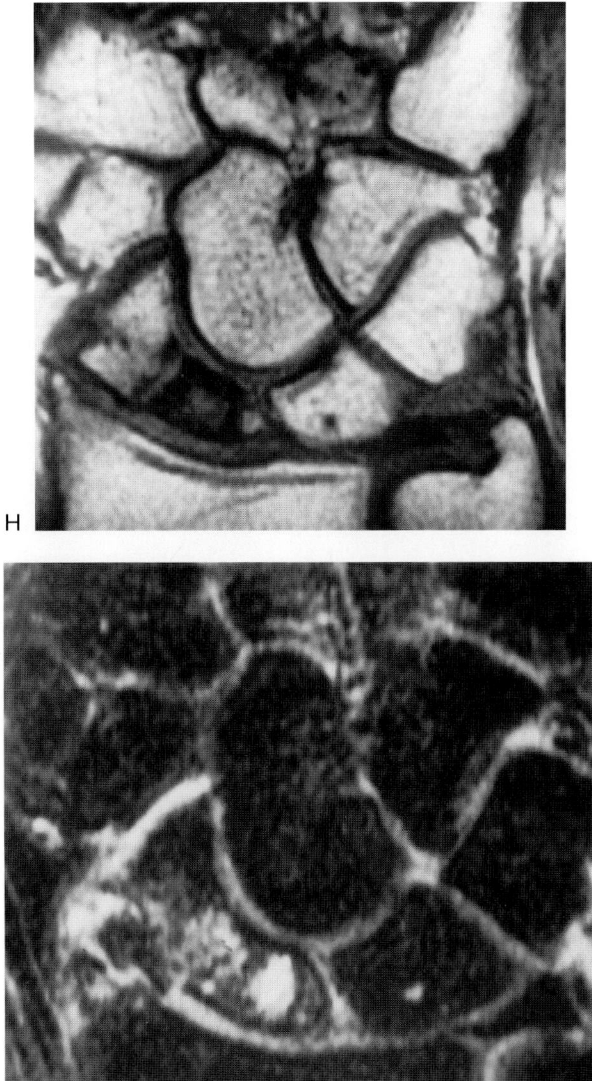

FIG. 4. H,I: Scaphoid osteonecrosis. The T1-weighted image (TR/TE 500/16 msec) **(H)** demonstrates a nondisplaced scaphoid waist fracture that has comminution of the distal pole and low signal throughout the proximal pole. The short tau inversion recovery image (TR/TE/TI 2,000/20/160) **(I)** shows high marrow signal surrounding the fracture site in the scaphoid, in the proximal pole fragment, and in the radial styloid, at the site of another nondisplaced fracture. A small cyst is also seen in the lunate.

CASE 5

History: A 33-year-old woman had chronic ulnar-sided wrist pain. What constellation of findings are illustrated in Fig. 5AB, and what if any therapeutic options are available to patients with early detection to prevent progressive soft tissue and bony complications?

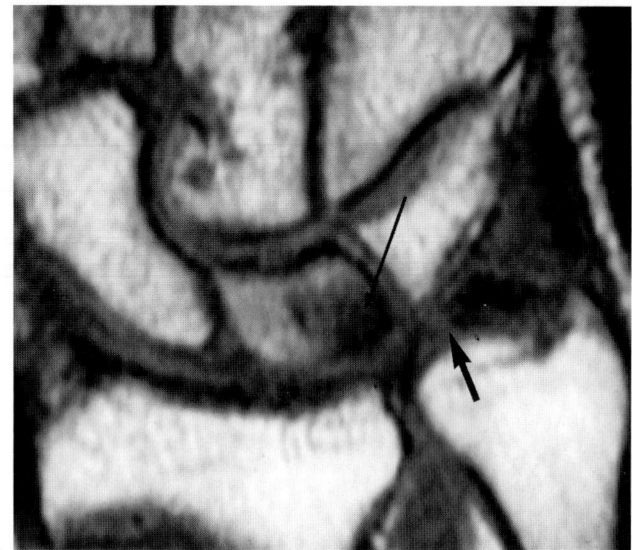

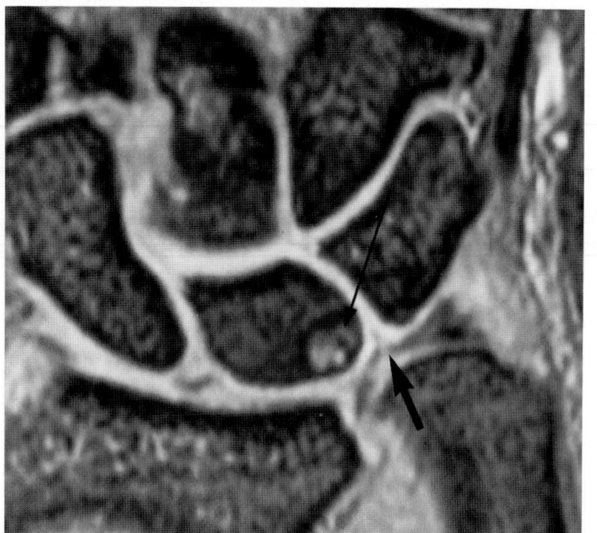

FIG. 5.

Findings: Figure 5A is a T1-weighted (TR/TE 400/16 msec) coronal MR image of the wrist demonstrating ulnar-positive variance with distal convex bowing of the triangular fibrocartilage complex (TFCC), which appears draped over the distal articular surface of the ulna. Intermediate signal intensity is seen extending through the central portion of the TFC (arrow), and low signal is illustrated within the proximal ulnar subchondral portion of the lunate (long arrow).

In Fig. 5B, a gradient recalled echo (GRE) (TR/TE/flip angle 40/14 msec/20) coronal MR image demonstrates discontinuity of the central aspect of the TFC (arrow) and a small amount of fluid within the distal radioulnar joint. Increased signal intensity is noted within the subchondral aspect of the lunate, corresponding to the low signal on the T1-weighted image (long arrow).

Diagnosis: Degenerative-type tear of the triangular fibrocartilage; ulnolunate abutment syndrome.

Discussion: The TFCC is interposed between the medial aspect of the distal radius and the base of the ulnar styloid. The TFCC is comprised of the TFC, the meniscal homologue, the dorsal and volar radioulnar ligaments, the ulnar collateral ligament, the ulnolunate and ulnotriquetral ligaments, and the tendon sheath of the extensor carpi ulnaris tendon. Biomechanically the TFCC functions to stabilize the ulnar aspect of the wrist and to absorb axial loading forces occurring secondary to flexion at the MCP and carpometacarpal joints, across the ulnar aspect of the radiocarpal joint (1).

The TFC is well evaluated on MR imaging. As with the intercarpal ligaments, the TFCC is best demonstrated on coronal images, especially with the use of three-dimensional GRE volumetric acquisitions with 1-mm slice thickness and 0.3 × 0.3 mm in-plane resolution (2–5). This resolution can be accomplished either utilizing an 8-cm field of view (FOV) and a 256 × 256 matrix or a 16-cm FOV and a 512 × 512 matrix. More recently, the authors have performed dedicated imaging of the TFC with a 4-cm FOV (Fig 5C,D). The normal TFC has classically been described as a homogeneous low-signal-intensity structure extending from the sigmoid notch of the radius to the base of the ulnar styloid process (6–9). With higher resolution imaging and using gradient-echo techniques, the TFC is seen as an inhomogeneous structure (4–5). The TFC appears to attach to the ulnar head and styloid by two different methods (5). In the high-resolution study of Totterman and Miller (4), the most common type of attachment consisted of two bands of tissue: one proximal and extending from the ulnar aspect of the TFC to the ulnar aspect of the head at the base of the styloid, and the other distal, connecting the ulnar aspect of the TFC to the tip of the styloid (Fig. 5D). The "cushion" function of the TFC between the ulnar head and carpal bones correlates well with the observation that the thickness of the TFC varies directly with the space available between the ulnar head and carpal bones (5). Negative ulnar variance is invariably associated with thicker TFCs and positive ulnar variance with thinner TFCs (5,10).

The meniscus homologue represents a band of tissue that extends from the sigmoid notch of the radius to the triquetrum and base of the fifth metacarpal. It has a common origin with the dorsal radioulnar ligament from the dorsal radius and borders the prestyolid recess ulnarly as it extends toward the triquetrum (Fig. 5D) (4,5). The opening into the prestyloid rescess may be visualized between the distal TFC attachment and meniscus homologue. The dorsal and volar radioulnar ligaments (Fig. 5C) arise from the dorsal and volar corner of the sigmoid notch of the radius and border the TFC dorsally and ventrally (4,5). They have a striated appearance on MR but, as a consequence of their oblique course, cannot be depicted along their entire course on a single coronal image (5). The ulnolunate and ulnotriquetral ligaments extend from the volar aspect of the volar radioulnar ligament to the lunate and triquetrum (Fig. 5C). Their size and extent are variable and most commonly appear on coronal MR images as a single inhomogeneous intermediate-signal-intensity structure (5).

Tears of the TFCC can occur secondary to chronic degenerative changes or as a sequela of traumatic injuries to the ulnar aspect of the wrist. The prevalence of TFC perforations in the elderly population is known to be high and is thought to be degenerative in nature (5). While tears found in younger patients are frequently symptomatic, numerous studies have attested to the fact that asymptomatic TFC tears and/or perforations are common (5,11,12). These observations further complicate interpretation of ulnar wrist pain and emphasize the need for close clinical correlation with the results of any imaging study of the region (5).

Palmer (13) classified traumatic lesions as 1A to D depending on location of the injury. Class 1A represents a 1- to 2-mm slit-like tear close to the radial attachment of the TFC. These small tears may be difficult to depict even on high-resolution studies (5). Class 1B represents traumatic avulsions of the ulnar attachment of the TFC (Fig. 5E). Class 1C lesions represent disruptions of the ulnolunate and ulnotriquetral ligaments (5,13). Class 1D lesions represent avulsions of the TFC from the sigmoid notch of the radius. Degenerative class 2 lesions are divided by Palmer (13) into five groups (A to E) depending on extent and location. With class 2A lesions, the TFC is degenerated, but intact and chondromalacic changes are noted in the lunate and triquetrum. Class 2C lesions represent central disc perforations that are ovoid in configuration. Class 2D lesions demonstrate further progression, with degenerative perforation of the lunotriquetral ligament. Class 2E lesions are characterized by generalized ulnocarpal arthrosis.

On MR, disruption of the low-signal-intensity TFC with a high-signal-intensity area is a relatively reliable sign for a tear or perforation (2–12). Signal within the central aspect of the TFC may be an asymptomatic variant, particularly on GRE sequences (11,12). Differential points in the diagnosis of a TFC tear include linearity of the abnormal signal, as well as the visualization of signal on both T1- and in particular on T2-weighted images. Visualization of a tear on T2-weighted images is a quite specific albeit insensitive indicator of a TFC tear (Fig. 5F,G). Chondromalacia (with or without accompanying subchondral edema or sclerosis) within the ulnar head may also be seen as secondary indirect features of TFCC tears, especially in those cases associated with ulnar impaction or abutment syndrome.

MR imaging can detect both the soft tissue and osseous changes seen in ulnocarpal impaction. The gross osseous relationships of the distal ulna and radius are well demonstrated on MR imaging, as are the possible secondary changes of TFCC and lunotriquetral tears, and degenerative ulnocarpal joint disease (14). Subchondral sclerosis and cyst formation associated with such degenerative disease are most frequent in the lunate, then in the triquetrum, and least commonly in the distal ulna. Cartilage abnormalities are frequently seen with MR and are best demonstrated with high-resolution GRE imaging techniques.

Ulnolunate abutment syndrome has a direct association with positive ulnar variance (14–18). Ulnar variance, or the radioulnar index, is a measure of the relative relationships of the distal articulating surfaces of the radius and ulna. Positive ulnar variance refers to the ulnar articular surface lying distal to that of the radius. Conversely, in cases of negative ulnar variance, the distal aspect of the ulna ends proximal to that of the radius. The importance of these structural relationships, whether developmental or post-traumatic in origin, lies in their association with clinically significant disorders of the wrist due to soft tissue changes and altered biomechanical forces.

Assessment of ulnar variance is greatly dependent on wrist and forearm positioning at the time of imaging. Suggested standard radiographic techniques compare the relative lengths of the distal ulna and radius with the elbow abducted 90° from the body and with 90° of elbow flexion, preventing pronation or supination of the forearm and wrist (15,16). Pronation and supination of the forearm cause the radius to cross over the ulna, shortening its apparent relative length by up to 3.5 mm. MR imaging of the wrist performed in the neutral anatomic position, however, is not possible with most MR core configurations. MR imaging of the wrist is commonly performed with the patient's arm extended over the head or at the patient's side.

Positioning in this manner invariably results in differing degrees of pronation or supination of the forearm, limiting the utility of MR imaging in the accurate assessment of ulnar variance.

With positive ulnar variance, greater than 2.5 mm, excessive compressive biomechanical forces are transmitted from the distal ulna to the lunate and triquetrum through the TFC. Such forces cause early degeneration and tears of the TFC. Subsequent rotational contact may occur between the ulna and the lunate and triquetrum, resulting in degenerative joint disease along the ulnolunate or triquetral articulations, as well as possible lunotriquetral ligament tears and secondary ligamentous instability. In the appropriate clinical setting therapeutic intervention in cases of ulnar impaction is surgical, with attempted ulnar shortening and ''unloading'' of excessive load across the ulnocarpal articulations (17–19).

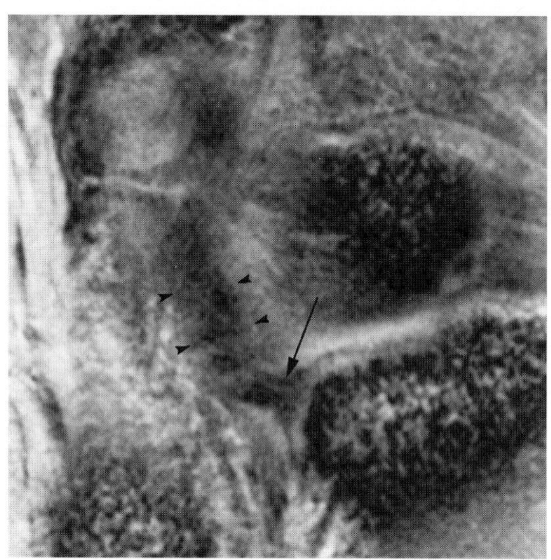

C

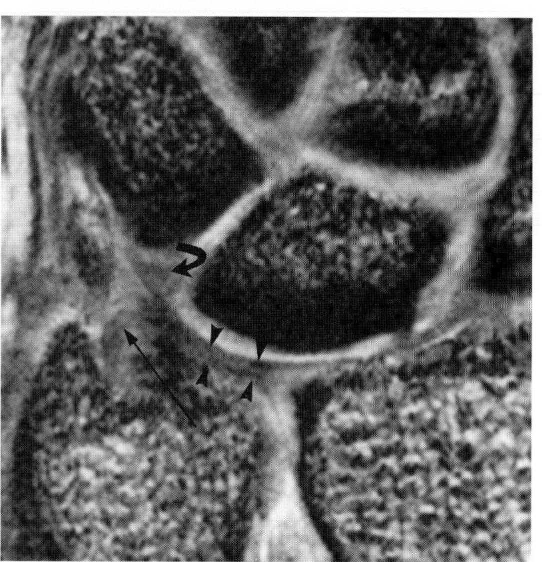

D

FIG. 5. C,D: Normal TFCC. Coronal 3D fast spoiled gradient-recalled acquisition in the steady state (GRASS) images (TR/TE/flip angle 70.9/19.6 msec/30). These images were acquired with a small (4 × 4 cm) field of view and rectangular matrix. **C.** On the most volar section the volar radioulnar ligament and its origin from the radius (*arrow*) is well demonstrated, as is the ulnotriquetral ligament (*arrowheads*) **D.** The TFC (*arrowheads*) attaches to radial articular cartilage but ends well short of the ulnar styloid. The proximal and distal attachments of the TFC to the base and tip of the ulnar styloid process are seen. The proximal attachment (*long arrow*) is formed from bands extending from the volar and dorsal radioulnar ligaments. The meniscus homologue and lunotriquetral ligament (*curved arrow*) are also demonstrable on this section.

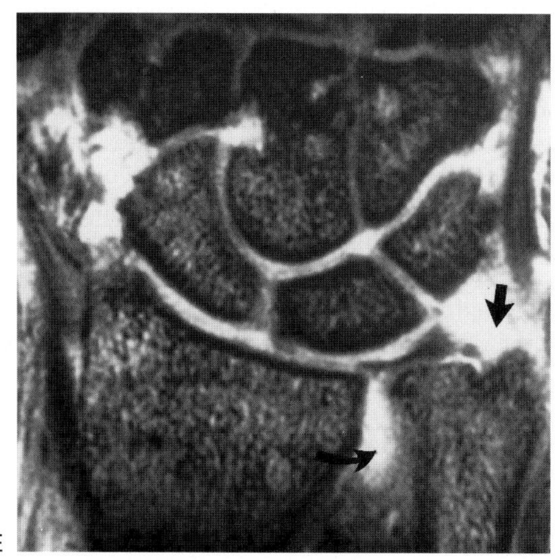

E

FIG. 5. E: Ulnar-sided TFC tear. A 35-year-old man experienced ulnar-sided wrist pain following trauma. Coronal T2*-weighted multiplanar GRASS image demonstrates full-thickness hyperintense signal along the ulnar side of the articular disc of the TFC (*arrow*) with displacement of the disc from the ulnar styloid. In addition, a moderate amount of fluid is seen within the distal radioulnar joint (*curved arrow*).

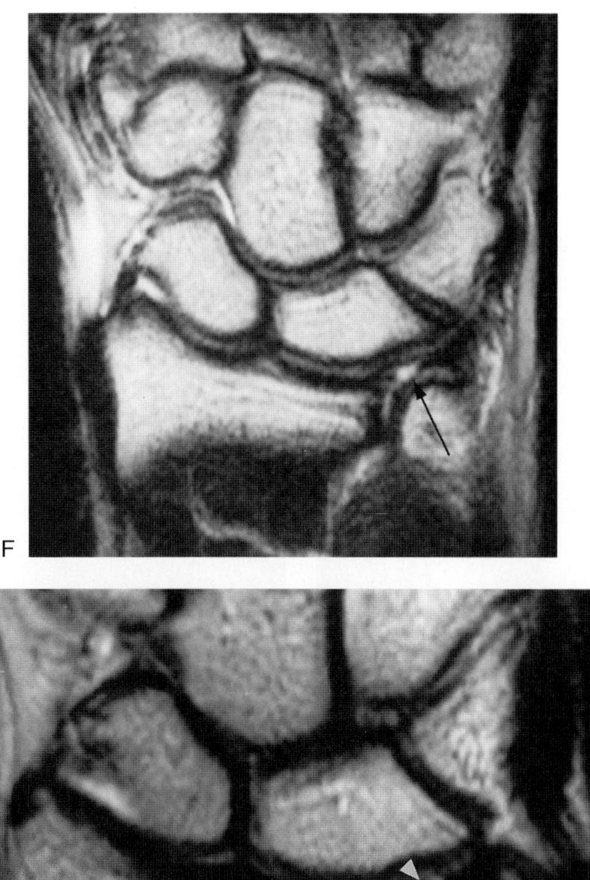

FIG. 5. F,G: Degenerative TFC tear demonstrated utilizing fast spin-echo imaging. A 45-year-old woman presented with ulnar-sided wrist pain. **F.** A T2-weighted (TR/TE 2,983/96 msec, ET 8) coronal fast spin-echo MR image of the wrist demonstrates high signal across a focal discontinuity of the central aspect of the TFC (*arrow*). **G.** On a T2-weighted (TR/TE 2,983/96 msec, ET 8) coronal fast spin-echo MR image of the carpus with fat saturation, the discontinuity of the TFC is again seen, with fluid traversing through the defect extending from the radiocarpal joint proximally into the distal radioulnar joint (*arrowheads*).

CASE 6

History: What are the abnormalities in this MR examination of a 78 year old female with bilateral chronic wrist pain and stiffness?

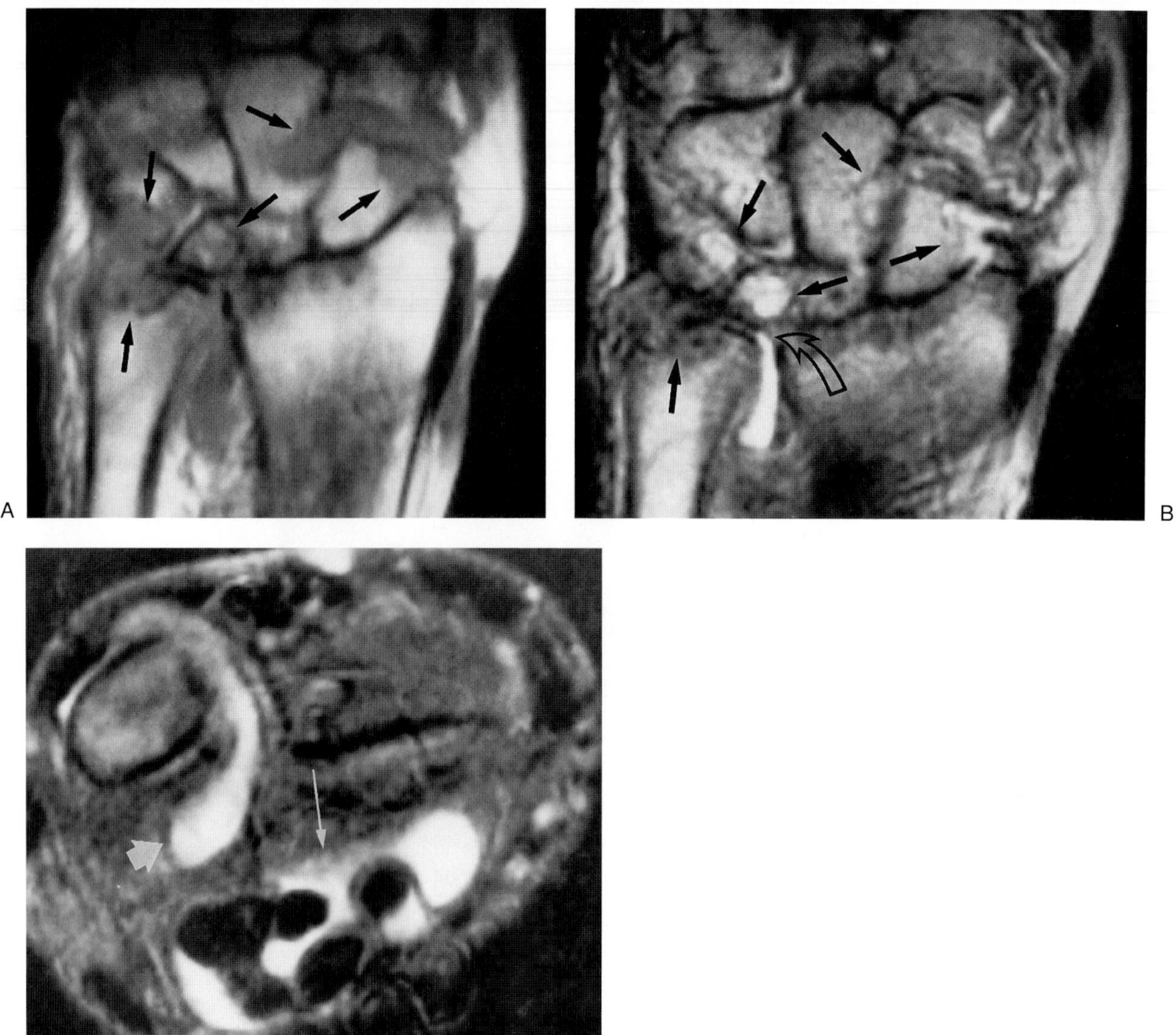

FIG. 6.

Findings: On the T1- and T2-weighted images (Fig. 6A: TR/TE 600/20 msec and Fig. 6B: TR/TE 2,500/95 msec), round cystic erosions (small arrows) are visible throughout the scaphoid, lunate, triquetrum, capitate, hamate, and ulnar styloid. The articular cartilage throughout all intercarpal and radiocarpal articulations is absent and there is subchondral sclerosis of the distal radius. A radioulnar joint effusion is communicating with the radiocarpal joint, consistent with a tear of the TFC (open arrow). On the axial, fat-saturated, T2-weighted image (Fig. 6C: TR/TE 8,300/84 msec), the radioulnar effusion is seen (short arrow), and there is fluid in the flexor digitorum tendon sheaths (long arrow).

Diagnosis: Rheumatoid arthritis with a secondary TFC tear and flexor tenosynovitis.

Discussion: The differential diagnosis for multiple cystic lesions in the wrist includes rheumatoid arthritis, juvenile rheumatoid arthritis, gout, calcium pyrophosphate deposition disease, hemophilia, amyloidosis, and pigmented villonodular synovitis. Distribution of the erosions, pancarpal articular cartilage loss, patient age, bilaterality, and the presence of tenosynovitis and a TFC tear are most consistent with rheumatoid arthritis.

In rheumatoid arthritis, inflammatory synovial pannus attacks the joints, destroying articular cartilage and fibrocartilage, subchondral bone, and the ''bare'' areas of bone at the margins of the cartilage. The resulting articular surface deformities, combined with ligamentous erosion, lead to joint instability, subluxation, and eventual dislocation. Rheumatoid arthritis in the wrist initially causes pancarpal articular cartilage loss and erosion of the ulnar styloid. The TFC and the carpal ligaments may be eroded. When pannus arises in the flexor and extensor tendon sheaths, it can cause tenosynovitis, tendonitis, and tendon ruptures. Mechanical erosion of tendons over the distorted radiocarpal and radioulnar joints also contributes to tendon ruptures. Late in the disease process, the carpal bones may undergo fibrous or bony ankylosis.

While the classical radiologic triad of rheumatoid arthritis has been described as the combination of erosions, joint space narrowing, and periarticular osteopenia, these changes are usually not visible on plain radiographs early in the course of the disease. Since the erosions of rheumatoid arthritis are irreversible, identifying the disease in its earliest stages is an important step toward maintaining maximum integrity and function of the joints.

Several studies have shown that rheumatoid arthritis can be diagnosed earlier with MR imaging than with other imaging modalities (1–3). The superior soft tissue contrast of MR allows the depiction of synovial hypertrophy and joint effusion that would not be evident on plain radiography but would represent the earliest findings of rheumatoid arthritis. It has also been demonstrated that MRI may depict more erosions of cartilage and bone than plain radiography (2,3).

On T1-weighted images, both the synovial pannus and the joint effusion will have low-to-intermediate signal characteristics, similar to skeletal muscle. On T2-weighted images, joint fluid will become high signal while the pannus may be of low,intermediate, or high signal intensity. The variability of the signal characteristics of the pannus are thought to be dependent on the degree of vascularity and the amount of fibrous tissue and hemosiderin deposited in the synovium. Active synovitis has been shown to be hypervascular and to have high signal on T2-weighted images, while burnt-out pannus is hypovascular and infiltrated with fibrin deposition (4,5).

When differentiation between fluid and synovial pannus is difficult, contrast enhancement can be helpful. Gadolinium enhancement has been studied in patients with rheumatoid arthritis, and the results have been encouraging (5–9). Reiser et al. (6) found that hypervascular pannus enhanced rapidly after bolus injection, with 123% signal increase seen at 1 minute on dynamic GRE sequences. On delayed T1-weighted images, the pannus demonstrated a signal increase of 131% over baseline. Synovial fluid enhanced only minimally, with an average 15% signal increase on similar images (6). These enhancement characteristics were also easily perceptible qualitatively (Fig. 6D,E).

These investigators also found that the cystic bone erosions in patients with rheumatoid arthritis demonstrated enhancement with contrast injection, indicating that they contain synovium and are not true cysts (7). However, as with the synovium throughout the rest of the body, enhancement may be variable. The signal characteristics and enhancement pattern of these cysts may be those of fibrous pannus in ''burnt-out'' areas of rheumatoid arthritis, appearing as low to intermediate signal on T2-weighted images without significant contrast enhancement.

MR imaging with gadolinium enhancement can also be used to follow the activity of rheumatoid arthritis (10). Clinical assessment and serum rheumatoid factor levels have been found to be relatively insensitive in evaluating the effectiveness of anti-rheumatoid drug therapy. The amount of enhancing synovium present within the joint and its vascularity on MR images can provide a more accurate measure of disease activity.

MR has not been found to be useful for predicting extensor tendon rupture in the wrist (11). While synovitis could be visualized within the tendon sheaths, there was no correlation between the degree of tendon sheath swelling and the clinical risk of tendon rupture. It should also be noted that the presence of proliferative synovium is a nonspecific finding and may be secondary to other inflammatory or infectious arthridites. All available clinical information and laboratory studies should be evaluated before the diagnosis of rheumatoid arthritis can be made.

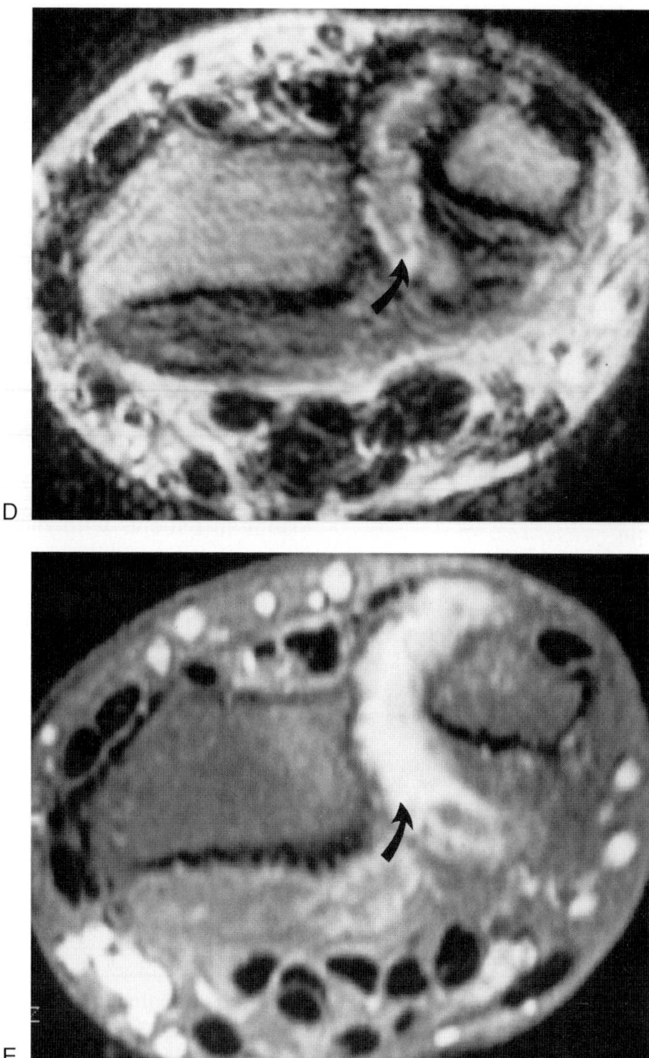

FIG. 6. D: On the T2-weighted axial image of a 53-year-old woman with rheumatoid arthritis, an effusion is seen in the radioulnar joint. **E:** Within the effusion is nodular, intermediate-signal tissue that enhances on the post-intravenous gadolinium-enhanced, fat-saturated, T1-weighted image [TR/TE 450/12 msec] image. This tissue represents hypervascular synovial pannus that is contained within and lining the joint capsule.

CASE 7

History: A 29-year-old male professional baseball player presented with complaints of persistent ulnar-sided wrist pain and symptoms of ulnar neuropathy. What are the findings demonstrated in Fig. 7A–C? Do the findings in Fig. 7A,B suggest a possible cause for the patient's ulnar neuropathy?

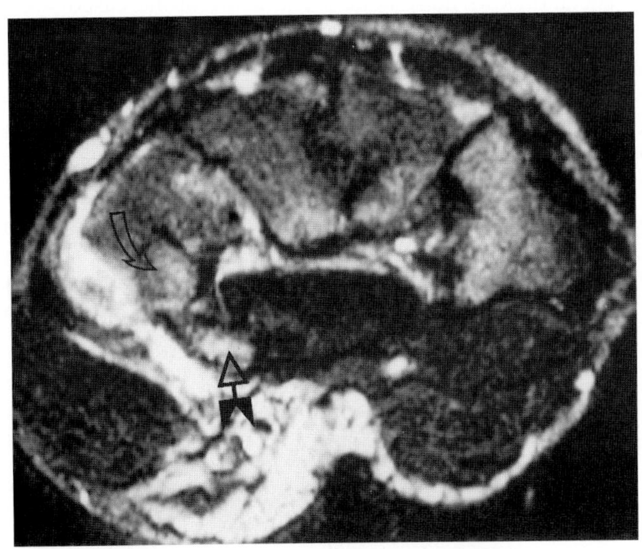

A

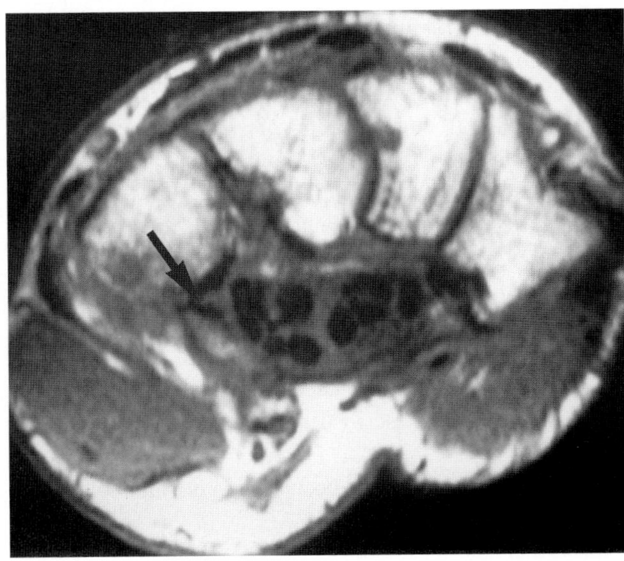

B

C

FIG. 7.

Findings: Figure 7A shows an axial T2-weighted (TR/TE 2,983/96 msec, ET 8) fast spin-echo MR image through the carpus at the level of the carpal tunnel. Heterogeneous high signal intensity is seen within the hook of the hamate and palmar aspect of the body of the hamate bone (open curved arrow). Extensive high signal intensity is demonstrated within the soft tissues along the ulnar palmar aspect of the wrist with extension into Guyon's canal surrounding the ulnar nerve (arrow).

Figure 7B is an axial T1-weighted (TR/TE 400/16 msec) MR image at the same location as Fig. 7A illustrating linear parallel low-signal bands across the base of the hook of the hamate (arrow). Intermediate to low signal is seen, corresponding to both the intra- and extraosseous regions of high T2-weighted signal in Fig. 7A.

Figure 7C is a sagittal T1-weighted (TR/TE 400/16 msec) MR image through the hamate, again demonstrating low-signal bands across the base of the hook of the hamate (arrow) with adjacent low-to-intermediate intraosseous signal.

Diagnosis: Stress fracture of the hook of the hamate.

Discussion: MR imaging has proved to be a highly sensitive and potentially cost-effective modality in the early evaluation of patients for possible occult fractures (1). By definition, such injuries are not detectable with conventional radiographs and may manifest as continued pain and/or loss of function following a single traumatic episode. A spectrum of injury exists following trauma including responses to altered biomechanics, bone bruises, occult fractures, and "true" fractures (in order of increasing severity). While bone bruises may be seen in the knee without significantly altering patient management, the use of the term *bone bruise* in the wrist may not be appropriate (2,3). It is our experience that such injuries seen in the bones of the hand and wrist, like those seen in the ankle and foot, may be the cause of significant pain and disability. The natural history of these injuries necessitates casting for adequate healing and such findings should be described and treated as occult fractures (4).

Stress fractures are those occurring in normal bone subject to minor repetitive injury usually as a result of new or different repeated loading or strenuous activity, as may be seen in athletic individuals. In contrast to occult or "true" fractures, patients with stress fractures have no history of a distinct traumatic episode. The hook of the hamate is a classic site of stress injuries in the wrist, typically occurring in tennis players, golfers, and baseball players. Theoretically such injuries occur secondary to accelerated bone remodeling at a site of increased stress, weakening the underlying bone and subjecting it to fracture with continued repetitive stress.

The utility of imaging in the management of occult and stress fractures lies in early detection to guide appropriate therapy, as well as to prevent further progression and complications of the osseous insult. MR imaging of these injuries includes findings of isolated focal or diffuse osseous marrow edema and hemorrhage manifest as areas of low T1- and high T2-weighted signal intensity. Fast STIR imaging is particularly sensitive in the detection of such marrow changes. Occasionally a distinct fracture line may be seen as a single or paired parallel intraosseous low-signal-intensity bands continuous with the bony cortex, with or without surrounding edema and hemorrhage (1,5) (Fig. 7D,E).

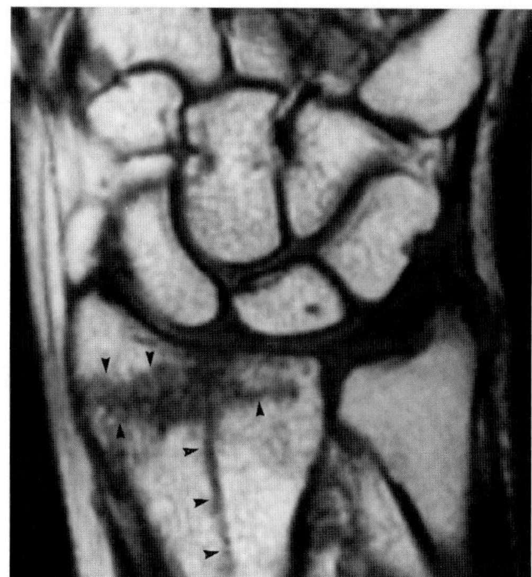

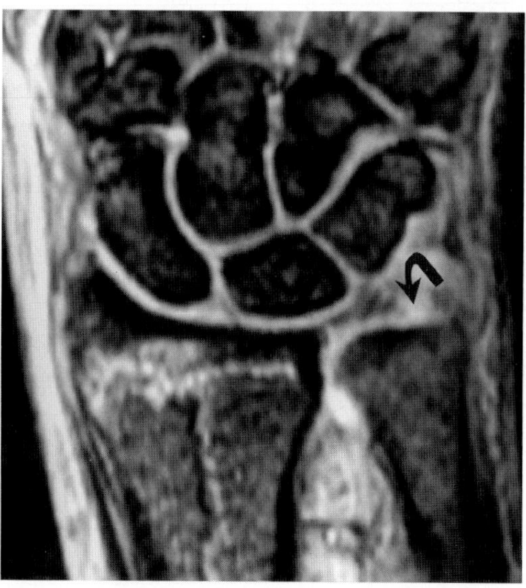

FIG. 7. D,E: Occult distal radial fracture. A 42-year-old advertising executive was referred for MR imaging by a hand surgeon to exclude a possible navicular fracture. Plain radiographs were normal. **D:** On the coronal T1-weighted spin-echo image, a striking band of low signal intensity is seen paralleling the articular border of the distal radius (*small arrowheads*). In addition, a second well-defined low-signal-intensity line is seen paralleling the long axis of the radius (*small arrowheads*). The navicular bone is normal. **E:** A coronal gradient-echo sequence again demonstrates the findings, which have increased signal intensity. In addition, fluid is seen within the inferior radioulnar joint and on this and adjacent images (not shown), there was evidence of avulsion of the ulnar attachment of the TFC (*curved arrow*).

CASE 8

History: A patient presents with chronic pain along the radial aspect of the wrist. What is the finding? Can it be clinically confirmed? What associated conditions may be seen with this disorder?

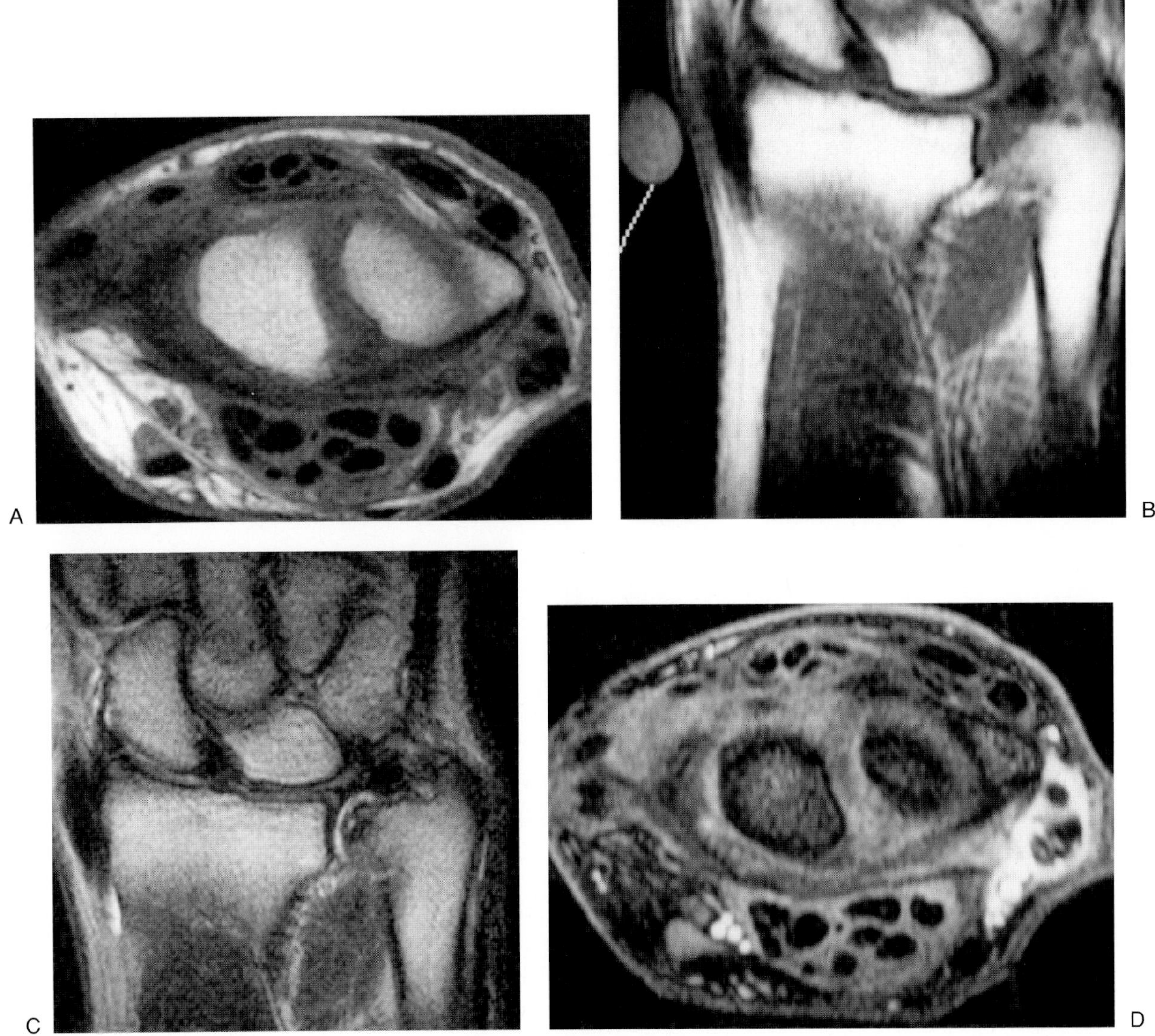

FIG. 8.

Findings: Figure 8A presents a T1-weighted axial image demonstrating a loss of adjacent fat between the first extensor compartment and the skin. In Fig. 8B, a coronal T1-weighted image demonstrates the same finding as well as poor definition of the margins of the tendon when compared with the adjacent infiltrated fat. In Fig. 8C, coronal T2-weighted imaging demonstrates a small amount of high signal around the tendon in the region of the tendon sheath. This signal is slightly lower than that typically seen with synovial fluid and has somewhat more ill-defined margins than those seen with synovial fluid. Figure 8D gives an axial T1-weighted fat-suppressed image following intravenously administered gadolinium; intense enhancement of the entire region of the synovial sheath of the first extensor compartment is demonstrated, with poor definition of the margin of the tendon sheath similar to that seen in Fig. 8B.

Diagnosis: De Quervain's stenosing tenosynovitis.

Discussion: This is an analogous disorder to peroneal stenosing tenosynovitis in the ankle and adhesive capsulitis, seen in the shoulder. De Quervain's disease is characterized by synovial hypertrophy and fibrosis. This is a sequela of chronic inflammation, and consequently rheumatoid arthritis is a risk factor for this disorder. The most frequent cause of de Quervain's disease is overuse injuries or as a sequela of trauma subsequent to hemorrhage within the synovial sheath with resultant hypertrophy and fibrosis.

De Quervain's disease is usually a relatively straightforward clinical diagnosis. Findings are consistent with synovitis, which appears on imaging as somewhat inflammatory. ''Nonspecific'' (simple) synovitis manifests on MR by fluid in tendon sheaths. It should be noted, however, that the appearance of fluid in tendon sheaths varies considerably, particularly in the first extensor compartment, where visualization of fluid is not definitive evidence of synovitis or de Quervain's disease. Synovitis of the II to VI extensor compartment is less variable and therefore more specific, but this also may be a normal variant. Younger patients as well as athletic patients without symptoms tend to have more visible synovial fluid. With this factor in mind, some type of inflammatory arthropathy should be suspected in any patient who demonstrates synovitis in II to VI extensor compartments. Although synovitis of the first compartment is also associated with inflammatory arthropathies, this is less frequent than in compartments II to VI (1).

Synovitis may be ''inflammatory'' or noninflammatory. The degree of enhancement may correlate with clinical symptomatology (1). Most cases of inflammatory synovitis demonstrate enhancement. Tendon sheath enhancement is less typical in noninflammatory synovitis. Synovial appearance is also an important diagnostic criteria for inflammatory synovitis. As demonstrated in this case, inflammatory synovitis may give slightly lower than fluid signal on T2-weighted imaging. This is a manifestation of synovial hypertrophy and/or fibrosis. In addition, rheumatoid arthritis leads to bleeding and hemosiderin deposition in the synovium. Ferritin deposition has been associated with disease activity.

In this case, loss of the adjacent fat pad is a rather characteristic finding for de Quervain's disease. This condition typically enhances profusely and is usually bright on T2-weighted images although there is variability. The perisynovial enhancement is also a frequent finding in stenosing tendon disorders (2).

CASE 9

History: A patient presents with marked soft tissue swelling of the hand and wrist. Are the findings consistent with the clinically suspected cellulitis or is osteomyelitis present?

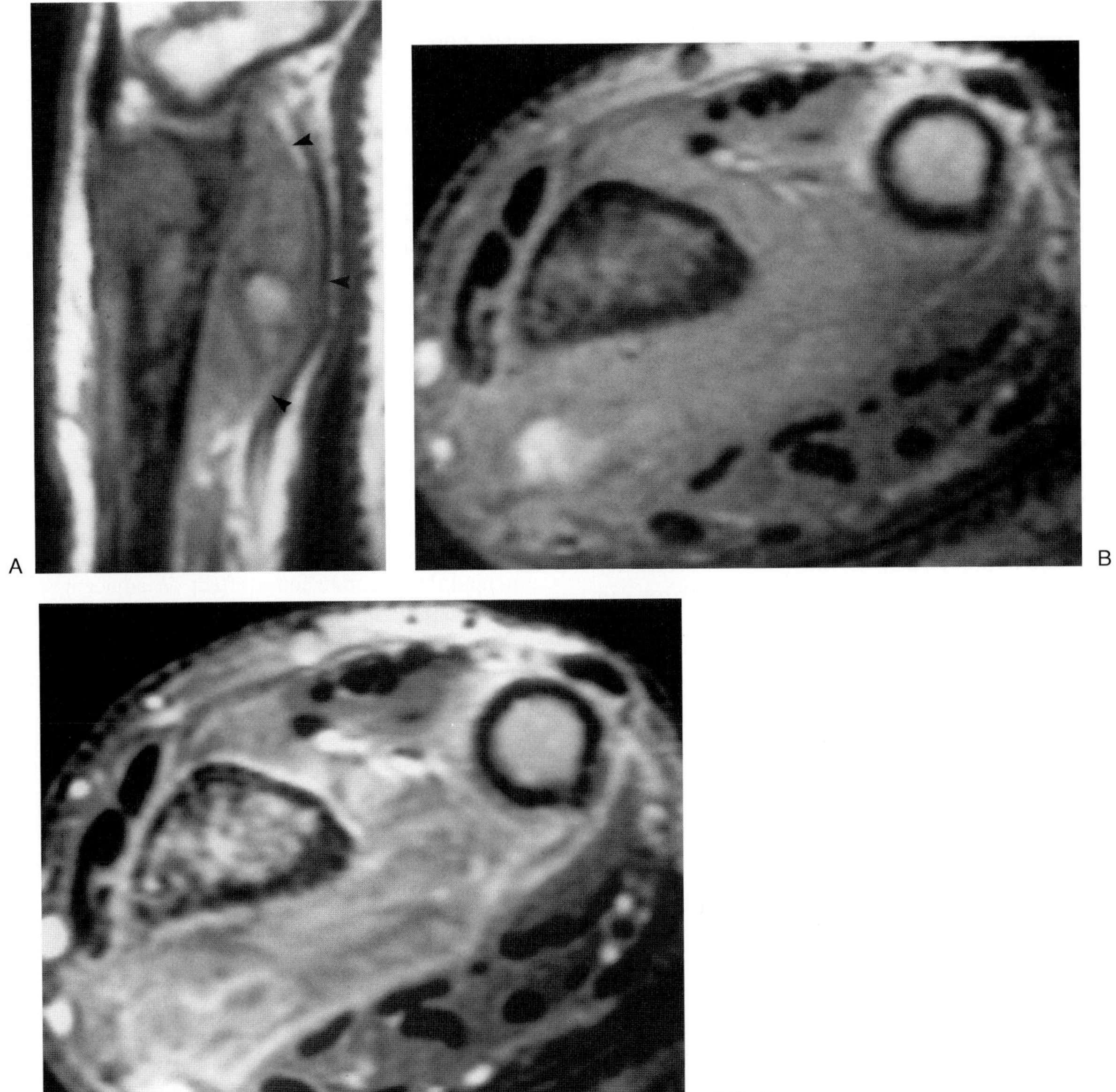

FIG. 9.

Findings: Figure 9A shows a sagittal T1-weighted image of the distal radius demonstrating diffuse marrow replacement with an adjacent predominantly subperiosteal mass (arrowheads). Figure 9B demonstrates a precontrast axial T1-weighted fat-suppressed image. There is low signal intensity within the radius and a large intermediate-signal-intensity mass extending along the volar aspect of wrist and through the interosseous membrane. In Fig. 9C, a postcontrast enhanced axial T1-weighted fat-suppressed image demonstrates, intra- and extrosseous heterogeneous enhancement with extension through the interosseous membrane and periosteal and marrow enhancement.

Diagnosis: Osteomyelitis with soft tissue extension.

Discussion: Infection in the hand and wrist is usually secondary to direct inoculation. The most susceptible areas are the tufts and the MCP joints. Infection typically begins in the soft tissues. It can then extend into adjacent joints or osseous structures (1).

In cases of septic arthritis the infection is usually hematogenous. Patients with inflammatory arthropathies, particularly rheumatoid arthritis, have an increased susceptibility to septic arthritis. Occasionally septic arthritis can occur secondary to spread of an adjacent soft tissue infection (2).

Early soft tissue infection has the appearance of edema. When it occurs from inoculation, it is typically seen first in the subcutaneous fat. Early infections demonstrate poorly defined margins that become well defined over time. The most frequent cause of hand infections is trauma, human bites, and drug abuse (3). Most are secondary to aerobic organisms (*Staphylococcus* in 35%) (3). Infections respect compartment/fascial boundaries and if they begin subcutaneously rarely extend into the muscle. Likewise, if they begin in muscles they only infrequently extend to the subcutaneous tissue or marrow. Nevertheless, there may be adjacent reactive changes such as edema in the subcutaneous fat with deep infections and muscle edema with superficial infections. This edema is sympathetic and is not a manifestation of infection in the adjacent area. Deep web space infections are rare, but they tend to spread within compartments (radial space, hypothenar space, web space, midpalmar space, thenar space). Infections from the thenar and hypothenar space may spread proximally because of communication with the space of Parona in the forearm (1).

Infection of muscle itself is uncommon. When seen, it may be a manifestation of neurologic abnormalities and subsequent unknown inoculations. Otherwise, an immunosuppressive disorder should be suspected such as the acquired immunodeficiency syndrome, diabetes, or immunosuppression from chemotherapy or associated with neoplasms elsewhere. These muscle infections may present as a suspected neoplasm. On imaging it may be difficult to differentiate them from neoplasms since infections may have infiltrative margins and demonstrate a high degree of biologic aggressiveness similar to malignant neoplasms.

Chronic infections may become encapsulated and form abscesses. When an abcess is present it will be high signal on T2-weighted imaging with rim enhancement on gadolinium-enhanced imaging. This rim is usually irregular and thick. If postcontrast imaging is delayed, the entire abscess may enhance so delayed images showing complete enhancement do not allow a confident exclusion of a neoplasm. Any infection about the nail whether encapsulated or not, may also spread into the adjacent bone because of the association of the periosteum with the nail bed (4). With chronicity, abscesses may become large and erode bone (5).

Infections may begin in the bone and spread to adjacent soft tissues (6). When they occur in the wrist, they tend to spread along tendon sheaths. When they occur in the tuft secondary to paronychia, the infection is isolated by a separate fascial sheath, deep to the subcutaneous fat. Because of mass affect, the infection may then block venous drainage, leading to soft tissue and bone infarction. This combination of infarction and infection is termed a felon (5).

Septic arthritis usually presents as fluid in joints. A small amount of fluid in the midcarpal and distal radioulnar joint is normal (7). Fluid in other locations and in excessive volume should be suspected to be pathologic. Synovial thickening may occur with septic arthritis. Synovial thickening on MR is demonstrated by visible synovium. The synovium demonstrates variable enhancement following contrast administration and may be low or high on nonenhanced T1- or T2-weighted images. Cartilage erosions are frequently seen, but high-resolution images are necessary to evaluate the cartilage.

Osteomyelitis usually, but not always, demonstrates low signal on T1-weighted images and almost always increases in intensity on T2-weighted images, particularly with fat suppression, and on STIR images. An adjacent soft tissue reaction is a nearly ubiquitous MR finding, as is the appearance of an adjacent soft tissue mass. Soft tissue enhancement can be differentiated from soft tissue mass because soft tissue masses have better defined margins and demonstrate mass effect on adjacent skin, tendons, and particularly musculature. Ulcers and sinus tracts are helpful signs to confirm the diagnosis of osteomyelitis (8). Sinus tracts are more common in the upper extremity than ulceration.

When contrast is administered, cases of osteomyelitis demonstrate marrow and soft tissue enhancement. If the images are delayed, sympathetic edema around the area of osteomyelitis may enhance. Sinus tracts and active (not scarred or chronic) ulcers also demonstrate enhancement. Periosteal reaction may also be recognized on MR imaging. The thickness of this reaction is typically related to its biologic aggressiveness. Periosteal reaction is usually seen in cases of osteomyelitis (9). Location is a helpful finding for differentiating osteomyelitis from other types of arthritic disorders; osteomyelitis is more common distally and neuropathic disease more common proximally. STIR images may not be helpful in the differentiation of osteomyelitis from septic arthritis since sympathetic marrow edema may be seen in uncomplicated septic arthritis. This sympathetic marrow edema may be recognized in arthritic disorders other than septic arthritis as well. When osteomyelitis spreads, it tends to do so within compartments from distal to proximal; therefore it is important to define the proximal extent of the infection. Abscesses with osteomyelitis are rare in adults and almost invariably occur in the soft tissues. Abscesses in children are more frequent and tend to occur intraosseously or more commonly periosteally.

CASE 10

History: A 35-year-old man presented with pain and swelling after a fall on his outstretched hand prior to his MR study. Based on the findings, what is your diagnosis?

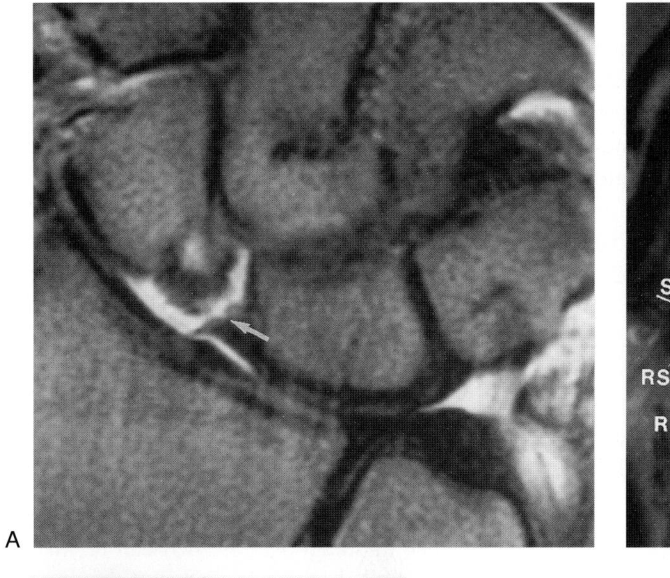

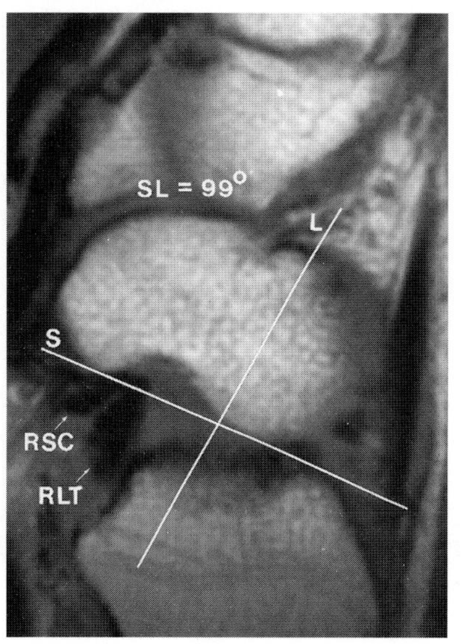

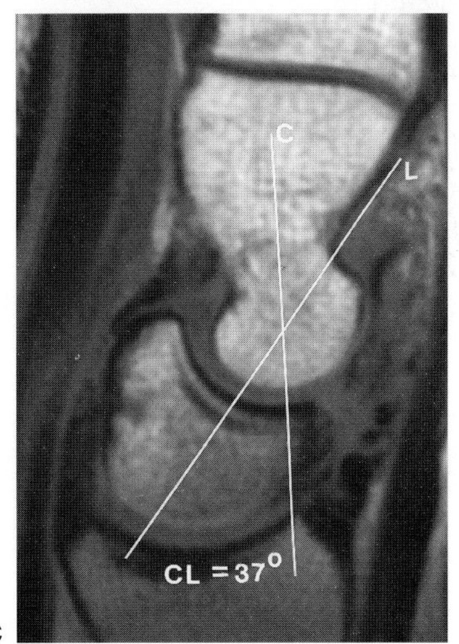

FIG. 10.

Findings: A coronal T2-weighted (TR/TE 2,000/80 msec) image demonstrates a wide separation between the scaphoid and lunate (Fig. 10A). There is apparent avulsion of the scapholunate ligament from its scaphoid attachment (arrow). Sagittal T1-weighted (TR/TE 500/20 msec) images at the level of the scaphoid (Fig. 10B) and the lunate (Fig. 10C) demonstrate an abnormal scapholunate (SL) angle of 99° and an abnormal capitate-lunate (CL) angle of 37°. The radioscaphocapitate ligament (RSC) and radiolunotriquetral ligament (RLT) are displaced volar to the waist of the scaphoid but are still intact.

Diagnosis: Scapholunate dissociation with dorsal intercalated segment instability (DISI).

Discussion: Scapholunate dissociation, which involves a tear of the scapholunate ligament, is the most common type of carpal instability. Carpal instability is defined as loss of the normal alignment of the carpal bones. The mechanism of injury for tearing the scapholunate ligament involves a fall on the outstretched hand with the blow directed at the hypothenar eminence. The resultant acute radial deviation severely compresses the space around the scaphoid, leading to extreme palmarflexion of the scaphoid; this force is compounded by the battering ram effect of the head of the capitate on the scapholunate joint (1). As the scaphoid and lunate are driven apart, the ligament between them tears.

As is shown in this case, scapholunate dissociation may result in a malalignment known as DISI. In DISI, the lunate is abnormally dorsiflexed with the wrist held in neutral position; if left untreated, this malalignment can gradually develop into carpal collapse and early degenerative disease. An understanding of how scapholunate dissociation leads to DISI can be gained by delving into the mechanics of the wrist, as described below.

Normally the wrist is under constant axial loading forces generated by the extrinsic muscles of the forearm. The wrist, of course, is made up of two carpal rows; the proximal carpal row is considered an "intercalated segment" (i.e., it is "inserted between" the radius and the distal carpal row). The position of this intercalated segment, especially that of the lunate, is somewhat passive in that it is determined by the positioning of the distal carpal row and the radius. Such a system would tend to collapse in a zigzag pattern under the constant loading forces acting across the wrist were it not for the presence of the scaphoid, which functions as a bridge across the proximal and distal carpal rows. The stability afforded by the scaphoid is critical because the radius, lunate, and capitate form the main "control column" for wrist flexion and extension and the inherent instability of the lunate endangers this normal functional architecture (2).

Besides its stabilizing bridge effect across the carpal rows, the scaphoid influences the positioning of the lunate in another way as well. This latter influence is best appreciated by understanding that the constant axial loading forces across the wrist affect proximal carpal bones in different ways. Because the articular surface of the distal radius is sloped volarly, a component of the axial force tends to displace the axis of rotation of the lunate volarly. In addition, because the lunate is thinner dorsally than volarly, axial forces tend to act more on the dorsal aspect of the lunate, thereby angulating it dorsally (3). Like the lunate, the triquetrum tends to dorsiflex under axial loading. This tendency is due to the hamate sliding down the inclined plane of the dorsal aspect of the triquetrum (4). Unlike the lunate and triquetrum, the scaphoid has a tendency to palmarflex under compressive force; this tendency helps to balance the dorsiflexion tendency of the lunate and triquetrum to keep the proximal carpal row most importantly, the lunate in neutral position. This balancing, or distributing of forces across the members of the proximal carpal row, is the function of the SL and lunotriquetral ligaments. Thus these ligaments are primary wrist stabilizers (Fig. 10); should these ligaments become disrupted, alignments of lower potential energy occur, leading to carpal instability (5). Specifically, when the SL ligament becomes torn, the palmarflexion influence of the scaphoid on the lunate is lost, thereby allowing the dominance of the dorsiflexion influence of the triquetrum, which is transmitted to the lunate by the lunotriquetral ligament. (As mentioned above, the lunate has an inherent dorsiflexion tendency as well.) The end result may be DISI (Fig. 10E).

As the sagittal images in Fig. 10B and 10C demonstrate, DISI is present in this case. As explained in Fig. 10F–H, the diagnosis of DISI can be made by determining the angle between the axes of the scaphoid and lunate and the angle between the axes of the capitate and lunate.

Why some scapholigament tears lead to DISI, and others do not, is controversial. Blatt (1) believes that the entire ligament must be injured, but Fisk (6) and Taleisnik (7) think that the RSC and long radiolunate ligaments must also be injured, along with the entire SL ligament before instability occurs. In this case, when the RSC and long radiolunate ligaments remain intact, Blatt's position is supported over that of Fisk and Taleisnik.

The treatment of tears of the SL ligament with associated DISI depends on how long standing the injury is. Acute tears of the ligament may be treated with suture approximation and temporary internal fixation of the scaphoid, but chronic SL dissociation may require reconstructive techniques, such as dorsal capsulodosis (1).

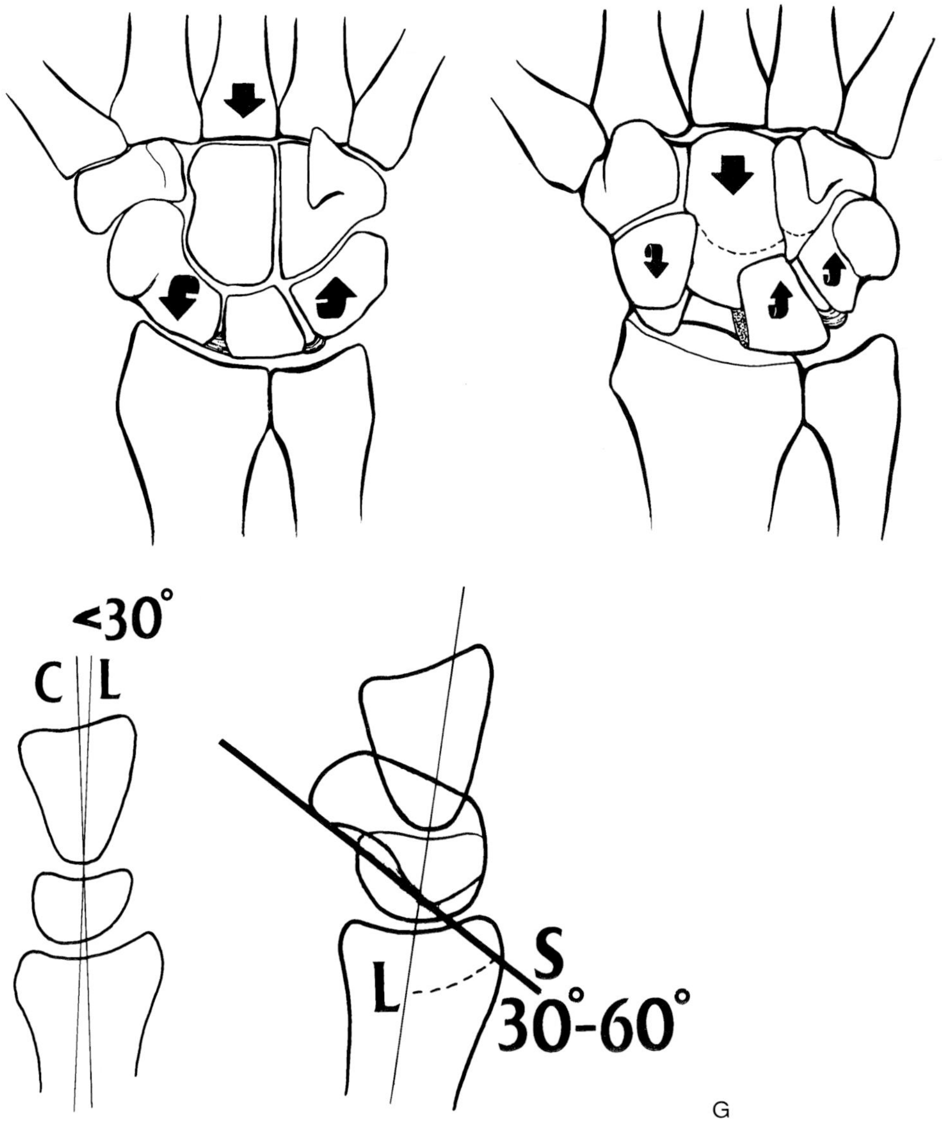

D

E

<30°
C L

S
L
30°-60°

F

G

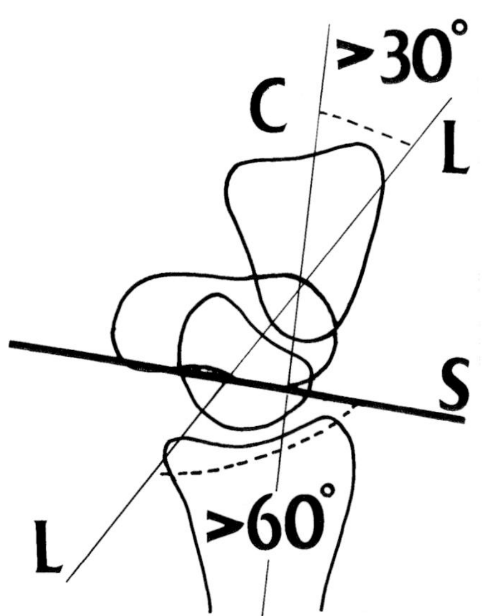

H

FIG. 10. D: Diagram shows that the integrity of the SL and lunotriquetral ligaments is critical for balancing out the palmarflexion tendency of the scaphoid and the dorsiflexion tendency of the triquetrum. (Adapted from Timins ME, Jahnke JP, Krah SF, et al. MR imaging of the major carpal stabilizing ligaments: normal anatomy and clinical examples. *Radiographics* 1995;15:575, with permission.) **E:** Scapholunate dissociation. Diagram shows how a torn SL ligament decouples the palmarflexion influence of the scaphoid on the lunate. The intact lunotriquetral ligament continues to transmit the dorsiflexion influence of the triquetrum to the lunate, leading to possible DISI. (Adapted from Timins ME, Jahnke JP, Krah SF, et al. MR imaging of the major carpal stabilizing ligaments: normal anatomy and clinical examples. *Radiographics* 1995;15:575, with permission.) **F–H:** Normal alignment and malalignment of the carpal bones. *C,* axis of capitate; *L,* axis of lunate; *S,* axis of scaphoid. **(F)** Diagram shows the capitate–lunate angle, which should be less than 30°. The axis of the capitate is drawn from the mid-point of its head to the center of its articular surface. The axis of the lunate is drawn perpendicular to a line through its distal poles. Both axes should be parallel to the long axis of the radius. **(G)** Diagram shows the SL angle, which should be between 30° and 60°. The axis of the scaphoid is drawn as a line connecting its proximal and distal volar convexities. **(H)** Diagram shows that, in DISI, the scaphoid is palmarflexed and the lunate is tipped dorsally. Therefore, the capitate–lunate angle is greater than 30° and the scapholunate angle is greater than 60°.

CASE 11

History: A 42-year-old woman presented with persistent ulnar-sided wrist pain after a fall on her outstretched hand. The injury occurred 3 weeks prior to the MR examination. What is your diagnosis?

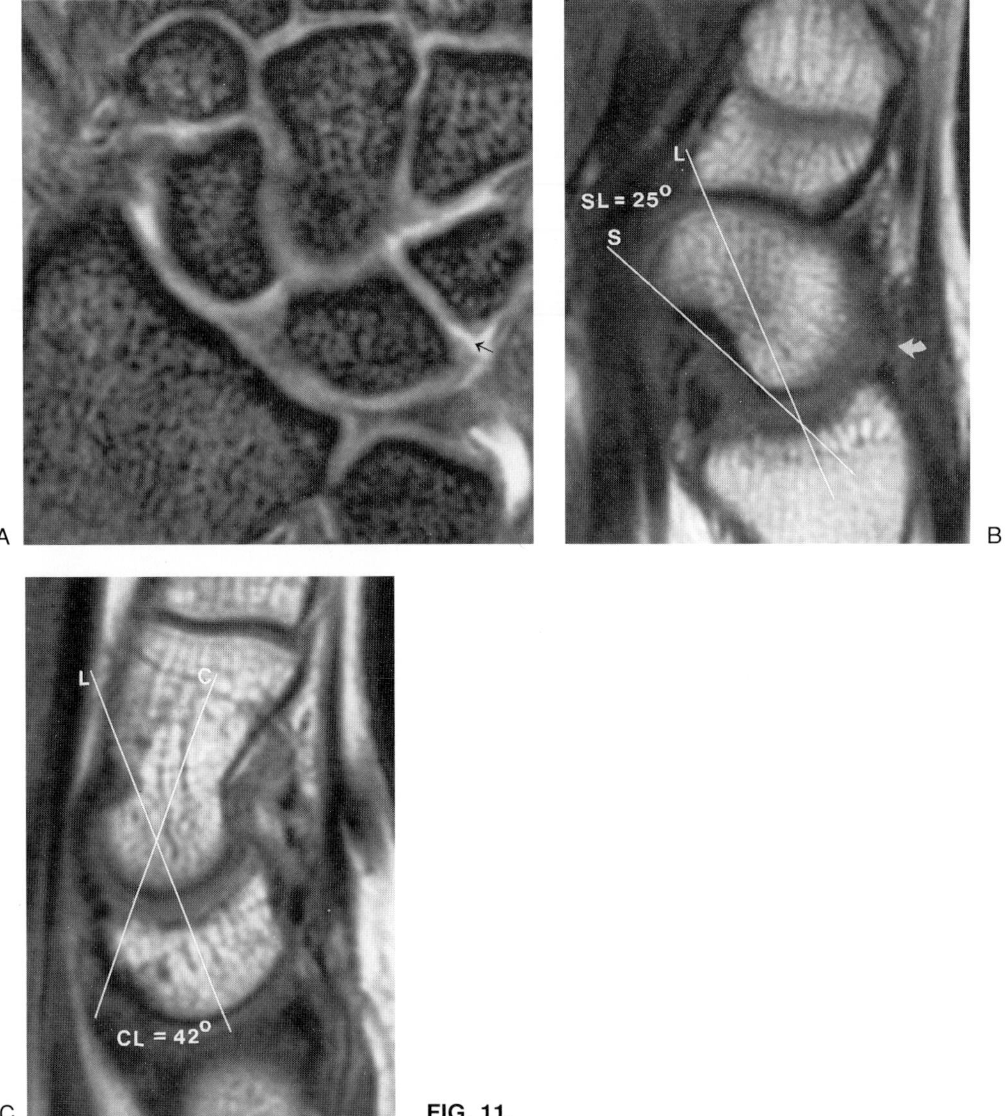

FIG. 11.

Findings: Coronal gradient-echo image (Fig. 11A) demonstrates nonvisualization of the lunotriquetral ligament (arrow). Sagittal T1-weighted (TR/TE 500/20 msec) images at the level of the scaphoid (Fig. 11B) and the lunate (Fig. 11C) demonstrate an abnormal SL angle of 25° and abnormal CL angle of 42°. An SL angle of less than 30° and a CL angle of more than 30° suggests volar intercalated segment instability (VISI). Indistinctness of the dorsal radiocarpal ligament (arrow in Fig. 11B) suggests injury to this structure, a finding considered by some investigators to be necessarily associated in VISI.

Diagnosis: Lunotriquetral ligament tear with VISI.

Discussion: Tear of the lunotriquetral ligament, that is, lunotriquetral dissociation, is the second most common ligamentous cause of carpal instability. (SL ligament tear is the most common.) Isolated tears of the lunotriquetral ligament may lead to ulnar-sided wrist pain, as in this case. In addition, lunotriquetral dissociation may result in a malalignment known as VISI, which we see here. In VISI the lunate is tipped volarly; like DISI, this instability can lead to early and accelerated degenerative joint disease because the lunate has lost its normally congruent articulation with the radius.

VISI is diagnosed in this case on the sagittal images by measuring the angle between the axes of the scaphoid and the lunate and the angle between the capitate and the lunate. The normal SL angle is between 30° and 60°, and the normal CL angle should be less than 30° as shown in Case 10, Fig. 10F–H. In VISI, as shown in this case and as schematically illustrated in Fig. 11D, the SL angle is less than 30°, and the CL angle is greater than 30°.

A complete tear of the lunotriquetral ligament should theoretically always result in a VISI malalignment, as shown in Fig. 11E. After all, the lunate should be under the influence of the palmarflexion tendency of the scaphoid (through the SL ligament) while dissociated from the dorsiflexion tendency of the triquetrum (because of the torn lunotriquetral ligament). Nevertheless, on the basis of a series of cadaver dissections and load studies, Viegas et al (1) showed that the dorsal radiocarpal ligament must also be torn at its scaphoid and lunate attachments for there to be enough dorsal space to allow the lunate to undergo palmar tilt. [The dorsal radiocarpal ligament connects the dorsal aspect of the distal radius with the dorsal aspects of the proximal carpal row (2).] Cooney et al. (3) have confirmed that VISI requires the dorsal radiocarpal ligament to be involved but specify that it is the triquetral attachment of this ligament that must be disrupted for VISI to develop. In any case, a lunotriquetral ligament tear with associated VISI requires involvment of both the lunotriquetral and dorsal radiocarpal ligaments.

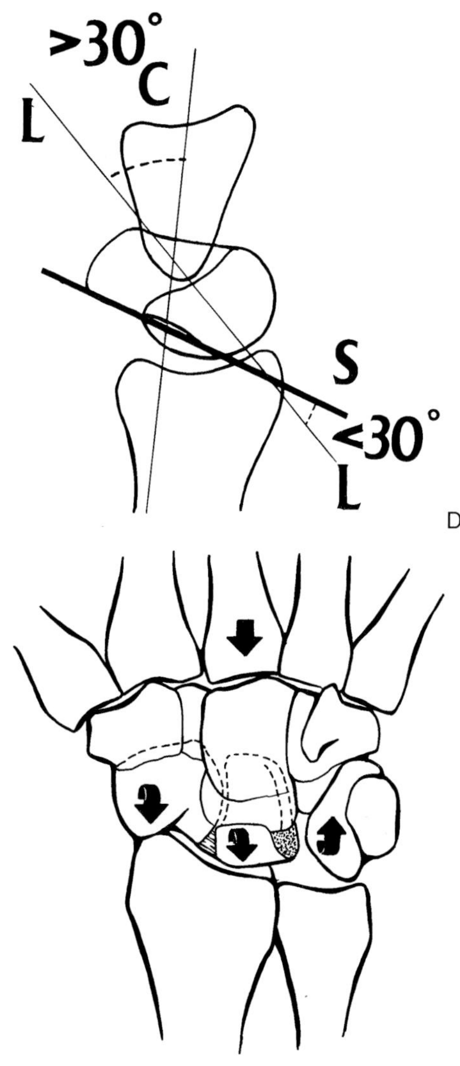

FIG. 11. D: Diagram shows that, in VISI, the scaphoid and lunate are both palmarflexed; the capitate–lunate angle is greater than 30°, and the scapholunate angle is less than 30°. **E:** Diagram shows how a torn lunotriquetral ligament could lead to VISI. When the triquetrum becomes dissociated from the lunate, the lunate no longer shares the dorsiflexion tendency of the triquetrum but instead palmarflexes with the scaphoid. However, it appears that the dorsal radiocarpal ligament must also be injured for VISI to occur. (Adapted from Cooney WP, Garcia-Elias M, Dobyns JH, Linscheid RL. Anatomy and mechanics of carpal instability. *Surg Rounds Orthop* 1989;3:15–24, with permission.)

CASE 12

History: A 42-year-old woman had undergone prior repair of a tendon laceration. The patient did not follow the prescribed rehabilitation program and attempted to lift a heavy object. Is the repair intact?

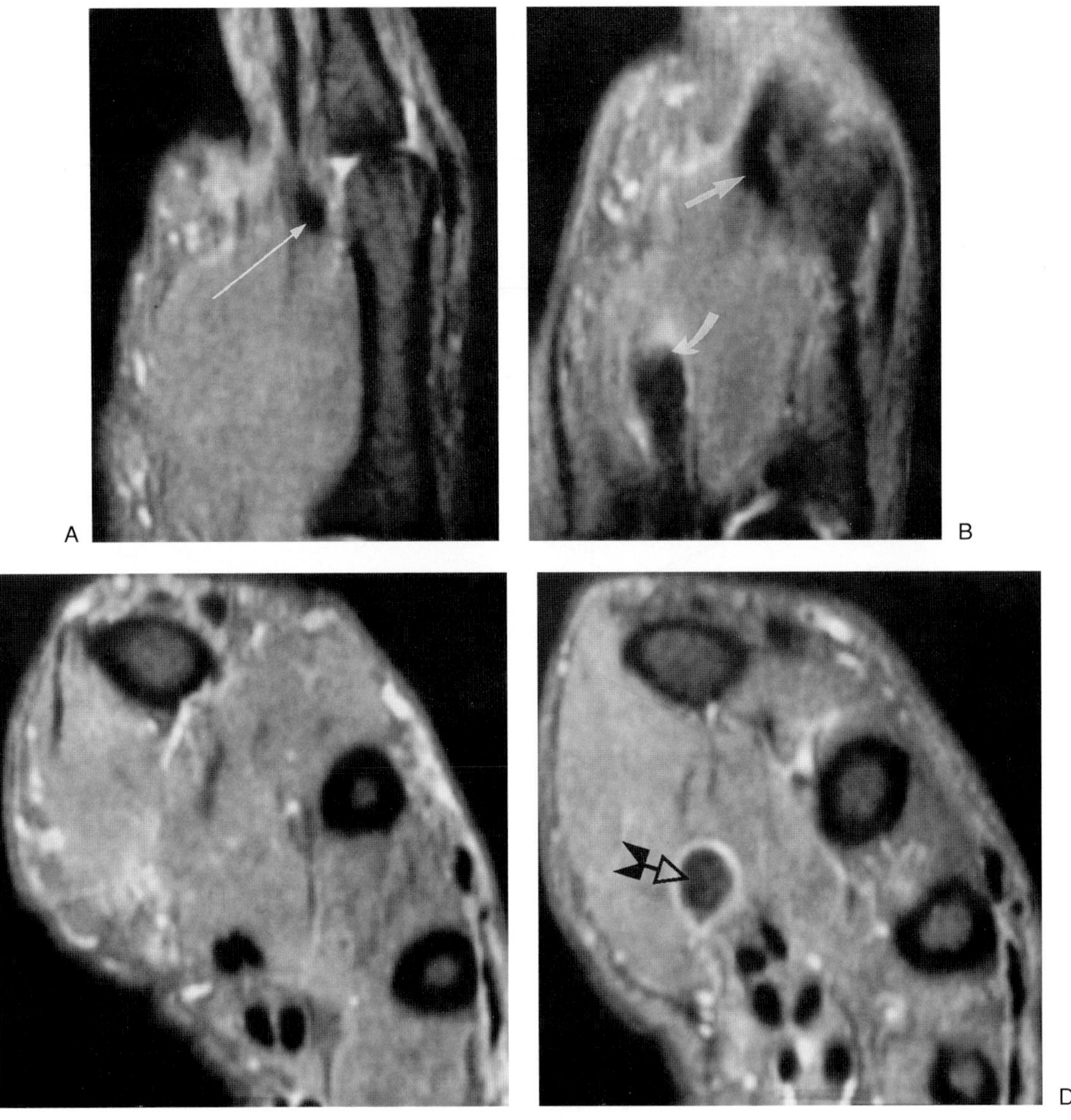

FIG. 12.

Findings: Fat-suppressed sagittal proton-density fast spin-echo image (TR/TE 3,500/32 msec) demonstrates an intact distal flexor tendon to the level of the MCP joint, where it abruptly terminates (Fig. 12A, arrow). Contiguous image demonstrates the gap between the distal (straight arrow) and proximal tendon (Fig. 12B, curved arrow). Axial fat-suppressed sagittal PD weighted fast spin-echo image (TR/TE 3,500/32 msec) through the level of the gap illustrated in (Fig. 12B) demonstrates complete absence of the tendon from its normally expected position. Axial fat-suppressed sagittal PD weighted fast spin-echo image (TR/TE 3,500/32 msec) through the proximal tendon demonstrates marked increased girth and surrounding high signal fluid (arrow).

Diagnosis: Re-tear of the deep flexor tendon of the index finger.

Discussion: In the digits, the flexor tendons are enclosed within sheaths lined by synovium (1,2). Five fibrous bands serve as sturdy annular pulleys that serve important biomechanical functions and assist in keeping the tendons closely applied to the phalanges (1,2). The A1 and A4 annular pulleys arise from the periosteum of the proximal half of the proximal phalanx and the midportion of the middle phalanx, respectively (1). The A1, A3, and A5 pulleys are joint pulleys that arise successively from the palmar plates of the MCP, proximal interphalangeal, and distal interphalangeal joints (1). Between the annular pulleys are three thin and pliable cruciate pulleys that collapse to allow full digital flexion (1,2). Loss of portions of the digital pulley system may significantly alter the normal integrated balance between the flexor (intrinsic) and extensor tendons (1). The A2 and A4 pulleys are the most important biomechanically; the loss of either may substantially diminish digital motion and power or may lead to flexion contractures (1).

The digital flexor tendons consist of both superficial (FDS) and deep (FDP) components. The FDS tendons usually arise from single muscle bundles and act independently (1,2). There is often a common muscle origin for several FDP tendons; as a consequence, there may be simultaneous flexion of multiple digits. The FDS tendons lie on the palmar side of the FDP tendons until they enter the A1 entrance of the digital sheath (at the MCP joint level) (1). Early in the proximal flexor tendon sheath, the FDS tendon divides and passes around the FDP tendon. The two portions of the FDS tendon reunite dorsally by means of fibers referred to as Camper's chiasma and terminate as they insert along the proximal half of the middle phalanx. The FDP tendons pass through the FDS bifurcation to insert into the proximal aspect of the distal phalanges. The FDP muscle serves as the primary digital flexor. The FDS and intrinsic muscles combine for forceful flexion (1).

Primary repair of flexor tendons severed in the digital sheath has become the accepted treatment of these injuries (3). This need not be accomplished on an emergency basis; studies have demonstrated equal or better results with delayed primary repair (3). Current opinion favors repair of both the FDP and FDS tendons rather than the FDP alone, as was once the preferred option (3). Lacerations on the palmar aspect of the fingers will almost always injure the FDP tendon before severing the FDS tendon (1). The absence of FDP function, however, does not exclude the possibility of a nearly complete FDS division. MR may allow both detection and characterization of the extent of tendon injury (Fig. 12E–G). Partial tears may be demonstrated as well as the degree of tendon retraction in complete lacerations, a finding that may aid in surgical planning.

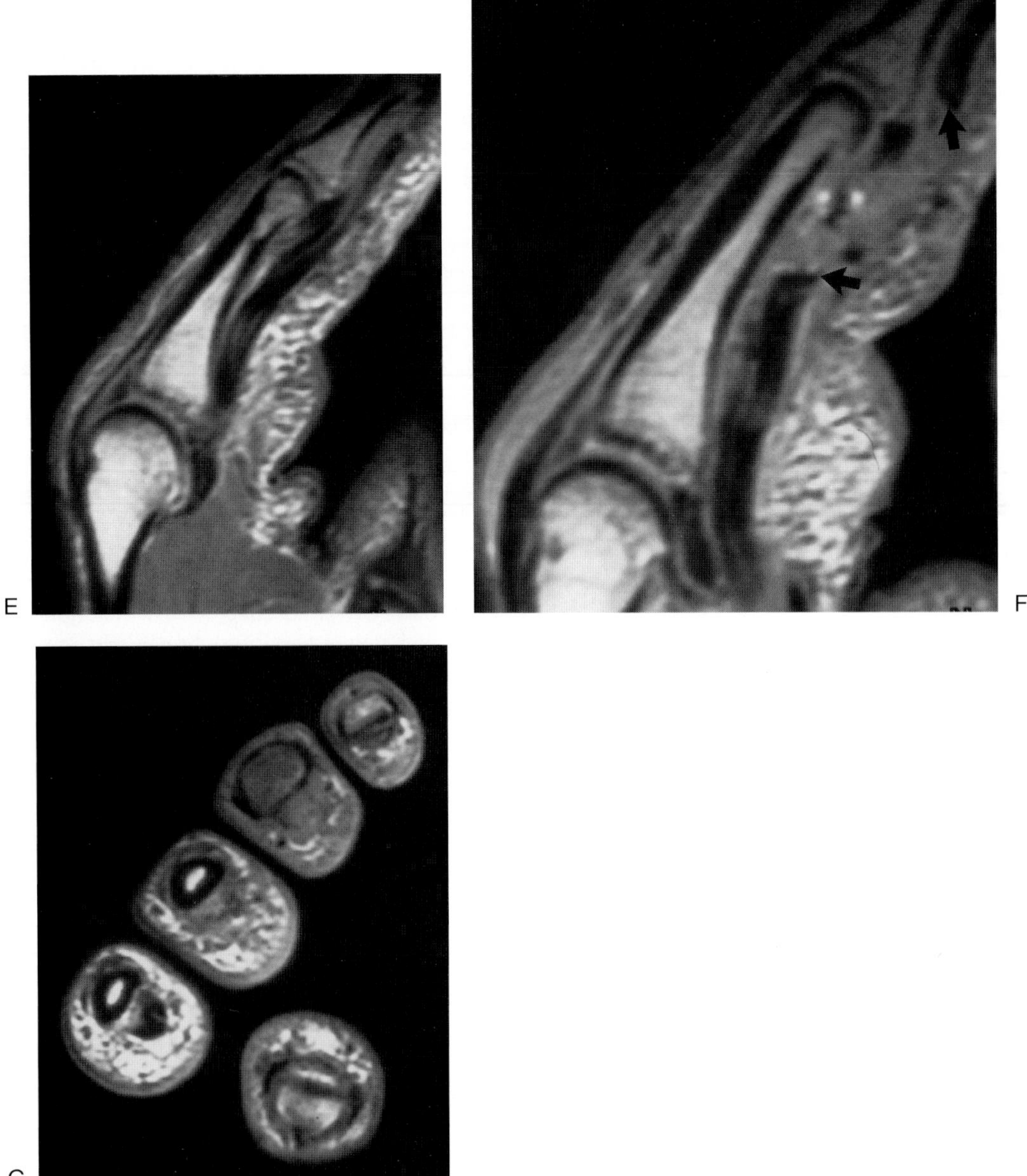

FIG. 12. E–G: A 23-year-old man with a knife injury to the palmar aspect of the hand. **E:** Sagittal proton-density image of the normal uninjured index finger demonstrates normal superficial and deep flexor tendons. **F:** Sagittal proton-density image of the adjacent long finger demonstrates discontinuity and retraction of the deep and superficial flexor tendons (*arrows*). **G:** Axial proton-density image demonstrates normal flexor tendons of the index finger with nonvisualization of the flexor tendons of the third through fifth digits. The tendons were completely lacerated at surgery. (Images courtesy of Mitchell Klein, M.D., Milwaukee, Wisconsin.)

CASE 13

History: A 42-year-old woman presented with a painless, slowly growing mass over the dorsoradial aspect of the wrist. What are the findings demonstrated in Fig 13A–C? What characteristics of the lesion illustrated suggest a possible diagnosis?

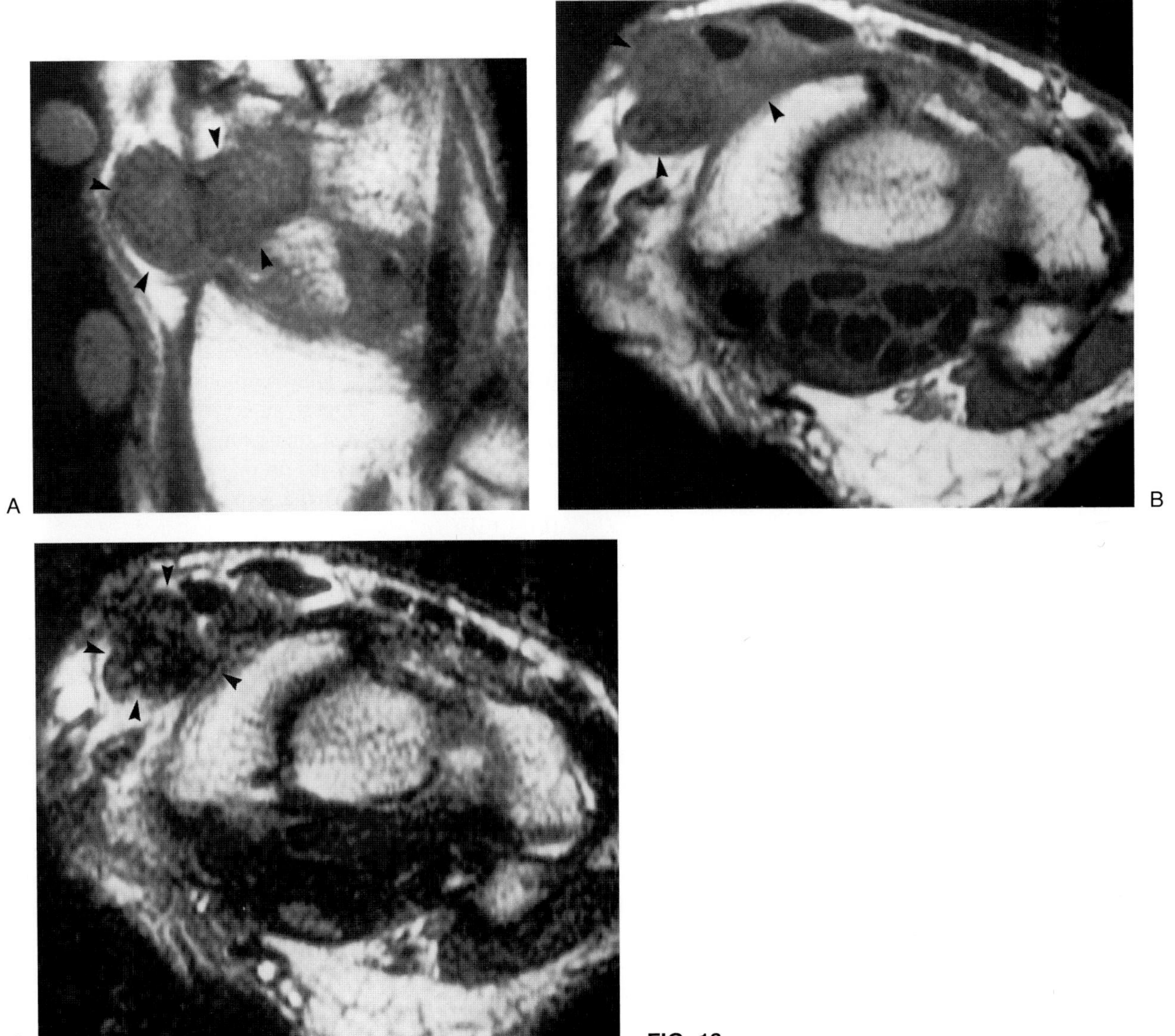

FIG. 13.

Findings: T1-weighted (TR/TE 500/16 msec) coronal MR image (Fig. 13A) obtained over the dorsal aspect of the wrist demonstrates a lobulated well-defined low-signal mass over the dorsoradial aspect of the proximal carpus (arrowheads, Fig. 13A). T1-weighted (TR/TE 500/20 msec) axial MR image through the level of the proximal carpal row, again illustrates a lobulated extraosseous mass with low signal intensity in close proximity to the extensor carpi radialis longus tendon (arrowheads, Fig. 13B). T2-weighted (TR/TE 200/80 msec) axial MR image at the same location as Fig. 13B shows persistent homogeneous low signal intensity within the mass (arrowheads, Fig. 13C). Incidental note is made of a small amount of fluid within the tendon sheath of the extensor carpi radialis longus and brevis.

Diagnosis: Giant cell tumor (GCT) of the tendon sheath.

Discussion: GCT of the tendon sheaths forms the localized extra-articular counterpart of pigmented villonodular synovitis (PVNS) (1–8). The tendon sheaths along the volar aspect of the hand represent one of the most common locations for this localized or ''nodular'' form of PVNS, which histologically is distinguished from PVNS by the presence of numerous multinucleated giant cells.

The appearance of PVNS and GCT on MR has been well described and while not pathognomonic, the diagnosis can be strongly suggested when typical features are present (Fig. 13D, E). Recognized MR features of PVNS include: (a) low signal intensity on both T1- and T2-weighted images (related to hemosiderin deposition and thick fibrous tissue); (b) the presence of fat signal within the mass reflecting clumps of lipid-laded macrophages; (c) hyperplastic synovium; and (d) the presence of bone erosions (in cases of PVNS) in those joints with restricted intracapsular space (e.g., hip) (1–8). The actual appearance of any individual lesion will vary, however, depending on the relative proportions of lipid, hemosiderin, fibrous stroma, septation, pannus formation, fluid, cyst formation (when present), and cellular elements (2,3). In addition, the degree of signal heterogeneity of the lesion will be affected by the pulse sequence utilized to image it and by the use of contrast material. The greatest heterogeneity of PVNS lesions has been observed with contrast-enhanced T1-weighted images and gradient-echo sequences (probably reflecting the sensitivity of this pulse sequence to susceptibility effects) (3).

Most soft tissue tumors involving the wrist are benign (1). In addition to GCT, the most common lesions include ganglion cysts (see Case 3-3), lipomas, granulomas secondary to embedded foreign bodies or chronic infections, and neural and vascular lesions such as schwannomas (see Case 5-22) (2), glomus tumors (see Case 3-15) (3), and hemangiomas (Fig. 13F).

Of the soft tissue masses affecting the wrist, ganglion cysts are the most common (1). They are commonly located over the dorsal aspect of the wrist and are composed of a fibrous encapsulated mass containing internal mucinous material. On MR imaging ganglions typically demonstrate homogeneous internal signal similar to that of fluid. Occasionally, however, complicated ganglion cysts are encountered demonstrating atypical imaging characteristics. In such instances differentiation from other soft tissue masses may not be possible. Important preoperative information includes the extent of the lesion and its relationship to adjacent soft tissue and bony structures. Ganglion cysts may compress adjacent neurovascular structures and in this manner produce symptoms of one of the mass lesions that may be responsible for carpal tunnel syndrome (see Case 3-3).

Lipomas are well delineated and characterized by MR imaging. In the hand they typically appear as well-defined lesions commonly within the thenar eminence, demonstrating signal characteristics similar to those of subcutaneous fat on all pulse sequences (9). Occasionally the lesions may demonstrate low-signal internal septations (see Case 10-1). Fibrolipomatous hamartomas, rare lesions of the median nerve, may also demonstrate foci of internal fatty signal suggestive of their diagnosis (see Case 2-2). Areas of high signal intensity representing fat may also be seen in soft tissue hemangiomas (9). Soft tissue hemangiomas may demonstrate poorly defined margins that can suggest an aggressive malignancy. A distinctive feature of hemangioma, however, is the relative lack of mass effect or mass effect that is disproportionately small for the expected size of the lesion (Fig. 13F).

Embedded foreign bodies are also common in the hand, and MR imaging may be useful for their detection. GRE imaging with long TE parameters can be used to magnify inherent magnetic susceptibility artifacts at the interface between the foreign material and native soft tissues. While metallic foreign bodies are most conspicuous with such imaging, any foreign material including plastic, glass, and wood may be detectable. Frequently the presence of foreign material may incite surrounding infectious or noninfectious inflammatory reactions, manifesting on MR imaging as irregular areas of increased soft tissue signal with heterogeneous enhancement, or peripheral rim-like enhancement in cases of abscess formation. In such instances a GRE sequence may be particularly helpful in the potential detection of a centrally located foreign body.

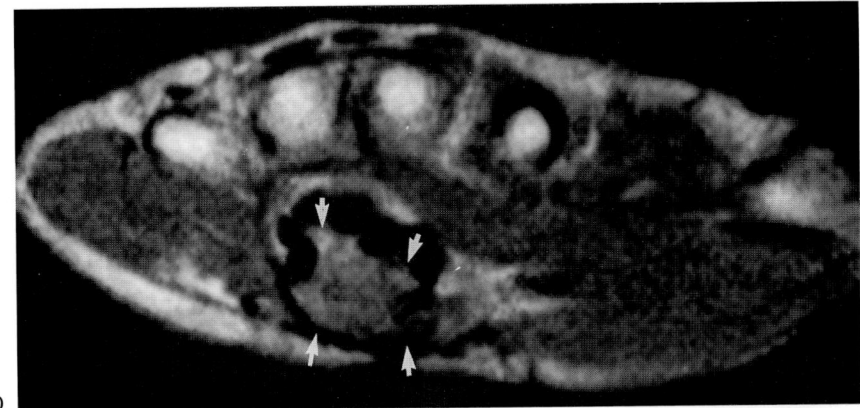
D

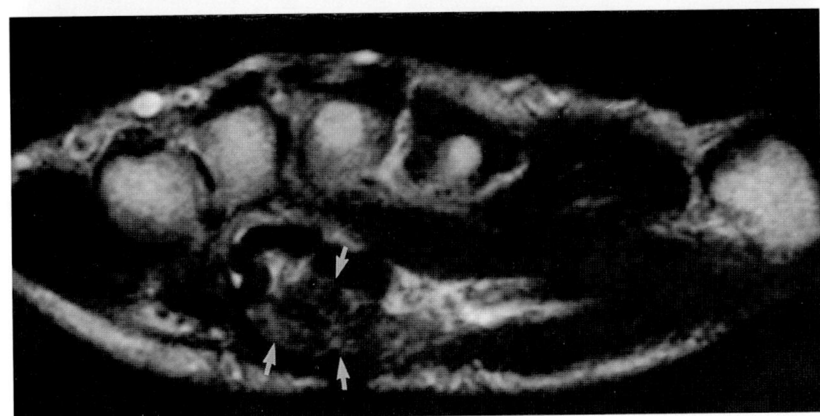
E

FIG. 13. D–E: Giant cell tumor of the tendon sheath. **D:** Proton-density (TR/TE 2,000/20 msec) sequence demonstrates a heterogeneous mass (*short arrows*) with areas of slightly augmented signal intensity seen along the volar flexor tendons of the hands just distal to the carpal tunnel. **E:** On the T2-weighted sequence (TR/TE 2,000/80 msec), the mass (*short arrows*) is of predominantly low signal intensity.

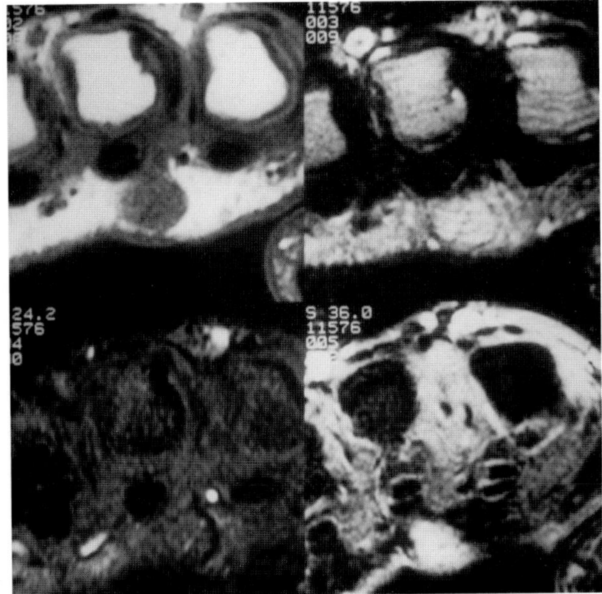
F

FIG. 13. F: Soft tissue hemangioma of the hand. Composite image. T1-weighted (TR/TE 517/15 msec) sequence (*upper left*) demonstrates a nodular mass located along the subcutaneous fat adjacent to the flexor tendon. On a fast spin-echo fat-suppressed image (TR/TE 3,766/92 msec) (*upper right*), the lesion increases in signal intensity and becomes isointense to the adjacent suboptimally suppressed fat. The T2-weighted images, however, demonstrate small areas of edema adjacent to the lesion. The margins of the lesion are better defined on the T1- than on the T2-weighted image. T1-weighted fat-suppressed images both before (*lower left*) and after (*lower right*) the intravenous administration of gadopentetate dimeglumine demonstrate marked enhancement of the lesion. The postcontrast study again demonstrates poorly defined margins without enhancement of adjacent edema. There is enhancement of the overlying skin, which was not demonstrated to yield abnormal signal on the conventional sequences without contrast.

CASE 14

History: A 37-year-old woman with, wrist pain several years after a lunate implant for Kienbock's disease. What is your diagnosis?

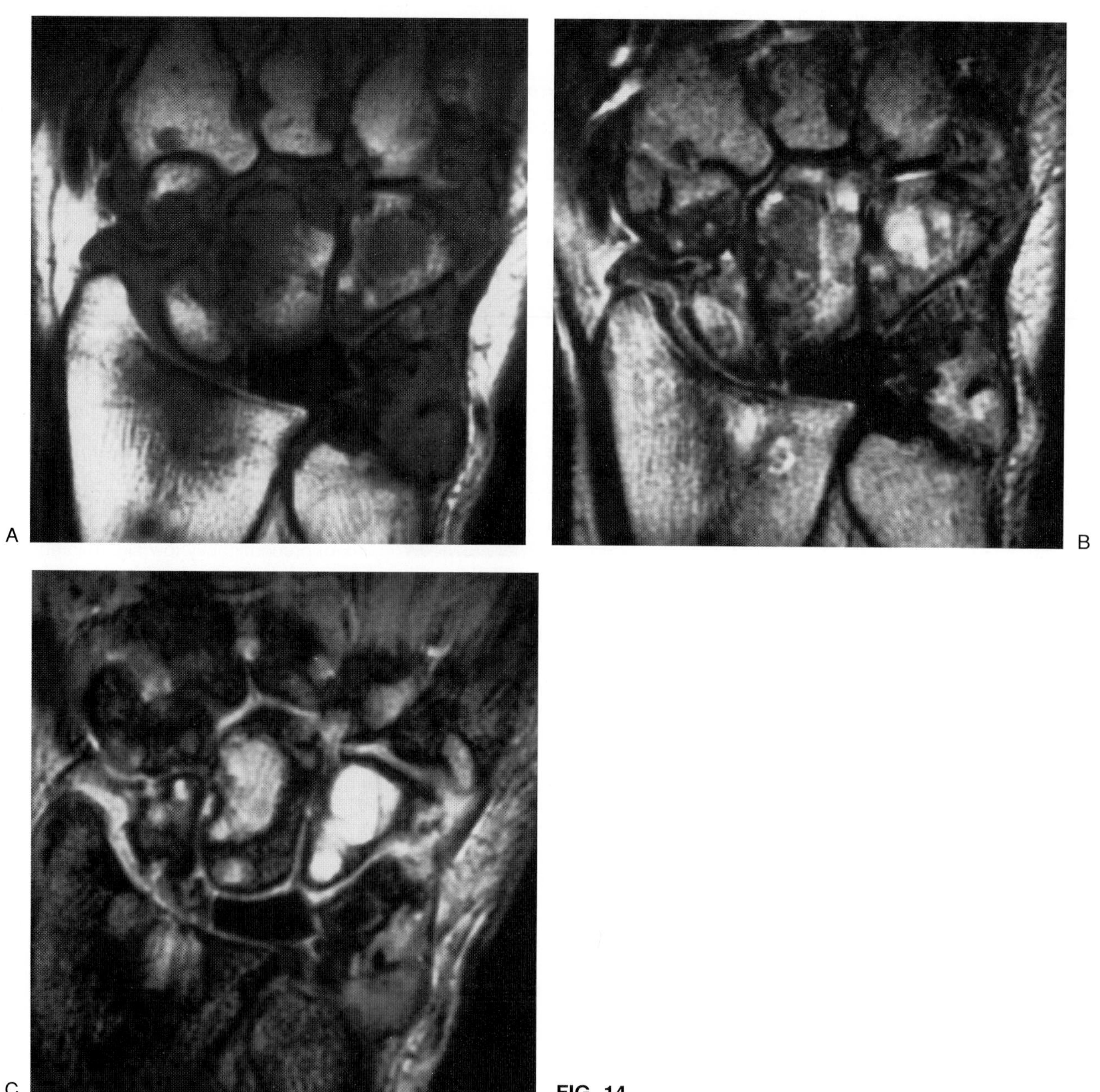

FIG. 14.

Findings: Coronal T1-weighted image demonstrates a signal void in the region of the lunate corresponding to a silicone implant (Fig. 14A). There is diffuse low signal seen throughout the carpus both intraosseously and distributed throughout the joint space. Coronal T2-weighted spin-echo sequence again demonstrates the silicone implant as a signal void (Fig. 14B). The intraosseous lesions demonstrate variable increase in signal intensity, as does abnormal signal within the joint (e.g., prestyloid recess). Coronal T2*-weighted gradient-echo image demonstrates striking high signal within several of the intraosseous lesions (Fig. 14C). The intra-articular process is of intermediate signal. (Images courtesy of Mitchell Klein, M.D., Milwaukee, Wisconsin)

Diagnosis: Silicone synovitis and synovial hyperplasia.

Discussion: Silicone rubber elastomer prostheses have been utilized for replacement of multiple joints of the hand as well as for carpal bone substitutes (1–4). Principal among the complications associated with these prostheses include fracture (reported in up to 50% of cases in one series) and synovitis (1–4). The synovitis is believed to represent a response to the shedding of silicone particles from the prostheses related to damage as a result of shear and compressive forces (1–4). This complication appears more common following carpal bone replacement than after finger joint replacement, a finding that may relate to the greater compressive forces at the level of the wrist (2,3).

The synovitis occurs months to years following the surgical procedure. The condition may be painful, although some patients remain asymptomatic despite evidence of radiographic changes. Particles resulting from fibrillation of the prosthesis become imbedded in the synovium with resultant synovial hypertrophy and multinuclear foreign body giant cell infiltration of synovium and bone (2,3). Bone changes may eventuate in cyst formation, collapse, and fracture. Surgical removal of the prosthesis is associated with resolution of the synovitis (4).

Radiographic changes are typical, with soft tissue swelling, initial preservation of the joint spaces, and well-circumscribed subchondral cyst formation (5) (Fig. 14D). Radiographic diagnosis is relatively straightforward, although in the differential diagnosis entities including PVNS, amyloidosis, tuberculosis, or fungal infection could be entertained (2). On MR, the prosthesis, as well as large fragments, is of low signal intensity on all pulse sequences (2). MR also allows direct depiction of synovial hypertrophy as well as enhanced demonstration of bony changes including cyst formation that may not be evident radiographically.

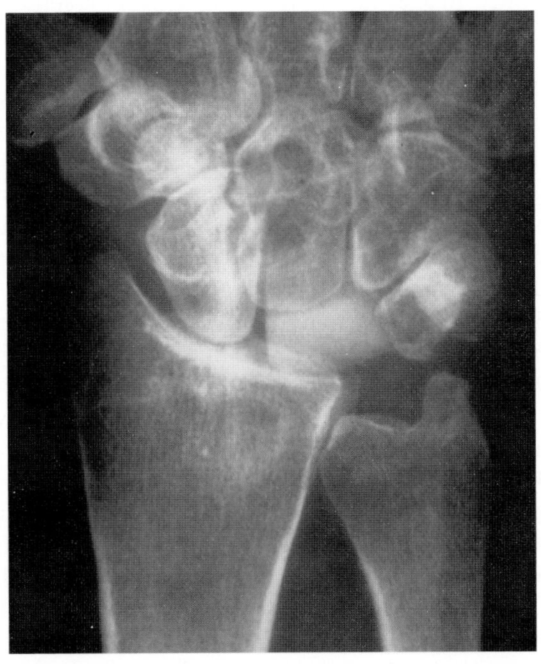

FIG. 14. D: Anteroposterior radiograph of the wrist demonstrates the radiodense lunate prosthesis. Multiple cysts are seen throughout the carpus.

CASE 15

History: A 42-year-old woman complained of pain in the region of the distal finger for the past 4 years. No mass was identified on physical examination. What is your diagnosis?

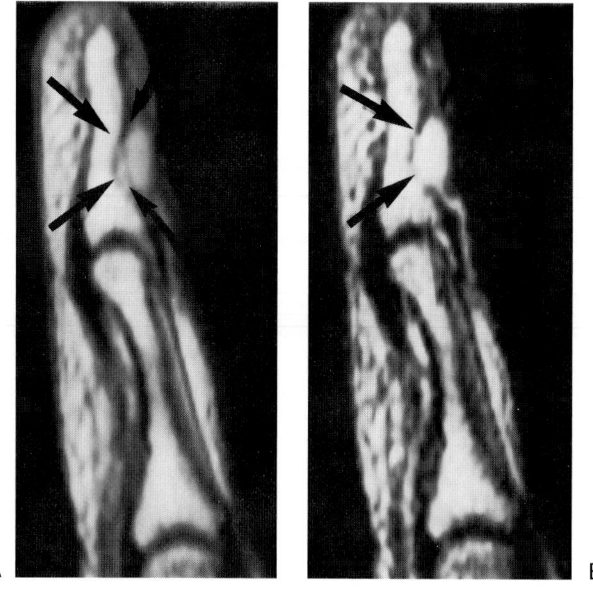

FIG. 15.

Findings: Sagittal PD weighted spin echo image (TR/TE 2,000/20 msec) demonstrates an elliptical mass of intermediate signal in a subungual location (arrows, Fig. 15A). There is a suggestion of slight scalloping of the underlying bone. On the T2-weighted image, the mass becomes increased in signal intensity (arrows, Fig. 15B).

Diagnosis: Glomus tumor (subungual).

Discussion: The normal glomus is a neuromyoarterial receptor that connects distal arterioles to terminal venules and functions to control circulation and body temperature (1,2). Glomus bodies are distributed throughout the body, contained within the deepest layer of the dermis (1). They are most highly concentrated, however, in the tips of the fingers (particularly beneath the nail bed), a finding accounting for the high incidence of tumors found in this location (13). Each glomus body is a tiny encapsulated oval organ measuring 300 μm long, and the nail beds of the fingers and toes contain 93 to 501 glomus bodies cm^2 (3,4).

Glomus tumors of the digit are more commonly encountered in women (3). While clinical findings are often indicative of the diagnosis, the classic triad of pain, point tenderness, and cold sensitivity may be seen in only a minority of patients. A blue or hemorrhagic nodule may be evident on physical examination in cases of subungual lesions but is an inconstant finding and early lesions may be occult to detection on physical examination (3). Symptoms typically are of long duration (average, 4–7 years) before definitive diagnosis is established (1,3,5,6). Analysis of clinical signs has facilitated correct diagnosis in 50% to 78% of cases (3,7).

Determination of precise tumor location is important to assist in preoperative planning and to avoid postoperative nail dystrophy. Prior to the application of MR for purposes of detection of glomus tumors, ultrasound was the only technique available for direct imaging of these lesions. While capable of detecting small lesions, the technique is highly operator dependent, and difficulties with small and flattened subungual lesions may be encountered (3,8). Normal glomus bodies may be detected with high-resolution MR techniques; they demonstrate high signal intensity on T2-weighted spin-echo sequences and enhance following the administration of gadopentetate dimeglumine (Fig. 15C,D) On MR, most glomus tumors are iso-or slightly hyperintense to the dermal layers of the nail bed on T1-weighted sequences and strongly hyperintense on T2-weighted images (3,9–15). Most glomus tumors are surrounded by a capsule, which may appear as a low-signal-intensity rim on T2-weighted and contrast-enhanced T1-weighted images (3). The capsule represents a secondary response of surrounding tissues and may be incomplete (3). In addition to primary detection, MR may be of particular value in the assessment of patients with persistent or recurrent symptoms following surgery. Recurrence and the need for repeat surgery has been reported in between 12% and 24% of cases (1,3,5). MR could be particularly valuable in detection of a lesion beyond an incomplete capsule or in detection of multiple tumors when present (3).

Most glomus tumors are composed of a mixture of three elementary components: vessels, glomus cells, and mucoid tissue (3) (Fig. 15E). While a wide variation in appearance is typical of glomus tumors at histologic examination, three predominant types based on cellular pattern can be recognized (vascular, myxoid, and solid) (3). The MR appearance of the lesions has been correlated with cellular subtype in the study by Luc-Drape and colleagues (3). The solid or cellular type demonstrated mildly augmented signal intensity on T2-weighted images. The administration of gadopentetate dimeglumine significantly assisted in its detection (3). With the mucoid type, the T2 relaxation times were markedly prolonged. The vascular type demonstrated marked enhancement (3).

Glomus tumors are most commonly encountered in a subungual location or within the lateral nail fold (3). Lesions located beneath the nail matrix near the nail root may be particularly difficult to detect (3). Tumors in this location compress the nail matrix. Under the matrix, the reticular dermis of the nail bed is homogeneous and enhances avidly with contrast (3). This may lower the conspicuity of small tumors, especially the solid subtype lesion (3). The ability to detect the normal glomus and small lesions (e.g. 1, mm) places significant demands on the imaging system. The order of magnitude of the thickness of the epidermis of the nail bed is 100 μm. Current clinical imagers allow spatial resolution on the order of about 300 μm (3,9,1). In part this relates to limitations in gradient strength. Echo-planar equipment allows the use of higher gradient strengths, which will allow for higher resolution examinations (3). The use of specialized coils including local gradient coils can give spatial resolution on the order of 100 to 200 μm, a level that will facilitate demonstration of quite small lesions (3,16).

Not all high-signal-intensity masses in the region of the nail bed are glomus tumors. Mucoid cysts and angiomas should be mainly considered in the differential diagnosis (3). These lesions are classically located in the proximal nail fold and are painless. In contrast to glomus tumors, mucoid cysts demonstrate only faint peripheral contrast enhancement (3). Angiomas of the nail bed may have similar signal characteristics to glomus tumors but typically are located more superficially in the papillary dermis and the epidermis (3).

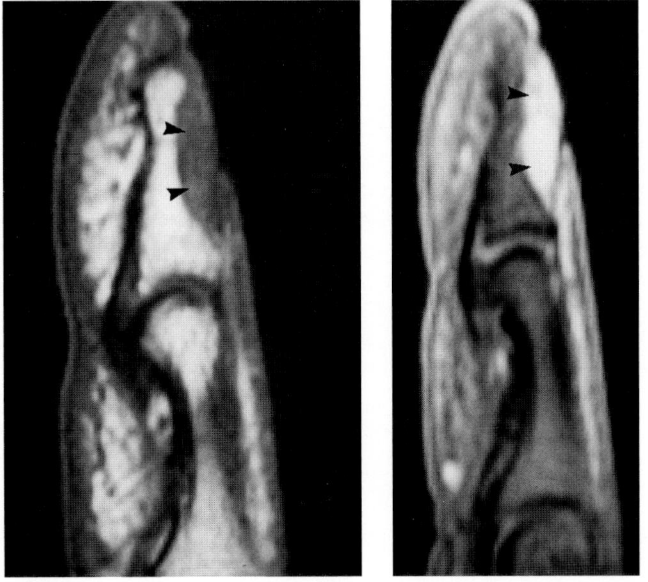

FIG. 15. C: Sagittal T1-weighted image (TR/TE 500/20 msec) image demonstrates an elliptical subungual mass (*arrowheads*). **D:** Following the administration of gadopentetate dimeglumine, a T1-weighted image demonstrates striking enhancement of the lesion (*arrowheads*).

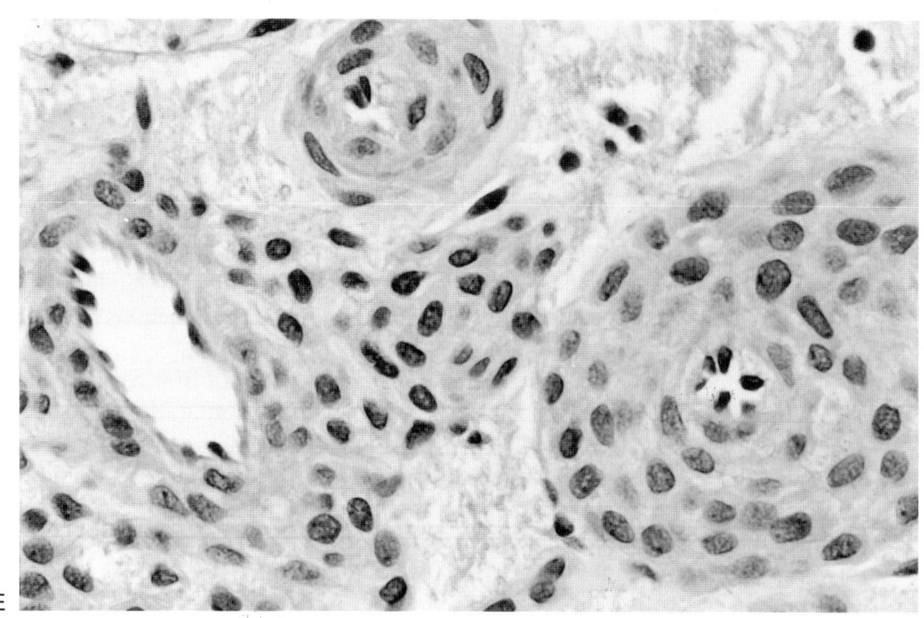

FIG. 15. E: Glomus tumor, microscopic findings. Study is from patient illustrated in the test case. High-power (300×) view demonstrating characteristic dilated vascular channels lined by flat endothelial cells and cuffed by bland epitheloid glomus cells.

CASE 16

History: A 38-year-old man with a history of previous minor wrist trauma presents with progressive wrist pain and swelling somewhat relieved by rest and immobilization. What are the findings demonstrated in Fig. 16A–C? Are these findings early or late manifestations of this process?

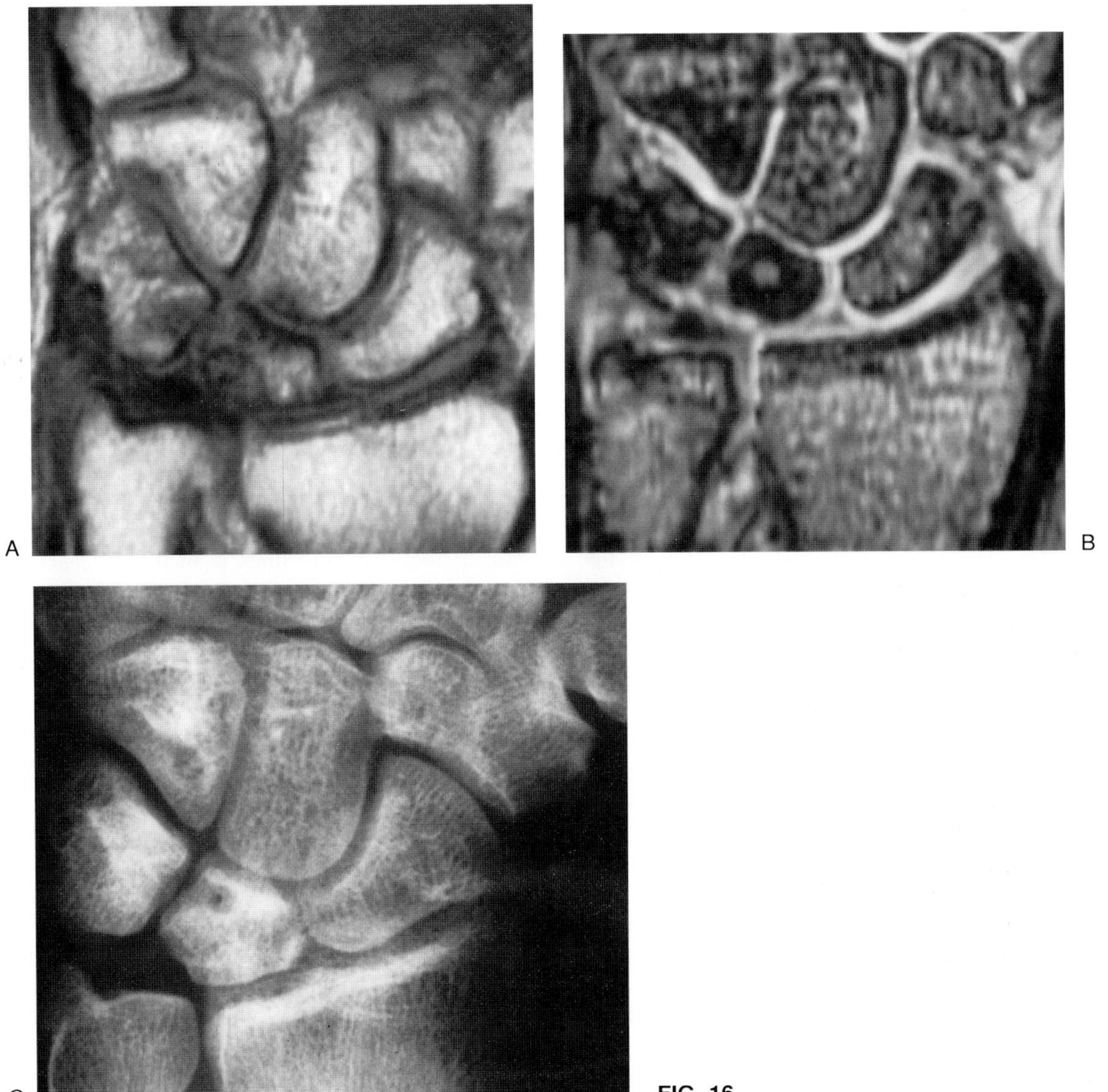

FIG. 16.

Findings: T1-weighted (TR/TE 600/20 msec) coronal MR image of the wrist shows heterogeneous low-signal replacement of the normal fatty marrow signal within most of the lunate (Fig. 16A). Zones of subchondral low signal intensity are demonstrated within the neighboring capitate and triquetrum. GRE (TR/TE 30/12, 20 msec) coronal MR image of the wrist illustrates corresponding low T2-weighted signal intensity throughout the carpal lunate aside from a small rounded focus of high signal intensity centrally (Fig. 16B). Conventional posteroanterior radiograph of the wrist demonstrates patchy internal sclerosis and a mildly abnormal contour to the lunate bone. A focal cyst is additionally appreciated within the mid- to distal aspect of the lunate (Fig. 16C).

Diagnosis: Avascular necrosis of the lunate (Kienbock's disease).

Discussion: Osteonecrosis of the lunate, or Kienbock's disease, is most commonly seen in patients in the third to fourth decades of life. Affected patients are more commonly male and are often involved in manual labor; the dominant extremity is usually affected. The condition often presents with mild pain, swelling, and loss of grip strength (1). Unlike other sites of osteonecrosis, lunate involvement is most often related to local causes and is rarely encountered as an affected site in more generalized conditions associated with osteonecrosis such as lupus (1).

The central position of the lunate exposes it to compressive forces from different directions, and the axial load is greater on its proximal convex surface than on any other bone in the proximal carpal row (1,2). In the presence of negative ulnar variance, the shift of force transmission is even greater (3). Also relatively unique to the lunate is the large proportion of its surface that is covered by articular cartilage. This latter characteristic of the lunate leaves only two small areas accessible to nutrient vessels. These two sites correspond with the sites of ligamentous insertions, and tears of these ligaments or avulsion of the bone surface on which they insert further predispose to impairment of the blood supply. Another consequence of the large proportion of coverage of the lunate by hyaline cartilage is the limited amount of nerve endings. This accounts for the relatively silent or minimally symptomatic clinical characteristics of the early stages of Kienbock's disease. The intraosseous arterial supply of the lunate is characterized by a series of anastomic channels. The blood supply to the proximal part of the bone is relatively poor and is accomplished mostly by terminal circulation without demonstrable arteriolar anastamoses (2).

Kienbock's disease is postulated to occur as a result of repetitive or minor injuries presumed to result in compression fracture and segmental occlusion of the intraosseous blood supply of the lunate (1). Less commonly, the development of Kienbock's disease may be related to a single major traumatic event. An additional contributory factor may be negative ulnar variance (greater than 2.5 mm), which, as previously discussed, increases force transmission along the convex proximal surface of the bone. A recent study, however, has challenged the validity of the association between negative ulnar variance and Kienbock disease (4).

Golimbu and colleagues (1) have offered a five-stage classification system for osteonecrosis of the lunate. In the earliest phase of the condition, stage 0, the abnormality depicted can be considered a stress fracture. A line of low signal, most often located close to the proximal part of the lunate and horizontal in orientation, is encountered. In stage 1, the contour of the lunate remains normal. Abnormal signal (bone edema pattern) is seen in 50% or less of the volume of the bone (1). In stage II disease, abnormal signal is seen throughout most, if not the entire marrow of the lunate. The bone is low in signal on T1-weighted images and of inhomogeneously low or mixed low and increased signal on T2-weighted images (1). The patchiness on T2-weighted images probably reflects a combination of osteonecrosis and edema and/or granulation tissue at the interface between necrotic and revascularized bone. The augmented signal on T2-weighted images has been regarded as a reflection of retained vascularity in some parts of the bone marrow and is considered a good prognostic sign on follow-up examinations (5). Conversely, homogeneous low signal on both T1- and T2-weighted images has been associated with a poorer prognosis (5). Within the necrotic bone area, lines of separation may be demonstrated between layers of bone marrow and can be regarded as the equivalent of the crescent sign described in the femoral head (1).

Stage III disease is characterized by structural collapse of the lunate in addition to the abnormalities seen in the bone marrow, as described for stage II. As a result of the reduction in vertical dimension of the lunate, there is a consequent shift of the carpal bones, with proximal migration of the capitate between the scaphoid and lunate (1). Stage III can be subdivided into stage IIIA, in which normal carpal relationships are maintained, and stage IIIB, in which subluxations of the scaphoid and triquetrum occur (1). In addition, the lunate in this stage may become fragmented, with distinct fracture lines separating the bone into several sections. The most common orientation of these fracture lines is in the coronal plane and therefore they are best visualized on sagittal images (1). In stage IV disease, there is concomitant degenerative disease involving the radiocarpal or intercarpal joints.

Synovitis is usually present in the advanced stages of Kienbock's disease and probably results from continuous stimulation of the synovium by osteochondral fragments and necrotic bone from the lunate. The increased thickness of the synovium may assume the appearance of proliferative pannus. On MR, this may be demonstrable as a thick layer (of intermediate or low signal on T1-weighted images) that enhances avidly following the intravenous administration of gadopentetate dimeglumine. Enhancement of the synovium is most evident within the first 2 to 3 minutes following contrast administration (6). After approximately 5 minutes, gadolinium begins to be actively transferred into the synovial fluid, producing an arthrographic effect that continues for 45 to 60 minutes (6). The synovitis present in Kienbock's disease is often detected more readily on the dorsal surface of the lunate and capitate, where the normal layers of soft tissues are usually thin.

The early stages of Kienbock's disease are treated conservatively with a trial of immobilization. Considerable controversy exists as to the best approach to stage II disease; some advocate revascularization through bone grafting and others propose correction of ulnar length discrepancies (1). In stage III disease, intercarpal fusion (capitate–hamate) is undertaken in an attempt to shift the pressure off the central axis of the carpus (1). Stage IV disease is generally approached with extensive synovectomy and pancarpal and radiocarpal fusion. As this procedure compromises all wrist motion, it is undertaken after all other possibilities have been exhausted, and its goal is pain reduction.

Chondromalacia of the lunate related to ulnolunate abutment (see Case 3–5) must be considered in the differential diagnosis (1). This condition is usually associated with positive ulnar variance, and erosive changes involving the ulnar head in addition to the lunate provide the clue to the true nature of this process and allow its differentiation from a localized form of osteonecrosis (1). Intraosseous ganglia could, potentially be confused with a localized form of Kienbock's disease, but it can be differentiated by location; Kienbock's disease is most often proximal, compared with ganglion cysts, which occur near the ligamentous insertion sites (Fig. 6D,E). Of the arthritic processes affecting the wrist, the one most likely to be confused with Kienbock's disease is that of calcium pyrophosphate deposition disease (1). This condition may lead to the development of SL advanced collapse, which could be confused with stage IIIB Kienbock's disease. The pancarpal involvement in calcium pyrophosphate deposition disease allows differentiation. The pannus-based inflammatory arthritides such as rheumatoid arthritis are usually well-established clinical entities by the time collapse of the lunate occurs and should not be confused with stage IV Kienbock's disease.

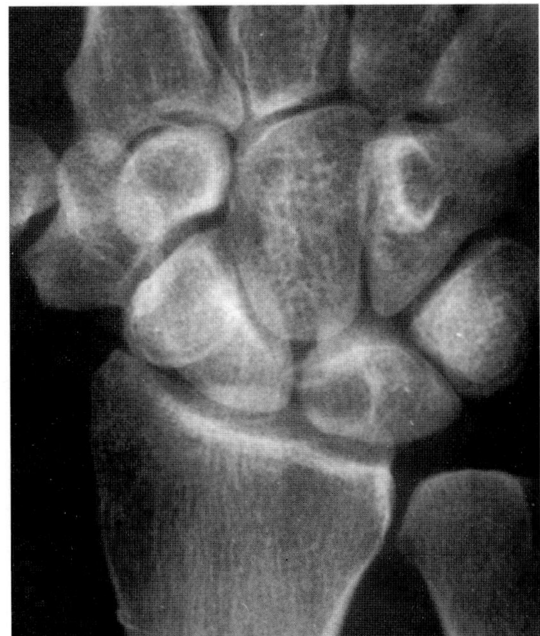
D

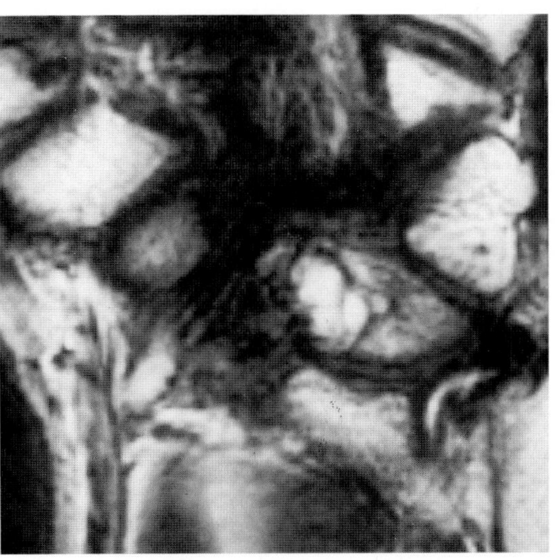
E

FIG. 16. D,E: Intraosseous ganglion cyst. **D:** Anteroposterior radiograph of the wrist demonstrates a well-defined "cystic" lesion with partially sclerotic margins along the radial aspect of the lunate. The location along either the radial or ulnar aspect of the wrist near a ligamentous insertion site is characteristic of this entity. **E:** Coronal T2-weighted fast spin-echo image demonstrates that the lesion is of high signal intensity.

REFERENCES

Introduction

1. Smith DK. MR imaging of normal and injured wrist ligaments. MRI Clin North Am 1995;3:229–248.
2. Totterman SM, Miller RJ. MR imaging of the triangular fibrocartilage complex. *MRI Clin North Am* 1995;3:213–228.
3. Kneeland JB. Technical considerations for MR imaging of the hand and wrist. *MRI Clin North Am* 1995;3:191–196.

Case 1

1. Campbell CS. Gamekeepers thumb. *J Bone Joint Surg* 1955;37: 148–149.
2. Newland CC. Gamekeepers thumb. *Orthop Clin of North Am* 1992; 23:41–48.
3. Stener B. Displacement of the ruptured ulnar collateral ligament of the metacarpal-phalangeal of the thumb. *J Bone Joint Surg [Br]* 1962;44:869–879.
4. Heyman P, Gelberman RH, Duncan K, Hipp JA. Injuries of the ulnar collateral ligament of the thumb metacarpophalangeal joint. *Clin Orthop* 1993;292:165–171.
5. O'Callaghan BI, Kohut G, Hoogewoud HM. Gamekeeper thumb: identification of the Stener lesion with US. *Radiology* 1994;192: 477–480.
6. Noszian IM, Dinkhauser LM, Orthner E, Straub GM, Csanady M. Ulnar collateral ligament: differentiation of displaced and nondisplaced tears with US. *Radiology* 1995;194:61–63.
7. Hinke DH, Erickson SJ, Chamoy L, Timins ME. Ulnar collateral ligament of the thumb: MR findings in cadavers, volunteers and patients with ligamentous injury (gamekeepers thumb). *AJR* 1994; 163:1431–1434.
8. Louis DS, Buckwalter KA. Magnetic resonance imaging of the collateral ligaments of the thumb. *J Hand Surg [Am]* 1989;14A: 739–741.
9. Spaeth JH, Abrams RA, Bock GW, et al. Gamekeeper's thumb: differentiation of non-displaced and displaced tears of the ulnar collateral ligament with MR imaging. *Radiology* 1993;188:553–556.
10. Hergan K, Mittler C, Oser W. Ulnar collateral ligament: differentiation of displaced and nondisplaced tears with US and MR imaging. *Radiology* 1995;194:65–71.

Case 2

1. Cavallaro MC, Taylor JAM, Gorman JD, et al. Imaging findings in a patient with fibrolipomatous hamartoma of the median nerve. *AJR* 1993;161:837–838.
2. Walker CW, Adams BD, Barnes CL, et al. Case report 667. *Skeletal Radiol* 1991;20:237–239.
3. Silverman TA, Enzinger FM. Fibrolipomatous hamartoma of nerve: a clinicopathologic analysis of 26 cases. *Am J Surg Pathol* 1985: 9:7–14.
4. Langa V, Posner MA, Steiner GE. Lipofibroma of the median nerve: a report of two cases. *J Hand Surg [Br]* 1987;12B:221–223.
5. Amadio PC, Reiman HM, Dobyns JH. Lipofibromatous hamartoma of nerve. *J Hand Surg [Am]* 1988;13A:67–75.
6. Enzinger FW, Weise SW. *Soft tissue tumors.* St. Louis: CV Mosby; 1988:332–334.
7. Jacob RA, Buchino JJ. Lipofibroma of the superficial branch of the radial nerve. *J Hand Surg [Am]* 1989;14A:704–706.

Case 3

1. Middleton WD, Kneeland JB, Kellman GM, et al. MR imaging of the carpal tunnel: normal anatomy and preliminary findings in the carpal tunnel syndrome. *AJR* 1987;148:307–316.
2. Reiser MF, Bongartz GP, Erlemann R, et al. Gadolinium-DTPA in rheumatoid arthritis and related diseases: first results with dynamic magnetic resonance imaging. *Skeletal Radiol* 1989;18:591–597.
3. Mesgarzadeh M, Schneck CD, Bonakdarpour A. Carpal tunnel: MR imaging. Part I. Normal anatomy. *Radiology* 1989;171:743–748.
4. Mesgarzadeh M, Schneck CD, Bonakdarpour A, Mitra A, Conaway D. Carpal tunnel: MR imaging. Part II: Carpal tunnel syndrome. *Radiology* 1989;171:749–754.
5. Healy C, Watson JD, Longstaff A, Campbell MJ. Magnetic resonance imaging of the carpal tunnel. *J Hand Surg [Br]* 1990;15B: 243–248.
6. Murphy RX, Chernofsky MA, Osborne MA, Wolson AH. Magnetic resonance imaging in the evaluation of persistent carpal tunnel syndrome. *J Hand Surg [Am]* 1993;18A:113–120.

Case 4

1. Weber ES, Chao EYS. An experimental approach to the mechanism of scaphoid waist fractures. *J Hand Surg* 1978;3:142–148.
2. O'Brien ET. Acute fractures and dislocations of the carpus. In: Lichtman DM, ed. *The wrist.* Philadelphia: WB Saunders; 1988: 129–159.
3. Mack GM, Lichtman DM. Scaphoid nonunion. In: Lichtman DM, ed. *The wrist.* Philadelphia: WB Saunders; 1988:293–328.
4. Gelberman RH, Menon J. The vascularity of the scaphoid bone. *J Hand Surg [Am]* 1980;5A:508–513.
5. Schweitzer M, Karasick D. MRI of the ankle and hindfoot. *Semin US CT MRI* 1994;15:410–422.
6. Pathria MN, Wiber JH, Yulish BS. MR imaging of fracture nonunion. *Radiology* 1988;169P:191.
7. Steinbach L, Fritz RC. The wrist and hand. In: Chan WP, Lang P, Genant HK, eds. *MRI of the musculoskeletal system.* Philadelphia: WB Saunders, 1994.
8. Mazet R, Hohl M. Conservative treatment of old fractures of the carpal scaphoid. *J Trauma* 1961;1:115–127.
9. Cristiani G, Cerofolini E, Squarzina PB, Zanasi S, et al. Evaluation of ischaemic necrosis of carpal bones by magnetic resonance imaging. *J Hand Surg [Br]* 1990;15B(2):249–255.
10. Gelberman RH, Menon J. The vascularity of the scaphoid bone. *J Hand Surg [Am]* 1980;5A:508–513.
11. Zlatkin MB, Greenan T. Magnetic resonance imaging of the wrist. *Semin US CT MR* 1990;11:267–287.
12. Ferlic DC, Norin PM. Idiopathic avascular necrosis of the scaphoid: Preiser's disease? *J Hand Surg [Am]* 1989;14A:13–16.
13. Reinus WR, Conway WF, Totty WG, et al. Carpal avascular necrosis: MR imaging. *Radiology* 1986;160:689–693.
14. Desser TS, McCarthy S, Trumble T. Scaphoid fractures and Kienbock's disease of the lunate: MR imaging with histopathologic correlation. *Magn Reson Imaging* 1990;8:357–361.
15. Totty WG, Murphy WA, Ganz WI, Kumar B, Daum WJ, Siegel BA. Magnetic resonance imaging of the normal and ischemic femoral head. *AJR* 1984;143:1273–1280.
16. Mitchell DG, Rao VM, Dalinka MK, et al. Femoral head avascular necrosis: correlation of MR imaging, radiographic staging, radionuclide imaging, and clinical findings. *Radiology* 1987;162:709–715.
17. Mitchell DG, Kundal DL, Steinberg ME, Kressel HY, Alavi A, Axel L. Avascular necrosis of the hip: comparison of MR, CT and scintigraphy. *AJR* 1986;147:67–71.
18. Cova M, Kang YS, Tsukamoto H, et al. Bone marrow perfusion evaluated with gadolinium-enhanced dynamic fast MR imaging in a dog model. *Radiology* 1991;179:533–539.
19. Tsukamoto H, Kang YS, Jones LC, et al. Evaluation of marrow perfusion in the femoral head by dynamic magnetic resonance imaging. Effect of venous occlusion in the dog model. *Invest Radiol* 1992;27:275–281.
20. Russe O. Fracture of the carpal navicular. Diagnosis, non-operative treatment, and operative treatment. *J Bone Joint Surg [Am]* 1960; 42A:759–768.

Case 5

1. Palmer AK, Werner FW. The triangular fibrocartilage complex of the wrist: anatomy and function. *J Hand Surg* 1981;6:153–162.
2. Greenan T, Zlatkin B. Magnetic resonance imaging of the wrist. *Semin US CT MR* 1990;11:267–287.
3. Skahen JR, Palmer AK, Levinson EM, et al. Magnetic resonance

imaging of the triangular fibrocartilage complex. *J Hand Surg* 1990; 15:552–557.
4. Totterman SS, Miller RJ. Triangular fibrocartilage complex: normal appearance on coronal three-dimensional gradient-recalled-echo MR images. *Radiology* 1995;195:521–527.
5. Totterman SS, Miller RJ. MR imaging of the triangular fibrocartilage complex. *MRI Clin North Am* 1995;3:213–228.
6. Cerofolini E, Luchetti R, Pederzini L, et al. MR evaluation of triangular fibrocartilage complex tears in the wrist: comparison with arthrography and arthroscopy. *J Comput Assist Tomogr* 1990;14: 963–967.
7. Kang HS, Kindynis P, Brahme SK, et al. Triangular fibrocartilage and intercarpal ligaments of the wrist: MR imaging. Cadaveric study with gross pathologic and histologic correlation. *Radiology* 1991; 181:401–404.
8. Schweitzer ME, Brahme SK, Hodler J, et al. Chronic wrist pain: spin-echo and short-term inversion recovery MR imaging and conventional and MR arthrography. *Radiology* 1992;182:205–211.
9. Golimbu CN, Firooznia H, Melone CP Jr, et al. Tears of the triangular fibrocartilage of the wrist: MR imaging. *Radiology* 1989;173: 731–733.
10. Oneson SR, Timins ME, Erickson SJ, et al. MR imaging diagnosis of triangular fibrocartilage pathology with arthroscopic correlation. *Radiology* 1994;193P:183.
11. Metz VM, Schratter M, Dock WI, et al. Age-associated changes of the triangular fibrocartilage of the wrist: evaluation of the diagnostic performance of MR imaging. *Radiology* 1992;184:217–220.
12. Sugimoto H, Shinozaki T, Ohsawa T. Triangular fibrocartilage in asymptomatic subjects: investigation of abnormal MR signal intensity. *Radiology* 1994;191:193–197.
13. Plamer AK: Triangular fibrocartilage complex lesions: a classification. *J Hand Surg* 1989;14:594–606.
14. Sullivan PP, Berquist TH. Magnetic resonance imaging of the hand, wrist and forearm: utility in patients with pain and dysfunction as a result of trauma. *Mayo Clin Proc* 1991;66:1217–1221.
15. Escobedo EM, Bergman GA, Hunter JC. MR imaging of ulnar impaction. *Skeletal Radiol* 1995;24:85–90.
16. Epner RA, Bower WH, Guilford WB. Ulnar variance: the effect of wrist positioning and roentgen filming technique. *J Hand Surg* 1982; 7:298–305.
17. Palmer AK, Glisson RR, Werner FW. Ulnar variance determination. *J Hand Surg* 1982;7:376–379.
18. Chun S, Palmer AK. The ulnar impaction syndrome: follow-up of ulnar shortening osteotomy. *J Hand Surg* 1993;18:46–53.
19. Friedman SL, Palmer AK. The ulnar impaction syndrome. *Hand Clin* 1991;7:295–310.

Case 6

1. Gilkeson G, Polisson R, Sinclair H, et al. Early detection of carpal erosions in patients with rheumatoid arthritis: a pilot study of magnetic resonance imaging. *J Rheumatol* 1988;15:1361–1366.
2. Foley-Nolan D, Stack JP, Ryan M, et al. Magnetic resonance imaging in the assessment of rheumatoid arthritis: a comparison with plain film radiographs. *Br J Rheumatol* 1991;30:101–106.
3. Beltran J, Caudill JL, Herman LA, et al. Rheumatoid arthritis: MR imaging manifestations. *Radiology* 1987;165:153–157.
4. Rominger MB, Bernreuter WK, Kenney PJ, et al. MR imaging of the hands in early rheumatoid arthritis: preliminary results. *Radiographics* 1993;13:37–46.
5. Konig H, Sieper J, Wolf KJ. Rheumatoid arthritis: evaluation of hypervascular and fibrous pannus with dynamic MR imaging enhanced with Gd-DTPA. *Radiology* 1990;176:473–477.
6. Reiser MF, Bongartz GP, Erlemann R, et al. Gadolinium-DTPA in rheumatoid arthritis and related diseases: first results with dynamic magnetic resonance imaging. *Skeletal Radiol* 1989;18:591–597.
7. Gubler FM, Algra PR, Maas M, Dijkstra PF, Falke TH. Gadolinium-DTPA enhanced magnetic resonance imaging of bone cysts in patients with rheumatoid arthritis. *Ann Rheum Dis* 1993;52:716–719.
8. Yanagawa A, Takano K, Nishioka K, Shimada J, Mizushima Y. Clinical staging and gadolinium-DTPA enhanced images of the wrist in rheumatoid arthritis. *J Rheumatol* 1993;20:781–784.
9. Jevtic V, Watt I, Rozman B, et al. Precontrast and postcontrast (Gd-DTPA) magnetic resonance imaging of hand joints in patients with RA. *Clin Radiol* 1993;48:176–181.
10. Kursonuglu-Brahme S, Riccio T, Weisman MH, et al. Rheumatoid knee: role of gadopentetate-enhanced MR imaging. *Radiology* 1990; 176:831–835.
11. Rubens DJ, Blebes JS, Totterman SM, Hooper MM. RA: evaluation of wrist extensor tendons with clinical examination versus MR imaging-a preliminary report. *Radiology* 1993;187:831–838.

Case 7

1. Yao L, Lee JK. Occult intraosseous fracture: detection with MR imaging. *Radiology* 1988;167:749–751.
2. Mink JH, Deutsch AL. Occult cartilage and bone injuries of the knee: detection, classification, and assessment with MR imaging. *Radiology* 1989;170:823–829.
3. Vellet AD, Marks PH, Fowler PJ, et al. Occult post-traumatic osteochondral lesions of the knee: prevalence, classification and short-term sequelae evaluated with MR imaging. *Radiology* 1991;178: 271–276.
4. Schweitzer ME, Karasick D. MRI of the ankle and hindfoot. *Semin US CT MR* 1994;15:420–422.
5. Lee JK, Yao L. Stress fractures: MR imaging. *Radiology* 1988;169: 217–220.

Case 8

1. Ruben DJ, Blebea JS, Totterman SMS, Hopper MM. Rheumatoid arthritis: evaluation of wrist extensor tendons with clinical examination versus MR imaging—preliminary report. *Radiology* 1993;187: 831–838.
2. Glajchen N, Schweitzer ME. MRI features in de Quervain's tenosynovitis of the wrist. *Skeletal Radiol* 1996;25:63–65.

Case 9

1. Brown DM, Young VL. Hand infections. Current concepts in therapy. *South Med J* 1993;86:56–66.
2. Glass KD. Factors related to the resolution of treated hand infections. *J Hand Surg* 1982;3:388.
3. Linscheid RL, Dobyns JH. Common and uncommon infections of the hand. *Orthop Clin North Am* 1975;6:1063.
4. Bolton H, Fowler PJ, Jepson RP. Natural history and treatment of pulp space infection and osteomyelitis of the terminal phalanx. *J Bone Joint Surg [Br]* 1949;31:499.
5. Lowden TJ. Infection in digital pulp space. Lancet 1951;1:196.
6. Waldvogel FA, Vasey H. Osteomyelitis: the past decade. *N Engl J Med* 1980;303:360.
7. Marks B, Schweitzer ME. Carpal fluid: normal appearance in vivo, effusion criteria and association. [work in progress]
8. Morrison WB, Schweitzer ME, Bock GW, et al. Diagnosis of osteomyelitis: utility of fat-suppressed contrast-enhanced MR imaging. *Radiology* 1993;189:251–257.
9. Schweitzer ME, Karasick D. MRI of the ankle and hindfoot. *Semin US CT MR* 1994;15:410–422.

Case 10

1. Blatt G. Scapholunate instability. In: Lichtman DM, ed. *The wrist and its disorders.* Philadelphia: WB Saunders; 1988:251–273.
2. Taleisnik J. The ligaments of the wrist. *J Hand Surg* 1976;1:110–118.
3. Reicher MA, Kellerhouse LE. Carpal instability. In: Reicher MA, Kellerhouse LE, eds. *MRI of the wrist and hand.* New York: Raven Press, 1990:69–85.
4. Mayfield JK, Johnson RP, Kilcoyne FR. The ligaments of the human wrist and their functional significance. *Anat Rec* 1976;186:417–428.
5. Watson HK, Black DM. Instabilities of the wrist. *Hand Clin* 1987; 3:103–111.
6. Fisk GR. Carpal instability and the fractured scaphoid. *Ann R Coll Surg Engl* 1970;46:63–76.

7. Taleisnik J. *The wrist.* New York: Churchill Livingstone: 1985: 239–280.

Case 11

1. Viegas SF, Patterson RM, Peterson PD, et al. Ulnar-sided perilunate instability: an anatomic and biomechanic study. *J Hand Surg [Am]* 1990;15:268–278.
2. Taleisnik J. *The wrist.* New York: Churchill Livingstone: 1985:13–38.
3. Cooney WP, Garcia-Elias M, Dobyns JH, Linscheid RL. Anatomy and mechanics of carpal instability. *Surg Rounds Orthop*1989;3: 15–24.

Case 12

1. Strickland JW. Flexor tendon injuries: treatment principles. *J Am Acad Orthop Surg* 1995;3:44–53.
2. Doyle JR. Anatomy of the finger flexor tendon sheath and pulley system. *J Hand Surg [Am]* 1988;13:473–484.
3. Strickland JW. Flexor tendon injuries: operative technique. *J Am Acad Orthop Surg* 1995;3:55–62.

Case 13

1. Miller TT, Potter HG, McCormack RR. Benign soft tissue masses of the wrist and hand: MRI appearances. *Skeletal Radiol* 1994;23: 327–332.
2. Jelinek JS, Kransdorf MJ, Utz JA, et al. Imaging of pigmented villonodular synovitis with emphasis on MR imaging. *AJR* 1989; 152:337–342.
3. Hughes TH, Sartoris DJ, Schweitzer ME, et al. Pigmented villonodular synovitis: MRI characteristics. *Skeletal Radiol* 1995;24:7–13.
4. Poletti SC, Gates HS, Martinez SM, et al. The use of magnetic resonance imaging in the diagnosis of pigmented villonodular synovitis. *Orthopedics* 1990;13:185–190.
5. Besette PR, Cooley PA, Johnson RP, et al. Gadolinium enhanced MRI of pigmented villonodular synovitis of the knee. *J Comput Assist Tomogr* 1992;16:992–994.
6. Kottal RA, Bogler JB, Matamoros A, et al. Pigmented villonodular synovitis: a report of imaging in two cases. *Radiology* 1987;163: 551–553.
7. Cotten A, Flipo RM, Chastanet P, et al. Pigmented villonodular synovitis of the hip: review of radiographic features in 58 patients. *Skeletal Radiol* 1995;24:1–7.
8. Binkovitz LA, Berquist TH, McLeod RA. Masses of the hand and wrist: detection and characterization with MR imaging. *AJR* 1990; 154:323–326.
9. Kransdorf MJ, Moser RPJ, Meis JM, et al. Fat containing soft-tissue masses of the extremities. *Radiographics* 1991;11:81–106.

Case 14

1. Jolly SL, Rerlic DC, Clayton ML, et al. Swanson silicone arthroplasty of the wrist in rheumatoid arthritis: a long term follow-up. *J Hand Surg [Am]* 1992;17:142.
2. Weissman BN. Imaging of joint replacement. In Resnick D, ed. *Diagnosis of bone and joint disorders.* 3rd ed. Philadelphia: WB Saunders; 1995:559–606.
3. Atkinson RE, Smith RJ: Silicone synovitis following silicone implant arthroplasty. *Hand Clin* 1986;2:291.
4. Smith RJ, Atkinson RE, Jupiter JB. Silicone synovitis of the wrist. *J Hand Surg* 1985;10:47–60.
5. Schneider HJ, Weiss MA, Stern PJ: Silicone induced erosive arthritis: radiologic features in seven cases. *AJR* 1987;148:923.

Case 15

1. Carroll RE, Berman AT. Glomus tumors of the hand. *J Bone Joint Surg [Am]* 1972;54A:691–703.
2. Schneider LH, Hunter JM, DePaula C. The glomus tumor. *Am Fam Physician* 1975;12:140–141.
3. Luc-Drape J, Idy-Peretti I, Goettmann S, et al. Subungual glomus tumors: evaluation with MR imaging. *Radiology* 1995;195:507–515.
4. Dawber RPR, Baran R. Structure, embryology, comparative anatomy and physiology of the nail. In: Baran R, Daqber RPR, eds. *Diseases of the nail and their management.* Oxford, England: Blackwell Scientific; 1984:1–23.
5. Rettig AC, Strickland JW. Glomus tumor of the digits. *J Hand Surg* 1977;2:261–265.
6. Kouskoukis CE. Subungual glomus tumor: a clinicopathological study. *J Dermatol Surg Oncol* 1983;9:294–296.
7. Gandon F, Legaillard P, Brueton R et al. Forty-eight glomus tumors of the hand: retrospective study and four-year follow-up. *Ann Hand Surg* 1992;11:401–405.
8. Fornage BD. Glomus tumors in the finger: diagnosis with US. *Radiology* 1988;167:183–185.
9. Kneeland JB, Middleton WD, Matloub HS et al. High resolution MR imaging of glomus tumor. *J Comput Assist Tomogr* 1987;11: 351–352.
10. Jablon M, Horowitz A, Bernstein DA. Magnetic resonance imaging of a glomus tumor of the fingertip. *J Hand Surg [Am]* 1990;15: 507–509.
11. Schneider LH, Bachow TB. Magnetic resonance imaging of a glomus tumor. *Orthop Pev* 1991;20:255–256.
12. Holzberg M. Glomus tumor of the nail: a ''red herring'' clarified by magnetic resonance imaging. *Arch Dermatol* 1992;128:160–162.
13. Idy-Peretti I, Cermakova E, Dion E, et al. Subungual glomus tumor: diagnosis based on high resolution MR images [letter]. *AJR* 1992; 159:1351.
14. Hou SM, Shih TTF, Lin MC. Magnetic resonance imaging of an obscure glomus tumor in the fingertip. *J Hand Surg [Br]* 1993;18: 482–483.
15. Matloub HS, Muoneke UN, Prevel CD, et al. Glomus tumor imaging: use of MRI for localization of occult lesions. *J Hand Surg [Am]* 1992;17:472–475.
16. Wong EC, Jesmanowicz A, Hyde JS. High resolution, short echo time MR Imaging of the fingers and wrist with a local gradient coil *Radiology* 1991;181:393–397.

Case 16

1. Golimbu CN, Firooznia H, Rafii M. Avascular necrosis of carpal bones. *MRI Clin North Am* 1995;3:281–303.
2. Taleisnik J: Kienbock's disease. In: *The wrist.* New York: Churchill Livingstone; 1985:858–873.
3. Gelberman RH, Salamon PB, Jurist JM, et al: Ulnar variance in Kienbock disease. *J Bone Joint Surg [Am]* 1975;57:674–676.
4. D'Hoore K, DeSmet L, Verelen K, et al: Negative ulnar variance is not a risk factor for Kienbock disease. *J Hand Surg [Am]* 1994; 19:229–231.
5. Imaeda T, Nakamura R, Miura T, et al: Magnetic resonance in Kienbock disease. *J Hand Surg [Br]* 1992;17:20–27.
6. Winalski CS, Aliabadi P, Wright RJ, et al. Enhancement of joint fluid with intravenously administered gadopentetate dimeglumine: technique, rationale and implications. *Radiology* 1993;187:179–185.

CHAPTER 4

The Hip

Mini N. Pathria, Andrew L. Deutsch, and Dennis M. Wilcox

The hip represents one of the first articulations to which magnetic resonance (MR) imaging was substantially applied, and many of the pioneering applications of musculoskeletal MR were first described in this joint. Among these applications include the use of MR for assessment of osteonecrosis, the first description of the phenomenon of transient bone marrow edema, and the initial determination of the cost-effective use of MR for occult fracture detection. This chapter reflects the importance of all these applications of MR for the assessment of bone and marrow, and these topics are considered in detail.

MR may also be applied to the evaluation of a number of diverse congenital and developmental anomalies that affect the hip. The use of non-radiography-based technique is obviously highly desirable in the pediatric population. The use of MR in the assessment of variety of pediatric conditions including developmental dysplasia, coxa vara, proximal focal femoral deficiency, Legg-Perthes disease, and slipped capital femoral epiphyses is reviewed.

Emerging applications of MR in the region of the hip are also considered. Most MR examinations of the hip performed today are done as quasi-MR pelvic examinations. The body coil is most frequently utilized and both hips evaluated simultaneously with the spin-echo and various fat-suppression methods [e.g., short tau inversion recovery (STIR)]. While assessment of the contralateral hip is desirable in several situations (such as osteonecrosis) higher spatial resolution is required for the assessment of a variety of nonosseous applications. The increasing availability of surface coils that can be readily used in assessment of the hip (such as flexible wrap around coils) has facilitated the practical performance of higher resolution studies directed at the area of maximal symptomatology. The use of surface coils is mandated for the detection of localized abnormalities including loose bodies and tears of the acetabular labrum. The emerging use of MR arthrography in the hip for assessment of these latter disorders is featured in several cases devoted to technique, applications, and pitfalls of this newly developed application.

As in the case of any articulation, the hip may be affected by a variety of non-site-specific inflammatory and neoplastic disorders. Several of these are considered in cases in this section, and these discussions are coordinated and cross-referenced to the treatment of these subjects in other sections of the text.

CASE 1

History: A 28-year-old woman first complained of pain within her left hip during the late third trimester of her pregnancy. The pain persisted post partum and had recently become incapacitating. The pain did not respond to rest or analgesics. What is the most likely diagnosis?

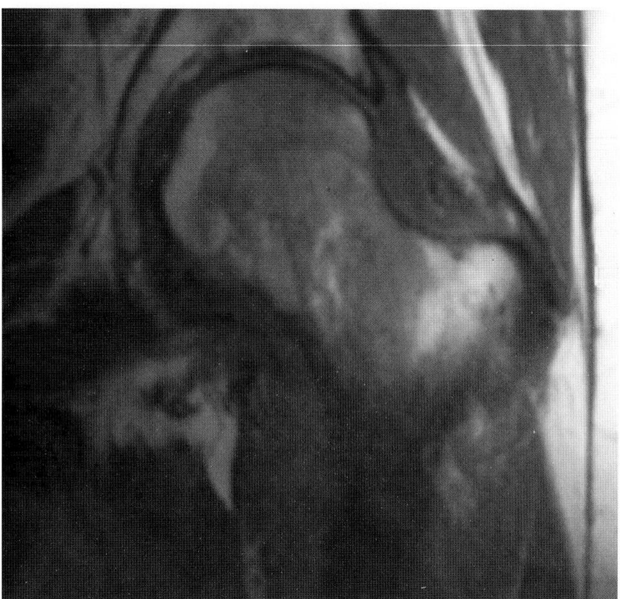
A

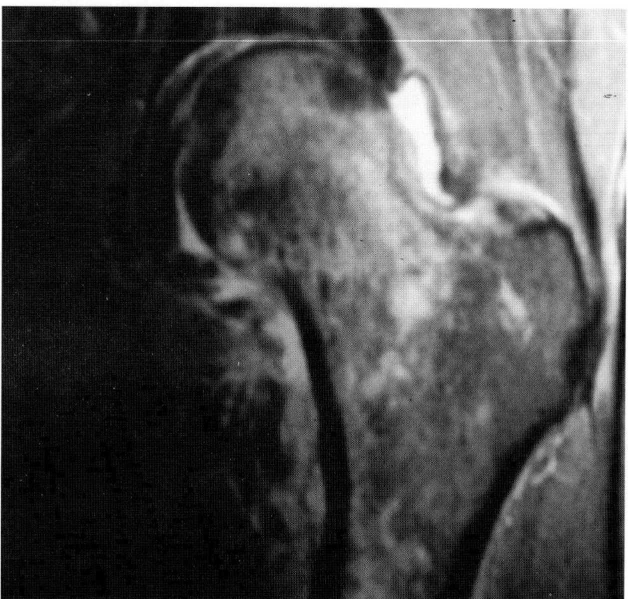
B

FIG. 1.

Findings: Coronal T1-weighted spin-echo [TR (repetion time)/TE (echo time) 600/20] image demonstrates diffuse loss of the normally expected high signal intensity within the femoral head and neck (Fig. 1A). The findings extend into the intertrocanteric region. Coronal fast spin-echo inversion recovery [TR/TE/inversion time (TI) 3,500/32/60 msec] image (Fig. 1B) demonstrates markedly increased signal intensity corresponding to the areas of decreased signal depicted in Fig. 1A. A small effusion is present.

Diagnosis: Transient osteoporosis (marrow edema) of the left hip.

Discussion: Edema of the bone marrow within the femoral head and neck is a common finding on MR imaging when imaging an adult with unexplained hip pain. Bone marrow edema refers to an increase in free water within the marrow space in the absence of significant cellular infiltration. Edema may be reactive to a known underlying disorder or may be present on an idiopathic basis. Reactive bone marrow edema is seen in the early stages of avascular necrosis, following trauma, subsequent to fractures, adjacent to osteomyelitis or chronic soft tissue infection, and at the margins of neoplastic tissue. In many cases, no apparent cause for bone marrow edema can be identified, and the marrow edema is considered idiopathic in origin.

Three related syndromes, known as transient osteoporosis of the hip (TOH), regional migratory osteoporosis, and transient bone marrow edema syndrome are associated with proximal femur bone marrow edema of uncertain etiology (1). All three have an identical MR appearance, although they differ in their radiographic findings and clinical course. While the precise mechanism of marrow edema in this group of disorders is uncertain, vascular, neural, and mechanical causes have all been implicated. These syndromes have some clinical similarities to reflex sympathetic dystrophy, but their predictably limited duration and benign clinical course suggest a distinct entity. Histologic findings in patients with idiopathic bone marrow edema are nonspecific; they include fat necrosis, bone resorption, granulation tissue, synovial inflammation, and edema (2). These self-limited conditions are treated by analgesics and supportive measures, including restricted weight bearing to prevent insufficiency fractures in osteopenic bone.

TOH was initially described in women in the third trimester of pregnancy, based on the radiographic demonstration of focal osteoporosis of the proximal femur (3).

It is now known that TOH more commonly affects previously healthy men in the third and fourth decades of life. Symptoms include an abrupt onset of progressive hip pain that is exacerbated by weight bearing, as well as increasing functional disability. The initial radiographs in TOH are normal. Several weeks after the onset of symptoms, osteoporosis of the femoral head develops, with less involvement of the femoral neck, trochanters, and acetabulum. The joint space is preserved, although there may be prominent loss of the subchondral cortex of the femoral head (1,3). Scintigraphy demonstrates positive findings before the radiographic detection of osteopenia. Marked homogeneous accumulation in the femoral head is typical; uptake decreases before the radiograph returns to normal. Resolution of symptoms and osteopenia is usually complete 6 to 8 months after the onset of symptoms (3).

The next related condition is regional migratory osteoporosis, a less common disorder in which multiple anatomic sites are affected sequentially (3). This condition occurs mostly in middle-aged men, who present with spontaneous local pain and swelling in one joint that resolves within 1 year without sequelae. The large joints of the lower extremity are most commonly involved. Recovery of function of the initial site is followed by the development of similar symptoms in other regions of the ipsilateral or contralateral extremity. Symptoms rarely recur in the same joint, and synchronous involvement of several regions is uncommon. When the initial episode of bone marrow edema involves the hip, subsequent episodes most commonly involve the contralateral hip or the ipsilateral knee, ankle, or foot (Fig. 1C–G).

The most recently described entity in this group of disorders has been designated the transient bone marrow edema syndrome (1). This newly recognized pattern has no conventional radiographic findings and is diagnosed solely on the basis of clinical findings and abnormalities on MR imaging. The clinical symptoms and the MR appearance of bone marrow edema are identical to those of TOH, but this form is not associated with radiographic osteopenia. This variant is presumably an earlier, milder, or previously unrecognized form of TOH (1).

The MR appearances of TOH, regional migratory osteoporosis involving the hip, and bone marrow edema are identical. A diffuse bone marrow edema pattern involves the femoral head, femoral neck, and sometimes the intertrochanteric region and/or the acetabulum (1,2,4). The edema appears as a homogenous, fairly well-marginated zone of diminished signal on the T1-weighted images. Increased signal within the femoral head and neck are evident on T2-weighted images, particularly if fat-suppression techniques are employed. The STIR sequence is highly sensitive to changes in water content and shows markedly increased signal in the region of the abnormal bone marrow. Because of its high sensitivity to changes in free water, MR can show alterations in the bone marrow signal within 48 hours after the onset of symptoms of TOH (5).

The areas of signal loss are often quite extensive and (rarely) may be associated with mild loss of signal in the acetabulum. These signal alterations may involve the entire femoral head or they may spare some portions, either anteriorly or posteriorly. The anatomic distribution of the marrow abnormalities may change on sequential MR studies. Hauzeur et al. (6) noted alterations over time in distribution of the regions of abnormal bone marrow signal in five of their eight patients with TOH (6). Marrow edema was initially observed in the anterior femoral head but subsequently migrated to the posterior femoral head as the anterior portion converted back to normal signal. The interface between the normal bone marrow and the edematous marrow is well defined, but there are no low-signal bands demarcating the interface. No low-signal band, ring, or double-line sign is seen in patients with idiopathic bone marrow edema. The cortex and osseous contour remain intact and the adjacent soft tissues are normal. A small joint effusion is usually present during the acute period.

The differential diagnosis of proximal femoral bone marrow edema includes early osteonecrosis without focal lesions, post-traumatic bone bruise, osteomyelitis, and infiltrative neoplasm. The hypointensity of bone marrow fat on T1-weighted MR sequences tends to be more extensive in TOH than in osteonecrosis. In the former condition, altered signal often extends from the femoral neck to the intertrochanteric ridge and occasionally involves the acetabulum, whereas signal alterations confined to the femoral head are more characteristic of osteonecrosis. Distinguishing TOH from osteonecrosis is not difficult if the latter condition demonstrates a zone of low signal intensity with a T2-weighted hyperintense signal margin just inside it—the double line sign. However, osteonecrosis may occasionally present as diffuse signal abnormality in the femoral head and neck; in these instances, the MR study may not allow differentiation of TOH from osteonecrosis (7). In such cases, the diffuse bone marrow changes may obscure a focal subchondral abnormality. Follow-up examination in patients with risk factors for osteonecrosis may be obtained. In transient bone marrow edema, the signal alterations disappear over a period of several months. In osteonecrosis, evolution of the infarction results in the development of the diagnostic low-signal bands in the area of the infarct. The altered bone marrow signal in TOH extends from the subchondral portion of the epiphysis into the metaphysis, which may be useful in distinguishing TOH from infiltrative neoplasm and osteomyelitis, which are usually located in the metaphysis. Osteomyelitis and neoplasm also often demonstrate associated soft tissue extension, a finding absent in TOH (4).

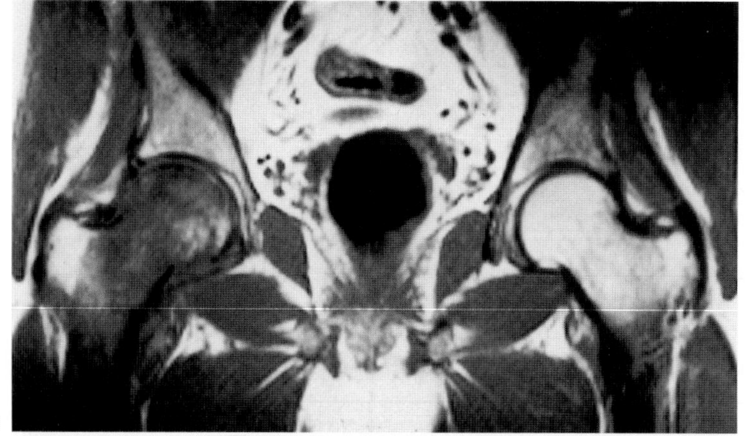
C

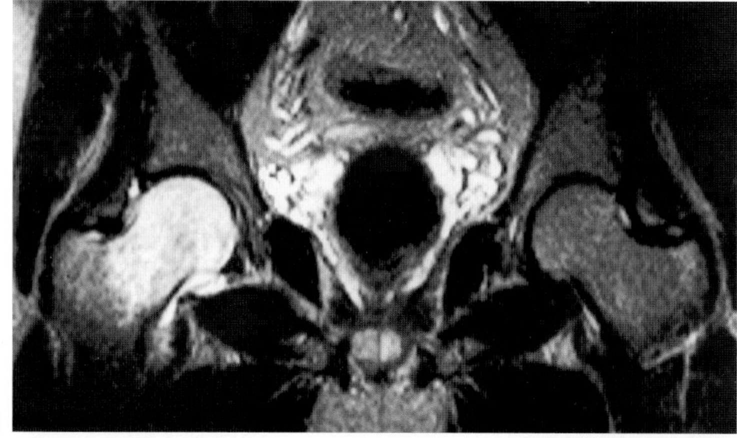
D

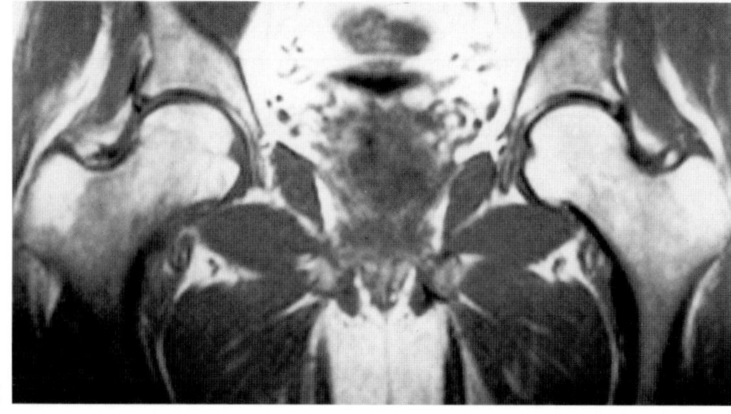
E

FIG. 1. C–G: MR images from three sequential examinations performed on a 42-year-old man illustrate the MR findings of regional migratory osteoporosis. The clinical presentation at the time of this man's initial MR study was consistent with transient osteoporosis of the right hip. Following complete resolution of symptoms, he presented again with identical symptoms involving the left hip, when the diagnosis of regional migratory osteoporosis was established. **C:** In September, a T1-weighted (TR/TE 422/15 msec) spin-echo coronal image of the pelvis shows homogeneous low signal in the right femoral head and neck. **D:** The T2-weighted (TR/TE 2,000/110 msec) spin-echo image at this time shows intense high signal in the corresponding region. **E:** In January, the bone marrow signal has reverted to a normal appearance on the T1-weighted (TR/TE 422/15 msec) MR image.

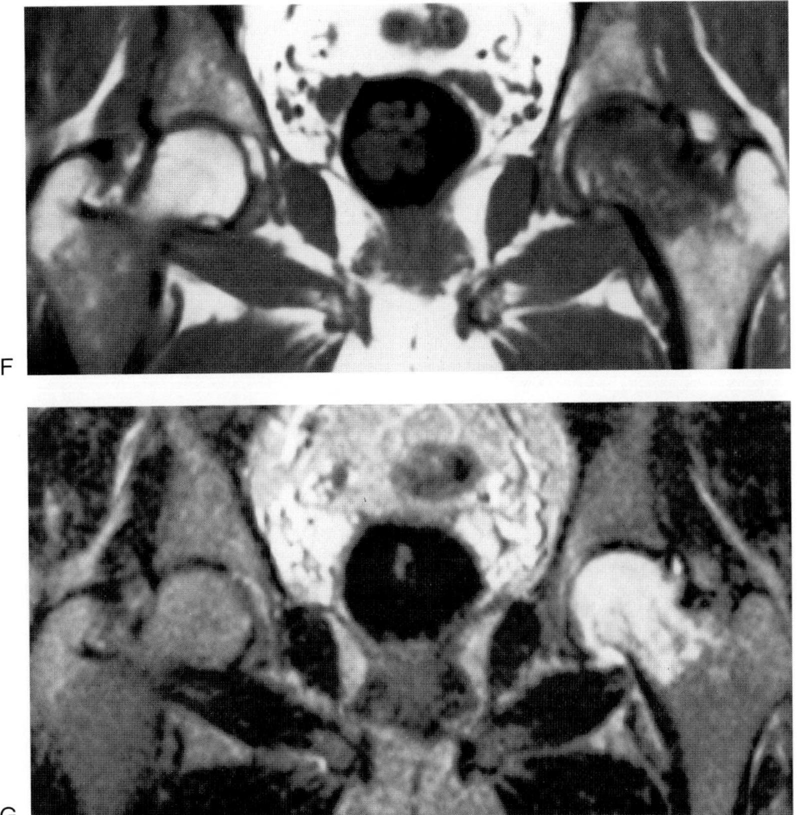

FIG. 1. F: In June, symptoms developed in the left hip. The T1-weighted (TR/TE 422/15 msec) spin-echo MR image shows abnormal low signal in the left femoral head and neck. The right hip remains normal. **G:** The T2-weighted (TR/TE 2,000/110 msec) spin-echo image shows high signal in the left femoral head and neck. No focal lesions are present.

CASE 2

History: A 13-year-old boy complained of proximal thigh pain that had been increasing over the past two months and was particularly severe at night. What is the most likely histology of the lesion in the right femur? What is the explanation for the soft tissue changes?

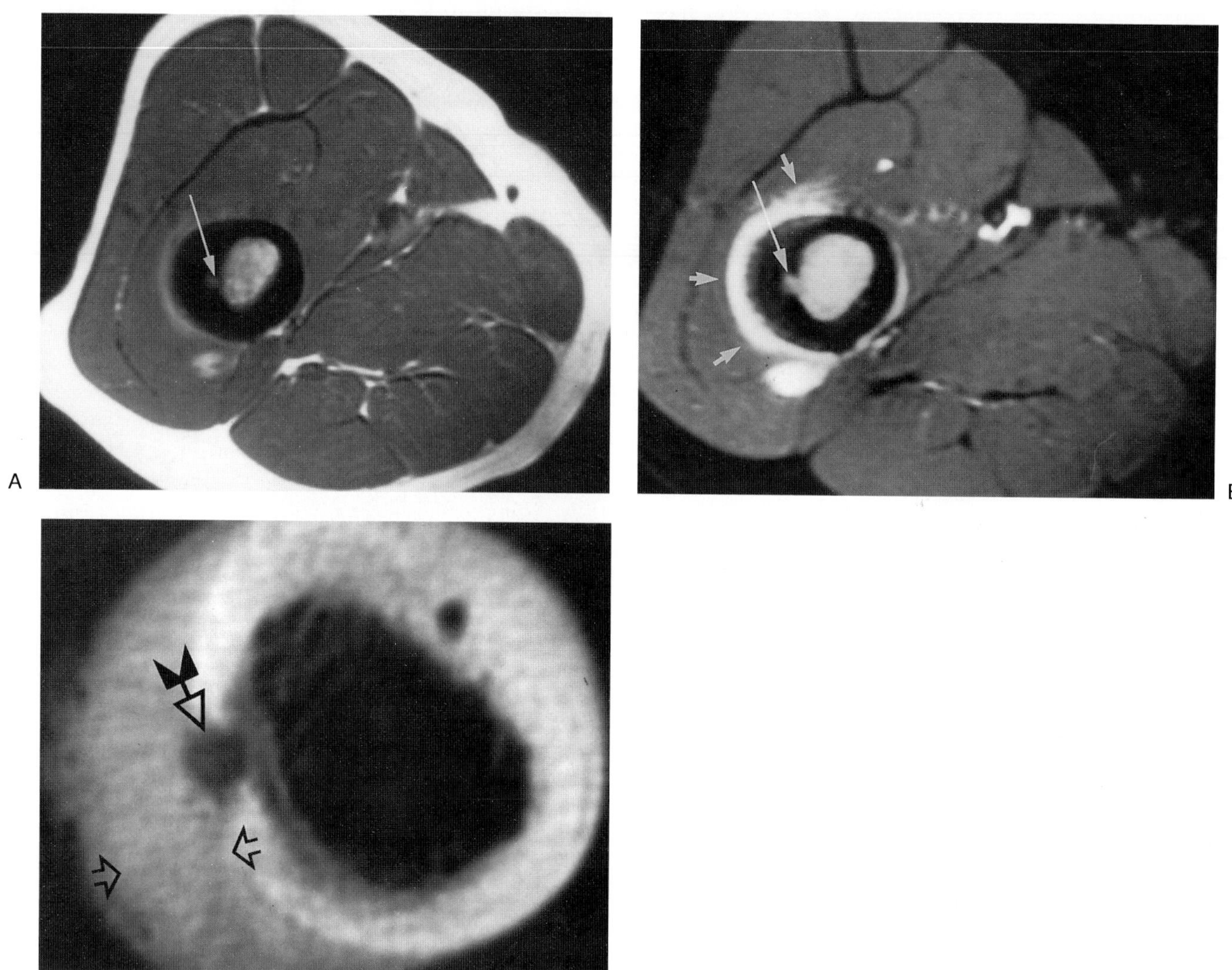

FIG. 2.

Findings: Axial proton-density (TR/TE 1,550/20 msec) image (Fig. 2A) through the femoral diaphysis demonstrates a small circular to ovoid focus of slightly increased signal intensity (long white arrow) within a markedly thickened femoral cortex (low signal intensity). There is a small rim of intermediate signal intensity partially encircling the medial femoral cortex. Axial inversion recovery fast spin-echo (TR/TE/TI 3,000/35/160 msec) sequence (Fig. 2B) demonstrates that the previously defined focus of signal alteration within the cortex has become markedly hyperintense (long white arrow). Additionally noted is a rim of high signal intensity most consistent with edema encircling the markedly thickened femoral cortex (small white arrows). Axial computed tomography (CT) scan (Fig. 2C) demonstrates a circular lesion (triangular arrow) contained with the markedly thickened cortex and periosteal new bone formation (between open black arrows). (Images courtesy of Scott Kingston, M.D., LaCanada, California.)

Diagnosis: Intracortical osteoid osteoma of the proximal femur.

Discussion: Osteoid osteomas are common benign

neoplasms of the skeleton, representing 11% of all benign bone tumors (1). The tumor typically affects individuals under the age of 30 years and occurs in males three times more frequently than females. Severe aching pain, usually worst at night and lessened by salicylates, is the classic clinical presentation. The tumor itself consists of a small (<1.5 cm) highly vascular nidus that is surrounded by dense nontumorous reactive bone. Treatment is aimed at surgically removing the intensely painful nidus. A cortical location of the nidus is most common, but subperiosteal and intramedullary sites also occur (1). In the femur, osteoid osteomas occur most commonly in the cortex of the intertrochanteric region and in an intracapsular subperiosteal location along the medial femoral neck.

Cortical lesions are most common and are the form easiest to diagnose with conventional radiography. On plain radiographs, the nidus of cortical tumors appears as a small round or oval radiolucent area, surrounded by periosteal and endosteal sclerotic bone. The nidus is well demonstrated with CT as a small radiolucent focus surrounded by medullary sclerosis and a thickened cortex. The nidus of osteoid osteomas is often less conspicuous on MR imaging than it is with CT, which remains the imaging modality of choice for detection of osteoid osteoma (1,2). In one series of nine patients with osteoid osteomas, CT was more accurate than MR in identifying the tumor nidus (2). When the nidus is evident by MR, it has a hypointense signal on T1-weighted imaging and a variable signal on T2-weighted sequences (Fig. 2D,E). The nidus may be hypointense on T2-weighted images if it contains abundant fibrosis or calcification. In most cases, the nidus itself demonstrates hyperintensity on T2-weighted images (2). Nidus enhancement is typically seen following the intravenous administration of gadolinium (2). The nidus, regardless of its internal signal, is surrounded by signal void from reactive sclerotic bone. There are no studies comparing high-resolution surface-coil MR with CT for detection of the nidus. High-resolution MR, using thin slices, small field-of-view, and high reconstruction matrices is recommended to facilitate recognition of a small nidus in a patient with a suspected diagnosis of osteoid osteoma.

Intramedullary edema is present in the marrow space contiguous with the lesion in up to two-thirds of patients (2). This edema is seen as inhomogeneous low signal on T1-weighted sequences and becomes more obvious as high signal on STIR or T2-weighted sequences. The inflammatory response incited by the tumor also leads to peritumoral edema in the soft tissue in the vicinity of the osteoid osteoma. The soft tissue changes typically consist of a thin rim of high signal parallel to the cortex on T2-weighted images. This type of perilesional edema was seen in half the patients with cortical lesions in one series (2). Less commonly, the inflammatory changes in the overlying soft tissues appear mass-like, suggesting a more aggressive process (3). Correlation with plain films and CT helps to avoid overinterpretation of the reactive edema surrounding an osteoid osteoma (2,4,5).

Intramedullary and subperiosteal osteoid osteomas tend to incite less osseous response than cortical lesions. The lack of significant bone sclerosis should not preclude the diagnosis of an osteoid osteoma, particularly in the right clinical setting. When osteoid-osteomas occur in an intracapsular location, they tend to be of the subperiosteal variety. Intracapsular lesions result in effusion and synovitis, manifest radiographically as juxta-articular osteoporosis and widening of the medial joint space.The clinical presentation often mimics other entities such as inflammatory arthritis, septic arthritis, and traumatic disorders. Also the amelioration of pain by salicylates is less apparent in cases of intracapsular osteoid osteomas. The joints most commonly affected include the hip, elbow, foot, and wrist (1). The femoral neck is intracapsular and therefore encased by synovium rather than periosteum. Without periosteal reaction, bone sclerosis on plain films is minimal, and the radiographs often suggest the diagnosis of inflammatory or septic arthritis. When present, a thick calcar from the intramedullary reaction strongly favors the diagnosis of an intracapsular osteoid osteoma.

Despite the paucity of radiographic findings, radionuclide scans demonstrate the nidus as a small focal area of intense accumulation. This scintigraphic finding is not specific; it can also be present with a fatigue fracture, Brodie's abscess, or other neoplasm. The bone scan, however, is useful in anatomically localizing the lesion and planning the CT study, which is the optimal method for detection of a small subperiosteal nidus. When a subperiosteal nidus is small, it may not be detected on MR, and only the presence of joint effusion, synovitis, and adjacent bone marrow and soft tissue edema may be appreciated. These MR findings can easily be misinterpreted as a more aggressive process, such as septic arthritis or malignancy. The marked inflammatory response in the bone marrow and in the soft tissues adjacent to an osteoid osteoma may lead to erroneous diagnosis.

Peritumoral edema is also commonly seen in association with osteoblastoma, a larger cousin of osteoid osteoma, and chondroblastoma, an uncommon benign cartilaginous neoplasm that typically presents in children and teenagers. Osteoblastoma is encountered most frequently in the spine, although the long bones of the lower extremities are also frequent sites for this tumor. Chondroblastoma occurs in the epiphyses of long tubular bones, particularly in the femur, humerus, and tibia. The MR appearance of osteoblastoma is nonspecific; it usually appears as an expansile inhomogeneous lesion with surrounding bone marrow edema (6). On MR imaging, chondroblastoma typically manifests as a well-defined, lobulated epiphyseal mass with signal intensity similar to muscle on T1-weighted sequences. T2-weighted images reveal a poorly defined internal lobular pattern with foci of low, intermediate, and high signal intensity (7). With both osteoblastoma and chondroblastoma, periostitis, extensive bone marrow edema, and adjacent soft tissue edema are frequently present (3,6,8). As is the case with osteoid osteoma, the extensive peritumoral edema and periostitis can make benign chondroblastoma appear aggressive on MR imaging.

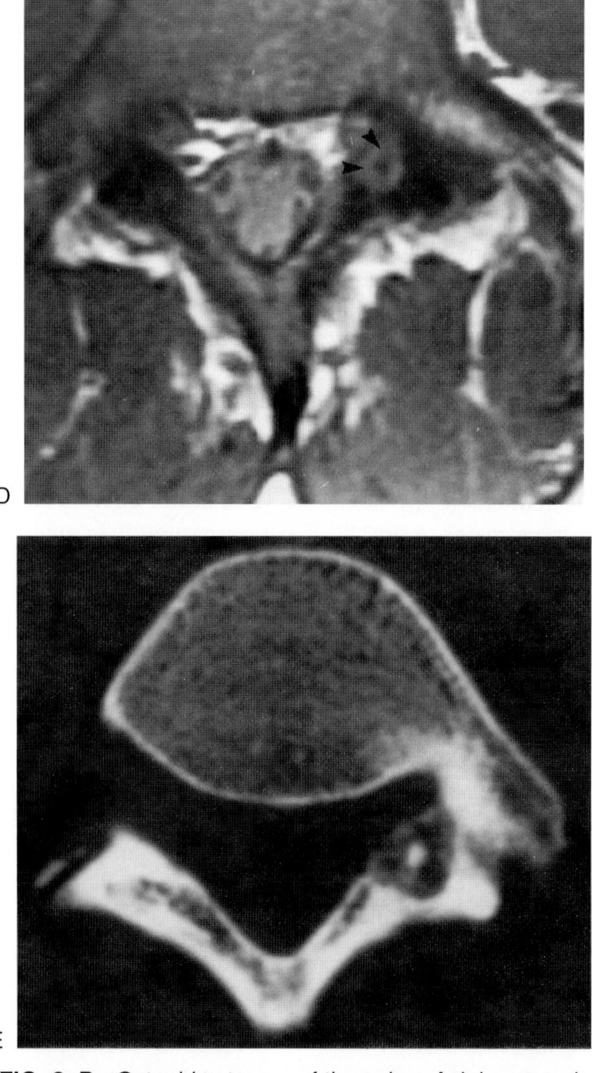

FIG. 2. D: Osteoid osteoma of the spine. Axial proton-density (TR/TE 2,000/20 msec) conventional spin-echo image demonstrates a circular intermediate-signal-intensity lesion representing the osteoid ostoma (*arrowheads*). There is a low-signal-intensity center representing the nidus. **E:** Corresponding axial CT section again demonstrates the lesion as well as the calcified nidus and correlates excellently with the MR.

CASE 3

History: A 4-year-old child underwent MR evaluation to characterize an abnormality of the proximal left lower extremity that had been present since birth. What is your diagnosis? How is this malformation characterized and what are the advantages of MR in assessing this condition?

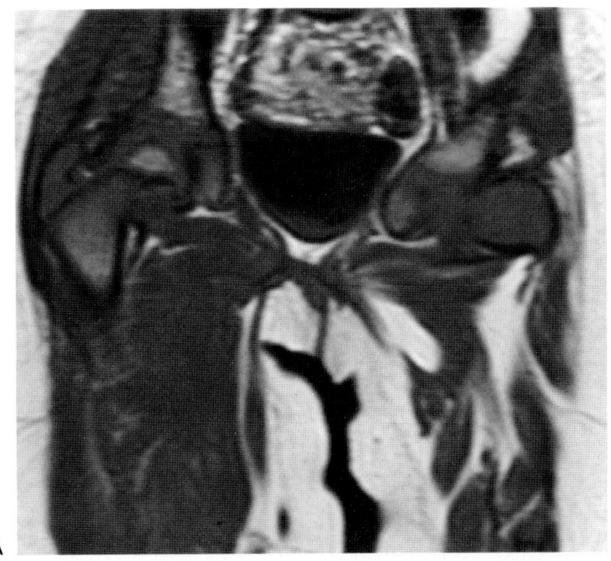

A

FIG. 3.

Findings: Coronal T1-weighted (TR/TE 800/11 msec) image (Fig. 3A) of the pelvis and proximal femora demonstrates absence of the femoral capital epiphysis on the left side. There is abundant nonossified femoral cartilage within the acetabulum and in the region of the greater trochanter. The left lower extremity musculature is atrophic.

Diagnosis: Proximal focal femoral deficiency.

Discussion: Proximal focal femoral deficiency (PFFD) is a congenital condition characterized by partial absence and shortening of the proximal portion of the femur (1–3). The manifestations of PFFD are variable, ranging from mild femoral shortening with congenital varus of the femoral neck to the most severe form, in which only a segment of the distal femur develops. PFFD is distinguished from both total femoral agenesis and congenital coxa vara by the shortening of the femoral neck.

The vast majority of PFFD cases are unilateral, and the distal femur is usually unaffected. Other congenital anomalies, particularly longitudinal deficiency of the ipsilateral fibula, may be present. Clinically, infants with PFFD demonstrate a short bulky thigh that is flexed and abducted. Treatment of PFFD depends on the stability of the hip, leg length discrepancy, femoral malrotation, and inadequacy of the proximal femoral musculature (4). Surgical procedures attempt to correct varus deformity, femoral pseudoarthrosis when present, and the leg length discrepancy.

A variety of classification systems have been suggested, based on the presence and location of the femoral head and neck. The Aitken classification designates four types based on radiographic findings (4). In the least severe group, class A, a femoral head is present and attached to the shaft, although a subtrochanteric pseudoarthrosis may subsequently develop through the gracile attachment site (Fig. 3B–C,D). In class B, there is a femoral head and an adequate acetabulum, but at maturity there is no connection between the femoral head and shaft. In class C, the acetabulum is severely dysplastic and the femoral head is absent. In class D, both the acetabulum and femoral head are absent. Accurate classification is essential for determining prognosis.

Selection of appropriate therapy has been difficult in infants due to difficulty in visualizing the unossified cartilaginous portions of the femur and acetabulum. In the past, arthrography has been utilized to determine the amount of tissue present and to evaluate hip stability. Now MR imaging can noninvasively demonstrate the presence or absence of a cartilaginous femoral head, the presence or absence of a cartilaginous connection in the femoral neck, and the size of these cartilaginous structures

(4). Fat-suppressed gradient-echo imaging has been demonstrated to be advantageous in imaging cartilaginous structures (5). Accurate early classification with MR eliminates a change in classification type as the child grows (4). The major differential diagnosis of the radiographic findings is developmental coxa vara (see Case 4–19). This entity may be familial, is bilateral in one-half of cases, presents later, and represents a true decrease in the neck shaft angle as opposed to the subtrochanteric varus that is present in PFFD.

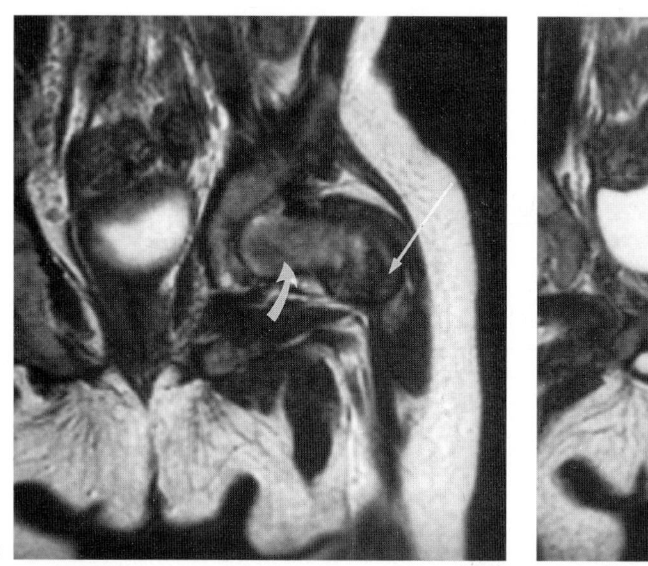
B

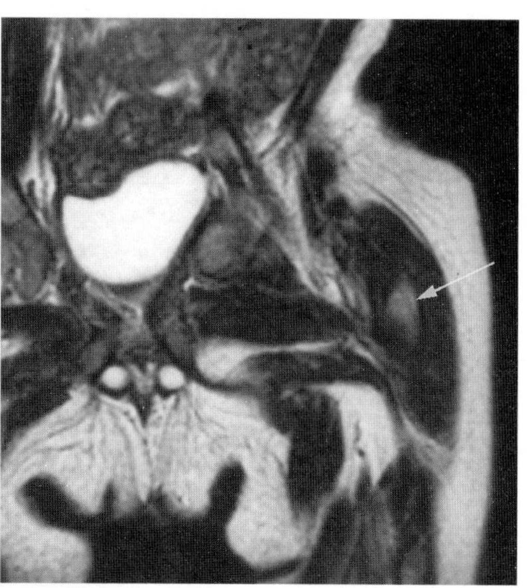
C

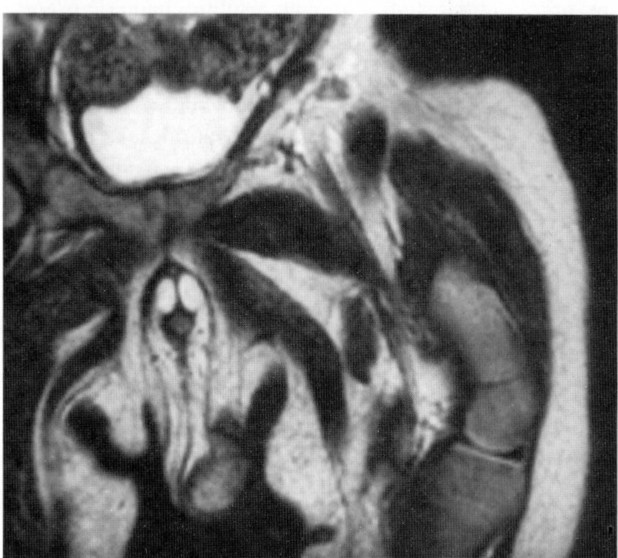
D

FIG. 3. B–D: Proximal focal femoral deficiency, type A. Series of coronal T1-weighted images (TR/TE 500/20 msec) from an MR examination on a 5-month-old boy. **B:** The femoral head (*curved arrow*) and acetabulum are developed. There is marked varus deformity and possible subtrochanteric pseudoarthrosis at the focally deficient attachment site (*long thin arrow*). **C:** The mid-shaft is present (*arrow*) but is incompletely visualized as the image obliquely sections the femoral shaft. **D:** The distal femur and knee are well demonstrated.

CASE 4

History: A 78-year-old woman was involved in a recent accident. She complained of severe left hip pain and refused to bear weight on her left leg. Radiographs of the pelvis and left hip were normal. An MR examination was obtained. What is the cause of her hip pain? What are the advantages of MR compared with scintigraphy for this diagnosis?

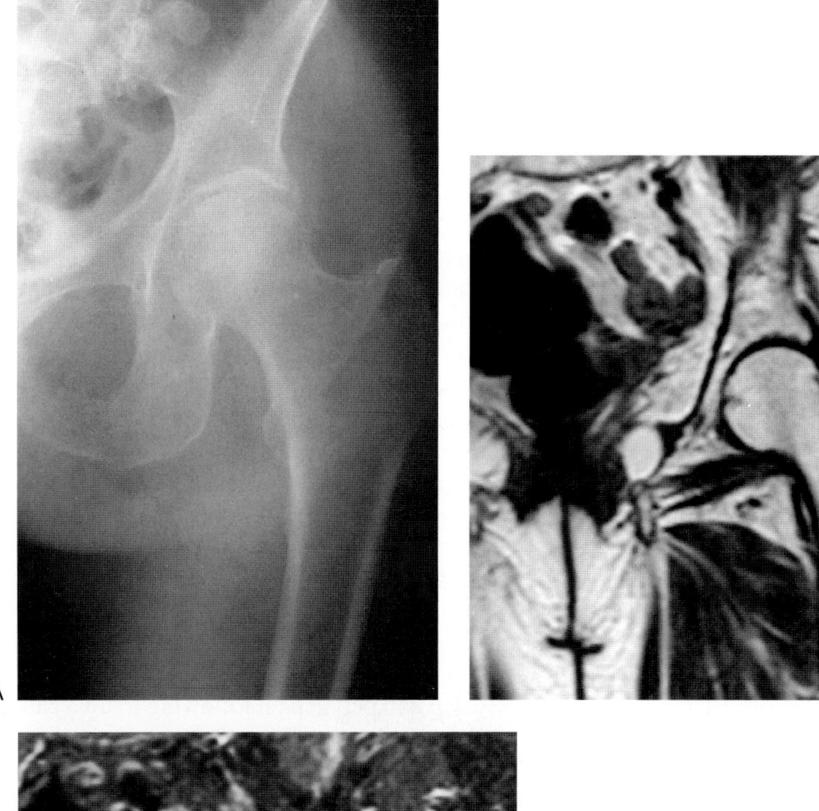
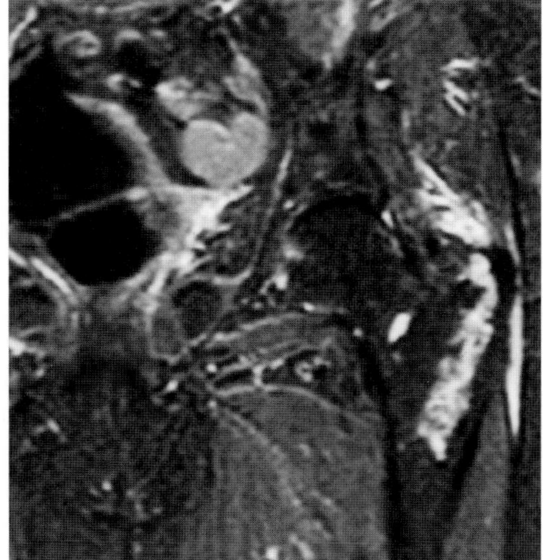

A B C FIG. 4.

Findings: The anteroposterior (AP) radiograph of the left hip (Fig. 4A) is normal. A T1-weighted (TR/TE 600/14 msec) spin-echo image of the pelvis (Fig. 4B) shows a serpiginous line of low signal extending obliquely from the greater trochanter of the left femur into the subtrochanteric region. A STIR (TR/TE/TI 3,000/19/150 msec) image (Fig. 4C) shows high signal in the corresponding area. There is increased signal in the adjacent musculature.

Diagnosis: Occult fracture of the proximal femur.

Discussion: Relatively little trauma is required to fracture the pelvis and proximal femur in the elderly, osteoporotic patient. Conventional radiography is adequate for diagnosis of displaced fractures, but many pelvic and hip fractures are undisplaced and have subtle or absent radiographic findings. In 1989, Deutsch et al. (1) initially described the use of MR for diagnosis of radiographically occult fractures of the proximal femur. Since that time, MR imaging has rapidly become the imaging modality of choice for detection of hip fractures in patients with normal or equivocal radiographs. Fractures in major weight-bearing areas, such as the femoral neck, intertrochanteric ridge, and acetabular roof, require prompt diagnosis to minimize patient morbidity and expense. Undisplaced fractures of the femoral neck, in particular, are at risk for displacement and secondary avascular necrosis if the patient resumes weight bearing. Fractures of the femoral neck and intertrochanteric region require either prolonged immobilization or early operative intervention for optimal results. The radiographic demonstration of undisplaced fractures of the pelvis and proximal femur is especially difficult in patients with severe osteopenia due to poor contrast on the radiograph. Patient obesity, overlying bowel gas, superimposed osteoarthrosis, and difficulties in positioning the acutely traumatized patient add further difficulty to plain film interpretation of the pelvis and hip (2).

Besides MR imaging, conventional tomography, CT, and scintigraphy have been employed in the patient with high clinical suspicion for fracture, particularly of the femoral neck. MR offers advantages compared with other modalities because it can be performed immediately, is noninvasive, and has higher diagnostic accuracy. Undisplaced fractures are difficult to identify with tomographic imaging modalities that rely on trabecular or cortical disruption for fracture detection. The axial plane used for CT also limits the detection of undisplaced fractures of the femoral neck, which are usually oriented transversely (2). Until recently, scintigraphy has been the most widely used modality for detecting radiographically occult fractures of the hip. In elderly patients, it is customary to wait 48 to 72 hours after the injury to minimize the rate of false-negative bone scans. Unlike scintigraphy, MR can be performed immediately following the injury, obviating the 48-to 72-hour delay considered necessary for scintigraphic abnormalities to develop in the elderly patient. MR consistently shows proximal femoral fractures imaged 24 hours after trauma (3). Evans et al. (4) compared MR with scintigraphy obtained at least 48 hours after the injury in patients with traumatic hip pain and normal radiographs (4). In their 37 patients, MR showed eight undisplaced proximal femoral fractures, whereas only six were diagnosed on radionuclide imaging. There were no false-positive or false-negative MR examinations, based on 3 months of clinical follow-up (4). Bone scans also do not provide the anatomic depiction of the fracture afforded by MR, such as the degree of angulation and displacement of the fracture site. A low-cost screening MR examination, consisting of one or two sequences, is cost effective when compared with either radionuclide imaging or CT (1,2),

The MR appearance of a traumatic fracture consists of an oblique or wavy linear band of low signal intensity in the bone marrow on T1-weighted spin-echo images (1,5) (Fig. 4D). The normal physeal scar also should not be misinterpreted as a fracture. The physeal scar is curved, very thin, and located more proximally than the typical fracture. The low-signal fracture line is surrounded by a poorly defined area of edema, which is of high signal on T2-weighted and STIR images. The fracture line itself may be obscured on T1-weighted images if there is extensive surrounding edema. On T2-weighted images, the fracture line appears of low signal, whereas the surrounding edema increases in signal intensity. On STIR sequences, hypointense intramedullary lines present within the hyperintense bone marrow, probably representing compressed trabeculae, are seen in only in half of the patients (6). In the remaining patients, the fracture line itself appears as a linear band of intramedullary hyperintense signal extending to the cortical margin (6). When the fracture line and the surrounding edema are both hyperintense, it is more difficult to ascertain the orientation of the fracture line itself. None of these sequences is sensitive for detecting diminished perfusion and early osteonecrosis due to an intracapsular fracture (7).

Since clinical localization of fractures in patients with hip and pelvic pain following trauma can be difficult, MR evaluation should include not only the affected hip, but also the adjacent soft tissues and pelvic girdle. A T1-weighted spin-echo coronal sequence of the entire pelvis is widely used as a screening study for occult fractures (1,2,4). STIR sequences can also be employed for initial screening because of their high sensitivity to the edema surrounding the fracture. However, the configuration of the fracture line may be difficult to define on the STIR sequence (Fig. 4E,F). At our institution, we obtain a fast STIR coronal sequence of the entire pelvis using the body coil to localize areas of bone trauma. Subsequently, smaller field of view (FOV) T1-weighted spin-echo or proton-density fast spin-echo images limited to the areas of abnormality detected on the STIR sequence are obtained using either the body coil or a surface coil, depending on the patient's habitus and the number of regions to be evaluated.

The role of MR in detection of fractures of the femoral neck has been emphasized in the literature, but MR is useful in other regions of the pelvis and proximal femur. Occult fractures of the acetabulum, sacrum, pubic rami, and intertrochanteric region are readily detected on the coronal MR images. Occult fractures of the acetabulum are less common than in the proximal femoral region. Acetabular fractures are caused by major trauma, such as a fall from a height or a motor vehicle accident, and are often associated with a posterior or central dislocation of the hip. CT is generally employed to determine the configuration of the acetabular fracture, to assess joint congruity, and to evaluate for intra-articular bodies. MR can demonstrate the fracture lines but is inferior to CT for identifying intra-articular bodies following an acetabular fracture (8). In patients with acetabular fractures, MR has equivalent sensitivity to CT for detection of associated osteochondral fractures of the femoral head. Even when the femur appears normal on CT, MR demonstrates subchondral edema of the femoral head in 65% of patients sustaining a fracture of the acetabulum (8). Acute traumatic fractures involving the pubic rami and the sacrum can also be evaluated with MR, although CT is usually sufficient in these regions.

Even when the radiograph shows a fracture, MR can be used to clarify the anatomic extent of the injury. We have employed MR in several patients who appeared to have isolated greater trochanteric fractures by plain film. In all cases, MR demonstrated a much more extensive fracture than could be appreciated radiographically. Most significantly, MR showed a complete intertrochanteric fracture in over half of our patients initially diagnosed as having limited trochanteric injury (Fig. 4G–I). This distinction is important because isolated greater trochanteric fractures are treated conservatively, whereas intertrochanteric fractures require internal fixation.

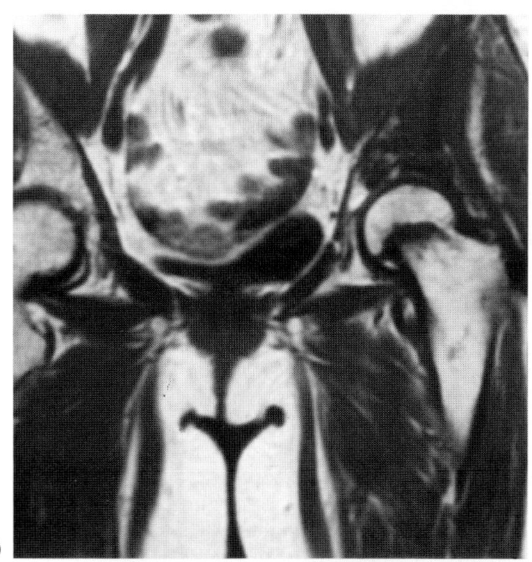

D

FIG. 4. D: An elderly woman with nonHodgkin's lymphoma complained of severe left hip pain after a fall. A T1-weighted (TR/TE 600/17 msec) spin-echo coronal image of the pelvis shows a thick band of low signal in the left femoral neck due to a radiographically occult fracture. The abnormal signal in the left acetabulum is due to lymphomatous involvement. No malignancy was present in the femoral neck on the biopsies obtained during operative pinning of the femoral neck fracture.

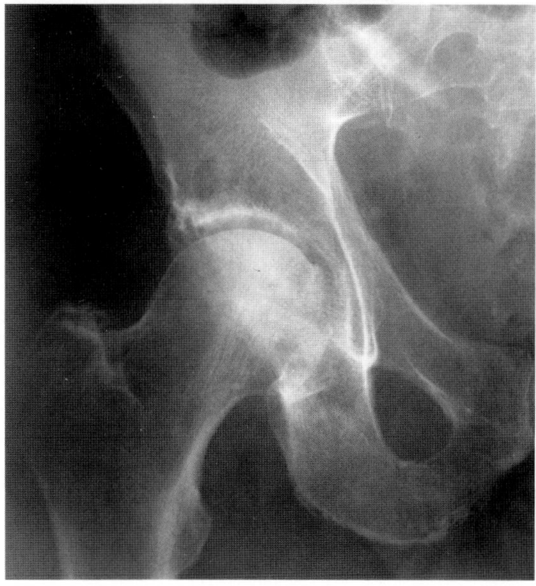

E

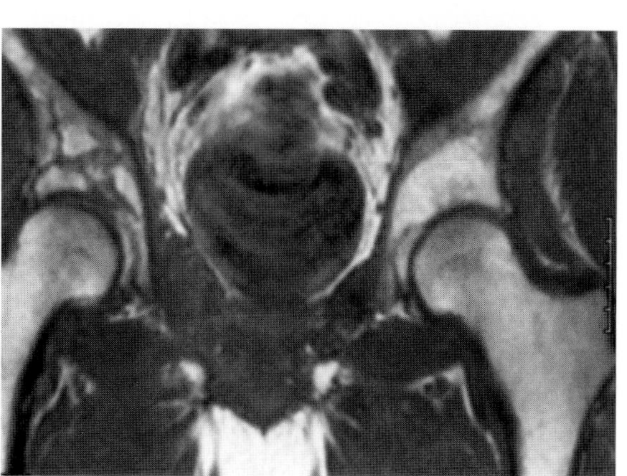

F

FIG. 4. E,F: This 78-year-old man fell and complained of right hip pain. (Courtesy of Steven Eilenberg, M.D., San Diego, California.) **E:** The AP radiograph of the right hip is normal. **F:** A T1-weighted (TR/TE 549/17 msec) spin-echo coronal MR of the pelvis shows multiple low-signal fracture lines involving the right acetabular roof and supraacetabular region.

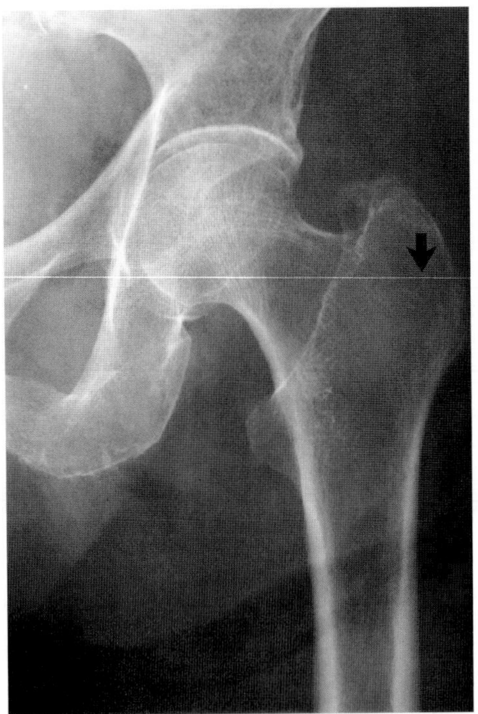

G

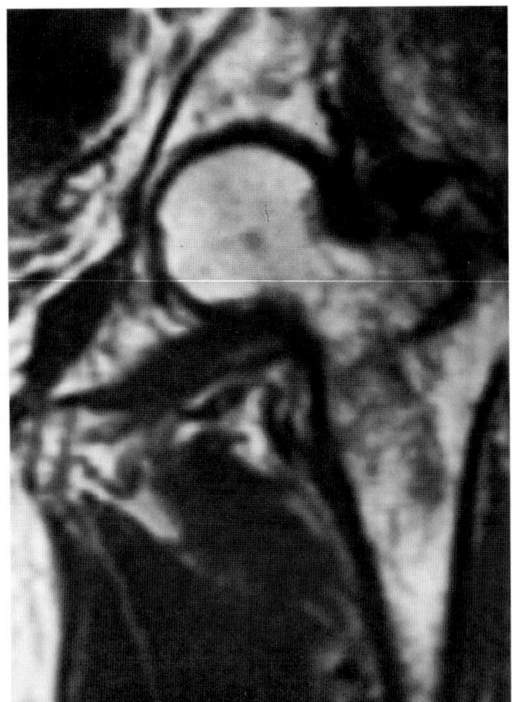

H

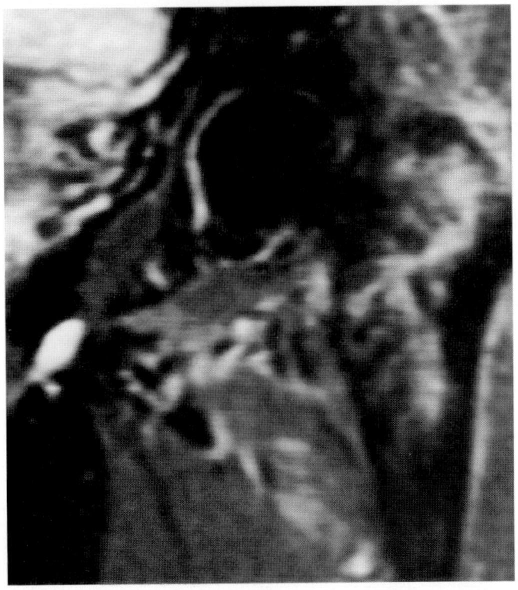

I

FIG. 4. G–I: This 70-year-old women developed severe left hip pain after a fall. Radiographs of the left hip showed a questionable fracture of the greater trochanter. MR examination demonstrated an occult intertrochanteric fracture. **G:** The AP radiograph of the left hip shows a subtle area of cortical disruption in the greater trochanter (*arrow*). **H:** The T1-weighted (TR/TE 450/15 msec) image of the left hip shows a linear low-signal fracture line crossing the intertrochanteric ridge. There is extensive edema in the intertrochanteric area resulting in a large area of abnormal medullary signal. **I:** The STIR image (TR/TE/TI 3,000/30/160 msec) shows the fracture line and the surrounding edema as irregular, linear regions of high signal. Note that a low-signal fracture line cannot be seen. There is edema in the muscles adjacent to the fracture line.

CASE 5

History: A 40-year-old man developed a slowly enlarging, painless mass in his left thigh following surgical repair of a deep laceration of the thigh resulting from a motorcycle accident. What is your diagnosis? Why is there a fluid–fluid level in the lesion?

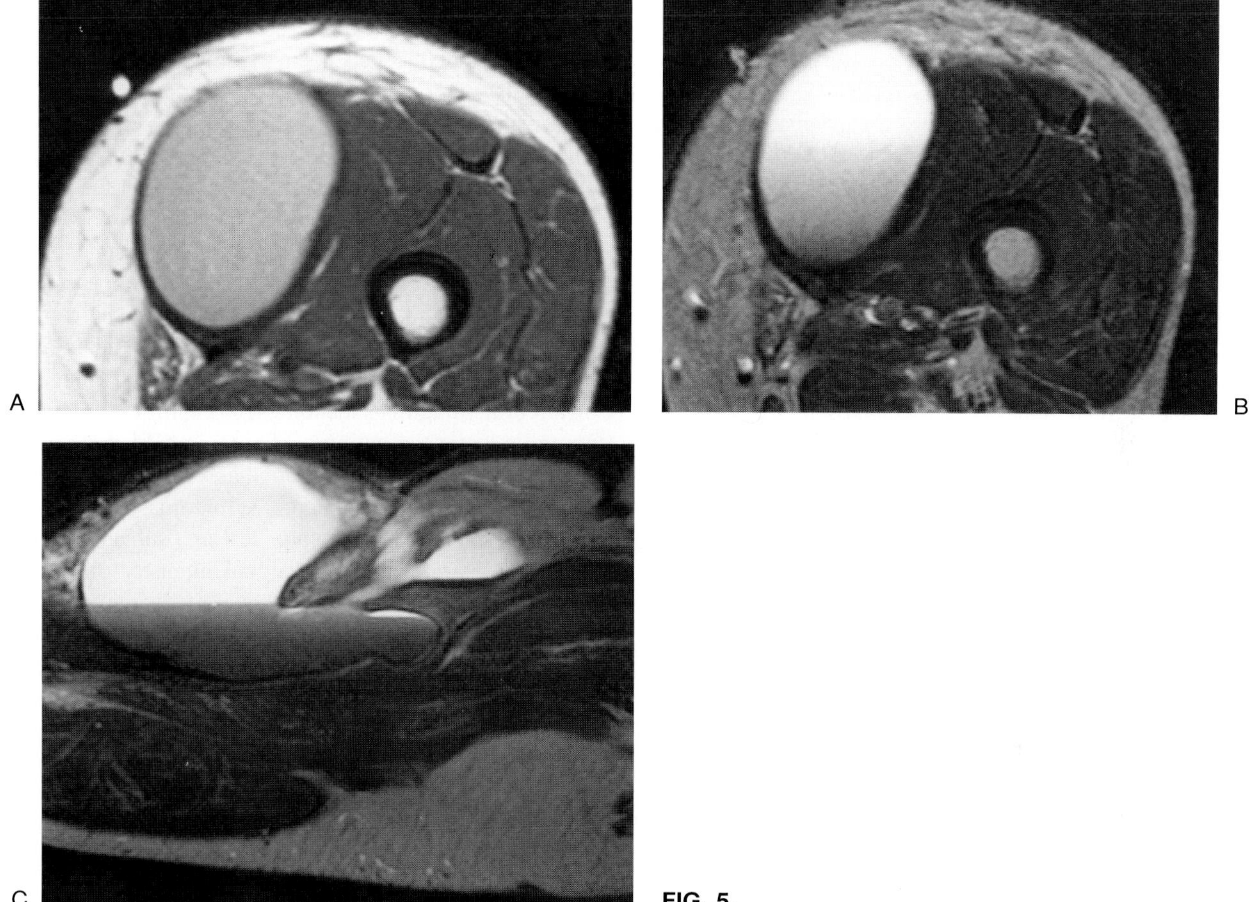

FIG. 5.

Findings: The T1-weighted (TR/TE 717/20 msec) axial image of the proximal thigh (Fig. 5A) shows a well-defined, homogeneous mass in the region of the left sartorius muscle, compressing the quadriceps muscle. The mass followed the course of the sartorius muscle on the adjacent slices. The T2-weighted (TR/TE 2,033/70 msec) axial image (Fig. 5B) shows high signal in the superior portion of the mass and low signal inferiorly, with a fluid–fluid level in the central portion of the lesion. The fluid–fluid level is more readily apparent on the T2-weighted (TR/TE 1,800/70 msec) sagittal image (Fig. 5C). (Courtesy of Steven Eilenberg, M.D., San Diego, California.)

Diagnosis: Large hematoma with fluid–fluid level.

Discussion: MR is an excellent technique for assessment of soft tissue injury because of the high soft tissue contrast afforded by a combination of different pulse sequences. While CT can detect soft tissue edema, parenchymal hemorrhage, and focal hematoma, far more accurate assessment of soft tissue injuries is afforded by MR. Injury to the muscles and musculotendinous junction is common in the large muscles of the pelvis and proximal thigh. Traumatic muscle injuries demonstrate a spectrum of findings (1,2). Muscle strains, which are minor injuries caused by microscopic tears of muscle fibers, are difficult to visualize on T1-weighted images. In this injury there is ill-defined hyperintense signal on T2-weighted and STIR images within the abnormal muscle due to a combination of hemorrhage, edema, and inflammation (1,3). In some patients with muscle strain, high signal intensity is most pronounced at the periphery of the injured muscles. This rim of high signal intensity is most pronounced 5 to 6 days after the injury (3). More extensive muscle tears exhibit diffuse mild hyperintensity within large regions of the muscle, as well as more focal hyperintense areas within the area of the muscle tear (Fig. 5D). The high signal intensity is most pronounced on fat-suppressed T2-weighted and STIR sequences. Associated perimuscular, subfascial, and soft tissue hemorrhage is frequently present in association with muscle tears (1). Complete tears of the muscles produce extensive signal abnormality as well as morphologic disruption of the contour of the muscle. Injuries of the myotendinous junction, a site frequently injured in athletes, have a poorer prognosis than injury limited to the muscle belly itself (4).

MR is also an excellent modality for identifying hemorrhage within the soft tissues. Hemorrhage may be traumatic, or due to excessive anticoagulation, or related to an underlying bleeding diathesis, or it may be secondary to an underlying vascular malformation or soft tissue neoplasm. Soft tissue hemorrhage can be parenchymal, where blood dissects through the stromal tissue, or it may form a discrete collection of blood known as a hematoma (5). Parenchymal hemorrhage appears as an ill-defined region of low signal on T1-weighted images and of high signal on T2-weighted images. As the parenchymal hemorrhage organizes, foci of hemosiderin interspersed within the muscle result in low-signal regions on all pulse sequences. Parenchymal hemorrhage typically does not contain significant concentrations of methemoglobin, so it does not usually appear bright on T1-weighted images, even in its subacute phase, (5).

Hematomas are well-defined masses that are typically confined to a single muscle or located in the subcutaneous tissues (2). The MR appearance of subcutaneous and intramuscular hematomas is highly variable depending on their chronicity. Hematomas are best evaluated using a combination of T1- and T2-weighted spin-echo sequences to characterize specifically the relative oxidation states of hemoglobin (2). Hyperacute blood is low signal on T1-weighted images and remains low signal on T2-weighted images. Acute hematomas are low or intermediate in signal on T1-weighted images, depending on the field strength of the MR unit, and high signal on T2-weighted images (Fig. 5E–G). The appearance of an acute hematoma is highly nonspecific and is difficult to differentiate from more aggressive processes. Subacute hematomas typically contain some regions that are intermediate or high signal on T1-weighted images due to the presence of methemoglobin (5). Fluid–fluid levels due to sedimentation of material in different phases of oxidation may also be present during this stage (1). The magnetic susceptibility effects of intact red blood cells containing deoxyhemoglobin layering at the bottom of the hematoma produce low-signal material within the dependent portions of the lesion. The noncellular plasma in the nondependent portions of the lesion and extracellular methemoglobin produce high signal. Chronic hematomas manifest a low signal rim due to the presence of hemosiderin-laden macrophages at the periphery of the hematoma. Hemosiderin is dark on all sequences but is particularly conspicuous on gradient-echo sequences.

Acute hematomas have a nonspecific appearance on MR imaging and are difficult to distinguish from neoplasms, abscesses, and cystic masses. Once the hematoma shows increased signal on the T1-weighted images, the differential diagnosis becomes more limited. The most common lesions showing high signal on T1-weighted images are fatty lesions, cystic masses containing highly proteinaceous fluid, subacute hematomas, and areas of slow blood flow. Unlike fatty lesions, subacute hematomas demonstrate increased signal intensity on both T1- and T2-weighted images and do not decrease in signal with selective fat-suppressed sequences. Proteinaceous fluid is rarely as high in signal as subacute blood on T1-weighted images. Intramuscular hemangiomas, which contain both fat and slowly flowing blood, show interspersed low-signal serpiginous channels due to enlarged, abnormal vessels. Hemorrhage within a pre-existing neoplasm can be difficult to differentiate from a complex hematoma on the basis of MR imaging (Fig. 5H) A history of a pre-existing mass or lack of any significant trauma should raise concern for an underlying local lesion with secondary hemorrhage.

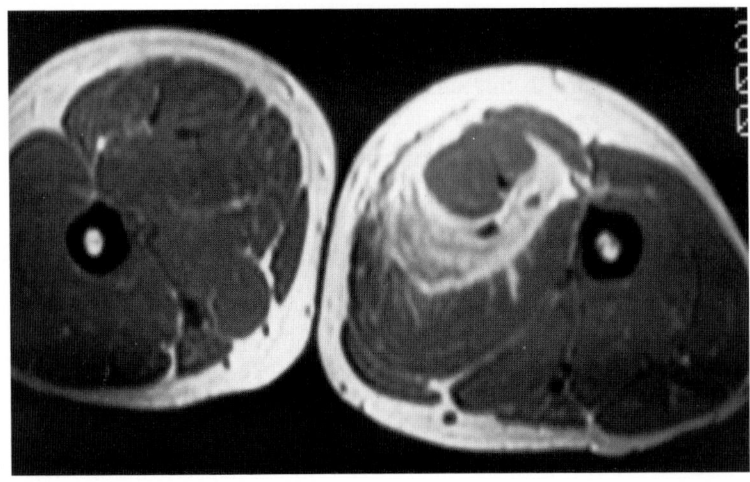

FIG. 5. D: An axial proton-density (TR/TE 2,000/25 msec) image of the proximal thighs in a patient with a large tear of the adductor muscles of the left thigh shows extensive high signal within the injured muscle.

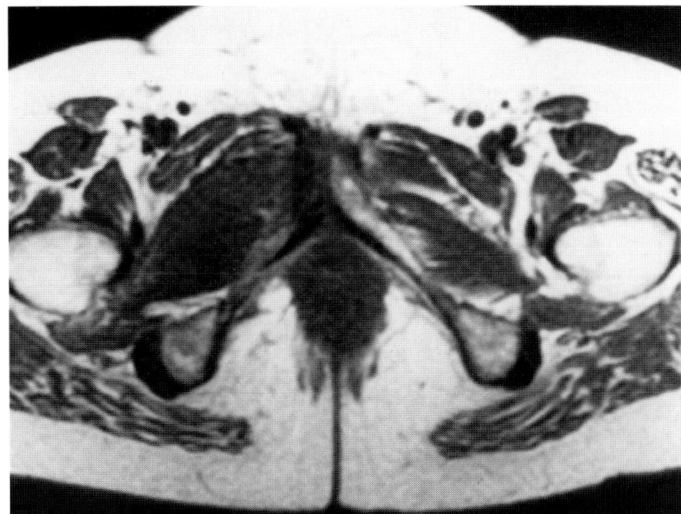

E

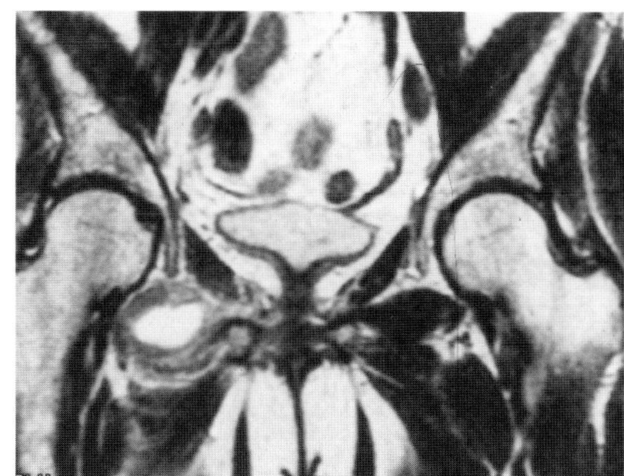

F

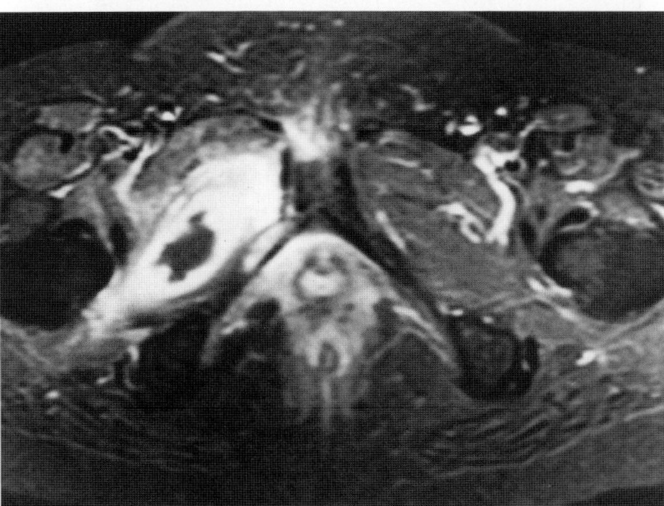

G

FIG. 5. E–G: An elderly woman complained of groin pain after a fall. Radiographs were normal so an MR was obtained to evaluate for occult fracture. (Courtesy of David Levey, M.D., Corpus Christi Texas **E:** An axial T1-weighted (TR/TE 600/17 msec) image shows low signal in the right parasymphyseal region due to an occult fracture. There is enlargement and loss of the normal intramuscular fat planes of the right obturator externus muscle. **F:** A coronal T2-weighted (TR/TE 3,000/84 msec) fast spin-echo coronal image shows a well-defined high-signal mass within the right obturator externus muscle. There is also mild hyperintensity of the muscle itself as it surrounds the focal mass. **G:** An axial fat-suppressed T1-weighted (TR/TE 600/17 msec) image obtained following gadolinium enhancement shows no enhancement of the hematoma. There is enhancement of the adjacent muscle and within the site of osseous fracture.

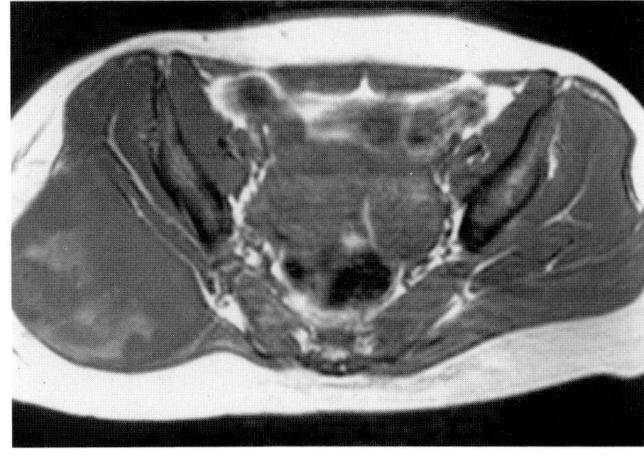

H

FIG. 5. H: A T1-weighted (TR/TE 817/20 msec) axial image of the pelvis in a patient with a large sarcoma within the right buttock shows hemorrhage within the tumor. There are irregular regions of high signal within the low-signal mass in the areas of hemorrhage.

CASE 6

History: A 32-year-old man presented with a 2-year history of back pain. Both lumbar spine (not illustrated) and sacroiliac joint examinations were performed. What are the findings and the most likely diagnosis? What is the role of MR in the evaluation of sacroiliac joint disease?

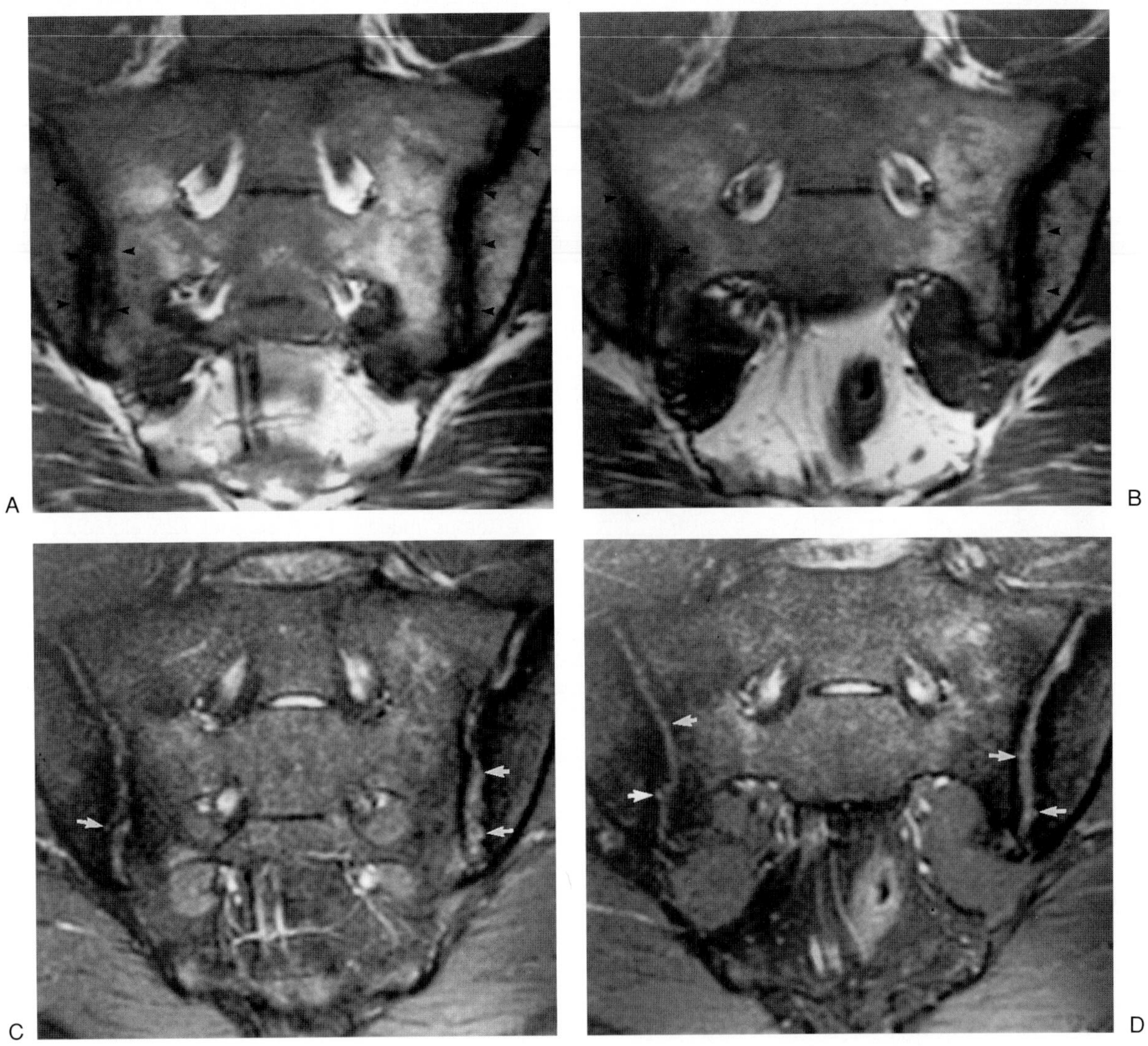

FIG. 6.

Findings: Sequential oblique coronal T1-weighted (TR/TE 600/20 msec) images demonstrate markedly decreased signal intensity involving the subchondral bone with findings predominating on the iliac side of the articulation (Fig. 6A,B, arrowheads). Oblique coronal fat-suppressed proton-density images (TR/TE 3,500/32 msec) graphically delineate the low signal intensity of the subchondral bone, increased signal intensity of the articular cartilage within the joint spaces, and localized bone erosions (Fig. 6C,D, arrows).

Diagnosis: Ankylosing spondylitis with bilateral sacroiliac joint involvement.

Discussion: MR has several potential advantages compared with CT for assessment of the sacroiliac joints (1,2–4). Included among these are: (a) the ability to image cartilage directly, (b) the ability to obtain the desirable true oblique coronal plane (CT may be limited by sacral orientation and allowable gantry angle tilt in many patients), and (c) the avoidance of ionizing radiation (15 to 20 mGy/CT examination). A few studies have compared

MR with CT for the assessment of sacroiliac disease, and several have suggested that MR may allow detectable changes of sacroiliitis prior to their depiction utilizing other diagnostic methods (1,2,4). No large-scale experience, however, presently exists for utilizing MR in this disorder, and the ultimate role of MR in the diagnostic algorithm remains unsettled.

MR of the sacroiliac joints is ideally performed in the true coronal plane with the angle obtained from a sagittal localizer sequence (1,4). In this plane, the fibrous component of the articulation is clearly distinguished from the synovial portion of the joint (1). The ligamentous component of the joint is posterior and obliquely oriented. It contains fat with an interspersed area of connective tissue and interosseous ligaments. The synovial portion is anterior and more vertical in orientation. The sacral articular surface is covered with hyaline cartilage that measures up to 4 mm in thickness (1). The iliac side is covered by a thinner layer of fibrocartilage measuring 1 to 2 mm. The articular cartilage of the articulation is thicker posteriorly than anteriorly and inferiorly. The appearance of the cartilage varies with the pulse sequence and spatial resolution utilized in performance of the examination. The normal subchondral bone is demonstrated as a well-defined low-signal-intensity interface between the articular cartilage and cancellous bone of the ilium and sacrum.

In patients with sacroiliitis, changes are invariably present both within the articular cartilage and in the adjacent subchondral bone. Histopathologic studies have demonstrated that early changes of inflammation in patients with ankylosing spondylitis are initially manifest in the subchondral bone (2,3). This observation correlates with MR findings of altered signal prominently observed along the subchondral borders of the sacroiliac joints. In patients with early disease, these signal changes may demonstrate an edema pattern with variably increased signal on T2-weighted and STIR sequences. Areas of fatty infiltration bordering subchondral bone, initially patchy and more confluent with increasing grade, have been described in association with sacroiliitis in several reports (1,2). These have an appearance similar to those of end-plate changes seen in association with degenerative disc disease of the lumbar spine. Low signal on all pulse sequences with MR has been correlated with sclerotic changes on CT (2).

Dynamic MR acquisitions utilizing gadopentetate dimeglumine contrast and rapid scan techniques have also been applied to the assessment of sacroiliitis (4,5). This technique may permit the demonstration of changes both within the joint space and in the subchondral bone in patients with sacroiliitis prior to their detectability by other techniques (4,5). The articular cartilage, joint capsule, and bone marrow of normal sacroiliac joints do not demonstrate enhancement following the intravenous administration of gadopentetate dimeglumine (5). Increased signal within the joint space can be demonstrated qualitatively and quantitatively in a patient with sacroiliitis (4,5). Initial experience suggests a correlation of degree of quantitative enhancement with the degree of patient symptomatology (5). Bone erosions seem to be more sensitively depicted utilizing contrast and appear as contrast-enhanced areas directly communicating with the joint space. Areas of localized or confluent bone marrow enhancement (juxtracortical osteitis) can be demonstrated in the adjacent bone marrow and are differentiated from erosions by a lack of direct communication with the joint. These areas of enhancement are particularly evident in patients with early disease, a finding that further supports the contention that the earliest changes in sacroiliitis are manifest in the subchondral bone adjacent to the joint (4,5).

MR has also been described in the assessment of septic arthritis and has compared favorably with the techniques most commonly used before for assessment of this problem, including CT and nuclear medicine techniques (5). In one series MR provided a diagnosis of septic arthritis in all cases; even in retrospect, a definite diagnosis of this condition could be made in only five of six joints by gallium citrate scans, in three of six joints by CT scans, and in one of six joints by ^{99m}Tc methylene diphosphonate bone scans (6). Reported MR findings in septic arthritis include fluid and/or inflammation in the sacroiliac joint space, in the bone marrow of the sacrum and/or ilium, and in the psoas muscle. In addition, fluid tracking posterior to the iliopsoas muscle has been described as a characteristic finding of septic arthritis (Fig. 6E,F).

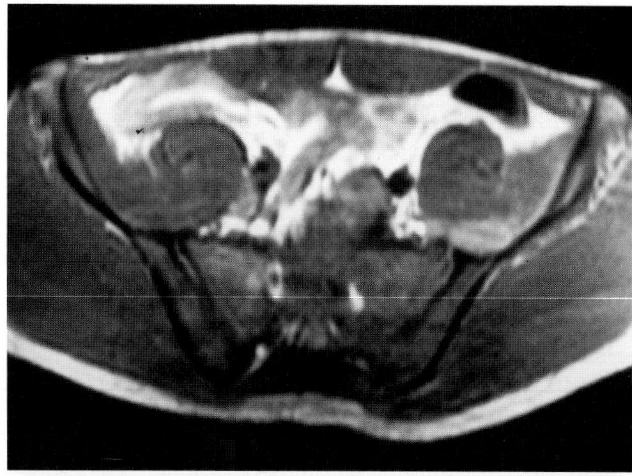

E

F

FIG. 6. E,F: Septic arthritis. A 30-year-old runner with no history of intravenous drug abuse presented with a 3-day history of left buttock pain, fever, and elevated white blood cell count. Proton-density **(E)** and T2-weighted **(F)** transverse images demonstrate fluid within the anterior aspect of the left sacroiliac joint. The fluid tracks posterior to the left iliopsoas muscle (*curved arrow*). Hyperintense signal is identified on the sacral side of the sacroiliac joint (*straight arrow*). From Klein MA, Winalski CS, Wax MR, Piwnica-Worms DR. MR imaging of septic sacroiliitis. *J Comput Assist Tomogr* 1991;15:126–132, with permission.)

CASE 7

History: This 39-year-old woman presented with right hip pain that had been progressively increasing for the past 18 months. She was an avid runner and the pain had developed suddenly after she had finished running a half-marathon. What might be the etiology of the lesion seen on the MR?

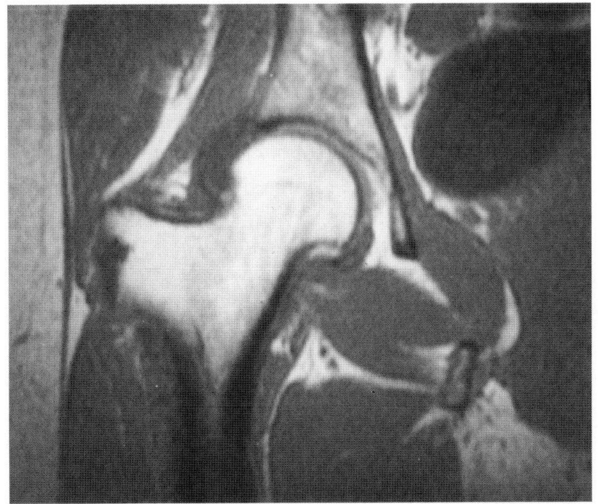
A

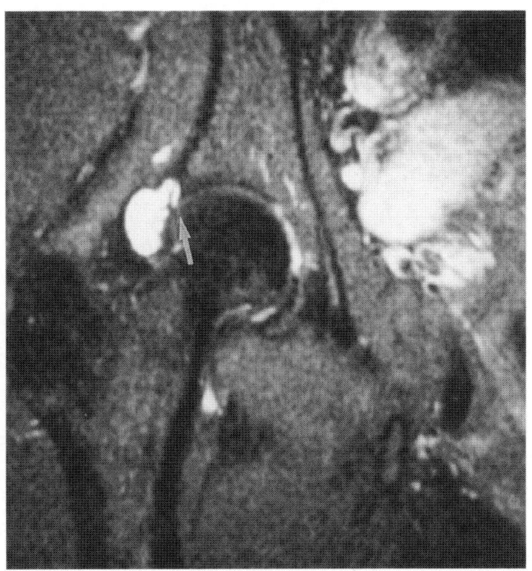
B

FIG. 7.

Findings: The coronal T1-weighted spin-echo (TR/TE 600/11 msec) image (Fig. 7A) shows an oval soft tissue density adjacent to the supra-acetabular region. This density is low signal and obliterates the fat plane between the hip joint and the gluteus minimus muscle. The normal low-signal labrum at the superolateral margin of the acetabulum is not visible. The adjacent bone is normal. The T2-weighted IRF spin-echo (TR/TE/TI 4,000/19/150 msec) image (Fig. 7B) shows a well-defined, lobulated mass of homogeneous high signal. A linear band of high signal is seen in the labrum adjacent to the lesion (arrow). There is no effusion within the joint.

Diagnosis: Periacetabular ganglion of the right hip, associated with tear of the superolateral acetabular labrum.

Discussion: A ganglion is a common benign cystic or myxomatous tumor-like lesion. The distinction between a ganglion and a soft tissue synovial cyst is somewhat controversial, although most pathologists distinguish between these two entities through the lack of a synovial cellular lining in ganglia and the mucinous character of their contents (1). Although the exact etiology of ganglia remains unknown, it is speculated that either trauma resulting in synovial herniation or degeneration of the joint capsule, tendon sheaths, or subchondral bone leads to their development (2). Ganglia are classically found attached with or in close proximity to the synovium of joint capsules and tendon sheaths, but these fluid-filled collections can be found in a variety of other less common anatomic regions. Purely intraosseous, subperiosteal, intramuscular, intratendinous, and subcutaneous locations have all been described (1,3).

Ganglia are most commonly found in the soft tissues adjacent to the wrists and knees. Involvement of the acetabular bone and periacetabular soft tissues is less common, but the hip region is not a infrequent site of involvement by both intraosseous and soft tissue ganglia. The etiology of periacetabular ganglia appears to be post-traumatic since most such ganglia are associated with acetabular labral tears. The weakest site of the acetabular labrum is the posterosuperior portion, the location of most tears (4). A high frequency of acetabular labral tears is found in patients with underlying acetabular dysplasia with uncovering of the lateral portion of the femoral head, presumably due to excessive stress on the posterolateral labrum. Post-traumatic labral tears are generally observed in high-performance athletes. Usually this latter group presents with an abrupt onset of hip pain during a sporting activity. Pain results from the enlarged edematous torn labrum stretching the joint capsule and may mimic sciatica as it radiates down the thigh (5). Labral tears may be managed conservatively with a period of non-weight bearing or by partial resection of the torn labrum (6). If untreated, labral tears may enlarge and result in clinical locking, secondary osteoarthrosis, and ganglia formation. It has been suggested that increased intra-articular pressure forces synovial fluid through a labral tear via a one-way valve mechanism, resulting in the formation of a periarticular ganglion. Eventually, the loss of congruity between the femoral head and acetabulum following a labral tear can lead to the development of premature osteoarthrosis (6).

Several recent reports have described the clinical and radiographic manifestations of periacetabular ganglion formation (2,4–7). Conventional radiographs of patients with a periacetabular ganglion are usually normal but in rare instances may reveal a soft tissue mass with well-defined extrinsic erosion of the supra-acetabular lateral cortex. When nitrogen gas fills the ganglion, a specific diagnosis can be rendered. (Fig. 7C) The presence of intralesional gas within the ganglion is only seen in association with degenerative joint disease of the hip (7). Its appearance should not be confused with gas in overlying bowel or with a soft tissue abscess. Gas within periacetabular ganglia appears to be common, even though gas within ganglia in other locations is unusual.

Arthrography or high-resolution arthroCT can usually demonstrate the labral tear as a linear collection of contrast within the triangular labrum. (Fig. 7D) Small acetabular labral tears may be difficult to detect since they may be obscured by the overlying contrast within the hip joint. The ganglion does not always fill following injection of the hip joint, possibly due to the presence of a one-way valve mechanism. Therefore, a normal hip arthrogram does not exclude the diagnosis of an acetabular labral tear or a periacetabular ganglion. Arthroscopy has been reported to be more sensitive for the detection of labral tears than arthrography (4). MR permits noninvasive diagnosis of labral tears and also allows visualization of both communicating and noncommunicating ganglia. However, the sensitivity of MR for the detection of labral tears has not yet been established.

Periacetabular ganglia associated with labral tears are well demonstrated by MR. The normal hip labrum is well seen on sagittal and axial MR images as a triangular area of signal void encircling the osseous borders of the acetabulum, analogous to the fibrous labrum of the shoulder joint. Nonvisualization, diffuse increase in signal intensity, linear bands of high signal within the labrum, or superolateral displacement of the labrum suggest the diagnosis of labral pathology. (Fig. 7E,F) In our limited experience, intra-articular fluid does not appear to be essential for identification of acetabular labral tears. With MR imaging, the contiguity of ganglia with the acetabular joint capsule and labrum is well depicted. Periacetabular ganglia are typically located posterosuperior to the hip joint, adjacent to the lateral cortex of the supracetabular ilium, although they may be seen in other locations adjacent to the labrum. Ganglia are usually homogenous low-signal masses on T1-weighted images, similar to articular fluid and muscle. Because of their proteinaceous contents, some ganglia are hyperintense relative to muscle on T1-weighted images. Ganglia are typically best seen on T2-weighted and STIR sequences because of their marked hyperintensity on these sequences (1,3). Like ganglia in other locations, periacetabular ganglia appear sharply marginated and may be spherical, elliptical, or lobulated. Thin linear or curvilinear internal septa are frequently present.

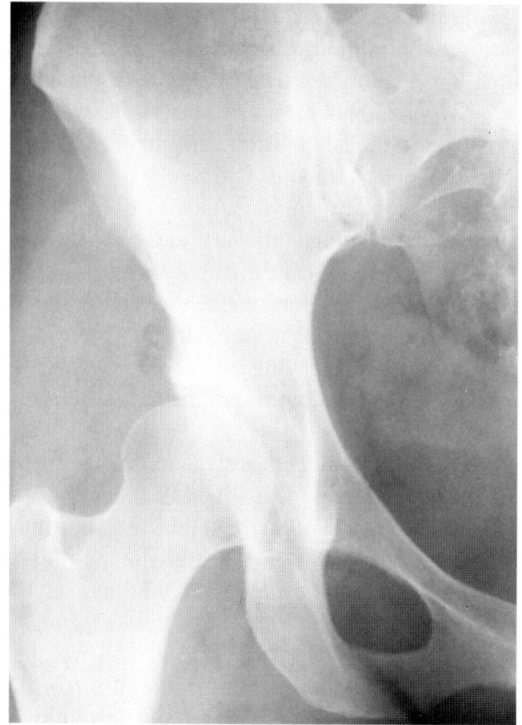

C

FIG. 7. C: A soft tissue radiograph of the right hip in a 32-year-old woman with mild acetabular dysplasia and uncovering of the right femoral head. There is a 2-cm mass containing multiple rounded gas collections in the periacetabular soft tissues, diagnostic of a periacetabular ganglion. Note the subtle extrinsic erosion of the supra-acetabular cortex.

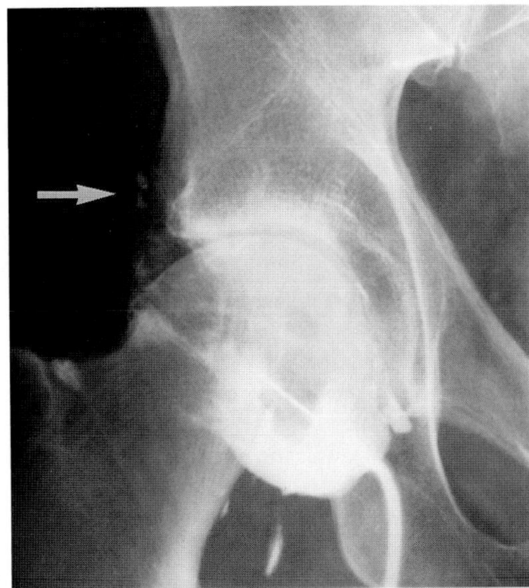

D

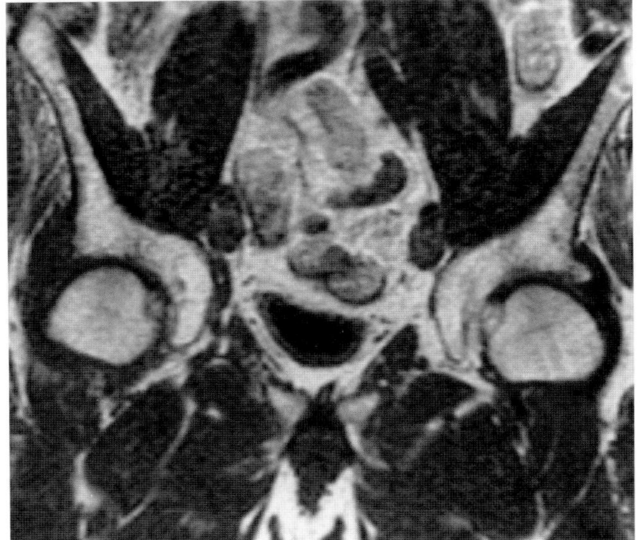

E

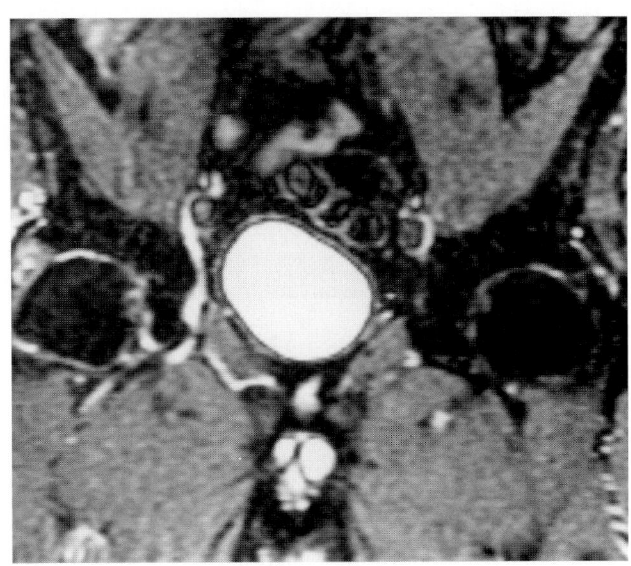

F

FIG. 7. D–F: A periacetabular ganglion associated with a labral tear in a 60-year-old physician who was a long-distance runner. He presented with vague right hip pain of several month's duration. Conventional radiographs were interpreted as normal, although, in retrospect, subtle cortical erosion could be identified in the supra-acetabular cortex. **D:** A single-contrast arthrogram shows a linear band of contrast extending across the labrum and puddling of small amounts of contrast in the periacetabular soft tissues (*arrow*). **E:** A T1-weighted spin-echo (TR/TE 300/11 msec) coronal image of the pelvis shows a low-signal mass adjacent to the right superior labrum. There is irregularity of the cortex of the iliac bone adjacent to the lesion. **F:** A fast multiplanar inversion recovery (FMPIR) (TR/TE/TI 4,000/19/160 msec) image shows a lobulated high-signal mass in the soft tissues and a small effusion within the right hip joint. The normal triangular low-signal labrum seen on the left side cannot be visualized. The right acetabular labrum is abnormally enlarged and shows diffuse increase in signal intensity, consistent with a labral tear.

CASE 8

History: This 6-month-old girl was noticed to have abnormal hips at birth. She had been treated conservatively but was not responding to her treatment regimen.

What is the diagnosis? What is the role of MR imaging in this entity?

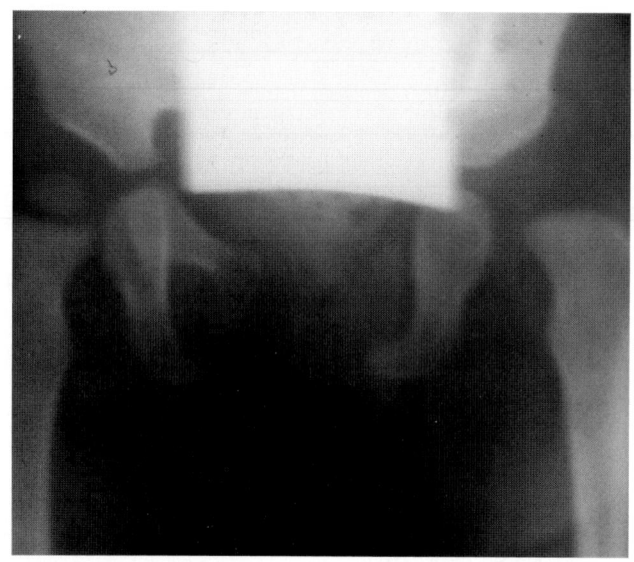

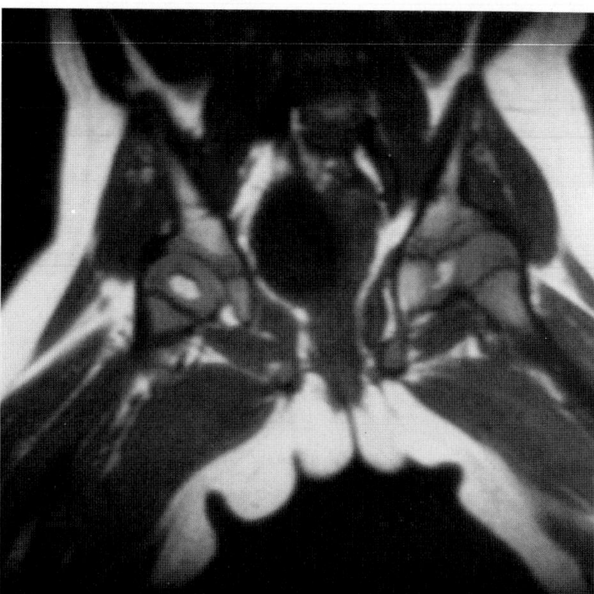

A B

FIG. 8.

Findings: The AP radiograph of the pelvis (Fig. 8A) shows steep acetabular roofs bilaterally. There is very mild subluxation of the right hip, with loss of alignment of Shenton's line and mild superolateral subluxation of the right femoral head. The left femoral head has not yet ossified. There is increased distance between the medial wall of the acetabulum and the medial femoral neck due to lateral subluxation. The T1-weighted (TR/TE 600/17 msec) MR image of the pelvis (Fig. 8B) shows absence of ossification of the left femoral head. The ossific nucleus of the right femur is well formed and has normal signal. The left femoral head is superolaterally subluxed, with lack of containment and poor coverage of the femoral head. There is thickening of the medial wall of the acetabulum. A hypertrophied fat pad is present in the left acetabular fossa. (Courtesy of Sevil Brahme, M.D., San Diego, California).

Diagnosis: Developmental dysplasia of the hips with subluxation of the left femoral head.

Discussion: Developmental dysplasia of the hip (DDH), previously known as congenital dislocation of the hip, is a common disorder of the infant that ranges in severity from minimal subluxation to frank dislocation of the hip joint. In the United States, DDH occurs in 0.5% to 1% of all live births (1). Typically, hip instability and dislocation develop in an otherwise healthy infant just before or after birth. Hip subluxation and dislocation result from the combination of tight maternal musculature (which restricts fetal motion in utero), often breech presentation, and hormonal influences (particularly laxity of the hip capsule from maternal estrogens). The left hip is affected more frequently than the right hip; bilateral involvement is present in approximately 20% of cases. If the dislocation is diagnosed in the newborn period, the femoral head can easily be reduced, and 95% of affected children will develop normal hips (2). The longer the hip is dislocated, the greater the development of adaptive changes, which increases the difficulty of treatment. The medial acetabulum becomes flattened and shallow without the stimulus of the femoral head, the muscles about the hip become contracted, and the labrum enlarges and may enfold into the joint, preventing reduction (3).

Typically, radiographs of a newborn with DDH appear normal at birth. The articular structures are cartilaginous in the neonate, and pathologic adaptations are not yet present. Physical examination using Ortolani and Barlow maneuvers is the most reliable clinical method of diagnosing DDH in the neonatal period (3). By 6 weeks of age, abnormal development of the acetabulum may be evident on radiography. The acetabular angle is increased compared with the normal angle (<30°), and the femoral head is abnormally positioned. In the infant, the exact position of the femoral head cannot be determined until the infant is 4 to 6 months of age because the femoral head is not ossified. In patients with DDH, ossification is typically even more delayed, so accurate assessment of the position of the femoral head may not be possible until the infant is 8 to 10 months old. In the first year of life, ultrasonography, performed with dynamic movement of the hip, is the imaging modality of choice for the diagnosis of DDH (2,4). Ultrasound allows accurate determination of the position of the cartilaginous femoral head within the acetabulum throughout its full range of motion (4). The acetabular cartilage is not as well seen with ultrasound, and femoral and acetabular version are difficult to determine (2). The technique is operator dependent and requires considerable experience on the part of the examiner.

MR plays a complementary role to conventional radiography, arthrography, CT, and ultrasound for evaluation of development dysplasia of the neonatal hip. MR is rarely necessary for diagnosis and assessment of uncomplicated cases of DDH. MR also requires that the infant be sedated and remain immobile for the duration of the study. Therefore, the role of MR imaging of individuals with DDH is limited to assessment of the 5% of patients who have a teratologic dislocation or a late presentation or those who fail standard treatment (2,5) (Fig. 8C) In patients with sufficient ossification to preclude satisfactory ultrasound imaging, MR can accurately depict the shape and position of both the ossified and cartilaginous portions of the femoral head. The viability of the ossific nucleus can be assessed by noting the signal of the marrow within the femoral head. Cartilage is well seen, so acetabular development and hip containment, factors that are important determinants of normal hip development, can be accurately evaluated (Fig. 8D,E). Unlike ultrasonography, MR is generally performed as a static examination in only one position. Dynamic instability may therefore be underestimated with MR examination.

Soft tissue impediments to hip reduction, which previously required hip arthrography for diagnosis, can be visualized by MR imaging (1,6). An enlarged ligamentum teres is a common impediment to reduction and can be identified more accurately with MR imaging than with arthrography (1). An inverted hypertrophied labrum appears as a triangular or rounded low-signal structure enfolded between the femoral head and the acetabular cartilage. In patients with iliopsoas tendon impingement, a thickened, rounded mass of low signal intensity, representing the hypertrophied tendon, is seen invaginating the anterior aspect of the joint capsule (2) (Fig. 8F,G). These soft tissue impediments to reduction are more difficult to visualize with CT and ultrasound (2). In patients with chronically subluxed hips, a hypertrophied pulvinar (acetabular fossa fat pad) can be seen as a mass with signal intensity identical to fat filling the central acetabular region.

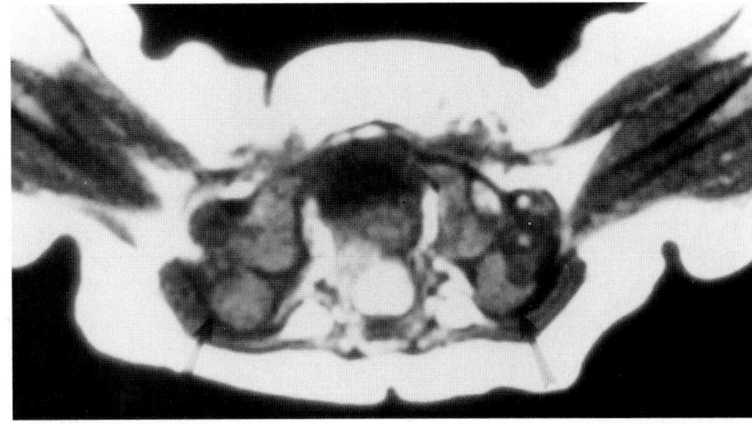

FIG. 8. C: An axial MR with the hips flexed and abducted shows persistent dislocation of both femoral heads, which are completely dislocated and lie posterior to the acetabuli. This child had failed treatment with a Pavlik harness.

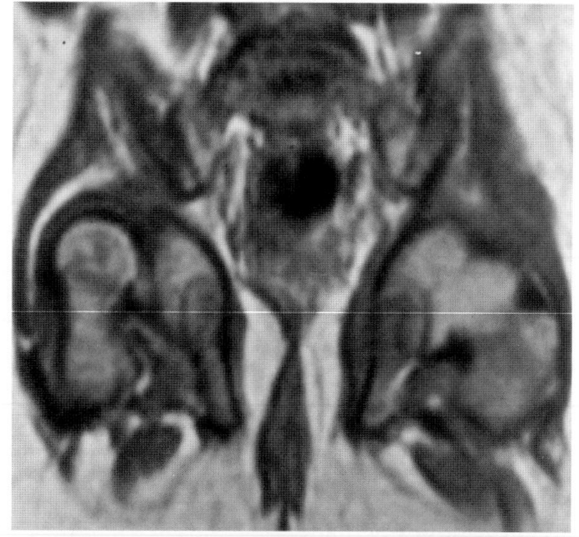

D

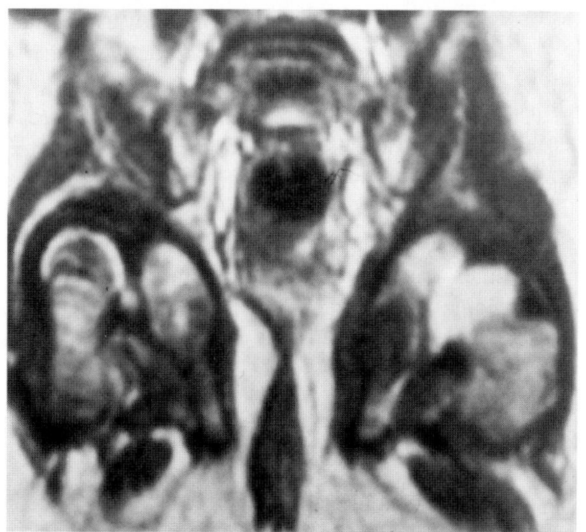

E

FIG. 8. D,E: This infant had failed conservative therapy for DDH of the right hip. (Courtesy of Sevil Brahme, M.D., San Diego, California). **D:** The T1-weighted (TR/TE 450/15 msec) image shows dislocation of the right femoral head. The acetabulum is hypoplastic and there is coxa valga. **E:** The T2-weighted (TR/TE 1,400/60 msec) image shows the dislocation. The cartilage is higher signal on this sequence. The dislocated femoral head is superolateral to the acetabular cartilage.

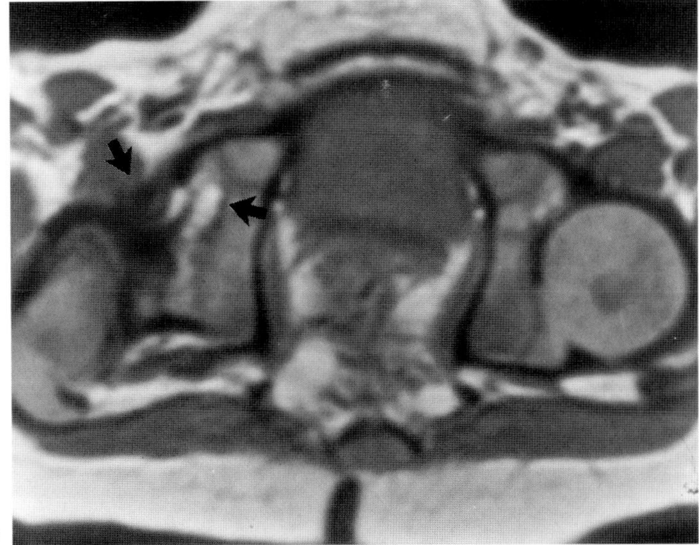

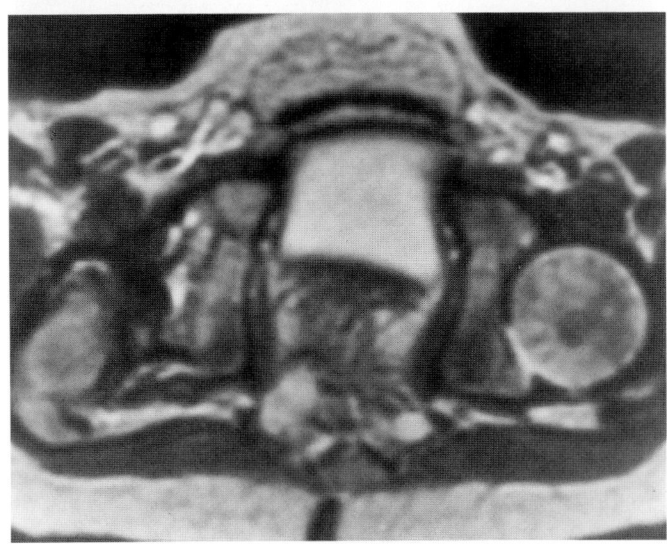

FIG. 8. F,G: The MR examination was obtained to determine the reason for failure to respond to therapy. (Courtesy of Sevil Brahme, M.D., San Diego, California). **F:** The T1-weighted (TR/TE 600/20 msec) axial image shows a normal left hip. The right acetabulum is poorly formed posteriorly and there is subluxation of the right femoral head. Fatty tissue is seen in the acetabular fossa, representing a hypertrophied pulvinar (*arrow*). There is invagination of the anterior capsule by a thickened iliopsoas tendon (*arrow*). **G:** The T2-weighted image (TR/TE 1,800/80 msec) shows that the thickened pulvinar maintains fat signal. The capsular invagination by the iliopsoas tendon is not as well seen in the absence of a joint effusion.

CASE 9

History: A 38-year-old woman presented with several years of right hip pain aggravated by walking. On physical examination, flexion and internal rotation of the hip reproduced the pain. What is the abnormality detected on this MR arthrogram?

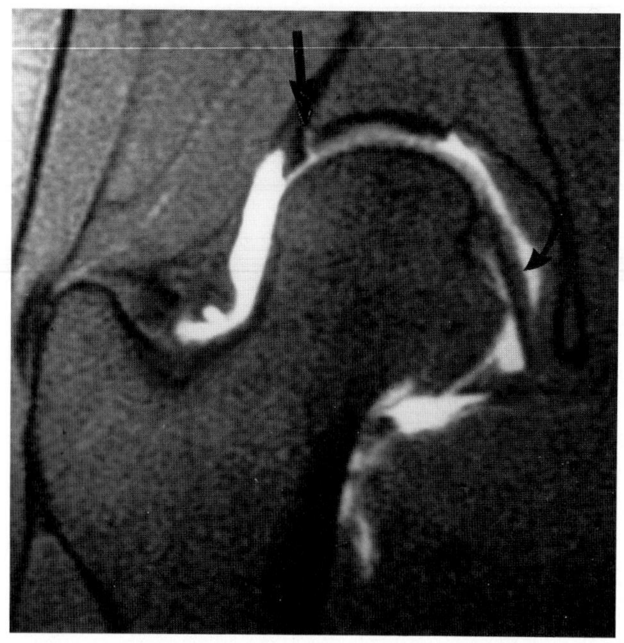

A

FIG. 9.

Findings: T1-weighted fat-suppressed coronal image (TR/TE 550/17 msec) of the right hip obtained following the intra-articular injection of dilute gadopentetate dimeglumine demonstrates an abnormal high signal intensity line at the base of the posterosuperior labrum (Fig. 9A). This line (arrow) is isointense with intra-articular contrast and also communicates with it. It has higher signal intensity than the articular cartilage. Also noted is the normal ligamentum teres (curved arrow). (Image courtesy of Mitchell Klein, M.D., Milwaukee, Wisconsin).

Diagnosis: Torn posterosuperior labrum.

Discussion: Acetabular labral tears represent an important but only recently recognized cause of mechanical hip pain in adult patients (1). Pain is initially experienced as one or more discrete episodes of sharp clicking pain precipitated by a pivoting or twisting motion. Early in the course of the condition, the pain typically resolves on the order of minutes (1). Later the pain may be continuous in nature and is frequently reported as groin pain. These tears may result from either an acute injury or chronic repetitive trauma (1–5). The trauma is frequently minor, such as a prior slipping or twisting episode. Hip abnormalities such as acetabular dysplasia or congenital hip dislocation may also predispose patients to labral tears (1–5). Passive hip flexion and internal rotation of the hip reproduce the pain.

Labral tears are most commonly encountered in younger individuals (1–5). The tears most frequently consist of a separation of the labrum from the adjacent articular cartilage of the articulating surface of the acetabulum. The anterosuperior labrum has been recently reported as the most common site of injury (1), although prior reports have emphasized a posterior location (4), and it is clear that labral tears may be identified in any location. Besides deepening the acetabulum, the intact labrum creates negative pressure in the hip, adding to its stability. A torn labrum reduces the negative pressure in the hip and decreases the acetabular depth; both factors contribute to hip instability. While it is possible to establish the diagnosis utilizing conventional MR techniques, the authors have found the assessment of the acetabular labrum on MR to be far improved utilizing MR arthrographic techniques (Fig. 9B)

Acetabular labral tears need to be distinguished from other possible causes of mechanical hip pain (1). Snapping of the iliopsoas tendon over the greater trochanter is reported as the most common cause of snapping pain around the hip (1). This condition may be difficult to distinguish from an acetabular labral tear. Arthrographic methods have been utilized in making this differentiation. If the hip joint does not communicate with the tendon sheath of the iliopsoas tendon, the patient with a painful snapping of the iliopsoas tendon should not experience relief from an intra-articular administration of marcaine and steroid at the time of arthrography (1). Both pigmented villonodular synovitis and synovial chondromatosis need to be considered in the differential diagnosis. Another entity that can cause mechanical pain of the hip and that can be confused with an acetabular labral tear is acute hemorrhage of the ligamentum teres (1).

In a manner analogous to the glenoid labrum in the shoulder, the hyaline cartilage of the acetabulum undercuts the acetabular labrum (6,7). The interface between the hyaline articular cartilage and the labrum can simulate a labral tear (Fig. 9C,D). On fat-suppressed T1-weighted sequences, the hyaline cartilage is typically less hyperintense than the diluted gadopentetate dimeglumine solution utilized for MR arthrography. This difference in intensity, however, may be quite subtle. If uncertainty remains, a fast spin-echo T2-weighted sequence can be obtained. On this sequence cartilage is intermediate in signal intensity, in contrast to fluid, which is hyperintense, thus facilitating differentiation.

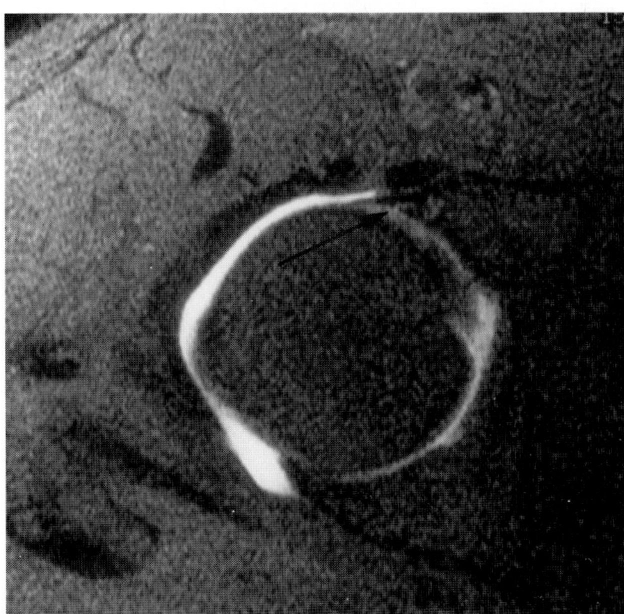
B

FIG. 9. B: Acetabular labral tear. Axial fat-suppressed T1-weighted axial (TR/TE 550/17 msec) images of the right hip after intra-articular injection of dilute gadopentetate dimeglumine demonstrates a high-signal-intensity line (*arrow*) within the substance of the anterior labrum, consistent with a labral tear. The patient had experienced the sudden onset of pain after abrupt hip flexion during a soccer match.

FIG. 9. C,D: Acetabular labral tear pitfall. **C:** Axial T1-weighted fatsuppressed image (TR/TE 550/17 msec) of the left hip obtained after intra-articular administration of dilute gadopentetate dimeglumine. **D:** Magnified view of the same area demonstrates a high-signal-intensity linear structure (*arrow*) identified along the base of the anterior acetabular labrum. The signal intensity of the linear structure is less than that of the intra-articular gadopentetate dimeglumine and is similar to that of the articular cartilage along the articulating surfaces of the acetabulum and femoral head. This represents normal undercutting of acetabular cartilage along the base of the acetabular labrum and should be distinguished from a labral tear. (Images courtesy of Mitchell Klein, M.D., Milwaukee, Wisconsin).

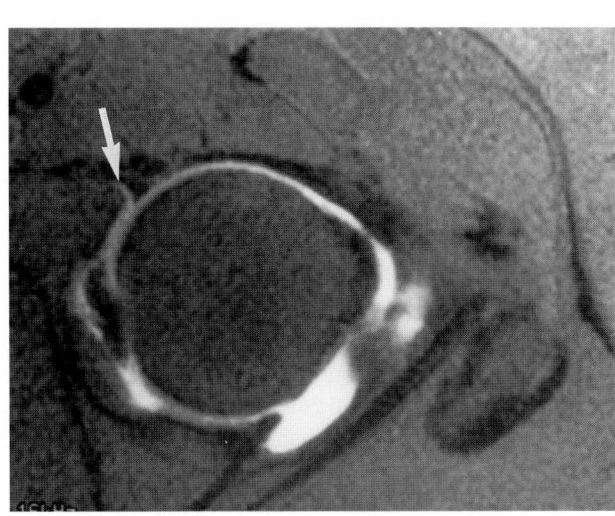
C

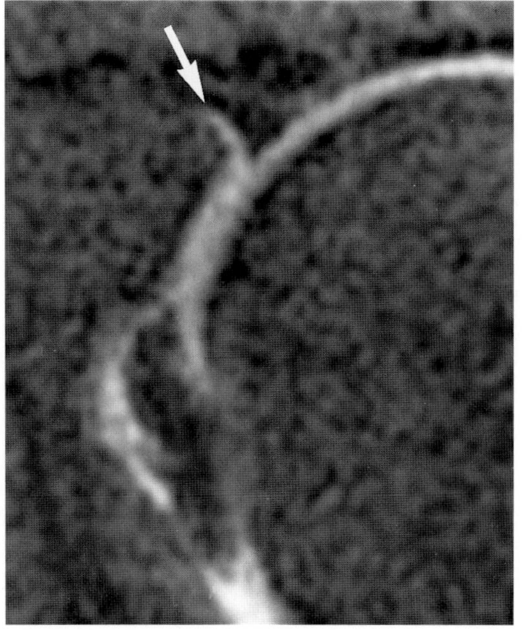
D

CASE 10

History: A 19-year-old male naval recruit was referred for evaluation of bilateral hip pain for the previous 3 weeks. He complained of exercise-induced pain with only mild discomfort while at rest. His pain was markedly restricting his performance during his physical training, which he had started 6 weeks before. He had no systemic complaints and had not experienced any single episode of major trauma. A radiograph of the pelvis at the time of presentation was normal (not illustrated). What is the cause of the MR abnormalities? What other imaging modality is commonly used to make this diagnosis?

A
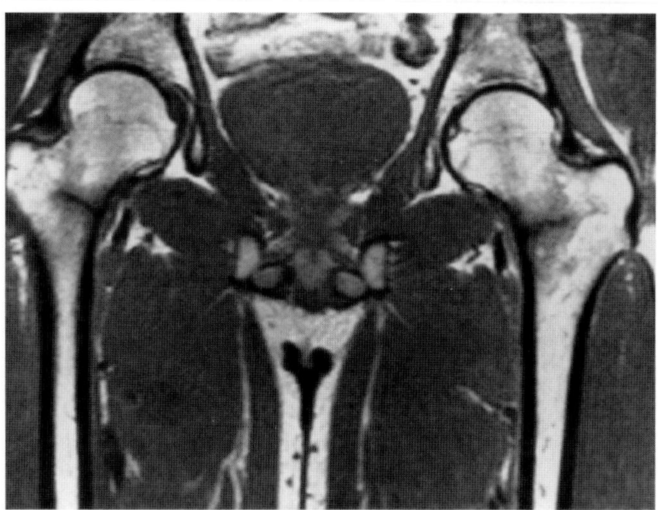

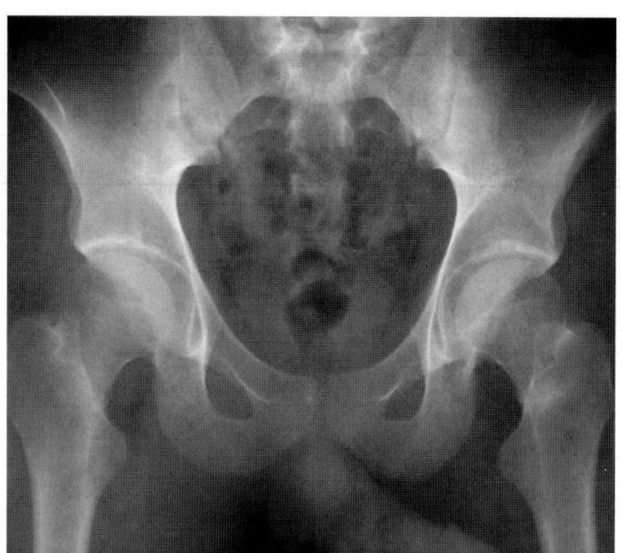
B

FIG. 10.

Findings: A proton-density (TR/TE 2,000/20 msec) spin-echo coronal MR image of the pelvis (Fig. 10A) shows linear bands of low signal extending into the medullary space from the medial femoral necks bilaterally. These linear bands are perpendicular to the cortex and major weight-bearing trabeculae. Both these linear bands are incomplete since they do not extend across the entire width of the femoral neck. A pelvic radiograph obtained 4 weeks after the MR (Fig. 10B) shows linear bands of sclerosis and periosteal proliferation at both sites.

Diagnosis: Bilateral fatigue fractures of the femoral necks.

Discussion: Fatigue fractures result from repetitive, prolonged muscular activity applied to bone of normal elasticity that has not accommodated itself to the repetitive load (1). Fatigue fractures are caused by repetitive microtrauma without any single individual episode of trauma sufficient to break the bone. Fatigue fractures can involve any bone but have a predilection for the weight-bearing lower extremity. The specific locations of fatigue fractures differ depending on the repetitive activity responsible for the osseous injury. Fatigue fractures of the femoral neck are seen in recreational and competitive runners, who also develop fatigue fractures of the proximal tibia and distal fibula (1). Ballet dancers are also prone to developing fatigue fractures of the femoral neck and femoral shaft (1). Femoral neck fatigue fractures are classified into transverse and compressive types. Transverse fatigue fractures of the lateral femoral neck usually develop in elderly patients and are at high risk of displacement. Displacement is less likely in compression-type fatigue fractures of the cancellous bone of the medial femoral neck (2).

The patient classically presents with pain associated with a particular activity; the pain is typically relieved by rest (1). Diagnosis at this stage depends on a thorough clinical history. The physical findings at the onset of symptoms are nonspecific. Early diagnosis at this stage, with avoidance of the offending activity until the fracture heals, affords the best long-term prognosis. In their early stage, fatigue fractures of the femoral neck are incomplete and undisplaced. Undetected fatigue fractures of the femoral neck can become complete and develop displacement with continued repetitive stress. Complications such as osteonecrosis, delayed union, and nonunion occur more frequently in the presence of such displacement (3). Early detection and evaluation of the extent of femoral fatigue fractures can be accomplished by several imaging modalities.

Radiographs are generally normal for the first few weeks following the onset of symptoms. Scintigraphy demonstrates increased accumulation at the fracture site days or weeks before radiographic findings are evident. In the right clinical setting, increased uptake on a bone scan in a location consistent with a fatigue fracture is sufficient to establish the diagnosis. CT is usually not helpful in early identification of the fracture line in the femoral neck because the orientation of the fracture tends to be transverse. With ongoing trauma, resorption at the site of the fracture may result in a radiographically visible cortical defect and intracortical lucent line. In most cases, however, the fracture line itself is obscured by adjacent reactive bone, so cortical disruption can not be identified. Instead, lamellar periosteal and endosteal sclerosis or hazy callous is seen at the site of the fracture. The reactive bone is generally focal, confined to a limited portion of the cortex, and does not involve the cortex circumferentially (1).

Like scintigraphy, MR is highly sensitive for the diagnosis of stress injury to bone, typically showing abnormalities several weeks prior to the development of radiographic alterations. MR has much higher spatial resolution than scintigraphy, allowing determination of the extent and orientation of the osseous fracture. Unlike scintigraphy, the MR appearance of a fatigue fracture is usually distinctive, enabling a specific diagnosis of fracture in most cases (4). Two MR patterns of fatigue fractures have been emphasized (5). The most common pattern is a band-like fracture line that is low signal on all sequences, surrounded by an ill-defined zone of edema. The metaphyseal and diaphyseal regions of the bone are most commonly involved. The fracture line is continuous with the cortex and extends into the intramedullary space oriented perpendicular to the cortex and the major weight-bearing trabeculae (5,6). The zone of surrounding edema within the medullary space can be very extensive, producing a large area of abnormal signal loss in the marrow on T1-weighted images (7). The fracture line itself may be obscured on the T1-weighted images by extensive surrounding low-signal edema (Fig. 10C,D) In our experience, proton-density images are excellent for demonstrating the presence, orientation, and extent of the fracture line. This sequence is very insensitive for bone marrow edema and should not be used as the primary sequence for identifying the presence of an intramedullary abnormality. On T2-weighted and STIR images, the fracture line remains low signal, and the surrounding marrow edema is seen as an amorphous region of increased signal (Fig. 10E,F) Increase in signal on the T2-weighted images becomes less prominent with increasing duration of symptoms. High signal may not be present if the patient is imaged more than 4 weeks after the onset of symptoms (6). Ill-defined regions of high signal in the juxtacortical and subperiosteal tissues, as well as longitudinal periosteal zones of hyperintensity, are present on T2-weighted and STIR sequences in the acute phase of the fracture (6,8).

The less common pattern is that of amorphous alterations of marrow signal without a clear-cut low signal fracture line (5). This pattern presumably represents an earlier stage in the evolution of the stress injury. The amorphous pattern shows ill-defined regions of low signal within the medullary space on T1-weighted images that increase in signal on T2-weighted or STIR images. This nonspecific pattern is difficult to distinguish from transient bone marrow edema, very early avascular necrosis, (AVN), osteomyelitis, and infiltrative neoplasms.

Other sites of fatigue fractures in the pelvis are injured less commonly than the femoral neck. Fatigue fractures of the sacrum are reported secondary to running and aerobics. Bowling and gymnastics can result in fatigue fractures of the pelvis and obturator ring (1). Avulsion injuries of the apophyses in the skeletally immature often present with a chronic history of pelvic pain. Apophyseal avulsions are usually due to a single episode of trauma; less commonly, they represent fatigue fractures due to chronic, repetitive overactivity such as running, jumping hurdles, and gymnastics. Common sites of apophyseal fatigue fractures include the anterior superior iliac apophysis at the site of sartorius muscle origin, the anterior inferior iliac apophysis at the site of the rectus femoris origin, and the ischial apophysis at the site of origin of the hamstrings (9) (Fig. 10G,H). Displacement of the apophysis, edema in the adjacent bone marrow, and soft tissue hemorrhage can be identified with MR imaging. These injuries are easily misinterpreted as representing neoplasms, particularly during the subacute healing phase, when considerable reactive bone is present at the site of injury. Knowledge of their characteristic locations and correlation with the clinical findings help to avoid misdiagnosis of fatigue fractures in these locations.

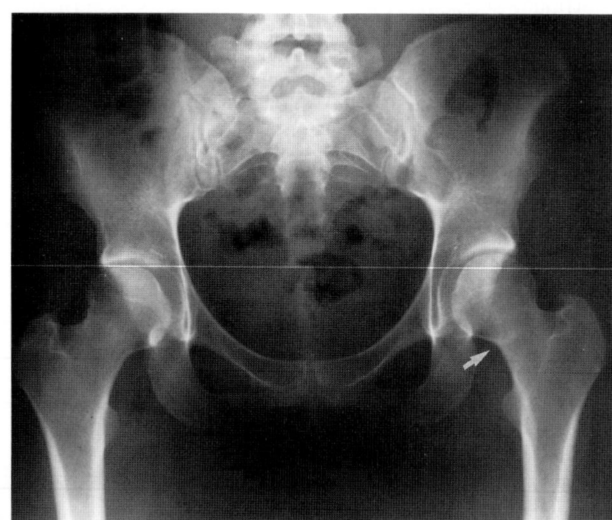

C

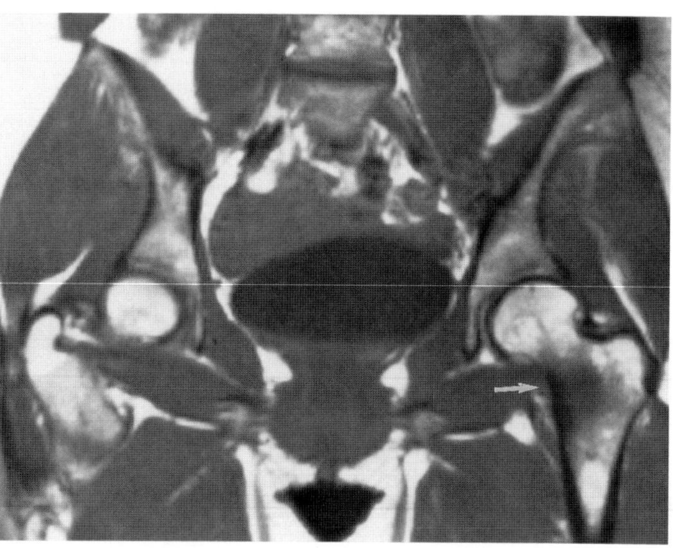

D

FIG. 10. C,D: A 25-year-old woman recently started a running program and developed pain in the left hip. **C:** The AP radiograph of the pelvis demonstrates a thin linear band of sclerosis in the medial left femoral neck *(arrow)*. This finding was not appreciated at the time of the initial reading. **D:** The coronal T1-weighted (TR/TE 500/15 msec) spin-echo image of the pelvis shows a poorly defined focal area of low signal in the marrow of the medial aspect of the left femoral neck. The transverse low-signal fracture line *(arrow)* is difficult to appreciate because of the surrounding edema.

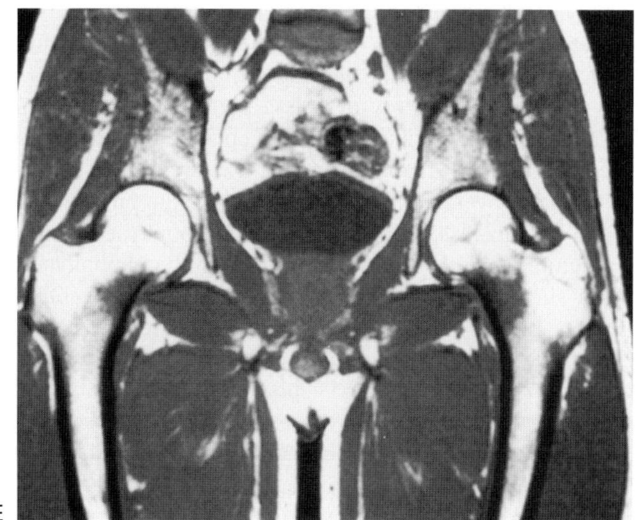

E

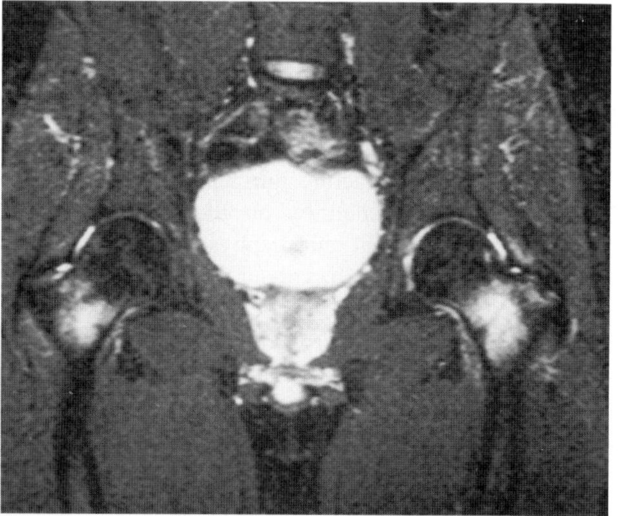

F

FIG. 10. E,F: This 20-year-old male military recruit complained of left hip pain. The right hip was asymptomatic. Radiographs were normal at the time of presentation. **E:** The T1-weighted (TR/TE 717/20 msec) coronal image shows abnormal signal in the bone marrow adjacent to the medial femoral necks bilaterally. No definite fracture line can be seen on this sequence. **F:** A STIR (TR/TE/TI 2,500/40/150 msec) image shows that the marrow edema in both femoral necks is more extensive than it appears on the T1-weighted sequence. There are incomplete fracture lines, seen as linear low-signal bands, arising from the medial femoral necks bilaterally. A small amount of joint fluid and mild increase in the soft tissue adjacent to the femoral necks can also be seen.

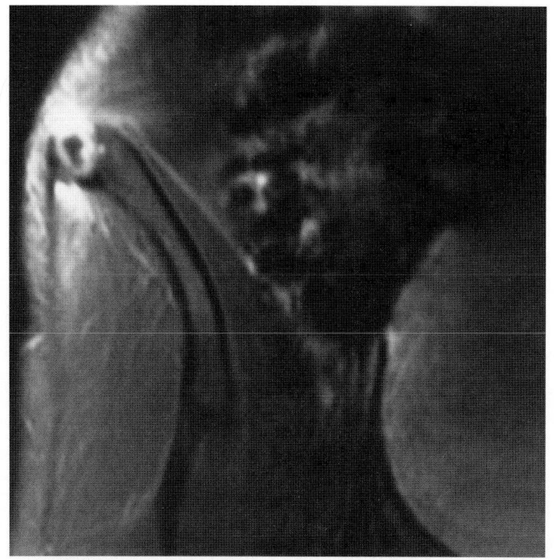

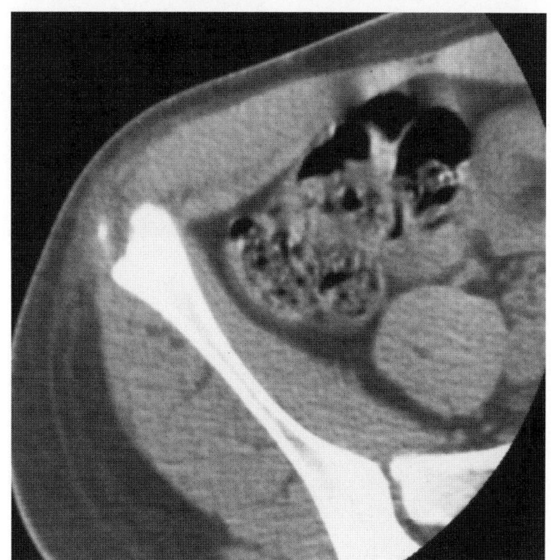

FIG. 10. G,H: A 13-year-old female gymnast complained of several weeks of pain in the right iliac crest. Outside radiographs were interpreted as showing a possible neoplasm, and the patient was referred to MR for tumor staging. (Courtesy of David Levey, M.D., Corpus Christi, Texas.) **G:** A coronal fat-suppressed T1-weighted (TR/TE 800/11 msec) spin-echo image obtained following intravenous administration of gadolinium shows an osseous avulsion of the anterior superior iliac spine. There is enhancement in the soft tissues adjacent to the fracture. **H:** An axial CT image confirms the presence of an avulsion fracture of the right anterior superior iliac spine.

CASE 11

History: This 13-year-old boy presented with intermittent left hip pain that had been progressively increasing in severity over the past 2 months. He had no significant prior medical history and denied recent trauma. What is the likely diagnosis? What are some risk factors associated with this condition?

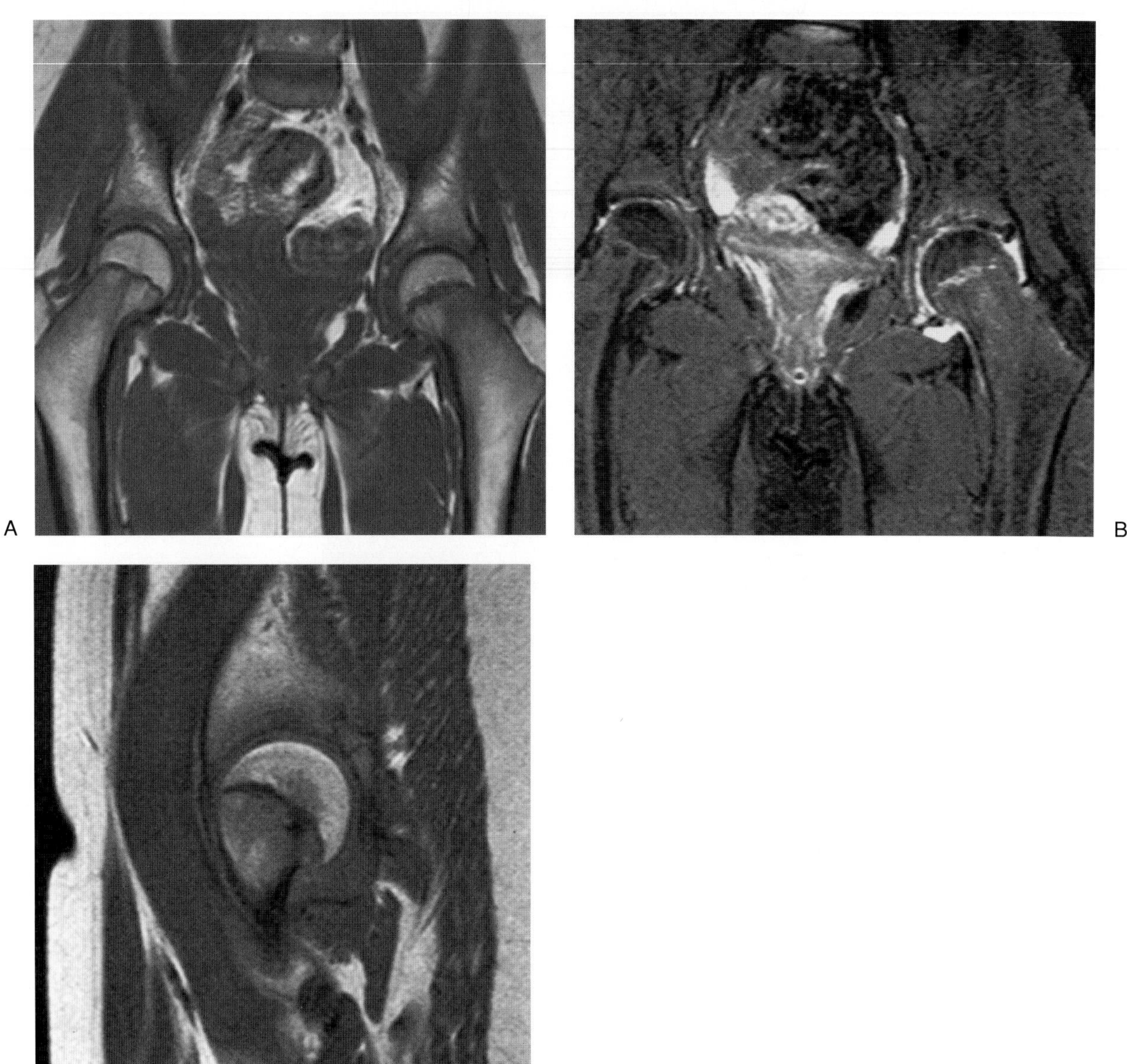

FIG. 11.

Findings: The T1-weighted (TR/TE 500/20 msec) spin-echo coronal image (Fig. 11A) of the pelvis shows diminution in vertical height of the left femoral head. The marrow signal within the femoral head is normal. The growth plate on the left side is wider than on the right and appears thickened and irregular in its mid-portion. There is subtle loss of marrow signal in the left femoral metaphysis adjacent to the physeal plate. The proton-density (TR/TE 2,000/30 msec) fat-suppressed spin-echo coronal image (Fig. 11B) demonstrates high signal within the left physeal plate and mild increase in signal of the marrow in the proximal metaphysis. There is a moderate-size joint effusion within the left hip joint. The sagittal T1-weighted (TR/TE 800/20 msec) image (Fig. 11C) shows posterior displacement and angulation of the epiphysis relative to the femoral neck. There is irregularity of the posterior metaphyseal cortex, consistent with subacute epiphyseal displacement. (Courtesy of Steven Eilenberg, M.D., San Diego, California.)

Diagnosis: Slipped capital femoral epiphysis.

Discussion: Slipped capital femoral epiphysis (SCFE) is a disruption of the open physeal plate of the adolescent hip resulting in displacement of the femoral epiphysis from the femoral neck. SCFE is characterized by medial, posterior, or posteromedial displacement of the proximal femoral epiphysis relative to the metaphysis. Although the etiology of this disorder is often uncertain, causal factors include acute trauma, mechanical stress due to repetitive trauma, prior irradiation, renal failure, obesity, and hormonal abnormalities (1). SCFE most commonly affects boys between the ages of 10 and 16 years and girls between the ages of 8 and 15 years. There is a definite male predominance. In boys, SCFE is bilateral in 25% to 33% of cases and involves the left hip twice as frequently as the right. In girls, the incidence of bilaterality is lower, and both hips are involved with equal frequency (1). Patients often give no history of hip pain, providing only a nonspecific history of aching thigh or knee pain exacerbated by activity. The physical examination is usually normal except for an antalgic gait and limited range of motion of the involved hip (1).

The physeal disruption may be acute or chronic at the time of presentation. Acute displacement is defined as symptoms of less than 3 weeks' duration and often presents following an episode of minor trauma. Despite the frequency history of minor trauma, SCFE is considered a distinct entity from the acute Salter 1 fracture of the proximal femoral physeal plate seen in childhood (1). Chronic insidious displacement is more common because the presence of a physeal slip is frequently not recognized for several months after the onset of symptoms. Once the physeal plate is disrupted, the physis is at risk of progressive slippage and ischemia as long as the growth plate remains open. In the long term, the slip results in retroversion and abnormal mechanics of the hip joint, which predispose the joint to premature degenerative osteoarthritis (2). Because of the risk of osteonecrosis and permanent deformity, surgical treatment (in situ fixation, epiphysiodesis, or osteotomy) is usually instituted immediately following the diagnosis of this entity (1).

Plain radiography is sufficient to establish the diagnosis of SCFE in most instances (3). Loss of vertical height of the physis, widening of the physeal plate, and indistinctness of the metaphyseal margin are early radiographic findings. On the AP view of the hip, a line tangent to the lateral femoral neck intersects less than 20% of the epiphysis. In severe cases, the line may completely fail to intersect any portion of the epiphysis. Since the slip is most frequently in a posterior or posteromedial direction, the true groin lateral view and the frog-leg lateral view are both more sensitive than the AP view for early detection of femoral head displacement. In very mild cases of SCFE, abnormalities of the physeal plate can be detected without identifiable epiphyseal displacement. In these mild cases, sometimes termed *preslips* MR or CT can be used to confirm the presence of physeal disruption and subtle epiphyseal slippage. Further imaging is usually not necessary in uncomplicated SCFE (3).

MR imaging in the acute postslip period shows signal changes within the disrupted physeal plate, associated with morphologic displacement of the femoral head. The normal growth plate is poorly seen on T2-weighted spin-echo and STIR images, intermediate in signal on T1-weighted images, and high signal on T2*-weighted gradient-recalled echo (GRE) images (4). Acute physeal plate injury is best seen on high resolution low flip angle GRE images, appearing as a linear low-signal cleavage plane through the normally high-signal physis (4). In acute slips, decreased marrow signal on T1-weighted images that increases in signal on T2-weighted images probably represents adjacent edematous changes in the epiphysis and adjacent metaphysis (3).

The major role of MR imaging in patients with SCFE is to evaluate the hip for complications related to the physeal displacement. The three most common complications associated with SCFE are osteonecrosis of the femoral head, acute chondrolysis, and secondary osteoarthritis (1,2,5). Osteonecrosis is uncommon in untreated SCFE but can develop as a complication following traction reduction or surgical treatment. The risk of osteonecrosis is particularly high following pin placement into the superior quadrant of the femoral head (1). The MR findings of osteonecrosis complicating SCFE are similar to those of AVN of the hip from other causes, usually manifesting as a ring or band of low signal demarcating the infarcted zone. Acute chondrolysis is another uncommon complication of SCFE, most often associated with surgical pin penetration into the joint space (1,5). A large effusion, regional osteoporosis, and diffuse loss of cartilage are typically present. We have no experience using MR for diagnosis of this disorder. Currently, joint aspiration is considered necessary in almost all cases of chondrolysis to exclude joint infection. In the future, MR may allow noninvasive differentiation between septic arthritis and acute chondrolysis of the hip.

CASE 12

History: A young man with persistent hip pain following a sprinting injury underwent evaluation with MR arthrography. What is the problem with this MR arthrogram?

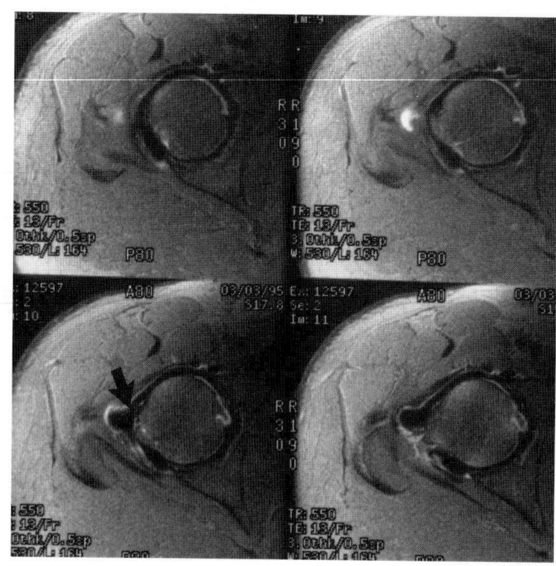

FIG. 12.

Findings: Figure 12A shows four contiguous images obtained in the axial plane using a fat-suppressed T1-weighted sequence after intra-articular injection of dilute gadopentetate dimeglumine. The fluid in the hip joint (arrow) is markedly decreased in signal intensity.

Diagnosis: Decreased signal intensity caused by a too highly concentrated solution of gadopentetate dimeglumine.

Discussion: MR arthrography (primarily of the shoulder and knee) has been the subject of several reports (1–6). The object of the technique is to utilize fluid to obtain joint distention and achieve high signal intensity within the fluid to provide contrast (e.g., arthrogram effect). Gadopentetate dimeglumine is utilized instead of saline to allow both of these objectives to be accomplished while taking advantage of several of the properties associated with T1- rather than T2-weighted images that would be required with saline alone (e.g., faster acquisition time, high signal-to-noise ratio). The optimal concentration of gadopentetate dimeglumine to be utilized is critical for obtaining the desired result (1–6).

Gadolinium, the metal in gadopentetate dimeglumine, has seven unpaired electrons. Unpaired electrons cause substances to be paramagnetic. Paramagnetic substances in turn cause shortening of both T1 and T2 relaxation times in nearby protons, although T1 shortening is the more important effect in the concentrations used most typically for MR imaging (7,8). A decrease in the T1 relaxation time caused by the paramagnetic agent results in increased signal intensity, whereas the T2 relaxation time shortening causes decreased signal intensity. At high concentrations of gadopentetate dimeglumine, the T2 shortening effect overwhelms the T1 shortening effect, and overall signal intensity decreases (7,8). The overall decreased signal intensity of the fluid seen in this case reflects the decreased T2 relaxation time dominating the T1 effect.

Anatomic delineation of the acetabular labrum and associated capsular structures of the hip is facilitated by administration of intra-articular contrast. For purposes of hip MR arthrography, the authors utilize a 2 mmol/L solution of gadopentetate dimeglumine. To obtain this dilute concentration, 10 ml of a 1:125 dilution of gadopentetate dimeglumine (0.5 mol/L) in normal saline is mixed with 5 ml each of 60% iodinated contrast material and 1% lidocaine. Epinephrine [0.3 ml (1:1,000)] may be added to this solution. This concentration of gadopentetate dimeglumine was adapted from shoulder MR arthrography (1–6). The injection is done under fluoroscopic guidance, and the presence of iodinated contrast material allows documentation of intra-articular administration.

MR arthrography facilitates demonstration of the intrinsic soft tissue structures of the hip joint. The bony hip itself is formed by the femoral head and acetabulum. The acetabulum has contributions from the ileum, ischium, and pubis. The surface of the acetabulum that articulates with the femoral head is C-shaped, with the acetabular notch directed inferiorly (9,10). The articular surfaces of the acetabulum and femoral head are covered with hyaline cartilage. The center of the acetabulum, the acetabular fossa, is filled with fat and fibrous tissue. A fibrocartilaginous labrum circumscribes the perimeter of the acetabulum (Fig. 12B) (9,10). The neurovascular structures of the acetabulum pass through a small foramen along the inferior aspect of the ligament, which is bounded externally by the transverse ligament. The ligamentum teres attaches the transverse ligament to the fovea of the femoral head. The zona orbicularis circumscribes in its mid-portion (Fig. 12A). A focal capsular thickening contributes to the maintenance of hip stability. The fibrocartilaginous labrum of the acetabulum deepens the acetabulum and, in concert with the transverse ligament, completely encircles the head of the femur (Fig. 12C). The joint capsule attaches along the periphery of the labra with more direct attachments to the labrum and transverse ligament anteriorly and inferiorly. It attaches along the intertrocanteric line of the femur anteriorly and posteriorly attaches along the mid- to distal femoral neck. The capsule is thickened by the iliofemoral, pupfemoral, and ischiofemoral ligaments (Fig. 12B).

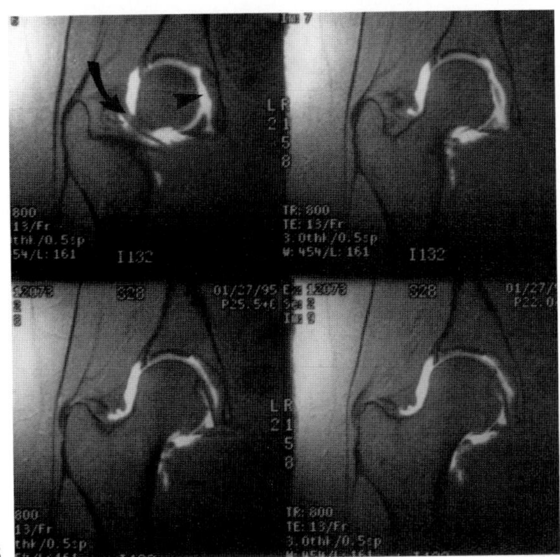

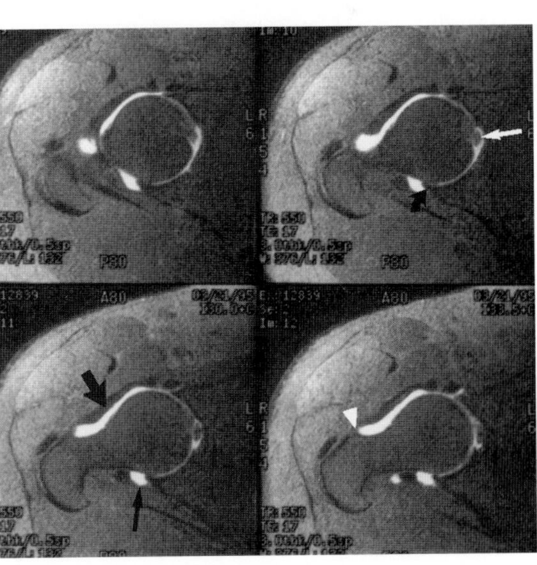

FIG. 12. B: Series of fat-suppressed T1-weighted coronal images after intra-articular administration of dilute gadopentetate dimeglumine demonstrates the zona orbicularis *(curved arrow)* and acetabular fossa *(arrowhead)*. **C:** Series of fat-suppressed T1-weighted axial images after intra-articular administration of dilute gadopentetate dimeglumine demonstrates the iliofemoral ligament *(thick arrow)*, ischiofemoral ligament *(thin arrow)*, anterior and posterior *(curved arrows)* acetabular labra, and ligamentum teres *(white arrow)*. Note that the anterior capsule *(white arrowhead)* inserts lateral to the posterior capsule. (Images courtesy of Mitchell Klein, M.D., Milwaukee, Wisconsin.)

CASE 13

History: This 67-year-old woman complained of pain in the right hip that had been increasing for the past 6 weeks. Radiographs of the right hip obtained 1 week after onset of symptoms were normal. These studies were obtained after failure of conservative therapy. What is the cause of the patient's pain? What test should be performed next?

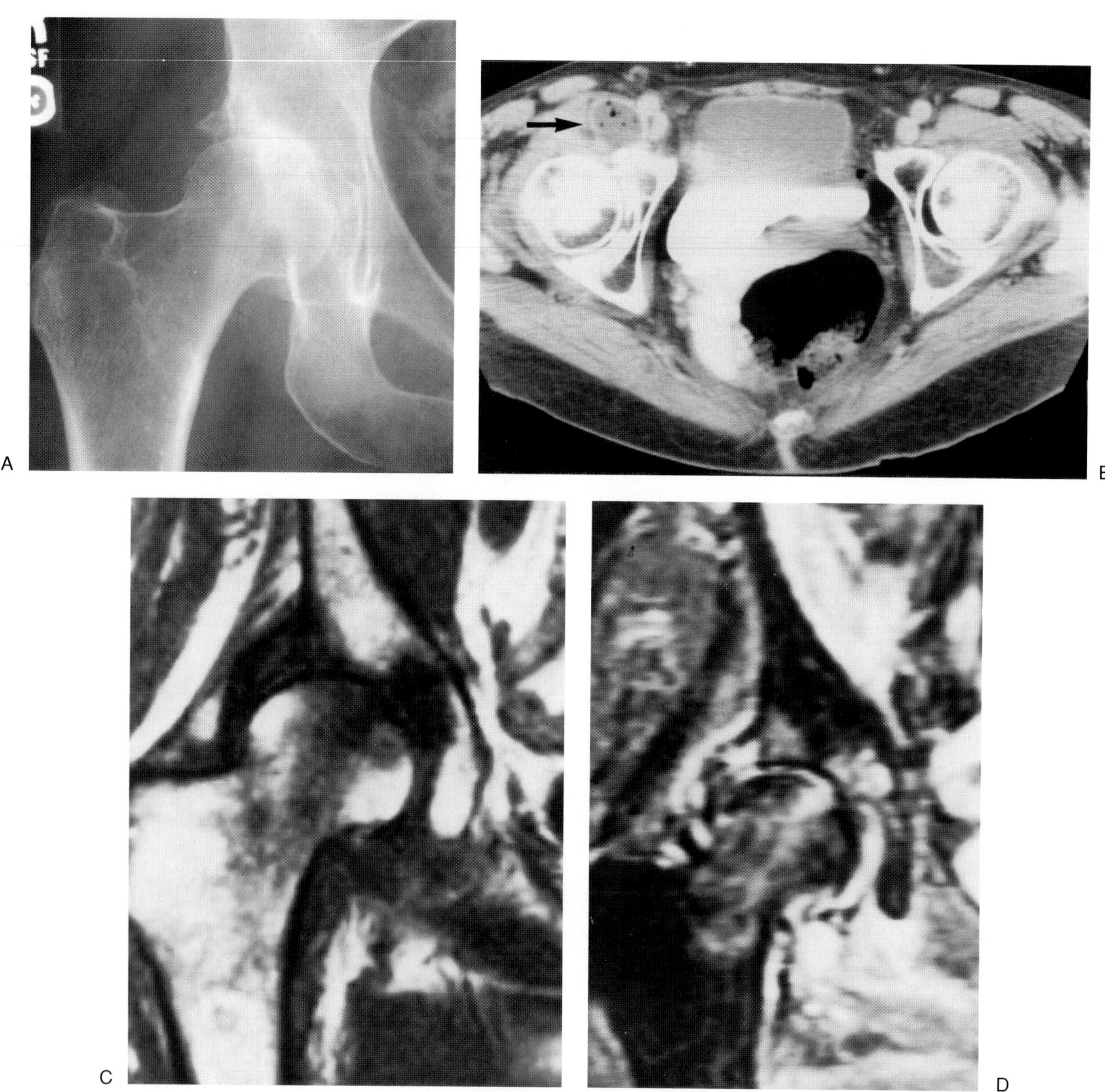

FIG. 13.

Findings: The radiograph of the hip (Fig. 13A) shows osteopenia, loss of joint space, and irregularity of the superolateral acetabular cortex. The axial CT scan (Fig. 13B) shows gas bubbles within a distended iliopsoas bursa (arrow). There is increased density of the bone marrow in the anterior acetabulum of the right hip. The T1-weighted (TR/TE 600/20 msec) coronal MR image of the right hip (Fig. 13C) shows large effusion obliterating the pulvinar fat pad and loss of marrow signal on both sides of the hip joint. The STIR (TR/TE/TI 2,700/30/160 msec) image (Fig. 13D) shows increased signal in the femoral head and acetabulum. There is a large effusion and abnormal increased signal in the soft tissues adjacent to the hip joint. Erosion of the cortex of the superolateral acetabulum is present.

Diagnosis: Septic arthritis of the right hip complicated by osteomyelitis.

Discussion: Infection of synovium-lined joints is a common disorder that requires immediate and accurate diagnosis to prevent permanent damage to the articulation. Synovial infection can involve tendon sheaths, bursae, or articulations. In the pelvic region, infection of the hip and sacroiliac joints is the most common form of synovial infection. Septic arthritis involves all age groups, although there is an increased frequency of this condition in the very young, the elderly, and immunocompromised patients. The clinical symptoms consist of fever, local pain, and a limited range of motion of the involved joint. An effusion is generally present, associated with regional hyperemia causing swelling and redness in the overlying soft tissues. The most common causative organism in all age groups combined is *Staphylococcus aureus.* In young children, *Haemophilus influenzae* is the most common cause of septic arthritis. Group D *Streptococcus* is an important cause in neonates and infants, especially in multifocal infection. Multifocal septic arthritis in young adults is suggestive of gonococcal arthritis. Lyme disease may also produce multifocal involvement. Nonpyogenic septic arthritis is less common and is usually caused by *Mycobacterium tuberculosis.*

The major mechanisms responsible for septic arthritis include hematogenous spread to the synovium, direct spread from adjacent osseus infection, and direct inoculation of the articulation, via either penetrating trauma or iatrogenic contamination of the joint (1). Septic arthritis due to hematogenous spread, the most common mechanism for synovial infection, typically affects the large central articulations, such as the shoulder, knee, hip, and sacroiliac joints. Predisposing conditions for hip, and sacroiliac infection include septicemia, intravenous drug abuse, gastrointestinal or genitourinary infection, pelvic trauma, pelvic surgery, and pregnancy. Infection of the hip carries a particularly poor prognosis because the joint is deep, and an effusion within this joint is difficult to detect clinically and radiographically, which may lead to a delay in diagnosis (2). Additionally, the relatively small volume of this major weight-bearing articulation leads to rapid erosion of cartilage and bone due to an early increase in intra-articular pressure.

Diagnostic imaging plays a relatively minor role in the initial diagnosis of septic arthritis, which is typically diagnosed on the basis of aspiration of a clinically symptomatic joint. The radiographic demonstration of any monoarticular erosive arthritis should be considered septic arthritis until proved otherwise. Plain film findings of early septic arthritis include soft tissue edema, joint effusion, and para-articular osteoporosis. At this stage, scintigraphic examination shows increased blood flow adjacent to the joint and prominent activity on the blood pool images on both sides of the articulation. The delayed bone scan images show diminution of activity, in contrast to osteomyelitis, in which continued accumulation of the activity is seen. With more advanced joint space infection, changes in the joint space and adjacent bones can be seen. Loss of cartilage leads to uniform narrowing of the joint space, particularly in patients with pyogenic infection. Loss of the subchondral cortex, marginal erosions, periostitis, and central subchondral regions of bone destruction become apparent as the infection goes on to involve the adjacent osseous structures (1). Septic arthritis secondary to tuberculosis and fungal diseases shows prominent osteoporosis, a slower rate of cartilage destruction, and less joint space narrowing than with pyogenic infection. CT is nonspecific but will demonstrate erosions of bone prior to plain films, particularly in areas of complex anatomy, such as the spine and pelvis. Eventually, untreated septic arthritis leads to joint destruction and adjacent osteomyelitis, resulting in fibrous or osseous ankylosis of the infected joint.

MR imaging of septic arthritis in its early stages often only shows a nonspecific effusion due to excess fluid within the synovial lined space. In the hip, large effusions may be associated with distension of the iliopsoas bursa, a rounded bursal cavity anterior to the hip joint that normally communicates with the joint in 10% to 15% of patients (3). Like all synovial fluid, the septic effusion is low signal on T1-weighted images and high signal on T2-weighted or STIR images. Unlike bland sterile effusions, septic effusions (and other inflammatory effusions) tend to obliterate the intra-articular fat pads and produce inflammation in the adjacent soft tissues (4). In more chronic infection, synovial hypertrophy and debris within the joint may be apparent. Synovial hypertrophy can usually be distinguished from effusion on T2-weighted images because the thickened synovium is characteristically irregular, inhomogeneous, and of lower signal than joint fluid. In questionable cases, gadolinium enhancement may be helpful in distinguishing joint fluid, which does not enhance on the early postinjection images, from the enhancing thickened synovium.

Diagnostic imaging of patients with septic arthritis is often performed for monitoring therapeutic response and for the early detection of complications. Complications associated with septic arthritis include epiphyseal destruction and growth disturbance in the skeletally immature individual, as well as secondary osteomyelitis in all age groups (1) (Fig. 13E). There is some controversy regarding the extent of osseous abnormality necessary before the patient is considered to have frank osteomyelitis. Small marginal erosions, even though they technically result in communication of the infected joint with the underlying bone marrow, are not considered to represent frank osteomyelitis by some clinicians because the therapeutic protocol and prognosis are not significantly altered. Others consider any cortical erosion in a patient with septic arthritis as a sign of osteomyelitis and alter the therapy protocol. Because of this controversy, it is difficult to state conclusively when a patient with septic arthritis should be considered to have osteomyelitis. At our institution, we do not diagnose osteomyelitis unless there is nonmarginal cortical destruction, periostitis, or medullary trabecular destruction.

With MR, it is often difficult to evaluate the adjacent bones for secondary osteomyelitis, particularly with pyogenic articular infection. Septic arthritis frequently results in signal changes in the adjacent bone marrow due to adjacent reactive edema. Abnormal decreased T1-signal intensity and increased STIR-signal intensity has been demonstrated in periarticular bone marrow without infectious involvement in 60% of patients with uncomplicated septic arthritis (5). Frank adjacent osteomyelitis can be confidently diagnosed if there is cortical destruction, but signal changes alone are nonspecific.

MR is an excellent method for evaluating septic sacroiliitis because it is difficult to visualize the sacroiliac joint adequately on conventional radiographs, and scintigraphy is often difficult to interpret because of the normal high activity present in the sacroiliac region. MR demonstrates an abnormal volume of intra-articular fluid within the sacroiliac joint and often shows additional abnormal fluid collections extending posterior to and within the iliopsoas muscle (Fig. 13F). Abnormal signal in the adjacent bone marrow within the sacrum and/or ilium is also well shown (6,7) (also see Case 4–6).

Septic arthritis secondary to adjacent osteomyelitis occurs most frequently in the infant hip. The metaphysis of the proximal femur lies within the capsule, and infection of the femur can spread into the adjacent hip joint. In the lax joints of an infant, the accumulation of joint fluid can produce marked widening of the joint space and pathologic subluxation. The tense effusion interferes with the vascular supply to the epiphysis. Epiphyseal destruction may lead to permanent joint deformity, growth disturbance, or AVN. The epiphysis often completely disappears following the episode, only to reconstitute in an irregular and hypoplastic form.

In the initial phase of septic arthritis, when only a joint effusion can be identified, the differential diagnosis is quite extensive and includes numerous process that can lead to articular irritation. Once joint space alterations and erosions become apparent, the differential diagnosis is limited to those entities that present as an erosive arthritis. Rheumatoid arthritis and juvenile chronic arthritis are erosive inflammatory arthritides that typically involve the hips in a bilateral and symmetric fashion (Fig. 13G,H). Monoarticular rheumatoid arthritis is difficult to distinguish from septic arthritis without joint aspiration. Rapidly destructive osteoarthritis of the hip is a recently described disorder of the elderly characterized by severe, rapidly progressive, atrophic bone destruction of the proximal femur and acetabulum (8,9). Complete destruction of the hip joint takes place over a period of months. The radiographic findings can mimic septic arthritis, rheumatoid arthritis, severe osteonecrosis, and neuroarthropathy. MR shows a large joint effusion, prominent cartilage loss, and bone resorption with normal marrow signal (8) (Fig. 13I,J). Pigmented villonodular synovitis (PVNS) and synovial osteochondromatosis show radiographic alterations that can be difficult to distinguish from chronic septic arthritis, although they can usually be differentiated on the basis of their clinical findings. On MR, PVNS and synovial osteochondromatosis tend to be associated with normal signal in the bone marrow and overlying soft tissues. Intra-articular PVNS is low signal on the T2-weighted images due to the paramagnetic effects of hemosiderin. In synovial osteochondromatosis, signal inhomogeneity of the intra-articular contents and rounded low-signal osteochondral bodies adherent to the synovium can be identified (Fig. 13K).

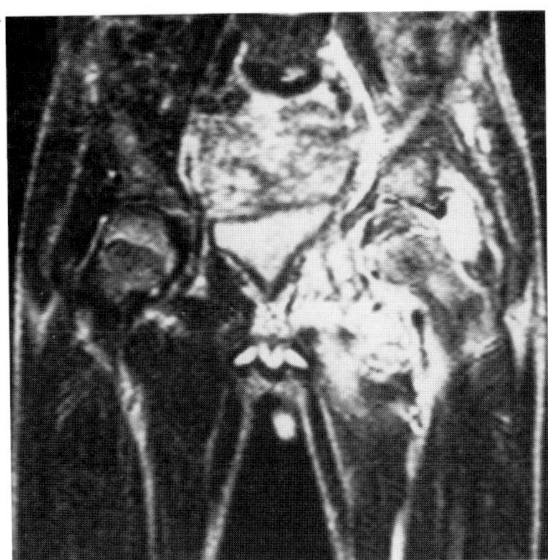

E

FIG. 13. E: A coronal STIR (TR/TE/TI 2,200/35/160) coronal image of the pelvis in a 12-year-old boy with septic arthritis of the left hip due to septic emboli from a valve replacement shows almost complete dissolution of the proximal femoral epiphysis. There is a large, inhomogeneous effusion that is higher signal than the bladder fluid, suggesting a chronic septic arthritis. There is loss of marrow signal within the left femur and erosion and fragmentation of the destroyed epiphysis.

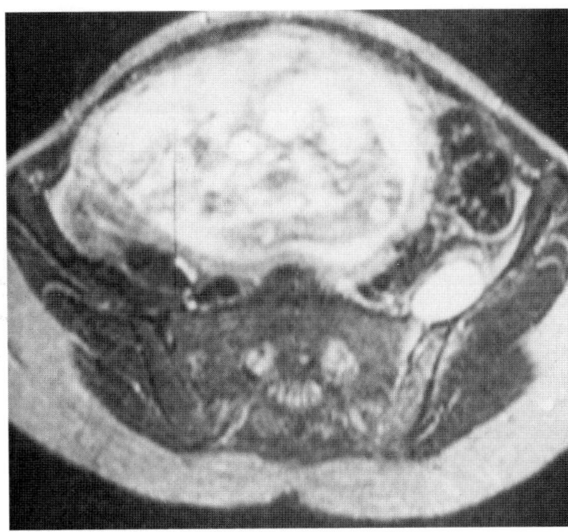

F

FIG. 13. F: An axial T2-weighted (TR/TE 2,000/70 msec) image of the pelvis in a 17-year-old pregnant woman with *S. aureus* septic sacroiliitis of the left sacroiliac joint demonstrates an effusion, increased signal in the adjoining bone marrow, and a rounded fluid collection anterior to the joint.

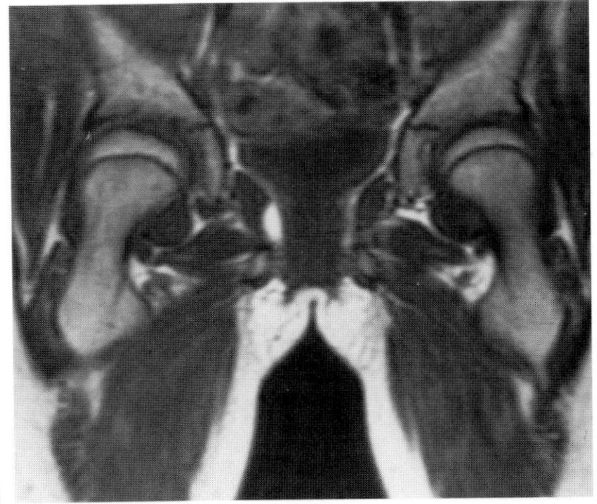

G

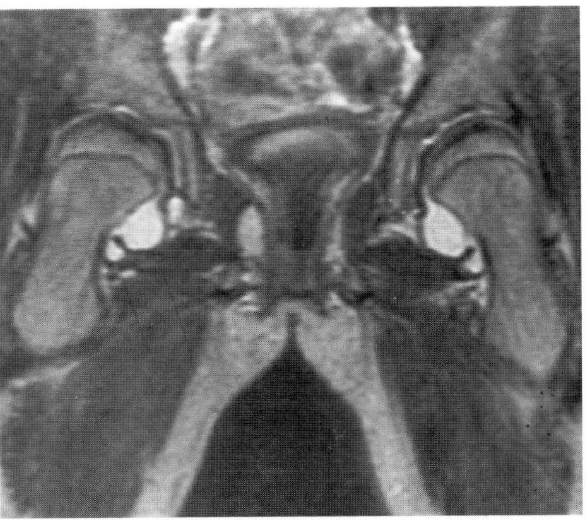

H

FIG. 13. G,H: MR images of both hips in a woman with juvenile rheumatoid arthritis are illustrated. **G:** The T1-weighted (TR/TE 400/20 msec) coronal image shows bilateral hip effusions, normal bone marrow signal, and abnormal morphology of the proximal femora due to chronic hyperemia and arthritis. There is coxa valga and diminution in size of the femoral heads bilaterally. **H:** The T2-weighted (TR/TE 2,000/70 msec) image shows the effusions within both hip joints. Note the symmetry of the process.

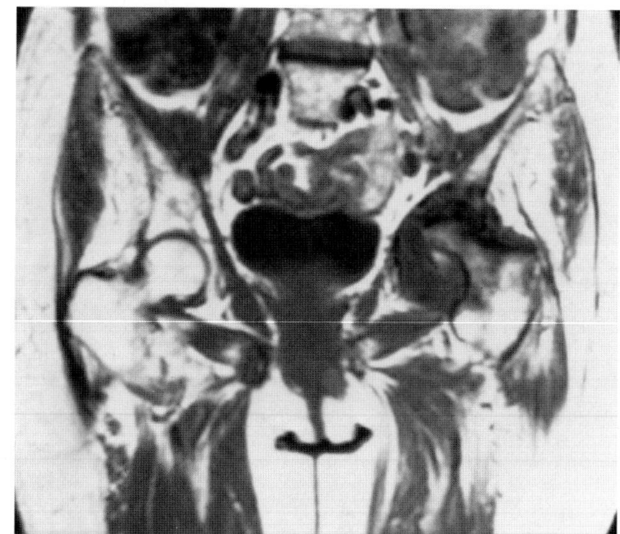

I

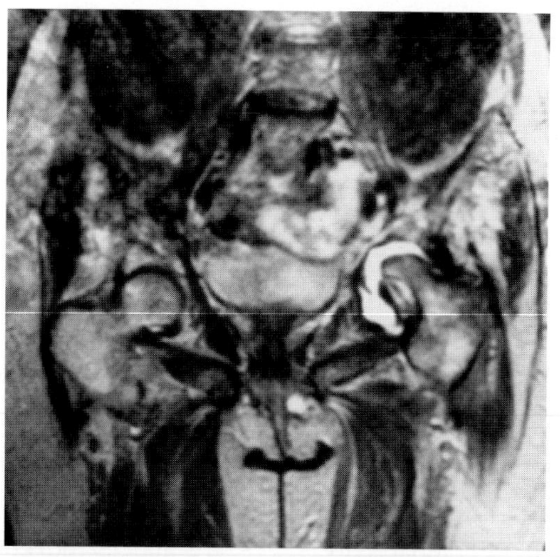

J

FIG. 13. I,J: This 66-year-old woman complained of increasingly severe left hip pain. Radiographs obtained over a period of 6 months showed rapid dissolution of the hip joint with atrophic destruction of the proximal femur. Extensive rheumatologic and laboratory workup was negative, and a joint aspiration revealed no growth or crystals. Histology at the time of joint replacement showed nonspecific synovitis and osteoarthritis, consistent with rapidly destructive osteoarthritis of the hip. **I:** A T1-weighted (TR/TE 517/14 msec) spin-echo coronal MR of the pelvis shows destruction of the femoral head and proximal neck, as well as erosion of the acetabulum. There is a large effusion within the joint as well as abnormal signal in the bone marrow on both sides of the joint. **J:** A T2-weighted (TR/TE 2,300/80 msec) spin-echo image also shows the atrophic destruction of the proximal femur. The bone marrow edema is not very pronounced on this sequence. The joint effusion appears homogeneous and the soft tissues are normal.

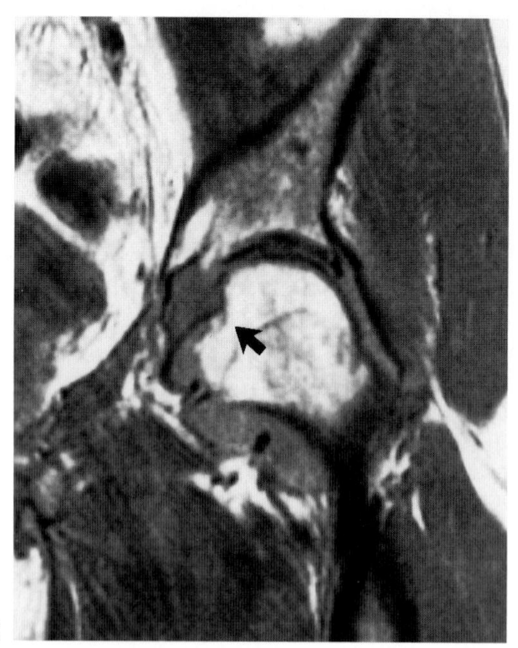

K

FIG. 13. K: A coronal proton-density (TR/TE 2,000/20 msec) MR image of the left hip in a 37-year-old man with idiopathic synovial osteochondromatosis shows a lobulated soft tissue density within the hip joint. There are at least three rounded low-signal densities within the joint, representing calcified osteochondral bodies. Note the enlargement of the capital fovea *(arrow)* and the adjacent erosion of the medial wall of the acetabulum. The acetabular fat pad (pulvinar) is invaded by the abnormal tissue. (Courtesy of G. Greenway, M.D., Dallas, Texas.)

CASE 14

History: This 69-year-old woman complained of pain in her back and both hips following a minor fall. She had undergone a mastectomy for breast cancer 14 years previously. Radiographs of the pelvis and lumbar spine showed osteoporosis but were otherwise normal. An MR of the pelvis was ordered. What is the cause of the abnormality on the MR examination? Should a biopsy be performed?

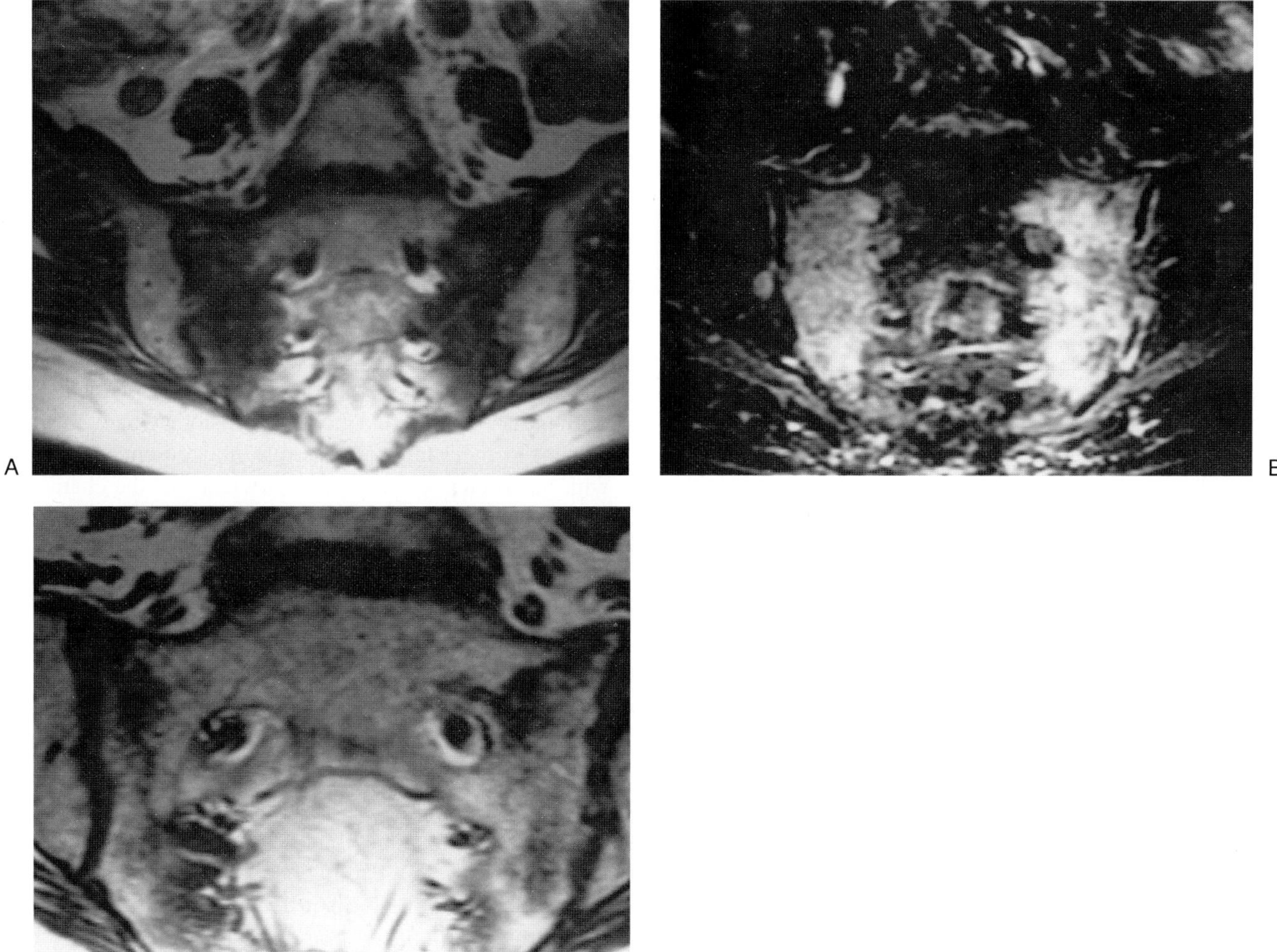

FIG. 14.

Findings: The axial T1-weighted (TR/TE 300/15 msec) spin-echo image of the sacrum (Fig. 14A) shows bilateral vertical bands of abnormal low signal in the sacral alae. The low signal regions are located adjacent to the sacral foramina and the sacroiliac joints. The STIR (TR/TE/TI 2,200/35/160 msec) image (Fig. 14B) shows high signal in the corresponding area. No low-signal bands are seen within these regions of abnormal signal. The T1-weighted (TR/TE 350/16 msec) gadolinium-enhanced image (Fig. 14C) shows enhancement of large portions of the abnormal marrow. Serpiginous low-signal bands persist in both sacral alae following contrast enhancement. (Courtesy of R. Kerr, M.D., Los Angeles, California).

Diagnosis: Bilateral insufficiency fractures of the sacrum.

Discussion: Insufficiency fractures develop when minor trauma or the usual activities of daily living stress bone that is insufficient in mineral or elastic resistance (1). An insufficiency fracture is one of the two subtypes of stress fractures, the other being fatigue fracture caused by repetitive loading of normal bone. Insufficiency fractures are most common in patients with senile or steroid-induced osteoporosis and in patients who have undergone pelvic irradiation for genitourinary malignancy (1,2). Other risk factors for the development of insufficiency fractures include osteomalacia, hyperparathyroidism, rheumatoid arthritis, diabetes mellitus, and resumption of weight bearing following immobilization (1). Insufficiency fractures predominate in the cancellous bone of the sacrum, pelvis, and lower extremities. Patients typically present with an acute onset of local pain at the site of the fracture with no antecedent trauma. This nonspecific history in an elderly patient often raises the suspicion of primary or secondary neoplasia. The spontaneous onset of pain is particularly worrisome in the previously irradiated patient, who is at high risk of metastatic disease (2). Less commonly, the presence of an insufficiency fracture is clinically unsuspected.

Insufficiency fractures of the sacrum are most commonly seen in women with postmenopausal osteoporosis and in patients who have undergone prior pelvis irradiation (2,3). Symptoms include low back pain or pain in the region of the sacroiliac joint, often with radiation of pain to the buttock, hip, and lower extremities. Sacral insufficiency fractures have a predictable location in the sacral alae, coursing parallel to the sacroiliac joint at the lateral margins of the ventral sacral foramina (3). Radiographs of the sacrum are insensitive for detecting these fractures due to the complex anatomy of the sacrum, the subtle nature of the radiographic abnormalities, and the presence of overlying soft tissues, vascular calcification, and bowel gas (3). In the acute phase, the radiographs are normal. In the subacute and chronic phase, radiographs do not show the fracture lines but do demonstrate reactive endosteal sclerosis at the fracture margins. Sclerosis may be unilateral or bilateral. Most often, the sclerosis is linearly oriented and runs vertically along the lateral edges of the sacral arcuate lines, parallel to the sacroiliac joints (3,4). The finding of widespread sacral sclerosis is often misleading and is misinterpreted as representing sclerotic metastatic disease.

Bone scanning is a sensitive method for early detection of sacral insufficiency fractures, becoming positive days or weeks earlier than radiographs (1). In the acute phase, a nonspecific mild generalized increase in uptake is present within the sacrum. This appearance progresses over a period of a few weeks to a characteristic pattern of increased accumulation in a well-marginated linear pattern in one or both sacral alae (1,5). Bilateral vertical fractures are often connected by a transverse bridge of radioactivity across the sacral body, resulting in an H-or butterfly-shaped pattern of uptake (5). CT is diagnostic for the presence of sacral insufficiency fractures. On CT, condensed sclerotic bone at the fracture margins runs vertically in the sacral alae, parallel to the sacroiliac joints. With appropriate windowing, disruption of the anterior sacral cortex at the lateral margins of the sacral foramina and linear intraosseous lucencies at the fracture sites can be identified (3).

The MR appearance of sacral insufficiency fractures, particularly in the acute phase, can be confusing, and the findings are easily misinterpreted as representing metastatic disease (2,6). In the acute phase, the fracture lines themselves are seldom identified on the MR examination. Wide bands of bone marrow edema at the fracture sites represent the typical MR finding in acute sacral insufficiency fractures. The edema is low signal on T1-weighted and high signal on T2-weighted, STIR, and gadolinium-enhanced sequences (1,2). These zones of marrow edema may be vertical or horizontal, or they may involve almost the entire sacrum (6). The extent of marrow abnormality is much larger than can be appreciated on CT scanning of these fractures (2). As the fracture evolves and reactive sclerosis develops adjacent to the fracture margins, linear bands of low signal can be seen within the wide areas of edema (6) (Fig. 14D). Once the low-signal linear fracture line is identified, a confident diagnosis of insufficiency fracture can be made. In the acute phase, it is difficult to distinguish insufficiency fractures from metastatic disease. Metastatic disease is often disseminated and frequently involves the spine and other portions of the pelvis. Soft tissue masses and bone expansion, which may be present in metastatic disease, are not present in patients with insufficiency fractures. Knowledge of the characteristic locations of these injuries is necessary to avoid misinterpretation of the MR. In questionable cases, CT is a useful modality to identify subtle cortical and trabecular disruption.

Insufficiency fractures of the pelvis involve multiple locations simultaneously in up to 50% to 85% of patients (3,7,8). The entire pelvis should be carefully evaluated when a seemingly isolated insufficiency fracture of the sacrum is encountered. Other common locations for insufficiency fractures in the pelvis include the pubic rami, parasymphysis, medial iliac wing, and supra-acetabular region (2,4,7,8). Pubic rami and parasymphyseal fractures are relatively simple to identify radiographically. Supra-acetabular insufficiency fractures, however, are very difficult to visualize on radiographs. These fractures, which are frequently bilateral, develop immediately superior to the acetabulum and appear as an arched band of subchondral sclerosis oriented parallel to the acetabular roof (8). The curvilinear low-signal fracture line is well seen on MR, as is the surrounding marrow edema. Marrow edema adjacent to insufficiency fractures of the supraacetabular region may be very extensive, particularly following radiation therapy (Fig. 14E–G). Unlike insufficiency fractures of the sacrum CT is rarely helpful for establishing the diagnosis of a supra-acetabular fracture, due to the horizontal orientation of the fracture line. As for insufficiency fractures, knowledge of their characteristic locations will avoid misdiagnosis of these common skeletal injuries.

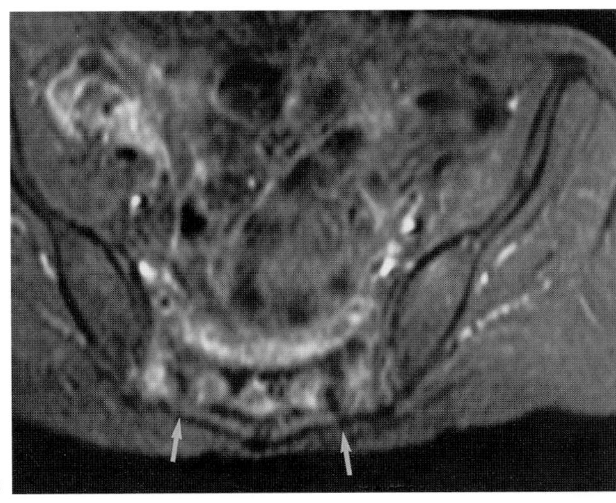

D

FIG. 14. D: A T1-weighted (TR/TE 600/10 msec) fat-suppressed spin-echo image of the sacrum following gadolinium enhancement shows diffuse enhancement of the entire bone in this patient with bilateral sacral insufficiency fractures. Note the linear bands of low signal in both sacral alae *(arrows)* at the sites of reactive sclerosis at the margins of the fractures.

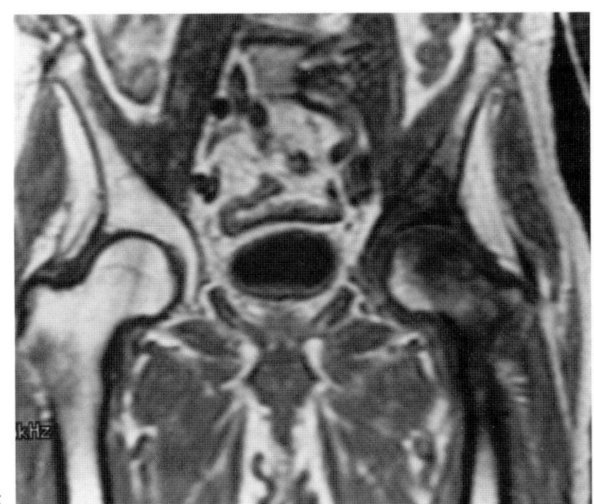

E

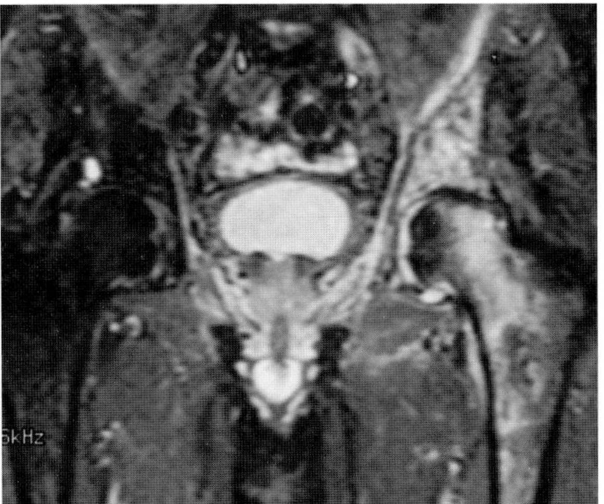

F

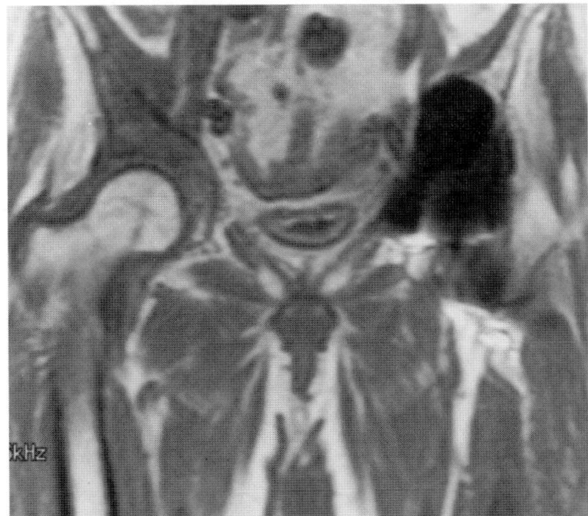

G

FIG. 14. E–G: This 80-year-old man had undergone pelvic irradiation for prostate cancer several years previously. He complained of severe left hip pain at the time of presentation. Radiographs of the left hip were normal, and he was referred for MR examination. **(E,F).** Biopsy of the acetabulum showed reactive bone and no tumor cells. The patient underwent a total hip replacement, and the resected specimens showed healing insufficiency fractures. One year later, he developed right hip pain. Radiographs and MR **(G)** at that time showed another supra-acetabular insufficiency fracture. **E:** A T1-weighted (TR/TE 549/10 msec) coronal image of the pelvis shows extensive loss of marrow signal in the left acetabulum and left femur caused by radiation-induced insufficiency fractures. **F:** A STIR (TR/TE/TI 2,700/40/160 msec) image shows even more extensive areas of high signal in the corresponding regions. **G:** T1-weighted MR (TR/TE 600/20 msec) image obtained 1 year later following left total hip arthroplasty shows a new zone of signal loss in the right supra-acetabular region due to another insufficiency fracture.

CASE 15

History: An 86-year-old man with mild right buttock pain was noted to have an abnormality on a pelvis radiograph. An MR examination of the pelvis was obtained. What is the most likely diagnosis? What are the specific MR findings in this patient that suggest this entity?

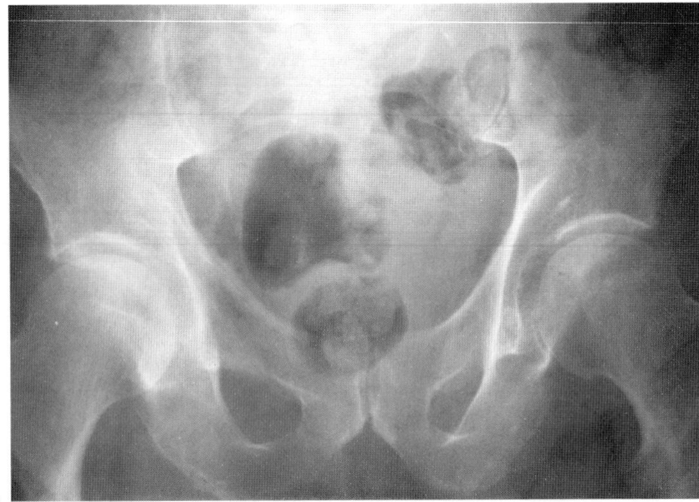

A

FIG. 15.

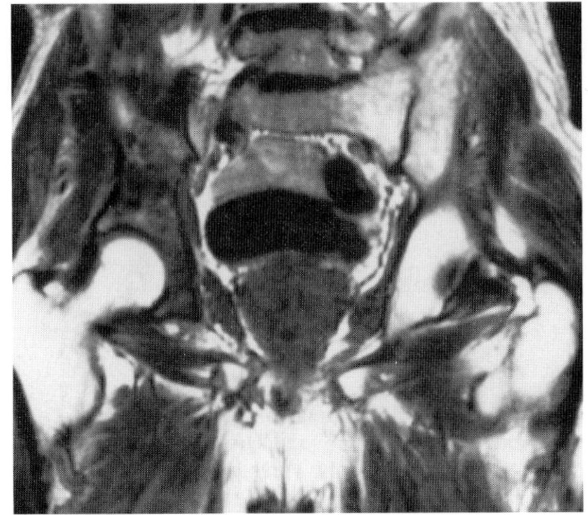

B C

Findings: The radiograph of the pelvis (Fig. 15A) shows mild sclerosis of the right iliac wing and thickening of the right iliopectineal cortex. The medial acetabular wall and posterior acetabular column also show mild cortical thickening when compared with the left side. The T1-weighted (TR/TE 716/19 msec) image of the posterior pelvis (Fig. 15B) shows low-signal marrow within the iliac bone. A few small islands of normal high-signal marrow are interspersed within the abnormal marrow. A more anterior image from the same series (Fig. 15C) shows irregular bands of low signal, with interspersed fatty marrow, within the acetabular roof. There is thickening of the cortex at the medial border of the acetabulum on this image.

Diagnosis: Paget's disease of the right ilium.

Discussion: Paget's disease is a common skeletal disorder of the middle-aged and elderly, affecting approximately 3% to 4% of the population over the age of 40 years (1). The disorder is of unknown etiology, although hereditary predisposition, inflammation, and chronic viral infection have been implicated (1). Paget disease is characterized by abnormal and excessive osseous remodeling, leading to disorganized osseous resorption and apposition, which produces thickened and disorganized osseous trabeculae. The most common sites of involvement are in the axial skeleton, particularly the innominate bone, sacrum, spine, and calvarium (1). The proximal long bones, particularly the femur, are also frequently involved. The condition is frequently asymptomatic and is discovered incidentally on radiographs obtained for another reason. In some patients, local pain and deformity develop at the site of involvement. The bone in Paget's disease is weak and prone to fracture with minimal trauma. Insufficiency fractures may also develop in the involved areas in the absence of any injury. Neurologic compromise due to bony hypertrophy may develop, particularly with craniofacial and spinal involvement. The most serious complication of Paget's disease is the development of Paget's sarcoma, a highly malignant neoplasm that develops in approximately 1% of patients with Paget's disease (1–3).

Paget's disease produces characteristic radiographic findings, allowing accurate diagnosis in most cases on the basis of the radiographic appearance alone. In the early active phase of Paget's disease, osteolysis may predominate, particularly in the skull and long bones. With maturation of the process, a combination of osteolysis and osteosclerosis is apparent. In this mixed phase, coarsening of the trabeculae, cortical thickening, bone expansion, and osteosclerosis are apparent. As the process becomes inactive, osteosclerosis predominates, and the changes in the cortex and trabeculae are readily apparent. Scintigraphy is positive during the active and mixed phases of Paget's disease, and scintigraphic changes may precede radiographic abnormalities (1). CT shows the cortical and trabecular alterations well, particularly in areas of complex anatomy like the pelvis and spine.

MR is rarely necessary in patients with Paget's disease of bone. Because of the wide variability in the appearance of the marrow in such patients, correlation with the radiographs is necessary to avoid misdiagnosis of this benign disorder, particularly in patients with active disease. Because of the high prevalence of this condition, Paget's disease is usually discovered as an incidental finding on MR imaging of the spine and pelvis. Knowledge of the spectrum of signal changes in the marrow in this common disorder is essential for accurate diagnosis. On MR images, the thickened cortex appears as a thick rim of signal void, distinct from the higher signal medullary cavity (4). An undulating contour of the endosteal cortex may be present. The appearance of the medullary cavity is quite variable. Patterns seen in Paget's disease include normal fatty marrow, focal areas of fat, sclerosis, or soft tissue material resembling granulation tissue (4). In active disease, intramedullary foci of low signal on T1-weighted images that increase to high signal on T2-weighted images can be seen in the medullary space. These nodular areas of abnormal marrow represent areas of fibrovascular tissue that histologically resembles granulation tissue (4). In uncomplicated active Paget's disease, these areas of fibrovascular tissue remain intracortical and show no soft tissue component. Their MR appearance is nonspecific, and correlation with the radiographs is essential to exclude metastatic disease, primary osseous sarcoma, and osteomyelitis, which may have identical signal characteristics (4). In long-standing, inactive Paget disease, the signal intensity of the marrow shows a return to a more normal marrow signal appearance. Prominent regions of persistent fat signal are apparent within the areas of thickened trabeculae. The fat signal may be even more intense and homogeneous in these areas of inactive Paget's disease than in the normal bone marrow (4,5). These cyst-like areas of fatty marrow are surrounded by the thickened cortical bone and interspersed with the thickened, irregular trabeculae characteristic of long-standing involvement. The osseous thickening results in diminution in volume of the medullary space (4). The finding of persistent fat signal in an area of osseous sclerosis and trabecular thickening is highly suggestive of Paget's disease because other processes producing reactive osteosclerosis, such as metastatic disease and chronic osteomyelitis, show loss of fat signal within the medullary space (5).

Osseous neoplasms that can complicate Paget's disease include metastatic disease, giant cell tumor, and sarcomatous transformation in the areas of involvement. A variety of different sarcomas, particularly osteosarcoma, fibrosarcoma, and malignant fibrous histiocytoma, may arise in the areas of Paget's disease. Any histologic type of sarcoma arising in patients with Paget's disease is usually referred to as Paget's sarcoma. The prognosis of these highly aggressive neoplasms is particularly poor, and 5-year survival is unusual (2,3). On conventional radiographs, sarcoma formation results in the development of large areas of osteolysis and cortical destruction. Less commonly, osteoblastic changes or mixed osteolysis and osteosclerosis may be present (3). Periostitis is not typically present, although tumor mineralization in the soft tissues may be apparent in some cases (3). A large soft tissue mass is usually seen (2). The extent of the soft tissue mass, which can be as large as 25 cm or more, is more clearly delineated with CT or MR imaging. MR is the modality of choice for defining the anatomic extent of the soft tissue component of Paget's sarcoma (Fig. 15D). The neoplastic tissue in most cases of Paget's sarcoma is low signal on T1-weighted images and high signal on T2-weighted images. In osteosclerotic lesions with extensive tumor mineralization, variable signal and inhomogeneity are present on the T2-weighted images.

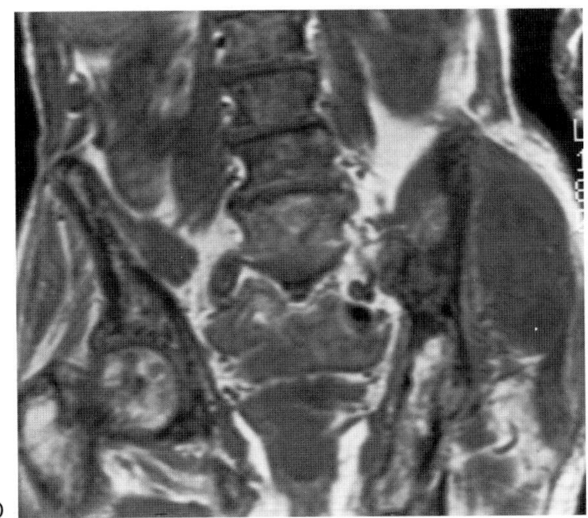

D

FIG. 15. D: A T1-weighted (TR/TE 400/10 msec) image of the pelvis in a patient with known widespread Paget's disease shows inhomogeneous marrow, cortical thickening, and prominent trabeculae throughout the skeleton. There is a large soft tissue mass in the left buttock that distorts the gluteal fat planes. Axial images showed that the mass arose from the left iliac bone. Biopsy confirmed the diagnosis of Paget's sarcoma.

CASE 16

History: A young dancer underwent MR arthrographic evaluation for chronic hip pain. What is the abnormality demonstrated?

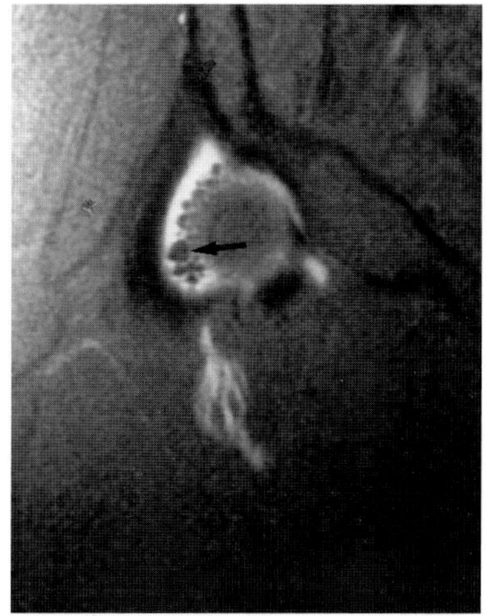

FIG. 16.

Findings: Coronal T1-weighted image (Fig. 16A) of the right hip after intra-articular injection of dilute gadopentetate dimeglumine demonstrates multiple low-signal-intensity fillings defects (arrow) within the right hip joint fluid. (Image courtesy of Mitchell Klein, M.D., Milwaukee, Wisconsin.)

Diagnosis: Intra-articular air simulating multiple loose bodies.

Discussion: Images of the right hip in other projections demonstrate a curvilinear low-signal-intensity focus within the nondependent aspect of the hip joint. These areas of low signal intensity appear to conform to the surface of the femoral head and demonstrate a more linear morphology than on the coronal projections (Fig. 16B,C). The nondependent nature of the filling defects and their morphology suggest the correct diagnosis that they represent the iatrogenic introduction of air at the time of injection and not loose chondral bodies within the joint. As the introduction of small quantities of air at the time of injection is not an uncommon event, it is important to be aware of this potential pitfall.

The diagnosis of chondral defects and loose osteocartilaginous bodies within the joint is facilitated either by the presence of sufficient joint fluid to provide for an arthrogram effect (T2-weighted images) or by the performance of MR arthrography (1–3). When evaluating patients with unilateral hip pain, it is imperative to utilize a surface coil to obtain a sufficient signal-to-noise ratio to support high-resolution techniques. The authors presently utilize a flexible wraparound coil for this purpose. Three-plane imaging using various combinations of T1, fast spin-echo, and fast inversion recovery sequences are employed. Most recently, the authors have increasingly found that fat-suppressed intermediate-weighted fast spin-echo images (e.g., TR more than 4,000/TE 30 msec) provide excellent contrast for depiction of both bone and soft tissue abnormalities (Fig. 16D–F). Chondral defects can be exceedingly subtle, and localized contour abnormalities along the articular surfaces highlighted by joint fluid must be sought (Fig. 16G,H). The signal characteristics of loose bodies are varied, reflecting whether they are entirely cartilaginous or variably ossified (1–3). Loose bodies may be quite subtle on routine MR examinations and again must be approached with a high degree of suspicion. Loose bodies may also be simulated by volume averaging of normal intra-articular structures including capsular ligaments as well as fat (Fig. 16I,J).

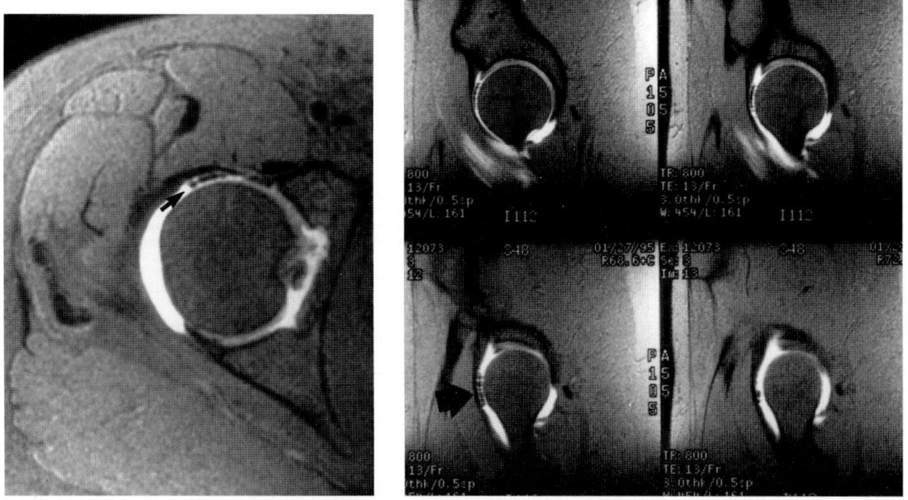

FIG. 16. B: Fat-suppressed axial T1-weighted image of the right hip after intra-articular injection of dilute gadopentetate dimeglumine demonstrates air *(arrow)* rather than loose bodies. **C:** Series of fat-suppressed sagittal T1-weighted images of the right hip after intra-articular injection of dilute gadopentetate dimeglumine again demonstrates the nondependent, linear air *(arrow).*

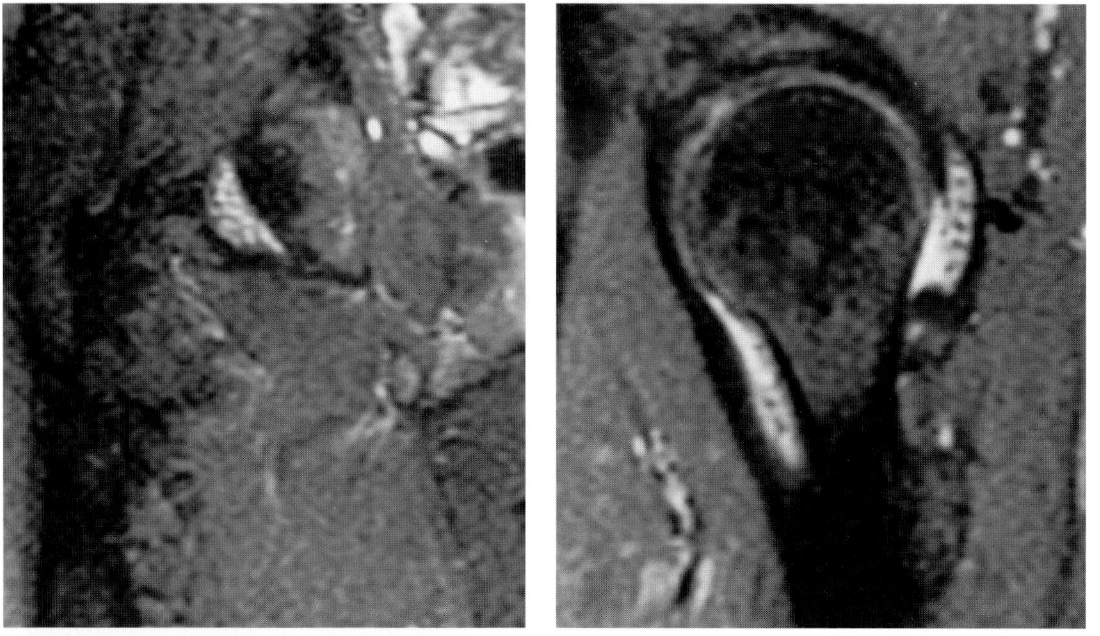

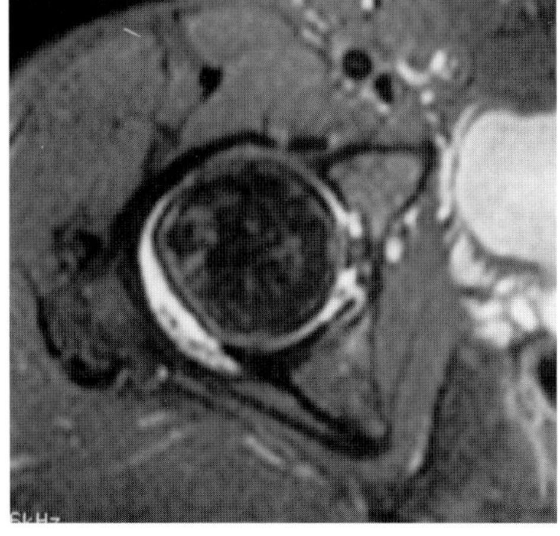

FIG. 16. D–F: Synovial chondromatosis. Coronal **(D)**, **(E)** sagittal, and axial fat-suppressed **(F)** proton-density (TR/TE 4,000/32 msec) images demonstrate multiple small low-signal-intensity filling defects contained within the synovial fluid of the right hip joint of a 22-year-old man. Biopsy revealed synovial chondromatosis.

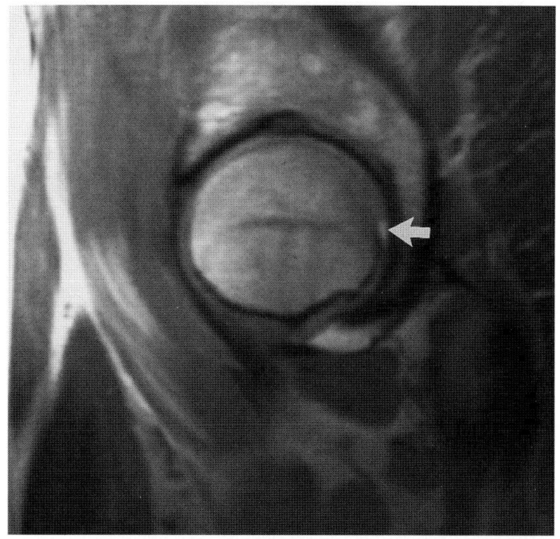

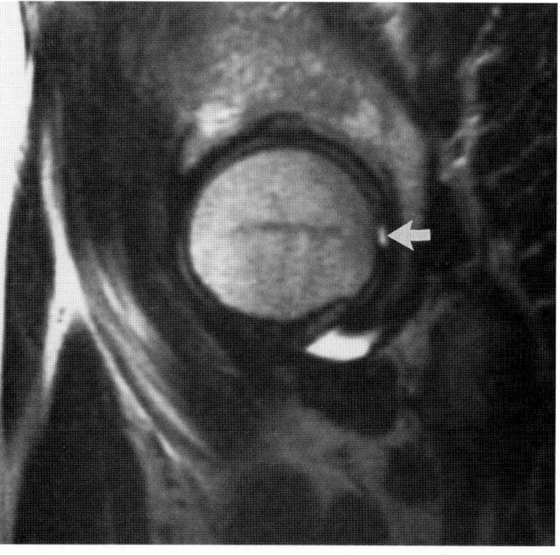

FIG. 16. G,H: Coronal proton-density-weighted **(G)** and T2-weighted **(H)** spin-echo images demonstrate localized area of slightly augmented signal **(G)** along the medial articular surface of the femoral head, reflecting a localized chondral defect *(arrow)*. High-signal-intensity joint fluid extends into the localized articular cartilage defect and is more conspicuous on the T2-weighted image *(arrow)*. (Images courtesy of Hollis Fritz, M.D., Minneapolis, Minnesota.)

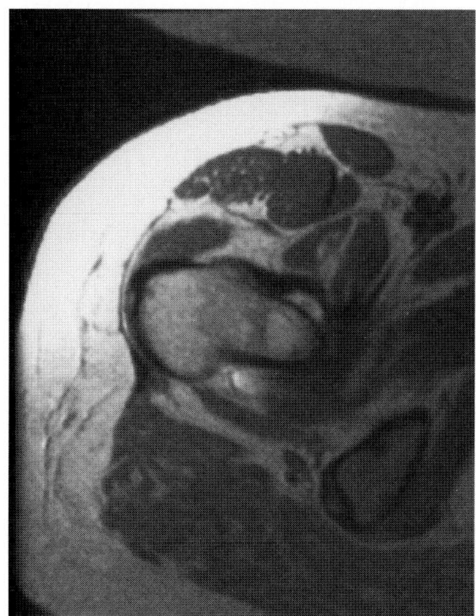

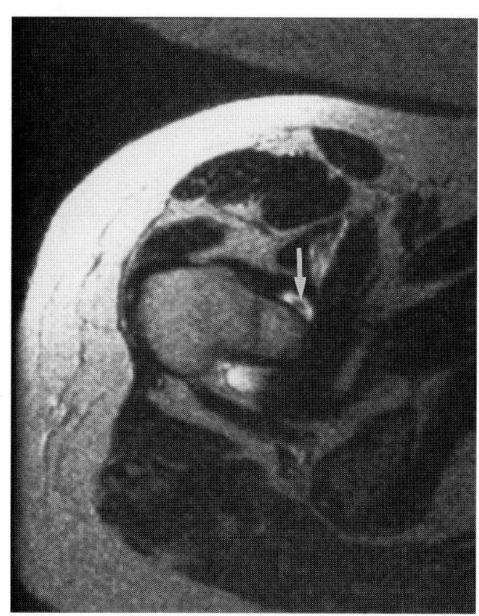

FIG. 16. I,J: Axial proton-density **(I)** and T2-weighted **(J)** spin-echo images obtained at a level corresponding to the inferior aspect of the hip joint. A subtle low-signal-intensity filling defect is identified within the joint fluid, reflecting a small loose body *(arrow)*. (Images courtesy of Hollis Fritz, M.D., Minneapolis, Minnesota.)

CASE 17

History: A 6-year-old child complained of recent onset of pain in the left hip. There was no history of trauma. Physical examination revealed a mild limp and limitation of motion of the left hip. What is the diagnosis based on the MR images? Does the child need surgery?

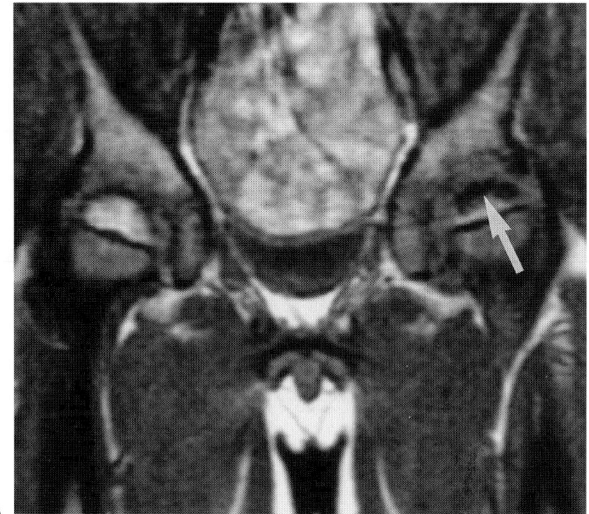
A

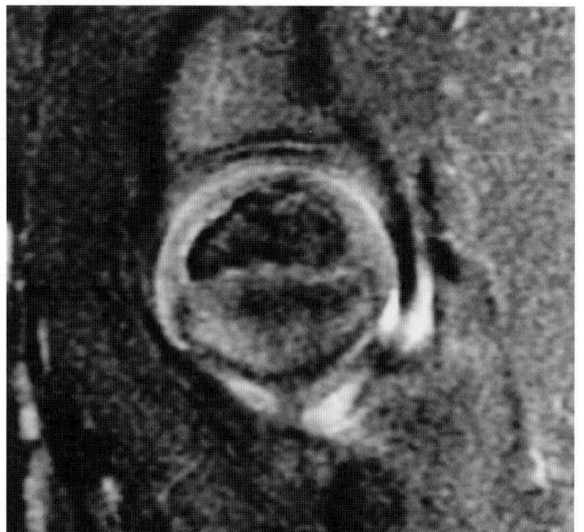
B

FIG. 17.

Findings: Coronal T1-weighted (TR/TE 600/17 msec) image (Fig. 17A) demonstrates diminution in signal intensity involving the left femoral capital epiphysis (arrow). Sagittal PAT suppressed proton-density fast spin-echo (TR/TE 4,000/36 msec) image (Fig. 17B) demonstrates that the articular surface of the hip maintains its smooth convex contour.

Diagnosis: Legg-Calvé-Perthes disease (LCPD) (idiopathic osteonecrosis) of the left hip.

Discussion: LCPD is the eponym applied to the common development of idiopathic osteonecrosis of the proximal femoral epiphysis in the child. LCPD can develop at any time during childhood but is most common between the ages of 4 and 7 years. This period is thought to be a time when the blood supply to the femoral head is relatively precarious (1,2). LCPD is bilateral in 10% to 20% of cases and affects boys more commonly than girls. The child typically presents with nonspecific hip pain, a mild limp, and a limited range of hip motion. The differential diagnosis for this clinical presentation includes transient synovitis of the hip, traumatic injury, and joint infection. Aspiration of the joint may be necessary to exclude infection in the child with a hip effusion and normal radiographs.

LCPD is a self-limited condition, and the necrotic femoral head is capable of significant reconstitution and remodeling, resulting in a generally favorable prognosis (2). Unfortunately, children with extensive necrosis and relatively advanced age may develop permanent femoral head deformity. This deformity leads to loss of congruity between the femoral head and acetabulum, with loss of containment of the femoral head. Nonoperative and operative treatment in LCPD is directed toward ensuring containment of the femoral head within the acetabulum during the healing process. Such containment is necessary to foster development of a normal, spherical femoral head and prevent the development of secondary osteoarthrosis (2).

The diagnosis of LCPD is usually established on the basis of clinical examination and conventional radiography, although radiographs may be normal early in the course of the disease. Early cases may only show a hip effusion, causing joint space widening and displacement or obscuration of the lateral capsular fat plane. At this stage, scintigraphy usually shows an area of deficient radionuclide accumulation in the femoral head. Quantification of the extent of necrosis is difficult with bone scanning, however, due to the limited spatial resolution of this technique. Later in the course of the disease, abnormalities in the size, density, and architecture of the femoral ossification center become radiographically apparent. Abnormal radiographic findings include diminution in size, lateral displacement, fracture and fragmentation, sclerosis, and flattening of the proximal femoral epiphysis. At this stage, arthrography has been used to evaluate the cartilaginous surfaces of the hip and acetabulum. Hip arthrography is an invasive procedure and often requires general anesthesia in this age group, so its use is limited (3). In more advanced disease, radiolucencies in the proximal metaphysis of the proximal femur, shortening and broadening of the femoral neck, and greater trochanteric overgrowth are apparent.

MR imaging allows early diagnosis of LCPD and can also determine the extent of involvement of the femoral head. The normal epiphysis in children contains fatty marrow with high signal intensity on T1-weighted images. Osteonecrosis initially involves the more central portions of the epiphysis, with relative sparing of the periphery (3). In acute LCPD, T1-weighted images reveal areas of hypointensity in the fatty marrow of the necrotic femoral epiphysis (Fig. 17C). This finding is nonspecific since any process altering the bone marrow produces loss of signal on T1-weighted images (1). As the infarct evolves, the signal becomes more heterogeneous on the T1-weighted images, presumably due to progressive irregular revascularization (1,4). The signal of the infarcted area is more heterogeneous on T2-weighted images. The sequestered dead bone demonstrates hypointense signal, whereas the adjacent repair tissue is hyperintense in signal on the T2-weighted sequences (4). The subchondral fracture line is well shown with MR, even when it is not visible with plain radiography (5). During the phase of active necrosis, hypertrophy of the synovial tissue may be identified (1). With revascularization, the signal of the femoral head returns to normal, although the morphologic abnormalities persist until remodeling is complete (1,4).

The most important role for MR in patients with LCPD is evaluation of femoroacetabular containment (1,3). Loss of containment, which is the major indication for surgery, is present in up to 25% of patients with LCPD. Conventional radiographs do not depict the contour of the cartilage, which is frequently much less deformed than the fragmented ossific nucleus would indicate. With MR imaging, the shape of the femoral head, its position within the acetabulum, and the degree of femoral head uncovering can be assessed in the absence of significant ossification (Fig. 17D,E). MR accurately depicts the shape and congruity of the cartilaginous components of the femoral head and the acetabulum, particularly on gradient-echo images with small flip angles (6). The T2*-weighted gradient-echo sequences most accurately distinguish intra-articular fluid from cartilage (6). In LCPD, thickening of both the acetabular and femoral head cartilage results in loss of femoral head containment. The mean increase in cartilage thickness compared with normal individuals varies from 1.8 to 3.9 mm, depending on the location within the joint (1). MR affords noninvasive assessment of cartilage thickness, joint congruity, and the status of the femoral head in patients with LCPD. MR will continue to play an important role in helping to select appropriate patients for surgical correction.

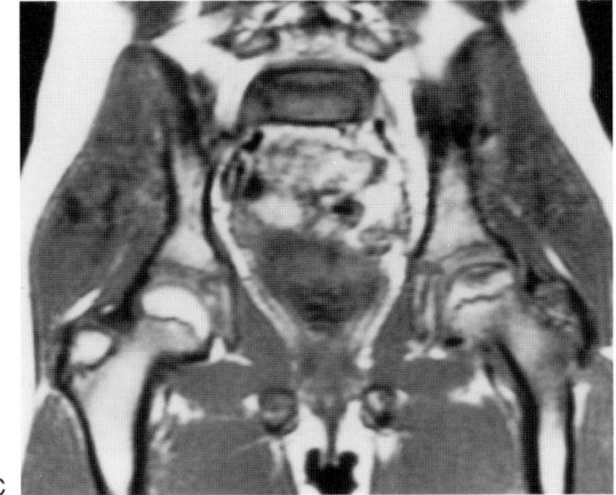

FIG. 17. C: A T1-weighted (TR/TE 700/20 msec) coronal image of the pelvis in a child with LCPD of the left hip shows a central area of necrosis. The shape of the femoral head appears normal.

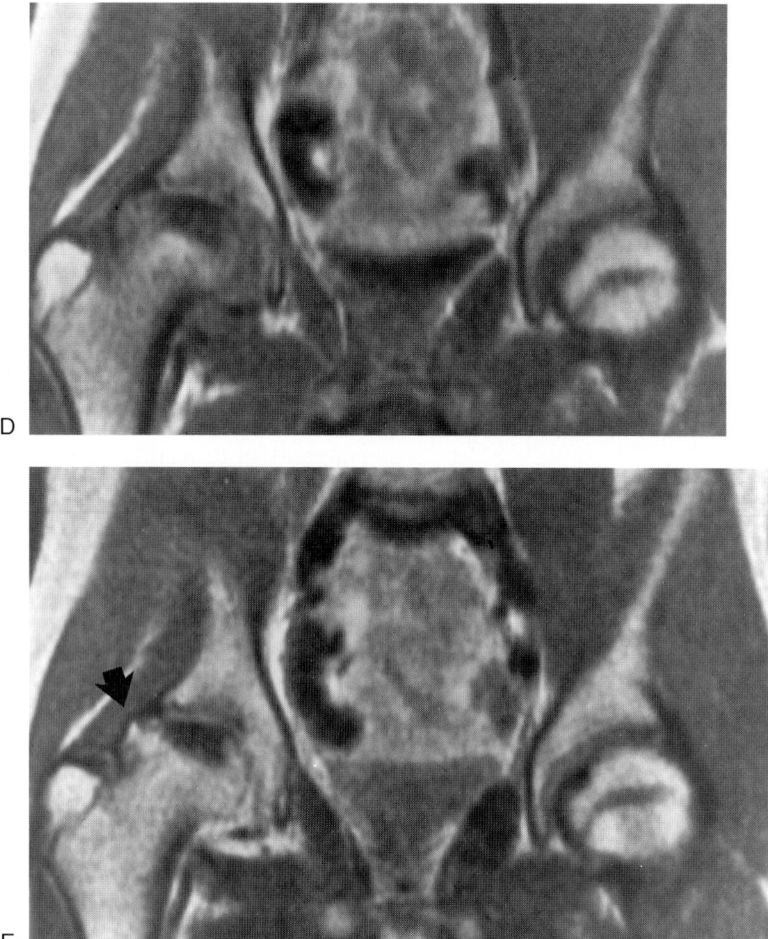

FIG. 17. D,E: Coronal images of the pelvis in a child with advanced LCPD of the left hip show loss of containment of the femoral head due to lateral cartilage overgrowth. (Courtesy of Sevil Brahme, M.D., San Diego, California.) **D:** The T1-weighted (TR/TE 600/20 msec) image shows loss of signal within the entire left femoral head. The ossification center is small, and there is excessive lateral cartilage that is not contained by the acetabulum. **E:** The proton-density (TR/TE 1,600/20 msec) shows the overgrowth of the lateral femoral head cartilage *(arrow)* to better advantage.

CASE 18

History: A 50-year-old asymptomatic woman was noted to have an abnormality of the femur on an abdominal radiograph. Following conventional radiography, an MR examination of the right femur was obtained. What is the best diagnosis based on the radiograph? Are the MR findings consistent with this disorder?

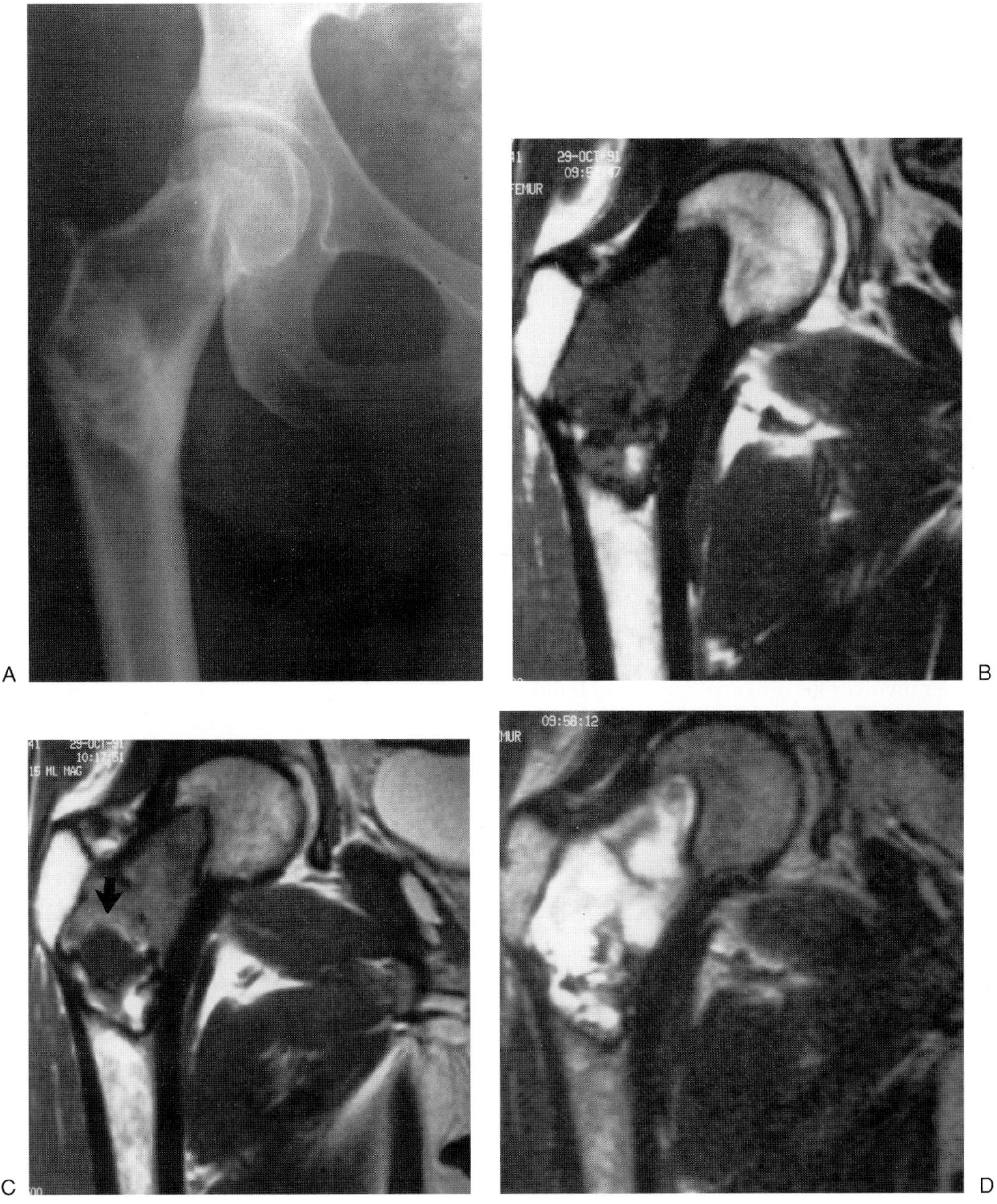

FIG. 18.

Findings: An AP radiograph (Fig. 18A) of the right femur shows a ''ground-glass'' area of osteolysis in the femoral neck and intertrochanteric and subtrochanteric regions. The lesion has a thick sclerotic border around its periphery and dense ossification in its inferior third. The T1-weighted (TR/TE 500/15 msec) image (Fig. 18B) shows a well-defined, low-signal region of marrow replacement corresponding to the radiographic abnormality. The inferior, ossified portions of the lesion show a few high-signal nodular areas, probably representing marrow formation in areas of mature ossification. Following gadolinium enhancement (Fig. 18C), there is minimal enhancement in the upper portions of the abnormal area. Inferiorly, there is a rounded area (arrow) that shows no enhancement. The T2-weighted (TR/TE 2,000/80 msec) image (Fig. 18D) shows inhomogeneous high signal throughout the lesion. Small foci of low signal persist, particularly inferiorly in the ossified portions. (Courtesy of Dr. Josef Kramer, M.D., Vienna, Austria.)

Diagnosis: Fibrous dysplasia of the femur.

Discussion: Fibrous dysplasia is an idiopathic, nonhereditary developmental abnormality of bone-forming mesenchyme resulting in abnormal differentiation and maturation of osteoblasts (1). The medullary cavity of the bones is replaced by an abnormal fibrous tissue that contains immature, poorly calcified trabeculae (2). Areas of mature ossification, cartilage formation, and necrosis are interspersed within the abnormal tissue. The process is monostotic in 70% to 80% of cases, polyostotic in 20% to 30%, and associated with an endocrinopathy in 2% to 3% (1). Any bone in the skeleton may be involved in this process. The most common sites of skeletal involvement are the rib, femur, tibia, mandible, calvarium, and humerus (1). The iliac bone is commonly involved in polyostotic fibrous dysplasia but is rarely a solitary site of involvement. Lesions may be discovered incidentally on radiographs or there may be clinical symptoms related to local deformity, pain, or the development of a fracture through the abnormal bone. Fractures can develop in regions of fibrous dysplasia following minimal trauma. Insufficiency-type fractures are also common, particularly when the disorder involves the proximal femur and femoral shaft. Extraskeletal manifestations include cutaneous and mucosal pigmentation and formation of soft tissue myxomata overlying the involved skeletal regions (3).

Fibrous dysplasia can usually be diagnosed on the basis of conventional radiographs. Intraosseous, centrally located, radiolucent lesions predominate in the diaphyseal and metadiaphyseal regions of the long bones. The lesions classically have a hazy appearance due to the presence of mineralization within the lesion, imparting a ''ground-glass'' appearance (1). The radiolucencies are well-defined, mildly expansile, and frequently surrounded by reactive sclerosis or a thickened, hypertrophied cortex. In the pelvis and proximal femur, the bone softening inherent to this process may lead to significant osseous deformities, such as acetabular protrusion, coxa vara, and curvature of the metadiaphyseal region. CT is helpful in defining the anatomic extent of the lesions and in demonstrating the hazy, amorphous calcifications within the areas of medullary abnormality (2). The Hounsfield units in fibrous dysplasia are usually in the range of 70 to 130, values considerably higher than are encountered in most lytic lesions of bone (2).

MR imaging of fibrous dysplasia shows a wide variability in the appearance of the lesions (4). On MR imaging, the lesions tend to be very well defined, and the sclerotic border present at the margin of most lesions is seen as a thick, low-signal band on all pulse sequences. Deformity of the underlying bone, particularly bowing of the long bones, is readily apparent. (Fig. 18E) The central, medullary regions of the bone have a more variable appearance. On T1-weighted images, the areas of fibrous dysplasia tend to be of inhomogeneous low or intermediate signal intensity, similar to muscle (4,5). The abnormal areas stand out clearly from the normally fatty high-signal marrow on the T1-weighted sequences. Lesions of fibrous dysplasia have greater signal heterogeneity on the T2-weighted images (4). Approximately half of cases show only low or intermediate signal on T2 weighting, findings suggestive of fibrous dysplasia. These lesions presumably consist solely of fibrous tissue and metaplastic bone spicules (4). In the other half of cases, the areas of fibrous dysplasia increase in signal on T2 weighting, although an admixture of high, intermediate, and low signal within the lesion is common. The areas of increased signal on T2-weighted images are thought to represent areas of hyaline cartilage, immature collagen matrix, and regions of cystic necrosis (4). In patients with high-signal lesions on T2 weighting, the signal characteristics are nonspecific, and a specific diagnosis of fibrous dysplasia cannot be established on the basis of the MR examination (4).

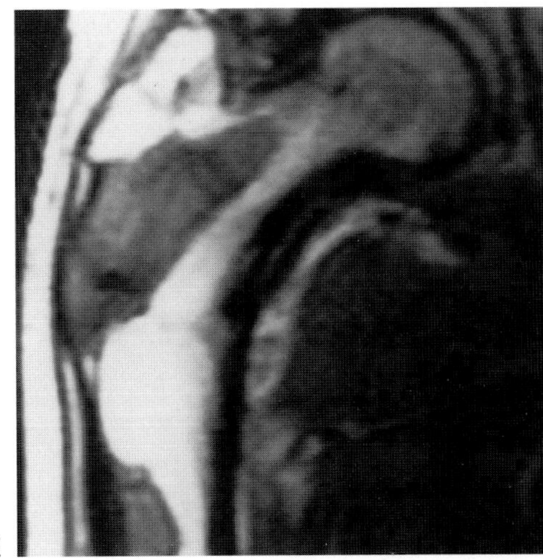

FIG. 18. E: A T1-weighted (TR/TE 400/20 msec) of the right femur in a patient with fibrous dysplasia shows low-signal lesions in the areas of medullary replacement. The femur shows coxa vara and lateral apex bowing deformities due to the bone softening caused by the lesions.

CASE 19

History: An 8-month-old infant underwent MR evaluation for evaluation of suspected hip abnormality. The child was otherwise well with no known associated conditions. What is the most likely diagnosis?

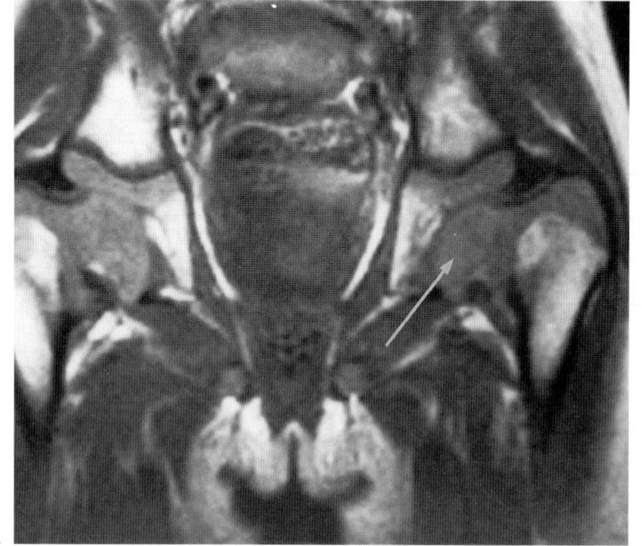
A

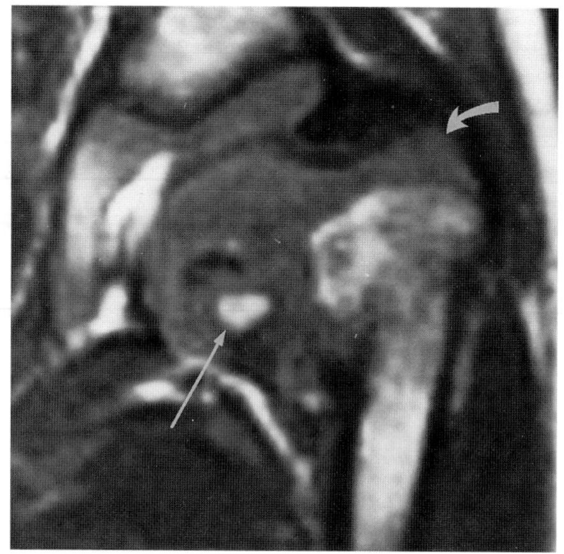
B

FIG. 19.

Findings: A T1-weighted coronal image (TR/TE 500/15 msec) of the pelvis demonstrates medial and apparent inferior displacement of the femoral capital epiphysis (long arrow) with a markedly acute femoral neck–shaft angle (Fig. 19A). A T1-weighted coronal image of the left hip graphically depicts the acute femoral neck–shaft relationship, which approaches a right angle (Fig. 19B). There is early ossification of the femoral capital epiphysis (arrow). The apophysis for the greater trochanter projects superior to the femoral head (arrow). (Images courtesy of J. Tehranzedah, M.D., Irvine, California.)

Diagnosis: Developmental coxa vara (bilateral).

Discussion: In the normal situation, the angle formed by the intersection of the femoral neck and shaft varies with age but is approximately 150° at birth and decreases to a value between 120° and 130° in the adult (1,2). The femoral neck valgus observed in the infant occurs secondary to a relatively increased growth of the medial aspect of the growth plate during the prenatal period. The change toward varus occurs during childhood and has been attributed to an acceleration of growth in the lateral portion of the subcapital growth plate (1,2). This process is also probably influenced by weight-bearing forces, since conditions seen with a lack of erect posture (e.g., cerebral palsy) are associated with persistent femoral neck valgus (2).

The term coxa vara refers to a neck–shaft angle of less than 120°. This condition (''true'' coxa vara) must be distinguished from apparent or ''functional'' coxa vara. Additionally, primary (developmental) coxa vara must be distinguished from coxa vara that occurs in association with a number of other conditions. Apparent (functional) coxa vara due to femoral neck shortening and relative trochanteric overgrowth is often confused with true coxa vara (1). In this condition, which often occurs secondary to growth plate dysfunction (trauma, osteonecrosis), the femoral neck is situated low on the shaft and the greater trochanter projects above it (1). ''True'' coxa vara, in which the neck–shaft angle is actually decreased toward a right angle, may be seen in association with a variety of conditions including PFFD, osteogenesis imperfecta, renal osteodystrophy, rickets, fibrous dysplasia, spondyloepiphyseal dysplasia, cleidocranial dysplasia, chrondrodysplasia, and spondylometaphyseal dysplasia (1,2). If no other primary problem or associated systemic diseases are present, the condition is termed idiopathic, developmental, or infantile coxa vara.

Developmental coxa vara usually becomes apparent in the first few years of life, frequently at the age that the child first begins to walk. The sexes are equally affected, and bilateral involvement is reported in one-third to one-half of cases (1,2). Developmental coxa vara appears to result from relatively greater growth in the lateral aspect of the growth plate, with a resulting tilt of the growth plate to an abnormally oblique angle (1). In this regard, abnormal transition from resting to proliferating cartilage cells in the growth plate medially has been reported (3). Increased shearing forces along this abnormally oblique growth plate likely result in progressive epiphyseal slips with resultant widening of the growth plate and progressive deformity. A metaphyseal triangular fragment is characteristically noted medially below the growth plate and probably represents an attempt to support the medially slipping growth plate (1). This fragment is typically seen in patients younger than 9 years of age; at older ages it merges with the shaft, remodeling thickens the medial cortex of the femoral neck, relative trochanteric overgrowth is noted, and secondary degenerative joint disease can appear (1,2).

MR provides graphic depiction of the altered femoral neck–shaft orientation. Widening of the growth plate with expansion of cartilage medially and distally between the capital epiphysis and metaphysis can be directly depicted. The diagnosis can be readily accomplished even prior to ossification of the capital epiphysis.

CASE 20

History: A 44-year-old man complained of 3 months of right hip pain. He had no significant medical history, denied trauma, and was not taking any medications. Radiographs of the right hip were normal. What is the diagnosis based on the MR examination? What are the treatment options for this patient?

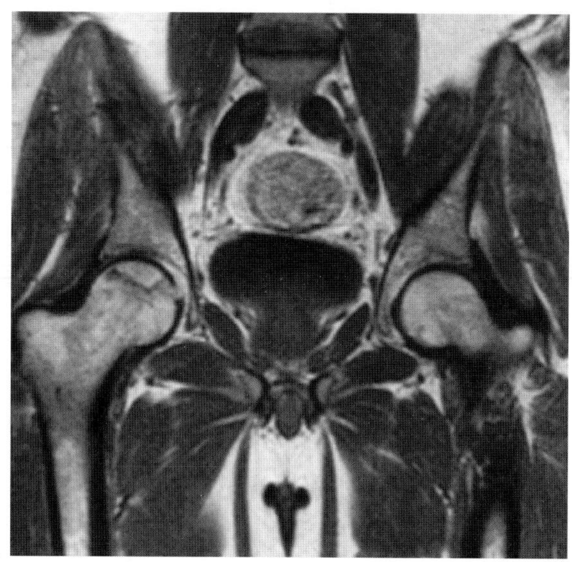

A

FIG. 20.

Findings: A T1-weighted (TR/TE 600/20 msec) coronal image of the pelvis (Fig. 20A) shows an angular band of low signal within the subchondral region of the right femoral head. The low-signal band surrounds an area of marrow with normal fatty signal. The contour of the femoral head is normal. No abnormality of the left femoral head is present.

Diagnosis: Idiopathic avascular necrosis of the right femoral head.

Discussion: The femoral head is the most frequent site of AVN, also known as osteonecrosis, aseptic necrosis, or ischemic necrosis of bone. There are numerous etiologies for this common condition. AVN following trauma is most commonly seen following displaced subcapital femoral neck fractures or dislocation of the hip (Fig. 20B). The injury to the posterior hip capsule leads to disruption of the blood supply to the femoral head. Although it is accepted that the final pathway for the development of AVN is also circulatory compromise to the blood supply of bone, the exact pathogenesis of nontraumatic AVN is uncertain (1). Some of the disparate conditions associated with nontraumatic AVN include corticosteroid use, renal transplantation, irradiation, alcohol consumption, pancreatitis, sickle cell disease, and dysbaric exposure. In significant proportion of patients with idiopathic AVN no known risk factor can be identified.

AVN is classified according to the modified Ficat classification system, which divides AVN into five successive stages (2). Stage 0 (the silent hip) is incidentally detected AVN on MR or radionuclide imaging. This stage is both preclinical and preradiographic. Prior to MR imaging, the incidence of clinically and radiographically occult AVN was difficult to estimate without invasive bone marrow pressure measurements and biopsy (3). MR has allowed large populations of high-risk patients to be examined noninvasively for early evidence of osteonecrosis. In one large study using a screening MR protocol, clinically occult AVN was present in 6% of asymptomatic renal transplantation patients who had also been treated with corticosteroids (3). In another large study of renal transplant patients, previously undiagnosed AVN was discovered in 7.6% of patients, with half these patients having bilateral disease (4). These and other studies suggest that subclinical AVN may be much more common than has been previously appreciated. At this time, the long-term prognosis for patients with asymptomatic osteonecrosis of the femoral head is unclear. The natural history of asymptomatic AVN may be relatively benign. In Mulliken et al.'s (4) series of 11 patients with stage 0 AVN, only 1 showed progression of disease over a 22-month follow-up period.

Progressive groin pain with radiation to the thigh and limitation of motion on physical examination characterize stage 1 disease. At this stage, radiographs remain normal or only show nonspecific indistinctness of the trabeculae. In stage 2, symptoms persist or worsen and radiographs demonstrate sclerosis, cystic changes, or mixed sclerosis and cystic changes. Stage 3 is characterized by the appearance of a crescentic subchondral fracture line, followed by subchondral collapse. Progression of the stage of subchondral collapse is believed to occur over a 3- to 5-year period (5). Radiographic osteoarthrosis marks the last phase, stage 4 (2).

Stages 0, 1, and 2 are early stages and are potentially reversible. There is limited understanding of the natural history of these early stages, and the value of early therapy in preventing progression is controversial (4). Early therapy for early-stage osteonecrosis usually consists of limited weight bearing, avoidance of known risk factors, and core decompression. Less commonly employed treatments for early AVN include osteotomies to minimize weight bearing by the necrotic segment and revascularization procedures utilizing bone grafts. The value of core decompression has been the subject of considerable controversy in the orthopedic literature (4). Advocates of the procedure report excellent results in stages 1 and 2 disease and recommend early diagnosis and treatment for optimal results (2). The more recent literature has shown higher failure and complication rates for core decompression (4). Beltran et al. (6) found that 47% of patients with stages 1 and 2 AVN of the femoral head developed collapse following core decompression surgery (6). The prognosis and likelihood of collapse following core decompression of AVN are closely related to the extent and site of femoral head involvement (6,7). Lesional extent can be accurately quantified by assessing the percentage of the weight-bearing femoral cortex involved on the coronal and sagittal images. This percentage is estimated by dividing the circumference of the involved median femoral cortex facing the acetabular roof by the acetabular weight-bearing area circumference (Fig 20C,D). When the lesion is entirely circumscribed and the cortical femoral head spared, the weight-bearing surface involvement is estimated as 0%. Collapse following core decompression develops in 43% of patients with 25% to 50% involvement and in 87% of patients with greater than 50% involvement of the femoral head (6). The usefulness of core decompression for small areas of osteonecrosis at low risk of postsurgical collapse has not yet been established. The clinical prognosis is good when less than 45% of the weight-bearing surface is necrotic, regardless of the type of therapy employed (8). Stages 3 and 4 are considered irreversible. The treatment options available for these late stages are unfortunately limited to surgical replacement arthroplasty and arthrodesis.

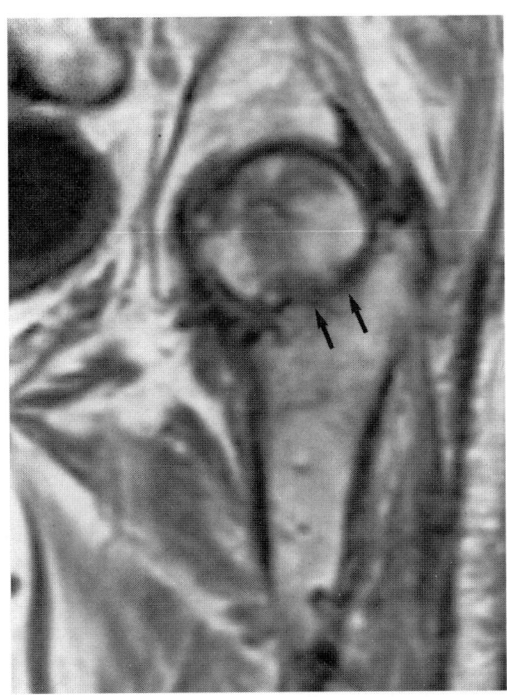

B

FIG. 20. B: This 49-year-old woman sustained a displaced subcapital fracture 7 months prior to the MR examination. She refused surgical treatment for her fracture and presented with persistent pain in the left hip. The T1-weighted (TR/TE 600/20 msec) coronal image of the left hip shows shortening and deformity of the femoral neck. The fracture line is still clearly visible *(arrows)*. There is AVN of the subchondral portion of the femoral head. Irregular lines of low signal are seen in the epiphyseal portion of the femoral head in the region of the infarct.

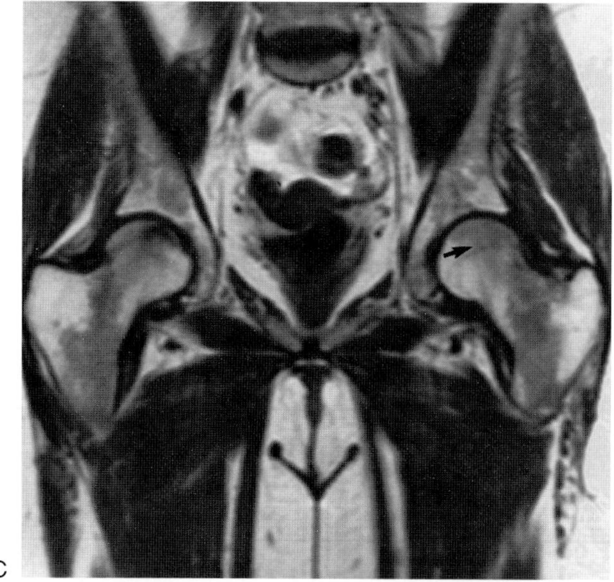

C

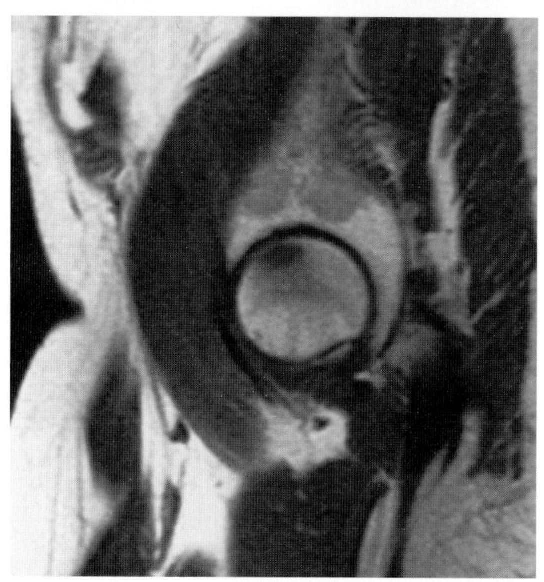

D

FIG. 20. C,D: A 38-year-old asymptomatic woman underwent a screening MR for AVN as part of a study of patients taking corticosteroids for asthma. The MR shows a small (less than 25% of the weight-bearing surface of the femoral head) lesion of osteonecrosis in the anterosuperior femoral head. **C:** The T1-weighted (TR/TE 467/11 msec) image shows a linear band of low signal in the left femoral head. There is normal hematopoietic marrow in the intertrochanteric regions bilaterally. The thick oblique band of intermediate signal in the left femoral head *(arrow)* is caused by the compressive trabeculae and is a normal finding. **D:** A T1-weighted (TR/TE 467/11 msec) sagittal image of the left hip shows the low-signal band of AVN in the femoral head more clearly.

CASE 21

History: A 12-year-old girl presented with a 3-day history of spontaneous onset of pain in the right thigh. Physical examination revealed local tenderness and a mild limp, but the patient had minimal systemic symptoms, a normal range of hip motion, and normal radiographs. At follow-up 1 week later, radiographs revealed permeative bone destruction in the proximal metaphysis of the right femur, and an MR examination was obtained. What is the differential diagnosis for this appearance? What is the cause of the intracortical abnormality?

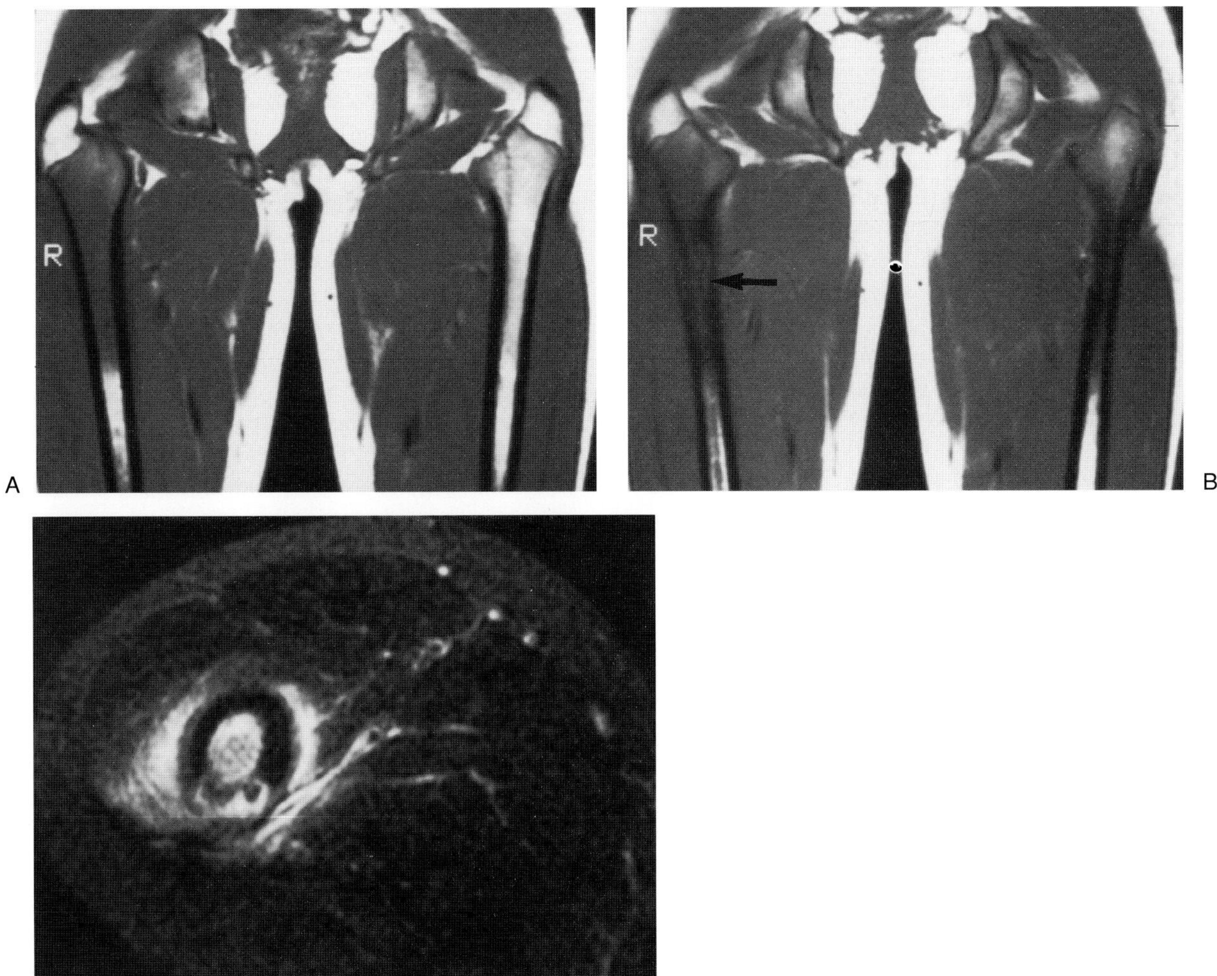

FIG. 21.

Findings: The T1-weighted (TR/TE 450/12 msec) spin-echo coronal image of the proximal femora (Fig. 21A) shows loss of the marrow fat signal in the right proximal femur. There is an extensive region of low-signal marrow involving the metaphysis and metadiaphyseal area. The greater trochanter and femoral head (not illustrated) were normal. A more posterior image from the same series (Fig. 21B) shows a serpiginous linear band of low signal (arrow) within the posterior femoral cortex. The abnormal-appearing marrow on the left side is an artifact due to volume averaging of the posterior cortex. An axial fat-suppressed T2-weighted (TR/TE 2,000/80 msec) image (Fig. 21C) through the proximal right femur shows very high signal within the medullary space. In addition, there is a rounded intracortical focus of high signal within the posterior femoral cortex. A 3- to 4-mm round low-signal density is seen lying within the cortical lesion. There is high signal in the soft tissues circumferentially surrounding the abnormal bone. (Courtesy of Guerdon Greenway, M.D., Dallas, Texas.)

Diagnosis: Osteomyelitis of the proximal right femur. Intracortical abscess with sequestrum formation.

Discussion: Infection of the musculoskeletal system is common in all age groups and can involve a number of different tissue planes (Table 1). Infection in various tissues can coexist or can develop sequentially, with osteomyelitis being the most serious. The major routes of spread of infection to bone are hematogenous dissemination, direct spread from adjacent soft tissue infection, and inoculation, either through a skin ulcer or due to traumatic or iatrogenic penetration of the bone (Fig. 21D). Infection due to hematogenous spread is most common and typically affects the axial skeleton, particularly in adults who have limited amounts of vascularized marrow in the peripheral skeleton. Imaging plays an important role in detection of osteomyelitis, in determining the intraosseous and extraosseous extent of the infection, in planning surgical debridement, and in monitoring antibiotic therapy (1). Therefore an optimal imaging modality should be capable of accurate diagnosis and assessment of the anatomic extent of bone marrow involvement. MR is a valuable imaging modality in the evaluation of osteomyelitis due to its high sensitivity to changes in water content in all tissue planes and its excellent spatial resolution.

The limitations of conventional radiographs for the detection of acute osteomyelitis are well known. Radiographic alterations are delayed in the acute phase by 1 or 2 weeks. This lag period limits the usefulness of radiographs because early diagnosis is necessary to begin antibiotic therapy before bone devitalization (1). Although three-phase ^{99m}T-labeled methylene diphosphonate bone scanning is more sensitive for detection of osteomyelitis than plain films, bone scans provide low anatomic detail and have low specificity in the presence of overlying cellulitis or other bone lesions such as fractures, surgical defects, or arthropathy (2,3). Leukocyte-labeled radionuclide scans allow greater specificity for infection in the presence of processes causing increased bone turnover but have very low spatial resolution and are technically more difficult to perform.

MR has been shown to be as sensitive as scintigraphy for detection of acute osteomyelitis and has higher specificity and more anatomic information than scintigraphy (2–4). Acute active osteomyelitis produces marrow inhomogeneity with poorly defined regions of loss of the normal high signal intensity of marrow fat on T1-weighted sequences and increased signal intensity on T2-weighted or STIR images (5). Chronic, indolent infections tend to be better defined, with geographic bone destruction. The diminished T1 signal of infected bone marrow in infants and young children may be difficult to identify because the normal hematopoietic bone marrow in this age group normally produces low T1 signal intensity. Similarly, osteomyelitis in adults is more difficult to identify when excessive hematopoiesis results in diffusely low signal marrow (Fig. 21E,F). Fat-suppressed T2-weighted spin-echo and STIR images are more sensitive for marrow infection in these situations. Fat-suppressed T1-weighted images following the intravenous administration of gadolinium are also more accurate than standard T1- and T2-weighted spin-echo sequences (3,6).

TABLE 1. *Infections of the musculoskeletal system*

Condition	Site of infection
Cellulitis	Skin and subcutaneous tissues
Pyomyositis	Muscle
Septic arthritis	Articulation
Infectious tenosynovitis	Tendon sheath
Infective periostitis	Periosteum
Osteitis	Cortical bone
Osteomyelitis	Bone marrow

Osteomyelitis can be complicated by the development of pubperiosteal and/or soft tissue abscesses. These areas of necrosis and liquefaction are resistant to systemic antibiotics and require surgical drainage. Soft tissue abscesses are usually round or oval mass lesions that displace adjacent soft tissues. They are typically surrounded by a thick, irregular wall and may contain gas. Abscesses typically demonstrate low signal on T1-weighted sequences and high signal on T2-weighted sequences. If the protein content is elevated in the fluid collection, the T1-weighted signal may be higher than the adjacent muscle. Focal fluid collections may be difficult to distinguish from soft tissue edema and vascularized inflammatory tissue, particularly on STIR sequences. The presence of focal disruption of fascial planes and mass effect suggest the presence of an abscess collection (7). Soft tissue abscesses can be distinguished from vascularized phlegmonous tissue most reliably by demonstrating failure of enhancement of the central portions of the lesion following the intravenous administration of gadolinium (6).

Alterations in the low-signal cortical bone are not seen in the acute phases of osteomyelitis, although periostitis may be present early in the course of the disease (7). Remodeling of bone characterizes chronic osteomyelitis. In chronic osteomyelitis, cortical changes dominate because the marrow signal changes diminish and become more inhomogeneous (5,6). Areas of chronic osteomyelitis are demarcated from normal bone by a low-signal rim, presumably due to surrounding fibrosis or bony sclerosis, in up to 93% of cases (4,7). An important feature of chronic active osteomyelitis is the presence of sequestra, which represent dense necrotic bone fragments isolated from the viable osseous tissue. Although sequestra are not as obvious with MR as with CT, they may be identified as linear bands of low signal on T1- and T2-weighted images, similar to cortical bone. The sequestrum is surrounded by necrotic soft tissue debris that is high signal on the T2-weighted images. Sinus tracts may also be present with chronic osteomyelitis. These cortical channels appear as linear areas of high signal intensity on T2-weighted images that extend from the medullary cavity into the soft tissues. An intramedullary (Brodie) abscess appears as a well-demarcated intraosseous lesion of low or intermediate signal intensity on T1-weighted images and high signal intensity on T2-weighted sequences. A rim of low-density sclerotic bone surrounds the abscess; this rim will enhance with intravenous administration of gadolinium (4,6).

Several noninfectious disorders are difficult to distinguish from osteomyelitis on MR imaging, leading to loss of specificity (7). Infiltrative neoplasms and eosinophilic granuloma may be indistinguishable from osteomyelitis on the basis of MR imaging, but the clinical history is usually helpful in distinguishing these entities from infection. Traumatic bone contusions and insufficiency fractures may show a nonspecific bone marrow edema pattern with signal characteristics identical to those of osteomyelitis. Transient osteoporosis of the hip can appear identical to acute osteomyelitis (see Case 4-1). Osteonecrosis in its early stages may also show bone marrow edema that resembles osteomyelitis, although a ring or band of low signal is usually apparent.

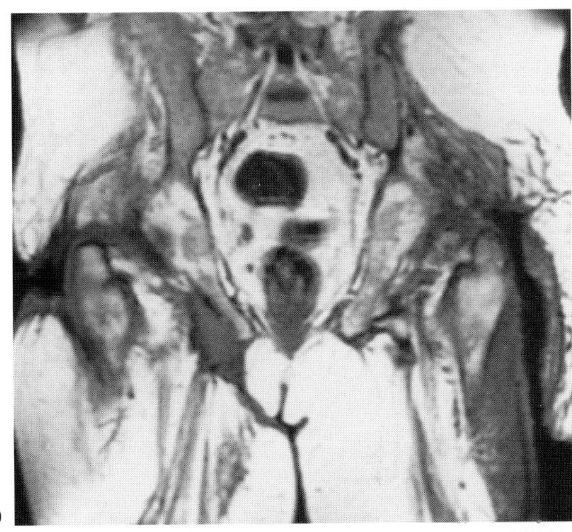

D

FIG. 21. D: A T1-weighted (TR/TE 667/12 msec) spin-echo coronal image of the pelvis in a paraplegic patient is illustrated. There are large decubitus ulcers filled with signal void from room air overlying both lateral femora and greater trochanters. There is loss of the normal high-signal fat within the greater trochanters bilaterally due to osteomyelitis, caused by direct implantation via the ulcers. There is nonseptic bursa formation adjacent to the ischium due to chronic supine positioning.

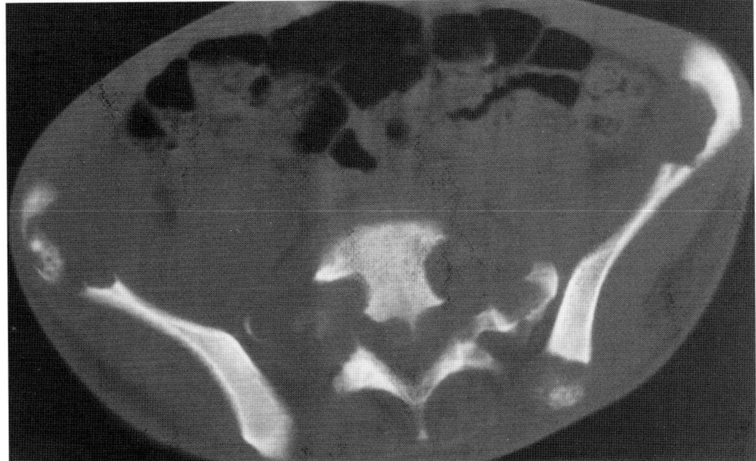

E

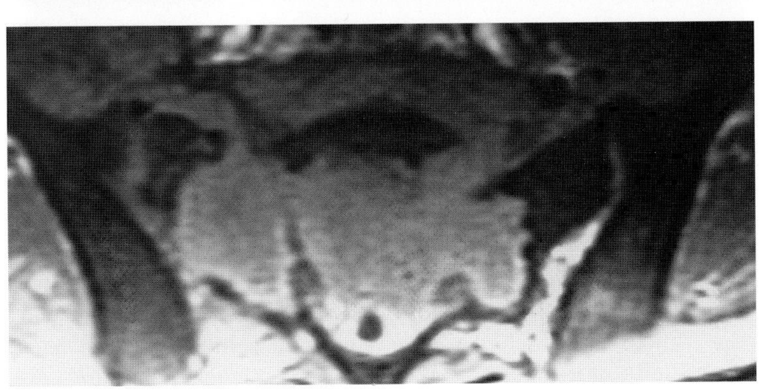

F

FIG. 21 E,F: A 20-year-old man with chronic, widely disseminated coccidiomycosis osteomyelitis shows multiple areas of bone destruction. **E:** An axial CT image of the pelvis shows well-defined osteolytic lesions in the sacrum and both iliac wings. **F:** An axial T1-weighted (TR/TE 600/20 msec) image of the sacrum shows well-defined osteolytic lesions in the sacrum. The normal bone marrow in this patient is low signal because of excessive hematopoiesis due to chronic hypoxemia resulting from severe chronic pulmonary coccidiomycosis infection. The areas of chronic osteomyelitis are intermediate in signal, probably due to accumulation of proteinaceous debris in the area.

CASE 22

History: A 48-year-old man with a remote history of lymphoma complained of bilateral hip pain. His lymphoma had been successfully treated with chemotherapy and corticosteroids. The pain in his hips had been present for over a year but had increased in severity in the past month. Radiographs of the pelvis were normal. What is the diagnosis based on the MR findings?

A

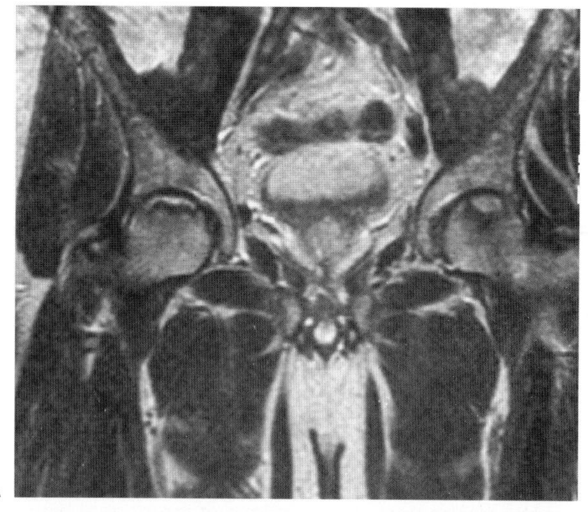

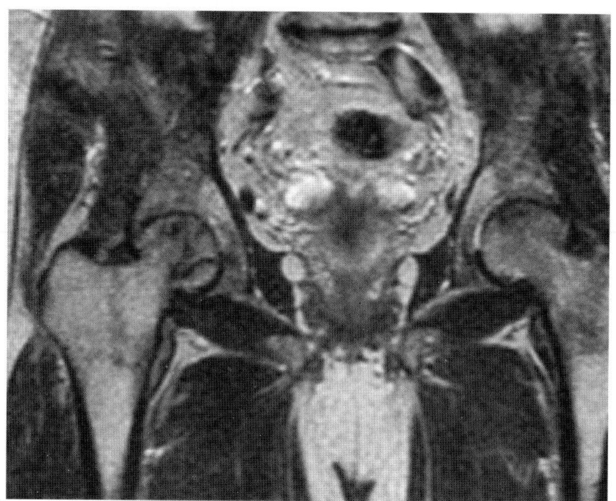

B

C

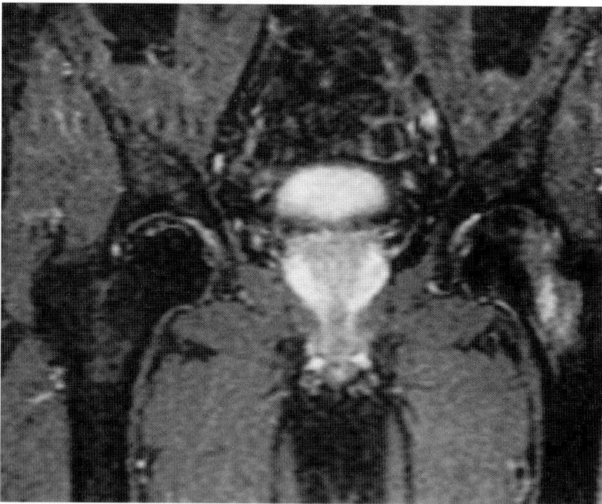

FIG. 22.

Findings: A T2-weighted (TR/TE 2,000/80 msec) spin-echo coronal image of the pelvis (Fig. 22A) shows low-signal bands in both femoral heads. A more posterior image from the same series (Fig. 22B) shows that the right femoral head lesion is extensive. The band in the left femoral head has almost disappeared at this level. The STIR image (TR/TE/TI 1,800/30/160 msec) (Fig. 22C) shows a linear band of high signal in the subchondral bone of the right femoral head. There is increased signal in the bone marrow in the femoral neck and intertrochanteric ridge of the left femur.

Diagnosis: AVN both femoral heads. Subchondral fracture of right femoral head.

Discussion: MR imaging has become the imaging modality of choice for the diagnosis of AVN of bone. The sensitivity of MR for AVN exceeds that of conventional radiography, CT, and radionuclide imaging (1,2). MR has demonstrated a 97% sensitivity and 85% specificity in differentiating AVN from other causes of hip pain (1). MR is particularly useful in early avascular necrosis, prior to the onset of radiographic and scintigraphic changes. Prior to MR, diagnosis at this early stage required invasive bone marrow pressure measurements or biopsy. Unlike bone marrow pressure measurements, MR is noninvasive and allows evaluation of both the symptomatic and contralateral hip. Evaluation of both hips is valuable because AVN is bilateral in up to 50% of cases. In patients with unsuspected bilateral AVN, the asymptomatic hip generally demonstrates an earlier stage of involvement, and joint-preserving therapy is more successful.

The MR appearance of AVN of the femoral head correlates with the histopathologic features present in various stages of osteonecrosis. Following the onset of vascular insufficiency, necrosis of all the elements of living bone develops over a period of a few days. Death of different cell lines occurs sequentially, with initial necrosis of hematopoietic cells, followed by fat cells, and finally osteocytes. In acute-onset AVN, the MR examination is typically normal quite early in the course of the disease. In very early AVN, only invasive bone marrow pressure measurements and bone marrow biopsy allow detection of osteonecrosis. Both these techniques are more sensitive than MR in detecting acute AVN (3,4). The reason for the normal appearance of the bone marrow in acute AVN is presumably lack of any significant reactive edema, hemorrhage, or bone response at the junction between the normal and abnormal marrow. Speer et al. (5) demonstrated that intracapsular femoral neck fractures complicated by avascularity of the femoral head show normal high-signal-intensity marrow fat for at least the first 48 hours after the injury (5). Dynamic contrast-enhanced MR shows lack of enhancement of the avascular marrow prior to the development of any abnormal MR findings on standard spin-echo and STIR sequences (6).

The earliest MR finding in AVN is nonspecific bone marrow edema, which appears as ill-defined areas of low signal on T1-weighted and intermediate-to-high signal intensity on T2-weighted images (7). This diffuse reactive edema is located in the femoral head and neck and often extends into the intertrochanteric area (Fig. 22D,E). The marked tissue response to osteonecrosis can result in the bone marrow edema pattern even in the presence of a very small infarct (8). It may not be possible to distinguish this phase from idiopathic transient bone marrow edema unless one can identify a focal abnormality in the femoral head (7). In AVN, high spatial resolution sagittal T2-weighted images usually show small hypointense subchondral crescent-shaped areas in the anterosuperior femoral head, allowing a correct diagnosis of AVN (8). At follow-up MR studies in 3 to 6 months, bone marrow pressure measurements or core biopsy may be necessary to differentiate early osteonecrosis from transient bone marrow edema in rare cases.

Over a period of days or weeks, reactive changes at the margins of the area of infarction become apparent. In early osteonecrosis, the central region of infarcted marrow maintains normal fat signal. The active reparative tissue at the interface between necrotic and normal bone consists of thickened trabecular bone and therefore produces a peripheral border of low signal intensity on both T1- and T2-weighted sequences (8). Vascularized granulation tissue just inside the reactive bone interface results in a second band-like area of intermediate-to-high signal intensity on T2-weighted images, producing the classic double-line sign of osteonecrosis (2,9). The reactive low-signal interface and double-line sign can be seen on the coronal images of the hip in most cases. Small FOV sagittal images are more sensitive, particularly for small anterior lesions (10). The double-line sign is present in 80% of cases of nontraumatic osteonecrosis of the femoral head and is virtually pathognomonic for AVN (9).

The granulation tissue and portions of the necrotic bone that have become revascularized demonstrate enhancement following intravenous administration of gadolinium-DTPA (11). The enhancing areas correspond to vascularized reparative tissue, whereas nonenhancing regions correspond to persistently necrotic bone marrow. The role of MR contrast enhancement in osteonecrosis has yet to be defined. Assessment of the weakened weight-bearing portion of the femoral head may be aided by defining the extent and location of the hypervascularized repair tissue (11).

With more advanced osteonecrosis and ongoing resorption of necrotic bone, subchondral fractures within the area of infarction and collapse of the articular surface may develop. Subchondral fractures that have not progressed to collapse can be difficult to recognize with MR. On T1-weighted images the fracture line may be visible as a low-signal band parallel to the subchondral cortex. On T2-weighted images, this line is replaced by a high-signal-intensity band, probably representing fluid or edema within the fracture cleft (8). Collapse of the articular surface results in loss of the normal spherical contour of the femoral head and incongruity between the femoral head and acetabulum (Fig. 22F–H). Small areas of collapse are often easier to appreciate on the sagittal images. When collapse is present, the infarcted marrow often shows evidence of bone collapse and fibrosis, manifest as low signal intensity on T1- and T2-weighted sequences (2).

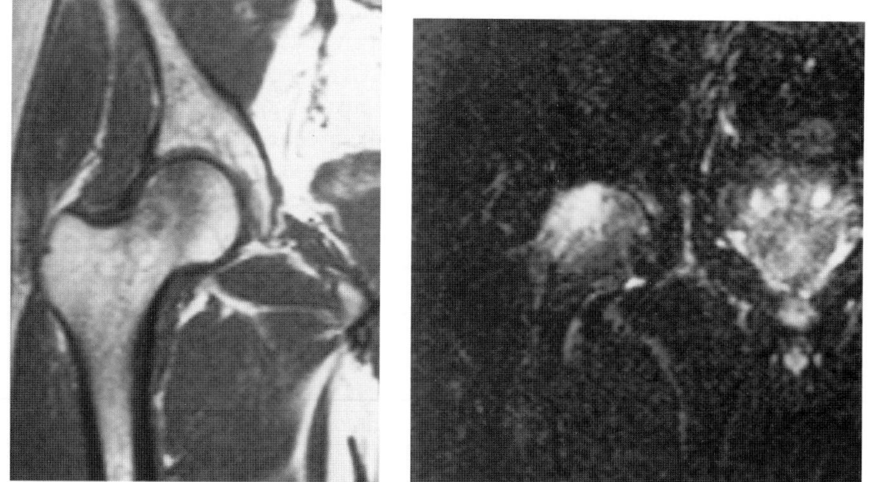

FIG. 22. D,E: This 50-year-old man had previously undergone core decompression of the left hip for AVN. He presented with right hip pain and normal radiographs. Following the MR study, a core decompression of the right femur was performed, which resulted in a decrease in his symptoms. The biopsy specimens from the core decompression showed avascular necrosis. (Courtesy of Steven Eilenberg, M.D., San Diego, California). **D:** A T1-weighted (TR/TE 540/15 msec) image of the right hip shows a geographic region of loss of bone marrow signal in the superolateral femoral head. No low-signal band could be seen on the coronal and sagittal images. **E:** A fat-suppressed heavily T2-weighted (TR/TE 6,700/90 msec) fast spin-echo image shows high signal in the corresponding region. No low signal lines, bands, or rings are apparent.

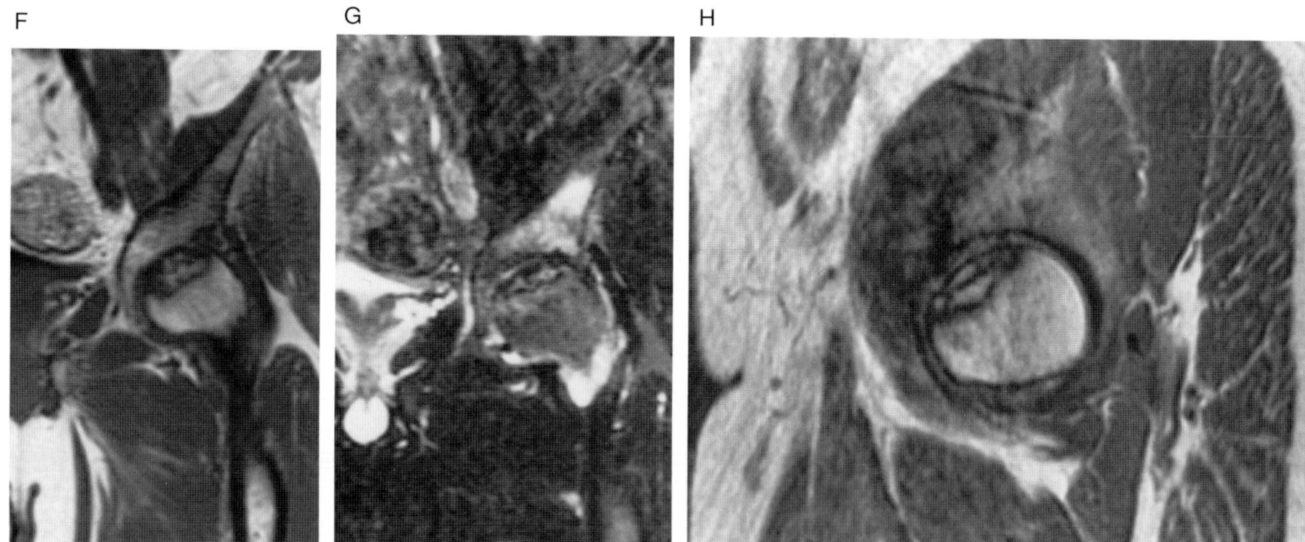

FIG. 22. F–H: A 48-year-old alcoholic complained of severe left hip pain that had been increasing in severity for the past 6 months. Radiographs of the pelvis showed mottled sclerosis of the left femoral head. **F:** A T1-weighted (TR/TE 600/12 msec) spin-echo coronal image of the left hip shows a linear band of low signal in the left femoral head. The left femoral head has lost its normal spherical contour. There is subtle decrease of marrow signal in the left acetabulum. The right femoral head is normal. **G:** The fat-suppressed T2-weighted (TR/TE 3,500/102 msec) fast spin-echo coronal image shows that the low-signal band persists. There is increased signal in the bone marrow surrounding the sclerotic band in the proximal femur and in the adjacent acetabulum. A small effusion is also present within the left hip joint. **H.** The sagittal T1-weighted image illustrates collapse of the anterosuperior portion of the femoral head.

CASE 23

History: A 40-year-old black woman complained of mild pain in the right hip. She had a long medical history due to an underlying hematologic disorder. Radiographs of the pelvis were normal. What are the abnormalities on the MR examination? What disease could produce these findings?

A 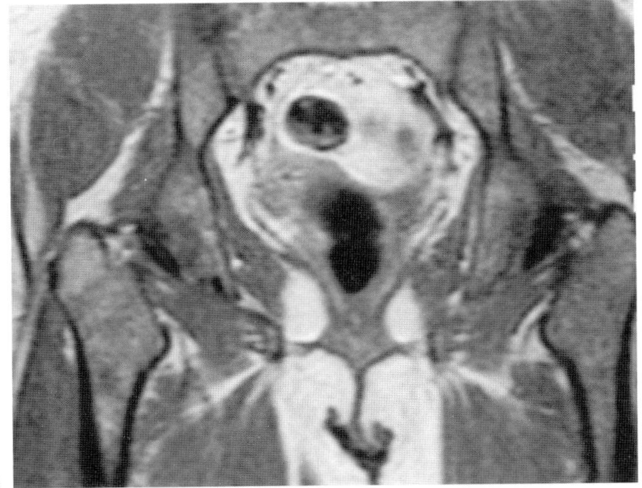

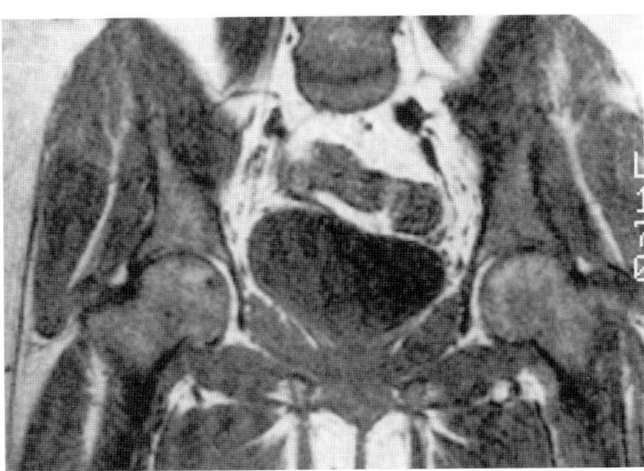 B

C 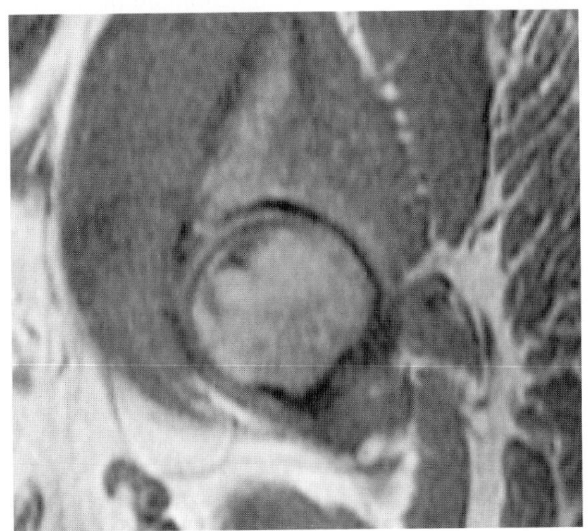

FIG. 23.

Findings: A T1-weighted (TR/TE 600/17 msec) coronal image (Fig. 23A) of the pelvis at the level of the greater trochanters shows diffuse low signal in the marrow with loss of the normal fatty marrow in the trochanters bilaterally. A more anterior image from the same series (Fig. 23B) shows homogeneous low signal in both femoral heads. There is a suggestion of a linear area of darker signal in the subchondral region of the right hip. A sagittal T1-weighted (TR/TE 600/17 msec) image of the right hip (Fig. 23C) shows an arcuate band of subchondral low signal in the anterosuperior portion of the femoral head.

Diagnosis: Sickle cell disease with expansion of hematopoietic marrow and avascular necrosis of the right femoral head.

Discussion: Sickle cell disease is due to an inherited defect in hemoglobin formation, which results in the formation of an abnormal β-hemoglobin. This abnormal hemoglobin is less soluble under conditions of low oxygen tension, resulting in polymerization of hemoglobin and sickling of red blood cells, which compromises microvascular flow (1). The abnormal gene is present in 8% of the black population in the United States, and 0.15% to 0.2% of blacks are homozygous for the gene defect; the homozygous state results in sickle cell disease (1). The chronic hemolysis in this disorder creates an abnormally high demand for hematopoiesis, resulting in hyperplastic marrow. The microvascular compromise results in widespread skeletal infarction. In addition, chronic hemolysis causes hemosiderin deposition within the bone marrow. Finally, the hypoperfusion of the bone marrow and the presence of bowel wall infarction combine to increase the risk of septicemia and osteomyelitis. The MR appearance of the bone in patients with sickle cell disease reflects all these pathologic alterations, although MR is employed primarily to detect osteonecrosis and osteomyelitis.

AVN of the bone is a common complication of sickle cell disease; in one large series of patients, 8.8% had AVN of the femoral head, with a mean age at diagnosis of 24.5 years (1). Infarcts are most common in the metadiaphyseal regions of the distal femur and the proximal tibia (2). Subchondral osteonecrosis within the epiphyses is also common, resulting in pain, limitation of movement, and/or serious disability in up to 50% of patients with radiographically evident osteonecrosis (1). The humeral and femoral heads are involved with approximately equal frequency. The radiographic and MR findings of AVN in sickle cell disease are identical to those seen in AVN from other causes (1). Early reports suggested that acute infarcts show low signal on T1-weighted images, increasing in signal on T2 weighting, presumably due to marrow edema, whereas chronic infarcts that have undergone fibrosis show low signal on both T1- and T2-weighted sequences (3). The distinction between acute and chronic infarcts is not always that clear-cut in patients with sickle cell disease, who typically have multiple infarcts of varying ages. Chronic infarcts can be complicated by liquefaction, cyst formation, and secondary infection, which also result in high signal on T2-weighted images. When infarction takes place in an area of expanded hematopoiesis, the infarct is more difficult to recognize because of lack of contrast between the infarcted marrow and the low-signal hematopoietic marrow (2). In this situation, AVN can only be diagnosed when there is a peripheral dark rim or a double-line sign around the infarcted segment or if there is subchondral collapse. In patients with epiphyseal hematopoiesis, diffuse homogeneous low signal in the epiphysis may be due to marrow expansion alone.

Expansion of the hematopoietic marrow tissue is seen in a variety of disorders associated with an increased need for hematopoiesis due to ineffective hematopoiesis, erythrocyte destruction, or failure of oxygenation. Examples of these disorders include sickle cell disease, spherocytosis, thalassemia, congenital heart disease, severe pulmonary disease, and diffuse marrow replacement processes (4,5). The increased demand for erythropoiesis encountered in these diseases results in persistent low-signal hematopoietic marrow in anatomic regions that should have converted to high-signal fatty marrow. Hematopoietic expansion is a symmetric generalized process that does not alter the morphology of the bone until the process is advanced, at which time expansion of the bone and extramedullary hematopoiesis may become evident. Recognition of hematopoietic tissue expansion requires knowledge of the normal appearance of bone marrow and the normal distribution of hematopoietic marrow at different ages. The three major constituents of marrow are mineralized osseous matrix, hematopoietic marrow, and fatty marrow. The MR appearance of marrow is largely determined by the ratio of hematopoietic to fatty marrow. In the child, 60% of the marrow is hematopoietic marrow, which is low signal on both T1- and T2-weighted sequences (5,6). In the adult, only 30% of the marrow retains myeloid tissue, and the remainder consists of fatty marrow, which is high signal on T1-weighted images and intermediate in signal on T2-weighted images. Newly formed ossification centers typically contain hypointense red marrow for a few months (7). The epiphyses and apophyses convert within the first year of their formation to completely fatty marrow, and then display marrow signal similar to subcutaneous fat (7). Conversion to fatty marrow progresses from the periphery centrally and in the long bones, from the diaphyses to the metaphyses (6). By puberty, all the long bones contain more fatty marrow than red marrow. In the mature adult, the only normal areas of hematopoietic marrow are in the axial skeleton and the metaphyses of the proximal long bones (6).

Expansion of the hematopoietic marrow is a common feature of sickle cell disease (3). On T1-weighted sequences, low-signal marrow is normally not seen in the epiphyses and apophyses. Hematopoiesis in these anatomic structures is always an abnormal finding except in the infant and young child. In adults, persistent hematopoietic marrow may be present in the peripheral appendicular skeleton or the diaphyses of the long bones (Fig. 23D) In sickle cell disease, the bone is rarely expanded, and extramedullary hematopoiesis is rare. Alterations in bone morphology are more commonly encountered in thalassemia. Secondary hemosiderosis is a well-recognized complication of sickle cell disease and further contributes to the diffuse loss of marrow signal in these patients. Hemosiderosis, related to chronic hemolysis and multiple transfusions, results in marked decrease in marrow signal, particularly on gradient-echo images, due to the dephasing caused by the paramagnetic effect of hemosiderin (3).

Expansion of hematopoietic marrow produces low signal marrow on both T1- and T2-weighted images. Other disorders, such as Gaucher's disease, myelofibrosis, osteopetrosis, sclerotic metastases, and leukemia may also produce diffusely abnormal low-signal marrow (4,5,8). Sclerotic metastases tend to be more focal and asymmetric than the other processes. Myelofibrosis produces dark bone marrow on all pulse sequences due to increases in the reticulin content of the fibrotic bone marrow (8) (Fig. 23E). Gaucher's disease is an inherited metabolic disorder resulting in accumulation of glucocerebrosides in the reticuloendothelial system (9). This abnormal accumulation produces low-signal marrow due to decrease in marrow fat as the marrow is replaced by progressive accumulation of glucocerebrosides. Gaucher's disease closely resembles sickle cell disease on MR imaging because both conditions are associated with diffusely abnormal marrow signal and superimposed osteonecrosis (Fig. 23F,G). Unlike patients with sickle cell disease, those with Gaucher's disease also develop abnormalities in bone remodeling, producing undertubulation of the long bones (Ehrlenmeyer flask deformity) (9).

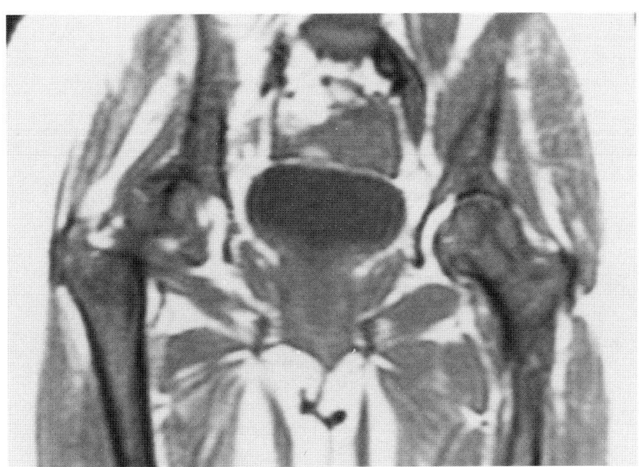

D

FIG. 23. D: A T1-weighted (TR/TE 600/20 msec) coronal image of the pelvis and proximal femora in a 27-year-old woman with sickle cell disease illustrates expansion of hematopoietic marrow into the epiphyses, apophyses, and diaphyses of the long bones. The marrow in the spine is darker than the disk, indicating some type of marrow infiltrative process. AVN is present in both femoral heads.

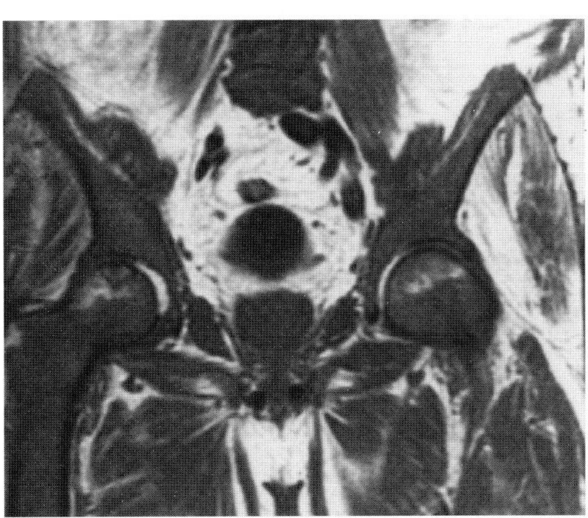

E

FIG. 23. E: A T1-weighted (TR/TE 500/17 msec) coronal image in a patient with myelofibrosis demonstrates very dark marrow throught the skeleton. Note the low signal within the epiphyses due to fibrosis of the normal fatty marrow in the absence of osteonecrosis.

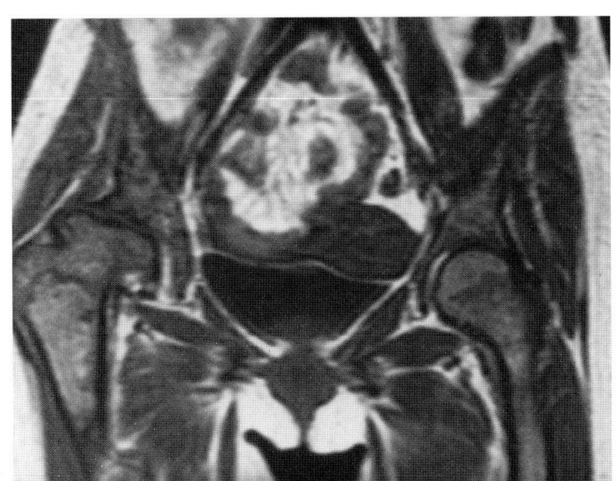

F

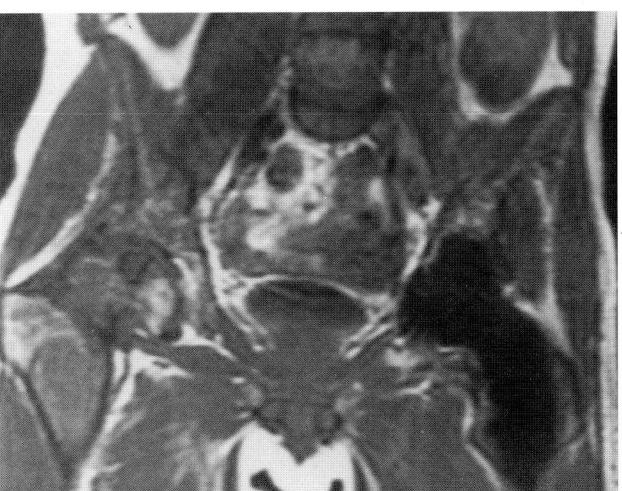

G

FIG. 23. F,G: Two different patients with Gaucher's disease show the marrow replacement, osteonecrosis, and bone deformities typical of this disorder. **F:** A T1-weighted (TR/TE 750/20 msec) image of the pelvis shows abnormally low-signal marrow in the skeleton. Note the replacement of the normal fatty marrow in both femoral heads. The right femoral head is flattened and deformed due to long-standing AVN. **G:** This T1-weighted (TR/TE 500/20 msec) image also shows widespread abnormal low signal in the bone marrow of the pelvis, right femur, and spine. There is a rounded subchondral area osteonecrosis in the right femoral head. There had previously been a left total hip replacement for osteonecrosis of the left femoral head. Note the signal void and lack of significant ferromagnetic artifact from the titanium prothesis.

REFERENCES

Case 1

1. Hayes CW, Conway WF, Daniel WW. MR imaging of bone marrow edema pattern: transient osteoporosis, transient bone marrow edema syndrome, or osteonecrosis. *Radiographics* 1993;13:1001–1011.
2. Potter H, Moran M, Schneider R, Bansal M, Sherman C, Markisz J. Magnetic resonance imaging in diagnosis of transient osteoporosis of the hip. *Clin Orthop* 1992;280:223–229.
3. Schapira D. Transient osteoporosis of the hip. *Semin Arthritis Rheum* 1992;22:98–105.
4. Bloem JL. Transient osteoporosis of the hip: MR imaging. *Radiology* 1988;167:753–755.
5. Daniel WW, Sanders PC, Alarcon GS. The early diagnosis of transient osteoporosis by magnetic resonance imaging. *J Bone Joint Surg* 1992;74:1262–1264.
6. Hauzeur J, Hanquinet S, Gevenois P, Appelboom T, Bentin J, Perlmutter N. Study of magnetic resonance imaging in transient osteoporosis of the hip. *J Rheumatol* 1991;18:1211–1217.
7. Turner DA, Templeton AC, Selzer PM, Rosenberg AG, Petasnick JP. Femoral capital osteonecrosis: MR finding of diffuse marrow abnormalities without focal lesions. *Radiology* 1989;17:135–140.

Case 2

1. Goldman AB, Schneider R, Pavlov H. Osteoid osteomas of the femoral neck: report of four cases evaluated with isotopic bone scanning, CT and MR imaging. *Radiology* 1993;186:227–232.
2. Assoun J, Richardi G, Railhac J, et al. Osteoid osteoma: MR imaging versus CT. *Radiology* 1994;191:217–223.
3. Woods ER, Martel W, Mandell SH, Crabbe JP. Reactive soft-tissue mass associated with osteoid osteoma: correlation of MR imaging findings with pathologic findings. *Radiology* 1993;186:221–225.
4. Sundaram M, McGuire MH. Computed tomography or magnetic resonance for evaluating the solitary tumor or tumor-like lesion of bone? *Skeletal Radiol* 1988;17:393–401.
5. Hayes CW, Conway WF, Sundaram M. Misleading aggressive MR imaging appearance of some benign musculoskeletal lesions. *Radiographics* 1992;12:1119–1134.
6. Crim JR, Mirra JM, Eckardt JJ, Seeger LL. Widespread inflammatory response to osteoblastoma: the flare phenomenon. *Radiology* 1990;177:835.
7. Weatherall PT, Maale GE, Mendelsohn DB, Sherry CS, Erdman WE, Pascoe HR. Chondroblastoma: classic and confusing appearance at MR imaging. *Radiology* 1994;190:467–474.
8. Brower AC, Moser RP, Kransdorf MJ. The frequency and diagnostic significance of periostitis in chondroblastoma. *AJR* 1990;154: 309–314.

Case 3

1. Ozonoff MB. *Pediatric orthopedic radiology.* 2nd ed. Philadelphia: WB Saunders; 1992:183–184.
2. Resnick D. Additional congenital or heritable anomalies and syndromes. In Resnick D, ed. *Diagnosis of bone and joint disorders.* 3rd ed. Philadelphia: WB Saunders; 1995:4283–4286.
3. Hillman JS, Mesgarzadeh M, Rvesz G, et al. Proximal femoral focal deficiency: radiologic analysis of 49 cases. *Radiology* 1989;165: 179.
4. Grogan DP, Lowve SM, Ogden JA. Congenital malformations of the lower extremities. *Orthop Clin North Am* 1987;18:537.
5. Gabriel H, Fitzgerald SW, Myers MT, et al. MR imaging of hip disorders. *Radiographics* 14:763–781.

Case 4

1. Deutsch AL, Mink JH, Waxman AD. Occult fractures of the proximal femur: MR imaging. *Radiology* 1989;170:113–116.
2. Quinn SF, McCarthy JL. Prospective evaluation of patients with suspected hip fracture and indeterminate radiographs: use of T1-weighted MR images. *Radiology* 1993;187:469–471.
3. Rizzo P, Gould ES, Lyden JP, Asnis S. Diagnosis of occult fractures about the hip. *J Bone Joint Surg [Am]* 1993;75A:395–401.
4. Evans PD, Wilson C, Lyons K. Comparison of MRI with bone scanning for suspected hip fracture in elderly patients. *J Bone Joint Surg [Br]* 1994;76B:158–159.
5. Yao L, Lee JK. Occult intraosseous fracture: detection with MR imaging. *Radiology* 1988;167:749–751.
6. Meyers SP, Wiener SN: Magnetic resonance imaging features of fractures using the short tau inversion recovery (STIR) sequence: correlation with radiographic findings. *Skeletal Radiol* 1991;20: 499–507.
7. Speer KP, Spritzer CE, Harrelson JM, Nuhley JA. Magnetic resonance imaging of the femoral head after acute intracapsular fracture of the femoral neck. *J Bone Joint Surg* 1990;72:98–103.
8. Potter HG, Montgomery KD, Heise CW, Helfet DL. MR imaging of acetabular fractures: value in detecting femoral head injury, intraarticular bone fragments and sciatic nerve injury. *AJR* 1994;163: 881–886.

Case 5

1. Greco A, McNamara MT, Escher MB, Trifilio G, Parienti J. Spin-echo and STIR MR imaging of sports-related muscle injuries at 1.5 T. *J Comput Assist Tomogr* 1991;15:994–999.
2. Steinbach LS, Fleckenstein JL, Mink JH. Magnetic resonance imaging of muscle injuries. *Orthopedics* 1994;17:991–999.
3. Fleckenstein JL, Weatherall PT, Parkey RW, Payne JA, Peshock RM. Sports related muscle injuries: evaluation with MR imaging. *Radiology* 1989;172:793–798.
4. Pomeranz SJ, Heidt RS. MR imaging in the prognostication of hamstring injury: work in progress. *Radiology* 1993;189:897–901.
5. Swensen SJ, Keller PL, Berquist TH, McLeod RA, Stephens DH. Magnetic resonance imaging of hemorrhage. *AJR* 1985;145:921–927.

Case 6

1. Murphey MD, Wetzel LH, Bramble JM, et al. Sacroiliitis: MR imaging findings. *Radiology* 1991;180:239–244.
2. Ahlstrom H, Nyman FR, Hallgren R. Magnetic resonance imaging of sacroiliac joint inflammation. *Arthritis Rheum* 1990;33:1763–1769.
3. Shichikawa K, Tsujimoto M, Nishioka J, et al. Histopathology of early sacroiliitis and enthesis in ankylosing spondylitis. In Ziff M, Cohen SB, eds. *Advances in inflammation research. The spondyloarthropathies.* New York: Raven Press; 1985.
4. Braun J, Bollow M, Eggens U, et al. Use of dynamic magnetic resonance imaging with fast imaging in the detection of early and advanced sacroiliitis in spondyloarthropathy patients. *Arthritis Rheum* 1994;37:1039–1045.
5. Bollow M, Braun J, Hamm B, et al. Early sacroiliitis in patients with spondyloarthropathy: evaluation with dynamic, gadolinium enhanced MR imaging. *Radiology* 1995;194:529–536.
6. Klein MA, Winalski CS, Wax MR, Piwnica-Worms DR. MR imaging of septic sacroilitis. *J Comput Assist Tomogr* 1991;15:126–132.

Case 7

1. Abdelwahab IF, Kenan S, Hermann G. Klein MJ, Lewis MM. Periosteal ganglia: CT and MR imaging features. *Radiology* 1993;188: 245–248.
2. Haller J, Resnick D, Greenway G, et al. Juxtaacetabular ganglionic (or synovial) cysts: CT and MR features. *J Comput Assist Tomogr* 1989;13:976–983.
3. Feldman F, Singson RD, Staron RB. Magnetic resonance imaging of para-articular and ectopic ganglia. *Skeletal Radiol* 1989;18:353–358.
4. Ikeda T, Awaya G, Suzuki S, Okada Y, Tada H. Torn acetabular

labrum in young patients: arthroscopic diagnosis and management. *J Bone Joint Surg [Br]* 1988;70B:13–16.
5. Ueo T, Hamabuchi M. Hip pain caused by cystic deformation of the labrum acetabulare. *Arthritis Rheum* 1984;27:947–950.
6. Dorrell JH, Catterall A. The torn acetabular labrum. *J Bone Joint Surg [Br]* 1986;68B:400–403.
7. Silver DAT, Cassar-Pullicino VN, Morrissey BM, Etherington RJ, McCall IW. Gas-containing ganglia of the hip. *Clin Radiol* 1992; 46:257–260.

Case 8

1. Guidera KJ, Einbecker ME, Berman CG, Ogden JA, Arrington JA, Murtagh R. Magnetic resonance imaging evaluation of congenital dislocation of the hips. *Clin Orthop* 1990;261:96–101.
2. Johnson ND, Wood BP, Jackman KV. Complex infantile and congenital hip dislocation: assessment with MR imaging. *Radiology* 1988;168:151–156.
3. Hensinger RN: Congenital dislocation of the hip: treatment in infancy to walking age. *Orthop Clin North Am* 1987;18:597.
4. Harcke HT, Grissom LE. Performing dynamic sonography of the infant hip. *AJR* 1990;155:837–844.
5. Gabriel H, Fitzgerald SW, Myers MT, Donaldson JS, Poznanski AK. MR imaging of hip disorders. *Radiographics* 1994;14:763–781.
6. Johnson ND, Wood BP, Noh KS, Jackman KV, Westesson P, Katzburg RW. MR imaging anatomy of the infant hip. *AJR* 1989;153: 127–133.

Case 9

1. Fitzgerald RH. Acetabular labral tears: diagnosis and treatment. *Clin Orthop* 1995;311:60–68.
2. Dorrel JH, Catteral A. The torn acetabulaar labrum. *J Bone Joint Surg [Br]* 1986;68B:400–403.
3. Carlidge IJ, Scott JHS. The inturned acetabular labrum and osteoarthrosis of the hip. *J Coll Surg Edinb* 1982;27:339–344.
4. Ikeda T, Awaya G, Suzuki, S, et al. Torn acetabular labrum in young patients. *J Bone Joint Surg [Am]* 1988;70A:13–16.
5. Suzuki S, Awaya G, Ikanda Y, et al Arthroscopic diagnosis of ruptured acetabular labrum. *Acta Orthop Scand* 1986;57:512–515.
6. Palmer WE, Brown JH, Rosenthal DE. Labral-ligamentous complex of the shoulder: evaluation with MR arthrography. *Radiology* 1994; 190:645–651.
7. Tirman PFJ, Bost FW, Garvin GJ, et al. Posterosuperior glenoid impingement of the shoulder: findings at MR imaging and MR arthrography with arthroscopic correlation. *Radiology* 1994;193: 431–436.

Case 10

1. Daffner RH, Pavlov H. Stress fractures: current concepts. *AJR* 1992; 159:245–252.
2. Meaney JEM, Carty H. Femoral stress fractures in children. *Skeletal Radiol* 1992;21:173–176.
3. Aro H, Dahlstrom S. Conservative management of distraction-type stress fractures of the femoral neck. *J Bone Joint Surg [Br]* 1986; 68B:65.
4. Bosch E, Pathria MN, Resnick D. Difficult-to-detect osseous injuries. *Physician Sportsmed* 1993;21:116–122.
5. Mink JH, Deutsch AL. Occult cartilage and bone injuries of the knee: detection, classification, and assessment with MR. *Radiology* 1989;170:823–829.
6. Lee JK, Yao L. Stress fractures: MR imaging. *Radiology* 1988;169: 217–220.
7. Stafford SA, Rosenthal DI, Gebhardt MC, Brady TJ, Scott JA. MRI in stress fracture. *AJR* 1986;147:553–556.
8. Meyers SP, Wiener SN. Magnetic resonance imaging features of fractures using the short tau inversion recovery (STIR) sequence: correlation with radiographic findings. *Skeletal Radiol* 1991;20: 499–507.
9. Fernbach SK, Wilkinson RH. Avulsion injuries of the pelvis and proximal femur. *AJR* 1981;137:581–584.

Case 11

1. Busch MT, Morrissy RT. Slipped capital femoral epiphysis. *Orthop Clin North Am* 1987;18:637.
2. Cooperman DR, Charles LM, Pathria MN, Latimer B, Thompson GH. Post-mortem description of slipped capital femoral epiphysis. *J Bone Joint Surg [Br]* 1992;74B:595–599.
3. Gabriel H, Fitzgerald SW, Myers MT, Donaldson JS, Poznanski AK. MR imaging of hip disorders. *Radiographics* 1994;14:763–781.
4. Rogers LF, Poznanski AK. Imaging of epiphyseal injuries. *Radiology* 1994;191:297–308.
5. Goldman AB, Schneider R, Martel W. Acute chondrolysis complicating slipped capital femoral epiphysis. *AJR* 1978;130:945–950.

Case 12

1. Kopa L, Funke M, Fischer U. MR arthrography of the shoulder with gadopentetate dimeglumine: influence of concentration, iodinated contrast material, and time on signal intensity. *AJR* 1994;163:621–623.
2. Chandnani VP, Yeakger TD, DeBerrdino T, et al. Glenoid labral tears: prospective evaluation with MR imaging, MR arthrography, and CT arthrography *AJR* 1993;161:1229–1235.
3. Flannigan B, D Kursunoglu-Brahme S, et al. MR arthrography of the shoulder: comparison with conventional MR imaging. *AJR* 1990;155:829–832.
4. Hajek PC, Baker LL, Sartoris DJ, et al. MR arthrography: anatomic-pathologic investigations. *Radiology* 1987;163:142–147.
5. Hajek PC, Sartoris DJ, Nemann CH. Potential contrast agents for MR arthrography: in vitro evaluation and practical observations. *AJR* 1987;149:97–104.
6. Palmer WE, Brown JH, Rosenthal DI. Labral-ligamentous complex of the shoulder; evaluation with MR arthrography. *Radiology* 1994; 190:645–651.
7. Jinkins JR, Robinson JW, Sisk L, et al. Proton relaxation enhancement associated with iodinated contrast agents in MR imaging of the CNS. *AJNR* 1992;13:19–27.
8. Brasch RC. New directions in the development of MR imaging contrast media. *Radiology* 1992;183:1–10.
9. Netter FN. *Atlas of human anatomy.* New Jersey: Ciba-Geigy Corporation; 1989.
10. Dorrel JH, Catteral A. The torn acetabular labrum. *J Bone Joint Surg [Br]* 1986;68B:400–403.

Case 13

1. Resnick D. Infectious arthritis. *Semin Roentgenol* 1982;17:49–58.
2. Lopez M, Sauerbrei E. Septic arthritis of the hip: sonographic and CT findings. *J Can Assoc Radiol* 1985;36:322–324.
3. Varma DGK, Richli WR, Charnsangavej C, Samuels BI, Kim EE, Wallace S. MR appearance of the distended iliopsoas bursa. *AJR* 1991;156:1025–1028.
4. Schweitzer ME, Falk A, Pathria MN, Brahme S, Hodler J, Resnick D. MR imaging of the knee: can changes in the intracapsular fat pads be used as a sign of synovial proliferation in the presence of an effusion? *AJR* 1993;160:823–826.
5. Erdman WA, Ramburro F, Jayson HT, Weatherall PT, Ferry KB, Peshock RM. Osteomyelitis: characteristics and pitfalls of diagnosis with MR imaging. *Radiology* 1991;180:533–539.
6. Klein MA, Winalski CS, Wax MR, Piwnica-Worms DR. MR imaging of septic sacroiliitis. *J Comput Assist Tomogr* 1991;15:126–132.
7. Savasegaran K, Saifuddin A, Coral A, Butt WP. Magnetic resonance imaging of septic sacroiliitis. *Skeletal Radiol* 1994;23:289–292.
8. Bock GW, Garcia A, Weisman MH, et al. Rapidly destructive hip disease: clinical and imaging abnormalities. *Radiology* 1993;186: 461–466.
9. Rosenberg ZS, Shankman S, Steiner GC, Kastenbaum DK, Norman

A, Lazansky MG. Rapid destructive osteoarthritis: clinical, radiographic, and pathologic features. *Radiology* 1992;182:213–216.

Case 14

1. Daffner RH, Pavlov H. Stress fractures: current concepts. *AJR* 1992;159:245–252.
2. Blomlie V, Lien HH, Iversen T, Winderen M, Tvera K. Radiation-induced insufficiency fractures of the sacrum: evaluation with MR imaging. *Radiology* 1993;188:241–244.
3. Cooper KL, Beabout JW, Swee RG. Insufficiency fractures of the sacrum. *Radiology* 1985;156:15–20.
4. DeSmet AA, Neff JR. Pubic and sacral insufficiency fractures: clinical course and radiologic findings. *AJR* 1985;145:601–606.
5. Schneider R, Yacovone J, Ghelman B. Unsuspected sacral fractures: detection by radionuclide bone scanning. *AJR* 1985;144:337–341.
6. Brahme SK, Cervilla V, Vint V, Cooper K, Kortman K, Resnick D. Magnetic resonance appearance of sacral insufficiency fractures. *Skeletal Radiol* 1990;19:489–493.
7. Abe H, Nakamura M, Takahashi S, Maruoka S, Ogawa Y, Kiyohiko S. Radiation-induced insufficiency fractures of the pelvis: evaluation with 99mTc-methylene diphosphonate scintigraphy. *AJR* 1992;158:599–602.
8. Cooper KL, Beabout JW, McLeod RA. Supraacetabular insufficiency fractures. *Radiology* 1985;157:15–17.

Case 15

1. Resnick D. Paget disease of bone: current status and a look back to 1943 and earlier. *AJR* 1988;150:249–256.
2. Smith J, Botet JF, Yeh SDJ. Bone sarcomas in Paget disease: a study of 85 patients. *Radiology* 1984;152:583–590.
3. Moore TE, King AR, Kathol MH, El-Khoury GY, Palmer R, Downey PR. Sarcoma in Paget disease of bone: clinical, radiologic, and pathologic features in 22 cases. *AJR* 1991;156:1199–1203.
4. Roberts MC, Kressel HY, Fallon MD, Zlatkin MB, Dalinka MK. Paget disease: MR imaging findings. *Radiology* 1989;173:341–345.
5. Kaufman GA, Sundaram M, McDonald DJ. Magnetic resonance imaging in symptomatic Paget's disease. *Skeletal Radiol* 1991;20:413–418.

Case 16

1. Quinn SF, Haberman JJ, Fitzgerald SW, et al. Evaluation of loose bodies in the elbow with MR imaging. *J Magn Reson Imaging* 1994;4:169–172.
2. Murphy BJ. MR imaging of the elbow. *Radiology* 1992;184:525–529.
3. Potter HG, Montogomery KD, Heise CW, et al. MR imaging of acetabular fractures: value in detecting femoral head injury, intraarticular fragments, and sciatic nerve injury. *AJR* 1994;163:881–886.

Case 17

1. Rush BH, Bramson RT, Ogden JA. Legg-Calvé-Perthes disease: detection of cartilaginous and synovial changes with MR imaging. *Radiology* 1988;167:473–476.
2. Thompson GH, Salter RB. Legg-Calvé-Perthes disease: current concepts and controversies. *Orthop Clin North Am* 1987;18:617.
3. Egund N, Wingstrand H. Legg-Calvé-Perthes disease: imaging with MR. *Radiology* 1991;179:89–92.
4. Toby EB, Koman LA, Bchtold RE. Magnetic resonance imaging of pediatric hip disorders. *J Pediatr Orthop* 1985;5:665–671.
5. Bos CFA, Bloem JL, Bloem RM. Sequential magnetic resonance imaging in Perthes' disease. *J Bone Joint Surg [Br]* 1991;73B:219–224.
6. Gabriel H, Fitzgerald SW, Myers MT, Donaldson JS, Poznanski AK. MR imaging of hip disorders. *Radiographics* 1994;14:763–781.

Case 18

1. Feldman F. Tuberous sclerosis, neurofibromatosis, and fibrous dysplasia. In Resnick D, ed. *Diagnosis of bone and joint disorders.* 3rd ed. Philadelphia: WB Saunders; 1994;4353–4395.
2. Daffner RH, Kirks DR, Gehweiler JA, Heaston DK. Computed tomography of fibrous dysplasia. *AJR* 1982;139:943–948.
3. Sundaram M, McDonald DJ, Merenda G. Intramuscular myxoma: a rare but important association with fibrous dysplasia of bone. *AJR* 1989;153:107–108.
4. Norris MA, Kaplan PA, Pathria MN, Greenway G. Fibrous dysplasia: magnetic resonance imaging appearance at 1.5 tesla. *Clin Imaging* 1990;14:211–215.
5. Glass-Royal MC, Nelson MC, Albert F, Lack EE, Bogumill GP. Case report 557. *Skeletal Radiol.* 1989;18:392–398.

Case 19

1. Ozonoff MB. *Pediatric orthopedic radiology.* Philadelphia: WB Saunders; 1992:228–229.
2. Resnick D. Additional congenital or heritable anomalies and syndromes. In Resnick D, ed. *Diagnosis of bone and joint disease.* (3rd ed.) Philadelphia: WB Saunders; 1995:4309–4310.
3. Bos CFA, Sakkers RJB, Bloem JL, et al. Histological, biochemical, and MRI studies of the growth plate in congenital coxa vara. *J Pediatr Orthop* 1989;9:660–666.

Case 20

1. Chang CC, Greenspan A, Gershwin ME. Osteonecrosis: current perspectives on pathogenesis and treatment. *Semin Arthritis Rheum* 1993;23:47–69.
2. Ficat RP. Idiopathic bone necrosis of the femoral head. *J Bone Joint Surg [Br]* 1985;67B:3–9.
3. Tervonen O, Mueller DM, Matteson EL, Velosa JA, Ginsburg WW, Ehman RL. Clinically occult avascular necrosis of the hip: prevalence in an asymptomatic population at risk. *Radiology* 1992;182:845–847.
4. Mulliken BD, Renfrew DL, Brand RA, Whitten CG. Prevalence of previously undetected osteonecrosis of the femoral head in renal transplant recipients. *Radiology* 1994;192:831–834.
5. Meyers MH. Osteonecrosis of the femoral head: pathogenesis and long-term results of treatment. *Clin Orthop* 1988;231:51–61.
6. Beltran J, Knight CT, Zueler WA, et al. Core decompression for avascular necrosis of the femoral head: correlation between long-term results and preoperative MR staging. *Radiology* 1990;175:533–536.
7. Ohzono K, Saito M, Takaoka K, et al. Natural history of nontraumatic avascular necrosis of the femoral head. *J Bone Joint Surg [Br]* 1991;73B:68–72.
8. Lafforgue P, Dahan E, Chagnaud C, Schiano A, Kasbarian M, Acquaviva P. Early-stage avascular necrosis of the femoral head: MR imaging for prognosis in 31 cases with at least 2 years of follow-up. *Radiology* 1993;187:199–204.

Case 21

1. Gold RH, Hawkins RA, Katz RD. Bacterial osteomyelitis: findings on plain radiography, CT, MR, and scintigraphy. *AJR* 1991;157:365–370.
2. Beltran J, McGhee RB, Shaffer PB, et al. Experimental infections of the musculoskeletal system: evaluation with MR imaging and Tc-99m and Ga-67 scintigraphy. *Radiology* 1988;167:167–172.
3. Morrison WB, Schweitzer ME, Bock GW, et al. Diagnosis of osteomyelitis: utility of fat-suppressed contrast-enhanced MR imaging. *Radiology* 1993;189:251–257.
4. Tang JSH, Gold RH, Bassett LW, Seeger LL. Musculoskeletal infection of the extremities: evaluation with MR imaging. *Radiology* 1988;166:205–209.
5. Quinn SF, Murray W, Clark RA, Cochran C. MR imaging of chronic osteomyelitis. *J Comput Assist Tomogr* 1988;12:113.
6. Dangman BC, Hoffer FA, Rand FF, O'Rourke EJ. Osteomyelitis

in children: gadolinium-enhanced MR imaging. *Radiology* 1992; 182:743–747.

7. Erdman WA, Ramburro F, Jayson HT, Weatherall PT, Ferry KB, Peshock RM. Osteomyelitis: characteristics and pitfalls of diagnosis with MR imaging. *Radiology* 1991;180:533–539.

Case 22

1. Glickstein MF, Burk DL, Schiebler ML, et al. Avascular necrosis versus other diseases of the hip: sensitivity of MR imaging. *Radiology* 1988;169:213–215.
2. Mitchell DG, Rao VM, Dalinka MK, et al. Femoral head avascular necrosis: correlation of MR imaging, radiographic staging, radionuclide imaging, and clinical findings. *Radiology* 1987;162:709–715.
3. Beltran J, Herman LJ, Burk JM, et al. Femoral head avascular necrosis: MR imaging with clinical-pathologic and radionuclide correlation. *Radiology* 1988;166:215–220.
4. Genez BM, Wilson MR, Houk RW, et al. Early osteonecrosis of the femoral head: detection in high-risk patients with MR imaging. *Radiology* 1988;168:521–524.
5. Speer KP, Spritzer CE, Harrelson JM, Nunley JA. Magnetic resonance imaging of the femoral head after acute intracapsular fracture of the femoral neck. *J Bone Joint Surg* 1990;72:98–103.
6. Nadel SN, Debatin JF, Richardson WJ, et al. Detection of acute avascular necrosis of the femoral head in dogs: dynamic contrast-enhanced MR imaging vs spin-echo and STIR sequences. *AJR* 1992; 159:1255–1261.
7. Turner DA, Templeton AC, Selzer PM, Rosenberg AG, Petasnick JP. Femoral capital osteonecrosis: MR finding of diffuse marrow abnormalities without focal lesions. *Radiology* 1989;171:135–140.
8. Vande Berg B, Malghem J, Labaisse MA, Noel HM, Maldague BE. MR imaging of avascular necrosis and transient marrow edema of the femoral head. *Radiographics* 1993;13:501–520.
9. Jergesen HE, Lang P, Moseley M, Genant HK. Histologic correlation in magnetic resonance imaging of femoral head osteonecrosis. *Clin Orthop Res* 1990;253:150–163.
10. Shuman WP, Castagno AA, Baron RL, Richardson ML. MR imaging of avascular necrosis of the femoral head: value of small field-of-view sagittal surface-coil images. *AJR* 1988;150:1073–1078.
11. Vande Berg B, Malghem J, Labaisse MA, Noel H, Maldague B. Avascular necrosis of the hip: comparison of contrast-enhanced and nonenhanced MR imaging with histologic correlation. *Radiology* 1992;182:445–450.

Case 23

1. Sebes JI. Diagnostic imaging of bone and joint abnormalities associated with sickle cell hemoglobinopathies. *AJR* 1989;152:1153–1159.
2. Rao VM, Mitchell DG, Rifkin MD, et al. Marrow infarction in sickle cell disease: correlation with marrow type and distribution by MRI. *Magn Reson Imaging* 1989;7:39–44.
3. Rao VM, Fishman M, Mitchell DG, et al. Painful sickle cell crisis: bone marrow patterns observed with MR imaging. *Radiology* 1986; 161:211–215.
4. Vogler JB, Murphey WA. Bone marrow imaging. *Radiology* 1988; 168:679–693.
5. Pathria MN, Isaacs P. Magnetic resonance imaging of bone marrow. *Curr Opin Radiol* 1992;4:21–31.
6. Moore SG, Dawson KL. Red and yellow marrow in the femur: age-related changes in appearance at MR imaging. *Radiology* 1990;175: 219–223.
7. Jaramillo D, Laor T, Hoffer FA, et al. Epiphyseal marrow in infancy: MR imaging. *Radiology* 1991;180:809–812.
8. Kaplan KR, Mitchell DG, Steiner RM, et al. Polycythemia vera and myelofibrosis: correlation of MR imaging, clinical, and laboratory findings. *Radiology* 1992;183:329–334.
9. Rosenthal DI, Barton NW, McKusick KA, et al. Quantitative imaging of Gaucher disease. *Radiology* 1992;185:841–845.

CHAPTER 5

The Spine

Andrew L. Deutsch, J. Randy Jinkins, Carl V. Bundschuh, Farid F. Shafaie, Jerrold H. Mink, Michael I. Rothman, and Greg H. Zoarski

Imaging of the lumbar spine represents one of the most common applications of magnetic resonance (MR). In many, if not most, centers, MR has replaced computed tomography (CT) as the primary means of spinal imaging, and both methods have relegated myelography to a secondary role in the evaluation of lumbar spine complaints. By virtue of the soft tissue contrast as well as the spatial resolution that can be achieved with present day MR imagers, the pathoanatomy of lumbar spine disorders has never been more clearly delineated. At the same time, however, it has become increasingly evident that direct correlation between gross anatomic findings (as depicted on MR images) and the clinical situation is often lacking. Defining the precise anatomic source of a patient's complaints on the basis of static imaging studies must be approached judiciously. Many of the cases in the present chapter explore the correlation between imaging and clinical findings, particularly with regard to lumbar disk disease. The prevalence of disk disease in the asymptomatic population and the natural history of conservatively managed disk disease are reviewed. Considerable attention is directed toward the postoperative spine in cases that explore the findings to be expected in successfully treated patients; thus we can establish the basis for imaging the patient with persistent or recurrent symptomatology following an interventional procedure. Considerable controversy remains regarding the terminology of lumbar disk disorders, and one standardized approach that appears to be gaining acceptance in radiologic and clinical circles is reviewed. Without attempting to be comprehensive, the chapter also reviews many other conditions that may be commonly encountered in studying patients with lumbar spine complaints.

The wide availability of phased-array coils has further enhanced the quality of lumbar spine MR imaging. The entire lumbar spine and region of the conus medullaris can be readily studied. Most of the contributors to this chapter presently employ fast spin-echo techniques for assessment of the lumbar spine and obtain both T1-weighted and proton-density as well as T2-weighted images. The editors present routine lumbar examination on a 1.5-T imager that uses the following parameters: sagittal T1-weighted fast spin echo [repetition time/echo time (TR/TE) 600/17 msec, 4 echo train length (ET), 256 × 256 matrix, 24-cm field of view (FOV), 2 excitations (NEX)]; sagittal T2-weighted fast spin echo (TR/TE 3,000/16/112 msec, 8 ET, 256 × 256 matrix, 3 mm skip 1 mm, 22-cm FOV); and axial T2-weighted fast spin echo (TR/TE 3,500/17/102 msec, 3 ET, 256 × 256 matrix, 4 mm skip 1.5 mm, 18 FOV, 2 NEX). For assessment of the postoperative spine, enhancement with gadopentetate dimeglumine coupled with fat-suppression techniques is utilized; the details of timing and optimization of contrast enhancement techniques are discussed in relevant cases.

CASE 1

History: A 34-year-old man presents with a history of back pain of 7 weeks' duration. The man works in the construction trade and attributes his problems to heavy lifting at work. What term would you use to describe the abnormality illustrated in Fig. 1A–D?

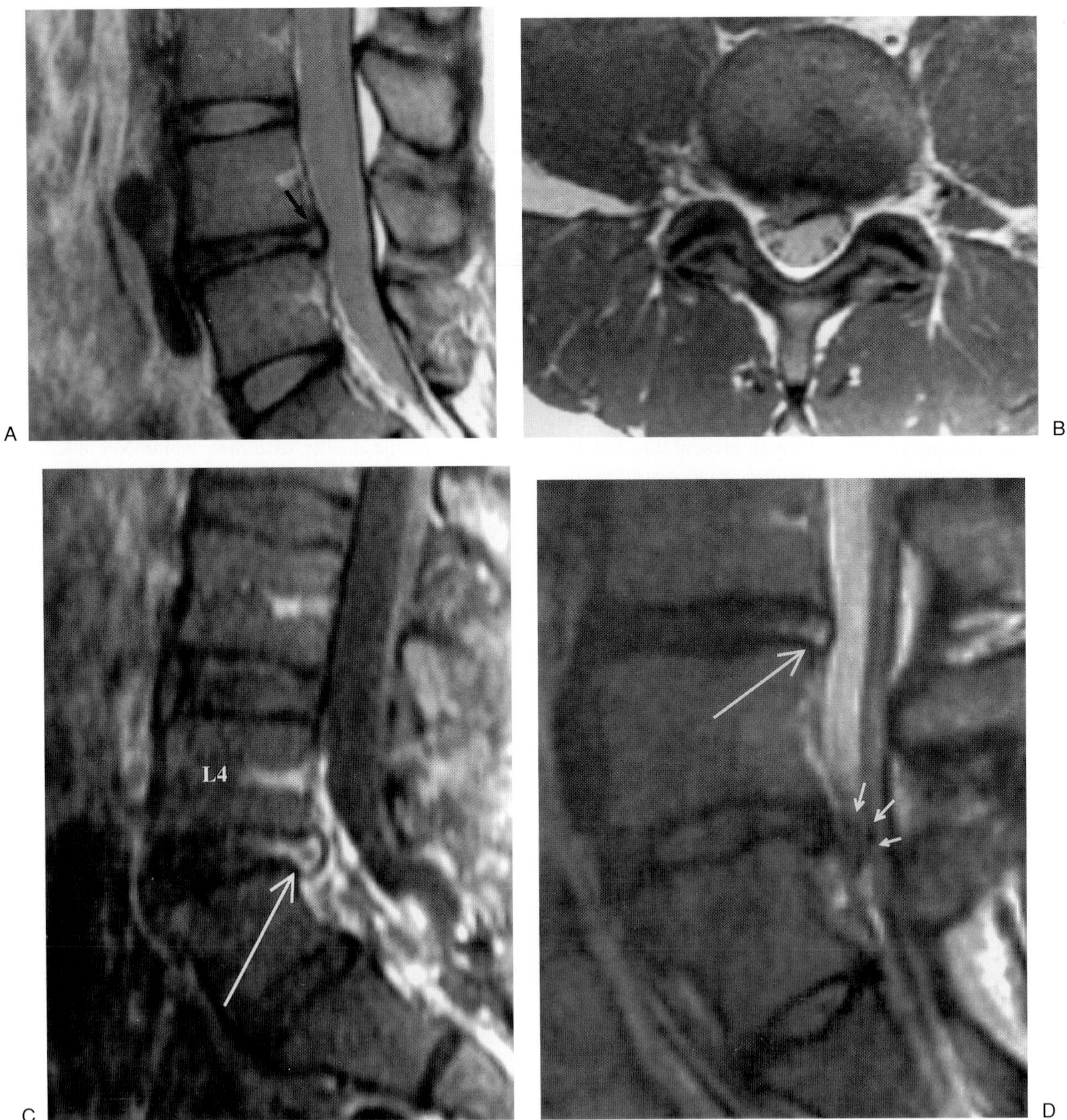

FIG. 1. A–D: Protrusion. **A:** Sagittal (TR/TE 3,000/18 msec) image shows that the L4-5 intervertebral disk space is narrowed. A small central/right paracentral protrusion *(arrow)* is bounded by the dark outer layer of the annulus/posterior longitudinal ligament complex. **B:** Axial (TR/TE 3,500/18 msec) image demonstrating that the right paracentral protrusion has a base along the disk that is greater than its anteroposterior dimension. **C:** In another patient, sagittal fat-saturated post-gadolinium (TR/TE 500/20 msec) image showing that the central protrusion of the L4-5 disc *(arrow)* enhances considerably following intravenous gadolinium injection. **D:** Sagittal (TR/TE 4,000/96 msec) image shows that a central protrusion at L3-4 *(arrow)* is bright on a T2-weighted image; most protrusions have the same signal intensity of the disk from which they arise *(dark)*. Incidentally noted is a lobulated extrusion at the L4-5 level *(small arrows)*.

Findings: A disk *protrusion* is defined as a *focal, asymmetric* condition in which a segment of the intervertebral disk contour extends beyond the margin of the adjacent vertebrae into the spinal canal (Fig. 1A–D). The base of a protrusion, that is, the dimension along the posterior margin of the disk, is greater than any other dimension of the protrusion (1). The finding that distinguishes a protrusion from an extrusion (see below) is the fact that, in a protrusion, at least some of the outermost anular fibers remain intact. On a sagittal MR, a sharply defined, uninterrupted dark rim defines the posterior margin of the protrusion (which, by definition, does not extend significantly above or below the adjacent end plates because of the integrity of the outer anulus). The nature of this dark line is not entirely certain; it represents the anulus and the posterior longitudinal ligament, although partial volume effect or chemical shift artifact involving the anterior margin of the dural sac with the anulus may contribute to this ''structure.'' The definition of a protrusion does not depend on the signal intensity of the protruded material (usually similar to the disk of origin on T2-weighted images), its histologic composition (nucleus, anulus, vertebral end-plate avulsion, or any combination of the foregoing), its location, or its size. Most, but certainly not all, protrusions and their disks of origin have decreased signal on T2-weighted images; some maintain a normal or even increased signal intensity (2) (Fig. 1A–D).

Diagnosis: Right paracentral protrusion (3 mm) of the L4-5 intervertebral disk.

Discussion: Even though great progress has been made in the radiologic assessment, surgical treatment, and nonsurgical therapy of degenerative spinal disorders, considerable misunderstanding and lack of uniformity with regard to the terminology used in describing these maladies continue to exist (3–5). The nomenclature committee of the North American Spine Society presented an exhibit on spinal terminology at the 1990 Annual Meeting of the American Academy of Orthopaedic Surgeons that incorporated a questionnaire (3). Eight highly detailed anatomic descriptions of different conditions of a lumbar disc as viewed through a laminotomy exposure were described. The surgeons were asked to write the diagnostic terms they would apply to each situation were they to encounter it at surgery. Astonishingly, the 49 spine surgeons who completed the questionnaire used a total of 53 different terms to describe the eight conditions! It is reasonable to conclude that the lack of a standardized glossary of terms, accepted and understood by surgeons, individuals rendering conservative spine care, radiologists, and personnel involved in the medicolegal arena, could potentially contribute to insufficient and/or inappropriate therapy and financial reward. Physicians and other medical professionals need to communicate with one another, a need accentuated by the recent demand by insurers for outcome data. How are we to gather such data if we are treating ''different conditions?''

A simple, limited, yet reproducible system has recently been proposed to ensure consistency in the reporting of imaging examinations of the lumbar spine (1,6,7).

The anatomy of the intervertebral disk is most graphically appreciated on T2-weighted sagittal MR studies utilizing a long TR (2,500–5,000 msec), long TE (80–120 msec) technique. The normal nucleus pulposus and the innermost layers of the anulus are bright and occupy the central two-thirds of the anteroposterior dimension of the vertebral bodies above and below. The outer layers of the normal anulus fibrosus are homogeneously dark. Typically, the normal annulus does not extend posteriorly beyond a line connecting the adjacent posterior corners of the vertebrae; however, a histologically normal anulus, even in children, may extend diffusely up to 2.5 mm beyond the vertebral margins (8).

Disk degeneration is a commonly used expression that implies irreversible structural and histologic alterations in the nucleus. Radiologists recognize a mildly degenerating disc as having a mild-to-moderate decrease in nuclear signal on T2-weighted images (Fig. 1E–G). The disk maintains a slightly diminished height. With increasing degrees of degeneration, the disk height is progressively reduced; the nuclear signal (T2-weighted) may be markedly diminished due to replacement of the hydrated nucleus pulposus with fibrous tissue. The disk signal may occasionally be paradoxically increased due to the appearance of fluid-filled cystic spaces within the disk space (2,8).

An *anular fissure* is a focal disruption of the fibers of the anulus (Fig. 1H,I). The fissure may be within the substance of the anulus and/or may involve its outer surface or that adjacent to the nucleus. Three distinct types of anular fissures have been described on pathologic material (concentric, transverse, radial), but because of its presumed relationship to degenerative disk disease, the radial fissure has received the most attention in the literature. It has been postulated by some authors that these radial fissures may produce ''diskogenic pain'' by noxiously stimulating nerve endings that grow into the fissures along with blood vessels and granulation tissue during attempted healing (9). This reactive tissue is believed to account for the increased signal observed on T2-weighted images and gadolinium-enhanced T1-weighted MR studies. The term *anular fissure* is generally reserved for those lesions identified as a focal signal alteration in an intervertebral disk with a normal contour, but anular fissures, generalized disk bulges, and disk protrusions frequently occur together (2,10–13).

A *generalized disk bulge, or bulging disk* is a circumferential, diffuse, symmetric extension of the anulus beyond the adjacent vertebral end plates (Fig. 1J,K). The anulus is best described as being ''lax.'' Bulging disks are generally associated with mild degrees of loss of disk height and frequently a decrease in the nuclear signal. The anulus itself is dark, as it normally is, unless there is a coexistent anular fissure. In an asymptomatic population, half of degenerated disks bulge, and half of disks that bulge are degenerated. By definition, the bulging disk extends *circumferentially* 3 or more mm beyond the adjacent vertebral end plates (4,8).

An *extrusion* is distinguished from a protrusion by the fact that an extrusion violates the entire anulus/posterior longitudinal ligament complex (Fig. 1E–G,L,M). The base of the disk abnormality against the disk of origin is narrower than the anteroposterior dimension of the extruding material itself (1). The lack of a restraining anulus allows disk material free access to the epidural space. In such instances, the disk may potentially completely separate from the disk of origin (see sequestration, below) and/or it may *potentially* migrate up or down the canal, away from the disk space of origin. On T2-weighted images, the solid dark rim defining the posterior margin of the anulus is interrupted or poorly defined, resulting in an indistinct and often lobulated or double disk contour. While extrusions usually have signal intensities similar to their disks of origin (low intensity), an inflammatory reaction, granulation tissue, and increased hydration may cause the extrusion to have increased signal on T2-weighted images and gadolinium-enhanced studies (14,15). A disk that migrates within the spinal canal but maintains its connection to the parent disk is termed an extrusion; migrated, but totally detached (from the parent disk) extrusions are termed *sequestrations.* These lesions have a particular import in that they might be overlooked by the operating surgeon, who, not being aware of the existence of the abnormality, operates through a microscopic incision with limited visibility (Fig. 1N,O).

It is often impossible to be specific when speaking of a disk disorder since its morphology may be unknown. Similarly, a variety of different types of disc abnormalities may be described in a discussion or paper, and a general term would be helpful to describe such situations. Using terms like ''disk disorder'' or ''disk extension beyond the interspace'' is awkward. Herniation may be an acceptable term in such instances. However, the word *herniation,* a generalized term, is probably the most frequently used but least understood word applied to disk abnormalities. To many individuals, the word ''herniation'' implies injury, an acute time course, pain, and often disability, especially to lay people such as judges, juries, lawyers, workers compensation adjusters, and arbitrators who are asked to examine the validity of a patient's claim for compensation. The word herniation is derived from a Greek word meaning ''to shoot or to sprout'' and is currently used in other medical contexts in which trauma is seldom a factor (hiatus hernia, tonsillar herniation). Many authors have suggested abandoning the term herniation because of the legal and compensation implications, yet this word has been used in so many scientific, clinical, and lay discussions of the past seven decades and is so firmly established in various lexicons as to make proscription impossible.

For formal reporting of imaging, surgical, or pathologic findings, when adequate data are available to describe specifically disk position as normal, bulging, protruded, or extruded, nonspecific terms like disk herniation *should not be used.* When, however, a nonspecific term is needed for informal discussion, when data are not available to allow for specificity, or when groups of terms containing a variety of conditions of displacement are discussed, disk herniation may be an acceptable term (7). *Disk herniation* is preferable to *herniated nucleus pulposus,* since, as noted above, it is not only the nucleus that extends beyond the disk space. Like all other descriptors of disk abnormality, the term disk herniation should not imply disability, cause, relation to injury, relation to pain, or need for treatment.

MR imaging is a extremely sensitive modality for the detection of intervertebral disk abnormalities and has a high correlation with findings at CT/myelogram examinations and with findings at surgery (15,16), yet disk abnormalities are frequently seen in patients who are, and have always been, completely asymptomatic with regard to the lumbar spine (1,2,17–21). However, an abnormal MR examination in a patient who either has no symptoms, or symptoms referable to another anatomic location, should not be identified as a *false-positive* MR study since the presence or absence of the lesion can only be verified at surgery, not by the clinical findings. *MR findings must, therefore, be considered to be gross pathoanatomic descriptions that nearly always accurately reflect the patient's anatomy, but that may or may not correlate with the patient's symptoms. MR studies are not functional or physiologic assessments of a patient's condition.* The results of the MR examination, and all imaging examinations of the spine, must be closely correlated with the history and physical evaluation to design a proper therapeutic course and achieve an optimal patient outcome.

The following terms are recommended to describe disk disorders of the lumbar spine:

Disk degeneration
Generalized disk bulge
Anular fissure
Protrusion
Extrusion
Sequestration

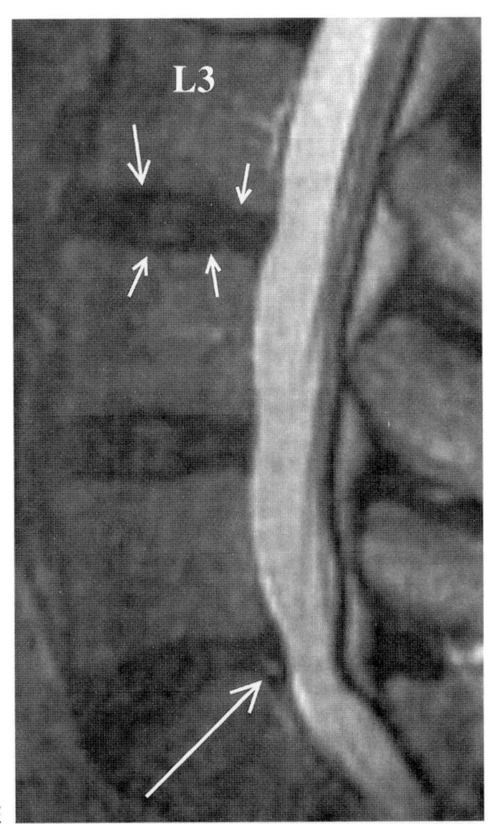

FIG. 1. **E–G:** Disk degeneration. Sagittal (TR/TE 3,000/102 msec) image. **E:** Sagittal (TR/TE 4,000/96 msec) image shows mild narrowing of the disk space and mild loss of disk signal at L3-4, indicative of a mild-to-moderate degree of disk degeneration. A small disk protrusion is seen at the L5-S1 level. **F:** The *arrow* defines a severely narrowed disk space with a bone-on-bone appearance. The remaining disk signal, if any, is low. The patient has a disk extrusion at the L3-4 level, displaying the typical bilobed appearance of such lesions. **G:** There is moderate degenerative disk disease at both L4-5 and L5-S1. In spite of severe disk space narrowing at L5-S1, a fluid-filled cleft *(arrow)* is seen within the disk.

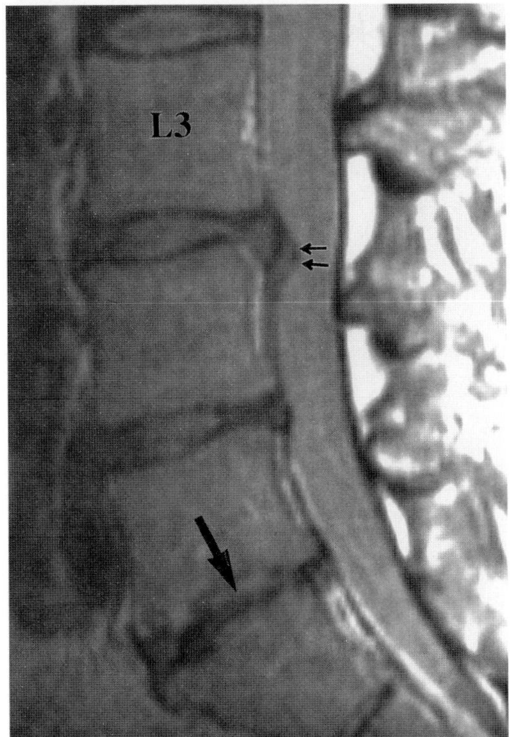

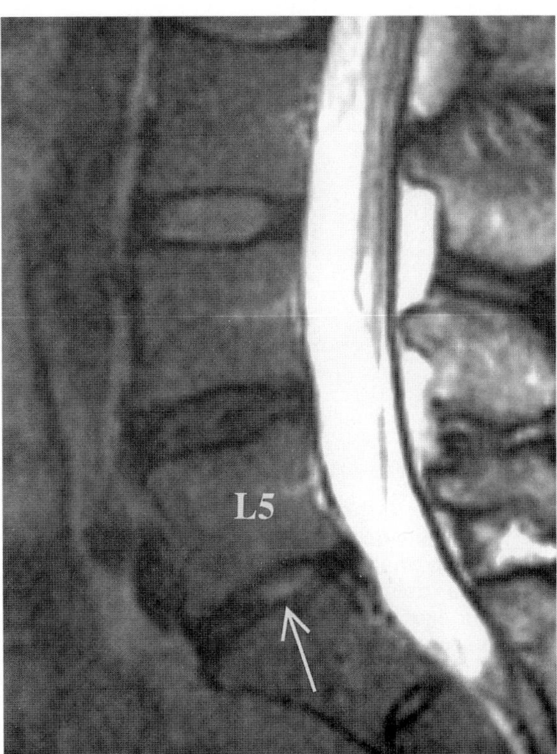

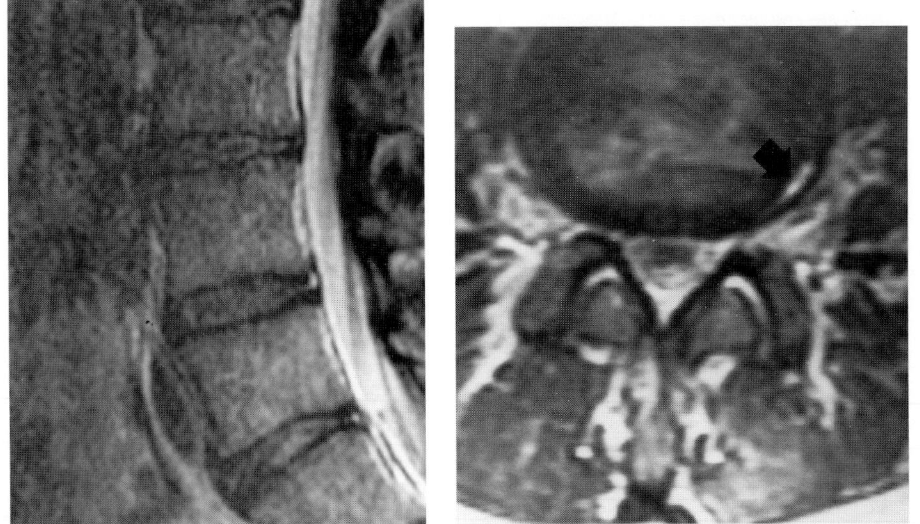

FIG. 1. H,I: Anular fissures. **H:** Sagittal (TR/TE 3,000/96 msec) image shows a focus of markedly increased signal in the posterior aspect of both the L4-5 and L5-S1 disks; there is no contour deformity of the posterior margin of the anulus. **I:** Axial (TR/TE 3,000/96 msec) image in another patient, showing a circumferential tear of the anulus *(arrow)* along the left side of the disk.

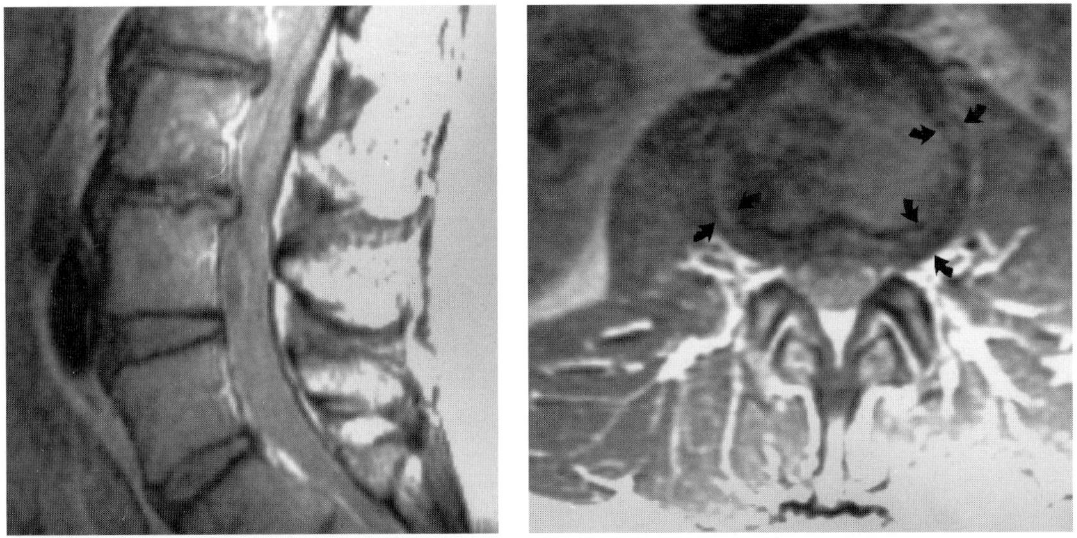

FIG. 1. J,K: Generalized disk bulge. Sagittal (TR/TE 3,000/17 msec) **(J)** and axial (TR/TE 3,500/17 msec) **(K)** images show marked narrowing of the L2-3 intervertebral disk space. The posterior margin of the disc extends into the spinal canal, much like a protrusion; however, the axial image **(K)** demonstrates a "double contour," revealing the circumferential nature of this abnormality. The lack of a *focal* contour alteration distinguishes a generalized disk bulge from a disk protrusion. Incidentally noted is the presence of a disk protrusion at the L3-4 level.

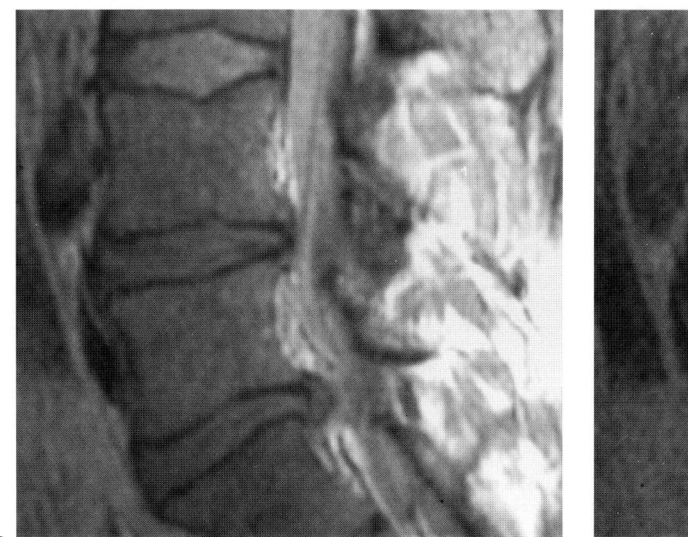
L

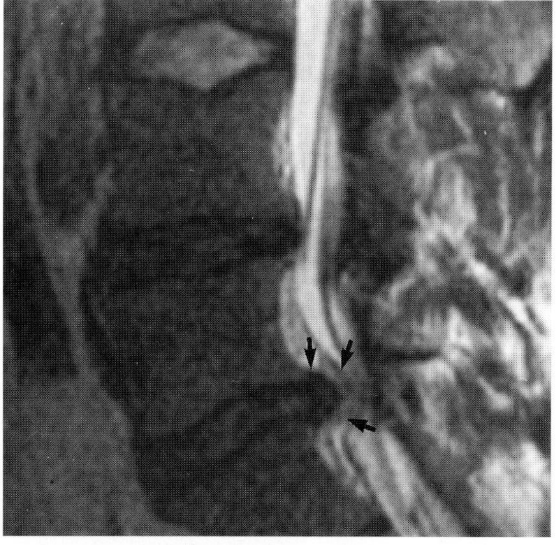
M

FIG. 1. L,M: Extrusion. Sagittal (TR/TE 3,000/17 msec) **(L)** and sagittal (TR/TE 3,000/102 msec) **(M)** demonstrate a lobulated mass associated with the posterior aspect of the L5-S1 disk. The posterior margin of the disk is less well defined than in a protrusion; the lobulated contour, best seen in **M,** is typical of an extrusion.

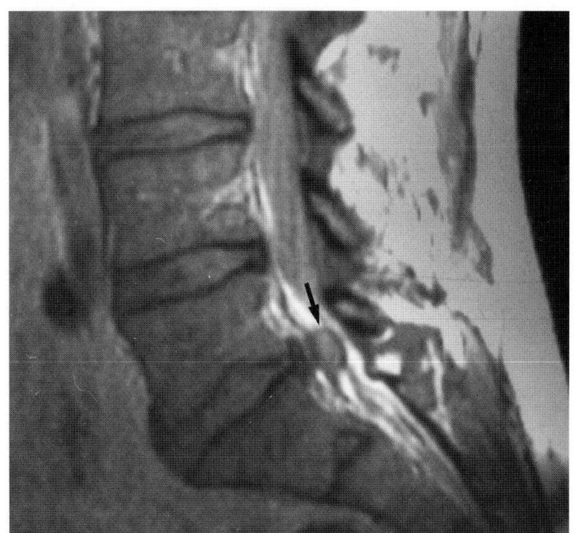
N

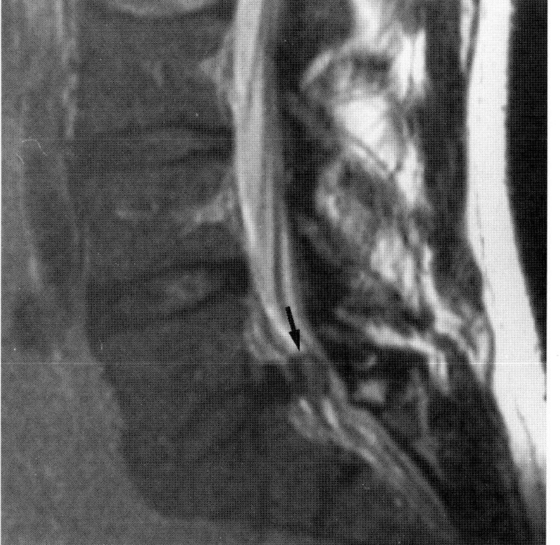
O

FIG. 1. N,O: Sequestration. Sagittal (TR/TE 3,000/17 msec) **(N)** and sagittal (TR/TE 3,000/102 msec) **(O)** images show a soft tissue mass *(arrow)* distinctly separated from the parent disk. Such sequestrations frequently migrate away from the intervertebral disk of origin.

CASE 2

History: You are asked to be an expert witness in a personal injury case. A 28-year-old man complains of nonradiating low back pain after a low-speed (10–15 mph) rear-end collision. Representative images from an MR examination performed following the accident are illustrated. What are the findings? What is the prevalence of similar findings on MR scans of the lumbar spine in asymptomatic individuals?

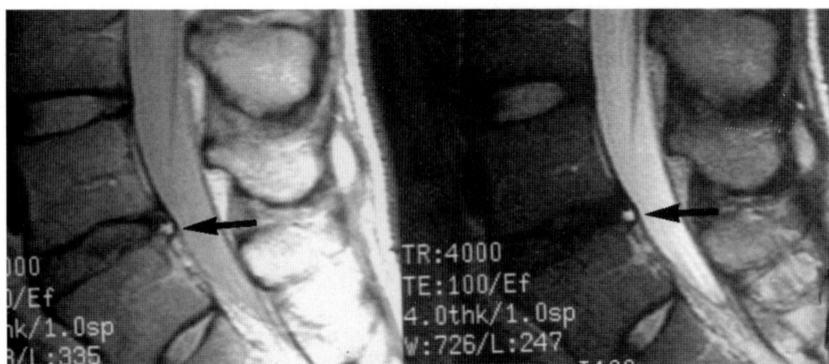

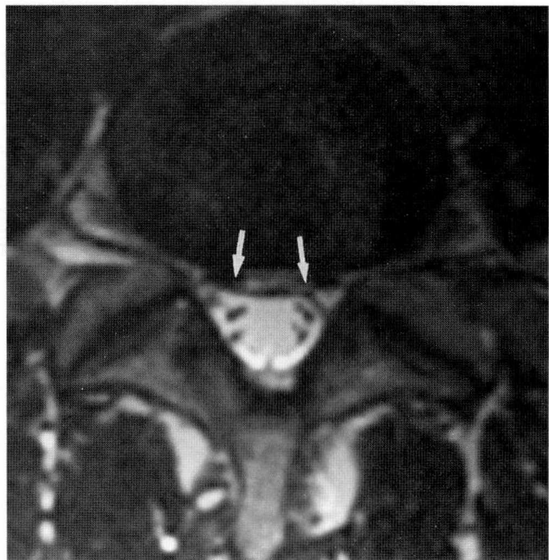

A B

FIG. 2.

Findings: Dual-echo (TR/TE 4,000/20,100) fast spin-echo midline sagittal MR images (Fig. 2A) show degenerative signal alteration as well as a mild focal central disk protrusion involving the L5-S1 intervertebral disk space. In addition, a small focus of high signal is seen at the posterior disk margin compatible with an anular fissure (arrows). T2-weighted (TR/TE 4,000/105 msec) axial fast spin-echo MR image (Fig. 2B) at the L5-S1 intervertebral disk space shows that the protrusion results in mild effacement of the ventral aspect of the thecal sac (arrows).

Diagnosis: Small central protusion of the L5-S1 disk. Because similar findings may be encountered in 27% of asymptomatic subjects on MR examinations, caution must be used in considering a finding such as this as the source of an individual's complaints, and particularly so in a legal or workers compensation setting.

Discussion: The relationship between degenerative disk disease and low back pain has not been firmly established. The presence of disk abnormalities in the asymptomatic population has long been known. In 1956, McRae (1) performed postmortem examinations of the entire spine in patients presumed to be free of symptoms; 39% showed a posterior protrusion. In patients without nerve root compression symptomatology, 24% have been shown to have defects involving the lumbosacral spine on myelograms (2). Wiesel et al. (3) examined CT scans of the lumbosacral spine in asymptomatic subjects and found that 36% had herniated disks. However, in this study only the L4-5 and L5-S1 disk levels were evaluated, thus presumably underestimating the prevalence of lumbar disk herniations. More recently, Boden, et al., (4) utilizing MR, found that 20% of asymptomatic subjects younger than 60 years had a herniated disk of the lumbosacral spine, a figure that increased to 36% for subjects 60 years or older. An MR study was recently performed in subjects without a history of low back pain or radiculopathy that classified disk abnormalities as bulge, protrusion, or extrusion. In this study 52% of the subjects showed a bulge at at least one level, 27% had a protrusion, and disk extrusions were rare, seen in only 1% of the subjects (5). Table 1 summarizes the prevalence of disk bulges, protrusions, and extrusions on the MR studies. The prevalence of bulging disks and protrusions was highest at L4-5 and L5-S1; there were few abnormalities at L1-2. Interestingly, in this study, although the prevalence of disk bulges increased with age, a significant relationship was not found between age and the presence of a disk protrusion.

In 1934, Mixter and Barr (6) were the first to associate sciatica with ''prolapse'' of the intervertebral disk. Since then, research efforts have focused on the intervertebral disk and the production of low back pain. However, it has been shown that mechanical compression can cause neurologic dysfunction and ischemia, leading to dysesthe-

TABLE 1. *Prevalence of bulges and protrusions by age, sex and level in 98 asymptomatic subjects (% with bulges/% with protrusions)*

Age	L1-2	L2-3	L3-4	L4-5	L5-S1	At least one level	
						Prevalence	No. of subjects
20–29							
M	0/0	0/0	19/0	25/13	6/0	25/13	8
F	0/0	0/0	4/0	13/13	21/13	29/25	12
30–39							
M	0/0	5/0	0/9	23/14	27/14	36/23	11
F	9/6	3/6	15/0	26/12	3/3	35/21	17
40–49							
M	3/3	9/0	22/3	34/22	25/19	66/38	16
F	0/0	0/0	29/0	43/7	36/14	50/21	7
50–59							
M	19/0	19/13	75/19	44/13	75/0	88/38	8
F	17/0	33/0	33/0	44/22	55/0	72/22	9
≥60							
M	7/0	21/0	43/7	50/14	43/29	71/29	7
F	0/0	33/50	50/0	67/33	50/0	100/67	3
Total	6/2	10/4	24/2	33/15	29/10	52/27	98

sia and even weakness in the muscles supplied by the nerve, but not necessarily pain (7). Thus it appears that the phenomenon of low back pain is far more complex than can be accounted for on the basis of simple mechanical compression of a nerve root by disk material. Another theory is that low back pain seen in the setting of degenerative disk disease is somehow related to an altered biochemical milieu. Such mechanisms are based on the direct biochemical and inflammatory effects of the contents of the nucleus pulposus on nerve fiber structure and function (8). The pathophysiology of such effects on the nucleus pulposus and nerve tissue have been demonstrated in animal models (8,9). High levels of phospholipase A_2 have been described in symptomatic lumbar disk herniations. This enzyme plays a critical role in the generation of inflammatory mediators such as prostoglandins, leukotrienes, and platelet-activating factor (10). Some have proposed that this mechanism may lead to low back pain and radiculopathy with disk degeneration and anular disruption, even in the absence of any extension of disk material into the spinal canal or neural channels (9).

The clinical significance of anular fissures, however, remains unknown. Hirsch and Schajowicz (11) postulated that nerves that grow into radial tears, along with blood vessels, may be the pain-producing factor. However, they were unable to draw this conclusion from subsequent histologic investigations. The high incidence of radial tears in cadavers, as well as on MR in asymptomatic subjects, suggests that they could well be incidental findings (12). Some have suggested that the common inciting event leading to disk desiccation is the mechanically caused annular defect, which in turn stimulates granulation tissue response, progressing to disk degeneration and scarring (13,14).

MR accurately depicts both the morphologic and the biochemical sequelae of disk degeneration. Caution must be used, however, before considering a particular finding the source of the patient's symptoms. Disk bulges and protrusions on MR scans in individuals with low back pain or even radiculopathy may be coincidental. Therefore, a patient's clinical situation must be carefully evaluated in conjunction with the results of the MR study.

CASE 3

History: A 30-year-old woman presented with low back and left leg pain. Selected images are demonstrated from two MR examinations obtained on the patient, at the time of presentation (Fig. 3A,B) and $1\frac{1}{2}$ years later as part of a study (Fig. 3C,D). The patient was asymptomatic at the time of the second examination. What probably occurred in the interval between the examinations?

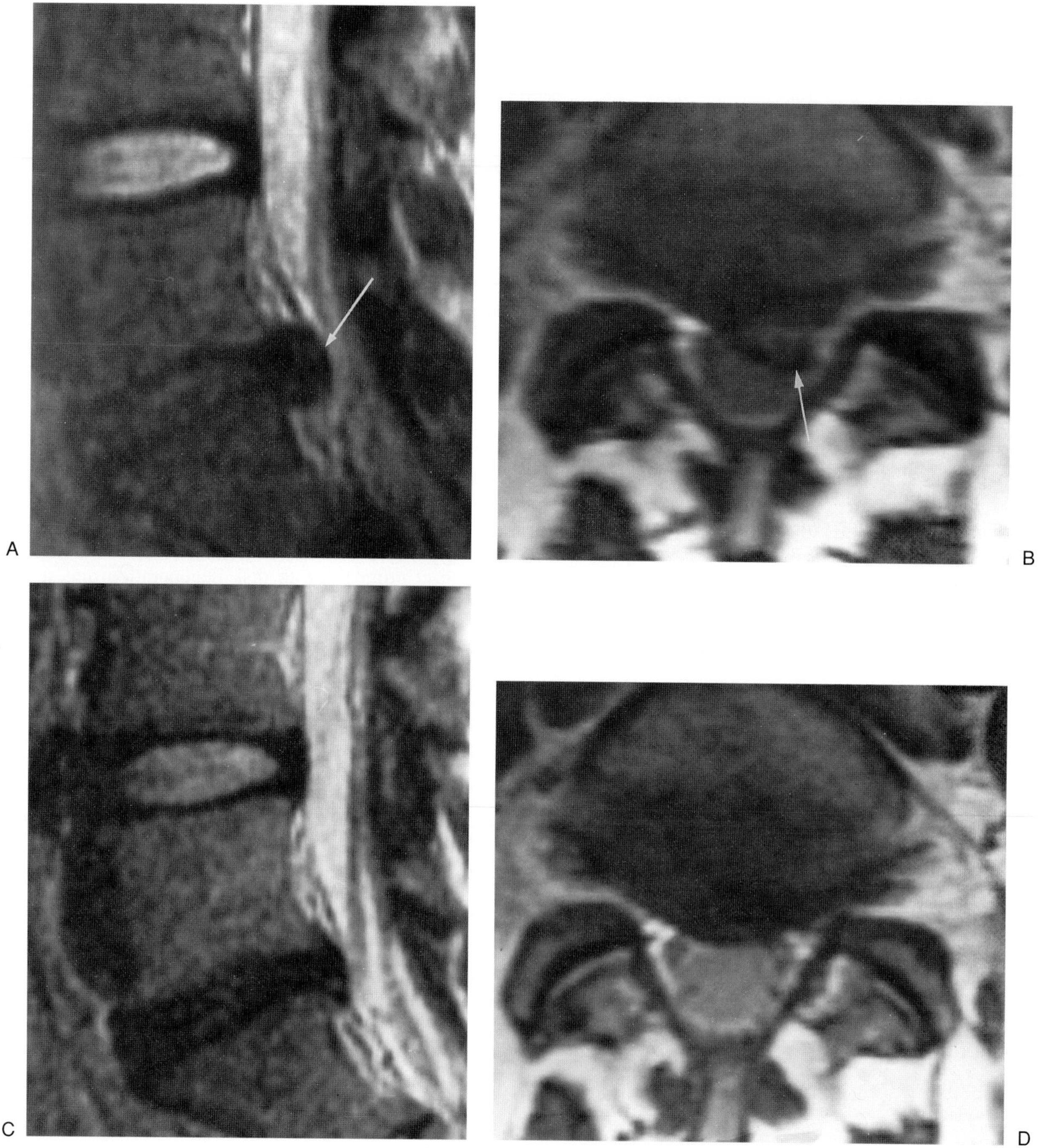

FIG. 3.

Findings: Sagittal (TR/TE 2,000/70 msec) image (Fig. 3A) demonstrates a central to left paracentral extrusion of the L5-S1 disk of approximately 7–8-mm (arrow). Axial (TR/TE 2,000/70 msec) image (Fig. 3B) demonstrates that the disk extends slightly to the left of midline (arrow). Sagittal T2-weighted fast spin-echo (TR/TE 3,000/108 msec eff) image (Fig. 3C) demonstrates diminution in disk signal intensity and disk height. Only a small disk protusion is present. Axial T2-weighted fast spin-echo image (TR/TE 3,000/108 msec, eff) image (Fig. 3D) demonstrates a small disk protusion.

Diagnosis: Spontaneous decrease in size of a lumbar disk extrusion with conservative management.

Discussion: The natural history of lumbar disk extrusions has been an area of considerable interest and some controversy. Several studies have now attested to the success of nonoperative management of lumbar disk extrusions (1–6). In one outcome study, Saal and Saal (1) reported a greater than 90% success rate in the nonoperative management of patients with disk extrusions and radiculopathy. Interest has also been directed toward studying the changes over time in appearance of lumbar disk extrusions managed nonoperatively (2–6). Multiple reports have documented the apparently spontaneous reduction in size of conservatively managed disk protusions and extrusions (2–6). In a recent prospective series utilizing MR, Bozzao et al. (3) reported on longitudinal follow-up studies of 69 patients with lumbar disk herniations. In this series, 63% of patients demonstrated a decrease in size (volumetric analysis utilizing both the sagittal and axial planes) of more than 30%; 48% of patients demonstrated a reduction of greater than 70% (3). In this series, as in the one reported by Saal et al. (2), a positive correlation was identified between size of the disk extrusion and amount of regression (i.e., the larger the initial disk abnormality, the greater the degree of reduction).

While regression in size of the disk extrusion has been correlated with successful clinical outcome, it is not clear that these events are directly related (2,7,8). The pain associated with nuclear extrusions may result as much from an inflammatory component as from neurocompression. Support for this concept comes from recent demonstrations of high levels of phopholipase A_2, a potent inflammatory enzyme, in human lumbar disk herniations (8). Support for the lesser importance of mass effect on pain also comes from follow-up studies in patients successfully treated with chemonucleolysis, automated percutaneous diskectomy, and surgical microdiskectomy, which have all documented successful clinical outcomes in the absence of documentable changes in the size of the extruded disk (9–11).

The mechanism by which disk extrusions are resorbed requires further investigation. Saal et al. (2) have suggested that separation of the nuclear material from the nutrient supply of the disk and exposure to the epidural vascular supply contribute to the process of disk resorption (12). Separation of the nuclear material from the parent disk may preclude production and replenishment of the hydrophilic proteoglycan in the disk and contribute to progressive dissecation. Additionally, cellular elements in the epidural space stimulated by the inflammatory response may promote phagocytosis and subsequent resorption (2).

In considering spontaneous regression of disk extrusions, it is important to differentiate this process from the phenomenon of spontaneous localized epidural hematoma (see Case 5–11). As Heithoff and colleagues (13,14) have suggested, this process may account for many of the previously reported spontaneous disk resorptions in the older myelographic literature. Spontaneous localized epidural hematomas demonstrate a characteristic appearance on MR as well as clinical course that should allow their differentiation from primary disk extrusions in most cases. Since localized epidural hematomas are typically associated with smaller underlying disk protrusions, there is clearly potential for overlap between the two groups.

CASE 4

History: A 40-year-old man presented with left-sided radiculopathy. What are the findings illustrated in Fig. 10B, how common are such findings, and what is their potential significance?

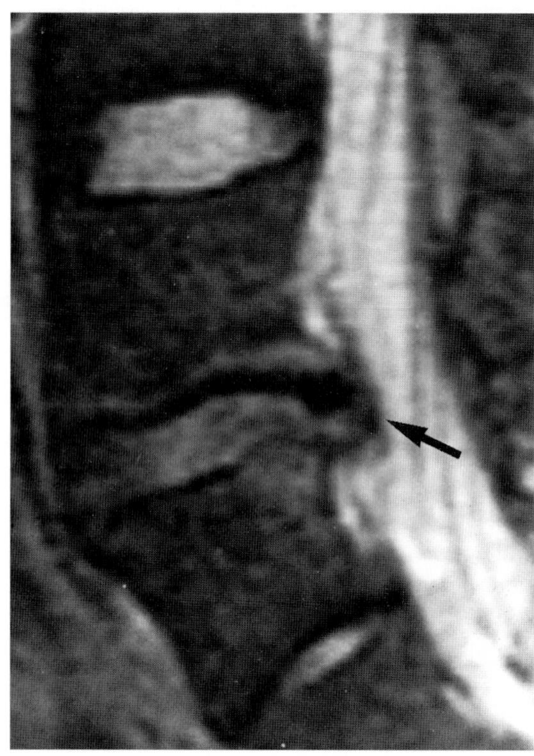

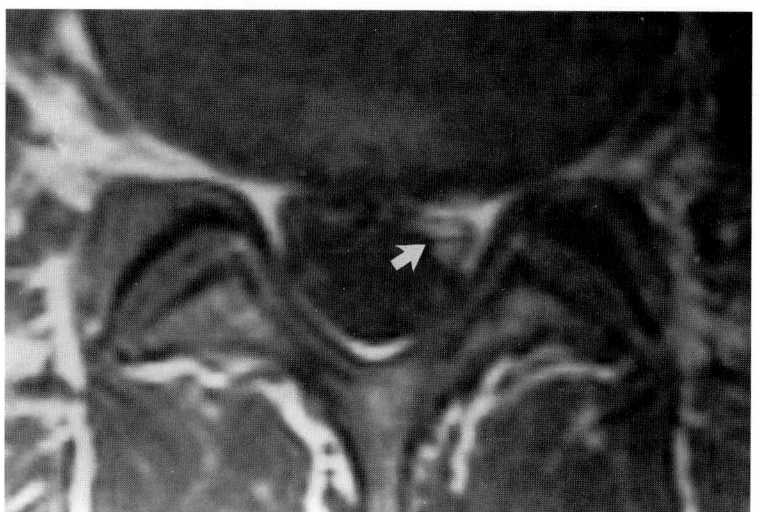

A B

FIG. 4.

Findings: T2-weighted (TR/TE 2,000/80 msec) conventional spin-echo MR image (Fig. 4A) acquired to the left of midline demonstrates a focal 6-mm extrusion of the L5-S1 disc. T1-weighted (TR/TE 600/20 msec) axial conventional spin-echo MR image (Fig. 4B) acquired following the administration of 0.1 mm/kg IV of gadopentetate dimeglumine at a level just cranial to the disk extrusion illustrated in Fig. 4A. There is evidence of localized high-signal intensity representing enhancement within at least two intrathecal nerve roots (arrow).

Diagnosis: Intrathecal nerve root enhancement (radiculitis) associated with lumbar intervertebral disk herniation.

Discussion: MR imaging has been remarkably valuable in the evaluation of spinal pain syndromes. It is well recognized, however, that morphologic abnormalities identified on imaging studies may not correlate directly with the presenting clinical syndrome (1). Frank disk protrusions and extrusions may be seen in the lumbosacral spine in asymptomatic individuals (see Case 5-2) (2). Because of the not infrequent dissociation of the gross morphologic pathoanatomy depicted and the presenting lumbosacral clinical syndrome, a noninvasive marker that could confirm clinically relevant and correlative underlying neural pathology would be of great value for identifying and confirming the specific cause, location, and extent of clinical signs and symptoms.

Enhancement of nerve roots in the lumbosacral spine after intravenous administration of gadopentetate dimeglumine has been observed in a number of benign conditions including disk disease and canal stenosis. In a report of 200 consecutive unselected and unoperated patients studied with 0.1 mmol/kg IV gadopentetate dimeglumine, Jinkins (3) showed an overall incidence of 5% enhancing nerve roots in the lumbosacral spine. If the incidence of focal disk protrusion was taken into account, the prevalence of pathologic enhancement was 21.2%. It is important to note that lumbosacral intrathecal nerve roots do not enhance normally on MR utilizing conventional spin-echo acquisitions at the above dose levels. This study also demonstrated an overall 1.5% incidence of idiopathic nerve root enhancement. This small group all had multiple, multilevel enhancing roots. While the etiology of this enhancement pattern is unknown, it probably represents a low-grade inflammatory response and thus an active arachnoiditis.

Nerve root enhancement represents actual pathology within nerve roots reflected by a breakdown in the blood–nerve barrier. The mechanism in the present context (e.g., illustrated case) is almost certainly mechanical, resulting from the disk protusion (4,5). The contribution of traumatic fibrosis to the picture of enhancement has also been considered (3). Multilevel neural enhancement remote from a compressive abnormality has been theoretically linked with wallerian axon degeneration (6–8).

The general correlation of enhancing nerve root cases with the radicular clinical syndrome was excellent in the series reported by Jinkins (1). For example, all patients with multiple enhancing nerve roots demonstrated a polyradiculopathy clinically. In addition, the individuals with multiple enhancing nerve roots associated with disk protrusions lend support to the theory that a protrusion at one level may cause clinical pathology within more than one root. Other relevant cases revealed that lateralizing bulges contralateral to the symptomatology, as well as midline disk protrusions, may on occasion be associated with enhancing nerve root (s) appropriate to the clinical syndrome.

The phenomenon of nerve root enhancement was shown in this study to be a marker for clinically relevant nerve root pathology in patients with a distinct lower extremity radiculopathy in the presence or absence of focal protrusive disk abnormalities. This information would not be available without the use of contrast enhancement and suggests that intravenous administration of gadopentetate dimeglumine may have a role in selected patients with lower extremity radiculopathy with equivocal or nonclinically correlative findings on preceding nonenhanced studies.

The use of contrast enhancement has also been investigated in a series of patients with central canal stenosis and suspected ''neurogenic claudication.'' Central stenosis of the lumbosacral canal may be either developmental or acquired, although most cases presenting in adulthood represent a combination of these factors (9,10). Preexistent ontogenic circumferential narrowing of the spine and/or degenerative changes of the intervertebral disks, facet joints, and related structures results in a constriction of the thecal sac and its contents. This may cause the clinical syndrome known as neurogenic claudication (11).

In a recent MR study, seven patients presenting with suspected neurogenic claudication demonstrated severe findings of central stenosis of the lumbar canal on nonenhanced MR acquisitions (12). Five of these patients (71.4%) demonstrated abnormal intrathecal nerve root enhancement on MR imaging at and extending craniad and/or caudad from the level(s) of severe spinal stenosis after intravenous administration of gadopentetate dimeglumine (dose 0.1 mmol/kg). Depending on the plane of imaging, the enhancement pattern was seen to be linear, curvilinear, and/or punctate in configuration (Fig. 4C–E). Since intrathecal nerve root enhancement is not a normal phenomenon at this dosage of contrast, the neural enhancement pattern in these patients with claudication was felt to be caused by traumatic disruption in the blood–nerve barrier resulting from the circumferential nerve root compression coupled with wallerian axonal neural degeneration (1,12,13). From a practical standpoint, the presence of intrathecal enhancement represents unequivocal underlying neural pathology associated with spinal disease of a stenotic nature. This information may be of clinical utility in a functional manner when dealing with ambiguous or equivocal cases and in particular when attempting to differentiate between neurogenic and vascular claudication.

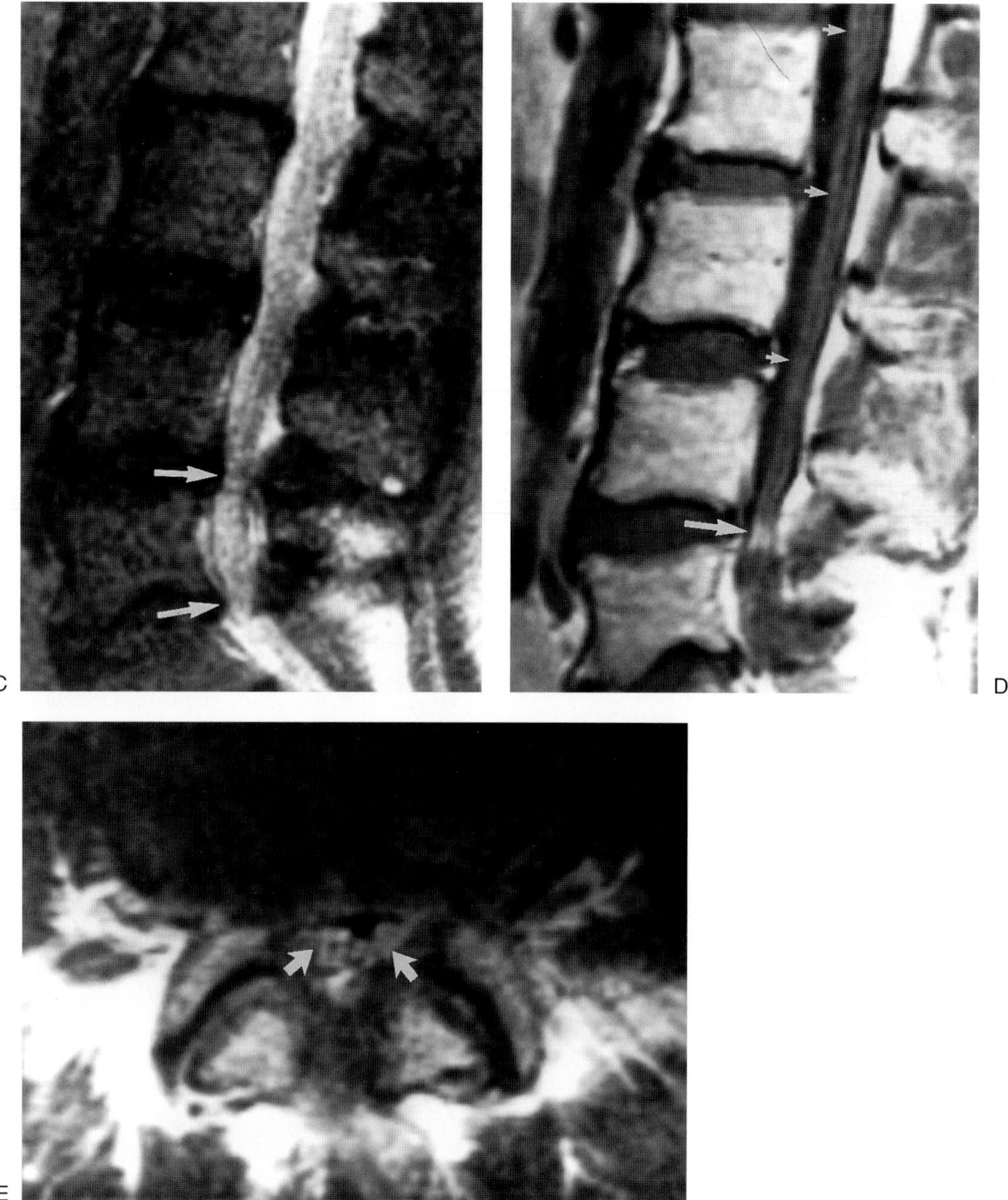

FIG. 4. C–E: Multiple, multilevel enhancing nerve roots in a previously unoperated 66-year-old man with low back pain and neurogenic claudication of 10 years' duration. The patient presented with a 4-month history of radiating radiculopathy to both lower extremities, bilateral lower extremity sensory paresthesias, sensory neuropathy, and motor weakness. **C:** T2-weighted (TR/TE 2000/80 msec) sagittal conventional spin-echo MR image demonstrates multilevel central stenosis of the spinal canal most severe at the L4-5 and L5-S1 levels *(arrows)*. **D:** T1-weighted (TR/TE 500/15 msec) sagittal conventional spin-echo MR image obtained after 0.1 mmol/kg IV gadolinium administration demonstrates focal intrathecal enhancement at L4-5 *(long arrow)*. Diffuse enhancement of the cauda equina is also identified above the levels of central spinal stenosis *(short arrows)*. **E:** T1-weighted (TR/TE 500/15 msec) axial conventional spin-echo MR image acquired at L4-5 after 0.1 mmol/kg gadolinium administration showing the generalized central canal stenosis as well as punctate areas of intrathecal enhancement *(arrows)*, indicating a diffuse breakdown in the blood–nerve barrier.

CASE 5

History: A 56-year-old man underwent lumbar disk surgery for an L5-S1 disk extrusion. Representative images from the preoperative study and from a postoperative MR obtained 16 months following surgery are demonstrated. Can you predict the outcome of the procedure from the postoperative image? What are the principal morphologic criteria for differentiating recurrent disk from postoperative scar? Is ''mass'' effect a feature commonly associated with postoperative scar?

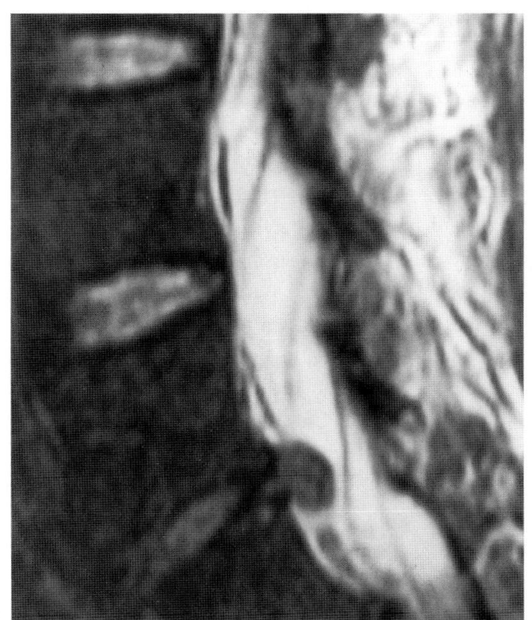

A

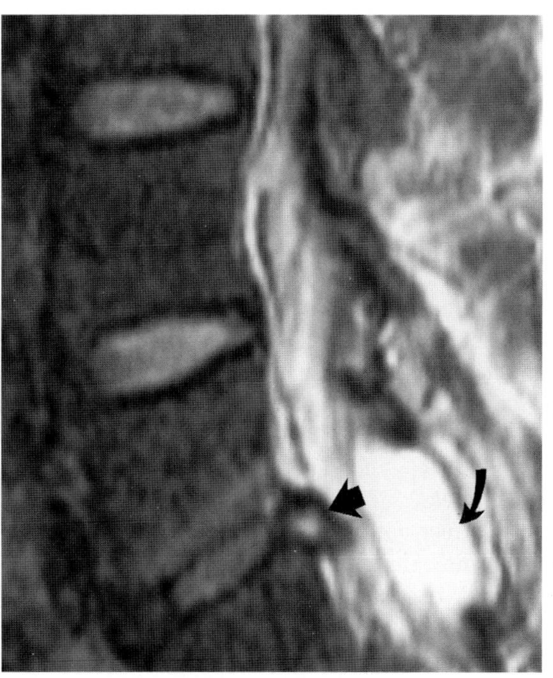

B

FIG. 5.

Findings: Sagittal T2-weighted (TR/TE 2,000/80 msec) spin-echo image (Fig. 5A) obtained preoperatively demonstrates a moderate to large central extrusion of the L5-S1 disk. Figure 5B shows a sagittal T2-weighted (TR/TE 2,000/80 msec) examination obtained 16 months postoperatively. There has been only a small interval decrease in size of the extradural defect paralleling the posterior disk margin. This material demonstrates a change in MR signal characteristics from the preoperative study with centrally increased signal intensity on the more heavily T2-weighted images (arrow). Incidentally noted is a small dorsal postoperative fluid collection (curved arrows). (From Deutsch AL, Howard M, Dawson EG, et al. Lumbar spine following successful surgical disectomy: magnetic resonance imaging features and implications. *Spine* 1993;18:1054–1060, with permission.)

Diagnosis: Localized ''mass-like'' epidural fibrosis. The patient was entirely asymptomatic, and the study was obtained as part of a longitudinal follow-up study.

Discussion: Fundamental to approaching analysis of the lumbar spine in patients with persistent or recurrent symptoms following prior surgical intervention is an appreciation of the findings to be expected normally after diskectomy. Recent studies have addressed this subject by studying groups of successfully treated postdiskectomy patients (1–4). These studies have provided observations that challenge both several of the previously established criteria for differentiating scar versus disk in the postoperative spine and the mechanisms responsible for pain production in patients with persistent or recurrent symptoms following diskectomy.

In the immediate postoperative period, increased anterior epidural soft tissue signal (presumably reflecting edema), and disruption of the posterior annular margin secondary to curettage of the disk combine to produce mass effect and an appearance that may closely simulate the preoperative disk herniation (1,2,5–7). The changes within the anterior epidural space gradually resolve over a period of 2 to 6 months after surgery. Ross and associates (1), in a study of patients undergoing lumbar surgery, performed MR scans on 15 patients preoperatively, in the immediate postoperative period (1 to 10 days), and in the ''late'' postoperative period (2 to 6 months). In the acute postoperative period, 69% of postdiskectomy patients demonstrated mass effect that mimicked preoperative findings (1). In 89% of the cases in which initial postoperative mass effect was noted, late postoperative images indicated an improvement in the appearance of the anterior epidural changes. In this series there was no correlation between the postoperative appearance of the disk and patient symptoms, and the authors concluded that the mass effect related to soft tissue edema precluded the effective use of MR for evaluation of the symptomatic patient in the immediate postoperative period, except for the detection of gross abnormalities and complications such as postoperative hemorrhage (1).

Boden and colleagues (2) studied 14 patients (15 levels) with successful results following lumbar diskectomy. Preoperative scans were compared with studies obtained 3 weeks, 3 months, and 6 months following surgery. In addition to routine spin-echo images, the authors performed contrast-enhanced examinations with gadopentetate dimeglumine (2). Intermediate signal intensity areas were uniformly seen at the operative site of the original disk herniation. In the immediate postoperative period, 66% of patients demonstrated significant mass effect upon the thecal sac (2). In the 3- and 6-month examinations, a decrease in the mass effect was evident, a finding corroborating the work of Ross et al. (1). In the early postoperative period (3 weeks), 38% of patients demonstrated extradural mass effect with peripheral enhancement using gadopentetate dimeglumine that was indistinguishable from recurrent or residual disk herniation surrounded by epidural scar. The authors concluded that despite the use of contrast material, persistent mass effect on the neural elements, demonstrating an enhancement pattern indistinguishable from recurrent disk material, can be frequently encountered in the first 3 to 6 months following surgery even on successfully treated patients (2). For this reason, the investigators urged caution with regard to using MR findings in this period for significant management decisions.

In an MR study of 36 lumbar intervertebral disk levels in asymptomatic postdiskectomy patients, Kahn et al. (3) demonstrated space-occupying features in 67% of patients in the early postoperative period (4 to 12 days) and in 13% of late (6 to 9 months) follow-up studies.

As a consequence of the extensive epidural changes in the immediate and early postoperative period, it appears that the use of MR is best limited to assessment of gross changes upon the dural tube and neural structures as well as significant complications such as postoperative hemorrhage, pseudomeningocele formation, and disk space infection rather than assessment of possible residual/recurrent disk herniation in the symptomatic patient.

Deutsch and colleagues (4) described persistent localized soft tissue intensity masses in apparent direct continuity with the operative disk level that produced varying degrees of mass effect in 60% of a group of 23 patients (26 levels) in the long term (at least 1 year) after successful lumbar disk surgery. All patients met rigorous criteria for a successful outcome. At only nine levels was there virtually complete resolution of the preoperative findings of a localized disk protrusion or extrusion (Fig. 5C,D). In 4 patients, the postoperative study demonstrated virtually no change in the morphology of the posterior disk contour, and in 13 patients persistent posterior contour abnormalities were present that contributed to continued mass effect on the thecal sac and or nerve roots (4) (Fig. 5E–H). Contrast-enhanced studies demonstrated uniform enhancement of the persistent contour abnormalities in 18 of 19 disk levels, a finding the authors attributed to the presence of fibrosis. These studies suggest that localized apparent posterior diskal contour abnormalities (morphologically simulating recurrent disk herniations and variably contributing to mass effect) may be commonly encountered in long-term follow-up imaging studies of successfully treated diskectomy patients.

The degree of mass effect demonstrated by the presumed residual scar in these asymptomatic patients is in part inconsistent with previous descriptions of the appearance of postoperative scarring and with many of the established criteria for differentiating between postoperative fibrosis and recurrent disk herniation on imaging studies (5–11). It has been well established that anterior epidural scar tends to demonstrate increased signal intensity compared with the intervertebral disk on T2-weighted sequences; this observation was again corroborated in the cases of presumed mass-like scar reported in the above series (4–6). Scar is also generally associated with a tendency to conform to existing space and to extend circumferentially around the dural sac (5,6). Scar may demonstrate irregular margins compared with the smooth polypoid configuration observed with recurrent disk herniation (12). Retraction of the sac toward the scar is also a reported differentiating feature of scar compared with recurrent disk. Recurrent disk may be associated with a low-signal-intensity rim, an observation that may further aid in distinguishing scar from disk (10,12). The presence of mass effect is generally a feature associated with recurrent/residual disk and not with scar. The occasional presence of significant mass effect associated with scar has, however, been recognized by other investigators (6) (Fig. 5E–H). Of all the morphologic criteria advanced for differentiating between scar and disk, the presence or absence of mass effect should be considered of secondary importance to other considerations (most prominently the presence or absence of contrast enhancement). Indeed, while MR imaging without intravenous contrast appears to be at least equivalent to contrast-enhanced CT in distinguishing scar from disk, the above considerations strongly suggest that simple morphologic analysis of lumbar disk contours on routine spin-echo MR imaging is at times misleading, an observation that underscores the necessity of contrast administration in postoperative patients (13). The dynamics of contrast enhancement, optimal techniques for administration and scanning, and interpretive criteria are discussed elsewhere in this text (see Cases 5-4, 5-6, 5-10, and 5-20).

The persistence of mass effect in unequivocally successfully treated patients also challenges many previously held conceptions regarding the role of mechanical compression of neural structures in the production of pain in symptomatic patients. The persistence of disk contour abnormalities and mass effect has been described in other groups of successfully treated lumbar disk herniation patients including those followed up after percutaneous diskectomy (16) (see Case 5-13), after chemonucleolysis (17), and after conservative management (18). Quite clearly, our understanding of the mechanisms that produce conscious pain remains incomplete. Mass effect alone is an insufficient explanation.

For pain to be perceived, a fundamental pathophysiologic change in the axonal membrane must occur. The axon theoretically becomes an ectopic source of neuroelectrical impulses, which are interpreted as pain and paresthesias within the distribution of the insulted axon (s) by the central nervous system. A variety of neurochemical mediators may be additional contributing factors (19). The role of epidural fibrosis itself in the possible generation of symptoms remains incompletely understood, as are the factors involved in development of epidural fibrosis. In addition to the direct trauma emanating from the initial disk protrusion and subsequent surgery, other stimuli may be responsible for inducing epidural inflammation. Byproducts of disc nutrition (i.e., lactic acid) or otherwise normal substances ordinarily contained within the intact intervertebral disk (i.e., glycoproteins) may hypothetically cause an inflammatory response following rupture of the disk into the epidural space (20,21). Alternatively, or possibly in addition to this response to chemical irritation, an allergic/autoimmune reaction to disk material may play a part in the epidural pathologic process. However, it seems unlikely that the spinal nerve, protected by the barrier formed by the dura, arachnoid, cerebrospinal fluid (CSF), and pia, would be intimately affected in most cases by such epidural reactions (22). Therefore, while epidural fibrosis may be a direct manifestation of these processes, actual pathology within the underlying nerve itself may not. Experimental evidence appears to support this contention (23).

While the process of epidural fibrosis appears ubiquitous following surgery and should be (reluctantly) incriminated as a pain source, it is theoretically possible that through CSF nutritional deprivation and tethering of the root causing traction on the nerve during somatic movement, epidural fibrosis may induce additional aberrant neuroelectrical potentials within already inflamed hypermechanosensitive axons (24–27). It is also theoretically possible that actual mechanical circumferential constriction of the underlying nerve root could be amplified by perineural scar (28). In assessing the possible role of epidural scar in the symptomatic patient, attention could be directed toward demonstration of any underlying nerve root enhancement that might suggest radiculitis. In a recent study of ten asymptomatic postoperative patients with varying degrees of epidural fibrosis, no abnormal radicular enhancement was identified on contrast MR studies (29) nor was any observed in a retrospective analysis of the study of Deutsch and colleagues (4). As such, the presence of distinct nerve root enhancement may be of significance in cases of epidural fibrosis and should be sought in analysis of MR studies of symptomatic postoperative patients. In the absence of demonstrable nerve root enhancement, the significance of epidural fibrosis in the symptomatic postoperative patient should remain questionable and is not presently verifiable on the basis of MR imaging. These observations further underscore the emerging importance of factors other than mechanical compression (such as neurogenic and non-neurogenic pain and inflammatory mediators) in the production of back pain (14,15,19) (see Cases 5-2, 5-3, 5-4, 5-13, and 5-20).

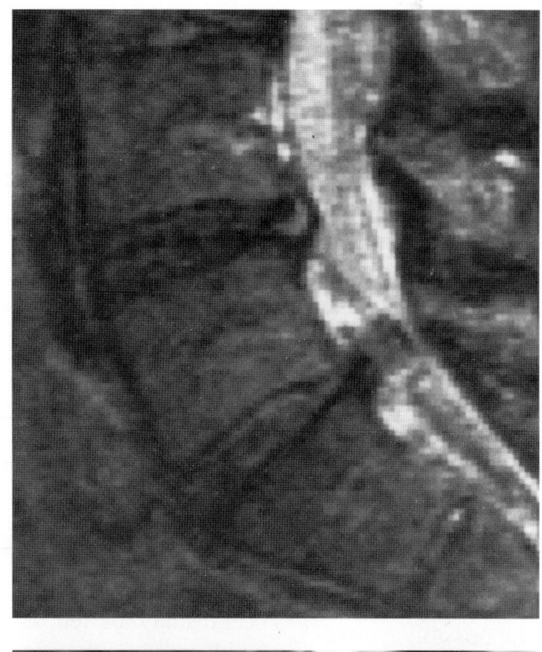

C

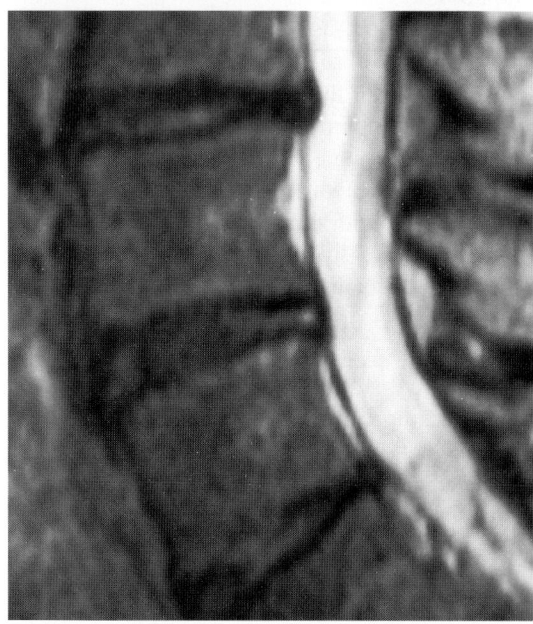

D

FIG. 5. C,D: Postoperative complete resolution of previous disk extrusion. **C:** Sagittal T2-weighted (TR/TE 2,000/80 msec) preoperative image demonstrates large L5-S1 disk extrusion *(arrow)*. **D:** Sagittal T2-weighted (TR/TE 2,000/80 msec) postoperative image demonstrates virtually complete resolution of the previously identified abnormality *(arrow)*.

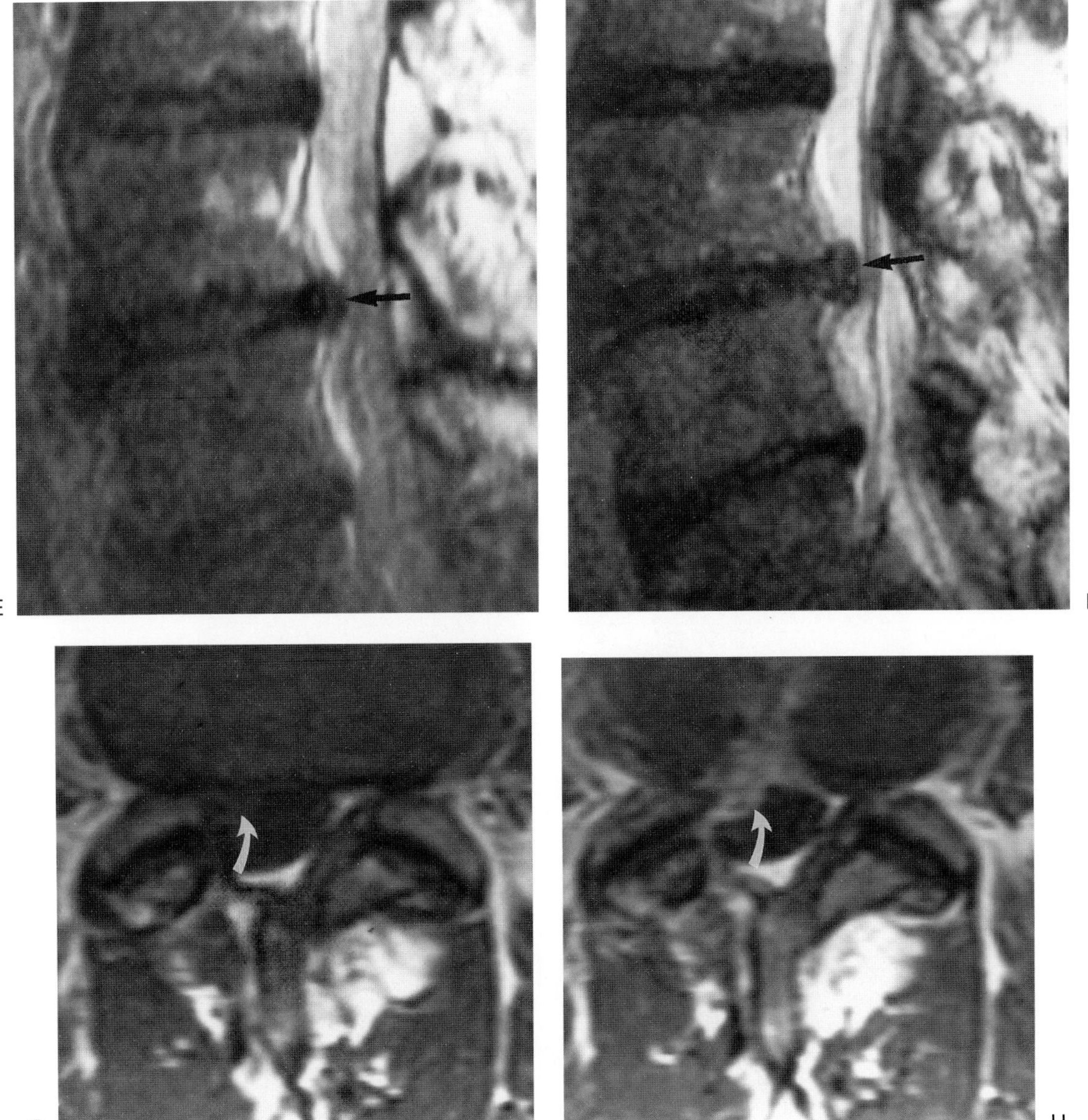

FIG. 5. E,H: Mass-like epidural fibrosis. **E:** Sagittal T2-weighted (TR/TE 2,000/80 msec) spin-echo image obtained preoperatively demonstrates a right paracentral disk protrusion. **F:** Sagittal T2-weighted (TR/TE 2,000/80 msec) spin-echo image obtained 15 months postoperatively on the same individual. The patient was asymptomatic and the study performed as part of longitudinal follow-up. A low-signal-intensity localized epidural mass is seen slightly to the right of midline *(arrow)*. There is mild impression upon the thecal sac. Axial T1-weighted (TR/TE 500/20 msec) image obtained before **(G)** and after **(H)** injection of gadopentetate dimeglumine. There is evidence of avid enhancement with contrast material of the right paracentral mass, which is seen to impress the right anterolateral aspect of the thecal sac *(curved arrows)*. The findings strongly suggest localized and asymptomatic epidural fibrosis with mass effect upon the thecal sac. (From Deutsch AL, Howard M, Dawson EG, et al. Lumbar spine following successful surgical disectomy: magnetic resonance imaging features and implications. Spine 1993;18:1054–1060, with permission.)

CASE 6

History: A 46-year-old woman with a history of a prior laminectomy and diskectomy presents with recurrent S1 radiculopathy for 3 months. Representative images from an MR examination performed are illustrated. What is the most likely diagnosis?

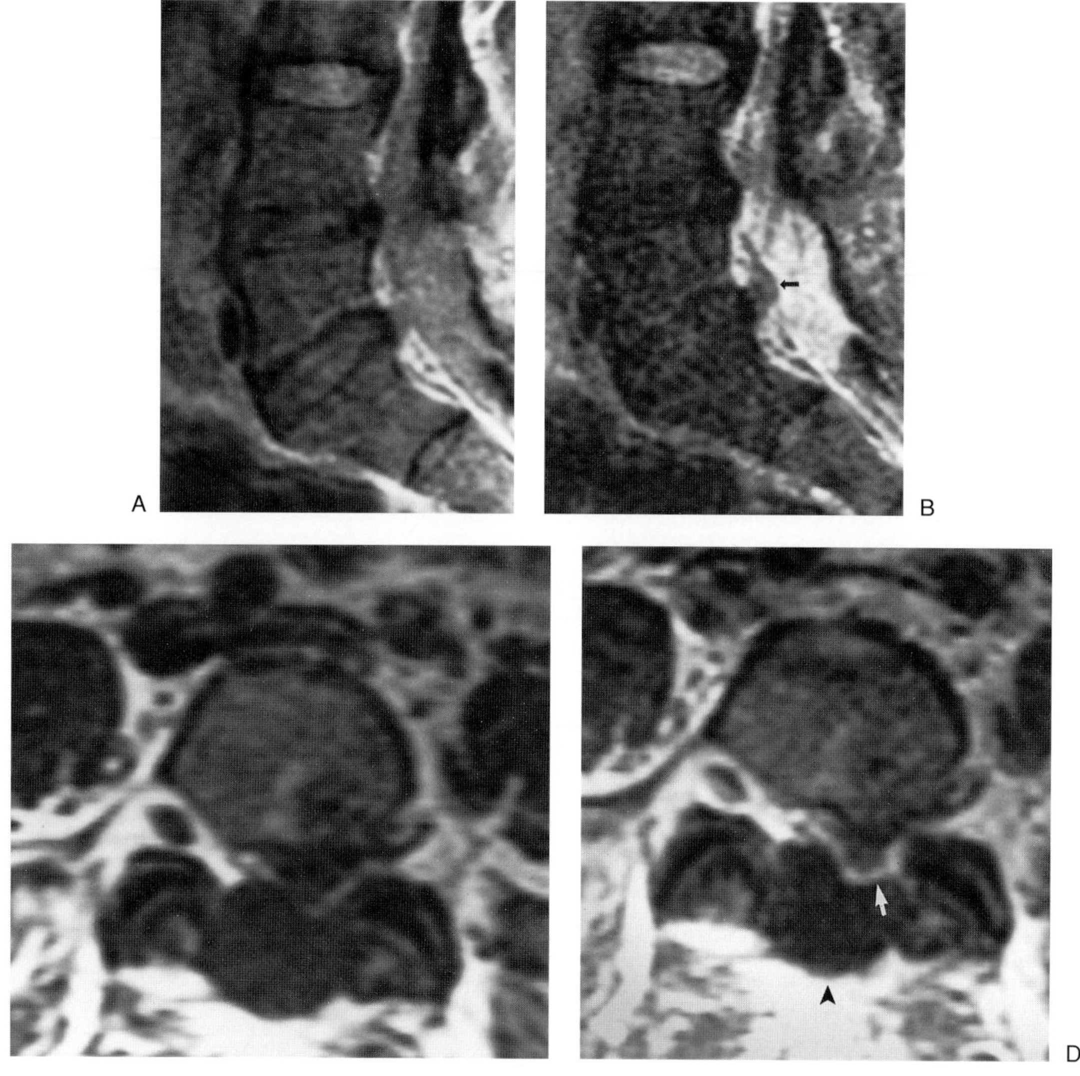

FIG. 6.

Findings: Sagittal proton-density (TR/TE 2,200/20 msec) and T2-weighted spin-echo (TR/TE 2,200/80 msec) images (Figs. 6A,B) demonstrate a localized extradural "mass" that is mildly hyperintense in comparison with cancellous bone, as seen on the T2-weighted sequence. The mass is juxtaposed to the intervertebral disk at the L5-S1 level (black arrow). Axial T1-weighted (TR/TE 600/20 msec) spin-echo images (Fig. 6C,D) before and after administration of gadopentetate dimeglumine demonstrate an extensive bilateral laminectomy and a mild posterior protrusion of the thecal sac (arrowhead). The aberrant soft tissue mass within the left paracentral epidural space demonstrates peripheral enhancement, suggesting peridiskal scar (white arrow). No central or homogeneous enhancement is identified.

Diagnosis: Recurrent free disk fragment.

Discussion: It has been estimated that 80% of the U.S. population will at some time experience significant back pain (1). In patients between 20 and 50 years of age, low back pain has been estimated to represent the most expensive component of health care (2). Despite significant advances in both diagnosis and therapeutic measures, less than satisfactory results are not uncommonly encountered following surgical intervention for lumbar disk disease; the term *failed back surgery syndrome* (FBSS) has been employed to encompass a syndrome characterized by intractable pain and various degrees of functional incapacitation following surgical removal of a herniated lumbar intervertebral disk (3). The most common causes of FBSS include epidural fibrosis, recurrent/residual disk herniation, herniation at another spinal level, and arachnoiditis, as well as lateral recess, foraminal, and/or central spinal stenosis (4). Other less common causes of persistent symptoms following surgical diskectomy include surgery inadvertently performed at the wrong side or wrong level, direct nerve injury at the time of surgery, chronic mechanical pain (e.g., facet joint disease), fusion failure, anterior disk protrusion, radiculitis, annular tears (type II), diskitis and epidural abscess, synovial cysts, epidural hematoma, facet joint fracture, and spondylolisthesis (4–7).

Progressive improvements in MR imaging have substantially advanced the critical analysis of the postoperative lumbar spine. Optimization of technique is critical to maximize the potential of the examination. Epidural scar has been demonstrated to enhance immediately after the injection of contrast material, although maximal enhancement is not observed until 5 to 6 minutes after contrast administration (8,9). This enhancement occurs regardless of the time that has elapsed since surgery and has been observed as long as 20 years after operation (9). The basis for differentiation of scar from disk utilizing contrast relates to nonenhancement of disk material on early images as a consequence of its avascular nature, in contrast to the immediate enhancement of scar tissue. It must be emphasized, however, that some disk herniations can be vascularized and enhance relatively early (see Case 5-10). Imaging within 2 minutes of contrast administration is desirable to optimize differentiation of scar from possible vascularized disk material. The use of fat suppression has also been demonstrated to facilitate the conspicuity of scar enhancement (10,11). Imaging beyond 20 to 30 minutes after contrast administration is of limited value, since disk material may enhance due to diffusion of contrast material into the disk from adjacent vascularized scar. In cases in which a combination of scar and disk material is present, the scar will enhance and the disk remain nonenhanced on early postcontrast administration images (Fig. 6E,F). It is again emphasized, however, that disk material will consistently enhance if sufficient time elapses to allow for diffusion of contrast material from the surrounding scar (Fig. 6G–J). The presence of a vascular supply, a pathway for contrast extension from the blood supply, and sufficient interstitial space to contain the contrast material are necessary for contrast material enhancement of any tissue (8,9). Histologic investigations have demonstrated that epidural scar possesses all these qualities, containing both abundant vascularity and extensive interstitial space. Contrast material may gain access to scar material via "leaky" junctional complexes between capillary endothelial cells or potentially through areas of endothelial discontinuity that permit its passage from the vascular lumen to capillary basement membrane (8,9).

The examination technique utilized differs among various institutions. It is emphasized that both pre- and postcontrast T1-weighted images must be obtained in the axial plane and that the patient not be moved between the acquisitions to allow for the most optimal comparison. The same window and level settings should be utilized for analysis of the pre- and postcontrast images to facilitate assessment of enhancement. As a consequence of the aforementioned considerations, the images obtained following contrast administration must be obtained without significant delay. Since the peak of contrast enhancement is not observed until approximately 5 minutes after injection, it is probably best if image acquisition is not completed within the first minute. Vascular pool images should be obtained initially in the axial plane. This can be accomplished utilizing either axial fat-suppressed short TR/TE fast spin-echo or fat-suppressed T1-weighted gradient-echo [i.e., spoiled gradient-recalled acquisition in the steady state (GRASS)] sequences, at 1.5 T within 2 minutes and 9 seconds of delivery of 0.1 mmol/kg of intravenous contrast material. "Delayed" enhanced spin-echo images can be acquired between 6 and 15 minutes after contrast material delivery. As previously empha-

TABLE 2. *Sample protocol for postoperative lumbar spine examination*

Protocol	TR/TE/matrix/NEX
1. Sagittal short TR/TE	300/16/256 × 192/4
2. Axial long TR/short and long TE (fast) level L1/L2–L5/S1	3,500/17,102/256 × 256/2
3. Axial short TR/TE Two disk spaces	600/11/256 × 192/2
4. Axial dynamically enhanced high-resolution (fast) Fat-suppressed, postcontrast One disk space Completed in 2 minutes of contrast delivery	500/17/512 × 256/4
5. Axial short TR/TE Postcontrast Two disk spaces	600/11/256 × 192/2.0
6. Sagittal short TR/TE Postcontrast	300/16/256 × 192/4

Remarks
1. Sagittal postcontrast sequence is obtained approximately 15–18 minutes and axial postcontrast sequence 8–12 minutes after contrast administration.
2. All sequences are conventional spin-echo unless otherwise noted.

sized, imaging after 20 to 30 minutes is not helpful because many herniations enhance in this time frame. A sagittal fat-suppressed T1-weighted image obtained within 20 minutes of contrast material injection can be useful to avoid partial volume averaging of scar and herniated disk and to highlight annular tears (particularly type II tears). A sample comprehensive protocol from one of our institutions for obtaining contrast-enhanced MR studies is listed in Table 2.

The lack of early contrast enhancement of disk herniations on T1-weighted images has been claimed to be the major criterion by which scar and disk are distinguished (12). In addition to the time of acquisition following contrast administration, the degree of enhancement of scar material is influenced by the "age" of the scar as well as its location. More recently developed scar (i.e., granulation tissue) enhances more intensely than established scar, and the extent of scar enhancement also differs by epidural location (less avid enhancement of posterior epidural scar) (13,14). The accuracy of contrast-enhanced MR imaging in the differentiation of herniated disk and epidural scar is 96% to 100% (12,15). This reflects an incremental improvement over noncontrast-enhanced studies, the diagnostic criteria for which are reviewed in Case 5-5. While the differentiation of scar versus disk represents the principal and most common indication for the use of gadopentetate dimeglumine in the postoperative spine, contrast enhancement may also be invaluable for assessment of spinal, leptomeningeal, and/or neural inflammation (infections or aseptic) or neural degeneration (16,17).

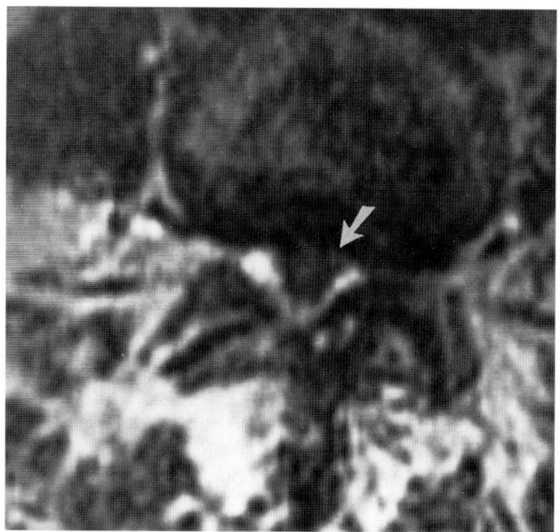

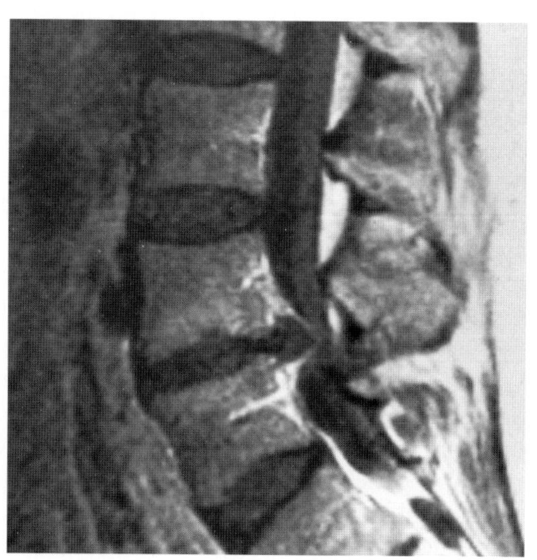

FIG. 6. E,F: Avascular disk. **E:** Axial T2-weighted (TR/TE 2,500/80 msec) spin-echo sequence demonstrates a central low-signal-intensity free fragment with marked mass effect on the adjacent dural tube *(arrow).* **F:** Sagittal contrast-enhanced T1-weighted (TR/TE 600/20 msec) spin-echo image demonstrates subtle enhancement of peridiskal scar with lack of enhancement of the fragment itself. Note enhancement of the epidural venous plexus. There is marked central canal stenosis.

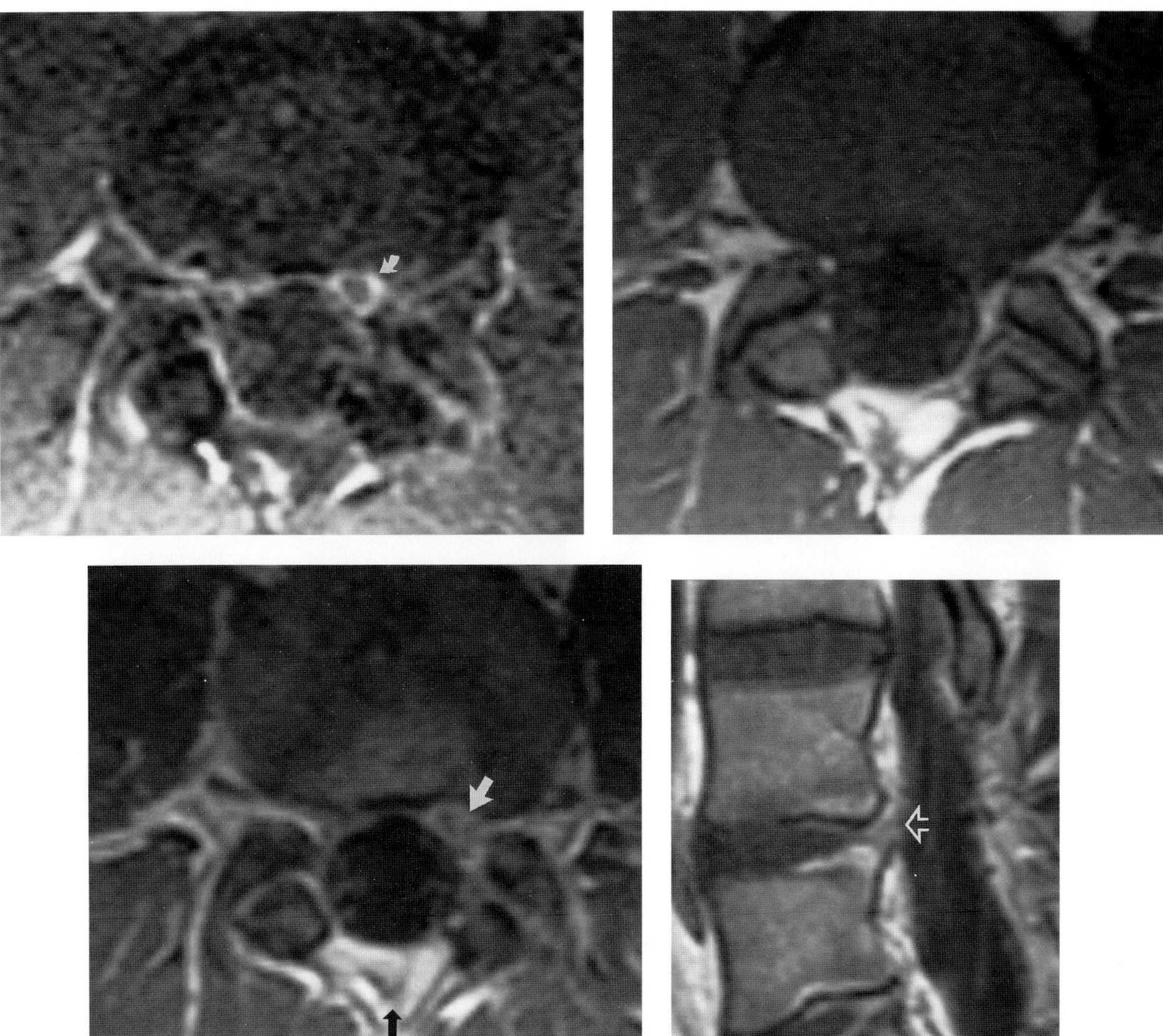

FIG. 6. G–J: Enhancement of a free disk fragment. Progressive enhancement of a disk herniation with time. On the initial image **(G)** obtained utilizing a fat-suppressed spoiled GRASS sequence (TR/TE 55/5 msec, 60° flip angle), there is evidence of a left paracentral free fragment at the L4-5 level with peripheral but no central enhancement *(white curved arrow)*. This sequence was obtained within 2 minutes and 9 seconds of the bolus administration of contrast material. On axial T1-weighted (TR/TE 600/11 msec) images obtained before **(H)** and within 10 minutes after contrast injection **(I)**, there is evidence of thickening of the rim of contrast enhancement *(white arrow)*. There is also evidence of a bilateral laminectomy *(black arrow)*. **J:** Postcontrast enhanced sagittal T1-weighted image (TR/TE 400/20 msec) image (obtained within 16 minutes of contrast delivery) demonstrates marked peripheral and moderate central enhancement of this free disk fragment *(white open arrow)*. This represents an example of seepage of contrast from the peridiskal scar into the nonvascularized herniated disk by diffusion.

CASE 7

History: A 57-year-old man presented with right-sided radicular symptoms and difficulty in walking. What is your diagnosis? What are possible treatment options for this condition?

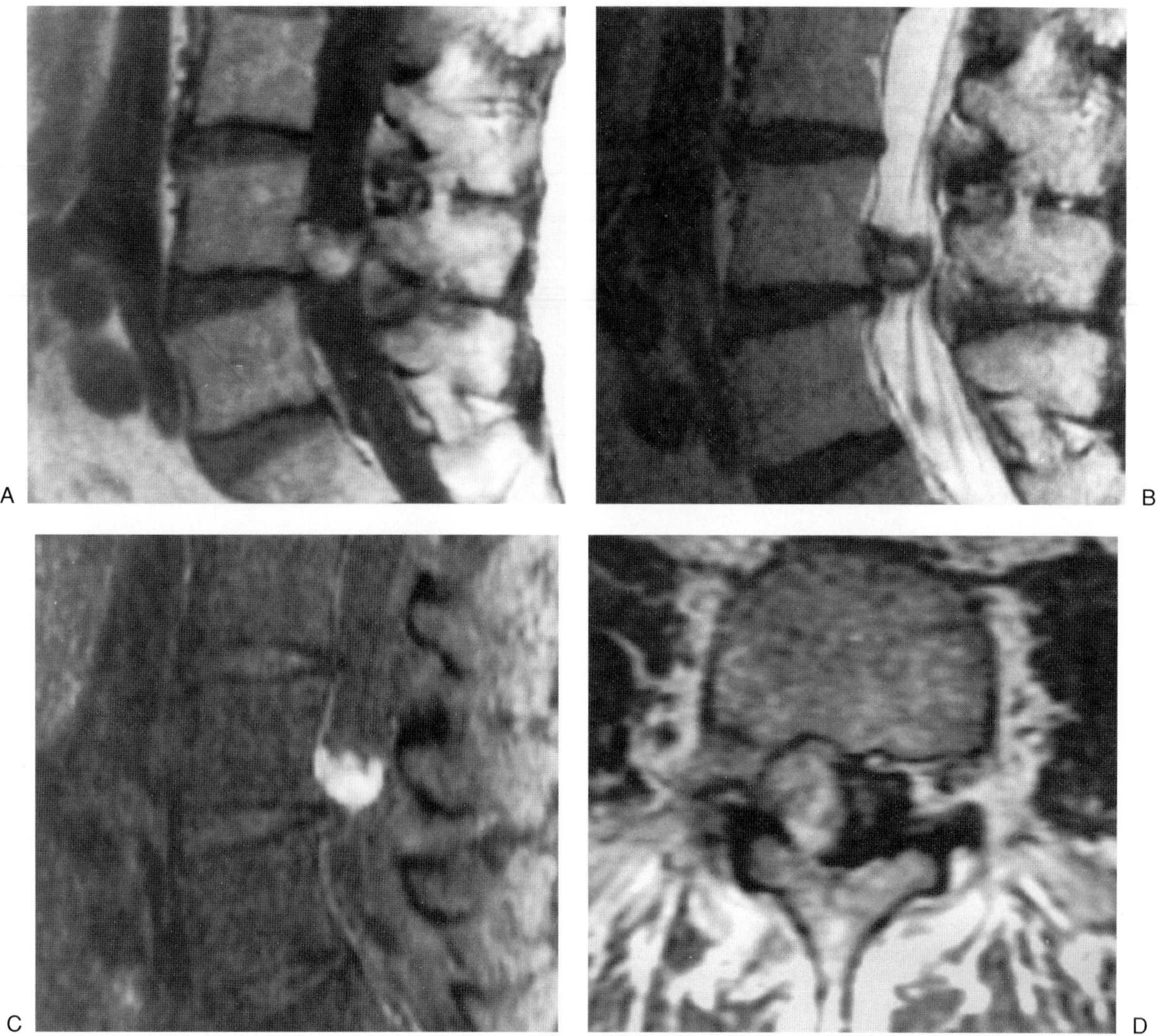

FIG. 7.

Findings: Sagittal T1-weighted fast spin-echo (TR/TE 600/16 msec) image (Fig. 7A) demonstrates a high-signal-intensity ''mass'' projecting slightly rostral to the L4-5 intervertebral disk space. Sagittal T2-weighted fast spin-echo (TR/TE 3,000/112 msec, eff) image (Fig. 7B) demonstrates that the ''mass'' diminishes in signal intensity within its periphery, with a smaller central area of continued increased signal intensity. Sagittal fat-suppressed T1-weighted fast spin-echo (TR/TE 700/16 msec) image (Fig. 7C) demonstrates that the mass remains of increased signal intensity. Axial T1-weighted fast spin-echo (TR/TE 500/14 msec, eff) image (Fig. 7D) demonstrates that the mass occupies the right side of the intervertebral canal. On a slightly more caudal section, the mass is in apparent continuity with the right L5 facet joint.

Diagnosis: Synovial cyst related to the right L4-5 facet joint. High signal within the cyst is presumably related to proteinaceous material or, less likely, subacute hemorrhage.

Discussion: Synovial cysts and synovial chondromas (localized noncystic focal thickening of the ligamentum flavum) are highly associated with degenerative spondylolisthesis (1). Synovial cysts most commonly arise at the medial margin of the degenerated facet joint. They have a characteristic appearance, with a flat base paralleling the ligamentum flavum at the medial margin of the zygo-apophyseal joint (2). The rounded medial aspect of the cyst projects in an anteromedial direction, in contrast to the dorsal extension of disk extrusions. Uncommonly, synovial cysts can vary in position and extend into the lateral aspect of the nerve root canal beneath the lateral attachment of the ligamentum flavum to the base of the superior articular facet. It is important to distinguish synovial cysts from synovial condromas, since the cysts are potentially amenable to percutaneous therapy with steroid injections. This can be performed either via the adjacent apophyseal joint or by direct injection (Fig. 7E–H).

On MR, synovial cysts typically appear as rounded or ovoid structures extending into the spinal canal. On T2-weighted sequences, the cyst may be defined by a thin, uniform, low-signal-intensity rim that may correlate with the dense perimeter of synovial cysts recognized on CT scanning (3,4). The center of the cyst may be heterogeneous or homogeneously increased in signal intensity on T2-weighted images. Dural and contralateral sac displacement is commonly observed. With the administration of gadopentetate dimeglumine, enhancement may be present along the margins of the cyst. Some cysts may demonstrate increased signal intensity on precontrast T1-weighted images, reflecting subacute blood products within the lesion (4). Most synovial cysts are located at the L4-5 level, which is also the most frequent site of degenerative spondylolisthesis.

Degenerative spondylolisthesis is a common and important cause of acquired central and subarticular recess stenosis (2,5–8). The condition is characterized by marked degenerative hypertrophic facet arthropathy, thickening of the ligamentum flavum, and erosive changes of the mesial aspect of the superior articular facet (SAP) at the involved level. It is the erosive change of the supportive SAP of the caudal vertebral body that facilitates ventral subluxation of the intact neural arch of the cephalic vertebral body, with resultant constriction of the central canal (2). The compression of the thecal sac occurs between the lamina and degenerated inferior articular process dorsally and the posterior aspect of the involved disk and caudal vertebral body ventrally (2). Diffuse bulging or protrusion of the disk in association with the hypertrophy of the ligamentum flavum is responsible for most of the neurocompression.

Degenerative spondylolisthesis occurs most commonly at the L4-5 level. The anterolisthesis rarely progresses beyond a grade I slip due to the intact neural arch. Bony lateral spinal stenosis is relatively unusual in these patients because the pedicles of L4 arise relatively more cephalad on the vertebral body than the pedicles of L5. As a consequence, diminution in disk height alone at the L4-5 level does not ordinarily produce cephalocaudal bony lateral exit stenosis (2). When bony exit stenosis does occur, it results from a rotary subluxation. The exit stenosis occurs on the side with anterior subluxation of the SAP that rotates into the intervertebral exit canal with the compression occurring between the cephalic margin of the hypertrophied SAP and the undersurface of the pedicle above (2). While osteophytic spurring causing neural impingement in the exit canal is unusual with degenerative spondylolisthesis, ''up-down'' stenosis secondary to cephalically oriented disk bulging or more focal protrusion accompanied by marked thickening of the ligamentum flavum at the anterior margin of the SAP as it attaches to the undersurface of the pedicle is common.

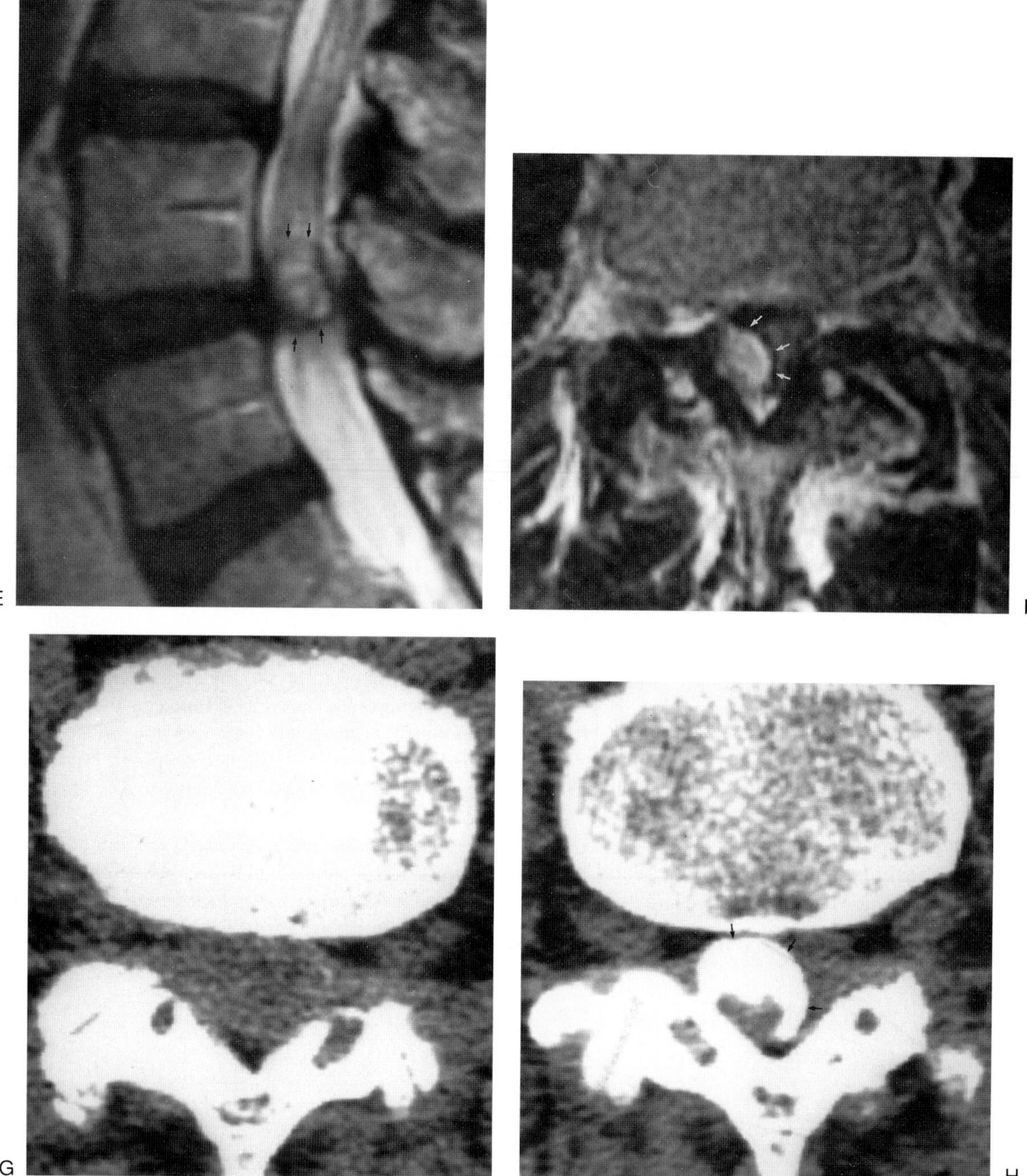

FIG. 7. E: Sagittal T2-weighted fast spin-echo (TR/TE 4,000/96 msec, eff) demonstrates a well-circumscribed circular to ovoid mass *(small arrows)* projecting slightly to the right of midline and apparently effacing the thecal sac at the L4-5 level. **F:** Axial T2-weighted fast spin-echo (TR/TE 3,000/108 msec, eff) demonstrates a focal high-signal-intensity mass *(small arrows)* projecting medial to the right L4-5 facet joint and markedly impressing the thecal sac, which is displaced to the left. **G:** Axial CT section obtained at level corresponding to Fig. 7F demonstrates degenerative changes of the facet joints bilaterally. The previously defined mass is more difficult to appreciate. **H:** Axial CT section obtained after cyst aspiration and corticosteroid injection demonstrates high attenuation contrast material which was injected into the cyst *(arrows)*. The injection was accomplished through the facet joint.

CASE 8

History: A 68-year-old physician underwent diskectomy at L5-S1. The patient developed increasing pain after surgery and 6 weeks later underwent the following imaging study to assess suspected diskitis. What are the imaging features associated with disk space infection? What are the potential difficulties in differentiating this complication from normal postoperative changes in the early period after surgery?

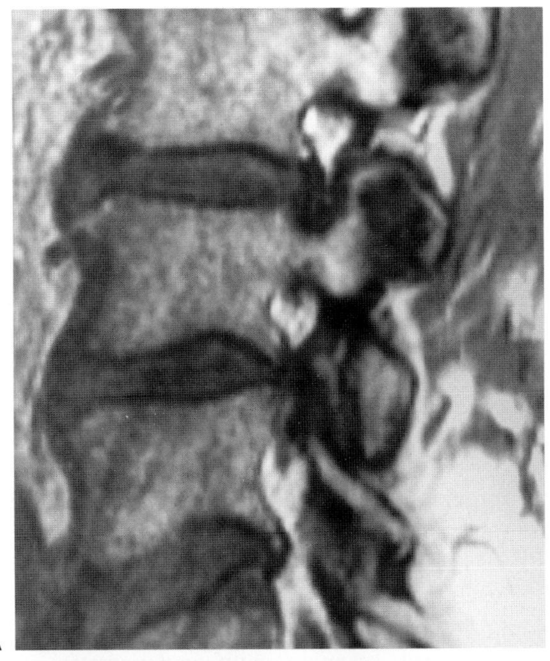

A

FIG. 8.

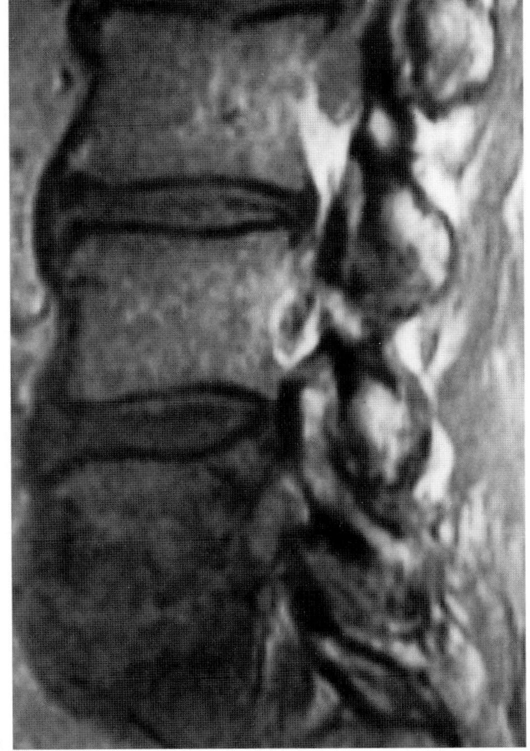

B

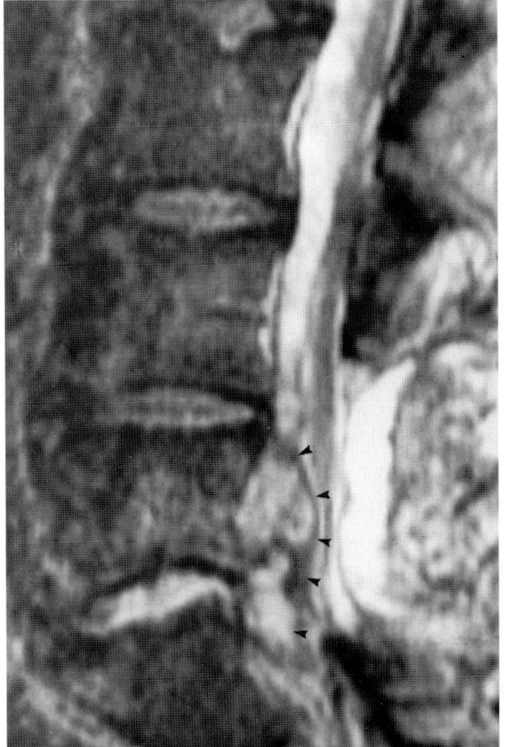

C

Findings: Sagittal proton-density (TR/TE 3,000/17 msec) preoperative study (Fig. 8A) to the left of midline demonstrates a minimal disk protrusion at the L5/S1 level. (The disk was larger to the right of midline.) The disk space is otherwise normal. Corresponding sagittal proton-density (TR/TE 3,000/17 msec) sequence (Fig. 8B) obtained 6 weeks postoperatively demonstrates markedly decreased signal within the inferior one-third of the L5 and superior one-half of the S1 vertebral bodies. There is marked diminution in disk height. Sagittal T2-weighted (TR/TE 3,000/108 msec) fast spin-echo image (Fig. 8C) to the right of Fig. 8B demonstrates marked increased signal intensity in the narrowed L5-S1 disk. The vertebral body end plates in this section appear grossly intact. There is an epidural soft mass extending rostrally from the disk space posterior to the L5 vertebral body (arrowheads).

Diagnosis: Infective spondylitis with epidural abscess.

Discussion: The MR appearance of pyogenic disk space infection is characteristic, and MR has become the preferred imaging modality for diagnosis of this disorder and detection of associated complications including vertebral body osteomyelitis and epidural abscess (1–5). Disk space infection demonstrates diminution in height and confluent diminution in signal intensity of the disk space and contiguous vertebral bodies on T1-weighted images and corresponding increased signal intensity of these structures on T2-weighted images (1–5). The signal alteration involving the vertebral end plates may precede the signal changes that develop in the disk, a finding consistent with the observed pathologic changes in this entity in which infection initially lodges in the bone and subsequently disseminates across the disk space into the adjacent vertebrae via transdiskal collateral vessels (6). The abnormal signal that develops in the intervertebral disk occurs in a nonanatomic distribution with loss of visualization of the internuclear cleft.

While the appearance of disk space infection and infective spondylitis is often characteristic, especially in the appropriate clinical setting, several potential difficulties in diagnosis may be encountered (1). The first of these potential problems relates to signal changes that may be encountered in vertebral body end plates related to degenerative disk disease (1). Initially described by Modic and associates (7,8) type 1 end-plate changes are characterized by band-like decreased signal parallelling the end plate on T1-weighted images, which increases in signal intensity on T2-weighted images (1). The changes may occur in approximately 4% of patients undergoing MR evaluation for disk disease. The histologic basis for these findings appears to relate to replacement of normal cellular marrow with fibrovascular marrow (7,8). Type 1 changes will characteristically demonstrate enhancement following administration of contrast material. Type 2 end-plate changes are characterized by band-like increased signal intensity on T1-weighted images that becomes only slightly increased in intensity on T2-weighted images. These changes, which may be seen in 16% of patients undergoing MR examination, relate to end-plate disruption with yellow marrow replacement in the adjacent vertebral body (7,8).

The presence of type 2 changes can potentially mask the presence of early signal changes in the end plates normally associated with early infection (1). Type 1 end-plates changes demonstrate a similar pattern of signal change as does disk space infection (e.g., low signal on T1, increased signal on T2). Important in the differential diagnosis is the signal intensity of the disk space itself. In degenerative disease, it is characteristically decreased in intensity, whereas with infection, the disk space should be of high signal intensity. When type 1 changes are seen in association with degenerative disks that demonstrate cystic changes, the differentiation of degenerative disease and early disk space infection can be extremely difficult (1). The integrity of the end plates can also be a helpful differential feature. Irregular or focally destroyed end plates obviously favor infection and should be specifically sought.

Infection occurring in the setting of recent spinal surgery may present particular diagnostic difficulties, since postoperative granulation tissue and alteration in normal tissue planes may closely mimic infection. An orderly progression of changes at the operative site has been demonstrated, and deviations from this sequence can be demonstrated by MR. Boden and colleagues (9) studied the usefulness of MR in distinguishing disk space infection from normally observed postoperative changes. Utilizing enhanced scanning with gadopentetate dimeglumine, these investigators found that the combination of disk space, end-plate, and annular enhancement was strongly associated with the presence of disk space infection (9) (Fig. 8D,E). Substantial overlap with normal patients was noted when enhancement of one or more of these components was not seen. It is also important to evaluate the abnormal disk space for the presence of increased T2-weighted signal intensity. This finding is seen with infection but typically not in the normal postoperative patient. The use of fat saturation with T1-weighted postcontrast sequences improves the conspicuity of abnormal enhancement in comparison with nonsuppressed images (10).

Infective spondylitis is the preferred term for a combined infection of the osseous vertebral structures and the disk space. Clinical signs and symptoms at presentation typically have been present for days to weeks depending on the particular organism and route of infection. Nonspecific complaints include fever, weight loss, weakness, and malaise, with symptoms of mechanical back pain, focal tenderness, and radiculopathy. Patients with contiguous spread to the epidural space, a recognized complication of infective spondylitis, may present with severe neurologic findings. MR has been demonstrated to localize the site of infection accurately and provides more information than CT with regard to defining extent of disease in the epidural space and paraspinal region.

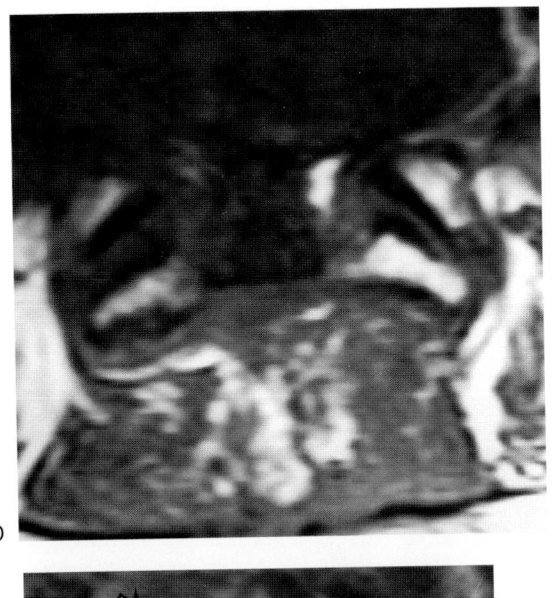

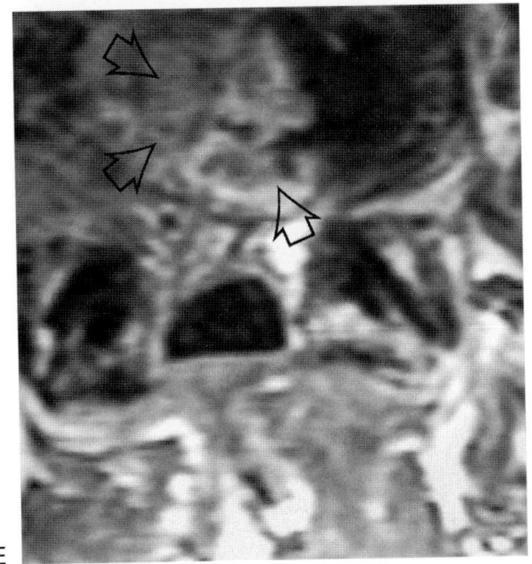

FIG. 8. D,E: T1-weighted (TR/TE 500/20 msec) images obtained before **(D)** and after the intravenous administration of gadopentetate dimeglumine. The study was accomplished on the same patient illustrated in Fig. 8 A–C. There is intense contrast enhancement throughout the entire intervertebral disk space in **E** *(arrows)*.

CASE 9

History: A 46-year-old man presents with a history of left leg pain and signs of left L5 nerve root impingement.

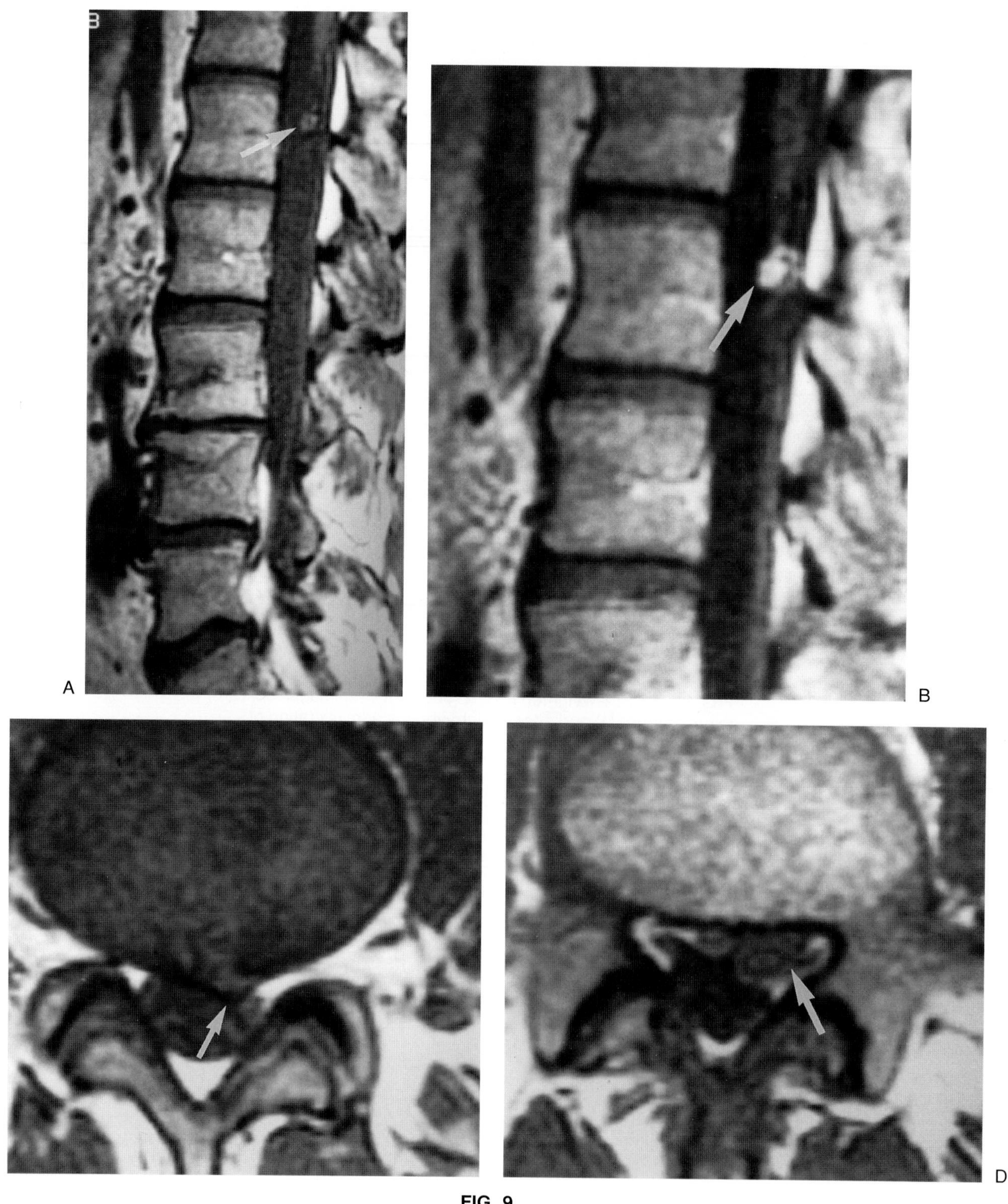

FIG. 9.

Findings: Sagittal T1-weighted (TR/TE 600/20 msec) sequence (Fig. 9A) of the lumbar spine demonstrates discogenic changes involving the lower lumbar spine and a subtle small circular to ovoid focus of increased signal intensity (arrow) projecting within the canal at the level of the distal conus. Detailed view of sagittal T1-weighted (TR/TE 600/20 msec) sequence (Fig. 9B) performed following the intravenous administration of gadopentetate dimeglumine demonstrates the avid contrast enhancement of the mass (arrow). Axial proton-density (TR/TE 1,000/20 msec) image (Fig. 9C) through the L4-5 intervertebral disk space demonstrates a localized left paracentral disk extrusion (arrow). Axial T1-weighted (TR/TE 600/20 msec) sequence (Fig. 9D) at the level of the superior lateral recess of L5 demonstrates the caudally migrated disk material (arrow). (Images courtesy of Michael Brandt-Zawadski, M.D., Newport Beach, California.)

Diagnosis: Ependymoma of the conus medullaris/filum terminale and left L4-5 paracentral disk extrusion with caudally migrated fragment.

Discussion: This case underscores the wide capability of MR to depict the entire gamut of possible pathoanatomic conditions involving the lumbar spine. Lesions involving the conus medullaris and filum terminale may be depicted as an incidental finding on an MR examination obtained for suspected lumbar disk disease, or, alternatively, a conus lesion may account for the patient's originally suspected lumbar pathology (e.g., back pain, radicular pain, unsteady gait, numbness, bladder or bowel dysfunction). Attention should be directed at the appearance of the conus during routine MR examination. In most individuals, the tip of the normal conus lies at or above L2-3 intervertebral disk level and is thus readily accessible to analysis on lumbar MR studies.

Ependymomas are the most common primary neoplasms of the conus medullaris and cauda equina and in one series represented 68% of all conus tumors (2). They occur in a population that is approximately 10 years older than that presenting with astrocystomas of the cord (early forties compared with early thirties) (3). In contrast to astrocytomas, the prevalence of ependymomas increases at progressively lower levels of the spine. One subtype of ependymoma, the myxopapillary type, is particularly common in the conus and represents approximately 30% of ependymomas. In a series from the Mayo Clinic, 65% of myxopapillary ependymomas were limited to the filum, 30% involved the filum and conus medullaris, and only 4% were located in the cervicothoracic cord (4). Myxopapillary ependymomas are often seen in younger adults and children and demonstrate a strong male predominance (3). Presenting signs and symptoms for cauda equina and filum terminale ependymomas include back and leg pain, altered gait, leg weakness, and bowel and bladder incontinence (4). Myxopapillary ependymomas may exude protein into the subarachnoid space, with resulting impaired CSF resorption and communicating hydrocephalus (4).

The MR appearance of ependymomas reflects their wide diversity. Ependymomas are typically hypointense or isointense on T1-weighted images and hyperintense on T2-weighted images. High protein concentration in the subarachnoid space may result in increased CSF signal on T1-weighted images and obscuration of nerve roots and spinal cord margins on T1- and T2-weighted images (4). The lesions may be well marginated and enhance intensely and homogenously with contrast, mimicking meningiomas; the latter entity is, however, far less common in the lumbar region. Alternatively, intramedullary lesions may grossly expand the cord and demonstrate areas of scattered mixed increased and decreased signal probably reflecting components of hemorrhage, cystic change, and calcification. Low-grade myxopapillary ependymomas of the filum terminale may enlarge to fill and expand the lumbar spinal canal (3). Such bulky tumors may extend through intervertebral foramina into the paraspinal region and may resemble lobulated schwannomas or neurofibromas (3). Blood products are frequently present within or adjacent to spinal ependymomas (3). Subarachnoid hemorrhage may occur, and repeated subarachnoid hemorrhage may result in superficial siderosis of the spinal cord and skull base manifest as a thin, low-signal-intensity lining caused by hemosiderin deposition on the surface of the affected structures (4).

In general, avid and homogeneous contrast enhancement is usually present within the solid components of spinal ependymomas (3). While uniform enhancement may also be seen within astrocytomas, in general, poorly enhancing tumors are more likely to be of astrocytic origin (3). In addition, in contrast to astrocytomas, ependymomas are usually encapsulated, with good demarcation from surrounding parenchyma. As a consequence, even extensive tumors can be removed with good preservation of function and excellent prognosis.

In the differential diagnosis, fatty tumors of the conus and filum are readily differentiated. Meningiomas are more commonly located dorsolaterally in the thoracic spine. Astrocytomas are rare at the level of the conus. Intradural metastatic lesions, particulary from lung, breast, and melanoma primaries, should be definite considerations for intramedullary lesions of the conus but should be readily differentiable on clinical grounds.

CASE 10

History: A 50-year-old patient with a history of a prior back surgery presented with progressive weakness and paresthesia for 6 months. What is your diagnosis?

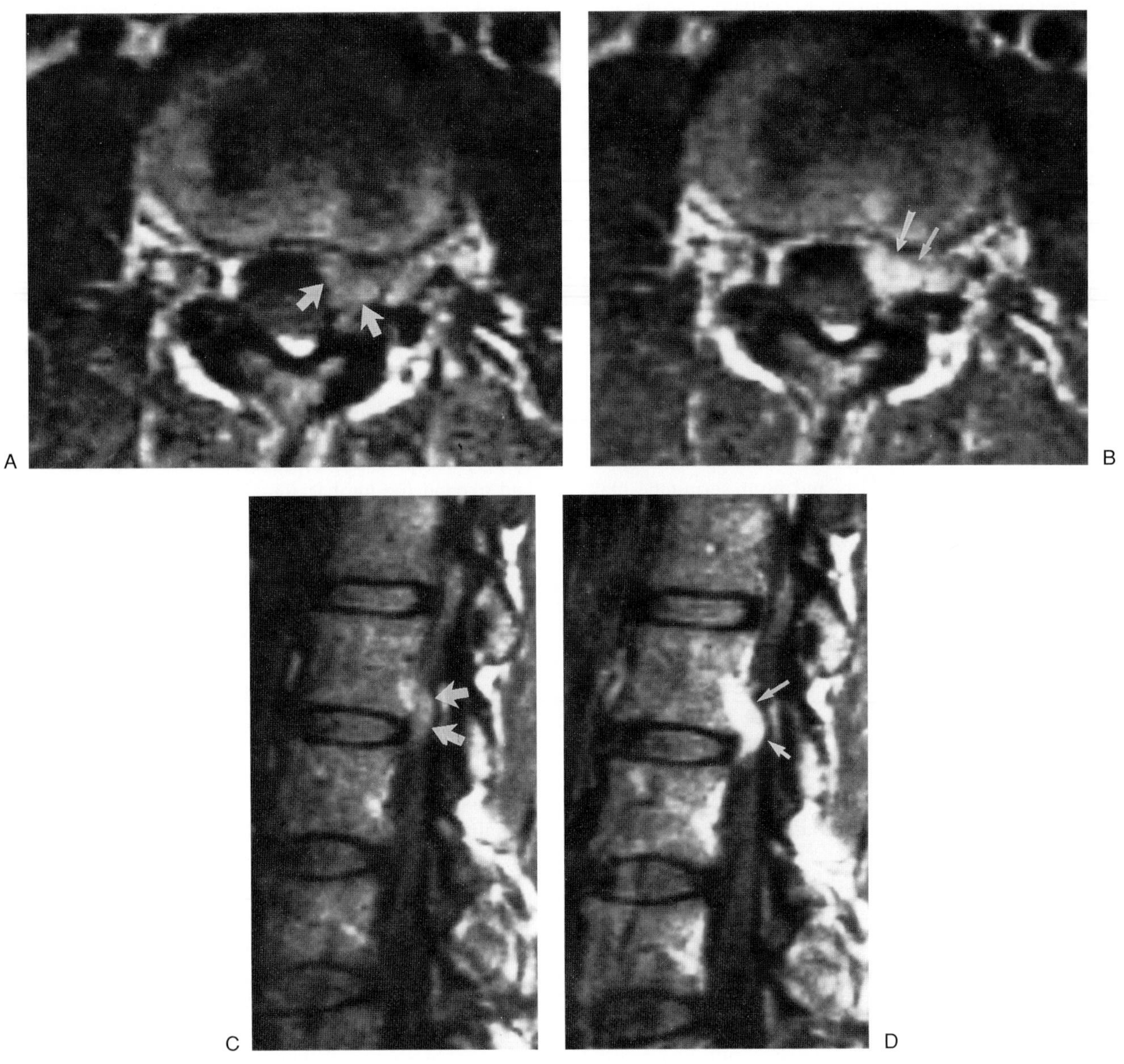

FIG. 10.

Findings: Axial (Fig. 10A,B) and sagittal (Fig. 10C,D) pre- and postcontrast-enhanced T1-weighted (TR/TE 750/35 msec) spin-echo images are demonstrated. The axial postcontrast-enhanced scans were obtained within 10 minutes of contrast administration and the sagittal sequence within 15 minutes following the injection of gadopentetate dimeglumine. On the precontrast images (Fig. 10A,C), there is evidence of a localized mass of increased signal intensity compared with CSF, located paracentrally to the left at the level of the L2-3 disk space (arrows). Following the administration of contrast material, there is avid and essentially homogeneous contrast enhancement of the mass (arrows in Fig. 10B,D). (From Shafaie F, Bundschuh C, Jinkins R. Post-operative lumbosacral spine. In: Edelman R, Hesselink J, eds. *Clinical magnetic resonance imaging.* 2nd ed. Philadelphia: WB Saunders; with permission.)

Diagnosis: Vascularized free disk fragment demonstrating avid contrast enhancement.

Discussion: The major criterion by which scar and disk are distinguished utilizing contrast MR with gadopentetate dimeglumine has been the lack of early contrast enhancement of disk herniations in comparison with the immediate enhancement of scar tissue (1). Unusual and early disk herniation enhancement therefore becomes problematic, as it may contribute to misdiagnosis of disk as scar. In one investigation it has been reported that up to 20% of disk herniations demonstrate "early" MR imaging contrast enhancement (i.e., within 20 minutes of contrast material infusion) (2). The time from injection of contrast material to the time in which the MR acquisitions are obtained (along with contrast dosage utilized) is an important variable to be considered when interpreting contrast-enhanced MR studies of the lumbar spine. Research in rabbits has shown that normal intervertebral disks enhance at doses greater than 0.3 mmol/kg at 1 to 2 hours (3).

Extracellular space size, degree of vascularity, and endothelial gap junction status are usually the major determinants of contrast enhancement in a given tissue. Neither iodinated contrast material nor gadolinium-DTPA, however, can easily penetrate the normal avascular disk matrix, and therefore rates of diffusion and partition coefficients between fluid compartments may also contribute to the extent of observed enhancement. In the case of herniated disk, however, vascular density is believed to be a more important factor with regard to early contrast or homogenous enhancement characteristics, whereas the size of the extracellular space and the status of the endothelial gap junction appear to be the more significant factors in enhancement of epidural scar (4).

A clear understanding of peridiskal fibrosis is important for understanding the dynamics of contrast enhancement. Peridiskal fibrosis is defined as loose, well-vascularized connective tissue that is often found intimately associated with and partially or completely surrounding a disk herniation. Even though it usually has sharp and definable boundaries with chondroid elements, peridiskal fibrosis is often labeled as part of the herniation by the surgeon, a practice that further contributes to difficulty in gaining precision in diagnosis. Peridiskal fibrosis probably represents a response to inflammatory agents released from herniated disks (e.g., phospholipase A_2 activity) as well as to disk antigens (5,6). It appears to be biologically different from epidural scar secondary to hemorrhage that is often associated with bone removal. Peridiskal fibrosis appears to be responsible for resorption of disk herniations and in particular noncontained disk extrusions (7).

Vascularized herniated disk, in addition to containing chondrocytes as identified utilizing periodic acid-Schiff stain, has a predominantly fibrocartilaginous matrix with granulation tissue in the center of the specimen or embedded within it (Fig. 10E). The various components are occasionally well intermixed. It is likely that a vascularized disk herniation is in a more advanced stage of disk destruction/resorption than is a herniated disk surrounded by peridiskal fibrosis and that the granulation tissue in both is biologically identical. The distinction with regard to imaging is important because a vascularized herniation is more likely to be diagnosed as scar: it can enhance heterogeneously within several minutes of contrast administration (Fig. 10F,G). After 5 minutes, the contrast enhancement characteristic may be relatively homogeneous. A small, nonvascularized disk herniation surrounded by exuberant peridiskal scar may enhance fairly homogeneously within 10 to 20 minutes (presumably secondary to diffusion), and therefore even avascular herniations could be misdiagnosed as scar if the details of contrast administration are not known (Fig. 10H–K). False-negative examinations on enhanced MR (with respect to missing the diagnosis of a vascularized or small nonvascularized disk herniation) are likely attributable to delays in obtaining the scans past the immediate postinjection period. In a small nonvascularized or large peripherally vascularized disk herniation, the contrast material gradually seeps into the center of the disk from the surrounding vasculature, occasionally masking its presence (8–10). In most cases, however, it appears that the center of the disk herniation will enhance less intensely than the periphery, if imaging is completed within 20 minutes. This peripheral vascularization explains the pattern of peripheral enhancement with subsequent delayed filling by diffusion. Possible patterns of enhancement that may be seen in association with disk herniations include thin or thick peripheral, patchy or linear central, and/or homogeneous (Fig. 10L).

Diskal material enhances by diffusion of extravascular contrast material assisted by highly vascular peridiskal fibrosis or vessels in the periphery of a herniated disk. It should be noted, however, that vascularity in the herniated disks has only been identified in certain phases of disk lesions (11). Some studies suggest that vascularity can be visualized in the periphery of extruded disks in patients with acute symptoms (i.e., 2 to 4 weeks) (3). Vessels have been identified in the center or diffusely throughout specimens in patients with longer term symptoms (i.e., 3 to 6 months) (2). These observations may be explained by the hypothesis that vascularization develops with time after herniation and grows from the periphery to the center of the specimen. While the pathophysiology remains to be completely elucidated, the stimulus for vascular proliferation and inflammation may be potent inflammatory agents arising from the intervertebral disk, an autoimmune response to degenerating cartilaginous fractions, or a combination thereof (5,12,13).

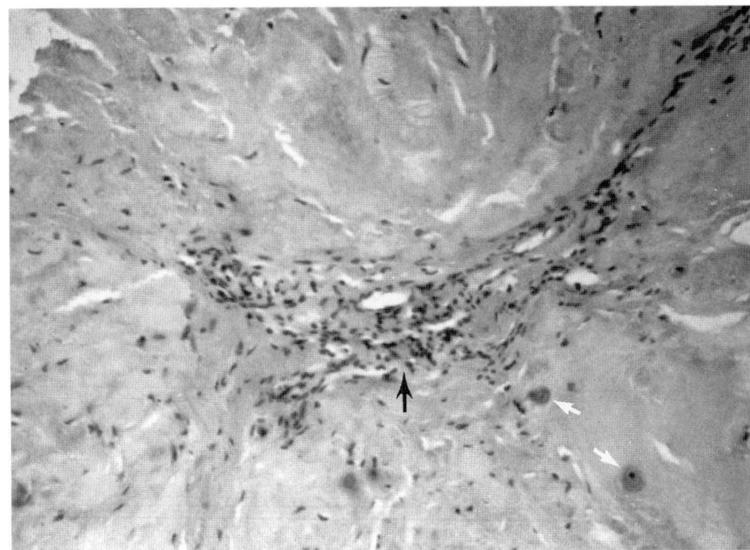

E

FIG. 10. E: Hematoxylin and eosin light photomicrograph demonstrates fibrocartilage within a central island of vascular granulation tissue *(black arrow)*. Several chondrocytes are indicated by white arrows. Original magnification ×100. (From Shafaie F, Bundschuh C, Jinkins R. Post-operative lumbosacral spine. In: Edelman R, and Hesselink J, eds. *Clinical magnetic resonance imaging.* 2nd ed. Philadelphia: WB Saunders; with permission.)

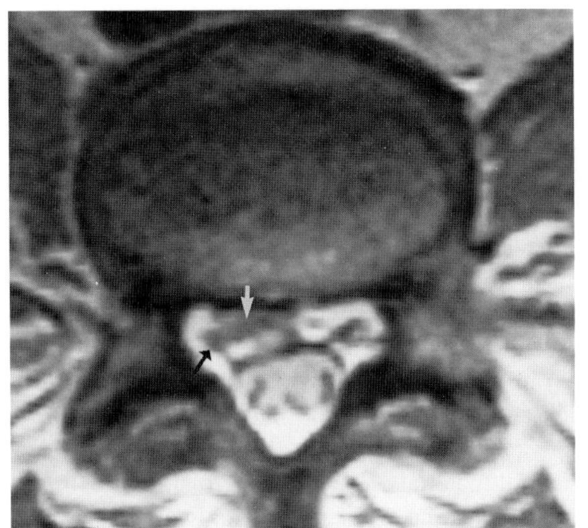

F

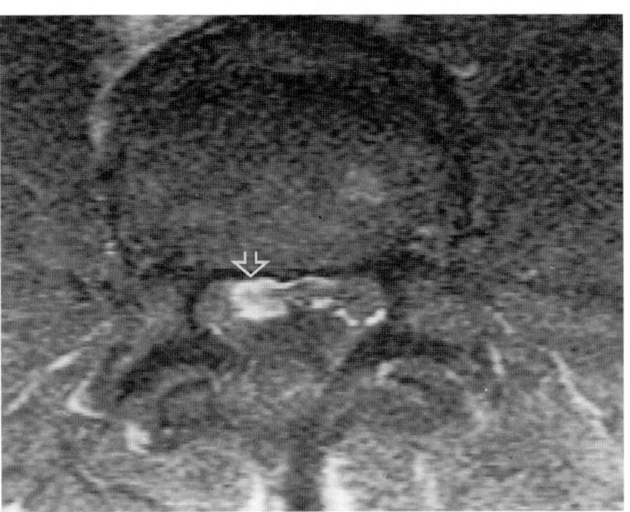

G

FIG. 10. F,G: Vascularized free disk fragment. **F:** Axial proton-density fast spin-echo (TR/TE 4,000/18 msec) sequence demonstrates a heterogeneously hyperintense mass within the anterior epidural space with mild mass effect on the dural tube. The mass *(white arrow)* is contiguous with the right L5 nerve root sleeve *(black arrow)* within the lateral recess. **G:** Dynamically enhanced high-resolution T1-weighted (TR/TE 600/18 msec) fat-suppressed fast spin-echo sequence (completed within 2 minutes of contrast administration) demonstrates the mass within the right anterior epidural space with marked peripheral enhancement and heterogeneous central enhancement *(open arrow)*. This enhancement pattern is typical of a vascularized herniated disk fragment.

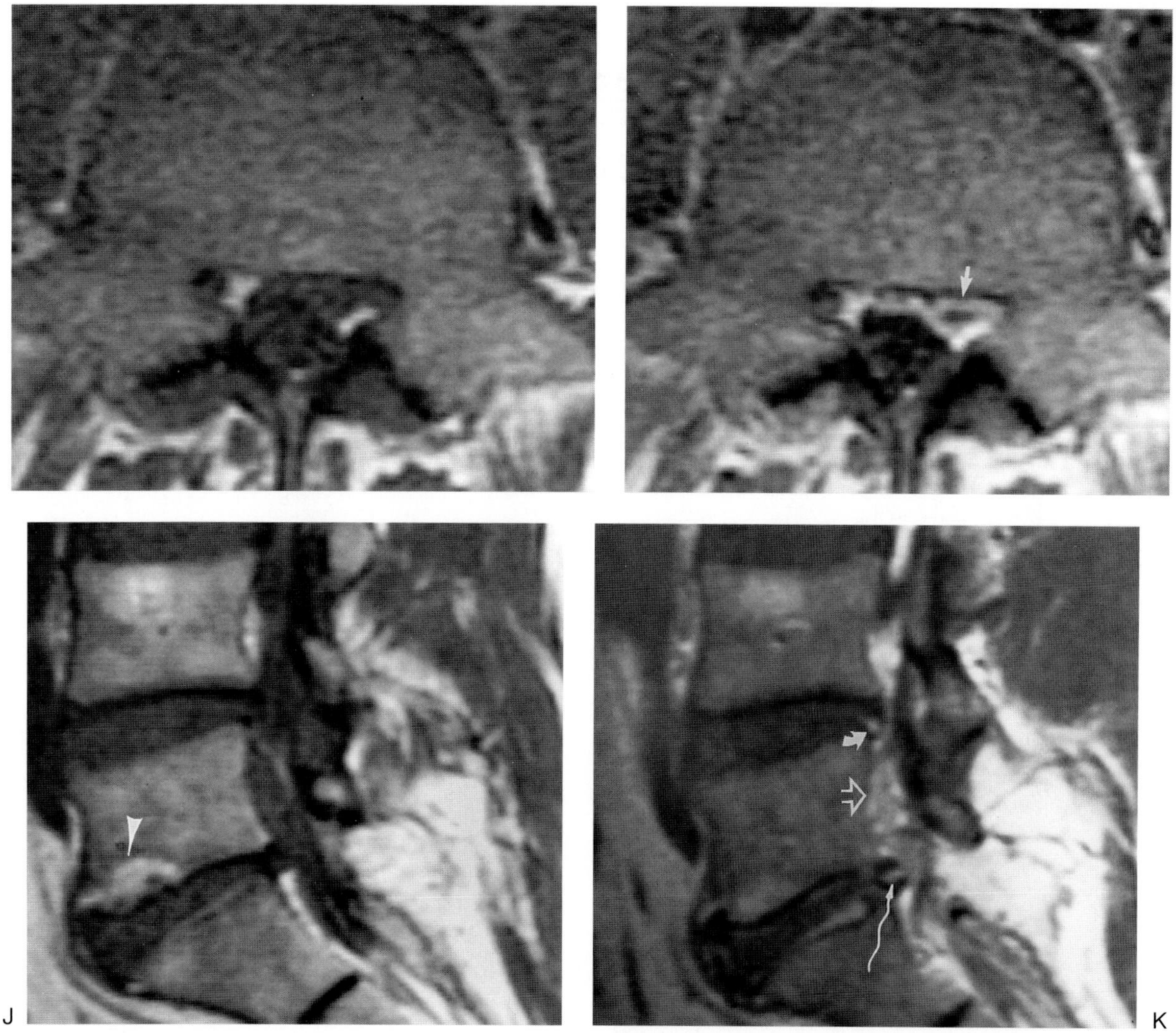

FIG. 10. H–K: Avascular herniated disk enhancement by diffusion. Axial T1-weighted (TR/TE 600/11 msec) spin-echo images before **(H)** and after **(I)** contrast and sagittal T1-weighted spin-echo images before **(J)** and after **(K)** are illustrated. The axial study **(I)** was accomplished within 15 minutes of contrast administration and the sagittal study **(K)** within 20 minutes of injection. On the axial images there is peripheral ring enhancement of the mass lesion within the left anterior epidural space and L5 lateral recess *(closed white arrow)*. On the sagittal images, this free fragment, which extends from the L5-S1 interspace nearly to the L4-5 interspace, demonstrates essentially homogeneous moderate enhancement and simulates epidural scar *(open arrow)*. Note the linear enhancement within the extrusion at the L5-S1 level *(long arrow)* as well as a type II annular tear involving the posterior aspect of the L4-5 intervertebral disk *(closed curved arrow)*. In addition, Modic type II changes involving the inferior aspect of the L5 vertebra *(arrowhead)* as well as anterior disk protrusion associated with pincer-type osteophytes are noted. The phenomenon of disk enhancement by diffusion is well illustrated here.

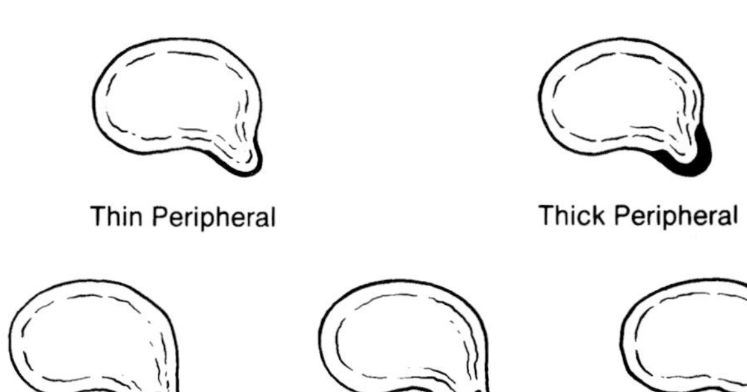

FIG. 10. L: Central and peripheral enhancement patterns of herniated disk fragments. The peripheral enhancement represents peridiscal fibrosis and/or vascularized peripheral herniated disk material. The central enhancement represents enhancement of a vascularized herniated disk (granulation tissue embedded within the center of the disk herniation). In the homogeneous form, there is either diffusion of contrast into the herniated disk or a fairly diffuse mixture of the herniated disk material and the vascularized granulation tissue.

CASE 11

History: A 46-year-old man presented with a 1-week history of severe back and left lower extremity pain. What is your diagnosis?

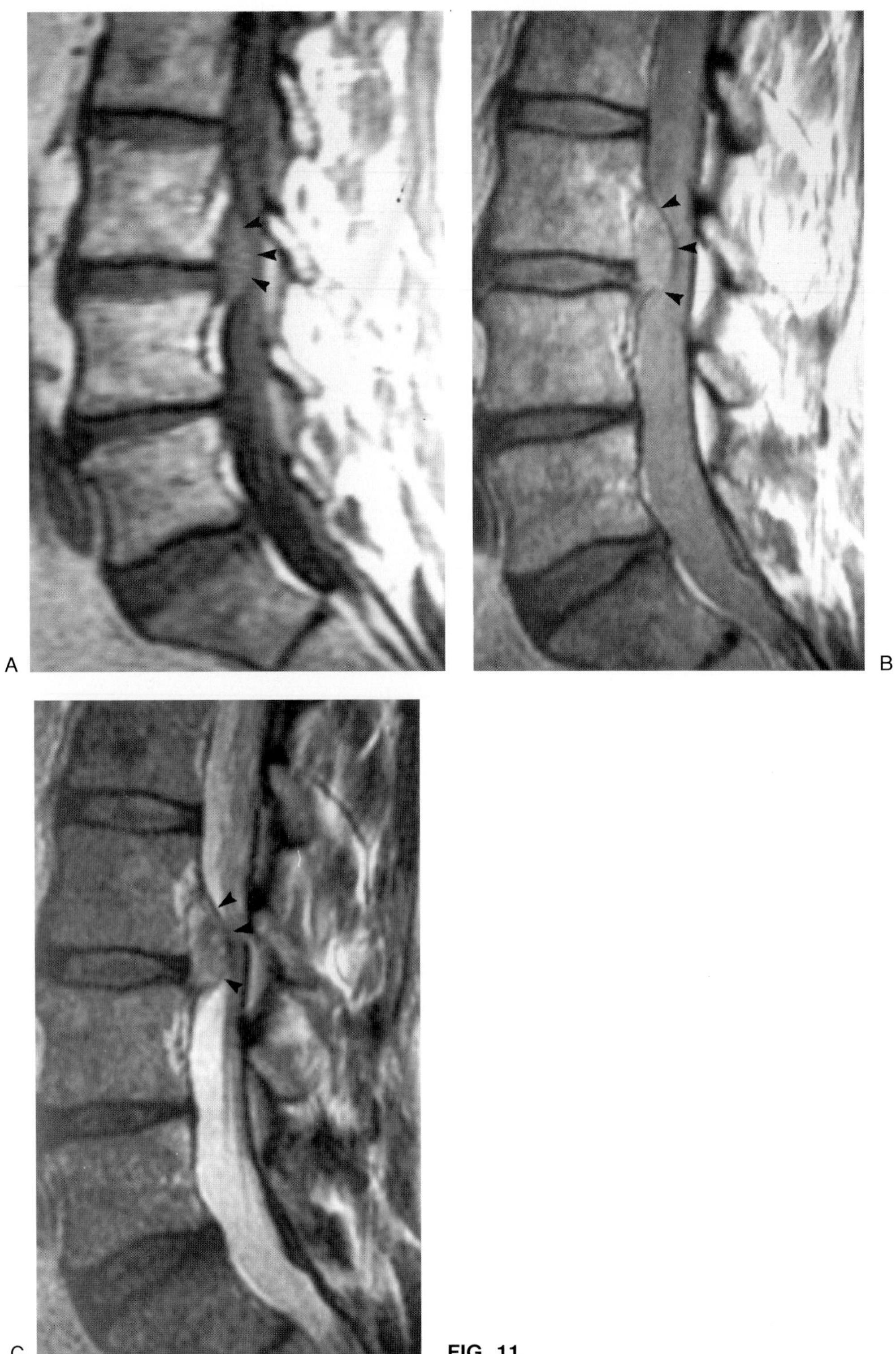

FIG. 11.

Findings: Sagittal T1-weighted (TR/TE 600/20 msec) image (Fig. 11A) demonstrates a well-defined elliptical "mass" of slightly increased signal intensity extending from the level of the L3-4 disk rostrally to the mid-vertebral level of L3 (arrowheads). Sagittal proton-density (TR/TE 3,000/16 msec) image (Fig. 11B) again demonstrates the mass (arrowheads) to be of slightly augmented signal intensity compared with CSF. Sagittal T2-weighted fast spin-echo image (Fig. 11C) reveals the mass to be of mixed signal intensity but predominantly lower than that of CSF. The mass extends from the level of the disk space to the basivertebral plexus level of L3.

Diagnosis: Lumbar localized epidural hematoma.

Discussion: Formerly considered an uncommon event, spontaneous localized epidural hematomas involving the lumbar spine have become increasingly encountered, particularly with the widespread use of MR imaging. Spontaneous epidural hematomas may manifest with clinical symptoms indistinguishable from those produced by acute disk protrusions and extrusions (1–5). In distinction to disk extrusions, symptoms related to epidural hematomas characteristically regress rapidly (e.g., days to weeks) (2). In the vast majority of cases, a small underlying disk protrusion or annular tear can be identified (1,2).

On MR, spontaneous epidural hematomas characteristically appear as ventral extradural masses (of varying sizes but frequently large) displacing the adjacent thecal sac and traversing nerve roots. The vast majority of hematomas are largest in anteroposterior diameter at the mid-vertebral body level immediately adjacent to the basivertebral plexus and taper as they extend to the level of the adjacent disk (1,2). In patients with underlying disk protrusions, cephalic extension of the hematoma is more commonly identified than caudal extension (approximately 70% extend superiorly) (1). Distinction of underlying disk from hematoma is readily accomplished in most cases, with the disk generally exhibiting lower signal intensity. The diskal component of the complex usually accounts for only a minor portion of the total extradural mass effect (typically less than 20%) (1). The signal intensity of the hematoma itself is variable, probably partly related to the age of the lesion, and findings suggestive of subacute blood are encountered in only a minority of cases. The hematoma may take months to resorb and often persists long after the symptoms have remitted.

Gundry and Heithoff (1) have hypothesized that the hematomas result from tearing of fragile epidural veins located along a fibrous membrane that has been described as lying between the posterior surface of the vertebral bodies and the posterior longitudinal ligament. The veins of Batson lie on the dorsal surface of this membrane and perforate through the membrane at multiple locations. Acute disk disruptions may result in shearing of these veins with bleeding contained within the perivascular space of the torn epidural vein, which extends to the basivertebral plexus. Containment of blood within this space creates the characteristic rounded mass effect at the mid-vertebral body level (1).

The possibility of a localized epidural hematoma must be strongly considered when any extradural mass is encountered that is rounded, appears largest at the level of the basivertebral plexus, and demonstrates tapered margins toward the adjacent disk space. While associated with underlying disk abnormalities, these are often minor, and symptoms frequently resolve rapidly with conservative management. It has been suggested that in the past spontaneous epidural hematoma may have accounted for many of the "disappearing" disks described in the myelographic and surgical literature (see Case 5-3). Differentiation of this entity from large disk extrusions with rostral or caudal migration should be possible in virtually all cases.

CASE 12

History: A 36-year-old man with chronic renal failure presented for evaluation of increasing low back pain. What are the principal differential considerations to account for the present findings?

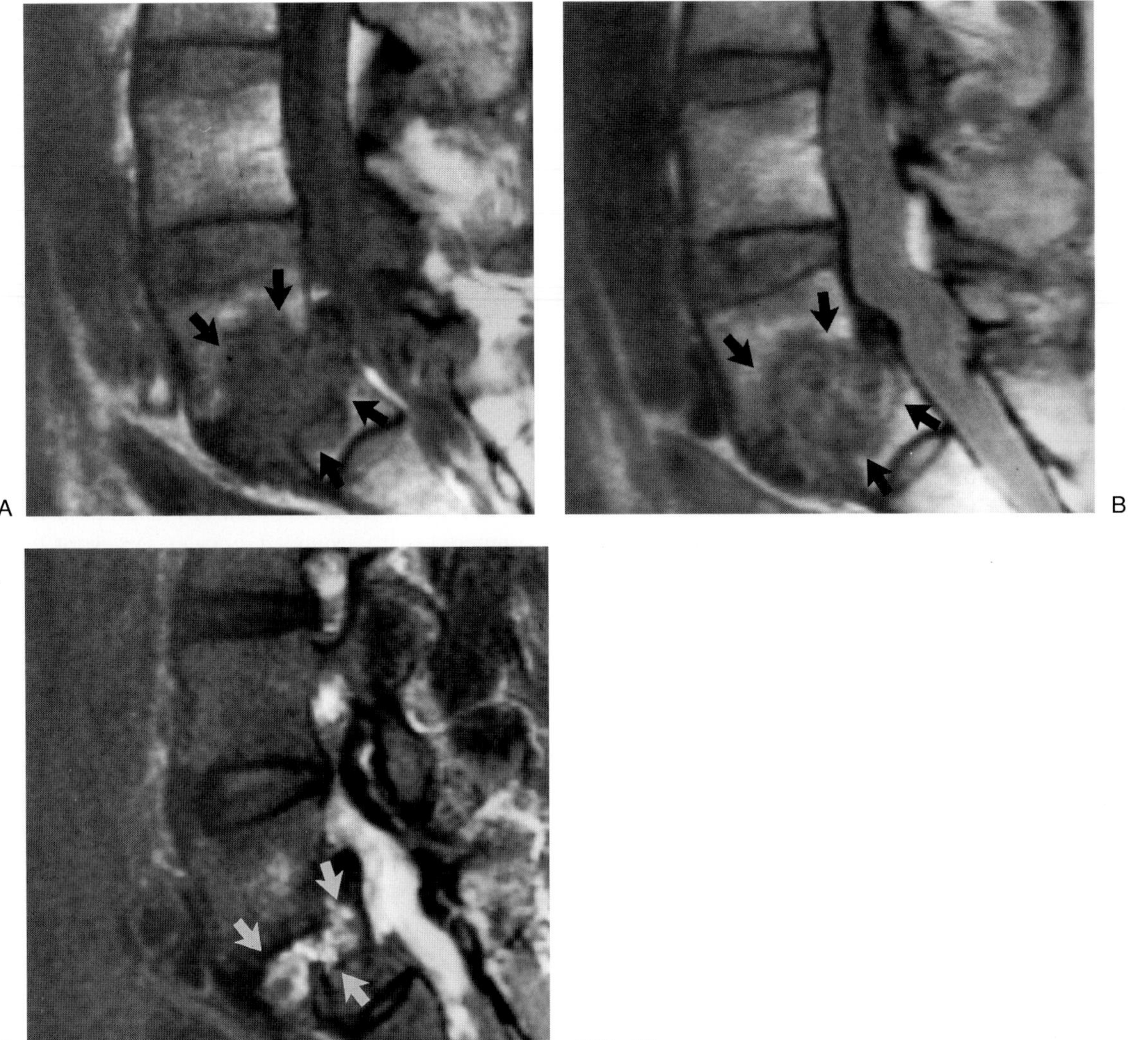

FIG. 12.

Findings: Sagittal T1-weighted (TR/TE 500/20 msec) image (Fig. 12A) through the lower lumbar spine demonstrates a strikingly abnormal appearance of the L5-S1 level characterized by destructive changes involving both end plates and extending to involve both vertebral bodies (arrows). There is a grade 1 spondylolisthesis of L5 in relation to S1. Sagittal proton-density (TR/TE 2,000/20 msec) image (Fig. 12B) demonstrates slight increased signal intensity of the process involving L5-S1 (arrows). Sagittal T2-weighted section (Fig. 12C) slightly to the left of Fig. 12A,B demonstrates markedly increased signal intensity at the level of the L5-S1 disk space (arrows). (Images courtesy of Roger Kerr, M.D., Los Angeles, California.)

Diagnosis: Spondyloarthropathy of renal dialysis.

Discussion: Dialysis-related spondyloarthropathy (DRSA) is a rapidly progressive destructive spondyloarthropathy that may be seen in approximately 5% to 15% of patients undergoing long-term hemodialysis (1–7). Symptomatic DRSA is rarely observed in patients undergoing dialysis for less than 2 years, and the incidence increases with the duration of hemodialysis up to a maximum of 10 years (1). Reported clinical symptoms include sitffness, pain, decreased range of motion, and radiculopathy. Cord compression may be seen in as many of 20% of cases and represents one of the most serious complications of DRSA (6). One-third of patients may be asymptomatic (8). The cervical spine is the most commonly affected site.

DRSA represents a component of a chronic progressive polyarthropathy known as dialysis-related arthropathy (DRA). DRA may affect both large and small joints and indeed it is unusual to encounter DRSA without concomitant DRA of the hands and wrists. DRA is characterized by periarticular cysts and erosions that communicate with the articular surface and may lead to its progressive destruction. DRSA is characterized initially by erosion of vertebral body end plates that may mimic early anklyosing spondylitis (1,3,7,8). Progressive disease is characterized by intervertebral disk space narrowing, subchondral geodes, erosive end-plates changes, vertebral collapse, erosive facet disease, and peridiskal calcification. Osteophytes and chondrocalcinosis are not typical of DRSA and help in distinguishing this condition from degenerative spondyloarthritis and calcium pyrophosphate deposition disease (1,8). Rapid progression over the time course of several months to a year is typical of DRSA, and advanced DRSA commonly involves multiple levels. Patients with clinically symptomatic DRSA may demonstrate no radiographic changes (1,9).

The pathogenesis of DRSA has been extensively studied. Initial reports attributed the condition to the deposition of calcium hydroxyapetite or calcium pyrophosphate within vertebral body end plates (1,3). Iron overload, elevated serum aluminum, secondary hyperparathyroidism, chronic heparin administration, and human lymphocyte antigen haplotype have also been suggested as possible etiologies (1,10,11). Most recently, deposition of an unusual type of amyloid, the major component of which is termed β_2-microglobulin, has been implicated as the cause DRSA (1,9,12). Patients undergoing either chronic hemodialysis or peritoneal dialysis manifest markedly increased plasma levels of β_2-microglobulin. These appear to be related to both poor clearance of this material by dialysis membranes and a possible immune response to the membranes themselves, leading to increased β_2-microglobulin production (1,9). Histologic studies have demonstrated that fibrils of β_2-microglobulin replace collagen fibers in articular cartilage, disk fibrocartilage, and joint capsules (1,9). β_2-Microglobulin deposits have been demonstrated in spine specimens, with vertebral body corner, end-plate, and facet erosive changes (1,13).

Multiple entities must be considered in the differential diagnosis of DRSA. Similar vertebral body end-plate changes may be seen in infectious diskitis, secondary hyperparathyroidism, neuropathic spine, and crystal deposition disease (particularly when associated with large geodes). The lack of extensive vertebral body osteophytes in the presence of marked disk space narrowing and reactive sclerosis favors DRSA over degenerative spondylitis, as does the lack of demonstrable chondrocalcinosis in the presence of extensive geode formation (1). Rapid progression is also characteristic of DRSA and is helpful in differentiation from etiologies other than possible infectious diskitis. Differentiation from infection represents the most difficult challenge. While the presence of high signal within the intervertebral disk space on T2-weighted images has been reported as a finding associated with infection rather than DRSA, as demonstrated in this case, this is not an invariable finding. Biopsy may ultimately be required to make this distinction with any degree of certainty.

CASE 13

History: Preoperative images and sequences from a study performed 1 year following percutaneous diskectomy are illustrated from two different patients. Which patient probably had the successful result? Can the images of the postoperative study be utilized to predict the outcome? What factors have been hypothesized to account for successful results in patients without discernable change in disk size following surgery? (Figure 13A,B shows the pre- and postoperative images of patient 1 and Fig. 13C,D the pre- and postoperative images of patient 2.)

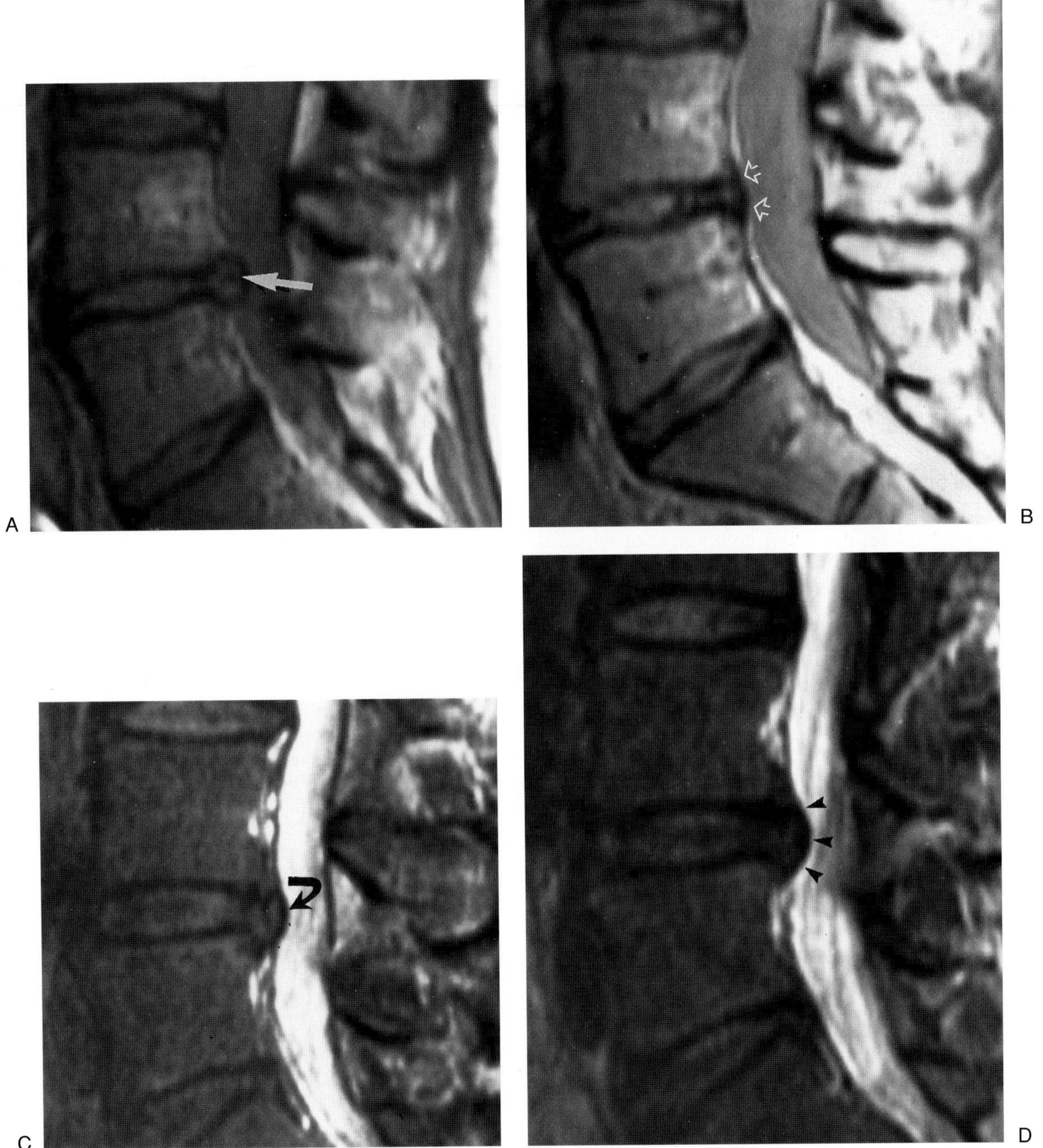

FIG. 13.

Findings: Preoperative sagittal proton-density (TR/TE 2,000/20 msec) image (Fig. 13A) demonstrates a 5- to 6-mm central protrusion of the L4-5 disk (arrow). The postoperative (TR/TE 2,000/20 msec) study (Fig. 13B) demonstrates marked reduction in size of the disk protrusion (small open arrows). Figure 13C is an image of a different patient (TR/TE 2,000/80 msec) demonstrating a contained 4- to 5-mm central protrusion of the L4-5 disk (curved arrow). Postprocedure T2-weighted (TR/TE 2,000/80 msec) study (Fig. 13D) demonstrates virtually no change to slight increase in size of the disk given slight differences in technique and magnification factors (arrowheads).

Diagnosis: Both patients reported highly successful results 1 year after percutaneous diskectomy, and the imaging studies were obtained as part of a longitudinal follow-up study. The appearance of the disk postoperatively is not predictive of procedure outcome.

Discussion: Percutaneous diskectomy has recently received considerable attention as a possible alternative method of treating patients with lumbar disk herniations. Two alternative techniques, one using a manual and the other an automated device were used (1–7). With the manual method, specially designed pituitary rongeurs are used to remove the nuclear material. The automated technique is performed with a pneumatically driven suction cutting probe. With both percutaneous methods, in contrast to open diskectomy, a central or paracentral disk protrusion is not approached directly. To ensure a safe approach to the disk, the automated probe or rongeur is passed anterolateral to the actual protrusion in most patients and is directed into the center of the disk. Most of the removal of the disk material occurs 1 cm anterior to the herniation, although attempts are made to explore the adjacent quadrants.

The reported rates of success of 50% to 85% (2,8,9) for percutaneous diskectomy are somewhat lower than those reported after laminotomy or laminectomy (85% to 95%) (10–12). Advocates of the percutaneous method cite advantages including performance on an outpatient basis, decreased epidural scar formation, avoidance of general anesthesia, preservation of spinal stability, and decreased cost compared with conventional diskectomy (7).

While considerable attention has been directed toward the assessment of symptomatic patients following unsuccessful disk surgery, relative little effort has been directed toward establishing the expected appearance of the disk following either successful conventional or percutaneous diskectomy. Hijikata (1) performed myelograms 2 to 3 weeks following percutaneous diskectomy in 136 patients and found no change in morphology of the disk. Kambin and Schaffer (3) reported no evidence of a herniated disk on CT scans of 16 of 22 patients 2 to 21 months following percutaneous diskectomy. Deutsch et al. (10) and Delamarter et al. (7) reported on a group of 30 patients in whom follow-up studies after percutaneous diskectomy were performed. On studies accomplished a mean of 14 months after the procedure, no changes in the morphology of the disc were noted in 24 patients (80%), irrespective of clinical outcome. Only 3 of 17 who had a successful result had a reduction of more than 2 mm in the size of the protruded disk segment. This study concluded that the outcome of the procedure could not be predicted based on the appearance of the disk on either the preoperative (provided that the patient met the inclusion criteria of a protrusion contained by the posterior longitudinal ligament) or postoperative studies and furthermore that lack of change in disk morphology was common in patients with successful results (7,10).

Three hypotheses have been advanced to explain successful results in patients who demonstrate no discernible change in the morphology of the disk following percutaneous diskectomy (7). The first possibility is that the intervention was too early and that these patients would have had improvement with nonoperative therapy. This possibility is in keeping with the reported natural history of disk herniations (see Case 5-3). This is a legitimate criticism of many studies in which nonoperative treatment has been attempted for only 6 weeks before operative intervention has been taken (7). In the series reported by Deutsch et al. (10) and Delamarter et al. (7), some patients had only a 3-month trial of nonoperative therapy, although the mean duration was 6 months (range, 3 to 24 months).

A second possibility to account for diminution in pain without morphologic disk change relates to a possible alteration in the chemical and humoral milieu in the area surrounding the herniated disk that cannot be appreciated on imaging studies (13–15). Several investigators have suggested that disk material in the epidural space releases chemical mediators of inflammation that produce a direct injury to the nerve root. The highest levels of phospholipase A_2 activity ever recorded in human tissue were found by Saal et al. (16) at the interface of a herniated extruded lumbar disk and the epidural space. Phospholipase A_2 is the enzyme that liberates arachidonic acid from cell membranes; this in turn affects the rate of production of eicosanoids (all prostaglandins and leukotrienes), which are important mediators of pain and inflammation. Other investigators have demonstrated that the dorsal ganglion has a role in the modulation of pain as it relates to each motion segment (17–19). Weinstein et al. (20) reported that neurochemical changes (substance P and vasoactive intestinal peptide) occur with manipulation of the disk and suggested that the pain sometimes associated with abnormal diskographic findings may be related in part to the chemical environment within the intervertebral disk and the sensitized state of its annular nociceptors. It is still not clear how fenestration of the annulus and nuclear lavage performed during a percutaneous diskectomy affect the neurochemical milieu of the disc.

The third possible explanation for the successful results of percutaneous diskectomy in the absence of morphologic changes in the disk is the placebo effect. While such an effect is difficult to exclude in any individual patient, the investigators reporting on these patients consider this an unlikely explanation for most cases. The mechanisms for the less commonly observed decrease in disk size following percutaneous diskectomy also remain to be established. One theory advanced to explain these findings is that removal of the center of the disk reduces the intradikcal pressure so that the herniated segment can fall back in place. This so-called implosion theory suggests that the contour of the disk would return to normal in a patient with a satisfactory clinical outcome (2). This phenomenon was demonstrated in only 2 of 16 patients with paracentral disk protrusions in the studies of Deutsch et al. (10) and Delamarter et al. (7). Indeed, one patient in their series had immediate clinical relief of symptoms following the procedure but demonstrated an actual increase in size of the disk herniation on a follow-up study obtained 12 months after percutaneous diskectomy. The only patient with a lateral foraminal herniation that was directly approached by the percutaneous method had marked reduction in size of the disk and relief of symptoms.

Another proposed mechanism is that of fenestration of the disk with concomitant diminution in intradiskal pressure. Hult (21) reported relief of chronic back and sciatic pain in 26 of 30 patients who had a fenestration of a retroperitoneal disk without a diskectomy. Hult (21) believed that the diversion of intradiskal pressure anteriorly into the defect created by the fenestration was a major factor in the alleviation of symptoms. Kambin and Brager (2) demonstrated a rapid decrease in intradiskal pressure and lateral fenestration of a disk in cadaveric specimens and stated that relief of intradiskal pressure was an important factor in the relief of sciatic pain.

Only patients with disk herniations that appear to be contained by the posterior longitudinal ligament are presently considered candidates for percutaneous diskectomy. MR is valuable in making this determination (22). Beyond this determination, imaging studies have not been demonstrated to be value in predicting the clinical outcome of the surgical procedure. No direct correlation between the size or the location of the herniation was useful in this regard (22). Additionally, follow-up MR studies have demonstrated that a lack of appreciable morphologic change in the disk in patients with successful clinical outcomes is common (7,10). This observation does not support the implosion theory of pain relief nor refute the fenestration theory. The lack of changes in the disk morphology appears to implicate a chemical or humoral effect rather than simple mechanical decompression.

CASE 14

History: A 42-year-old man who complained of low back pain underwent MR examination as part of a workers compensation claim evaluation. The initial examination diagnosed a central disk herniation at the L4-5 level. The study was of poor technical quality, and a repeat examination at another at another institution was performed. What is your diagnosis?

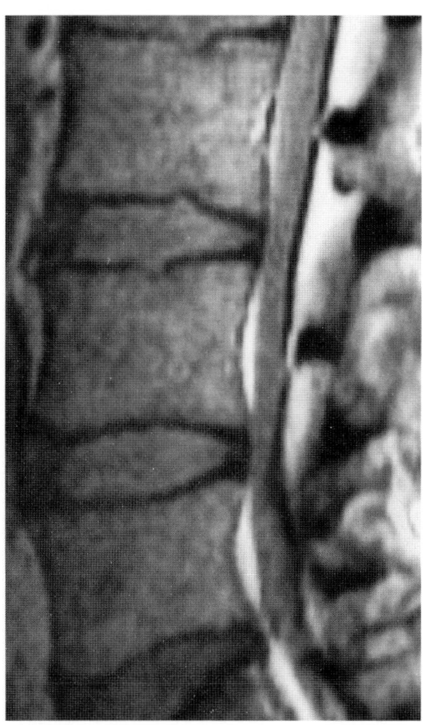

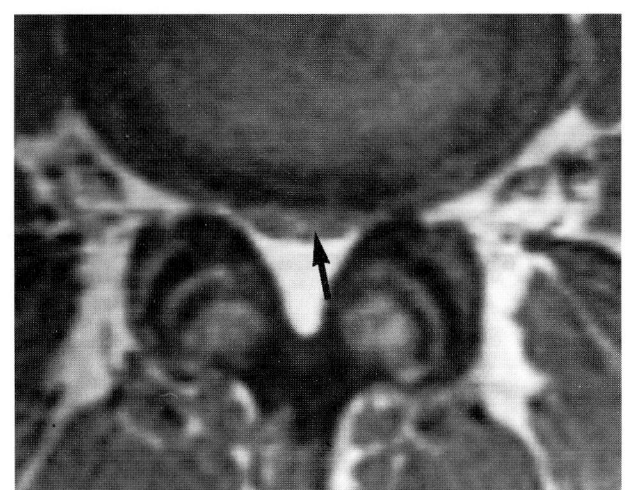

A B

FIG. 14.

Findings: Axial proton-density (TR/TE 2,000/20 msec) image (Fig. 14A) at the L4-5 level showed marked attenuation of the thecal sac (arrow), which appears compressed between the abundant epidural fat posteriorly and the intervertebral disk anteriorly. There is mild hypertrophy of the facet joints. Sagittal proton-density (TR/TE 2,000/20 msec) image (Fig. 14B) of the lumbar spine demonstrates no evidence of a disk protrusion. There is abundant epidural fat both anterior and posterior to the thecal sac.

Diagnosis: Epidural lipomatosis.

Discussion: On the axial section the compressed thecal sac superficially simulates a central disk protrusion. Indeed, this diagnosis was initially made at another institution. Recognition, however, that there is no other structure to account for the thecal sac, coupled with the lack of any disk abnormality on the sagittal sequence, makes this original diagnosis untenable.

While classically considered to represent a ''rare'' condition, increased epidural fat contributing to varying degrees of spinal cord and cauda equina compression are in actuality not uncommonly encountered on imaging examinations (MR and CT). The condition has been most commonly associated with iatrogenic Cushing's syndrome/chronic intake of exogenous corticosteroids (1–5). It may, however, be seen in the absence of an association with steroid usage and has been reported in association with obesity and other more diverse conditions including hypothyroidism, and juvenile rheumatoid arthritis, as well as after chemotherapy (4). The dorsal aspect of the spinal cord is most commonly involved, usually along its entire extent. Adults in all age groups may be affected and present with progressive signs and symptoms of spinal cord or cauda equina compression. Presenting symptoms range from nonspecific back pain to radiculopathy and/or frank myelopathy secondary to cord compression (1–3). Most patients demonstrate slowly progressive findings, although a subgroup has been reported in which acute irreversible paraplegia has been observed (4).

Diminution in epidural lipomatosis has been demonstrated following dietary control of obesity as well as diminution or withdrawal of exogenous steroid therapy

(1,2). Surgical intervention with decompression and debulking has been undertaken in advanced cases. At surgery, histologically normal unencapulated fat has been reported (5). When the dural sac is deformed by epidural tissue demonstrating short T1 values, the possibility of hematoma should be considered along with lipomatosis. A scan utilizing fat-suppression techniques will allow definitive evaluation when uncertainty exists. Angiolipomas and hemangiomas are rarely considered in the differential diagnosis.

CASE 15

History: A 28-year-old man was referred for follow-up MR examination after lumbar fusion. What is the effect on MR image quality of the various types of spinal instrumentation?

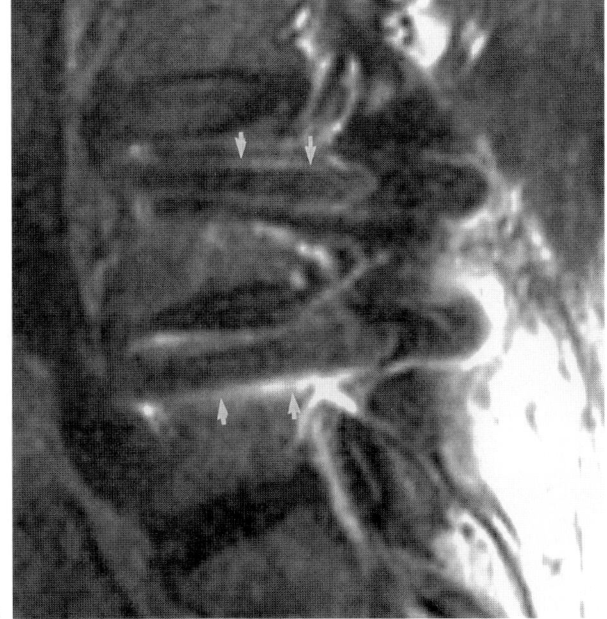

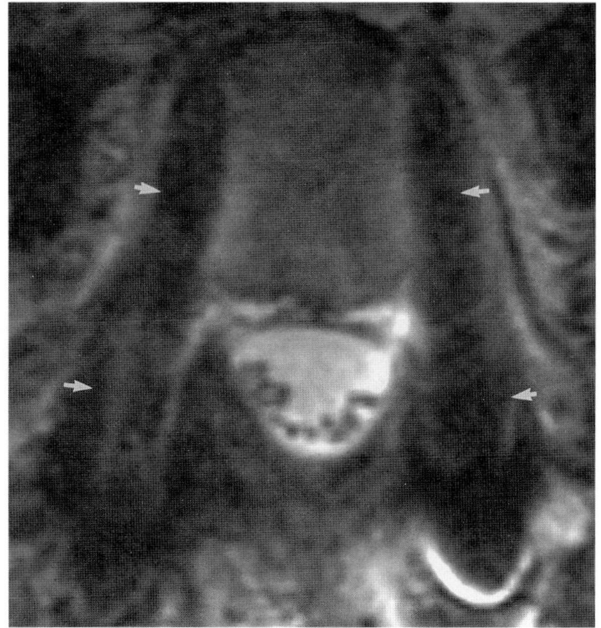

FIG. 15.

Findings: Sagittal T2-weighted fast spin-echo (TR/TE 3,000/102 msec) image (Fig. 15A) demonstrates low-signal-intensity bands with high-signal-intensity borders corresponding to the pedicle screws at the level of the L4 and L5 pedicles (small arrows). Axial T2-weighted fast spin-echo (TR/TE 3,000/102 msec) image (Fig. 15B) again demonstrates the low-signal-intensity bands corresponding to the pedicle screws (small arrows). There is no evidence of a recurrent disk protrusion.

Diagnosis: Only moderate image degradation in a patient with spinal fixation utilizing a titanium-based pedicle screw and rod system.

Discussion: Spinal instrumentation is utilized in conjunction with arthrodesis to stabilize and maintain alignment until fusion occurs (1–4). Fusion remains the goal in almost all cases, since the instrumentation itself inevitably fails (2). Assessment of the spine after instrumentation has presented significant challenges for both CT and MR because of image degradation secondary to artifact. The introduction of titanium-based systems, as illustrated in this case, allows for remarkable improvement in the assessment of spinal structures with MR even in the presence of instrumentation.

A large number of techniques have been employed for lumbar spine fixation and fusion. These can be divided into anterior and posterior approaches and by the method utilized for attachment to the spine (2). For posterior fixation, rods and plates are most commonly used to provide rigid linkage (2,4). The primary methods for attaching these devices include: (a) sublaminar, interspinous, and/or foraminal wires and cables; (b) laminar and pedicle hooks; and (c) pedicle screws (2). Pedicle screws were first introduced in 1969 as a means of obtaining segmental vertebral purchase for spinal instrumentation (2). The screws are angled medially to follow the long axis of the pedicle and continue into the vertebral body. The screws should be centered within the pedicle to avoid injury to adjacent neural structures (Fig. 15C). The screws have a shallow cancellous thread pattern to maximize their strength (1,2). Depending on the system, rods or plates are attached to the screws with eyebolts, clamps, or set screws. Pedicle screw systems provide three-dimensional rotational control, which makes them useful for correcting deformities (2).

In contrast to hooks, the screws need to be attached to only one segment above and below the level of concern to provide rigid fixation. This reduces the number of levels that may need to be involved in an arthrodesis and allows for maintenance of lumbar lordosis and avoidance of flat back syndrome (e.g., loss of normal lordosis, requiring the patient to flex at the knees to stand erect, which results in abnormal gait, fatigue, and pain) (2).

Pedicle screw strength is diminished when the screw is inserted too shallowly or utilized in osteopenic bone (2). Use of a screw too large in diameter may result in fracture. Nerve root injury may result from penetration of the screw beyond the cortex of the pedicle (2,4). Screws may be malpositioned at the time of placement, or can subsequently bend, break, or loosen from stress or because of bone resorption (2). Conventional radiographs remain the mainstay of implant evaluation. CT evaluation, while degraded by metallic image artifact, can provide adequate information. CT examination is facilitated by increasing the technique and by oversampling (2). Reformatted images are typically less affected by beam hardening artifacts than direct axial images (2). Titanium-based systems, such as the one utilized in this case, allow for high-quality MR evaluation of the postinstrumentation spine. This facilitates direct visualization of the central canal and foramen and neural structures, as well as optimal assessment of both immediate (i.e., hematoma, infection, neural injury) and longer term complications (i.e., disk protrusion, recurrent stenosis).

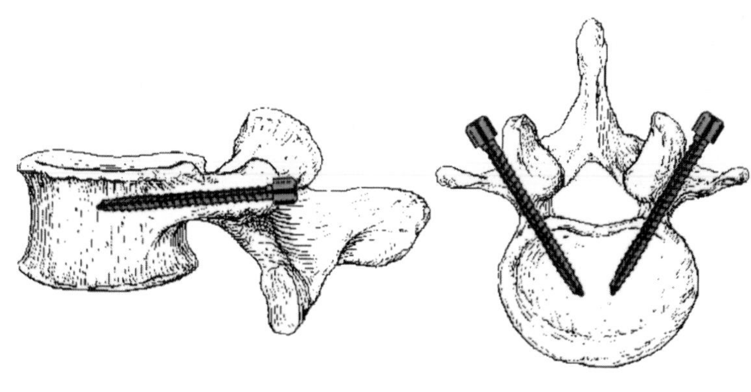

FIG. 15. C: Diagramatic representation of the optimal placement of pedicle screws. Note the orientation parallel to the plane of the pedicles. From Slone RM, McEnery KW, Bridwell KH, Montgomery WJ. Principles and imaging of spinal instrumentation. *Radiol Clin North Am* 1995;33:189–212, with permission.)

CASE 16

History: A 27-year-old man underwent MR evaluation for assessment of low back pain and increasing radiculitis.

What is your diagnosis? What factors contribute to nerve root entrapment with this entity?

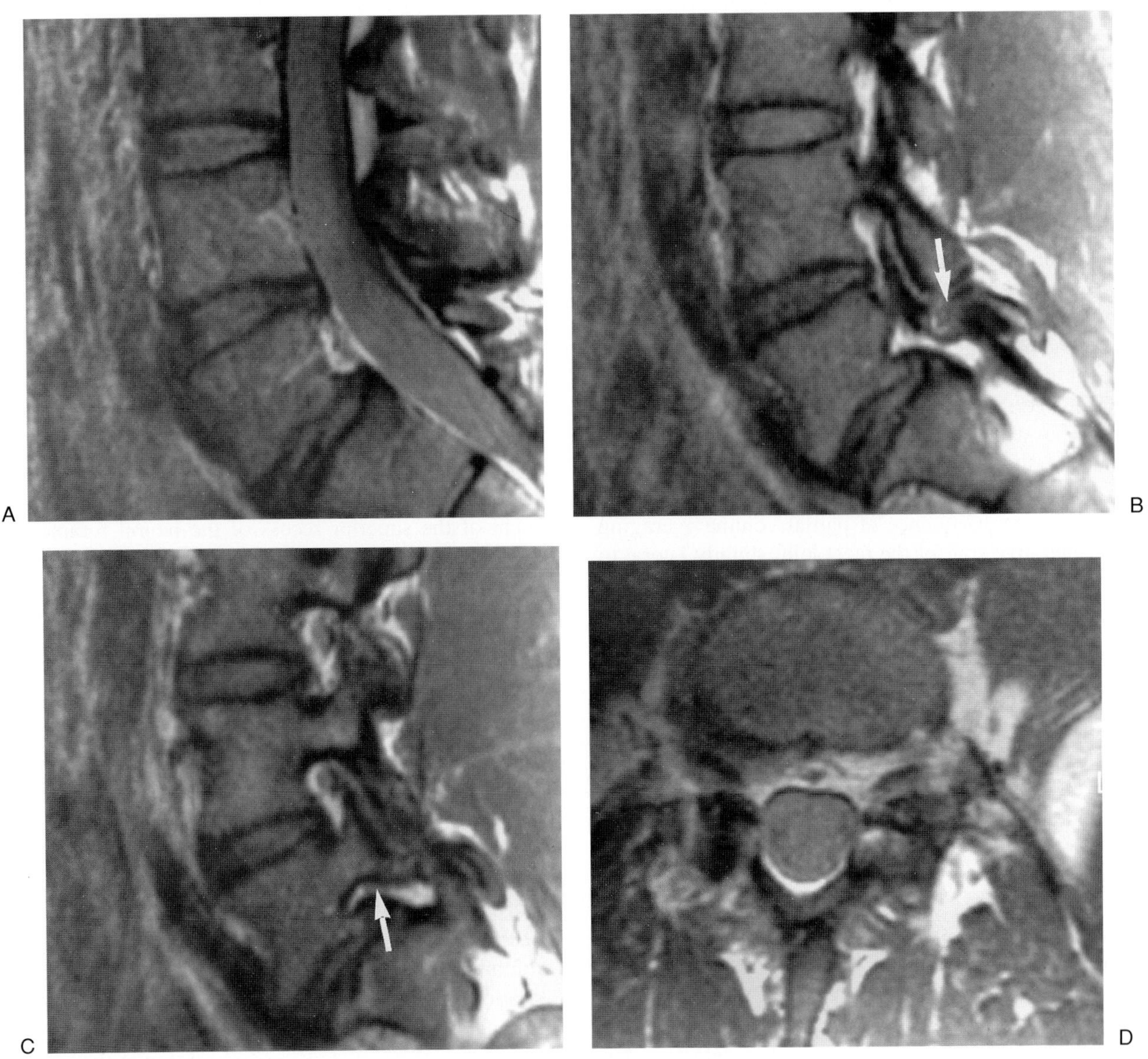

FIG. 16.

Findings: Sagittal proton-density (TR/TE 4,000/32 msec) image (Fig. 16A) in the midline demonstrates grade 1 spondylolisthesis of L5 in relation to S1. Parasagittal proton-density (TR/TE 4,000/32 msec) image (Fig. 16B) demonstrates a spondylitic defect within the L5 pars interarticularis (arrow). Parasagittal proton-density (TR/TE 4,000/32 msec) image (Fig. 13C) demonstrates ''up-down'' stenosis of the exit foramen with entrapment of the exiting L5 nerve root (arrow). Axial proton-density (TR/TE 4,000/32 msec) image (Fig. 16D) through the level of the exit foramen.

Diagnosis: Spondylolysis with spondylolisthesis and associated lumbosacral nerve root entrapment.

Discussion: Spondylolysis is defined as a bone defect in the pars interarticularis (1–6). Its cause has not been determined with certainty but in most cases it is probably related to repeated stress, trauma, or a combination of both (1,2). The estimated prevalence of defects in the pars interarticularis is on the order of 6% in adults (1,2). The rate of occurrence during childhood is somewhat lower, thus accounting for the overall 5% prevalence in the population as a whole (1,2). In a large survey of skeletons, defects were found significantly more often in males than in females (3:1). At the L5-S1 level, the defect almost always occurs bilaterally. It has been suggested that the unilateral spondylolysis reflects a healing of a defect that was originally bilateral (1).

The affected vertebral body is essentially divided into two segments; an anterosuperior (ventrocranial), consisting of the vertebral body with the pedicles, transverse processes, and superior interarticular facets, and a posterior-inferior (dorsocaudal) segment consisting of the inferior interarticular facet, the laminae, and the spinous process (4). The sagittal plane has been considered the best for evaluating the entire pars interarticularis, as the obliquity of the pars in this plane is minimal (4). An interruption of the continuity or intensity of the marrow signal of the pars interarticularis in cases of spondylolysis can be well depicted on sagittal MR images. It is emphasized, however, that low signal intensity relative to adjacent marrow within the pars interarticularis may be seen in other conditions including acquired benign or malignant sclerosis within the neck of the pars interarticularis itself and as a consequence of partial volume averaging of a degenerative spur of the facet joint slightly lateral to the pars interarticularis (4).

Segmental anteroposterior elongation of the spinal canal at the level of the defect in the pars interarticularis is seen in virtually all patients who have spondylolysis (4). Elongation of the canal can be caused both by anterior dislocation of the vertebral body from the neural arch associated with spondylolisthesis and/or by posterior displacement of the neural arch relative to the arches of the vertebrae above and below. This is demonstrable on midline sagittal MR images, which can be utilized as one of the more sensitive indirect indicators of the presence of spondylolysis (Fig. 16E).

Actual distraction of the dislocated bony elements at the pars interarticularis, or a step-off pattern, can be a specific indicator of spondylolysis (4). This finding is occasionally well shown on sagittal MR images and allows unequivocal demonstration of the defect in the pars interarticularis (Fig. 16F). As often occurs, the degree of subluxation may be bilaterally unequal at the pars interarticularis, resulting in rotational dislocation of the neural arch, or segmental rotatory scoliosis, or both. In the absence of gross subluxation at the pars interarticularis, detection of a simple linear interruption in marrow signal may suggest the presence of spondylolysis. This finding reflects a nondisplaced or stress fracture of the pars interarticularis (4). If the defect is bilateral, little spondylolisthesis of the involved vertebral body or retroluxation of the neural arch may occur. In this setting, the primary abnormality may be overlooked on imaging examinations.

In the presence of spondylolisthesis, the orientation of the foramen becomes more horizontal rather than vertical oblique, and the foramen assumes a more bilobed appearance. This appearance has previously been described as a flattening on CT examinations performed with multiplanar reconstruction (1). The neural foramina are well depicted on direct sagittal MR, which allows for assessment of foraminal encroachment originating from any radial direction. Axial MR images are frequently misleading or indeterminate with regard to assessment of foraminal encroachment, as is well illustrated in the above case. Obliteration or nearly complete loss of the high-signal-intensity fat that normally borders on and often surrounds the spinal nerve root on T1-weighted images suggests actual entrapment of the nerve root within the neural foramen.

Foraminal encroachment in patients with spondylolysis has a characteristic configuration, with specific encroachment of the superior recess of the neural foramen (Fig. 16G). With forward slippage of the craniad vertebral body (e.g., L5), there is consonant posterior displacement of the caudad vertebral body (e.g., S1). As a result, the leading posterosuperior corner of the caudal vertebral body and the contiguous bulging disk protrude into the superior recess of the neural foramen. This process takes place without a true herniation of the intervertebral disk and with only relatively minor degrees of spondylolisthesis (4–6). Such encroachment may be clinically significant, as this is the usual anatomic location of the exiting spinal nerve and dorsal root ganglion.

Some patients with spondylolysis have radicular signs and symptoms without frank circumferential entrapment of the nerve root within the superior recess of the neural foramen. In these cases, sagittal MR may allow demonstration of a form of encroachment in which the root is impinged between the respective bony elements of the spondylolisthesis in a pincer-like configuration. As a consequence, depending on the spatial location of the nerve root within its respective neural foramen relative to the components of the spondylolisthesis, early direct impingement without gross entrapment may be a clinically significant diagnostic feature on MR images.

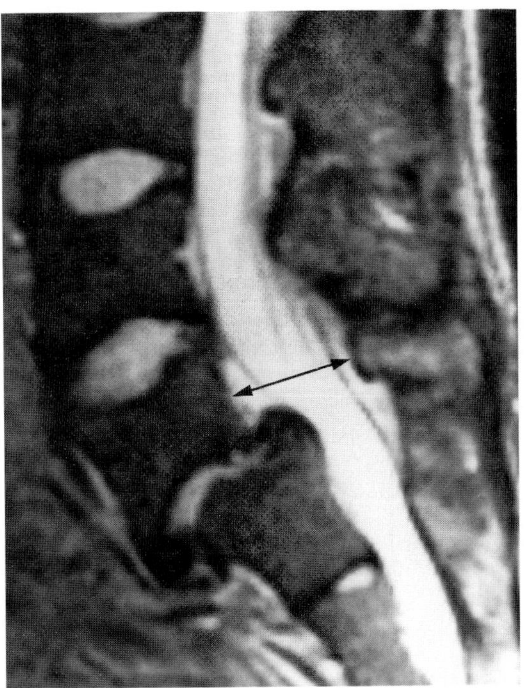

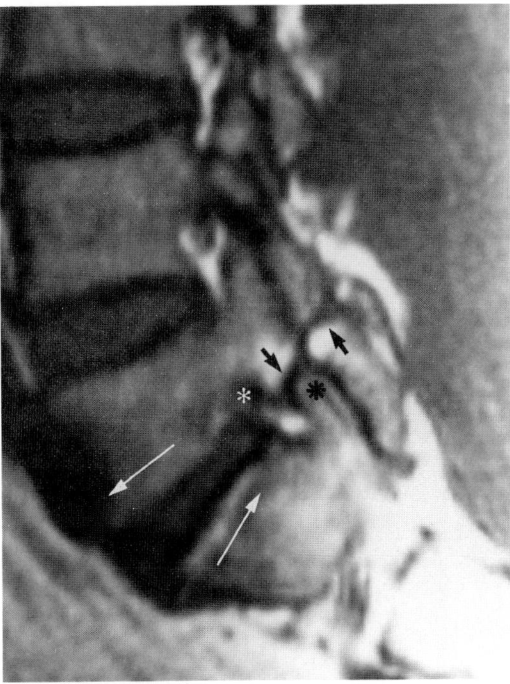

FIG. 16. E: Sagittal T2-weighted (TR/TE 2,500/80 msec) conventional spin-echo image in the midline demonstrates retroluxation of the neural arch of L5, producing malalignment of spinous process posteriorly relative to adjacent levels, and spondylolisthesis at L5-S1. Together these phenomena produce an elongation of the anteroposterior diameter of the spinal canal *(double-headed arrow)* at L5. **F:** Sagittal T1-weighted (TR/TE 600/20 msec) conventional spin-echo MR image of the left of midline shows step-off or subluxation *(black arrows)* of involved bony elements at the level of spondylolysis. Interposition *(black asterisk)* of the superior articular facet of S1 has occurred. Note relatively angulated spondylolisthetic translation of L5 on S1 *(white arrows)* and encroachment on the superior recess of the neural foramen with associated entrapment of the spinal nerve *(white asterisk)*. Compare this with normal neural foramina at craniad levels.

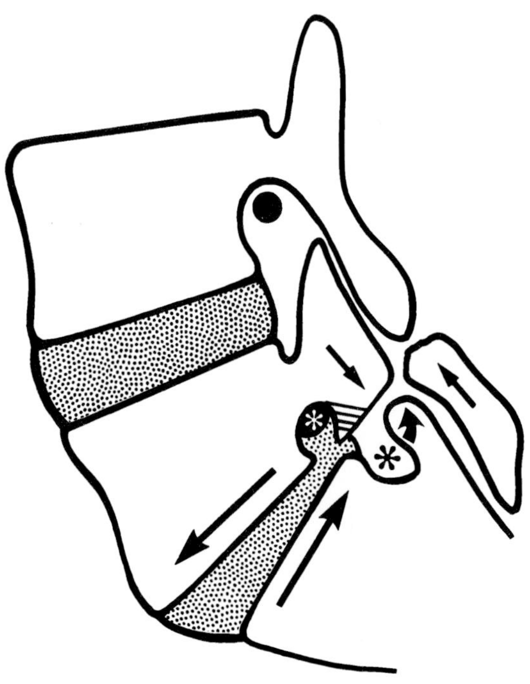

FIG. 16. G: Cutaway schematic drawing shows the pathologic anatomy of spondylolysis and spondylolisthesis at L5-S1. Cranial, angulated spondylolisthetic translation of S1 relative to L5 *(long straight arrows)* results in encroachment on the superior recess of the neural foramen and entrapment of the L5 spinal nerve *(white asterisk)*. In many instances, the inferior recess of the neural foramen *(black asterisk)* remains patent. In this case of spondylolysis, note associated dislocation or step-off of distracted bony elements *(short straight arrows)* and interposition of the superior articular facet of S1 *(curved arrow)*. Some of the foraminal stenosis may be contributed to in certain instances by adjacent hypertrophic pseudoarthrosis induced by the spondylolysis *(hatched area) (solid black circle/oval,* spinal nerves exiting the neural foramina; *stippled areas,* intervertebral disk). (From Jinkins JR, Matthes JC, Sener RN, Venkatappan S, Rauch R. Spondylolysis, spondylolisthesis and associated nerve root entrapment in the lumbosacral spine: MR evaluation *AJR* 1992;159:799–803, with permission.)

CASE 17

History: A 36-year-old man underwent MR examination of the spine. There was a questionable history of ankylosing spondylitis. What are the potential causes and the significance of the findings illustrated? What mechanisms account for the observed signal changes within the intervertebral disk?

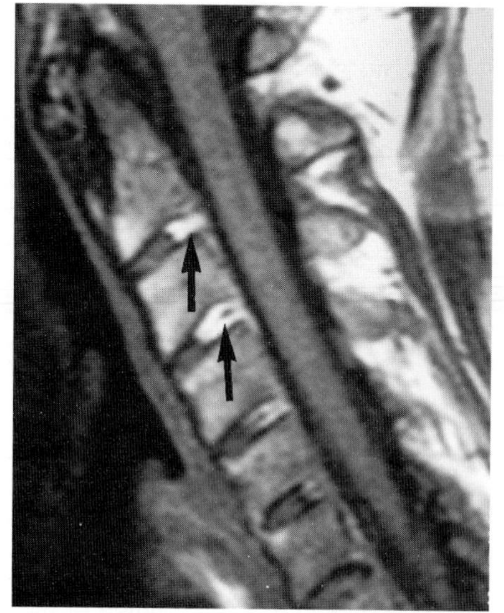

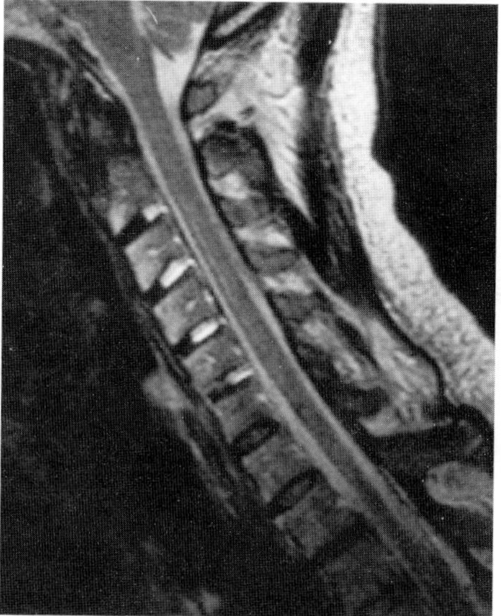

FIG. 17.

Findings: Sagittal T1-weighted (TR/TE 600/13 msec) image (Fig. 17A) demonstrates mild increased signal intensity within the intervertebral disk (arrow). Sagittal T2-weighted (TR/TE 3,300/108 msec) spin-echo image (Fig. 17B) demonstrates persistent high signal intensity within the disks. (Courtesy of Jeffrey Ross, M.D., Cleveland, Ohio.)

Diagnosis: Hyperintense disk on T1-weighted image.

Discussion: While traditionally described as low to intermediate in signal intensity on T1-weighted sequences, increased signal within the intervertebral disk on T1-weighted spin-echo sequences is not uncommonly encountered and has now been described by several groups (1–3). This increased signal appears to be related in most instances to mineralization (calcification) within the intervertebral disk (1–5). The phenomenon of mineralization accounting for increased signal intensity on T1-weighted spin-echo and gradient-echo sequences is at variance with the classic MR literature, which has associated calcification with diminished signal intensity on these sequences. Henkelman and co-workers (4), however, have attributed T1 shortening associated with mineralization to a surface relaxation phenomenon caused by the binding of a layer of water molecules to the crystal surface of calcium. The adsorbed water molecules have rotational and translational frequencies that more closely approximate the Larmor frequency than those of ''free'' water and as a consequence ''relax'' more rapidly (3,4). Other studies have been reported that also lend support to this hypothesis, and it appears that this phenomenon becomes more likely as the surface area of the calcium particles increases (3,6,7).

The effect of calcium on signal intensity, however, appears to be concentration dependent (3,4,8). Initially at relatively low concentrations of calcium, the signal intensity on T1-weighted spin-echo sequences increases with increasing calcium concentration. As the concentration of calcium further increases, the effects of the associated shortening of the T2 relaxation time traditionally recognized with calcium begin to dominate, and signal intensity decreases. The T2 shortening effect of calcium appears to predominate over the T1 shortening effect when the concentration of calcium exceeds 30% to 40% by weight (4). As a consequence, signal intensity on T1-weighted images may be increased with relatively low concentrations of calcium, become isointense as the effects of simultaneously shortened T1 and T2 relaxation times move into equilibrium, and decrease when the T2 and proton-density changes predominate (3,5).

Correlative studies have demonstrated high signal within intervertebral disks in which calcium could be demonstrated on either plain radiographs or CT (3). It has been suggested, however, that the above-described mechanism may not sufficiently account for all cases of high intervertebral disk mechanism (3,5). In some cases, markedly hyperintense disks have been associated with extremely opaque rather than the expected mild-to-moderate calcifications as depicted by CT (3). Bankert and associates (3) have suggested that ossification rather than calcification may account for such cases. In this situation, the increased signal is related to fat within the marrow in distinction from the proposed mechanism of surface relaxation in cases of mineralization. Neither mechanism, however, definitively explains the presence of hyperintense signal on T1-weighted images in the absence of demonstrable disk calcification (3,5). It is possible that in such cases the calcium concentration within the disk is sufficient to contribute to increased signal on T1-weighted images (accounted for by the surface relaxation phenomenon) but is not high enough to show radiographically (3,5,8). This explanation would suggest that MR is more sensitive to low levels of calcification than CT (8). It remains possible, however, that other mechanisms are responsible for the observed signal changes and that other materials such as trace elements or other macromolecules, rather than calcium, may be responsible for the phenomenon observed (3,5,8).

Increased signal intensity within soft tissue on T1-weighted spin-echo sequences may be encountered in a variety of conditions (3,5). Lipid-containing materials including endogenous fat, posterior pituitary neurosecretory granules, and oil-based myelographic contrast agents are all characterized by high signal on T1-weighted images secondary to short T1 relaxation times (3,5). Conditions associated with proteinaceous material including craniopharyngiomas, colloid cysts, pituitary tumors, bronchogenic cysts, and inspissated paranasal secretions have all been associated with a pattern of increased signal intensity on T1-weighted images (3,5). The paramagnetic effects of subacute hemorrhage, melanin, and chelated gadolinium contrast agents also are associated with increased signal on T1-weighted spin-echo images (3,5). Few of these possible causes of increased signal on T1-weighted sequences tend to involve the intervertebral disk (3,5) (Fig. 17C).

The list of possible causes of disk calcification is also extensive (3,5). The most common causes include include degenerative, post-traumatic, and postfusion changes (3,5). Less commonly, disk calcification may be associated with calcium pyrophosphate dihydrate deposition disease, gout, hemochromatosis, hyperparathyroidism, acromegaly, amyloidosis, vitamin D intoxication, ochronosis due to alkaptonuria, homocystinuria, and infection (3,5). Intervertebral disk calcification is also associated with conditions characterized by limited spinal mobility including ankylosing spondylitis, Klipppel-Feil syndrome, juvenile rheumatoid arthritis, diffuse idiopathic skeletal hyperostosis, and paraplegia (3,5). In most instances, these less common causes of disk calcification should be clinically obvious or readily distinguishable by other radiographic features. In the absence of a clinical history highly suggestive of one of the rarer causes of disk calcification, the presence of high signal intensity within an intervertebral disk on a T1-weighted image as an isolated finding in an otherwise normal-appearing spine or in a spine with other findings consistent with degenerative changes can reasonably be attributed to degenerative change, probably related to intervertebral disk calcification (5). In most patients, no clinical significance should be attributed to this finding (5).

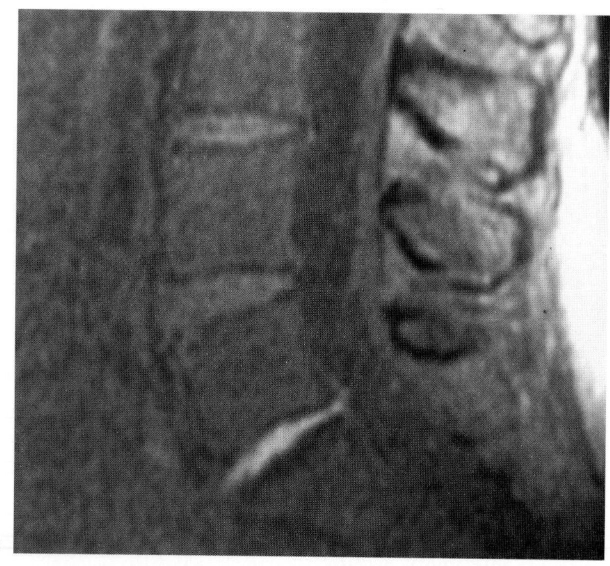

C

FIG. 17. C: MR diskogram. Sagittal fat-suppressed T1-weighted (TR/TE 600/20 msec) image demonstrates uniform increased signal intensity within the L5-S1 disk. The patient was imaged immediately following the performance of a diskogram with the injection of diluted gadopentetate dimeglumine. The case represents one of the more unusual causes of increased signal within disks on T1-weighted images.

CASE 18

History: This 50-year-old patient with a history of bilateral laminectomies complained of several months of leg weakness and paresthesia. The MR study was obtained for evaluation of possible recurrent herniated disk. What are the findings illustrated in Fig. 18A? What is your diagnosis? What classification system has been employed in the description of this disorder?

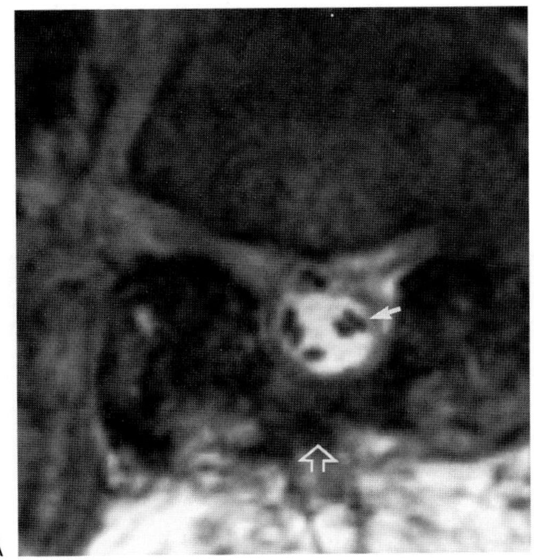

FIG. 18.

Findings: Axial T2-weighted fast spin-echo (TR/TE 4,000/108 msec) image (Fig. 18A) demonstrates clumping of the nerve roots of the cauda equina (arrow). The patient is status post bilateral laminectomy with hypointense scar at the laminectomy site (open arrow).

Diagnosis: Adhesive arachnoiditis at L4-5 (type 1).

Discussion: In addition to recurrent/residual disk herniation and scar formation, other causes may be responsible for recurrence of pain after back surgery. Characteristic imaging findings along with the patient's clinical course may assist in establishing the diagnosis.

Arachnoiditis is responsible for persistent signs and symptoms in 6% to 16% of postsurgical patients (1). All three layers of the meninges are commonly involved in this relatively avascular inflammatory process that has usually been attributed to a complication of preoperative myelography or a mechanical insult during surgery (2). Three patterns of arachnoiditis can be identified by high-resolution MR examinations (3,4). The first pattern is that of central adhesion of nerve roots within the thecal sac into a central clump of soft tissue signal. The normal ''feathery'' pattern of nerve roots is replaced by roots clumped into one or more cords. This pattern may be best appreciated on axial T1-weighted MR images (3,4) (Figs. 18B–E).

The second pattern of arachanoiditis has been termed the *empty sac* sign and reflects adhesion of the nerve roots to the meninges (3,4). On MR the signal intensity of CSF is homogeneous, and the nerve roots can be identified attached to the meninges, which are in turn attached to the thecal sac. With CT myelography the nerve roots themselves may not be resolved (3,4). In the third pattern, the arachnoiditis essentially becomes an inflammatory mass that may fill the thecal sac. This ''end-stage'' arachanoiditis may result in a myelographic block (Fig. 18F). Both MR and CT may demonstrate nonspecific-appearing soft tissue masses that may mimic intradural mass lesions (Fig. 18G).

While the three patterns have been described separately, combinations are commonly seen; in particular, central and peripheral clumping are often identified together (3). Following the administration of gadopentetate dimeglumine, there is typically little enhancement, which does not appear to correlate with clinical or radiographic severity (4). This relative lack of enhancement may be of value in the differential diagnosis of arachanoiditis, particulary in the inflammatory mass stage, which might simulate possible neoplasm (4). The clinical history and relative lack of enhancement of the mass would favor arachanoiditis in this setting. In addition, the relative lack of enhancement may be a valuable sign in differentiating arachanoiditis from leptomeningeal spread of tumor (see Case 5-19). In addition, leptomeningeal tumor may demonstrate a more irregular morphologic pattern (4). Myelography and CT-myelography may depict even subtle changes indicative of arachnoiditis (5,6). However, subtle changes of arachnoiditis (i.e., absence of nerve root sleeve filling) may mimic the presence of a herniated disk. A 92% sensitivity, a 100% specificity, and a 99% accuracy has been reported for diagnosis of moderate-to-severe arachnoiditis by MRI (3).

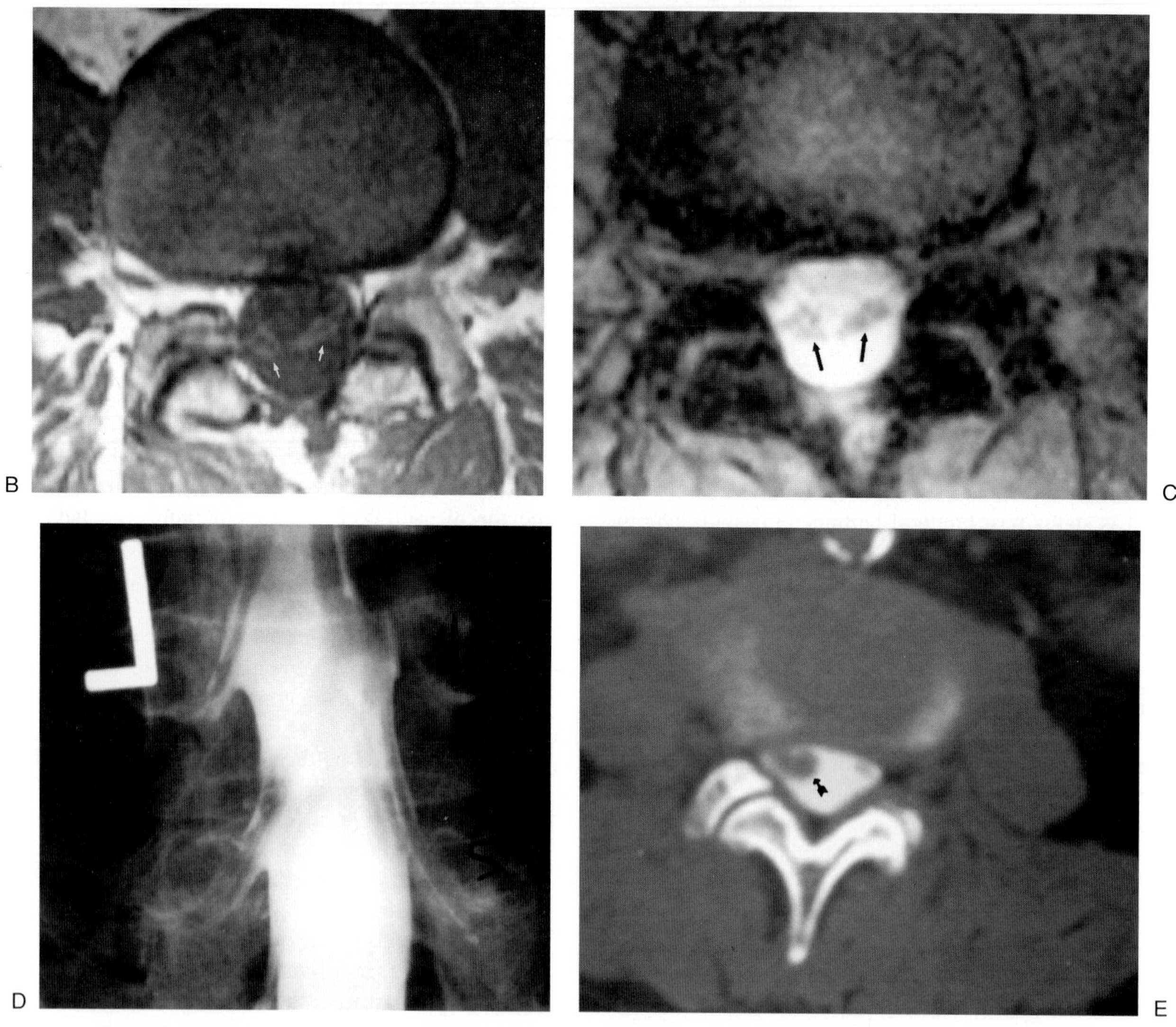

FIG. 18. B–E: Axial T1-weighted MR image demonstrates that the nerve roots are clumped into two centrally located bundles or hemicords in this postoperative patient with recurrent symptoms *(arrows)*. This appearance is often well appreciated on T1-weighted images. **C:** Axial gradient-echo MR image also demonstrates well the central clumping consistent with adhesive arachnoiditis *(arrows)*. **D:** Lumbar myelogram demonstrates central clumping of the right L4 and L5 nerve roots. **E:** Axial CT myelogram better demonstrates the central clumping of nerve roots *(arrow)*, a finding consistent with adhesive arachanoiditis.

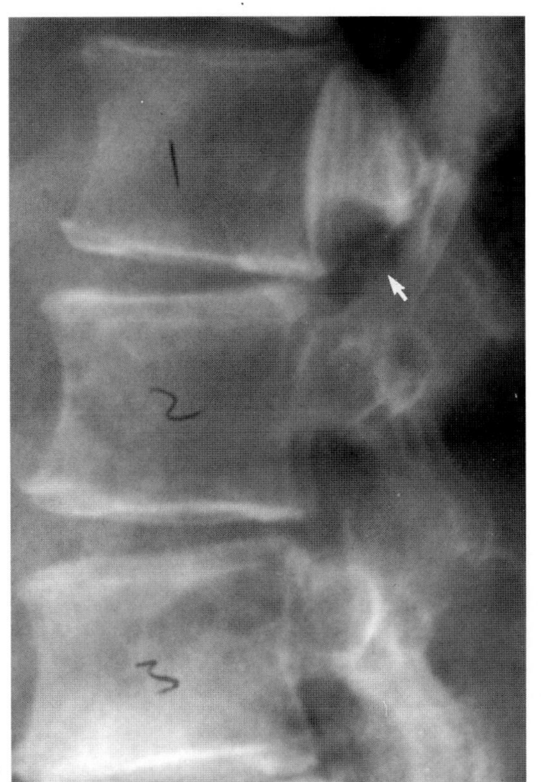

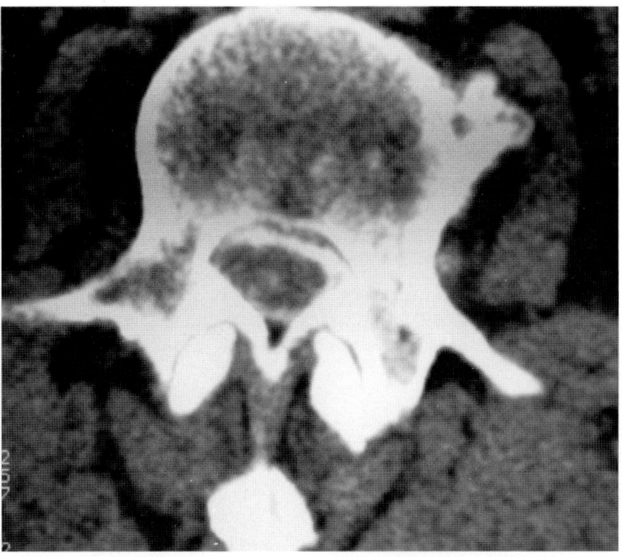

F G

FIG. 18. F,G: Severe arachnoiditis (type 3). **F:** Lateral myelogram demonstrates a complete block to flow of contrast material at the L1-2 level *(short arrow).* **G:** CT myelogram demonstrates an apparent intradural "mass" caused by advanced arachnoiditis. (Courtesy of M. Basham, M.D.)

CASE 19

History: A 66-year-old man had undergone prior lumbar disk surgery (at the L5-S1 level). The patient remained symptomatic following surgery and a repeat MR was performed. Images are presented from his preoperative and postoperative studies. What is the significance of the findings illustrated on the postoperative study?

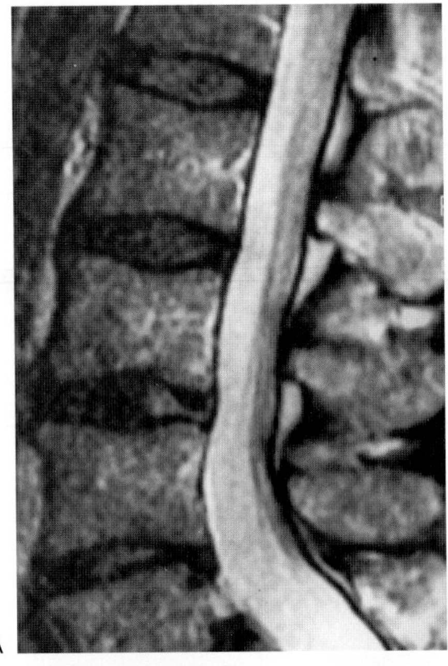
A

FIG. 19.

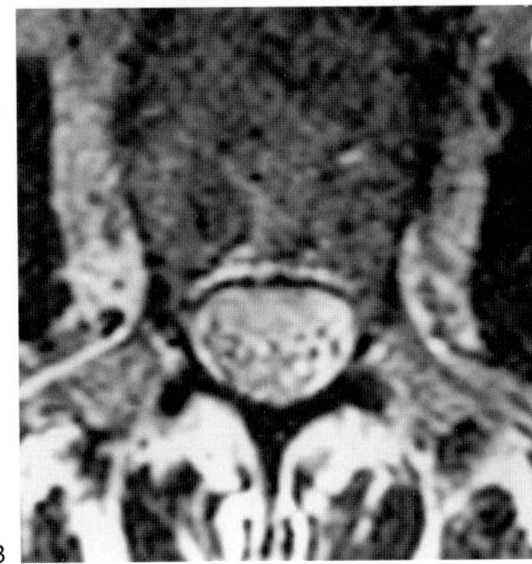
B

C

Findings: Sagittal preoperative T2-weighted fast spin-echo (TR/TE 3,000/96 msec, eff) image (Fig. 19A) demonstrates multiple levels demonstrating diminution of disk height and signal intensity. The upper lumbar levels are unremarkable. Axial preoperative T2-weighted fast spin-echo (TR/TE 3,500/102 msec, eff) image (Fig. 19B) through the L2 level is normal. Sagittal fat-suppressed T1-weighted (TR/TE 700/16 msec) image (Fig. 19C) following the administration of gadopentetate dimeglumine. A small focus of intrathecal signal enhancement is noted at the L2 level that is remote from the operative level (arrow).

Diagnosis: Leptomeningeal spread secondary to melanoma. The patient had a prior history of melanoma. The CSF was examined twice. On the second "tap," 25 ml of CSF was removed, yielding two melanoma cells on cytologic analysis.

Discussion: Tumor involvement of the leptomeningeal

membranes is seen as a complication of both primary intracranial neoplasms (more common in children) as well as secondary to metastatic involvement (1–4). The systemic tumors most commonly associated with leptomeningeal spread include breast, lung, melanoma, and lymphoma (1–3). Anatomic-pathologic studies suggest direct tumor extension through cranial and spinal foramina via perivascular and perineural lymphatics as one of the primary mechanisms of spread in this type of tumor (1–3). Hematogenous extension and seeding via the choroid have also been implicated as possible routes.

In spinal involvement, leptomeningeal tumor most often involves the lumbosacral spine, probably due to gravitational effects (1–3). Lesions may, however, be seen distributed throughout the spine, and in the cervical and thoracic region are frequently dorsal in position, a finding that is probably accounted for by the natural flow of CSF (e.g., travels from the brain dorsal to the cord and returns ventral to the cord) (3). Subarachnoid spread of tumor may be asymptomatic, although presentation with multiple symptoms, representing concurrent involvement of multiple locations along the neuroaxis, is not uncommon (1–3). Multiple cranial and spinal nerve deficits, headache and changing mental status, neck, back, and leg pain, and gait disturbances are commonly reported symptoms (2).

Analysis of CSF cytology is reported to be the most sensitive method for detection of leptomeningeal metastases (3). Prior to the introduction of MR, myelography was considered the imaging standard for detection of intradural extramedullary spread of tumor. Findings on myelographic examination include nodular filling defects, thickened adherent nerve roots, scalloping of the subarachanoid membrane, and irregularity and constriction of the thecal sac (3,4). Early experience with noncontrast MRI suggested a significantly lower sensitivity compared with myelography, with a high number of false-negative examinations (4). Indeed, in the present case, a ''screening'' noncontrast MR examination was considered within normal limits. Nonenhanced MR may demonstrate a ''ground-glass'' or mildly heterogeneous pattern of the subarachnoid space, but the findings are often subtle, and images must be approached with a high degree of suspicion (2,4–6). More recent experience with contrast-enhanced MR, however, has proved considerably more sensitive for detection of subarachnoid tumor (5–7). Tumor spread may demonstrate a variety of different appearances. Tumor growth may be quite localized and result in multiple discrete nodules within the subarachanoid space or, alternatively, as one or more discrete masses (6). In other instances, tumor may be more diffuse and coat or ''frost'' the neural structures (6). The diffuse coating pattern is reported to be more common with ''drop'' metastases from intracranial neoplasms, especially medulloblastoma and ependymoma, although intracranial tumors may be associated with any pattern including a coarser intradural nodularity (6). A superficial enhancement pattern along the cord margins is not specific for malignant disease and may be encountered in inflammatory disorders such as sarcoidosis or cytomegalovirus myelitis (6). In still other cases, more extensive involvement can result in an apparently amorphous or featureless appearance of the intradural compartment related to the presence of cells, protein, and contrast within the spinal fluid (6). As in the present case, with severe involvement, enhancement of the entire thecal sac may be seen (Fig. 19D,E).

As indicated previously, cytologic examination of the CSF may remains the most sensitive method for detection of leptomeningeal tumor involvement. Present experience suggests that contrast-enhanced CT and MR are approximately equivalent for detection of leptomeningeal spinal tumor with the noninvasive nature of MR representing an advantage (3). MR appears to be the better method if diffuse coating of the neural structures is suspected, since myelography is less sensitive to subtle changes in cord caliber (3). Minimal focal involvement may be missed by both methods and meticulous technique and high clinical suspicion are required for early detection. The findings of focal tumor are important since patients may be treated with both a combination of intrathecal chemotherapy and localized radiation directed at the site of local tumor involvement. The overall prognosis with leptomeningeal spread of tumor remains poor.

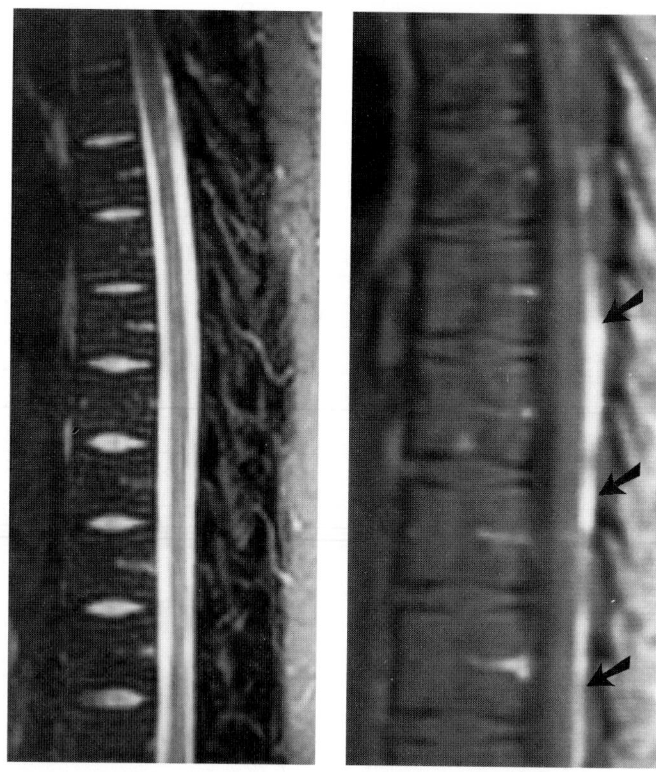

FIG. 19. D,E: Leptomeningeal carcinomatosis. A 26-year-old man with a known history of melanoma presented with multilevel neurologic complaints. A screening MR examination was performed. **D:** A representative T2-weighted fast spin-echo image (TR/TE 3,000/112 msec) from this examination. The cord appears of normal caliber. No intrinsic or extrinsic lesion is seen. The patient returned 1 week later with increasing symptomatology. **E:** A repeat examination with gadopentetate dimeglumine was performed. A fat-suppressed sagittal T1-weighted fast spin-echo image (TR/TE 600/16 msec, eff) demonstrates diffuse linear and somewhat plaque-like enhancement reflective of meningeal tumor *(arrows)*.

CASE 20

History: A 42-year-old man underwent follow-up MR 3 months after lumbar diskectomy. The patient continued to complain of anterior thigh pain; symptoms had not been relieved by the prior surgical procedure. What is your diagnosis? Figure 20A,B represents the patient's preoperative examination, and Fig. 20C–E the postoperative study.

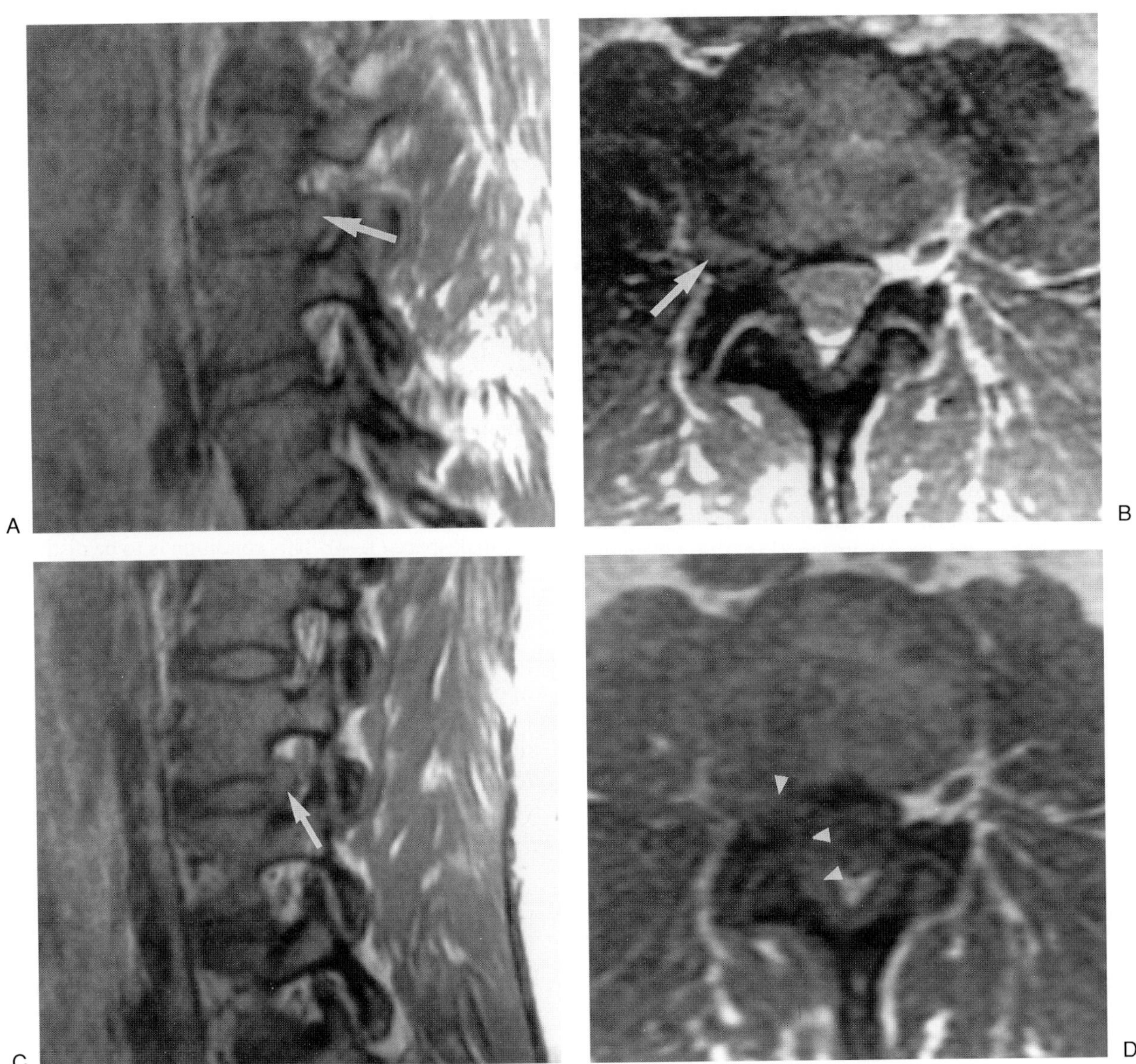

FIG. 20.

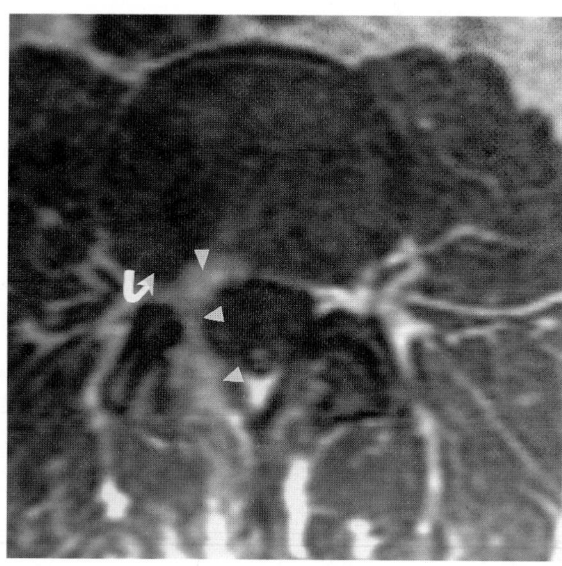

FIG. 20. (*continued.*)

Findings: Sagittal proton-density fast spin-echo (TR/TE 3,000/16 msec, eff) image (Fig. 20A) demonstrates a well-defined intermediate-signal-intensity mass extending into the right L3-4 exit foramen (arrow). Axial proton-density fast spin-echo (TR/TE 3,000/16 msec) image (Fig. 20B) graphically demonstrates the large foraminal disk extrusion (arrow). Sagittal proton-density fast spin-echo (TR/TE 3,000/16 msec, eff) image (Fig. 20C) demonstrates that the extrusion is essentially unchanged (arrow). Axial T1-weighted (TR/TE 600/20 msec) demonstrates intermediate signal intensity replacing the normal high-signal-intensity fat along the right anteromedial aspect of the thecal sac and extending slightly toward the medial aspect of the exit foramen (arrowheads). Axial T1-weighted (TR/TE 600/20 msec) image (Fig. 20E) at the same level following administration of gadopentetate dimeglumine reveals avid enhancement reflective of scar along the course of the conventional microdiskectomy route paralleling the lateral aspect of the thecal sac (arrowheads). A component of the foraminal disk extrusion demonstrates no enhancement (curved arrow).

Diagnosis: Normal enhancement along the conventional microdiskectomy route. No enhancement of a ''mass,'' reflecting a persistent foraminal disc extrusion, is seen.

Discussion: This case graphically illustrates one of the causes of the failed back surgery syndrome, namely, persistent disk extrusion secondary to faulty surgical technique (see also Cases 5-5, 5-6, 5-10, 5-18, and 5-21). The case underscores the need for precision in interpretation and communication of imaging findings to ensure that the abnormalities identified and their ramifications are completely understood by all involved in the affected individual's care. In foraminal extrusions, virtually the entire fragment is within the foramen, and only a small tail of disk material lies in the usual posterolateral quadrant of the disk space (1–4). This condition is to be distinguished from the more common lateral disk protrusion, which generally is contained, in contrast to the extruded nature of foraminal disks (1). Affected individuals typically experience far more significant leg pain than back pain. Spontaneous onset without trauma is characteristic (1). Involvement of two nerve roots with motor, sensory, and reflex deficits (e.g. double root syndrome) is common (e.g., L4 and L5 roots at the L4-5 level). In patients with compression of the L2, L3, or L4 nerve roots, intense anterior thigh pain is typically reported.

The foramen cannot be adequately decompressed from a conventional posterior surgical approach without sacrifice of the inferior facet, which may lead to instability. A conventional posterior approach, as performed in the present case, failed to gain access to the extruded disk. Current microsurgical technique employs a lateral paraspinal muscle splitting exposure. This allows access to the intertransverse ligament, which is resected, exposing the nerve root and extruded disk. This approach requires the removal of the superior tip of the facet joint at the affected level (e.g., superior tip of L5 superior articular facet at the L4-5 level) to gain access into the disk space.

CASE 21

History: A 42-year-old man presents with right S1 distribution radicular symptoms 6 years after intervertebral disk surgery at the L5-S1 level. What is the significance of the demonstrated findings?

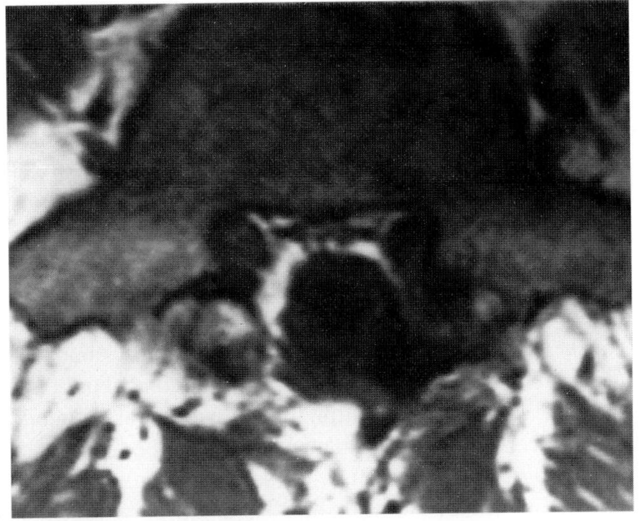

A

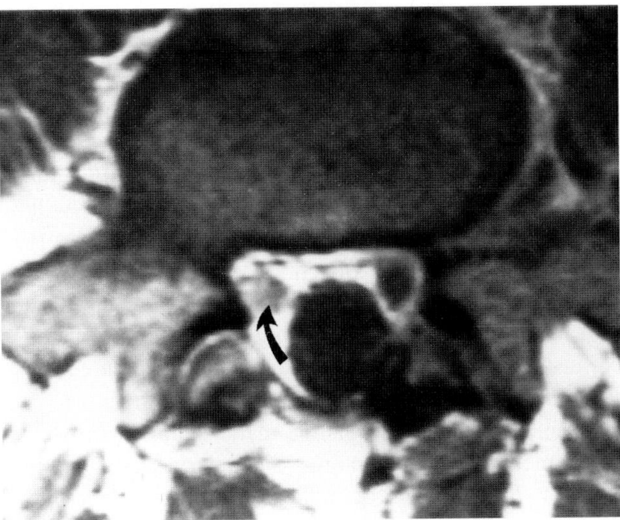

B

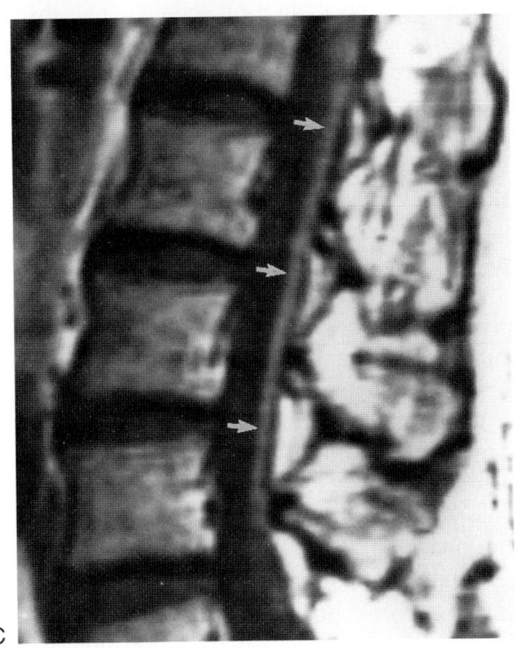

C

FIG. 21.

Findings: Precontrast axial T1-weighted (TR/TE 600/20 msec) conventional spin-echo MR image (Fig. 21A), acquired through the L5-S1 level. There is evidence of a bilateral laminectomy. No evidence of recurrent disk is appreciated. Gadopentetate dimeglumine-enhanced axial T1-weighted (TR/TE 600/20 msec) conventional spin-echo MR image (Fig. 21B) acquired at the same level as Fig. 20A reveals unilateral enhancement of the right S1 nerve root (arrow). Postcontrast enhanced sagittal T1-weighted (TR/TE 600/20 msec) conventional spin-echo MR image (Fig. 21C) demonstrates multilevel enhancement of the S1 nerve root (arrows) previously demonstrated in Fig. 20B. The enhancement extends craniad to the level of the conus medullaris.

Diagnosis: Nerve root enhancement (radiculitis) occurring 6 years following disk surgery.

Discussion: The lumbosacral postsurgical syndrome consists of signs and symptoms that recur following operative treatment for disk disease. The literature has described in detail the two most important differential diagnoses frequently seen in association with this syndrome: (a) recurrent disk herniation (see Case 5-6) and (b) epidural scarring (see Case 5-5) (1). The clinical significance of recurrent disk protrusion/extrusion is obvious; however, the precise relationship of scarring to the clinical syndrome has never been explained. Additional changes responsible for failure of therapy in the chronic postoperative period include osseous foraminal or lateral recess stenosis (Fig. 21D,E), residual disk protrusion (see Case 5-20), and chronic adhesive arachanoiditis (see Case 5-18), as well as a large category comprising ''indeterminate'' phenomenon (e.g., local vertebrogenic or remote musculoskeletal pathology and idiopathic causes).

A recent retrospective MR study of 120 patients with recurrent symptomatology following lumbar disk surgery was performed utilizing conventional spin-echo MR before and after IV gadopentetate dimeglumine (0.1 mmol/kg) (2). An additional ten asymptomatic subjects, who were at least 6 months postoperative, were evaluated using the same imaging protocol.

Overall, 21.6% of the symptomatic subjects showed enhancement of one or more spinal nerve roots. This enhancement was focal or multisegmental and involved single or multiple nerve roots. The abnormal neural enhancement was associated with otherwise isolated epidural fibrosis in 88.5% and with residual/recurrent disk protrusion in the remaining 11.5%. Presumably the latter circumstance was related to direct mechanical trauma that resulted in the observed blood–nerve barrier disruption and neural enhancement (3,4). Conversely, multilevel neural enhancement was probably linked to wallerian axon degeneration/regeneration (5). Some cases of multiple enhancing nerve roots may have been caused by generalized low-grade aseptic inflammation (i.e., arachnoiditis). There were no clinical findings to suggest active infection or neoplastic involvement in any individual in this study, and no other responsible agent could be identified.

In other studies, asymptomatic neural enhancement on gadolinium-enhanced MR may occur during the first 6 months following surgery (6). After this period, the enhancement is seen to resolve. Persistent enhancement beyond 6 to 8 months in symptomatic postoperative patients had a very high correlation (95.7%) with the presenting clinical syndrome in the study of Jinkins and colleagues (2). It should also be noted that 21.7% of the patients in this study demonstrated faint, additional nerve root enhancement that did not have a clear clinical correlation. Ten patients in the asymptomatic subgroup of Jinkins et al.'s (2) study manifested degrees of epidural fibrosis but no abnormal radicular enhancement. Such epidural fibrosis is a regular feature of lumbosacral disk surgery, but it is not necessarily linked with signs and symptoms.

Laboratory experimentation has demonstrated that a definite blood–nerve barrier exists (7,8). Any nonspecific pathologic insult or acquired neural alteration (e.g., ischemia, trauma, inflammation, active demyelination) involving roots or nerves can potentially disrupt this barrier. Enhancement of nerve roots occurs secondary to the pathologic exit of the intravenously administered agent into the extravascular space (i.e., endoneurium) of the nerve. The presence of enhancement depictable on MR studies probably relates in part to the extent of injury as well as the timing of the examination (i.e., enhancement likely represents a transient phenomenon).

While nerve root enhancement is abnormal (at a contrast dose of 0.1 mmol/kg), spinal dorsal root ganglia demonstrate intense enhancement, since there is little or no blood–nerve barrier in relation to these structures. As a caveat to the preceding observations concerning abnormal root enhancement, it should be recognized that at or below the S1 level (as well as within the lumbar neural foramen), any perceived neural enhancement may be caused by physiologic enhancement of dorsal root ganglia and thus not be abnormal. Additionally, it should be appreciated that true nerve root enhancement itself should not be absolutely equated with radiculopathy. Contrast enhancement and radiculopathy represent different pathophysiologic processes. While these may be temporally and spatially related and may correlate reasonably well with patient signs and symptoms in many cases, it must also be recognized that these two processes may become dissociated in any neural structure at any given time. Nerve root enhancement may also indicate a more severely damaged nerve than otherwise might be expected or a nerve root in a reparative phase. The precise relationship between nerve root enhancement, pathophysiologic alteration, and prognosis remains a subject for further study. Regardless of these considerations, the presence of intrathecal nerve root enhancement signifies unequivocal underlying neural/perineural pathology (e.g., direct/focal trauma, remote wallerian degeneration) in patients presenting with radicular signs and symptoms of either obvious or ambiguous etiology. The use of contrast-enhanced examinations may be of particular value in a number of settings including: (a) clinically reliable evidence of radiculopathy in the absence of correlative anatomic abnormalities on routine studies (e.g., no recurrent disk or stenosis) and (b) discordant anatomic abnormalities and clinical findings (e.g., pain syndrome remote from anatomic level). In such settings, contrast-enhanced studies may demonstrate the true nature and location of the pathologic neural alteration responsible for the clinical syndrome.

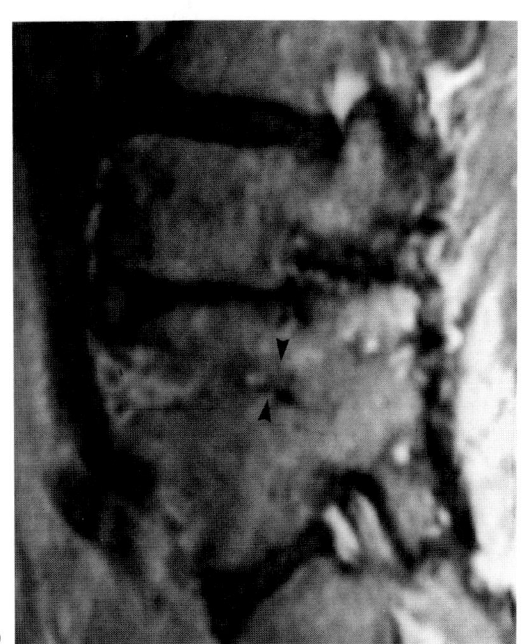

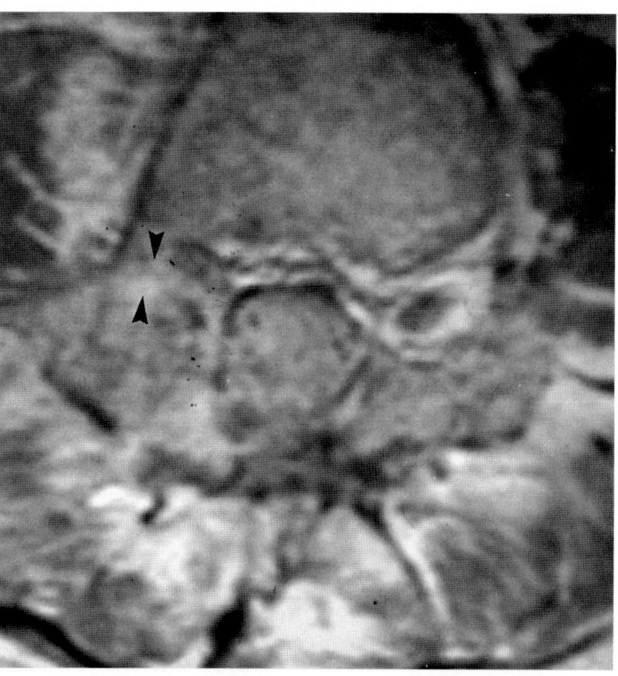

FIG. 21. D,E: Exit foraminal obliteration after fusion. These images demonstrate an extremely unusual cause of exit foramen stenosis following surgery and fusion. In the most common case, existing foraminal stenosis is not adequately addressed at the time of surgery, leading to an unsuccessful result. In the present case, marked fusion overgrowth obliterated the foramen. **D:** Right parasagittal proton-density image (TR/TE 3,300/23 msec) demonstrates complete osseous overgrowth of the right L4-5 exit foramen (*arrowheads*). **E:** Axial proton-density image (TR/TE 3,300/23 msec) again demonstrates loss of the exit foramen (*arrowheads*).

CASE 22

History: A 52-year-old man presented with a history of several months of right radicular symptoms. What is the most likely diagnosis?

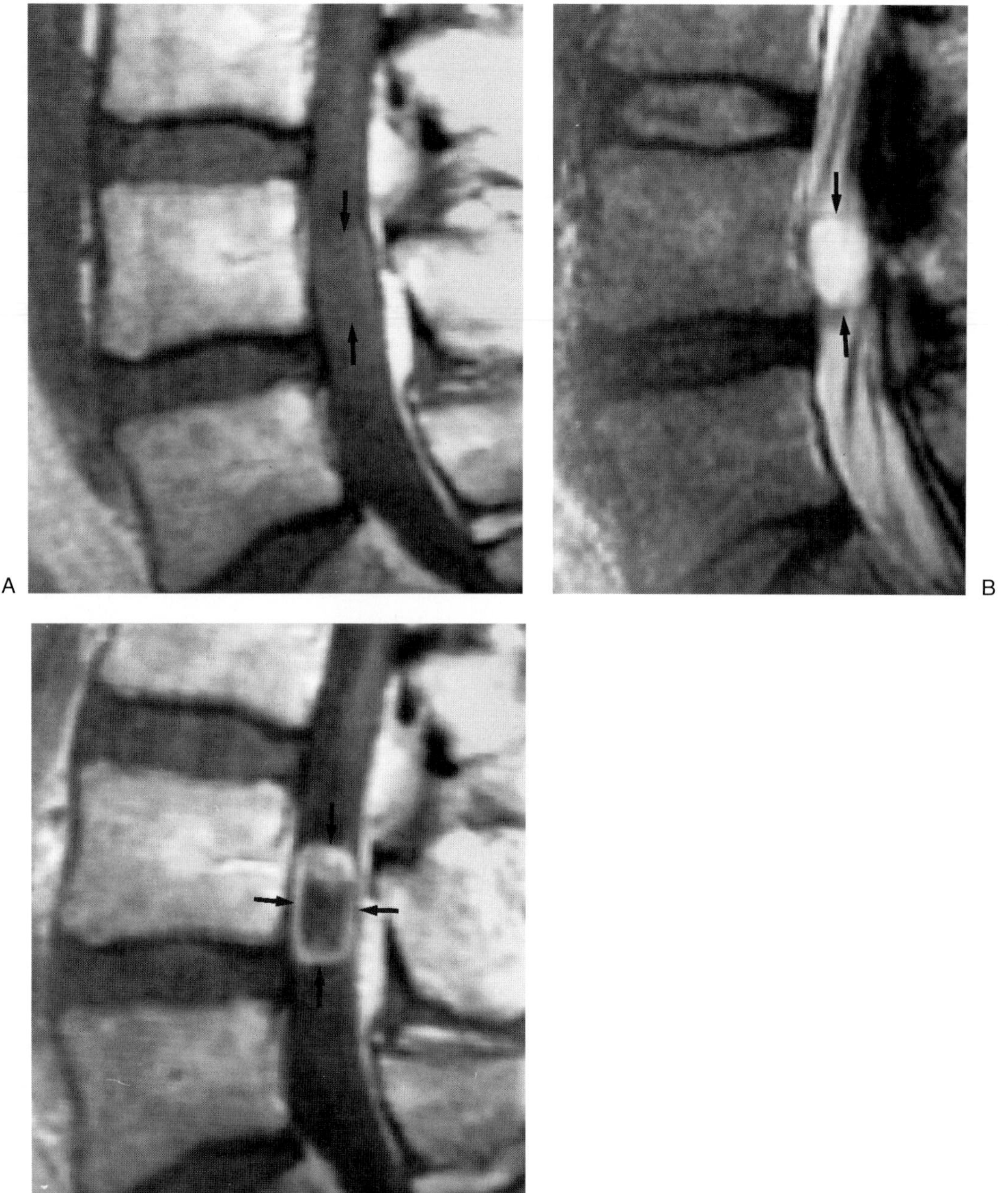

FIG. 22.

Findings: Sagittal T1-weighted (TR/TE 500/20 msec) image (Fig. 22A) demonstrates subtle increased signal intensity posterior to the L4 vertebral body within the thecal sac (arrows). Sagittal T2-weighted (TR/TE 2,000/80 msec) image (Fig. 22B) demonstrates a well-circumscribed circular to ovoid mass within the thecal sac at the L4 level (arrows). Sagittal T1-weighted (TR/TE 500/20 msec) image (Fig. 22C) obtained following the administration of gadopentetate dimeglumine demonstrates peripheral enhancement of the mass (arrows).

Diagnosis: Intradural schwannoma.

Discussion: Schwannomas and neurofibromas are the

most common primary tumors of spinal nerves (1–7). Since many of the imaging characteristics of these lesions are similar, they are often classified together as nerve sheath or nerve root tumors, although both clinically and pathologically they are distinct entities (1,2). Schwannomas occur sporadically and are almost always solitary; they take origin from the focal proliferation of Schwann cells along the perimeter of a nerve and arise adjacent to and displace the dorsal sensory root (1,2,4,6). Multiple spinal schwannomas may be seen in patients with neurofibromatosis type 2 (NF2), a condition that results from a chromosome 22 defect and shows autosomal dominant inheritance (3).

Schwannomas represent eccentric masses that compress and displace otherwise uninvolved axons (1,2 4,6). Neurofibromas, in contrast, are composed of Schwann cells and fibroblasts, which results in symmetric enlargement that surrounds individual axons (3). Neurofibromas are almost always multiple and are associated with NF1. NF1 occurs with an incidence of 1 in 4,000; it results from a chromosome 17 defect, demonstrates an autosomal dominant inheritance in approximately 50% of cases, and occurs sporadically in the remainder (3). Most spinal nerve root tumors are entirely contained within the dura. Approximately 40% are extradural or demonstrate both intradural and extradural components (3). Such dumbbell-shaped tumors typically involve and expand intervertebral foramen (Fig. 22D–F). Expansion is a clue to the long-standing presence of a benign mass (2). Schwannomas are typically isointense or mildly hypointense on T1-weighted images and may resemble an enlarged root sleeve on non-contrast-enhanced studies (1,2,4,6). Spinal schwannomas typically appear heterogeneously hyperintense on T2-weighted spin-echo images. It has been suggested that the high signal intensity may relate to increased free water within the tumor, possibly reflecting the predominance of Antoni ''B'' zones, which are characterized by a relatively loose arrangement of cells (1,2,6). Spinal schwannomas may demonstrate central areas of decreased signal intensity on T2-weighted images. This hypointensity may relate to collagen deposition, hemorrhage, and/or densely packed schwann cells. Cystic or necrotic degeneration is not uncommon within schwannomas as they increase in size, and small or large areas of hemorrhage are commonly found within schwannomas on histologic examination (1,2,4,6,8) (Fig. 22G–I).

Spinal schwannomas frequently contain nonenhancing areas. These probably correlate with the areas of necrosis, hemorrhage, and cystic degeneration previously described. Peripheral contrast enhancement is typical of schwannomas (Fig. 22J–L). The T2-weighted image characteristics and contrast enhancement pattern are helpful in distinguishing schwannomas from meningiomas (1,2). Neurofibromas, however, occur in similar locations and frequently have similar features on T1 and T2 images. Neurofibromas typically enhance more uniformly than schwannomas, but enhancement pattern alone cannot allow confident differentiation of schwannomas from neurofibromas (1,6). When a solitary nerve sheath tumor is encountered, it should probably be diagnosed as a schwannoma rather than a neurofibroma, although this rule is not invariable (1).

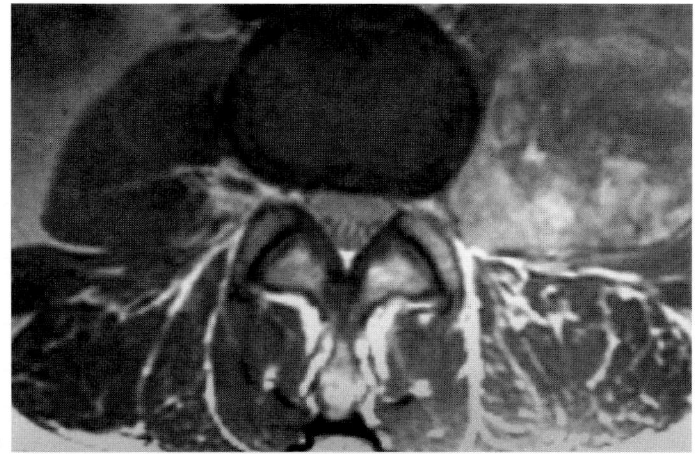
D

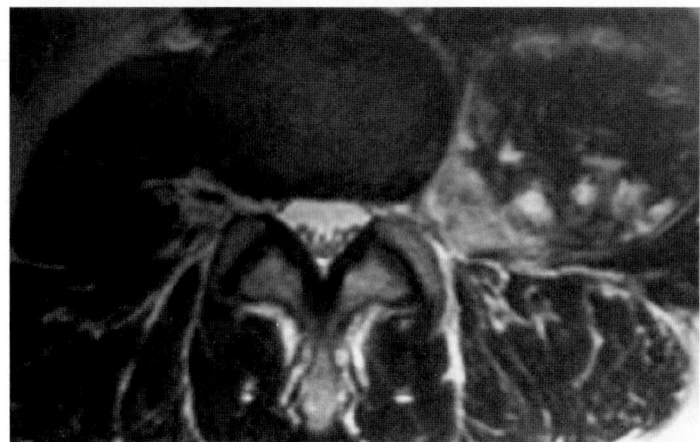
E

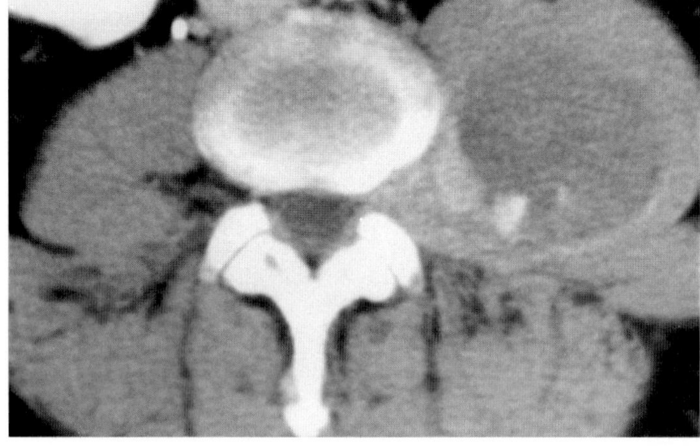
F

FIG. 22. D–F: Dumbbell-shaped schwannoma extending out of the left L3-4 neuroforamen in a 68-year-old man who presented with back/flank pain. **D:** Axial proton-density (TR/TE 4,000/17 msec) fast spin-echo image demonstrates a well-defined mass extending out from the intervertebral foramen and displacing the left psoas muscle. The mass is of mixed signal intensity with areas of moderately increased signal intensity. **E:** Axial T2-weighted (TR/TE 4,000/102 msec) fast spin-echo image demonstrates that much of the mass remains of relatively low signal intensity, with at most slight increase in intensity of the previously defined areas of increased signal. **F:** Axial contrast-enhanced CT demonstrates slight enhancement in regions of previously defined increased signal on MR.

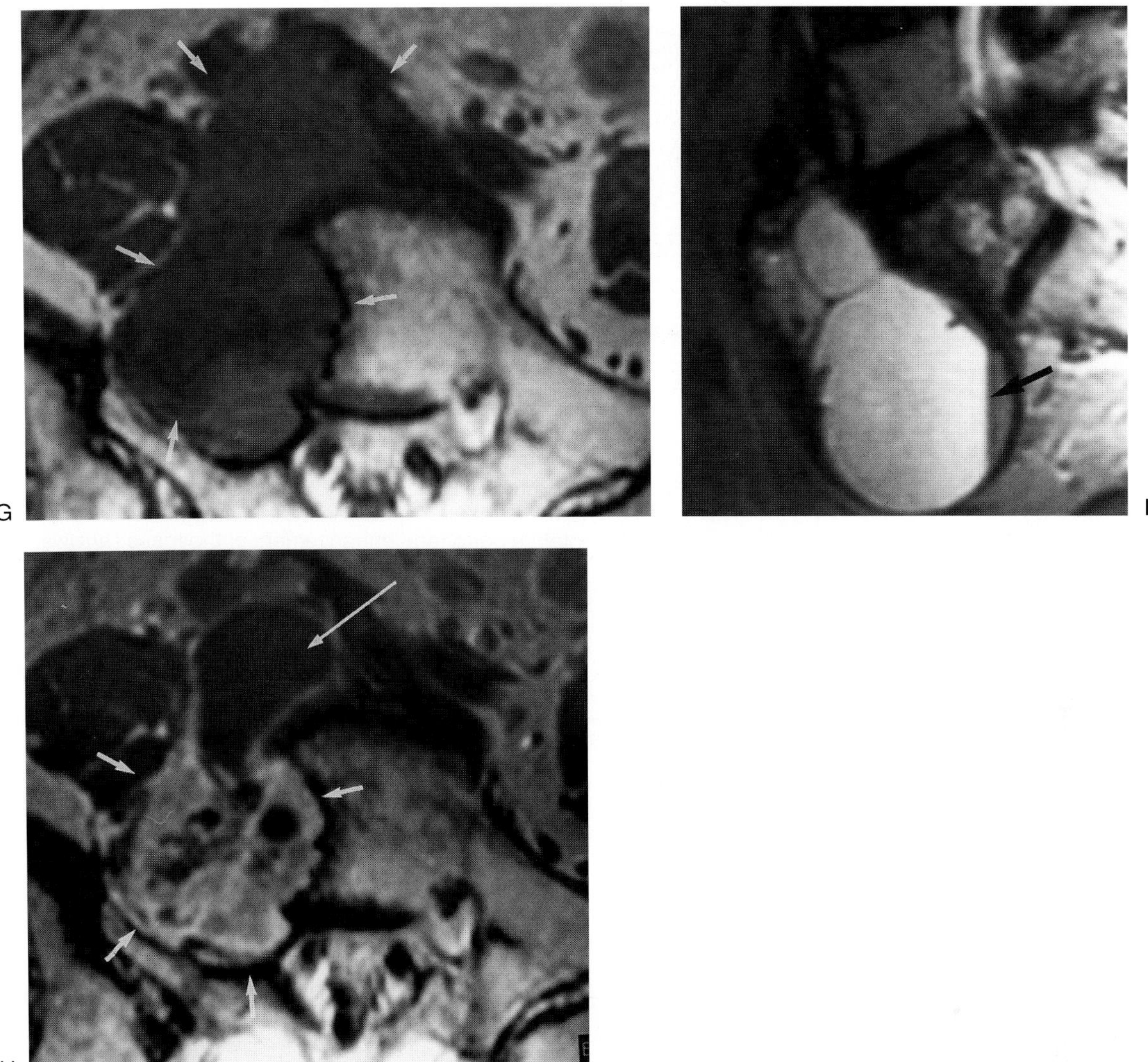

FIG. 22. G–I: Large sacral schwannoma with multiple areas of cystic degeneration in an 82-year-old woman who presented with a several-month history of lower back and leg pain. **G:** Axial T1-weighted image (TR/TE 800/16 msec) through the upper sacrum demonstrates a bilobed mass of low-to-intermediate signal intensity that completely obliterates the right S1 sacral foramen and extends anteriorly into the pelvis (*arrows*). **H:** On a T1-weighted image (TR/TE 800/16 msec) obtained following the administration of gadopentetate dimeglumine, the mass (*small arrows*) demonstrates a complex enhancement pattern with central areas of nonenhancement. Additionally, the anterior component of the "bilobed" mass does not enhance with contrast (*long arrow*). **I:** Sagittal T2-weighted fast spin-echo image (TR/TE 4,000/102 msec) through the more anterior component of the mass demonstrates high signal intensity throughout much of the lesion with an apparent fluid–fluid level (*arrow*). (Courtesy of Michael Brandt-Zawadski, M.D., Newport Beach, California.)

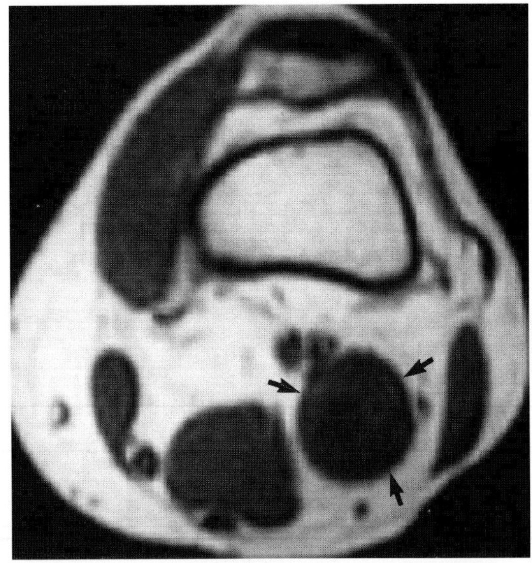

J

FIG. 22. J–L: Peripheral schwannoma in a 39-year-old attorney who complained of pain and a palpable mass along the posterior aspect of her knee. Physical examination elicited a positive percussion test. **J:** Axial T1-weighted image (TR/TE 500/20 msec) demonstrates a low-signal-intensity mass adjacent to the popliteal neurovascular bundle (*arrows*). **K:** Axial fat-suppressed T2-weighted fast spin-echo (TR/TE 4,600/104 msec) image demonstrates a "target" appearance with a central core of extremely high signal intensity (*arrow*) surrounded by a rim of slightly less intense but still augmented signal intensity. **L:** Fat-suppressed T1-weighted (TR/TE 500/20 msec) image following the administration of gadopentetate dimeglumine. There is intense peripheral enhancement of the mass (*arrows*).

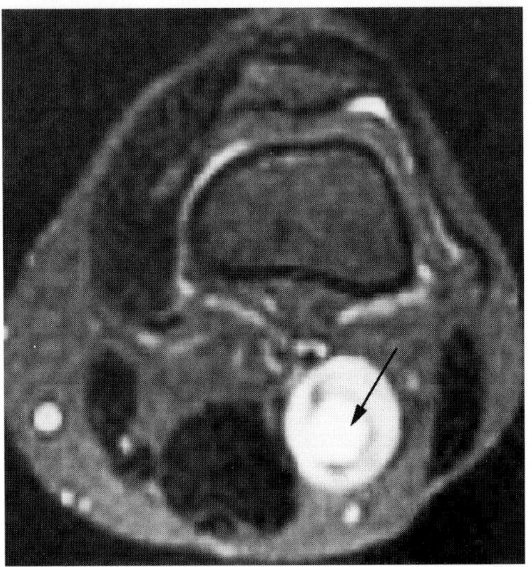

K

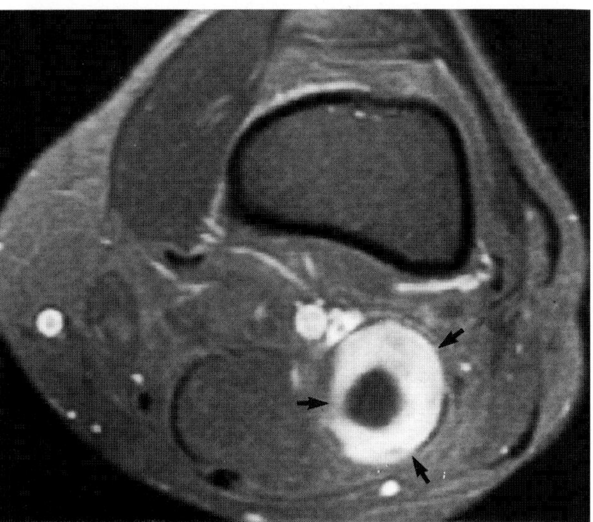

L

REFERENCES

Case 1

1. Jensen M, Brant-Zawadski M, Obuchowski N. Magnetic resonance imaging of the lumbar spine in people without back pain. *N Engl J Med* 1995;**331:**69–73.
2. Herzog RJ. Magnetic resonance imaging of the spine. In: Frymoyer, JW, ed. *The adult spine: principles and practice.* New York: Raven Press, 1991.
3. Fardon D, Pinkerton S, Balderston R. Terms used for diagnosis by English speaking spine surgeons. *Spine* 1993;**18:**274–277.
4. Boden SD, Davis DO, Dina TS. Abnormal magnetic-resonance scans of the lumbar spine in asymptomatic subjects: a prospective investigation. *J Bone Joint Surg* 1990;**72:**403–408.
5. Fardon, DF. Disorders of the spine: A coding system for diagnoses. *North American Spine Society.* Philadelphia: Hanley & Belfus, 1991.
6. Mink J. Terminology of disc disorders of the lumbar spine. California Managed Imaging Medical Group, Inc., 1993.
7. Fardon D. Position of the nomenclature committee of the North American Spine Society. 1995. (Personal communication.)
8. Yu S, Haughton VM, Sether IA. Anulus fibrosis in bulging intervertebral disks. *Radiology* 1988;**169:**761–763.
9. McCarron RF, Wimpee MW, Hudkins PG. The inflammatory effect of nucleus pulposus: a possible element in the pathogenesis of low-back pain. *Spine* 1987;**12:**760–764.
10. Park WM, McCall IW, O'Brien JP. Fissuring of the posterior anulus fibrosus in the lumbar spine. *Br J Radiol* 1979;**52:**382–387.
11. Ross JS, Modic MT, Masaryk TJ. Tears of the anulus fibrosus: assessment with Gd-DTPA-enhanced MR imaging. *AJNR* 1989;**10:** 1251–1254.
12. Yu S, Sether LA, Ho PSP. Tears of the anulus fibrosus: correlation between MR and pathologic findings in cadavers. *AJNR* 1988;**9:** 367–370.
13. Hirsch C, Schajowicz F. Studies on structural changes in the lumbar anulus fibrosus. *Acta Orthop Scand* 1952;**22:**184–223.
14. Masaryk TJ, Ross JS, Modic MT. High-resolution MR imaging of sequestered lumbar intervertebral disks. *AJR* 1988;**150:**1155–1162.
15. Modic M. Degenerative disorders of the spine. In: Modic M, Masaryk T, Ross J, eds. *Magnetic resonance imaging of the spine.* Chicago: Year Book Medical Publishers; 1989.
16. Modic M, Masaryk T, Boymphrey F. Lumbar herniated disc disease and canal stenosis: prospective evaluation by surface coil MR, CT, and myelography. *AJNR* 1986;**7:**709–717.
17. Hitselberger WE, Witten RM. Abnormal myelograms in asymptomatic patients. *J Neurosurg*1968;**28:**204–206.
18. McRae DL. Asymptomatic intervertebral disc protrusions. *Acta Radiol* 1956;**46:**9–27.
19. Wiesel SW, Tsourmas N, Feffer HL. A study of computer-assisted tomography. I. The incidence of positive CAT scans in an asymp-

tomatic group of patients. *Spine* 1984;**9:**549–551.
20. Powell MC, Wilson M, Szpryt P. Prevalence of lumbar disc degeneration observed by magnetic resonance in symptomless women. *Lancet* 1986;**2:**1366–1367.
21. Weinreb JC, Wolbarsht LB, Cohen JM. Prevalence of lumbosacral intervertebral disk abnormalities on MR images in pregnant and asymptomatic nonpregnant women. *Radiology* 1989;**170:**125–128.

Case 2

1. McRae D. Asymptomatic intervertebral disc protrusions. Acta Radiol 1956;46:9–27.
2. Hitselberger WF, Witten RM. Abnormal myelograms in asymptomatic patients. J Neurosurg 1968;28:2–6.
3. Wiesel S, Tsourmas N, Feffer H, Citrin C, Patronas N. A study of computer-assisted tomography. 1. The incidence of positive CAT scans in an asymptomatic group of patients. *Spine* 1984;9:549–551.
4. Boden SD, Davis D, Dina T, Patronas N, Wiesel S. Abnormal magnetic resonance scans of the lumbar spine in asymptomatic subjects. *J Bone Joint Surg* [*Am*] 1990;72A:403–408.
5. Jensen M, Brant-Zawadzki M, Obuchowski N, Modic M, Malkasian D, Ross J. Magnetic resonance imaging of the lumbar spine in people without back pain. *N Engl J Med* 1994;331:69–73.
6. Mixter WJ, Barr JS. Rupture of the intervertebral disc with involvement of the spinal canal. *N Engl J Med* 1934;211:210–215.
7. Vanharanta H. Etiology, epidemiology, and natural history of lumbar disk disease. *Spine* 1989;3:1–12.
8. Olmarker K, Rydevik B, Nordborg C. Autologous nucleus pulposus induces neurophysiologic and histologic changes in porcine cauda equina nerve roots. *Spine* 1993;18:1425–1432.
9. McCarron R, Wimpee M, Hudkins P, Laros G. The inflammatory effect of nucleus pulposus. A possible element in the pathogenesis of low-back pain. *Spine* 1987;12:760–764.
10. Franson R, Saal JS, Saal JA. Human disc phospholipase A_2 is inflammatory. *Spine* 1992;17:S129–S132.
11. Hirsch C, Schajowicz F. Studies on structural changes in the lumbar annulus fibrosus. *Acta Orthop Scand* 1953;22:184–231.
12. Kieffer SA, Stadian EM, Mohandas A, Peterson HO. Discographic-anatomical correlation to developmental changes with age in the intervertebral disc. *Acta Radiol* 1969;9:733–739.
13. Lipson SJ, Muir H. Experimental intervertebral disc degeneration—morphologic and proteoglycan changes over time. *Arthritis Rheum* 1981;24:12–20.
14. Lipson SJ, Muir H. Proteoglycans in experimental intervertebral disc degeneration. *Spine* 1981;6:194–210.

Case 3

1. Saal JA, Saal JS. The nonoperative treatment of herniated nucleus pulposus with radiculopathy: an outcome study. *Spine* 1989;14:431–437.
2. Saal JA, Saal JS, Herzog RJ. The natural history of lumbar intervertebral disc extrusions treated nonoperatively. *Spine* 1990;15:683–687.
3. Bozzao A, Gallucci M, Masciocchi C, et al. Lumbar disk herniation: MR imaging assessment of natural history in patients treated without surgery. *Radiology* 1992;185:135–141.
4. Pardatscher K, Fiore Dl, Barbiero A. The natural history of lumbar disc herniations assessed by a CT follow-up study. *Neuroradiology* 1991;33[Suppl]:84.
5. Teplick JG, Haskin ME. Spontaneous regression of herniated nucleus pulposus. *AJNR* 1985;6:331–335.
6. Guinto FC, Hashim H, Stumer M. CT demonstration of disc regression after conservative therapy. *AJNR* 1984;5:632–637.
7. McCarron RF, Wimpee MW, Hudgins PG, et al. The inflammatory efect of nucleus pulposus: a possible element in the pathogenesis of low back pain. *Spine* 1987;12:760–764.
8. Saal JS, Franson RC, Dobrow R, et al. High levels of inflammatory phospholipase A_2 activity in lumbar disc herniations. *Spine*.
9. Kornings JG, Williams FJB, Deutman R. The effects of chemonucleolysis as demonstrated by computerized tomography. *J Bone Joint Surg* [*Br*] 1984;66:417–421.
10. Deutsch AL, Howard M, Dawson EG. Lumbar spine following successful surgical discectomy. Magnetic resonance imaging features and implications. *Spine* 1993;18:1054–1060.
11. Deutsch AL, Mink JH, Goldstein T et al. MR imaging in percutaneous lumbar discectomy. *Radiology* 1989;173(P):544.
12. Lindblom K, Hultqvist G. Absorption of protruded disc tissue. *J Bone Joint Surg* [*Am*] 1950;32:557–560.
13. Gundry CR, Heithoff KB. Epidural hematoma of the lumbar spine: 18 surgically confirmed cases. *Radiology* 1993;187:427–431.
14. Heithoff KB, Amster JL. The spine. In Mink JH, Deutsch AL, eds. *MRI of the musculoskeletal system: a teaching file.* New York: Raven Press;1990:119–123.

Case 4

1. Jinkins JR. Magnetic resonance imaging of benign nerve root enhancement in the unoperated and postoperative lumbosacral spine. *Neuroimaging Clin North Am* 1993;3:525–541.
2. Boden SD, Davis DO, Dina TS, et al. Abnormal magnetic resonance scans of the lumbar spine in asymptomatic subjects. *J Bone Joint Surg* [*Am*] 1990;72A:403–408.
3. Jinkins JR. MR of enhancing nerve roots in the unoperated lumbosacral spine. *AJNR* 1993;14:193–202.
4. Rydevik B, Lundborg G. Permeability of intraneural microvessels and perineurium following acute, graded experimental nerve compression. *Scand J Plast Reconstr Surg* 1977;11:179–187.
5. Kobayashi S, Yoshizawa H, Hachiya Y, et al. Vasogenic edema induced by compression injury to the spinal nerve root. Distribution of intravenously injected protein tracers and gadolinium enhanced magnetic resonance imaging. *Spine* 1993;18:1410–1424.
6. Sparrow JR, Kiernan JA. Endoneurial vascular permeability in degenerating and regenerating peripheral nerves. *Acta Neuropathol* 1981;53:181–188.
7. Weerasuriya A, Rapoport SE, Taylor RE. Perineurial permeability increases during Wallerian degeneration. *Brain Res* 1980;10:633–637.
8. Breber RK, Williams AL, Daniels DL, et al. Contrast enhancement in spinal MR imaging. *AJNR* 1989;10:633–637.
9. Kirkaldy-Willis WH, Paine KWE, Cauchoix J, et al. Lumbar spinal stenosis. *Clin Orthop* 1974;99:30–50.
10. Moreland LW, Lopez Mendez A, Alarcon GS. Spinal stenosis. A comprehensive review of the literature. *Semin Arthritis Rheum* 1989;19:127–149.
11. Salibi BS. Neurogenic intermittent claudication and stenosis of the lumbar canal. *Surg Neurol* 1976;5:269–272.
12. Jinkins JR. Gd DTPA enhanced MR of the lumbosacral spine in patients with claudication. *J Comput Assist Tomogr* 1993;17:555–562.
13. Watanabe R, Parke WW. Vascular and neural pathology of lumbosacral spinal stenosis. *J Neurosurg* 1986;64:64–70.

Case 5

1. Ross JS, Masaryk TJ, Modic MT, et al. Lumbar spine: postoperative assessment with surface coil MR imaging. *Radiology* 1987;164:851–860.
2. Boden SD, Davis DO, Dina TS, et al. Contrast enhanced MR imaging performed after successful lumbar disk surgery: prospective study. *Radiology* 1992;182:59–64.
3. Kahn T, Roosen N, Messign AM, et al. Follow-up studies with Gd DTPA enhanced MR imaging of the asymptomatic postdiskectomy lumbar spine. *Radiology* 1990;177(P):233.
4. Deutsch AL, Howard M, Dawson EG, et al. Lumbar spine following successful surgical disectomy: magnetic resonance imaging features and implications. *Spine* 1993;18:1054–1060.
5. Ross JS. Magnetic resonance assessment of the postoperative spine: degenerative disc disease. *Radiol Clin North Am* 1991;29:793.
6. Ross JS. Post-operative spine. In Forbes G, Quencer RM, Harnsberger HR, eds. *Syllabus: special course in neuroradiology: clinical approach and management for diagnostic imaging.* Presented at the 80th Scientific Assembly and Annual Meeting of the Radiological Society of North America, November, 1994. RSNA Publications.
7. Dina TS, Boden SD, Davis DO. Lumbar spine after surgery for herniated disk: imaging findings in the early postoperative period. *AJR* 1995;164:665.

8. Ross JS, Masaryk TJ, Schrader M. MR imaging of the postoperative lumbar spine: assessment with gadopentetate dimeglumine. *AJR* 1990;155:867.
9. Bundschuh CV, Stein L, Slusser JH, et al. Distinguishing between scar and recurrent herniated disk in postoperative patients: values of contrast-enhanced CT and MR imaging. *AJNR* 1990;11:949.
10. Hueftle MG, Modic MT, Ross JS, et al. Lumbar spine: postoperative MR imaging with Gd-DTPA. *Radiology* 1988;167:817.
11. Bundschuh CV. Imaging of the postoperative lumbosacral spine. *Neuroimaging Clin North Am* 1993;3:499.
12. Hochhauser L, Kieffer Sa, Cacayorin ED, et al. Recurrent postdiskectomy low back pain: MR-surgical correlation. *Am J Roentgenol* 1988;151(4):755–760. .
13. Bundschuh CV, Modic MT, Ross JS, et al. Epidural fibrosis and recurrent disc herniation in the lumbar spine: assessment with magnetic resonance. *AJNR* 1988;9:169–178.
14. Jinkins JR. The pathoanatomic basis of somatic and autonomic syndromes originating in the lumbosacral spine. *Neuroimaging Clin North Am* 1993;3:443.
15. Jinkins JR. Magnetic resonance imaging of benign nerve root enhancement in the unoperated and postoperative lumbosacral spine. *Neuroimaging Clin North Am* 1993;3:525.
16. Deutsch AL, Mink JH, Goldstein T. MR imaging in percutaneous lumbar disckectomy. *Radiology* 1989;173(P):544.
17. Masaryk TJ, Boumphrey F, et al. Effects of chemonucleolysis demonstrated by MR imaging. *J Comput Assist Tomogr* 1986;10:917–923.
18. Bozzao A, Gallucci M, Masciocchi C, et al. Lumbar disc herniation: MR imaging assessment of natural history in patients treated without surgery. *Radiology* 1992;185:135–141.
19. Saal JJ, Franson RC, Dobrow R, et al. High levels of inflammatory phospholipase A_2 activity in lumbar disc herniations. *Spine* 1990;15:674.
20. Haughton VM, Nguyen CM, Ho KC. Effect of experimental root sheath compression on dura. *Invest Radiol* 1989;24:204–205.
21. Marshall LL, Trethewie ER, Curtain CC. Chemical radiculitis. A clinical, physiological and immunological study. *Clin Orthop* 1977;129:61–67.
22. Rydevik B, Brown MD, Lundborg G. Pathoanatomy and pathophysiology of nerve root compression. *Spine* 1984;9:7–15.
23. Rydevik B, Brown MD, Ehira T, et al. Effects of graded compression and nucleus pulposus on nerve tissue; an experimental study in rabbits. *Acta Orthop Scand* 1983;54:670–671.
24. Howe JF, Loeser JD, Calvin WH. Mechanosensitivity of dorsal root ganglia and chronically injured axons; a physiological basis for the radicular pain of nerve root compression. *Pain* 1977;3:25–41.
25. Smyth MJ, Wright V. Sciatica and the intervertebral disc: an experimental study. *J Bone Joint Surg [Br]* 1958;40(B):1401–1418.
26. Rydevik B, Holm S, Brown MD. Nutrition of spinal nerve roots; the role of diffusion from the cerebrospinal fluid. *Trans Orthop Res Soc* 1984;9:276.
27. Breig A, Marions O. Biomechanics of the lumbosacral nerve roots. *Acta Radiol* 1963;1:1141–1160.
28. Ransford AO, Harries BJ. Localized arachanoiditis complicating lumbar disc lesions. *J Bone Joint Surg [Br]* 1972;54B:656–665.
29. Jinkins JR, Osborn AG, Garret D, Hunt S, Story JL. Spinal nerve enhancement with Gd-DTPA: MR correlation with the postoperative lumbosacral spine. *AJNR* 1993;14:383–394.

Case 6

1. Rish BL. A critique of the surgical management of lumbar disc disease in a private neurosurgical practice. *Spine* 1984;9:500.
2. Ross JS. Magnetic resonance assessment of the postoperative spine: degenerative disc disease. *Radiol Clin North Am* 1991;29:793.
3. Burton CV. Lumbosacral arachnoiditis. *Spine* 1978;3:24.
4. Burton CV, Kirkaldy-Willis WH, Yong-Hing K, et al. Causes of failure of surgery on the lumbar spine. *Clin Orthop* 1981;157:191.
5. Jinkins JR. The pathoanatomic basis of somatic and autonomic syndromes originating in the lumbosacral spine. *Neuroimaging Clin North Am* 1993;3:443.
6. Jinkins JR, Garrett D, Osborn AG, et al. Spinal nerve radiculitis: relationship to the lumbosacral post surgical syndrome. Presented at the Ninth Annual Meeting of the Society of Magnetic Resonance Imaging, Chicago, 1991.
7. Ross JS, Modic MT, Masaryk TJ. Tears of the annulus fibrosus: assessment with Gd-DTPA-enhanced MR imaging. *AJNR* 1989;10:1251.
8. Ross JS, Delamarter R, Heuftle MG, et al. Gadolinium-DTPA-enhanced MR imaging of the postoperative lumbar spines: time course and mechanisms of enhancement. *AJNR* 1989;10:37.
9. Ross JS. Post-operative spine. In Forbes G, Quencer RM, Harnsberger HR, eds. *Syllabus: special course in neuroradiology: clinical approach and management for diagnostic imaging.* Presented at the 80th Scientific Assembly and Annual Meeting of the Radiological Society of North America, November, 1994. RSNA Publications.
10. Mirowitz SA, Shady KL. Gadopentatate dimeglumine-enhanced MR imaging of the postoperative lumbar spine: comparison of fat-suppressed and conventional T1 weighted images. *AJR* 1992;159:385.
11. Bobman SA, Atlas SW, Listerund J, et al. Postoperative lumbar spine; contrast enhanced chemical shift MR imaging. *Radiology* 1991;179:557.
12. Ross JS, Masaryk TJ, Schrader M. MR imaging of the postoperative lumbar spine: assessment with gadopentetate dimeglumine. *AJR* 1990;155:867.
13. Bundschuh CV, Stein L, Slusser JH, et al. Distinguishing between scar and recurrent herniated disk in postoperative patients: values of contrast-enhanced CT and MR imaging. *AJNR* 1990;11:949.
14. Ross, JS, Blaser S, Masaryk TJ, et al. Gd-DTPA enhancement of posterior epidural scar: an experimental model. *AJNR* 1989;10:1083.
15. Hueftle MG, Modic MT, Ross JS, et al. Lumbar spine: postoperative MR imaging with Gd-DTPA. *Radiology* 1988;167:817.
16. Jinkins JR. Magnetic resonance imaging of benign nerve root enhancement in the unoperated and postoperative lumbosacral spine. *Neuroimaging Clin North Am* 1993;3:525.
17. Bangert BA, Ross JS. Arachnoiditis affecting the lumbosacral spine. *Neuroimaging Clin North Am* 1993;3:517.

Case 7

1. Liu SS, Williams KD, Drayer BP, et al. Synovial cysts of the lumbosacral spine: diagnosis by MR imaging. *AJR* 1990;154:163–167.
2. Heithoff KB, Armster JL. The spine. In Mink JH, Deutsch AL, eds. *MRI of the musculoskeletal system: a teaching file.* New York: Raven Press;1990:146–149.
3. Silbergleit R, Gebarski SS, Brunberg JA, et al. Lumbar synovial cysts: correlation of myelographic, CT, MR, and pathologic findings. *AJRN* 1990;11:777–779.
4. Yock DH Jr. *Magnetic resonance imaging of CNS disease: a teaching file.* St. Louis: CV Mosby; 1994:532–533.
5. Herzog RJ. Magnetic resonance imaging of the spine. In: Frymoyer JW, ed. *The adult spine: principles and practice.* New York: Raven Press; 1991:457–507.
6. MacNab I. Spondylolisthesis with an intact neural arch, the so-called pseudo-spondylolisthesis. *J Bone Joint Surg [Br]* 1950;32:325–333.
7. Inoue SI, Watanabe T, Goto S, et al. Degenerative spondylolisthesis. *Clin Orthop* 1988;227:90–98.
8. Sato K, Wakamatsu E, Yoshizumi A, et al. The configuration of the laminas and facet joints in degenerative spondylolisthesis; a clinicoradiologic study *Spine* 1989;14:1265–1271.

Case 8

1. Ross JS. Post-operative spine. In Forbes G, Quencer RM, Harnsberger HR, eds. *Syllabus: special course in neuroradiology: clinical approach and management for diagnostic imaging.* Presented at the 80th Scientific Assembly and Annual Meeting of the Radiological Society of North America, November, 1994. RSNA Publications.
2. Modic MT, Feiglin DH, Piraino DW, et al. Vertebral osteomyelitis: assessment using MR. *Radiology* 1985;157:157–166.
3. Sharif HS. Role of MR imaging in the management in the management of spinal infections. *AJR* 1992;158:1333–1345.
4. Shariff HS, Clark DC, Aabed MY, et al. Granulomatous spinal infections: MR imaging. *Radiology* 1990;177:101–107.
5. Smith AS, Weinstein MA, Mizushima A, et al. MR imaging charac-

teristics of tuberculous spondylitis vs vertebral osteomyelitis. *AJNR* 1989;10:619–625.
6. Wiley Am Trueta J. The vascular anatomy of the spine and its relationship to pyogenic vertebral osteomyelitis. *J Bone J. Surg [Br]* 1941;4:796–809.
7. Modic MT, Steinberg PM, Ross JS, et al. Degenerative disc disease: assessment of changes in vertebral body marrow with magnetic resonance imaging. *Radiology* 1988;166:193–199.
8. Modic MT, Masaryk TH, Ross JS, et al. Imaging of degenerative disk disease. *Radiology* 1988;168:177–186.
9. Boden SD, Davis DO, Dina TS, et al. Post-operative diskitis: distinguishing early MR imaging findings from normal postoperative changes. *Radiology* 1992;184:765–771.
10. Mirowitz SA, Shady KL. Gadopentetate dimeglumine enhanced MR imaging of the postoperative lumbar spine: comparison of fat-suppressed and conventional T1 weighted images. *AJR* 1992;159: 385–389.

Case 9

1. James CCM, Lasman LP. *Spinal dysraphism: spina bifida oculta.* New York: Appleton Century Croft; 1972.
2. Kernohan JW, Woltman HW, Adson AW. Gliomas arising from the region of the cauda equina. *Arch Neurol Psychiatry* 1933;29: 287–307.
3. Yock DH. *Magnetic resonance imaging of CNS disease: a teaching file.* St. Louis: CV Mosby; 1995:586–594.
4. Rapoport RJ, Flander AE, Tartaglino LM. Intradural extramedullary causes of myelopathy. *Semin US CT MRI* 1994;15:189–225.
5. Sonneland PR, Scheithauer BW, Onofrio BM. Myxopapillary ependymoma, a clinicopathologic and immunocytochemical study of 77 cases. *Cancer* 1985;56:883–893.

Case 10

1. Ross JS, Masaryk TJ, Schrader M. MR imaging of the postoperative lumbar spine: assessment with gadopentetate dimeglumine. *AJR* 1990;155:867.
2. Bundschuh CV, Mittman B, Ladaga LE, et al. Correlation of patient symptoms, histology, and MR imaging characteristics of lumbar herniated discs. Presented at the Ninth Annual Meeting of the Society of Magnetic Resonance Imaging, Chicago, 1991.
3. Dina TS, Boden SD, Davis DO. Lumbar spine after surgery for herniated disk: imaging findings in the early postoperative period. *AJR* 1995;164:665.
4. Bundschuh CV, Stein L, Slusser JH, et al. Distinguishing between scar and recurrent herniated disk in postoperative patients: values of contrast-enhanced CT and MR imaging. *AJNR* 1990;11:949.
5. Saal JJ, Franson RC, Dobrow R, et al. High levels of inflammatory phospholipase A_2 activity in lumbar disc herniations. *Spine* 1990; 15:674.
6. Naylor A, Happey F, Turner RL, et al. Enzymic and immunological activity in the intervertebral disc. *Orthop Clin North Am* 1975;6: 51.
7. Djukic S, Lang P, Morris J. The postoperative spine: magnetic resonance imaging. *Ortho Clin of North America* 1990;21:603.
8. Ross JS, Delamarter R, Heuftle MG, et al. Gadolinium-DTPA-enhanced MR imaging of the postoperative lumbar spines: time course and mechanisms of enhancement. *AJNR AM J Neuroradiol* 1989;10:37.
9. Bundschuh CV. Imaging of the postoperative lumbosacral spine. *Neuroimag Clin of North America* 1993;3:499.
10. Hueftle MG, Modic MT, Ross JS, et al. Lumbar spine: postoperative MR imaging with Gd-DTPA. *Radiology* 1988;167:817.
11. Branemark P-I, Ekholm R, Lundskog J, et al. Tissue response to chymopapain in different concentration. Animal investigations on microvascular effects. *Clin Orthop Rel Res* 1969;67:52.
12. Bobechko WP, Hirsch C. Auto-immune response to nucleus pulposus in the rabbit. *J Bone Joint Surg* 1965;47B:574.
13. McCarron RF, Wimpee MW, Hudkins PG, et al. The inflammatory effect of nucleus pulposus. A possible element in the pathogenesis of low-back pain. *Spine* 1987;12:760.

Case 11

1. Gundry CR, Heithoff KB. Epidural hematoma of the lumbar spine: 18 surgically confirmed cases. *Radiology* 1993;187:427–431.
2. Heithoff KB, Amster JL. The spine. In: Mink JH, Deutsch AL, eds. *MRI of the musculoskeletal system: a teaching file.* New York: Raven Press; 1990:119–123.
3. Boyd HR, Pear BL. Chronic spontaneous epidural hematoma. *J Neurosurg* 1972;36:239–242.
4. Levitan LH, Wiens CW. Chronic lumbar extradural hematoma: CT findings. *Radiology* 1983;148:707–708.
5. Rothfus WE, Chedid MK, Deeb ZL, et al. MR imaging in diagnosis of spontaneous spinal epidural hematomas. *J Comput Assist Tomogr* 1987;11:851–854.

Case 12

1. Stolpen AH. Case of the season: spondyloarthropathy of renal dialysis. *Semin Roentgenol* 1993;28:96–100.
2. Fiocchi O, Bedani PL, Orzincolo C, et al. Radiological features of dialysis amyloid spondyloarthropathy. *Int J Artif Organs* 1989;12: 216–222.
3. Kaplan P, Resnick D, Murphey M, et al. Destructive noninfectious spondyloarthropathy in hemodialysis patients: a report of four cases. *Radiology* 1987;162:241–244.
4. Weiske R, Munding M, Schneider HW. Destructive non-infectious spondyloarthropathy in chronic renal insufficiency. *ROFO Fortschr Geb Rontgenstr Nuklearmed* 1988;149:129–135.
5. Stabler A, Kroner G, Seiderer M. MRI in dialysis associated destructive spondyloarthropathy of the atlantoaxial region. *ROFO Fortschr Geb Rontgenstr Neuen Bildgeb Verfahr* 1991;154:469–474.
6. Davidson GS, Montanera WJ, Fleming JE, et al. Amyloid destructive spondyloarthropathy causing cord compression; related to chronic renal failure and dialysis. *Neurosurgery* 1993;33:519–522.
7. Kerr R, Bjorkengren A, Bielecki DK, et al. Destructive spondyloarthropathy in hemodialysis patients. *Skeletal Radiol* 1988;17: 176–180.
8. Bindi P, Chandard J. Destructive spondyloarthropathy in dialysis patients: an overview. *Nephron* 1990;55:104–109.
9. Sethi D, Morgan TC, Brwon EA, et al. Dialysis arthropathy: a clinical, biochemical, radiological and histological study of 36 patients. *Q J Med* 1990;77:1061–1082.
10. Naidich JB, Mossey RT, McHeffrey-Atkinson B, et al. A spondyloarthropathy from long-term hemodialysis. *Radiology* 1988;167: 761–764.
11. Cary NR, Serthi D, Brown EA, et al. Dialysis arthropathy; amyloid or iron. *BMJ* 1986;293:1392–1394.
12. Brown EA, Arnold IR, Gopwer PE. Dialysis arthropathy: complication of long term treatment with hemodialysis. *BMJ* 1986;292:163–166.
13. Athanasou UN, Ayers D, Rainey AJ, et al. Joint and systemic distribution of dialysis amyloid. *Q J Med* 1991;78:205–214.

Case 13

1. Hijikata S. Percutaneous nucleotomy. A new concept technique and 12 years experience. *Clin Orthop* 1989;238:9–23.
2. Kambin P, Brager MD. Percutaneous posterolateral discectomy. Anatomy and mechanism. *Clin Orthop* 1987;223:145–154.
3. Kambin P, Schaffer JL. Percutaneous lumbar discectomy. Review of 100 patients and current practice. *Clin Orthop* 1989;238:24–34.
4. Morris J, Onik G. Percutaneous nucleotomy. In: White AH, Rothman RH, Ray CD, eds. *Lumbar spine surgery: techniques and complications.* St. Louis: CV Mosby; 1987:186–191.
5. Onik G, Helms CA, Ginsburg L, et al. Percutaneous lumbar diskectomy using a new aspiration probe. *AJR* 1985;144:1137–1140.
6. Onik G, Helms CA, Ginsburg L, et al. Percutaneous lumbar diskectomy using a new aspiration probe: porcine and cadaver model. *Radiology* 1985;155:251–252.
7. Delamarter RB, Howard MW, Goldstein T, et al. Percutaneous lumbar discectomy. *J Bone Joint Surg [Am]* 1995;77A:578–584.
8. Goldstein TB, Mink JH, Dawson EG. Early experience with automated percutaneous lumbar discectomy in the treatment of lumbar

disc herniation. *Clin Orthop* 1989;238:77–82.
9. Onik G, Maroon J, Helms C, et al. Automated percutaneous discectomy: initial patient experience. Work in progress. *Radiology* 1987; 162:129–132.
10. Deutsch AL, Mink JH, Goldstein T, et al. MR imaging in percutaneous lumbar discectomy. *Radiology* 1989;173(P):43.
11. Maroon J, Allen RC. Retrospective study of of 1054 automated percutaneous lumbar discectomy cases: a 20 month clinical followup at 35 U.S. centers. *J Neurol Orthop Med Surg* 1989;10:335–337.
12. Onik G, Mooney V, Maroon JC, et al. Automated percutaneous discectomy: a prospective multi-institutional study. *Neurosurgery* 1990;26:228–238.
13. Marshall LL, Trethewie ER. Chemical irritation of nerve root in disc prolapse. *Lancet* 1973;2:320.
14. Marshall LL, Trethewie ER, Curtain CC. Chemical radiculitis. A clinical, physiological and immunological study. *Clin Orthop* 1977; 129:61–67.
15. Nathan C, Tsunawaki S. Secretion of toxic oxygen products by marcrophages, regulatory cytokines and their effects on the oxidase. Read at Rheumatology and Clinical Immunology: An Advanced Course, San Francisco, California, 1986.
16. Saal JS, Franson RC, Dobrow R, et al. High levels of inflammatory phospholipase 2 activity in lumbar disc herniations. *Spine* 1990;15: 674–678.
17. Korkala O, Gronblad M, Liesi P, et al. Immunohistochemical demonstration of nociceptors in the ligamentous structures of the lumbar spine. *Spine* 1985;10:156–157.
18. Weinstein J. Mechanisms of spinal pain. The dorsal root ganglion and its role as a mediator of low back pain. *Spine* 1896;11:999–1001.
19. Weinstein J, Pope M, Schmidt R, et al. Neuropharmacologic efects of vibration on the dorsal root ganglion. An animal model. *Spine* 1988;13:521–525.
20. Weinstein J, Claverie W, Gibson S. The pain of discography. *Spine* 1988;13:1344–1348.
21. Hult L. Retroperitoneal disc fenestration in low back pain and sciatica. A preliminary report. *Acta Orthop Scand* 1951;20:342–348.
22. Mink JH. Imaging evaluation of the candidiate for percutaneous lumbar discectomy. *Clin Orthop* 1989;238:83–91.

Case 14

1. Haddad SF, Hitchon PW, Godersky JC. Idiopathic and glucocorticoid induced spinal epidural lipomatosis. *J Neurosurg* 1991;74:38–42.
2. Roy-Camille R, Mazel C, Husson JL. Symptomatic spinal epidural lipomatosis induced by a long term steroid treatment. Review of the literature and report of two additional cases. *Spine* 1991;16: 1365–1371.
3. Quint DJ, Boulos RS, Sanders WP, et al. Epidural lipomatosis. *Radiology* 1988;169:485–490.
4. Kaplan JG, Barasch E, Hirschfeld A, et al. Spinal epidural lipomatosis; a serious complication of iatrogenic Cushing's syndrome. *Neurology* 1989;39:1031–1034.
5. Toshniwal PK, Glick RP. Spinal epidural lipomatosis: report of a case secondary to hypothyroidism and review of the literature. *J Neurol* 1987;234:172–176.

Case 15

1. Slone RM, McEnery KW, Bridwell KH, Montgomery WJ. Principles and imaging of spinal instrumentation. *Radiol Clin North Am* 1995;33:189–212.
2. Slone RM, McEnery KW, Bridwell KH, Montgomery WJ. Fixation techniques and instrumentation used in the thoracic, lumbar, and lumbosacral spine. *Radiol Clin North Am* 1995;33:233–265.
3. Slone RM, Montgomery WJ, MacMillan M. Spinal fixation: Part 3: Complication of spinal instrumentation. *Radiographics* 1993;13: 797–816.
4. Slone RM, MacMillan M, Montgomery WJ, et al. Spinal fixation: Part 2: Fixation techniques and hardware for the thoracic and lumbosacral spine. *Radiographics* 1993;13:521–543.

Case 16

1. Rothman SLG, Glenn WV. CT multiplanar reconstruction in 253 cases of lumbar spondylolysis. *AJNR* 1984;5:81–90.
2. Grenier N, Kressel HY, Schiebler ML, Grossman RI. Isthmic spondylolysis of the lumbar spine: MR imaging at 1.5T *Radiology* 1989; 170:489–493.
3. Teplick JG, Laffey PA, Berman A, Haskin ME. Diagnosis and evaluation of spondylolisthesis and or spondylolysis on axial CT. *AJNR* 1986;7:479–491.
5. Rauch RA, Jinkins JR. Lumbosacral spondylolisthesis associated with spondylolysis. *Neuroimaging Clin North Am* 1993;3:543–555.
6. Jinkins JR, Rauch RA. Magnetic resonance imaging of entrapment of lumbar nerve roots in spondylolytic spondylolisthesis. *J Bone Joint Surg [Am]* 1994;76:1643–1648.

Case 17

1. Major NM, Helms CA, Genant HK. Calcification demonstrated as high signal intensity on T1 weighted MR images of the disks of the lumbar spine. *Radiology* 1993;189:494–496.
2. Williams AL. Lumbar disk disease; imaging update. *Radiology* 1994;191:884 (abst).
3. Bankert BA, Modic MT, Ross JS, et al. Hyperintense discs on T1 weighted MR: correlation with calcification. *Radiology* 1995;195: 437–443.
4. Henkelman RM, Watts MF, Kucharczyk W. High signal intensity in MR images of calcified brain tissue. *Radiology* 1991;179:199–206.
5. Quint DJ. Hyperintense disks on T1-weighted MR images: are they important? *Radiology* 1995;195:325–326.
6. Hanus F, Gillis P. Relaxation of water adsorbed on the surface of silica powder. *J Magn Reson Imaging* 1984;59:437–445.
7. Gamsu G, deGeer G, Cann C, et al. A preliminary study of MRI quantification of simulated calcified pulmonary nodules. *Invest Radiol* 1987;22:853–858.
8. Kucharczyk W, Henkelman RM. Visibility of calcium on MR and CT; can MR show calcium that CT cannot? *AJNR* 1994;15:1145–1148.

Case 18

1. Burton CV, Kirkaldy-Willis WH, Yong-Hing K, et al. Causes of failure of surgery on the lumbar spine. *Clin Orthop* 1981;157:191.
2. Quencer RM, Tenner M, Rothman L. The postoperative myelogram. Radiographic evaluation of arachnoiditis and dural/arachnoidal tears. *Radiology* 1977;123:667.
3. Ross JS, Masaryk TJ, Modic MT, et al. MR imaging of lumbar arachnoiditis. *AJNR* 1987;8:885.
4. Ross JS. Post-operative spine. In: Forbes G, Quencer RM, Harnsberger HR, eds. *Syllabus: special course in neuroradiology: clinical approach and management for diagnostic imaging.* Presented at the 80th Scientific Assembly and Annual Meeting of the Radiological Society of North America, November, 1994. RSNA Publications.
5. Castan P, Bourbotte G, Herail JP, et al. Follow-up and postoperative radiculography. *J Neuroradiol* 1977;4:49.
6. Meyer JD, Latchow RE, Roppolo HM, et al. Computed tomography and myelography of the postoperative lumbar spine. *AJNR* 1982;3: 223.

Case 19

1. Grain GO, Karr JP. Diffuse leptomenigeal carcinomatosis: clinical and pathological characteristics. *Neurology* 1955;5:706–722.
2. Heithoff K, Armster J. *The spine.* In: Mink JH, Deutsch AL, eds. *MRI of the musculoskeletal system: a teaching file.* New York: Raven Press; 1990:176–178.
3. Sze G, Twohig M. Neoplastic disease of the spine and spinal cord. In Atlas S, ed. *Magnetic resonance imaging of the brain and spine.* New York: Raven Press; 1991:950–952.
4. Krol G, Sze G, Malkin M, et al. MR of cranial and spinal meningeal carcinomatosis: comparison with CT and myelography. *AJNR* 1988; 9:709–714.

5. Sze G. MR of the spine: contrast agents. *Radiol Clin North Am* 1988;26:1009–1024.
6. Yock DH. Tumors of the spinal canal. In: Yock DH, ed. *Magnetic resonance imaging of CNS disease: a teaching file.* St. Louis: CV Mosby; 1995:548–597.
7. Sze G, Abramson A, Krol G, et al. Gadolinium DTPA/dimeglumine in the MR evaluation of intradural extramedullary spinal disease. *AJNR* 1988;9:153–163.

Case 20

1. McCulloch JA. Microsurgical approach to the foraminal HNP.
2. Abdullah AF, Wolber PGH, Warfield JR, et al. Surgical management of extreme lateral lumbar disc herniations: review of 138 cases. *Neurosurgery* 1988;22:648–653.
3. Abdullah AF, Ditto E, Byrd E, et al. Extreme lateral lumbar disc herniations. *J Neurosurg* 1974;41:229–234.
4. Gado M, Patel J, Hodges FJ. Lateral disc herniations into the lumbar intervertebral foramen. *AJNR* 1983;4:598–600.

Case 21

1. Byrd SE, Cohn ML, Biggers SL, et al. The radiographic evaluation of the symptomatic postoperative lumbar spine patient. *Spine* 1985; 10:652–611.
2. Jinkins JR, Osborn AG, Garrett D, et al. Spinal nerve enhancement with Gd-DTPA: MR correlation with the postoperative lumbosacral spine *AJNR* 1993;14:383–394.
3. Fukuhara N, Kumamoto T, Nakazawa T. Blood-nerve barrier: effect of ligation of the peripheral nerve. *Exp Neurol* 1979;63:573–582.
4. Kobayashi S, Yoshizawa H, Hachiya Y, et al. Vasogenic edema induced by compression injury to the spinal nerve root: distribution of intravenously injected protein tracers and gadolinium enhanced magnetic resonance imaging. *Spine* 1993;18:1410–1424.
5. Jinkins JR. Magnetic resonance imaging of benign nerve root enhancement in the unoperated and postoperative lumbosacral spine. *Neuroimaging Clin North Am* 1993;3:525–541.
6. Boden SD, Davis DO, Dina TS, et al. Contrast enhanced MR imaging performed after successful lumbar disc surgery; prospective study. *Radiology* 1992;182:59–64.
7. Olsson Y. Vascular permeability in the peripheral nervous system. In: Dyck PJ, Thomas PK, Lambert EH, et al., eds. *Peripheral neuropathy.* Philadelphia: WB Saunders; 1984:579–597.
8. Sjostrand J, Rydevik B, Lundborg G, et al. Impairment of intraneural microcirculation, blood-nerve barrier and axonal transport in experimental nerve ischemia and compression. In: Korr IM, ed. *The Neurobiologic mechanisms in manipulative therapy.* New York: Plenum; 1978:337–355.

Case 22

1. Rapoport RJ, Flander AE, Tartaglino LM. Intradural extramedullary causes of myelopathy. *Semin US CT MRI* 1994;15:189–225.
2. Yock DH Jr. *Magnetic resonance imaging of CNS disease: a teaching file.* St. Louis: CV Mosby; 1994:574–579.
3. Egelhoff JC, Bates DJ, Ross JS, et al. Spinal MR findings in neurofibromatosis types 1 and 2. *AJNR* 1992;13:1071–1077.
4. Friedman DP, Tartaglino LM, Flanders AE. Intradural schwannomas of the spine: MR findings with emphasis on contrast-enhanced characteristics. *AJR* 1992;158:1347–1350.
5. Demachi H, Takashima T, Kadoya M, et al. MR imaging of spinal neurinomas with pathological correlation. 1990;14:250–254.
6. Verstraete KL, Achten E, DeSchepper A, et al. Nerve sheath tumors: evaluation with CT and MR imaging. *J Belge Radiol* 1992;75:311–320.
7. Cerofolini E, Landi A, DeSantis G, et al. MR of benign peripheral nerve sheath tumors. *J Comput Assist Tomogr* 1991;15:593–597.
8. Shu HH, Mirowitz SA, Wippold FJ: Neurofibromatosis: MR imaging findings involving the head and spine. *AJR* 1993;160:159–164.

CHAPTER 6

The Knee

Andrew L. Deutsch, Jerrold H. Mink, Arthur A. DeSmet, Michael J. Tuite, David W. Stoller, Charles G. Peterfy, and Michael P. Recht

MR has had its greatest impact on evaluation of the knee. Multiple studies have attested to the high accuracy of the technique for the assessment of both meniscal and ligamentous injuries, an accuracy competitive with the more invasive and costly method of arthroscopy. As is now widely appreciated, the technique has the ability to demonstrate the complete spectrum of possible bone and soft tissue pathology, intra- or extraarticular, that may affect this articulation.

Despite these considerable achievements, the future role of MR imaging of the knee will greatly depend on the establishment of its cost-effectiveness and its effect on patient outcome (1). These studies have only begun to be undertaken. In an early study in the orthopedic literature, Boden et al. (2) concluded on the basis of a cost analysis formula that MR imaging was not cost-effective compared with diagnostic arthroscopy. This study, however, was not an outcome examination and had multiple limitations including a very selected study population and no defined criteria for avoiding arthroscopy based on the MRI data (1). In a more recent outcome study by Ruwe and colleagues (3) of the cost effectiveness of MRI, 51% of patients with knee complaints considered sufficient to warrant a diagnostic arthroscopy, were successfully managed nonoperatively based on the results of MR examinations. In this study of 103 patients, the use of MR resulted in a cost savings of over $100,000 (based on an MR global cost of $1000). The cost of surgery in all patients was balanced against the cost of MR imaging in all patients plus the cost of surgery required based on the result of the MR imaging as well as the cost of surgery of any patient with a negative MR imaging that went on to have a positive arthroscopy. Costs not addressed in this study included those related to lost work time, physical therapy, and surgical complications—factors that all could have significantly added to the cost of surgical management (1).

The finding of Ruwe and colleagues that many patients with significant knee complaints could be successfully managed without arthroscopy gains further support from the findings of Hede and coworkers (4), who found that 14 of 36 (39%) of patients decided not to proceed with knee arthroscopy while on a surgical waiting list. Boeree and associates (5) have suggested that the accuracy of MR should allow more appropriate selection of patients for whom arthroscopy would be beneficial. The authors postulate that this would eliminate the need for arthroscopy in one third to one half of those considered on clinical grounds to have internal derangement.

The marked reduction in charges for MR imaging examinations being seen in many regions of the country will also likely significantly impact the analysis of its cost-effectiveness. The development of low-cost extremity scanners may also have a major impact on the use of MR should their diagnostic accuracy be found comparable to that achievable with whole body scanners. Continued research in this area will dramatically affect the practice of musculoskeletal MR in the coming years.

CASE 1

History: A 28-year-old man underwent MR evaluation for assessment of an acute knee injury with hemarthrosis. What is your diagnosis?

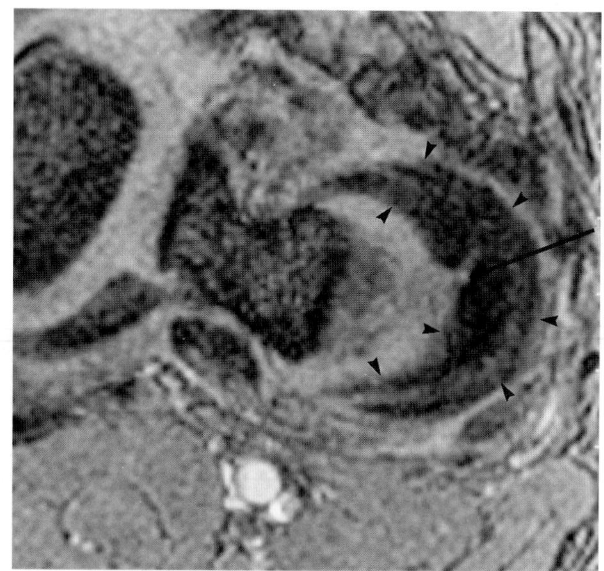

A

FIG. 1.

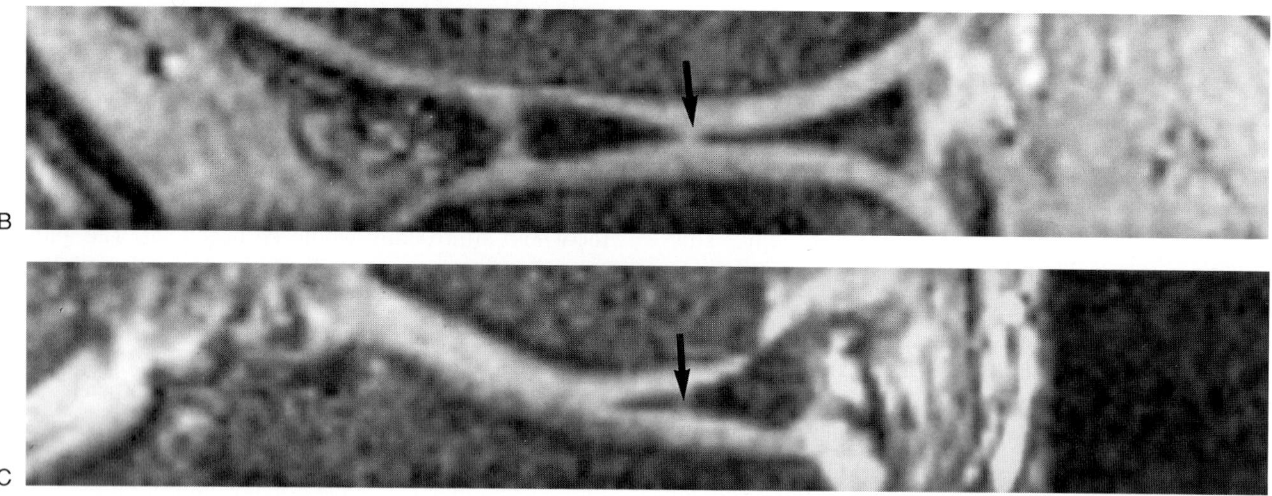

B

C

Findings: In Fig. 1A the axial image from a three-dimensional (3D) volume acquisition (TR/TE/theta 31/15/11) demonstrates a small line of increased signal extending from the free edge of the lateral meniscus at the junction of the midzone and anterior horn (arrow). The meniscus is visualized as a semilunar structure (arrowheads). In Fig. 1B the sagittal reformatted sequence from the same acquisition demonstrates a small focus of signal at the junction of the mid-zone and anterior horn (arrow). In Fig. 1C the coronal reformation of the same data set also demonstrates the abnormality along the inner margin of the meniscus (arrow).

Diagnosis: Radial tear of the free edge of the lateral meniscus.

Discussion: Since the inception of the clinical use of MR for the assessment of meniscal abnormalities, two-dimensional Fourier transform spin echo methods have predominated in clinical practice. Multiple investigations have attested to the high diagnostic accuracy of knee MR primarily utilizing conventional spin-echo sequences obtained in the sagittal and coronal planes and utilizing short echo times (1–10). It remains apparent, however, that despite the overall high accuracy of conventional knee MR there are limitations to these methods, and that some meniscal abnormalities cannot be demonstrated either prospectively or retrospectively utilizing these sequences and their relatively thick sections and spacing e.g., 4-mm section thickness, 1-mm interslice gap (10,11). In particular, small tears involving the meniscal free edge as well as vertical tears perpendicular to the margin of the meniscus may be difficult to detect. In a recent series that analyzed error patterns in the MR evaluation of meniscal tears, the largest category of errors related to false-negative examinations involving the posterior horn of the lateral meniscus (10). The findings of this study were also in agreement with those in a recent review that described a preponderance of false-negative errors (especially involving the posterior horn of the lateral meniscus) (2) (also see Case 6-12). These findings contrast, however, with earlier studies that demonstrated the majority of errors in mensical evaluation to consist of false-positive studies (3–5). The apparent trend toward a reduction in false-positive diagnoses suggested by these recent series likely reflects increased interpreter experience with normal variants and expected postoperative findings as well as improvements in arthroscopic meniscal evaluation (10). Of particular interest, the false-negative errors in the study by Justice and Quinn (10) were resistant to reduction on retrospective analysis. These findings suggest that the diagnostic accuracy of MR, while quite high, may not be improved without significant changes in technique.

Both two-dimensional (2DFT) and three-dimensional Fourier transform (3DFT) gradient echo techniques have also been applied to meniscal evaluation. In general, gradient echo techniques demonstrate enhanced depiction of intrameniscal signal (12–19). This has resulted in a reported increased sensitivity compared with spin echo techniques for the detection of meniscal tears but unfortunately at the cost of lower specificity. In addition, these methods provide gradient echo type contrast, necessitating the use of additional spin echo sequences to provide comprehensive evaluation of nonmeniscal abnormalities. Specialized 2DFT gradient echo techniques such as radial acquisition of images has also been investigated (12). Despite theoretical advantages of this approach, which provides images acquired in multiple planes that are rotated around the center of curvature of the meniscus being studied, no significant difference in accuracy between this method and conventional spin echo techniques has been demonstrated (12). Combined interpretation of conventional sagittal images with 2D radial gradient echo images provided greater accuracies than either technique in isolation in one study (12). The use of surface rendering with 2D and 3D techniques has also been investigated and appears to increase sensitivity to detection of both radial and horizontal posterior horn tears (13).

Several studies have reported on the use of 3DFT sequences for meniscal evaluation (14–19). These sequences allow for both thin contiguous sections as well as the potential for multiplanar reconstruction. Results of 3DFT gradient echo meniscal imaging have been reported for primary acquisitions in both the sagittal and axial planes. In the present case, a 3DFT technique was utilized that provided 28 contiguous 0.7-mm-thick sections acquired in the axial plane. Reported experience with this technique, while limited, has suggested extremely high sensitivity and specificity (18,19). Axial imaging is particularly valuable for assessment of radial tears involving the free edge of the meniscus, as in the present case. With the increasing accessibility to and trend toward direct interpretation of images from workstations, the use of techniques such as this may assume greater practicability in the coming years. Quite clearly, while the overall accuracy of MR of meniscal evaluation is quite high, advances in technique, such as illustrated in this case, will likely be required to advance beyond the current level of interpretive performance.

Radial meniscal tears are oriented perpendicular to the long circumferential axis of the meniscus. They frequently result from acute trauma, may be partial or full thickness, and may occur anywhere along the meniscus. Depending on their location, radial tears (full thickness) may split the meniscus into separate components, each of which is attached to the tibia at only one end. The loss of meniscal integrity destroys the ability of the meniscus to distribute hoop stresses (centrifugally oriented forces generated in the meniscus by weight bearing) by transecting the circumferentially oriented collagen fibers that function to distribute these forces. Simultaneously, the loss of meniscal attachment to the tibia at two ends allows the normal centrifugal pull of the meniscocapsular attachments to displace the meniscus peripherally. These two events underlie the accelerated rate of degenerative articular cartilage changes within the knee that commonly are observed with this type of meniscal tear.

Radial tears may demonstrate a variety of appearances on MR images. Axial images, as previously discussed, are ideal for their depiction and demonstrate a higher sensitivity for detection of partial free edge radial tears than standard techniques (18,19). On axial images, radial tears in any location are well depicted as a localized meniscal defect. Utilizing standard sagittal and coronal imaging planes, the most readily recognized radial tears are full-thickness disruptions oriented perpendicular to the meniscal long axis. Most commonly, these tears are depicted as the complete "absence" of meniscus on at least one of a series of contiguous sections, with the images adjacent to the tear appearing normal or nearly normal (20). Radial tears located along the "curved" portion of the meniscus (with respect to the routine sagittal and coronal imaging planes) will appear as a series of partial-thickness tears progressing from the apex to the periphery of the meniscus on contiguous sections.

The most subtle form of radial tear to depict with conventional MR imaging are those tears that occur adjacent to the tibial attachments of the menisci. These tears separate the meniscus from one of its tibial attachments and result in a nearly normal appearing meniscus on the side of the tear away from the center of the joint. Diagnosis is facilitated by knowledge of the normally expected anatomy of the meniscal attachments. On routine sagittal MR images, the lack of visualization of meniscus medial to the tibial attachment of the posterior cruciate ligament should be strongly suggestive of a radial tear (20). This observation can be corroborated by inspection of coronal images in which the posterior medial meniscus should extend across the entire medial tibial plateau on at least one image. Failure of the meniscus to cover the extreme lateral aspect of the medial tibial plateau is confirmatory of a radial tear through the attachment. Coronal and axial images are the most helpful planes for depiction of radial tears located centrally at the attachment of the posterior horn of the lateral meniscus to the tibia (20,21). Failure of the meniscus to extend to the most medial aspect of the lateral tibial plateau on at least one image should be viewed with suspicion. Radial tears of the meniscus at the level of the tibial attachment may lead to peripheral subluxation of the involved meniscus and the presence of meniscal subluxation beyond the margins of the tibia should elicit a search of this type of meniscal tear. This finding, however, is far from pathognomonic, being seen in association with severe degenerative changes of the meniscus and in patients of large body habitus.

CASE 2

History: Two different patients suffered similar injuries. When were they injured relative to one another, and who will have the most severe disability?

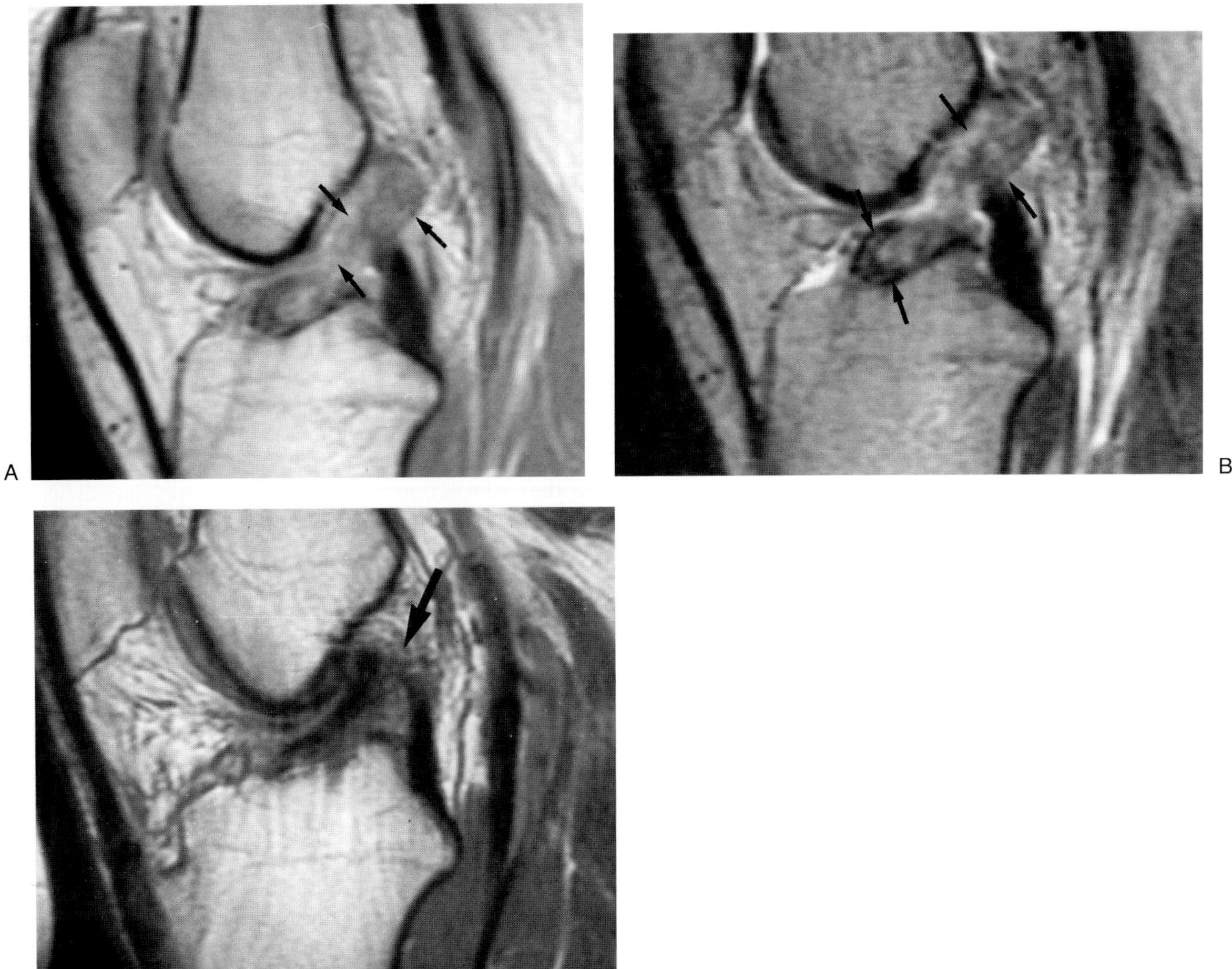

FIG. 2.

Findings: Figure 2A shows sagittal MR, TR/TE 2,300/20 and Fig. 2B shows sagittal (TR/TE 2,300/80 msec) image. The anterior cruciate ligament (arrows) is diffusely widened throughout its length, and demonstrates a marked inhomogeneity of signal intensity. Such interstitial tears may either be partial or complete. Figure 2C shows. Sagittal (TR/TE 2,800/20 msec) image. The ACL has reattached itself to the lateral aspect of the PCL (arrow is at point of attachment). When such reattachment occurs, the arthroscopist may mistakenly assume that the ACL is intact.

Diagnosis: Both patients have suffered tears of the anterior cruciate ligament (ACL). The patient in Fig. 2A,B suffered his injury 3 days prior and had a stable knee at the examination after injury; the patient in Fig. 2C was injured last year and has a mild degree of instability on clinical testing. One cannot tell from this study alone whether or not the patient in Fig. 2A,B will eventually have instability equal to the other patient. Many patients who suffer seemingly partial tears ultimately have significant instability.

Discussion: The classic mechanism of injury to the ACL is a change of direction during deceleration with the foot fixed in position and subsequent contact with another person/object, usually from the lateral side of the knee. This is the well-known American football injury.

Recently however, ACL injuries have been occurring more frequently in noncontact sports. These injuries occur when the body and femur rotate externally about a fixed tibia with the leg in near extension (1). Alpine skiing has produced a virtual epidemic of ACL injuries as a result of changes in equipment, technique, and terrain.

Acute ruptures of the ACL may be defined as those that come to MR examination within 1 to 2 weeks of the injury. Several different MR patterns of acute ruptures share a combination of characteristics including loss of continuity, mass effect, retraction of the torn ends, waviness or concavity of the anterior margin of the ACL, and increased signal within the ligament (2,3).

An interstitial tear is one in which the injury pattern as seen on an MR examination is over virtually the entire length of the ACL (Fig. 2A,B). The ligament is diffusely widened. This tear may appear to the arthroscopist to be only a partial injury; several remaining fibers, and the hematoma hold the ligament in place. At surgery, however, such ligaments may be, or quickly become, functionally complete. This pattern of injury occurs in nearly 50% of all ACL tears (4).

Focal mass-like ruptures most typically affect the proximal end of the ACL (3). The proximal mass has a cloud-like character on MR with signal characteristics of edematous soft tissue intermingled with the torn dark ligament fibers (5). Another common MR pattern of an acute ACL tear is a focal angular deformity of the midsegment of the ACL in which there is rather little mass effect (Fig. 2D).

The precise site and degree of an ACL tear is more readily identified in the subacute (2–8 weeks), rather than in the acute phase. The proximal and distal ends will have retracted slightly, the hematoma will have partially resolved, and bright joint fluid will intravasate into the defect. Subacute tears of the ACL are often manifest by the presence of an intermediate to dark mass within the joint due to retraction of the torn ligament and the development of a synovial cap over the torn end. During the subacute period, the mass-like appearance of the torn ACL may masquerade as one of several lesions. The "double posterior cruciate ligament (PCL)" sign describes a curvilinear dark structure anterior and parallel to the normal PCL that represents the displaced fragment of a bucket handle tear. This appearance can also be simulated by the distal segment of an ACL that has torn proximally and has fallen into the joint. Alternatively, the proximal end of an ACL that is completely torn distally may become displaced behind the PCL, where it mimics a large loose body situated in the posterior capsule (6).

Occasionally, it may be difficult to assess the ACL, especially its femoral attachment, when there is hemorrhage into the joint, edema in the notch, other intraarticular pathology, and/or a narrow intercondylar notch. In such cases, it is extremely helpful to review the axial sequences, especially if the examination was performed utilizing inversion recovery or fat-saturated techniques (7) (Fig. 2E–H). The attachment of the ACL to the medial wall of the lateral femoral condyle is always evident where it is identified as an ovoid, dark structure closely applied to the medial condylar wall. It can normally be followed distally on three or four images, depending on the imaging intervals. The distal-most ligament, however, is not well imaged on the axial sequence because of partial volume effect with the intercondylar eminence, the sharp anterior angulation of the distal ligament, morphology of the separate bands of the ACL, and the normal increase in signal intensity that occurs in the ACL, especially in elderly patients (8).

The MR image of the chronically torn ACL is dependent on the degree of atrophy of the ligament and the degree of instability that supervenes. In many cases, the ACL itself is not visualized; the lateral intercondylar notch is empty, or a tiny nubbin of soft tissue representing a retracted, ischemic ACL remnant may be identified on the tibial plateau. In such cases, the ACL may be difficult to find at surgery. On the coronal sequence, the bald medial face of the lateral femoral condyle, corresponding to the arthroscopically described "empty lateral wall," is evident (9) (Fig. 2I).

The chronically torn ACL may occasionally be present as an intact band of low signal in the lateral aspect of the intercondylar notch simulating a virgin ligament (5). The structure, however, typically has a low-lying axis, and the proximal-most segment of the ACL (that portion cephalad to the PCL), is usually never visualized. In fact, the proximal end of a torn ACL may secondarily reattach itself to the lateral aspect of the PCL. (At one time, surgeons intentionally reattached the ACL to the PCL in the hope of establishing some element of stability and parasitizing some new blood supply.) The ACL, residing in the same synovial envelope with the PCL, falls slightly following complete rupture. It finds a secondary blood supply from the PCL to which it adheres, and the scarring that invariably occurs has a low signal intensity identical to that of the normal ACL. An ACL that has attached itself to the PCL is commonly mistaken for an intact ligament both on MR images and remarkably even at arthroscopy (5,9) (Fig. 2C). Such reattached ligaments may provide some stability, further leading to arthroscopic confusion.

Avulsion of the attachment of the ACL occurs almost exclusively at the tibia since the tibial end of the ligament is stronger than the proximal end (Fig. 2J–M). Fewer than 5% of ACL ruptures occur in adults as a result of avulsion, but the relative strength of the ACL compared with that of the adjacent bone and physeal plate explain why most ACL injuries in children are avulsive or result in a physeal fracture. Avulsion injuries are often overlooked on MR because of the small size of the fragment, the relative lack of edema at the avulsion site, and the similar signal of the bony intercondylar eminence to the fat pad into which it is displaced.

Numerous recent articles have addressed the role of MR in the management of acute knee injuries, and in particular, its value and accuracy in assessment of the integrity of the ACL (3,5,10–20). These studies have been performed using a variety of MR imagers and imaging protocols and have been conducted by examiners of differing skill levels. Eleven recent studies, utilizing *high field strength* imagers, have examined a total of 1,661 patients. In these reports, the sensitivity of MR for the detection of a complete ACL tear has varied from 92% to 100% (6 of the 11 studies reported 100% sensitivity), the specificity from 82% to 100%, and the accuracy from 92% to 100%. Since arthroscopy is not a true gold standard, that is, it is not 100% accurate, it is best to refer to the values given above as "agreement rates of MR and arthroscopy" (9,21,22). The negative predictive value of MR, that is, the percentage of normal MR's that do not have a tear at arthroscopy, ranged between 89% and 100% in those studies for which the data were available (four studies reported 100%). It must be concluded that MR of the knee, performed on a high field strength MR scanner, is an extremely accurate method of detecting complete ACL tears.

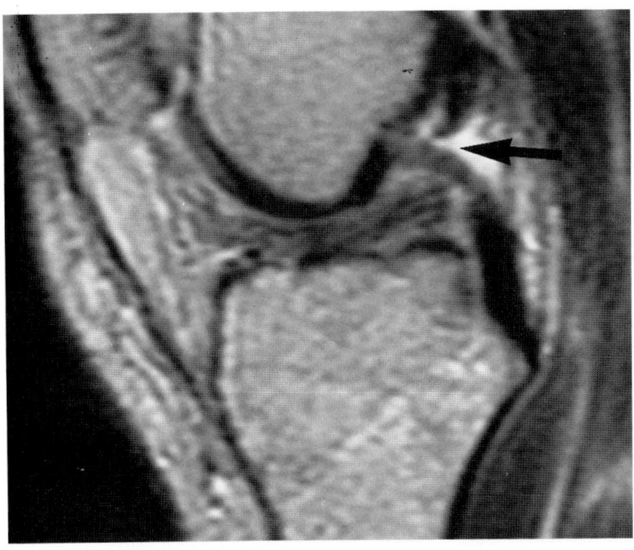
D

FIG. 2. D: Acute angular tear of the anterior cruciate ligament. Sagittal (TR/TE 2,000/80 msec) image. An acute tear of the midsubstance of the ACL has occurred in this professional football player. The distal ACL has a low-lying axis. The tear through the midsubstance of the ligament *(arrow)* can be seen by virtue of the fact that joint fluid fills the gap.

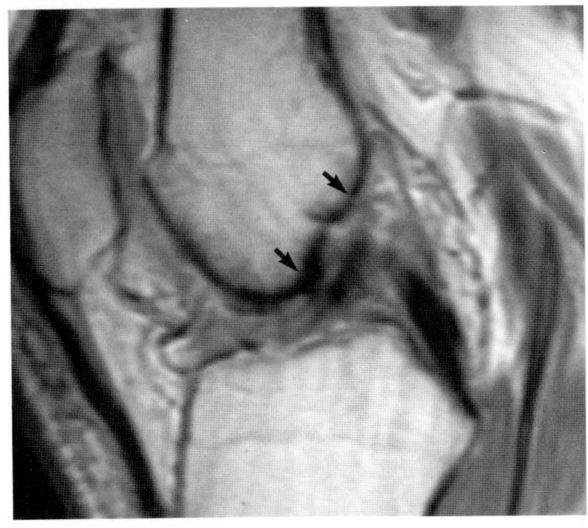
E

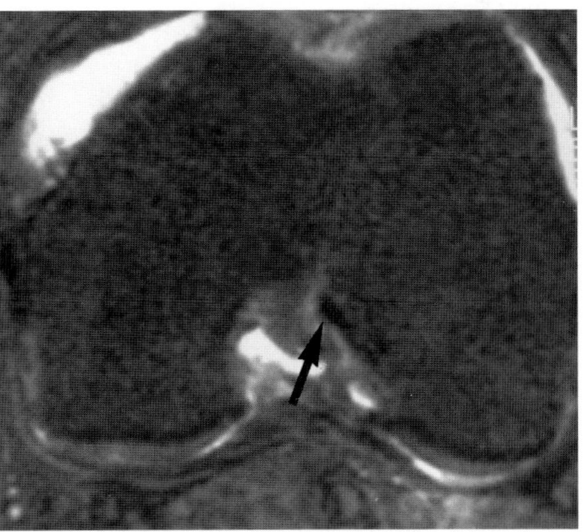
F

FIG. 2. E–H: Normal ACL: use of axial images. **(E)** Sagittal (TR/TE 2,300/20) image; **(F–H)** axial (TR/TE/TI 2,800/36/130) image. Although this sagittal image is directly through the lateral intercondylar notch as evidenced by identification of the intercondylar roof *(small arrows),* only the midportion of the anterior cruciate ligament could be clearly identified and the integrity of the ligament was in doubt. The three axial images demonstrate that the ACL *(arrow)* is intact. Axial images are extremely valuable in assessment of the proximal ACL.

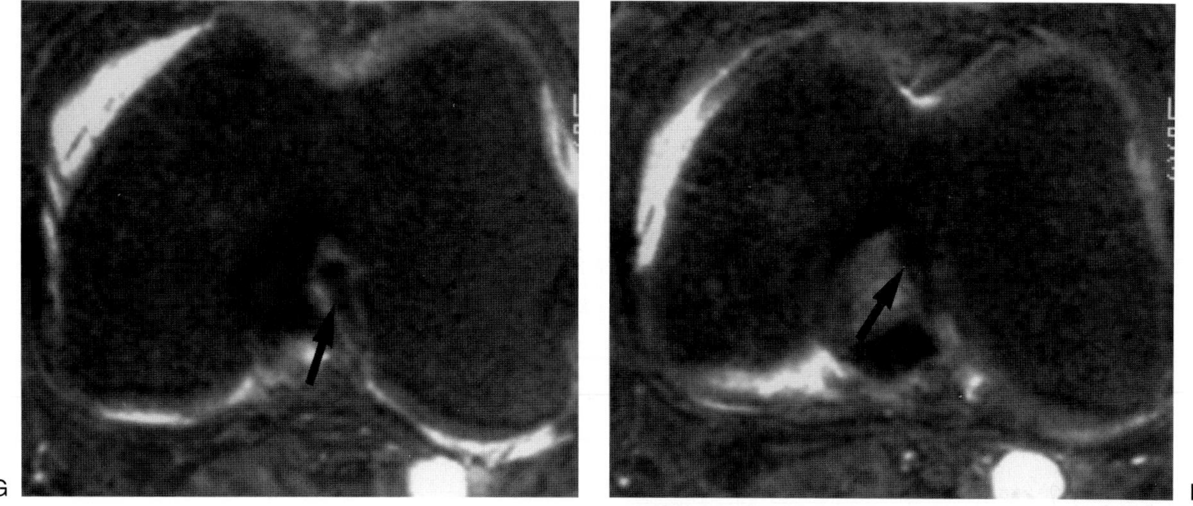

FIG. 2. G,H. (*continued.*)

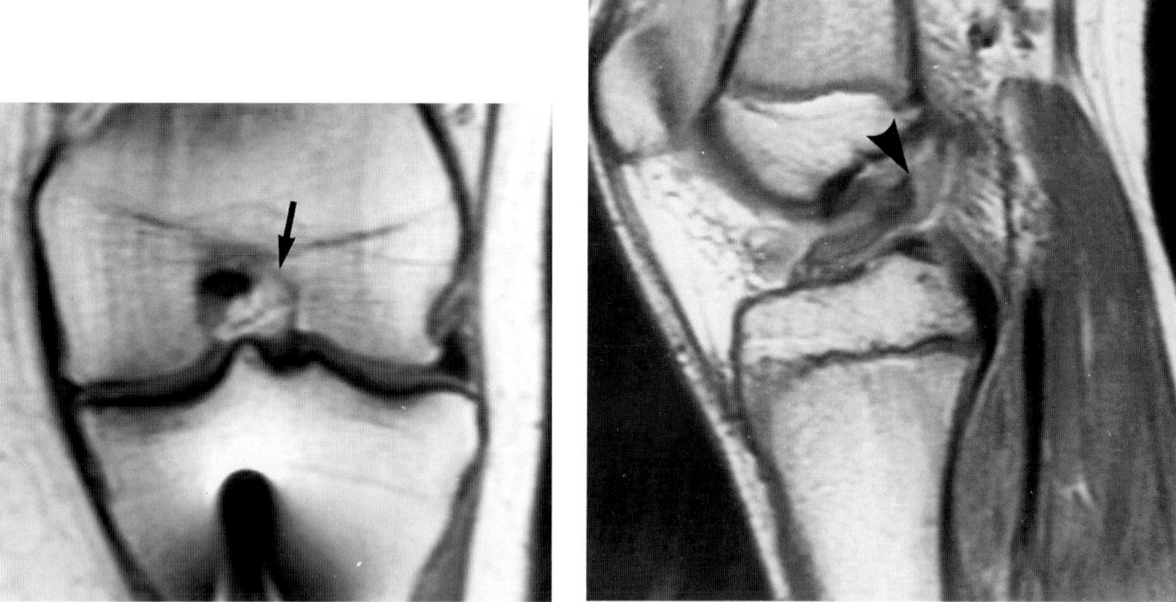

FIG. 2. I: Chronic tear of the ACL. Coronal (TR/TE 600/20 msec) image. These images demonstrate complete ACL absence in this patient injured 3 years prior. The patient has an "empty lateral wall" sign *(arrow)*. The metallic artifact is from a previous patellar tendon procedure. **J:** Detachment of the proximal ACL; sagittal (TR/TE 2,000/20) image. This 15-year-old suffered an ACL tear during a soccer game. The femoral end of the ACL *(arrowhead)* has become detached from the femur with a small flake of bone which was not identified on the MR. The fragment was much too small to be reattached at the time of surgery.

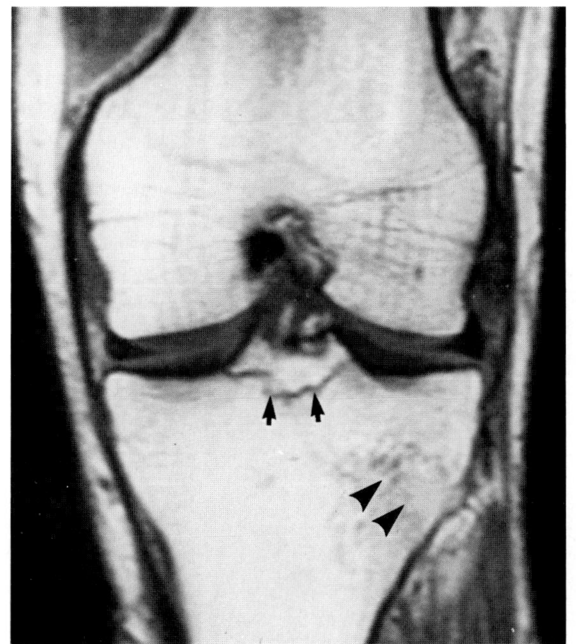
K

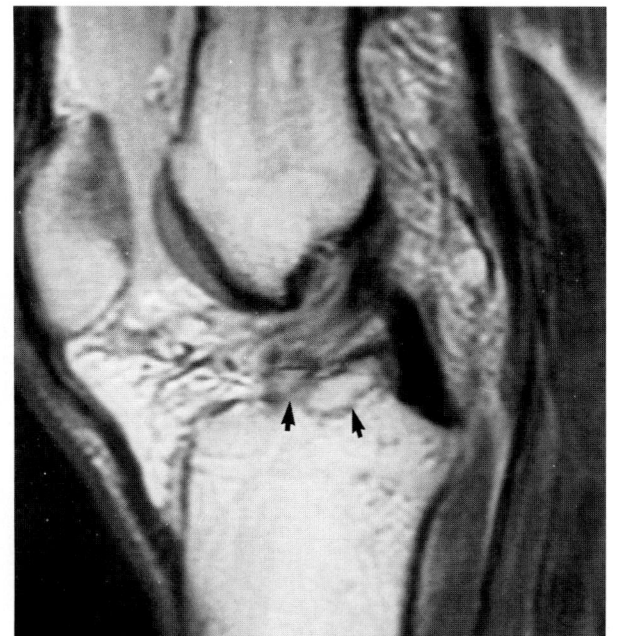
L

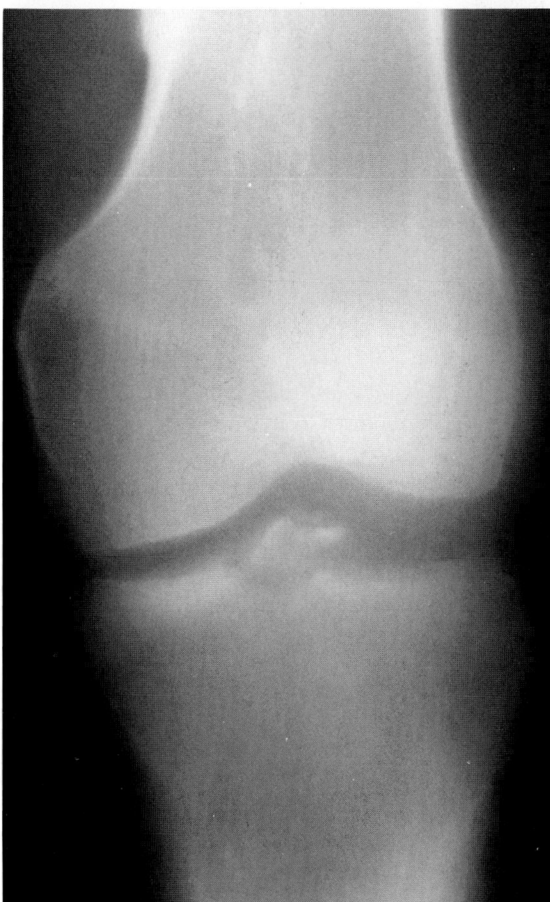
M

FIG. 2. K–M: Avulsion ruptures of the ACL. K: Coronal (TR/TE 2,000/20) image. **L:** Sagittal (TR/TE 2,000/20) image. Both images demonstrate a minimally displaced avulsion fracture of the insertion of the ACL *(arrows).* There is surprisingly little bone marrow edema associated with the fracture. A medullary contusion *(arrowheads)* is present in the lateral proximal tibia. **M:** This anteroposterior tomogram demonstrates the avulsion of the intercondylar eminence which is displaced more than on the MR which was performed 1 week earlier.

CASE 3

History: This 10-year-old girl had a fall and suffered a knee injury. She was taken to an orthopedist who took conventional radiographs, which demonstrated a Salter-Harris II fracture of the distal femur. What is the diagnosis and what is the value of an MR in this case?

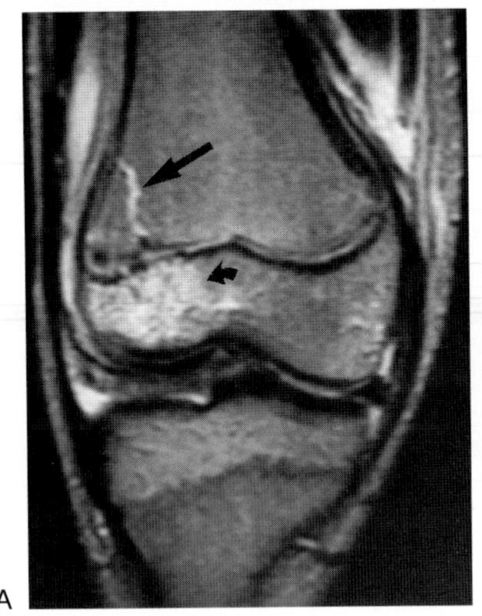

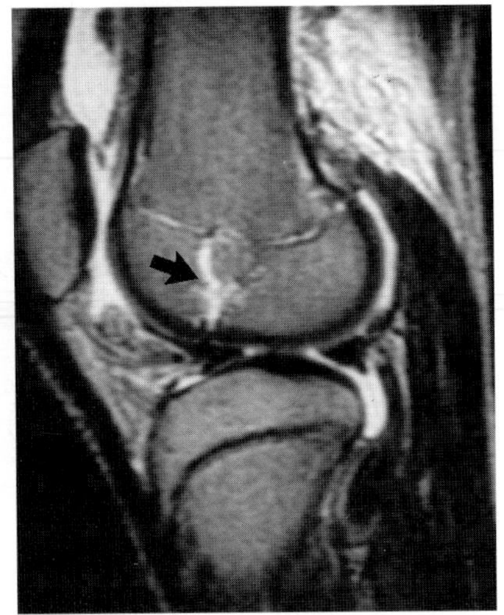

FIG. 3.

Findings: Figure 3A is a coronal (TR/TE 2,200/80) image. Figure 3B is a sagittal (TR/TE 2,300/80 msec) image. This 13-year-old girl was known to have suffered a Salter II fracture (arrow in A) which was visible on the conventional radiograph. The physician suspected that the lateral growth plate was slightly widened. The coronal MR examination demonstrates edema within the lateral femoral condyle (small arrow); however, the sagittal image (B) confirms the linear nature of the abnormality. A Salter-Harris IV fracture implies injury to the growth plate, and subsequently a less-favorable prognosis than a Salter II fracture.

Diagnosis: Salter-Harris IV fracture. MR has proven itself to have a valuable role in diagnosing radiographically occult pediatric fractures, reclassifying them into different Salter-Harris groups, and in assessing the potential complications of such injuries. For the most part, reclassification into a more serious injury than originally suspected is the rule.

Discussion: Utilizing MR in the pediatric patient population has achieved considerable popularity. The lack of ionizing radiation, the relatively lesser need for intravenous contrast than with computed tomography (CT), the multiplanar capability of the method, the often greater reluctance of physician and parent to "wait and see" or "do a little operation"—all contribute to the increasing use of MR in this age group (1). In the musculoskeletal system, the ability of MR to noninvasively assess the knee and avoid arthroscopy has been a particularly appealing application. In children, physical examination may be more stressful for both the patient and the examining physician! In adults, arthroscopically undetectable lesions such as stress fractures, trabecular microfractures, muscle injuries, and osteochondral fractures frequently account for the patient's pain; judiciously utilized MR can avoid needless surgery. The pediatric skeleton suffers many of the same injuries, but provides an even greater degree of difficulty for the clinician since it is significantly in cartilage, and therefore detection of a fracture by conventional radiography is markedly more difficult.

Numerous publications have addressed the sensitivity of MR in the detection of radiographically occult lesions of the musculoskeletal system (2–10) (Fig. 3C–E). Its ability to directly and noninvasively assess articular cartilage, fibrocartilage, and synovial surfaces makes it an ideal tool for examining the acutely injured patient. While MR has certainly improved the radiologists ability to diagnose meniscal tears, it has most revolutionized current concepts of radiographically occult, intraosseous injuries. MR permits direct evaluation of the bone marrow compartment, allowing the examiner to visualize the edema and hemorrhage that typically accompany trauma.

The Salter-Harris classification describes growth plate fractures according to the direction and extent of the fracture line as seen on plain radiographs (11,12). The prognosis of these injuries is dependent not only on the type of fracture but also on the age of the patient, the epiphysis involved, the intensity of the trauma, and the size and location of any fibrous or bony physeal bar that may result (13). Salter-Harris II (metaphyseal) fractures are the most frequently encountered type, followed, in order, by types III (epiphyseal), IV (combined metaphyseal and epiphyseal), and I (transphyseal). The wrist, distal tibia, elbow, and the distal femur are common sites of growth plate injuries. These lesions have potential lifelong effects since interruption of these growth centers may lead to disturbances of growth and limb shortening (5).

Improved detection of the adjacent epiphyseal and/or metaphyseal components of growth plate fractures may change the radiologic Salter-Harris classification (14). MR is particularly useful in detecting an epiphyseal component of a radiographically suspected Salter-Harris II fracture (particularly when the epiphysis is nonossified), in detecting metaphyseal extension of an apparent Salter III fracture, and in reclassifying a type I lesion as a II, III, or IV (Fig. 3A,B and F,G).

Redefining a less ominous II or III fracture into the more worrisome IV category carries significant prognostic implications (see below) since the likelihood of formation of a bony or fibrous bar across the physis increases significantly. Jaramillo et al. (14) examined a small number of patients who had suffered fractures near the growth centers. They found that MR imaging changed the radiologic Salter-Harris classification in half of their 12 cases. In another study group, the classification as judged by MR changed from the injury radiographs in three of four patients (15). Exclusion of extension of suspect fractures is just as important as detecting occult lesions (Fig. 3H,I).

The normal growth plate (physis) has a dual blood supply, each with a different function. The epiphyseal vessels are responsible for the nutrition of the chondrocytes and are necessary for their growth and maturation from the zone of resting cartilage, through the proliferating and hypertrophied stages. The metaphyseal vessels come into contact with the hypertrophied cartilage in the zone of provisional calcification where the matrix is being mineralized, the terminal chondrocytes are dying, and blood-borne osteoblastic precursors are being deposited. The physis receives its nutrition by diffusion from the epiphyseal vessels, but the epiphyseal and metaphyseal vessels are normally never in contact with one another. Injury to either vascular system can result in growth disturbance. Destruction of the epiphyseal vessels leads to death of the growth cartilage and formation of a bony bridge by means of a vascular connection through the physis (16). Injury to the metaphyseal system results in persistence of the activity of the chondrocytes of the hypertrophic zone, leading to widening of the growth plate. With continued longitudinal growth, cartilage rests may become incorporated into the metaphysis, resulting in linear or ovoid areas of decreased mineralization of the metaphysis. When the circulation is restored, the aberrant cartilage is gradually ossified (17).

A Salter-Harris IV injury implies that the metaphyseal and epiphyseal circulations come into contact (5). In general, vertical interruption of the growth plate may be associated with subsequent growth abnormality, but horizontal fracture (type I) of the physis is not (14). The pathologic process following injury can be serially studied by utilizing MR with gadolinium augmentation (5). The normal growth plate is seen on MR as a relatively high signal intensity band on long TR or T2*-imaging sequences. Normally, the cartilage plate interacts with the metaphyseal and epiphyseal vessels to produce endochondral ossification (progressive ossification of a cartilage precursor) within the metaphysis (1). Following creation of a traumatic defect in the physis, transphyseal bone formation without evidence of cartilage repair occurs. Gadolinium enhancement indicates the development of vascularity through the plate 1 week postinjury, followed by formation of a bony bridge by 3 weeks. Areas of interruption of the growth plate are manifest as zones of decreased signal intensity within the otherwise bright epiphyseal cartilage (T2* images).

MR has the potential to depict fibrous bridges that by their nature are not demonstrable by conventional radiographs or CT. The physis can adapt to small violations of its integrity. Central bars that cause less than 2 cm of limb shortening are well tolerated, but peripheral lesions causing angular deformity are frequently treated by early excision of the bony bridge (14). The multiplanar capabilities of MR permit optimal mapping and cross-referencing of abnormalities prior to surgical revision.

Abnormal growth recovery lines are easily detected with MR as dark lines across a bright background of fatty marrow within the metaphysis on short TR sequences. These lines, when oblique, tilted, or curved with respect to the physis, indicate an asymmetry in longitudinal bone growth and suggest the presence of a bar (18,19). Posttraumatic extension of growth cartilage into the metaphysis has little likelihood of causing growth abnormalities but should alert the examiner to the possibility of previous growth plate injury (20) (Figs. 3J–L).

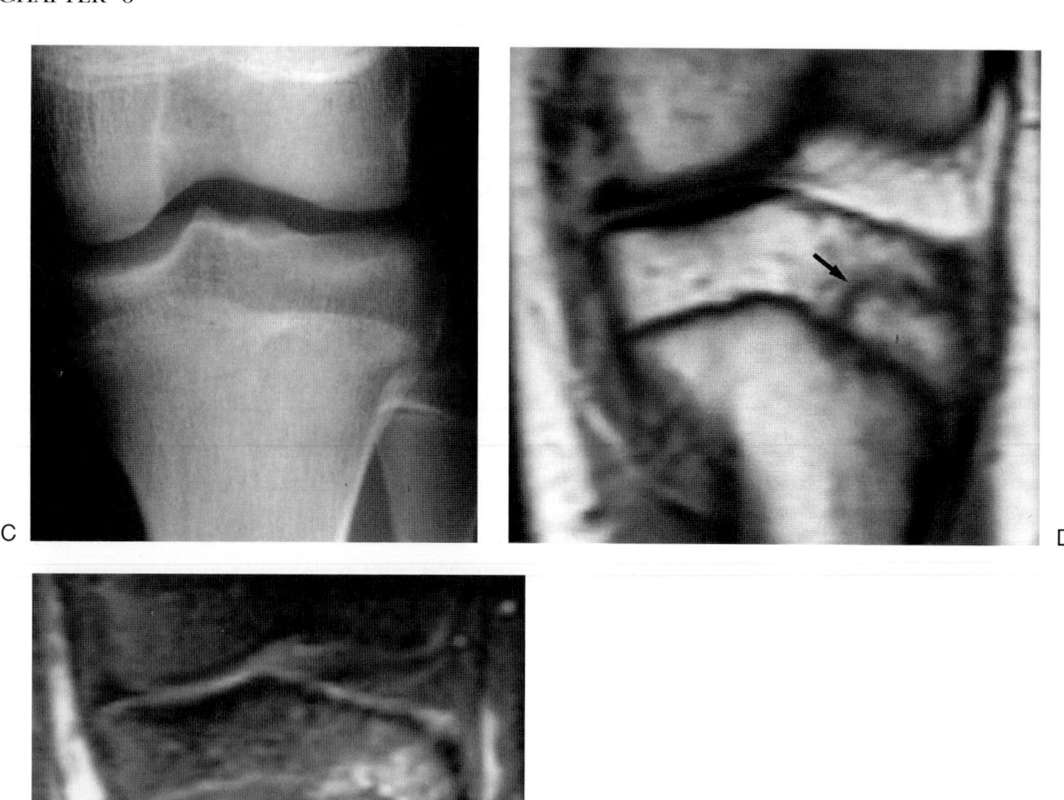

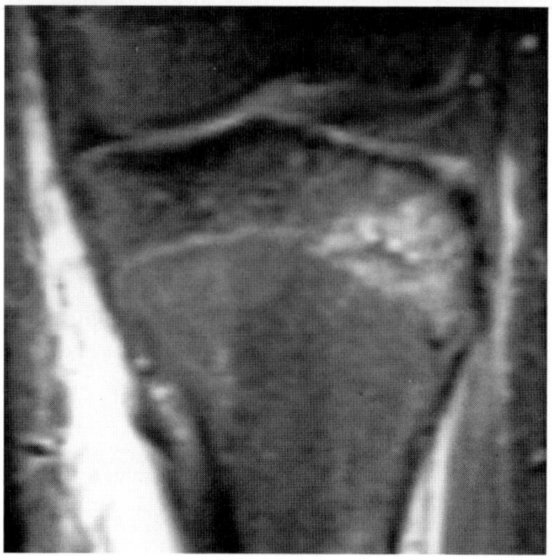

E

FIG. 3. C–E: Detection of radiographically occult fracture of the epiphysis. **C:** Conventional radiograph; **D:** coronal (TR/TE 2,000/20 msec) image; **E:** coronal (TR/TE/TI 2,500/20/130) image. This 11-year-old girl suffered a fall and had conventional radiographic examination, which was normal. A linear epiphyseal fracture (Salter III) *(arrow)* is manifest by low signal on the proton density image, and by edema on the inversion recovery.

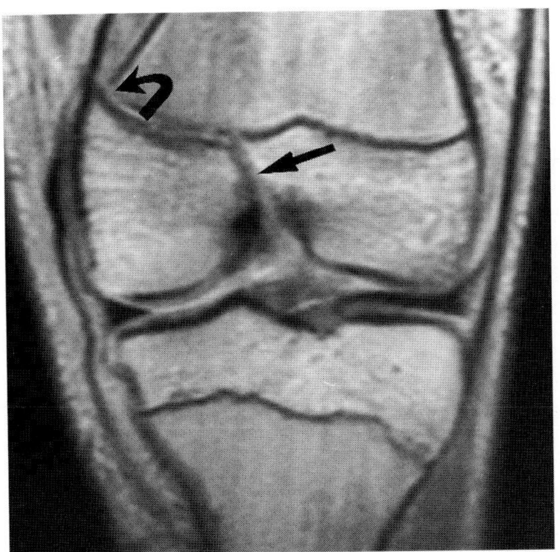

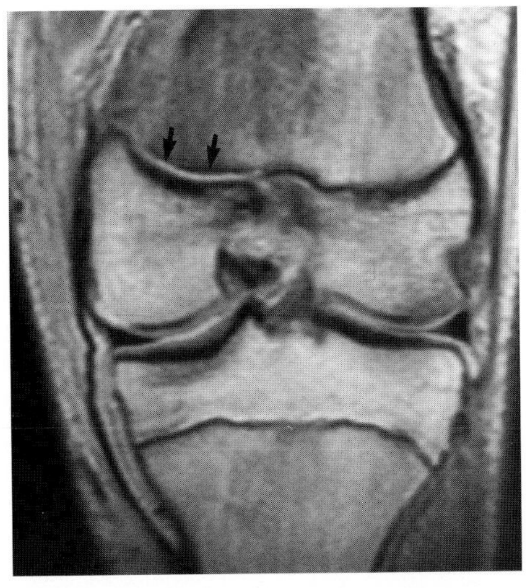

FIG. 3. F,G: Salter III fracture. Coronal MR, TR/TE 22,000/20 msec. This 12-year-old boy suffered a fall. Conventional radiography demonstrated a minor degree of incongruity of the metaphysis and epiphysis along the medial aspect of the distal femur but no distinct fracture line was seen. The MR examination clearly identified the incongruity *(curved arrow),* the Salter III fracture *(arrow),* and subtle widening and increased signal of the growth plate **(G)** *(small arrows).*

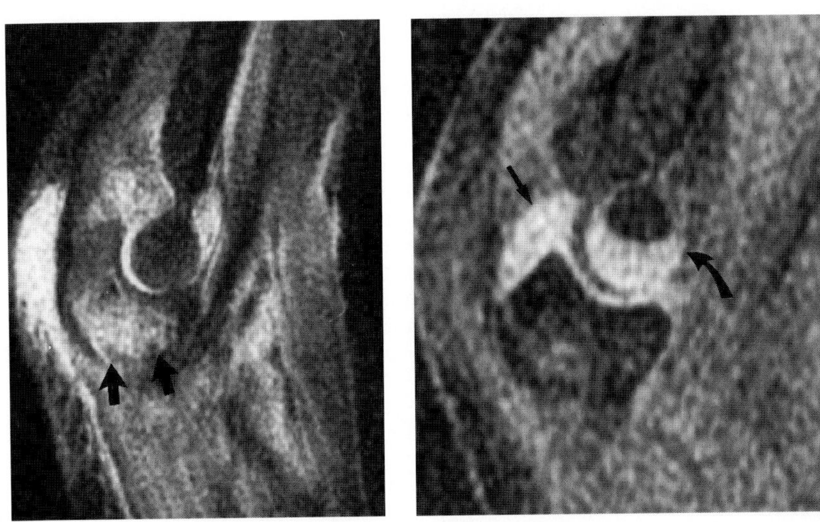

FIG. 3. H,I: Salter II fracture. H: Sagittal (TR/TE/TI 2,300/23/130) image; I: Sagittal (TR/TE 27/8 msec) image (volume). This 4-year-old girl suffered a metaphyseal injury (arrows) manifest as edema in the proximal ulna. She was unable to full straighten her elbow, and the clinician was concerned about an epiphyseal separation. The gradient sequence **(I),** optimized for visualization of cartilage, demonstrates the bright olecranon *(arrow)* and trochlear growth centers *(curved arrow).* While the fracture in the ulna is not seen on the latter imaging sequence, the growth centers are seen to be in normal position. The patient made and prompt and uneventful recovery, and difficult and painful (for both patient and physician!!) elbow arthrography was avoided.

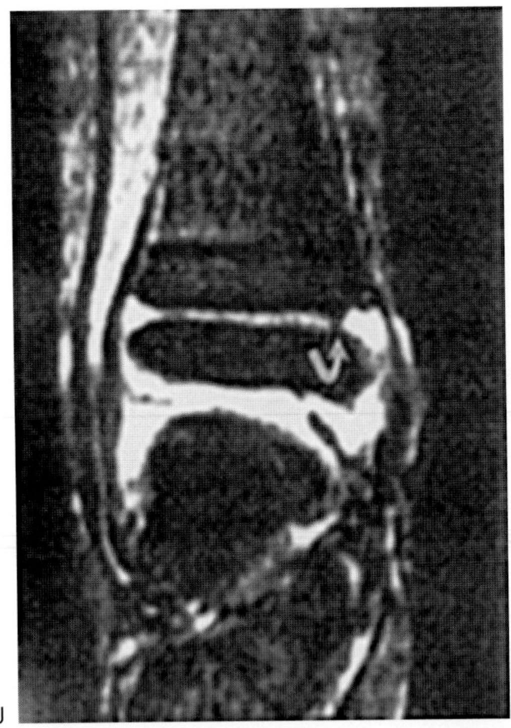
J

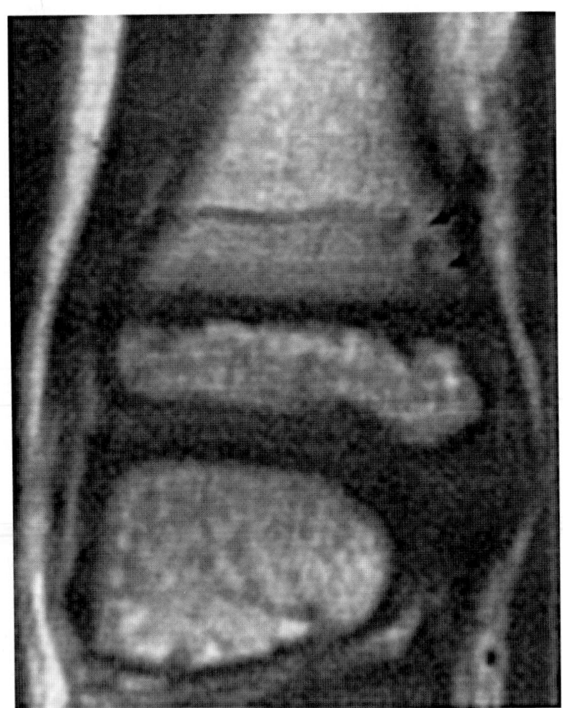
K

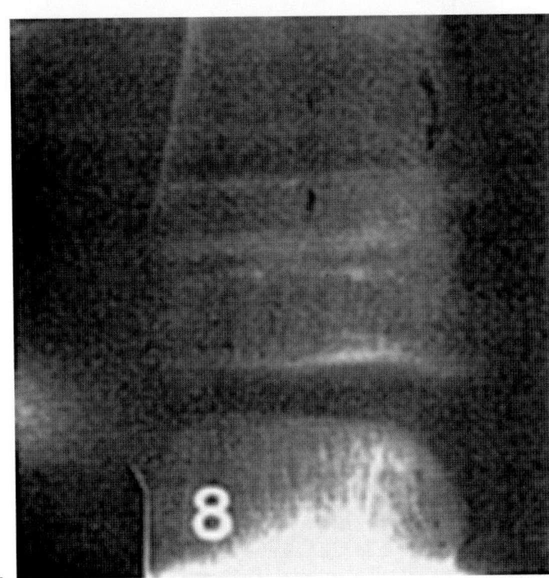

L

FIG. 3. J–L: Growth plate injury. **J:** Coronal gradient echo (GE/TR/TE 10/100/25 msec). A 5-year-old girl sustained a Salter IV fracture of the distal tibia. This scan was performed 15 months after the initial injury. There is focal interruption of the medial growth plate manifested by a low signal intensity focus *(arrow)*. At surgery, a fibrous bridge was found in the region of interruption. **K:** Coronal TR/TE 500/20 msec. There is focal curving of a growth recovery line medially *(arrows)*. More lateral growth recovery line is not affected. **L:** Anteroposterior tomogram obtained near to the time of the MRI examination. The localized curving of the growth recovery line is less apparent than on the MR *(arrow)*. The fibrous bridge is not detectable. (Courtesy of Diego Jaramillo, M.D. Boston, MA.) (From Jaramillo D, Hoffer FA, Shapiro F, et al. MR imaging of fractures of the growth plate. *AJR* 1990;155:1261–1265, with permission.)

CASE 4

History: This 40-year-old man presented with intermittent locking of his right knee after a weekend jog. Pain was referred to the medial joint line. Is the meniscus torn? If so, what primary tear pattern is present? Could this information be used to predict surface tear morphology as would be visualized if arthroscopy were performed (Fig. 4A)?

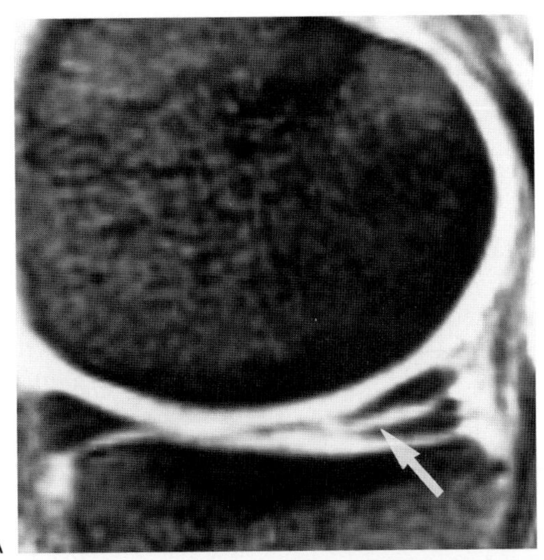

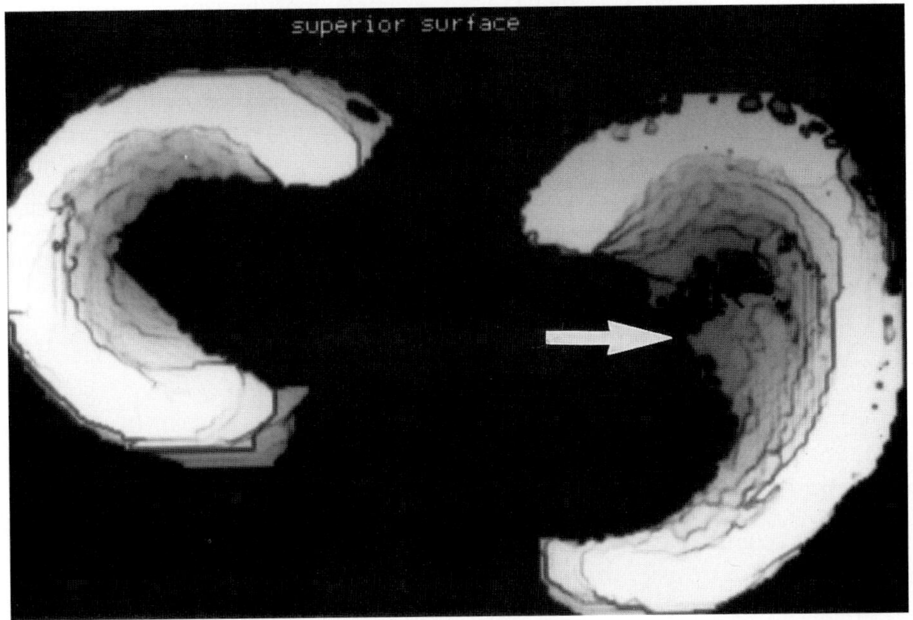

FIG. 4. A: A 2D gradient echo sagittal image (TR/TE/°, 600/15 msec/15°) through the medial meniscus demonstrating tear of the posterior horn of the medial meniscus with both superior and inferior surface extension *(arrow)*. **B:** A 3D rendering performed from a 0.7-mm axial gradient echo sequence demonstrating the morphology of an oblique (flap) tear pattern of the meniscus *(arrow)*.

Findings: In Fig. 4A 2D T2* sagittal images demonstrate a primary horizontal tear with complex morphology in communication with both the superior and inferior meniscal surfaces Fig. 4B. 3D rendering of the meniscus as generated by 0.7-mm (submillimeter) axial gradient echo imaging displays the circumferential meniscal morphology of a flap tear also known as an oblique tear.

Diagnosis: Flap tear of the medial meniscus.

Discussion: An oblique tear or flap tear represents a composite of a longitudinal and transverse or radial tear.

The tear starts on the free edge of the meniscus and curves obliquely into the meniscal fibrocartilage. These tears are often referred to as flap tears. A parrot-beak tear usually describes an oblique tear with a smaller horizontal component or is commonly used to also refer to radial tears. On cross-sectional imaging, flap tears or oblique tears demonstrate either a primary vertical or horizontal tear pattern. On sagittal sections without thin section axial images or a 3D composite image of the meniscus, it is difficult to differentiate between oblique and longitudinal tear morphology. However, when grade 3 signal intensity is seen to extend to the superior and inferior surface of the meniscus on separate sagittal images, oblique morphology may be inferred. Flap tears may generate an anterior or posterior based flap of the meniscus. Flap tears are treated with arthroscopic resection of the flap and contouring of the meniscus tissue to a stable rim through the remaining horizontal component.

As viewed from the surface of the meniscus at arthroscopy and relative to its circumference, three basic tear patterns are found:

1. Longitudinal.
2. Transverse or radial.
3. Oblique or flap tear.

Vertical tears extend to the meniscus surface, either as a longitudinal, transverse, or oblique (flap) tears. Horizontal tears are either longitudinal or oblique, unless they remain on the plane of the middle periphery and collagen bundle and extend to the meniscal apex as degenerative cleavage tears with approximately equal size, superior and inferior leaves.

A tear seen from the surface of the meniscus as longitudinal may either be vertical or horizontal on sagittal images (Fig. 4C,D). Peripheral vertical tears are successfully treated with primary meniscal repair, whereas horizontal tears that extend into avascular fibrocartilage are often treated with partial meniscectomy.

A displaced longitudinal tear of the meniscus, usually the medial meniscus, is called a bucket-handle tear because the separated central fragment resembles the handle of a bucket (Fig. 4E). In complex bucket-handle tears, 3DFT axial images show the relationship of the displaced tear to the remaining meniscus in a single section. The lateral meniscus may also be the site of a bucket-handle tear in which the body of the lateral meniscus is displaced into the intercondylar notch.

A transverse or radial tear is, by definition, a vertical tear perpendicular to the free edge of the meniscus. No horizontal vector or flap generation is present in radial tears. On sagittal images, the only evidence of a radial tear may be increased signal intensity on one or two peripheral sections, since the sagittal plane sections through the meniscus are perpendicular to the tear pattern (Fig. 4F,G,H).

Using cross-sectional anatomy of the meniscus as demonstrated on sagittal sections, meniscal tears can be classified into two primary tear planes, vertical and horizontal (Fig. 4I,J). However, because most meniscal tears are not exclusively perpendicular or parallel to the tibial plateau surface, tears classified as vertical may have secondary tear patterns (i.e., horizontal or vertical, respectively). For example, most horizontal tears have a secondary vertical vector associated with extension of the tear to the inferior surface of the meniscus. An accurate description of the morphology and location of a tear is particularly useful in making the choice between primary meniscal or partial meniscectomy. The 3D composite images spatially disarticulated from the knee, processed from 3DFT T2*-weighted images are particularly useful in demonstrating the internal tear patterns and displaying the circumferential morphology of the meniscus.

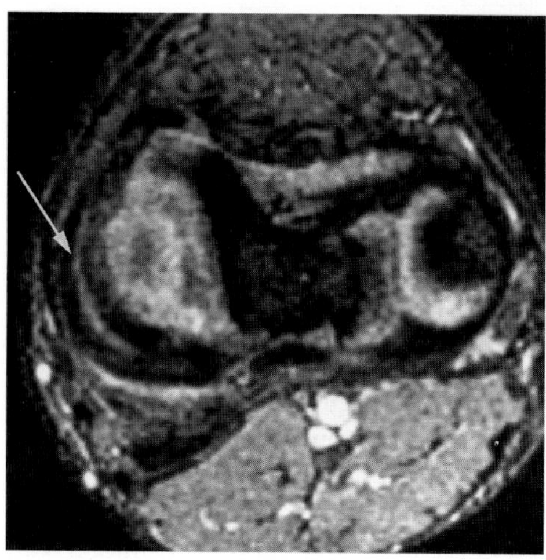

FIG. 4. C: A 0.7-mm gradient echo axial image (TR/TE/°, 55/15 msec/11°) demonstrating high signal intensity paralleling the borders of the medial meniscus in a longitudinal tear pattern *(arrow).*

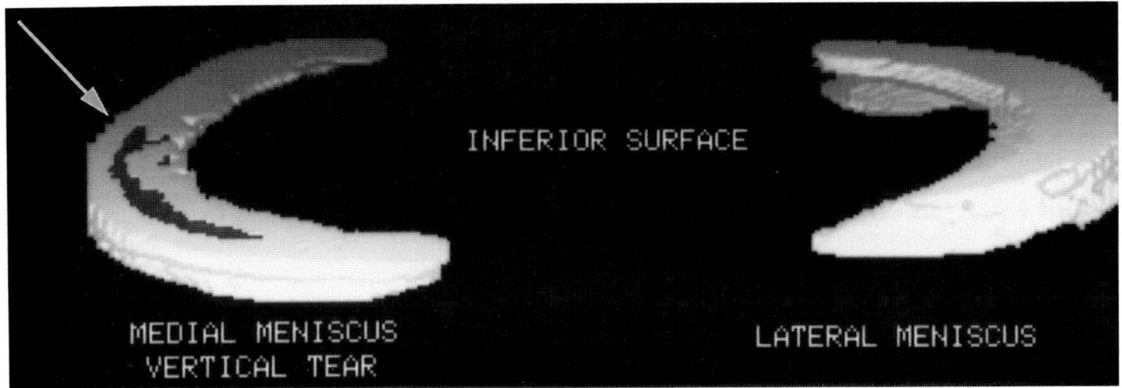

D

FIG. 4. D: A 3D rendering of the longitudinal tear of the posterior horn of the medial meniscus with peripheral location *(arrow)*.

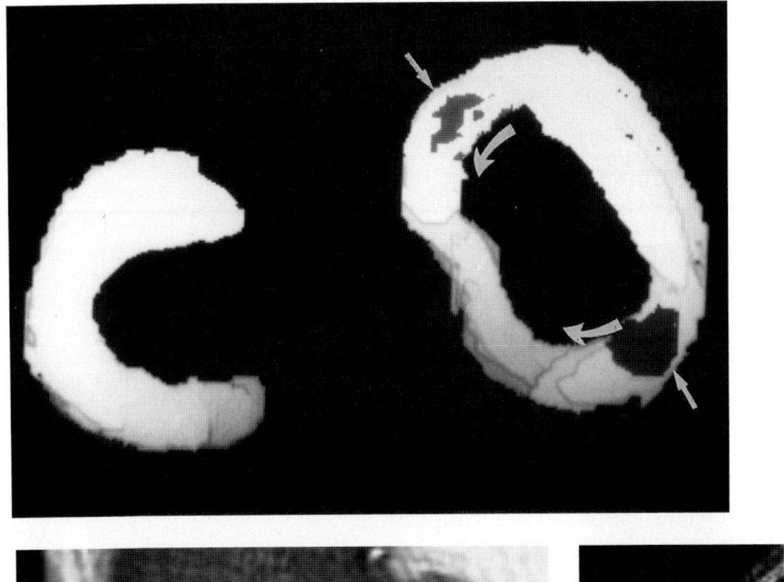

E

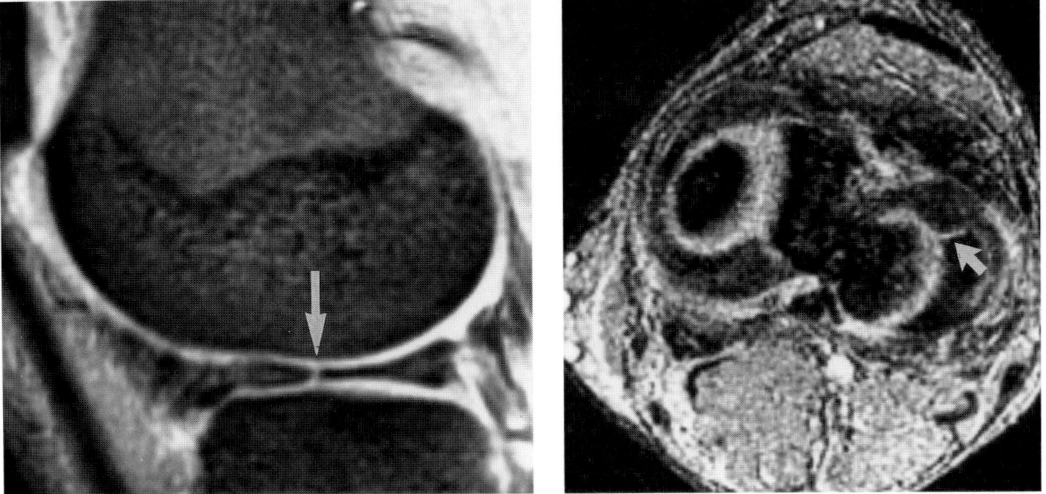

F G

FIG. 4. E: A 3D rendering of medial meniscus bucket handle tear demonstrating the displaced portion of the meniscus *(curved arrows)* and tear points adjacent to the anterior horn and posterior horn of the medial meniscus *(straight arrows)*. Intact lateral meniscus is demonstrated. **F:** A 2D gradient echo sagittal image (TR/TE/°, 600/15 msec/15°) demonstrating grade 3 signal in the anterior horn/body junction of the lateral meniscus representing a peripheral radial tear *(arrow)*. **G:** A 0.7-mm gradient echo axial 3DFT acquisition image (TR/TE/°, 55/15 msec/11°) demonstrating free edge vertical tear of the anterior horn/body junction of the lateral meniscus *(arrow)*.

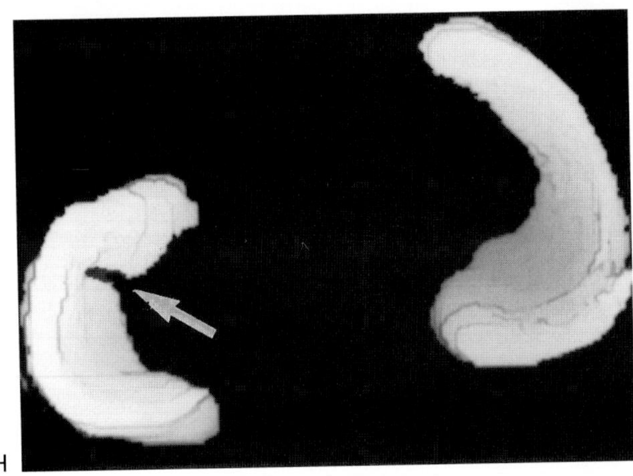

FIG. 4. H: A 3D rendering demonstrating the relationship of the radial tear to the free edge of the three-dimensionally displayed lateral meniscus *(arrow).*

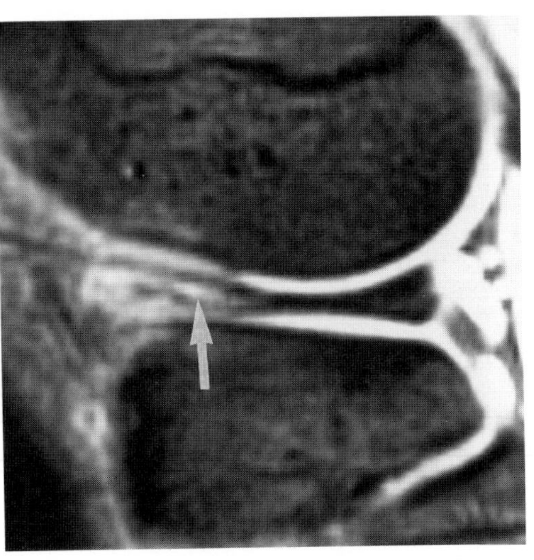

FIG. 4. I: A 2D gradient echo sagittal image (TR/TE/°, 600/15 msec/15°) demonstrating horizontal tear decompressing into a small meniscal cyst in the anterior horn of the lateral meniscus *(arrow).*

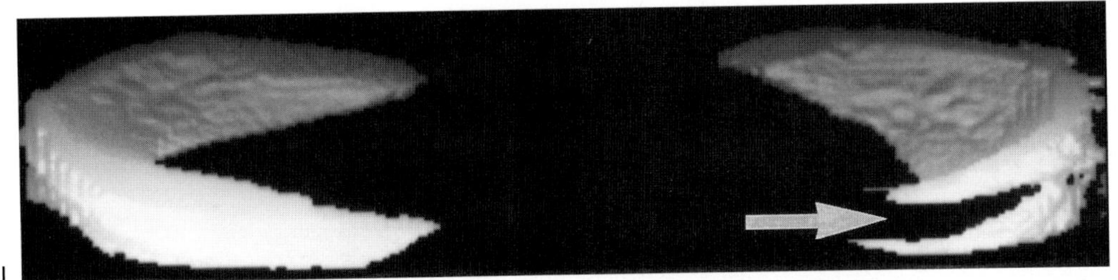

FIG. 4. J: A 3D rendering demonstrating division of the anterior horn of the lateral meniscus into a superior and inferior leaf in a horizontal cleavage tear *(arrow).*

CASE 5

History: A 58-year-old woman underwent MR examination of the knee for assessment of possible internal derangement. What is the most likely diagnosis for the marrow disorder depicted?

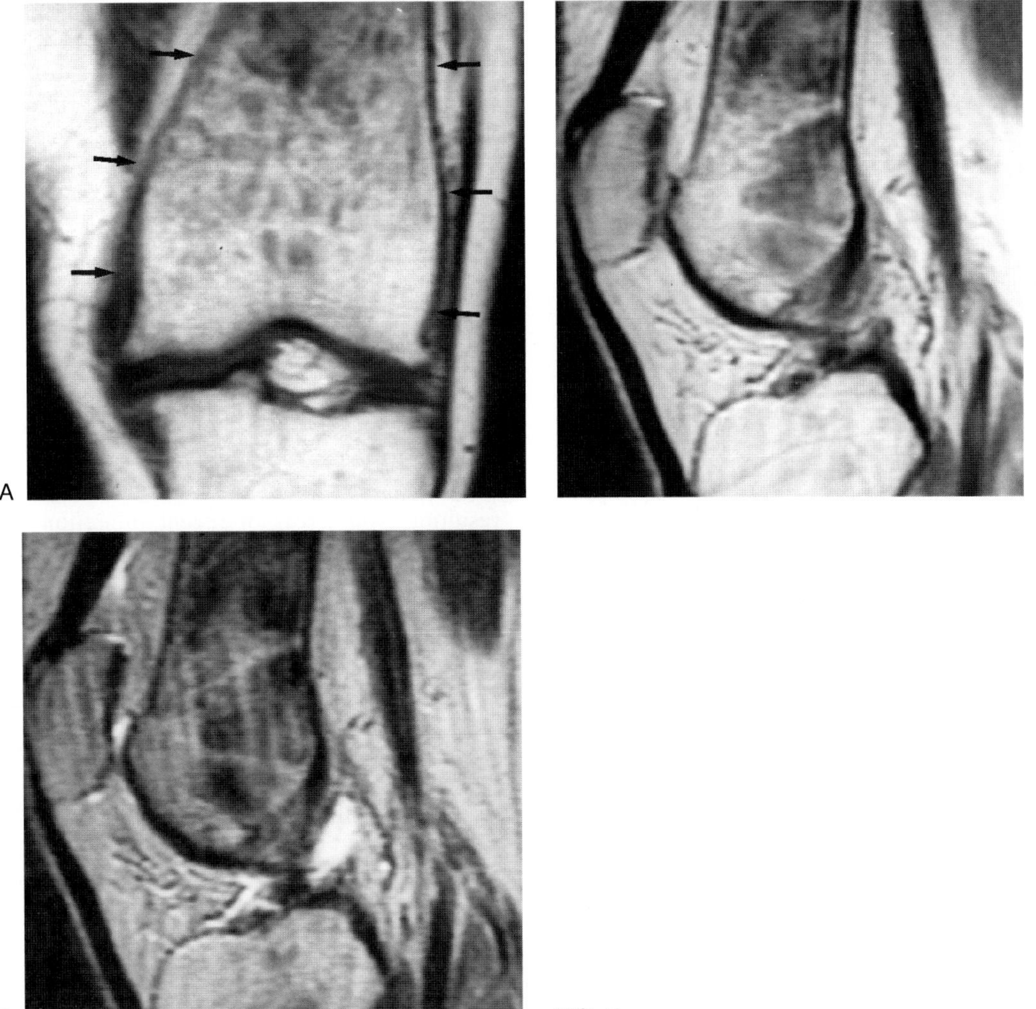

FIG. 5.

Findings: In Fig. 5A coronal T1-weighted (TR/TE 600/30 msec) image of the distal femur demonstrates expansion of the contour of the distal femur with convex outer margins (arrows). There are multiple areas of intermediate signal intensity dispersed throughout the marrow space. In Fig. 5B sagittal proton density (TR/TE 2,300/20 msec) image demonstrates the diffuse signal abnormality throughout the distal femur. In Fig. 5C sagittal T2-weighted (TR/TE 2,300/80 msec) image does not demonstrate any significant augmentation in signal intensity within the process infiltrating the marrow space.

Diagnosis: Type 1 gaucher disease.

Discussion: The clinical and imaging manifestations of Gaucher disease result from the accumulation of enlarged lipid laden histiocytes (Gaucher cells) that are distributed in a variable fashion in the organs and tissues of the reticuloendothelial system (1,2). The cells, which represent the hallmark of the disease, are characterized by an abnormal accumulation of a lipid material (glycosylceramide) that results from a deficit of a specific enzyme (glucocerebroside hydrolase or β-glucosidase) (1,2). Gaucher disease is an autosomal recessive disorder that affects most commonly Ashkenazi Jews in whom the incidence is ten times that of the non-Jewish population. The disease may affect both men and women and may present at any time, although it is particularly frequent in childhood and early adult life.

Gaucher disease has been divided into three clinical forms. Common to all forms are the recessive inheritance, deficient acid β-glucosidase activity, and characteristic Gaucher cells within the bone marrow (1). Type 1 disease or ''adult'' Gaucher disease is the most common and demonstrates a spectrum of clinical manifestations (protuberant abdomen, episodic pain in the arms, legs, and back, fever, respiratory distress and pneumonia, diffuse yellow-brown pigmentation on the lower legs and face, and bleeding diathesis). Bone involvement is common (1). Type 2 is a rare, fatal (average survival 1 year), neurodegenerative disorder and becomes manifest shortly after birth. This form of the disease demonstrates no ethnic predilection (1). Type 3, ''juvenile'' Gaucher disease, is an uncommon form characterized by the development of hepatosplenomegaly within the first few years of life and the development of neurologic (convulsions, hypertonicity) and skeletal manifestations during childhood (1).

Gaucher disease represents a classic example of skeletal ''marrow-packing'' disorder. The accumulation of Gaucher cells within the marrow results in cellular necrosis, fibrous proliferation, and resorption of spongy trabeculae (1,2). The radiographic correlates of these pathologic changes include (a) increased radiolucency of bone with endosteal scalloping and thinning, (b) coarsened trabecular pattern, (c) localized geographic or moth-eaten areas of bone destruction that may simulate the appearance of myeloma or metastasis, and (d) reactive sclerosis (often accompanying osteonecrosis) (1). Modeling deformities, particularly in the appendicular skeleton, represent one of the most characteristic manifestations of Gaucher disease (1,2). Expansion of the contour of the long tubular bones is most characteristic in the lower ends of both femurs and the resulting convex osseous margin, termed an Erlenmeyer flask deformity, is quite suggestive of this disease, particularly when seen in association with epiphyseal osteonecrosis (1). The Erlenmeyer flask deformity, however, is not specific for Gaucher disease and may be seen in other disorders, particularly marrow packing diseases, including (a) Niemann-Pick disease, (b) anemias, (c) fibrous dysplasia, (d) osteopetrosis, (e) metaphyseal dysplasia, and (f) heavy metal poisoning (1). Other osseous manifestations of Gaucher disease include (a) fractures (particularly of veterbral bodies), (b) osteonecrosis (particularly epiphyses and diaphyses), (c) and infection (increased incidence) (1).

The signal intensity of the involved marrow in Gaucher disease is typically decreased on both T1- and T2-weighted conventional spin echo images (1,3–8). This signal intensity pattern may be homogeneous or inhomogeneous (with islands of interposed preserved fatty marrow), progresses from proximal to distal sites in the appendicular skeleton, and generally spares the epiphyses unless extensive bone involvement is present (3,4). The T1-weighted signal changes in Gaucher disease are similar to those of other marrow packing disorders on MR. The generally persistent low signal intensity on T2-weighted images is more helpful in differentiating Gaucher disease from other disorders of marrow infiltration such as metastases, certain tumors, and infections, which may typically, although not invariably, be associated with increased signal intensity on T2-weighted images (1,3–8).

MR has been investigated in the assessment of disease activity in Gaucher disease. In the study of Hermann and coworkers (7), areas of increased signal intensity on T2-weighted images had a high correlation with acute bone pain in patients with this disease. These investigators suggested that the increased signal may reflect edema resulting from areas of recent infarction related to intramedullary extravascular compression caused by the accumulation of Gaucher cells (7). Osteomyelitis and or intraosseous subperiosteal hemorrhage complicating Gaucher disease may also manifest as areas of increased signal intensity on T2-weighted spin echo sequences (6). The use of MR has also been investigated as a method for evaluating the extent of marrow involvement in Gaucher disease. Johnson and associates (8) correlated the findings from quantitative chemical shift imaging of the lumbar spine (modified Dixon) with quantitative analysis of marrow triglycerides and glucocerebrosides and MR determined quantitative splenic volume. Bulk T1 values were significantly longer in areas of infiltrated marrow, reflecting an overall decrease in marrow fat. Glucocerebroside concentrations were higher in diseased marrow and correlated inversely with triglyceride concentrations. The extent of marrow infiltration as determined by fat fraction measurements correlated with the disease severity measured by splenic enlargement (8). With the recent implementation of enzyme replacement therapy for Gaucher disease, the role of MR in evaluation of this disorder may well increase as a consequence of its ability to noninvasively assess both disease extent and activity (7).

CASE 6

History: A 32-year-old man was injured while playing soccer. What is the nature of the injury? What are the mechanisms involved in production of injury to this structure? Can you reliably grade this injury?

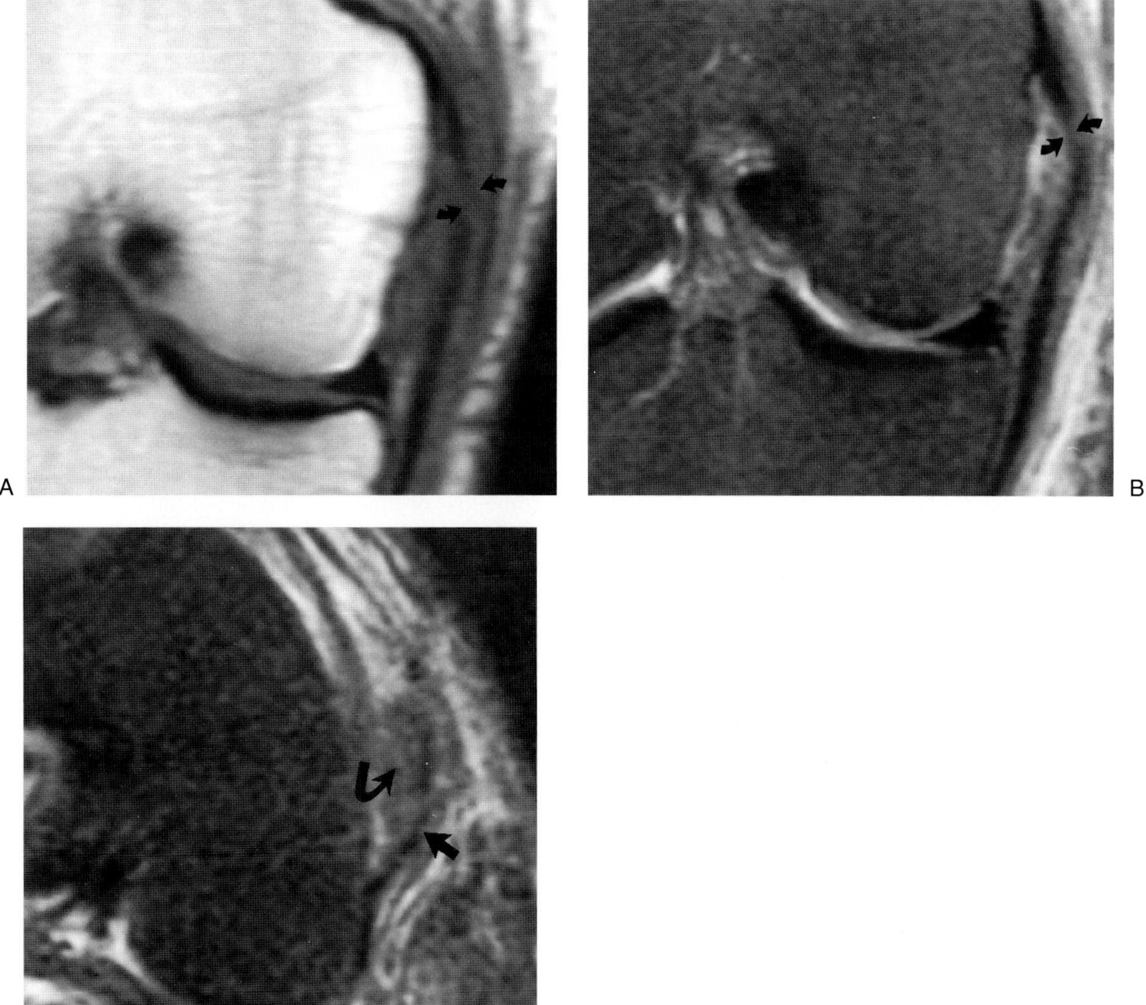

FIG. 6.

Findings: In Fig. 6A coronal T1-weighted (TR/TE 600/20 msec) imaging demonstrates intermediate weighted signals both superficial and deep to the tibial collateral ligament (TCL). The ligament appears markedly attenuated near its femoral attachment (small curved arrows). In Fig. 6B fat-suppressed proton density fast spin echo image (TR/TE 4,000/30 msec) demonstrates increase in intensity to the signal presumably representing edema and hemorrhage superficial and deep to the tibial collateral ligament. The focally markedly attenuated region of the TCL is better defined (small curved arrows). In Fig. 6C axial inversion recovery fast spin echo (TR/TE/TI 3,000/30/130) is obtained through the region of maximal attenuation of the TCL. The TCL is displaced from the femur and has a serpentine configuration (short arrow). A presumed localized acute soft tissue hematoma is present (curved arrow).

Diagnosis: Extensive grade 2 to 3 sprain of the tibial collateral ligament.

Discussion: The medial collateral ligament of the knee is composed of both superficial and deep components (1–6). The TCL represents the superficial layer of the medial collateral ligament (MCL) complex (1–6). The TCL arises from the medial epicondyle of the femur 5 cm above the joint line and inserts on the tibial metaphysis 6 to 7 cm below the joint line just deep to the pes anserinus tendons. The TCL is approximately 1.5 cm wide and most of the fibers are vertically oriented. One segment of the TCL, however, has a slightly more obliquely posterior course extending from the posterior aspect of the TCL to cover the anterior portion of the semimembranosus tendon and to eventually insert on the joint line. This oblique segment is referred to the posterior oblique ligament. The TCL represents the principal component of the MCL complex that is visualized at MR imaging. It should be appreciated, however, that it represents the superficial component of the complex and not the MCL itself. The TCL is the strongest part of the MCL and is infrequently torn.

Just deep to the TCL, the joint capsule of the knee, which is otherwise a relatively thin and redundant structure, demonstrates a focal condensation that is known as the deep medial ligament, the deep layer of the medial collateral ligament, or the medial capsular ligament. The medial capsular ligament is firmly adherent to the medial meniscus. The proximal and distal ends of the medial capsular ligament are known as the meniscofemoral and meniscotibial ligaments. This deep component of the MCL is relatively weak and is the first component to tear. As such, there is a direct relationship between MCL injuries and meniscocapsular separations (7). A relatively constant bursa can be identified within the fibrofatty tissue that separates the TCL from the medial capsular ligament (5,7).

The function of the TCL reflects its division into parallel and oblique fiber segments (1,2). The femoral attachment of the parallel fibers is arranged around the axis of flexion of the knee joint so that its tension remains constant throughout the range of motion; the oblique fibers, however, become lax with increasing degrees of knee flexion. The TCL acts as the primary restraint against valgus stress although biomechanical studies suggest that its function is only crucial when the knee is in flexion. As such, injury to the MCL occurs when a valgus stress is applied to a flexed knee (2). If the TCL is sectioned, approximately 4 to 5 mm of opening to valgus stress at the knee can be provoked; if only the medial capsular ligament is cut, there is not significant valgus laxity. As such, the deep fibers are demonstrated to have only a minor role in maintaining knee stability. As the deep fibers tear first, and the TCL is only infrequently torn, it follows that most MCL tears are not associated with instability (8). The MCL also serves to prevent external rotation of the tibia on the femur and to provide a secondary restraint to anterior tibial translations in the ACL-deficient knee.

The extent of a knee injury is determined in part by the nature of the interface between the foot and surface on which it is operating at the time of injury (1). As such, isolated TCL tears are common in hockey players because the interface between the skate and ice is forgiving, allowing for a pure valgus stress without the associated rotary deceleration of the lower limb that commonly occurs for example at the interface between ski and snow (1). In situations in which a shoe is in contact with turf, as in football, TCL tears are caused by severe valgus stress with contact (clipping injury) or with valgus stress and external rotation without contact (1). If the knee is flexed 90° at the time of injury, the valgus stresses are borne by the medial capsular ligament fibers, which tear first; more severe forces result in disruption of the TLC and anterior cruciate ligament (ACL) (2). While the injury may result in peripheral meniscocapsular detachment, tears within the substance of the meniscus are uncommon unless a rotary component accompanies the valgus stresses (2). Indeed, while O'Donohue's ''unhappy triad'' classically includes a tear of the TLC, the ACL, and the medial meniscus, it is in fact the lateral meniscus that is more frequently torn with acute anterior cruciate injuries (1).

The severity of a ligamentous injury may be graded according to the amount of ligamentous disruption (2,9). A grade I sprain is predominantly a microscopic tearing or stretching of the ligament and results in no loss of ligament restraining function (Fig. 6D,E). In a grade 2 sprain, disruption of a portion of the ligament results in some loss of function, and in a grade 3 sprain, or complete rupture, anatomic continuity of the ligament is absent and function is lost. Clinically the degree of injury is graded according to the amount of medial joint space opening with valgus stress applied at 30° of flexion. In a first-degree sprain, there is tenderness and pain at the site of injury but there is a firm end point and there is no identifiable laxity. In second-degree injuries, up to 5 mm of excessive laxity is found, but the end point remains firm. The middle third of the medial capsule is probably intact. In third-degree tears, the end point is soft. If 10 mm of laxity is found, the TCL, capsular ligament, and one or both cruciates are usually torn.

On MR, the medial collateral ligament can be well evaluated in both the coronal and axial planes (1,10). The normal TCL is visualized as a thin dark stripe paralleling and in many cases inseparable from the dark medial cortices of both the femur and the tibia. The femoral attachment is usually more readily identifiable than is the distal tibial insertion. On more anterior sections, the TCL, which is only 1.5 cm wide, blends with the superficial layer of the MCL, which represents an extension of the deep fascia that covers the quadriceps. The patellar retinaculum, which is invariably displaced from the underlying bone, should not be confused with a displaced and or torn TCL.

In grade I MCL injuries, in which there is only microscopic tearing of the ligament, the predominant alterations occur in the periligamentous tissues. On MR, the thin dark stripe of the TCL is maintained, but periligamentous edema is present (1,10). More extensive MCL injuries (e.g., grade 2) are characterized by intrinsic signal alteration and or morphologic disruption (ligamentous widening and striated appearance) accompanied by fluid within the MCL bursa, fascial edema, and loss of demarcation from adjacent fat (1,8,10). In severe injuries, the TCL is variably thickened, discontinuous, and often serpiginous. The distance between the subcutaneous fat and the medullary fat of the osseous structures is increased due to deep soft tissue edema and/or hemorrhage between the TCL and capsular ligament (1).

It has not been established that MR can reliably distinguish between grade 2 and 3 injuries (1,8,10). As current orthopedic practice, however, favors conservative management of *isolated* MCL injuries (including complete disruptions, e.g., grade 3), the precise grade of MCL disruption is of limited clinical relevance (11–14). What is critical with regard to management, however, is the presence of concomitant injuries, particularly to the anterior cruciate ligament. It is the presence or absence of these findings that determines theraputic options. The results of nonoperative management of complex injuries (nonisolated MCL) were judged as unsatisfactory in 79% of patients in one study (2). As such, it is not critical to precisely differentiate between advanced MCL injuries on MR studies and when uncertainty exists regarding advanced grade, they can be effectively classified as combined grade 2/3 injuries. Effort should be concentrated at detection of significant concomitant injury (e.g., mensicocapsular separation, ACL disruption, and lateral meniscal tears) that might alter management. In cases in which surgical intervention is undertaken, MR can also potentially be of value in the assessment of the integrity of the reconstruction, although experience in this area is quite limited (Fig. 6F–I).

There has been a relative lack of critical studies regarding the accuracy of MR in the assessment of MCL injuries. This may in part relate to the infrequent surgical treatment of these injuries and as such an imperfect standard for comparison of results. Two recent studies, however, have both demonstrated MR–clinical disagreement regarding the classification of higher grade MCL injuries (8,10). We believe that this further substantiates the approach of grouping these grades for purposes of MR description. Of interest, in one recent study, Schweitzer and colleagues (8) reported on a high incidence of medial compartment bone contusions in association with MCL injuries. While lateral bone bruises can be accounted for by valgus stresses, the medial location is more difficult to explain. It is possible that these areas relate not to contusions but rather reflect a manifestation of microavulsions (8). The previously reported high incidence of lateral bone contusions with MCL injuries may in part reflect bone injuries secondary to associated ACL injuries (8).

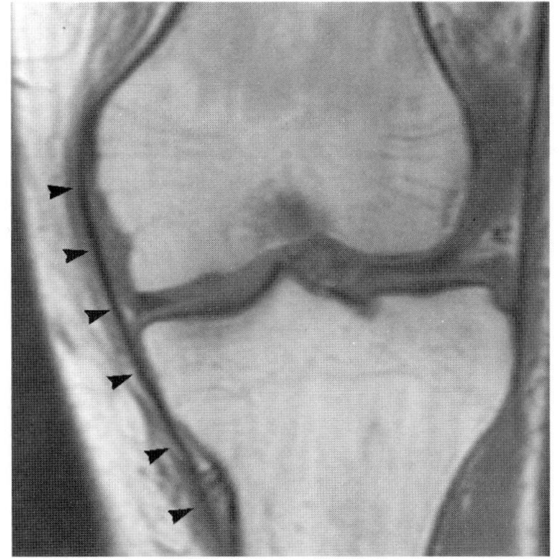

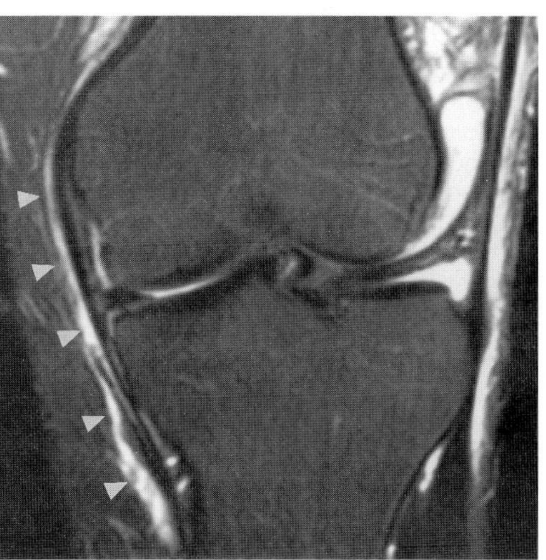

FIG. 6. D,E: Grade 1 TCL sprain. **D:** Coronal T1-weighted (TR/TE 600/20 msec) image demonstrates the TCL to be intact *(arrowheads).* There is intermediate signal intensity both superficial and deep to the TCL. **E:** Coronal fat suppressed PD (TR/TEeff 800/32) image better demonstrates the high signal fluid superficial to the grossly intact TCL *(arrowheads).*

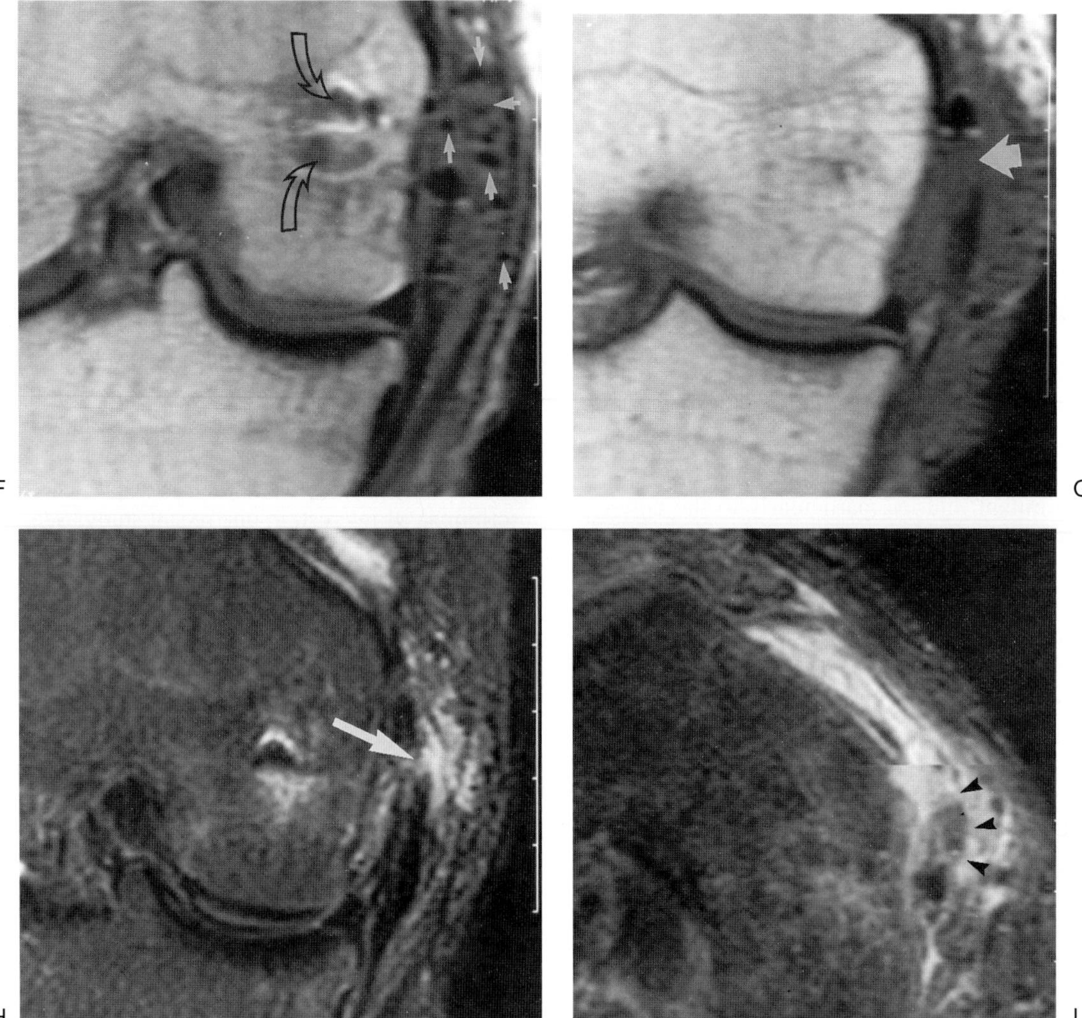

FIG. 6. F–I: Re-rupture of a recent TCL repair. A 32-year-old man had undergone a repair of a prior acute TCL injury. During rehabilitation, the patient experienced a reinjury with increased pain and swelling at the surgical site. The patient was sent for evaluation of the status of the repair. **F:** Coronal T1-weighted (TR/TE 600/16 msec) image demonstrates marked signal alteration along the medial aspect of the knee in the expected course of the TCL. There are several discrete foci of low signal intensity identified representing suture material and micrometallic artifact *(arrows)* and areas of signal alteration within the medial femoral condyle representing the site of placement of Mytex anchors *(open curved arrows).* The distal TCL is visualized although the proximal and mid-TCL are entirely obscured. **G:** Slightly more anterior coronal T1-weighted image from the same series demonstrates apparent discontinuity at the approximate level of the proximal third of the TCL *(arrow).* **H:** Coronal fat-suppressed T2-weighted fast spin echo (TR/TE 2,500/102 ef) demonstrates the site of ligamentous discontinuity *(arrow)* and surrounding superficial edema. **I:** Axial inversion recovery fast spin echo (TR/TE/TI 800/28eff/130) demonstrates a small localized hematoma at the site of tendon disruption *(arrowheads).*

CASE 7

History: These young women suffered similar injuries of their knees. What is the diagnosis? What is the prognosis?

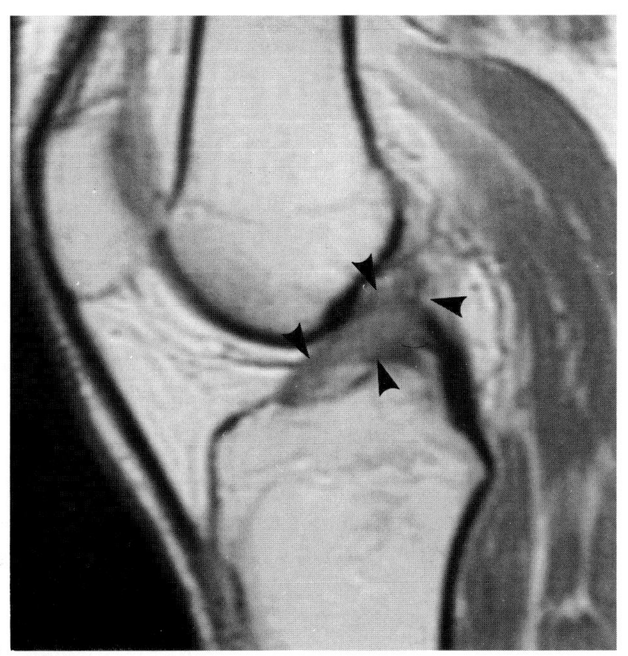
A

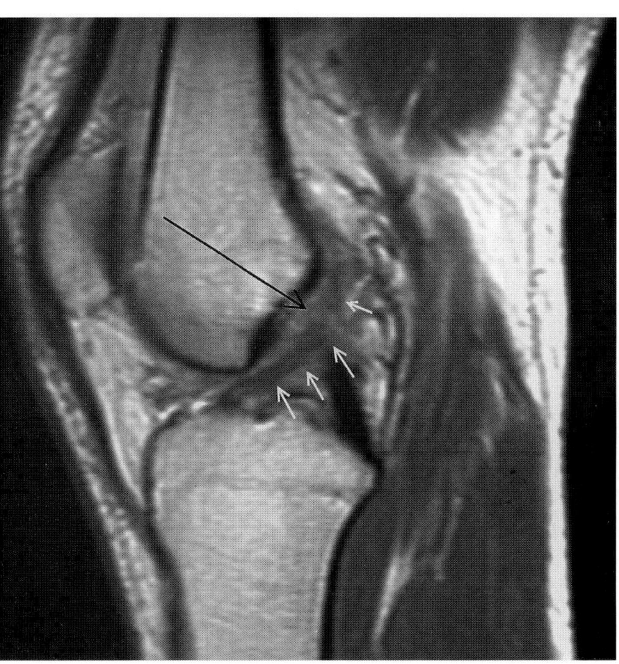
B

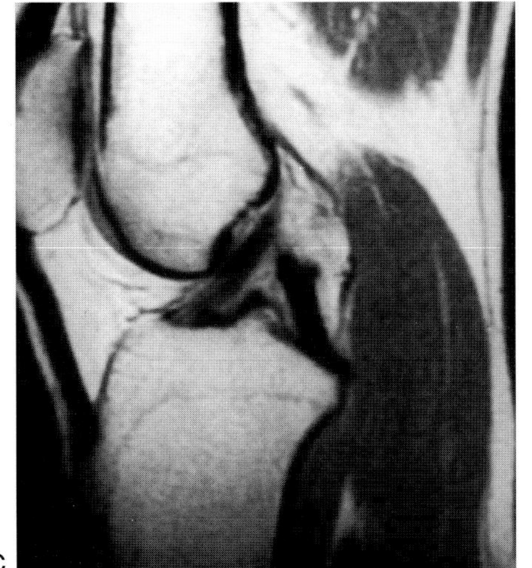
C

FIG. 7. C.

Findings: Figure 7A: Complete tear of the ACL. Figure 7B: Partial tear of the ACL. Sagittal (TR/TE 2,300/20 msec) image. In another patient (B,C), this 46-year-old competitive runner injured her ACL during a fall. The axis of the ACL is low and the ligamentous margins are poorly defined (small arrows). Clinical testing revealed a mild degree of discrepancy between her normal and abnormal knees; she was thought to have at least a partial tear of the ACL.

Diagnosis: Figure 7A is a complete tear of the ACL; Fig. 7B is a partial tear of the ACL. Figure 7C is the follow-up to Fig. 7B. It may be impossible to state with certainty the degree of disruption of an ACL tear in the acute situation.

Discussion: Partial ruptures of the ACL account for 15% to 17% of such injuries (1), but the distinction between a partial tear and a complete tear by MR may be extremely difficult. The diagnostic problem is compounded by the fact that partial tears diagnosed by arthroscopy may rapidly progress to completeness (2); there is rather little anatomic/pathologic correlation and there is a tendency of arthroscopy to underestimate the degree of damage to the ACL (1,3); the degree of disruption of the ACL at the time of arthroscopy does not correlate with the eventual outcome (3–12); partial ACL tears frequently result in instability (13). Of 35 patients with proven partial ACL tears, 17% had fair results and 31% poor results (14).

The diagnosis of a partial tear may be made when a discrete focus of high signal is present within the substance of an ACL in which the majority of the fibers are clearly seen to be intact from femur to tibia (although this pattern is quite similar to the interstitial type of complete tear) (Fig. 7C) (1). If a partial tear is truly present, it is the anteromedial band that is torn, and the tear is most fully appreciated at the distal end of the ligament. Unfortunately, the fat interposed between the diverging fascicles may produce increased signal in association with the anteromedial band (15). The knee suffering a partial tear and one that is stable at the time of initial examination may become unstable over time because of progression of the injury and compromise of the vascular pedicle. On the other hand, the edema, hemorrhage, and synovial reaction that surround the ACL immediately following injury may completely obscure normal, intact ACL fibers and lead the radiologist to suggest that a complete interstitial tear is present. Even a dramatically abnormal appearance of the ACL may become normal at a later date. Yao et al. (1) found that a low-lying axis may be seen in as many as 48% of partial tears (although his criterion for partial tear was the arthroscopic appearance alone; see above). After 4 to 6 weeks, the synovitis accompanying an ACL injury resolves somewhat, and the devascularized segments of the ACL (if present) will retract and the true status of ACL integrity will be evident (Fig. 7D–I).

One should carefully examine the remainder of the MR for secondary signs of ACL disruption, many of which indicate instability. These signs, however, only provide a limited ability to distinguish between partial and complete tears, and while the specificity of these signs is reasonably good the sensitivity is not (1,16). The distinction of a partial from a complete ACL rupture will remain problematic as half or more of detected partial tears will be indistinguishable from complete ruptures (1).

The conventional secondary signs of ACL disruption do not reliably distinguish between complete and partial tears. In knees suffering partial tears, 48% suffer contusional injuries of bone, while 87% of knees with complete tears have this common marker; 43% of partial and 55% of complete tears have an abnormal PCL index, and 48% of partial, and 52% of complete tears have an abnormal ACL axis. On the other hand, an uncovered lateral meniscus was never seen in a knee with a partial tear, and is the best discriminator between these two grades of injury (1). In those cases in which there are findings that could be consistent with a partial ACL tear, but in which some uncertainty exists, it may be prudent to report that the patient has "at least a partial, but possibly/probably a complete ACL tear. Reexamination at 6 to 12 weeks would likely reveal the extent of the tear."

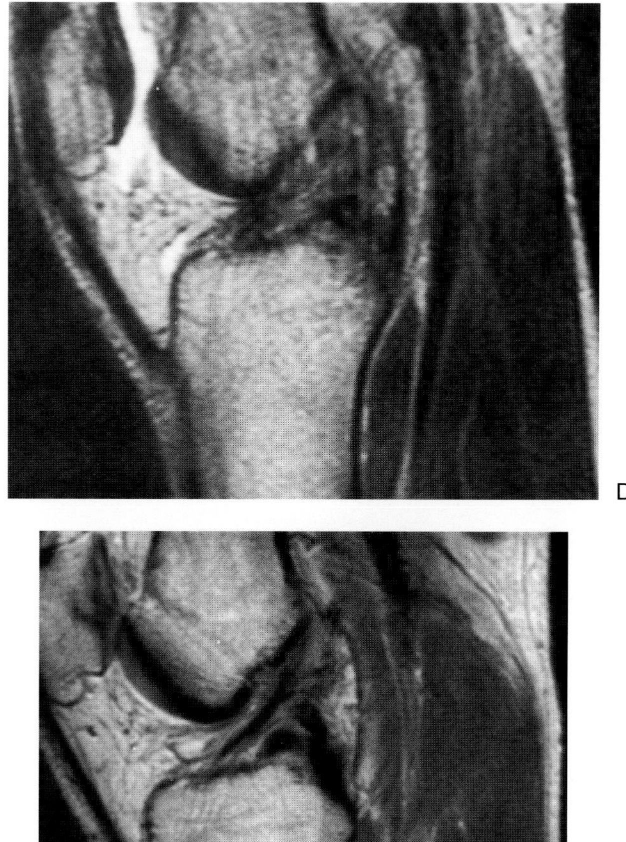

FIG. 7. D,E: Longitudinal study of a partial ACL tear. **D:** Sagittal (TR/TE 2,000/80 msec) image. Diffuse increase signal intensity is seen in the proximal ACL, but the ligament appears intact. **E:** Sagittal (TR/TE 2,300/20 msec) image. This image made 3 years after that in D demonstrates a thin and slightly serpiginous anterior cruciate ligament. Clinical testing demonstrated a mild degree of instability.

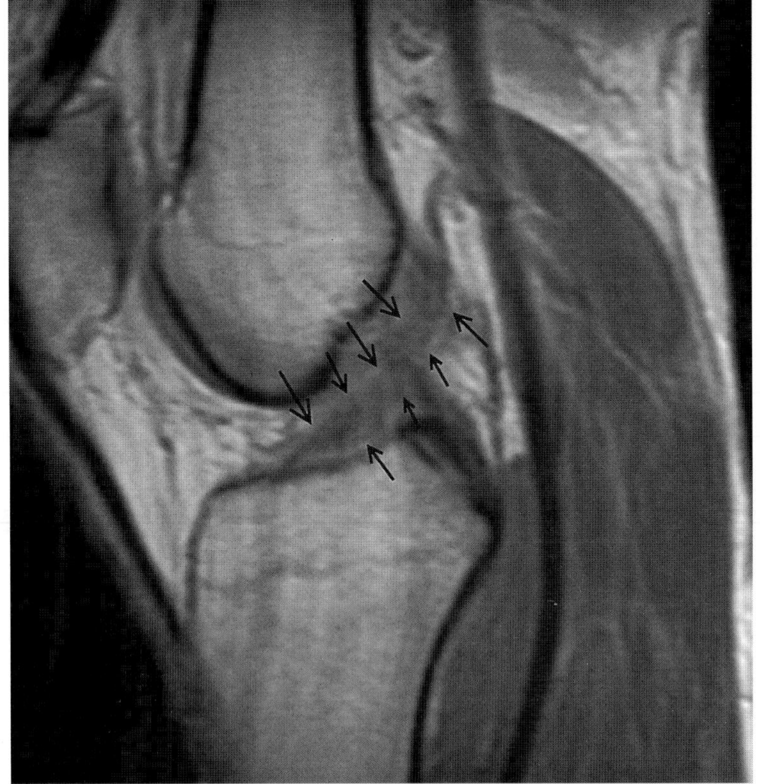
F

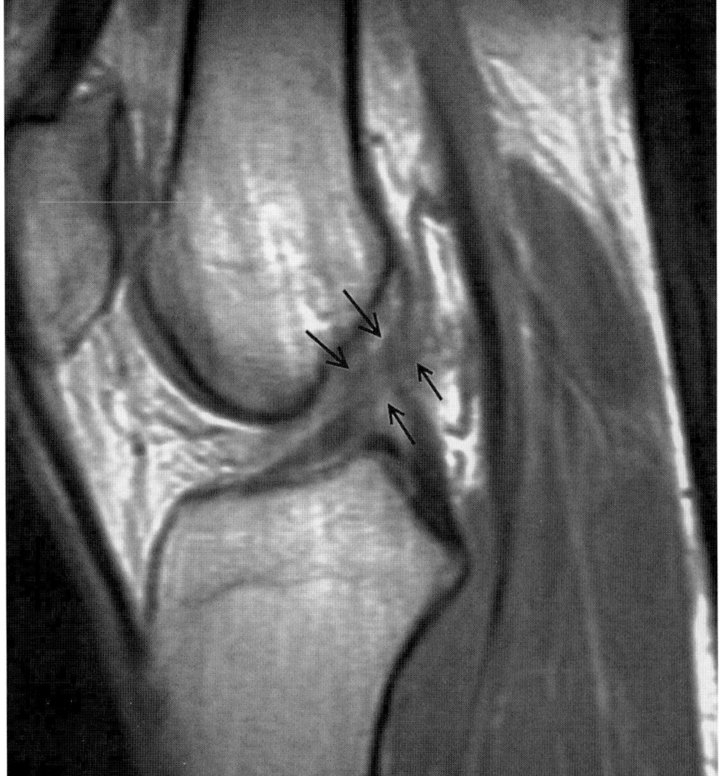
G

FIG. 7. F,G: Partial tear of the ACL. Sagittal MR TR/TE 2,300/20 msec. **F** is made 6 weeks before G. **(F)** The ACL *(arrows)* is seen to be lax and mildly concave. The ligament demonstrates a diffuse increase in signal. **G:** The distal end of the ligament is again dark and sharply defined, but the proximal end has not changed in appearance. Clinical examinations at the time of both MR demonstrated mild instability.

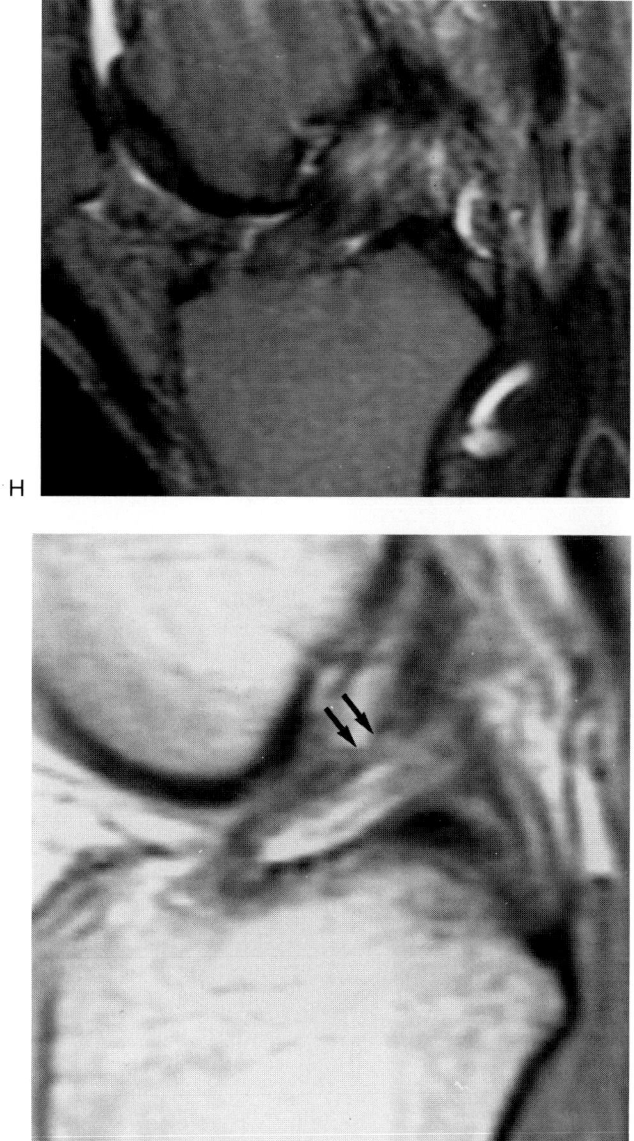

FIG. 7. H: Partial tear of the ACL. Sagittal (TR/TE 2,300/80 msec) image. While the ACL appears intact, there is clearly a focus of increased signal in its midsegment. Eight weeks later, the patient's knee remained stable. **I:** Partial tear of the ACL. Sagittal (TR/TE 2,000/20) image. The mid-portion of the ACL demonstrates a slight angulation *(arrows)* at the site of a surgically demonstrated partial ACL tear.

CASE 8

History: A 63-year-old man underwent MR evaluation to exclude internal derangement of the knee. What is your diagnosis?

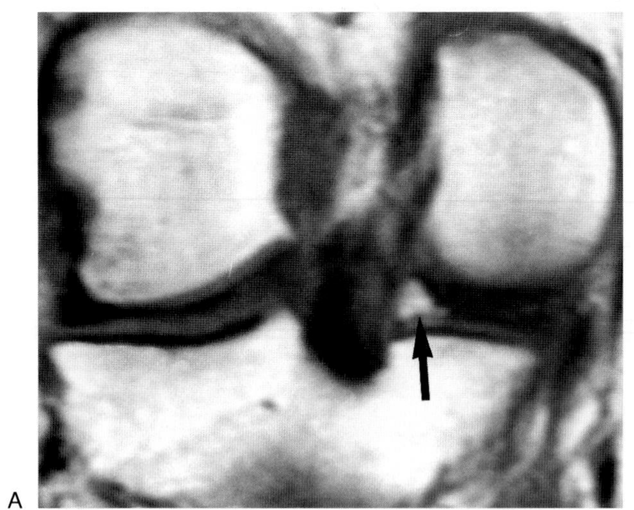

A

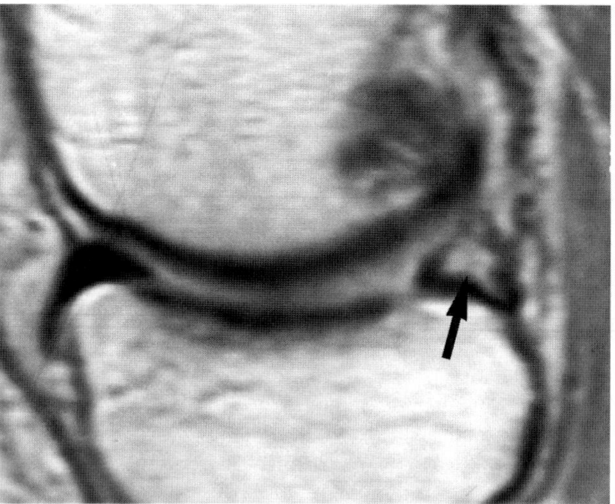

B

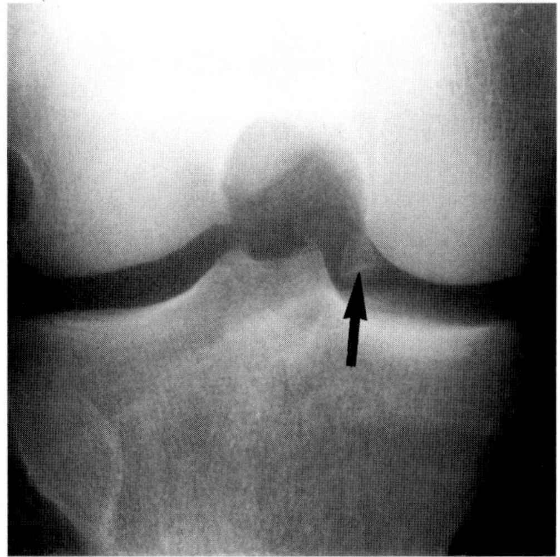

C

FIG. 8.

Findings: In Fig. 8A the coronal T1-weighted (TR/TE 600/15 msec) image demonstrates a triangular-shaped focus of increased signal intensity projecting at the posterior-medial aspect of the medial meniscus (arrow). In Fig. 8B the sagittal proton density (TR/TE 2,100/20 msec) sequence obtained quite centrally through the medial meniscus again demonstrates the high signal intensity focus within the meniscus (arrow). In Fig. 8C AP radiograph demonstrates triangular-shaped area of ossification corresponding to the findings depicted in A and B.

Diagnosis: Meniscal ossicle.

Discussion: Focal ossification within the meniscus of the human knee is an uncommonly reported phenomenon, although such ossicles may commonly be found in animals such as rodents (1–3). Meniscal ossicles are most frequently found in the posterior horn of the medial meniscus (4,5). The vast majority of reported patients have been men ranging in age from 16 to 40 years with a mean age of 25 years. Histologically, meniscal ossicles represent mature lamellar and cancellous bone containing fatty bone marrow surrounded by hyaline cartilage (1,3). The overwhelming majority of reported patients with meniscal ossicles have been symptomatic (1). Clinically, intermittent pain is the predominant symptom and the condition is associated with tears (typically longitudinal). The condition is not associated with knee locking, a symptom that might be expected with an intraarticular osteocartilaginous body, an entity with which meniscal ossicles are often confused on plain radiographs (1).

The pathogenesis of meniscal ossicles remains debated with four predominant theories: (1) a degenerative basis related to impaired meniscal nutrition, resulting in replacement of areas of mucoid degeneration with bone; (2) a posttraumatic basis with metaplasia and development of heterotopic ossification within the meniscus; (3) the meniscal ossicles represent vestigial structures, given their more common occurrence in lower class animals; (4) ossicles result from avulsion of the tibia at the site of meniscal attachment.

On plain radiographs, the ossicle most commonly appears as a well-corticated density in the posteromedial aspect of the knee joint. Differential diagnostic considerations include (a) loose intraarticular body, (b) osteochondral lesion, (c) avulsion of the meniscal attachment, (d) avulsion of the semimembranosus or popliteus tendon insertion, and (e) chondrocalcinosis (1,2,6,7) On MR, meniscal ossicles appear as signal foci contained within the meniscus that demonstrate characteristics identical to those of medullary bone (1). Their intrameniscal location allows definitive differentiation from loose intraarticular bodies. The appropriate treatment course is controversial, and there are advocates of both surgical and conservative regimens.

CASE 9

History: Is there a meniscal tear in this 15-year-old girl, who has had a recent injury? What is your level of certainty?

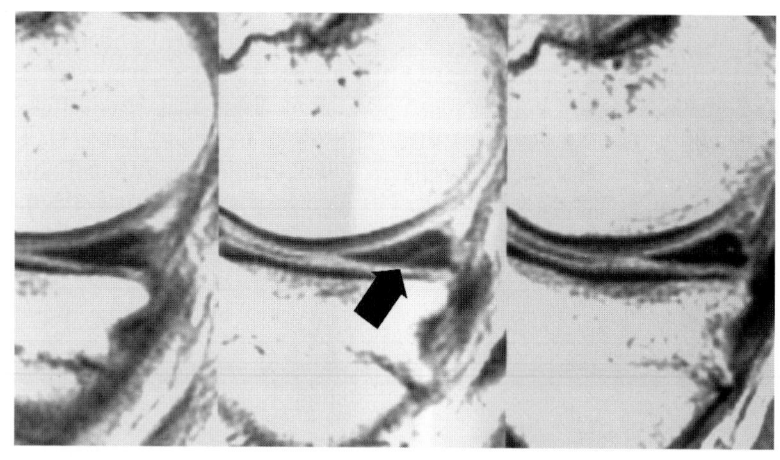

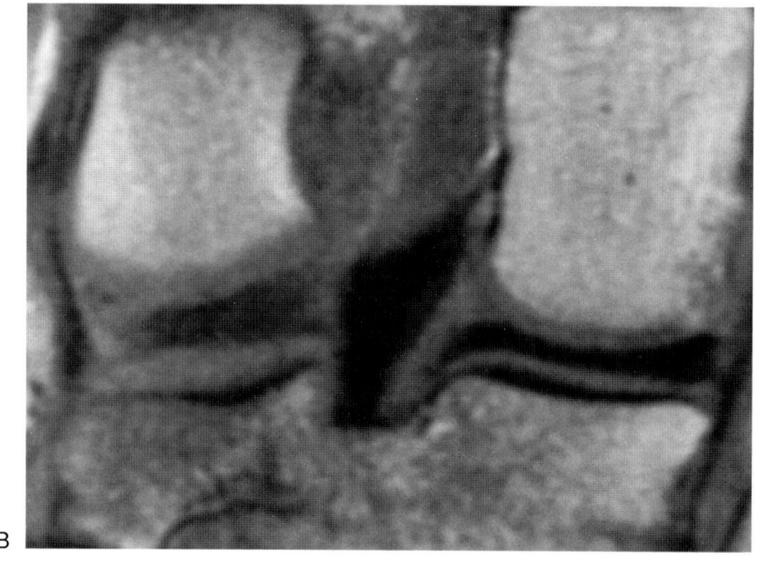

FIG. 9.

Findings: There is a high signal intensity line that extends to the inferior articular surface (arrow) on a single image. Fig. 9A: Three consecutive sagittal MR TR/TE 2,000/16 msec. Fig. 9B: Coronal MR TR/TE 600/16 msec.

Diagnosis: No meniscal tear identified at arthroscopy.

Discussion: Meniscal high signal intensity is most commonly classified using the system originally described by Lotysch et al. (1). In this classification, a grade 3 meniscal abnormality is defined as increased intrameniscal signal that is contiguous with an articular surface, and this is felt to indicate a meniscal tear (1). There are two important features to analyze when evaluating a meniscal signal that appears to contact an articular surface. The first is whether the high signal definitely, or only possibly, contacts the articular surface. The second is to determine the number of slices where the signal is contiguous with the surface.

Crues et al. (2) quickly recognized that false positives were common in menisci with signal only possibly contacting the surface, and therefore clarified the definition of grade 3 signal to include only unequivocal extension. There have been several reports since then that have evaluated the prevalence of tears in menisci with possible surface contact. Kaplan et al. (3) reported that no tears were identified at arthroscopy in 13 cases that had equivocal surface extension. DeSmet et al. (4) found that 13% of 15 medial and 6% of 18 lateral menisci with possible surface extension were torn at surgery, which was not significantly different than the 7% to 8% MR false positive rate for menisci with no signal contacting the surface. Unequivocal contact, or focal distortion of a meniscus, is now considered the MR criterion for a meniscal tear, and the positive predictive value using this definition has been demonstrated to be 86% to 98% (2,4,5).

The second important feature is the number of images that demonstrate intrameniscal signal contacting an articular surface. Several authors have shown that small radial and parrot-beak tears may only be seen on a single image (5–7). The meniscus may appear torn on an MR image in one plane, however, but appear completely normal in the orthogonal plane (7). DeSmet et al. (4) specifically evaluated the positive predictive value of identifying grade 3 signal on one versus multiple images using a 3-mm slice thickness and a 1.5-mm interslice gap. These authors found that 56% of medial and 30% of lateral menisci with grade 3 signal on only one image were torn at surgery, compared with 90% to 94% when such signal was present on more than one image.

There are several potential reasons why grade 3 signal may not correlate with a meniscal tear at arthroscopy. The discrepancy may be due to one of the anatomic pitfalls, such as the transverse meniscal ligament, which were a source of false positives in early MR studies (6,7). Motion and intraarticular gas have also been shown to simulate tears (7). Truncation artifact, which is linear high signal that parallels the meniscal-hyaline cartilage interface, can also appear as surfacing high signal and therefore be mistaken for a tear (7). The obliquely oriented grade 3 signal in our case was probably not due to any of these artifacts or pitfalls.

Another cause for grade 3 signal not correlating with a tear at arthroscopy involves tears that surface inferiorly in the outer third of the posterior meniscus (8,9). The orthopedist cannot directly visualize this area through an arthroscope, and therefore must use probing an compression to evaluate for a tear. This arthroscopic technique is probably not as accurate as direct visual inspection for the detection of tears (8,10,11). The accuracy of arthroscopy is reported to be above 95%, especially when performed by an experienced orthopedic surgeon (8,11). The arthroscopy in our patient was performed by an experienced knee subspecialist, and the area of the meniscus where the grade 3 signal was observed on MR was specifically stated to be normal in the surgical report.

Several articles have also described menisci with high signal that contacted an articular surface on MR but represented intrasubstance tears at arthroscopy (9,12). Stoller et al. (12) demonstrated that intrasubstance tears can have histologic separation of fibers at the meniscal surface without macroscopic evidence of a tear. Adsorption of water molecules on collagen in the area of the separated fibers may be the cause of the high signal seen on short TE MR images (13).

Finally, tears that have healed can cause a false-positive grade 3 signal within a meniscus. In patients with meniscal tears that have been treated conservatively and where a second-look arthroscopy has documented that the meniscus has healed, grade 3 signal on MR has been demonstrated to persist for years after the acute injury (14,15). Healed tears in an area that cannot be directly visually examined may be especially difficult to identify at arthroscopy (8). We believe that this patient likely had a healed tear so that the meniscus appeared normal at arthroscopy.

In summary, the full answer to the questions regarding this case are that the signal possibly represents a medial meniscal tear, with a likelihood of about 50%. If a single image showing signal contacting the surface is seen in the lateral meniscus, then the likelihood would be about 30%.

CASE 10

History: A 22-year-old man presented with anterior knee pain. What are the findings in Figure 1A and B?

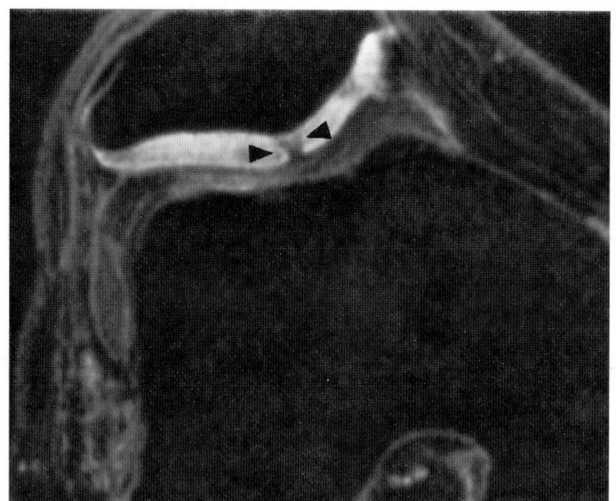
A

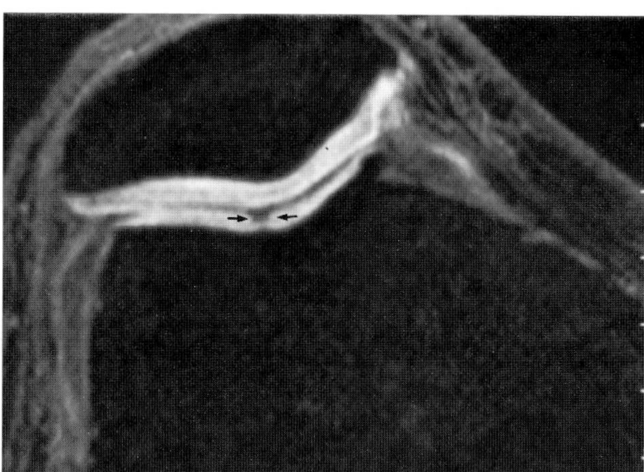
B

FIG. 10.

Findings: Figure 10A is a transaxial fat suppressed (FS) FLASH 3D (TR/TE/FA = 47/10/45) image of the knee demonstrating a partial-thickness defect involving the median ridge of the patella (arrowheads). Figure 10B is a transaxial FS 3D FLASH image demonstrating a partial thickness lesion of the medial trochlea (arrow).

Diagnosis: Localized chondral defect medial patella facet.

Discussion: Noninvasive diagnosis of cartilage abnormalities is assuming greater importance with the recent development of surgical and pharmacologic methods for the treatment and prevention of chondral degeneration and injury (1–5). MR appears to be the ideal modality by virtue of its excellent soft tissue contrast, its ability to allow direct visualization of hyaline cartilage, and its multiplanar capabilities. Despite its theoretical advantages, however, MR has not yet achieved widespread use for the detection of chondral abnormalities. Both the MR appearance of normal articular cartilage as well as the pulse sequences that optimally display cartilage defects are not fully defined. Recent studies, however, have begun to clarify these issues.

Although early studies of articular cartilage depicted a structure of uniform signal intensity, most recent studies have depicted cartilage as a trilaminar structure (Fig. 10C,D); there is a superficial lamina of high signal intensity, a middle lamina of intermediate to low signal intensity, and a deep lamina of intermediate to high signal intensity. A fourth layer of low signal intensity separates articular cartilage from subchondral bone marrow. This multilaminar appearance is felt to relate to the complex histologic structure of articular cartilage. There are four histologic layers of articular cartilage (Fig. 10E); the superficial or tangential zone, the transitional or intermediate zone, the deep radial zone, and the zone of calcified cartilage. These layers differ in the size and shape of their cells, the size and orientation of collagen fibers, and in their content of water and proteoglycan.

Using electron microscopy, Rubenstein et al. (6) correlated the different lamina on the MR images with the histologic layers of cartilage as follows: the superficial lamina was thought to represent the tangential and transitional chondral layers and the upper part of the radial layer. The middle lamina was believed to represent the bulk of the radial layer, and the deep lamina was thought to correlate with the lower part of the radial zone. The zones of calcified cartilage and bony cortex were believed to form the region of low signal intensity that separated cartilage from the subjacent bone marrow.

In addition to the uncertainty regarding the MR appearance of normal articular cartilage, there also has been uncertainty regarding the optimal pulse sequence for the detection of chondral abnormalities. A multitude of pulse sequences including T1, proton density, T2-weighted spin echo, 2D and 3D gradient echo, inversion recovery, and fat suppressed (FS) sequences have been used to assess articular cartilage abnormalities with varied and often disappointing results. More recently, a T1-weighted FS 3D gradient echo sequence with radio frequency spoiling (SPGR, FLASH) has been shown to be accurate in the detection and grading of articular cartilage abnormalities of the patellofemoral joint (7). In preliminary clinical studies of the knee, this sequence has also performed well in the detection of chondral defects (8). On this sequence articular cartilage has high signal intensity, joint fluid has low to intermediate signal intensity, and bone marrow has low signal intensity. This leads to high contrast between cartilage vs. fluid and cartilage vs. bone marrow. The ability of this sequence to detect early cartilage lesions before actual surface defects occur is still uncertain as is its ability to detect lesions in smaller joints with thinner articular cartilage. Despite these possible limitations, this sequence appears to offer an excellent method for the noninvasive evaluation of articular cartilage. Examples of additional articular cartilage abnormalities of the knee are illustrated in Fig. 10F and G.

One final issue regarding the detection of cartilage defects concerns the system used to grade the lesions. There are a number of possible classification schemes, but the one currently favored at our institution is the Noyes system (9). The Noyes system is based on four separate variables; the integrity of the articular surface, the extent (depth) of involvement, the location of the lesion, and the diameter of the lesion. Table 10A illustrates the Noyes classification scheme.

Although the MR evaluation of articular cartilage remains an ongoing area of intense research, great progress has been made. It now appears clear that there is a multilaminar appearance of normal articular cartilage that is related to its histologic structure. In addition, although larger clinical studies need to be performed, a FS 3D FLASH(SPGR) sequence appears highly accurate in detecting chondral defects.

TABLE 1. *Noyes classification of articular cartilage lesions*

Surface description	Extent of involvement	Diameter (mm)	Location
Grade 1: cartilage surface intact	A. Definite softening with some resilience remaining B. Extensive softening with loss of resilence (deformation)	<10 <15 <20 <25 >25	Patella Proximal 1/3 Middle 1/3 Distal 1/3 Odd facet Middle facet Lateral facet Trochlea
Grade 2: cartilage surface damaged; cracks, fissures, fibrillation, or fragmentation	A. <1/2 thickness B. >1/2 thickness		Medial femoral condyle a. anterior 1/3 b. middle 1/3 c. posterior 1/3 Lateral femoral condyle a. anterior 1/3 b. middle 1/3 c. posterior 1/3
Grade 3: bone exposed	A. Bone surface intact B. Bone surface cavitation		Medial tibial condyle anterior 1/3 middle 1/3 posterior 1/3 Lateral tibial condyle anterior 1/3 middle 1/3 posterior 1/3

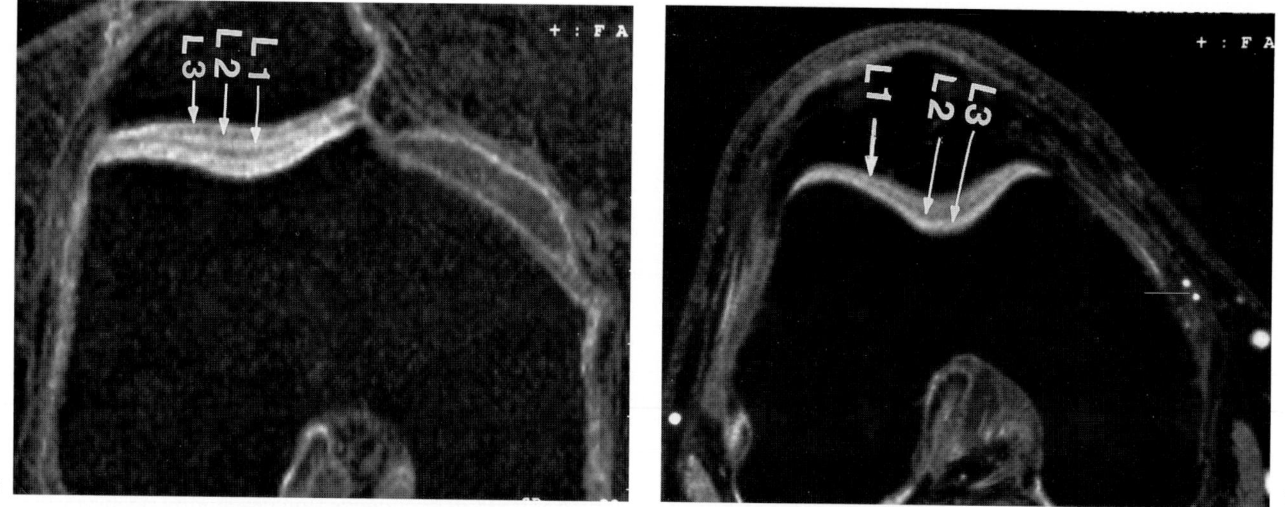

FIG. 10. C,D: Transaxial FS 3D SPGR (TR/TE/TI 58/11/60) images of two volunteers show trilaminar appearance of patellar and trochlear cartilage. Superficial lamina (L1) has high signal intensity, middle lamina (L2) has intermediate signal intensity, and deep lamina (L3) has high signal intensity.

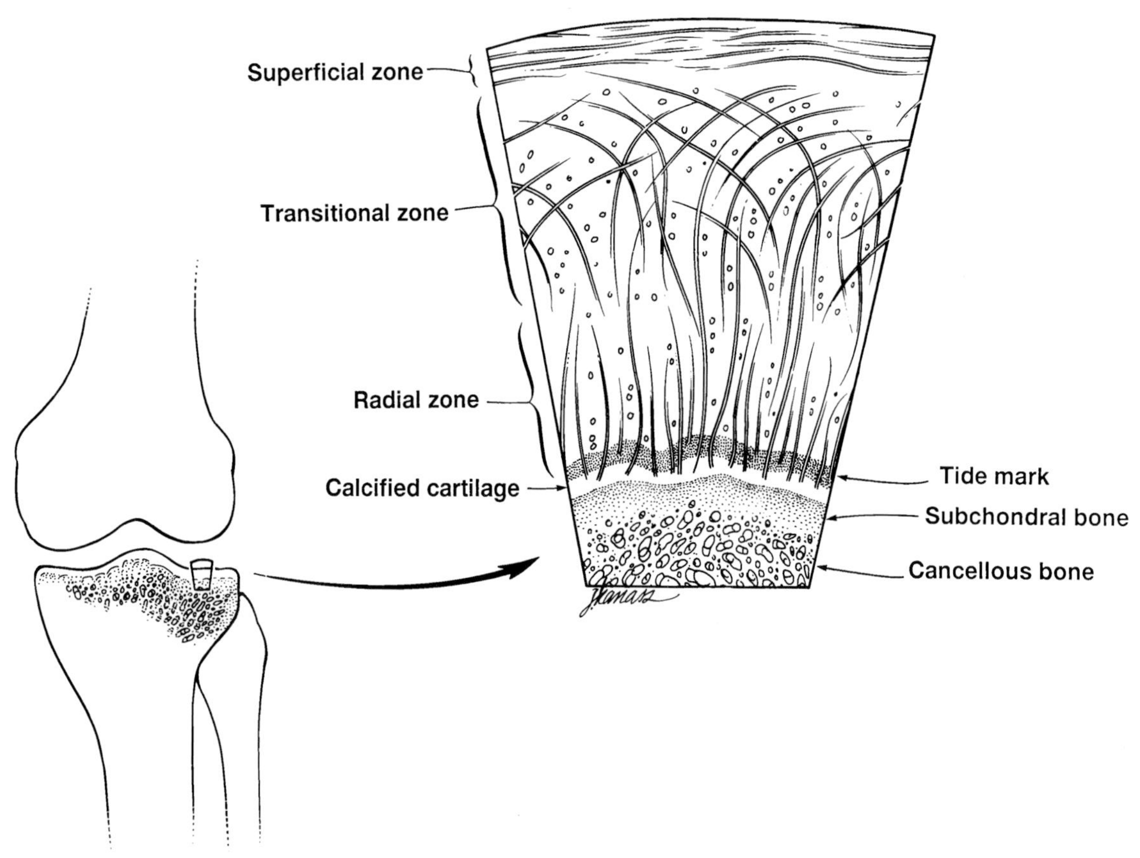

FIG. 10. E: Drawing shows different histologic layers of articular cartilage. (Figures C, D, and E From Recht MP, Resnick D. MR imaging of articular cartilage: current status and future directions. *AJR* 1994;163:283–290, with permission.)

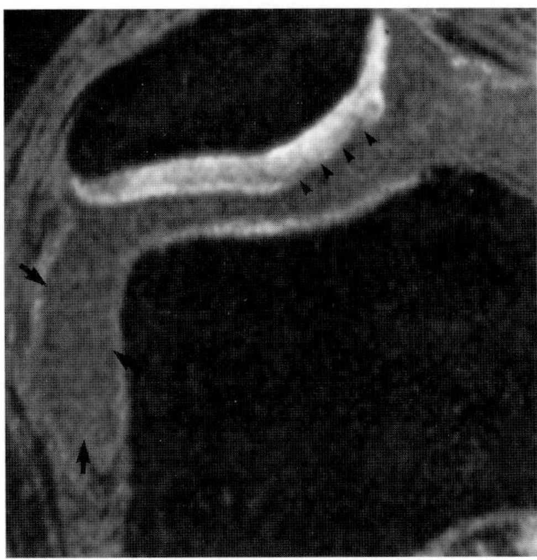

F

FIG. 10. F: Lesion of medial patellar facet *(arrowheads),* which involves less than half the thickness of the articular cartilage and is therefore a 2A lesion according to the Noyes system. Note that there is a joint effusion that is of low signal intensity *(arrows).*

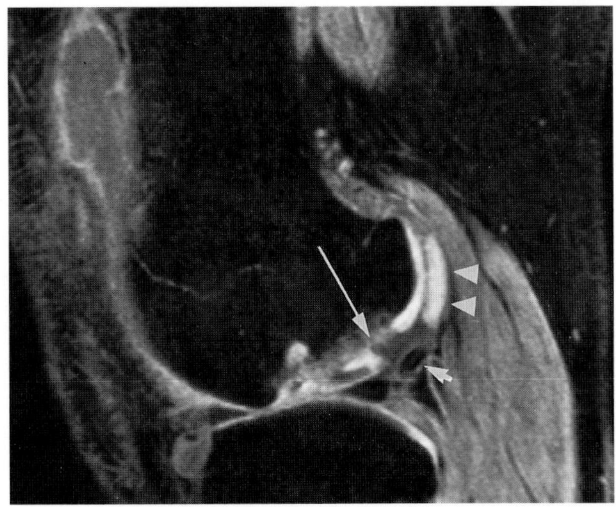

G

FIG. 10. G: Sagittal FS FLASH (TR/TE/TI 47/10/45) image demonstrating a large osteochondral defect involving the lateral femoral condyle *(large arrow).* There is a detached cartilage fragment *(arrowhead)* as well as a detached bony fragment *(small white arrow).*

CASE 11

History: These two patients have the same diagnosis. What is it, and what are the diagnostic signs that lead one to the correct analysis?

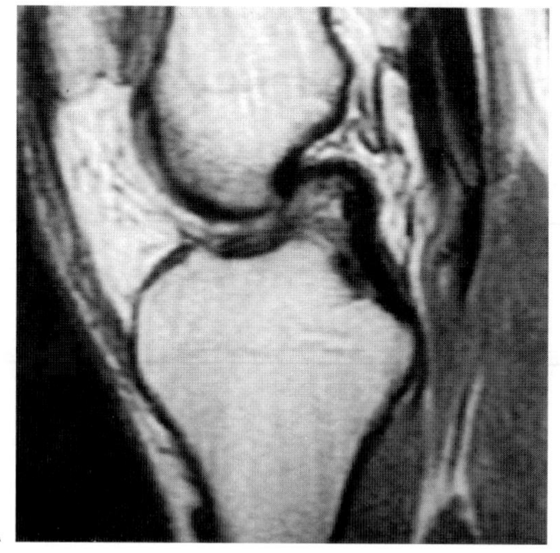

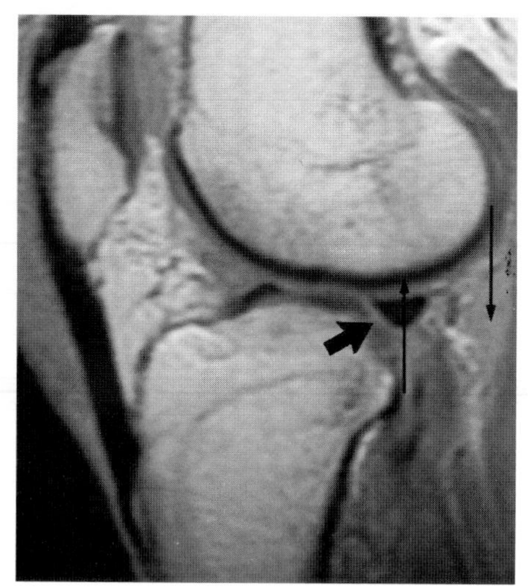

FIG. 11.

Findings: Figure 11A shows buckled PCL. Sagittal (TR/TE 2,300/20 msec) image. The PCL has assumed a buckled appearance due to the presence of a tear of the ACL and a subsequent shift of the tibia forward relative to the femur. Figure 11B shows tibial shift. There is gross tibial subluxation relative to the femur; the posterior aspect of the femur is nearly 2 cm behind the tibia (long arrows). The undersurface of the posterior horn of the lateral meniscus is uncovered secondary to the tibial shift, and there is a fracture of the posterior lateral tibial plateau (arrow).

Diagnosis: Tear of the anterior cruciate ligament. Although these images do not demonstrate the ligament disruption itself (the primary sign), each image reveals one of the so-called secondary signs that lead one to the correct diagnosis. Most of the secondary signs are a reflection of the instability that occurs following an ACL tear.

Discussion: The most accurate method of diagnosing a tear of the anterior cruciate is by means of primary inspection of the ligament. ACL tears are most frequently manifest by diffuse ligamentous widening, indistinct ligamentous contours, irregular or serpiginous anterior margin, interruption of the ligament, mass-like appearance of the proximal end, absence of the ligament, and sharp angulation of the mid-ligament (1). Occasionally, however, intercondylar edema and/or a narrow intercondylar notch, as well as partial volume effect, make diagnosis of an ACL tear difficult. A number of secondary signs of ACL disruption have been defined in an attempt to assist the uncertain examiner.

The most commonly encountered secondary sign of ACL disruption is the presence of an osseous or osteochondral injury of femur or tibia. A fracture of the posterior-lateral tibial plateau is extremely common in patients with an ACL tear (Fig. 11B) (2–15), and some believe that such a lesion is virtually diagnostic of a cruciate ligament injury (2).

The ACL is most commonly torn as a result of a forward shift of the femur relative to the tibia, which is commonly fixed to the ground; the body and the femur rotate externally over the tibia. Most skiing and football injuries occur via this mechanism. Bony injuries result from this transient dislocation; there is impaction of the posterior lateral tibia against the mid- or anterior lateral femoral condyle (Fig. 11C). This subluxation usually reduces immediately, but a more subtle degree of distortion of alignment may persist.

If the tibia remains minimally subluxed, the distance between the femoral attachment of the PCL, and its insertion on the tibia, is shortened (Fig. 11A). Subsequently, the posterior cruciate ligament buckles and assumes a curved, serpiginous, or ''sea-horse'' configuration (16). Since there is some element of variation in the shape of the PCL, several authors have attempted to quantify the degree of distortion. The *PCL curvature value* is described by measuring the distance between the anteriormost tibial and femoral insertion points of the ligament (line *y*), and a line extending perpendicular from the first line to the anterior most aspect of the mid-PCL (line *x*) (14) (Fig. 11D). Line *x* is then divided by line *y*. Normal values are 0.27, while values in patients with proven ACL tears have a mean values of 0.40 or more.

The *posterior cruciate ligament angle* is defined as the angle between a line through the central portion of the tibial insertion of the PCL and a line drawn through the central portion of its femoral insertion. Normal values are variable, but values below 105° define a ligament that is buckled and angulated. This sign has a higher sensitivity but lower specificity than the presence of bone injuries (13,17) (Fig. 11E).

The *PCL sign* has been described as a useful ancillary sign of ACL tear with instability (18). A line is drawn adjacent and parallel to the posterior margin of the distal portion of the PCL and is extended proximally toward the tibia. A positive PCL line does not intersect the medullary cavity of the femur. A positive line correctly predicted ACL disruption in 19/22 patients.

Uncovering of the undersurface of the posterior horn of the lateral meniscus has been described as a sign of tibial subluxation relative to the femur (16). The lateral meniscus moves with the femur, resulting in posterior displacement of the posterior horn. In the normal knee, the posterior edge of the posterior horn of the lateral meniscus lies on the rim of the lateral tibial plateau. In a knee in which an ACL tear has resulted in instability, the tibia remains in a relatively forward position and the posterior horn of the lateral meniscus lies posterior to the rim. The *uncovered lateral meniscus sign* is said to be positive if a line drawn along the posteriormost cortical margin of the lateral tibial plateau on sagittal images intersects any part of the posterior horn of the lateral meniscus (14,17) (Fig. 11B). This sign has a high specificity, but rather low sensitivity for the detection of ACL tears.

It is useful to standardize the method by which measurements of tibial shift are made (Fig. 11F). Two lines are constructed directly on the MR image, both of which are perpendicular to the bottom of the image frame (15). The first line parallels the posterior aspect of the lateral (or medial) femoral condyle, and the other the posterior aspect of the lateral tibial plateau. These lines are made on the image made through the midsagittal lateral femoral condyle. (While the same measurements can be made using the medial femoral condyle and plateau, the lateral sided measurement has been found to be a better discriminant between normal and abnormal). If 5 mm is used as the discriminator between normal and torn ACL, the differentiation is measured with a sensitivity of 58%, a specificity of 93%, and an overall accuracy of 68%. With a 7-mm cutoff, the sensitivity is 38%, the specificity is 100%, and the overall accuracy is 58%. Sixty percent of patients with ACL tears had an average shift of 3.6 mm, but this degree of instability was never seen in patients without ACL tears (13).

The posterior cruciate ligament and the fibular collateral ligament (FCL) both arise from the femur at a position relatively more anterior than their distal, tibial insertions. Because of their posterior-inferior angular course, neither one is identified on a single coronal image. In the presence of a tear of the anterior cruciate ligament, however, the tibia may be free to glide forward. In such circumstances, the tibial insertions of both the PCL and the FCL may come to lie directly distal to their respective femoral origins. In such cases, the entire course of either/both ligaments may be seen on a single image frame (Fig. 11G).

Several other signs have been described as occurring in association with tears of the ACL. The forward shift of the tibia that so commonly occurs with ACL disruption is thought to tear a portion of the posterior lateral capsule, permitting joint fluid to escape along the course of the popliteus tendon and muscle (1,16). The extravasated fluid dissects into the popliteus muscle, giving an appearance similar to a simple muscle strain (Fig. 11H). Rarely, the popliteus itself may be primarily injured. When the posterolateral capsule is torn, the root of the lateral meniscus may also be disrupted resulting in a tear of the posterior lateral meniscus (1).

Assessment of the presence of secondary signs may help in distinguishing whether or not the injury to the ACL is partial or complete. Uncovering of the lateral meniscus is never seen in partial tears, but occurs frequently in complete tears (52%); injury to the popliteus and posterior capsule rarely occurs in partial tears but is seen in 32% of complete tears. Remarkably, trabecular injury, a highly sensitive sign to the presence of an ACL injury, is not a very good discriminator between partial and complete tears because patients with partial tears have trabecular microfractures 48% of the time (1).

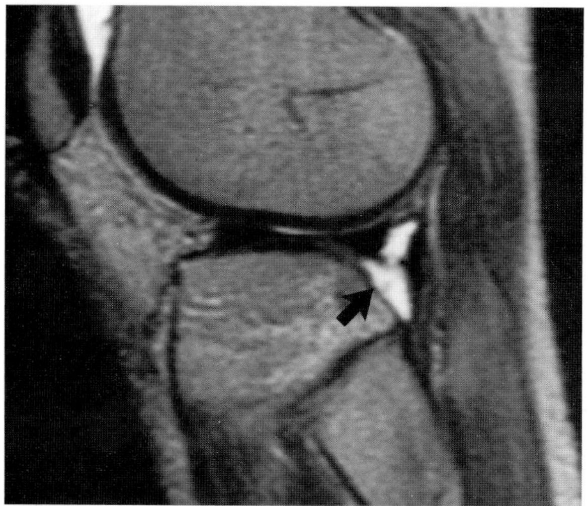
C

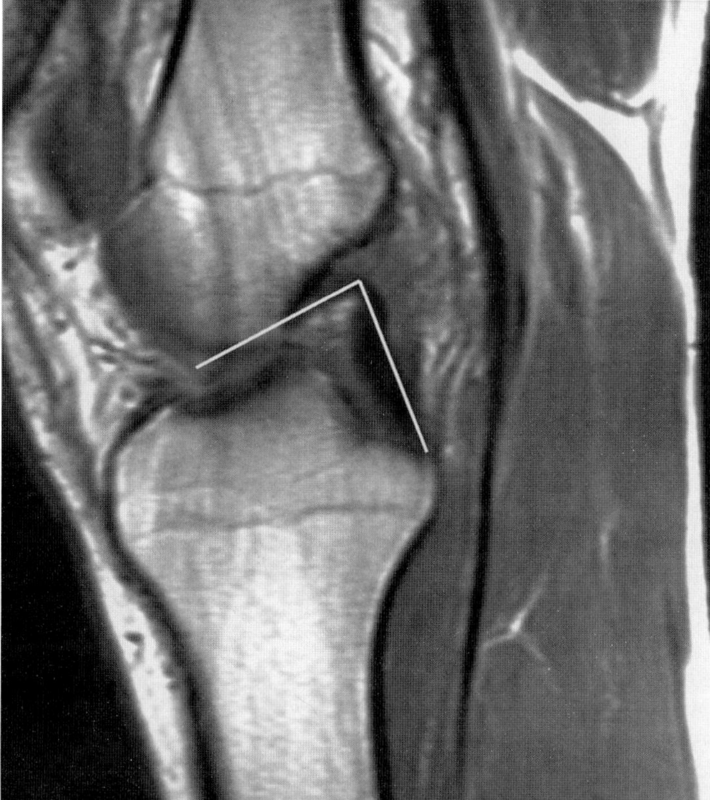
E

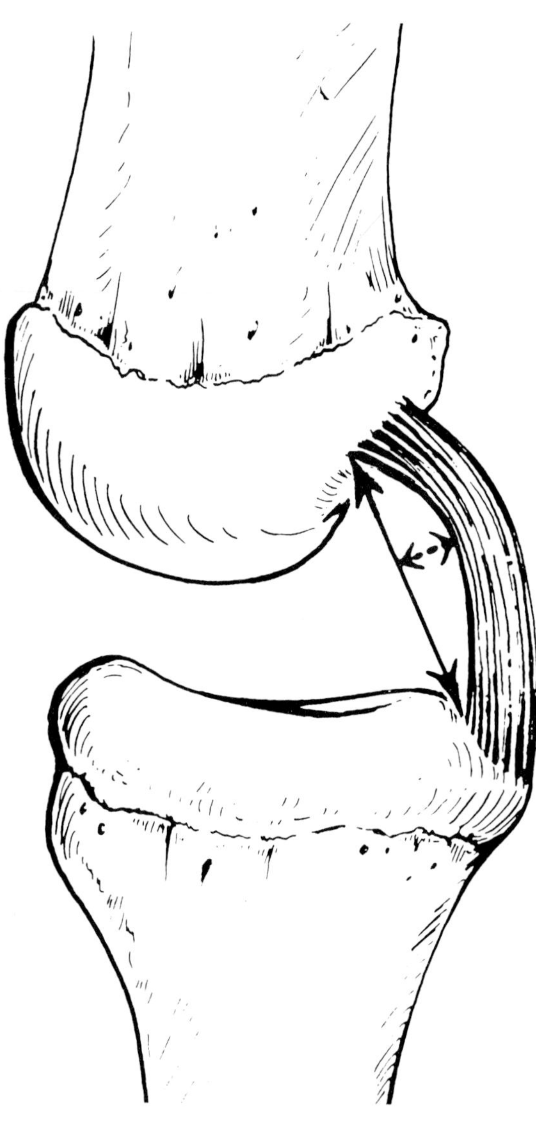
D

FIG. 11. C: Uncovering of the lateral meniscus/posterolateral tibial plateau fracture. Sagittal (TR/TE 2,300/20 image). Because of the impaction fracture of the posterolateral tibial plateau *(arrow)*, the undersurface of the posterior horn of the lateral meniscus is uncovered. **D:** PCL curvature. The distance between the anteriormost tibial and femoral insertion points of the PCL is indicated by the solid line (line *y*). At the point of greatest curvature, a second line, *x (dashed line)*, is drawn to the anterior surface of the PCL, perpendicular to line *y*. The PCL curvature value is defined as the length of line *x* divided by the length of line *y*. (From Tung GA, Davis LM, Wiggins ME. Tears of the anterior cruciate ligament: primary and secondary signs at MR imaging. *Radiology* 1993;188:661–667, with permission.) **E:** PCL angle. Sagittal MR TR/TE 2300/20 msec. Lines are drawn through the distal and proximal portions of the PCL and the angle between them is measured. In this instance, the angle is approximately 90°.

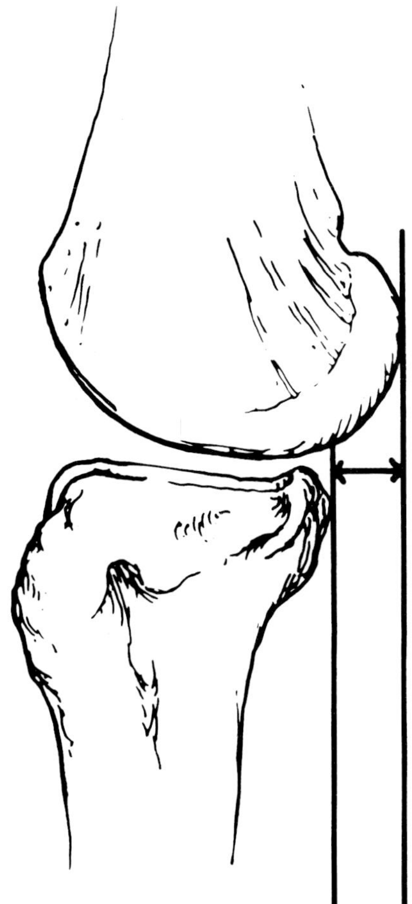
F

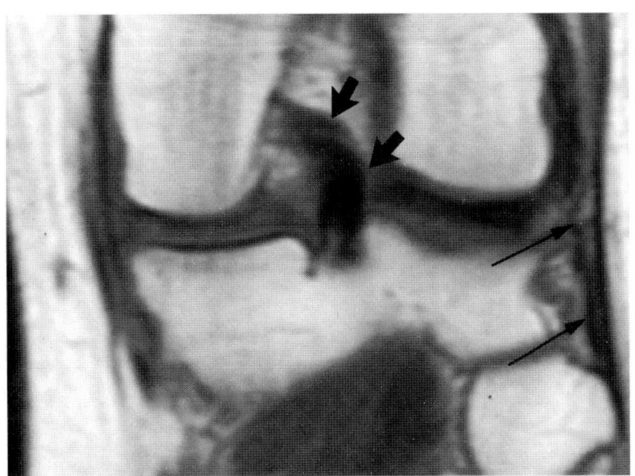
G

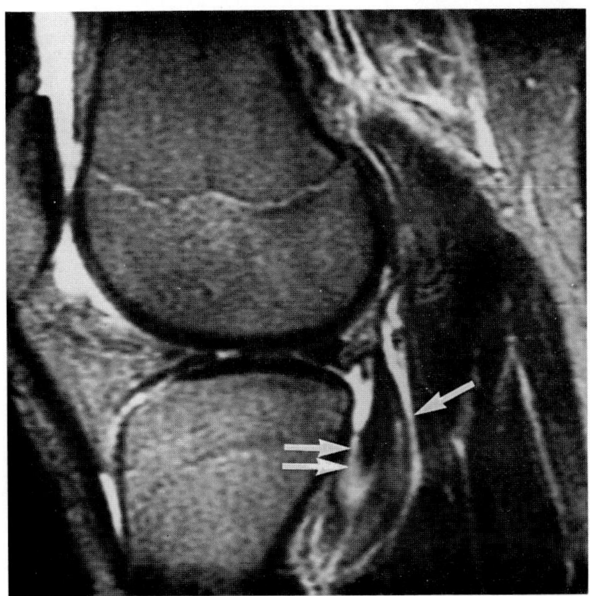
H

FIG. 11. F: Tibial shift measurement. The degree of anterior tibial translation is measured with regard to a plane perpendicular to the long axis of the MR image. The degree of shift is the distance between a line along the posterior aspect of the lateral femoral condyle and the posterior lip of the posterior lateral tibial plateau. (From Vahey TN, Hunt JE, Shelbourne KW. Anterior translocation of the tibia at MR imaging: a secondary sign of anterior cruciate ligament tear. *Radiology* 1993;187:817–819, with permission.) **G:** Visualization of the PCL and the FCL. Both the PCL *(arrows)* and the FCL *(long arrows)* are identified on a single coronal image. Both ligaments normally have insertions on the tibia that are more posterior than their femoral origins. In knees with ACL disruption, the origins and insertions of both ligaments are one above the other. **H:** Capsular rupture at the popliteus; sagittal (TR/TE 2,000/80) image. During ACL rupture, the posterolateral capsule may tear, permitting joint fluid *(double arrows)* to extravasate along the popliteus muscle/tendon *(single arrow)*. Such extravasation implies a tear of the attachment of the arcuate complex.

CASE 12

History: What is your diagnosis in this 15-year-old girl with a recent basketball injury? What is the accuracy of MR for meniscal tears in the presence of an anterior cruciate ligament (ACL) tear?

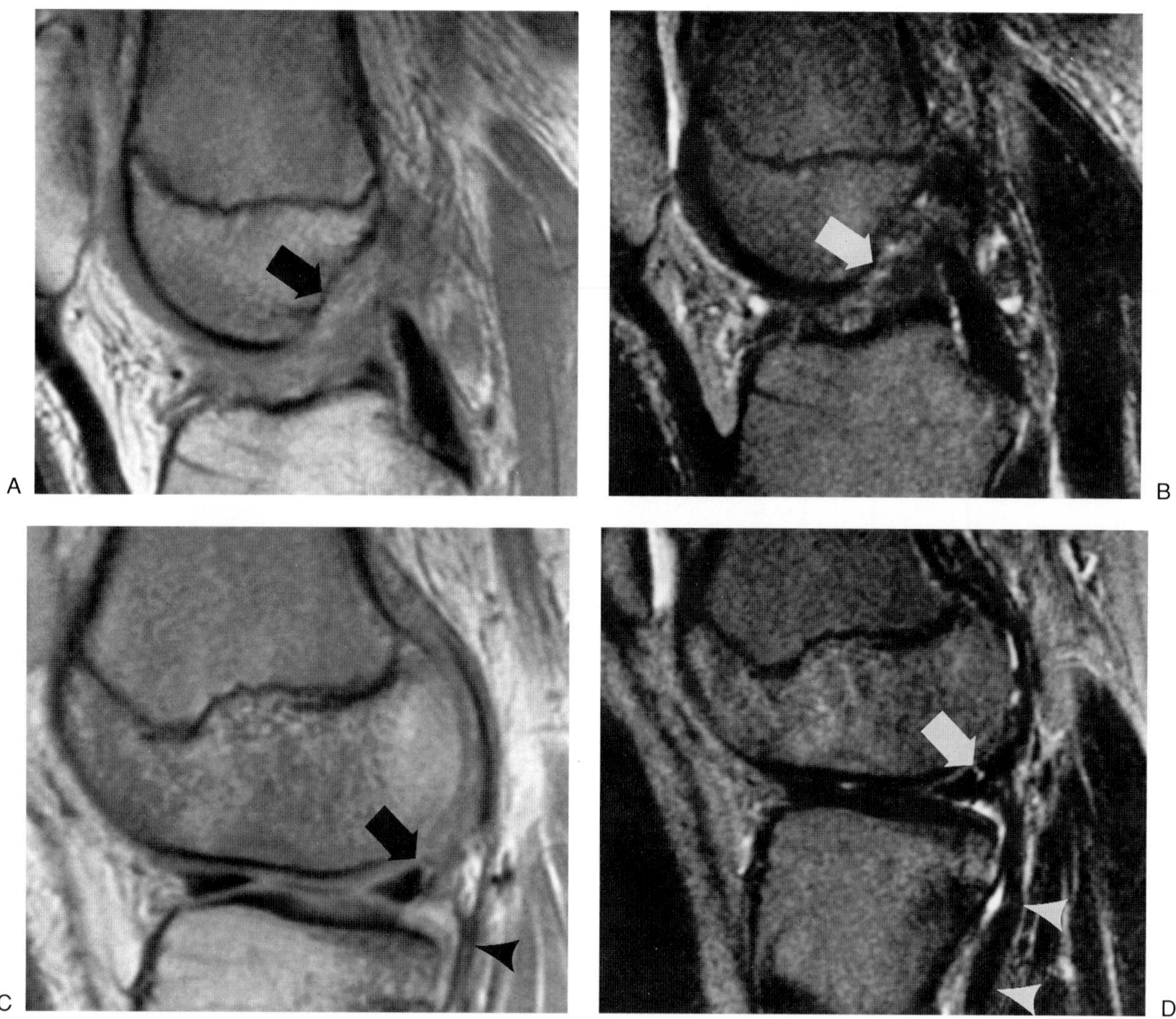

FIG. 12.

Findings: Figure 12A,B: Sagittal MR TR/TE 2000/16,90 msec and Fig. 12C,D: Lateral parasagittal MR TR/TE 2000/16,90 msec. There is increased signal in the intercondylar notch on both the proton density (Fig. 12A) and T2-weighted (Fig. 12B) images (arrows), with loss of the normal low signal ACL fibers. On a more lateral pair of images (Fig. 12C,D), there is high signal that extends to the superior surface of the posterior horn of the lateral meniscus (arrows). Note the location of the popliteus muscle and tendon (arrowheads).

Diagnosis: Complete tear of the anterior cruciate ligament and a peripheral tear of the posterior horn of the lateral meniscus.

Discussion: The association between meniscal tears and ACL tears is well documented, with various articles reporting a 46% to 80% incidence of meniscal tears in patients with torn anterior cruciate ligaments (1–4). Both medial and lateral meniscal tears are associated with a cruciate tear (2,4). Several authors have reported a higher incidence of lateral than medial meniscal tears, with Mink (3) indicating that lateral tears are seen in 66% of patients with ACL tears, compared with medial tears in 43% of patients (3,5). Even in articles on ACL tears that report a higher incidence of medial meniscal tears, the ratio of lateral to medial tears is still higher than in a comparable population of patients with meniscal tears but intact ACLs (2,4). Clearly, the mechanism of injury that produces an ACL tear also causes lateral meniscal tears.

There are also distinct patterns to the types of meniscal tears seen in patients with ACL tears. One of the more common types is a peripheral tear in the posterior third of either the medial or lateral meniscus (4,6). These tears are believed to result from the femoral condyles rolling over the posterior rim of the tibia at the time of the ACL disruption, the same mechanism that causes the posterior tibial bone contusions commonly seen on MR (2,7).

These patterns of injury are important to the radiologist because they affect the accuracy of the MR interpretation. Mesgarazadeh et al. (8) reported that most of their missed lateral meniscal tears were in the posterior horn. DeSmet and Graf (4) reported a sensitivity of only 65% for tears in the posterior third of the lateral meniscus, many of which were in patients with a torn ACL. When analyzed by the type of lateral meniscal tear, the sensitivity for peripheral tears was found to be 50%, compared with 97% for oblique tears and 80% overall. In their series, MR was more accurate for the detection of posterior horn tears in the medical meniscus than the lateral meniscus.

There are several reasons why these lateral tears may be more difficult to identify on MR. The radius of curvature of the lateral meniscus is smaller than the medial meniscus, and the posterior horn angles superiorly toward its midline attachment to the tibia. This may cause volume averaging of the meniscus when imaged in the usual sagittal and coronal planes (4). The orientation of the fibrocartilage fibers also results in magic-angle effects in the posterior horn of the lateral meniscus, which leads to increased signal that can obscure a tear (9). Popliteal artery pulsation artifact may also obscure tears in this region (4). Anatomic features such as the popliteus tendon and the origin of the meniscofemoral ligaments also make this area somewhat difficult to interpret (3,4).

A reasonable question might be, What is the significance of missing these posterior peripheral tears in the lateral meniscus? Most of these patients elect to have an ACL reconstruction, and these tears are not in an area that is difficult for the orthopedist to identify arthroscopically. Although most orthopedic surgeons now wait several weeks before performing an ACL reconstruction, these nondisplaced peripheral tears are not considered an indication to perform early arthroscopic meniscal surgery. Because the outer third of the meniscus is vascularized, many of these tears may heal spontaneously, and the others can be repaired at the time of the ACL reconstruction (2,10). DeSmet and Graf (4) reported that fewer of their meniscal tears missed on MR, many of which had an associated ACL tear, required surgical intervention than tears that were not missed. The patient illustrated in this case had a single suture placed across the tear during surgery.

In summary, the posterior horn of both the medial and lateral menisci should be carefully evaluated on MR in patients with a torn ACL (2,4). Peripheral tears in the posterior horn of the lateral meniscus are especially difficult to identify on MR because of the magic-angle effect and other anatomic features of this area. Although failure to identify these tears are a common cause of false-negative MR studies in patients with torn ACLs, they may not significantly change the clinical outcome.

CASE 13

History: A 16-year-old boy sustained an acute injury to the knee while running. He was seen by a university orthopedist who diagnosed an ACL tear and recommended surgery. The MRI examination was ordered by another orthopedist seeing the patient for a second opinion. What is your diagnosis?

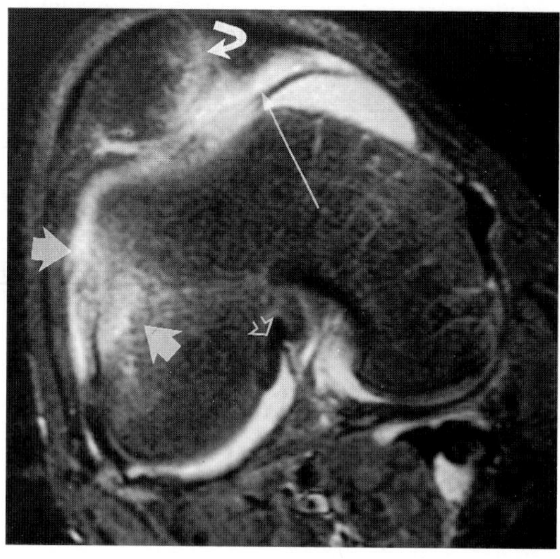

FIG. 13.

Findings: In Fig. 13A an axial scan through the patella demonstrates high signal intensity along the anterolateral femoral condyle (arrow) and medial facet of the patella (curved arrow). Also note a prominent but not thickened medial patellar plica (long arrow), and the proximal attachment of the ACL (open arrow), which was demonstrated to be intact.

Diagnosis: Prior lateral patellar dislocation.

Discussion: The natural history of most acute patellar dislocations is for the patella to relocate spontaneously prior to patient's clinical presentation (1–5). The patient is often unaware that an actual dislocation has occurred. Physical findings vary widely and the diagnosis may not be considered initially. Many patients present with the patella relocated and a tense hemarthrosis. In this setting, patellar dislocation may be confused clinically with anterior cruciate ligament disruption. In a clinical series of patients undergoing surgery for suspected ACL disruption, 10% of cases were ultimately determined to represent prior transient patellar dislocations (5). In an MR series of 22 patients, less than half had a preimaging diagnosis of suspected prior transient patellar dislocation, a finding that further underscores the difficulty in clinical diagnosis (1). In some patients, tenderness or a defect in the vastus medialis obliquus (VMO) muscle may be demonstrable or a medial retinacular hematoma may be present (4). Other patients may have only mild swelling, which is predominantly periarticular. In less extensive injuries, no palpable medial defect may be present. Lateral patellar hypermobility accompanied by apprehension and pain can often be demonstrated.

Lateral patellar dislocation results in demonstrable injury to both soft tissue and bony structures that can be well demonstrated utilizing MR. The medial capsule and retinaculum, along with the medial patelloepicondylar and patellotibial ligaments may be ruptured. In addition to these static supporting structures, the VMO, which contributes dynamic medial patellar support, may be ruptured at one of several sites: (a) at the level of its insertion into the patella, (b) an interstitial rupture in its midportion, or (c) a tear from its origin at the adductor tubercle. Articular cartilage injury is common and occurs typically along the medial edge of the patella and lateral femoral trochlea as the patella impacts at the time of spontaneous reduction. Loose purely chondral or osteochondral fragments may result.

The diagnosis of prior lateral patellar dislocation is established on MR by recognition of a combination of injuries that typically occur as a consequence of the dislocation and subsequent relocation of the patellar (1–3). In the author's series (1) and in the one reported by Virolainen et al. (2), the three most common consistent findings were hemarthrosis, disruption of the medial retinaculum, and contusional injury of the lateral femoral condyle.

Contusional injuries involving the medial facet of the patella were common but less consistently observed than those involving the lateral femoral condyle. Identification of this constellation of findings should lead to the diagnosis of prior patellar dislocation (Fig. 13B,C). Injuries to other structures within the knee, particularly medial collateral ligament injuries and medial meniscal tears, may occur with patellar dislocation, but they do not specifically add to the constellation of findings that suggest the diagnosis.

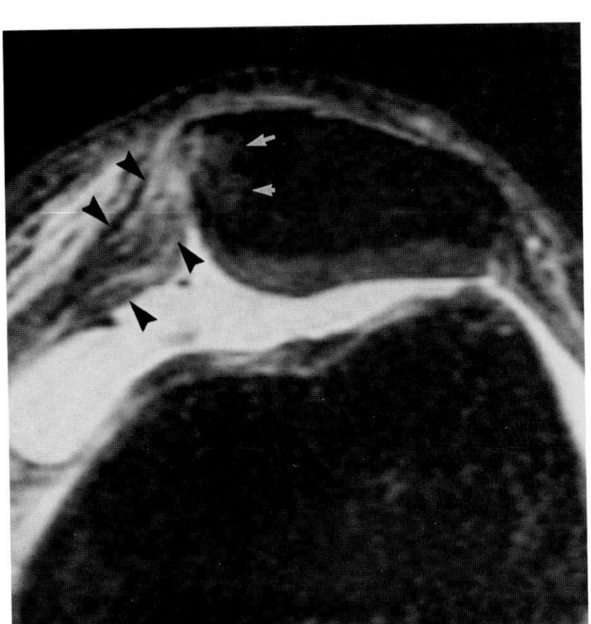

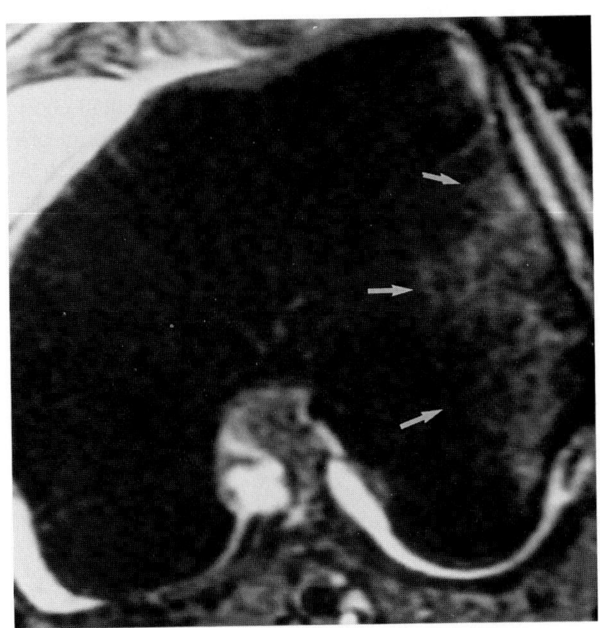

FIG. 13. B: Axial (STIR) sequence (TR/TE/TI 2,200/35 msec/60) demonstrates high signal intensity consistent with edema and or hemorrhage within the subchondral bone of the medial facet of the patella *(small arrows),* and extensive disruption of the medial retinaculum *(arrowheads).* **C:** On an image obtained more inferiorly, a corresponding site of edema is seen within the subchondral and cancellous bone of the lateral femoral condyle *(arrows).* This combination of findings strongly suggests a prior patellar dislocation. (From Deutsch AL, Shellock FG. The extensor mechanism and patellofemoral joint. In: Mink JH, Reicher M, Crues JV, Deutsch AL, eds. *MRI of the knee, 2nd edition.* New York: Raven Press 1994:218–220, with permission.)

CASE 14

History: An MR examination is performed to evaluate for possible meniscal tear. What is your diagnosis?

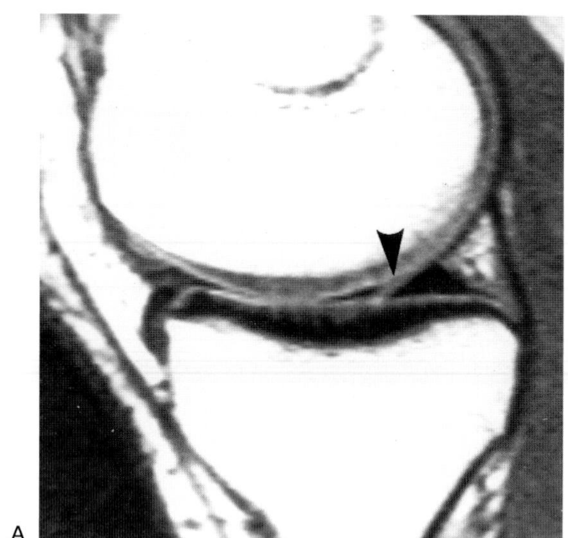

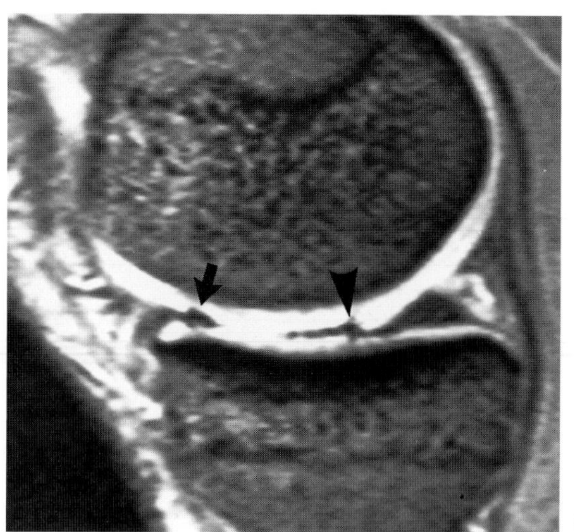

FIG. 14.

Findings: In Fig. 14A a sagittal T1-weighted image (TR/TE 800/20 msec) demonstrates an oblique line of high signal (arrowhead) that extends across the posterior horn of the medial meniscus and intersects both meniscal articular surfaces. In Fig. 14B, on a sagittal gradient echo sequence the oblique line demonstrates a ''blooming effect'' (arrowhead) both in the posterior horn and to a lesser degree adjacent to the anterior horn (arrow). (Reproduced with permission (From Shogry MEC, Pope TL. Vacuum phenomenon simulating mensical or cartilaginous injury of the knee at MR imaging. *Radiology* 1991;180:513–515, with permission.)

Diagnosis: Pseudotear of the medial meniscus secondary to intraarticular gas.

Discussion: High signal linear intrameniscal foci simulating meniscal tears may result on the basis of one of several factors (1). Intraarticular gas, either idiopathic (vacuum phenomenon) or iatrogenic (arthrography), may produce a high signal artifact within the meniscus on T1-weighted images, which can simulate a meniscal tear (2). This pitfall relates to susceptibility artifact, which occurs when adjacent tissues vary greatly and abruptly in magnetic susceptibility. Local field inhomogeneties produced by intrinsic magnetic field gradients generated in such regions result in areas of high and low signal intensity similar to, but less predictable than, those seen with chemical shift artifact. Bone and air have less susceptibility, and ferromagnetic metals have more susceptibility than soft tissues, and it is at boundaries between these tissues where artifacts occur. If gas within the knee joint collects in a triangular configuration between the femoral or tibial articular cartilage and the meniscus, it may appear as a signal void separated from the low signal meniscus by a thin, high signal line, an appearance highly suggestive of a meniscal tear (1,2). Gradient-recalled images in patients with intraarticular gas distributed in a linear or triangular configuration demonstrate artifactual blooming of the signal void, resulting in dark and bright streaks as a consequence of misregistration of spatial information. In these cases, the findings suggest meniscal and/or cartilaginous disorders. Many of the patients with this reported artifact have had intraarticular gas present on a conventional radiograph, a finding that may help in differential diagnosis. It is the blooming effect on the gradient echo sequences that should alert the examiner to this pitfall.

Meniscal tears may also be simulated by truncation artifacts, which can produce high signal lines within the meniscus (3) (Fig. 14C,D). These artifacts result from the use of Fourier-transform methods to construct MR images of high contrast boundaries, such as between the spinal cord and cerebrospinal fluid, between cortical bone and medullary fat, or between articular cartilage and the menisci (1,3). When an attempt is made to describe such boundaries, an inaccurate representation may occur. These truncation artifacts are a series of high and low signal intensity lines adjacent to, and parallel to, a high contrast boundary. When a high intensity artifact is projected over a low signal intensity structure such as the meniscus, a pseudotear of the meniscus may result.

Conventional knee MR may be performed with either a 128 or 256 acquisition matrix in the phase-encoded direction, which can be either anteroposterior or superoinferior. Truncation artifacts are most evident on those studies in which 128 phase-encoding steps are oriented in a superoinferior direction. If such boundary artifacts occur, they can be minimized by increasing the phase-encoding matrix to 192 or 256, and or by redirecting a 128-step phase-encoding direction to anteroposterior rather than superoinferior. Recognition of the characteristics of truncation artifacts will help avoid mistaking them for meniscal tears. The artifacts are relatively subtle and have a uniform thickness. They are parallel to and two pixels distant from the high contrast boundary from which they originate. Careful inspection may reveal that these lines extend beyond the margins of the menisci. The artifact does in fact extend across the entire image but is generally recognized only when projected over a low signal intensity structure such as the meniscus. In our experience, truncation artifacts have been only rarely confused with meniscal tears.

Another possible pitfall in meniscal interpretation relates to ghosting artifacts related to knee motion simulating mensical tears (1,4). Motion of the knee during image acquisition results in high signal intensity replications of marrow within the distal femoral condyle and the hyaline cartilage covering it. Motion artifact, when present, may serve as a source of intrameniscal signal (4). All motion artifacts propagate along the phase-encoded direction. In the knee, motion artifacts may lead to curvilinear foci of relative increased signal intensity that overlie the menisci and simulate meniscal tears. This problem is exacerbated by the frequent reorientation of the phase and frequency encoding gradients undertaken in an attempt to redirect pulsation artifacts arising from the popliteal vessels away from the intercondylar notch, thus improving visualization of the cruciate ligaments (1,4).

When an apparent meniscal tear is demonstrated on an MR examination, it is important to carefully review the entire series of images to exclude motion-related artifacts. These are manifested as a series of alternating high and low signal intensity curvilinear bands that may be seen overlying the femoral condyles or tibial plateau (4). Additional clues to motion-related meniscal pseudotears include a relatively thin signal alteration that corresponds to the contour of the femoral condyle and observation of signal intensity alterations that may project beyond the confines of the meniscus. When motion artifacts are suspected, it is imperative to repeat the examination to exclude the possibility of a pseudotear (4).

C

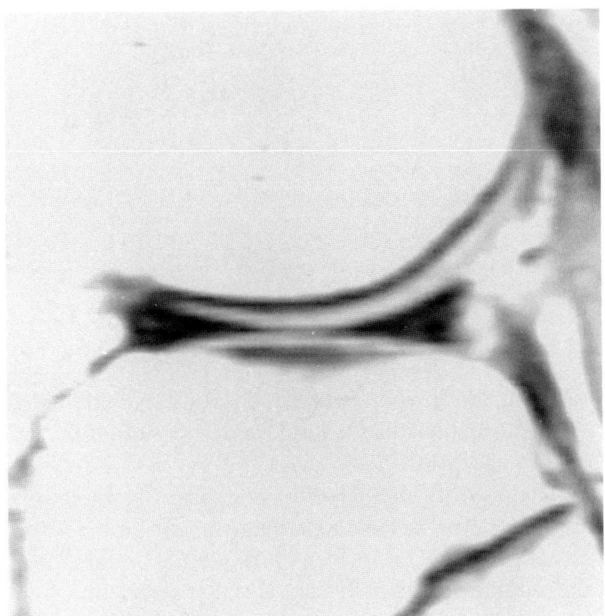

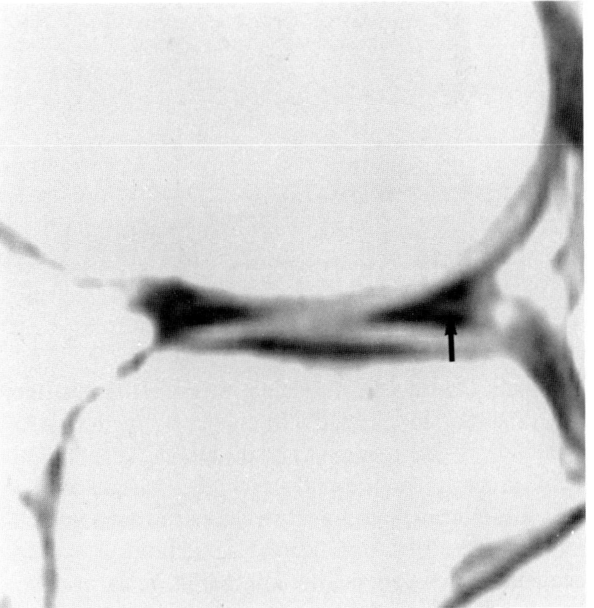

D

FIG. 14. C,D: Truncation artifact. Sagittal proton density weighted images. **C:** The phase-encoded direction is anterior-posterior, and the posterior horn of the lateral meniscus does not demonstrate any significant linear signal. **D:** The phase-encoded direction has been changed to superior-inferior and only 128 phase-encoded steps were utilized. A faint truncation artifact *(arrow)* is not evident. (From Mink JH. Pitfalls in interpretation. In: Mink JH, Reicher MA, Crues JVIII, Deutsch AL, eds. In: *MRI of the knee, 2nd edition.* New York: Raven Press, 1993; with permission.)

CASE 15

History: This professional football player stumbled on artificial turf and claims to have immediately felt pain in his knee. Is this an acute injury, a chronic abnormality, or is it really abnormal at all?

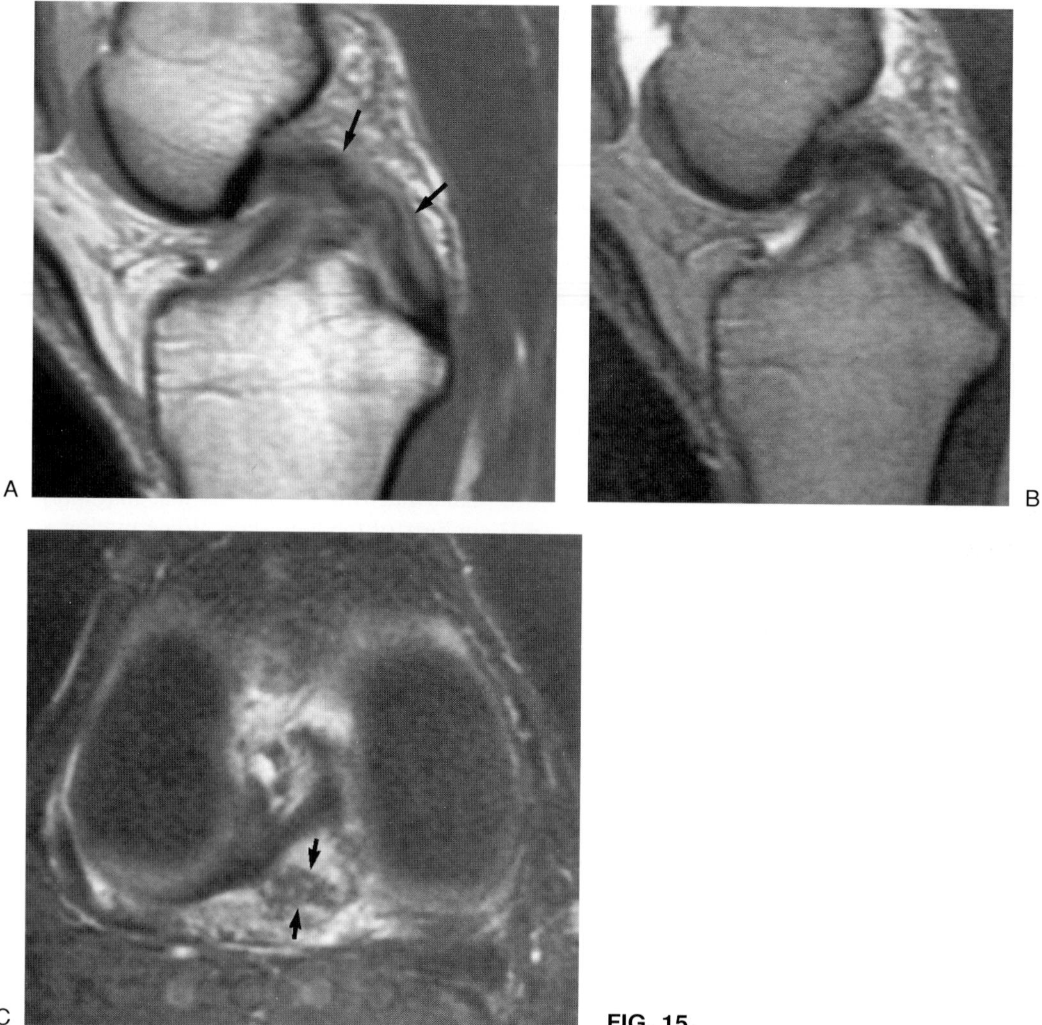

FIG. 15.

Findings: Figure 15A,B,C is an interstitial tear of the PCL. A: Sagittal (TR/TE 2300/20 image; B: Sagittal (TR/TE 2,300/80 msec) image; C: axial (TR/TE/TI 2,800/36 msec/130) image. The normally dark PCL is replaced by serpiginous, intermediate signal structure (arrows), which does not demonstrate significant brightening on a long TE sequence. This patient was injured 24 hours prior in a football game.

Diagnosis: Acute interstitial-type tear of the posterior cruciate ligament. There is a minor increase in signal on the long TE images, but the inversion recovery images (Fig. 15C) reveal intrasubstance edema and hemorrhage. Lack of intrasubstance edema/hemorrhage, especially on T2-weighted image, certainly does not exclude an acute tear. The patient had a mild degree of instability at the time of the MR.

Discussion: The posterior cruciate ligament (PCL) originates from a fan-shaped attachment on the posterior aspect of the lateral surface of the medial femoral condyle and courses posteriorly and distally across the joint to attach to the extreme posterior intercondylar region of the tibia 1 cm below the tibial articular surface (1). Similar in basic structure to the ACL, the PCL is composed of individual fascicles that are grouped into two major fiber bundles. The anterolateral group, which composes most of the PCL, and the smaller posteromedial group, are named for their attachment locations on both the femur (anterior and posterior), and the tibia (medial and lateral). The fibers twist on their longitudinal axis in a clockwise direction during flexion and unwind during extension; this action is analogous to, but directly opposite that of, the ACL.

The meniscofemoral femoral ligament (MFL), a fibrous cord that connects the posterior horn of the lateral meniscus to the medial femoral condyle, has been called the "third cruciate ligament" because of its relationship to the origin of the PCL from the femur.

The MFL is composed of two components, the anterior limb, known as the ligament of Humphry, and the posterior limb, called the ligament of Wrisberg. The MFL functions to pull the posterior horn of the lateral meniscus anteriorly and medially during flexion, balancing the action of the popliteus muscle whose fibers attach to the posterolateral meniscus. While both branches of the MFL may exist in the same knee, there are considerable variations in their presence and size. Various anatomic dissections have found that the branches of the meniscofemoral ligaments may be found in 80% to 100% of knees; the anterior is found in 34% and the posterior in 60% of the specimens. Either of the two branches of the meniscofemoral ligament may be present in up to 58% of MR examinations, but ready visualization of both is uncommon (11%) (2).

Ruptures of the PCL are thought to compose 3% to 20% of knee ligament injuries, but the true rate is probably higher given the fact that acute tears often go undiagnosed (3). Since the PCL is a large strong structure, it is ruptured only following forces and injuries of great magnitude, and therefore it is common that other intraarticular injuries simultaneously occur (1,4). In fact, only 30% of PCL injuries are isolated (5). Most commonly, the medial collateral ligament (50%), the medial meniscus (30%), the articular cartilaginous surfaces (12%), the ACL (65%), the posterior capsule, and the fibular collateral ligament are simultaneously damaged, and the peroneal nerve and popliteal artery may also be injured (4,6–9). If the PCL injury is not associated with other ligamentous instability, the affected knee will probably will remain symptom-free, although the incidence of both chondral defects and meniscal tears increases with time (9,10).

Approximately half of PCL tears occur during athletic endeavors, and the remainder occur during traffic and industrial accidents. The PCL is injured by one of three mechanisms: (1) hyperextension of the knee, as occurs in football; (2) hyperflexion of the knee; (3) posterior displacement of the tibia on the femur with the knee in flexion (11,12). The latter injury, resulting from the high-velocity forces that are present during deceleration actions in automobile accidents, is the most common mode of injury and is often associated with a telltale ecchymosis or abrasion to the pretibial region.

The PCL, by virtue of its size and low signal intensity on all spin echo imaging sequences, is perhaps the most easily identified structure within the knee on MR examinations. The entire course of the PCL is readily visualized on a single sagittal image on 95% of properly performed knee MR examinations. The midportion of the PCL appears slightly thickened, often due to close approximation of the ligaments of Humphry and/or Wrisberg (Fig. 15D). The ligaments are often more readily visualized in patients with a torn PCL. Occasionally, a displaced fragment from a bucket-handle tear of the medial, or rarely a lateral meniscal tear may lie adjacent to the PCL and create the appearance of a "double PCL" (Fig. 15E,F).

Conventional spin echo sequences have been the standard protocol used in assessment of suspected internal derangement disorders of the knee, but some centers utilize a variety of gradient recalled (GRE) sequences as their primary imaging mode. Areas of high signal intensity on GRE sequences have been identified in the normal PCL on images in which the ligament lies at an angle of 45° to 85° to the axis of the main magnetic field (13). It is at this angle ("magic angle") that the effective T2 of the PCL fibers is the longest, and signal alterations have been seen at this critical point in a number of different ligamentous structures. In patients with high signal foci in the midsubstance of the PCL on GRE studies, spin echo images, and arthroscopic assessment are often normal, and such isolated "abnormalities" must be assessed with great caution (Fig. 15G).

The MR morphology of the normal PCL is dependent on the degree of flexion of the knee and the integrity of the ACL and the secondary restraints. When the knee is in extension or a minimal degree of flexion, the PCL has a gently convex posterior margin; if the knee is flexed more than 10°, the ligament becomes taut and perhaps slightly thinner than in the extended position (14). Minimal degrees of buckling, or a slightly degree of accentuated convexity of the proximal end of the PCL are normal. Disruption of either the PCL (or the ACL) must be suspected when the proximal PCL assumes a kinked, sharply buckled, S-shaped or so-called sea-horse appearance (15). Such descriptions describe a PCL that is (a) stretched, as a result of an intrasubstance tear; (b) foreshortened, as a result of a distal avulsion or distal substance tear of the PCL; or (c) kinked, as a result of a complete tear of the ACL, with instability and a shift of the tibia anteriorly (Fig. 15H,I).

Acute tears of the PCL can be divided into two broad groups, those occurring as a result of an interstitial tear and those occurring as a result of avulsion (Fig. 15J). Nearly all substance tears of the PCL demonstrate morphologic characteristics described for an interstitial tear of the ACL. A long segment, or perhaps the entire length of the PCL is diffusely widened, but the torn PCL has less of a mass-like character than does the torn ACL. The signal on T1-weighted image is inhomogeneously increased; on long TR/TE sequences, there are small areas within the ligament that increase in signal intensity, due to the presence of intraligamentous edema, but it is unusual to identify coalescent areas of joint fluid traversing the PCL. Avulsive ruptures typically affect the tibial attachment of the PCL (Fig. 15K–M). If the knee has suffered a hyperextension type of injury, ''kissing contusions'' of the anterior tibial plateaus and the anterior femoral condyles will be seen (12). Contusional lesions of the anterior tibia is often seen in patients with ''dashboard'' injuries.

Since most PCL tears occur in the substance of the ligament, it is difficult, by MR, to precisely determine which PCL tears are partial and which are complete; it is probably impossible to attempt to determine future stability (Fig. 15N,O). There is not a statistically significant difference between patients with complete tears and those with partial tears with regard to thickness, margination, or signal intensity of the PCL, but MR images in patients with complete tears are more likely to show focal areas of ligamentous discontinuity (16). Such discontinuities, however, are rare.

Since most PCL tears occur interstitially, the ''continuity'' of the ligament is maintained. Spontaneous repair of the torn PCL occurs over the torn ligament and the interposed hematoma, and dense fibrous scar may replace the normal PCL, unlike the ACL in which vascular interruption and ligamentous atrophy are common (Fig. 15P,Q). Both fibrous scar and the normal PCL have low signal character at all imaging sequences, and the anatomically ''intact'' but functionally torn PCL may be a source of interpretive error (15) (Fig. 15H).

Since PCL ruptures are uncommon, frequently escape clinical (and arthroscopic) detection, and if suspected, are often treated conservatively, it is quite difficult to determine the prevalence of such injuries and the efficiency of MR in assessing them. While the PCL is a very large structure, it is not easily assessed at the time of arthroscopy (2,4,7,17–21). The inferior two-thirds of the PCL are poorly seen from an anterior approach because of the presence of large amounts of synovium and subsynovial fat, the presence of an intact ACL, and the fact that the PCL attachment is well below the articular surface of the tibia and is therefore difficult to identify at arthroscopy, especially from an anterior approach (1). Ideally, the PCL should be viewed by means of posteromedial and posterolateral arthroscopic portals. Occasionally, the synovial envelope of the PCL needs to be opened to find the tear predicted by MR (7). Occasionally, the patient with a PCL tear may have a taut, intact ligament of Humphry and the arthroscopist may mistake it for an intact PCL (1).

Several recent studies have specifically addressed the ability of MR to assess the integrity of the PCL (22,23). In a recent study, there were 190 intact and 13 torn PCL. Of the 11 patients with surgically proven tears, the MR was positive in all 11; of the 190 normal ligaments, there were no false positives and there were no false negatives (24). In another group, there were 11 patients with surgically confirmed tears of the PCL; four of these tears were clinically unsuspected (14).

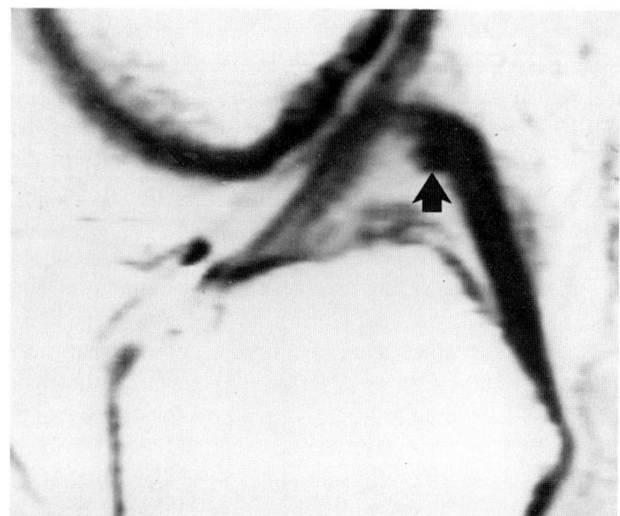
D

FIG. 15. D: Double PCL. Sagittal (TR/TE 2,000/20 msec) image. The anterior limb of the meniscal femoral ligament *(arrow)* is identified as a small rounded mass immediately anterior from the posterior cruciate ligament.

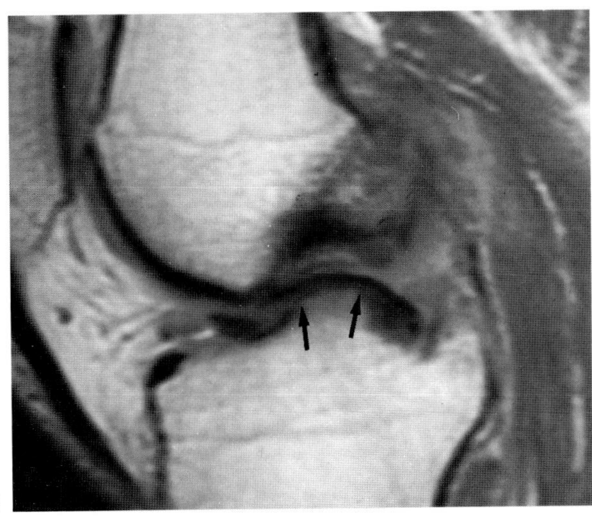

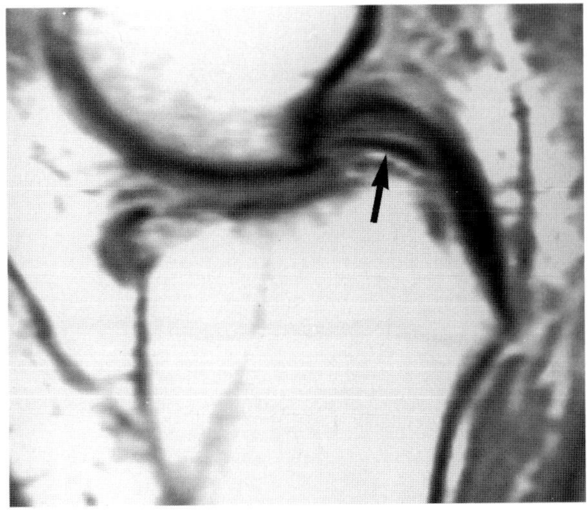

FIG. 15. E,F: Double PCL. Sagittal (TR/TE 2,200/20 msec) images. A low signal band *(arrows)* is seen paralleling the posterior cruciate ligament. In this case, the displaced fragment from a bucket handle tear is responsible for the appearance. **F:** In this case, the double PCL is mimicked by a prominent ligament of Humphry *(arrow).*

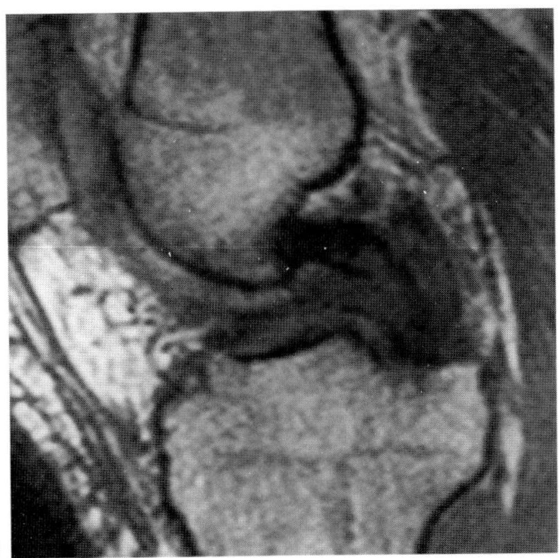

FIG. 15. G: Magic angle effect. There is an increase in signal intensity in the midportion of this otherwise normal PCL. The midsegment of the PCL frequently lies at 55° to the main magnetic field and is subject to such magic angle effects.

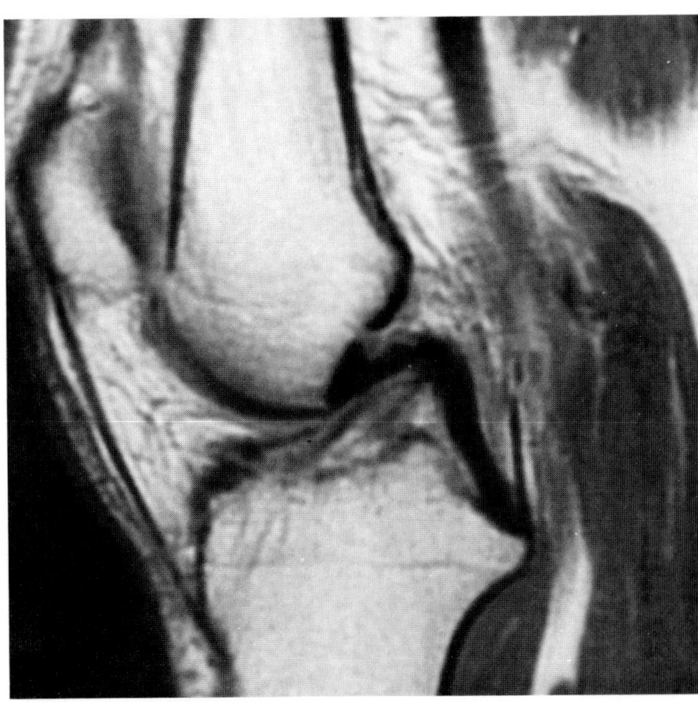

FIG. 15. H: Abnormal morphology of the PCL. Sagittal (TR/TE 200/20 msec) image. At arthroscopy, this PCL was clinically insufficient although the MR signal and contour is normal. The midportion of the PCL is somewhat redundant.

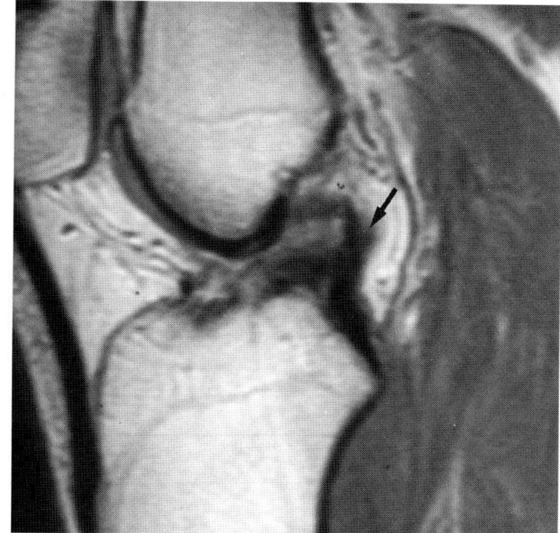

FIG. 15. I: Buckled PCL. Sagittal (TR/TE 2,100/20 msec) image. This patient has a chronic tear of the anterior cruciate ligament, which has reattached itself to the lateral aspect of the PCL *(arrow)*. Because of a shift of the tibia, the posterior cruciate ligament, which is structurally normal, has assumed a serpiginous, or seahorse configuration.

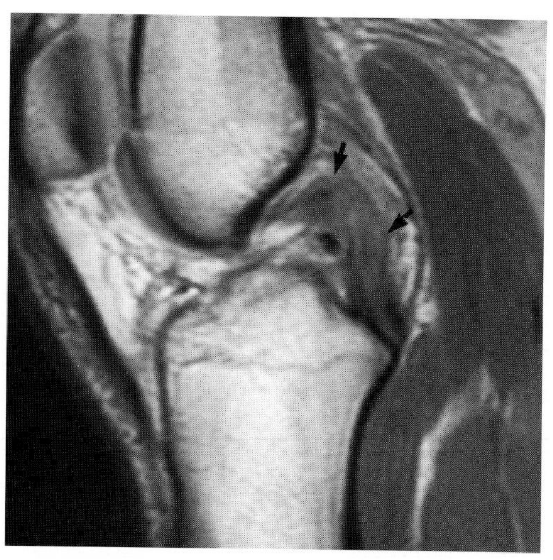

Fig. 15. J: Acute tear of the PCL. Sagittal (TR/TE 2,300/20 msec) image. The PCL has an increase in signal intensity, and demonstrates marked widening and poor definition of its margins *(arrows)*. A rather prominent ligament of Humphry lies just anterior to the torn ACL.

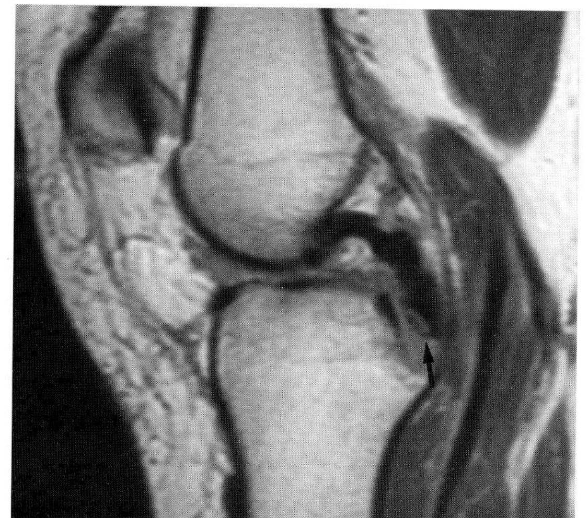

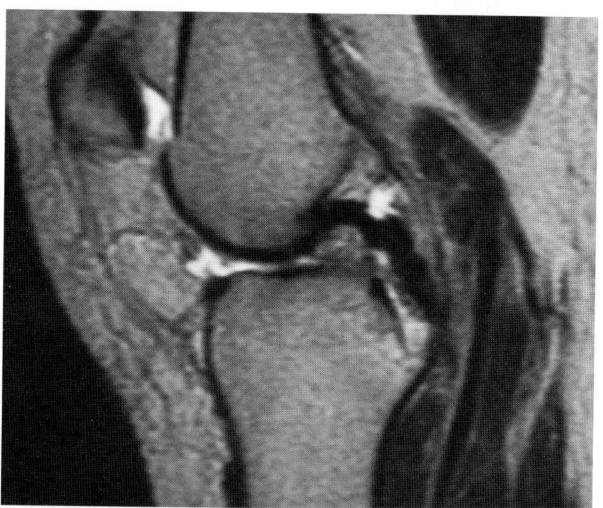

FIG. 15. K,L: Avulsion of the posterior cruciate ligament insertion. **(K)** Sagittal (TR/TE 2,300/25 msec); **L:** sagittal (TR/TE 2,300/80 msec) image. The distal end of posterior cruciate ligament has been avulsed from the tibia via a small flake of bone *(arrows)*.

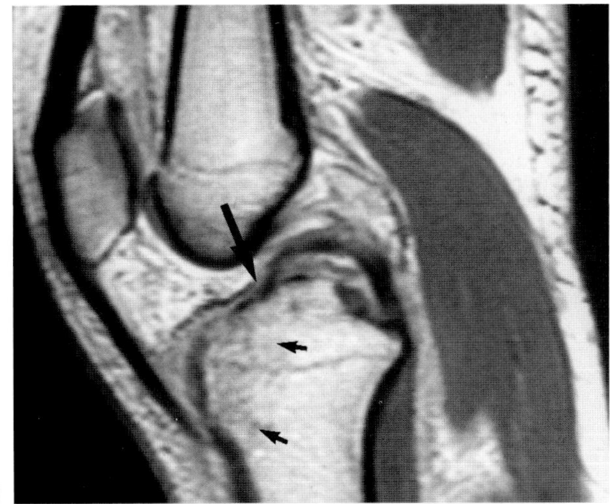

M

FIG. 15. M: Avulsion of the PCL. Sagittal (TR/TE 2,300/20 msec) image. A poorly defined contusional injury of the tibia *(small arrows)* is identified in this patient who struck his leg against a dashboard. The posterior cruciate ligament has become detached from its femoral origin, and its proximal end *(arrow)* lies on the tibial plateau. Anterior tibial contusions are seen in those patients who suffer typical dashboard injuries, and are a clue to the diagnosis. N,O: Partial tear of the PCL.

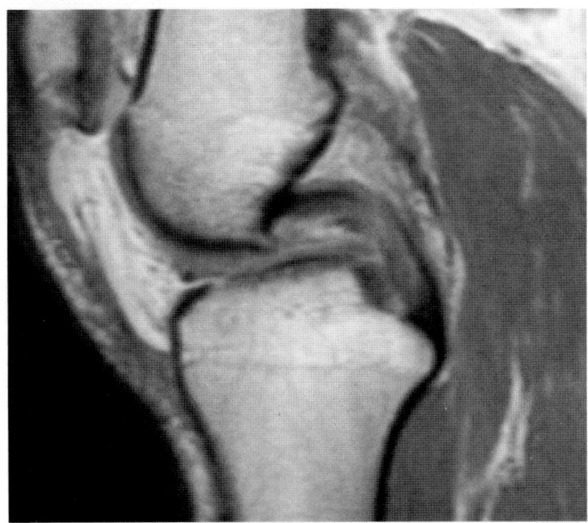

N

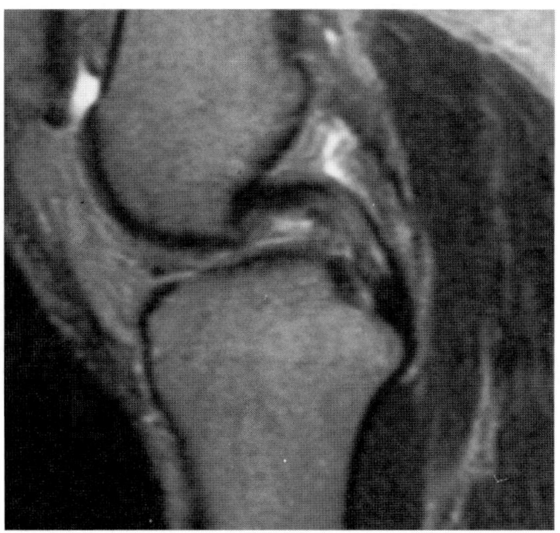

O

FIG. 15. N: Sagittal (TR/TE 2,300/20) image; **O:** Sagittal (TR/TE 23,00/80 msec) image. There is an increase in signal intensity of the PCL, which becomes minimally brighter on the long TR sequence. The ligament is not significantly widened. The patient had no instability at the time.

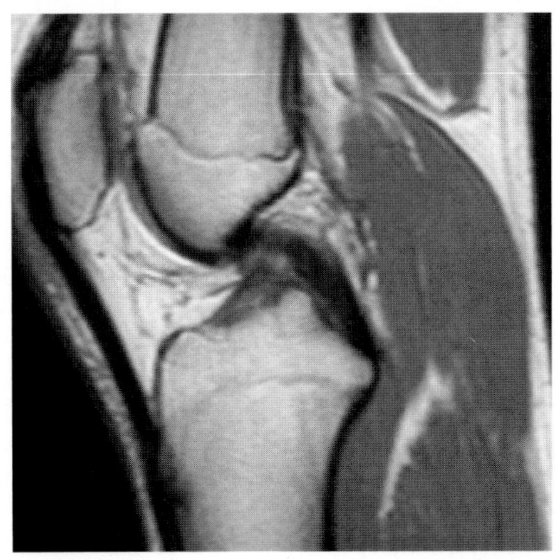

P

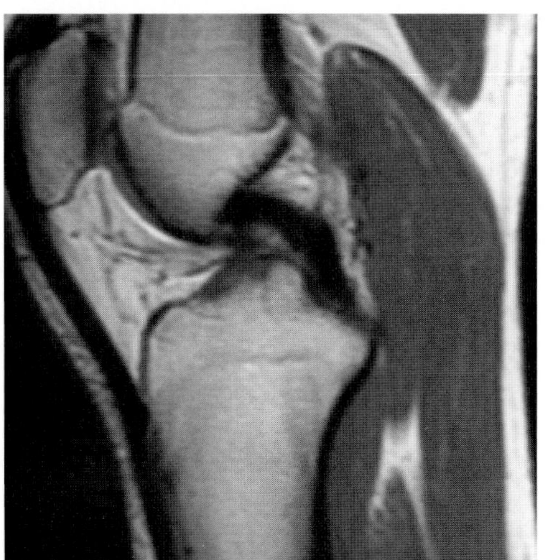

Q

FIG. 15. P,Q: Healing of a torn PCL. Sagittal (TR/TE 2,400/20 msec) image. Approximately 6 weeks before, the patient had an arthroscopically confirmed tear of the posterior cruciate ligament but had a stable knee to examination. On the first image **(P)** the ligament has a slightly increased signal intensity, and its margins are poorly defined. Two months later **(Q)** the PCL is dark and sharply defined from its surrounding tissues. The proximal end of the ligament is thicker and darker than it was in P. The knee is stable at the time of Q.

CASE 16

History: A 42-year-old man status post–ACL reconstruction undergoes MR evaluation for increasing knee symptomatology. Two representative images from the examination are illustrated. How would you assess the position of the femoral tunnel with regard to graft isometry?

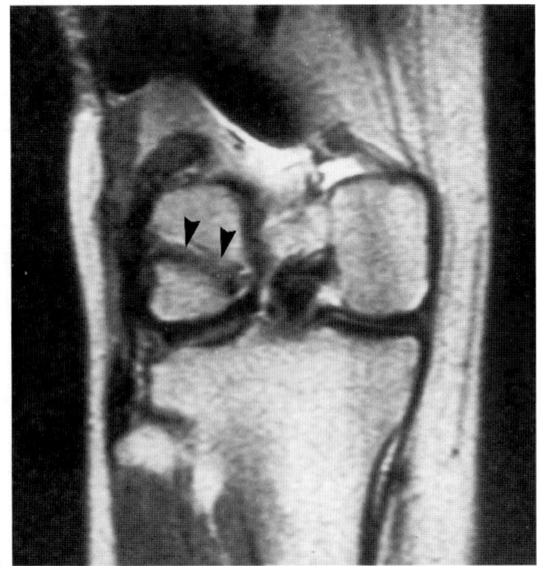

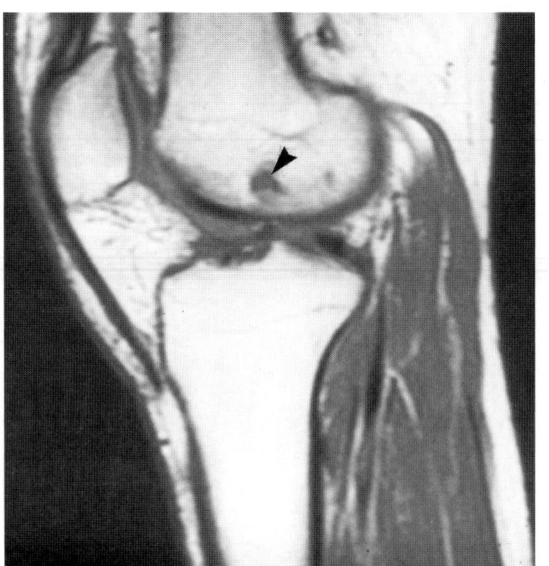

FIG. 16.

Findings: In Fig. 16A a coronal T2-weighted SE (TR/TE 2,000/80 msec) sequence demonstrates an inferiorly placed femoral tunnel that appears to open at the approximately 9 o'clock position (arrowheads). In Fig. 16B, on the sagittal PD image (TR/TE 2,000/20 msec), the femoral tunnel is seen to be quite low and far anterior in position. (From Deutsch AL, Mink JH. The Postoperative knee. In: Mink JH, Reicher MA, Crues JV III, Deutsch AL, eds. *MRI of the knee, 2nd edition.* New York: Raven Press 1993;270–292, with permission.)

Diagnosis: Abnormal and nonisometric femoral tunnel positions.

Discussion: Critical to the longevity and stability of an intraarticular ACL reconstruction is adherence to the surgical principles of isometry and impingement free movement. Isometry refers to the ability of a graft to maintain a constant length and tension during flexion and extension of the knee (1–5). Lack of length and tension change is critical since multiple studies have shown that the native ACL has a normal length change of 2 mm (5%) beyond which ligamentous damage begins to occur (3–5). Overloading the graft will result in limitation of motion and graft failure due to stretching of the neoligament. Underloading of the graft will not restore knee stability, which was the original goal of the procedure. While the achievement of isometric tracking is dependent on the interrelationship of the positions of the femoral and the tibial sites, moderate variations in the tibial attachment site of the graft only minimally affect isometry (2).

Tunnels to accommodate the graft are drilled into the femur and the tibia at sites that have been identified as being isometric (5–8). These tunnels may be seen on conventional radiographs by visualization of their thin, sclerotic edges which often are visible by 4 months postimplantation (9) (Fig. 16C,D). Since MR is a tomographic technique, it has enhanced capabilities to optimally assess tunnel position (5). As noted above, the femoral tunnel site is largely responsible for the isometry of the graft. The isometric position of the femoral tunnel can be visualized on the sagittal MR (or the lateral radiograph) as the intersection of two lines: the posterior femoral cortex and the posterior edge of the intercondylar roof (the posterior edge of the intercondylar roof corresponds nearly precisely to the posterior edge of the physeal scar) (Fig. 16E,F). The femoral tunnel opening on the coronal MR is at either the 11 o'clock or 1 o'clock position, depending on whether one is dealing with the right or left knee (5).

On the sagittal MR, the impingement-free tibial tunnel opens distally below the tibial tubercle and courses posterosuperiorly to exit the tibial articular surface immediately anterior to the anterior tibial spine (5). The location of the center of a well-placed tibial tunnel is 43% of the sagittal distance from the front of the tibial plateau; those grafts that demonstrate roof impingement have tibial tunnels that are only 30% of the distance from the anterior to the posterior edges of the tibial plateau (10). On the coronal MR images, the tibial tunnel opening into the joint should be centered on the intercondylar eminence (5). The position of the tibial tunnel in large part determines whether the graft will suffer from impingement. Since roof impingement is difficult to diagnose either clinically or at the time of surgery, the criteria for diagnosis are radiographic. Roof impingement is said to be present when a significant portion of the tibial tunnel lies anterior to the inferiorly projected slope of the intercondylar roof (5). It is critical to recognize that the position of the bone plug, the securing staples, or the interference fit screws is not the site of the exit/entrance of the graft into the bony tunnels (5).

While a wide variety of autologous and allogeneous materials have been utilized for ACL reconstruction, the bone patellar tendon bone (BPTB), hamstring tendons, and Achilles tendon allografts are among the most commonly employed (5). Preliminary evidence suggests that the natural history of these grafts is identical, and in the authors experience the MR appearance differs only in the surgical changes occurring at the donor site (5). Several studies have focused on the MR appearance of the normal BPTB and hamstring grafts (Fig. 16F,G). The immediate postinsertion signal of a graft is identical to the signal of the posterior cruciate ligament; the harvesting, preparing, sizing, and insertion of the graft do no result in an appreciable signal increase (9,11). The portions of the graft that lie within the tunnels also demonstrate low signal. Ninety percent of BPTB grafts in clinically stable knees during the first year postimplantation appear as solid, low signal bands traversing the joint (5). If clinical stability is assumed to represent the gold standard, MR achieves a sensitivity of 90%, a specificity of 100%, and an accuracy of 90% in determining the status of the neoligament (12).

Ten percent of normal grafts as judged by clinical stability alone (without reference to imaging assessment) demonstrate an intermediate signal intensity (13) (Fig. 16G,H). Occasionally, the increase in signal intensity is so significant that the graft may appear discontinuous (9,10,13). This potential for increased signal may make assessment of graft integrity difficult in some cases. It is to be emphasized that an increase in graft signal, even to the extent that the graft is very poorly delimited, is not necessarily indicative of a rerupture (5). Indeed, rerupture without a secondary traumatic event is unusual. There are two potential explanations for an increase in graft signal in clinically stable knees in which second-look arthroscopy reveals an intact graft: (1) normal temporal changes due to physiologic ligamentization; and (2) graft impingement (9,13) (see case 5-22). In patients in whom sequential MR studies are performed within 4 months of reconstruction, between 4 and 8 months, and again after 9 months postoperatively, MR demonstrates that there is a tendency for some grafts to acquire an increase in signal intensity of a moderate to a marked degree by the second MR examination, and a trend to diminish this newly acquired signal by the time of the last examination (14,15).

It is possible that this appearance can be explained by the process of ligamentization (5). Initially after harvest, the graft is avascular. By approximately 6 weeks, the graft becomes completely enveloped in a vascular synovial sheath arising from the vessels of the infrapatellar fat pad and the adjacent synovium (16). The synovial membrane, which engulfs the ACL reconstruction, has an intermediate signal character on T1 and proton density sequences, and can potentially hamper graft visualization (5,16). Ischemic necrosis occurs in the central core of the graft followed by revascularization progressing from the endosteal vessels of the bony tunnels and from the fat pad (16). Revascularization of the graft is complete by 20 weeks (16). The central ischemic necrosis and peripheral revascularization may sufficiently alter the relaxation characteristics of the graft to cause it to appear isointense with the intraarticular soft tissues (5). Because of the partial volume effect of the intermediate signal synovial sheath with the dark ligament, the graft might appear smaller than it actually is (5). If these processes of synovial reaction, graft edema, and revascularization are extensive, it is possible for a totally normal graft to be difficult to depict, thereby accounting for the reported inability of some examiners to identify the neoligament in a clinically stable knee (5). The authors have examined multiple stable grafts over the first year postimplantation. At least one half of the grafts demonstrated a mild diffuse increase in signal intensity (T1-weighted images) that developed over the first 3 months. The signal increase is diffuse, imparting an overall gray or occasionally a striped appearance to the graft (5).

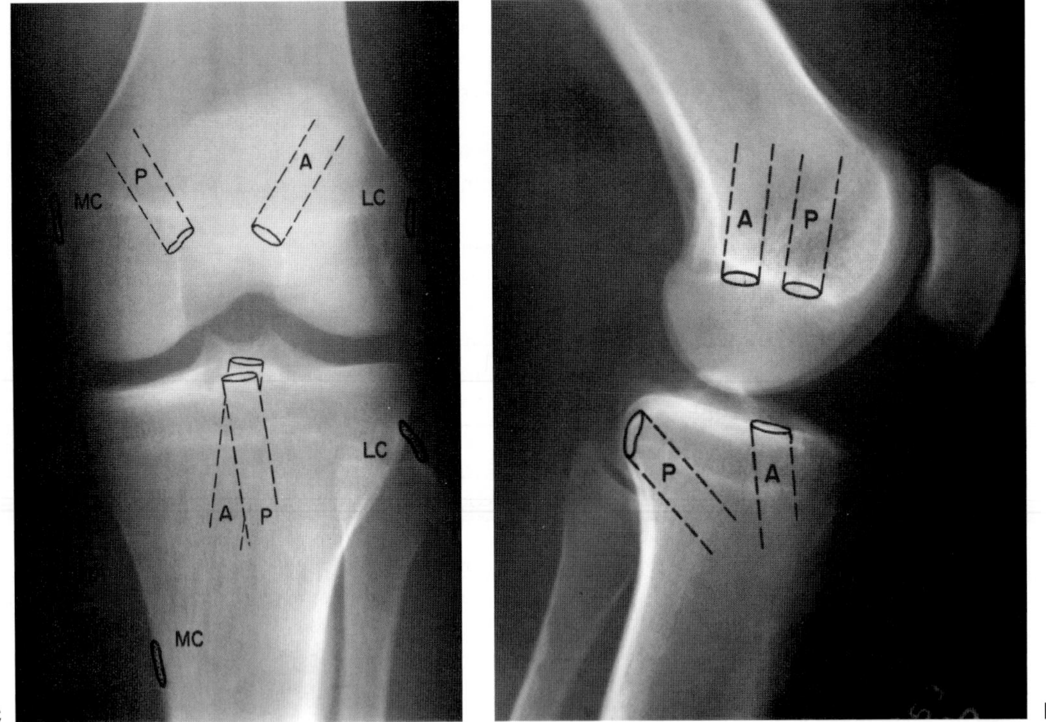

FIG. 16. C,D: Normal tunnels. The position of the anterior (A) and posterior (P) cruciate tunnels are depicted on these radiographs. On the anteroposterior radiograph **(C),** the attachment sites of both the medial collateral ligament (MC) and lateral collateral ligament (LC) are identified.

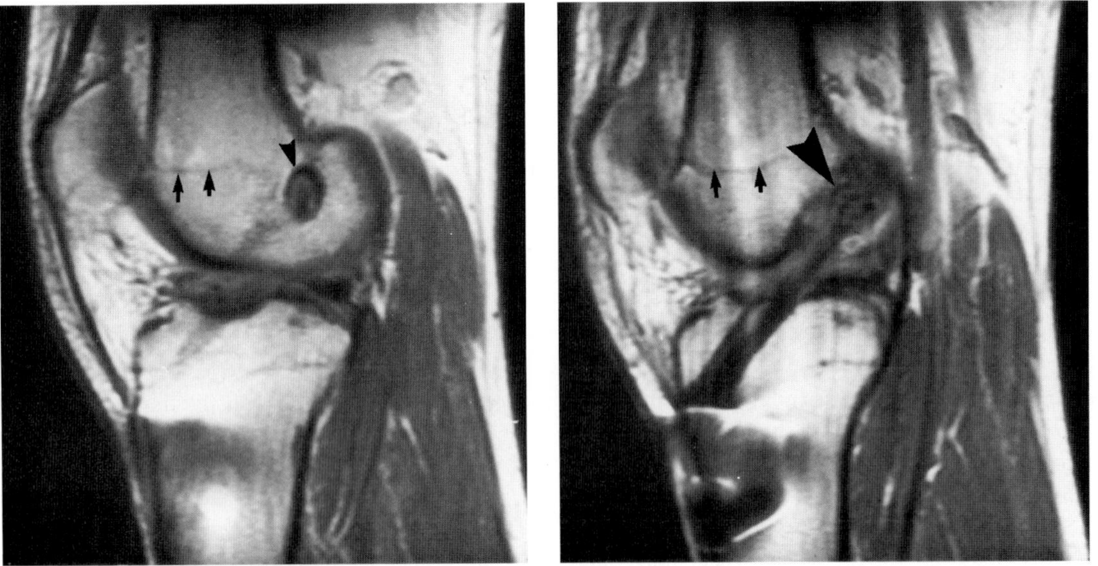

FIG. 16. E,F: Normal femoral tunnel position. Adjacent 3-mm sagittal T1-weighted spin echo sequences (TR/TE 800/20 msec) demonstrating the isometric position of the femoral tunnel *(arrowheads)* at the intersection of the line described by the posterior femoral cortex and the physeal scar *(arrows).* (Figures C–F from Deutsch AL, Mink JH. *The Postoperative knee.* In: Mink JH, Reicher MA, Crues JV III, Deutsch AL, eds. *MRI of the knee, 2nd edition.* New York: Raven Press, 1993;270–292, with permission.)

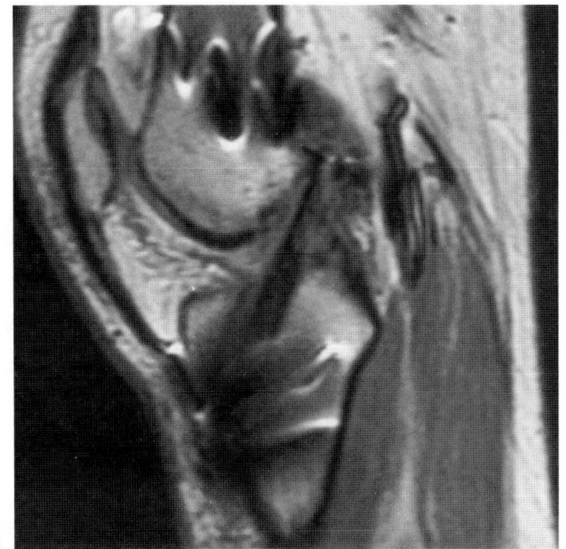

G

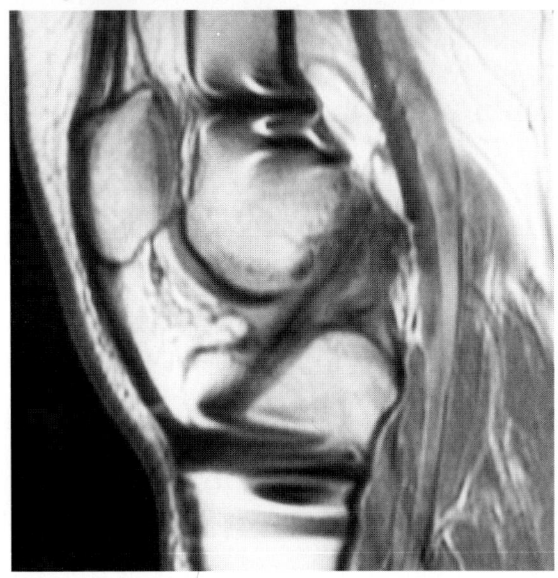

H

FIG. 16. G,H: Normal temporal changes in autografts. **G:** Sagittal proton density weighted image of a hamstring graft illustrates two principles of ACL reconstructive surgery: (a) the high, posterior origin of the graft in the joint indicates that the graft has been placed isometrically and (b) the entire tibial tunnel lies posterior to the projected slope of the intercondylar roof ensuring impingement free motion. The graft is seen as solid, dark band with only minimal intratendinous signal inhomogeneities. **H:** Obtained five months after G, the graft appears wider, with a somewhat linear signal increase within the graft. The patient was totally asymptomatic and the knee stable to examination. (From Deutsch AL, Mink JH. *The Postoperative knee.* In: Mink JH, Reicher MA, Crues JV III, Deutsch AL, eds. *MRI of the knee, 2nd edition.* New York: Raven Press, 1993;270–292, with permission.)

CASE 17

History: A 30-year-old man with history of knee pain. What is the most likely diagnosis for the findings illustrated and what is the significance of the findings?

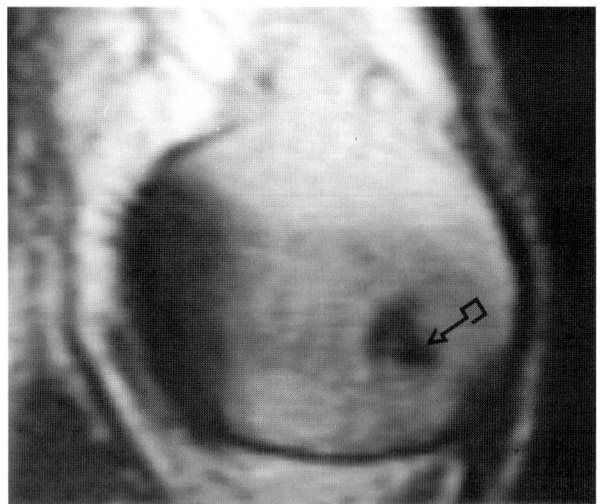

A

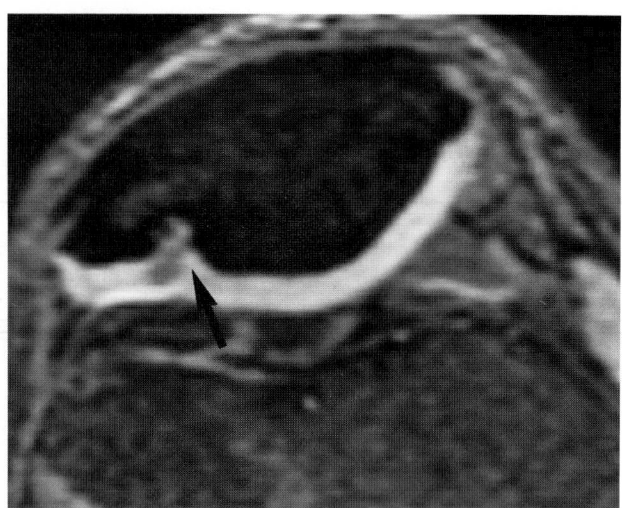

B

FIG. 17.

Findings: In Fig. 17A, a coronal T1-weighted (TR/TE 600/15 msec) sequence through the patella demonstrates a circular to ovoid focus of low signal intensity within the upper and outer quadrant (arrow). In Fig. 17B, an axial fat-suppressed spoiled GRASS (TR/TE/theta 60/10/60) demonstrates the dorsal defect along the lateral aspect of the patella (arrow). The overlying articular cartilage appears grossly intact and appears to extend into the defect within the subchondral bone plate.

Diagnosis: Dorsal defect of the patella (DDP).

Discussion: DDP is a relatively uncommon entity variably estimated to occur in from 0.3% to 1.0% of the population (1–6). Radiographically, the lesion is recognized as a relatively well-defined radiolucent lesion located within the superolateral aspect of the patella (1,6). It is most commonly encountered in the second and third decades. The etiology of the condition is not known although it is thought most likely to reflect an anomaly of ossification (1–6). Initial descriptions of this entity reported intact overlying articular cartilage, although cartilage abnormalities have subsequently been described and may be associated with symptoms (4–6). The natural history of DDP is that it does not progress in size and generally heals spontaneously leaving an irregular sclerotic zone in the patella (6). Curettage of the lesion has been reported in symptomatic patients with good results, although in general DDP is considered innocuous and generally merits no further intervention. The radiologic differential diagnosis includes osteochondral fractures, osteochondritis dissecans, chondromalacia patella, bone tumors, intraosseous ganglia, and Brodie abscess (6).

On MR imaging, DDP is typically seen, as in the present case, as a localized lesion with apparent disruption of the subchondral bone plate at the site of the lesion with ingrowth of articular cartilage (3,6). The articular cartilage overlying the lesion may appear of increased girth, and foci of increased signal intensity may be seen within the cartilage extending into the lesion on T2-weighted sequences (3,6). The sclerotic zone defining the lesion on plain radiographs is seen as an area of low signal intensity on all pulse sequences. DDP is differentiated on MR from simple chondromalacia by the defect in the subchondral bone plate that allows for extension of the overlying cartilage into the defect (6). Concurrent chondromalacia of the overlying thickened articular cartilage may account for the signal changes previously described as well as for symptoms in patients with DDP.

CASE 18

History: The patient experienced a twisting injury at the time of an automobile accident. Conventional radiographs were interpreted as normal. What is your diagnosis?

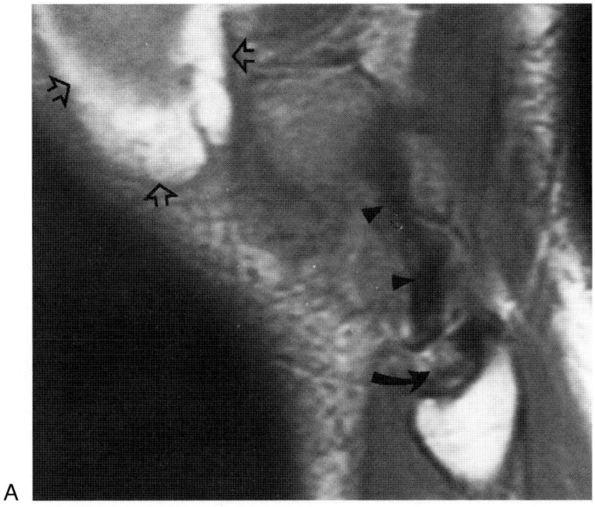

FIG. 18.

Findings: In Fig. 18A a sagittal T2-weighted image (TR/TE 2,000/80 msec) demonstrates a serpiginous contour and separation of the distal FCL from its fibular attachment (arrow). There is a complex appearance to fluid within the lateral suprapatellar bursa, suggesting acute hemarthrosis.

Diagnosis: Avulsion of the fibular collateral ligament.

Discussion: The major structures composing the lateral supporting ligaments include the iliotibial band (the distal tendon of the tensor fascia lata), the fibular collateral ligament (FCL), the popliteus tendon, the capsule, and biceps femoris tendon (1). The FCL is the primary constraint to lateral joint opening, with varus stress and functions to limit hyperextension as well. Injuries to the lateral compartment are far less common than those of the medial and often are unrecognized because signs of injury may be subtle (1). Significant injuries of the lateral complex may be disabling because these structures are under maximal tension when the leg is extended during walking.

The normal FCL arises from the lateral epicondyle just above the groove for the popliteus tendon; it extends in a posterior, inferior, and lateral direction on its course to its attachment to the fibular head (2). Just prior to its insertion, the FCL joins with the tendon of the biceps femoris where they intertwine as a conjoint tendon (2). Both are cord-like, dark structures at all imaging parameters. Because of its slight lateral and posterior-directed course, the FCL is only fortuitously seen on sagittal or coronal images, but is more reliably found on coronal sequences. The biceps tendon may be found on either coronal or sagittal images (Fig. 18B–D).

The iliotibial band is a thin, dark, vertically oriented stripe arising as a thickening of the fascia lata and coursing distally in the bright subcutaneous fat of the lateral thigh to its insertion on the lateral tibia just below the plateau anteriorly (Gerdy tubercle). It is identified 100% of the time on coronal sequences, but is uncommonly found in the sagittal plane (Fig. 18E,F).

Complete ruptures of the FCL occur either in midsubstance or as an avulsion from the fibular head following a contact or noncontact hyperextension varus stress (2). Rupture on occasion may be associated with a "pop" at the time of injury, but frequently little or no swelling is detectable. A significant effusion indicates the presence of an associated capsular or cruciate ligament injury since the FCL is a completely extraarticular structure. Lateral ligamentous disruptions may occur following a ski injury in which the tips of the ski cross, permitting internal rotation and varus torques to act at the knee. Lateral disruptions are often the result of more severe trauma than normally occurs in athletic injuries, as evidenced by the high incidence of motor vehicle accidents as the mechanism of injury (3). Large varus forces will result in disruption of the iliotibial band (ITB), the popliteus tendon, and the biceps femoris tendon. Complete lateral compartment disruption, essentially a knee dislocation, may be associated with peroneal nerve injury, vascular compromise, and long-term disability, all of which serve to underscore the severity of the injury (3).

Injury to the lateral supporting structures share several characteristics (2,4). They occur almost exclusively at the distal (tibial and fibular) attachments. The ITB and the FCL may be injured by means of avulsion; identification of their different insertion sites will allow for correct diagnosis. Acute substance tears of any of the lateral structures result in interruption of the contours, a wavy appearance, and localized fluid within/around them, but most commonly extensive disorganization of the surrounding soft tissues with areas of both increased and decreased signal are found (Fig. 18G,H). In chronic injuries, complete absence, a serpiginous contour, and focal areas of abnormal signal, within the ligament are common. Extensive varus stresses may result in medial compartment trabecular microfracture.

Two additional disorders of the lateral supporting structures that, while less common, may be a source of clinical symptoms. Runners may develop pain over the posterolateral aspect of the knee which has been ascribed to popliteus tendonitis or ruptures (5,6). The popliteus helps prevent forward displacement of the femur on the tibia and helps maintain internal rotation of the tibia. Running downhill produces stress on the tendon and subsequently pain. The diagnosis of tendinitis or a rupture of the popliteus is quite problematic clinically. MR may allow definitive evaluation of popliteus rupture (Fig. 18I). The iliotibial band friction syndrome presents with pain over the lateral femoral condyle in patients in training programs that include walking and running (7). Pain is maximal when the knee is flexed; the symptoms are thought to be due to the movement of the band from a position anterior to the condyle in full extension to posterior in full flexion. Localized tenderness can be elicited in a small area over the lateral femoral condyle when the knee is held in 30° of flexion.

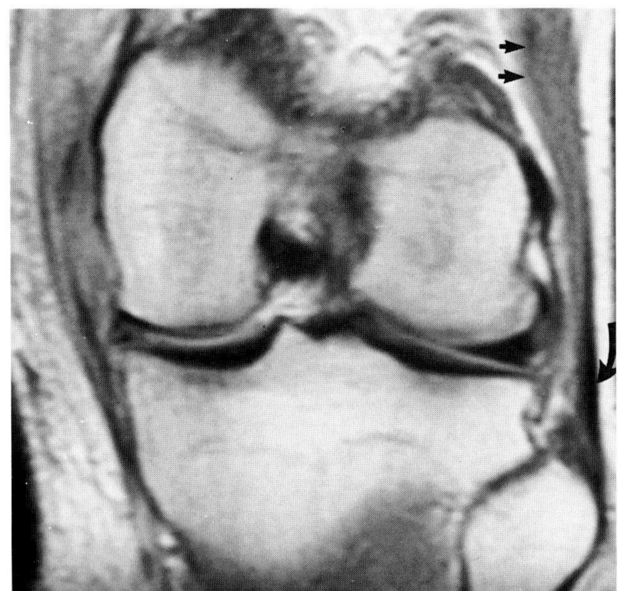
B

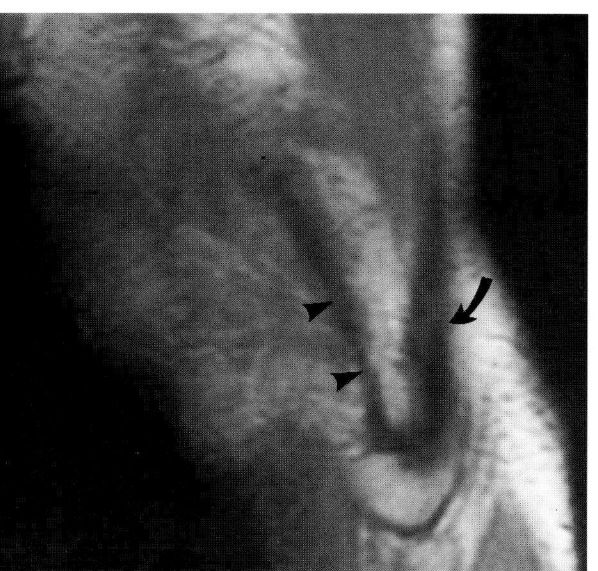
D

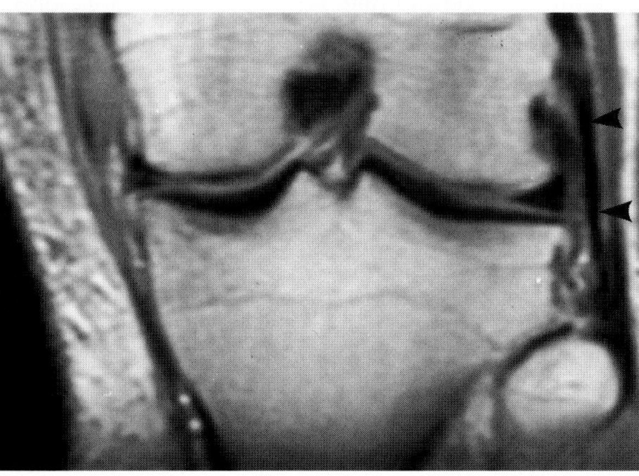
C

FIG. 18. B–D: Normal FCL and biceps tendon. Consecutive coronal PD weighted images (TR/TE 2,000/20 msec). **B,C:** The biceps muscle *(arrows),* tendon *(curved arrow),* and FCL *(arrowheads)* are shown as they course to comprise the conjoint tendon; one of the main lateral supports of the knee. Sagittal PD-weighted image (TR/TE 2,000/20 msec). **D:** Both the FCL *(arrowheads)* and the biceps tendon *(curved arrow)* are shown as they extend to attach onto to fibular head.

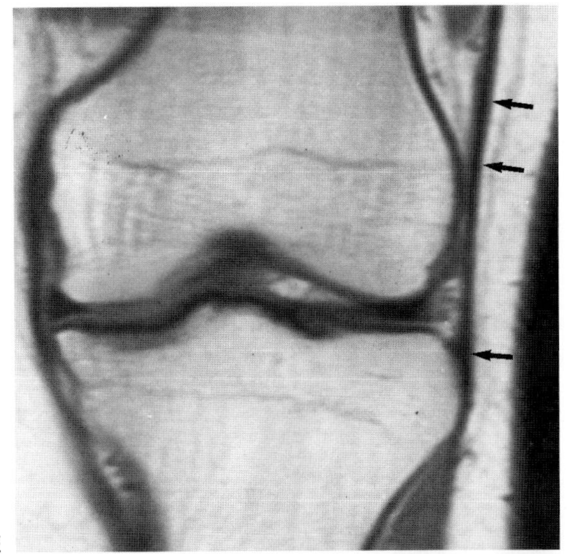

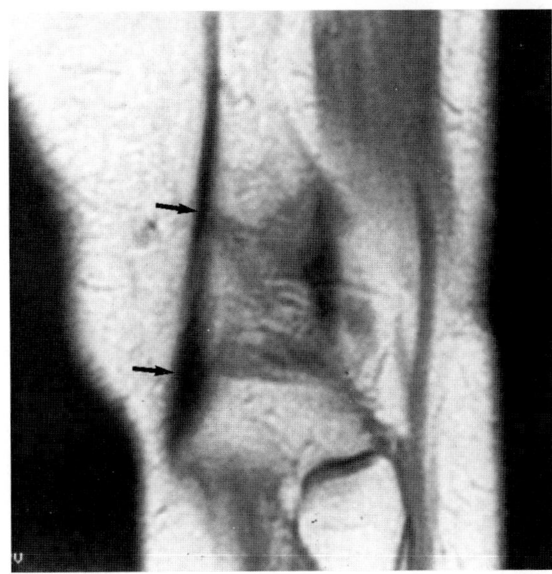

FIG. 18. E,F: Normal iliotibial band. **E:** Coronal PD-weighted image (TR/TE 2,000/20 msec) demonstrates the iliotibial band *(arrows)* inserting onto the anterolateral tibia at the Gerdy tubercle. **F:** On a sagittal PD image (TR/TE 2,000/20 msec), the iliotibial band *(arrows)* is seen to insert well anterior to the FCL and the middle third of the capsular ligament. It is generally unusual to depict the iliotibial band on routine sagittal images. (From Mink JH. The cruciate and collateral ligaments. In: Mink JH, Reicher MA, Crues JV III, Deutsch AL, eds. *MRI of the knee, 2nd ed.* New York: Raven Press, 1993;179–185, with permission.)

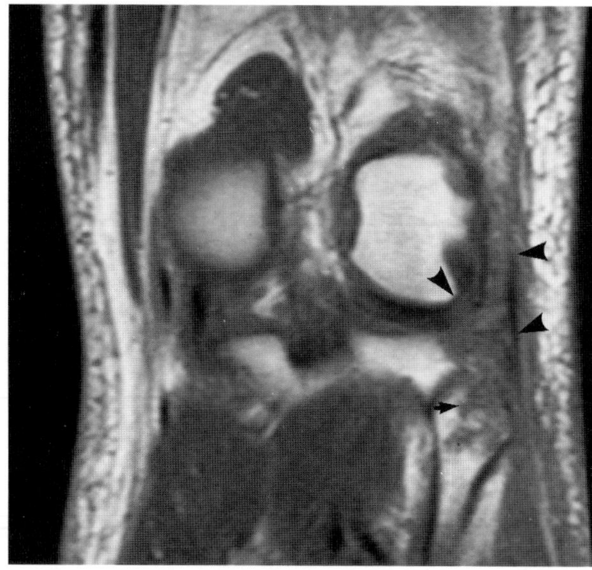

G

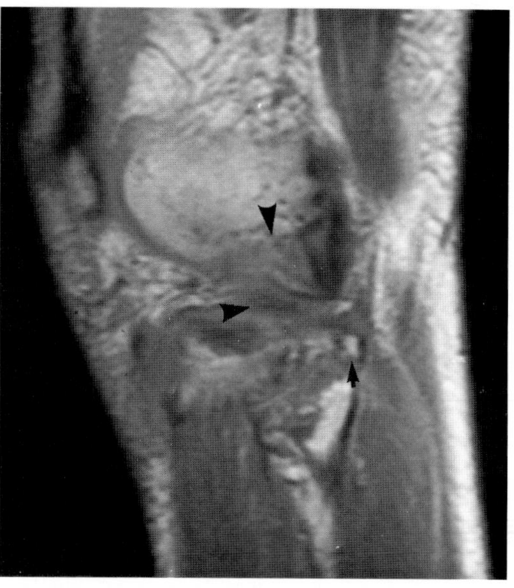

H

FIG. 18. G,H: (G:) Coronal T1-weighted (TR/TE 600/20 msec) image demonstrates signal alteration within the proximal fibula consistent with an occult fracture *(small arrow)*. There is increased intermediate signal intensity in the region of the distal attachment of the fibular collateral ligament (FCL) and its attachment to the fibular head cannot be defined. **H:** PD-weighted image (TR/TE 2,000/20 msec) demonstrates complete disorganization of the soft tissues in the expected position of the FCL and biceps tendon.

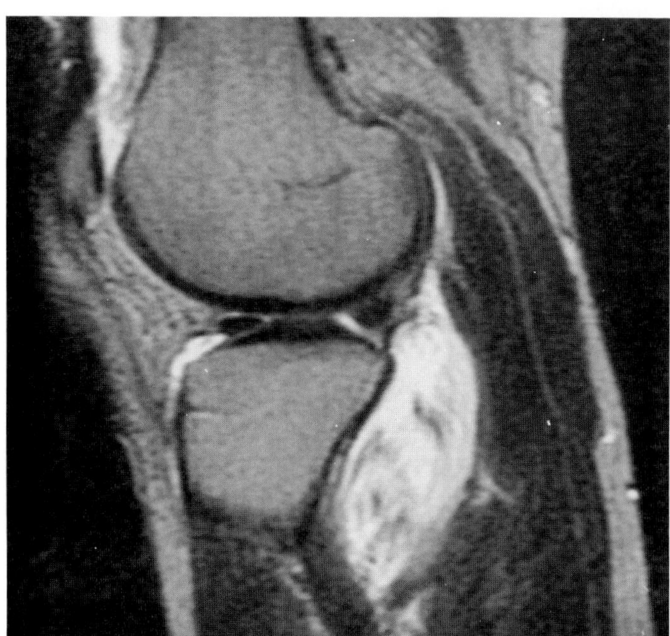

I

FIG. 18. I: Presumed popliteus muscle/tendon rupture. Sagittal T2-weighted image (TR/TE 2,000/80 msec) demonstrates complete replacement of the normal popliteus muscle by high signal intensity fluid and possible interstitial hemorrhage. The patient experienced sudden pain along the posterolateral knee while running downhill. (From Mink JH. The cruciate and collateral ligaments. In: Mink JH, Reicher MA, Crues JV III, Deutsch AL, eds. *MRI of the knee, 2nd ed.* New York: Raven Press, 1993;179–185, with permission.)

CASE 19

History: A 45-year-old man with history of prior partial medial menisectomy underwent MR evaluation for recurrent symptoms following a new injury. What is your diagnosis and level of confidence? What modifications in diagnostic criteria are necessitated by the postoperative state?

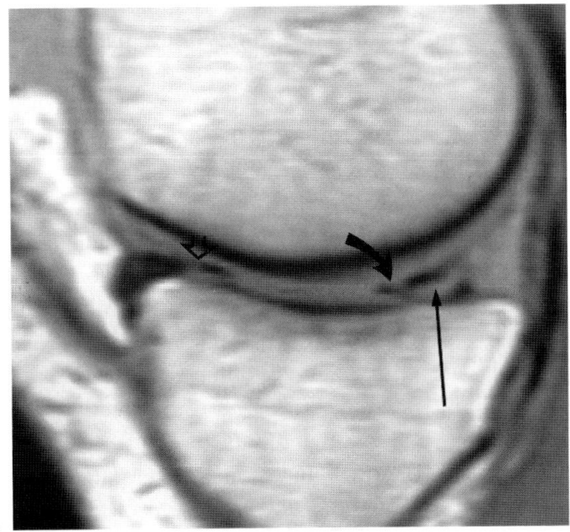
A

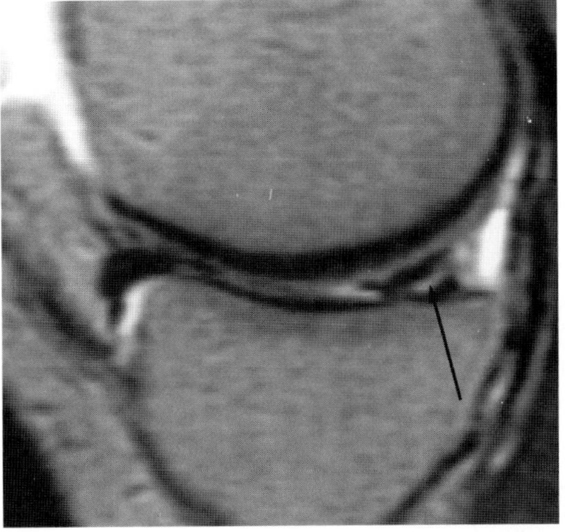
B

FIG. 19.

Findings: In Fig. 19A a sagittal proton density (TR/TE 2,300/20 msec) image demonstrates blunting of the apex of the posterior horn (curved arrow), a finding that may be seen normally following partial menisectomy. The finding in the present case, however, appears more extensive and irregular than would normally be expected and as such suggests the presence of a flap or displaced fragment. This suspicion is also supported by the presence of abnormal signal within the anterior horn, suggesting anterior extension of the tear (short open arrow). Additionally there is an oblique line of intrameniscal signal extending to the tibial articular surface (long arrow) (Fig. 19B). Corresponding T2-weighted (TR/TE 2,300/80 msec) image demonstrates the line of signal within the posterior horn to become of increased intensity (long arrow). This finding strongly suggests the presence of a retear even in the absence of the other morphologic alterations.

Diagnosis: Complex retear of the posterior horn of the medial meniscus with displaced fragment. The diagnosis in this case is established both on the basis of abnormal morphology and abnormal signal within the posterior horn remnant. The diagnosis should be made with a high degree of confidence in this case.

Discussion: Assessment of the meniscus following partial menisectomy is often problematic. The appearance of the meniscus may be highly variable, reflecting the techniques utilized and options chosen by the surgeon at the time of resection. In approaching the postoperative meniscus, our group as well as others have found it valuable to divide these menisci into three groups according to the apparent size of the remnant (1,2).

Utilizing this schema, group 1 menisci are characterized by menisci having undergone relatively minor prior resection (arbitrarily considered to be less than 25%) and appear nearly normal on MR scans. The tips of these menisci are often blunt and they are focally smaller than normal but otherwise are not remarkable. Group 1 menisci represent an important group because the standard criteria utilized for assessment of the nonoperated meniscus (e.g., grade 3 signal and altered meniscal morphology) can be applied with a degree of accuracy approaching that of the virgin meniscus (approximately 90%) (2–6) (Fig. 19C,D).

Group 2 menisci present greater interpretive challenges. These meniscal remnants are quite variable in size (resection between 25% and 75%) and morphology and the presence of grade 3 signal is not a reliable finding for a definite retear. One of the factors that contributes to the diagnostic difficulty observed in this group is a phenomenon that has been termed intrameniscal signal conversion (Fig. 19E–H). The line of signal typically seen with MR in horizontal and oblique tears is invariably greater than the recognized extent of the tear as depicted at arthroscopy. When the unstable portion of the meniscus is resected and the meniscus recontoured, it is theoretically possible to create from the remaining intrasubstance (grade 2) signal a new appearance that could mimic a grade 3 tear. While the presence of apparent grade 3 signal is not in and of itself a reliable predictor of a retear, the presence of increased signal within the focus of apparent grade 3 signal on T2-weighted images is a valuable sign of a retear. The high predictive value of hyperintensity along the course of the grade 3 signal was first established in assessment of the virgin meniscus (4). This presumably reflects fluid tracking into the actual tear. While highly specific, this finding is not sensitive (40%) (5,6). The overall accuracy for assessment of group 2 menisci approaches 65% (2,3,5,6).

Group 3 menisci are characterized by small remnants of unresected meniscus (>75% meniscal resection). Assessment of meniscal stability in this group is highly problematic with an accuracy approaching 50% (2,3). MR has proven of value in this group for assessment of nonmeniscal abnormalities including articular cartilage defects, osteonecrosis, and loose intraarticular osteocartilaginous bodies (Fig. 19I,J).

The difficulties presented by signal conversion and other equivocal changes in postoperative menisci have stimulated interest in MR arthrography as a means to accomplish a more accurate examination of meniscal stability. For MR arthrography, gadopentetate dimeglumine can be diluted at a ratio of 1:100 with saline and 20 to 30 cc injected intraarticularly (2,3,5,6). We presently perform fat-suppressed sagittal and coronal T1-weighted sequences following the injection of contrast. The principal diagnostic criteria for meniscal retears are similar to those established for conventional radiographic arthrography and include contrast imbibition into the tear and altered meniscal morphology (Fig. 19K,L). The addition of contrast material significantly facilitates assessment of meniscal morphology, particularly in the presence of diffuse increased intrameniscal signal (Fig. 19M,N). In the experience reported to date, MR arthrography has significantly improved assessment of both group 2 and 3 menisci. With MR arthrography, accuracy for group 2 menisici improved from 65% to 87% in a recently reported series (5,6). For group 1 menisci, which are well evaluated with conventional criteria, MR arthrography provided no significant incremental improvement in accuracy (5,6).

Our present approach toward the symptomatic postoperative knee is to perform a conventional MR examination. If a definite diagnosis can be established (e.g., meniscal retear, osteonecrosis), or if the study is considered normal with a group 1 meniscus, no further evaluation is performed. If no explanation for the patient's symptoms has been established, and the menisci are group 2 or 3, we frequently proceed to MR arthographic examination.

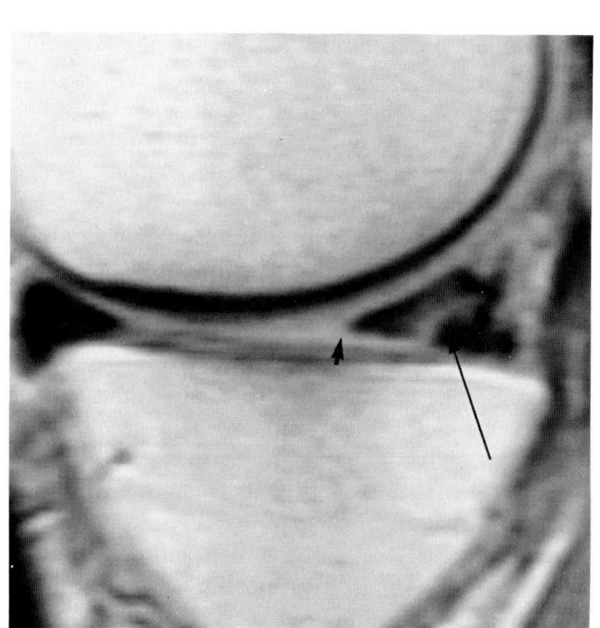

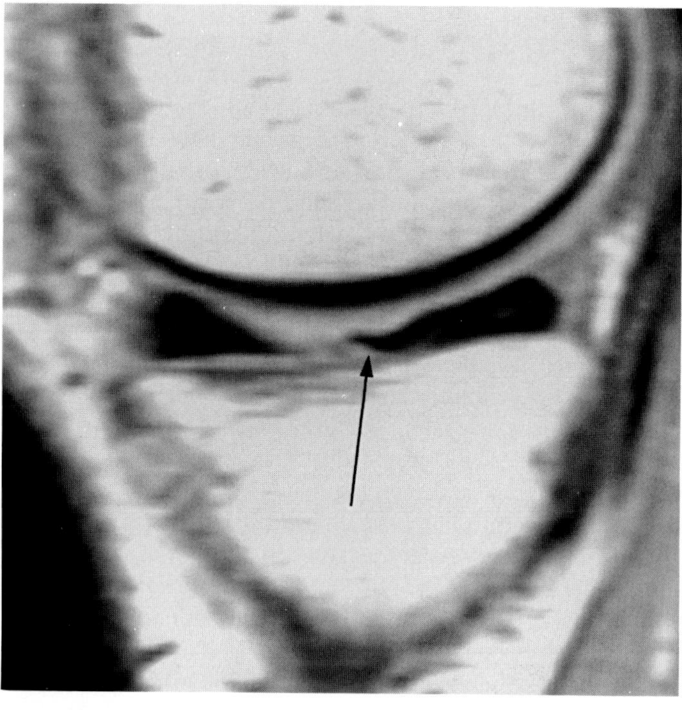

FIG. 19. C,D: Demonstration of re-tears within group 1 menisci. **C:** Sagittal intermediate-weighted image of the medial meniscus demonstrates the meniscus to be minimally decreased in size with a slightly blunted tip to the posterior horn; findings consistent with limited prior meniscal resection. A curvilinear line of signal is seen traversing the posterior horn *(long arrow).* The criteria of grade 3 signal can be applied to postoperative menisci such as this one with nearly the same confidence as for the virgin meniscus. **D:** Sagittal intermediate weighted image in a different patient presenting with new symptomatology following prior partial meniscectomy. Instead of a smoothly rounded tip as may normally be seen postoperatively, the apex of the meniscus demonstrates a clearly abnormal contour *(arrow).* A flap tear was identified at subsequent arthroscopy. Morphological criteria can also be utilized in assessment of group 1 menisci.

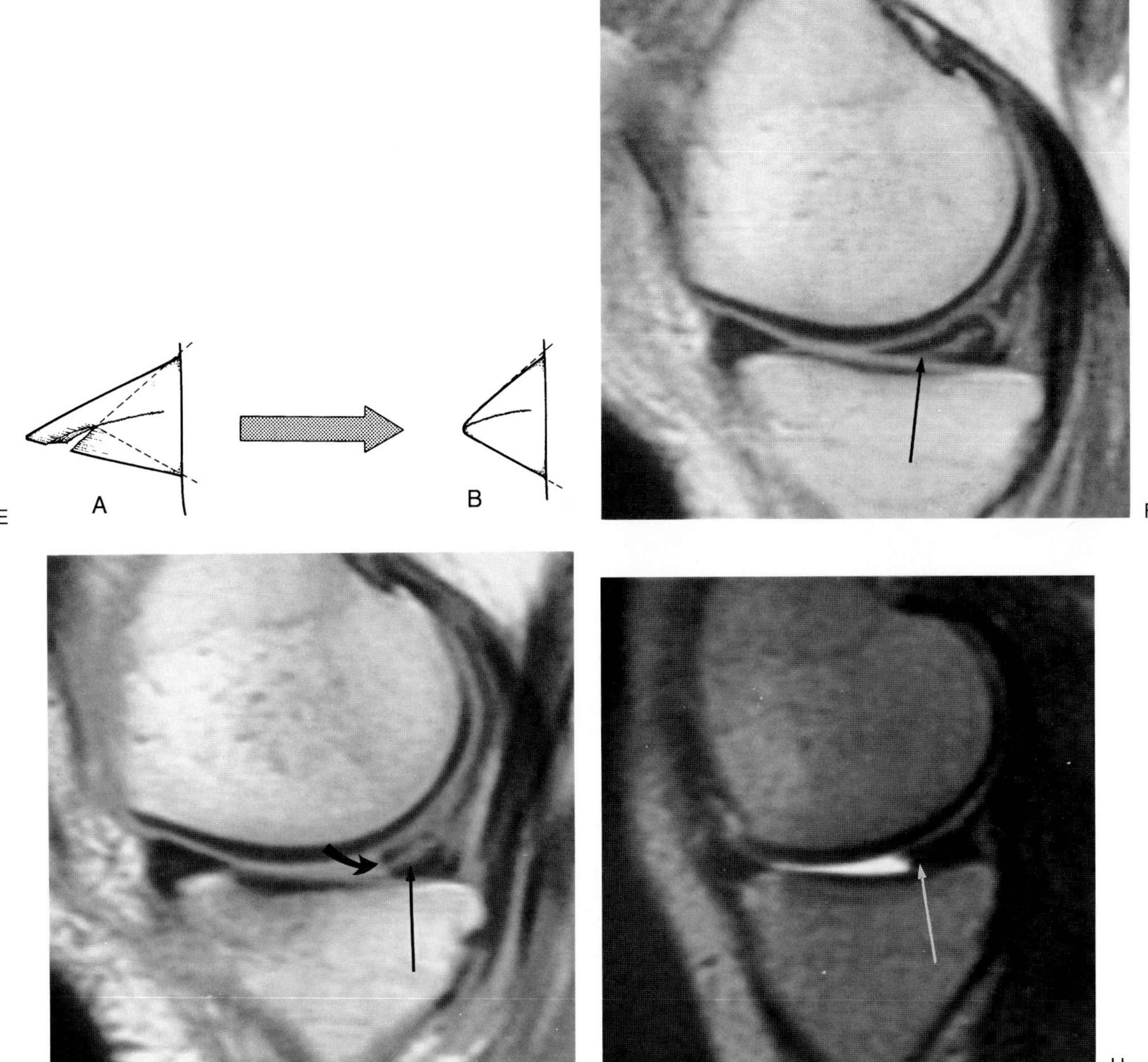

FIG. 19. E–H: Intrameniscal signal conversion. **E:** Schematic diagram demonstrating the potential conversion of intrasubstance grade 2 signal to mimic a grade 3 tear. The extent of the arthroscopically demonstrable mensical tear is less than the degree of abnormal signal seen on the MR (A). Following partial mensical resection *(dotted lines),* the postoperative meniscus (B), which is clinically stable, could contain intrasubstance signal that now appears to extend toward the articular surface and could thus be confused with grade 3 signal indicative of a meniscal tear. **F:** Sagittal PD-weighted (TR/TE 2,000/20 msec) preoperative MR demonstrating a nondisplaced horizontal tear of the posterior horn of the medial meniscus *(arrow).* **G:** Sagittal PD-weighted (TR/TE 2,000/20 msec) postoperative study (group 2 meniscus) demonstrates blunting of the meniscal tip *(curved arrow)* and intramensical signal extending to the recontoured meniscal margin *(long arrow)* in a manner directly analogous to that illustrated in the schematic diagram. The meniscus was proven stable at follow-up arthroscopy. **H:** Sagittal T2-weighted image. The existing synovial fluid provides a natural contrast medium and highlights the meniscal contour. No high signal intensity fluid is seen extending into the previously identified line to suggest fluid tracking within a meniscal tear *(arrow).* Indeed, the intrameniscal signal is not demonstrable on this sequence. Such findings can be of assistance in excluding retear of the meniscal remnant with increased confidence. (From Deutsch AL, Mink JH. The Post-operative knee. In Mink JH, Reicher M, Crues JV, Deutsch AL, eds. *MRI of the knee, 2nd ed.* New York: Raven Press, 1993;237–290, with permission.)

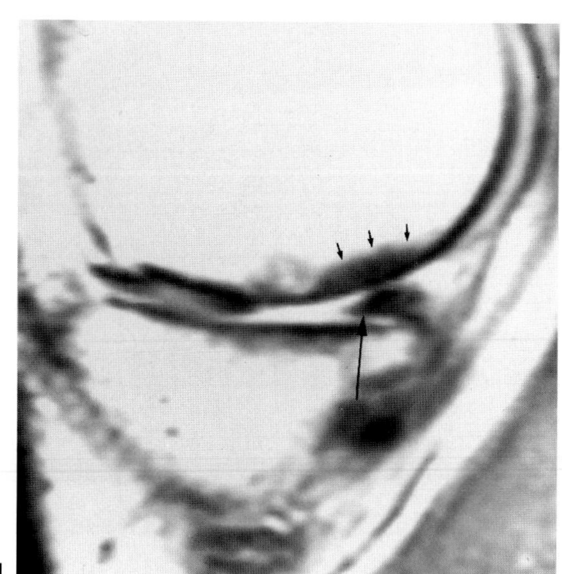

I

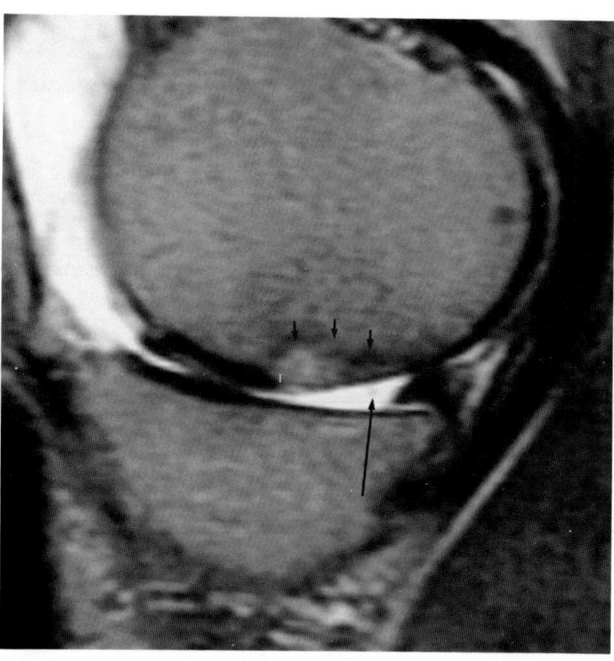

J

FIG. 19. I,J: Group 3 postoperative meniscus. **I:** Sagittal PD-weighted (TR/TE 2,000/20 msec) MR demonstrates a small posterior horn remnant *(arrow).* No significant intrameniscal signal is noted. Advanced loss of articular cartilage with changes extending into subchondral bone typical for the postoperative knee are well depicted. **J:** Sagittal T2-weighted image (TR/TE 2,000/80 msec). The arthrogram effect is valuable in assessing meniscal contour as well as in depicting the articular cartilage loss and degenerative changes *(small arrows).* Additionally, a small loose body can be faintly seen as a filling defect within the joint fluid *(arrow).* The patient underwent arthroscopy for an MR demonstrable bucket handle tear of the lateral meniscus. Inspection of the medial meniscus demonstrated a small central nondisplaced tear of the posterior horn remnant. The tear was considered significant enough to warrant further partial resection. This case illustrates the degree of difficulty that may be encountered in distinguishing retorn and stable menisci. (From Deutsch AL, Mink JH. The Postoperative knee. In Mink JH, Reicher M, Crues JV, Deutsch AL, eds. *MRI of the knee, 2nd ed.* New York: Raven Press, 1993;237–290, with permission.)

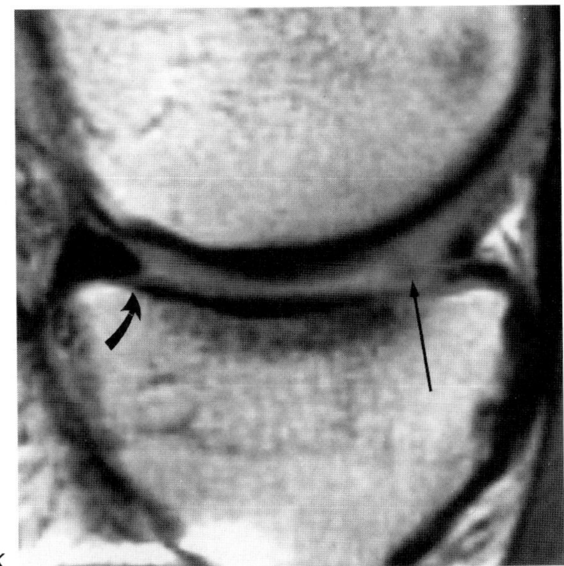

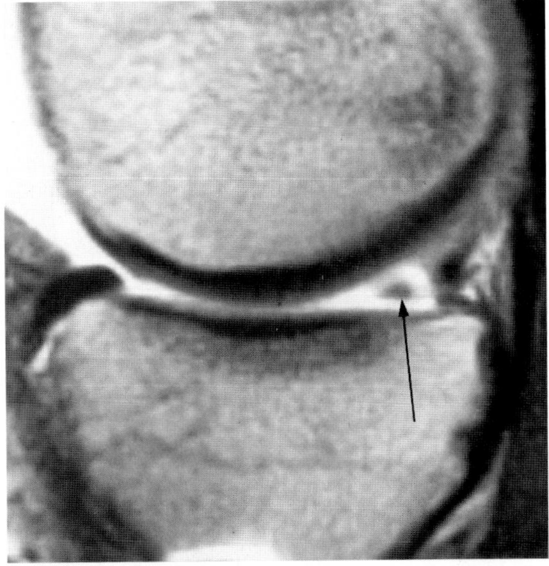

FIG. 19. K,L: MR arthrographic depiction of a meniscal retear. **K:** Sagittal T1-weighted image (TR/TE 500/20 msec) in a patient status post–prior partial meniscectomy. The tip of the anterior horn is mildly blunted consistent with prior surgery *(curved arrow)*. The posterior horn remnant is irregular with the suggestion of a small centrally displaced flap *(arrow)*. **L:** Sagittal T1-weighted image following the intraarticular administration of a dilute solution of gadopentetate dimeglumine. The depiction of the centrally displaced fragment is markedly improved *(arrow)*. (With permission from ref. 2.)

FIG. 19. M: MR arthrography sagittal PD image (TR/TE 2,000/20 msec) in a patient status post–prior partial menisectomy demonstrates a blunt mensical tip as well as equivocal intramensical signal.

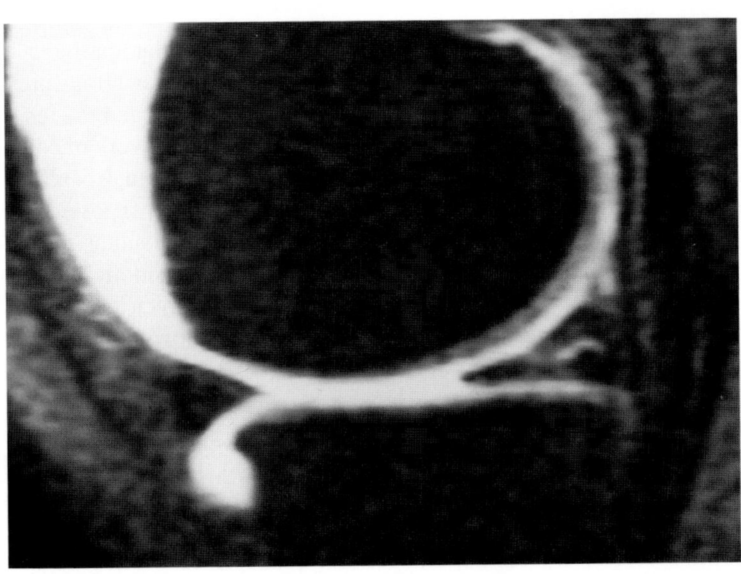

FIG. 19. N: Sagittal fat-suppressed image obtained following the intraarticular administration of a dilute solution of gadopentetate dimeglumine demonstrates increased signal intensity to the previously demonstrated intramensical signal reflecting contrast imbibition into the substance of the meniscus. (From Deutsch AL, Mink JH. The Postoperative knee. In: Mink JH, Reicher M, Crues JV, Deutsch AL, eds. *MRI of the knee, 2nd ed.* New York: Raven Press, 1993;237–290, with permission.)

CASE 20

History: A 23-year-old woman complained of anterior knee pain following lateral dislocation of the patella. What are the findings in Fig. 20B, and what are their potential clinical implications?

A 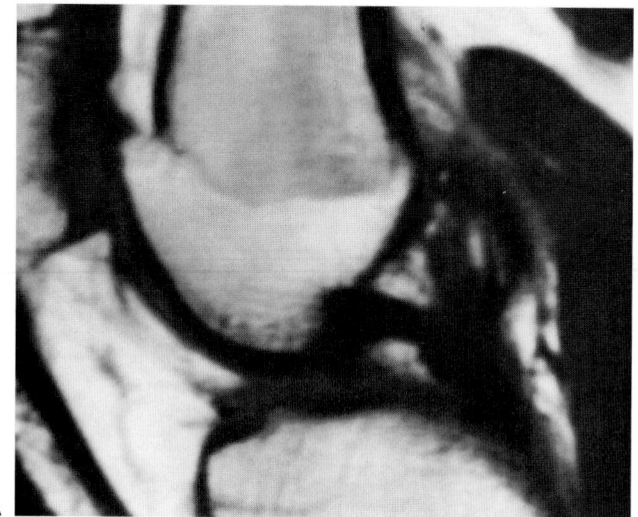

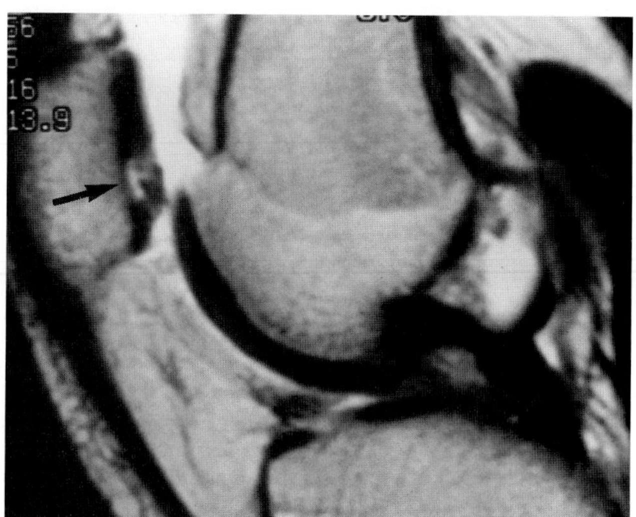 B

FIG. 20.

Findings: In Fig. 20A a T1-weighted (TR/TE 500/11 msec) midsagittal conventional spin echo image through the knee shows a normal-appearing patellofemoral joint with no evidence of cartilage bone abnormality. In Fig. 20B a T2-weighted (TR/TE 2500/102 msec) fast spin echo image, however, shows linear high signal in the patellar cartilage paralleling the cartilage-bone interface (arrow).

Diagnosis: Localized articular cartilage defect (posttraumatic).

Discussion: The integrity of articular cartilage is essential to the normal function of diarthrodial joints (1). However, since cartilage does not contain any nerves, significant damage may develop in relative silence or with only vague and misleading clinical findings. MR imaging is uniquely suited to evaluating articular cartilage and is unparalleled in its ability to detect and characterize early abnormalities in this important tissue (2).

The MR imaging appearance of articular cartilage reflects its complex histologic and biochemical composition (3). It is useful to conceptualize cartilage as being composed of two phases: a solid phase containing chondrocytes, aggregated proteoglycans, and fibrous collagen, and a fluid phase composed of water and ions in solution. The fluid phase constitutes approximately 70% of the weight fraction of normal cartilage and provides a high proton density for generating MR signal. The solid matrix, particularly collagen, tends to constrain water protons and thus shorten T2 relaxation time. Therefore, despite its high proton density, normal articular cartilage shows a uniformly low signal intensity on T2-weighted images. Magnetization transfer between the water protons and the solid matrix (primarily collagen) also contributes to signal loss in normal cartilage (4,5). Magnetization transfer routinely occurs in this tissue during multislice imaging, but can be particularly noticeable when fast spin echo technique is used (6).

The degree of hydration of the articular cartilage is determined by the balance between the ''swelling pressure'' exerted by hydrophilic proteoglycans and the resistance to expansion imposed by the dense network of collagen in cartilage (3). Damage to the collagen matrix allows aggregated proteoglycans to expand and expose more negatively charged moieties that attract counter ions—mostly sodium—and thus draw water osmotically into the cartilage (7,8). Proteoglycan loss, which usually accompanies collagen degradation, also allows residual proteoglycans to swell and thus increase the water content of cartilage.

In general, therefore, matrix damage is associated with increased proton density, decreased T2 relaxation, and decreased magnetization transfer in cartilage. These factors combine to increase the signal intensity of cartilage on T2-weighted images (9). Fast spin echo imaging provides heavy T2 weighting and high spatial resolution in only a fraction of the time required to acquire conventional spin echo images, and is therefore an extremely useful sequence for evaluating the articular cartilage during routine clinical imaging (2).

In osteoarthritis, changes usually begin superficially with fraying and fibrillation of the articular surface (10). Surface damage alone causes cartilage softening (chondromalacia) but does not necessarily progress to complete cartilage loss and clinical osteoarthritis (11). Full-thickness cartilage loss requires damage to the deep layers of cartilage; however, this may arise without involvement of the articular surface (12). Isolated injury to the deep cartilage may occur following a subfracture impact to the cartilage surface (13). Rapid, excessive loading of the cartilage produces severe shear forces at the cartilage-bone interface that can cause horizontal cracks to develop in this region without necessarily affecting the articular surface (13–15). In extreme cases, the cartilage may actually separate from the subchondral bone and subsequently break down over time. This may be accompanied by focal ingrowth of the subchondral bone (i.e., central osteophyte formation) (Fig. 20C). The thicker the cartilage, the greater the magnitude of the deep shear forces generated and therefore the higher risk of damage to the matrix (16). This may explain why the patellar cartilage is more frequently damaged than the corresponding femoral cartilage during dislocation injuries (12,16).

Consistent with these pathologic changes, traumatized articular cartilage often shows a deep linear pattern of high signal intensity paralleling the subchondral cortex on T2-weighted MR images (2) (Fig. 20A,B,D). In its early stages, this form of cartilage damage may not be evident during inspection of only the articular surface by arthroscopy, and the apparent preservation of the overlying cartilage on the MR images may lead one to underestimate the potential significance of these otherwise innocuous-appearing signal abnormalities. It is therefore important to recognize such areas on MR images as lesions that may be arthroscopically occult and yet have serious implications to the future integrity of the joint.

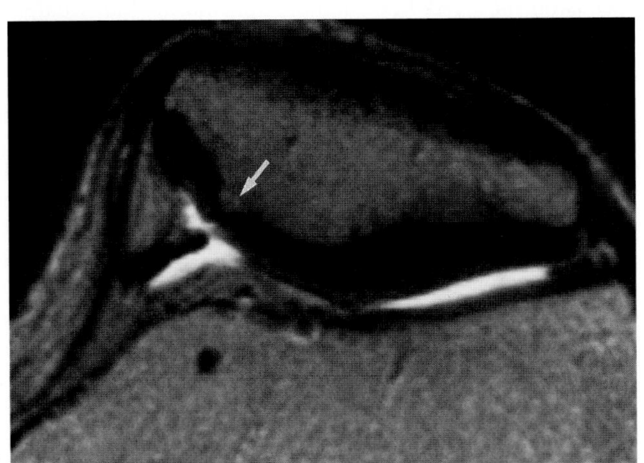

FIG. 20. C: T2-weighted (TR/TE 2,500/92 msec), fat-suppressed, axial fast spin echo image of the patella in a patient who remotely injured his knee during free-style skiing and now complains of anterior pain. Focal cartilage loss with central osteophyte formation *(arrow)* is present at the medial facet of the patella.

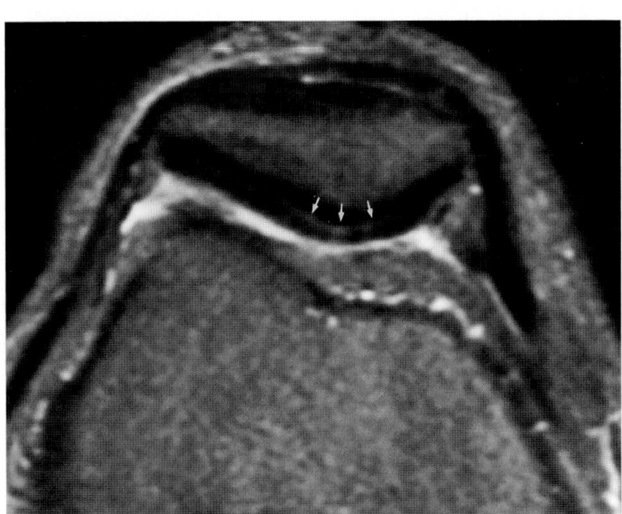

FIG. 20. D: T2-weighted (TR/TE 2,500/92 msec), fat-suppressed, axial fast spin echo image of the patella in a patient who suffered a direct impact to the front of the bent knee shows linear high signal intensity paralleling the subchondral surface *(arrows)*. The overlying cartilage appears otherwise normal.

CASE 21

History: A 26-year-old man underwent MR evaluation following a traumatic injury to the knee. There was clinical concern with regard to the integrity of the ACL. What is your diagnosis?

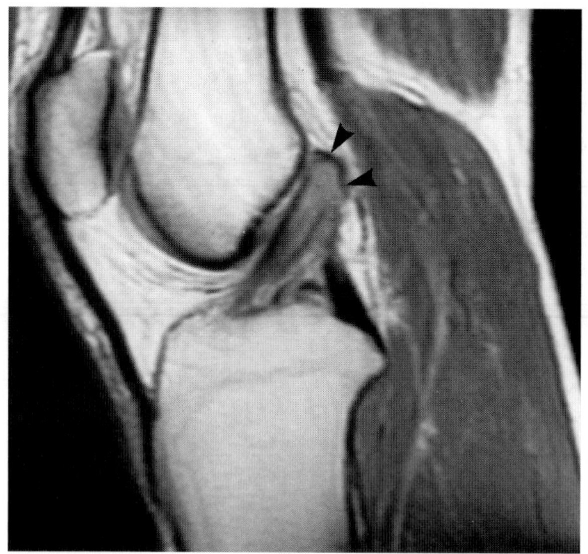

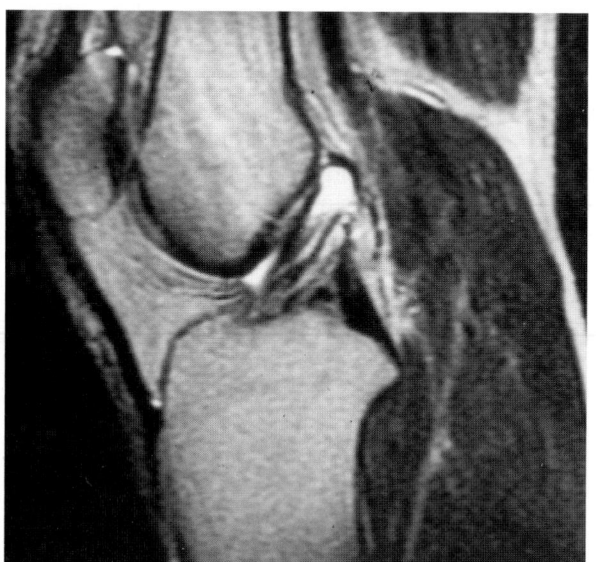

FIG. 21.

Findings: In Fig. 21A a sagittal PD image (TR/TE 2,000/20 msec) demonstrates intermediate signal intensity partially obscuring visualization of the ACL in the intercondylar notch. In Fig. 21B the ovoid-shaped signal becomes hyperintense on the T2-weighted image.

Diagnosis: Ganglion cyst associated with the anterior cruciate ligament.

Discussion: While previously considered to be rare lesions, intraarticular ganglion cysts arising from the cruciate ligaments, are not uncommonly encountered on MR imaging of the knee (1–7). They may arise from either the anterior (ACL) or posterior cruciate ligament (PCL) and depending on the site of origin typically demonstrate a distinctive appearance (1). Ganglia associated with the ACL characteristically demonstrate a fusiform appearance extending along the course of and interspersed within the fibers of the ligament. The ACL can usually be seen within the mass and this appearance should be readily distinguishable from an acute ACL disruption. Ganglia associated with the PCL typically demonstrate a well-defined multilocular appearance along the surface of the ligament. They may extend from a ligamentous attachment distally posterior to the joint (Fig. 21C,D).

In common with other ganglion cysts, the lesions typically demonstrate signal that is nearly isointense to muscle on T1-weighted images (believed to be related to the high protein concentration within the cyst fluid that results in relative T1 shortening) (7). On T2-weighted images, the cysts are usually homogeneously hyperintense. On occasion the cysts may demonstrate intraosseous extension (1). Symptoms associated with reported cases include pain and limitation of motion (1).

The pathogenesis of ganglion cysts remains uncertain. Multiple theories have been suggested and include herniation of synovium into surrounding tissue, displacement of synovium during early development, and posttraumatic degeneration of connective tissue, originating from proliferating pluripotential mesenchymal cells (1,6). The cysts are filled with clear mucinous fluid and the presence of a synovial cell lining is variable. While most commonly seen near or attached to tendon sheaths and joint capsules, they may occur within tendons, bone, as well as intramuscularly (see related cases 3-3 and 4-7).

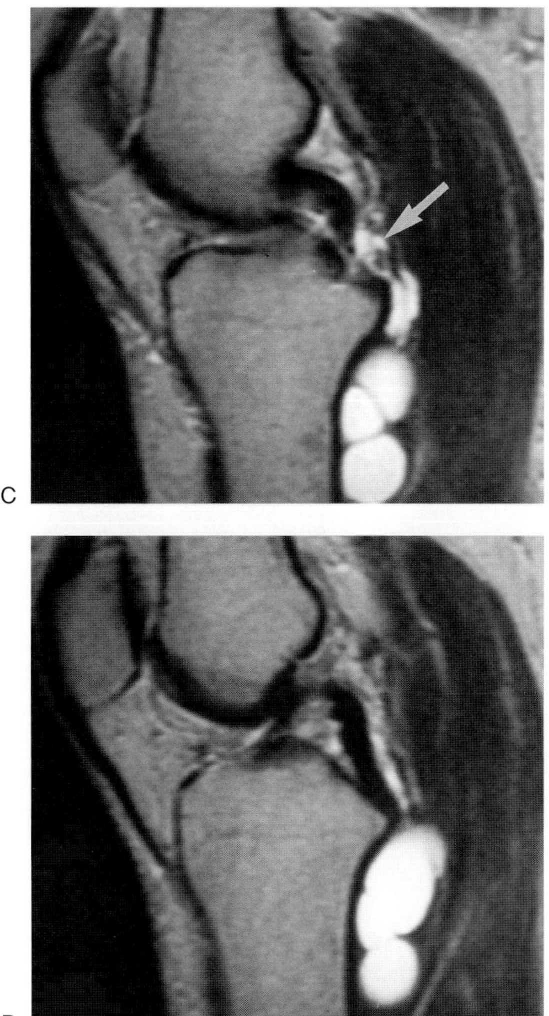

FIG. 21. C,D: Ganglion cyst related to the PCL. Contiguous sagittal T2-weighted (TR/TE 2,000/80 msec) demonstrated a well-defined high signal intensity mass tracking posterior to the distal tibia. The attachment of the mass to the PCL can be demonstrated *(arrow).* Detection of the attachment site of ganglion cysts is important as failure to resect this component of the mass is a reported cause of cyst recurrence.

CASE 22

History: A 36-year-old man is being evaluated 2 years following ACL reconstruction. What is your assessment of the status of the reconstruction?

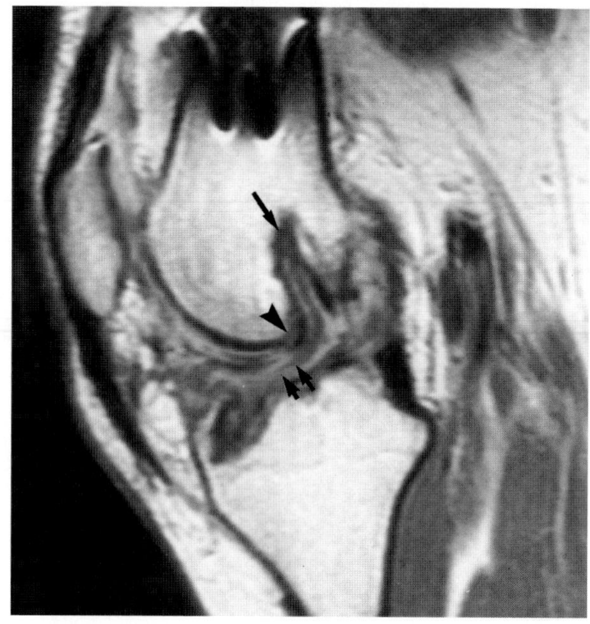

FIG. 22.

Findings: In Fig. 22A a sagittal proton density weighted image demonstrates a grossly intact graft. The graft, however, appears to acutely curve to pass under a spur along the anterior inferior aspect of the intercondylar roof (arrowhead). The distal end of the intraarticular portion of the graft (double arrows) demonstrates increased signal intensity. The femoral tunnel opens low and relatively anterior on the femoral condyle in a nonisometric position. (From Deutsch AL, Mink JH. The Postoperative Knee. In: Mink JH, Reicher MA, Crues JV III, Deutsch AL, eds. *MRI of the knee, 2nd edition.* New York: Raven Press 1993;270–292, with permission.)

Diagnosis: Grossly intact ACL reconstruction demonstrating increased intragraft signal and evidence of roof impingement.

Discussion: Impingement refers to the interference with the smooth, frictionless motion of the graft (1–4). While occasionally impingement is secondary to fibrous or fibrocartilaginous scar tissue, most cases are due to contact of the graft with a rough or protuberant bony surface. Impingement may not be associated with restricted joint motion, pain, or instability, but abnormal contact of the graft may lead to excessive wear on the graft, and ultimately to its failure. Impingement most commonly occurs within and at the exit points of the grafts from the bony tunnels, the side wall, and the roof of the intercondylar notch (1,3).

Clinically unrecognized roof impingement has been identified as a cause for a regionalized increase in graft signal (1,2). Howell (1,2,4) has described the MR changes in the graft that result from occult roof impingement. In these cases, the tibial tunnel is placed too far anteriorly, and the anteroinferior edge of the roof impacts the superior surface of the graft during full extension. Two groups of patients were followed by means of sequential MR studies; one group had and the other group did not have imaging evidence of roof impingement. Those without roof impingement had grafts that demonstrated low signal intensity throughout the period of assessment (up to 1 year). In contrast, those with radiographic evidence of impingement developed regionalized signal increases in the distal two-thirds of the intraarticular segment of the graft. This localized increase in signal is thought to be secondary to entrapment of the graft in extension between the tibial plateau below and the inferior lip of the intercondylar roof above (3,4).

Sequential studies on patients with the presumptive diagnosis of roof impingement, made by the finding of increased signal (T1-weighted images) in the distal graft, suggest that these changes become well established during the first 3 months following surgery. It is during this time period that the patient attempts full extension, the position that would most likely result in graft impingement (1). The regionalized signal increase tends to remain constant over at least 3 years (4). While frequently the signal increases are mild, they are occasionally so severe as to suggest that the graft is severely attenuated or in fact discontinuous. Resection of the offending anteroinferior lip of the roof or replacement of the tibal tunnel in proper position has been shown to relieve the impingement, and the graft in such cases returns to a normal, low signal state within 12 weeks after delayed roofplasty (1,2). Therefore, increased signal within ACL grafts may not be a normal part of the ligamentization process, but rather a result of clinically occult roof impingement (3). Since the clinical diagnosis of impingement may be difficult, MR may be the first indicator that a problem exists. The potential long-term consequences (e.g., rerupture, progressive instability) of untreated roof impingement are not known at this time. Preliminary data in animals suggest that an increase in signal intensity of the graft corresponds to an increase in the water concentration and a reduction in strength.

Contrast-enhanced MR with gadopentetate dimeglumine facilitates study of the physiologic events occurring around the graft (1,2,4–7). As previously discussed (see case 6–16), the graft becomes cloaked by a vascular envelope 10 to 20 weeks after implantation (1,2,4,6,7). It is this period of revascularization by means of vessels from synovial folds, endosteal vessels, and remnants of the native torn ACL that ultimately reestablishes the strength of the graft. Enhancement following intravenous administration of gadopentetate dimeglumine occurs most commonly within the synovial membrane around the graft (Fig. 22B,C). During the first 4 months postimplantation only minor degrees of signal enhancement characteristically occur in the synovial envelope. By 7 months, however, approximately one-half of the grafts demonstrate significant synovial enhancement (7). Rarely does the synovial membrane demonstrate moderate signal enhancement after 7 months. In those patients who underwent gadolinium (Gd)-enhanced studies both early and late in the first postoperative year, the degree of membrane enhancement decreased in 45%, remained unchanged (minimal) in 45%, and increased in 10%. This temporal pattern of perigraft enhancement correlates extremely well with the known physiologic events occurring in the knee, e.g., synovial proliferation, vascular ingrowth, reestablishment of vascularization, and eventual subsidence of the inflammatory reaction of the host. Graphic demonstration of these physiologic events by MR might possibly be valuable in designing and timing of rehabilitation programs.

While Gd enhancement of the synovial envelope is physiologic, focal intraligamentous enhancement is abnormal and must suggest the diagnosis of unrecognized roof impingement (Fig. 22D,E). The abnormal signal decreases somewhat over 2 to 3 years. The specificity or sensitivity of intraligamentous signal increase is not known, but the authors have not identified such enhancement in otherwise normal grafts.

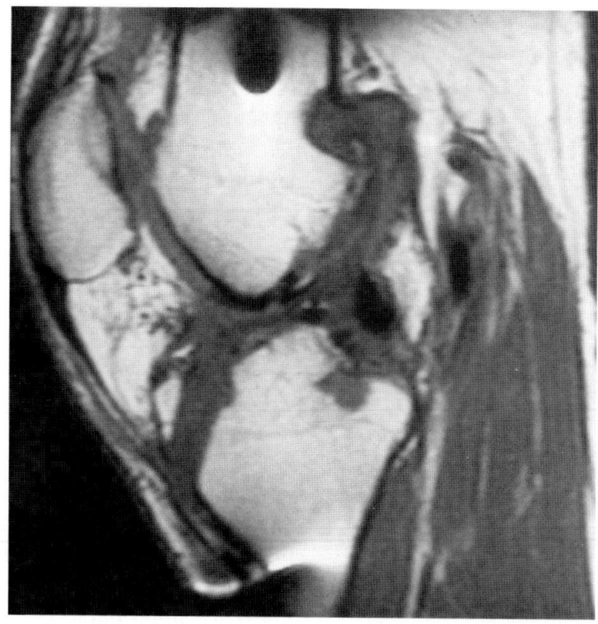

B

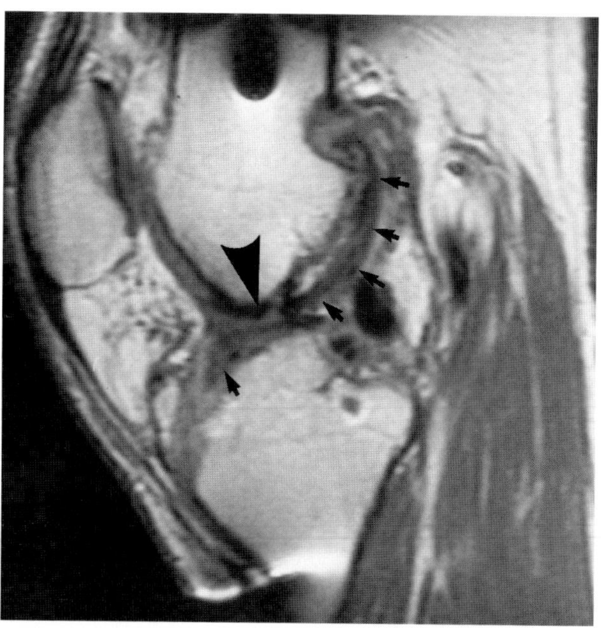

C

FIG. 22. B,C: Enhancement of the synovial membrane with gadopentetate dimeglunine. **B:** Sagittal T1-weighted MR demonstrates a tubular soft tissue mass in the expected position of the ACL graft. The dark graft itself is difficult to identify. **C:** Following the intravenous administration of gadopentetate dimeglumine (Gd) there is enhancement of the envelope and the graft *(small arrows)* is seen to be intact. There is, however, evidence of roof impingement *(large arrowhead)* and the graft appears questionably focally discontinuous at this point.

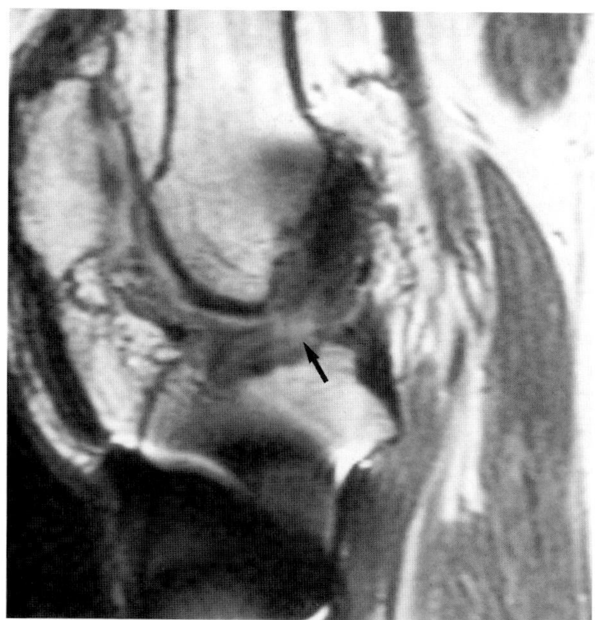

D

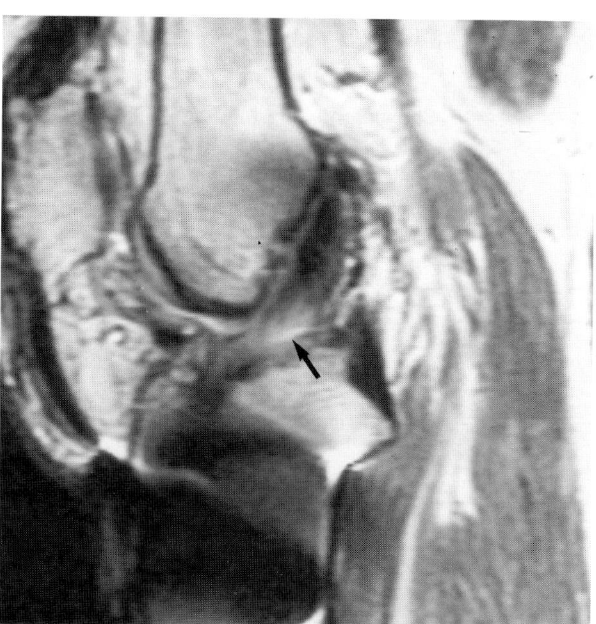

E

FIG. 22. D,E: Graft impingement assessed with gadopentetate dimeglumine. **D:** Sagittal T1-weighted image demonstrates a focal increase in signal *(arrow)* with the distal graft. **E:** This increase in signal is more evident following the intravenous administration of gadopentetate dimeglumine. The entire graft is more readily identified since the synovial membrane has enhanced. The tibial tunnel and distal graft lie anterior to the projected slope of the intercondylar roof. (From Deutsch AL, Mink JH. The Post-operative knee. In: Mink JH, Reicher MA, Crues JV III, Deutsch AL, eds. *MRI of the knee, 2nd edition.* New York: Raven Press, 1993;270–292, with permission.)

CASE 23

History: A 46-year-old man complained of pain aggravated by motion and knee swelling that had increased over the past several months.

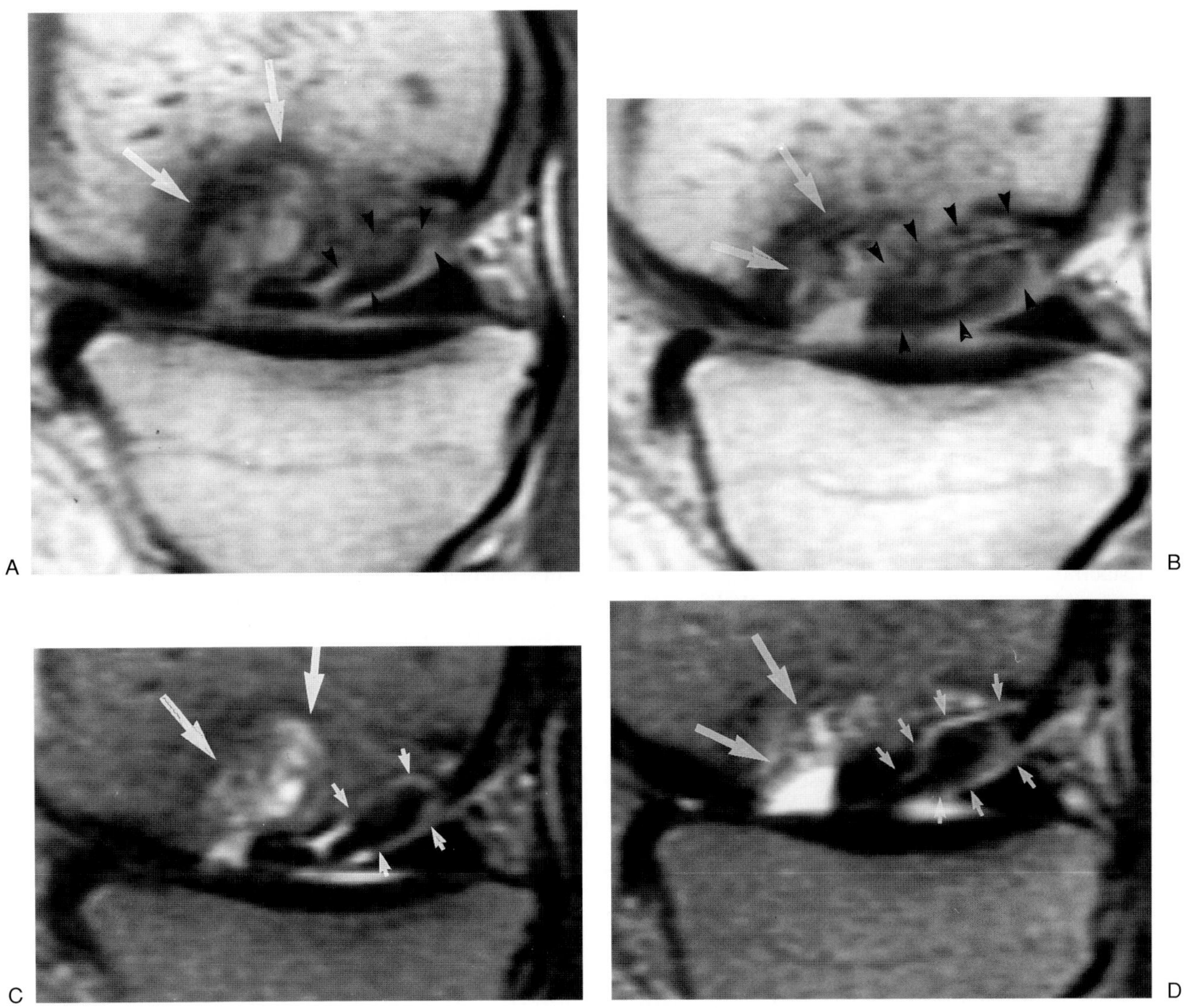

FIG. 23.

Findings: In Fig. 23A,B contiguous sagittal PD (TR/TE 2,000/20 msec) images through the central third of the medial femoral condyle demonstrate a large elliptical-shaped relatively low signal intensity fragment (arrowheads). Inhomogeneous intermediate signal intensity is seen within the contiguous cancellous bone of the femur (long arrows). In Fig. 23C,D corresponding T2-weighted images demonstrate the fragment to be encircled by high signal intensity synovial fluid (short arrows). The signal within the underlying bed of femur becomes moderately increased in intensity (long arrows).

Diagnosis: Osteochondritis dissecans with a large unstable fragment.

Discussion: The term *osteochondritis dissecans* (OCD) refers to the fragmentation and possible separation of a portion of the articular surface of the affected joint (1–4). The most commonly affected site is the condylar surface of the distal femur. Men are affected more commonly than women and the average age of onset of symptoms is 15 to 20 years, although the range is highly variable (1,2). The condition is generally believed to be traumatic in etiology and to result from an osteochondral impaction fracture, although a familial history has been noted in some cases, particularly those involving the knee (1). The relative insensitivity of cancellous bone to pain is believed to account for the absence of acute symptoms and signs and contributes to difficulty in determining the time of initial injury with any degree of precision. A history of significant trauma, however, can be elicited in up to 50% of cases. Within the knee, the medial femoral condyle is affected in approximately 85% of cases with the classic location (in approximately 70% of cases) along the inner (lateral) aspect of the medial condyle (3).

At present there is no uniformly accepted classification system for osteochondritis lesions (see case 6-10). Arthroscopic inspection has confirmed the existence of a wide spectrum of abnormalities ranging from ballotable fragments with intact overlying articular cartilage, to cartilage disruption with displaced fragments (1,2,4). Considerable orthopedic interest has focused on the determination of the "stability" of the fragment. Those fragments with intact cartilage may be amenable to surgical fixation while loose fragment with cartilage violation are frequently removed (1,2,4). Multiple imaging techniques, including conventional radiographs, arthrography, CT, scintigraphy, and most recently MR have been employed in attempting to make the determination of fragment stability.

Mesgarazdeh and colleagues (5) compared plain radiographs, radionuclide bone scans, and MR in the assessment of osteochondritic lesions of the femoral condyles utilizing arthroscopy as the reference standard. On conventional radiographs, larger lesions and those associated with sclerotic margins (greater than 3 mm) were generally loose. Significant activity on all phases of scintigraphic evaluation (e.g., flow, blood pool, delayed static) was also associated with loose fragments. On MR, loose fragments were identified when directly displaced from their underlying beds and by the presence of fluid between the fragment and underlying bone. DeSmet and associates (6,7) further focused attention on the MR signal intensity characteristics of the zone between the lesion and parent bone. High signal intensity, reflecting either granulation tissue or fluid, was found to represent a highly associated (but not invariable) finding of loose fragments. Fluid encircling a fragment and or focal cystic areas beneath the fragment were the most highly correlative findings for loose fragments. In the work of DeSmet, the absence of high signal between the fragment and underlying bone was a reliable sign of lesion stability.

In a further attempt to assess lesion stability and the integrity of the overlying articular cartilage, the use of gadopentetate dimeglumine, administered intravenously as well as intraarticularly, has been evaluated (8–10). With intravenous administration, enhancement of the interface between the fragment and underlying bone has been associated with loose fragments and the absence of such enhancement with stable lesions. Additionally, the delayed diffusion of gadopentetate dimeglumine into synovial fluid can provide for a noninvasive arthrogram effect. The administration of a dilute solution of gadopentetate dimeglumine intraarticularly has also been reported with improved accuracy in the assessment of the overlying articular cartilage (see case 7-18). MR arthrography, in particular, can help in the differentiation of fluid and granulation at the interface between fragment and parent bone, a determination that may be difficult to make with conventional MR alone. The determination that the signal changes observed are reflective of fluid tracking through a cartilaginous defect into the base of a fragment provides definitive determination of an unstable fragment. Fat-suppressed techniques may further enhance this determination and should be employed in determination of cartilage integrity (11).

CASE 24

History: A 45-year-old man complained of pain and limitation of extension 1 week following repair of a bucket-handle tear of the posterior horn of the medial meniscus. He was referred to assess the integrity of the repair. What is your determination?

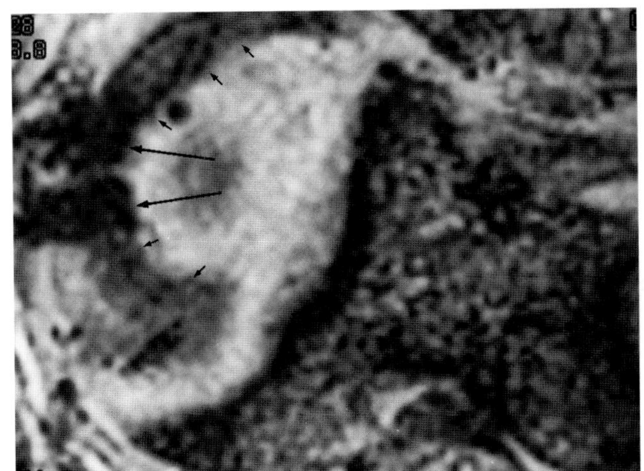

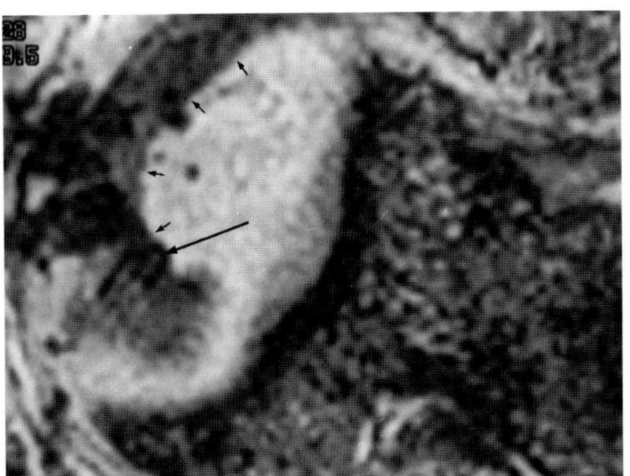

A B

FIG. 24.

Findings: In Fig. 24A,B two consecutive axial 0.7-mm sections from a gradient volume acquisition (TR/TE/theta 31/13/15). In Fig. 24A the meniscus (small arrows) is seen as a semilunar structure. Two bands of low signal seen traversing the meniscus (long arrows) reflect suture material. In Fig. 24B contiguous section demonstrates a third suture traversing the junction of the midzone and posterior horn. The meniscus (small arrows) is grossly intact.

Diagnosis: Intact meniscus following recent meniscal repair.

Discussion: This case illustrates the versatility of MR and one of the ways in which the technique may be of value in the noninvasive assessment of patients having undergone meniscal preservation surgery. The concept of meniscal preservation and repair of meniscal tears as an alternative to partial menisectomy has emerged as an area of great interest in recent years and has been given impetus from an enhanced appreciation of the potentially deleterious effects of total and even partial menisectomy (1–4). Tears that occur through the outer vascularized one-third of the meniscus represent the best candidates for repair, although techniques have been developed to extend the possibility of meniscal repair to those located more centrally (4). In general, full-thickness longitudinal tears contained within the periphery of the meniscus (''red-red'' and ''red-white'') are the best candidates for repair (2,3). The tears should be longer than 5 to 10 mm in length and the central component should be of good quality. Short vertical tears, stable partial-thickness tears involving less than 50% of meniscal height, and shallow radial tears of less than 3 mm can be freshened with a shaver or rasper to facilitate healing and can be managed without sutures.

Meniscal repair can be accomplished by either open surgical methods or by means of arthroscopic techniques, and advocates of both approaches exist (1–4). Advocates of open repair suggest that meniscal rim and capsular preparation are better and placement of vertically oriented sutures to reattach the capsular bed to the entire height of the mensical rim is easier (2). There are two principal arthroscopic techniques for meniscal repair: the ''inside-out'' and the ''outside-in'' method. With the inside out technique, long flexible needles are passed through a curved cannula and placed across the meniscal tear, allowing placement of a horizontal mattress suture that is tied subcutaneously. With this technique, special attention must be directed toward the neurovascular bundle, which may be damaged by the blind passage of the needles. Concerns about possible neurovascular injury complications have led to development of limited open approaches in which an open incision permits visualization of the vital structures during needle passage and suture ligation, while still allowing arthroscopic guidance of suture placement with the meniscus to be accomplished. Analysis of multiple reports on the success of meniscal repair suggest that approximately 75% will heal completely, 15% will heal partially, and 5% to 10% will fail to heal (3). Factors contributing to the success of the procedure include (a) acute tears, (b) peripheral tears (within 5 mm of the meniscocapsular junction), (c) stable knees (intact ACL), (d) patient age (younger is better), and (e) use of synovial abrasion or exogenous blood clot (3).

As in the present case, MR can be of great value in confirming the morphologic integrity of previously repaired menisci. Depending on the suture material utilized and MR technique, direct visualization of the sutures can be accomplished. Displaced flaps and fragments can be readily recognized (Fig. 24C,D). More problematic, however, can be assessment of signal changes within previously repaired menisci as well as those that have been conservatively managed without suture repair. Several studies have now attested to the fact that persistent signal can be seen within successfully repaired menisci and that this signal can potentially mimic re-tears (5–8). Indeed, the mere presence of grade 3 signal within a previously repaired meniscus (on T1 or intermediate weighted sequences) cannot be reliably utilized as an MR criteria for nonhealing or re-tear (5–8). It has been additionally suggested that the persistence of grade 3 signal in previously torn but presently healed menisci may represent an explanation for some apparent false-positive MR examinations (5,8). As in the knee status prior partial menisectomy, the presence of increased signal on T2-weighted images corresponding to the grade 3 signal has a significantly greater positive predictive value for re-tear (6). Longer-term follow-up of menisci with persistent signal following repair or conservative management has demonstrated a progressive modulation and diminution in the degree of intrameniscal signal (8,9) (Fig. 24E,F). Experimental studies have corroborated the persistence in signal intensity at the site of meniscal repairs despite the modulation of fibrovascular scar tissue into fibrocartilage (Fig. 24G–K). MR arthrographic techniques may be of additional value in selected cases but have not to date been the subject of a critical review with regard to previously repaired menisci.

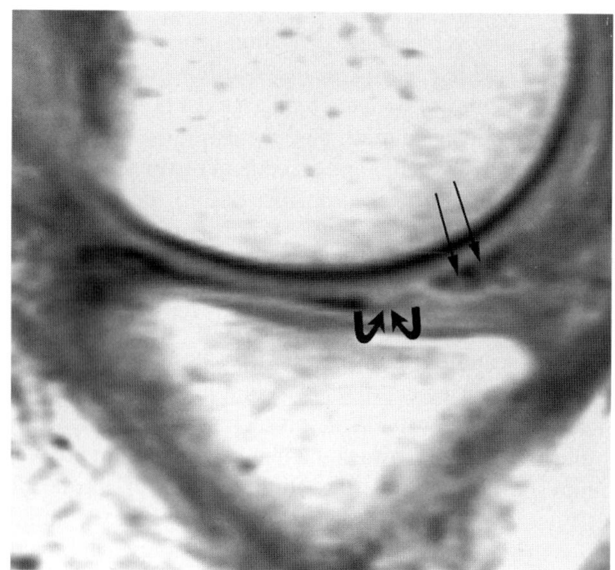

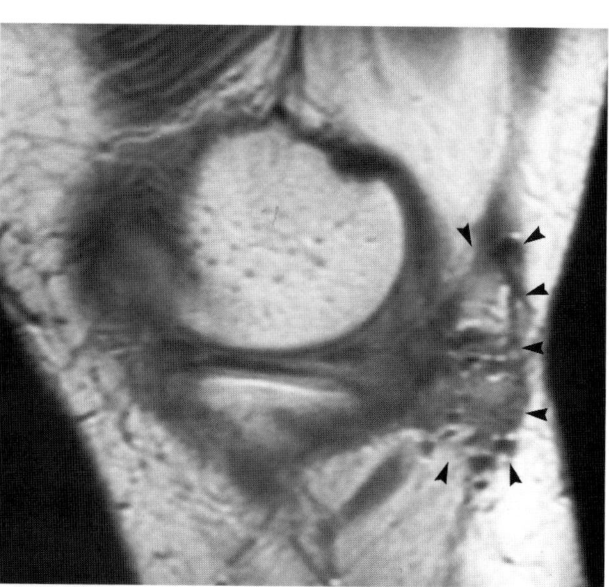

FIG. 24. C,D: Failed meniscal repair. **C:** Sagittal proton density–weighted image demonstrates extensive micrometallic artifact seen in and around the posteromedial corner of the knee *(arrowheads)*. The extent of the findings is quite unusual in the author's experience. **D:** Sagittal proton density weighted image demonstrates wide displacement of the patient's repaired meniscus *(curved arrows)*. Sutures are seen coursing through the peripheral fragment *(long arrows)*.

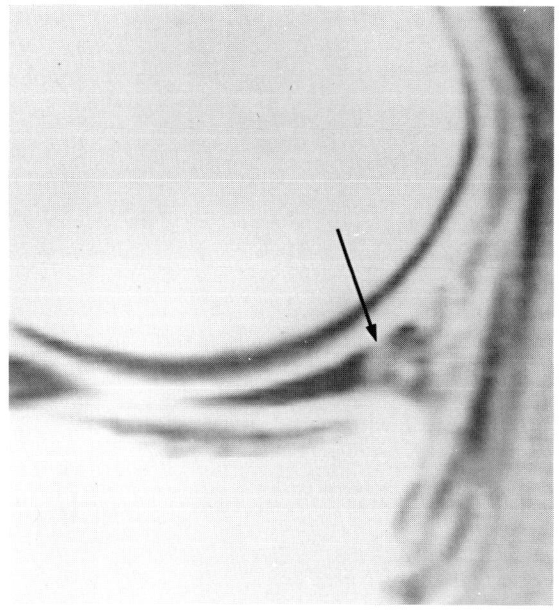

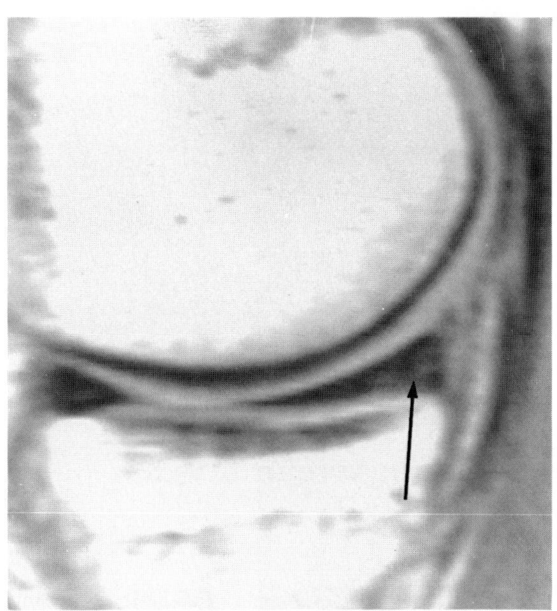

FIG. 24. E,F: Modulation of signal following conservative management of a peripheral meniscal tear. **E:** Sagittal proton density–weighted image demonstrates the typical appearance of a vertical tear through the "red-white" junction of the medial meniscus *(arrow).* The tear was nondisplaced. **F:** Sagittal proton density–weighted image performed as a follow-up 3 years later demonstrates marked diminution in the previously defined area of signal alteration. The patient was asymptomatic and considered clinically stable at the time of the examination. (From Deutsch AL, Mink JH. The Post-operative knee. In: Mink JH, Reicher M, Crues JV, Deutsch AL, eds. *MRI of the knee, 2nd ed.* New York: Raven Press, 1993;240–251, with permission.

G
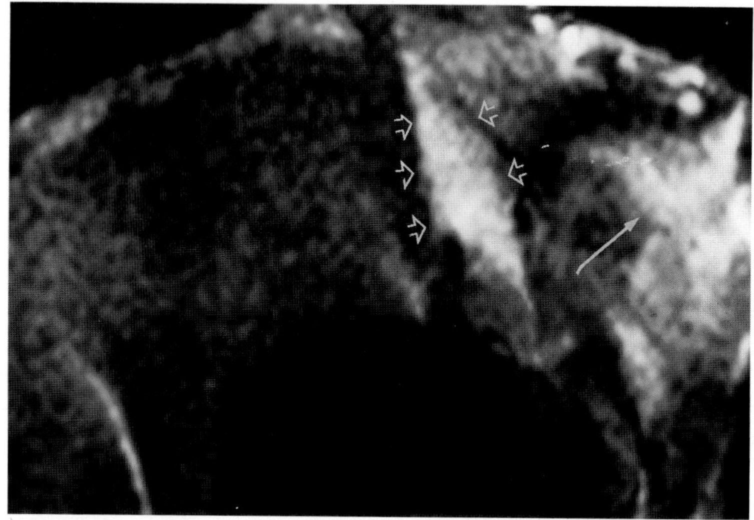

H
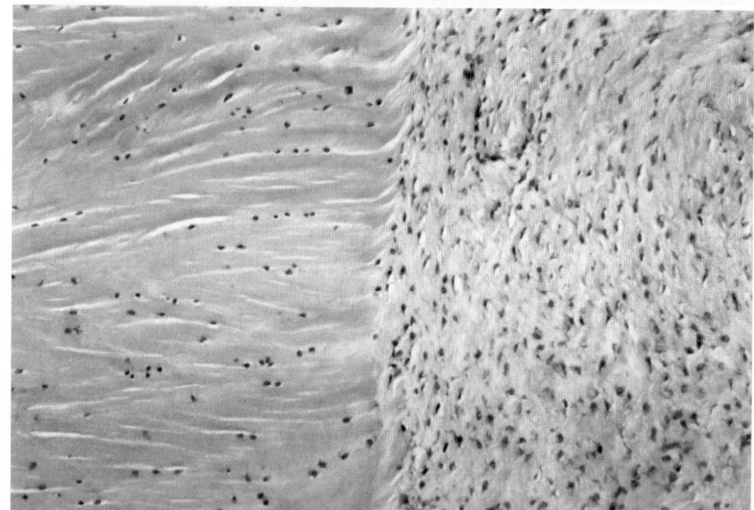

I
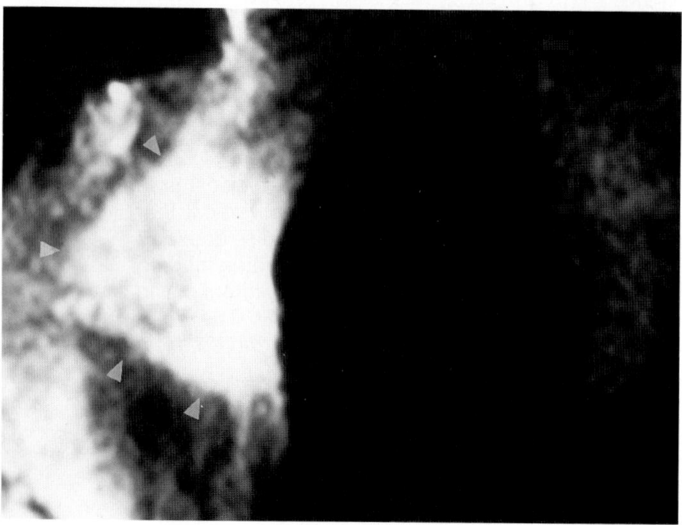

FIG. 24. G–K: Investigational study of meniscal healing in dogs. **G:** MR examination of a meniscal specimen harvested 8 weeks following repair of a complete radial transection. Increased signal at the repair site is readily demonstrable *(arrowheads).* The perimeniscal (PM) tissue attached to the meniscus also demonstrates increased signal. **H:** Photomicrograph of the meniscus in G demonstrates the junction between the normal meniscal tissue and the highly cellular, fibrovascular scar tissue (hematoxylin and eosin, original magnification ×100). **I:** MR examination of meniscus harvested 26 weeks following meniscal repair. Increased signal at the site of the repair *(arrowheads)* appears more diffuse and remains of increased signal even though the repair tissue has matured into fibrocartilage.

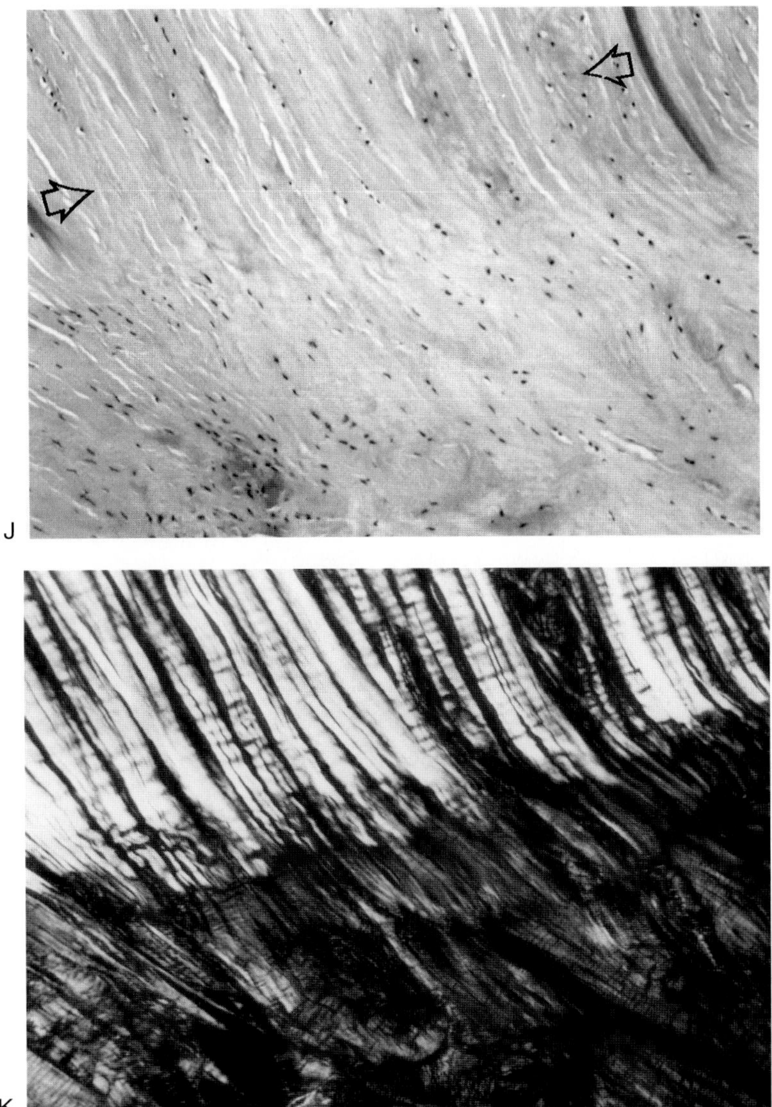

FIG. 24. J: Photomicrograph of the meniscus shown in I demonstrates the histologic appearance of the junction between the repair tissue (R) and the normal meniscus (N) (H&E ×100). **K:** Polarized photomicrograph of the same specimen graphically demonstrates that while the repair tissue can be histologically classified as fibrocartilage, it is still at this early stage obviously different from normal meniscal tissue. (From Arnoczky SP, Cooper TG, Stadelmier DM, et al. Magnetic resonance signals in healing menisci: an experimental study in dogs. *Arthroscopy* 1994;10(5):522–557, with permission.)

CASE 25

History: A 32-year-old man had undergone repair of a patellar tendon rupture approximately 5 months previously. The patient complained of continued discomfort and weakness. What is your diagnosis?

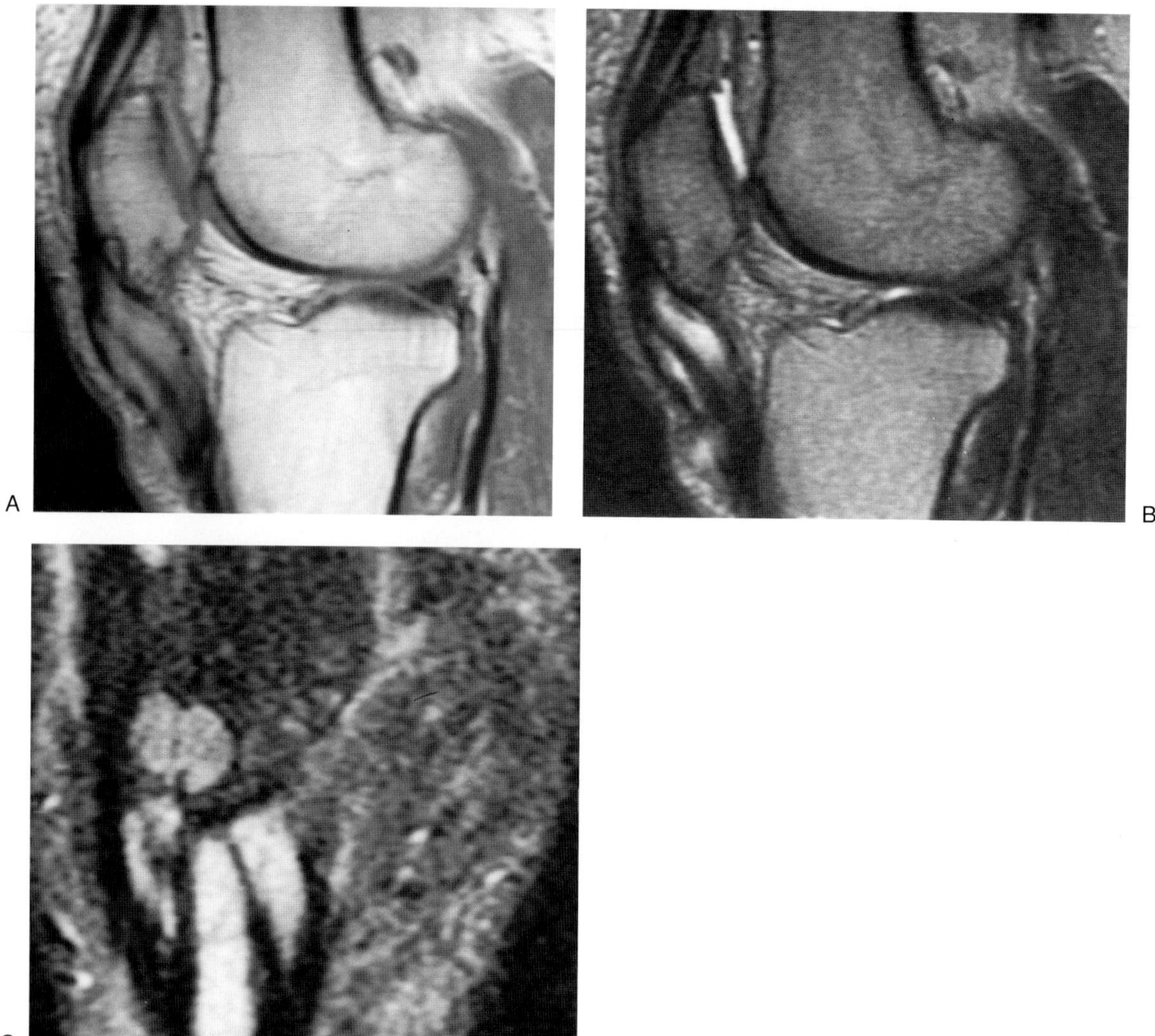

FIG. 25.

Findings: In Fig. 25A a sagittal PD (TR/TE 2,000/20 msec) image demonstrates increased girth and prominent band-like areas of increased signal intensity within the patellar tendon. In Fig. 25B a sagittal T2-weighted (TR/TE 2,000/80 msec) image demonstrates the band-like areas to become hyperintense. In Fig. 25C a coronal fast inversion recovery (TR/TE/TI 2,800/35/130) image also depicts striking band-like areas of increased signal intensity occupying much of the patellar tendon.

Diagnosis: Failed patellar tendon repair.

Discussion: Rupture of the patellar tendon is the least common cause of extensor mechanism disruption (1). In contrast to patients with quadriceps rupture, patients with patellar tendon disruption are typically younger, frequently male, and often involved in sports or active recreational activities (2–4). The most common site of rupture is at the junction of the tendon with the lower pole of the patella. Midsubstance patellar tendon tears are less common and may be seen as a result of a single episode of severe trauma likely related to forced knee flexion against a contracted quadriceps mechanism (3). Tears at the level of the lower pole of the patellar predominate in younger individuals.

The patellar tendon may be predisposed to rupture as

a consequence of its exposure to repetitive microtrauma that results in traumatic disruption of individual tendon fibers (4). Healing in this situation occurs by means of organization of edematous fluid and hemorrhage into mucoid material. Insufficient healing related to microtrauma may lead to progressive loss in tendon strength and eventual rupture. This possible etiology may best account for tears seen in patients with systemic disorders (e.g., systemic lupus erythematosus) or patients on corticosteroids, as well as athletes with repetitive submaximal trauma (3). A second theory to account for tendon ruptures proposes that excessive loads can be sufficient to rupture otherwise normal tendons. This theory would best account for ruptures seen following motor vehicle accidents or falls with large hyperflexion forces against the contracting quadriceps (3). In studying a weight lifter whose infrapatellar tendon ruptured while undergoing kinetic analysis, Zernicke and Jobe (5) found that the tendon ruptured after stress greater than 17.5 times body weight. In studies of the strength of the patellar tendon as an anterior cruciate ligament substitute, Noyes and colleagues (6) reported the strength of the central third of the patellar tendon as 175% of an intact ACL. By extrapolation, the strength of the intact infrapatellar tendon would be 525% of the average ACL strength. Given the strength of the patellar tendon, it is understandable why transverse patellar fractures are seen more commonly as a result of hyperflexion injuries than patellar tendon disruptions.

Patellar tendinitis is believed to be the result of microtears of the tendinous fibers and is believed to predispose to tendon rupture (4). This condition is common in athletes such as volleyball and basketball players who participate in sports that expose the patellar tendon to repetitive violent contraction of the quadriceps musculature. This action results in shearing of fibers within the tendon, the tissue response to which results in ingrowth of bone, synovium, or cartilage within the tear (4). The findings most commonly involve the proximal patellar tendon. This condition has been referred to as ''jumpers knee'' and has the potential to progress to substantial mucoid degeneration of the tendon (Fig. 25D,E).

Abnormalities of the patellar tendon are characterized by increased girth and increased signal within the tendon, both of which need to be distinguished from the normal appearance of the tendon. The normal patellar tendon is characterized by relative uniform thickness with mild proximal and distal expansions close to the sites of attachment to bone (7,8). The upper limit of normal thickness for the tendon has been stated as 7 mm (7), although recent studies have suggested a wide variation in width of the tendon in asymptomatic subjects (8). The leading edge of the tendon is usually straight. Mild undulation of the tendon may be associated with either injury to the extensor mechanism or other abnormalities in the knee joint, and a wavy appearance to the patellar tendon should prompt the search for other abnormalities (9). As with other tendons, the patellar tendon is subject to angle-dependent signal (''magic angle'') that may contribute to the appearance of increased intratendinous signal particularly on short TE sequences (e.g., T1-weighted and gradient echo sequences). Increased signal on gradient echo images is commonly encountered in the proximal and distal ends of the patellar tendons of asymptomatic individuals (8).

In the present case, the patient presented with persistent symptoms 5 months following repair of a prior patellar tendon tear. Striking bands of increased signal intensity were evident extending through much of the course of the tendon. Information regarding tendon healing can be gained from the study of portions of patellar tendons that remain following harvest of the central third of the tendon for purposes of ACL reconstruction procedures (10,11). Immediately following the procedure, the remaining tendon appears quite abnormal with 87% of tendons studied revealing a diffuse increase in signal for 6 months. By 12 months, asymptomatic patellar tendons return to their normally expected low signal intensity appearance, or demonstrate only a linear increase in signal at the harvest site. Thickening of the remaining patellar tendon is seen immediately following harvesting and persists even in asymptomatic patients. At 12 months, symptomatic tendons always measure greater than 10 mm in anteroposterior dimension. Persistent signal at the donor site and thickening of the tendon 3 months following the procedure correlate with residual clinical symptoms (10). The extent of the signal alteration demonstrated in the present case was considered beyond that to be reasonably expected at this stage postrepair, and the failure of the repair was confirmed at surgery.

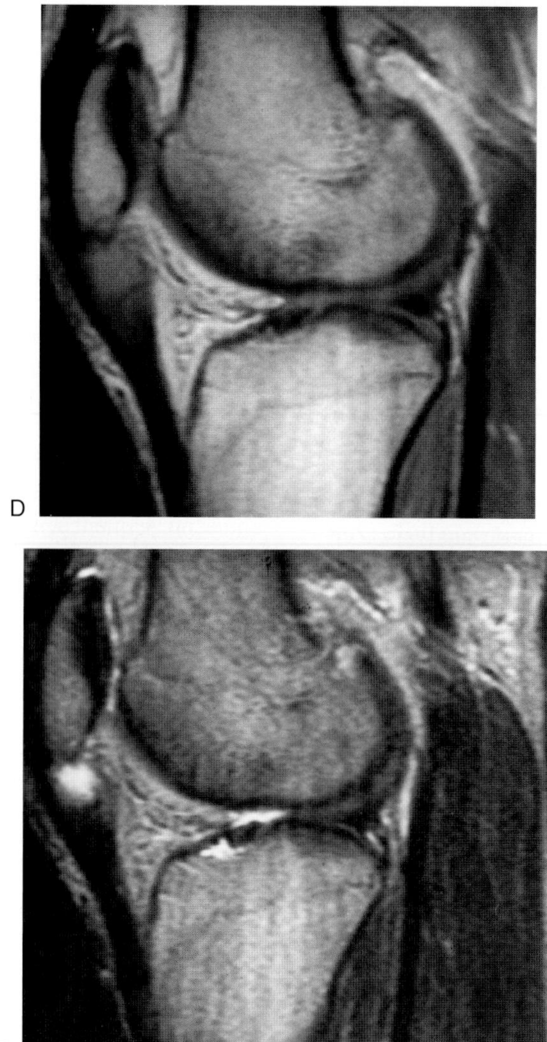

FIG. 25. D,E: Jumper's knee. A 28-year-old professional basketball player. **D:** Sagittal PD (TR/TE 2,000/20 msec) image demonstrates increased girth of the proximal patellar tendon with moderately increased signal intensity. **E:** Sagittal T2-weighted image (TR/TE 2,000/80 msec) demonstrates the previously defined signal to become markedly hyperintense. At surgery the findings reflected a combination of marked myxoid degeneration and partial disruption of the proximal tendon.

CASE 26

History: A 24-year-old woman with anterior knee pain that limits her activities. Conventional tangential x-ray of the patella was normal (Fig. 26A). A dynamic patellofemoral motion study was requested. What is your diagnosis?

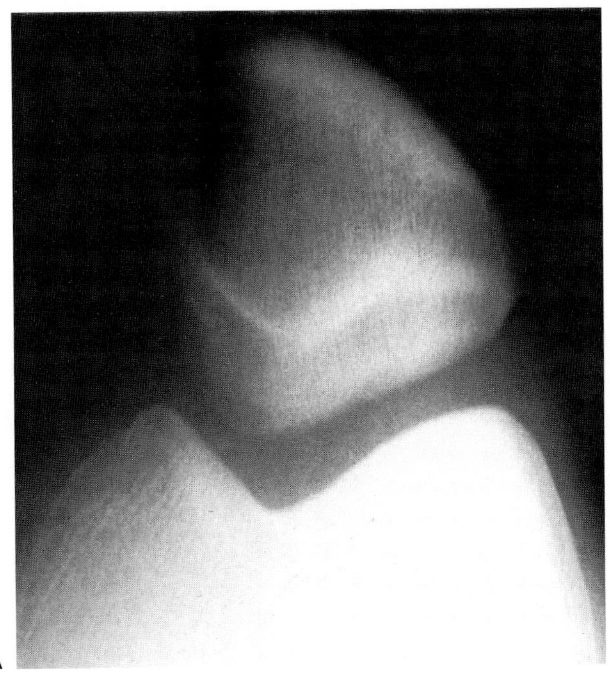

FIG. 26.

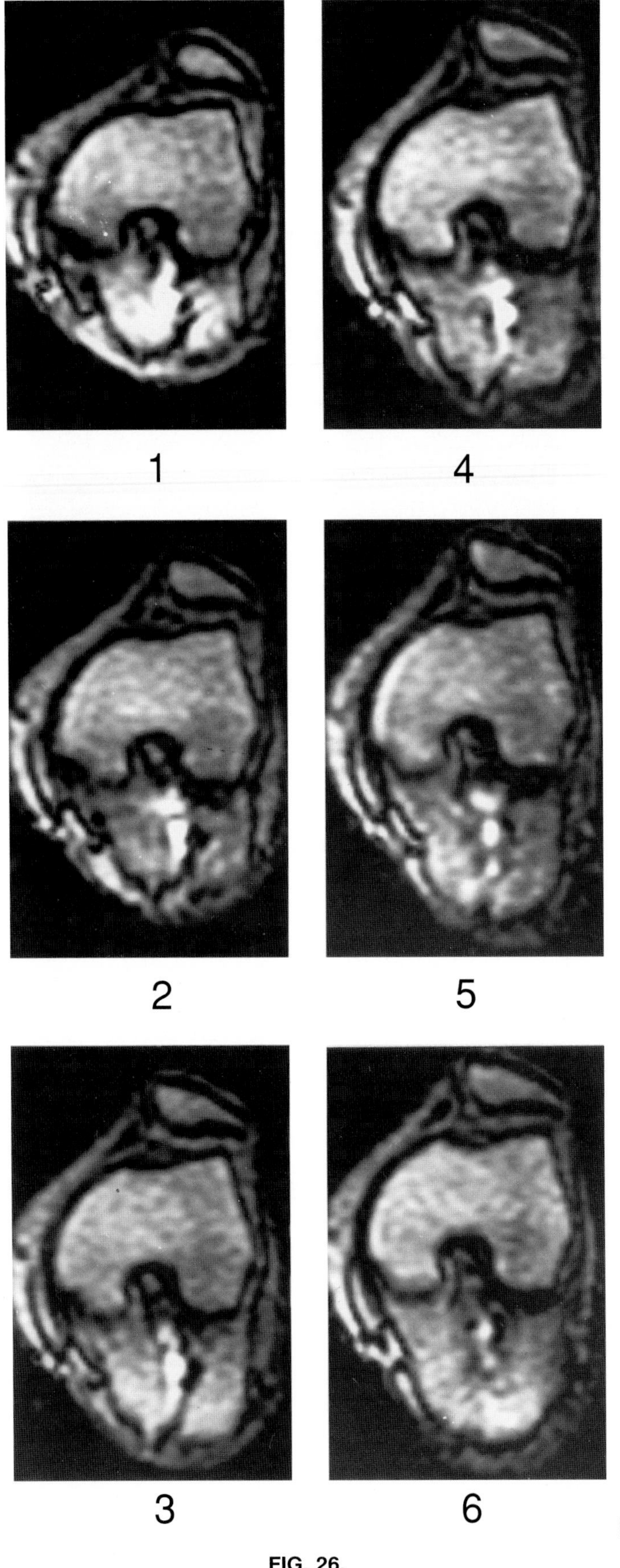

FIG. 26.

Findings: In Fig. 26A a conventional tangential x-ray of the patella was obtained at approximately 45° of flexion. There is no evidence of patella subluxation. In Fig. 26B axial plane images of the patellofemoral joint were obtained utilizing a spoiled GRASS technique while the patient actively moved her knee from extension (1) to approximately 30° of flexion (2). On the initial images there is evidence of significant lateral subluxation of the patella, a finding that could not be appreciated on the conventional radiographic examination. With increasing degrees of knee flexion, the patella becomes increasingly centered, although it remains slightly lateralized. (From Deutsch AL, Shellock FG. Extensor mechanism and patellofemoral joint. In: Mink JH, Reicher MA, Crues JVIII, Deutsch AL eds. *MRI of the knee, 2nd edition.* New York: Raven Press, 1993, with permission.)

Diagnosis: Lateral patellar subluxation demonstrated utilizing kinematic MR examination.

Discussion: The term *kinematics* refers to the science of motion that is applied to the study of movement of a body without reference to force or mass (1,3). Kinematic MR imaging was first utilized for the evaluation of the patellofemoral joint in 1988 and multiple studies have attested to the value of this method (3–11). Over the years, with the introduction of newer scanning techniques, the evaluation has undergone several refinements.

It is to be emphasized that abnormal patellar alignment and tracking occur during the earliest portion of the range of motion of the patellofemoral joint as the patella engages the femoral trochlear groove (1,3,4,5). As flexion of the joint increases, the patella moves deeper into the femoral trochlear groove and displacement is less likely to occur because the groove functions to buttress and stabilize the patella. As a consequence, diagnostic techniques, principally conventional tangential radiography, that cannot demonstrate the initial degrees of flexion, are limited in their ability to demonstrate early and subtle patellofemoral joint derangement. The major advantage of the application of cross-sectional imaging techniques (first CT and later MR) to the evaluation of the patellofemoral joint has been their ability to depict the early degrees of knee flexion.

The initial work with kinematic MR employed a specially designed positioning device that facilitated obtaining sections at each of three or four locations through the femoral groove chosen to depict the path of the patella during flexion (1,3,6,7). While static images were obtained, it was considered more practical to display the data in a simulated real-time format for purposes of qualitative analysis. The development of faster MR imaging techniques (i.e., ultrafast spoiled GRASS, turbo FLASH, echo planar imaging) has allowed for studies of patellofemoral motion to be performed during active movement of the knee (8–10). Utilizing these newer sequences, MR images are typically obtained at the rate of one image per second or faster. Active movement kinematic MR allows study of the influence of activated muscle and other related soft tissue supporting structures on patellofemoral motion. In addition, in contrast to the previously employed methods, no sophisticated positioning device is required to flex or position the joint incrementally and overall examination time is considerably reduced.

The kinematic MR evaluation of the patellofemoral joint may be further refined to include the study of active movement against an externally applied load (8–10). Patients with symptoms attributable to the patellofemoral joint typically experience symptoms during activity. Studying the patient against a physiologic load enhances demonstration of abnormal patellofemoral motion and allows for evaluation of the effect of quadriceps contraction. The authors have performed active movement-loaded kinematic MR utilizing a positioning device that incorporates a mechanism that allows for an adjustable resistance to be applied to the patellofemoral joint in the sagittal plane (8,9). The patient is examined prone and the study is directed in particular to evaluation of the influence of the extensor mechanism on joint motion. The device is designed to facilitate bilateral simultaneous evaluation as the knees move through a range of motion from approximately 45° of flexion to extension. A force of 30 foot-lbs/sec is applied during the active movement. The application of resistance to stress the patellofemoral joint during kinematic MR has been shown to elicit patellar malalignment and tracking abnormalities that may not have been observed during unloaded examinations. Preliminary results by Brown et al. (10) utilizing a different loaded active movement technique further corroborate the observation that tracking abnormalities are more evident with the loaded than the unloaded examination.

Another reported active movement technique of performing kinematic MR imaging of the patellofemoral joint uses a special nonferromagnetic positioning device that incorporates a trigger system that senses movement (12). The patellar alignment and tracking may then be assessed during active movement; this is similar to how cardiac-gated MR studies are performed. With this technique, the patient is placed supine on the positioning device and a single patellofemoral joint is flexed and extended repeatedly, while gradient echo images are obtained using a circular, receive only, surface coil. The results obtained using motion triggered kinematic MR imaging of the patellofemoral joint support the importance of using an active movement technique for their evaluation.

Normal alignment and tracking of the patella is considered to be present when the median ridge of the patella is positioned in the femoral trochlear groove as it travels in a vertical plane during flexion of the knee (1,3,4,5). There should be no significant transverse displacement of either patella facet and the patella appears "centered" in the groove. Any deviation of this normal pattern of patellar alignment or tracking seen on one or more slice locations at 5° of flexion or greater on the kinematic study is regarded as abnormal. Four principal patterns of patellar malalignment utilizing MR have been described.

Lateral subluxation of the patella is the most common form of patellar maltracking (1,3,4,5). MR may facilitate demonstration of associated articular cartilage abnormalities that may result from chronic hyperpressure related to contact stress of the laterally subluxed patella. One of the more unique contributions to study of patella tracking dynamics has been the observation of a redundant lateral retinaculum associated with some cases of lateral patellar subluxation. As release of the lateral retinaculum is the most commonly performed surgical procedure for suspected lateral patellar subluxation, this observation has important implications. Redundancy of the lateral retinaculum suggests insufficiency of the medial stabilizing structures, e.g., vastus medialis obliqus (VMO), as opposed to excessive ''tightness'' of the lateral retinaculum as the underlying etiology of the patellar malalignment (1,3).

Lateral patellar tilt or excessive lateral pressure syndrome (ELPS) is another commonly encountered category of patellar malalignment characterized clinically by anterior knee pain and radiologically by tilting of the patella with functional patellar lateralization commonly with a dominant lateral patellar facet (5) (Fig. 26C). A small amount of lateral displacement of the patella may or may not be present during joint flexion, as increasing tension from one or more overlying taut soft tissue structures tilt the patella in a lateral fashion. The underlying hyperpressure found in ELPS can produce significant destruction of the articular cartilage. ELPS is usually treated effectively by surgical release of the lateral retinaculum.

Medial subluxation of the patella is distinguished by medial displacement of the median patellar ridge relative to the femoral trochlear groove (Fig. 26D). Medial subluxation has been reported as a complication of prior surgical realignment procedures for presumed lateral subluxation and has more recently been reported in patients without prior history of surgical intervention (1,3,13). A variety of causative mechanisms are suspected to be responsible for medial subluxation of the patella that exist either individually or in combination, including (a) an insufficient lateral retinaculum, (b) overtaut medial retinaculum, (c) abnormal patellofemoral anatomy, and (d) unbalanced quadriceps muscles. In addition, excessive internal rotation of the lower extremities is a common clinical finding with medial subluxation of the patella. The presence of medial subluxation of the patella is especially important to identify and to distinguish from lateral or tilting forms of patellar subluxation because various surgical stabilization techniques can further increase medial subluxation of the patella, and not only lead to failure of the procedure but also potentially exacerbate the patient's symptoms. In addition, most physical rehabilitation techniques, including McConnel taping and patellofemoral bracing procedures are usually implemented to correct lateral subluxation of the patella and need to be modified when treating patients with medial subluxation of the patella.

Lateral to medial subluxation of the patella is a pattern of abnormal patella tracking whereby the patella is in a slightly laterally subluxated position during the early increments of knee flexion, moves across the femoral trochlear groove or femoral trochlear as flexion continues, and displaces medially at the higher increments of flexion. This type of abnormal patellar tracking is commonly exhibited by the patellofemoral joint that tends to have patella alta, a poorly developed femoral trochlear groove, and/or misshapen patella. No mechanical restraint is provided by the osseous anatomy and patellofemoral instability results when these structural abnormalities exist. As with other malalignment disorders, kinematic MR is particularly well suited to identifying this unusual type of patellar tracking.

MR may also be utilized to assess the effect of therapeutic endeavors undertaken in attempts to correct patellar tracking abnormalities. In addition to a wide variety of surgical procedures including lateral release and anterior tibial tubercle transfer procedures, a number of nonsurgical techniques including taping and bracing have been applied. MR can provide information regarding the apparent efficacy of these efforts (Fig. 26E,F).

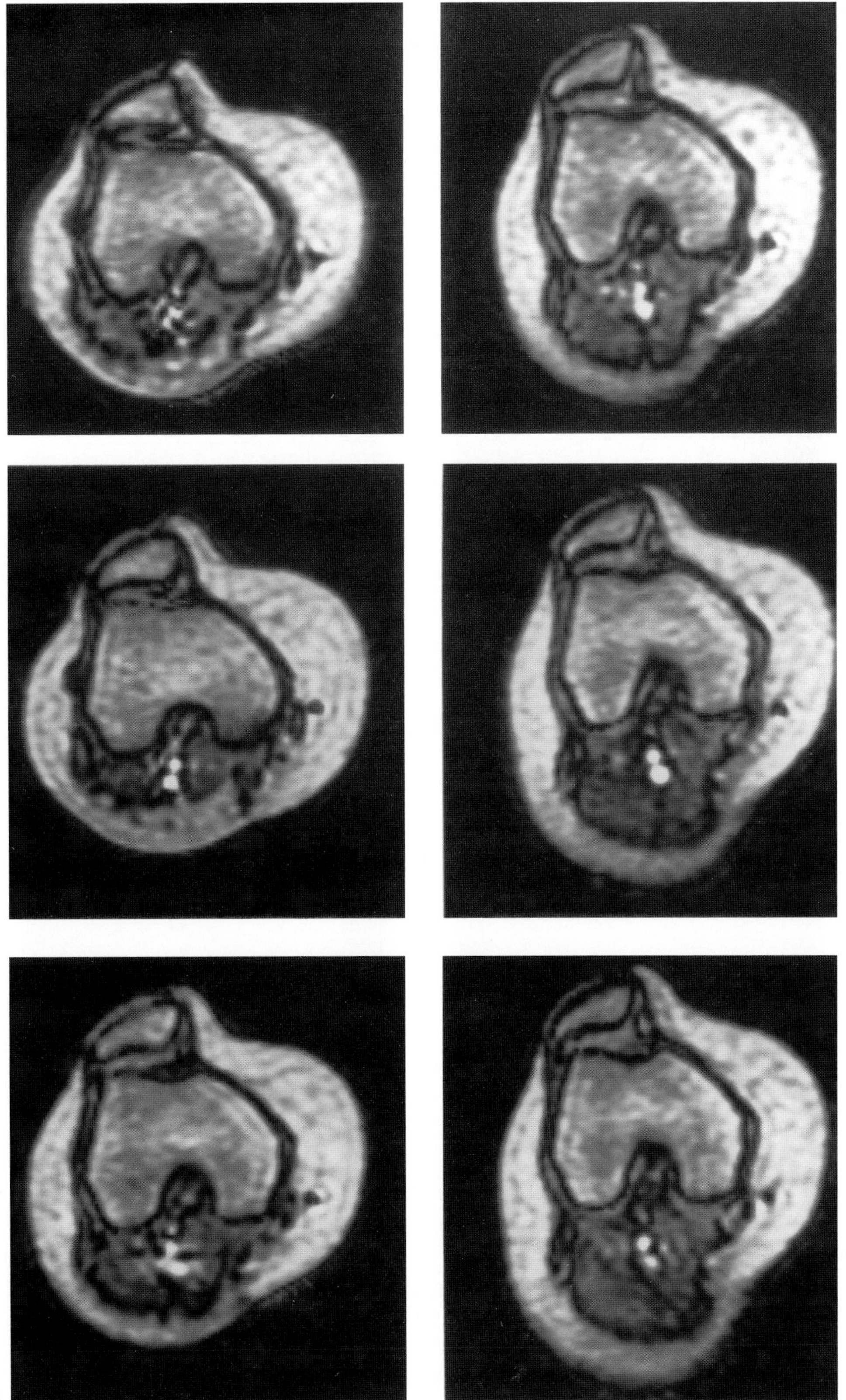

FIG. 26. C: Excessive lateral pressure syndrome. Series of axial spoiled GRASS images performed during active motion. Note the marked patellar tilting in the presence of only mild lateral subluxation. The degree of lateral tilting increases at the higher degrees of flexion. (From Deutsch AL, Shellock FG. Extensor mechanism and patellofemoral joint. In: Mink JH, Reicher MA, Crues JVIII, Deutsch AL, eds. *MRI of the knee, 2nd edition.* New York: Raven Press, 1993;220–233, with permission.)

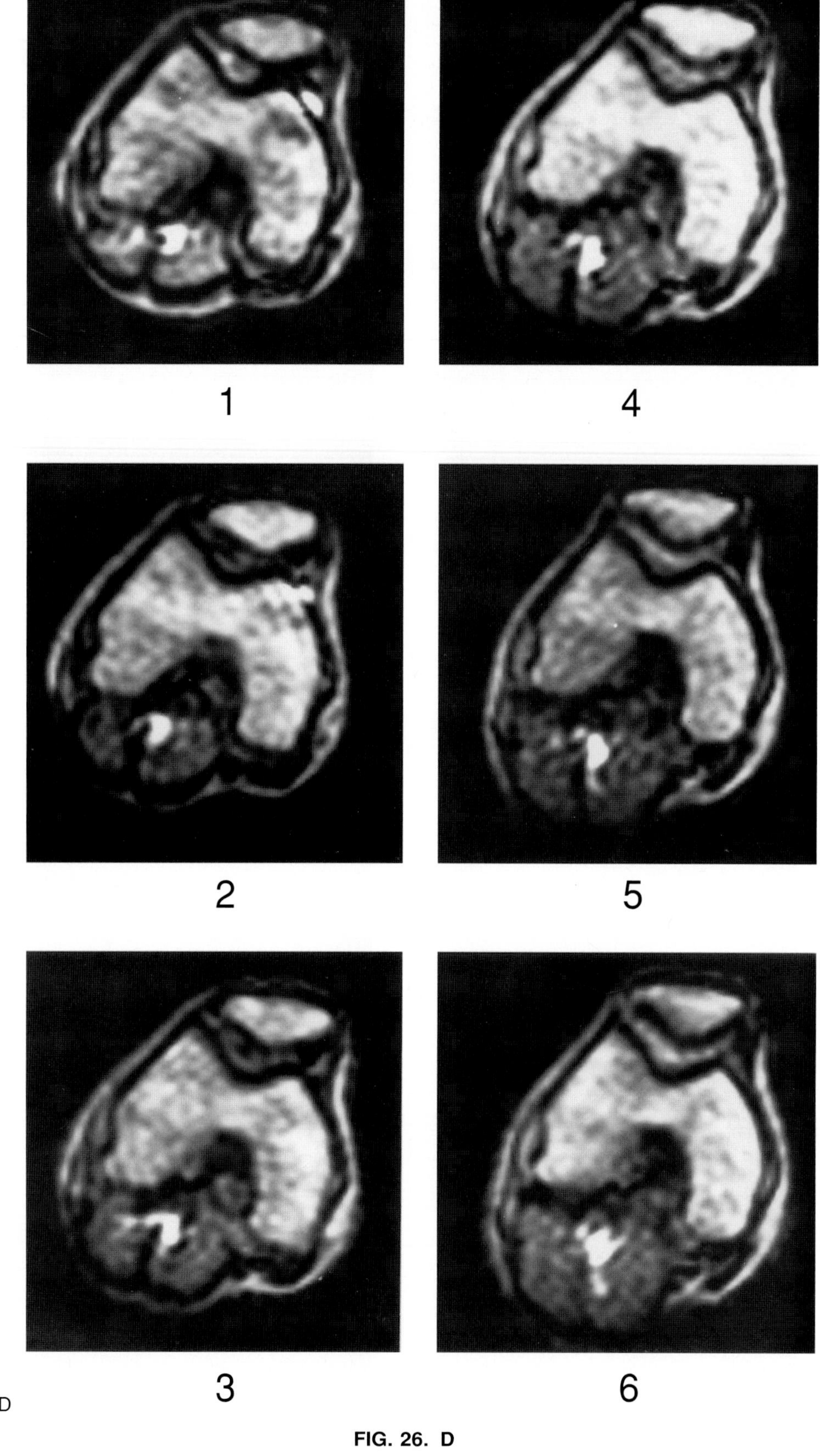

FIG. 26. D

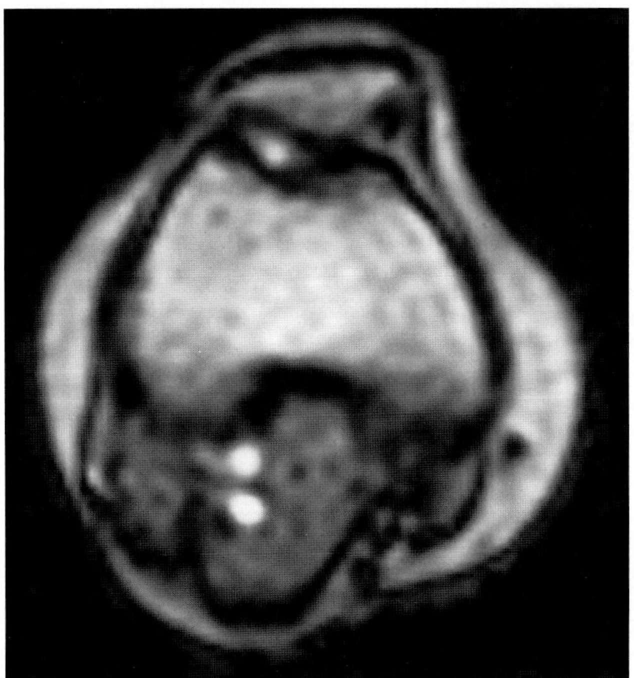

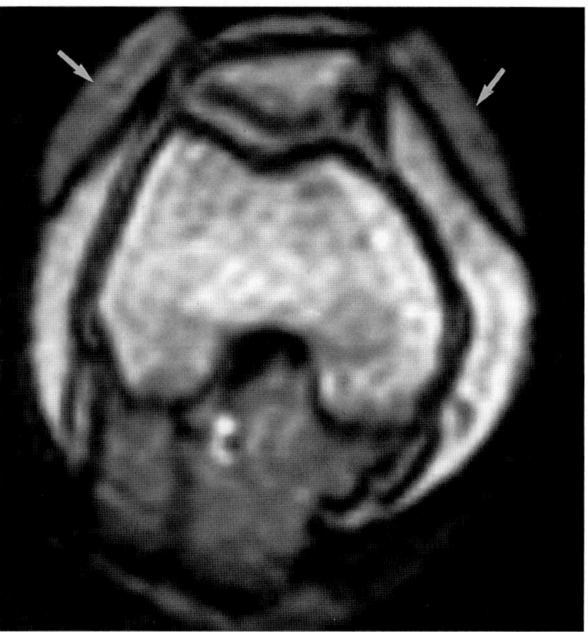

FIG. 26. E,F: Dynamic examination pre- and postbracing. **E:** Selected axial image (TR/TE 7/3 msec) from an SPGR sequence obtained at approximately 30° of active flexion demonstrates medial displacement of the apex of the patella. **F:** Following the placement of a specialized brace *(arrows),* a repeat MR study again performed during active motion demonstrates increased centering of the patella. Such MR-compatible devices allow direct monitoring of the potential effects of different therapeutic regimens. (From Deutsch AL, Shellock FG. Extensor mechanism and patellofemoral joint. In: Mink JH, Reicher MA, Crues JVIII, Deutsch AL, eds. *MRI of the knee, 2nd edition.* New York: Raven Press, 1993;220–233, with permission.)

FIG. 26. D: Medial subluxation. Axial (TR/TE 7/3 msec) image. Series of SPGR images obtained during active motion of the knee from extension to 45° of flexion. There is evidence of medial subluxation of the patella in a patient without prior surgery. Note that there is centralization of the patella by the higher increments of flexion. The findings of medial subluxation were not evident on a conventional tangential radiograph of the patella. (From Deutsch AL, Shellock FG. Extensor mechanism and patellofemoral joint. In: Mink JH, Reicher MA, Crues JVIII, Deutsch AL, eds. *MRI of the Knee, 2nd edition.* New York: Raven Press, 1993;220–233, with permission.)

CASE 27

History: A 48-year-old man presented with localized swelling and mild tenderness over his proximal tibia. There was an uncertain history of prior trauma.

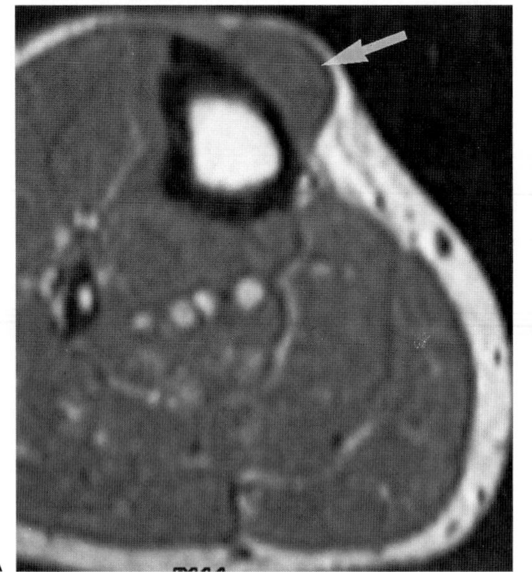

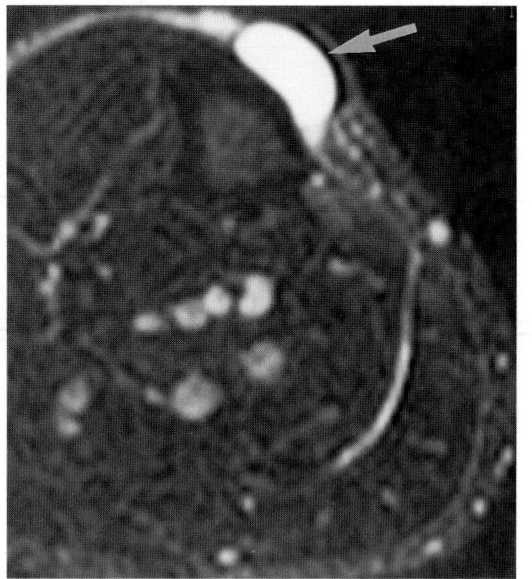

FIG. 27

Findings: In Fig. 27A an axial fast spin echo T1-weighted image (TR/TE 600/17 msec) obtained through the proximal tibia demonstrates a an ovoid-shaped mass (arrow) that is contiguous with the cortical surface of the anteromedial aspect of the tibia. In Fig. 27B an axial fat suppressed T2-weighted (TR/TE 4,000/108 msec) fast spin echo image demonstrates the mass to become of uniform high signal intensity (arrow).

Diagnosis: Periosteal ganglion.

Discussion: Ganglia are common cystic soft tissue–like lesions that are usually encountered attached to a tendon sheath, particularly in the hands, wrists, and feet (1–5). Less commonly, ganglia can be located within muscle and bone (intraosseous ganglion) (4). On rare occasions, at least according to the reported literature, ganglia may also arise within the periosteum (1–3). In this location they have a quite characteristic appearance on MR, and under appropriate clinical circumstances a specific diagnosis can be strongly suggested.

In common with their soft tissue counterparts, periosteal ganglia are cystic structures with fluid contents ranging in consistency from thin to viscid and mucinous (1). As in the present case, the lesions are more commonly encountered in males and the proximal shaft of the tibia is a favored location, particularly in proximity to the pes anserinus (1). All prior reports of periosteal ganglia indicate involvement of long tubular bones with no reports of axial skeletal involvement or of flat bone involvement. The radiographic appearance of the lesions varies. There may be no discernible bone involvement or different degrees of cortical erosion, scalloping, and reactive bone formation (1–3). Cortical erosion with scalloping is the hallmark of radiographic diagnosis and may be associated with thick spicules of reactive periosteal bone extending from the scalloped area. These radiographic appearances are nonspecific and can mimic those of other benign surface lesions of bone including periosteal chondroma and subperiosteal aneurysmal bone cyst.

On MR, periosteal ganglia are seen as sharply defined masses adjacent to the cortical bone that are isointense to surrounding muscle on T1-weighted sequences and demonstrate uniform high signal intensity on T2-weighted sequences (1). No evidence of surrounding marrow edema or inflammatory changes are seen. This latter sign may be of value in distinguishing these lesions from other surface lesions and periosteal/cortical processes including subperiosteal abscess and hematoma. A periosteal chondroma, particularly if it lacked matrix calcification and a calcified rim, could closely resemble the MR appearance of a periosteal ganglia (1). These lesions are commonly seen in a younger age group than the patients reported with periosteal ganglia. Treatment of periosteal ganglia consists of surgical excision including removal of a margin of normal periosteum along with the lesion itself. Local recurrence is not uncommon. While previously considered a rare lesion, the increasing use of MR for evaluation of soft tissue tumors may facilitate increased awareness and diagnosis of these lesions.

CASE 28

History: A 68-year-old woman complained of knee pain for the past 6 weeks. The patient recalls a sudden onset of the pain as she was walking down a flight of stairs. There was no history of trauma. What is your diagnosis?

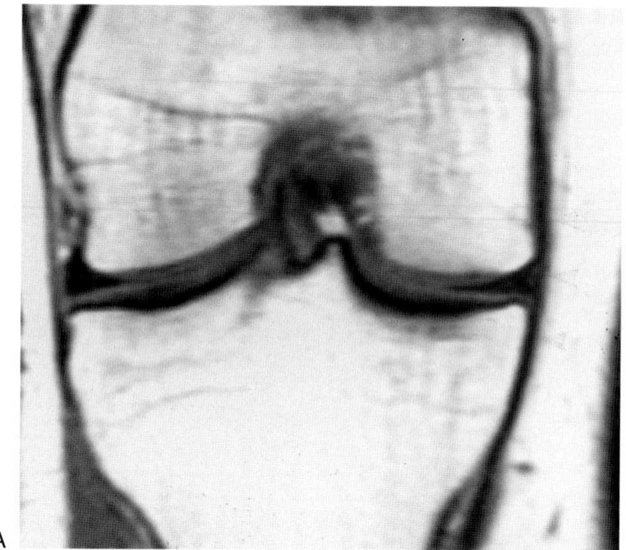

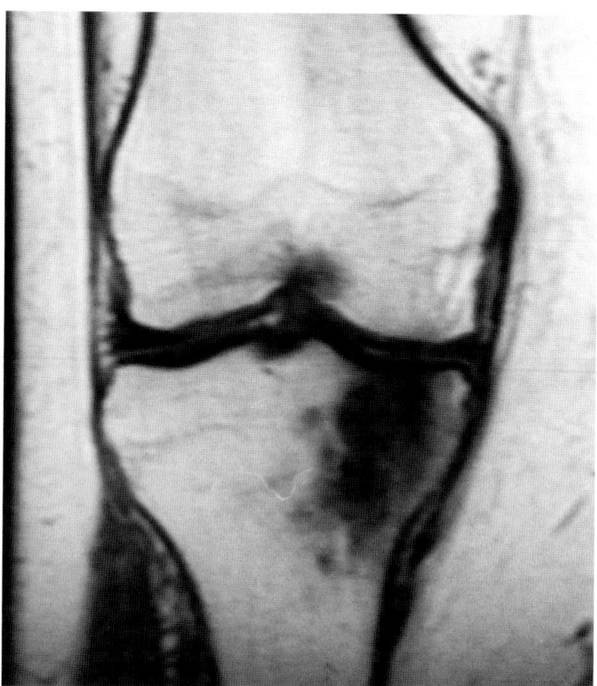

FIG. 28

Findings: Figure 28A is a coronal T1-weighted (TR/TE 800/20 msec) image obtained 2 years prior to the study demonstrated in Fig. B. The initial study is normal. In Fig. 28B a coronal T1-weighted (TR/TE 800/20 msec) image demonstrates the development of marked low signal intensity with the proximal medial tibial meta-epiphysis.

Diagnosis: Spontaneous osteonecrosis of the medial tibial plateau.

Discussion: Spontaneous osteonecrosis occurring about the knee is a well-recognized clinical entity that classically occurs in patients over the age of 60 and demonstrates a predilection for the weight-bearing surface of the medial femoral condyle (1–10). Involvement of the tibial plateau, as in the present case, is far less common, but has been reported. Women are affected three times more often than men, unilateral involvement predominates, and patients typically relate the abrupt onset of pain that characteristically persists at rest and is often worse at night. Physical examination may reveal local tenderness in the medial femoral condyle, restricted range of motion, and effusion. Progression of disease is accompanied by subchondral collapse, fragmentation of the articular cartilage, and progressive osteoarthritis. In fact, it is postulated that many cases of medial tibiofemoral osteoarthritis are preceded by spontaneous osteonecrosis. Spontaneous healing has been reported when the affected area is small and weight bearing is avoided (1–4). The prognosis is unfavorable if the lesion is larger than 5 cm or involves more than 40% of the width of the condyle (11).

The cause and pathogenesis of this disorder remain unknown. There is no association with alcohol or steroid use. Vascular insufficiency leading to infarction with elevation of intraosseous pressure has been proposed as has possible traumatic insult leading to microfractures in the subchondral bone plate with overlying osseous and cartilaginous collapse (1–3,8,11). Meniscal tears have been reported in high association with this condition and may play a role in the pathogenesis. The appearance of apparent spontaneous osteonecrosis following meniscal surgery also supports a possible pathogenetic relationship between this condition and meniscal alterations (5).

Prior to the widespread use of MR, scintigraphic methods were commonly employed to diagnose spontaneous osteonecrosis before the development of overt radiographic abnormalities. On MR, spontaneous osteonecrosis is characterized by low signal intensity of the affected area on T1-weighted images and a surrounding area of increased signal intensity on T2-weighted and short inversion time inversion recovery sequences. Additional MR findings include the presence of cystic lesions containing fluid, bone collapse with buckling of the articular cartilage, and tears of degeneration of the adjacent meniscus (1,2,4). Accurate measurement of the lesion, facilitated by the tomographic nature of MR, may contribute prognostic information.

Entities to be included in the differential diagnosis of this condition include osteonecrosis from other causes, osteochondritis dissecans, calcium pyrophosphate dihydrate (CPPD) deposition disease, transient marrow edema syndrome, neuropathic arthropathy, and osteoarthritis (1,2). Osteonecrosis secondary to systemic causes, including steroid use, differs from spontaneous osteonecrosis in several respects. There may be multiple sites of involvement, larger areas may be affected, and the lateral femoral condyle is commonly involved (1,2). Osteochondritis dissecans affects younger patients, the pain is insidious in onset, and the lesion does not classically involve the weight-bearing aspect of the medial femoral condyle (1). CPPD may contribute to flattening of the femoral condyle but is frequently associated with chondrocalcinosis within the knee or at another site (2). Stress fractures are typically encountered in younger individuals and occur at some distance from the articular surface. Osteoarthritis is characterized by the insidious onset of pain that is relieved by rest, whereas spontaneous osteonecrosis frequently results in abrupt onset of pain that persists at rest (1). Osteoarthritis often begins on the tibial side of the joint, whereas spontaneous osteonecrosis begins on the femoral side. In osteoarthritis, cartilaginous erosions occur early; in spontaneous osteonecrosis erosions occur late.

Spontaneous ischemic necrosis has been described in the medial portion of the tibial plateau (1,2). The condition occurs in the same age group and with a similar presentation to spontaneous osteonecrosis of the femoral condyle. Osteonecrosis of the tibia has been postulated to be caused by stress microfractures of the plateau that result in elevation of the intraosseous pressure due either to intrusion of synovial fluid or lack of sufficiently rigid subchondral bone to support the load. MR is particularly valuable for lesion detection, assessment of the overlying articular cartilage, and differentiation from other conditions including medial meniscal tear, pes anserinus bursitis, and spontaneous osteonecrosis of the medial femoral condyle.

CASE 29

History: What is the lesion depicted by the arrow, and what is its significance? What other bony abnormalities might one see in this knee (Fig. 29A–C)?

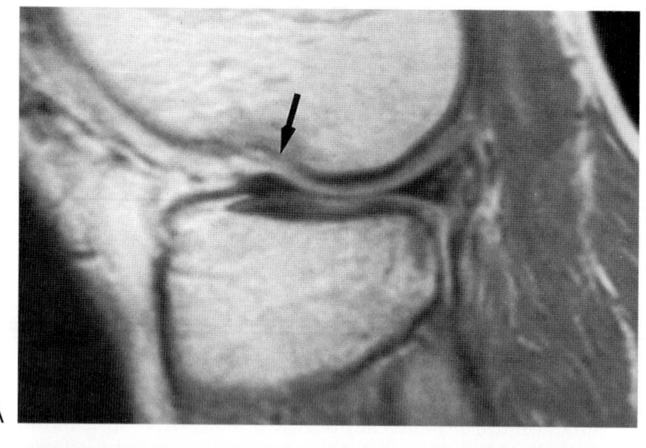
A

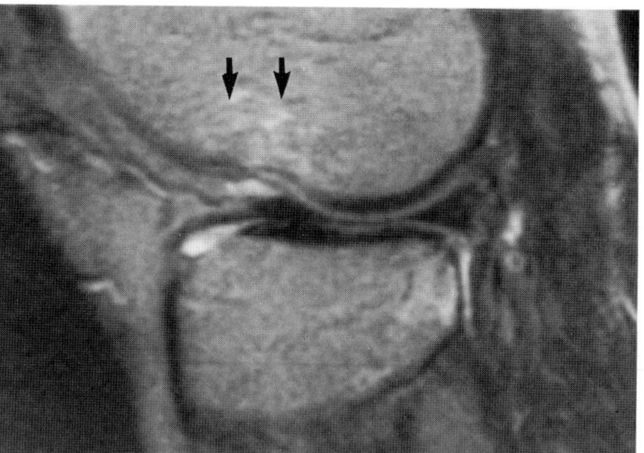
B

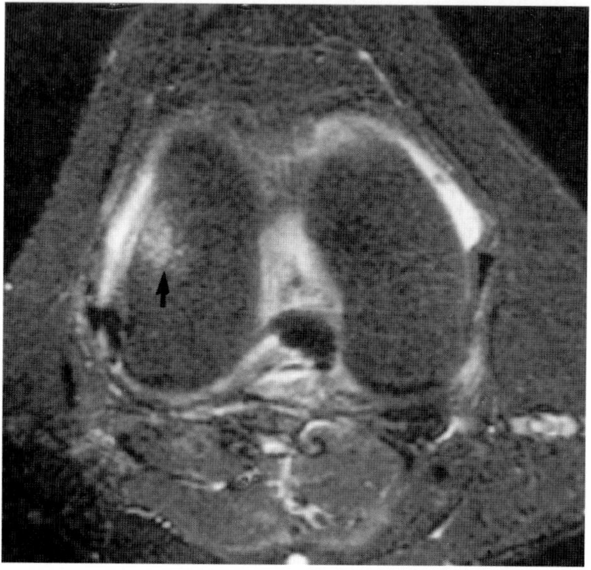
C

FIG. 29.

Findings: Figure 29A–C demonstrates impaction fracture of the lateral femoral condyle and the posterior lateral tibial plateau (''kissing lesions''). **A:** Sagittal (TR/TE 2,000/20 msec) image; **B:** sagittal (TR/TE 2,000/80 msec) image; **C:** axial (TR/TE/TI 2,800/35 msec/160) image. There is an impaction injury of the LFC (arrows) directly over the anterior horn of the lateral meniscus. The condylopatellar sulcus is too deep, and there is marrow edema adjacent to the lesion (small arrows) and along the posterior lateral tibial plateau. The marrow edema is more graphically seen on the STIR image than on the T2-weighted sequence.

Diagnosis: Osteochondral impaction injury of the lateral femoral condyle.

Discussion: The lesion is one of the several osseous and osteochondral lesions that occur in association with tears of the anterior cruciate ligament (ACL). In this instance, there is also an impaction injury of the posterior lateral tibial plateau. Many of the bony injuries occur as a result of impaction of one bony surface on another. Other osseous lesions have an avulsive etiology, resulting from the action of the joint capsule, ligaments, or tendon on bone.

The anterior cruciate ligament (ACL) is most commonly torn when the leg is nearly completely extended, and the femur externally rotates over a fixed tibia (1–4). The excursion is normally limited by the ACL, but when the forces exceed that of the ligament, the tibia is free to move forward. The tibia glides forward until the posterior lip of the lateral tibial plateau (LTP) approaches the middle or anterior margin of the lateral femoral condyle (LFC). At this point, the posterior aspect of the LTP impacts against the LFC (Fig. 29D–F). Along the anterior and middle aspects of the LFC, it is important to distinguish the normal shallow condylopatellar sulcus from an impaction injury. This normal sulcus, which accommodates the anterior horn of the lateral meniscus during full extension, is not more than 1.5 mm deep (5); a greater degree of depth is usually a result of a transchondral fracture. If rotation of the tibia is minimal, but the degree of direct anterior subluxation of the tibia is greater, the posterior medial tibial plateau (MTP) will also impact the medial femoral condyle (Fig. 29G,H). The resulting medial tibial plateau injury is usually less extensive than is the lateral. A less common mode of disruption of the ACL occurs when the knee is subjected to severe hyperextension with the leg internally rotated. In such instances, the anterior LTP and LFC, as well as the medial tibial and femoral surfaces, come into direct contact, producing a "kissing" impaction injury in each bone (1). All indirect type trabecular microfractures are not always secondary to ruptures of the ACL. Posttraumatic regions of bone marrow edema are frequently seen in the medial femoral condyle and medial tibial metaphysis in patients who have incurred injuries to the medial collateral ligamentous complex but in whom the ACL is normal (6).

Certain bony lesions are so common that some authors have suggested that the presence of one or more of these abnormalities is virtually proof of an ACL tear (1,4,7,8). As many as 85% of patients with proven ACL tears had bony injuries; 25% of the patients had more than one lesion (7). Eighty percent of the abnormalities were in the lateral compartment, involving the femur or tibia, but 21% of patients also had lesions of the medial tibial plateau. In another study, 94% of patients had abnormalities of the posterior tibia, 91% had signal alterations in the LFC, and 100% of patients had signal alteration in either the tibia or femur (8). The tibial lesion is nearly always found at the posterior lip of the plateau, and the femoral lesion is either in the middle or anterior thirds, depending on the degree of knee flexion at the time of injury (8,9).

The MR appearance of each of these variously located impaction injuries is quite similar. Impaction lesions are most frequently characterized by a large area of edema and presumed hemorrhage within the adjacent medullary space, occurring at a variable distance from the dark cortical bony plate. This edema is usually identified as an area of decreased signal on short TR/TE sequences, and possibly increased signal on long TR/TE sequences. The T1-weighted images are more efficient in demonstrating cortical lesions, while the T2 images were more likely to reveal signal abnormalities in the medullary bone (8). However, the co-mingling of the T2 signals of medullary bony fat, hemorrhage, edema, and normal hematopoetic marrow may make detection of signal alteration difficult (Fig. 29I,J). Several authors have advocated the use of inversion recovery or fat-saturation techniques to more accurately and graphically depict these occult injuries (3,10,11). In one series comparing standard T1 and T2 sequences, fast spin echo T2 sequences, and fat-saturated fast spin echo techniques in the detection of posttraumatic lesions of bone, conventional spin echo imaging depicted only 16/26 such lesions, and standard fast spin echo techniques revealed only 20 of the 26 lesions uncovered by the fat saturation technique. The T1-weighted images did not demonstrate four of the abnormalities (10).

A variety of terms have been adopted by various authors to describe posttraumatic hemorrhagic, edematous lesions of bone. The first term applied to these lesions was *bone bruise* (2). The term referred to a wide variety of types of *acute* posttraumatic lesion. It was originally believed that these injuries represented a "new category of incomplete fracture that spares the cortex" (12), but given that trabecular fractures are well-known occurrences in man and animals, it is likely that these fractures are "newly imaged" (3). Subsequently, various authors progressively subcategorized these abnormalities. Lynch et al. (13) proposed their own system of nomenclature (types 1, 2, 3) as did Yao and Lee (12) (occult intraosseous fracture), Mink and Deutsch (2) (bone bruise, osteochondral fracture, stress fracture, true fracture), and Vellet et al. (14) (geographic, reticular, linear).

The data on the prognosis of these posttraumatic osseous lesions is varied and still incompletely known, due in no small part to the variations in terminology. Posttraumatic osseous lesions that are exclusively meta-epiphyseal in location and that do not significantly involve the overlying cortex typically resolve by 6 weeks to 3 months without sequelae if there is no further stress placed across the lesion (3,9,12–15) (Fig. 29K,L). In one study, none of the patients who suffered a meta-epiphyseal injury at the time of ACL tear had MR evidence of the lesion at reexamination 9 weeks later (16). It has been assumed that these lesions result from abnormal compressive forces being delivered to the joint surface and that the force is absorbed by the medullary bone. A pattern of linear and globular edema in the meta-epiphyseal or meta-diaphyseal bone marrow occurs. The injury itself initially leads to alteration in meshwork of the medullary bone. During repair of the microfractures, the supporting trabecular meshwork is further weakened and each remaining intact trabecula is subjected to increased stress. The loss of compliance of the trabeculae lead to greater peak forces being transmitted to the overlying subchondral plate (17). Massive numbers of new fractures could occur in the face of only a minor degree of new trauma. It is possible that this process could lead to coalescence of neighboring microfractures resulting in macroscopically visible fracture and eventual collapse of the subchondral plate (3,10). Based on animal work that suggests repairing bone is in its most weakened state at 4 weeks posttrauma, it is reasonable to afford weight-bearing protection for 6 weeks, at which time complete, uncomplicated resolution is to be expected, but long-term follow-up is still lacking (17,18).

There is a second group of lesions that appear, by MR, identical to the meta-epiphyseal lesions described above except for the fact that they extend to the immediate subcortical region (Fig. 29M–R). These abnormalities are also invisible to the arthroscopist in the acute phase as are the lesions above, but many of *these* lesions have sequelae. In patients with subcortical lesions, there may be histologic evidence of injury to the cortical plate and, as such, they may induce a different physiologic response in the host (3,13). While the cartilage may be initially deformed by the injury, young cartilage is extremely resilient and elastic and may rebound back into normal position (14), and its surface is usually normal on arthroscopic inspection. Subchondral plate microfracture results in histologic alteration in the contour of the overlying articular cartilage. The location of the injury implies that the cortical plate and cartilage surfaces have not simply transmitted energy into the medullary space, but rather that the cortical plate has already failed and that the overlying cartilage has suffered microscopic damage. The articular surface is intolerant of even minor degrees of increased shear force that follow loss of joint congruity, and the cartilage becomes subject to progressive fissuring. Short-term osteochondral sequelae, including cartilage thinning or defects, osteosclerosis, and cortical impaction, were seen in 67% of 21 patients who had subcortical trabecular microfractures (14). Chondrocytic death, release of streptolysin and other enzymes, and disruption of the supporting collagen network may lead to delayed changes of osteoarthrosis and abnormal cartilage function (17,19). Therefore, in this group of patients, the acute appearance may vary substantially from the subacute or chronic appearance even without further trauma.

In an attempt to standardize terminology, to promote the ability of radiologists to communicate with each other as well as with their clinical colleagues, and to provide some sense of prognosis of these various lesions, it is recommended that the term *trabecular microfracture* be adopted to describe the appearance of acute posttraumatic abnormalities of bone that are characterized by medullary bone edema in the metaphysis and/or epiphysis but in which the articular surface is initially macroscopically normal. This term identifies the presumed basic pathology which, for all of the above described lesions, is probably the same. A qualifier regarding the location of the lesion relative to the cortical plate should be specified in the report since the patient's prognosis appears to be determined by the site of injury.

MR images that demonstrate macroscopic alterations of the articular surface and/or subchondral bone on examinations performed in the immediate posttrauma are best termed *chondral fracture, osteochondral fracture,* or *transchondral fracture* (Fig. 29S). In some cases, the injury is manifest by a subtle alteration in the convex contour of the femoral condyle; in others, there is an obvious defect in the articular cartilage. Acute transchondral fracture is readily diagnosed by MR because most patients suffering such injury have an effusion, resulting in an arthrographic effect within the joint. Laceration of the articular surface appears as an increased signal intensity within the cartilage structure, probably secondary to intravasation of joint fluid (3). If the fracture involves the full depth of the articular surface, the underlying bone may reveal an edematous, hemorrhagic response.

ACL-associated fractures may occur through the mechanism of avulsion. The Segond fracture, an avulsion of the meniscotibial insertion of the middle third of the lateral joint capsule, occurs with internal rotation and varus stress and carries an extremely high association with ACL tears (1,20) (Fig. 29T,V). This small osseous fragment is quite constant; it is elliptical in shape, measuring 10 × 3 mm with its proximal border 4 mm distal to the joint line. On MR, the lesion is manifest by (a) a linear flake fracture of the tibial rim (only 25% of the lesions are so identified because of their small size, similar in signal character to the periosseous fat in which they lie, and their propensity to become edematous and of low signal on T1 images); and (b) focal marrow edema at the expected site of avulsion.

Hyperextension of the knee may result in avulsion of the capsule and the arcuate ligament from the posterior LTP, and the fibular head; such avulsion may occur with a small fleck of bone. Similarly, external rotation and abduction with the knee flexed will put stress on the central tendon of the semimembranosus tendon, resulting in a small avulsion of the posterior medial tibial plateau (21).

Avulsion of the attachment of the ACL occurs almost exclusively at the tibia since the tibial end of the ligament is stronger than the proximal end (Fig. 29V). Less than 5% of ACL ruptures occur in adults as a result of avulsion, but the relative strength of the ACL compared with that of the adjacent bone and physeal plate explain why most ACL injuries in children are avulsive or result in a physeal fracture. It is essential to rapidly diagnose ACL avulsion; the fragment may block extension and preclude normal knee function and reattachment of the bony fragment has a high rate of success in restoring ACL function. Remarkably, these serious, but easily repairable injuries may be somewhat difficult to fully appreciate on MR due to several factors. The lack of displacement of the fragment is constant. The avulsed fragment, composed of medullary bone, has signal characteristics identical to that of the fat in the intercondylar notch and may therefore be difficult to identify. Finally, bone marrow edema in the bed of the avulsed fragment is minimal to absent.

Cystic lesions are occasionally found along the intercondylar eminence, adjacent to the attachments of either cruciate ligament. These rounded, occasionally multiple cysts occur in 1% of otherwise normal patients and are thought to be incidental findings rather than intraosseous ganglia or sequelae of ACL injury (22).

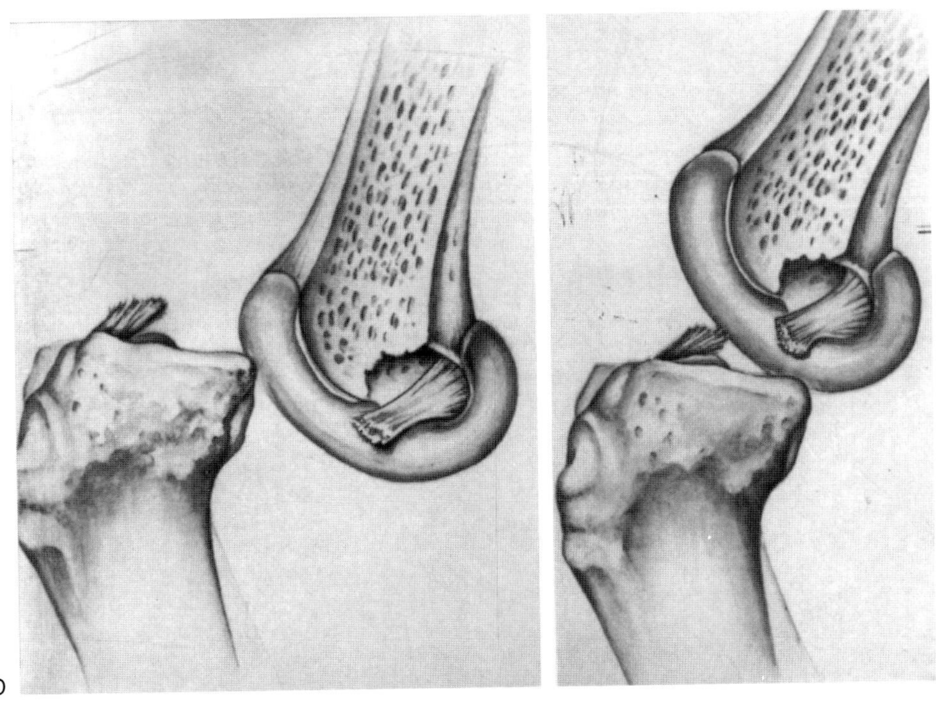
D

FIG. 29. D: Bony injuries associated with ACL tears. The variable location of femoral impaction sites of the lateral femoral condyle in patients with ACL disruptions depends upon the degree of knee flexion at the time of injury. If the knee is flexed 20° or less, the posterior lateral tibial plateau contacts the more anterior aspect of the femoral condyle; if the knee is flexed to a greater degree, impaction occurs over the mid–weight-bearing surface.

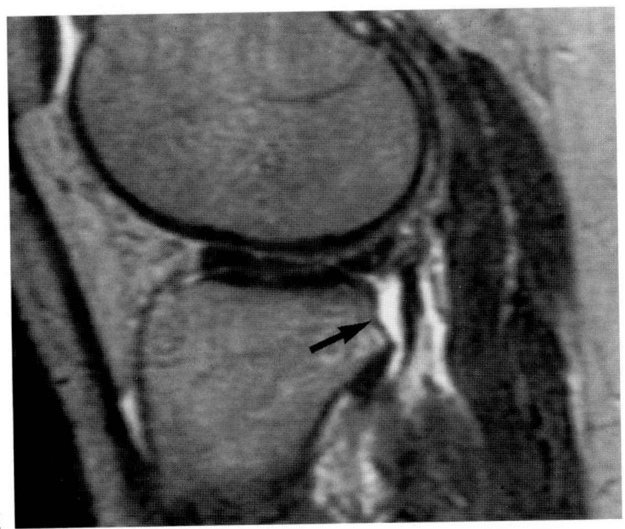
E

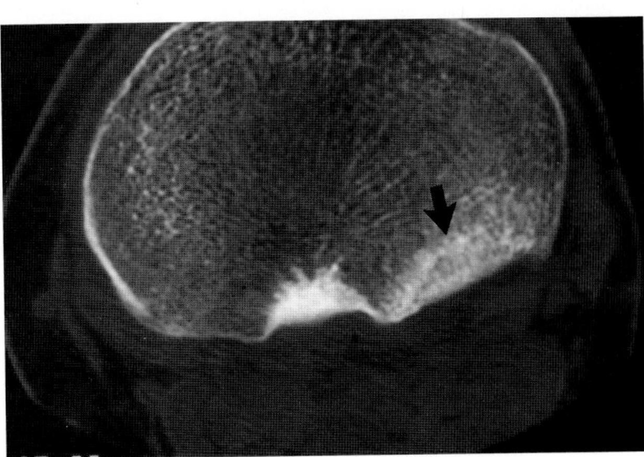
F

FIG. 29. E,F: Fracture of the posterolateral tibial plateau. **E:** Sagittal (TR/TE 2,300/80 msec) image; **F:** CT scan. This is the classic impaction fracture seen in patients who suffer tibial subluxation during ACL rupture. There is flattening of the posterior aspect of the lateral tibial plateau *(arrow)* with a usually minor degree of involvement of the articular surface.

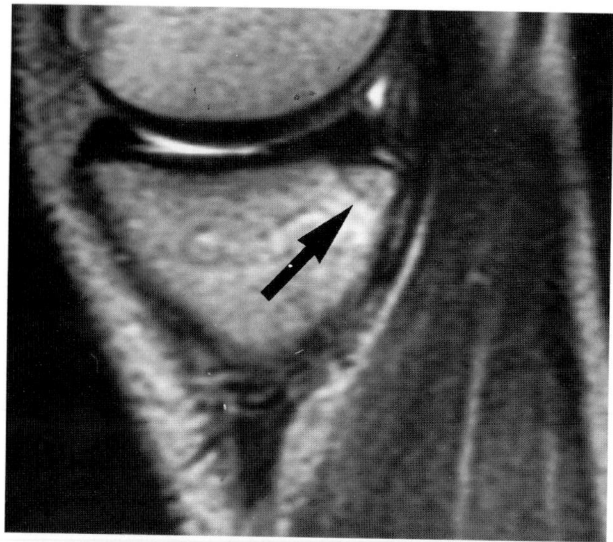

G

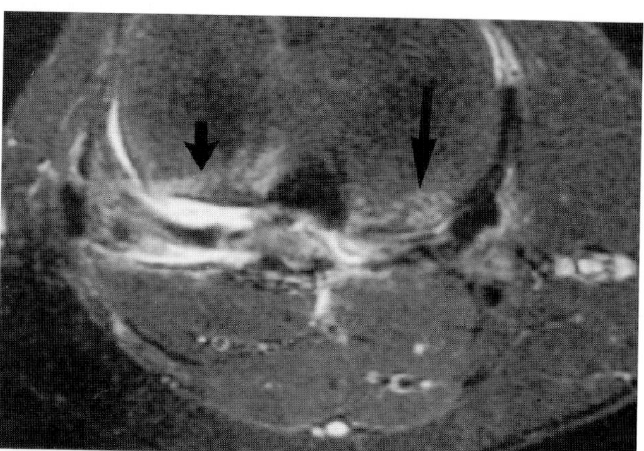

H

FIG. 29. G,H: Impaction fracture of the medial tibial plateau. **G:** Sagittal (TR/TE 2,000/80 msec) image; **H:** Axial (TR/TE/TI 2800/35/160) image. Both images demonstrate an injury to the medial tibial plateau *(large arrow)*, better seen on the axial STIR image. An impaction fracture of the lateral tibial plateau is also seen *(small arrows)*.

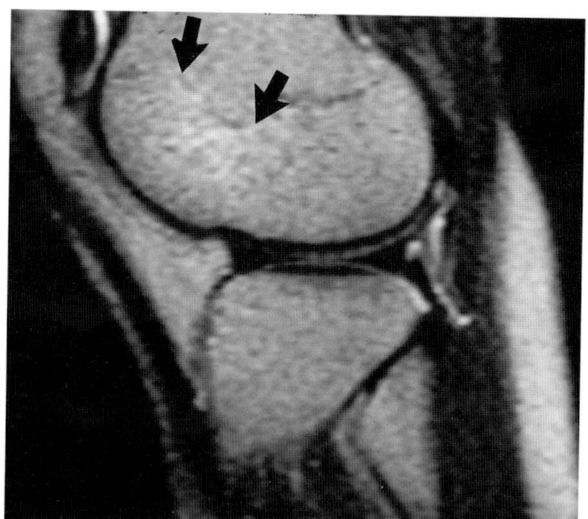

I

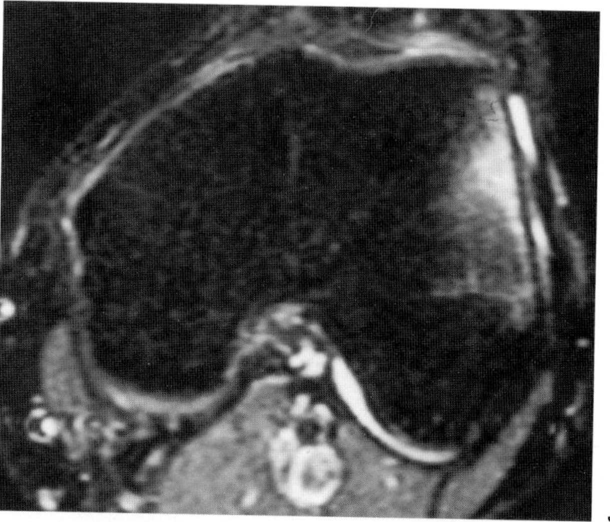

J

FIG. 29. I,J: Use of STIR imaging. **I:** Sagittal (TR/TE 2,200/80 msec) image; **J:** Axial (TR/TE/TI 2,200/35/160 msec) image. The patient recently suffered a lateral dislocation of the patellar. The conventional T2-weighted image in **I** demonstrates a minor degree of inhomogeneity of the marrow signal *(arrows)* of the anterolateral femoral condyle but the finding is so subtle as to be readily overlooked. In **J**, however, the inversion recovery study very graphically depicts the edema. Fat suppression techniques, such as inversion recovery or fat saturation, are extremely valuable in assessing the marrow space, which normally contains tissues with a variety of signal intensities.

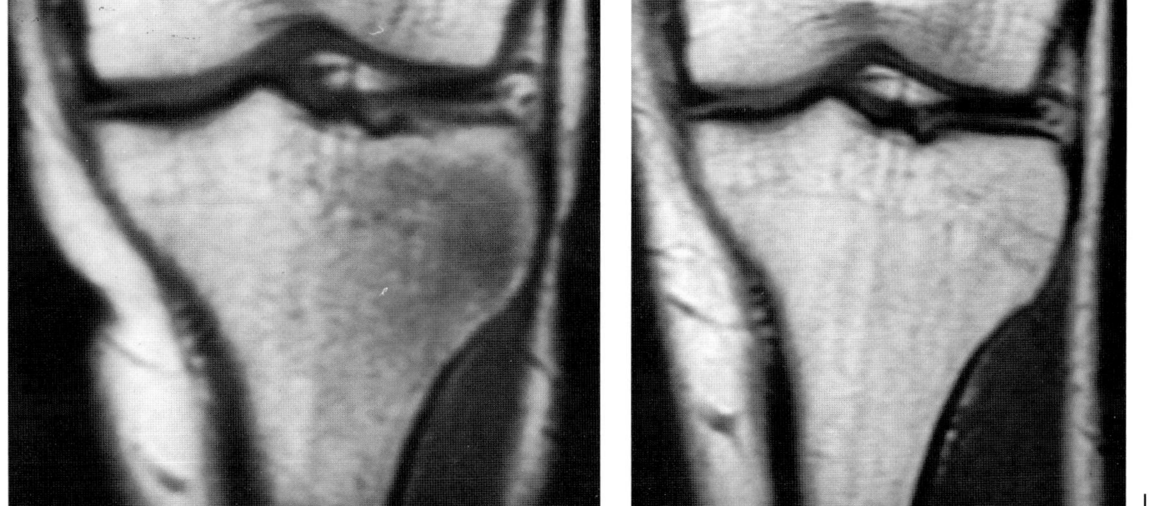

FIG. 29. K,L: Resolution of a metaphyseal trabecular microfracture. Coronal (TR/TE 600/20 msec). **L** is made 6 weeks after **K.** The initial examination demonstrates an extensive pattern of bone marrow edema in the lateral tibial metaphysis but there is no involvement of the immediate subcortical bone. The image in **L** demonstrates complete resolution; the patient was asymptomatic.

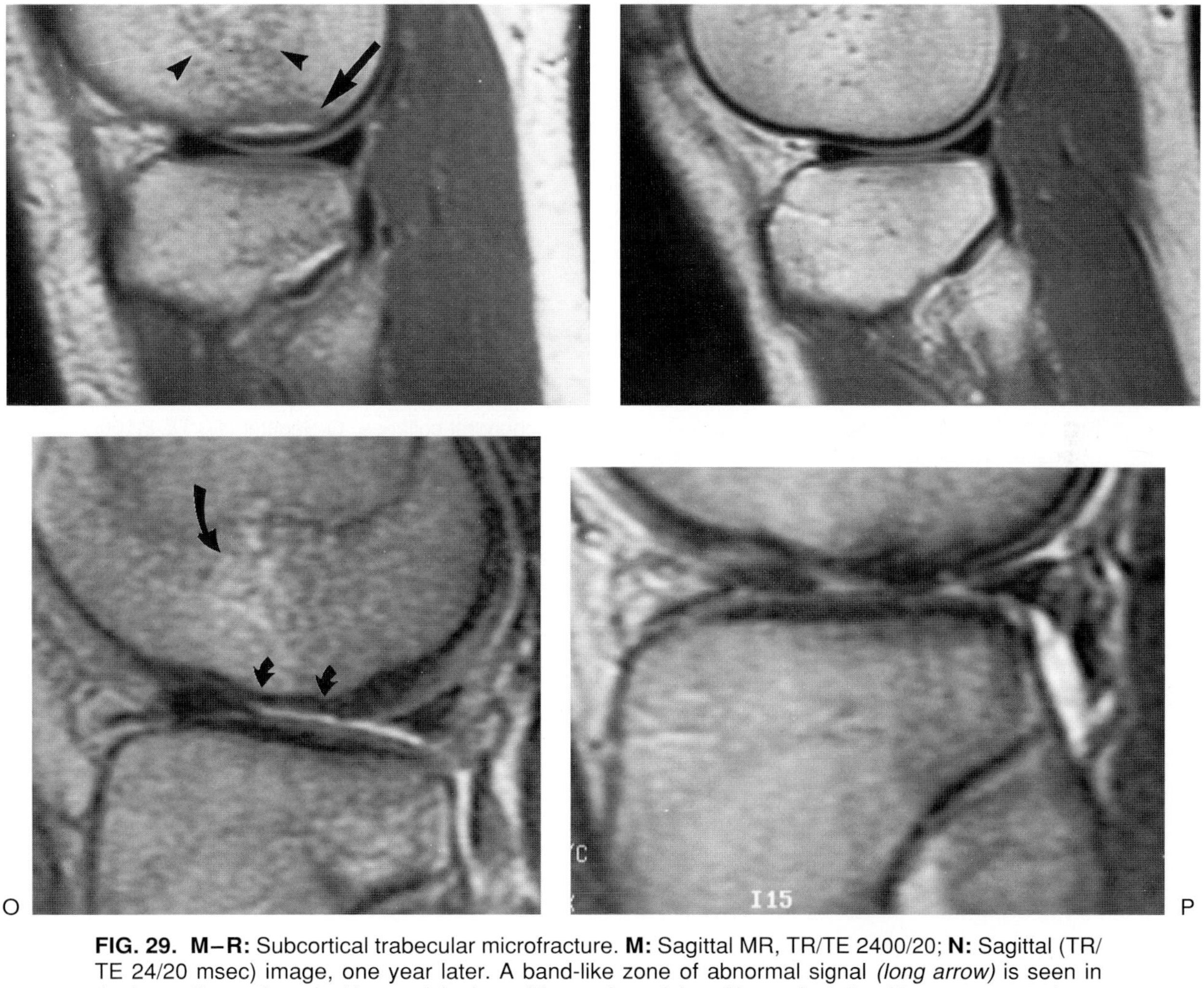

M N

O P

FIG. 29. M–R: Subcortical trabecular microfracture. **M:** Sagittal MR, TR/TE 2400/20; **N:** Sagittal (TR/TE 24/20 msec) image, one year later. A band-like zone of abnormal signal *(long arrow)* is seen in the immediate subcortical bone of the lateral femoral condyle, with poorly defined bone marrow edema in the epiphysis *(short arrows).* There is no defect in the contour of the cartilage or the cortical bone plate. The image in **N** demonstrates complete resolution of the lesion. Complete resolution of subcortical lesions is unusual. **O,P:** Sagittal (TR/TE 2,300/80 msec) image. **P** is 18 months following **O**. At the time of **O,** the patient suffered an acute ACL tear, and had a subcortical lesion with perhaps a minor degree of involvement of the articular surface *(arrows).* Edema *(long curved arrow)* is seen in the typical location in the meta-epiphysis. In **P,** there are areas of collapse, sclerosis and articular incongruity.

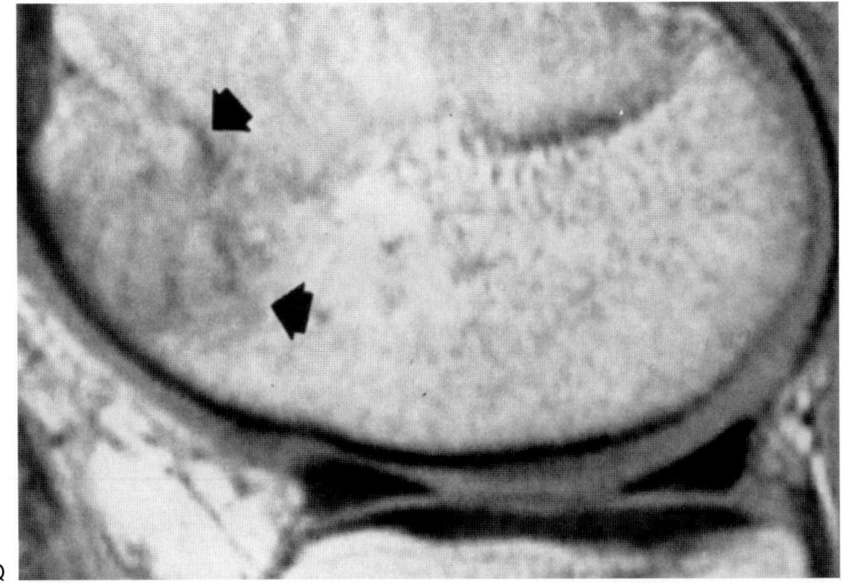

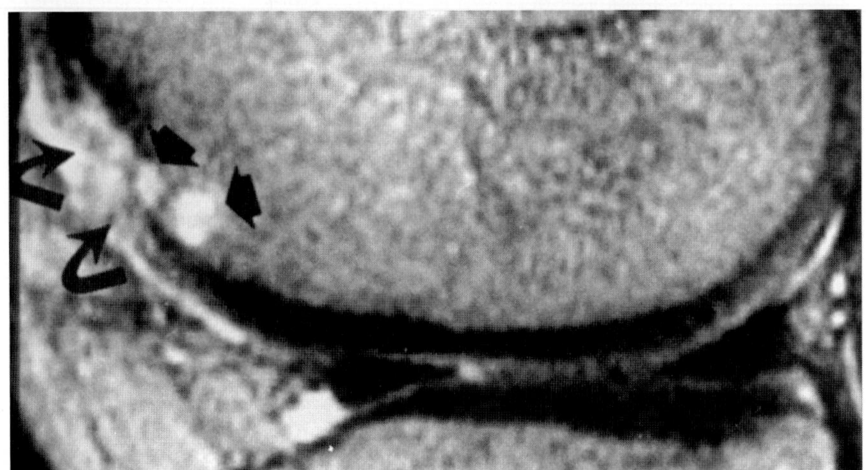

FIG. 29. Q,R: Trabecular injury with progression to subchondral collapse. **Q** Sagittal (TR/TE 2,100/80 msec); **R:** Sagittal (TR/TE 2,100/80 msec), 6 months following **Q.** This 28-year-old professional basketball player sustained a fall on his flexed knee. Multiple foci of abnormal signal *(arrows)* are seen in the medullary bone of the anterior lateral femur, immediately deep to the subchondral bone. **R:** The patient's pain worsened. A follow-up scan demonstrates violation of the subchondral bone and overlying cartilage *(curved arrows).* Small cyst-like regions are present within the medullary bone *(arrows).* (Figures Q and R from Blum GM, Tirman PFJ, Crues JVIII. Osseous and cartilaginous trauma. In: Mink JH, Reicher MA, Crues JVIII, Deutsch AL, eds. *MRI of the knee, 2nd edition.* New York: Raven Press, 1993;311, with permission.)

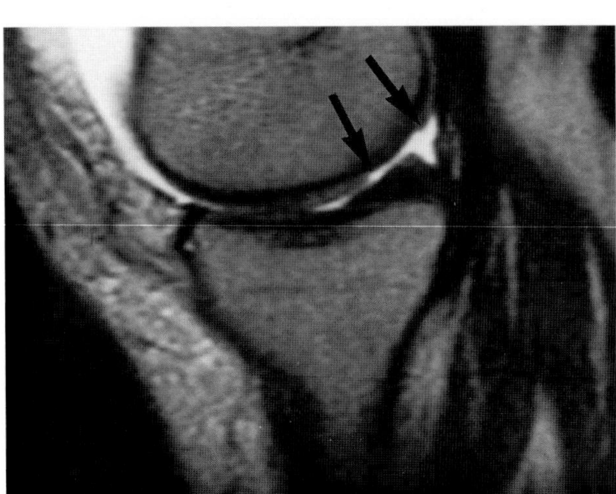

FIG. 29. S: Chondral fracture. Sagittal (TR/TE 2,200/80 msec) image. The patient fell, but had little pain in her knee. Several hours later, she developed an effusion and developed locking of the knee. The image demonstrates a defect in the articular cartilage covering the posteromedial femoral condyle *(arrows).* The lesion is identified by a change in contour of the cartilaginous surface and by a focal accumulation of joint fluid over the posteromedial meniscus.

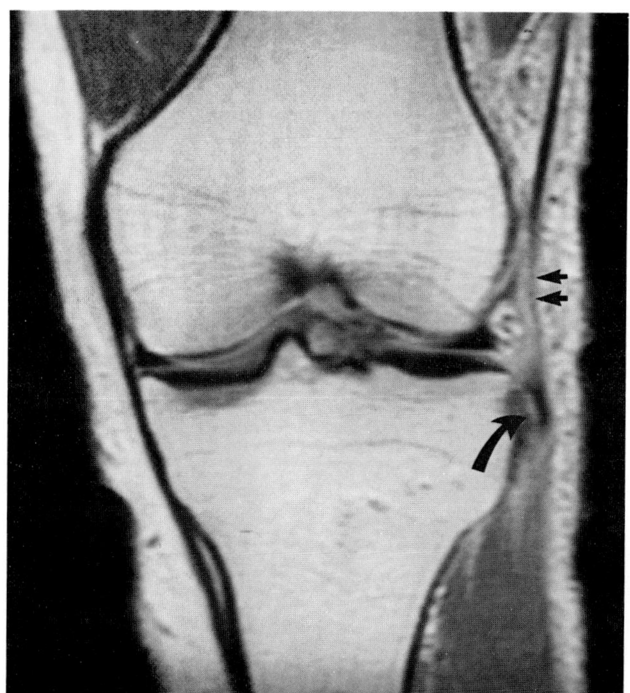

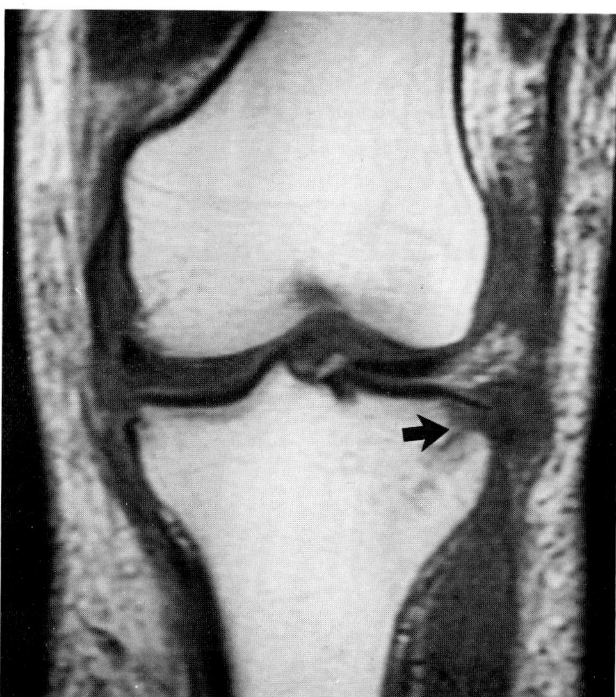

FIG. 29. T,U: Segond fractures. Coronal (TR/TE 2,000/20 msec). **T:** The middle third of the lateral capsular ligament has avulsed a small fragment of bone *(arrow)* from the tibial epiphysis. The distal iliotibial band *(arrows)* is only partially seen. **U:** In another patient, edema of the medullary bone *(arrow)* is seen at the site of avulsion of a very small Segond fracture, which could not be seen on the MR.

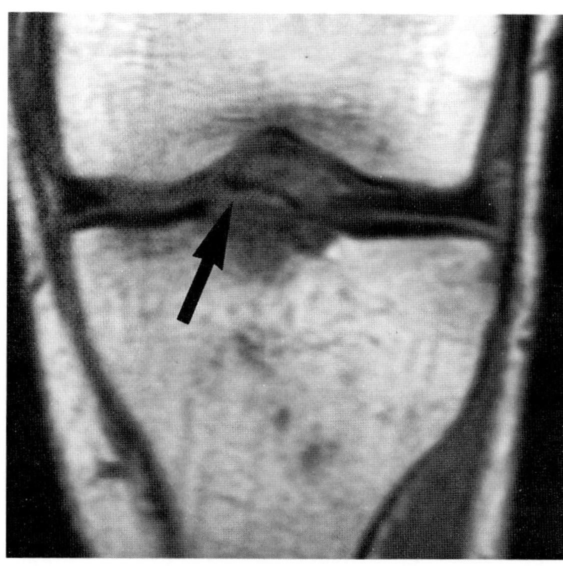

FIG. 29. V: Avulsion fracture of the insertion of the ACL. Coronal (TR/TE 2,000/20 msec) image. The image demonstrates a minimally displaced avulsion fracture of the insertion of the ACL *(large arrow)*. There is a minor degree of bone marrow edema (trabecular microfracture) associated with the fracture.

CASE 30

History: A 28-year-old man presents with history of intermittent knee locking. How would you characterize the findings relating to the posterior horn of the lateral meniscus?

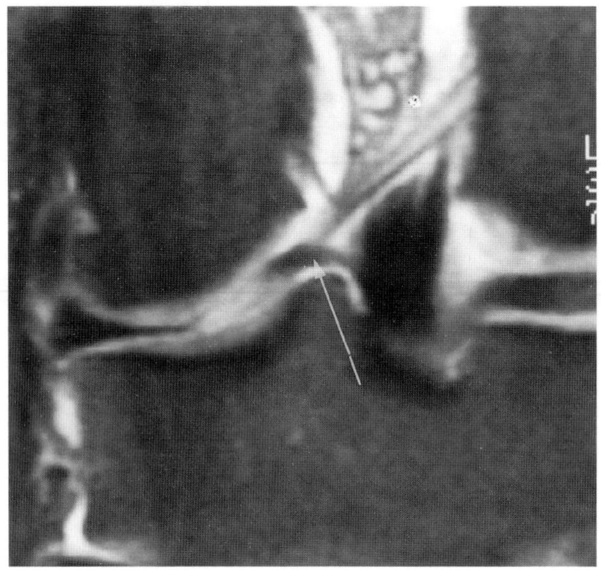

A

FIG. 30.

Findings: Figure 30A is a coronal fat-suppressed PD weighted image (TR/TE 4,000/17 msec) fast spin echo image through the posterior horn of the lateral meniscus. There is an apparent centrally displaced meniscal fragment (arrow). What would represent the peripheral fragment, however, demonstrates a sharp triangular configuration (arrowheads), a finding highly unlikely to be seen with a displaced bucket handle tear.

Diagnosis: Pseudotear of the posterior horn of the lateral meniscus.

Discussion: There are several areas of possible interpretive difficulty with regard to analysis of meniscal abnormalities involving the posterior horn of the lateral meniscus with conventional MR (1). One such problem relates to the distinction between pseudo and true flap tears involving the posterior horn as illustrated in the above case. The diagnosis of a displaced mensical tear is made by the recognition of abnormal mensical morphology and the detection of a component of the meniscus in an abnormal location, most commonly within the intercondylar notch. On conventional coronal MR images, however, the normal posterior horn of the lateral meniscus may simulate the appearance of a fragment projecting into the intercondylar notch posteriorly. The appearance results because the imaging plane, which is a straight line, intersects two nonadjacent points on the semicircular meniscus (1). The more lateral fragment represents the portion of the meniscus at the junction of the posterior horn with the midzone; the more central fragment is the extreme central portion of the meniscus and/or the mensical attachment to the intercondylar notch and the posterior cruciate ligament (see Fig. 30B,C). This central pseudofragment may be seen on one or occasionally two images if the scan is performed at a 3-mm or 4-mm section thickness. The possibility of mistaking the partially sectioned posterior horn/central attachment for a free meniscal fragment is much less common on the medial side of the knee because the medial meniscus is more comma-shaped than the lateral meniscus; the likelihood of acquiring an image that obliquely sections two noncontiguous segments of the medial posterior horn is much less likely. Excessive external rotation of the knee at the time of image acquisition accentuates the problem. This pitfall can be avoided by assessing the degree of internal or external rotation of the knee, by carefully examining the sagittal sections to ascertain that the posterior horn has a totally normal configuration, and by finding the central fragment on only one or occasionally two successive images (1).

The most commonly encountered area of interpretive difficulty with regard to the posterior horn of the lateral meniscus relates to the complex anatomy in and around the region of the popliteus tendon and tendon sheath as it violates the meniscocapsular junction (Fig. 30D–G).

This pseudotear of the posterior horn, which is now well recognized, may be encountered in 27% of routine conventional MR examinations (1,2). Several features serve to distinguish this pseudotear from a real tear of the posterior horn. The pseudotear invariably courses in a 30° to 45° anterosuperior to posteroinferior direction and the ''peripheral fragment,'' which is in fact the popliteus tendon, can be followed from the popliteal fossa above the joint line to its attachment on the posterior tibia below the joint line. The width of the pseudotear should only be on the order of 1 mm, it is seen on the most peripheral image through the ''bow-tie'' of the lateral meniscus, and rarely is seen on more than one image. True tears in comparison have a much less constant appearance.

A third pseudotear of the posterior horn of the lateral meniscus relates to the meniscofemoral ligament (MFL) and may be encountered on sagittal images quite near the intercondylar notch. The MFL represents a cord-like attachment of the lateral meniscus to the medial femoral condyle and been termed the *third posterior cruciate ligament* because of its relationship to the origin of the PCL (3,4). Anatomically, two distinct branches can be identified. The more anterior is known as the ligament of Humphry, which courses in an oblique orientation from the posterior lateral meniscus anterior to the PCL and toward its medial femoral condylar insertion (1). It is typically one third the diameter of the PCL. The posterior branch of the MFL, also known as the ligament of Wrisberg, has a similar orientation but courses immediately posterior to the PCL and is the larger of the two branches, approximating one half the cross-sectional diameter of the PCL. The MFL functions to pull the posterior horn of the lateral meniscus anteriorly and medially during flexion, balancing the action of the popliteus muscle whose fibers attach to the posterior and lateral aspects of the meniscus. (1)

While both branches of the MFL may exist, there are considerable variations in their frequency and size. In one MR study, the ligament of Humphry was visualized 33% of the time as a small ovoid low signal focus just anterior to the PCL, or more commonly in our experience, as a discrete low signal bulge inseparable from the concave surface of the PCL. The ligament of Wrisberg, visualized in 33% of MR cases, is virtually always identified as a discrete structure behind the PCL. Both ligaments are visualized in only 3% of MR studies. Either of these two low signal foci may be mistaken for free meniscal or osteochondral fragments (2,5,6) (Fig. 30H). Errors of interpretation will be avoided if one carefully follows the ''lesion'' on consecutive sagittal images.

When the origin of the MFL parallels and closely approximates the posterior margin of the lateral meniscus, a pseudotear is produced by the presence of fat interposed between the meniscal attachment and the meniscofemoral ligament. As the meniscus blends imperceptibly with its attachment, it loses its normal triangular configuration, becomes more rhomboid, and or may have a blunted tip. It is at this point that the meniscus may demonstrate a line of increased signal running from its superior to its inferior surface; the lesion might well be classified as a grade 3 signal because of its extension to an articular surface. This line is either nearly vertical in orientation (16%) or more commonly slightly oblique (84%), extending from superoanteriorly to inferoposteriorly (3). While the abnormal morphology of the meniscus at this point is often sufficient evidence that the plane of section is quite central and that one is no longer imaging the fibrocartilaginous meniscus (the image plane actually intersects the more fibrous meniscal attachment), one occasionally encounters a nearly triangular meniscal attachment.

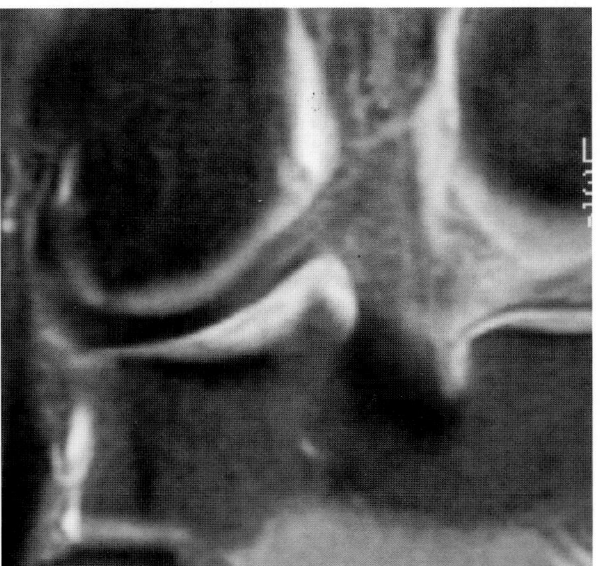
B

FIG. 30. B,C: Normal coronal anatomy of the posterior horn of the lateral meniscus. Coronal fat suppressed images obtained from same series as **A,B:** 5 mm posterior to **A; C:** 10 mm anterior. The most posterior section **B** intersects the extreme posterior margin of the lateral meniscus; the meniscal tissue has a bow-tie appearance quite analogous to the image depicted on sagittal sequences through the midzone.

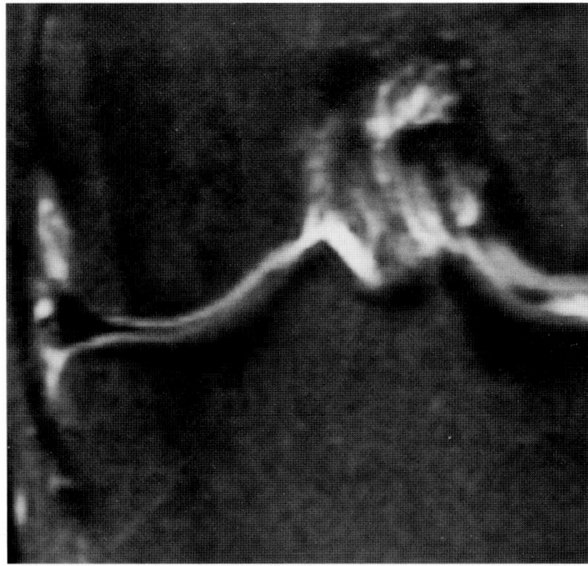

C

FIG. 30. C: The image plane has moved anteriorly into the midsubstance of the meniscus, which demonstrates a characteristic triangular configuration. In **A** the imaging plane "crossed" the free edge and intersected two noncontiguous portions of the posterior horn. It is on this image (between B and C) that confusion with a free meniscal fragment may occur.

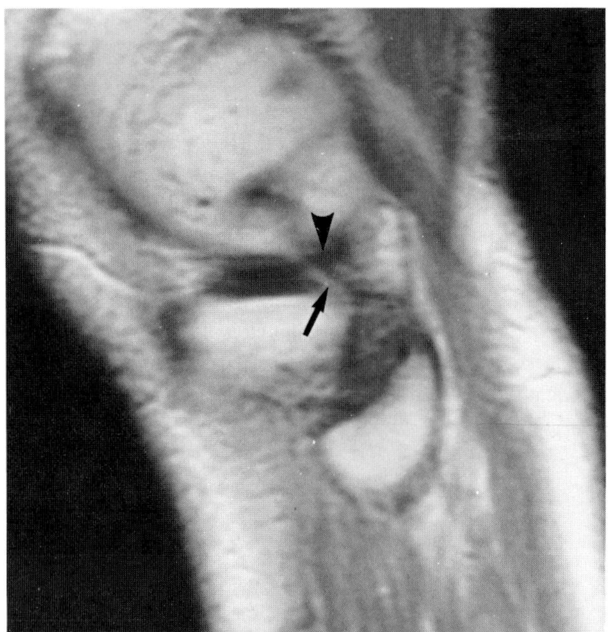

D

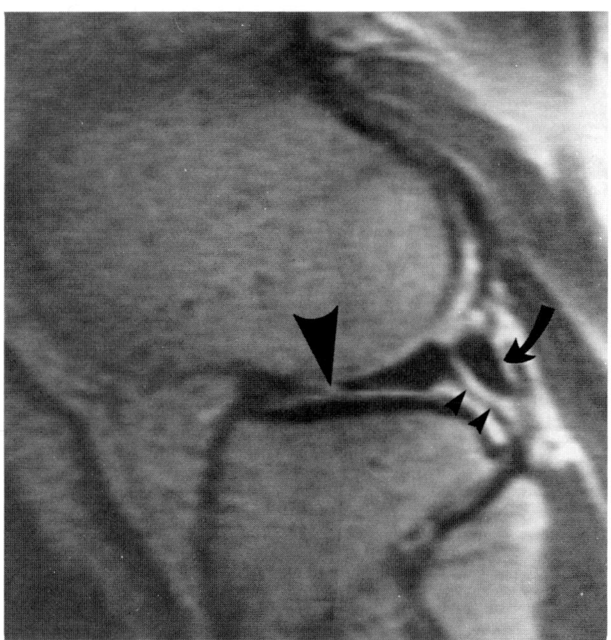

E

FIG. 30. D–G: Pseudotear of the posterolateral meniscus. **D:** Sagittal PD (TR/TE 2,000/20 msec) image demonstrates a linear focus of high signal *(arrow)* extending across the posterior horn of the lateral meniscus at approximately 30° angle from the vertical. The location and appearance is overwhelming suggestive of a pseudotear caused by the popliteus tendon. **E,F:** Sagittal T2-weighted images from a different patient with a radial tear of the midzone of the lateral meniscus *(large arrowhead).* **E:** The popliteus tendon *(arrow)* is violating the superior capsular margin and the superior fascicle is absent. The inferior fascicle *(arrowheads)* is the only peripheral attachment.

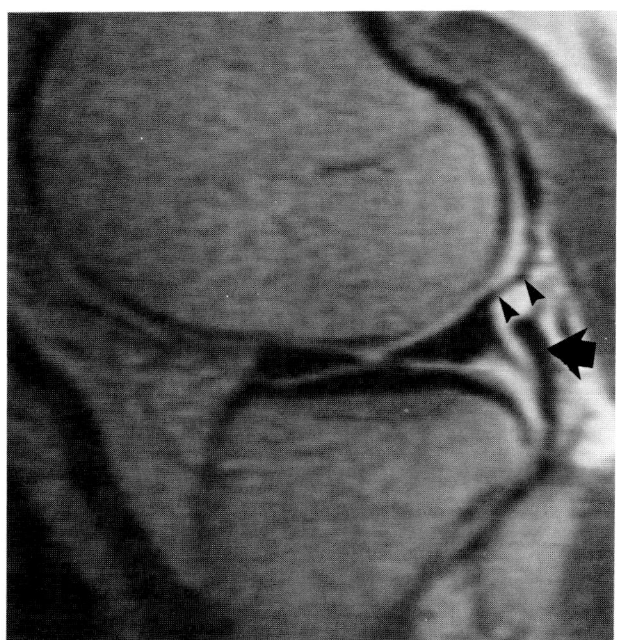

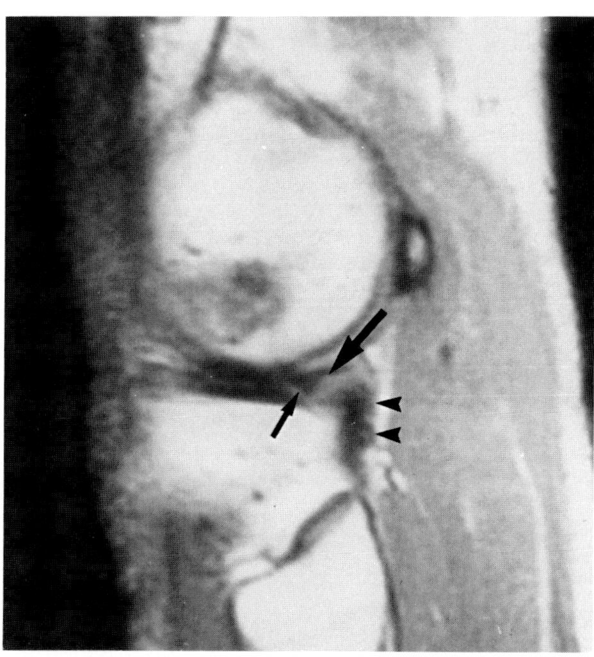

FIG. 30. F: The superior fascicle has reestablished itself *(arrowheads)* and the inferior fascicle appears torn as the popliteus tendon interposes itself between the meniscus and the capsule. In both E and F the popliteus bursa contains fluid and is clearly delineated anterior to the popliteus tendon. **G:** Sagittal T1-weighted (TR/TE 800/20 msec) image of a patient with a true tear of the posterior horn of the lateral meniscus. The oblique line of signal *(arrow)* extending across the posterior horn suggests the presence of an oblique pseudotear. The section, however, is not the most peripheral section where one would expect to encounter the pseudotear; in fact, the actual position of the popliteus tendon sheath *(large arrow)* is posterior to the surgically proven tear. The popliteus tendon *(arrowheads)* has traversed the meniscus and lies below the tibial plateau. (From Mink JH. Pitfalls in interpretation. In: Mink JH, Reicher MA, Crues JV III, Deutsch AL, eds. *MRI of the knee, 2nd edition.* New York: Raven Press, 1993;433–439, with permission.)

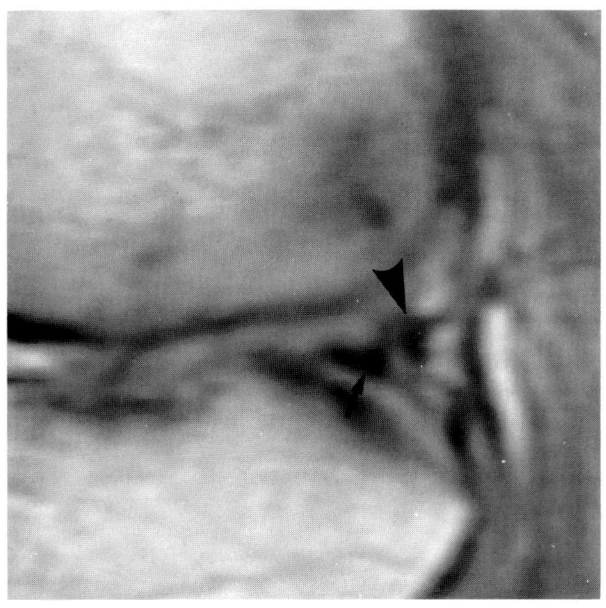

FIG. 30. H: Pseudotear of the lateral meniscus. Sagittal T1-weighted (TR/TE 800/20 msec) image accomplished near the central attachment of the posterior horn of the lateral meniscus. Two low signal intensity structures resemble portions of a torn meniscus. The origin of the ligament of Wrisberg *(arrowhead)* from the meniscal attachment *(arrow)* accounts for the appearance. Following the course of these structures on sequential images assists in establishing the correct diagnosis. (From Mink JH. Pitfalls in interpretation. In: Mink JH, Reicher MA, Crues JV III, Deutsch AL, eds. *MRI of the knee, 2nd edition.* New York: Raven Press, 1993;433–439, with permission.)

REFERENCES

Introduction

1. Ruwe PA, McCarthy S. Cost-effectiveness of magnetic resonance imaging. In: Mink JH, Crues JV, Reicher MA, Deutsch AL, eds. *MRI of the knee,* 2nd ed. New York: Raven Press, 1993.
2. Boden SD, Labaopoulos PA, Vailas SL. MR scanning of the acutely injured knee: sensitive, but is it cost effective? *Arthroscopy* 1990; 6:306–310.
3. Ruwe P, McCarthy S, Wright J, et al. Does MR imaging effectively replace diagnostic arthroscopy? *Radiology* 1992;183:335–339.
4. Hede A, Hempel-Poulson S, Jensen JS. Symptoms and level of sports activity in patients awaiting arthroscopy for meniscal lesions of the knee. *J Bone Joint Surg* 1990;72A:550–552.
5. Boeree NR, Watkinson AF, Ackroyd CE, et al. Magnetic resonance imaging of meniscal injuries of the knee. *J Bone Joint Surg* 1991; 73B:452–457.

Case 1

1. Reicher MA, Hartzman S, Duckwiler GR, et al. Meniscal injuries: detection with MR imaging. *Radiology* 1986;59:753–757.
2. Stoller DW, Martin C, Crues JV, et al. Meniscal tears: pathologic correlation with MR imaging. *Radiology* 1987;163:731–735.
3. Crues JV, Mink J, Levy TL, et al. Meniscal tears of the knee: accuracy of MR imaging. *Radiology* 1987;164:445–448.
4. Fischer SP, Fox JM, DelPizzo W, et al. Accuracy of diagnoses from magnetic resonance imaging of the knee. *J Bone Joint Surg* 1991; 73A:2–10.
5. Raunest J, Oberle K, Loehnert J, et al. The clinical value of magnetic resonance imaging in the evaluation of meniscal disorders. *J Bone Joint Surg* 1991;73A:11–16.
6. Glashow JL, Katz R, Schneider M, et al. Double blind assessment of the value of magnetic resonance imaging in the diagnosis of anterior cruciate and meniscal lesions. *J Bone Joint Surg* 1989; 71A:113–119.
7. Mink JH, Levy T, Crues JV. Tears of the anterior cruciate ligament and menisci of the knee: MR imaging evaluation. *Radiology* 1988; 167:769–774.
8. Firooznia H, Golimbu C, Rafi M. MR imaging of the menisci: fundamental of anatomy and pathology. *Magn Reson Imaging Clin North Am* 1994;2:325–347.
9. Bonamo J, Saperstein A. Contemporary magnetic resonance imaging of the knee: the orthopedic surgeon's perspective. *Magn Reson Imaging Clin North Am* 1994;2:481–495.
10. Justice WW, Quinn SF. Error patterns in the MR imaging evaluation of menisci of the knee. *Radiology* 1995;196:617–621.
11. DeSmet A, Graf B. Meniscal tears missed on MR imaging: relationship to meniscal tear patterns and anterior cruciate ligament tars. *AJR* 1994;162:1419–1423.
12. Quinn SF, Brown RE, Szumonwski J. Menisci of the knee: radial MR imaging correlated with arthroscopy in 259 patients. *Radiology* 1992;185:577–580.
13. Disler DG, Kattaapuram SV, Chew FS, et al. Meniscal tears of the knee: preliminary comparison of three dimensional MR reconstruction with two dimensional MR imaging and arthroscopy. *AJR* 1993; 160:343–345.
14. Tyrell RL, Gluckert K, Paatria M, et al. Fast three dimensional MR imaging of the knee: comparison with arthroscopy. *Radiology* 1988; 166:865–872.
15. Solomon SL, Totty WG, Lee JOT. MR imaging of the knee: comparison of three dimensional FISP and two dimensional spin echo pulse sequences. *Radiology* 1989;173:739–742.
16. Spritzer CE, Vogler JB, Garret WE, et al. MR imaging of the knee: preliminary results with a 3DFT GRASS pulse sequence. *AJR* 1988; 150:597–603.
17. Reeder JD, Matz SO, Becker L, et al. MR imaging of the knee in the sagittal projection: comparison of three dimensional gradient echo and spin echo sequences. *AJR* 1989;153:537–540.
18. Harms SE, Flamig DP, Fisher CF, et al. New method for fast MR imaging of the knee. *Radiology* 1989;173:743–750.
19. Araki Y, Ootani F, Tsukaguchi I, et al. MR diagnosis of meniscal tears of the knee: value of axila three dimensional Fourier transformation GRASS images. *AJR* 1992;158:587–590.
20. Tuckman GA, Miller WJ, Remo JW, Fritts HW, Rozansky MI. Radial tears of the menisci: MR findings. *AJR* 1994;163:395–400.
21. Crues JV, Stoller DW. The menisci. In: Mink JH, Reicher MA, Crues JV, Deutsch AL, eds. *MRI of the knee,* 2nd ed. New York: Raven Press, 1993:96–115.
22. Metcalf RW. Arthroscopic meniscal surgery. In: McGinty JB, ed. *Operative arthroscopy.* New York: Raven Press, 1991;227–236.

Case 2

1. Noyes FR, Bassett RW, Grood ES. Arthroscopy in acute traumatic hemarthrosis of the knee. *J Bone Joint Surg* 1980;62A:687–695.
2. Stoller DW. *Magnetic resonance imaging in orthopaedics and rheumatology.* Philadelphia: JB Lippincott, 1989.
3. Mink JH, Levy T, Crues JV III. Tears of the anterior cruciate ligament and menisci of the knee: MR imaging evaluation. *Radiology* 1988;167:769–774.
4. Barry KP, Mesgarzadeh M, Moyer R. Patterns and accuracy of diagnosis of anterior cruciate ligament tears with MR imaging. *Radiology* 1991;181(P):303.
5. Vahey TN, Broome DR, Kayes KJ. Acute and chronic tears of the anterior cruciate ligament: differential features at MR imaging. *Radiology* 1991;181:251–253.
6. Mink JH, et al. *Magnetic resonance imaging of the knee,* 2nd ed. New York: Raven Press, 1993.
7. Remer EM, Fitzgerald SW, Friedman H. Anterior cruciate ligament injury: MR imaging diagnosis and patterns of injury. *RadioGraphics* 1992;12:901–915.
8. Hodler J, Haghighi P, Trudell D. The cruciate ligaments of the knee: correlation between MR appearance and gross and histologic findings in cadaveric specimens. *AJR* 1992;159:357–360.
9. Bach BRJ, Warren RF. "Empty wall" and "vertical strut" signs of ACL insufficiency. *Arthroscopy* 1989;5:137–140.
10. Reiser M, Rupp N, Pfandner K. Imaging of cruciate ligament injuries using MR tomography. *ROFO* 1986;145:193–198.
11. Boeree NR, Watkinson AF, Ackroyd CE. Magnetic resonance imaging of meniscal and cruciate injuries of the knee. *J Bone Joint Surg* 1991;73B:452–457.
12. Glashow J, Katz R, Schneider M. Double-blind assessment of the value of magnetic resonance imaging in the diagnosis of anterior cruciate and meniscal lesions. *J Bone Joint Surg* 1989;71:113–119.
13. Fischer SP, Fox JM, Del Pizzo W. Accuracy of diagnoses from magnetic resonance imaging of the knee. *J Bone Joint Surg* 1991;73A:1–10.
14. Jackson DW, Jennings LD, Maywood RM. Magnetic resonance imaging of the knee. *Am J Sports Med* 1989;16:28–34.
15. Lee J, Yao L, Phelps C. Anterior cruciate ligament tears: MR imaging compared with arthroscopy and clinical tests. *Radiology* 1988; 166:861–864.
16. Christel P, Roger B, Witvoet J. Validity and diagnostic value of magnetic resonance imaging in traumatic meniscoligamentous pathology of the knee. Prospective study of 22 knees. *Rev Chir Orthop* 1988;74:402–412.
17. Jung T, Rodriguez M, Augustiny N. 1.5 T MRI arthrography and arthroscopy in the evaluation of knee lesions. *ROFO* 1988;148:390–393.
18. Fritz R, Helms C, Genant H. Frequency and significance of equivocal meniscal and anterior cruciate ligament tears in MR imaging of the knee. *Radiology* 1990;177(P):225.
19. Jacobs B, Applegate GR, Recht M. Comparison of frequency-selective fat saturation with spin-echo MR images in the detection of knee pathology. *J Magn Reson Imaging* 1991;1:176.
20. Crues J, Mink J, Levy T. Meniscal tears of the knee: accuracy of MR imaging. *Radiology* 1987;164:445–448.
21. Odensten M, Lysolm J, Gillquist J. The course of partial anterior cruciate ligament ruptures. *Am J Sports Med* 1985;13:183–186.
22. Casscells SW. *Arthroscopy: diagnostic and surgical practice.* Philadelphia: Lea & Febiger, 1984.

Case 3

1. Zobel MS, Borrello JA, Siegel MJ. Pediatric knee MR imaging: pattern of injuries in the immature skeleton. *Radiology* 1994;190:397–401.
2. Deutsch A, Mink J, Waxman A. Occult fractures of the proximal femur. *Radiology* 1989;170:113–116.
3. Vahey TN, Broome DR, Kayes KJ. Acute and chronic tears of the anterior cruciate ligament: differential features at MR imaging. *Radiology* 1991;181:251–253.
4. Graf BK, Cook DA, De Smet AA. "Bone bruises" on magnetic resonance imaging evaluation of anterior cruciate ligament injuries. *Am J Sports Med* 1993;21:220–223.
5. Jaramillo D, Shapiro F, Hoffer FA. Posttraumatic growth-plate abnormalities: MR imaging of bony-bridge formation in rabbits. *Radiology* 1990;175:767–773.
6. Kapelov SR, Teresi LM, Bradley WG. Bone contusions of the knee: increased lesion detection with fast spin-echo MR imaging with spectroscopic fat saturation. *Radiology* 1993;189:901–904.
7. Rosen MA, Jackson DW, Berger PE. Occult osseous lesions documented by magnetic resonance imaging associated with anterior cruciate ligament ruptures. *Arthroscopy* 1991;7:45–51.
8. Mink JH, et al. *Magnetic resonance imaging of the knee,* 2nd ed. New York: Raven Press, 1993.
9. Vellet AD, Marks PH, Fowler PJ. Occult posttraumatic osteochondral lesions of the knee: prevalence, classification, and short-term sequelae evaluated with MR imaging. *Radiology* 1991;178:271–276.
10. Spindler KP, Schils JP, Bergfeld JA. Prospective study of osseous, articular, and meniscal lesions in recent anterior cruciate ligament tears by magnetic resonance imaging and arthroscopy. *Am J Sports Med* 1993;21:551–557.
11. Rogers L. The radiography of epiphyseal injuries. *AJR* 1970;96: 289–299.
12. Ogden J. Injury to the growth mechanisms of the immature skeleton. *Skeletal Radiol* 1981;6:237–253.
13. Shapiro F. Epiphyseal disorders. *N Engl J Med* 1987;317:1702–1710.
14. Jaramillo D, Hoffer FA, Shapiro F. MR imaging of fractures of the growth plate. *AJR* 1990;155:1261–1265.
15. Smith BG, Rand F, Jaramillo D. Early MR imaging of lower-extremity physeal fracture-separations: a preliminary report. *J Pediatr Orthop* 1994;14:526–533.
16. Trueta J, Amato V. The vascular contribution to osteogenesis. III. Changes in the growth cartilage caused by experimentally induced ischaemia. *J Bone Joint Surg* 1960;42B:571–587.
17. Brighton C. The growth plate. *Orthop Clin North Am* 1984;15: 571–595.
18. Hynes D, O'Brien T. Growth disturbance lines after injury of the distal tibial physis. *J Bone Joint Surg* 1988;70B:231–233.
19. O'Brien T, Millis M, Griffin P. The early identification and classification of growth disturbances of the proximal end of the femur. *J Bone Joint Surg* 1986;68A:970–980.
20. Laor T, Jaramillo D. Metaphyseal abnormalities in children: pathophysiology and radiologic appearance. *AJR* 1993;161:1029–1036.

Case 5

1. Resnick D. Lipidoses, histiocytoses, and hyperlipoporteinemias. In: Resnick D, ed *Diagnosis of bone and joint disorders,* 3rd ed. Philadelphia: WB Saunders, 1995;2190–2203.
2. Beutler E. Gaucher's disease. *N Engl J Med* 1991;325:1354–1360.
3. Rosenthal DI, Scott JA, Barranger J, et al. Evaluation of Gaucher disease using magnetic resonance imaging. *J Bone Joint Surg* 1986; 68:802–806.
4. Lanir A, Hadar H, Cohen I, et al. Gaucher disease: assessment with MR imaging. *Radiology* 1986;161:239–244.
5. Cremin BJ, Davey H, Goldblatt J. Skeletal complication of type 1 Gaucher disease. The magnetic resonance features. *Clin Radiol* 1990;41:244–249.
6. Horev G, Kornreich L, Hadar H, et al. Hemorrhage associated with "bone crisis" in Gaucher's disease identified by magnetic resonance imaging. *Skeletal Radiol* 1991;20:479–482.
7. Hermann G, Shapiro R, Abdelwahab IF, et al: MR imaging in adults with Gaucher disease type 1: Evaluation of marrow involvement and disease activity. *Skeletal Radiol* 1993;22:247–251.
8. Johnson LA, Hoppel BE, Gerard EL, et al. Quantitative chemical shift imaging of vertebral bone marrow in patients with Gaucher disease. *Radiology* 1992;182:451–456.

Case 6

1. Mink JH. The cruciate and collateral ligaments. In: Mink JH, Reicher MA, Crues JV, Deutsch AL, eds. *MRI of the knee,* 2nd ed. New York: Raven Press, 1993;171–175.
2. Nicholas J, Hershman E, eds. *The lower extremity and spine in sports medicine.* St Louis: CV Mosby, 1986.
3. Hughston J, Andrews J, Cross M, et al. Classification of knee ligament instabilities. Part I: The medial compartment and cruciate ligaments. *J Bone Joint Surg* 1976;58:159–172.
4. Warren LF, Marshall JL. The supporting structures and layers on the medial side of the knee. An anatomic analysis. *J Bone Joint Surg* 1979;61A:56–62.
5. Brantigan OC, Voshell AF. The tibial collateral ligament: its function, its bursae,and its relationship to the medial meniscus. *J Bone Joint Surg* 1943;25:121–131.
6. Andrish JT. Ligament injuries of the knee. *Orthop Clin North Am* 1985;16(2):273–284.
7. Lee JK, Yao L. Tibial collateral ligament bursa: MR imaging. *Radiology* 1991;178:855–857.
8. Schweitzer ME, Tran D, Deely DM, et al. Medial collateral ligament injuries: evaluation of multiple signs, prevalence and location of associated bone bruises, and assessment with MR imaging. *Radiology* 1995;194:825–829.
9. Howe J, Johnson RJ. Knee injuries in skiing. *Orthop Clin North Am* 1985;16(2):303–313.
10. Yao L, Dungan D, Seeger LL. MR imaging of tibial collateral ligament injury: comparison with clinical examination. *Skeletal Radiol* 1994;23:521–524.
11. Jones RE, Henley MB, Francis P. Nonoperative management of isolated grade III collateral ligament injury in high school football players. *Clin Orthop* 1986;213:137–140.
12. Indelicato PA. Non-operative treatment of complete tears of the medial collateral ligament. *J Bone Joint Surg* 1983;65A:323–327.
13. Kannus P. Long term results of conservatively treated medial collateral ligament injuries of the knee. *Clin Orthop* 1988;226:103–112.
14. Fetto JF, Marshall JL. Medial collateral ligament injuries of the knee: a rationale for treatment. *Clin Orthop* 1978;132:206–218.

Case 7

1. Yao L, Gentilli A, Petrus L. Partial ACL tear: an MR diagnosis? *Skeletal Radiol* 1995;24:247–251.
2. Finsterbush A, Frankl U, Matan Y. Secondary damage to the knee after isolated injury of the anterior cruciate ligament. *Am J Sports Med* 1990;18:475–479.
3. Buckley SL, Barrack RL, Alexander H. The natural history of conservatively treated partial anterior cruciate ligament tears. *Am J Sports Med* 1989;17:221–225.
4. Noyes FR, Bassett RW, Grood ES. Arthroscopy in acute traumatic hemarthrosis of the knee. *J Bone Joint Surg* 1980;62A:687–695.
5. Fruensgaard S, Johannsen HV. Incomplete ruptures of the anterior cruciate ligament. *J Bone Joint Surg* 1989;71B:526–530.
6. Sandberg R, Balkfors B. Partial ruptures of the anterior cruciate ligament. *Clin Orthop* 1987;220:176–178.
7. Odensten M, Lysolm J, Gillquist J. The course of partial anterior cruciate ligament ruptures. *Am J Sports Med* 1985;13:183–186.
8. Glashow J, Katz R, Schneider M. Double-blind assessment of the value of magnetic resonance imaging in the diagnosis of anterior cruciate and meniscal lesions. *J Bone Joint Surg* 1989;71:113–119.
9. Stoller DW. *Magnetic resonance imaging in orthopaedics and rheumatology.* Philadelphia: JB Lippincott, 1989.
10. Yao L, Lee JK. Avulsion of the posteromedial tibial plateau by the semimembranosus tendon: diagnosis with MR imaging. *Radiology* 1989;172:513–514.
11. Noyes F, De Lucas J, Torvik P. Biomechanics of anterior cruciate ligament failure; an analysis of strain rate sensitivity and mechanisms of failure in primates. *J Bone Joint Surg* 1994;56A:236–253.
12. Daniel D, Akeson W, O'Connor J. *Knee ligaments. Structure, function, injury, and repair.* New York: Raven Press, 1991.
13. Noyes FR, Mooar LA, Moorman CT. Partial tears of the anterior

cruciate ligament. Progression to complete ligament deficiency. *J Bone Joint Surg* 1989;71:825–833.
14. Barrack RL, Buckley SL, Bruckner JD. Partial versus complete acute anterior cruciate ligament tears. The results of nonoperative treatment. *J Bone Joint Surg* 1990;72B:622–624.
15. Hodler J, Haghighi P, Trudell D. The cruciate ligaments of the knee: correlation between MR appearance and gross and histologic findings in cadaveric specimens. *AJR* 1992;159:357–360.
16. McCauley TR, Moses M, Kier R. MR diagnosis of tears of anterior cruciate ligament of the knee: importance of ancillary findings. *AJR* 1994;162:115–119.

Case 8

1. Yu J, Resnick D. Meniscal ossicle: MR imaging appearances in three patients. *Skeletal Radiol* 1994;23:637–639.
2. Harris HA. Calcification and ossification in the semilunar cartilage. *Lancet* 1934;1:1114–1116.
3. Pedersen HE. The ossicles of the semilunar cartilages of rodents. *Anat Rec* 1949;105:107.
4. Bernstein RM, Olsson HE, Spitzer RM, et al. Ossicle of the meniscus *AJR* 1976;127:785–788.
5. Glass RS, Barnes WM, Kells DU, et al. Ossicles of the knee. Menisci: report of seven cases. *Clin Orthop* 1975;111:163–171.
6. Yao J, Yao L. Magnetic resonance imaging of a symptomatic meniscal ossicle. *Clin Orthop* 1993;293:225–228.
7. Yao L, Lee J. Avulsion of the posteromedial tibial plateau by the semimembranosus tendon. Diagnosis with MR imaging. *Radiology* 1989;172:513.

Case 9

1. Lotysch M, Mink J, Crues JV, Schwartz SA. Magnetic resonance imaging in the detection of meniscal injuries. *Magn Reson Imaging* 1986;4:185.
2. Crues JV, Mink J, Levy TL, Lotysch M, Stoller DW. Meniscal tears of the knee: accuracy of MR imaging. *Radiology* 1987;164:445–448.
3. Kaplan PA, Nelson NL, Garvin KL, Brown DE. MR of the knee: the significance of high signal in the meniscus that does not clearly extend to the surface. *AJR* 1991;156:333–336.
4. DeSmet AA, Norris MA, Yandow DR, Quintana FA, Graf BK, Keene JS. MR diagnosis of meniscal tears of the knee: importance of high signal in the meniscus that extends to the surface. *AJR* 1993;161:101–107.
5. Davis SJ, Teresi LM, Bradley WG, Burke JW. The "notch" sign: meniscal contour deformities as indicators of tear in MR imaging of the knee. *J Comput Assist Tomogr* 1990;14:975–980.
6. Mesgarzadeh M, Moyer R, Leder DS, et al. MR imaging of the knee: expanded classification and pitfalls to interpretation of meniscal tears. *Radiographics* 1993;13:489–500.
7. Mink JH, Reicher MA, Crues JV, Deutsch AL. In: *MRI of the knee,* 2nd ed. New York; Raven Press, 1993;433–462.
8. Quinn SF, Brown TF. Meniscal tears diagnosed with MR imaging versus arthroscopy: how reliable a standard is arthroscopy? *Radiology* 1991;181:843–847.
9. Burk DL, Kanal E, Brunberg JA, Johnstone GF, Swensen HE, Wolf GL. 1.5T surface-coil MRI of the knee. *AJR* 1986;147:293–300.
10. Sisk DT. Arthroscopy of the knee and ankle. In: Crenshaw AH, ed. *Campbell's operative orthopaedics,* 7th ed. St. Louis: Mosby, 1987; 2547–2608.
11. DeHaven KE, Collins HR. Diagnosis of internal derangement of the knee: the role of arthroscopy. *J Bone Joint Surg* 1975;57A:802–810.
12. Stoller DW, Martin C, Crues JV, Kaplan L, Mink JH. Meniscal tears: pathologic correlation with MR imaging. *Radiology* 1987;163:731–735.
13. Mink JH, Levy T, Crues JV. Tears of the anterior cruciate ligament and menisci of the knee: MR imaging evaluation. *Radiology* 1988; 167:769–774.
14. Deutsch AL, Mink JH, Fox JM, et al. Peripheral meniscal tears: MR findings after conservative treatment or arthroscopic repair. *Radiology* 1990;176:485–488.
15. Weiss CB, Lundberg M, Hamberg P, DeHaven KE, Gillquist J. Non-operative treatment of meniscal tears. *J Bone Joint Surg* 1989; 71A:811–822.

Case 10

1. Convery FR, Meyers MH, Akerson WH. Fresh osteochondral allografting of the femoral condyle. *CORE* 1991;273:139–145.
2. Bert JM. Role of abrasion arthroplasty and debridement in the management of osteoarthritis of the knee. *Rheum Dis Clin North Am* 1993;19:725–740.
3. Itay S, Abramovici A, Nevo Z. Use of cultured embryonal chick epiphyseal chondrocytes as grafts for defects in chick articular cartilage. *Clin Orthop* 1987;220:284–303.
4. Ghosh P, Wells C, Smith M, Hutadilok N. Chondroprotection, myth or reality: an experimental approach. *Semin Arthritis Rheum* 1990; 19:3–9.
5. Howell DS, Altman RD. Cartilage repair and conservation in osteoarthritis. *Rheum Dis Clin North Am* 1993;19:713–724.
6. Rubenstein JD, Kim JK, Morava-Protzner I, Stanchev PL, Henkelman RM. Effects of collage orientation of MR imaging characteristics of bovine articular cartilage. *Radiology* 1993;188:219–226.
7. Recht MP, Kramer J, Marcelis S, Pathria MN, Trudell D, Haghighi P, Sartoris DJ, Resnick D. Abnormalities of articular cartilage in the knee: analysis of available MR techniques. *Radiology* 1993; 187:473–478.
8. Disler DG, McCauley TR, Fuchs MD, Wirth CR. Fat-suppressed 3D SPGR imaging of knee hyaline cartilage with arthroscopic correlation (Abst.) *AJR* 1995;164(3):40.
9. Noyes FR, Stabler CL. A system for grading articular cartilage lesions at arthroscopy. *Am J Sports Med* 1989;17(4):505–513.
10. Recht MP, Resnick D. MR imaging of articular cartilage: current status and future directions. *AJR* 1994;163:283–290.

Case 11

1. Yao L, Gentilli A, Petrus L. Partial ACL tear: an MR diagnosis? *Skeletal Radiol* 1995;24:247–251.
2. Kaplan PA, Walker CW, Kilcoyne RF. Occult fracture patterns of the knee associated with anterior cruciate ligament tears: assessment with MR imaging. *Radiology* 1992;183:835–838.
3. Cobby MJ, Schweitzer ME, Resnick D. The deep lateral femoral notch: an indirect sign of a torn anterior cruciate ligament. *Radiology* 1992;184:855–858.
4. Graf BK, Cook DA, De Smet AA. "Bone bruises" on magnetic resonance imaging evaluation of anterior cruciate ligament injuries. *Am J Sports Med* 1993;21:220–223.
5. Kapelov SR, Teresi LM, Bradley WG. Bone contusions of the knee: increased lesion detection with fast spin-echo MR imaging with spectroscopic fat saturation. *Radiology* 1993;189:901–904.
6. Lynch TCP, Crues JV III, Morgan FW. Bone abnormalities of the knee: prevalence and significance at MR imaging. *Radiology* 1989; 171:761–766.
7. Mink JH, Deutsch AL. Occult cartilage and bone injuries of the knee: detection, classification, and assessment with MR imaging. *Radiology* 1989;170:823–829.
8. Murphy BJ, Smith RL, Uribe JW. Bone signal abnormalities in the posterolateral tibia and lateral femoral condyle in complete tears of the anterior cruciate ligament: a specific sign? *Radiology* 1992;182: 221–224.
9. Rosen MA, Jackson DW, Berger PE. Occult osseous lesions documented by magnetic resonance imaging associated with anterior cruciate ligament ruptures. *Arthroscopy* 1991;7:45–51.
10. Stallenberg B, Gevenois PA, Sintzoff SA Jr. Fracture of the posterior aspect of the lateral tibial plateau: radiographic sign of anterior cruciate ligament tear. *Radiology* 1993;187:821–825.
11. Tervonen OGS, Stuart MJ. Traumatic trabecular lesions observed on MR imaging of the knee. *Acta Radiol* 1991;32:389–392.
12. Vellet AD, Marks PH, Fowler PJ. Occult posttraumatic osteochondral lesions of the knee: prevalence, classification, and short-term sequelae evaluated with MR imaging. *Radiology* 1991;178:271–276.
13. Moses M, Kier R, McCauley M. Anterior cruciate ligament injury: indirect sing at MR imaging. *Radiology* 1992;185P:146–147.
14. Tung GA, Davis LM, Wiggins ME. Tears of the anterior cruciate ligament: primary and secondary signs at MR imaging. *Radiology* 1993;188:661–667.
15. Vahey TN, Hunt JE, Shelbourne KW. Anterior translocation of the

tibia at MR imaging: a secondary sign of anterior cruciate ligament tear. *Radiology* 1993;187:817–819.
16. Mink JH, et al. *Magnetic resonance imaging of the knee,* 2nd ed. New York: Raven Press, 1993.
17. McCauley TR, Moses M, Kier R. MR diagnosis of tears of anterior cruciate ligament of the knee: importance of ancillary findings. *AJR* 1994;162:115–119.
18. Schweitzer ME, Cervilla V, Kursunoglu-Brahme S. The PCL line: an indirect sign of anterior cruciate ligament injury. *Clin Imaging* 1992;16:43–48.

Case 12

1. DeHaven KE. Diagnosis of acute knee injuries with hemarthrosis. *Am J Sports Med* 1980;8:9–14.
2. Cerabona F, Sherman MF, Bonamo JR, Sklar J. Patterns of meniscal injury with acute anterior cruciate ligament tears. *Am J Sports Med* 1988;16:603–609.
3. Mink JH, Reicher MA, Crues JV, Deutsch AL. In: *MRI of the knee,* 2nd ed. New York: Raven Press, 1993;142,433–439.
4. DeSmet AA, Graf BK. Meniscal tears missed on MR imaging: relationship to meniscal tear patterns and anterior cruciate ligament tears. *AJR* 1994;162:905–911.
5. Palmer I. On the injuries to the ligaments of the knee: a clinical study. *Acta Chir Scand Suppl* 1938;53:1–282.
6. Poehling GG, Ruch DS, Chabon SJ. The landscape of meniscal injuries. *Clin Sports Med* 1990;9:539–549.
7. McCauley TR, Moses M, Kier R, Lynch JK, Barton JW, Jokl P. MR diagnosis of tears of anterior cruciate ligament of the knee: importance of ancillary findings. *AJR* 1994;162:115–119.
8. Mesgarzadeh M, Moyer R, Leder DS, et al. MR imaging of the knee: expanded classification and pitfalls to interpretation of meniscal tears. *Radiographics* 1993;13:489–500.
9. Peterfy CG, Janzen DL, Tirman PFJ, van Dijke CF, Pollack M, Genant HK. "Magic-angle" phenomenon: a cause of increased signal in the normal lateral meniscus on short-TE MR images of the knee. *AJR* 1994;163:149–154.
10. Weiss CB, Lundberg M, Hamberg P, DeHaven KE, Gillquist J. Non-operative treatment of meniscal tears. *J Bone Joint Surg* 1989; 71A:811–822.

Case 13

1. Lance E, Deutsch AL, Mink JH. Prior lateral patellar dislocation: MR imaging findings. *Radiology* 1993;189:905–907.
2. Virolainen H, Visuri T, Kuusela T. Acute dislocation of the patella: MR findings. *Radiology* 1993;189:243–246.
3. Deutsch AL, Shellock FG. The extensor mechanism and patellofemoral joint. In: Mink JH, Reicher M, Crues JV, Deutsch AL, eds. *MRI of the knee,* 2nd ed. New York: Raven Press, 1994;218–220.
4. Hawkins RJ, Bell RH, Anisette G. Acute patellar dislocations. *Am J Sports Med* 1986;14(2):117–120.
5. Feagin JA. Subluxation of the patella versus ACL injury. In: Feagin JA, ed. *The crucial ligaments.* New York: Churchill-Livingstone, 1988.

Case 14

1. Mink JH. Pitfalls in interpretation. In: Mink JH, Reicher MA, Crues JV III, Deutsch AL, eds. *MRI of the knee,* 2nd ed. New York: Raven Press, 1993;450–452.
2. Shogry MEC, Pope TL. Vacuum phenomenon simulating mensical or cartilaginous injury of the knee at MR imaging. *Radiology* 1991; 180:513–515.
3. Turner DA, Rappaport ME, Erwin WD, et al. Truncation artifact: a potential pitfall in MR imaging of the menisci of the knee. *Radiology* 1991;179:629–633.
4. Mirowitz SA. Motion artifact as a pitfall in diagnosis of meniscal tear on gradient reoriented MRI of the knee. *J Comput Assist Tomogr* 1994;18(2):279–282.

Case 15

1. Feagin J. *The crucial ligaments.* New York: Churchill-Livingstone, 1988.
2. Van Dommelen B, Fowler P. Anatomy of the posterior cruciate ligament. A review. *Am J Sports Med* 1989;17:24–29.
3. Miller MD, Johnson DL, Harner CD. Posterior cruciate ligament injuries. *Orthop Rev* 1993;22:1201–1210.
4. Kennedy J. The posterior cruciate ligament. *J Trauma* 1967;7:367–377.
5. Sonin AH, Fitzgerald SW, Friedman H. Posterior cruciate ligament injury: MR imaging diagnosis and patterns of injury. *Radiology* 1994;190:455–458.
6. Delee J, Riley M, Rockwood C. Acute straight lateral instability of the knee. *Am J Sports Med* 1983;11:404–410.
7. Hughston J, Andrews J, Cross M. Classification of knee ligament instabilities. Part I: the medial compartment and cruciate ligaments. *J Bone Joint Surg* 1976;58:159–172.
8. Torisu T. Isolated avulsion fracture of the tibial attachment of the posterior cruciate ligament. *J Bone Joint Surg* 1977;59:68–72.
9. Geissler WB, Whipple TL. Intraarticular abnormalities in association with posterior cruciate ligament injuries. *Am J Sports Med* 1993;21:846–849.
10. Torg JS, Barton TM, Pavlov H. Natural history of the posterior cruciate ligament-deficient knee. *Clin Orthop* 1989;246:208–216.
11. Fowler PJ, Messieh SS. Isolated posterior cruciate ligament injuries in athletes. *Am J Sports Med* 1987;15:553–557.
12. Sonin AH, Fitzgerald SW, Hoff FL. MR imaging of the posterior cruciate ligament: normal, abnormal, and associated injury patterns. *RadioGraphics* 1995;15:551–561.
13. Oleaga L, Kressel H. High signal intensity in the posterior cruciate ligament on sagittal gradient-echo images: normal variant. *Radiology* 1990;177(P):195.
14. Grover JS, Bassett LW, Gross ML. Posterior cruciate ligament: MR imaging. *Radiology* 1990;174:527–530.
15. Stoller DW. *Magnetic resonance imaging in orthopaedics and rheumatology.* Philadelphia: JB Lippincott, 1989.
16. Patten RM, Richardson ML, Zink-Brody G. Complete vs partial-thickness tears of the posterior cruciate ligament: MR findings. *J Comput Assist Tomogr* 1994;18:793–799.
17. Casscells SW. *Arthroscopy: diagnostic and surgical practice.* Philadelphia: Lea & Febiger, 1984.
18. Jackson RW, Dandy DJ. *Arthroscopy of the knee.* New York: Grune & Stratton, 1976.
19. Cross MJ, Fracs MB, Powell JF. Long-term follow up of posterior cruciate ligament rupture: a study of 116 cases. *Am J Sports Med* 1984;12:292–297.
20. Yerys P. Arthroscopic posterior cruciate ligament repair. *Arthroscopy* 1991;7:111–114.
21. Loos WC, Fox JM, Blazina ME. Acute posterior cruciate ligament injuries. *Am J Sports Med* 1981;9:86–92.
22. Jackson DW, Jennings LD, Maywood RM. Magnetic resonance imaging of the knee. *Am J Sports Med* 1989;16:28–34.
23. Fischer SP, Fox JM, Del Pizzo W. Accuracy of diagnoses from magnetic resonance imaging of the knee. *J Bone Joint Surg* 1991; 73A:1–10.
24. Gross ML, Grover JS, Bassett LW. Magnetic resonance imaging of the posterior cruciate ligament. *Am J Sports Med* 1992;20:732–737.

Case 16

1. Johnson RJ, Eriksson E, Haggmark T, et al. Five to ten year follow-up evaluation after reconstruction of the anterior cruciate ligament. *Clin Orthop* 1984;183:122–140.
2. Melhorn JM, Henning CE. The relationship of the femoral attachment site to the isometric tracking of the anterior cruciate graft. *Am J Sports Med* 1987;15:539–542.
3. Jackson DW, Drez D. The anterior cruciate deficient knee. *New concepts in ligament repair.* St. Louis: CV Mosby, 1987.
4. Penner DA, Daniel DM, Wood P, et al. A study of anterior cruciate ligament graft placement and isometry. *Am J Sports Med* 1988;16: 238–243.
5. Deutsch AL, Mink JH. The post-operative knee. In: Mink JH, Reicher MA, Crues JV III, Deutsch AL, eds. *MRI of the knee,* 2nd ed. New York: Raven Press, 1993;270–292.
6. Manaster BJ, Remley K, Newman AP, et al. Knee ligament reconstruction: plain film analysis. *AJR* 1988;150:337–342.
7. Daniel D, Akerson W, O'Connor J. *Knee ligaments: structure, function, injury, and repair.* New York: Raven Press, 1990.

8. Feagin J. *The crucial ligaments.* New York: Churchill-Livingstone, 1988.
9. Howell SM. Serial magnetic resonance imaging of hamstring anterior cruciate ligament autografts during the first year of implantation. *Am J Sports Med* 1991;19:42–47.
10. Howell SM, Berns GS, Farley TE. Unimpinged and impinged anterior cruciate ligament grafts; MR signal intensity measurements. *Radiology* 1991;179:639–643.
11. Howell SM. Pixel signal intensity measurements unimpinged and impinged anterior cruciate ligament grafts: a prospective and reconstructive MR study. *Radiology* 1990;177:295.
12. Rak KM, Gillogly SD, Schaefer RA, et al. Anterior cruciate ligament reconstruction: evaluation with MR imaging. *Radiology* 1991;178:553–556.
13. Schultz TK, Black KP, Kneeland BJ, et al. Magnetic resonance imaging evaluation of an autogenous patellar tendon graft following anterior cruciate ligament reconstruction. In: Anaheim, CA: American Academy of Orthopedic Surgeons, 1991.
14. Bachmann G, Sens MA, Rauber K. MR imaging in follow-up of ligament reconstruction. In: 77th Scientific Assembly and Annual Meeting, Radiological Society of North America, Chicago, 1991.
15. Tosch US, Sander B, Lais ES, et al. MR follow-up of Gd DTPA enhanced autologous patellar ligament repair. In: 77th Scientific Assembly and Annual Meeting Radiological Society of North America, Chicago, 1991.
16. Arnoczky SP, Garvin GB, Marshall JL. Anterior cruciate ligament replacement using patellar tendon. *J Bone Joint Surg* 1982;64A:217–224.

Case 17

1. Goergen TG, Resnick D, Greenway G, et al. Dorsal defect of the patella (DDP): a characteristic radiographic lesion. *Radiology* 1979;130:333.
2. Johnson JF, Brogdon BG. Dorsal defect of the patella: incidence and distribution. *AJR* 1982;130:339.
3. Ho VB, Kransdorf MJ, Jelinek JS, et al. Dorsal defect of the patella: MR features. *J Comput Assist Tomogr* 1991;15:474.
4. Hunter LY, Hensinger RN. Dorsal defect of the patella with cartilaginous involvement. *Clin Orthop Rel Res* 1979;143:131.
5. Van Holsbeeck M, Vandamme B, Marchal G, et al. Dorsal defect of the patella; concept of its origin and relationship with bipartite and multipartite patella. *Skeletal Radiol* 1987;16:304.
6. Monu JUV, DeSmet AA. Case report 789. *Skeletal Radiol* 1993;22:528–531.

Case 18

1. Nicholas J, Hershman E, eds. *The lower extremity and spine in sports medicine.* St Louis: CV Mosby, 1986.
2. Mink JH. The cruciate and collateral ligaments. In: Mink JH, Reicher MA, Crues JV III, Deutsch AL, eds. *MRI of the knee,* 2nd ed. New York: Raven Press, 1993;179–185.
3. Delee J, Riley M, Rockwood C. Acute straight lateral instability of the knee. *Am J Sports Med* 1983;11:404–410.
4. Seebacher JR, Inglis AE, Marshall JL, et al. The structures of the posterolateral aspect of the knee. *J Bone Joint Surg* 1982;64:536–545.
5. Mayfield GW. Popliteus synovitis. *Am J Sports Med* 1977;5(1):31–36.
6. Rose DJ, Parisien JS. Popliteus tendon rupture. *Clin Orthop* 1988;226:113–117.
7. Renne JW. The iliotibial band friction syndrome. *J Bone Joint Surg* 1975;57A(8):1110–1111.

Case 19

1. Smith DK, Totty WG. The knee after partial menisectomy: MR imaging features. *Radiology* 1990;176:141–144.
2. Deutsch AL, Mink JH. The post-operative knee. In: Mink JH, Reicher M, Crues JV, Deutsch AL, eds. *MRI of the knee,* 2nd ed. New York: Raven Press, 1993;237–290.
3. Deutsch AL, Mink JH, Fox JM, et al. The post-operative knee. *Magn Reson Q* 1992;8:23–54.
4. Mink JH, Levy T, Crues JV. Tear of the anterior cruciate ligament and menisci of the knee: MR imaging evaluation. *Radiology* 1988;167:769–774.
5. Tolin BS, Fox JM, DelPizo, et al. MR arthrography of the post-operative meniscus in the knee. Presented at American Orthopedic Society for Sports Medicine, 18th Annual Meeting, 1992.
6. Applegate GR, Flannigan BD, Tolin BS, et al. MR diagnosis of recurrent tears in the knee: value of intraarticular contrast material. *AJR* 1993;161:821–825.

Case 20

1. Mankin HJ, Brandt KD, Shulman LE. Workshop on etiopathogenesis of osteoarthritis. *J Rheum* 1986;13:1130–1159.
2. Peterfy CG, Genant HK. Emerging applications of magnetic resonance imaging for evaluating the articular cartilage. *Radiol Clin North Am* 1996; (in press).
3. Mow VC, Ratcliffe A, Poole AR. Cartilage and diarthrodial joints as paradigms for hierarchical materials and structures. *Biomaterials* 1992;13:67–97.
4. Peterfy CG, Majumdar S, Lang P, van Dijke CF, Sack K, Genant H. MR imaging of the arthritic knee: improved discrimination of cartilage, synovium and effusion with pulsed saturation transfer and fat-suppressed T1-weighted sequences. *Radiology* 1994;191:413–419.
5. Kim DK, Ceckler TL, Hascall VC, Calabro A, Balaban RS. Analysis of water-macromolecule proton magnetization transfer in articular cartilage. *Magn Reson Med* 1993;29:211–215.
6. Melki PS, Mulkern RV. Magnetization transfer effects in multislice RARE sequences. *Magn Reson Med* 1992;24:189–195.
7. Mow V, Fithian D, Kelly M. Fundamentals of articular cartilage and meniscus biomechanics. In: Ewing J, ed. *Articular cartilage and knee joint function.* New York: Raven Press, 1990;1–18.
8. Buckwalter J, Rosenberg L, Hunziker E. Articular cartilage: composition, structure, response to injury, and methods of facilitating repairs. In: Ewing J, ed. *Articular cartilage and knee joint function.* New York: Raven Press, 1990;19–54.
9. Broderick LS, Turner DA, Renfrew DL, Schnitzer TJ, Huff JP, Harris C. Severity of articular cartilage abnormality in patients with osteoarthritis: evaluation with fast spin-echo MR vs arthroscopy. *AJR* 1994;162:99–103.
10. Dodge GR, Poole AR. Immunohistochemical detection and immunochemical analysis of type II collagen degradation in human normal, rheumatoid and osteoarthritic articular cartilages and in explants of bovine articular cartilage cultured with interleukin 1. *J Clin Invest* 1989;83:647–661.
11. Mankin H. The response of articular cartilage to mechanical injury. *J Bone Joint Surg* 1982;64A:460–466.
12. Donohue J, Buss D, Oegema TJ, Thompson R. The effects of indirect blunt trauma on adult canine articular cartilage. *J Bone Joint Surg* 1983;65A:948–957.
13. Armstrong CG, Mow VC, Wirth CR. Biomechanics of impact-induced microdamage to the articular cartilage: a possible genesis for chondromalacia. In: Finerman G, ed. *AAOS symposium on sports medicine: the knee.* St. Louis: WB Saunders, 1985;70–84.
14. Vener MJ, Thompson RCJ, Lewis JL, Oegema TR. Subchondral damage after acute transarticular loading: an in vitro model of joint injury. *J Orthop Res* 1992;10:759–769.
15. Shahriaree H. Chondromalacia. *Contemp Orthop* 1985;11:27–39.
16. Armstrong C. An analysis of the stresses in a thin layer of articular cartilage in a synovial joint. *N Engl J Med* 1986;15:55–61.

Case 21

1. Hecht MP, Applegate G, Kaplan P, et al. The MR appearance of cruciate ganglion cysts: a report of 16 cases. *Skeletal Radiol* 1994;23:597–600.
2. Bromley JW, Cohen P. Ganglion of the posterior cruciate ligament. Report of a case. *J Bone Joint Surg* 1965;47A:1247.
3. Garcia A, Hodler J, Vaughn L, et al. Case report 677. *Skeletal Radiol* 1991;20:373–375.
4. McLaren DB, Buckwalter KA, Vahey TN. The prevalence and significance of cyst-like changes at the cruciate ligament attachments in the knee. *Skeletal Radiol* 1992;21:365–369.
5. Lattes R. *Tumors of the soft tissues.* Washington, DC: Armed Forces Institute of Pathology, 1982.

6. Kissane JM. *Anderson's pathology,* vol 2, 8th ed. St. Louis: Mosby, 1985;1842.
7. Burk DL, Dalinka MK, Kanal E, et al. Meniscal and ganglion cysts of the knee: MR evaluation. *AJR* 1988;150:331–336.

Case 22

1. Howell SM. Serial magnetic resonance imaging of hamstring anterior cruciate ligament autografts during the first year of implantation. *Am J Sports Med* 1991;19:42–47.
2. Howell SM. Pixel signal intensity measurements unimpinged and impinged anterior cruciate ligament grafts: a prospective and reconstructive MR study. *Radiology* 1990;177:295.
3. Deutsch AL, Mink JH. The post-operative knee. In: Mink JH, Reicher MA, Crues JV III, Deutsch AL, eds. *MRI of the knee,* 2nd ed. New York: Raven Press, 1993;270–292.
4. Howell SM, Berns GS, Farley TE. Unimpinged and impinged anterior cruciate ligament grafts: MR signal intensity measurements. *Radiology* 1991;179:639–643.
5. Penner DA, Daniel DM, Wood P, et al. An introductory study of anterior cruciate ligament graft placement and isometry. *Am J Sports Med* 1988;16:238–243.
6. Schaefer RS, Gillogly SD, Rak KM, et al. Evaluation of patellar tendon autograft revascularization using Gadolinium-DTPA enhanced magnetic resonance imaging. (In press).
7. Tosch US, Sander B, Lais ES, et al. MR follow-up of Gd DTPA enhanced autologous patellar ligament repair. In: 77th Scientific Assembly and Annual Meeting Radiological Society of North America, Chicago, 1991.

Case 23

1. Resnick D. Physical injury: concepts and terminology. In: Resnick D, ed. *Diagnosis of bone and joint disorders.* Philadelphia: WB Saunders, 1995;2561–2692.
2. Pappas AM. *Osteochondrosis dissecans.* 1981;158:59–69.
3. Aichroth P. Osteochondritis of the knee. A clinical study. *J Bone Joint Surg* 1971;53B:440.
4. Nelson DW, DiPaola J, Colville M, et al. Osteochondritis dissecans of the talus and knee: prospective comparison of MR and arthroscopic classifications. *J Comput Assist Tomogr* 1990;14(5):804–808.
5. Mesgarzadeh M, Sapega AA, Bonakdarpour A, et al. Osteochondritis dissecans: Analysis of mechanical stability with radiography, scintigraphy, and MR imaging. *Radiology* 1987;165:775–780.
6. De Smet AA, Fisher DR, Graf BK, et al. Osteochondritis dissecans of the knee: value of MR imaging in determining lesion stability and the presence of articular cartilage defects. *AJR* 1990;155:549.
7. De Smet AA, Fisher DR, Bernstein MI, et al. Value of MR imaging in staging osteochondral lesions of the talus (osteochondritis dissecans): results in 14 patients. *AJR* 1990;154:555–558.
8. Adam G, Buhne M, Prescher A, et al. Stability of osteochondral fragments of the femoral condyle: magnetic resonance imaging with histopathologic correlation in an animal model. *Skeletal Radiol* 1991;20:601.
9. Kramer J, Stiglbauer R, Engel A, et al. MR contrast arthrography (MRA) in osteochondritis dissecans. *J Comput Assist Tomogr* 1992; 16(2):254–260.
10. Deutsch AL. Osteochondral injuries of the talar dome. In: Deutsch AL, Mink JH, Kerr R, eds. *MRI of the foot and ankle.* New York: Raven Press, 1993;118–134.
11. Peterfy CG, Majumdar S, Lang P, et al. MR imaging of the arthritic knee: improved discrimination of cartilage, synovium, and effusion with pulsed saturation transfer and fat suppressed T1-weighted techniques. *Radiology* 1994;191:413–419.

Case 24

1. Henning CE. Current status of meniscus salvage. *Clin Sports Med* 1990;9:567–576.
2. DeHaven KE, Sebastianelli WJ. Open meniscus repair. Indications, technique, and results. *Clin Sports Med* 1990;9:577–587.
3. Cooper DE, Arnoczky SP, Warren RF. Arthroscopic meniscal repair. *Clin Sports Med* 1990;9:589–607.
4. Arnoczky SP, McDevitt RF, Warren J, et al. Meniscal repair using an exogenous fibrin clot. An experimental study in the dog. *Orthop Trans* 1986;10:327.
5. Deutsch AL, Mink JH, Fox JM, et al. Peripheral meniscal tears: MR findings after conservative treatment or arthroscopic repair. *Radiology* 1990;176:485–488.
6. Farley TE, Howell SM, Love KF, et al. Meniscal tears: MR and arthrographic findings after arthroscopic repair. *Radiology* 1991; 180:517–522.
7. Kent R, Pope CF, Dillon EH, et al. MR imaging of the repaired meniscus: 12 month follow-up. *Magn Reson Imaging* 1991;9(3):379–388.
8. Deutsch AL, Mink JH. The post operative knee. In: Mink JH, Reicher M, Crues JV, Deutsch AL, eds. *MRI of the knee,* 2nd ed. New York: Raven Press, 1993;240–251.
9. Arnoczky SP, Cooper TG, Stadelmier DM, et al. Magnetic resonance signals in healing menisci: an experimental study in dogs. *Arthroscopy* 1994;10(5):522–557.

Case 25

1. Nance EPJ, Kaye JJ. Injuries of the quadriceps mechanism. *Radiology* 1982;142:301–307.
2. Kricum R, Dricum ME, Arangio GA, et al. Patellar tendon rupture with underlying systemic disease. *AJR* 1980;135(4).
3. Burgess RC, Guise ER. Infrapatellar tendon ruptures. *Orthopedics* 1985;8:362–364.
4. Yu JS, Petersilge C, Sartoris DJ, et al. MR imaging of injuries of the extensor mechanism of the knee. *Radiographics* 1994;14:541–551.
5. Zernicke RF, Jobe FW. Human patellar tendon rupture: a kinetic analysis. *J Bone Joint Surg* 1977;59A:179–183.
6. Noyes RF, Butler DL, Paulos LE, et al. Intra articular cruciate reconstruction. 1: Perspectives on graft strength, vascularization, and immediate motion after replacement. *Clin Orthop* 1983;172:71–77.
7. El-Khoury GY, Wira RL, Berbaum KS, et al. MR imaging of patellar tendinitis. *Radiology* 1992;184:849.
8. Reiff DB, Heenan SD, Heron CW. MRI appearances of the asymptomatic patellar tendon on gradient echo imaging. *Skeletal Radiol* 1995;24:123–126.
9. Berlin RC, Levinsohn EM, Chrisman H. The wrinkled patellar tendon: an indication of abnormality in the extensor mechanism of the knee. *Skeletal Radiol* 1991;20:181.
10. Gillogly SD, Schaefer RA, Rak KM, et al. The accuracy of magnetic resonance in assessment of patellar tendon autograft anterior cruciate ligament reconstruction. In: Anaheim, CA: American Academy of Orthopedic Surgeons, 1991.
11. Schultz TK, Black KP, Kneeland BJ, et al. Magnetic resonance imaging evaluation of an autogenous patellar tendon graft following anterior cruciate reconstruction. In: Anaheim, CA: American Academy of Orthopedic Surgeons, 1991.

Case 26

1. Deutsch AL, Shellock FG. Extensor mechanism and patellofemoral joint. In: Mink JH, Reicher MA, Crues JV III, Deutsch AL, eds. *MRI of the knee,* 2nd ed. New York: Raven Press 1993;220–233.
2. Shellock FG, Mink JH, Deutsch AL. Kinematic MRI of the joints: techniques and clinical applications. *Magn Reson Q* 1991;7:104–135.
3. Deutsch AL, Shellock FG, Mink JH. Imaging of the patellofemoral joint: emphasis on advanced techniques. In: Fox J, DelPizzo W, eds. *The patellofemoral joint.* New York: McGraw-Hill, 1993.
4. Molar NB, Koreas B, Juries GA. Patellofemoral incongruent in and instability of the patella. *Acta Orthop Scand* 1986;57:232–234.
5. Ficat RP, Hungerford DS. *Disorders of the patellofemoral joint.* Baltimore: Williams & Wilkins, 1977.
6. Shellock FG, Mink JH, Fox JM. Patellofemoral joint: kinematic MR imaging to assess tracking abnormalities. *Radiology* 1988;172:799–804.
7. Shellock FG, Mink JH, Deutsch AL, et al. Evaluation of patellar tracking abnormalities using kinematic MR imaging. Clinical experience in 130 patients. *Radiology* 1989;172:799–804.
8. Shellock FG, Mink JH, Deutsch AL, et al. Patellofemoral joint: identification of abnormalities using active movement, "unloaded" vs "loaded" kinematic MR imaging techniques. *Radiology* 1993; 188:575–578.

9. Shellock FG, Foo TKF, Deutsch AL, et al. Patellofemoral joint: evaluation during active flexion with ultrafast spoiled GRASS MR Imaging. *Radiology* 1991;180:581–585.
10. Brown SM, Muroff LR, Bradley WG, et al. Ultrafast kinematic MR imaging of the knee: increased sensitivity with a quadriceps loading device. *J Magn Reson Imaging* Suppl 1993;S29.
11. Shellock FG, Mink JH, Deutsch AL, et al. Evaluation in patients with persistent symptoms after lateral retinacular release by kinematic magnetic imaging of the patellofemoral joint. *Arthroscopy* 1990;6:226–234.
12. Brossman J, Muhle C, Schroder C, et al. Patellar tracking patterns during active and passive knee extension: Evaluation with motion-triggered cine MR imaging. *Radiology* 1993;187:205–212.
13. Shellock FG, Fox JM, Deutsch AL, et al. Medial subluxation of the patella: radiologic and physical findings. *Radiology* 1991;181(P):179.

Case 27

1. Abdelwahab IF, Denan S, Hermann G, et al. Periosteal ganglia: CT and MR imaging features. *Radiology* 1993;188:245–248.
2. McCarthy EF, Matz S, Steiner GC, et al. Periosteal ganglion; a cause of cortical bone erosion. *Skeletal Radiol* 1983;10:243–246.
3. Grange WF. Subperiosteal ganglion: a case report. *J Bone Joint Surg* 1978;60B:124–125.
4. Feldman F, Johnston A. Intraosseous ganglion. *AJR* 1973;118:328–343.
5. Feldman F, Singson RD, Staron RB. Magnetic resonance imaging of para-articular and ectopic ganglia. *Skeletal Radiol* 1989;18:353–358.

Case 28

1. Reicher MA. Knee joint disorders. In: Mink JH, Reicher MA, Crues JV III, Deutsch AL, eds. *MRI of the knee,* 2nd ed. New York: Raven Press, 1993;369–372.
2. Resnick D. Osteonecrosis: diagnostic techniques, specific situations, and complications. In: Resnick D, ed. *Diagnosis of bone and joint disorders,* 3rd ed. Philadelphia: WB Saunders, 1995;3539–3542.
3. Ahlback S, Bauer GCH, Bohne WH. Spontaneous osteonecrosis of the knee. *Arthritis Rheum* 1968;11:705–753.
4. Bjorkengren AG, Al Rowaih A, Lindstrand A, et al. Spontaneous osteonecrosis of the knee: p value of MR imaging in determining prognosis. *AJR* 1990;154:331–336.
5. Kursunoglu-Brahme S, Fox JM, Ferkel RD, et al. Osteonecrosis of the knee after arthroscopic surgery: diagnosis with MR imaging. *Radiology* 1991;178:851–853.
7. Lotke PA, Ecker ML. Osteonecrosis like syndrome of the medial tibial plateau. *Clin Orthop Rel Res* 1983;1765:148–153.
8. Norman A, Baker ND. Spontaneous osteonecrosis of the knee and medial meniscal tears. *Radiology* 1978;129:653–656.
9. Rozing PM, Insall J, Bohne W. Spontaneous osteonecrosis of the knee. *J Bone Joint Surg* 1980;62(A):2–7.
10. Pollack MS, Dalinka MK, Dressel HY, et al. Magnetic resonance imaging in the evaluation of suspected osteonecrosis of the knee. *Skeletal Radiol* 1987;16:121–127.
11. Aglietti P, Insall JN, Buzzi R, et al. Idiopathic osteonecrosis of the knee etiology, prognosis, and treatment. *J Bone Joint Surg* 1983; 65:588–597.

Case 29

1. Kaplan PA, Walker CW, Kilcoyne RF. Occult fracture patterns of the knee associated with anterior cruciate ligament tears: assessment with MR imaging. *Radiology* 1992;183:835–838.
2. Mink JH, Deutsch AL. Occult cartilage and bone injuries of the knee: detection, classification, and assessment with MR imaging. *Radiology* 1989;170:823–829.
3. Mink JH, et al. *Magnetic resonance imaging of the knee,* 2nd ed. New York: Raven Press, 1993.
4. Stallenberg B, Gevenois PA, Sintzoff SA, Jr. Fracture of the posterior aspect of the lateral tibial plateau: radiographic sign of anterior cruciate ligament tear. *Radiology* 1993;187:821–825.
5. Cobby MJ, Schweitzer ME, Resnick D. The deep lateral femoral notch: an indirect sign of a torn anterior cruciate ligament. *Radiology* 1992;184:855–858.
6. Schweitzer M, Tran D, Deely D. Medial collateral ligament injuries. *Radiology* 1995;194:825–829.
7. Rosen MA, Jackson DW, Berger PE. Occult osseous lesions documented by magnetic resonance imaging associated with anterior cruciate ligament ruptures. *Arthroscopy* 1991;7:45–51.
8. Murphy BJ, Smith RL, Uribe JW. Bone signal abnormalities in the posterolateral tibia and lateral femoral condyle in complete tears of the anterior cruciate ligament: a specific sign? *Radiology* 1992;182: 221–224.
9. Graf BK, Cook DA, De Smet AA. ''Bone bruises'' on magnetic resonance imaging evaluation of anterior cruciate ligament injuries. *Am J Sports Med* 1993;21:220–223.
10. Kapelov SR, Teresi LM, Bradley WG. Bone contusions of the knee: increased lesion detection with fast spin-echo MR imaging with spectroscopic fat saturation. *Radiology* 1993;189:901–904.
11. Tervonen OGS, Stuart MJ. Traumatic trabecular lesions observed on MR imaging of the knee. *Acta Radiol* 1991;32:389–392.
12. Yao L, Lee JK. Occult intraosseous fracture: detection with MR imaging. *Radiology* 1988;167:749–751.
13. Lynch TCP, Crues JV III, Morgan FW. Bone abnormalities of the knee: prevalence and significance at MR imaging. *Radiology* 1989; 171:761–766.
14. Vellet AD, Marks PH, Fowler PJ. Occult posttraumatic osteochondral lesions of the knee: prevalence, classification, and short-term sequelae evaluated with MR imaging. *Radiology* 1991;178:271–276.
15. Berger PE, Ofstein RA, Jackson DW. MRI demonstration of radiographically occult fractures: What have we been missing? *RadioGraphics* 1989;9:407–436.
16. Tung GA, Davis LM, Wiggins ME. Tears of the anterior cruciate ligament: primary and secondary signs at MR imaging. *Radiology* 1993;188:661–667.
17. Radin E, Pugh J, Rose R. Response of joints to impact loading. III. Relationship between trabecular microfractures and cartilage degenerations. *J Biomech* 1973;6:51–75.
18. Milgram J. Injury to articular cartilage joint surfaces. II. Displaced fracture of underlying bone. *Clin Orthop* 1986;206:236–247.
19. Repo R, Finlay J. Survival of articular cartilage after controlled impact. *J Bone Joint Surg* 1977;59A:1068–1076.
20. Weber WN, Neumann CH, Barakos JA. Lateral tibial rim (Segond) fractures: MR imaging characteristics. *Radiology* 1991;180:731–734.
21. Yao L, Lee JK. Avulsion of the posteromedial tibial plateau by the semimembranosus tendon: diagnosis with MR imaging. *Radiology* 1989;172:513–514.
22. McLaren DB, Buckwalter KA. The prevalence and significance of cyst-like changes at the cruciate ligament attachments in the knee. *Skeletal Radiol* 1992;21:365–369.

Case 30

1. Mink JH. Pitfalls in interpretation. In: Mink JH, Reicher MA, Crues JV III, Deutsch AL, eds. *MRI of the knee,* 2nd ed. New York: Raven Press, 1993;433–439.
2. Watanabe AT, Carter BC, Teitelbaum GP, et al. Common pitfalls in magnetic resonance imaging of the knee. *J Bone Joint Surg* 1989; 71A(6):857–862.
3. Vahey TN, Bennett HT, Arrington LE, et al. MR imaging of the knee: pseudotear of the lateral mensicus caused by the meniscofemoral ligament. *AJR* 1990;154:1237–1239.
4. Van Dommelen B, Fowler P. Anatomy of the posterior cruciate ligament. A review. *Am J Sports Med* 1989;17(1):24–29.
5. Carpenter WA. Meniscofemoral ligament simulating a tear of the lateral meniscus: MR features. *J Comput Assist Tomogr* 1990;14(6): 1033–1034.
6. Watanabe AT, Carter BC, Teitelbaum GP, et al. Normal variations in MR imaging of the knee: appearance and frequency. *AJR* 1989; 153:341–344.

CHAPTER 7

The Foot and Ankle

Andrew L. Deutsch, Mitchell A. Klein, and Jerrold H. Mink

Although one of the later articulations to which MR imaging was substantially applied, assessment of the ankle and foot has become one of the well-established musculoskeletal applications of MR. One of the most common indications for MR assessment of this region is the evaluation of tendon disorders. The content of this chapter reflects this interest with several cases depicting the gamut of Achilles tendon injuries, assessment of tendon repairs with MR, assessment of the posterior tibial tendon utilizing three-dimensional gradient echo techniques, and unique considerations including split tears of the peroneus brevis tendon, hemorrhagic tears of the subcutaneous tibialis tendon, and tendon lacerations.

Ankle sprains represent common injuries. MR is well suited to the assessment of both the medial and lateral collateral ligamentous complexes. The anatomy of these structures and the use of specialized techniques including three-dimensional gradient acquisitions for optimal assessment of the components of the deltoid complex are illustrated. While acute ankle injuries are uncommonly referred for MR evaluation (with the common exception of elite athletes), the sequelae of significant prior ankle sprains represent common indications for MR imaging. The use of MR for the assessment of chronic lateral impingement syndrome, the tarsal sinus syndrome, and osteochondral injuries of the talar dome (including MR arthrography) is extensively reviewed.

A number of entrapment syndromes occur in the foot and ankle. The anatomy of the tarsal tunnel and role of MR in the evaluation of suspected tarsal tunnel syndrome is illustrated. Morton neuroma represents an entrapment neuropathy of the interdigital nerve and the use of fat-suppressed contrast-enhanced techniques for the depiction of this entity are highlighted.

Bone abnormalities are common in this weight-bearing articulation. The role of MR in the assessment of occult bone injuries affecting this region is reviewed. Soft tissue infection and osteomyelitis are a frequent concern in the foot and ankle, particularly in diabetic patients. The role of MR and the use of dynamic contrast-enhanced examinations for detection of osteomyelitis is considered.

The authors approach the design of studies related to the foot by concentrating on the area of interest and performing small field-of-view acquisitions. Studies of the foot/ankle can be effectively divided into examinations of the ankle/hindfoot, midfoot, and forefoot. We have chosen this anatomic/imaging approach for two reasons: (1) the disease processes that affect each zone are different (e.g., osteochondritis dissecans and tendon ruptures occur in the hindfoot; infectious processes often predominate in the toes and forefoot); and (2) the imaging demands for each region are different and the coils and patient positioning may need to be adjusted for each site. The most common examination that we perform is that of the ankle/hindfoot. Utilizing a 12- to 14-cm field of view, the examination includes the distal tibial metaphysis above, the Achilles tendon posteriorly, the plantar soft tissues inferiorly, and the navicular or cuneiforms distally (depending on body size and the precise position of the patient). The midfoot examination includes most of the talus and calcaneus, the cuneiforms, and the metatarsals. The forefoot examination includes the phalanges, the metatarsals, and portions of the tarsals.

For examinations of the hindfoot and midfoot, the patient is positioned supine and the foot and ankle are positioned within the surface coil so that the medial malleolus is in the center of the coil. The foot is allowed to lie in a relaxed position; most patients will choose a position that places the foot in 10° to 20° of plantar flexion, and 10° to 30° of external rotation. Permitting the patient to choose a comfortable position helps ensure that motion artifact will be minimized. Specialized positioning, such as purposeful plantar flexion of the foot, can be performed but is rarely necessary. Examination of the forefoot has been considerably facilitated by the availability of flexible surface coils. Newly developed gradient coils allow extremely high resolution studies of the digits to be accomplished.

The authors presently use a combination of fast spin echo, inversion recovery, spoiled gradient echo, and three-dimensional gradient echo acquisitions depending on the specific area of concern and indication for the study. Fast spin echo T1 and T2 (often with fat suppression) and inversion recovery techniques are primarily utilized for general assessment and particularly for tendon disorders. Proton density–weighted fat-suppressed techniques can be useful for delineation of both bone and soft tissue edema while maintaining good anatomic delineation. Three-dimensional (3D) gradient echo techniques can be extremely useful for assessment of areas of complex anatomy including the distal posterior tibial tendon and deltoid ligament complex (Fig. A). In the absence of a 3D acquisition, at least two planes and frequently three orthogonal planes are acquired through the area of interest. Talar dome injuries require both true sagittal and coronal images. Fat suppression is commonly employed but can be problematic in the foot and ankle (see Marrow Disorders). The use of gadopentetate dimeglumine both intraarticularly and intravenously can be particularly useful for several applications including osteochondral injuries (intraarticular contrast) and osteomyelitis (intravenous contrast). These and other considerations are addressed in further detail in the teaching cases.

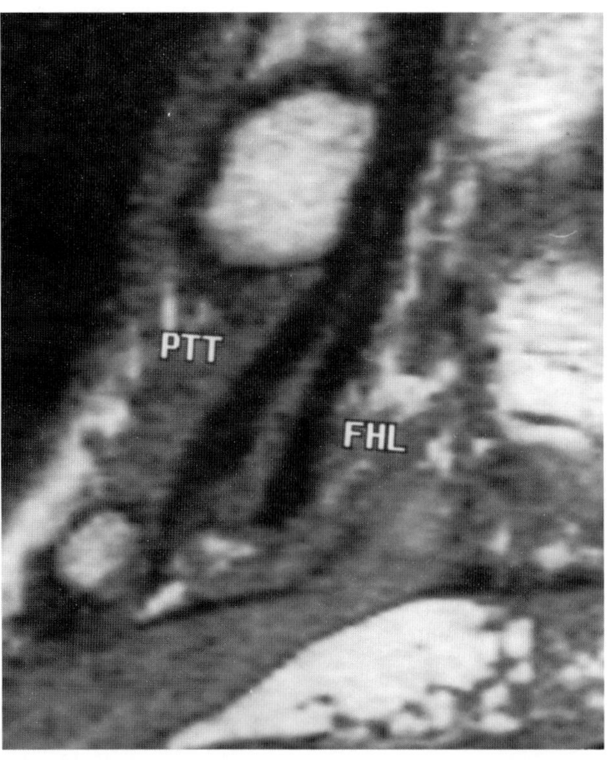

FIG. A: Oblique sagittal reformatted gradient echo image (TR/TE/° 50/10/45°) demonstrating the course of distal posterior tibial (PTT) and flexor hallucis longus (FHL) tendons. The 3D acquisition allows delineation of the entire course of the tendon and obviates the need for additional pulse sequences obtained in orthogonal planes.

CASE 1

History: A 50-year-old woman presents with chronic medial ankle pain and acquired flat foot deformity. What abnormality is depicted? In what way are the findings atypical?

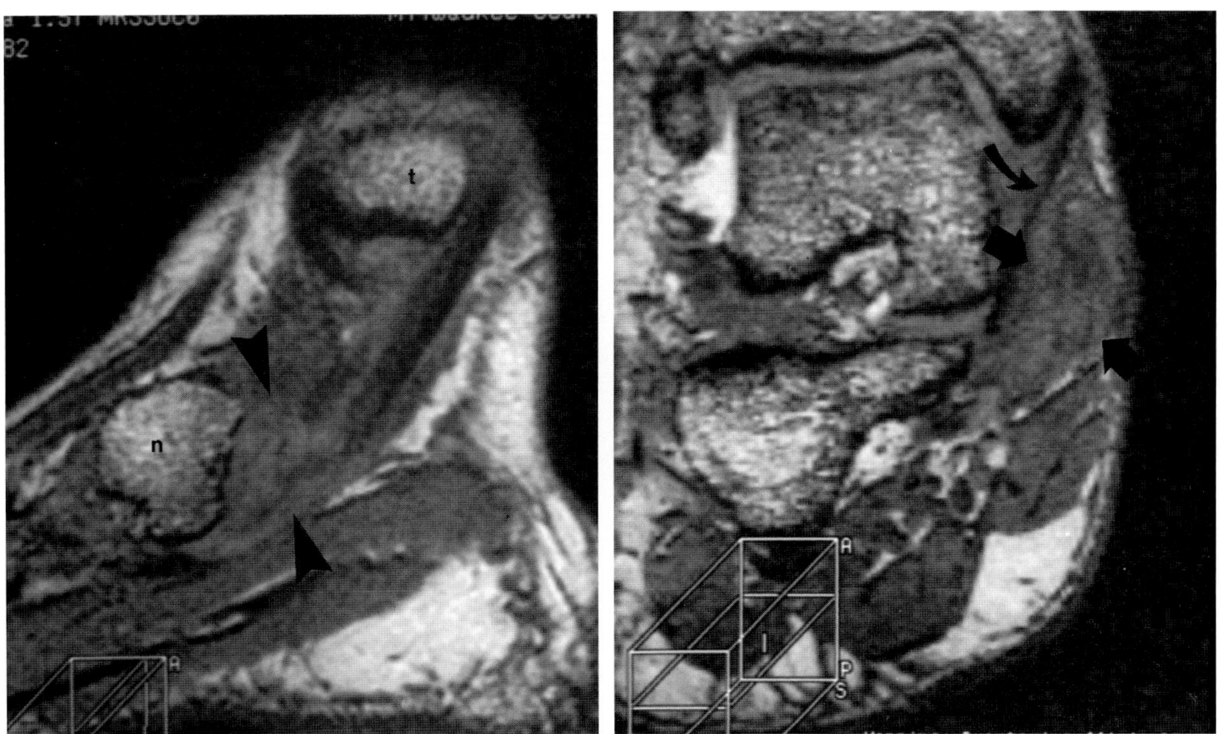

FIG. 1.

Findings: Figure 1A,B: Reformatted images from a three-dimensional gradient echo acquisition (TR/TE/° 30/10c/70).

Figure 1A: Oblique sagittal image. There is abnormal signal intensity and evidence of gross discontinuity of the distal posterior tibial tendon (PTT) (arrowheads) near its attachment site to the navicular bone (n). Proximally the intact portion of the PTT is visualized as a linear low signal intensity structure. T, medial malleous.

Figure 1B: Coronal oblique image. There is abnormal signal intensity in and around the expected location of the posterior tibial tendon (arrows). Deep to the tendon an intact tibiocalcaneal ligament is well depicted (curved arrow).

Diagnosis: Distal posterior tibial tendon disruption.

Discussion: The posterior tibial tendon (PTT) is one of the main stabilizers of the hindfoot and is responsible for inversion and plantar flexion (1–5). The PTT assists the spring ligament in supporting the talar head. The muscle belly arises from the proximal and middle thirds of the tibia and fibula; the muscle tendon junction is several centimeters above the medial malleolus under which the tendon turns on its course to insert following branching onto the navicular, the cuneiforms, and the bases of the second, third, and fourth metatarsals (Fig. 1C).

The oblique course of the PTT precludes imaging of the entire distal tendon on a true sagittal image in most cases (3,6). While conventional spin echo sequences are commonly employed in the evaluation of the PTT, utilization of a reformatted three-dimensional gradient echo acquisition facilitates visualization of the entire length of the distal tendon on a single image (Fig. 1D) (6). The normal PTT is depicted on MR as a homogeneously low signal intensity structure (3,6). In cross section it is round to ovoid in configuration and typically is twice the cross-sectional area of the adjacent flexor digitorum longus tendon. Just proximal to its insertion onto the navicular, the tendon may appear to "widen" and increased signal within the tendon may be demonstrated on short TE sequences simulating a tear (3,6). This intermediate signal should not be mistaken for a tear and represents a manifestation of the angle dependent signal (see Case 23) (6,7).

Parasagittal reformatted images from a three-dimensional gradient echo acquisition allow the long axis of an obliquely oriented tendon such as the PTT to be depicted along its long axis and are helpful in distinguishing a torn tendon from a normal tendon with angle-dependent signal (6). If necessary, reimaging with the patient's ankle in plantar flexion will eliminate this angle-dependent signal, or the three-dimensional gradient echo acquisition can initially be obtained with the patient's ankle in plantar flexion to eliminate this effect. In addition to allowing differentiation of distal PTT tears from normal angle-dependent signal, other normal variations of the posterior tibial tendon, such as branching, can be discerned utilizing the three-dimensional (3D) gradient echo acquisition. The 3D technique allows differentiation of normal branching from longitudinal tears that may be difficult with orthogonal spin echo images. An additional benefit of the 3D gradient echo acquisition includes easier evaluation of the extent of tears, allowing the radiologist to see the entire tear on one image rather than having to integrate the extent by mental summation from multiple images (Fig. 1E).

The typical presentation of a posterior tibial tendon tear is that of chronic medial ankle pain in a middle-aged woman (1,2). More recently, PTT tears on an acute basis have been increasingly recognized in young athletes engaged in sports requiring rapid changes in direction such as soccer, tennis, and basketball as well as in athletes, such as dancers and gymnasts, who point their toes excessively (4). As indicated previously, the PTT is a major stabilizer of the foot and rupture of the PTT allows for hindfoot valgus and forefoot abduction with resultant acquired flatfoot (2). Posterior tibial tendon tears usually occur in a small hypovascular zone located just below the medial malleolus. Distal attachment site tears, however, as illustrated in the present case, are less common although not rare.

Three patterns of posterior tendon ruptures have been described utilizing computed tomography (CT) and MR (1–3). These patterns correlate well with surgical subtypes familiar to foot and ankles surgeons. In the type 1 tear, longitudinal splits develop within the fiber bundles that compose the tendon (Fig. 1E). As a consequence of these longitudinal splits, hemorrhage and subsequently granulation and scar tissue develop and result in an increase in tendon girth such that it may be as great as four to five times the cross-sectional area of the adjacent flexor digitorum longus tendon (8). The normally ovoid tendon assumes a much more rounded configuration. On occasion, the splits are extremely long and result in apparent division of the tendon into two ''subtendons'' (8). In such cases, images in the axial plane may reveal four tendons (e.g., the anterior half of the PTT, the posterior half of the PTT, the flexor digitorum, and the flexor hallucis longus tendon) instead of the normal three along the medial side of the ankle. A similar pattern of rupture may affect the peroneus brevis tendon (see Case 10). Type I ruptures commonly demonstrate abnormal foci of signal intensity that are often best appreciated on heavily T2-weighted and short inversion time inversion recovery sequences. As a consequence of the increased girth of the tendon, this tear pattern can be considered to represent a ''hypertrophic'' pattern and is readily recognizable on MR images (3,8).

In contrast to the ''hypertrophic'' type I tear, the type II rupture is ''atrophic'' (8). Such tears are more severe, but are still partial tears in which the tendon undergoes such severe longitudinal splitting that it becomes attenuated to one-half to one-third of its normal size (3,8). This reduction in size is not circumferential; the dominant change in contour is a decrease in width of the tendon with a relative preservation of the anteroposterior dimension. The tendon then assumes a long ovoid configuration, a finding that may be appreciated on axial scans by careful comparison of the tendon diameter just above and below the suspected site of pathology. Recognition of type II tears is also facilitated on reconstructed 3D gradient echo acquisitions.

Type III tears of the PTT are complete ruptures with retraction of the proximal tendon and development of an obvious gap at the site of rupture. This pattern of tendon rupture is generally readily recognizable on MR. On occasion, the proximal and distal edges of the tear may retract and give a pseudo-widened appearance (3,8,9). It is important to carefully examine the entire length of the tendon to avoid mistaking the pseudo-widened segment of a type III tear for a type I ''hypertrophic'' rupture.

MR has a reported sensitivity of 95%, a specificity of 100% and an overall accuracy of 96% in determining the degree of rupture of a torn PTT (3). In the study of Rosenberg and associates (3) errors in interpretation were generally ones of underestimating the severity of the rupture in which partial volume effect of the tendon with the cortical bone of the tibia has occurred. Schweitzer and coworkers (9) have recently reported on the utility of using secondary signs related to abnormal biomechanics for detection of complete PTT tears (9). Failure of the long axis of the talus to bisect the navicular bone (i.e., normally the extension of a line representing the long axis of the talus should intersect the middle one-third of the navicular) was a reasonably sensitive and highly specific sign of complete PTT rupture. Prominence of the medial navicular tubercle (termed the cornuate navicular tubercle and believed to represent late fusion of an accessory navicular bone) was also a sensitive and reasonably specific (89%/75%) secondary sign of a PTT tear (9).

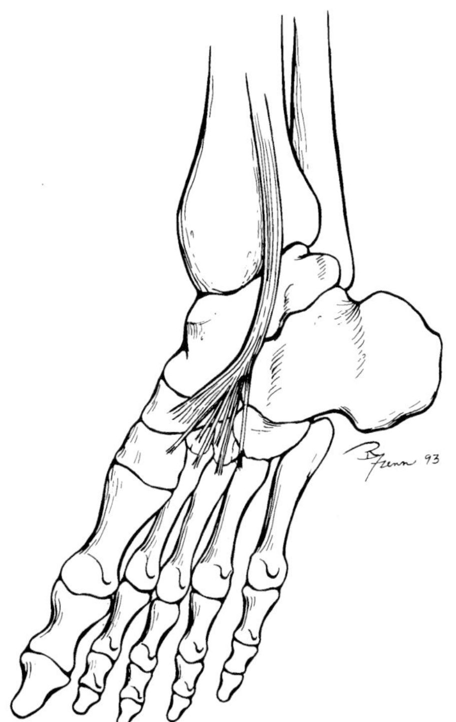

C

FIG. 1. C: Anatomic illustration of the posteromedial aspect of the ankle demonstrating the course and distal attachment sites of the posterior tibial tendon. The tibialis posterior muscle inserts along the posterior diaphysis of the tibia and fibula as well as the interosseous membrane. Its tendon travels posterior to the medial malleolus with the major attachment site on the medial navicular bone. Minor attachment sites include the medial, intermediate, and lateral cuneiform bone, as well as the second, third, and fourth metatarsal bones. (From Klein MA. Reformatted three-dimensional Fourier transform gradient-recalled echo MR Imaging of the ankle: spectrum of normal and abnormal findings. *AJR* 1993;161:831–836, with permission.)

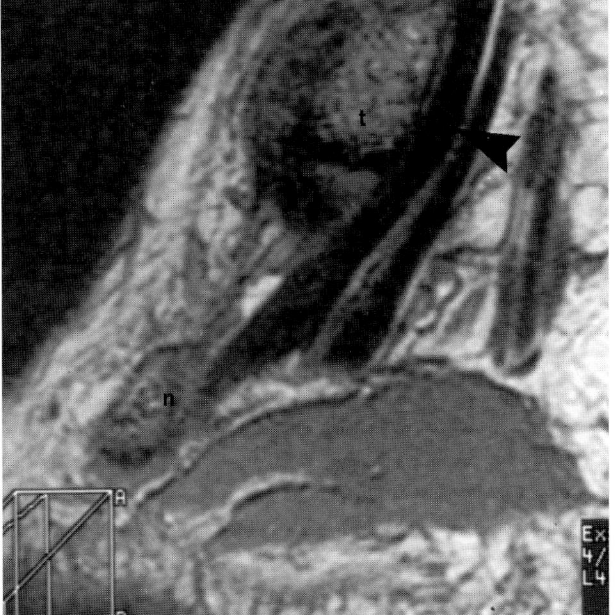

D

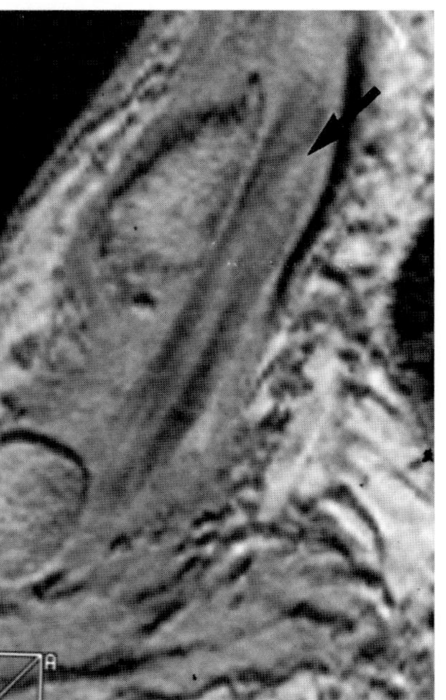

E

FIG. 1. D: Normal posterior tibial tendon. Oblique sagittal reformatted image from a three-dimensional gradient recalled echo acquisition in the steady state (TR/TE/° 30/10/70) demonstrates the entire length of the distal posterior tibial tendon *(arrowhead)* from the level of the medial malleolus (t) to its major distal attachment site on the medial navicular bone (n). The normal PTT demonstrates homogeneous low signal intensity without focal contour change or discontinuity. Focal thickening can be simulated by branches to minor attachment sites and by volume averaging with adjacent structures (spring ligament, tibionavicular ligament). **E:** Torn posterior tibial tendon. Sagittal oblique reformatted gradient echo image along the long axis of the distal posterior tibial tendon demonstrates a longitudinal tear in the posterior tibial tendon *(arrow).* The extent of the tear from the medial malleolus to the medial navicular bone is well demonstrated in this projection. (From Klein MA. Reformatted three-dimensional Fourier transform gradient-recalled echo MR Imaging of the ankle: spectrum of normal and abnormal findings. *AJR* 1993;161:831–836 with permission).

CASE 2

History: A 42-year-old woman presented with a several-month history of lateral ankle swelling. What is your diagnosis? What are the typical MR appearances of this lesion?

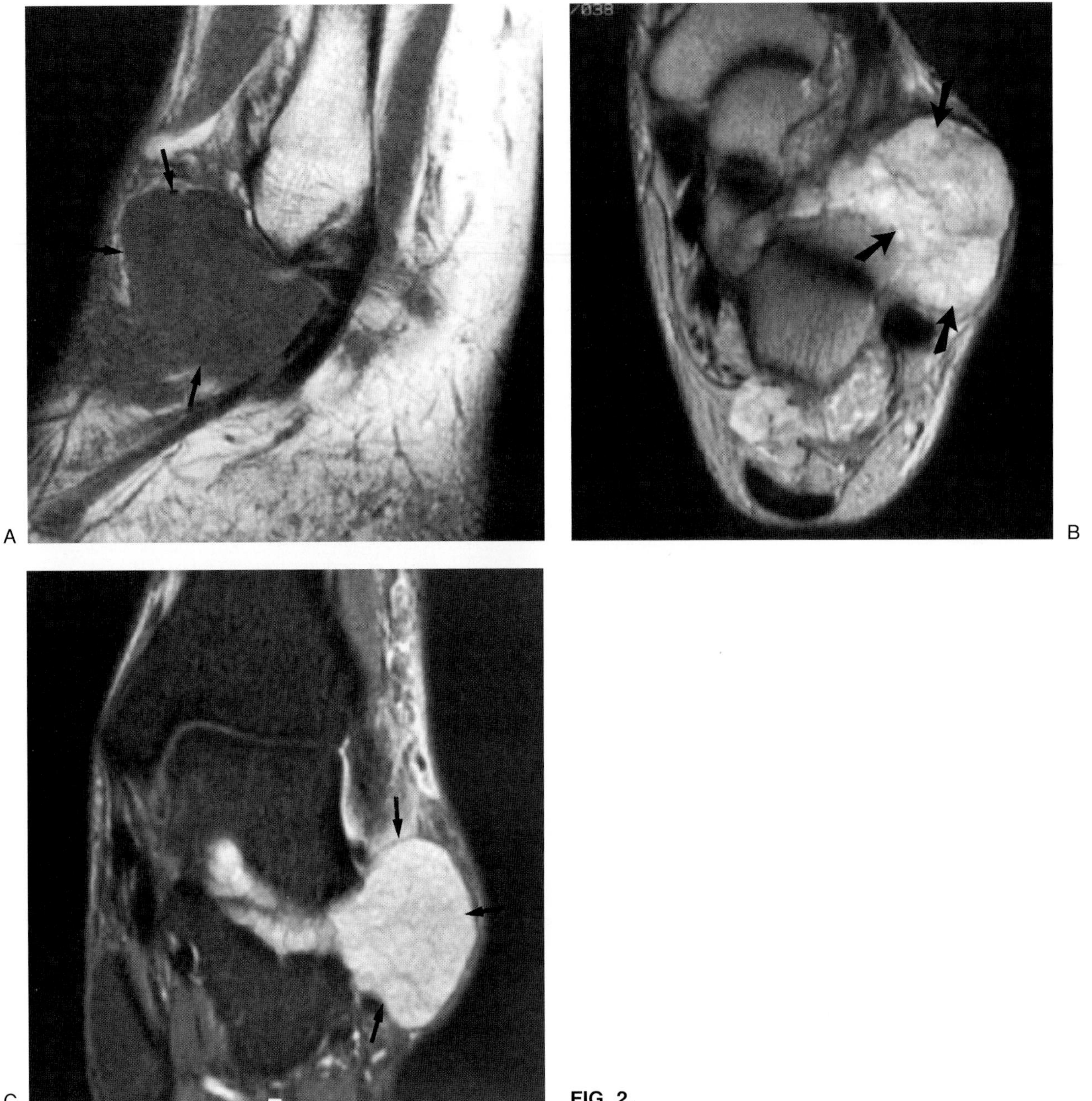

FIG. 2.

Findings: In Fig. 2A the sagittal T1-weighted (TR/TE 500/20 msec) image demonstrates a lobular low signal intensity mass projecting along the lateral aspect of the ankle anterior and inferior to the fibula (arrows). In Fig. 2B the axial T2-weighted fast spin echo (TR3/TE 3,800/95ef) reveals the mass to demonstrate high signal intensity and to extend out from the tarsal sinus (arrows) as well as the posterior ankle joint (long arrow). In Fig. 2C the coronal inversion recovery fast spin echo (TR/TE/TI 2,800/23/130) graphically depicts the lobular high signal intensity mass extending out from the tarsal sinus (arrows).

Diagnosis: Idiopathic synovial chondromatosis.

Discussion: Idiopathic synovial chondromatosis (ISO) is an uncommon condition characterized by cartilage formation by the synovial membrane (1–9). The cause is unknown but is thought most likely to represent a metaplasia or neoplasia (1,2). ISO is almost invariably monoarticular with typically a chronic progressive course (1,3,4). The knee, hip, and elbow represent the most commonly affected articulations (1,3,4). ISO is more common in men than woman (2×) and is most commonly encountered in the third to fifth decades. Patients typically present with a long (often several-year) history of joint pain, swelling, and limitation of motion (1,3,4). Joint effusion is rare but when present may be bloody. Intraarticular loose body formation is common and may contribute to mechanical destruction of the articular cartilage (1,3,4). Therapy is surgical and is directed toward loose body removal and resection of the involved synovium. Focal recurrence following surgery is not uncommon and may necessitate multiple surgical procedures (1,3,4).

Histologic findings include hyperplastic synovium with foci of cartilage metaplasia (2). These foci form cartilage nodules that grow (typically to a size of 2 to 3 cm) and project into the joint via a delicate pedicle (1,2). Disruption of the pedicle results in loose body formation. The cartilaginous nodules frequently calcify and enchondral bone formation may occur in the presence of an intact blood supply. The cartilaginous nodules may extend through the joint capsule into the surrounding soft tissues.

Radiographically, ISO is classically characterized by multiple intra/juxtaarticular calcified or ossified bodies that tend to be uniform in size, ranging from a few millimeters to several centimeters in diameter (1,5–9). The joint space is typically preserved initially, although secondary osteoarthritis may contribute to late joint space narrowing (1). Approximately one-third of patients with ISO demonstrate no calcification or ossification, a feature that contributes to diagnostic difficulty with conventional radiography (1,5). Noncalcified nodules may contribute to erosions of adjacent bone and apparent joint space widening. ISO is differentiated from multiple loose bodies associated with osteoarthritis, osteochondritis dissecans, osteochondral fractures, and neuropathic arthropathy by the radiographic features of large numbers of intraarticular bodies, their uniformity in size, and preservation of joint space (1,5).

Several reports of the appearance of ISO have appeared in the literature (5–9). The MR appearance of ISO is dependent on the pulse sequence employed as well as the presence and extent of calcification/ossification. In the largest reported experience, Kramer and colleagues (5) described three distinct MR patterns in ISO. The type A pattern consisted of lobulated homogeneous intraarticular signal that was isointense to slightly hyperintense to muscle on T1-weighted images and hyperintense on T2-weighted images. The signal pattern is similar to that of fluid but ISO is distinguished by its mass-like properties and the presence of internal septa. The type A pattern, as illustrated in the present case, is not associated with radiographically demonstrable calcifications and the diagnosis can be strongly entertained prior to the development of radiographic findings and without the use of contrast material. The type B pattern is similar to type A with the exception of focal areas of signal void/diminution, which correspond to foci of mineralization on radiographs or CT (5). This pattern was the most commonly encountered in the review of Kramer and coworkers. The use of gradient echo pulse sequences facilitates detection of areas of mineralization secondary to magnetic susceptibility effects. The type C pattern was characterized by features of A and B as well as additional foci of peripheral low signal surrounding central areas of signal consistent with fat (5). These correlated with radiographically demonstrable ossified bodies. Following the administration of gadopentetate dimeglumine, ISO is characterized by signal enhancement that may be inhomogeneous (5). While not a common condition, the MR findings associated with ISO may be sufficiently characteristic to allow the diagnosis to be strongly suggested.

CASE 3

History: Two reformatted images of the medial ankle from a three-dimensional gradient echo acquisition are demonstrated. The insert shows image location. What is the normal structure indicated by the arrow in Fig. 3A and arrowhead in Fig. 3B? What are the anatomic components of the medial collateral ligament complex of the ankle? What MR techniques can be utilized to optimize their depiction?

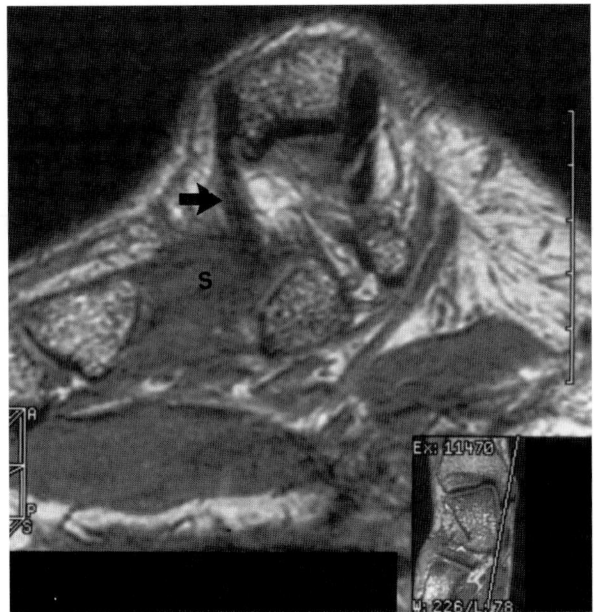

A

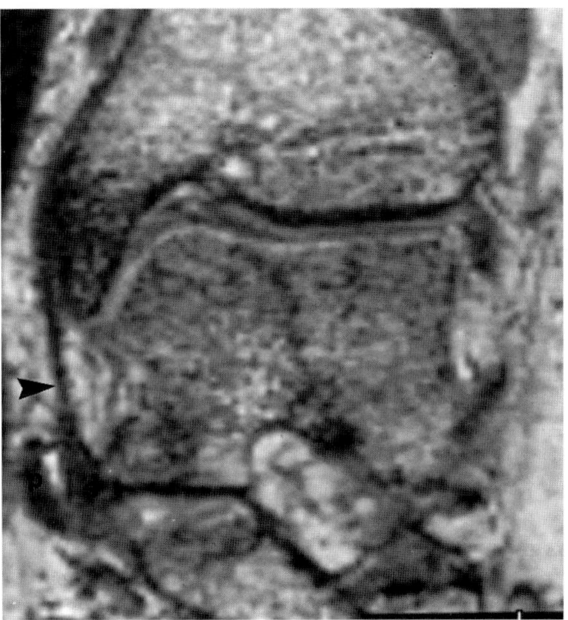

B

FIG. 3.

Findings: Figure 3A is a sagittal reformatted image from a three-dimensional gradient echo acquisition (TR/TE/° 30/10/70). A triangular low signal intensity structure (arrow) is identified coursing from the anterior colliculus of the medial malleolus to the thick superior plantar calcaneonavicular ligament (s). Figure 3B is a coronal oblique reformatted image from a three-dimensional gradient echo acquisition (TR/TE/° 30/10/70) that demonstrates the homogeneously low signal intensity structure coursing from the medial malleolus to the superior aspect of the plantar calcaneonavicular ligament (s). The medial collateral ligament is deep to the posterior tibial tendon. (p) (Figure A from Klein MA. MR imaging of the ankle: normal and abnormal findings in the medial collateral ligament. *AJR* 1994;162:377–383, with permission.)

Diagnosis: Normal tibiospring ligament.

Discussion: The medial collateral ligament (MCL), also known as the deltoid ligament, is an extremely strong ligament and demonstrates a delta shape with a narrow apical attachment superiorly to the medial malleolus and a broad fan-like distal insertion onto the talus, navicular, spring ligament, and the calcaneus. The MCL is generally considered to be composed of both superficial and deep layers although significant differences exist in the literature regarding the individual components of the ligament and descriptions of them (Fig. 3C) (1–8).

The superficial component attaches to the anterior aspect of the medial malleous (anterior colliculus) (4). The superficial layer is generally considered to be composed of the tibiocalcaneal, tibiospring, and tibionavicular ligaments (1–4). The deep component of the MCL attaches to the posterior colliculus and intercollicular groove.

The tibocalcaneal portion attaches the anterior colliculus of the medial malleolus to the posteror portion of the sustentaculum tali of the calcaneus. It extends across both the ankle and anterior subtalar joints (7). In the investigation of Klein (4), the tibiocalcaneal ligament was of uniform but variable width and had homogeneously low signal intensity on T2*-weighted images.

The tibionavicular component follows an anteroinferior course from the anterior colliculus of the medial malleolus to the medial navicular bone crossing both the ankle and talonavicular joints (4,7). It shares a common insertion with the posterior tibial tendon and the superior portion of the plantar calcaneonavicular ligament. On MR studies the ligament had a variable width and in the majority of ankles demonstrated a homogeneous low signal intensity appearance (4).

The tibiospring ligament, as illustrated in the present case, courses downward from the anterior colliculus of the medial malleolus to attach to the superior aspect of the spring ligament just anterior to the tibialcalcaneal ligament. Unlike the other MCL ligaments, it does not have two bony attachments (4). Most commonly it represents the second largest component of the MCL and in the majority of cases on MR it demonstrates homogeneous low signal intensity. While weaker than the posterior tibiotalar ligament, the tibiospring ligament is stronger than either the tibiocalcaneal or tibionavicular ligaments (4). Anatomic studies have demonstrated the tibiospring ligament to have a strength three to four times that of the tibiocalcaneal and tibionavicular ligaments (9). Superficial to the tibiospring ligament is the posterior tibial tendon. Usually the tibiospring ligament is of low signal intensity, although it can have diffuse intermediate signal. The intermediate signal is unlikely to result entirely from an angle-dependent signal increase since the tibiospring ligament is rarely oriented 45° to 65° to the stationary magnetic field, which is necessary for this effect to occur.

The posterior and anterior tibiotalar ligaments compose the deep layer of the deltoid. The posterior tibiotalar ligament is the thickest and strongest portion of the MCL and courses posterolaterally from the medial malleolus to attach to a large area on the posterior talar body. The tibiotalar ligament prevents the talus from being displaced laterally against the lateral malleolus (3). In instances in which a distal fibular fracture is present, displacement of the lateral malleolar fragment by the talus is precluded if the tibiotalar ligament is intact. In the study of Klein (4), the posterior tibiotalar ligament always appeared as a uniformly thick band of low to intermediate signal intensity with or without identifiable individual fibers.

The anterior tibiotalar ligament is the only component of the MCL that is inconsistently present (4). The anterior tibiotalar ligament attaches the medial malleolus to the talar neck. When present, it is thin and has either low or intermediate signal on T2*-weighted images.

The spring ligament, also known as the plantar calcaneonavicular ligament, extends from the anteromedial calcaneus to the medial plantar aspect of the navicular bone (7). The spring ligament is a specialized fibrocartilaginous structure that bridges the calcaneonavicular gap and supports the medial arch and the head of the talus (1). The medial border of this ligament curves superiorly to become continuous with the tibiospring portion of the deltoid (6). If the deltoid ligament is sectioned, the spring ligament sags, the talar head descends in a plantar direction, and the calcaneus slips into valgus, all of which contribute to a flatfoot deformity (1).

As a consequence of the their oblique course, full visualization of all the components of the MCL cannot be obtained with standard spin echo orthogonal imaging planes (4). Schneck et al. (7) were able to demonstrate the courses of the individual components of the MCL by fixing the ankle in various degrees of flexion and extension. In this work, the tibionavicular and anterior tibiotalar components of the deltoid were best depicted in coronal images obtained with the foot in maximum plantar flexion (40° to 50°) (6). Optimal visualization of the tibiospring component of the superficial deltoid was accomplished on coronal images obtained with the foot in 20° of dorsiflexion (6). The use of three-dimensional volume acquisitions and thin section multiplanar reformations obviates the need for specialized foot positioning while accomplishing full-length ligament visualization (Fig. 3D). The technique can also allow depiction of abnormalities involving the ligaments including ligamentous disruptions (Fig. 3E).

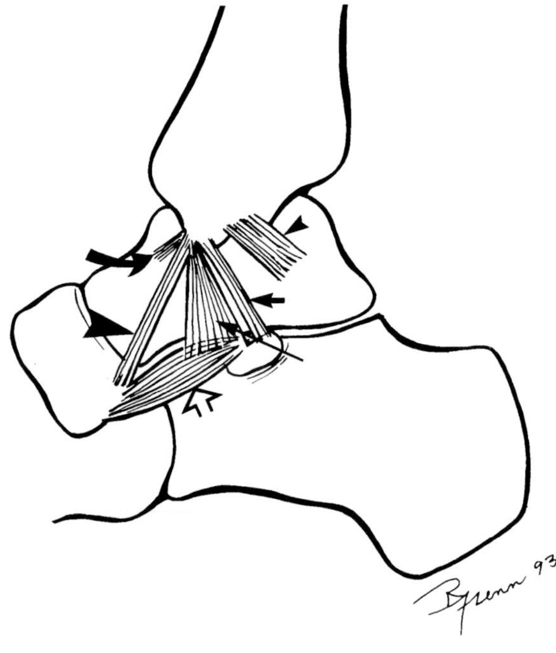

FIG. 3. C: Anatomic illustration of the medial collateral ligament. The medial collateral ligament attaches the medial malleolus to the talar body, sustentaculum tali, plantar calcaneonavicular ligament, medial navicular bone, and the talar neck. The deep components of the medial collateral ligament are the posterior tibiotalar ligament *(small arrowhead)* and the anterior tibiotalar ligament *(curved arrow).* The superficial components of the medial collateral ligament consist of the tibiocalcaneal ligament *(small straight arrow),* tibiospring ligament *(long straight arrow),* and the tibionavicular ligament *(large arrowhead).* The tibiospring ligament attaches inferiorly to the superior aspect of the plantar calcaneonavicular ligament *(open arrow).* This latter structure courses from the sustentaculum tali to the medial navicular bone. The tibionavicular ligament, plantar calcaneonavicular ligament, and the posterior tibial tendon all share an attachment site on the medial navicular bone. (From Klein MA. MR imaging of the ankle: normal and abnormal findings in the medial collateral ligament. *AJR* 1994;162:377–383, with permission).

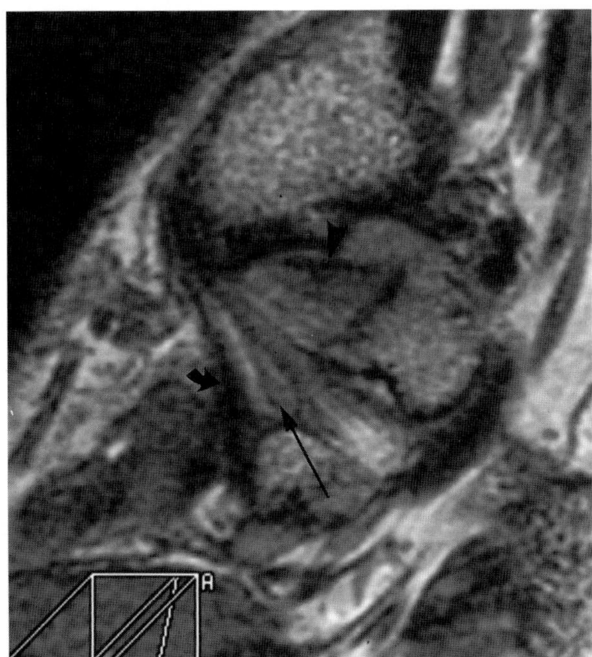

FIG. 3. D: Oblique sagittal image of the posteromedial ankle reformatted from a three-dimensional gradient echo acquisition (TR/TE/° 30/10 msec/70°). By obtaining a volume acquisition, the user can retrospectively reconstruct any obliquity to best demonstrate the desired structure. This obliquity demonstrates the entire course of several normal components of the medial collateral ligament. Shown are the tibiospring ligament *(curved arrow),* tibiocalcaneal ligament *(long arrow),* and the posterior tibiotalar ligament *(arrowhead).*

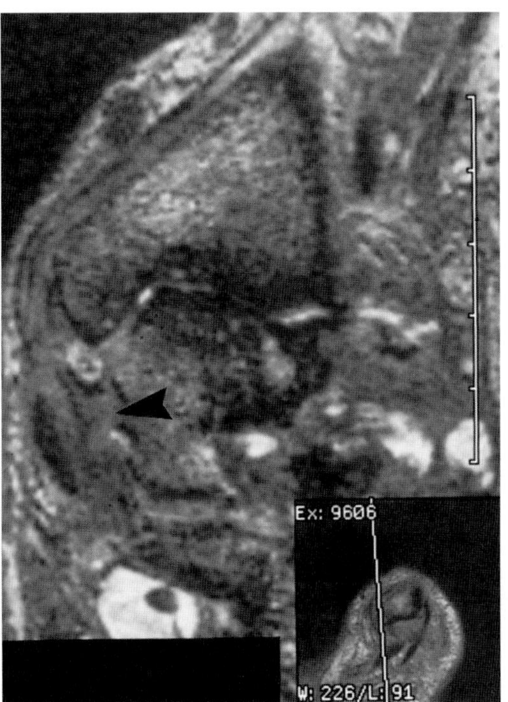

FIG. 3. E: Oblique coronal image of the posteromedial ankle reformatted from a three-dimensional gradient echo acquisition (TR/TE/° 30/10 msec/70°). There is gross discontinuity of the tibiospring ligament with associated hemorrhage and edema displacing the adjacent posterior tibial tendon.

CASE 4

History: A 55-year-old woman complained of increasing pain and difficulty in walking. What is the abnormality depicted on the MR examination? How common is this type of injury?

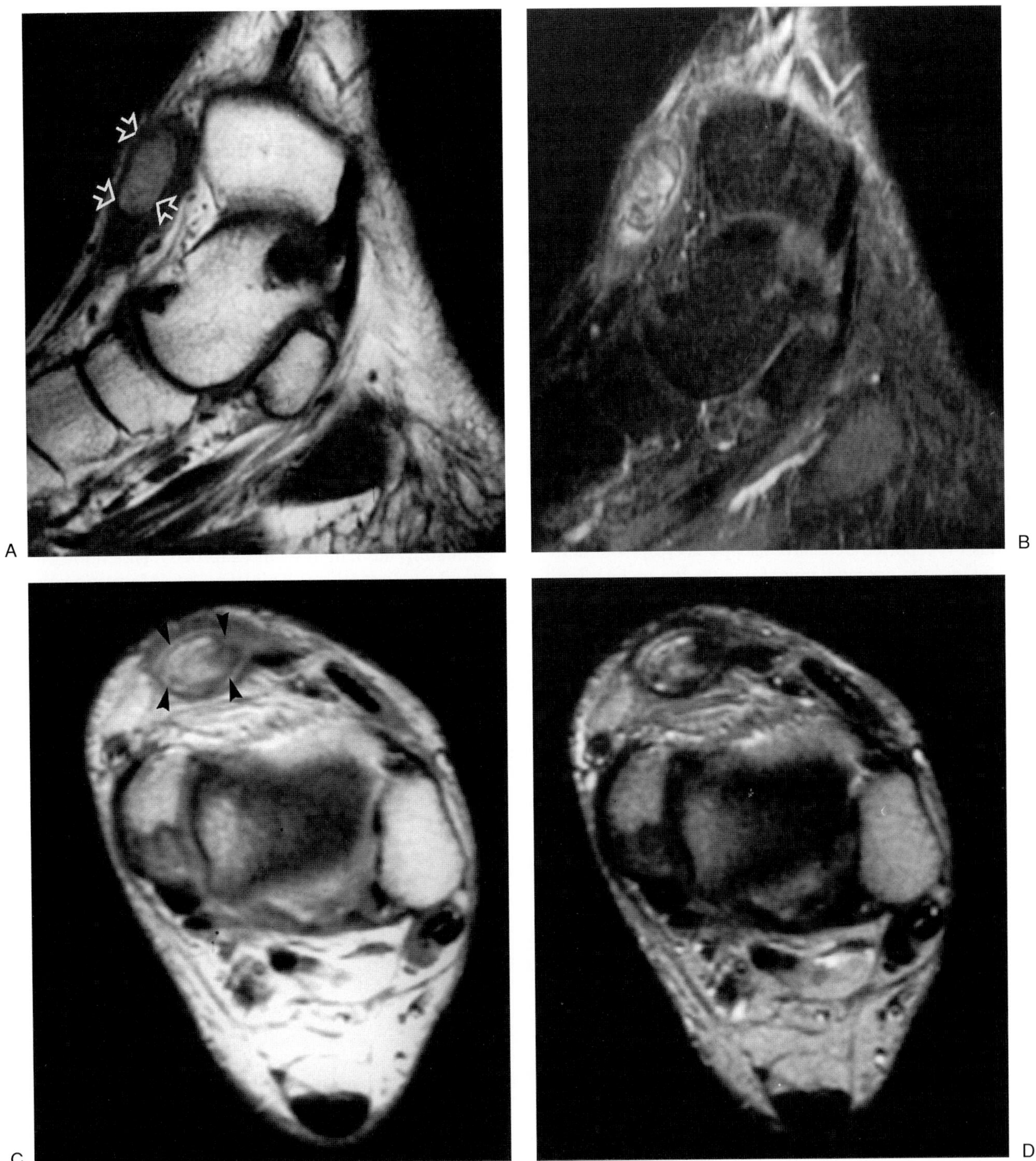

FIG. 4.

Findings: In Fig. 4A the sagittal T1-weighted fast spin echo (TR/TE 600/13eff) demonstrates marked increased girth and increased signal intensity within the tibialis anterior tendon (arrows). In Fig. 4B the sagittal inversion recovery fast spin echo (TR/TE/TI 3,300/30/130) demonstrates increased signal intensity within the focally thickened tendon segment. In Fig. 4C the axial PD (TR/TE 4,000/32 msec) image demonstrates localized increased girth and signal within the tibialis anterior tendon (arrowheads). In Fig. 4D the axial fat-suppressed T2-weighted fast spin echo (TR/TE 4,000/102eff) image demonstrates focally increased girth and signal within the tibialis anterior tendon.

Diagnosis: Tear of the subcutaneous segment of the tibialis anterior tendon with hemorrhagic component.

Discussion: Spontaneous rupture of the subcutaneous tibialis anterior tendon is an uncommon occurrence (1–5). Indeed, only 25 cases have been reported in the literature, although this likely underestimates the true prevalence of this entity (1). The condition is most commonly encountered in men in their fifth to seventh decade (1). Athletes who may be subject to forced plantar flexion of the foot (e.g., runners, soccer players) can stress and inflame the anterior tibial tendon.

Spontaneous tibialis tendon ruptures appear to occur through abnormal tendons (1). This contention is supported clinically by the rare occurrence in younger age groups as well as by gross and histological evidence. Papidus (4) reported extensive aseptic necrosis and hemorrhage into the necrotic area of one tendon on histological examination, and Moberg (5) described ''mucoid'' degeneration. Other investigators have described gross ''fibrillation'' and hemorrhage of the tendon at surgery (2,3).

The tibialis anterior muscle arises from the proximal two-thirds of the tibia and interosseous membrane and becomes tendinous at the level of the distal tibial metaphysis. Slightly more caudally, the tendon acquires a synovial sheath. The tendon traverses through three fibrous tunnels (retinacula) that prevent the tendon from displacing away from the subjacent skeletal elements during dorsiflexion. Most ruptures occur between the superior and inferior retinacula; the tough inferior edge of the superior retinaculum has been implicated in tendon abradement, which may predispose to rupture. The tendon provides 80% of the dorsiflexion power of the foot and is second only to the Achilles tendon in strength.

The optimal approach to treatment of tibialis anterior tendon ruptures remains controversial. Slapping gaits and flatfeet are among the reported complications in untreated patients (2,3). Full recovery, however, has also been reported with conservative (i.e., nonoperative) management. In groups treated by operative repair, full recovery is closely associated with acute as opposed to delayed repairs (1). This underscores the need for prompt and accurate diagnosis of this relatively uncommon entity.

MR may also be useful for assessment of muscle/tendon unit lacerations (Fig. 4E,F). Muscle/tendon unit lacerations result from penetrating injury. In this clinical setting, MR is capable of determining tendon integrity, characterizing the extent of injury if one is present, and evaluating the length of the gap between tendon ends in complete disruptions. This type of information can be especially valuable in preoperative decision making and surgical planning.

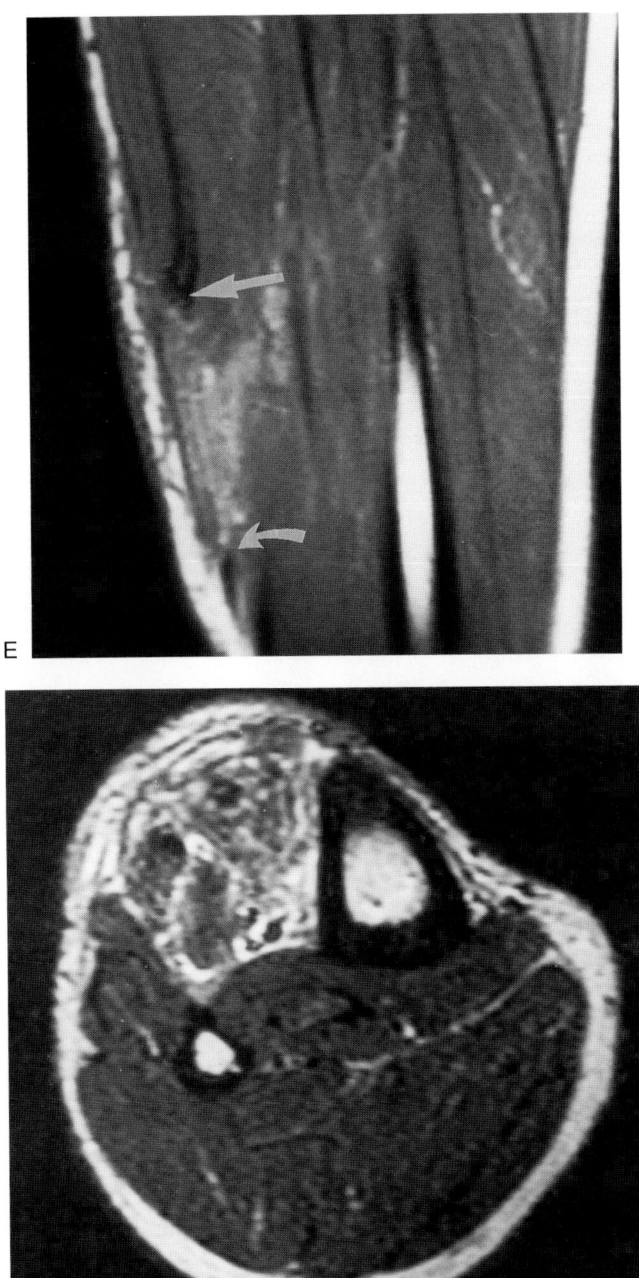

FIG. 4. E,F: Laceration of the tibialis anterior muscle/tendon unit. **E:** Sagittal T1-weighted image (TR/TE 600/20 msec) demonstrates focal disruption of the tibialis anterior tendon *(arrow)* at the site of laceration. There is surrounding signal alteration with the muscle consistent with hemorrhage. A component of the distal tendon can be seen *(curved arrow)*. **F:** Axial T2-weighted (TR/TE 2,000/80 msec) demonstrates marked and diffuse increased signal throughout the tibialis anterior muscle related to extensive edema.

CASE 5

History: A 24-year-old professional football player experienced a severe inversion injury during a game. The study was performed within several hours of the injury. What is your diagnosis?

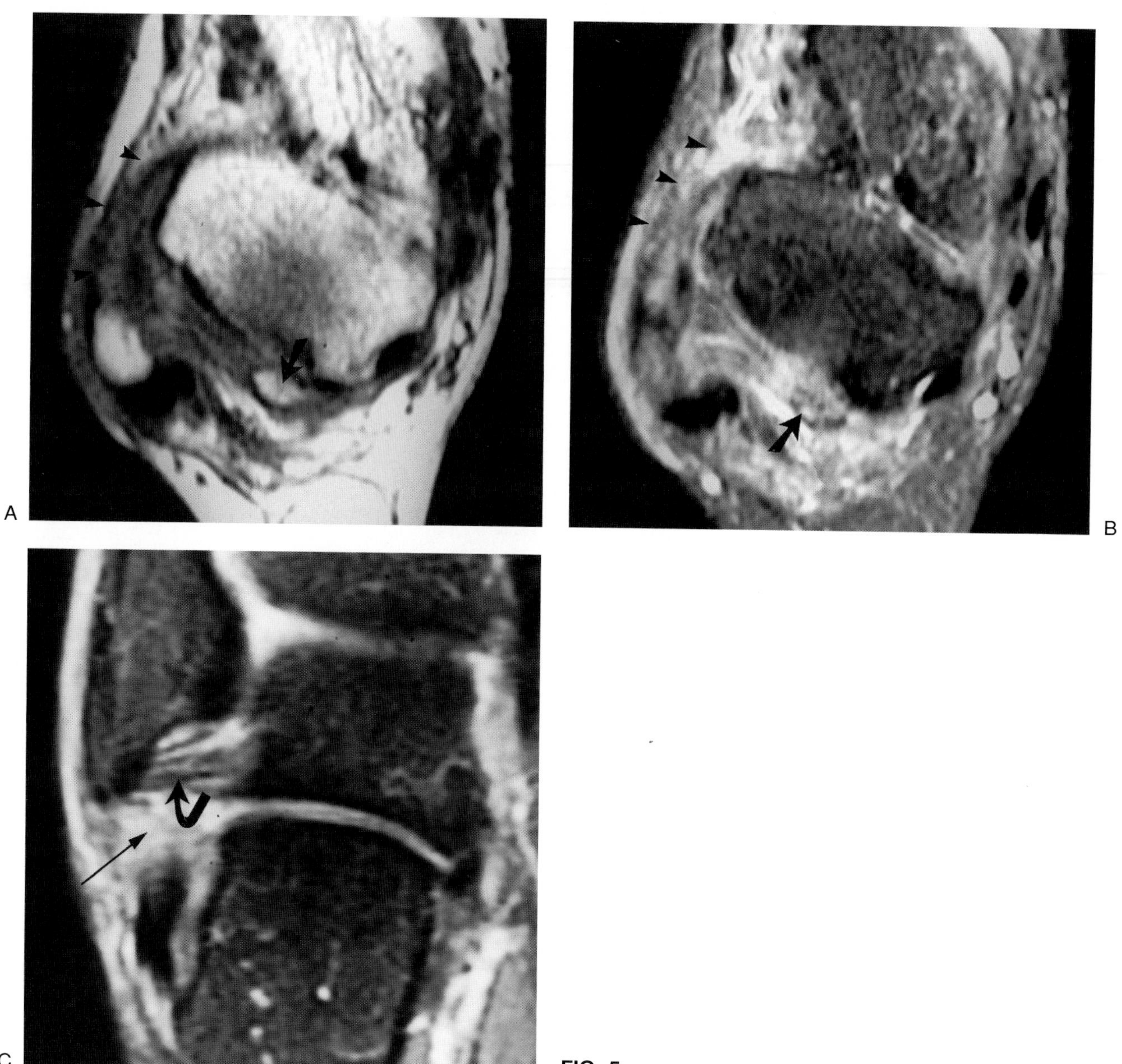

FIG. 5.

Findings: In Fig. 5A the axial T1-weighted (TR/TE 500/18 msec) image demonstrates increased intermediate signal intensity and poor definition in the expected region of the anterior talofibular ligament (ATAF) (arrowheads). A small ossific fragment is seen in the region of the posterior tubercle of the talus (arrow). In Fig. 5B the axial fat-suppressed proton density (TR/TE 4,000/36 msec) image demonstrates diffuse increased signal in the region of the ATAF (arrowheads). Increased signal is also identifed in the region of the posterior tubercle fragment and posterior talofibular ligament (PTAF). In Fig. 5C the coronal fat-suppressed proton density (TR/TE 4,000/36 msec) sequence demonstrates increased signal and nonvisualization of the calcaneofibular ligament (long arrow). The PTAF is well seen and appears grossly normal (curved arrow).

Diagnosis: Acute tear of the anterior talofibular ligament, calaneofibular ligament, and avulsion fracture of the posterior tubercle of the talus at the site of attachment of the posterior talofibular ligament.

Discussion: Inversion injuries of the ankle are among the most common injuries to the musculoskeletal system (1,2). Indeed, it has been estimated that there are as many as 23,000 ankle sprains per day in the United States. Such injuries may constitute up to 10% of the visits to emergency rooms in Scandinavia, and 85% of all ankle injuries (3). Inversion injuries occur primarily as a result of participation in sports and the majority of patients are under 35 years of age (4). Following inversion, a variety of traumatic lesions may occur singly or in combination including ligamentous sprain, ankle fractures, osteochondral lesions of the talar dome, and peroneal tendon dislocation.

Ligamentous sprains are classified clinically as first-, second-, or third-degree injuries, although it is impossible to quantify precisely the true extent of damage (3,5–8). First-degree sprains are characterized by local pain, point tenderness, minimal swelling, and demonstrable swelling on stress radiography. These are thought to correlate with collagenous microtears. Second-degree sprains are more severe injuries with swelling and hemorrhage. This injury most closely correlates with a partial ligament tear. The clinical hallmark is some loss of function; instability and abnormal motion may result. In third-degree sprains, there is complete loss of ligamentous integrity with abnormal motion and a soft or indistinct end point to stress testing. There is marked edema and tenderness, yet surprisingly the pain may be less than that of a second-degree injury. It is thought that with an incomplete tear, tension on partially torn fiber ends produces the pain (5). During the first week following a ligament tear, a hematoma, fibroblastic proliferation, and a mass of inflammatory tissue form at the site of the disruption. During the third to the sixth week healing begins by increasing collagenization and reorganization into a more organized ligament-like histology. The ultimate outcome is dependent on multiple factors including, but not limited to, proximity of the torn tendon edges to each other, local blood supply, and local environmental factors (5).

The lateral collateral ligament is a complex structure composed of the ATAF ligament, the PTAF ligament, and the calcaneofibular (CF) ligaments (Fig. 5D–F). The ATAF is 20 mm long and 2 to 3 mm thick. It arises from the anterior margin of the lateral malleolus and runs anteriorly and medially toward its attachment to the talus. The ATAF, the weakest and the most vulnerable of all of the ankle ligaments, serves to limit internal rotation and inversion and to prevent the talus from slipping forward from under the tibia (9). In its course to the talus, the ATAF crosses a ridge of the talus body just behind the neck of the talus. It is this ridge that is in part responsible for a tear of the ATAF (1). The ATAF tears in its midsubstance 45% of the time, and by talar avulsion 50% of the time (2). On MR examinations, the ATAF is nearly always identified on a single axial section if the foot is in a neutral or slightly dorsiflexed position because the ligament has a nearly horizontal course (10–15). As a consequence the ATAF is the most consistently and easily identified ligament on ankle MR studies. In patients who have experienced an acute complete tear of the ATAF, complete ligament disruption is readily evident on all MR sections (Fig. 5G,H). A poorly defined mass representing the proximal and/or distal extremities of the torn ligament may be demonstrated. Partial tears of the ATAF can be characterized by MR. Close inspection of images may reveal disruption of the upper section of the ATAF with the more inferior component of the ligament appearing grossly intact. This observation is consistent with the usual mechanism of injury in lateral ankle sprains, which includes both plantar flexion and inversion. This places increased strain on the upper portion of the ATAF. Three-dimensional volume acquisitions with the ability to reconstruct thin sections show promise in the quantification of ankle ligament tears.

The calcaneofibular ligament (CF) descends nearly vertically from the lateral malleolus to attach to the lateral calcaneus (Fig. 5D–F). The primary function of the CF ligament is to provide the primary static restraint against inversion of the calcaneus. The CF is situated just lateral to the calcaneus and just medial to the peroneal tendons in their common sheath. Unlike the PTAF and ATAF, the CF ligament is not intimately related to the talocrural joint. Cordlike extracapsular ligaments like the CF have a poorer blood supply than those that are flat and intracapsular. When round ligaments are torn, they tend to retract and often fail to unite even after extended periods of immobilization. Therefore, the arthrographic diagnosis of a CF tear can be made long after the traumatic event but the opposite is true for the ATAF. Demonstration of the normal CF on MR is complicated by its nonorthogonal course, its small size (2 to 3 mm thick), and partial volume effect of the low signal intensity ligament between the dark peroneal tendons and low signal intensity calcaneal cortex. The ability to image the CF in the axial plane can be improved by maximal plantar flexion of the foot (14). Alternatively, newly introduced fast three-dimensional sequences allow enhanced demonstration of the ligament utilizing oblique axial reformations. Tears of the CF can be reliably diagnosed when it is thickened (up to 5 mm) and assumes a serpiginous character. Findings associated with a CF tear include peroneal retinacular thickening and peroneal tendon abnormalities such as tenosynovitis and subluxation. Isolated tears of the CF are quite unusual, while CF tears in association with tears of the ATAF are common, reportedly occurring in combination 40% of the time that the ATAF tears (9).

The third component of the lateral collateral ligament complex, the PTAF, is generally considered the strongest of the ankle ligaments and disruption occurs only when the talus is dislocated without an associated fibular fracture. The PTAF arises from the medial aspect of the distal fibula at approximately the same level as the CF, courses nearly horizontally, and attaches to the posterior lateral talar tubercle, a landmark that defines the lateral margin of the groove for the flexor hallucis longus (Fig. 5D–F). On axial MR images, the normal PTAF can nearly always be visualized as a series of fibers running from the malleolar fossa of the fibula to the posterior lateral talar tubercle. The MR image of a parallel arrangement of individual fibers separated by fibrofatty tissue is similar to the appearance of the anterior cruciate ligament of the knee; the appearance must not be mistaken for a pathologic process. On sagittal images on which it is seen, the PTAF must not be mistaken for a loose body situated just behind the talus. In patients without joint effusion, the PTAF is closely applied to the posterior aspect of the talus and the joint line where it may simulate a loose body. If the joint is distended, however, the pseudo-loose body will be seen to be extracapsular in location (1). As a consequence of the strength of this ligament, an avulsion fracture of the posterior tubercle attachment may occur, as in the above illustrated case, prior to disruption of the ligament itself.

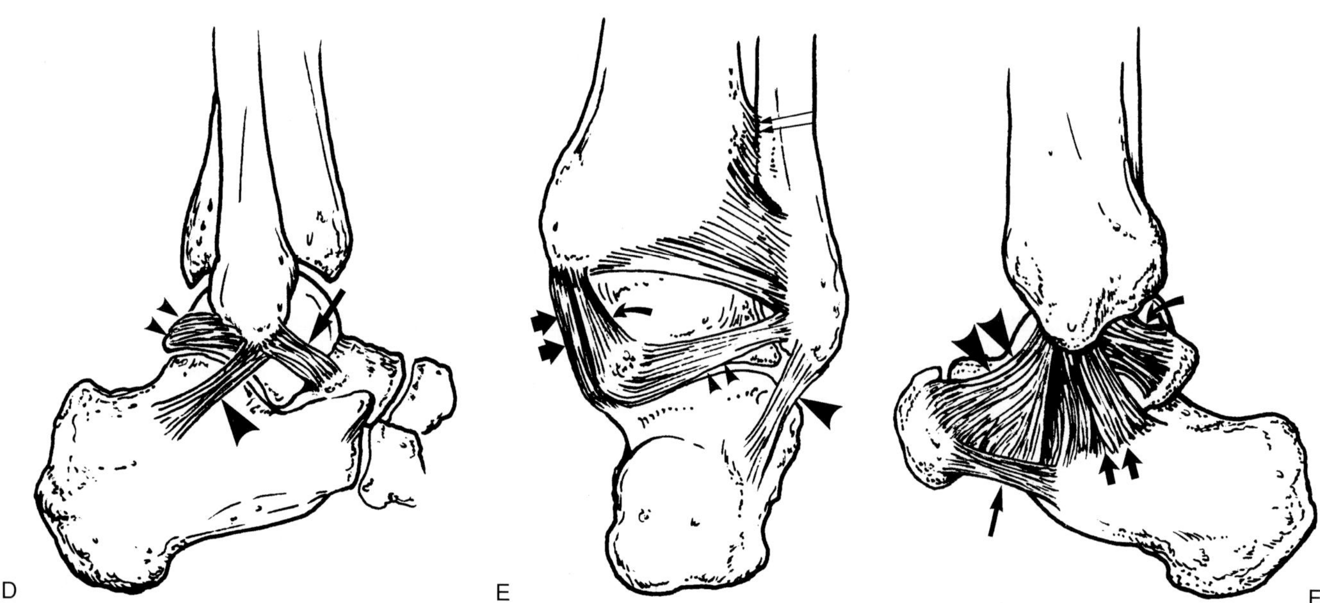

FIG. 5. D–F: Normal ligamentous anatomy of the ankle. These diagrams demonstrate the normal relationships of the components of the lateral collateral ligament, the medial collateral ligament, and the syndesmotic complex. ATAF, *arrow*; PTAF, *small arrowheads*; CF, *large arrowhead*; tibiotalar, *curved arrow*; tibiocalcaneal ligament, *double arrow*; tibionavicular, *double large arrowhead*; tibiospring, *long arrow*; ATIF, *double long arrow.* (From Mink JH. Ligaments of the ankle. In: Deutsch AL, Mink JH, Kerr R, eds. *MRI of the foot and ankle.* New York: Raven Press, 1992;173–178, with permission.)

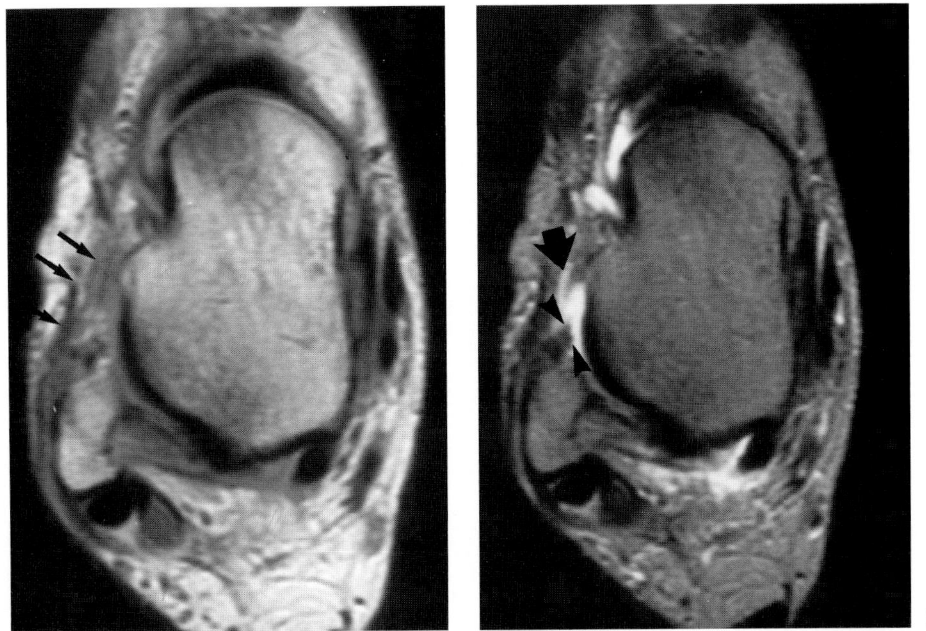

FIG. 5. G: Axial PD (TR/TE 2,000/20 msec) demonstrates disorganization of the soft tissues along the lateral aspect of the ankle with apparent discontinuity of the anterior talofibular ligament (ATAF) *(arrows)*. **H:** On the corresponding T2-weighted SE image (TR/TE 2,000/80 msec) the lateral extremity of the ATAF *(arrowheads)* is discontinuous and joint fluid *(large arrow)* is seen extravasating through the ligamentous defect. (From Mink JH. Ligaments of the ankle. In: Deutsch AL, Mink JH, Kerr R, eds. *MRI of the foot and ankle*. New York: Raven Press, 1992;173–178, with permission.)

CASE 6

History: A 12-year-old boy, the son of a staff physician, presents with ankle pain and a "destructive" lesion on plain radiography. Is this lesion most likely benign or malignant? What are your principal differential considerations?

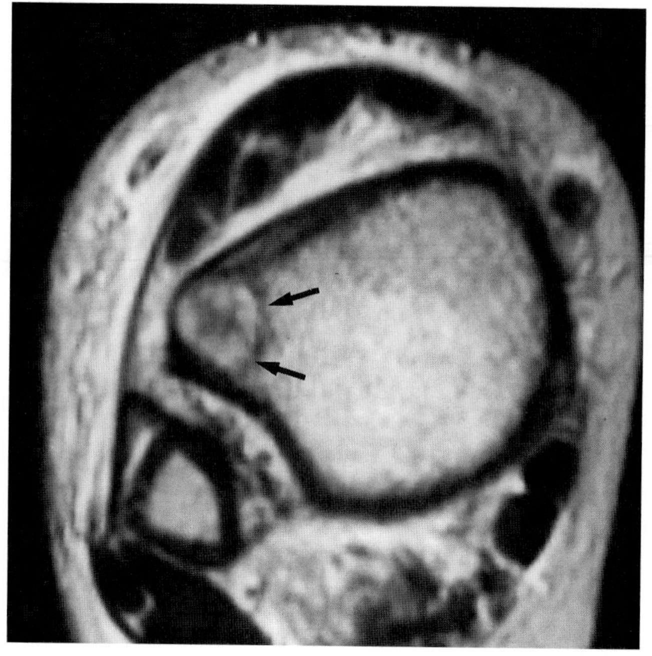

A

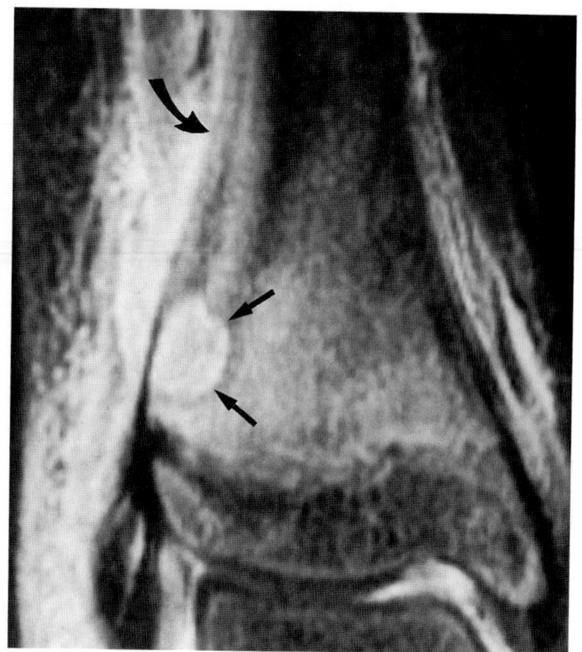

B

FIG. 6.

Findings: In Fig. 6A the axial T2-weighted fast spin echo image (TR/TE 4,000/108 msec) demonstrates a well-demarcated circular mass eccentrically located within the distal lateral tibia (arrows). In Fig. 6B the coronal inversion recovery fast spin echo (TR/TE/TI 2,800/23 msec/130) demonstrates the mass to become of uniform high signal intensity (arrow). There is extensive surrounding reactive edema. Periosteal new bone is seen along the lateral tibia metaphysis (curved arrow) and there is extensive reactive edema within the soft tissues and adjacent bone.

Diagnosis: Solid aneursymal bone cyst.

Discussion: First described by Sanerkin and coworkers (1) in 1983, the "solid" variant of aneurysmal bone cyst (ABC) has been increasingly recognized and accounts for approximately 5% to 7.5% of all aneursymal bone cysts (1–4). This entity presents a significant challenge particularly to the pathologist and can be confused histologically with osteosacrcoma (2). Indeed, the initial diagnosis rendered at the time of biopsy was that of a high-grade osteosarcoma. As this was so discordant with the well-circumscribed and relatively benign appearance of the lesion both radiographically as well as on MR, the biopsy was reviewed and the diagnosis of solid ABC rendered. The solid ABC demonstrates identical histologic characteristics to conventional ABC but lacks the typical blood-filled spaces and the cyst walls (2,3).

There appear to be no differences in the clinical or radiologic presentation in patients with the solid variant in contrast to conventional ABC (2,3). The vast majority of patients are younger than 20 years old and have pain and or swelling (3). Long bone lesions account for more than half of all cases and are usually eccentric and metaphyseal (3). Epiphyseal lesions are usually intramedullary and associated with chondroblastoma or giant cell tumor (3). Laminated periosteal reaction is common. Four classic radiologic stages have been described (3). In the initial phase, the lesion is depicted as a well-defined area of osteolysis with discrete periosteal reaction (3). In the active phase is progressive bone destruction with the development of the typical "blown-out" appearance (3). In the stabilization phase, there is the development of the "soap bubble" appearance, which results from maturation of the bony shell (3). In the healing phase, progressive calcification and ossification transform the lesion into a dense bony mass (3). MR imaging typically demonstrates a well-defined lesion with lobulated contours (3). Internal septa with multiple fluid-fluid levels may be seen, although adjacent loculi may demonstrate markedly differing signal characteristics (5). Fluid-fluid levels may demonstrate T1 shortening effects secondary to methemoglobin (e.g., increased signal intensity on T1-weighted images) and are more commonly demonstrated on MR than CT (3,5). The lesion may demonstrate a well-defined rim of decreased signal intensity that may be related to fibrous tissue.

Treatment is directed at surgical extirpation (2,3). Appropriate treatment requires recognition that ABC is often associated with a preexisting lesion. Treatment is directed toward the more aggressive component in cases where a preexisting lesion can be recognized (3). If no other lesion can be identified, most cases are treated with curettage and bone grafting. Recurrence is not uncommon and typically occurs within 2 years.

CASE 7

History: A 66-year-old man sustained acute ankle pain while running. What is your diagnosis?

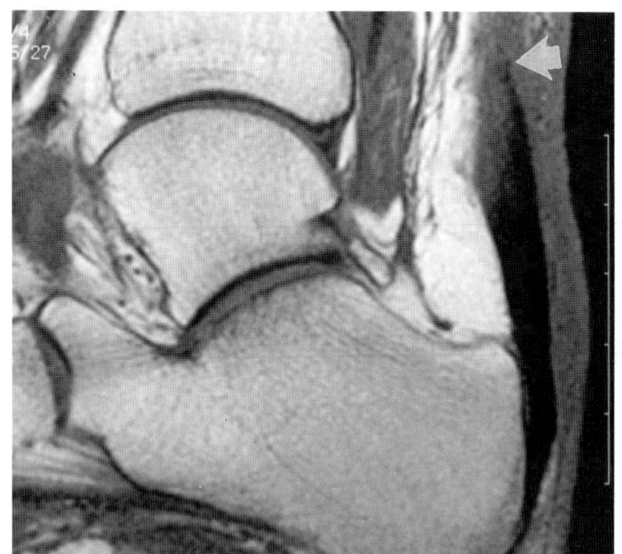
A

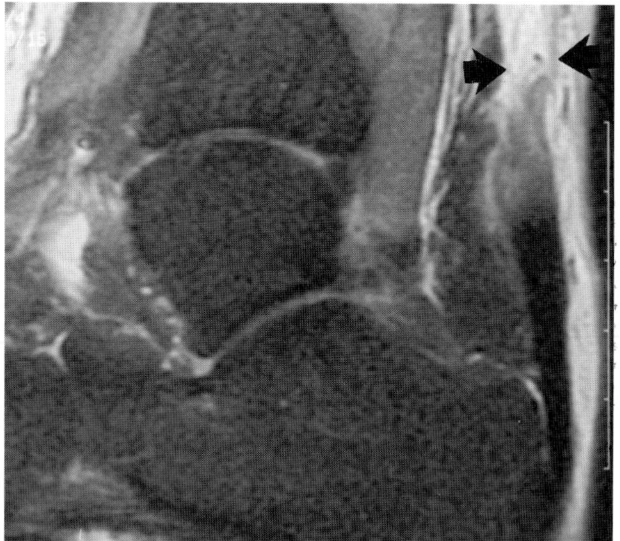
B

FIG. 7.

Findings: In Fig. 7A the sagittal T1-weighted fast spin echo (TR/TE 600/13eff) demonstrates loss of continuity of the Achilles tendon approximately 5 cm proximal to its calcaneal insertion (arrow). The tendon distally is slightly thickened but otherwise normal. In Fig. 7B the sagittal inversion recovery fast spin echo (TR/TE/TI 3,500/32 msec/130) demonstrates high signal intensity reflecting edema and hemorrhage within the gap between the two ends of the Achilles tendon and a ''mop-end'' appearance to the end of the distal tendon (arrows).

Diagnosis: Acute Achilles tendon disruption.

Discussion: The Achilles tendon is the largest and strongest tendon in the body and is capable of withstanding forces five to six times body weight during running or jumping (1). It is formed by the confluence of the individual tendons of the gastrocnemius and soleus muscles and in adults is about 10 to 15 cm in length and demonstrates a uniform thickness of 4 to 7 mm. Approximately 6 cm from its calcaneal insertion, the fibers of the Achilles tendon take a spiral twist. Ruptures most commonly occur at a level just cephalad to the twist, which represents a vascular watershed area (2–9).

It is generally agreed that a ''normal'' Achilles is not subject to rupture, and that some element of tendinosis precedes a complete tear (6). Achilles ruptures are far more common in men, favor the left side, and occur most often between the third and fifth decade. Ruptures may be seen in professional athletes of any age. Systemic and local disorders including rheumatoid arthritis, gout, hyperparathyroidism, chronic renal failure, diabetes, and steroid injections may weaken the tendon, but in most cases no predisposing factor is identified (4–9). Recurrent stress contributes to the development of numerous microfractures of the collagen fibrils of the tendon, which coalesce over time. Ultimately, tendinous repair is unable to keep pace with degeneration and failure finally occurs. Complete rupture typically occurs during push off, landing with the knee extended or the foot in dorsiflexion, or by direct trauma.

Complete ruptures of the tendon are manifest on MR by gross increase in tendon girth (often up to three to four times normal), tendinous discontinuity, and large areas of increased signal intensity on T2-weighted spin echo and STIR sequences within the gap (presumably representing edema and hemorrhage) (4–9). With complete ruptures, the proximal and distal ends of the tendon retract, the remaining tendon assumes a thicker appearance, and the torn tendon edges assume a mop-end appearance. Sagittal T2-weighted fast spin echo sequences with fat saturation and inversion recovery fast spin echo sequences are most commonly employed to assess the size of the gap and the condition and orientation of the torn fiber ends.

Differentiation of a complete tear from a partial tear (Fig. 7C,D) or from a rupture at the musculotendinous junction (Fig. 7E,F) may be clinically difficult. Ruptures at the musculotendinous junction, commonly referred to as tennis leg, are actually tears of the medial head of the gastrocnemius muscle with the tendon remaining intact. Such lesions, since they are largely injuries to muscle and tend to remain in continuity, are often treated nonoperatively. As the therapy for each condition is different, correct diagnosis is imperative, and MR can provide critical management information. While complete Achilles tendon tears are typically associated with excruciating, unremitting pain, the lesion may be misdiagnosed in up to 25% of patients and delayed diagnosis is not uncommon. The marked edema and hemorrhage that accompany the rupture make physical examination difficult, and plantar flexion of the affected ankle is often preserved by action of the posterior tibial tendon, the peroneals, and the flexor hallucis longus.

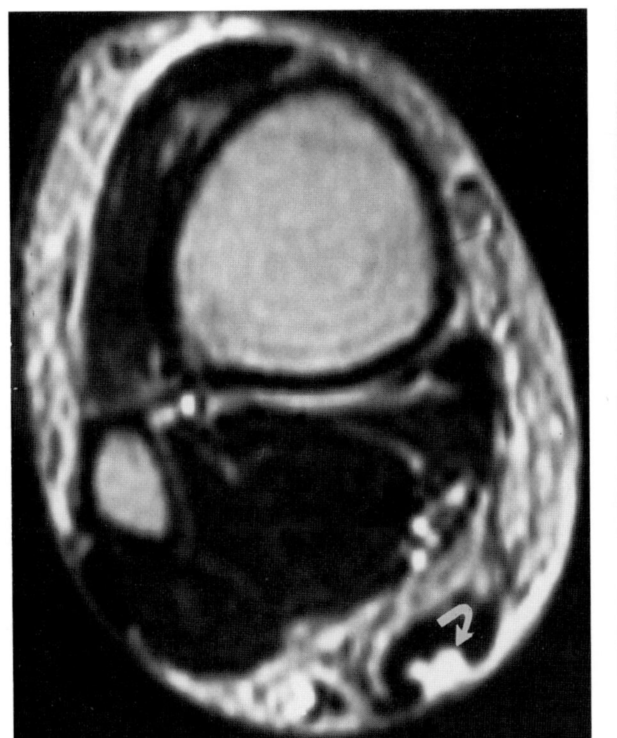
C

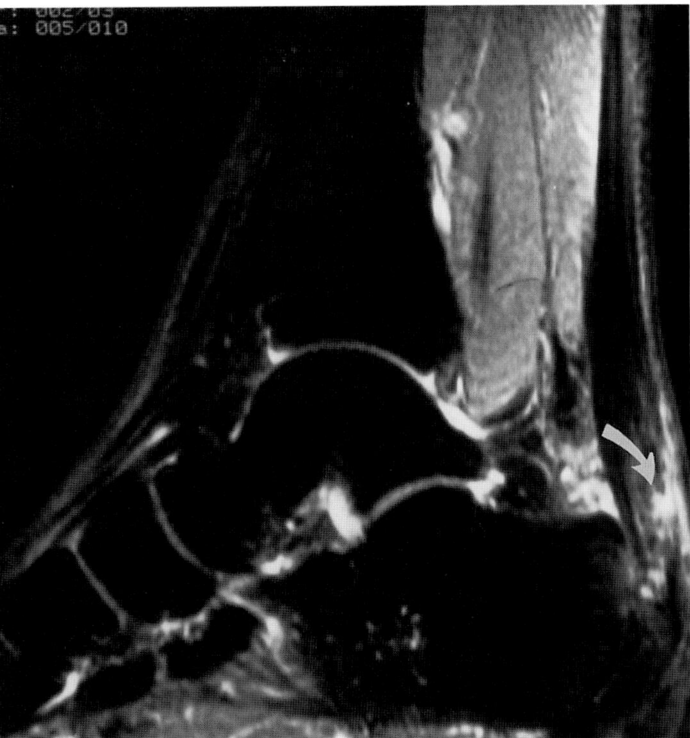

D

FIG. 7. C,D: Acute partial tear of the Achilles tendon. **C:** Axial T2-weighted sequence (TR/TE 2,000/80 msec) demonstrates the size and position of the partial tear *(curved arrow).* **D:** Sagittal (STIR) (TR/TE/TI 2,200/30 msec/160) sequence demonstrates bulbous enlargement of the Achilles tendon, a finding consistent with chronic tendinosis. There is a focus of high signal intensity along the distal dorsal aspect *(curved arrow)* with surrounding edema anteriorly and posteriorly.

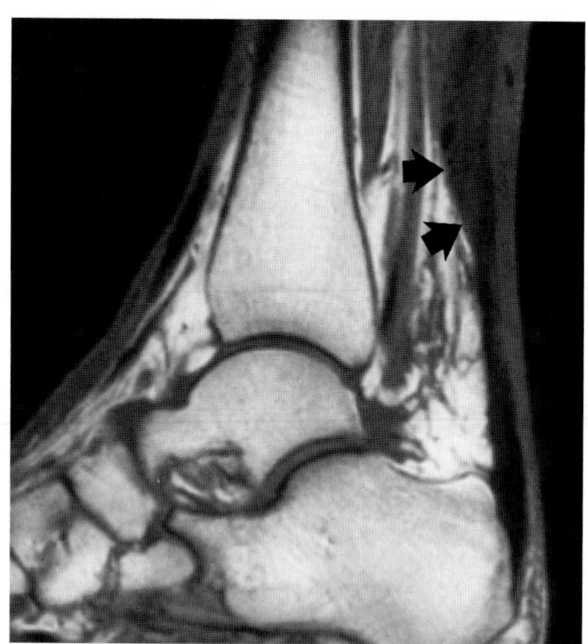
E

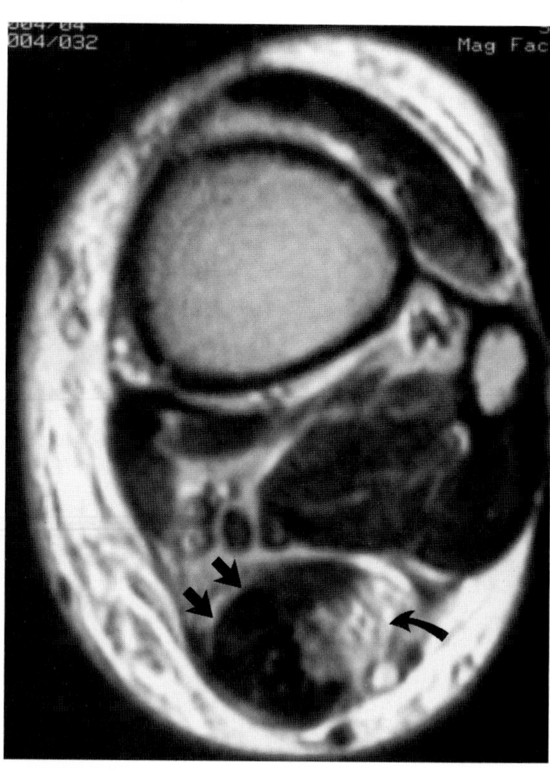

F

FIG. 7. E,F: Rupture of the musculotendinous junction of the Achilles ("tennis leg"). **E:** Sagittal T1-weighted image (TR/TE 500/20 msec) demonstrates site of abnormality *(arrows)* to be at the musculotendinous junction of the Achilles. **F:** Axial T2-weighted SE image (TR/TE 2,000/80 msec) demonstrates the pathologic process to represent a tear of the distal muscle. In this image, the tendon is somewhat medial *(arrows)* and is essentially intact, while the muscular portion *(curved arrow)* demonstrates disruption and edema of its internal architecture. (From Mink JH. Tendons. In: Deutsch AL, Mink JH, Kerr R, eds. *MRI of the foot and ankle.* New York: Raven Press, 1992, with permission.)

CASE 8

History: A 28-year-old professional athlete complained of chronic foot pain. What is your diagnosis? What are the relative advantages of MR in comparison to CT for establishing this diagnosis?

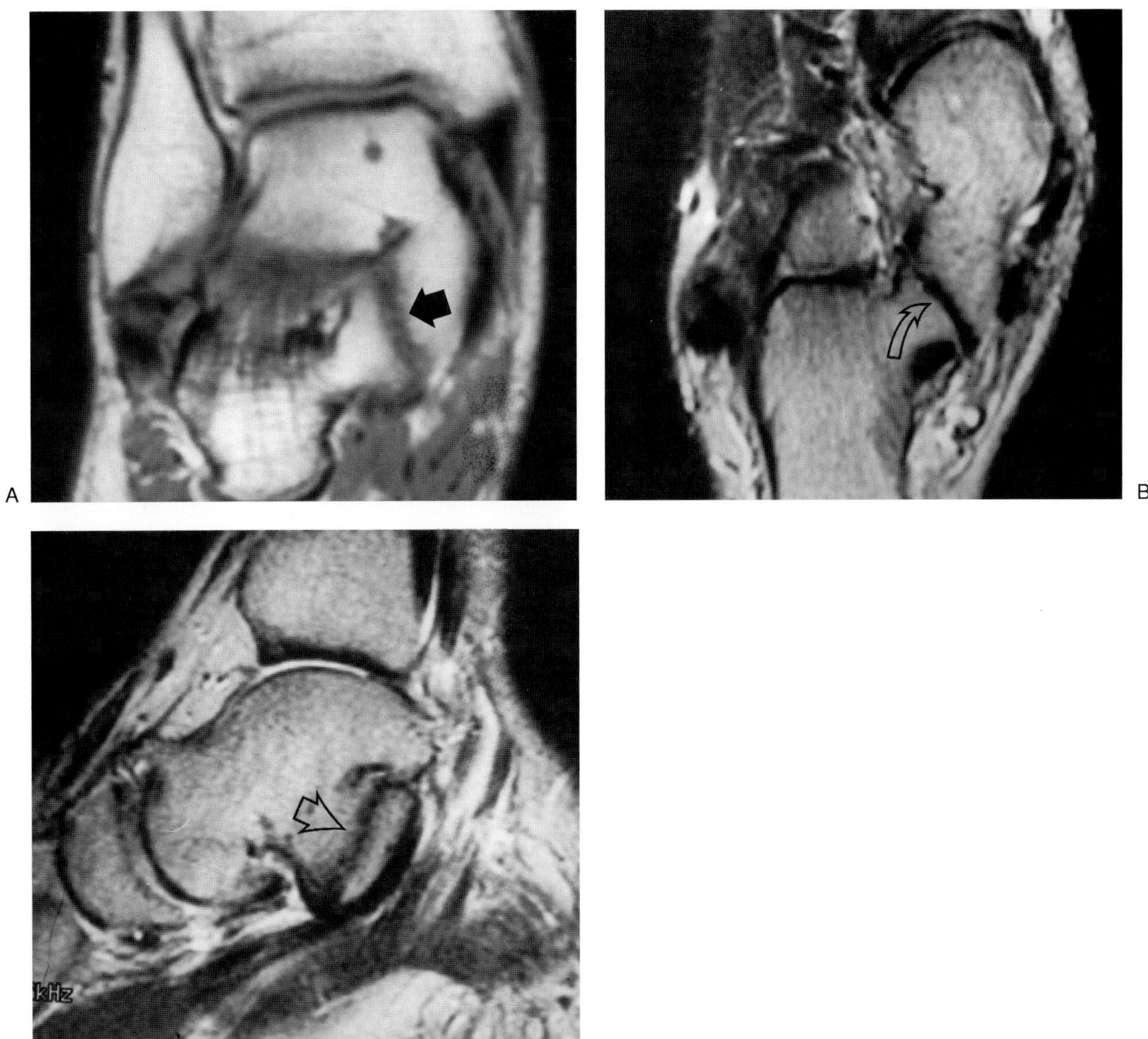

FIG. 8.

Findings: In Fig. 8A the coronal T1-weighted (TR/TE 500/20 msec) image through the level of the sustentaculum talus demonstrates marked narrowing of the joint space between the sustentaculum and the talus (short black arrow). In Fig. 8B the axial T2-weighted spin echo image (TR/TE 2000/80 msec) through the middle facet of the subtalar joint demonstrates marked narrowing of the expected joint space between the sustentaculum and the talus (open curved arrow). The interface remains of low signal intensity. In Fig. 8C the sagittal T2-weighted spin echo image through the middle facet again demonstrates the narrowed joint space without evidence of osseous continuity or of localized high signal to suggest fluid (short open arrow). (Images courtesy of Scott Kingston, M.D. LaCanada, California.)

Diagnosis: Tarsal coalition, fibrous type.

Discussion: *Tarsal coalition* is a term used to describe a symptomatic condition of the foot that results from an abnormal union between two or more tarsal bones. The union, which may be fibrous, cartilaginous, or osseous, results in restricted motion of the subtalar joint (1–3). Clinical complaints are uncommon before the second decade, presumably because the coalition is initially fibrous or cartilaginous and severe limitation of tarsal motion does not occur until ossific union develops. Foot pain is the usual complaint and a history of trauma is often present, suggesting that a ligamentous injury or possible fracture through the coalition may have occurred. Painful rigid flatfoot is a common clinical presentation although this complaint can also been seen in entities such as juvenile rheumatoid arthritis, posttrauma, and with septic arthritis. The three most common symptoms are pain, valgus deformity, and subtalar or talar stiffness. The two most common types of coalitions are those occurring between the calcaneus and talus (60%) and between the calcaneus and navicular (30%) (1–4). Ossification of calcaneonavicular coalitions occur at 8 to 12 years of age and that of talocalcaneal at 12 to 16 years (2,4). Congenital coalitions are bilateral in nearly 50% of cases.

The talus and calcaneus articulate by means of three facets (anterior, middle, and posterior) (1–7). The tarsal sinus separates the talocalcaneal joint into two components: (1) the posterolateral (anatomic) subtalar joint; and (2) the anteriormedial talocalcaneonavicular joint, which incorporates the anterior and middle facets (7). With normal gait, the calcaneus slides forward relative to the talus. The subtalar joints externally rotate to accommodate the internal rotation of the tibia. With restricted talar motion, and in particular with talocalcaneal coalitions, the calcaneus assumes a valgus position (2,7). This leads to shortening of the peroneal muscles. Instability of the talonavicular joint then results, with the talus forced underneath the navicular. Talar beaking, a characteristic radiographic findings of talocalcaneal coalition, results from traction on the periosteum on the superior neck of the talus (2).

Accurate detection and characterization of tarsal coalition has important implications for management. A coalition involving less than half of the joint area may be approached with surgical resection. More extensive coalitions and those demonstrating degenerative change may require arthrodesis. A number of diagnostic methods have been applied in detection and assessment of possible coalitions (1–7). Conventional radiographs are most useful for detection of osseous calcaneonavicular coalition (2). Conventional radiographs of the foot, including specialized calcaneal (Harris) projections, often fail to directly depict talocalcaneal coalitions (2). A number of changes that occur as a result of altered talar motion may be demonstrable. Among these the most common and suggestive is that of talar beaking (2). This finding, which appears as a steeply marginated dorsal projection at the talonavicular joint, must be distinguished from a normal dorsal ridge related to the insertion of the joint capsule and talonavicular ligament insertion that is set back from the talonavicular joint (2). Scintigraphic methods have been employed and may demonstrate augmented uptake in the subtalar and talonavicular regions. Radionuclide studies, however, lack both specificity and the ability to define anatomic detail (5).

CT has revolutionized the diagnosis of tarsal coalition and has become, since its application to assessment of this condition was first described, the technique of choice (1,2,4,6). The study is performed with the patient supine, knees flexed, with plantar surfaces flat on the table. Contiguous thin dorsoventral plane sections are accomplished through both feet simultaneously and are ideal for assessment of talocalcaneal coalitions, which occur through the middle facet of the subtalar joint. Oblique sagittal reformations can be accomplished for assessment of possible calcaneonavicular coalitions. Complete osseous coalition is readily recognized as loss of the normal joint space between the sustentaculum and the talus with bony continuity extending from one bone to the other. Cartilaginous coalitions are somewhat more difficult to define as the joint space remains present but the sustentaculum is abnormal (usually hypertrophied) (1,2,4). The joint space is usually narrow, and obliquely oriented, slanting downward in a medial direction. Subchondral cystic changes are seen involving the articulation. CT criterion for fibrous coalitions include subtle articular narrowing and cystic joint irregularity (Fig. 8D–H).

More recently, MR has been applied to assessment of tarsal coalitions and several investigators have noted advantages, particularly for detection of nonosseous coalitions (3,6,7). With MR, cartilaginous coalitions demonstrate joint space narrowing with signal intensity changes of cartilage or fluid between the two affected bones (7). Fibrous coalitions are diagnosed when areas of intermediate to low signal intensity are demonstrated bridging the two involved bones. In a recent series, Wechsler and colleagues (7) found MR better than CT for detection of fibrous coalitions and advocated the use of MR as the primary method of evaluation in suspected coalition.

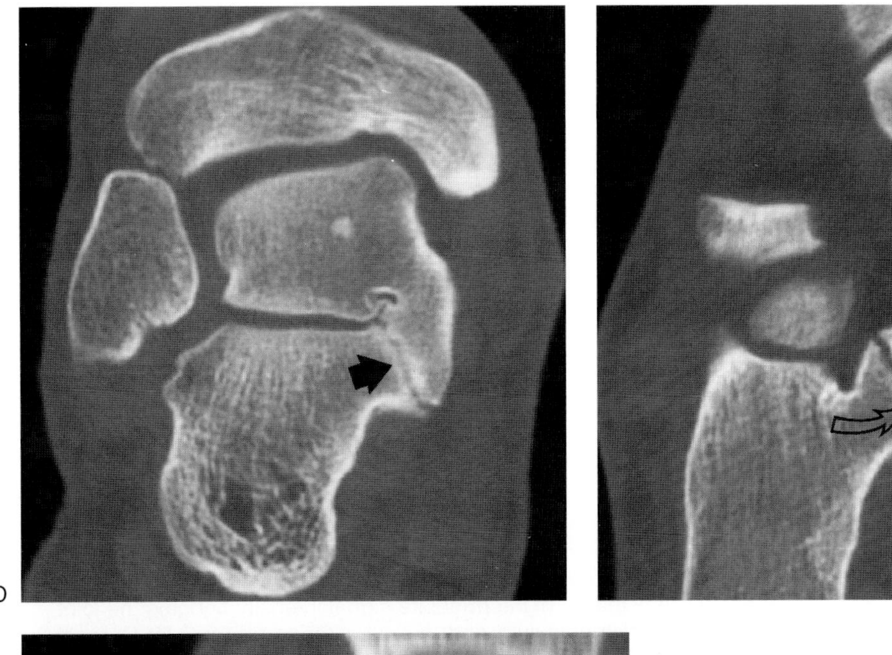

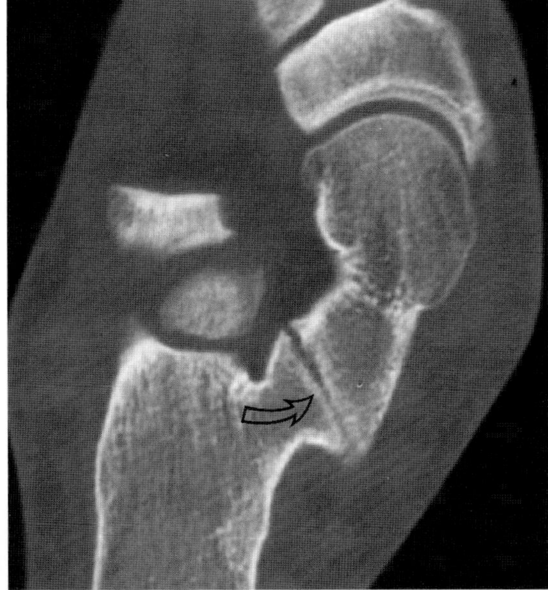

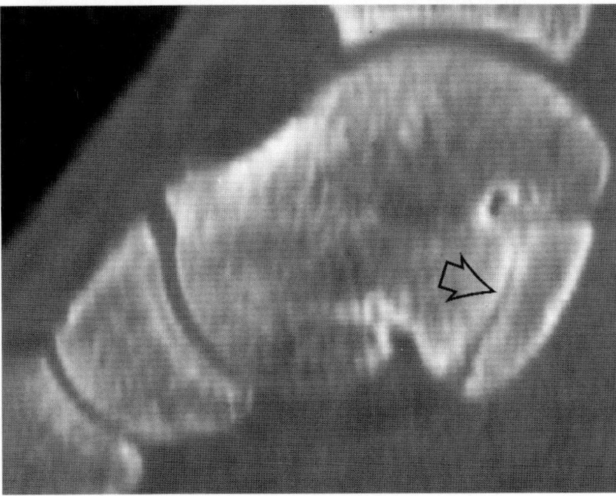

FIG. 8. D–F: Fibrous coalition; correlative CT sections. **D:** Coronal CT section obtained through the middle facet demonstrates marked narrowing of the joint space with a serpentine appearance of the subchondral bone plate. No bony continuity is identified *(short black arrow).* **E:** Axial CT section demonstrates the fibrous coalition at the level of the middle facet *(open curved arrow).* **F:** Sagittal reconstruction correlates well with the findings depicted in C. Again note the nonosseous coalition *(open short arrow).*

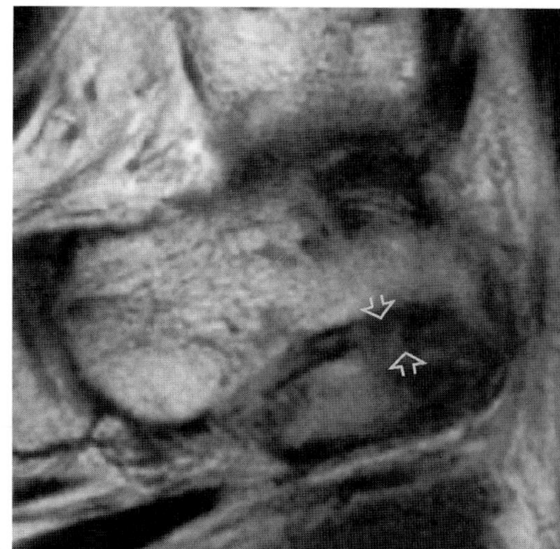

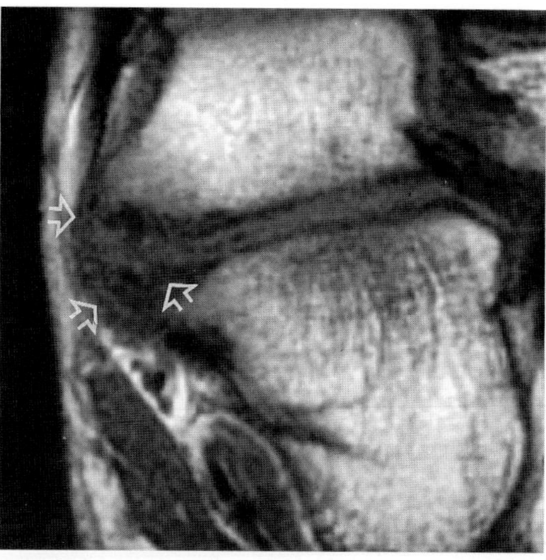

G H

FIG. 8. G,H: Cartilaginous coalition. A 10-year-old boy presenting with a protuberance along the medial aspect of the ankle. **G:** Sagittal T1-weighted (TR/TE 500/15 msec) image through the medial aspect of the ankle demonstrates a localized "mass" of intermediate signal intensity bridging the joint space between the posterior aspect of the sustentaculum and the talus *(open arrows).* **H:** Coronal T1-weighted image (TR5/TE 500/15 msec) also demonstrates the intermediate signal intensity "mass" between the talus and calcaneus representing a cartilaginous coalition *(open arrows).*

CASE 9

History: A 38-year-old man had suffered two prior significant lateral ankle sprain injuries. An avid tennis player, the patient had undergone a lateral collateral ligament reconstruction to help stabilize his ankle. The MR examination was obtained 14 months postsurgery at a time when the patient was complaining of increasing lateral ankle pain. What is your diagnosis? What are the sequelae of serious ankle sprains?

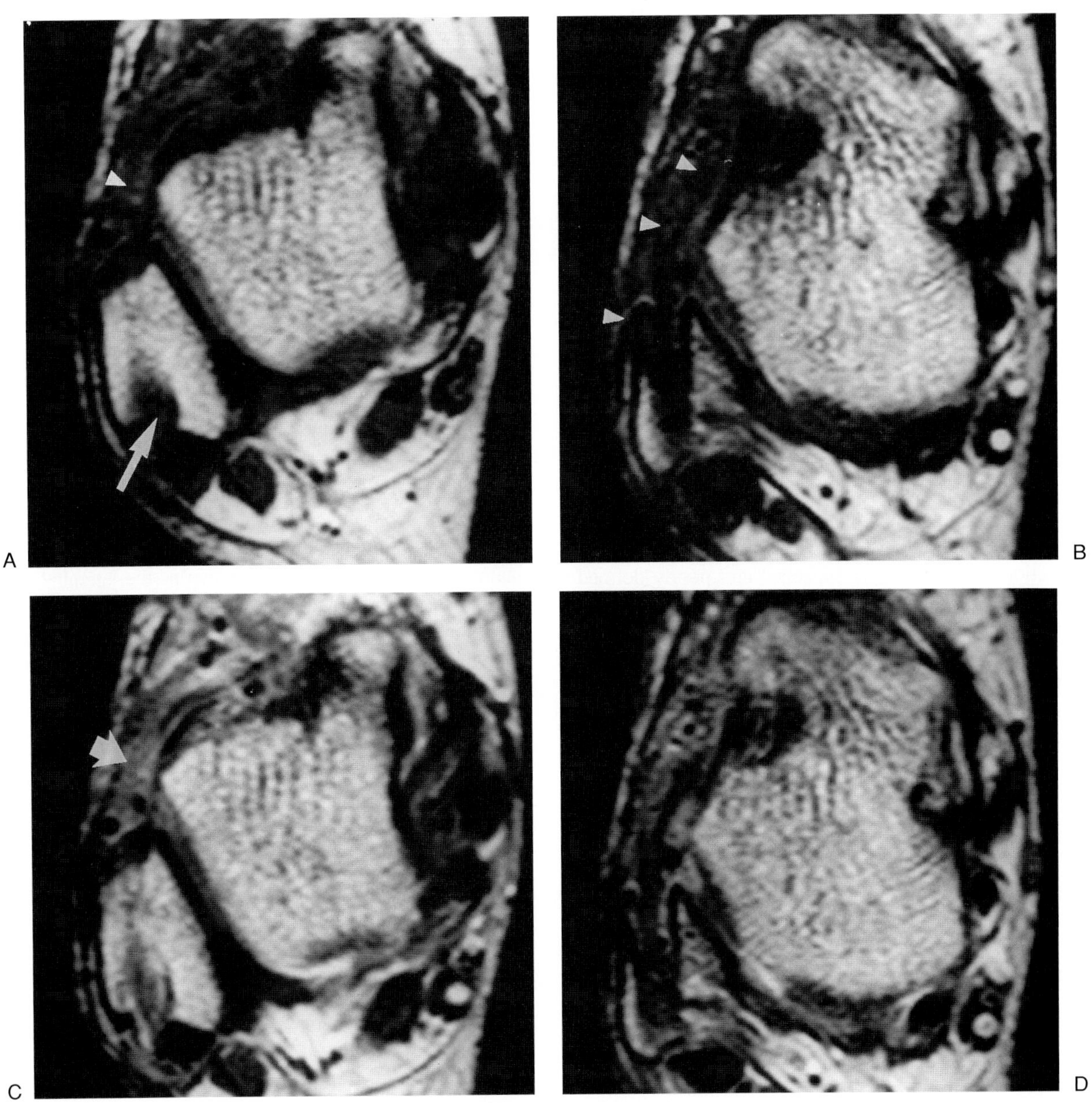

FIG. 9.

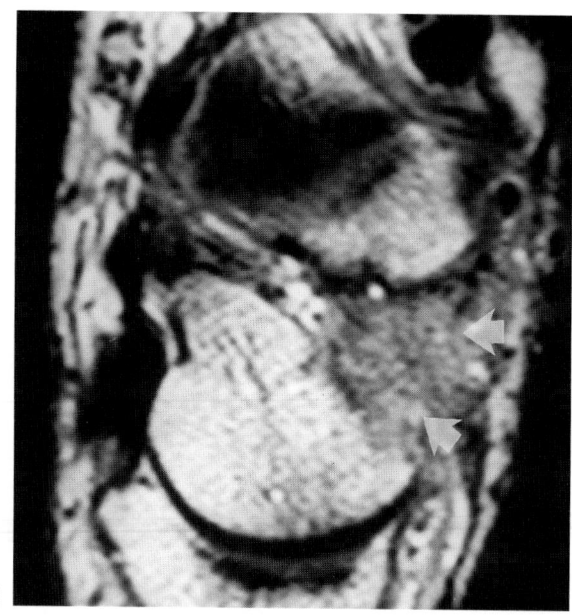

FIG. 9.

Findings: Figure 9A–D: Axial T1-weighted (TR/TE 517/20 msec) images prior to (A,B) and following (C,D) the intravenous administration of gadopentetate dimeglumine. In Fig. 9A there is somewhat poorly defined band-like intermediate signal intensity extending across the lateral gutter of the ankle (arrowheads). A small tunnel is seen within the fibula (arrow). In Fig. 9B, on a slightly more inferior section, the reconstruction (arrowheads) is thickened and poorly defined as it extends along the expected course of the anterior talofibular ligament. In Fig. 9C, on a postcontrast section, there is enhancement of the material within the lateral gutter (arrow). In Fig. 9D there is enhancement along the walls of the tunnel through the fibula and along the course of the reconstruction. In Fig. 9E, on a slightly more inferior section, there is extensive enhancing soft tissue mass extending into the tarsal sinus (arrow).

Diagnosis: Anterolateral ankle impingement syndrome and tarsal sinus syndrome secondary to scar formation and synovitis status post–reconstructive surgery.

Discussion: Ankle sprains are not always self-limited. In one study of high school athletes, 50% of those suffering acute lateral ligament tears reported residual symptoms from their injuries; 15% felt that their injury compromised their playing performance (1). The long-term consequences of severe lateral ankle ligament sprains include instability, weakness, pain, osteochondral fracture (osteochondritis dissecans), loose body formation, the sinus tarsi syndrome, and the anterolateral ankle impingement syndrome. The latter term refers to a group of patients who suffer from chronic lateral ankle pain, the etiology of which remains elusive even after the use of such conventional techniques as stress radiography, high-resolution computed tomography, and single photon emission computed tomography (1–3). Affected patients are typically young and active and 94% have sustained a previous lateral ankle sprain.

Following an inversion injury to the ankle, and at least a partial tear of the anterior talofibular ligament (ATAF), intraarticular hemorrhage may occur. Repetitive motion is believed to contribute to inflammation of the ligamentous ends with eventual synovial hyperplasia extending into the lateral talofibular space (also known as the lateral ''gutter'') (Fig. 9F). The synovial reaction may extend in a cephalic direction to involve the anterior tibiofibular joint and the mass of synovium may become entrapped between the talus, tibia, and fibula. In some cases, a ''meniscoid'' lesion of scar tissue forms in the anterolateral aspect of the joint space (4). This mass of tissue consists of hyalinized connective tissue that assumes a ligamentous contour. In more advanced cases, the hypertrophied synovium may erode the anterolateral talar dome. An additional purported cause of the anterolateral ankle impingement is hypertrophy of Bassett's ligament (3). This structure is thought to be a separate inferior fascicle of the anterior tibiofibular ligament. As illustrated in this case, lateral ankle impingement syndrome may arise secondary to marked synovitis post surgery. Histologically, moderate synovial hyperplasia with subsynovial capillary proliferation is identified (5,6). Arthroscopic synovectomy and resection of the offending lesion has been reported to improve symptoms in the majority of cases (5,6).

On MR, thickening and deformity of the anterior talofibular ligament is invariably present. Intermediate signal intensity material, representing scar tissue within the lateral gutter, may be demonstrated on both axial and sagittal sections. Protusion of the torn ends of the ATAF into the joint space may impede the uniform distribution of joint fluid from ''front to back'' within the lateral gutter (Fig. 9G,H). The findings may be more evident with the foot imaged in dorsiflexion or on ''kinematic'' examination.

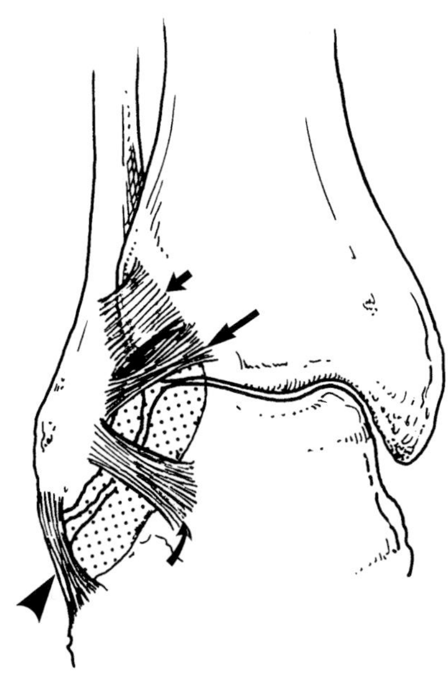

FIG. 9. F: Anterolateral ankle impingement syndrome. This diagram defines the boundaries of the lateral gutter *(shaded area),* the site at which ligamentous injury and reactive synovitis produce the "meniscoid" lesion typical of the impingement syndrome. ATIF, *arrow;* Bassett ligament, *long arrow;* ATAF, *curved arrow;* CF, *large arrowhead.*

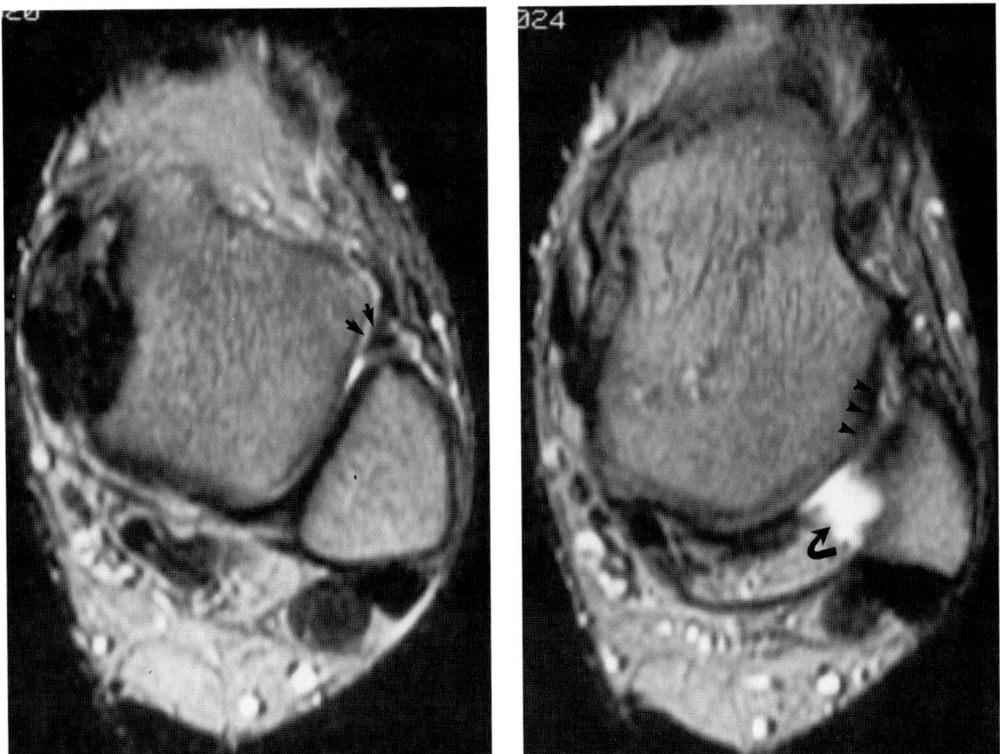

FIG. 9. G,H: Impingement syndrome. Axial T2-weighted spin echo (TR/TE 2,000/80 msec) images in a patient with chronic pain following multiple ankle sprains. **G:** The ATIF *(arrows)* is thickened and irregular, and a portion of the ATIF protrudes into the anterior aspect of the talofibular joint *(arrowhead).* **H:** At the level of the anterior talofibular ligament, a poorly defined soft tissue density *(arrowheads)* is seen in the talofibular joint space; the joint fluid *(curved arrow)* that is present is all posterior. This constellation of findings suggests the diagnosis of the anterolateral joint impingement syndrome. (From Mink JH. Ligaments of the ankle. In: Deutsch AL, Mink JH, Kerr R, eds. *MRI of the foot and ankle.* New York: Raven Press, 1992;178–185, with permission.)

CASE 10

History: A 32-year-old dancer underwent MR evaluation for lateral ankle pain of several months duration. What is your diagnosis? Is this a common pattern of injury?

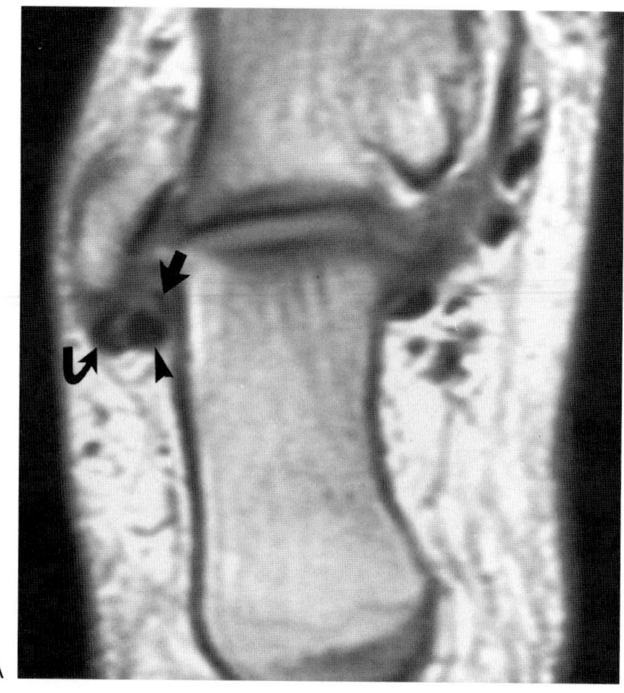

FIG. 10.

Findings: In Fig. 10A the oblique axial proton density MR image (TR/TE 2,000/20 msec) demonstrates two apparent ovoid-shaped structures along the medial (arrow) and lateral (arrow) aspect of the peroneus longus (PL) tendon (arrowhead). A ''normal'' peroneus brevis (PB) tendon is not visualized. (From Mink JH. Tendons. Deutsch AL, Mink JH, Kerr R, eds. In: *MR of the foot and ankle.* New York: Raven Press, 1993;160–171, with permission.)

Diagnosis: Longitudinal split tear of the peroneus brevis tendon.

Discussion: The peroneal muscles pronate and evert the foot, are primary lateral dynamic stabilizers of the ankle, and participate in the second stage of walking and pushoff (1–3). The peroneal tendons reside in a fibro-osseous tunnel that is formed by the lateral malleolus anteriorly, the superior peroneal retinaculum posteriorly and laterally, and the posterior talofibular and calcaneofibular ligaments medially. The peroneal retinaculum functions to prevent the peroneal tendons from bowstringing over the lateral malleolus during muscle contraction (3). Below the malleolus, the tendons diverge form one another and have separate sheaths; the PB inserts on the base of the fifth metatarsal while the longus passes posterior to the peroneal tubercle, curves beneath the calcaneus, and inserts on the base of the first metatarsal and medial cuneiform (1,3).

Mechanical stress on the peroneus brevis where it abuts the medial malleolus may result in decreased vascularity and longitudinal splitting, and eventually result in disruption (4). A longitudinal split of the peroneus brevis tendon occurs when, with forced dorsiflexion, the PL is driven into the posterior aspect of the PB, creating a cleft that never heals (4). It is important, however, to distinguish this type of tendon rupture from a bifurcated insertion of the peroneus brevis; the latter is a normal variant. Partial rupture of the peroneus brevis is commonly associated with abnormal intratendinous signal as well as tenosynovitis. It is critical to distinguish between artifactually increased angle-dependent signal (''magic angle'') and true intratendinous signal reflective of tendinosis and interstitial type tears (see Case 23).

Peroneal tenosynovitis may be idiopathic or occur as a complication of prior calcaneal fracture with resultant entrapment of the peroneal tendons between a laterally displaced fragment and the distal fibula. Mechanically induced tenosynovitis occurs at the points of direction change such as at the fibular malleolus, the peroneal tubercle, or the cuboid in the case of the peroneus longus (1). Peroneal tenosynovitis results in pain, swelling, tenderness, and occasionally involuntary inversion of the foot. On MR, the distention of the peroneal tendon sheath can be well demonstrated as can the peroneal tendons themselves, which are graphically contrasted against the joint fluid (Fig. 10B). When considering the diagnosis of possible tenosynovitis, it is important to realize that the depiction of some degree of fluid is common within many of the ankle tendons (exceptions include the extensor tendons) (9). Indeed in one study that addressed this issue, the degree of fluid detectable within the tendon sheaths of asymptomatic and symptomatic patients on MR was not appreciably different (9). Fluid is most commonly depicted proximally and posteriorly in ankle tendons, a finding that may reflect a manifestation of gravitational effects related to ankle positioning for MR evaluation. Large volumes of fluid are most commonly encountered in the flexor hallucis longus tendon sheath and are relatively rare in other ankle tendons (9). As such, while small amounts of MR detectable fluid are likely physiologic, significant amounts of fluid should raise the possibility of tenosynovitis or, in the setting of acute ankle trauma with a joint effusion, a possible tear of the CF ligament.

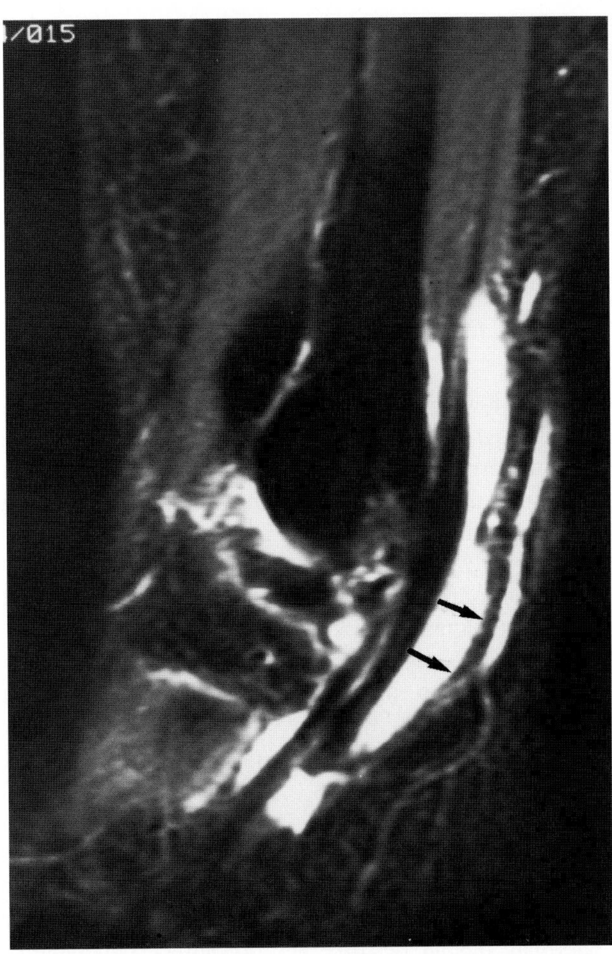

FIG. 10. B: Peroneal tenosynovitis. Sagittal STIR (TR/TE/TI 2,200/35 msec/160) sequence demonstrates marked distention of the peroneal tendon sheath *(arrows)*. The contained tendons appear grossly normal. (From Mink JH. Tendons. In: Deutsch AL, Mink JH, Kerr R, eds. *MRI of the foot and ankle.* New York: Raven Press, 1993;160–171, with permission.)

CASE 11

History: A 32-year-old woman presents with a several-month history of increasing heel pain and a palpable thickening along the plantar aspect of the calcaneus. What is your diagnosis?

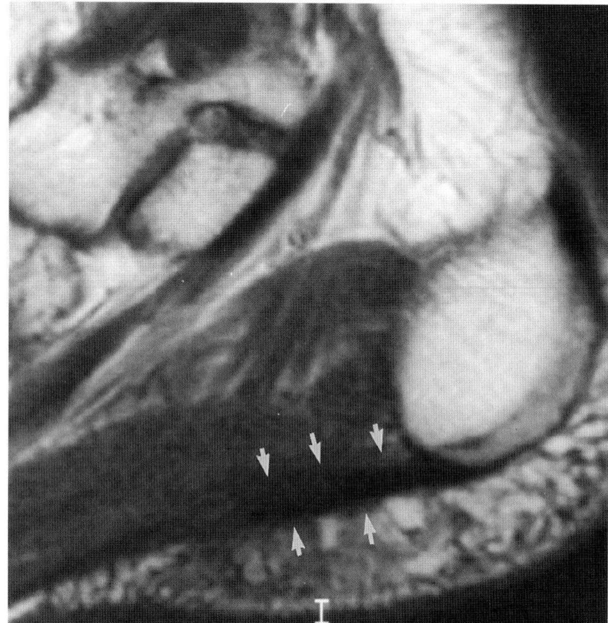

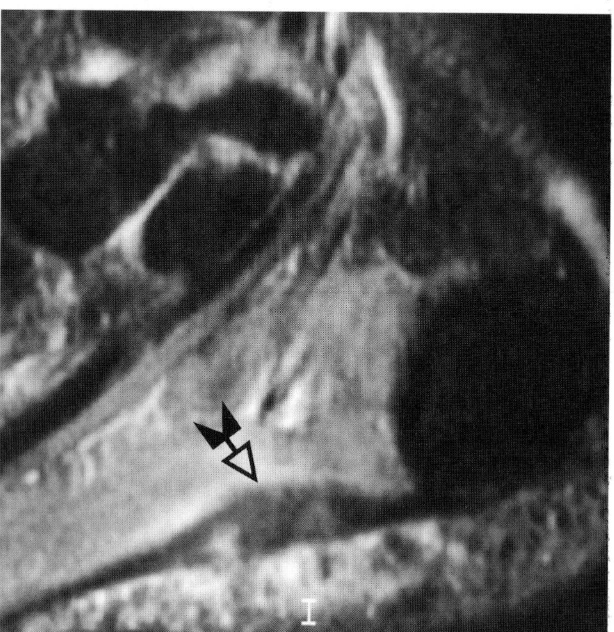

FIG. 11.

Findings: In Fig. 11A the sagittal T1-weighted (TR/TE 500/20 msec) image obtained through the hindfoot demonstrates marked thickening of the proximal attachment of the plantar fascia (small white arrows). In Fig. 11B the inversion recovery fast spin echo (TR/TE/TI 3,300/30 msec/130) image better delineates the marked focal thickening of the plantar fascia (black arrow). There is increased signal both within and surrounding the thickened plantar fascia.

Diagnosis: Plantar heel pain secondary to plantar fasciitis.

Discussion: Plantar heel pain is a common problem and is frequently seen both in the middle-aged and elderly population as well as in younger athletes, particularly runners (1–4). The basis for the pain may relate to any of the structures in this region and MR may be well utilized to specifically characterize the pathoanatomy in clinically confusing cases. Patients with plantar fasciitis have pain and tenderness in the region of the calcaneal tuberosity where the central and thickest component of the plantar fascia takes origin. The pain is often immediately present in the morning, may gradually subside with walking, and often is intensifed by athletic activity (5–8). The condition may be associated with obesity and is bilateral in 10% to 30% of cases. Heel pain related to plantar fasciitis is to be distinguished from other conditions that may produce symptoms in this region. Medial heel pain most often relates to disorders of the posterior tibial tendon or tarsal tunnel. Likewise, lateral heel pain may relate to peroneal tendon dysfunction. Posterior heel pain frequently relates to Achilles tendon disorders or retrocalcaneal bursitis.

The plantar fascia is a multilayered fibrous aponeurosis that has a medial, central, and lateral component. The central portion is largest and originates from the medial calcaneal tuberosity. The deep layer of the aponeurosis divides into five tracts distally that insert on the proximal phalanges. With extension of the metatarsophalangeal joints during gait, the plantar fascia is elongated and tightens and serves to facilitate elevation of the longitudinal arch in a manner analogous to a windlass (9). This function subjects the plantar fascia to a tensile force and places a traction stress at its origin. The pathogenesis of plantar fasciitis is believed to relate to both the previously discussed mechanical strain, which is accentuated by a pronation or cavus foot deformity, as well as to repetitive microtrauma. In both situations, microtears occur in the origin of the plantar fascia and elicit a local inflammatory reaction (10–12). Biopsy of patients with chronic plantar fasciitis demonstrates findings of collagen degeneration with collagen necrosis, angiofibroblastic hyperplasia, chondroid metaplasia, and matrix calcification (8,9,13). This process is believed to begin at the origin of the plantar fascia from the medial tuberosity and, in the chronic phase, to extend distally along the course of the plantar fascia (14). A similar inflammatory process at the origin of the flexor digitorum brevis muscle probably leads to the development of a plantar calcaneal spur (2).

In the acute phase, inflammation is localized to the origin of the plantar fascia and maximal tenderness is elicited at this site (8,14). In the chronic phase, pain extends distally along the entire course of the fascia and the fascia may appear thickened or nodular on physical examination. Patients with this chronic form are maximally tender along the midportion of the plantar fascia and symptoms are exacerbated by dorsiflexion of the toes. Morning pain and stiffness are common (8). Most patients with plantar heel pain respond to conservative management. Patients with recalcitrant pain may respond to surgical release of the origin of the plantar fascia with decompression of the first branch of the lateral plantar nerve. In one study this resulted in complete relief of symptoms in 82% of patients (10).

Conventional radiography is only of limited value in the evaluation of patients with plantar heel pain (15). Several studies have documented that the demonstration of a plantar spur is not necessarily a significant finding. A heel spur may be seen in 16% (16) to 65% (7) of nonpainful heels, and in one study only 10% of patients with heel spurs were symptomatic (17). The demonstration of sclerosis or periosteal thickening in the region of the medial calcaneal tuberosity may be seen in symptomatic as well as asymptomatic individuals and has been considered a normal variant (7). Bone scintigraphy has also been utilized in the assessment of patients with plantar heel pain. In one study, increased radionuclide activity was demonstrated on delayed images and on immediate blood pool images of 21% of patients with heel pain (18). The findings likely reflect periosteal inflammation accompanying an inflammatory or degenerative enthesopathy at the origin of the plantar fascia. Both conventional radiography and bone scintigraphy are limited by their inability to specifically delineate the soft tissue structures of the heel.

On MRI, the plantar fascia appears as a 3 to 4 mm thick, well-defined, low signal intensity structure on sagittal and coronal images (1). In symptomatic feet, the plantar fascia is thickened to 7 to 8 mm, and, instead of appearing black, reveals intermediate signal on T1-weighted and proton density images. Regions of variably increased signal intensity are observed within the thickened fascia on T2-weighted or STIR sequences (1,19). These abnormalities predominate within the proximal portion of the plantar fascia. High signal intensity is often seen in the adjacent soft tissue and the plantar fat pad. Focal increased signal intensity may be observed within the calcaneus, adjacent to the origin of the plantar fascia, presumably reflecting a localized mechanical or inflammatory reaction. The soft tissue and intraosseous regions of high signal intensity are often more prominent on STIR sequences. Relatively focal or diffuse patterns of involvement of the plantar fascia may also be observed, beginning several centimeters distal to its origin. With MR, a relatively small field of view (i.e., 10 to 12 cm) should be employed to optimize anatomic definition while still effectively screening the region for other abnormalities. The sagittal and coronal planes best demonstrate the structures of interest. We currently employ both fat-suppressed fast spin echo and inversion recovery fast spin echo sequences for assessment of plantar heel pain.

CASE 12

History: A 68-year-old woman presented to the emergency room with a 6-day history of heel pain. Radiographs obtained at the time of the visit were normal. The patient continued to complain of pain and was subsequently seen by an orthopedic surgeon and an MR examination undertaken. What is your diagnosis?

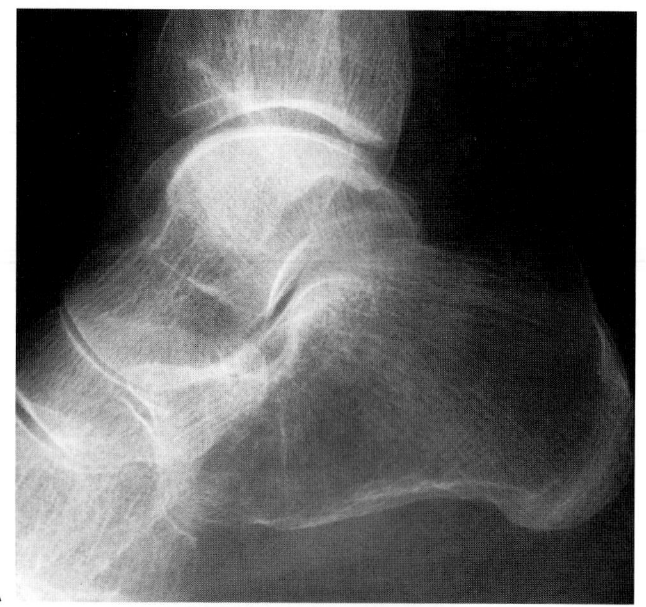
A

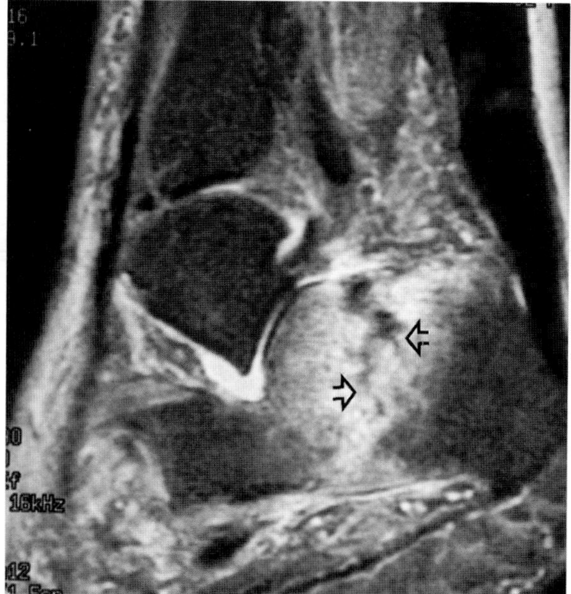
B

C

FIG. 12.

Findings: On Fig. 12A, a lateral x-ray of the ankle, there is mild osteopenia. No evidence of a fracture is identified. On Fig. 12B, a sagittal inversion recovery FSE (TR/TE/TI 3,300/30 msec/130) image, a serpentine line of low signal intensity is seen within the body of the calcaneus posterior to the posterior facet of the subtalar joint (arrowheads). There is striking surrounding high signal intensity reflective of reactive edema. On Fig. 12C, an axial T2-weighted fast spin echo image (TR/TE 3,500/108 msec), the serpentine line traversing the body of the calcaneus is vividly depicted.

Diagnosis: Calcaneal stress fracture.

Discussion: The case represents a graphic example of the ability of MR to depict abnormalities of bone that are not demonstrable on conventional radiography. The high contrast sensitivity of MR allowing depiction of marrow edema, coupled with sufficient resolution capabilities to allow definition of a fracture line, have allowed MR to emerge as the premier technique for skeletal trauma detection that is not initially demonstrable by conventional radiographic techniques.

The term *stress fracture* refers to the failure of the skeleton to withstand submaximal forces acting over time. Two types of stress fractures have been identified: (1) fatigue fractures, such as have been classically described in military recruits and runners in which normal bone is subjected to repeated abnormal stresses, and (2) insufficiency fractures, in which normal stresses are applied to a bone whose skeletal composition reflects deficient elastic resistance (e.g., Paget disease, osteoporosis) (1,2). Fatigue fractures are particularly common in the lower extremities in athletes, joggers, and dancers. The activity precipitating the fracture is typically new or different for the patient, is strenuous, and is repeated with a frequency that ultimately results in failure of the bone and the production of symptoms. Fatigue and insufficiency fractures may also complicate certain lower extremity surgical procedures that result in altered stress or an imbalance of muscular forces on normal or insufficient bone. A typical example can be found in the foot following bunion surgery; the second metatarsal bears an increased load and may suffer a stress fracture (3).

A stress fracture should be considered as the final chapter in a series of events (4). Preceding the actual fracture are a number of abnormal pathologic alterations that are associated with pain, and which place the patient at risk for the development of a fracture. The term *stress response* refers to certain prefailure events occurring at the cellular level that result in structural weakening. When a bone is stressed, gradual and progressive resorption of circumferential lamellar bone and its subsequent replacement by dense osteonal bone are identified (5). This period is characterized by local hyperemia, edema, and osteoclastic activity during which time the stressed bone undergoes remodeling. Because of this process, a vulnerable period exists following a stressful occurrence in which the cortical bone is less capable of withstanding further stress and in which foci of osseous resorption may be transformed into sites of microfracture. The stress fracture begins as a small cortical crack that can progress as the stress continues. This progression is characterized by the appearance of subcortical infraction in front of the main crack in the bone. Understanding the above sequence of events provides the basis for the development of conditioning programs for athletes. Stress events in bone are more likely to occur when there has been an increase in the strength of the acting muscle over a relatively short time as the concomitant increase in strength of bone lags behind that of muscle. If the more rapidly developing muscular stress is slowed down, the sequence is interrupted so that new bone formation can ''catch up'' with the increased muscular demand and a state of increased bone strength is achieved. Increasing physical activity under controlled circumstances can result in osseous hypertrophy without microfracture.

Among the tarsal bones, fatigue and insufficiency fractures most commonly involve the calcaneus and the shafts of the metatarsals. In the tarsal navicular, stress fractures have been reported in active individuals, especially basketball players and runners (6,7). Characteristically, this fracture is oriented in the sagittal plane and is located in the central one-third of the bone. It may extend through the navicular for a variable depth, and is typically difficult to image on conventional radiographs. Scintigraphy and computed tomography in the anatomic anteroposterior plane have been advocated for detection of these injuries (6).

MR, by virtue of its enhanced soft tissue contrast resolution, may also permit preradiographic detection of stress injuries of bone (1,2,8–10). Prefracture stress responses, as previously discussed, are most readily demonstrated on MR in areas in which bone marrow is overwhelmingly fatty (11,12). This is a distinct advantage for MR in imaging of the foot and lower extremity, which are overwhelming composed of fatty marrow except in the youngest of individuals. On MR, stress responses appear as globular foci of decreased signal on T1-weighted sequences that are graphically depicted against a background of fatty (bright) marrow. The signal variably increases in intensity on T2-weighted sequences. Stress responses are exceedingly well demonstrated on short inversion time inversion recovery sequences (STIR), in which they are depicted as areas of high signal intensity against the suppressed dark background of fatty marrow. While the detection of stress responses in fatty marrow is relatively straightforward, demonstrating the abnormalities in hematopoietic marrow can be more difficult. STIR sequences have been shown to be of considerable value in this setting (12).

Amorphous type stress fractures are best termed stress responses (2,11–14). Stress responses are differentiated from bone bruises predominantly by the clinical history, as the appearance on MR may be quite similar. The authors recently reviewed 11 instances of stress response in bone, diagnosed by history and resolution by both clinical and MR criteria. Five lesions occurred in the femoral neck, two at the knee, two involved the ankle, and two involved the foot. All lesions had identical MR findings: inhomogeneous, poorly defined signal decreases on T1-weighted sequences that became minimally and inhomogeneous brighter on T2-weighted sequences. All lesions demonstrated dramatically increased signal intensity on STIR sequences. In none of the patients were the initial or follow-up conventional radiographs abnormal. Bone scans obtained in seven patients were positive in five and negative in two.

The classic appearance of a stress fracture on MR is a linear zone of decreased signal intensity on T1-weighted sequences that is surrounded by a broad and more poorly defined area which is of lower signal intensity than the linear component. Typically, the linear component remains dark on T2-weighted sequences, but the surrounding zone of presumed hemorrhagic and nonhemorrhagic edema becomes inhomogeneously brighter. The linear components may be short and straight, or long and serpiginous, and are typically perpendicular to the adjacent cortex (1,2,8,11,14). Most stress fractures are metaphyseal or epiphyseal in location. It is the linear component that distinguishes a fracture from a stress response (Fig. 12D). The linear component of the fracture demonstrable on MR also provides increased specificity in diagnosis as compared with radionuclide bone scans. While initial reports emphasized the changes of bone stress and fracture within cancellous bone, with increasing experience and high-resolution images, changes in cortical bone and the periosteum can be demonstrated (15) (Fig. 12E,F). Simple periosteal reaction is most evident as a linear low signal intensity stripe paralleling the cortical margin of the underlying bone (15). The elevated periosteum may be separated from the cortex by material that increases in signal intensity on T2-weighted and STIR sequences and likely reflects reactive edema.

The comparative sensitivities of MR and scintigraphy for detection of stress fracture has not been the subject of a critical study. Recent assessments of MR versus scintigraphy for the detection of *other* marrow replacement processes such as metastases have demonstrated an increased sensitivity for MR. The greater specificity of MR as compared with scintigraphy would suggest that MR may well be the preferred method of evaluation.

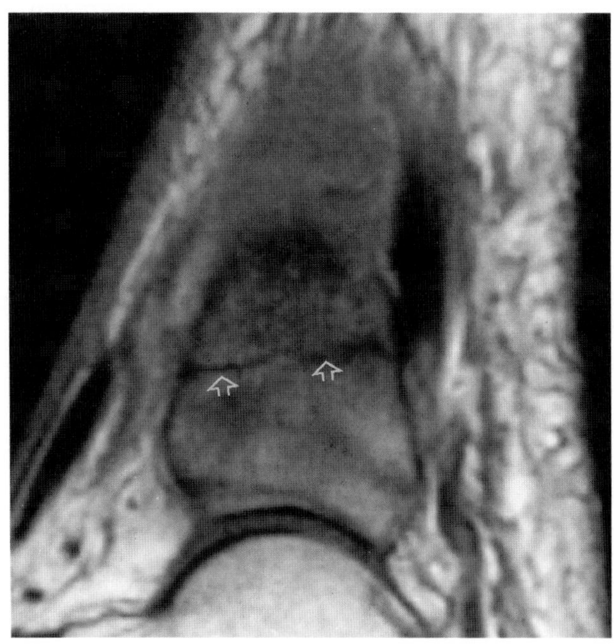

FIG. 12. D: Sagittal T1-weighted (TR/TE 500/20 msec) image of an ankle of a 63-year-old woman referred for evaluation of ankle pain for the past 6 weeks. There is striking signal alteration within the distal tibia consistent with marrow edema and a well-defined low signal intensity line reflecting the fracture *(open arrows)*. Prior radiographs had been reported as normal and the findings were clinically unsuspected.

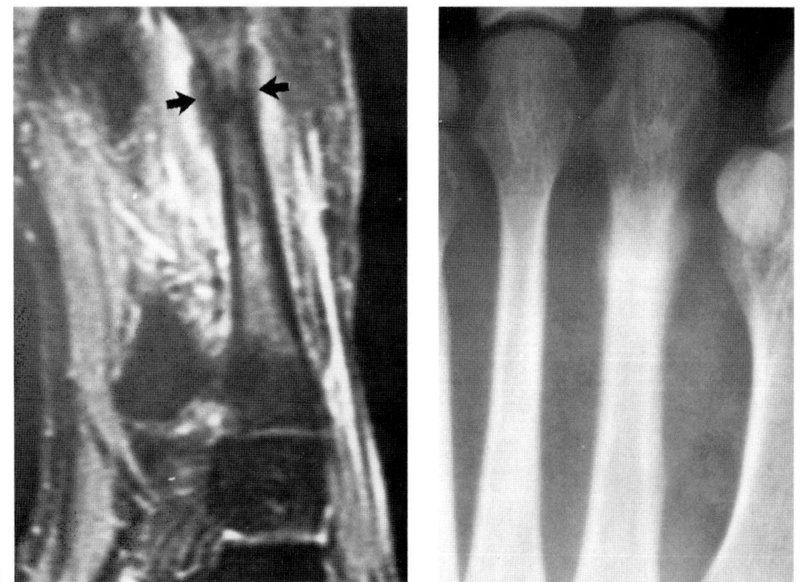

FIG. 12. E: Coronal inversion recovery fast spin echo image (TR/TE/TI 3,500/32 msec/130) demonstrating a stress fracture of the second metatarsal with developing callous *(arrows).* Note the surrounding high signal intensity soft tissue edema. **F:** Correlative AP radiograph demonstrating healing stress fracture.

CASE 13

History: A 46-year-old man complained of ''deep'' pain along the lateral aspect of his ankle, which continued for 2 months following a severe ankle sprain. What are the findings and likely diagnosis? What is the relationship of lateral collateral ligament complex injuries to this entity?

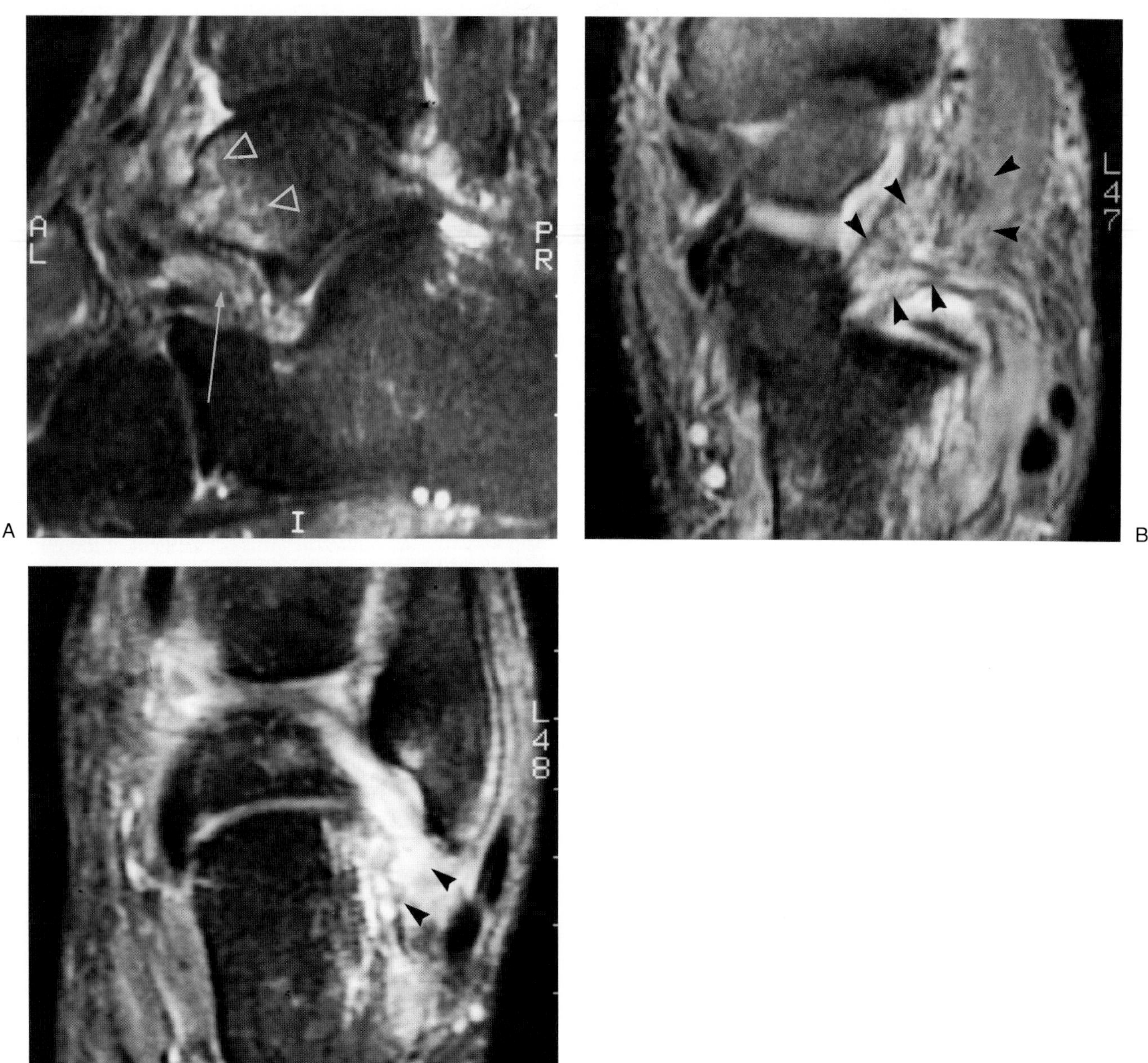

FIG. 13.

Findings: In Fig. 13 a sagittal inversion recovery fast spin echo (TR/TE/TI 2,800/20/130) image demonstrates increased signal intensity within the medial aspect of the tarsal sinus and nonvisualization of the cervical ligament (arrow). Bone marrow edema is seen within the talus (arrowheads). In Fig. 13B an axial fat-suppressed proton density (TR/TE 4,000/36 msec) image demonstrates marked increased signal and complete obliteration of anatomic landmarks within the lateral aspect of the tarsal sinus (arrowheads). In Fig. 13C a coronal fat-suppressed proton density (TR/TE 4,000/36 msec) image demonstrates marked increased signal and nonvisualization of the calcaneofibular ligament (arrowheads).

Diagnosis: Tarsal sinus syndrome secondary to prior subtalar ligament and lateral collateral ligament complex injury. The case illustrates the similar findings that may be encountered in tarsal sinus syndrome and acute subtalar and LCL sprains.

Discussion: The tarsal canal and sinus are the names given to segments of the space between the talus and calcaneus (1–5). This large channel extends anterolaterally to posteromedially, opening just behind the sustentaculum. The term *sinus* refers to the larger lateral opening, and the canal is its medial, more narrow continuation. Within the tarsal sinus run the artery of the tarsal canal, nerve endings, a small bursa, the cervical (lateral talocalcaneal) ligament, the talocalcaneal interosseous ligament (IOL), and the medial fibers of the inferior root of the extensor retinaculum. The exact role and significance of these ligaments remains controversial (6–8). It appears that the cervical ligament contributes to the limitation of inversion and that the inferior extensor retinaculum does not function as a significant stabilizer. Experimental section of the subtalar ligaments does produce minor instability of the hindfoot (increases dorsiflexion significantly), but patients who have clinical injuries to these ligaments rarely have an objective hindfoot abnormality on physicial examination (8). The major complaint of such patients, however, is a feeling of instability and weakness, as well as pain over the sinus tarsi. Injection of a local anesthetic relieves or significantly diminishes the pain. This constellation of symptoms is known as the sinus tarsi syndrome. The symptom complex results from an inversion injury to the ankle in 70% of cases. This explains the frequent association of the sinus tarsi syndrome with tears of the components of the lateral collateral ligament. Patients with space-occupying masses within the tarsal sinus may also present with similar symptoms. Treatment of the sinus tarsi syndrome consists of anti-inflammatory medications, strengthening exercises for the peroneals, and immobilization, and most patients respond to such measures. Exploration of the sinus with synovectomy is undertaken in patients not responding to the above measures. When surgical excision is accomplished, this is limited to the superficial component of the sinus to avoid damage to the vascular supply of the talus (9,10).

The ligaments of the normal subtalar articulation are best seen on sagittal MR images, but can on occasion be identified in any of the three orthogonal planes (1–3,7) (Fig. 13D–G). Volume acquisitions with multiplanar reformations facilitate assessment and identification of the subtalar ligaments. On the sagittal images, the inferior root of the extensor retinaculum appears as a thin, occasionally fasciculated long band seen on those sections through the extreme lateral aspect of the tarsal sinus. Its entire course from the calcaneus to the extensor tendons can be identified on a single section in 100% of cases (3). The cervical ligament, lying in the anterior portion of the capacious sinus tarsi, can be readily identified in 70% to 90% of normal ankles, depending on the imaging sequence and imaging plane chosen. The coronal and sagittal planes are usually the most valuable. The cervical ligament is seen as a single band or as several identifiable fascicles that appear as discrete structures. The interosseous ligament is the only one of the three subtalar ligaments to lie partially within the tarsal canal. Its calcaneal attachment is lateral, and it courses medially and superiorly to attach to the talus just behind the sustentacular joint. Although it is a broad, bandlike structure, it is the least reliably identified subtalar ligament (3). The IOL can be identified in 80% of high-resolution conventional spin echo sagittal and 40% of coronal spin echo pulse sequences. Visualization is improved with volume acquisitions and operator-directed reformatted images.

Patients with acute subtalar sprains and those suffering from the sinus tarsi syndrome manifest similar MR findings (1,2) (Fig. 13A–C). In both groups, images through the sinus tarsi and tarsal canal demonstrate poor definition of the soft tissue anatomy with replacement of the normal fat signal within the canal, edema of the sinus tarsi structures (particularly well depicted on STIR sequences), and lack of visualization of the cervical and talocalcaneal interosseous ligaments. It is emphasized that these findings lack specificity. In patients with acute ligament tears, associated edema within the lateral soft tissues as well as injuries to the lateral collateral ligament complex may be seen. Additionally, in some patients presenting with complaints considered consistent with the tarsal sinus syndrome, we have identified discrete fluid collections thought most consistent with ganglion cysts (Fig. 13H,I).

As previously discussed, associated lesions of the lateral collateral ligament (LCL) complex (most frequently the calcaneofibular ligament) are identified. In one reported investigation, 79% of patients with sinus tarsi syndrome had LCL injuries and 39% of patients with LCL disruptions had abnormal signa intensity within the tarsal sinus and canal (2). Tears of the posterior tibial tendon are also commonly associated with with sinus tarsi syndrome—79% in the series reported by Klein et al. (2). As such, detailed evaluation of these structures should be undertaken when evaluating patients for suspected tarsal sinus syndrome.

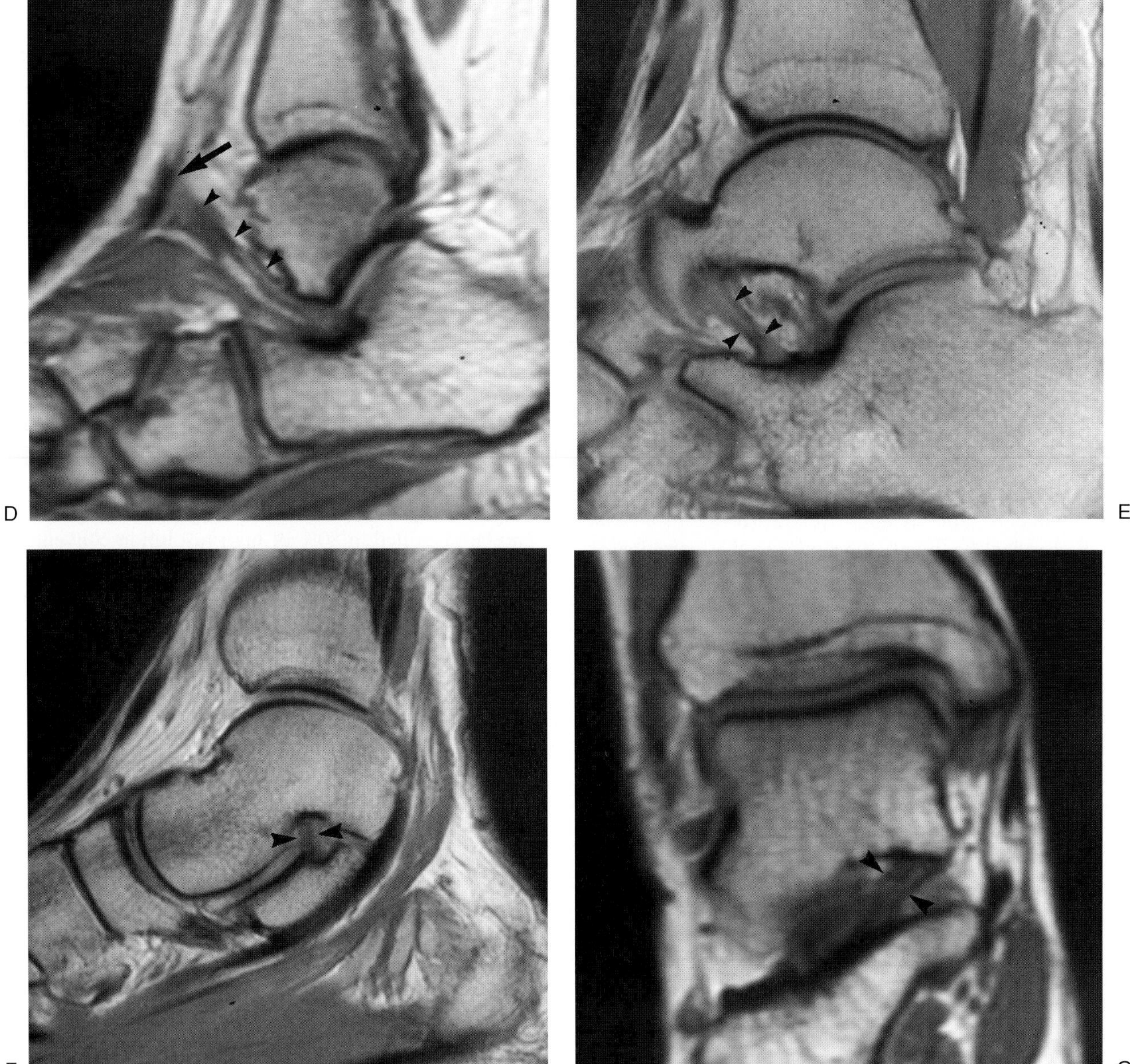

FIG. 13. D–G: Normal subtalar ligaments. **D,E:** Sagittal PD (TR/TE 2,000/20 msec) images and **(G)** coronal PD (TR/TE 2,000/20 msec) image. The subtalar ligamentous complex is composed of the medial root of the extensor retinaculum, the cervical ligament, and the talocalcaneal interosseous ligament. The extreme lateral sagittal **(D)** demonstrates the medial limb of the inferior extensor retinaculum *(arrowheads)* extending from the extensor tendons *(arrow)* to its tarsal sinus insertion. Just medial **(E)** a well-defined cervical ligament *(arrowheads)* can be seen extending from the talus to the calcaneus. Further medially **(F,G)** the talocalcaneal interosseous ligament *(arrowheads)* can be readily identified just posterior and medial to the sustentaculum. (From Mink JH. Ligaments of the ankle. In: Deutsch AL, Mink JH, Kerr R, eds. *MRI of the foot and ankle.* 1992;192–194, with permission.)

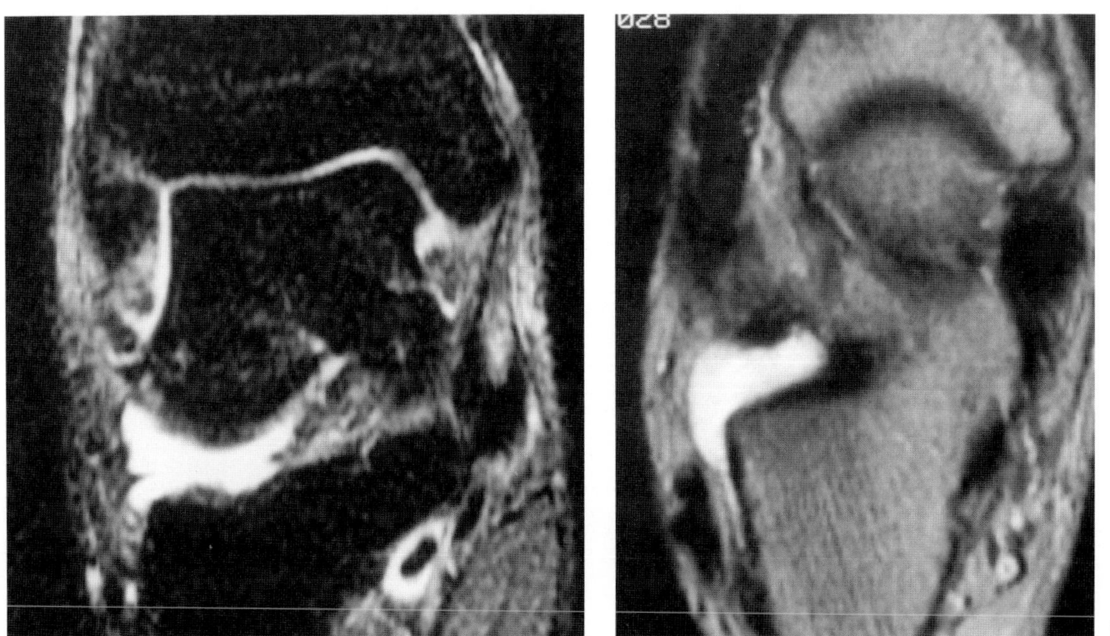

FIG. 13. H,I: Ganglion cyst within the tarsal sinus. **H:** Coronal inversion recovery fast spin echo (TR/TE/TI 3,500/30 msec/130) demonstrates a high signal intensity mass within the lateral aspect of the tarsal sinus. **I:** Axial T2-weighted fast spin echo (TR/TE 4,000/108 msec) again well demonstrates the well-defined mass, which proved to represent a ganglion cyst.

CASE 14

History: A 28-year-old woman presented with a several-month history of medial ankle pain. What are the findings and what classification system is utilized to describe this entity? What is the significance of the findings?

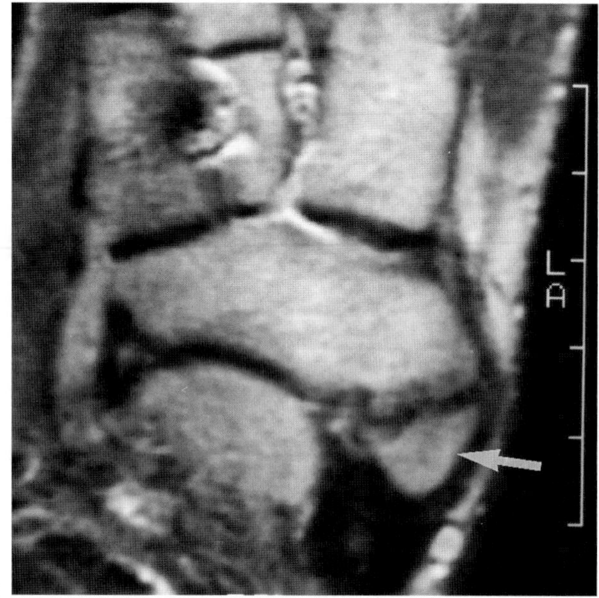

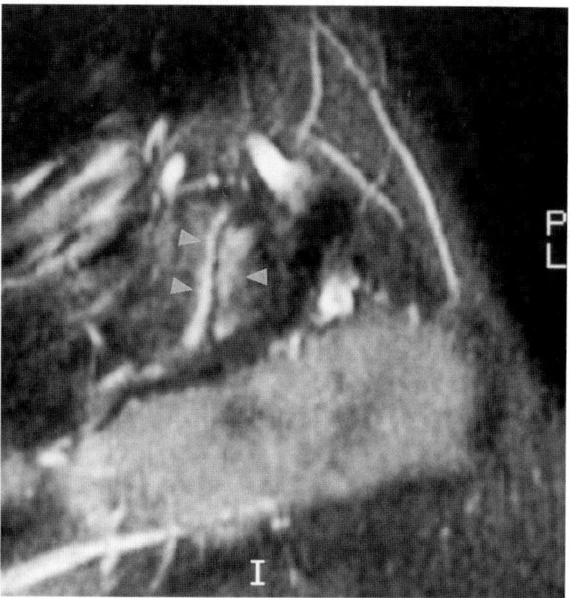

A B

FIG. 14.

Findings: In Fig. 14A the coronal T2 fast spin echo (TR/TE 4,000/108ef) image demonstrates a triangular-shaped ossific mass (arrow) adjacent to the navicular but separated by a low signal intensity line. In Fig. 14B the extreme sagittal inversion recovery fast spin echo image demonstrates high signal intensity along the interfaces between the triangular-shaped fragment and adjacent navicular (arrowheads). The fragment itself also demonstrates mild increased signal suggesting edema.

Diagnosis: Symptomatic type 2 accessory navicular with increased signal at the synchondrosis and within the secondary ossification center likely reflecting stress changes.

Discussion: Three variants of an accessory navicular have been described (1–5). The type 1 accessory navicular represents a sesamoid bone (os tibialis externum) contained within the distal PTT. It is typically round or oval in shape, 2 to 6 mm in diameter, and located within 5 mm medial and posterior to the medial aspect of the navicular. It is embedded within the distal tendon and does not have a cartilaginous connection to the navicular tuberosity. It may account for up to 30% of accessory navicular bones and is usually asymptomatic (4,5).

The type 2 variant is triangular or heart shaped in configuration with its base separated 1 to 2 mm from the posterior aspect of the navicular (1,3,6). It accounts for 50% to 60% of all navicular ossicles (4). As a true secondary ossification center, it is connected to the navicular by a 1 to 2 mm thick layer of either fibrocartilage or hyaline cartilage (4,6). Fusion of the accessory ossicle to the navicular is seen in 20% to 57% of cases (4). When fused, the resulting prominent navicular tuberosity is termed a cornuate navicular and this represents the third accessory navicular variant. The type 2 is the variant most likely to produce symptoms. The type 3 accessory navicular bone (cornuate navicular) may become symptomatic related to bunion formation over the osseous protuberance (2).

The basis for pain with symptomatic type 2 accessory navicular bones is thought to relate to shearing stress and tension across the synchondrosis as a result of forces related to the posterior tibial tendon (2). In the presence of an accessory navicular, the bulk of the PTT may insert onto it. This contributes to straightening of the distal tendon curve and results in the PTT acting as an adductor. The altered biomechanics may lead to pain in the region of the navicular as well as to a painful flatfoot. Previous investigators have described the use of radionuclide bone scans in the assessment of the symptomatic accessory navicular and have reported augmented activity at the region of the accessory navicular (1,5). More recently, Miller and associates (2) reported on the demonstration of bone marrow edema within symptomatic type 2 accessory navicular bones. With both the radionuclide and MR methods, in patients in whom the condition was bilateral, only the symptomatic side demonstrated increased activity (bone scan) and signal (MR).

Histologic analysis in two of the reported MR cases with bone marrow edema within a type 2 accessory navicular demonstrated reactive new bone formation in both cases and the presence of osteonecrosis in one (2). In addition to bone marrow edema, increased signal intensity involving the synchondrosis itself may be demonstrated on MR studies (Fig. 14C,D). This correlates well with established histologic studies of resected specimens in symptomatic patients, which have revealed proliferative changes in the cartilage of the synchondrosis at the cartilage-bone interface (4,5).

Patients with symptomatic accessory navicular bones that remain refractory to conservative management may require resection of the accessory bone and intervening cartilage with reattachment of the PTT for relief of pain. MR may be of value in these cases for both establishing the basis of the patients' symptoms as well as for evaluation of adjacent potential pathology and in particular for assessment of the status of the posterior tibial tendon (6).

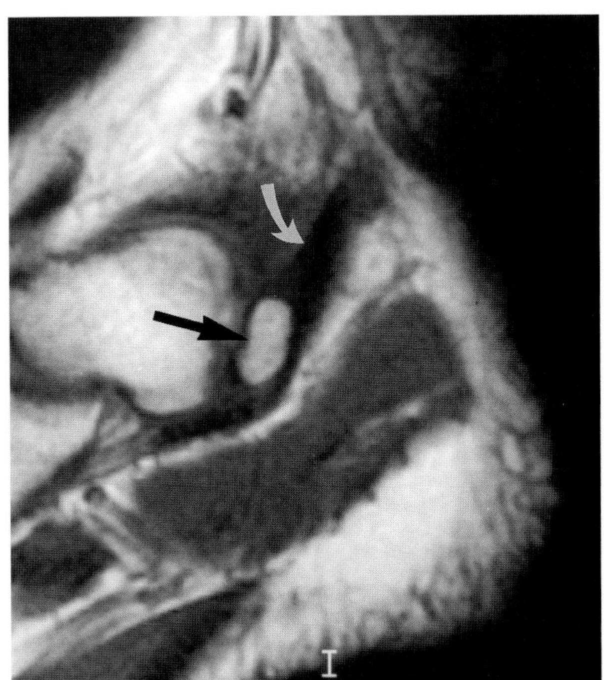

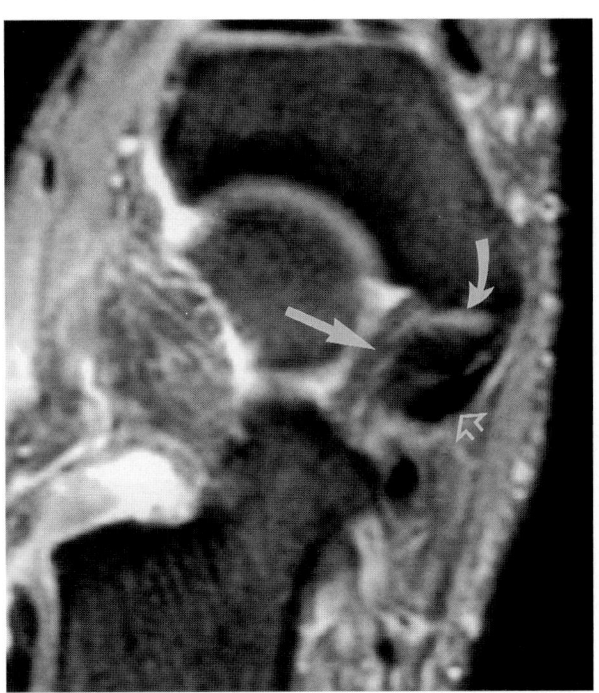

FIG. 14. C,D: Symptomatic type 2 accessory navicular with increased signal involving the synchondrosis. **C:** Sagittal T1-weighted (TR/TE 500/20 msec) image through the medial aspect of the ankle demonstrates a well-circumscribed accessory navicular bone *(black arrow)*, which provides the site of attachment for the posterior tibial tendon *(curved white arrow.)* **D:** Axial fat-suppressed proton density fast spin echo (TR/TE 4,000/30 msec) image demonstrates high signal intensity at the level of the synchondrosis *(curved white arrow)* between the navicular and the accessory navicular. Note excellent depiction of the plantar calcaneonavicular ligament *(straight white arrow)*. Posterior tibial tendon, *short open arrow.*

CASE 15

History: A 68-year-old man underwent repair of a complete Achilles tendon rupture 4 months previously. The patient was on chronic corticosteroid therapy. What is your evaluation of the status of the repair?

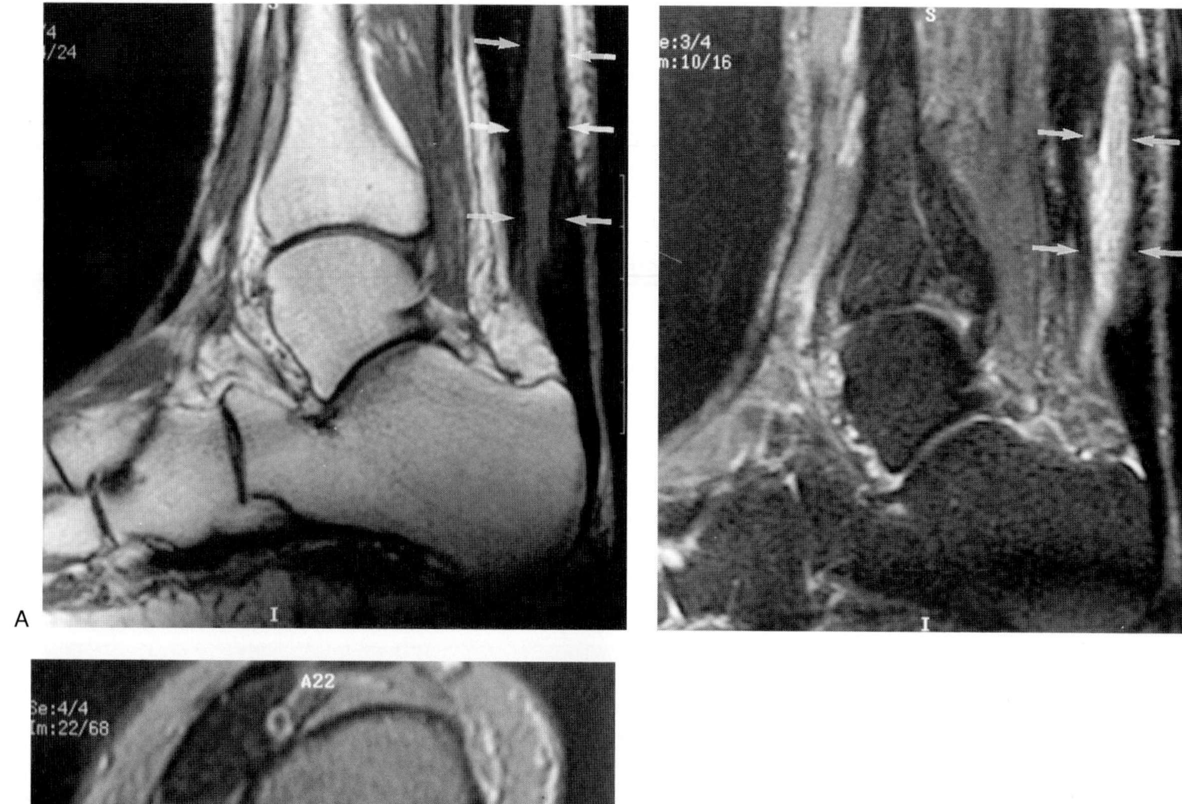

FIG. 15.

Findings: In Fig. 15A a sagittal T1-weighted (TR/TE 500/20 msec) image demonstrates a longitudinal intermediate signal intensity band coursing through the Achilles tendon (arrows). In Fig. 15B a sagittal inversion recovery fast spin echo (TR/TE/TI 3,300/30 msec/130) image demonstrates the band of signal to have markedly increased in signal intensity (arrows). In Fig. 15C an axial fat-suppressed T2-weighted fast spin echo image demonstrates the band of signal to extend across the entire course of the Achilles tendon (curved arrows), effectively separating it into two components. The tendon itself is grossly thickened.

Diagnosis: Failed Achilles tendon repair.

Discussion: The optimal treatment of complete Achilles tendon ruptures remains controversial with advocates of both conservative and operative management. In one large prospective study, Nistor (1) concluded that nonoperative management offered several advantages over surgical repair, including avoidance of anesthesia and surgical complications such as skin slough and infection. In a critical study of open versus closed therapy, Inglis et al. (2) concluded that patients treated surgically were more satisfied with their outcome. Strength, endurance, and power of the nonoperated group was only 70% of the surgically repaired group. No reruptures occurred in the surgical group, whereas the nonsurgical group experienced a nearly 30% rerupture rate. With nonsurgical repair, normal continuity of the tendon may not always be achieved and this can result in tendon lengthening and bridging of the gap with scar tissue of relatively low tensile strength. Surgical repair most frequently involves direct reapproximation of the torn tendon edges using either suture alone or suture reinforced with tendinous or fascial grafts; the gap may, however, be so large as to necessitate the interposition of a tendon substitute such as the flexor hallucis longus, the fascia lata, or the plantaris (3). More recently, Achilles tendon substitutes using glutaraldehyde, polylactic acid, or reconstituted type I collagen prostheses have been used (4,5). While no absolute consensus exists, many authorities reserve conservative management for the elderly, sedentary, or debilitated, with all others undergoing percutaneous or open surgical repair (6,7).

Several reports have focused on the process of tendon healing as manifest on MR (8,9). In one study of ruptured tendons treated either conservatively or with open repair, all tendons had intratendinous fluid spaces on studies performed at 3 and 6 months. All patients demonstrated progressive decrease in signal intensity over time. By 12 months, all tendons demonstrated uniformly low signal, indicative of dense collagen scar. In addition, all tendons demonstrated a distinctly widened contour. The widening may be focal at the site of repair, or may be diffuse and dramatic. Other MR studies have focused on the appearance of healing of the patellar tendon following harvesting for grafts utilized in anterior cruciate ligament reconstructions (9,10). This work has also substantiated a distinctly abnormal early appearance to the residual tendon, with 87% of tendons demonstrating a diffuse increase in signal on both T1- and T2-weighted sequences that commonly persisted for 6 months. By 12 months, asymptomatic patellar tendons returned to their normal low signal intensity appearance or demonstrated only linear increased signal at the site of graft harvest. Immediate and persistent thickening of the patellar tendon after harvesting was uniformly observed, with asymptomatic tendons measuring greater than 10 mm in dimension (normal: 7–8 mm).

In our experience with Achilles tendon repairs (11), MR examinations performed within the first weeks postoperatively uniformly appear dramatically abnormal, and indeed the degree of persistent edema and developing granulation tissue typically contribute to an appearance that could mimic apparent tendon discontinuity (Fig. 15D–F). As a consequence, the integrity of the tendon repair may be extremely difficult to evaluate utilizing MR in the very early postoperative period. At approximately 6 weeks postrepair, continuous areas of intratendinous signal change (increased signal on proton density and T2-weighted images) become apparent within repaired tendon. This presumably relates to immature scar tissue. This intratendinous signal, which is typically centered at the repair site, persists for months following the repair (often most extensive at 3 months), and typically progressively decreases in size in patients with successful outcomes. By approximately 6 months, in patients with good results, the previously demonstrated globular intratendinous signal modulates to an appearance that is either predominantly signalless or to one in which fine longitudinal streaks of signal may be seen (Fig. 15F). In addition, during the normal healing process, the cross-sectional area of the repaired tendon also significantly increases. This change in girth appears to progressively increase within the first several months postrepair and may involve the entire tendon well above and below the repair site. Tendon thickening persists well beyond the time that intratendinous signal resolves. An appearance such as demonstrated in the patient in the present test case, with striking intratendinous signal at 4 months or greater postrepair, is in our experience distinctly abnormal, suggests abnormal healing, and has correlated with poor results. In patients treated with polylactic acid implants, grafts may demonstrate streaks of intermediate signal diffusely throughout the tendon, and virtually the entire tendon is thickened in a fusiform manner (Fig. 15G,H).

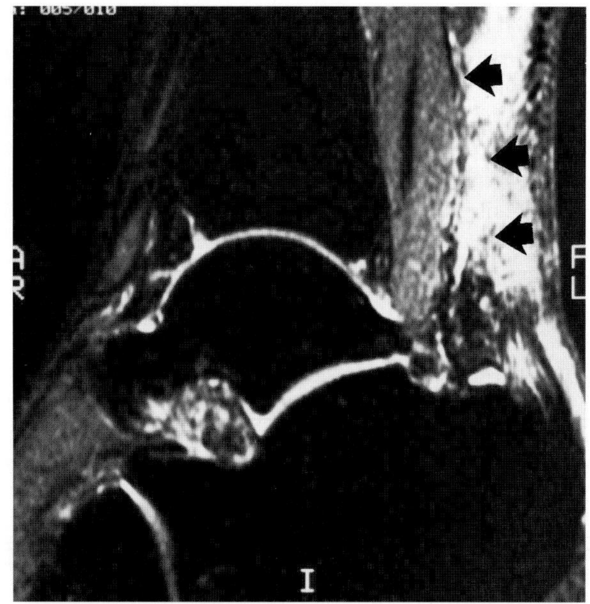
D

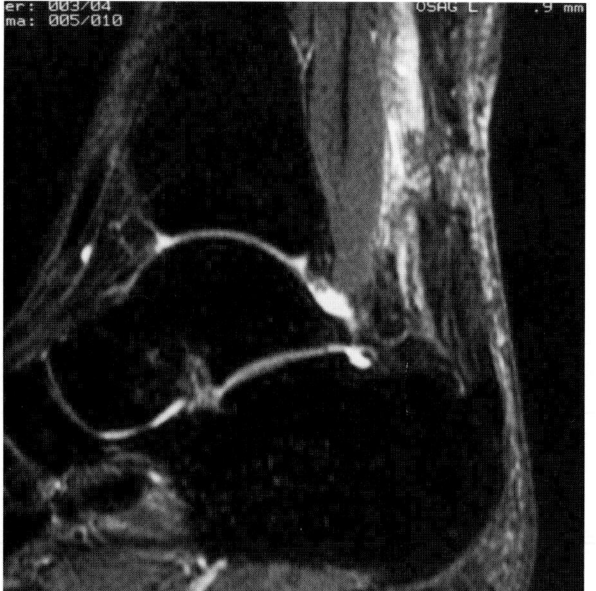
E

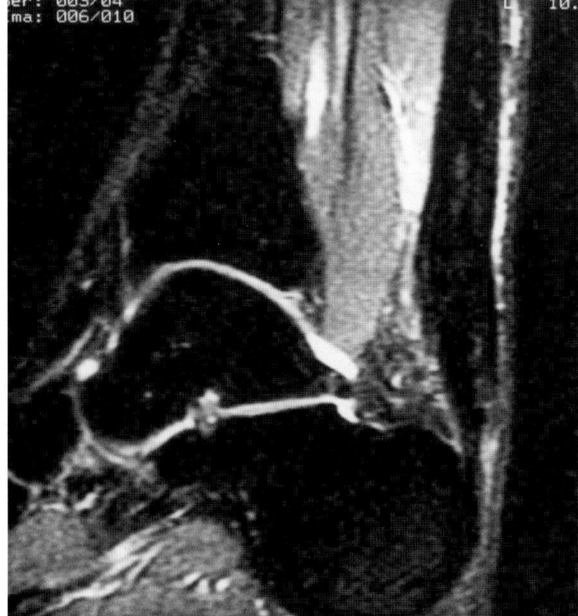
F

FIG. 15. D–F: Healing of a surgically repaired Achilles rupture. Sagittal STIR (TR/TE/TI 3,500/35 msec/160) sequences. **D:** Performed 1 month following a repair of a complete rupture of the Achilles tendon. The image demonstrates intra- and peritendinous edema at the site of the rupture *(arrow).* The tendon, in spite of the fact that it was surgically restored, appears remarkably narrow and discontinuous at the site of repair. **E:** Performed 1 month after D, the site of the repair appears thicker, and the degree of intratendinous signal has decreased. **F:** The intratendinous signal has virtually resolved and it is difficult to identify the site of the rupture. The entire Achilles tendon is markedly widened throughout its length. (From Mink JH. Tendons. In: Deutsch AL, Mink JH, Kerr R, eds. *MRI of the foot and ankle.* New York: Raven Press, 1992, with permission.)

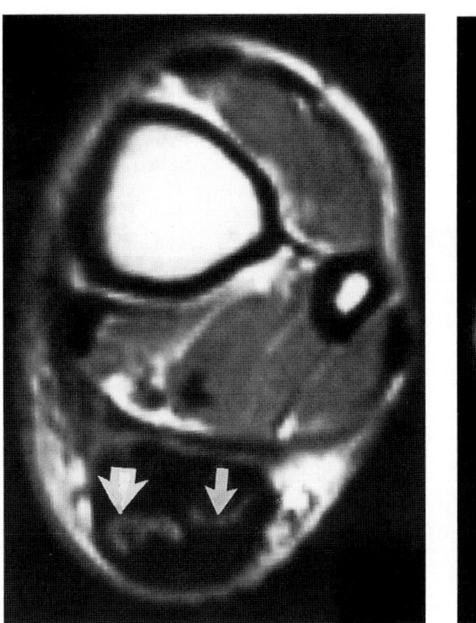

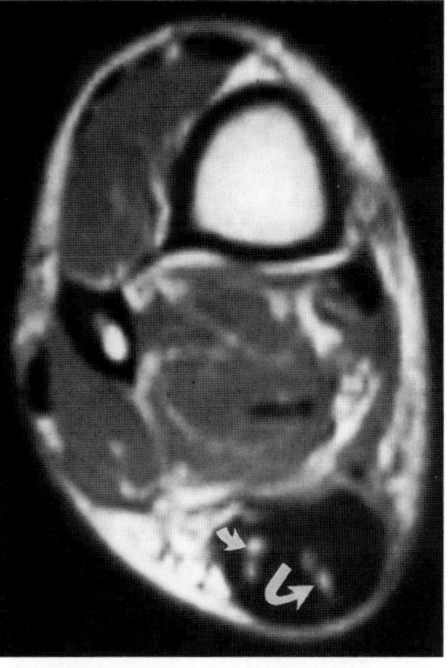

FIG. 15. G,H: Achilles tendon repair with polylactic acid implant. Axial proton density weighted images from two different patients. **G:** Performed 6 months after surgical repair, there are two well-defined bandlike clusters of high signal foci *(arrows)* corresponding to strands of polylactic acid covered with maturing collagenogenic tissue. **H:** Performed 12 months following implantation, this implant is manifest by a group of sharply defined punctate foci *(arrows)* corresponding to the strands covered by a healed tendon. In both images, the repaired tendon remains markedly thickened. (Case courtesy of M. Liem, M.D. From Mink JH. Tendons. In: Deutsch AL, Mink JH, Kerr R, eds. *MRI of the foot and ankle.* New York Raven Press: 1992, with permission.)

CASE 16

History: A previously healthy 82-year-old woman presented with a 3-week history of a painful swollen third toe. The clinical impression was that of probable osteomyelitis. What is your diagnosis?

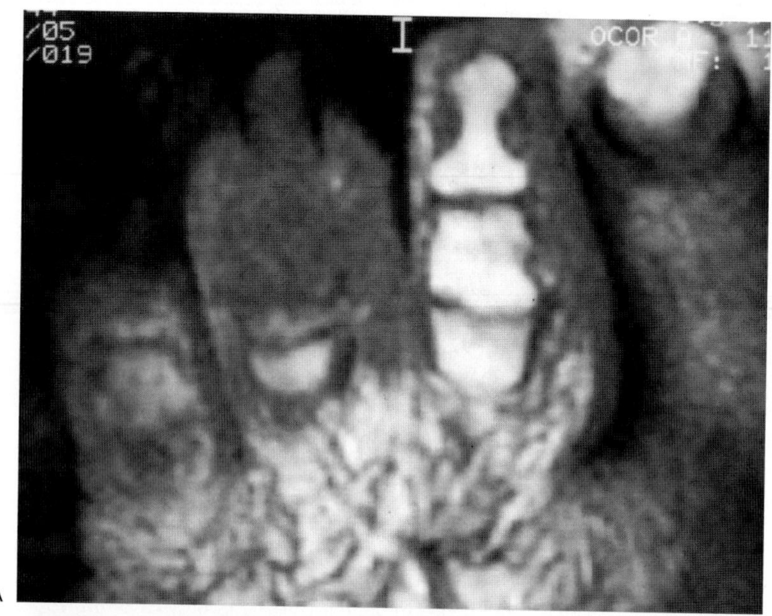

A

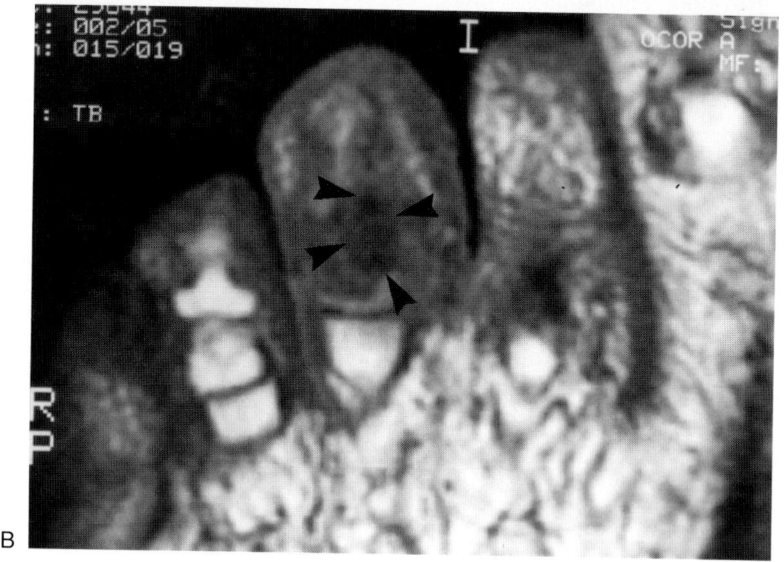

B

FIG. 16.

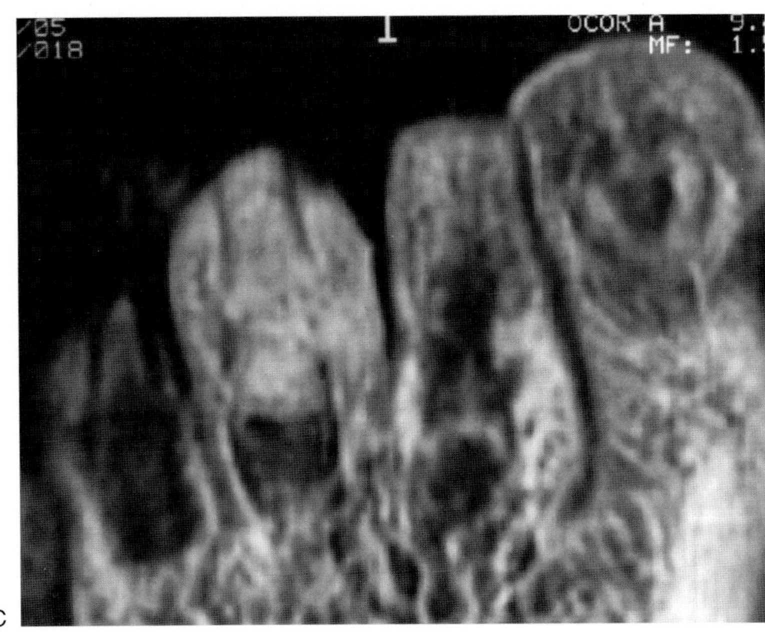

C

FIG. 16.

Findings: In Fig. 16A a coronal T1-weighted SE image (TR/TE 600/20 msec) demonstrates replacement of the normal high signal intensity marrow fat of the distal and middle phalanges of the third toe by an intermediate signal intensity process. In Fig. 16B, on a section contiguous with Fig. 16A, a more circumscribed, relatively low signal intensity lesion is seen within the middle phalanx (arrowheads). In Fig. 16C a coronal STIR (TR/TE/TI 3,500/35 msec/160) image demonstrates marked increased signal intensity within the involved portions of both the middle and distal phalanges. Increased signal is also identified within the contiguous soft tissues.

Diagnosis: Acute tophaceous gout.

Discussion: This case is unusual in several respects including the site of involvement as well as the presentation of tophaceous gout as the initial manifestation of the disease in this patient. Gout may be divided into two principal categories: (a) idiopathic, representing the vast majority of cases; and (b) gout associated with known clinical disorders or enzymatic defects, representing the basis for the hyperuricemia that is the biochemical hallmark of the disease (1). Idiopathic gout is far more common in men than women (20:1). While variable, gout most commonly presents in the fifth decade in men and in the postmenopausal period in women (1). Idiopathic gout may be divided into several clinical stages. Many patients have hyperuricemia for prolonged periods in the absence of signs or symptoms. Renal calculi or acute arthritis will occur in approximately 20% of these individuals and correlate with higher levels of serum urate and uric acid excretion. Urolithiasis or articular attacks of gout define the end of the stage of asymptomatic hyperuricemia. Acute gouty arthritis is typically mono- or oligoarticular. Gout has a predilection for the joints of the lower extremity with the first metatarsophalangeal (MTP) joint, representing the most common site of initial involvement (1). The onset and severity of arthritis in acute gout is often dramatic, and the clinical findings, as demonstrated in this case, may simulate those of a septic arthritis. Pain, tenderness, and swelling occur within hours and may persist for weeks (1). Prior to the advent of effective therapy for hyperuricemia, 50% to 60% of patients with gout developed clinical or radiographic evidence of deposits of monosodium urate (tophi). Chronic gouty arthritis is presently seen in fewer than half of patients who experience recurrent acute attacks (1). Visible tophaceous deposits most typically do not occur for years after the initial attack, with the average duration approximately 12 years (1). Tophi most commonly occur in the synovium and subchondral bone.

Radiographic abnormalities may be seen in the initial attacks of gout and consist of soft tissue swelling related to the presence of synovial inflammation, capsular distention, and surrounding soft tissue edema (1). As the attack subsides, the radiographic abnormalities usually disappear. Routine radiographic examination may reveal no abnormalities in a considerable percentage of patients with clinical evidence of gout and symptoms spanning many years (1). After years of intermittent episodic arthritis, chronic tophaceous gout may lead to permanent radiographically demonstrable abnormalities. Eccentric or asymmetric nodular soft tissue prominence accompanies soft tissue deposition of urates. Tophaceous deposits account for the bone erosions seen in gouty arthritis, and these may be intraarticular, paraarticular, or be located at a considerable distance from the joint (1). The joint space may be remarkably well preserved until late in the disease, a distinctive feature of this arthritis that relates to the relative integrity of cartilage adjacent to areas of extensive cartilaginous and osseous destruction. Subchondral lytic bone lesions may reveal communication with the joint cavity or appear as discrete cystic lesions apparently unrelated to the joint. Intraarticular erosions usually begin in the marginal areas of the joint and proceed centrally (1). Paraarticular erosions are eccentric in location, and frequently occur beneath soft tissue nodules (1).

The reported MR experience in assessment of gouty arthritis is quite limited (1–3). In common with other articular and musculoskeletal disorders, MR is capable of graphically demonstrating both soft tissue and bone involvement. The reports to date suggest that tophi may have a variable appearance on T2-weighted spin echo sequences, demonstrating increased signal intensity as in the present case, or intermediate to low signal intensity as in other reported experiences that have depicted soft tissue tophi (1,2). In this latter setting, gouty tophi would have to be included in the differential diagnosis of non–high signal intensity (on T2-weighted SE sequences) articular and paraarticular processes such as rheumatoid arthritis, pigmented villonodular synovitis, chronic infection, and amyloidosis. The case reported by Ruiz et al. (2) also featured an unusual presentation of tophaceous gout mimicking possible neoplasm in a patient without prior history of gouty arthritis. As such, the possibility of gouty tophi mimicking either infection or neoplasm, and presenting in patients without prior history of long-standing gout, should be considered in the differential diagnosis of articular and paraarticular masses on MR.

CASE 17

History: A 28-year-old man cut his foot on a porcelain soap dish in the shower. He presented 2 weeks after the injury with limitation of flexion. What is your diagnosis?

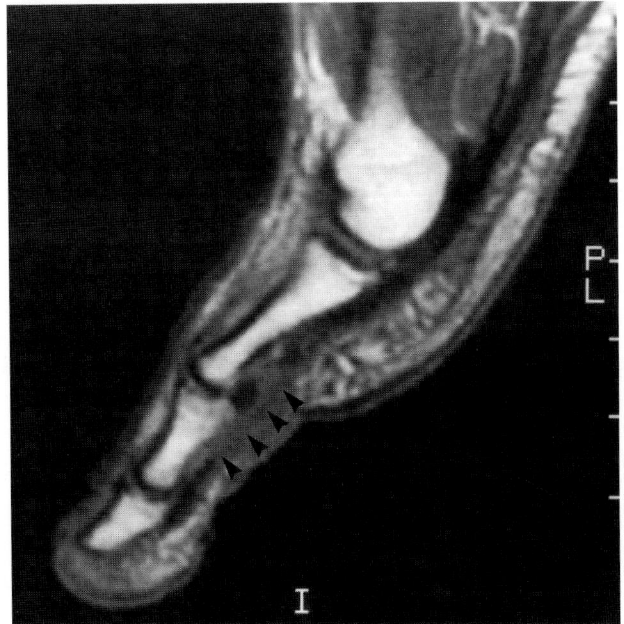

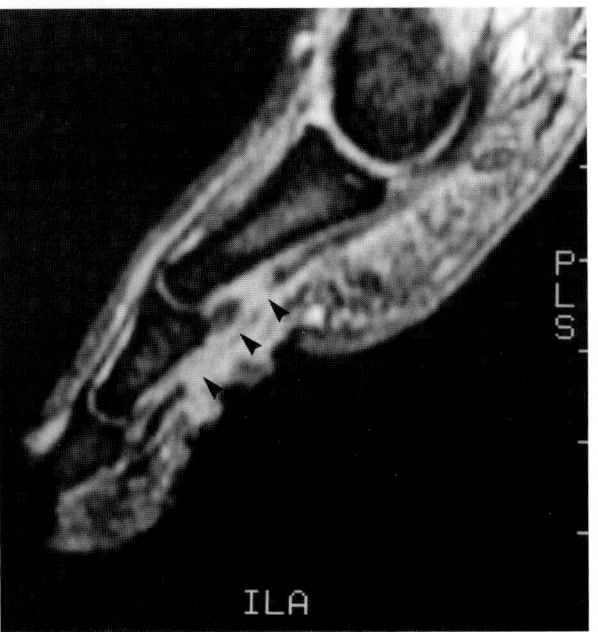

A B

FIG. 17.

Findings: In Fig. 17A a sagittal T1-weighted (TR/TE 500/21 msec) image demonstrates increased signal and poor definition of the long flexor tendon at the level of the proximal interphalangeal joint (small arrowheads). In Fig. 17B a sagittal gradient echo (TR/TE/TI 250/15 msec/20) image demonstrates a complete gap within the long flexor tendon (arrowheads).

Diagnosis: Complete laceration of the long flexor tendon of the third toe.

Discussion: MR may be of value for both the detection and characterization of tendon lacerations. This applies to both acute traumatic events as well as to the assessment of possible postoperative complications (Fig. 17C,D). The terminations of the tendons of the long and short flexor muscles are contained in bony/aponeurotic canals similar to those encountered in the fingers. The mode of division of the flexor digitorum brevis and that of its attachments to the phalanges also parallels that seen in the flexor digitorum superficialis in the hand. In lacerations extending through the full thickness of the tendon, the extent of tendon retraction can be directly visualized aiding in preoperative planning. The management approach to partial thickness tears remains controversial (1,2). While prior studies have advocated conservative management of partial thickness tears, recent studies have suggested that partial thickness tears involving greater than 60% of the cross-sectional area be repaired (3). The possibility of entrapment, rupture, and triggering of unrepaired partial tendon lacerations has been reported (1).

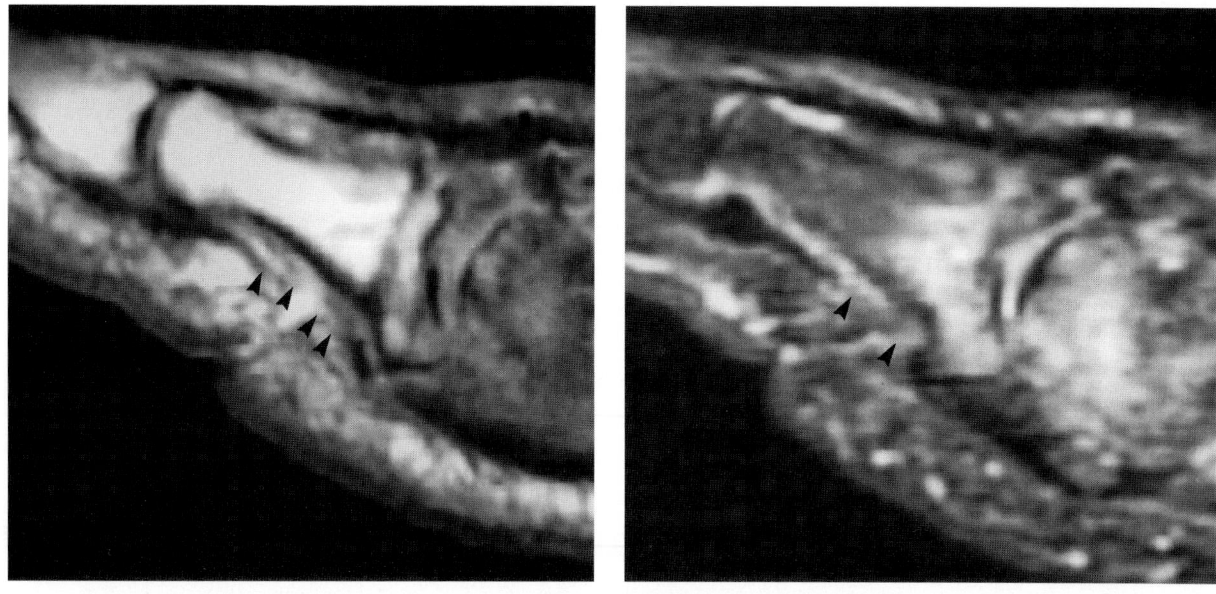

FIG. 17. C,D: Postoperative tendon laceration. **C:** Sagittal T1-weighted (TR/TE 300/16 msec) image demonstrates evidence of a prior osteotomy involving the first metatarsal with micrometallic artifact and decreased signal intensity of the first metatarsal head. There is localized discontinuity of the long flexor tendon at the level of the MTP joint with the gap in the tendon well demonstrated *(arrowheads)*. **D:** Sagittal fast inversion recovery (TR/TE/TI 2,800/23 msec/130) image demonstrates marked increased signal within the first metatarsal head as well as the focal discontinuity within the tendon *(arrowheads)*.

CASE 18

History: A 38-year-old avid tennis player complained of chronic ankle pain that significantly limited her ability to participate in sport and recreational activity. What is the nature of the abnormality depicted on the following images? How are such injuries classified?

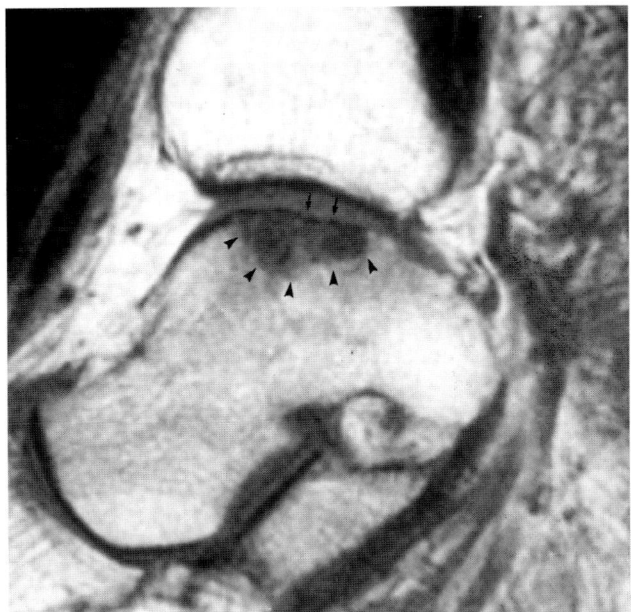

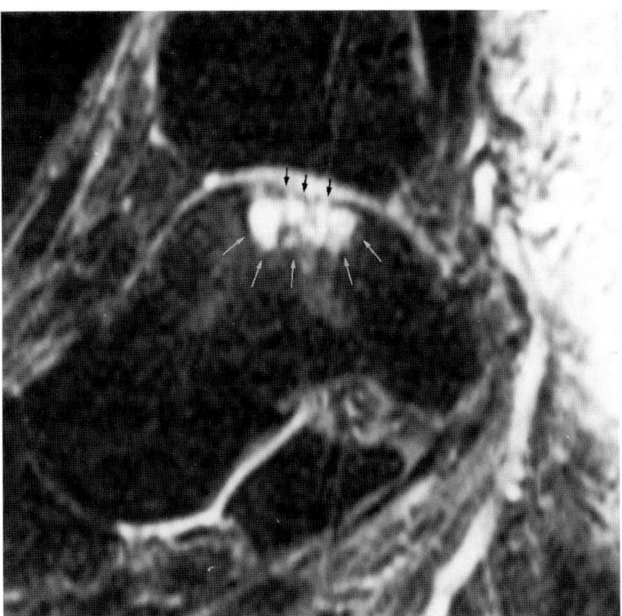

FIG. 18.

Findings: In Fig. 18A the sagittal T1-weighted image (TR/TE 500/30 msec) demonstrates a large ovoid-shaped low signal intensity region extending into the subchondral bone of the talar dome (small arrowheads). There is a subtle suggestion of flattening and early collapse of the articular surface. In Fig. 18B the sagittal STIR (TR/TE/TI 2,200/35 msec/160) image demonstrates multiple small hyperintense cyst like areas within the lesion (small white arrows). There is loss of the normal smooth convexity of the talar dome with slight undulation of the subchondral bone plate (small black arrows). There is little surrounding reactive edema consistent with the chronic nature of the patient's complaints. (From Deutsch AL. Osteochondral injuries of the talar dome. In: Deutsch AL, Mink JH, Kerr R, eds. *MRI of the foot and ankle.* New York: Raven Press, 1993;118–134, with permission.)

Diagnosis: Osteochondritis dissecans with unstable fragment and early collapse.

Discussion: The term *osteochondritis dissecans* (OCD) refers to an acquired lesion of the articular cartilage and subchondral bone of a diarthodial joint (2–9). These lesions are typically the result of either a single episode or multiple repetitive episodes of trauma. They frequently represent one of the long-term sequelae of severe ankle sprains. The talar dome is one of the most common sites of such injuries, and specifically the middle one-third of the lateral border and the posterior one-third of the medial border are most often affected (4). Medial lesions are typically described as being deeper and more cup-shaped and the lateral lesions shallower and more wafer-like. Talar dome osteochondral injuries are more common in men and present more often in the second to fourth decade but may be commonly encountered in both sexes and at a wide age range (3).

These injuries are generally believed to be caused by shearing, rotary, or tangentially aligned impaction forces and in the ankle commonly represent the result of a significant ligamentous sprain injury (1,2). Acute injuries can damage the subchondral bone without damage to the overlying articular cartilage (1,3). Alternatively, shearing injuries may result in fragments that consist of only cartilage (purely chondral fragments) or cartilage and underlying subchondral bone (osteochondral fragments) (Fig. 18C,D). The depth of the fracture line, which generally parallels the articular surface, determines the cartilaginous and osseous components of the lesions. Following injury, the osteochondral fragment may be deprived of its blood supply as a consequence of disruption of the subchondral surface. The overlying articular cartilage may remain viable by receiving sustenance from the synovial fluid. Subsequent healing and revascularization of the osteochondral lesion depends on both the stability of the fragment within its bony crater and the integrity of the overlying articular cartilage (10). The cartilage that overlies the osteochondritic defect may appear normal in the early stages of the condition; in later stages, it may variably appear flattened, fissured, discolored, or fibrillated (3). The detached portion of the articular surface (cartilage and/or bone) can remain in situ, become slightly displaced, or become loose within the joint. A ''loose'' fragment can continue to increase in size by gaining nourishment from synovial fluid or may attach to the synovial lining at a distant site and eventually be resorbed. The defect site is typically filled with dense fibrous tissue, and eburnated bone underlies the defect (3). Pathologically, the primary changes are observed in the bone with the cartilage affected secondarily (3). Hemorrhage at the site of the defect develops into a fibrin clot that may eventually be modulated into fibrovascular repair tissue that gradually increases in cellularity and eventually revascularizes the segment deprived of its blood supply. If there is failure of healing, the zone between the fragment and the cartilage evolves into a dense fibrous tissue somewhat analogous to the tissue characteristics of a fracture in the stage of delayed or nonunion (3). The bony bed presents a similar response—the early-healing defect demonstrating an active cellular response, early vasculariziation, and osteogenesis, whereas the nonhealing defect develops dense avascular eburnated bone. This leads to lack of subchondral support, subchondral cyst formation, secondary articular deformity, and subsequent degenerative changes (3).

Multiple approaches to management of osteochondral dome injuries have been advanced, and none has received universal approbation. If the lesion is mechanically stable and the overlying cartilage intact, conservative measures including a period of non–weight bearing may be sufficient to induce healing (6). Alternatively, the lesions may be percutaneously drilled in an attempt to facilitate revascularization. Larger fragments may also be pinned in an attempt to promote healing. Advances in arthroscopic techniques have allowed a more aggressive approach to these lesions. Advanced-stage lesions with unstable fragments and articular cartilage violations are commonly approached with curettage and drilling. The bony crater is debrided and any loose fragments removed. The various treatment options available underscore the need for an accurate lesion staging system. Most work today has used the system initially described by Berndt and Harty (2) based on their cadaver observations and extrapolated to plain radiography. This system classifies lesions into four stages (Fig. 18E). Stage 1 represents a small area of compression of subchondral bone resulting in microscopic damage to bony trabeculae. No lesion may be evident on superficial visual inspection. Stage 2 represents a partially detached osteochondral fragment with fissuring typically evident on plain radiographs. In stage 3, the fragment is completely displaced from its underlying bed, but remains in anatomic position. In stage 4, the detached fragment has become inverted in the fracture bed or lies displaced elsewhere in the joint. A significant limitation of this system, which is based on plain radiography, is the lack of inclusion of the status of the overlying articular cartilage. Pritsch and colleagues (9) proposed a staging system based on arthroscopic observation (Fig. 18F). Grade 1 lesions represented a small area of compression of subchondral bone with firm, intact, and shiny articular cartilage evident at arthroscopy. Grade 2 lesions represented partially detached osteochondral fragments with intact but softened cartilage, and grade 3 lesions represented partially detached osteochondral fragments that remained in the crater and demonstrated significantly frayed cartilage. When the cartilage is intact at arthroscopy, the fragment interface with underlying bone cannot be directly evaluated and in some circumstances the lesion itself cannot be demonstrated (3). The ability of MR to detect the subchondral component as well as to potentially characterize the status of the articular cartilage is a significant potential advantage that contributes to the appeal of applying this technique to the staging of osteochondral lesions.

Talar dome OCD lesions often have residual effects that long outlast the initial injury. To properly manage talar dome lesions, the abnormality must be detected and then characterized with regard to fragment stability and integrity of the overlying cartilage. On conventional radiographs, osteochondritic defects typically appear as a radiolucent defect involving the subchondral bone plate and a variable amount of underlying cancellous bone. One or more central bone fragments may be seen within the defect that is otherwise filled with noncalcified fibrocartilage. The fragments may remain ununited or heal and unite with radiolucent fibrocartilage or solid bridging bone. As ''healed'' lesions may appear as radiolucent areas, plain films are limited in their ability to provide information regarding lesion stability. Purely chondral defects remain radiographically occult. MR is utilized for detection of radiographically occult lesions and for characterization of known symptomatic lesions (1,10–15). Attention is directed toward assessment of the integrity of the articular cartilage as well as the status (stability) of the underlying fragment. Osteochondral fragments are reported to demonstrate variable signal intensity ranging from decreased to increased in comparison with that of remaining subchondral bone. The signal intensity of the fragment has been reported to have no correlation with lesion stability (11). Osteochondral fragments may be distinguished from their underlying bed of cancellous bone by means of an interface of low to intermediate signal intensity on T1- and proton density–weighted images.

The interface is composed of either fibrous or fibrovascular tissue along with a variable degree of bordering eburnated subchondral bone. Partially attached (unstable) lesions may demonstrate increased signal intensity at the interface on T2, T2*, and short inversion time inversion recovery sequences (11) (Fig. 18G,H). The presence of a high signal intensity interface between the fragment and underlying subchondral bone has been a reliable but not invariable indicator of fragment instability (1,11). A high signal intensity interface between the fragment and underlying subchondral bone can be seen with intact or disrupted overlying articular cartilage. Additionally, the presence of focal areas of high signal intensity on T2-weighted images, presumably reflective of cysts or localized granulation tissue, has been associated with unstable lesions (11). Conversely, the lack of a high signal intensity interface has been considered to indicate a stable or healed lesion, although experience with MR imaging in a large number of surgically staged stable lesions is limited.

Assessment of the talar dome articular cartilage is of significant clinical import. Optimal assessment requires high spatial resolution which places significant demands on many conventional MR pulse sequences with regard to their ability to maintain sufficient contrast to noise to support high-resolution imaging (1,16). The thinness of the articular cartilage of the talar dome (2 mm) places even further demands on the imaging techniques, as compared with the relatively thick articular cartilage of the patella (6–8 mm), which has served as the model for most investigational studies of MR imaging of articular cartilage. There is no present consensus regarding optimal pulse sequence selection for assessment of the talar dome. Most investigators currently continue to use some combination of conventional and fast spin echo techniques to obtain T1- and T2-weighted images. By allowing multiple acquisitions to be obtained within a reasonable imaging time, T2-weighted fast SE imaging can provide sufficient signal to noise to support a high in-plane resolution matrix (16). However, as with other commercial two-dimensional sequences in which sections are selected by means of individual excitation, the minimum section thickness with fast SE imaging is currently 3 mm. Therefore, even with favorable contrast properties, two-dimensional sequences offer limited scope for high spatial resolution (16). More recently introduced fast three-dimensional sequences, often combined with fat suppression, can combine high in-plane resolution with thin sections and offer promise with regard to articular cartilage assessment.

As a consequence of the present difficulty encountered in confidently assessing the integrity of the overlying articular cartilage in ostechondritic lesions, several investigators have advocated the use of intraarticular contrast material to enhance the MR examination in a manner analogous to its use in conventional and CT arthrography (1,17). Either saline or gadopentetate dimeglumine can be utilized. After injection, T1-weighted images, often with fat saturation, are obtained for cartilage integrity assessment. Extension of contrast material between the fragment and underlying subchondral bone is definitive for violation of the overlying articular cartilage (Fig. 18I,J). Gadopentetate dimeglumine can also be administered intravenously. Enhancement of the interface between fragment and bone has been reported as a reliable finding for fragment instability. In addition, delayed scans (15 to 45 minutes) post–contrast administration allow for an arthrogram type effect related to diffusion of the contrast material into the joint space. The quality of this latter application is in part predicated upon sufficient synovial fluid within the joint.

MR provides an excellent method to study the natural history of osteochondral injuries. In the author's experience with MR imaging of acute ankle injuries, radiographically occult talar dome injuries are commonly encountered. The clinical significance of these injuries and the incidence with which these progress to symptomatic OCD lesions remains unknown. Typically, however, the lesions are initially manifest as a localized bone marrow edema pattern within the talar dome. Over time, the marrow edema pattern subsides, and in lesions that do not resolve, a more defined and demarcated focus can be identified bordering the subchondral bone plate. Progressive change of the overlying articular cartilage may be demonstrable as well as progressive collapse of the underlying subchondral bone. Loose osteochondral fragments may be demonstrable if large enough or if there is sufficient synovial fluid to provide an arthrogram effect. Otherwise this determination is frequently better accomplished utilizing either MR or CT arthrographic techniques.

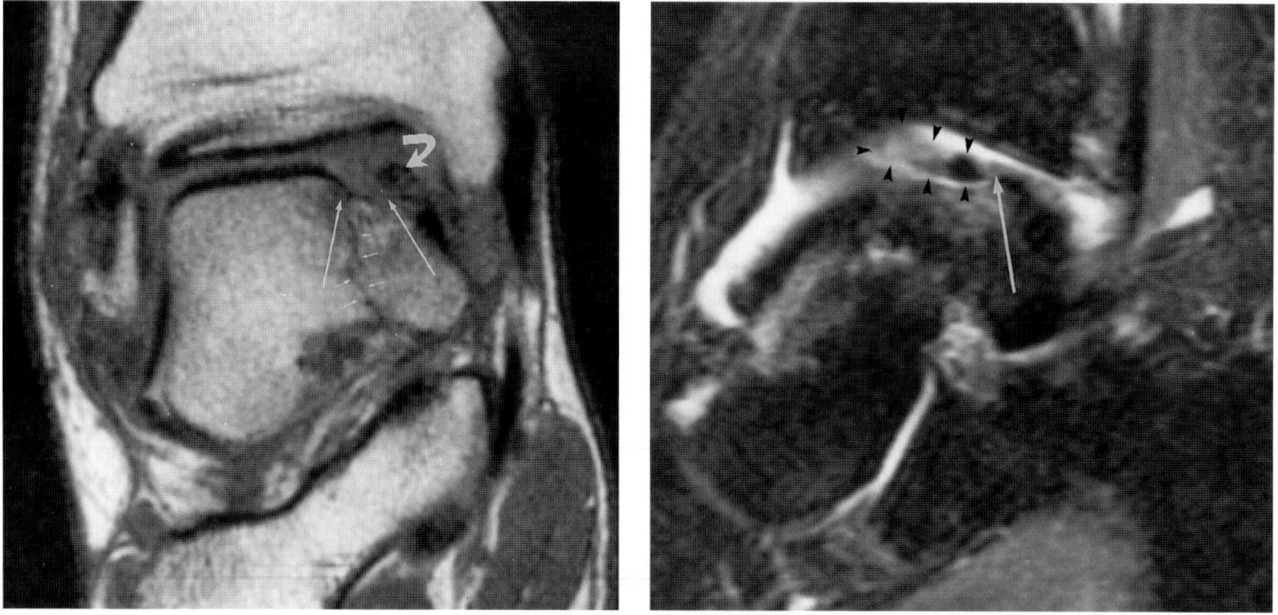

FIG. 18. C,D: Acute osteochondral injury. **C:** Coronal T1-weighted (TR/TE 500/20 msec) image extending through the medial third of the talar dome *(small arrows).* There is marked depression (11 mm) of the medial fragment, a finding well characterized by the tomographic capability of MR. A circular to ovoid slightly decreased signal intensity osteochondral fragment is seen within the medial joint space *(curved arrow).* **D:** Sagittal STIR sequence (TR/TE/TI 2,200/35 msec/160) well delineates the fragment *(small arrowheads)* and distinguishes the intermediate signal cartilaginous component (anterior two-thirds) and the low signal osseous/subchondral component (posterior one-third of fragment). A high signal intensity cleavage plane presumably representing synovial fluid separates the osteochondral fragment from its bony bed throughout its extent except for a small "hinge" of intact articular cartilage (*long arrow*).

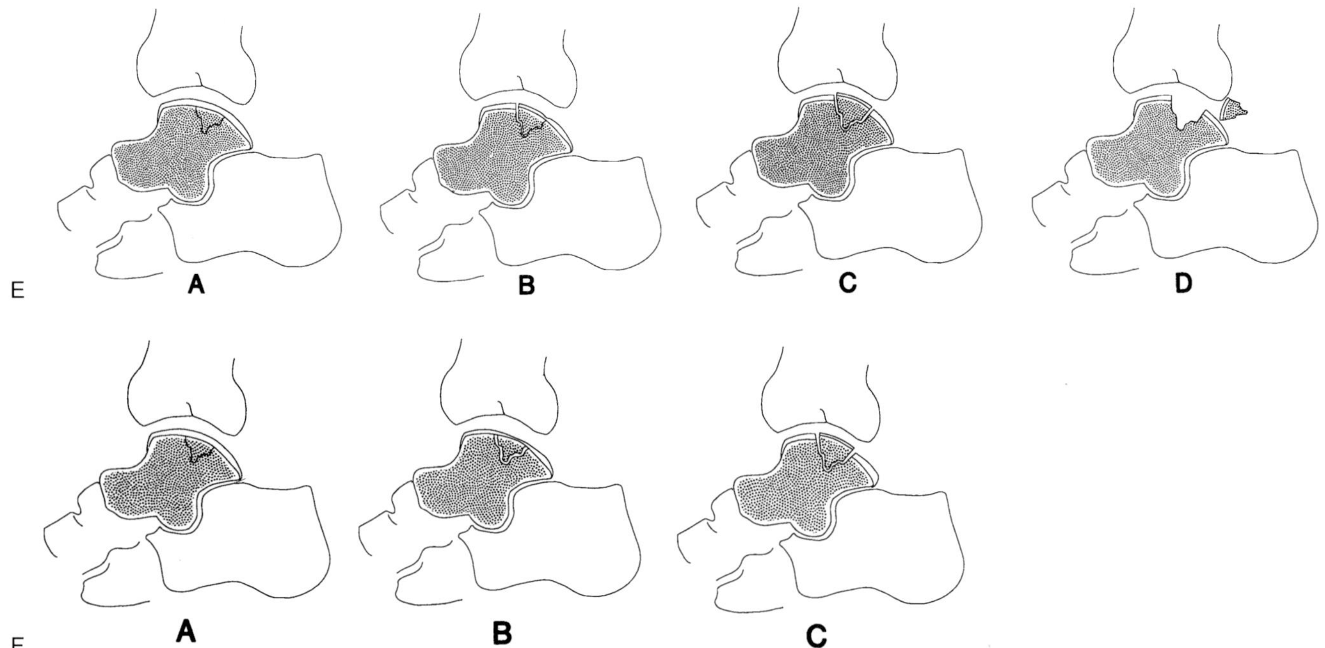

FIG. 18. E: Schematic representation of the different stages of osteochondritis dissecans as described by Berndt and Harty. 1: The injury is entirely confined to subchondral bone without significant findings at the articular surface. 2: The fracture has extended through the articular cartilage and the fragment remains largely in placed hinged by intact overlying cartilage. 3: The fragment is unattached and mildly displaced. 4: The fragment has become completely displaced and lies freely in the joint. **F:** Diagrammatic representation of staging of osteochondritis dissecans after Pritsch. 1: The lesion is entirely subchondral and no surface manifestation will be evident arthroscopically. 2: A discrete fragment is present but remains covered by intact but attenuated cartilage. 3: The fragment is partially displaced with complete disruption of the overlying articular cartilage. (From Deutsch AL. Osteochondral injuries of the talar dome. In: Deutsch AL, Mink JH, Kerr R, eds. *MRI of the foot and ankle.* New York: Raven Press 1993;118–134, with permission.)

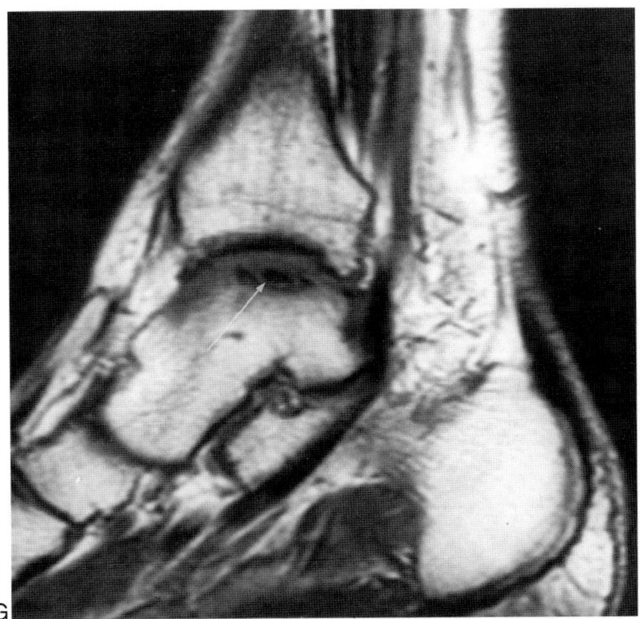

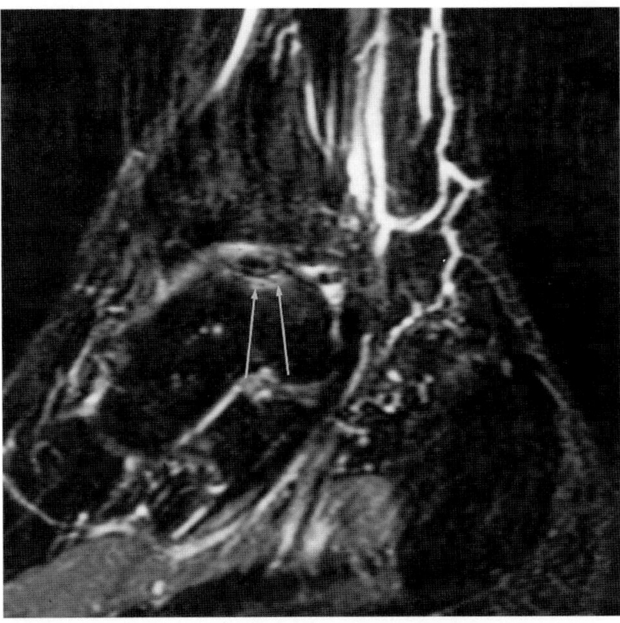

FIG. 18. G,H: Osteochondritis dissecans. **G:** Sagittal T1-weighted (TR/TE 500/20 msec) image demonstrates a discrete ovoid-shaped low signal intensity osteochondral fragment involving the middle third of the medial aspect of the talar dome *(arrow).* The articular cartilage is difficult to assess but is not grossly disrupted. **H:** Sagittal STIR (TR/TE/TI 2,200/35/160) sequence reveals a band of high signal intensity at the interface between the fragment and underlying subchondral bone, findings suggesting a partially attached but unstable fragment *(long arrows).* At surgery, the articular cartilage was grossly intact but on probing the fragment was readily ballotable.

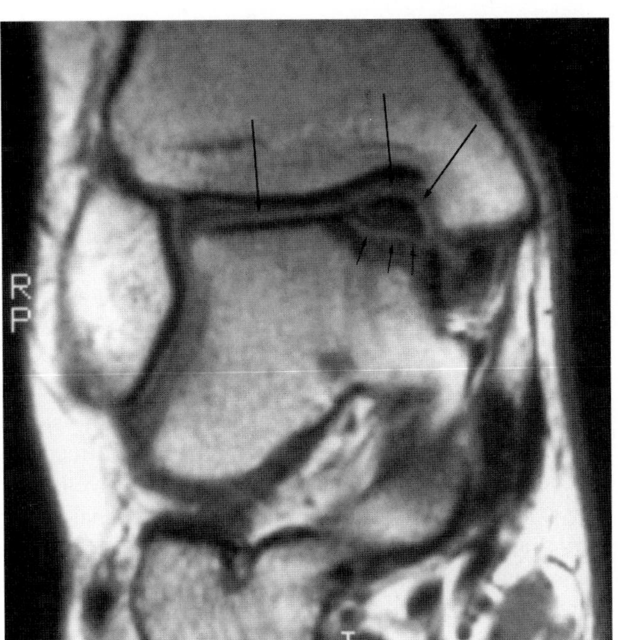

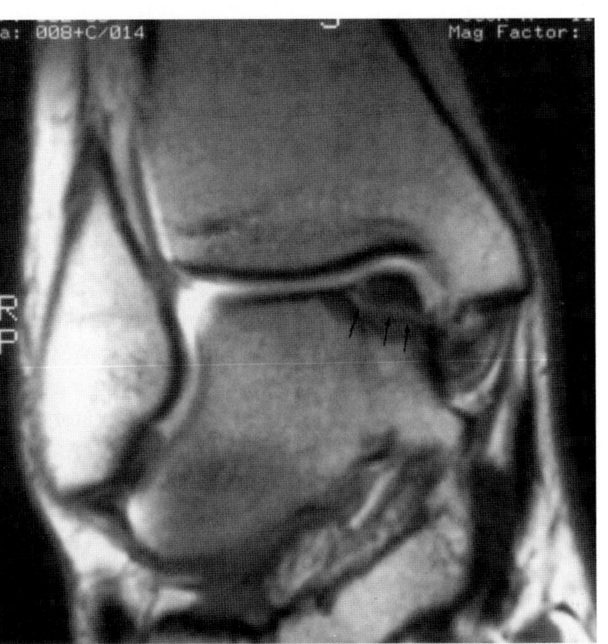

FIG. 18. I,J: Osteochondritis dissecans. MR arthrography. **I:** Coronal T1-weighted (TR/TE 300/20 msec) image demonstrates a well-defined osteochondral lesion involving the medial aspect of the talar dome. The fragment is of slightly decreased signal intensity compared with the remainder of the talus. The interface between the fragment and underlying subchondral bone is well delineated and is of slightly increased signal intensity *(small arrows).* The overlying articular cartilage appears grossly intact *(large arrow).* **J:** Coronal T1-weighted (TR/TE 300/20) image following the intraarticular administration of dilute gadopentetate dimeglumine. The contrast demonstrates high signal intensity and the articular surface is well defined and intact. No contrast extends into the interface between the fragment and underlying subchondral bone *(small arrows).* (From Deutsch AL. Osteochondral injuries of the talar dome. In: Deutsch AL, Mink JH, Kerr R, eds. *MRI of the foot and ankle.* New York: Raven Press, 1993;118–134, with permission.)

CASE 19

History: Two sequential imaging examinations obtained 3 weeks apart are presented. The patient presented 1 week following a puncture injury to the foot and the initial study was obtained (Fig. 19A). What is your diagnosis? What may have happened between examinations to account for the findings on the second examination (Fig. 19B). What is your diagnosis at this time?

A 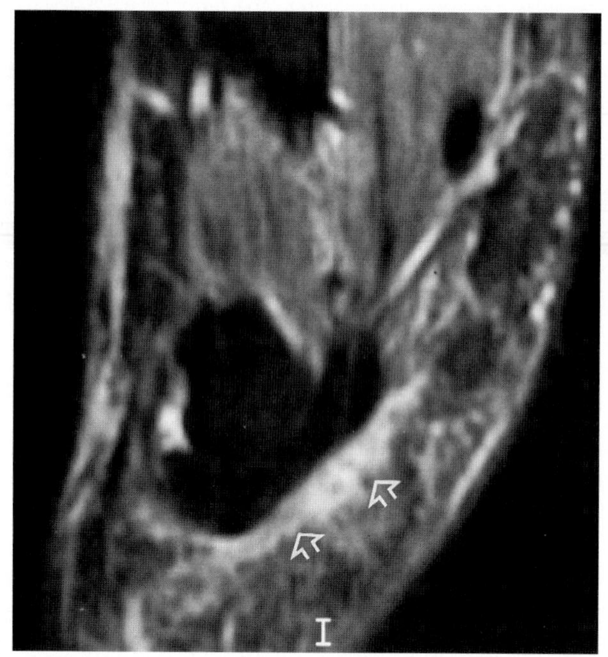

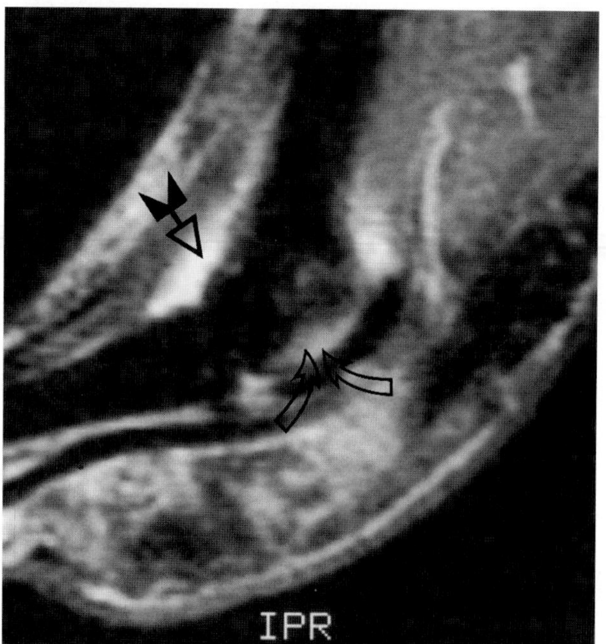 B

FIG. 19.

Findings: In Fig. 19A the sagittal inversion recovery fast spin echo (TR/TE/TI 3,500/30/130) image demonstrates increased signal within the soft tissues along the plantar aspect of second metatarsal head (open arrows). In Fig. 19B the sagittal inversion recovery fast spin echo (TR/TE/TI 3,500/30/130) is obtained approximately 3 weeks following Fig. 19A. There has been a significant increase in the degree of soft tissue changes involving the plantar aspect of the foot characterized by increased girth and diffuse edema. A moderate-sized joint effusion is present (triangle arrow) and increased signal is noted to have developed within the volar aspect of the metatarsal head involving both cortical and medullary bone (open curved arrows).

Diagnosis: Interval development of osteomyelitis involving the second metatarsal head. The patient was initially diagnosed as having soft tissue infection without involvement of bone. Antibiotic therapy was initiated but patient compliance was poor. The patient returned with increased swelling of her foot at which time the second MR was obtained.

Discussion: Multiple studies have addressed the utility of MR imaging in the evaluation of suspected osteomyelitis (1–9). The results of these studies, several of which have compared MR to more traditional scintigraphic methods, have been generally favorable. As a weight-bearing articulation, the foot presents several unique challenges with regard to the imaging assessment of osteomyelitis. One problem in particular, that has affected the specificity of both radionuclide techniques as well as MR, has been the differentiation between bone signal changes (MR) and activity (radionuclide) related to infection, and those that might result from chronic stress and microfracture (10–12). In this regard, the penultimate challenge is often presented in the patient with diabetes mellitus and neuroarthropathy.

Osteomyelitis is characterized by alterations in marrow signal intensity often associated with surrounding soft tissue changes depending the route of infection. In the foot, osteomyelitis most commonly results from extension of an adjacent soft tissue infection or occurs secondary to direct implantation from a traumatic source. In this setting, the combination of tissue characterization and anatomic detail provided by MR represents a significant advantage over the lower spatial/anatomic resolution of the scintigraphic techniques. MR has the potential to differentiate the sequential pathoanatomic stages of direct extension osteomyelitis including cellulitis, soft tissue abscess, infective periostitis, and frank osteomyelitis. Bone scintigraphy interpretation, in contrast, is entirely dependent on the biodistribution of technetium (Tc) 99m MDP, which is distributed based on capillary flow and incorporated into bone undergoing remodeling for whatever cause (e.g., infection, posttrauma, neuropathic, postoperative) (10–12).

MR marrow signal alterations seen in routine spin echo MR imaging are nonspecific. Decreased signal on T1-weighted images and increased signal on T2-weighted images may be seen in a spectrum of noninfectious conditions and may lead to false positive diagnoses, particularly in the complex clinical settings often encountered in the foot (1,3,6). Fat-suppressed contrast-enhanced techniques have been reported to yield significantly higher specificity for the diagnosis of osteomyelitis compared with nonenhanced MR, and higher sensitivity and specificity compared to three-phase bone scintigraphy (1). According to this work, patients with osteomyelitis typically demonstrate a similar degree of enhancement within involved soft tissue and marrow (Fig. 19C,D). In the series reported by Morrison et al. (1), 30 of 31 patients meeting the criteria for osteomyelitis demonstrated uniformly intense soft tissue and marrow enhancement. The one case in this series demonstrating disparate soft tissue and bone enhancement resulted in a false-positive diagnosis of osteomyelitis in a patient with cellulitis and neuroarthropathy (1).

According to Morrison et al. (1) the basis for uniform soft tissue and bone enhancement can be explained by the biodistribution of gadopentetate dimeglumine and the pathophysiology of musculoskeletal infection. Gadopentetate dimeglumine is deposited in the extracellular fluid compartment in areas with altered vascular permeability (13–15). Increased vascular permeability in seen in both bone and soft tissue compartments in infection (1). The relatively uniform enhancement observed in cases of osteomyelitis could be explained by a similar degree of vascular permeability occurring in both the bone and soft tissue compartments involved with osteomyelitis (1). Disparate soft tissue and bone enhancement might suggest two different mechanisms of contrast enhancement and thus two different processes (e.g., infection and neuroarthropathy).

Bone and soft tissue abscesses may demonstrate a high-intensity enhancing rim around a lower signal intensity central focus representing necrotic or devitalized tissue (1,13,14). In Morrison et al.'s (1) series, rim enhancement within bone was associated with clinical criteria for osteomyelitis in all cases. There was also a high association between rim-enhancing soft tissue abscesses and osteomyelitis in the adjacent bone. The rim effect seen with gadolinium increased the conspicuity of infected material and aided surgical and biopsy planning.

In the author's experience, the use of MR for the diagnosis of osteomyelitis has remained problematic. We have been most confident, as in the test case, utilizing MR to exclude osteomyelitis in the setting of cellulitis and contiguous soft tissue infection. For these purposes we utilize a combination of T1-weighted fast spin echo, fast STIR, and most recently fat-suppressed, proton density–weighted fast spin echo techniques. Absence of marrow edema in this setting allows us significant confidence in excluding the presence of osteomyelitis. Many cases sent for referral are, however, far more complicated and typically involve diabetic patients with coexisting neuropathic changes as well as previously operated-on patients. In this setting, marrow edema is frequently present, but not necessarily indicative of the presence of infection or its extent. Our experience with gadopentetate-enhanced studies, utilizing both dynamic perfusion techniques as well as standard fat-suppressed T1 spin echo imaging, has been quite variable and certainly has not been as consistent as that reported by Morrsion et al (16). It is our experience that close attention must be directed toward evaluation of the adjacent soft tissues (e.g., sinus tracts) as well as to the presence of periosteal and cortical bone changes. In our experience, these latter findings are critical to accurate diagnosis. We are reluctant to offer the diagnosis of osteomyelitis on the basis of marrow signal changes in the absence of definite cortical bone changes in the setting of nonhematogenous osteomyelitis (e.g., contiguous site infection).

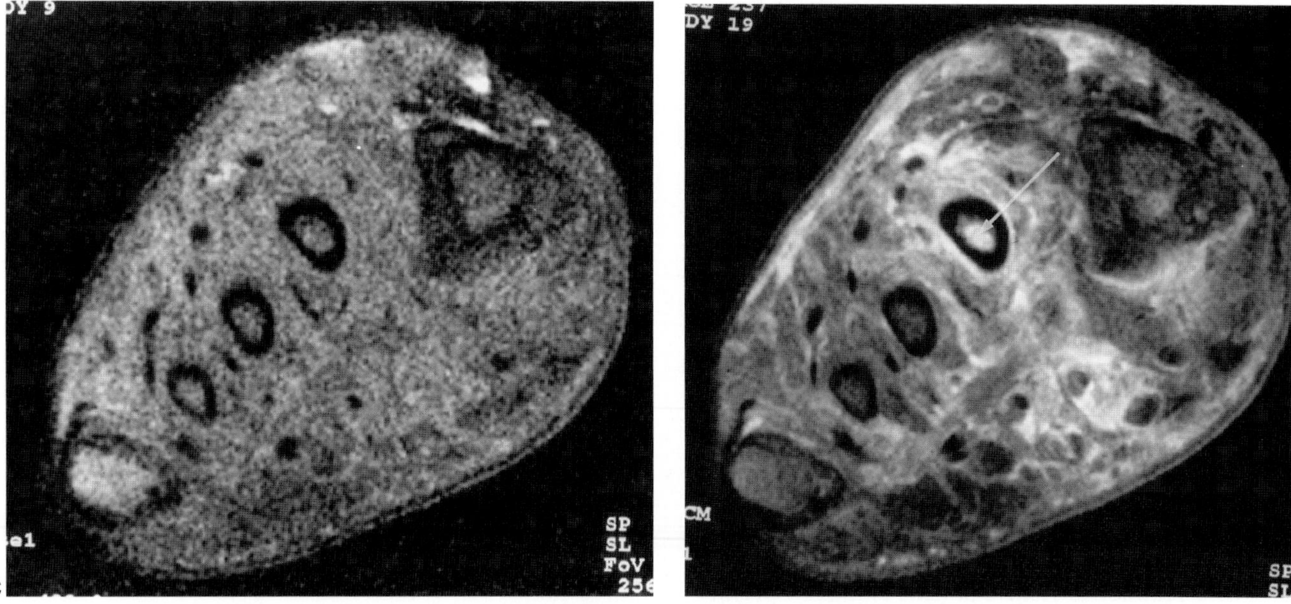

FIG. 19. C,D: Contrast enhancement in biopsy-proven osteomyelitis. A 69-year-old woman presented with a swollen foot and suspicion of cellulitis and possible osteomyelitis. Conventional radiographs were within normal limits. **C:** Axial T1-weighted (TR/TE 350/10 msec) image with fat suppression obtained prior to the administration of gadopentetate dimeglumine. There is increased girth and poor definition of the soft tissues along the dorsum of the foot. **D:** Axial T1-weighted fat-suppressed (TR/TE 350/10 msec) image obtained post–contrast administration. There is enhancement within the second metatarsal that is equal to minimally increased in intensity compared with the degree of enhancement of surrounding soft tissue *(arrow).*

CASE 20

History: An 82-year-old woman presents with pain and paresthesias along the medial aspect of the foot. Physical examination demonstrated a positive percussion sign.

What is your diagnosis? What is the role of MR in the assessment of these patients?

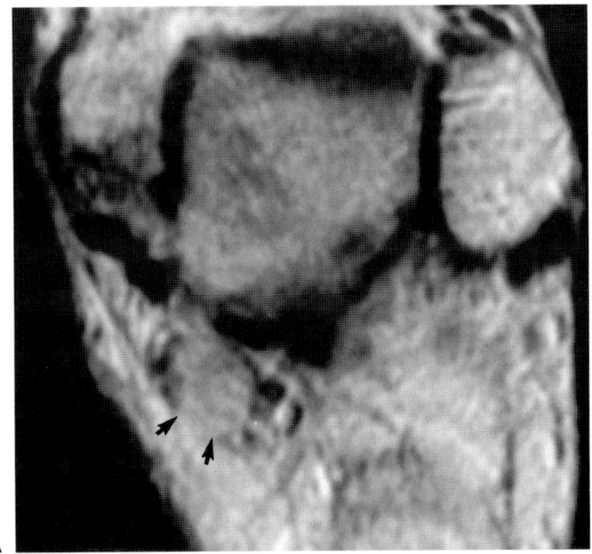

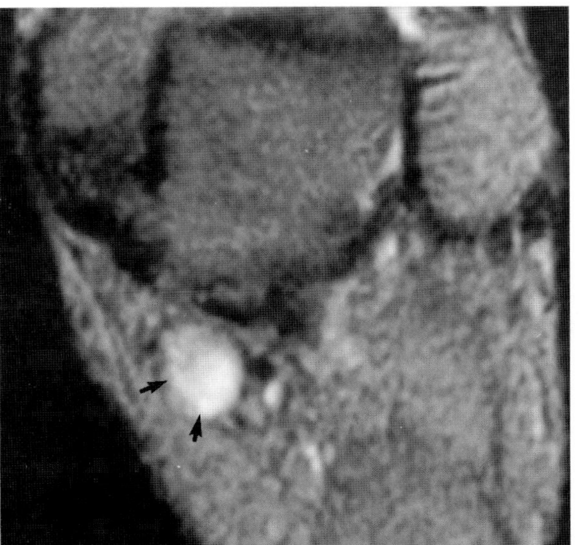

FIG. 20.

Findings: In Fig. 20A an axial proton density (PD) fast spin echo (TR/TE 3,500/32 msec) image demonstrates a mass to be of intermediate signal intensity (arrows). In Fig. 20B an axial T2-weighted fast spin echo (TR/TE 3,500/115 msec) image demonstrates the mass to increase in signal intensity (arrows).

Diagnosis: Tarsal tunnel syndrome secondary to a schwannoma.

Discussion: The tarsal tunnel is a passageway through which the medial ankle tendons and posterior tibial neurovascular bundle pass. While its precise proximal and distal borders are often difficult to define, in general, the tarsal tunnel extends from the level of the medial malleolus to the tarsal navicular bone. It occupies the medial-posterior aspect of the ankle and extends into the medial-plantar aspect of the foot. The tunnel has an osseous floor, which, from proximal to distal, is formed by the tibia, sustentaculum of the talus, and medial wall of the calcaneus (Fig. 20C). The roof is formed by the deep fascia of the leg, the flexor retinaculum, and distally by the abductor hallucis muscle.

The tarsal tunnel is compartmentalized by several deep fibrous septa that extend from the undersurface of the flexor retinaculum to the medial malleolus. These septa run between the tendons and the neurovascular bundle and enclose them in separate small tunnels. Some of these septa are attached to the neurovascular bundle, thereby causing it to be relatively immobile and quite vulnerable to traction forces or space-occupying lesions. Consequently, even mild compression or traction forces may elicit the sensory symptoms of tarsal tunnel syndrome.

From anterior to posterior, the contents of the tarsal tunnel include the posterior tibial tendon, flexor digitorum longus tendon, posterior tibial neurovascular bundle, and flexor hallucis longus tendon. A variety of branching patterns of the posterior tibial nerve have been described. Bifurcation into the medial and lateral plantar nerves usually occurs beneath the flexor retinaculum but rarely occurs proximal to it. The medial calcaneal nerve originates with equal frequency beneath or proximal to the flexor retinaculum. It may originate from the posterior tibial nerve or the lateral plantar nerve and may divide into multiple branches. Within the distal aspect of the tarsal tunnel, upper and lower chambers are formed for the medial and lateral neurovascular bundles, respectively. The medial plantar nerve is always in close proximity to the flexor hallucis longus tendon, coursing along its posteromedial aspect. The medial plantar nerve provides sensory innervation to most of the plantar surface of the foot and innervates the abductor hallucis, flexor digitorum brevis, flexor hallucis brevis, and first lumbrical muscles. The lateral plantar nerve provides sensory innervation to the lateral plantar aspect of the foot and innervates the remaining intrinsic muscles of the foot.

The tarsal tunnel syndrome represents an entrapment neuropathy of the posterior tibial nerve or of its branches and is one of several important entrapment neuropathies that have been described in the foot and ankle. It is characterized by burning pain and paresthesias along the plantar surface of the foot and in the toes. Pain may also localize to the medial plantar aspect of the ankle or radiate proximally into the calf. Sensory deficits predominate and muscle weakness is a late and infrequent finding. The symptoms are commonly exacerbated by weight bearing. The principal physical finding is a positive percussion sign (distal paresthesia produced by percussion over the affected portion of the nerve). The nerve may also be tender at and proximal to the site of compression.

The diagnosis of tarsal tunnel syndrome may be difficult as the physical findings may be poorly localized. Electromyography and nerve conduction studies may have variable results with reported sensitivities ranging from 65% to 90%. A normal nerve conduction study does not exclude the diagnosis. Definitive treatment of patients with tarsal tunnel syndrome is surgical release of the flexor retinaculum and removal of any compressing mass lesion. The results of surgical decompression have been quite variable with a significant number of patients reporting little or no relief postoperatively. The presence of an identifiable mass lesion contributing to neurocompression has a high predictive value for a successful surgical outcome. This underscores the value of preoperative MRI in the assessment of patients with suspected tarsal tunnel syndrome. The most common causes of tarsal tunnel syndrome include ganglion cysts, neurilemomas, tumors, venous varicosities, hemangiomas, and posttraumatic deformities with fibrous scarring (Fig. 20D,E). All of these conditions can be well demonstrated by MRI. Optimal assessment of the tarsal tunnel may require acquisition of images in the sagittal, coronal, and axial planes or alternatively three-dimensional techniques with multiplanar reconstruction. The authors presently utilize fast spin echo and fast inversion recovery pulse sequences. Small field of view (10–12 cm) surface coil imaging is advocated to provide the requisite anatomic detail.

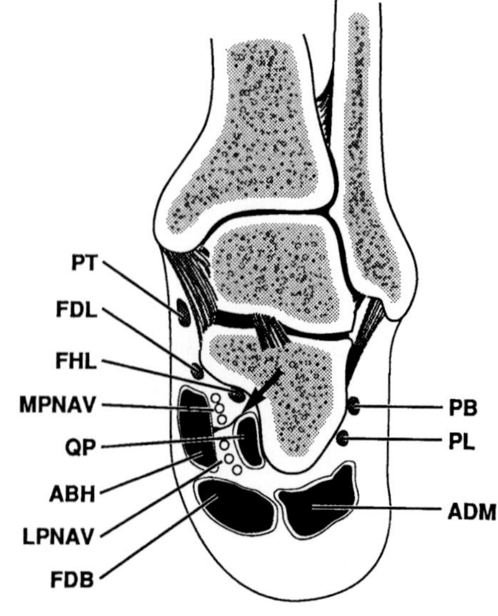

FIG. 20. C: Coronal diagram through the distal aspect of the tarsal tunnel demonstrates the medial and lateral neurovascular bundles separated by the transverse interfascicular septum *(arrow).* PT, posterior tibial tendon; FDL, flexor digitorum tendon; FHL, flexor hallucis tendon; MPNAV, medial plantar neurovascular bundle; QP, quadratus plantae muscle; ABH, abductor hallucis muscle; LPNAV, lateral plantar neurovascular bundle; FDB, flexor digitorum brevis muscle. (From Kerr R. Spectrum of disorders. In: Deutsch AL, Mink JH, and Kerr R, eds. *MRI of the foot and ankle.* New York: Raven Press, 1992;345–351, with permission.)

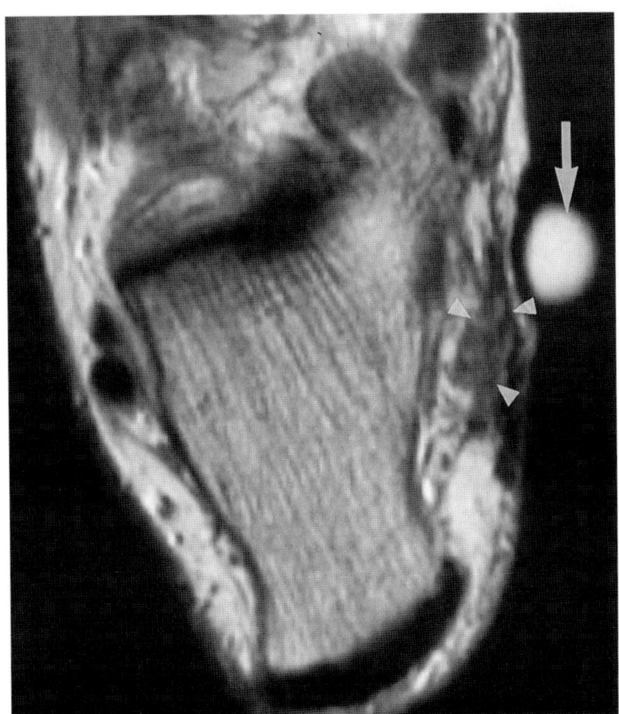

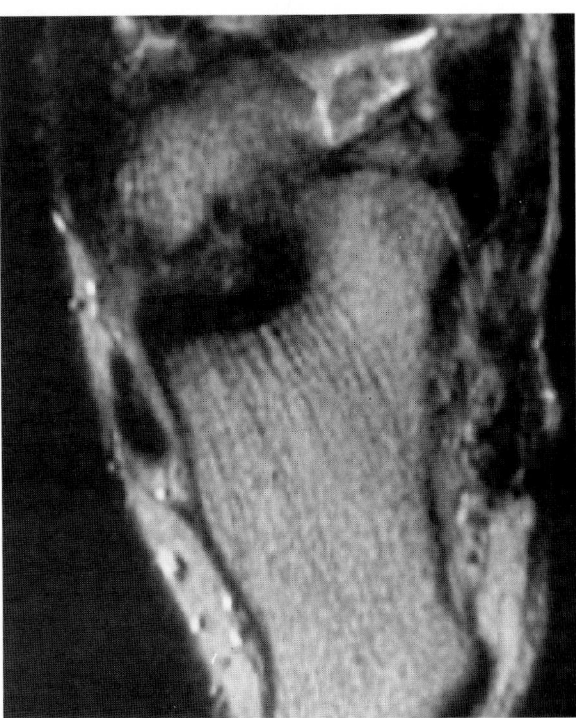

FIG. 20. D,E: Recurrent tarsal tunnel syndrome secondary to extensive scarring. A 41-year-old man presented with recurrent symptoms and a palpable mass $1\frac{1}{2}$ years status post–tarsal tunnel release. **D:** Axial T1-weighted (TR/TE 500/15 msec) image through the distal tarsal tunnel demonstrates an irregular collection of low to intermediate signal intensity replacing the normal fat within the distal tarsal tunnel *(arrowheads)*. This correlated with the level of a palpable abnormality indicated by an oil marker *(arrow)*. **E:** Axial T2-weighted (TR/TE 3,800/102 msec) image. The findings remained of low signal and essentially unchanged. On re-exploration, extensive scar and adhesions were identified.

CASE 21

History: A 3-year-old boy presented with a 2-week history of pain and swelling involving his midfoot. The patient was referred for MR imaging to evaluate for possible osteomyelitis.

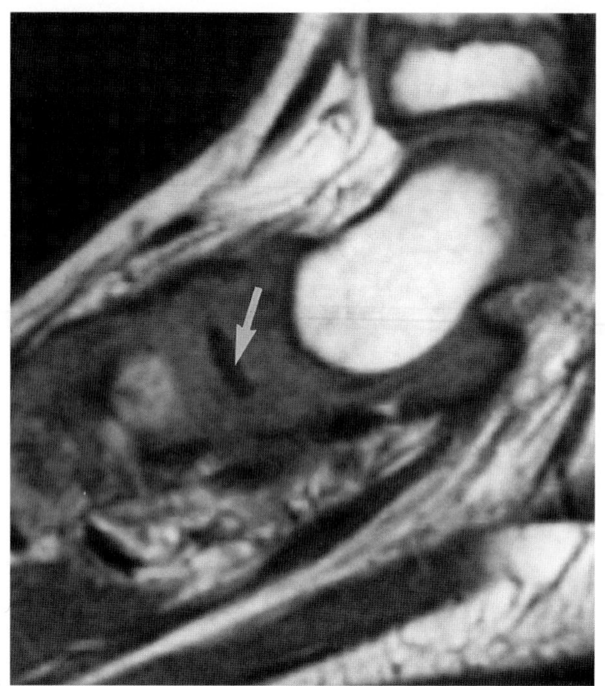
A

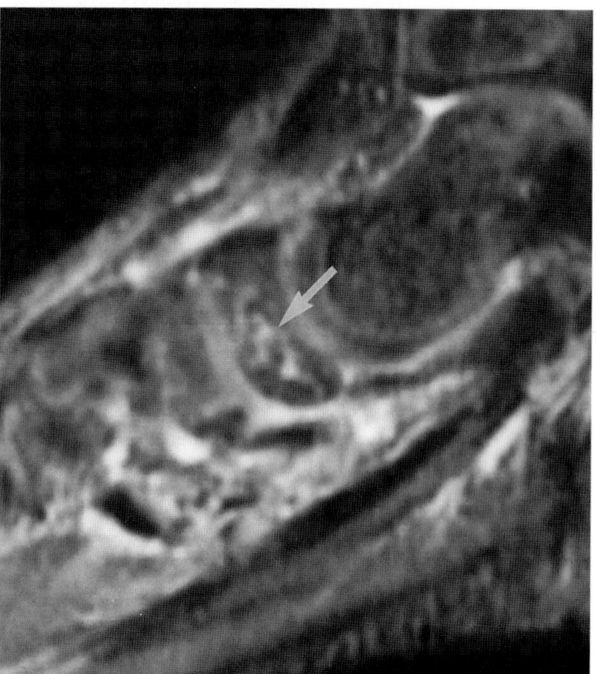
B

FIG. 21.

Findings: In Fig. 21A, on the sagittal T1-weighted sequence (TR/TE 500/20 msec), the ossification center for the tarsal navicular is ''wafer-like'' and of markedly low signal intensity. The other visualized tarsal ossification centers are of uniform high signal intensity surrounded by cartilage which appears of uniform intermediate to low signal intensity on this pulse sequence. In Fig. 21B, on the fat-suppressed sagittal fast spin echo T2-weighted sequence (TR/TE 4,000/102 msec), the surrounding cartilaginous components are better defined, demonstrating intermediate signal intensity. Areas of increased signal are seen within the ossification center for the navicular (arrow). The ''space'' for the navicular is preserved as the cartilaginous component remains normal in size.

Diagnosis: Kohler's disease.

Discussion: Kohler's disease is one of a number of conditions that has traditionally been classified as an osteochondrosis. These disorders, while now appreciated to represent a more heterogeneous group of unrelated conditions, share certain features in common including (a) predilection for the immature skeleton; (b) involvement of an epiphysis, apophysis, or epiphysoid bone; and (c) a radiographic picture characterized by fragmentation, collapse, sclerosis, and frequently reossification with reconstitution of the bony contour (1). Originally, it was believed that all of these conditions derived from a primary impairment of vascular supply that resulted in osteonecrosis (2). It is now apparent that osteonecrosis is not present on histologic examination in some of the osteochondroses (e.g., Blount's tibia vara), and that in some osteonecrosis is not a primary event but rather secondary to a fracture or traumatic event (e.g., Kienböck's disease of the lunate). Other conditions originally designated as osteochondroses, appear now to simply represent normal variations in ossification, e.g., Sever's disease of the calcaneus (1,2).

Kohler's disease is considered relatively rare, although its exact incidence is difficult to determine because many cases may not be of sufficient magnitude to require diagnostic workup. The disorder is more common in boys (4–6:1 ratio) and presentation is most commonly between the ages of 3 and 7 (1). Clinical manifestations include local pain, tenderness, swelling, and decreased range of motion (3). A history of trauma can be elicited in approximately 35% of cases and the condition has been reported to occur simultaneously with Legg-Calvé-Perthes disease (1). Radiographic findings include soft tissue swelling as well as increased radiodensity and fragmentation of multiple ossific nuclei. The bone may appear diminished in size and flattened or wafer-like in appearance (1). As in the present case, the interosseous space between the navicular and adjacent bones may be normal, reflecting the integrity of the cartilage surrounding the ossification center. Over a period of 2 to 4 years, the bone may return to an essentially normal appearance (1,3).

It is the self-limited and apparently reversible nature of the condition that has led some authorities to question whether it is really a "disease" or rather a manifestation of altered enchondral ossification (1,4). There is a marked paucity of histologic data available to help settle the controversy. A thickened zone of cartilage about the ossific nucleus has been reported to suggest a disturbance of enchondral ossification. The ossific nucleus has demonstrated signs of necrosis involving both spongiosa and marrow (1,4). The presence of symptoms, demonstrable changes in a previously normal navicular, and radiographic findings suggestive of osteonecrosis are utilized in support of a diagnosis of Kohler's disease.

CASE 22

History: Images from two patients with history of eversion injuries to the ankle are illustrated. What is the nature of the injury and what structures are involved? What is the normal anatomy of this ligamentous complex?

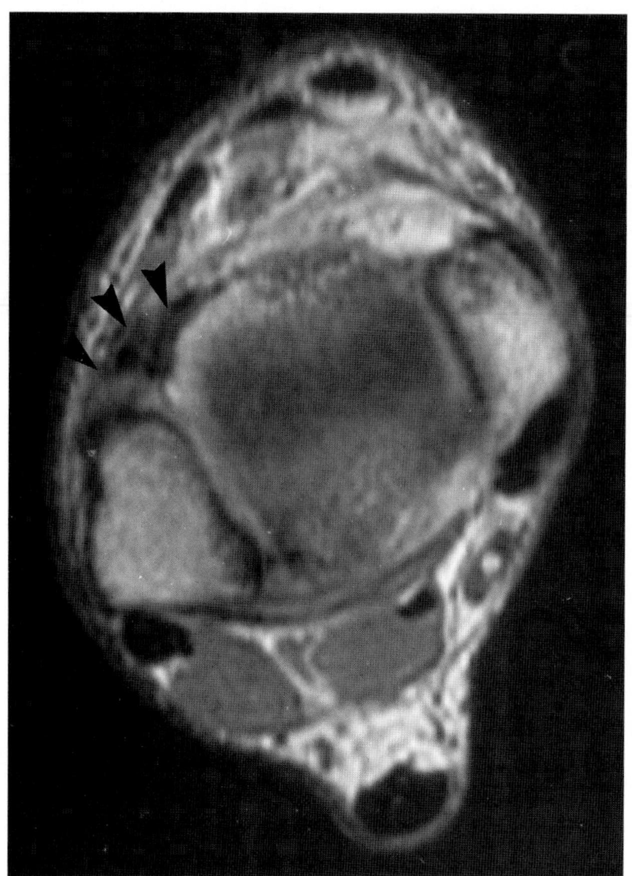

A

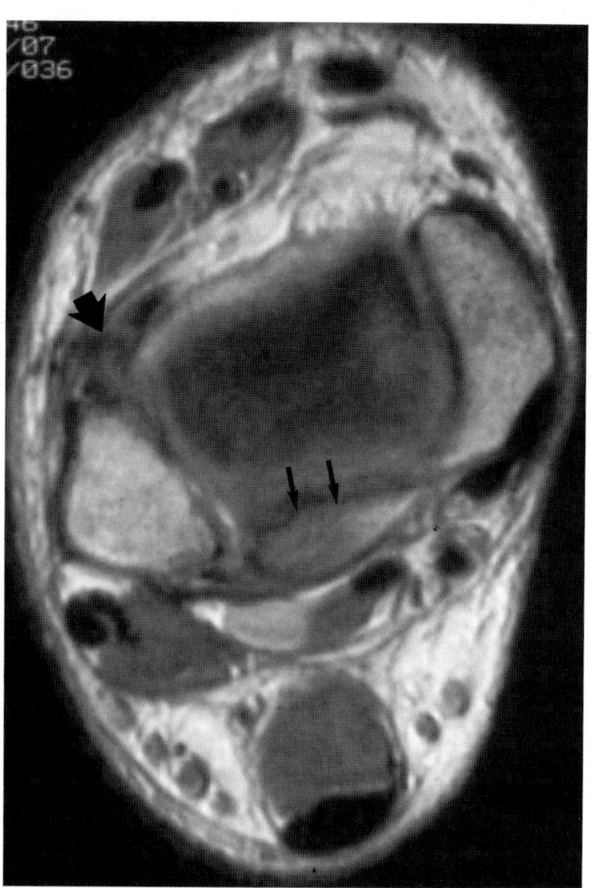

B

FIG. 22.

Findings: In Fig. 22A an axial PD (TR/TE 2,000/20 msec) image from a patient status post "ankle sprain" one month previously demonstrates marked thickening and deformity of the anterior tibiofibular ligament (ATIF) (arrowheads). There is only a minor degree of subcutaneous edema consistent with the subacute nature of the injury. In Fig. 22B, in a different patient who experienced an eversion injury of the ankle, PD axial image demonstrates marked thickening and distortion of the tissues in the expected position of the ATIF (large arrow). The PTIF is intact, but the patient has suffered a nondisplaced posterior malleolar fracture (arrows). (From Mink JH. Ligaments of the ankle. In: Deutsch AL, Mink JH, Kerr R, eds. *MRI of the Foot and Ankle.* New York: Raven Press, 1992;185–189, with permission.)

Diagnosis: Syndesmotic complex injury with disruption of the anterior tibiofibular ligament.

Discussion: The syndesmotic ligament complex is composed of the anterior tibiofibular ligament (ATIF), the posterior tibiofibular ligament (PTIF), and the interosseous ligament and membrane (1–3). The ATIF is a short band-like structure that extends from an origin on the anterior tibial tubercle and follows a caudal, lateral, and slightly posterior course to insert onto the fibula. The PTIF runs from the posterior malleolus of the tibia to insert on the fibula following a slightly downward path. The PTIF is stronger than its anterior counterpart and some authorities consider its lowest fibers to be a separate structure (the transverse tibiofibular ligament) that serves to deepen the ankle joint socket posteriorly (3,4). The interosseous ligament and membrane represent a series of short fibers running from a ridge on the tibia in a downward and lateral direction to a similar ridge on the fibula (4). The syndesmotic complex defines the uppermost extent of the ankle joint and extends 2 to 6 cm above the plafond. Isolated syndesmotic tears are rare except in the elite skier (5).

Injuries to the syndesmotic complex result from two major mechanisms (1,4). In the first, referred to as the anterior type, the talus laterally rotates in the mortise to impinge upon the fibula; the ATIF and interosseous ligament rupture, and the syndesmotic "books open" with the PTIF as the hinge. Either the medial malleolus fractures or the deltoid ligament ruptures. In the second type, a direct abduction force of the talus against the fibula results in a fracture and disruption of the ATIF, PTIF, and interosseous ligament. The deltoid ligament is either torn or the medial malleolus fractures. In all significant injuries of the syndesmosis, a fibular fracture is found. The fracture may occur at any point from the top of the inferior tibiofibular joint to the head of the fibula (Maisonneuve fracture). On conventional radiographs, rupture of the syndesmosis is not simply seen as diastasis of the distal tibiofibular articulation. The diagnosis can be confidently made if the space between the medial malleolus and the medial talus, or between the lateral talar margin and the medial fibula, measures greater than 5 mm. Major syndesmotic disruptions can occur and spontaneously reduce before the radiographic examination can be performed. Stress radiography can be employed if a syndesmotic rupture is suspected but not evident on conventional radiographs.

The syndesmotic complex is frequently assessed on MRI utilizing axial sections (1,6–9). The interosseous membrane is seen as a linear band of dark signal connecting the distal tibia and fibula. The most caudal extent is the interosseous ligament. In some individuals this structure is quite substantial, whereas in others it is virtually absent. The membrane ends at the level of the metaphyseal flare of the tibia. Just above the plafond, a thin dark band representing the ATIF arises from the anterolateral aspect of the tibia (Fig. 22C–E). The ATIF has a caudal and oblique course and is therefore difficult to identify on a single axial image. Volume acquisitions with reformatting may assist in demonstrating the ATIF on a single oblique axial image. The PTIF is demonstrated at approximately the same axial level as the ATIF as a triangular band with its base on the fibula and its apex on the posterior distal tibia and posterior malleolus (Fig. 22F,G). The superior margin of the PTIF may have a variable appearance (irregular and frayed) likely due to a fanlike or flared insertion of the ligament and partial volume effect of the ligament with the surrounding fat (8). On sagittal images, the PTIF projects at the posterior margin of the edge of the posterior malleolus. The dark signal, rounded appearance, and proximity to the joint may result in either the PTIF or the PTAF being mistaken for a loose body (1).

On axial MR sections, there may on occasion be confusion in identification and differentiation of the tibiofibular and talofibular ligaments particularly posteriorly (1,9). The reason for this confusion is related to the fact that the talus is invariably on the image in which the fibular attachment of the tibiofibular ligament is present. Two osseous landmarks can be utilized to correctly identify which group of ligaments one is observing when viewing MR images in the axial plane. At the level of its dome, the talus has a nearly rectangular configuration; this is the level at which the tibiofibular ligaments are observed. The talofibular group is identified more distally where the talus has a more elongated shape and a portion of the sinus tarsi is usually visible. The most dependable landmark to help distinguish the tibiofibular from the talofibular ligaments is the fibula (9). The tibiofibular ligaments are identified at the level of the distal fibular shaft; at this level the fibula has a flattened medial border. The talofibular ligaments are found at the level of the malleolar fossa, a prominent and rather deep indentation along the medial border of the lateral malleolus (9).

Disruptions of either the ATIF or PTIF are most commonly manifest by structural thickening and irregularity (1,7). Because the ATIF is not universally identified on axial sections, as a consequence of its steep oblique course, its absence cannot be utilized as a totally reliable sign of injury. Oblique axial reformations from three-dimensional volume acquisitions can give greater confidence with regard to the presence of an intact ligament. Rupture of the PTIF is uncommon and frequently injury to the ankle joint will be manifest by an avulsion fracture of the posterior malleous rather than a PTIF rupture. Chronic syndesmotic injuries are manifest by heterotopic calcification in the interosseous membrane and at the sites of ligamentous avulsion (Fig. 22H,I).

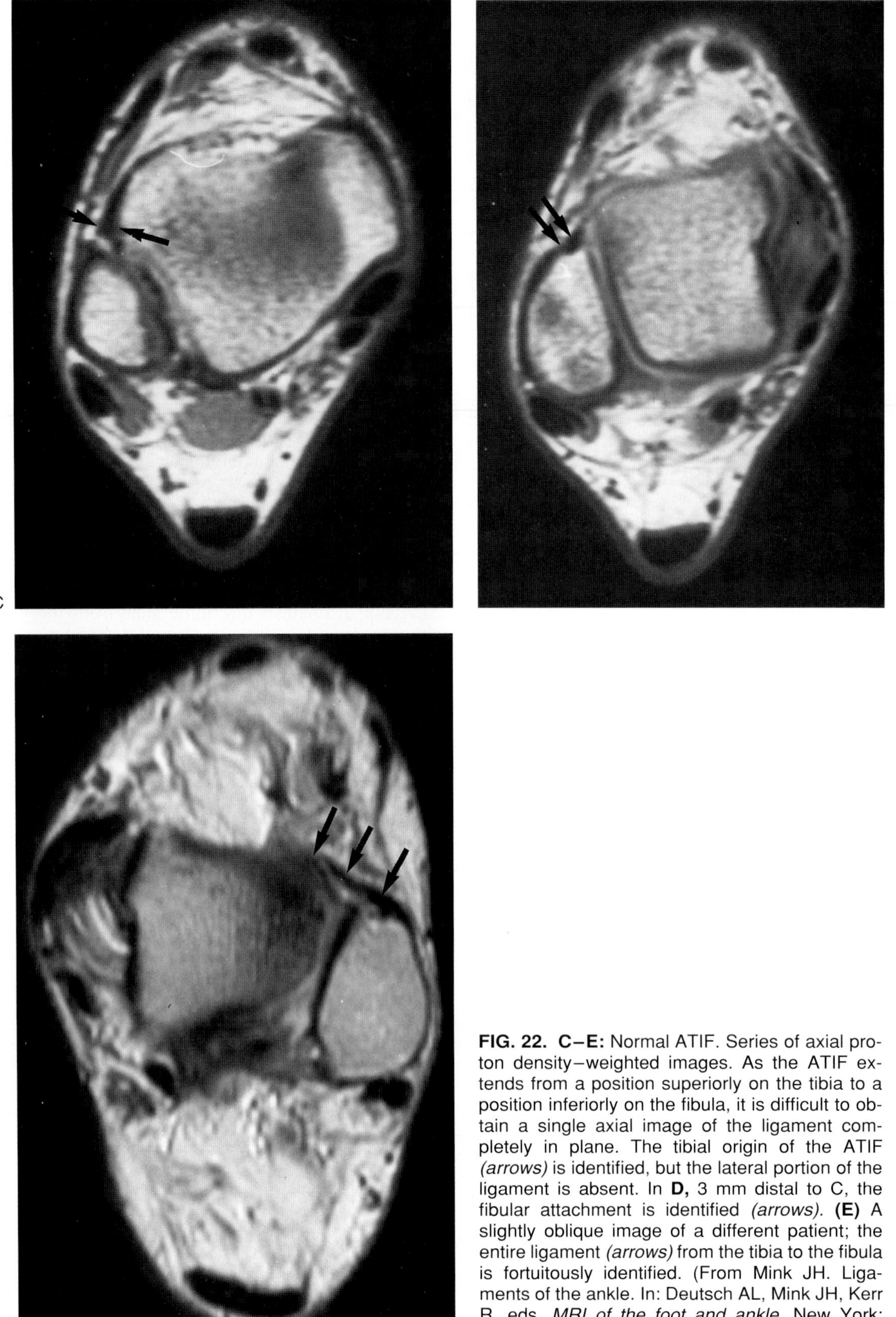

FIG. 22. C–E: Normal ATIF. Series of axial proton density–weighted images. As the ATIF extends from a position superiorly on the tibia to a position inferiorly on the fibula, it is difficult to obtain a single axial image of the ligament completely in plane. The tibial origin of the ATIF *(arrows)* is identified, but the lateral portion of the ligament is absent. In **D,** 3 mm distal to C, the fibular attachment is identified *(arrows).* **(E)** A slightly oblique image of a different patient; the entire ligament *(arrows)* from the tibia to the fibula is fortuitously identified. (From Mink JH. Ligaments of the ankle. In: Deutsch AL, Mink JH, Kerr R, eds. *MRI of the foot and ankle.* New York: Raven Press, 1992;185–189, with permission.)

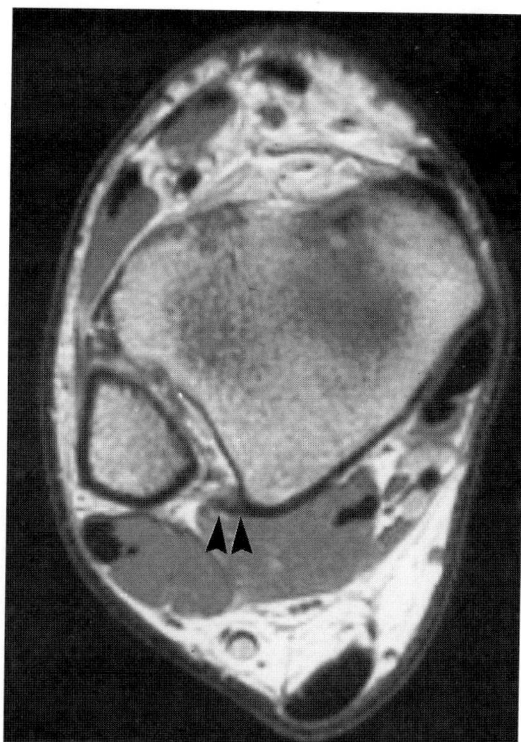

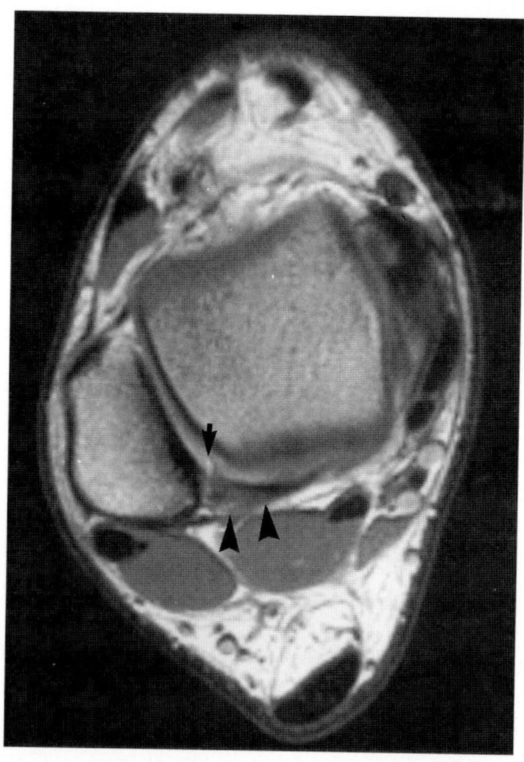

FIG. 22. F,G: Normal PTIF. Axial intermediate weighted MR images demonstrate the PTIF, in a manner similar to the ATIF, has a slightly caudal course as it runs from the tibia to the fibula. **F:** The proximal-most tibial origin of the PTIF *(arrowheads)* is seen arising from the posterior tibia/posterior malleolus. In **G,** 5 mm distal to F, the triangular, fan-shaped configuration of the PTIF *(arrowheads)* is best appreciated on this axial image. The PTIF is normally characterized by increased signal "streaks" arranged in a transverse direction. The small bright streak *(arrow)* seen over the PTIF is an artifact on the film. Inasmuch as most of the PTIF is seen on an image in which the talus is present and the tibia is not, the PTIF has been mistakenly identified as the PTAF. (From Mink JH. Ligaments of the ankle. In: Deutsch AL, Mink JH, Kerr R, eds. *MRI of the foot and ankle.* New York: Raven Press, 1992;185–189, with permission.)

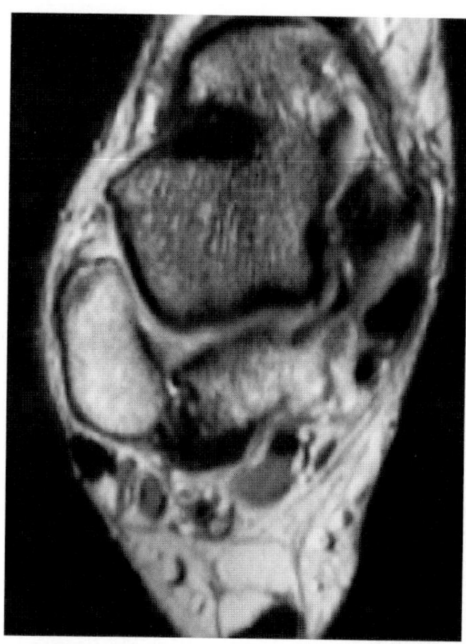

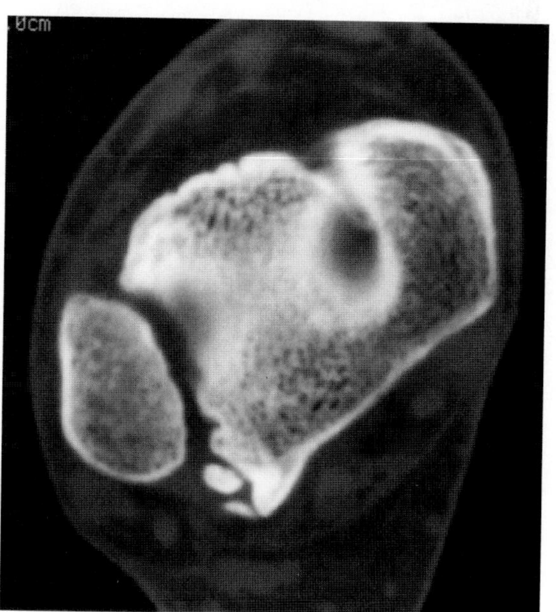

FIG. 22. H,I. H: Chronic syndesmotic injury with heterotopic bone formation. Axial proton density (TR/TE 2,000/20 msec) image of a professional football player demonstrates low signal intensity in the region of the PTIF. **I:** Axial CT demonstrates extensive proliferative bone formation corresponding to the area of signal alteration seen on MR.

CASE 23

History: A 36-year-old man underwent MR evaluation for pain and limitation in motion of the third toe. What is the significance of the signal identified in the long flexor tendon?

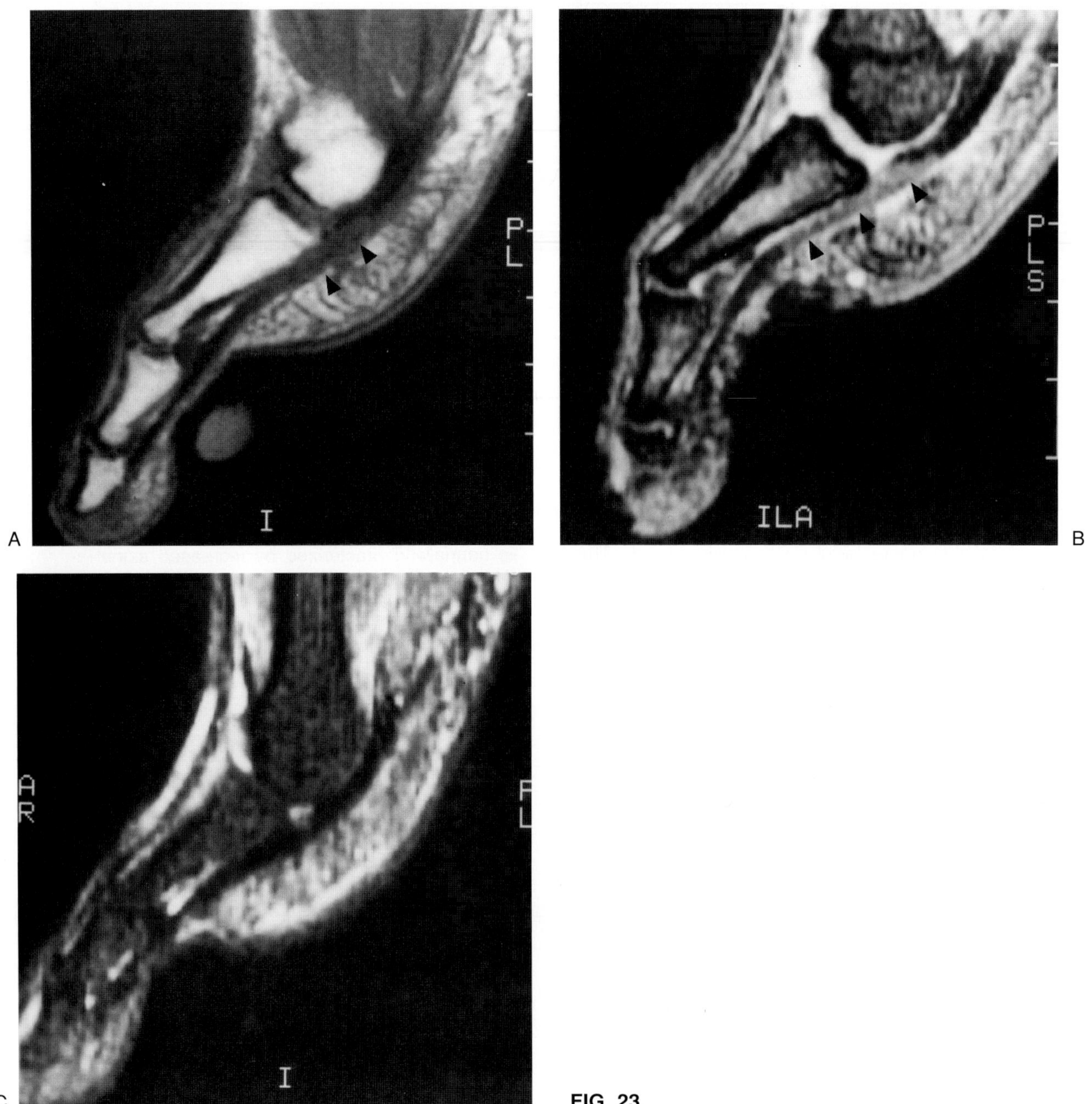

FIG. 23.

Findings: In Fig. 23A the sagittal T1-weighted (TR/TE 500/21 msec) image through the third toe demonstrates increased signal within the long flexor tendon at the level of the MCP joint (arrowheads). In Fig. 23B the sagittal gradient echo (TR/TE/° 250/15/20) sequence also demonstrates focal increased signal within the long flexor tendon at the level of the MCP joint (arrowheads). The tendon appears otherwise morphologically normal. In Fig. 23C the sagittal inversion recovery fast spin echo (TR/TE/TI 2,600/21/130) demonstrates a normal appearance to the long flexor tendon. The is no increased signal, no evidence of tenosynovitis, and no increased tendon girth.

Diagnosis: Increased intratendinous signal related to angle-dependent signal (''magic angle'' phenomenon).

Discussion: Tendons are principally composed of collagen and elastin. Collagen provides the structural strength and elastin the stretch properties. Collagen microfibrils are composed of tropocollagen, which is a protein consisting of three polypeptide chains arranged in a triple helix (1,2). The microfibrils are organized into fibers that are embedded into an amorphous ground substance. The fibers are arranged in parallel bundles resulting in a highly ordered structure (2). As a consequence of this highly ordered structure, the motion of water molecules binding to the collagenous tissue is greatly restricted (2). This arrangement results in greatly enhanced dipole-dipole (internuclear) interactions that can result in substantial shortening of the T2 relaxation time and contributes to the low signal intensity exhibited by tendons on MR images (2).

MR signal intensity is primarily determined by intrinsic factors (e.g., T1 and T2 relaxation times, proton density, flow, susceptibility effects) as well as extrinsic operator selectable parameters (e.g., repetition time, echo time, flip angle) (2). The observed signal intensity of highly structured tissues such as tendons, however, is dependent on yet another factor: the orientation of the tissue in relation to the constant magnetic induction field (Bo) (1–3). The physical basis for this relates to the influence on the magnitude of the previously discussed dipole-dipole interactions of the angle between the dipole-dipole vector and the constant magnetic induction field (Bo). When the angle between the vector joining the proton dipoles bound to the collagen triple helix and the constant magnetic induction field equals approximately 55°, the dipolar interaction virtually disappears and signal intensity is maximal (1–3).

This phenomenon of angle-dependent signal intensity within tendinous structures was first demonstrated by Berendsen (4) utilizing dehydrated tendons, and subsequently was studied by Fullerton et al. (5) and Peto et al. (6) utilizing fully hydrated tendons from recently sacrificed animals. These workers demonstrated augmentation of T2 relaxation times on the order of 100 times at the angle of 55° with negligible effect on the T1 relaxation time. Erickson and colleagues (1) first demonstrated that tendon orientation significantly affected signal intensity during routine clinical short TE sequences at high field strength. Angle-dependent signal effects are primarily manifested on T1-weighted spin echo and gradient echo sequences because the TE values of these sequences most closely approximates the augmented observed T2 relaxation times related to tendon orientation (1,2). The tendon demonstrates maximum signal intensity when oriented at the ''magic angle'' of 55° in relation to Bo. Intermediate signal is seen when the tendon is oriented within 10° of the magic angle, and tendons demonstrate minimal signal when oriented greater than 10° from the magic angle (1–3). Erickson and colleagues (1–3) have additionally noted that tendon signal intensity is independent of the prescribed scan plane.

When confronting intratendinous signal that may be contributed by the magic angle phenomenon, attention should be directed toward evaluation of tendon morphology as well as the appearance of the signal on T2-weighted sequences. Magic angle effects are commonly demonstrated in multiple tendons at the level of the malleoli where their course approximates 55° in relation to the main magnetic field with the patient positioned supine. Demonstration of signal in only one of two similarly coursing tendons such as the peroneus brevis and longus tendons should make the interpreter question whether the signal can indeed be attributed to magic angle. Signal that increases in intensity on T2-weighted images cannot be ascribed to magic angle. Tendon contour and morphology are important features to evaluate with regard to diagnosing abnormalities. Attention should be directed toward assessment of both increased girth as well as localized attenuation. These features may be subtle but the only objective findings in many derangements of the ankle tendons and particularly the posterior tibial tendon. In selected cases, when doubt still exists about the nature of intratendinous signal, the foot can be repositioned (e.g., plantar flexed) and scanning repeated. This should allow definitive determination regarding whether observed intratendinous signal relates to the magic angle phenomenon or is representative of an intrinsic tendon disorder.

CASE 24

History: A 42-year-old woman complained of distal metatarsalgia with pain radiating distally through the toes. What is your diagnosis?

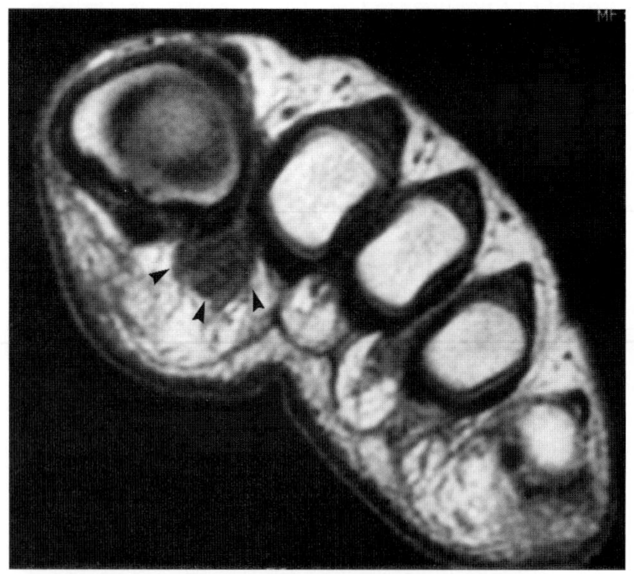
A

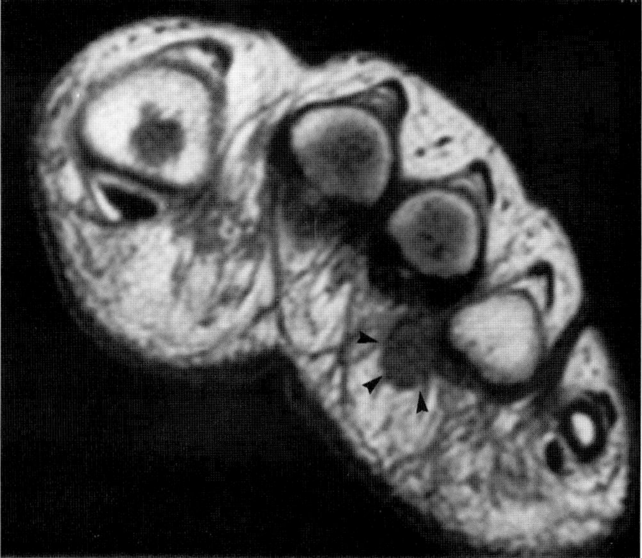
B

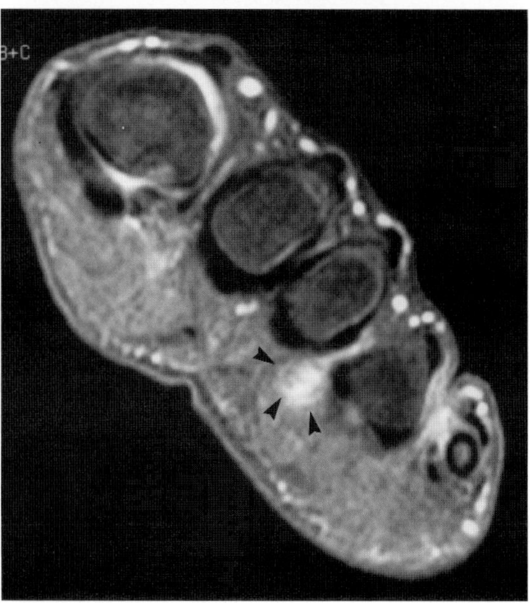

C

FIG. 24.

Findings: In Fig. 24A the axial T1-weighted (TR/TE 550/30 msec) image through the forefoot demonstrates a low signal intensity circular to ovoid mass projecting between the first and second metatarsals (arrowheads). In Fig. 24B the axial T1-weighted (TR/TE 550/30 msec) image obtained slightly more distally demonstrates a second low signal intensity mass between the third and fourth metatarsals (arrowheads). In Fig. 24C the axial T1-weighted (TR/TE 550/30 msec) fat-suppressed image obtained following the administration of gadopentetate dimeglumine demonstrates avid contrast enhancement of the previously defined mass between the third and fourth metatarsal heads (arrows). (Images courtesy of Dr. Gary Hinson, Kansas City.)

Diagnosis: Synchronous interdigital neuromas (Morton neuroma).

Discussion: The interdigital neuroma is not a tumor but rather a degenerative, fibrosing process occurring in and about the plantar digital nerve. It is the most common nerve entrapment in the foot and ankle. Interdigital neuroma is most frequent between the ages of 25 and 50 years and 80% of cases occur in women. Interdigital neuroma is most commonly unilateral and solitary, but bilaterality has been reported in up to 27% of patients and multiple synchronous lesions, as in the present case, may uncommonly be encountered. Patients typically complain of pain at the distal metatarsal interspace. The pain may radiate distally and transversely through the toes. A palpable mass is often not detected and, in some cases, the pain is more poorly localized. The pain may have a more throbbing, burning, or ''electric'' character, and is typically made worse with ambulation and relieved by removing the shoe and massaging the affected area.

The interdigital neuroma most often arises at the third intermetatarsal space between the metatarsophalangeal joints and plantar to the deep transverse metatarsal ligament. The etiology of interdigital neuroma is unproved but it is currently considered as an entrapment neuropathy that develops secondary to repetitive trauma and fibrous degeneration of the nerve. It has been postulated that dorsiflexion of the toes, a cavus arch, and walking in high-heeled shoes cause the interdigital nerve to be caught under the distal edge of the transverse ligament (Fig. 24D). A distended intermetatarsal bursa may also be contributory. The third nerve is the most vulnerable to compression because it is larger, formed by a conjunction of the medial and lateral plantar nerves, and is relatively fixed in position (Fig. 24E). Repeated trauma produces inflammation and fibrosis of the nerve and of the digital vessels. Surgery is often performed when conservative measures fail. Minimal or no improvement has been reported postoperatively in 13% to 14% of patients and residual tenderness and numbness were identified in roughly two-thirds of patients following surgical excision. These results, however, refer to series in which no definitive preoperative diagnosis was established, and it is unclear to what extent clinical failures represent cases in which no definite neuroma was demonstrated at surgery.

The MR diagnosis of interdigital neuroma is best accomplished in the coronal or axial plane using a small (10–12 cm) field of view (6–8). On spin echo sequences, the lesion demonstrates low to intermediate signal intensity on T1-weighted and proton density images and low signal intensity on T2-weighted sequences, a finding most likely reflecting the dense fibrous composition of the lesion. The lesions are better demonstrated on T1-weighted images than on T2-weighted sequences where they often appear isointense or slightly hypointense to surrounding fat. Fat-suppressed gadolinium-enhanced sequences allow enhanced visualization of neuromas as compared to conventional MR images and in our experience provide the best assessment for interdigital neuroma (6,8).

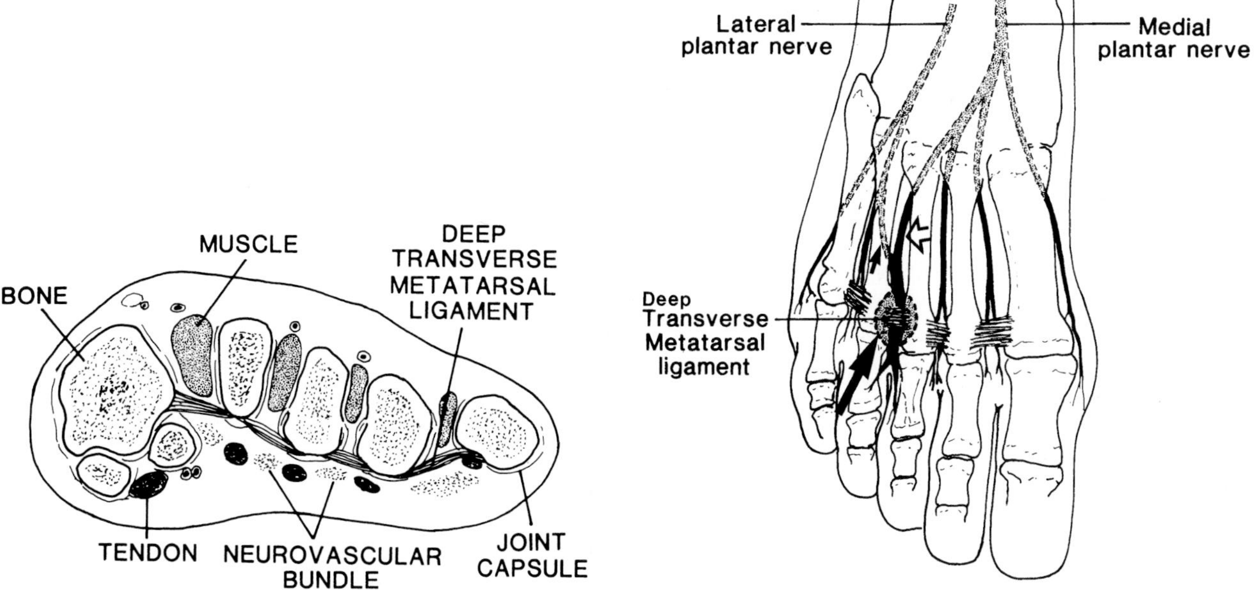

FIG. 24. D: Coronal diagram at the level of the metatarsophalangeal joints. The deep transverse metatarsal ligament blends with the planar metarsophalangeal joint capsules and forms a septum between the metatarsal heads and the plantar compartment. With dorsiflexion of the toes, the digital nerve may be compressed between the ligament and the plantar surface. **E:** Anatomic basis of interdigital neuroma at third interspace. The third digital branch of the medial plantar nerve (open arrow) and a communicating branch of the common digital nerve *(small arrow)* merge to form the third digital nerve. Increased size and lack of mobility predispose this nerve to the development of interdigital neuroma *(large arrow)* under the deep transverse metatarsal ligament. (From Kerr R. Spectrum of disorders. In: Deutsch AL, Mink JH, Kerr R, eds. *MRI of the foot and ankle.* New York: Raven Press, 1992;345–365, with permission.)

REFERENCES

Case 1

1. Funk DA, Cass JR, Johnson K. Acquired adult flatfoot secondary to posterior tibial tendon pathology. *J Bone Joint Surg* 1986;68-A:95–102.
2. Johnson KA. Tibialis posterior tendon rupture. *Clin Orthop Rel Res* 1983;177:140–147.
3. Rosenberg ZS, Cheung Y, Jahss MH, et al. Rupture of the posterior tibial tendon: CT and MR imaging with surgical correlation. *Radiology* 1988;169:229–235.
4. Frey C, Shereff M. Tendon injuries about the ankle in athletes. *Clin Sports Med* 1988;7:103–117.
5. Frey C, Shereff M, Greenidge N. Vascularity of the posterior tibial tendon. *J Bone J Surg* 1990;72-A:84–88.
6. Klein MA. Reformatted three-dimensional Fourier transform gradient-recalled echo MR imaging of the ankle: spectrum of normal and abnormal findings. *AJR* 1993;161:831–836.
7. Fullerton GD, Cameron IL, Ord VA. Orientation of tendons in the magnetic field and its effect on T2 relaxation times. *Radiology* 1985; 155:433–435.
8. Mink JH. Tendons. In: Deutsch AL, Mink JH, Kerr R, eds. *MRI of the foot and ankle.* New York: Raven Press, 1992;150–158.
9. Schweitzer ME, Caccese R, Karasick D, et al. Posterior tibial tendon tears: utility of secondary signs for MR imaging diagnosis. *Radiology* 1993;188:655–659.

Case 2

1. Resnick D. Tumors and tumor-like lesions in or about joints. In: Resnick D, ed. *Diagnosis of bone and joint disease,* 3rd ed. Philadelphia: WB Saunders, 1995;3957–3959.
2. Milgram JW. Synovial osteochondromatosis. A histopathological study of thirty cases. *J Bone Joint Surg* 1977;59:792–801.
3. Murphy FB, Dahlin DC, Sullivan R. Articular synovial chondromatosis. *J Bone Joint Surg [Am]* 1962;44:77–86.
4. Mussey RD, Henderson MS. Osteochondromatosis. *J Bone Joint Surg [Am]* 1949;31:619–624.
5. Kramer J, Recht M, Deely DM, et al. MR appearance of idiopathic synovial chondromatosis. *J Comput Assist Tomogr* 1993;17(5):772–776.
6. Burnstein MI, Fisher DR, Yandow DR, et al. Case report 502. *Skel Radiol* 1988;17:458–461.
7. Tuckman G, Wirth CZ. Synovial osteochondromatosis of the shoulder: MR findings. *J Comput Assist Tomogr* 1989;13:360–361.
8. Blandino A, Salvi L, Chrico G, et al. Synovial osteochondromatosis of the ankle: MR findings. *Clin Imaging* 1992;16:34–36.
9. Sanchez RB, Quinn SF. MRI of inflammatory synovial processes. *Magn Reson Imaging* 1989;7:529–540.

Case 3

1. Mink JH. Ligaments of the ankle. In: Deutsch AL, Mink JH, Kerr R, eds. *MRI of the foot and ankle.* New York: Raven Press, 1992; 173–178.
2. Kelikian H, Kelikian A. *Disorders of the ankle.* Philadelphia: WB Saunders, 1985.
3. Perlman M, Leveille D, DeLeonibus J, et al. Inversion lateral ankle trauma: differential diagnosis, review of the literature, and prospective study. *J Foot Surg* 1987;26(2):95–133.
4. Klein, MA. MR imaging of the ankle: normal and abnormal findings in the medial collateral ligament. *AJR* 1994;162:377–383.
5. Kneeland JB, Macrandar S, Middleton WE, et al. MR imaging of the normal ankle: correlation with anatomic sections. *AJR* 1988;151: 117–123.
6. Kier R, Dietz M, McCarthy S, et al. MR imaging of the normal ligaments and tendons of the ankle. *J Comput Assist Tomogr* 1991; 15:477–482.
7. Schneck CD, Mesgarzadeh M, Bonakdarpour A, et al. MR imaging of the most commonly injured ankle ligaments. Part 1. Normal anatomy. *Radiology* 1992;184:499–506.
8. Schneck CD, Mesgarzadeh M, Bonakdarpour A. MR imaging of the most commonly injured ankle ligaments. Part II. Ligament injuries. *Radiology* 1992;184:507–512.
9. Siegler S, Block J, Schneck CD. The mechanical characteristics of the collateral ligaments of the human ankle joint. *Foot Ankle* 1988; 8:234–242.

Case 4

1. Bernstein RM. Spontaneous rupture of the tibialis anterior tendon. *Am J Orthop* 1995;24(4):354–356.
2. Burman MS. Subcutaneous rupture of the tendon of the tibialis anticus. *Ann Surg* 1934;100:368–372.
2. Barnett T, Hammond NL. Anterior tibial tendon rupture. *Contemp Orthop* 1991;23(4):365–367.
3. Mensor MC, Ordway GL. Traumatic subcutaneous rupture of the tibialis anterior tendon. *J Bone Joint Surg [Am]* 1953;35(30):675–680.
4. Lapidus PW. Indirect subcutaneous rupture of the anterior tibial tendon. *Bull Hosp Joint Dis Orthop Inst* 1941;2:119–127.
5. Moberg E. Subcutaneous rupture of the tendon of the tibialis anterior muscle. *Acta Chir Scand* 1947;95:455–460.

Case 5

1. Mink JH. Ligaments of the ankle. In: Deutsch AL, Mink JH, Kerr R, eds. *MRI of the foot and ankle.* New York: Raven Press, 1992; 173–178.
2. Perlman M, Leveille D, DeLeonibus J, et al. Inversion lateral ankle trauma: differential diagnosis, review of the literature, and prospective study. *J Foot Surg* 1987;26(2):95–133.
3. Kannus P, Renstrom P. Treatment for acute tears of the lateral ligaments of the ankle. *J Bone Joint Surg* 1991;73A:305–312.
4. Smith R, Reischl S. Treatment of ankle sprains in young athletes. *Am J Sports Med* 1986;14(6):465–471.
5. Andrish JT. Ligament injuries of the knee. *Orthop Clin* 1985;16(2): 273–284.
6. Berquist T, Johnson K. Trauma. In: Berquist TH, ed. *Radiology of the foot and ankle.* New York: Raven Press, 1989.
7. Hamilton WG. Foot and ankle injuries in dancers. *Clin Sports Med* 1988;7(1):143–173.
8. Nicholas JA, Hershman EB. *The lower extremity and spine in sports medicine.* St. Louis: CV Mosby, 1986.
9. Kelikian H, Kelikian A. *Disorders of the ankle.* Philadelphia: WB Saunders, 1985.
10. Beltran J, Munchow AM, Khabiri H, et al. Ligaments of the lateral aspect of the ankle and sinus tarsi: an MR imaging study. *Radiology* 1990;177:455–458.
11. Erickson SJ, Smith JW, Ruiz ME, et al. MR imaging of the lateral collateral ligament of the ankle. *AJR* 1991;156:131–136.
12. Kneeland JB, Macrandar S, Middleton WE, et al. MR imaging of the normal ankle: correlation with anatomic sections. *AJR* 1988; 151:117–123.
13. Kier R, Dietz M, McCarthy S, et al. MR imaging of the normal ligaments and tendons of the ankle. *J Comput Assist Tomogr* 1991; 15:477–482.
14. Schneck CD, Mesgarzadeh M, Bonakdarpour A, et al. MR imaging of the most commonly injured ankle ligaments. Part 1. Normal anatomy. *Radiology* 1992;184:499–506.
15. Schneck CD, Mesgarzadeh M, Bonakdarpour A. MR Imaging of the most commonly injured ankle ligaments. Part II. Ligament injuries. *Radiology* 1992;184:507–512.

Case 6

1. Sanerkin NG, Mott MG, Roylnace J. An unusual intraosseous lesion with fibroblastic, osteoblastic, aneursymal and fibromysositic elements: ''solid'' variant of aneurysmal bone cyst. *Cancer* 1983;51: 2278–2288.
2. Mirra JM. Cysts and cystlike lesions. In: Mirra JM, ed. *Bone tumors: clinical, radiologic, and pathologic correlations.* Philadelphia: Lea and Febiger, 1989;1306–1309.

3. Kransdorf MJ, Sweet DE. Aneursymal bone cyst: concept, controversy, clinical presentation, and imaging. *AJR* 1995;164:573–580.
4. Bertoni F, Bacchini R, Capanna R, et al. Solid variant of aneurysmal bone cyst. *Cancer* 1993;71:729–734.
5. Munk PL, Helms CA, Holt RG, et al. MR imaging of aneurysmal bone cysts. *AJR* 1989;153:99–101.

Case 7

1. Hamilton WG. Foot and ankle injuries in dancers. *Clin Sports Med* 1988;7(1):143–173.
2. James S, Bates B, Ostering L. Injuries to runners. *Am J Sports Med* 1978;6:40–50.
3. Clement P, Taunton J, Smart G. Achilles tendinitis and peritendinitis: etiology and treatment. *Am J Sports Med* 1984;12:179–184.
4. Quinn SF, Murray WT, Clark RA, et al. Achilles tendon: MR imaging at 1.5 T. *Radiology* 1987;164:767–770.
5. Keene JS, Lash EG, Fisher DR, et al. Magnetic resonance imaging of Achilles tendon ruptures. *Am J Sports Med* 1989;17(3):333–337.
6. Mink JH. Tendons. In: Deutsch AL, Mink JH, Kerr R, eds. *MRI of the foot and ankle.* New York: Raven Press, 1992.
7. Mink JH, Deutsch AL, Kerr R. Tendon injuries of the lower extremity: magnetic resonance assessment. *Top Magn Reson Imaging* 1991;3(4):23–38.
8. Kier R, Dietz MJ, McCarthy SM, et al. MR imaging of the normal ligaments and tendons of the ankle. *J Comput Assist Tomogr* 1991; 15(3):477–482.
9. Schweitzer ME. Magnetic resonance imaging of the foot and ankle. *Magn Reson Q* 1993;9(4):214–234.

Case 8

1. Deutsch AL, Resnick D, Campbell G. Computed tomography and bone scintigraphy in the evaluation of tarsal coalition. *Radiology* 1982;144:137–140.
2. Ozonoff MB. The foot. In: Ozonoff MB, ed. *Pediatric orthopedic radiology.* Philadelphia: WB Saunders, 1992;440–446.
3. Pachuda NM, Lasday SD, Jay RM. Tarsal coalition: etiology, diagnosis, and treatment. *J Foot Surg* 1990;29:474–488.
4. Wechsler RJ, Karasick D, Schweitzer ME. Computed tomography of talocalcaneal coalition: imaging techniques. *Skel Radiol* 1992; 21:353–358.
5. Goldman AB, Pavlov H, Schneider R. Radionuclide bone scanning in subtalar coalitions: differential considerations. *AJR* 1982;138: 427–432.
6. Munk PL, Vellet AD, Levin MF, et al. Current status of magnetic resonance imaging of the ankle and hindfoot. *Can Assoc Radiol J* 1992;43:19–30.
7. Wechsler RJ, Schweitzer ME, Deely DM. Tarsal coalition: depiction and characterization with CT and MR imaging. *Radiology* 1994; 193:447–452.

Case 9

1. Mink JH. Ligaments of the ankle. In: Deutsch AL, Mink JH, Kerr R, eds. *MRI of the foot and ankle.* New York: Raven Press, 1992; 178–185.
2. Ferkel R, Fischer S. Progress in ankle arthroscopy. *Clin Orthop Rel Res* 1988;240:210–220.
3. Bassett F, Gates H, Billys J, et al. Talar impingement by the anteroinferior tibiofibular ligament. *J Bone Joint Surg* 1990;72A:55–59.
4. McCarrol J, Schrader J, Shelbourne K, et al. Meniscoid lesions of the ankle in soccer players. *Am J Sports Med* 1987;15:255–257.
5. Martin D, Baker C, Curl W, et al. Operative ankle arthroscopy. Long term follow-up. *Am J Sports Med* 1989;17(1):16–23.
6. Martin D, Curl W, Baker C. Arthroscopic treatment of chronic synovitis of the ankle. *Arthroscopy* 1989;5(2):110–114.

Case 10

1. Mink JH. Tendons. In: Deutsch AL, Mink JH, Kerr R, eds. *MRI of the foot and ankle.* New York: Raven Press, 1993;160–171.
2. Kelikin H. Disruption and dislocation of some tendon. In: Kelikin H, ed. *Disorders of the ankle.* Philadelphia: WB Saunders, 1987; 759–791.
3. Edwards M. The relations of the peroneal tendons to the fibula, calcaneus, and cuboideum. *Am J Anat* 1928;42:213–253.
4. Eckert WR, Davis EA. Acute rupture of the peroneal retinaculum. *J Bone Joint Surg* [*Am*] 1976;58(5):670–673.
5. Munk RL, Davis PH. Longitudinal rupture of the peroneus brevis tendon. *J Trauma* 1976;16:803–806.
6. Erickson SJ, Cox IH, Hyde JS, Carrera GF, Strandt JA, Estdowski LD. Effect of tendon orientation on MR imaging signal intensity: a manifestation of the "magic angle" phenomenon. *Radiology* 1991;181:389–392.
7. Erickson SJ, Prost RW, Timins ME. The "magic angle" effect: background physics and clinical relevance. *Radiology* 1993;188: 23–25.
8. Link SC, Erickson SJ, Timins ME. MR imaging of the ankle and foot: normal structures and anatomic variants that may simulate disease. *AJR* 1993;161:607–612.
9. Schweitzer ME, van Leersum M, Ehrlich SS, et al. Fluid in normal and abnormal ankle joints: amount and distribution as seen on MR images. *AJR* 1994;162:111–114.

Case 11

1. Kerr R. Spectrum of disorders. In: Deutsch AL, Mink JH, Kerr R, eds. *MRI of the foot and ankle.* New York: Raven Press, 1992; 355–358.
2. Pfeffer GB, Baxter DE. Surgery of the adult heel. In: Jahss MH, ed. *Disorders of the foot and ankle.* Philadelphia: WB Saunders, 1991;1396–1416.
3. Sundberg SB, Johnson KA. Painful conditions of the heel. In: Jahss MH, ed. *Disorders of the foot and ankle.* Philadelphia: WB Saunders, 1991;1382–1395.
4. DuVries HL. Heel spur (clacanela spur). *Arch Surg* 1957;74:536–542.
5. Schepsis AA, Leach RE, Corzyca J. Plantar fasciitis. Etiology, treatment, surgical results, and review of the literature. *Clin Orthop Rel Res* 1991;266:185–196.
6. Williams RL, Smibert JG, Cox R, et al. Imaging study of the painful heel syndrome. *Foot Ankle* 1987;7:345–349.
7. Clancy WG Jr. Tendinitis and plantar fasciitis in runners. *Orthopedics* 1983;6:217–233.
8. Hicks JH. The mechanics of the foot. II. The plantar aponeurosis and the arch. *J Anat* 1954;88:25–30.
9. Baxter DE, Thigpen CM. Heel pain operative results. *Foot Ankle* 1984;5:16–25.
10. Furey JG. Plantar fasciitis. The painful heel syndrome. *J Bone Joint Surg* 1975;57A:672–673.
11. Graham CE. Painful heel syndrome: rational of diagnosis and treatment. *Foot Ankle* 1983;3:261–267.
12. Snider MP, Clancy WG Jr, McBeath AA. Plantar fascia release for chronic plantar fasciitis in runners. *Am J Sports Med* 1983;11:215–219.
13. Kwong PK, Kay D, Voner RT, et al. Plantar fasciitis: mechanics and pathomechanics of treatment. *Clin Sports Med* 1988;7:119–126.
14. Amis J, Jenning L, Graham D, et al. Painful heel syndrome: radiographic and treatment assessment. *Foot Ankle* 1988;9:91–95.
15. Tanz SS. Heel pain. *Clin Orthop* 1963;28:169–177.
16. Rubin G, Witton M. Plantar calcaneal spurs. *Am J Orthop* 1963;5: 38–41.
17. Sewell JR, Black CM, Chapman AH, et al. Quantititive scintigraphy in diagnosis and management of plantar fasciitis: concise communication. *J Nucl Med* 1980;21:633–636.
18. Berkowitz JF, Kier R, Rudicel S. Plantar fasciitis: MR imaging. *Radiology* 1991;179:665–667.

Case 12

1. Mink JH, Deutsch AL. Occult cartilage and bone injuries of the knee: detection, classification, and assessment with MR imaging. *Radiology* 1989;170:823–829.
2. Stafford SA, Rosenthal DI, Gebhardt MC, et al. MRI in stress fracture. *AJR* 1986;147:553–556.
3. Unger EC, Summers TB. Bone marrow. *Top Magn Reson Imaging* 1989;1(4):31–52.
4. Greaney RB, Gerber FH, Laughlin RL. Distribution and natural history of stress fractures in US Marines. *Radiology* 1983;146:339–346.
5. Roub LW, Gummerman LW, Hanley ENJ. Bone stress: radionuclide imaging perspective. *Radiology* 1979;132–141.
6. Pavlov H, Torg JS, Freiberger RH. Tarsal navicular stress fractures: radiographic evaluation. *Radiology* 1983;148:641–645.
7. Torg JS, Pavlov H, Cooley LH, et al. Stress fractures of the tarsal navicular. A retrospective review of twenty-one cases. *J Bone Joint Surg* 1982;63A:700–712.
8. Yao L, Lee JK. Occult intraosseous fracture: detection with MR imaging. *Radiology* 1988;167:749–751.
9. Berger PE, Ofstein RA, Jackson DW, et al. MRI demonstration of radiographically occult fractures: What have we been missing? *Radiographics* 1989;9:407.
10. Lee JK, Yao L. Stress fractures: MR imaging. *Radiology* 1988;169:217.
11. Deutsch AD, Mink JH. Magnetic resonance imaging of musculoskeletal injuries. *Radiol Clin North Am* 1989;27:983–1002.
12. Deutsch AL, Mink JH, Waxman AD. Occult fractures of the proximal femur: MR imaging. *Radiology* 1989;170:113–116.
13. Simon JH, Szumowski J. Chemical shift imaging with paramagnetic contrast material enhancement for improved lesion depiction. *Radiology* 1989;171:538–543.
14. Mink JH, Deutsch AL, ed. *MRI of the musculoskeletal system. A teaching file.* New York: Raven Press, 1990.
15. Greenfield GB, Warren DL, Clark RA. MR imaging of periosteal and cortical changes of bone. *Radiographics* 1991;11:611–623.

Case 13

1. Mink JH. Ligaments of the ankle. In: Deutsch AL, Mink JH, Kerr R, eds. *MRI of the foot and ankle.* New York: Raven Press, 1992; 192–194.
2. Klein MA, Spreitzer AM. MR imaging of the tarsal sinus and canal: normal anatomy, pathologic findings and features of the sinus tarsi sundrome. *Radiology* 1993;186:233–240.
3. Beltran J, Munchow AM, Khabiri H, et al. Ligaments of the lateral aspect of the ankle and sinus tarsi: an MR imaging study. *Radiology* 1990;177:455–458.
4. Rosenberg AS, Cheung Y. Diagnostic imaging of the ankle and foot. In: Jahss MH, ed. *Disorders of the foot and ankle.* Philadelphia: WB Saunders, 1991;P109–154.
5. Resnick D. Radiology of the talocalcaneal articulations. *Radiology* 1974;111:581–586.
6. Cahill DR. The anatomy and function of the contents of the human tarsal sinus and canal. *Anat Rec* 1965;153:1–18.
7. Kier R, Dietz M, McCarthy S, et al. MR imaging of the normal ligaments and tendons of the ankle. *J Comput Assist Tomogr* 1991; 15:477–482.
8. Kjaersgaard-Anderson P, Wethelund JO, Helmig P, et al. The stabilizing effect of the ligamentous structures in the sinus and canalis tarsi on movement in the hindfoot. *Am J Sports Med* 1988;16(5): 512–516.
9. Reinherz RP, Sink CA, Krell B. Exploration into the pathologic sinus tarsi. *J Foot Surg* 1989;28(2):137–140.
10. Clanton T. Instability of the subtalar joint. *Orthop Clin North Am* 1989;20:583–591.

Case 14

1. Lawson JP, Ogden JA, Sella EJ, et al. The painful accessory navicular. *Skel Radiol* 1984;12:250–262.
2. Miller TT, Staron RB, Feldman F, et al. The symptomatic accessory tarsal navicular bone: assessment with MR imaging. *Radiology* 1995;195:849–853.
3. Zadek I, Gold AM. The accessory tarsal navicular scaphoid. *J Bone Joint Surg [Am]* 1948;30:957–968.
4. Mygrind HB. The accessory tarsal scaphoid: clinical features and treatment. *Acta Orthop Scand* 1953;23:142–151.
5. Sella EJ, Lawson JP, Ogden JA. The accessory navicular synchondrosis. *Clin Orthop Rel Res* 1986;209:280–285.
6. Lawson JP. Clinically significant radiologic anatomic variants of the skeleton. *AJR* 1994;163:249–255.

Case 15

1. Nistor L. Nonsurgical treatment of Achilles tendon ruptures. *J Bone Joint Surg [Am]* 1981;63:394.
2. Inglis AE, Scott N, Sculco TP, et al. Rupture of the tendon Achilles. An objective assessment of surgical and non-surgical treatment. *J Bone Joint Surg [Am]* 1976;58:990–993.
3. Mann R, Holmes GB, Seale KS, et al. Chronic rupture of the Achilles tendon: a new technique of repair. *J Bone Joint Surg [Am]* 1991; 73:214–217.
4. Kato YP, Dunn MG, Zawadsky JP, et al. Regeneration of Achilles tendon with a collagen tendon prosthesis. *J Bone Joint Surg [Am]* 1991;73(4):561–574.
5. Liem MD, Zegel HG, Balduini FC, et al. Repair of Achilles tendon ruptures with a polylactic acid implant: assessment with MR imaging. *AJR* 1991;156:768–773.
6. Bradley JP, Tibone JE. Percutaneous and open surgical repairs of Achilles tendon ruptures. *Am J Sports Med* 1990;18(2):118–196.
7. Ma GW, Griffith TG. Percutaneous repair of acute closed ruptured Achilles tendon. *Clin Orthop* 1977;128:247–255.
8. Dillon E, Pope C, Barber V, et al. Achilles tendon healing: 12 month follow-up with MR imaging. *Radiology* 1990;177:306.
9. Gillogly SD, Schaefer RA, Rak KM, et al. The accuracy of magnetic resonance imaging in assessment of patellar tendon autograft anterior cruciate ligament reconstruction. In: Anaheim, CA: American Academy of Orthopedic Surgeons, 1991.
10. Totterman SM, Adams MJ, Langhans MJ, et al. MR imaging of knee after reconstruction of anterior cruciate ligament. 77th Scientific Assembly and Annual Meeting Radiological Society of North America, Chicago, 1991.
11. Mink JH, Deutsch AL, Kerr R. Tendon injuries of the lower extremity. *Top Magn Reson Imaging* 1991;3(4):23–38.
12. Mink JH. Tendons. In: Deutsch AL, Mink JH, Kerr R, eds. *MRI of the foot and ankle.* New York: Raven Press, 1992.

Case 16

1. Resnick D. Gouty arthritis. In: Resnick D, ed. *Diagnosis of bone and joint disorders,* 3rd ed. Philadelphia: WB Saunders, 1995; 1511–1539.
2. Ruiz ME, Erickson SJ, Carrera GF, et al. Monoarticular gout following trauma: MR appearance. *J Comput Assist Tomogr* 1993;17(1): 151–153.
3. Image Interpretation Session. RSNA 1991. *Radiographics* 1991;11: 150–152.

Case 17

1. Strickland JW. Flexor tendon injuries: treatment principles. *Am J Orthop Surg* 1995;3(1):44–54.
2. Strickland JW. Flexor tendon injuries: operative technique. *J Am Acad Orthop Surg* 1995;3:55–62.
3. Bishop AT, Cooney WP III, Wood MB. Treatment of partial flexor tendon lacerations: the effect of tenorrhaphy and early protected mobilization. *J Trauma* 1986;26:301–312.

Case 18

1. Deutsch AL. Osteochondral injuries of the talar dome. In: Deutsch AL, Mink JH, Kerr R, eds. *MRI of the foot and ankle.* New York: Raven Press, 1993;118–134.

2. Berndt AL, Harty M. Transchondral fractures (osteochondritis dissecans) of the talus. *J Bone Joint Surg* 1959;41A:988–1020.
3. Pappas AM. Osteochondrosis dissecans. *Clin Orthop* 1981;158:59–69.
4. Flick AB, Gould N. Osteochondritis dissecans of the talus (transchondral fractures of the talus): review of the literature and new surgical approach for medial dome lesions. *Foot Ankle* 1985;5(4): 165–185.
5. Shelton ML, Pedowitz WJ. Injuries to the talar dome, subtalar joint, and mid foot. In: Jahss MH, ed. *Disorders of the foot and ankle. Medical and surgical management.* Philadelphia: WB Saunders, 1991;2274–2292.
6. Bauer M, Jonsson K, Linden B. Osteochondritis dissecans of the ankle. A 20-year follow-up study. *J Bone Joint Surg* 1987;69(1): 93–96.
7. Canale ST, Belding RH. Osteochondral lesions of the talus. *J Bone Joint Surg [Am]* 1980;62:97–102.
8. Lindholm TS, Osterman K, Vankka E. Osteochondritis dissecans of elbow, ankle, and hip: a comparison survey. *Clin Orthop* 1980; 148:245–253.
9. Pritsch M, Horoshovski H, Farine I. Arthroscopic treatment of osteochondral lesions of the talus. *J Bone Joint Surg [Am]* 1986;68: 862–865.
10. Nelson DW, DiPaola J, Colville M, et al. Osteochondritis dissecans of the talus and knee: prospective comparison of MR and arthroscopic classifications. *J Comput Assist Tomogr* 1990;14(5):804–808.
11. De Smet AA, Fisher DR, Bernstein MI, et al. Value of MR imaging in staging osteochondral lesions of the talus (osteochondritis dissecans): results in 14 patients. *AJR* 1990;154:555–558.
12. Yulish BS, Mulopulos GP, Goodfellow DB, et al. MR imaging of osteochondral lesion of talus. *J Comput Assist Tomogr* 1987;11(2): 296–301.
13. Anderson IF, Chichton KJ, Grattan-Smith T, et al. Osteochondral fractures of the dome of the talus (abstr). *Radiology* 1990;174:902.
14. Mesgarzadeh M, Sapega AA, Bonakdarpour A, et al. Osteochondritis dissecans: analysis of mechanical stability with radiography, scintigraphy, and MR imaging. *Radiology* 1987;165:775–780.
15. De Smet AA, Fisher DR, Graf BK, et al. Osteochondritis dissecans of the knee: value of MR imaging in determining lesion stability and the presence of articular cartilege defects. *AJR* 1990;155:549.
16. Peterfy CG, Majumdar S, Lang P, et al. MR imaging of the arthritic knee: improved discrimination of cartilage, synovium, and effusion with pulsed saturation transfer and fat suppressed T1-weighted techniques. *Radiology* 1994;191:413–419.
17. Kramer J, Stiglbauer R, Engel A, et al. MR contrast arthrography (MRA) in osteochondritis dissecans. *J Comput Assist Tomogr* 1992; 16(2):254–260.
18. Adam G, Buhne M, Pescher A, et al. Stability of osteochondral fragments of the femoral condyle: magnetic resonance imaging with histopathologic correlation in an animal model. *Skel Radiol* 1991; 20:601.

Case 19

1. Morrison WB, Schweitzer ME, Bock GW, et al. Diagnosis of osteomyelitis: utility of fat-suppressed contrast-enhanced MR imaging. *Radiology* 1993;189:251–257.
2. Totty WG. Radiographic evaluation of osteomyelitis using magnetic resonance imaging. *Orthop Rev* 1989;18:587–592.
3. Erdman WA, Tamburro F, Jayson HT, et al. Osteomyelitis: characteristics and pitfalls of diagnosis with MR imaging. *Radiology* 1991; 180:533–539.
4. Tang JSH, Gold RH, Bassett LW, et al. Musculoskeletal infection of the extremities: evaluation with MR imaging. *Radiology* 1988; 166:205–209.
5. Quinn SF, Murray W, Clark RA. MR imaging of chronic osteomyelitis. *J Comput Assist Tomogr* 1988;12:113–117.
6. Mason MD, Zlatkin MB, Esterhai JL, et al. Chronic complicated osteomyelits of the lower extremity: evaluation with MR imaging. *Radiology* 1989;173:355–359.
7. Chandnani VP, Beltran J, Morris CS, et al. Acute experimental osteomyelitis and abscesses: detection with MR imaging versus CT. *Radiology* 1990;174:233–236.
8. Yuh WTC, Corson JD, Baraniewski HM, et al. Osteomyelitis of the foot in diabetic patients: evaluation with plain film, 99mTC MDP bone scintigraphy, and MR imaging. *AJR* 1989;152:795–800.
9. Unger E, Moldofsky P, Gatenby R, et al. Diagnosis of osteomyelitis by MR imaging. *AJR* 1988;150:605–610.
10. Mauer AH, Chen DCP, Camargo EE, et al. Utility of three-phase skeletal scintigraphy in suspected osteomyelitis: concise communication. *J Nucl Med* 1981;22:941–949.
11. Shafer RB, Edeburn CF. Can the three phase bone scan differentiate osteomyelitis from metabolic or metastatic bone disease. *Clin Nucl Med* 1984;9:373–377.
12. Gupta NC, Prezio JA. Radionuclide imaging in osteomyelitis. *Semin Nucl Med* 1988;18:287–299.
13. Post MJD, Sze G, Quencer RM, et al. Gadolinium enhanced MR in spinal infection. *J Comput Assist Tomogr* 1990;14:721–729.
14. Beltran J, Chandnani V, McGhee RA. Gadopentetate dimeglumine-enhanced MR imaging of the musculoskeletal system. *AJR* 1991; 156:457–466.
15. Ross JS, Delamarter R, Hueftle MG, et al. Gadolinium DTPA enhanced MR imaging of the postoperative lumbar spine: time course and mechanism of enhancement. *AJRN* 1989;10:37–46.
16. Morrison WB, Schweitzer ME, Wapner KL, et al. Osteomyelitis in feet of diabetics: clinical accuracy, surgical utility, and cost-effectiveness of MR imaging. *Radiology* 1995;196:557–564.

Case 20

1. Kerr R. Spectrum of disorders. In: Deutsch AL, Mink JH, Kerr R, eds. *MRI of the foot and ankle.* New York: Raven Press, 1992; 345–351.
2. Kerr R, Frey C. MR imaging in tarsal tunnel syndrome. *J Comput Assist Tomogr* 1991;15:280–286.
3. Sarrafian SK. *Anatomy of the Foot and Ankle. Descriptive, Topographic, Functional.* Philadelphia: JB Lippincott, 1983;118–128.
4. Dellon AL, Mackinnon SE. Tibial nerve branching in the tarsal tunnel. *Arch Neurol* 1984;41:645–646.
5. Havel PE, Ebraheim NA, Clark SE, et al. Tibial nerve branching in the tarsal tunnel. *Foot Ankle* 1988;9:117–119.
6. Zeiss J, Fenton P, Ebraheim N, et al. Normal magnetic resonance anatomy of the tarsal tunnel. *Foot Ankle* 1990;10:214–218.
7. Edwards WG, Lincoln CR, Bassett FH III, et al. The tarsal tunnel syndrome. Diagnosis and treatment. *JAMA* 1969;207:716–720.
8. Janecki CJ, Dovberg JL. Tarsal tunnel syndrome caused by neurilemoma of the medial plantar nerve. *J Bone Joint Surg* 1977;59A: 127–128.
9. Keck C. The tarsal tunnel syndrome. *J Bone Joint Surg* 1962;44A: 180–182.
10. Sinscheid RL, Burton RC, Fredericks EJ. Tarsal-tunnel syndrome. *South Med J* 1970;63:1313–1332.
11. Radin EL. Tarsal tunnel syndrome. *Clin Orthop* 1983;181:167–170.

Case 21

1. Resnick D. Osteochondroses. In: Resnick D, ed. *Diagnosis of bone and joint disorders,* 3rd ed. Philadelphia: WB Saunders, 1995; 3559–3606.
2. Siffert RS. Classification of the osteochondroses. *Clin Orthop* 1981; 158:10.
3. Karp MG. Kohler's disease of the tarsal scaphoid. An end-result study. *J Bone Joint Surg* 1937;19:84.
4. Kidner FC, Muro F. Kohlers disease of the tarsal scaphod or os naviculare pedis retardatum. *JAMA* 1924;83:1650.

Case 22

1. Mink JH. Ligaments of the ankle. In: Deutsch AL, Mink JH, Kerr R, eds. *MRI of the foot and ankle.* New York: Raven Press, 1992; 185–189.
2. Kelikian H, Kelikian A. *Disorders of the ankle.* Philadelphia: WB Saunders, 1985.

3. Sclafani S. Ligamentous injury of the lower tibiofibular syndesmosis: radiographic evidence. *Radiology* 1985;56:21–27.
4. Monk C. Injuries to the tibo-fibular ligaments. *J Bone Joint Surg* 1969;51B:330–337.
5. Fritschy D. An unusual ankle injury in top skiers. *Am J Sports Med* 1989;17(2):282.
6. Schneck CD, Mesgarzadeh M, Bonakdarpour A, et al. MR imaging of the most commonly injured ankle ligaments. Part I. Normal Anatomy. *Radiology* 1992;184:499–506.
7. Schneck CD, Mesgarzadeh M, Bonakdarpour A. MR imaging of the most commonly injured ankle ligaments. Part II. Ligament injuries. *Radiology* 1992;184:507–512.
8. Noto A, Cheung Y, Rosenberg Z, et al. MR imaging of the ankle: normal variants. *Radiology* 1989;170:121–124.
9. Erickson SJ, Smith JW, Ruiz ME, et al. MR imaging of the lateral collateral ligament of the ankle. *AJR* 1991;156:131–136.

Case 23

1. Erickson SJ, Cox IH, Hyde JS, Carrera GF, Strandt JA, Estdowski LD. Effect of tendon orientation on MR imaging signal intensity: a manifestation of the "magic angle" phenomenon. *Radiology* 1991;181:389–392.
2. Erickson SJ, Prost RW, Timins ME. The "magic angle" effect: background physics and clinical relevance. *Radiology* 1993;188: 23–25.
3. Link SC, Erickson SJ, Timins ME. MR imaging of the ankle and foot: normal structures and anatomic variants that may simulate disease. *AJR* 1993;161:607–612.
4. Berendsen HJC. Nuclear magnetic resonance study of collagen hydration. *J Chem Phys* 1962;36:3297–3305.
5. Fullerton GD, Cameron IL, Ord VA. Orientation of tendons in the magnetic field and its effect on T2 relaxation times. *Radiology* 1985; 155:433–435.
6. Peto S, Gilllis P, Henri VP. Structure and dynamics of water in tendon from NMR relaxation measurements. *Biophys J* 1990;57: 71–84.

Case 24

1. Kerr R. Spectrum of disorders. In: Deutsch AL, Mink JH, Kerr R, eds. *MRI of the foot and ankle.* New York: Raven Press, 1992; 345–365.
2. Addante JB, Peicott PS, Wong KY, et al. Interdigital neuroma. Results of surgical excision of 152 neuromas. *J Am Podiartr Med Assoc* 1986;76:493–495.
3. Bradley N, Miller WA, Evans JP. Plantar neuroma: analysis of results following surgical excision in 145 patients. *South Med J* 1976;69:853–854.
4. Reed RJ, Bliss BO. Morton neuroma. *Arch Pathol* 1973;95:123–129.
5. Alexander IJ, Johnson KA, Parr JW. Morton neuroma: a review of recent concepts. *Orthopedics* 1987;10:103–106.
6. Terk MR, Kwong PK, Suthar M, et al. Morton neuroma: evaluation with MR imaging performed with contrast enhancement and fat suppression. *Radiology* 1993;189:239–241.
7. Erickson SJ, Canale PB, Carrera GF. Interdigital neuroma; high resolution MR imaging with a solenoid coil. *Radiology* 1991;181: 833–836.
8. Unger HR, Mattoso PQ, Drusen MJ. Gadopentetate enhanced magnetic resonance imaging with fat saturation in the evaluation of Morton neuroma. *J Foot Surg* 1992;31(3):244–246.

CHAPTER 8

Muscle

James L. Fleckenstein and Andrew L. Deutsch

As a consequence of its sensitivity to detect changes in muscle water and fat content, as well as its ability to distinguish between individual deep and superficial muscles, MR imaging has contributed significant insights into the study of skeletal muscle over the last 10 years (1). Both the scientific as well as the pragmatic clinical information now available suggest a significant continuing role for MR in the investigation of muscle disorders.

Muscle is closely linked to the nervous system and many neurologic diseases have associated muscle pathology (1). The locomotor system, which propels us, can be divided into a proximal limb, in which volition initiates motion, and a distal limb, which powers translocation through space. The proximal limb, or neuromuscular unit, begins in the cortex and is transmitted via corticospinal tracts to anterior horn cells and via peripheral nerves to the motor end plates on muscle fibers (1). Each motor axon innervates multiple muscle fibers, ranging from a few in ocular muscles to hundreds in the quadriceps femoris muscle. The distal limb of the locomotor system, termed the musculoskeletal unit, proceeds from the muscle fiber to the myotendinous junction, tendon, and bone. MR imaging may offer insights into processes involving both limbs of the system.

MR imaging has been extensively applied both to the depiction of normal and deranged anatomy as well as to the demonstration and characterization of tissue abnormalities. MR can effectively differentiate between muscle, fat, and edema (1–4). Fat is detected as high signal on T1-weighted sequences as a consequence of its short T1 relaxation time. Increased tissue water leads to increased proton density, T1, and T2 relaxation times. The prolongation of T1 may be demonstrable as decreased signal intensity on T1-weighted images, but these images must be heavily T1 weighted to counteract the relative increase in signal intensity secondary to T2 and spin density (1). A potential pitfall may arise in the attempt to discriminate between muscle edema and fat on T2-weighted sequences (1). The T2 of fat is approximately 60 msec, twice that of normal skeletal muscle (30 msec). When there is enough edema to cause the muscle T2 time to exceed 60 msec, edematous muscle can be distinguished from fat. When the edema is only mild to moderate, however, so that the T2 is increased above the normal 30 msec but to less than 60 msec, the edematous changes may be indistinguishable from mild fatty change if only routine spin echo pulse sequences are employed.

The detection of muscle edema is enhanced by the use of fat-suppression techniques (1–4). The short tau inversion recovery (STIR) sequence has the advantage that edema-associated increases in muscle T1 and T2 relaxation times are generally additive to signal intensity, as is increased spin density (1–3). This accounts for the great conspicuity of edema utilizing this sequence. Widely available fast STIR sequences can effectively reduce scan times while maintaining sensitivity to edema and are nearly ideal for the surveillance of skeletal muscle. Additional operator-dependent choices influencing image quality include coil size and design, number of excitations, voxel size, and spatial resolution. The added signal to noise afforded by small coils allows higher spatial resolution images to be obtained without significant time penalty. Such high-resolution images may be necessary to demonstrate fine detail in subtle lesions (1). In situations in which tissue characterization is more important than fine anatomic detail, large voxels can be utilized to maximize signal to noise. The combination of small coils and large voxels reduces the number of excitations and phase-encoding steps and provides the potential for relatively fast, economical imaging for tissue characterization.

Specialized techniques, including exercise enhancement MR, allow the investigation of dynamic muscle physiology and pathophysiology (5–9). Anatomic and functional applications of exercise enhancement include verification of muscle boundaries and assessment of intersubject variations in normal anatomy and usage. As an example, exercise-enhanced MR has been able to identify finger-specific components of the deep and superficial finger flexors, an accomplishment that had previously not been attainable by dissection or any other technique (8,9). These MR-demonstrable changes appear to depend on

TABLE 1. *Neuromuscular disorders*

Primary muscle diseases
- Metabolic myopathies
  - Inherited—enzyme defects (e.g., McArdle's disease)
- Acquired—endocrinopathies (Cushing's syndrome, hypothyroidism), electrolyte disturbances, toxic/drug induced
- Inflammatory muscle disease
  - Infection (pyomyositis, viral polymyositis)
  - Idiopathic (polymyositis, dermatomyositis)
  - Collagen vascular diseases
- Muscular diseases with myotonia
  - Myotonia congenita
  - Dystrophic myotonia
  - Paramyotonia congenita
- Muscular dystophies
  - X linked (Duchenne, Becker)
  - Autosomal recessive (limb-girdle, childhood, congenital)
  - Fascioscapulohumeral
  - Distal
  - Ocular
  - Oculopharyngeal

Primary neurogenic disorders
- Diseases of motor neurons (MN)
  - Upper MN—primary lateral sclerosis
  - Upper and lower MN—amyotrophic lateral sclerosis
  - Lower MN—spinal muscular atrophy
  - Spinal cord injury
  - Peripheral nerve injury
- Diseases of the neuromuscular junction
  - Myasthenia gravis
  - Lambert-Eaton syndrome
  - Toxic (botulism, tick paralysis)

effects of intracellular (low level exercise) and extracellular water (maximal exercise) content (10,11). The exercise-induced changes also correlate with high rates of glycogenolysis, and lactate production and blood flow is not required for the effects to occur. It is likely that the relative hyperosmolarity of exercised muscle, produced by products of glycogenolysis, draws water out of the vasculature by diffusion, whether or not the blood is moving (12).

Neuromuscular etiologies of locomotor dysfunction can be classified on the basis of whether the primary lesion lies within the nerve itself, the neuromuscular junction, or primarily within the muscle (Table 1) The response to these wide varieties of conditions is limited. Denervation changes, necrosis, alterations in fiber size (i.e., atrophy or hypertrophy), and mesenchymal tissue changes (e.g., fatty replacement and fibrosis) are important histopathologic findings in these disorders (73). Alteration in muscle fiber size may be difficult to detect clinically. In some conditions (e.g., dystrophies) hypertrophic and atrophic muscles may be juxtaposed, so that no net alteration in muscle bulk is apparent on clinical examination (1). MR can be of particular value in making these type of determinations.

In the differential diagnosis of neuromuscular lesions, it is frequently important to determine if the pattern of involvement is focal, multifocal, or diffuse; proximal, distal, or both; and symmetric or asymmetric (14–18). Some diseases such as Duchenne's muscular dystrophy, classically progress from proximal to distal, in a symmetric fashion, with highly focal atrophy and hypertrophy involving entire muscles. In dystrophies, certain thigh muscles (i.e., gracilis and sartorius) are reliably spared or are hypertrophic until late in the disease (15–17). In myotonic dystrophy, focal hypertrophy and atrophy also occur, but the progression tends to be from distal to proximal muscle involvement. Chronic polymyositis is similarly characterized by proximal and symmetric atrophy and highly focal involvement has been reported. The reason for the highly focal involvement and sparing of muscles in these diseases is unknown (1). Interestingly, compensatory hypertrophy does not often occur in polymyositis and this feature may prove to have diagnostic specificity and allow it to be distinguished from the dystrophies. Highly asymmetric atrophy and hypertrophy is unusual but has been reported in poliomyelitis. Diffuse changes in muscle bulk may occur in endocrinopathies (13,19) Diffuse muscle atrophy may be seen in Cushing's syndrome and muscle hypertrophy has been reported in hypothyroidism (19).

In primary neurogenic disease, early histopathologic findings are characterized by denervation atrophy and, when reinvervation has occurred, by fiber angulation and fiber type grouping (19). MR is unique among imaging techniques in its ability to identify the early phase of denervation before fatty deposition. This denervation has been attributed to fiber atrophy with a resultant increase in extracellular water and prolongation of proton T1 and T2 relaxation times. These characteristic are in contrast to the end-stage appearance of denervation, in which T1 shortening from fat deposition is the dominant MR alteration (see Case 1). Fiber necrosis is common to many

TABLE 2. *Musculoskeletal disorders*

Trauma
- Exertional
  - Strains and ruptures
  - Contusion
  - Delayed-onset muscle soreness
- Other trauma
  - Crush injury
  - Burns
  - Iatrogenic trauma

Mass lesions arising in muscle
- Myositis ossificans
- Vascular malformations
- Neoplasms
- Hematomas
- Bacterial myositis

Congenital disorders
- Complex limb deformities
  - Hemimelias
  - Arthrogryposis multiplex congenita

conditions affecting skeletal muscle (19) Although present in dystrophies, polymyositis, and other conditions, necrosis is perhaps best associated with inherited glycolytic muscle enzyme deficiencies (such as McArdle's disease) in which relatively trivial exercise results in muscle contracture and rhabdomyolysis (19) (see Case 2). Inflammatory conditions, such as idiopathic polymyositis and dermatomyositis (see Case 10) also show necrosis histopathologically; however, inflammatory cell infiltration of the muscle is typically present on biopsy specimens (1,19). Fatty infiltration of the muscle appears to be a common result if the disease does not abate and the ability of MR to distinguish between reversible and irreversible damage is of considerable therapeutic importance (18).

MR can also effectively localize and characterize pathology involving the musculotendinous unit (Table 2). Exertional trauma including acute strains and contusions (see Case 5), delayed-onset muscle soreness (see Case 7), and chronic overuse syndromes can be elegantly depicted. The sequelae of muscle injury including compartment syndromes and myositis ossificans can be evaluated (see Case 11). Continuing experience with MR has facilitated use of the technique in characterizing a variety of soft tissue abnormalities (also see Chapter 10). Muscle masses including hematomas (see Case 11) and infectious and idiopathic myositis (see Case 6) can also be assessed. The cases selected for this section provide an overview of the application of MR to the detection and characterization of abnormalities involving the system of locomotion.

CASE 1

History: A 64-year-old newly diagnosed diabetic man presented with pain and weakness in the left thigh of several months' duration.

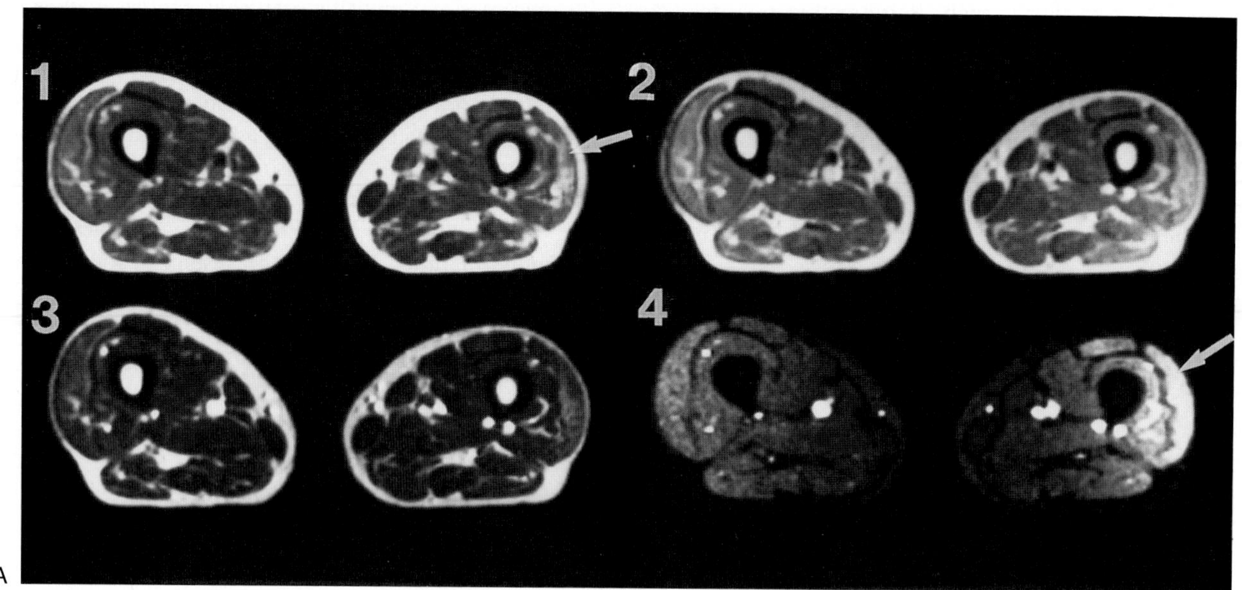

FIG. 1.

Findings: (1) Axial T1-weighted (TR/TE 500/30); (2) proton density (TR/TE 2,000/20); and (3) moderately T2-weighted (TR/TE 2,000/60) images show increased signal intensity within the left vastus lateralis, which is also slightly atrophic (arrows). Although fat likely contributes to the signal alteration, as judged by T1-weighted image (1), the highly edema-sensitive short tau inversion recovery (STIR) (TR/TE/TI 1500/30 msec/100) sequence (4) indicates that edema-like change is associated. Electromyography revealed denervation in the abnormal muscle. MR of the spine and pelvis showed only normal findings. Symptoms resolved over the subsequent 2 years with no additional treatment.

Diagnosis: Diabetic amyotrophy.

Discussion: Denervation of skeletal muscle results in a characteristic appearance on MR, appearing similar to edema in its early phases and fat in the late phases (1,2). The patient in the case example shows a mixture of both early and late denervation by showing both edema-like change (increased signal intensity on STIR) and fatty deterioration (increased signal intensity on T1-weighted image). These characteristics of denervated muscle are nonspecific, however, and are not frequently readily distinguished from similar changes that occur in other neuromuscular processes, especially myopathies. Only by combining the clinical data with the morphologic findings can specific diagnoses result.

In the patient presented here, the concurrence of diabetes mellitus and thigh pain and atrophy that are unilateral, together with absence of a compressive cause of radiculopathy, supported MR evidence of denervation of the vastus lateralis. Electromyography (EMG) was confirmatory of denervation. Spontaneous muscle infarction, another diabetic complication, typically presents with acute onset of severe muscle pain and swelling in patients with end-stage diabetes (3) (see Case 9). The MR features of atrophy and absence of mass effect provide persuasive evidence against an infarct or, for that matter, any mass lesion. Compression of the muscle's nerve supply could not be excluded until additional imaging tests were done to exclude pathologic involvement of the proximal nerve supply to the involved muscle (Fig. 1B).

MR studies of neuropathic muscle dysfunction are limited primarily to compressive and traumatic causes (1,2). In an initial study that established a broad time frame of signal changes in complete nerve transections, MR was reported to have a limited accuracy in detecting denervation-related alterations during the first few weeks following complete nerve transections due to the mild degree of signal changes that occur early on and to the occurrence of similar signal changes being caused by nearby wound edema (1). On the other hand, denervation occurring between 1 month and 1 year before the MR scan was performed was readily visible on edema-sensitive sequences, such as STIR, due to prolongation of proton T1 and T2 relaxation times (Fig. 1C).

A subsequent study more enthusiastically promoted the use of STIR as a technique to detect denervation earlier than EMG (2). That study emphasized the ability of MR to distinguish neuropraxia (myelin sheath damage with intact axon) from the more immediately serious condition of axonotemesis (complete transection of axons) by virtue of early onset of signal changes in the latter but only normal findings in the former. This conclusion requires further corroboration before it is relied upon to make surgical decisions, since wound edema may cause a low-level increase of muscle signal intensity that simulates early denervation (1).

While edema-like changes dominate the appearance of muscle in the first year of denervation (Fig. 1B), fatty change of the muscle is observed in more long-standing denervation (Fig. 1C). Other than serving to document a long-standing process, the occurrence of fat deposition is highly nonspecific, and is not helpful in securing a diagnosis of a neuropathic process. However, in the evaluation of neuromuscular diseases, MR provides information not only relating to gross mesenchymal variations of muscle, but it also imparts structural details that enhance the impact of MR on the clinical evaluation of patients. Such structural data include the distribution of involved and uninvolved muscles.

The distribution of deteriorated muscles in a given patient provides a clue as to the etiology and distribution of the disease but does not yield a specific diagnosis. Interestingly, however, the distribution of abnormal muscles is occasionally especially useful. Some causes of amyotrophy (deterioration of cord nerve cell bodies) have such a characteristic pattern as to suggest a specific disease. The best example of this is the nearly random distribution of attack of the anterior horn cells by the poliomyelitis virus. Although polio is now eradicated, survivors of the illness frequently present with a second wave of muscle complaints more than two decades after the initial illness. This post-poliomyelitis syndrome has provided the opportunity to see the long-term response to a single bout of a widespread spinal cord insult (Fig. 1D). The apparently random distribution of muscle atrophy and concurrent hypertrophy of spared motor units is in stark contrast to the majority of other neuromuscular diseases, which tend to be remarkably symmetrical (Fig. 1E,F).

A peculiar distribution of muscle involvement is also seen in certain myopathies and this selectivity can be helpful in distinguishing between myopathies and neuropathies. Among the diseases that present with symmetrical muscle deterioration, focal sparing of the gracilis and sartorius muscles is typical of muscular dystrophies (4) (Fig. 1E) and congenital myopathies (5), while selective involvement of these same muscles is typical of mitochondrial myopathies (6).

In contrast, many inherited neuropathies, including spinal muscular atrophies (SMA), tend to involve all muscles in a limb to a similar degree (Fig. 1F) (7). Unfortunately, the degree of similarity of involvement is a subjective determination, and in practice using MR to distinguish between neuropathies and other causes of muscle atrophy is difficult, with few exceptions (8,9). Also, since many neuropathies involve only one or a few nerves, focal muscle involvement occurs frequently (10). In fact, when the distribution of muscle atrophy conforms to that expected by disruption of a single nerve or group of nerves, some inference as to a likely site of the causative lesion can be suggested (e.g., nerve, plexus, etc.).

It is prudent to be aware of the MR appearance of denervated skeletal muscle. This is not so much because MR has replaced EMG as a means of mapping out motor nerve lesions. It has not. The overriding reason today is that when muscle atrophy or muscle edema-like change is identified on MRI, a compulsive search for underlying nerve lesions should commence. The muscle lesions may be relatively large while the causative nerve problem may be relatively small and focal.

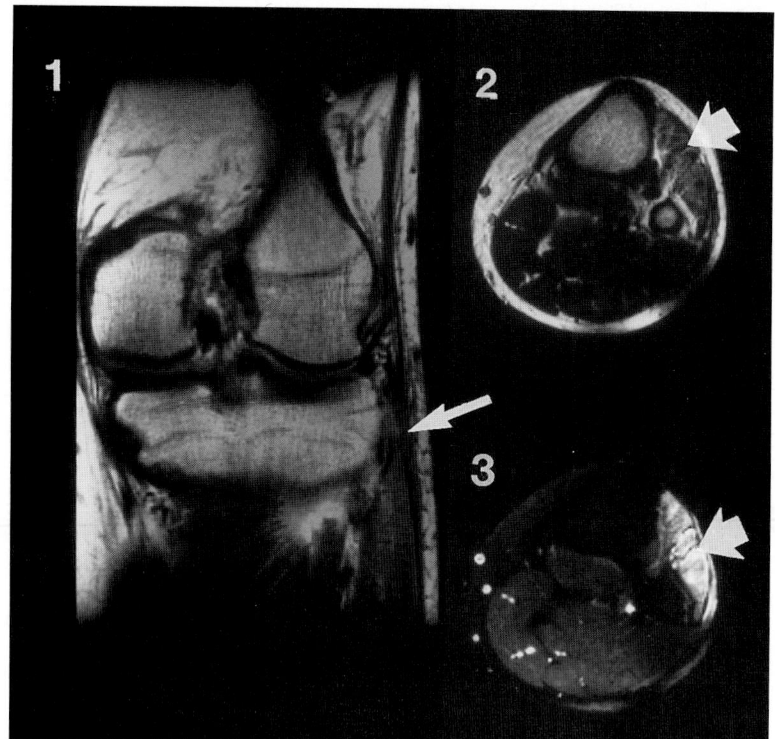

FIG. 1. B: MR imaging of subacute peripheral neuropathy: edema-like change. Lateral collateral ligament injury and subsequent scar formation resulted in compression of common peroneal nerve (not shown). Note that SI of denervated anterior leg muscles mimics edema, being normal on coronal T1-weighted image (*arrow,* 1, TR/TE 500/40 msec) and increased on axial T2-weighted (*arrow,* 2), and STIR (*arrow,* 3) images. (From Fleckenstein JL, Watumull D, Conner K, et al. Denervated human skeletal muscle: MRI evaluation. Radiology 1993;187:213–218, with permission.)

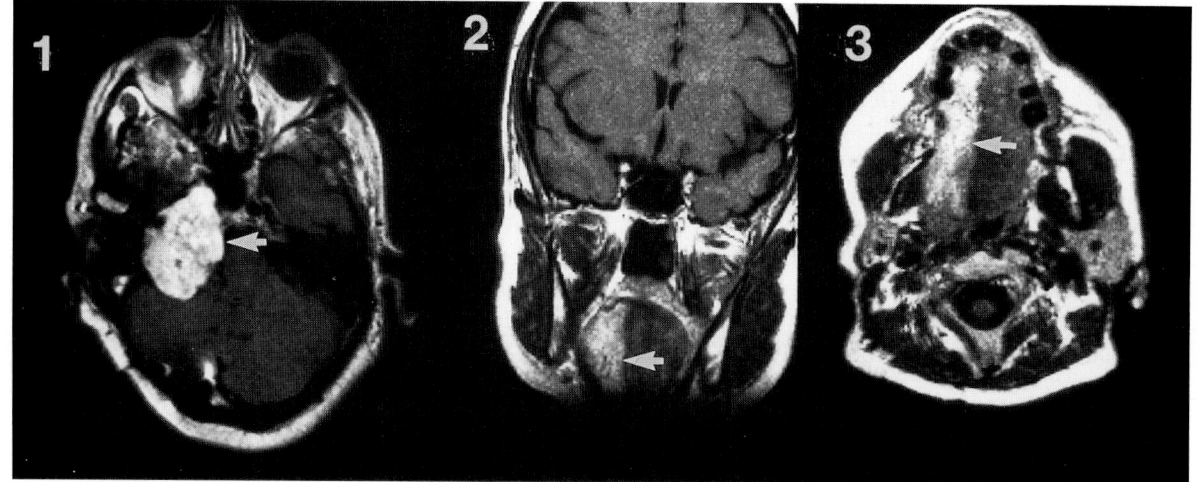

FIG. 1. C: MR of chronic peripheral neuropathy. Axial, postgadolinium T1-weighted image near skull base shows enhancing neuroma (*arrow,* TR/TE 500/40 msec, 1). Compression by the tumor of the ipsilateral hypoglossal nerve caused end-stage fatty atrophy of the ipsilateral hemisphere of the tongue, as shown on non-contrast T1-weighted coronal (*arrow,* TR/TE 500/40 msec, 2) and axial images (*arrow,* TR/TE 500/40 msec, 3).

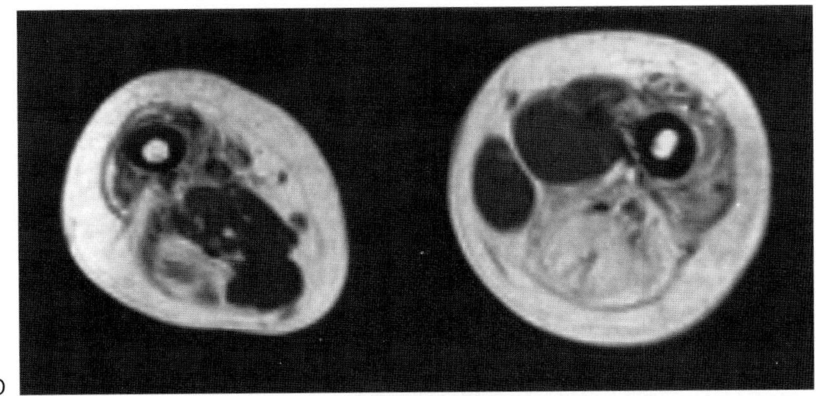

FIG. 1. D: Characteristic distribution of muscle atrophy: postpoliomyelitis syndrome. Axial T1-weighted image (TR/TE 500/30 msec) of thighs in a middle-aged man with recent progressive limitation in activities shows highly asymmetrical atrophy and hypertrophy of muscles.

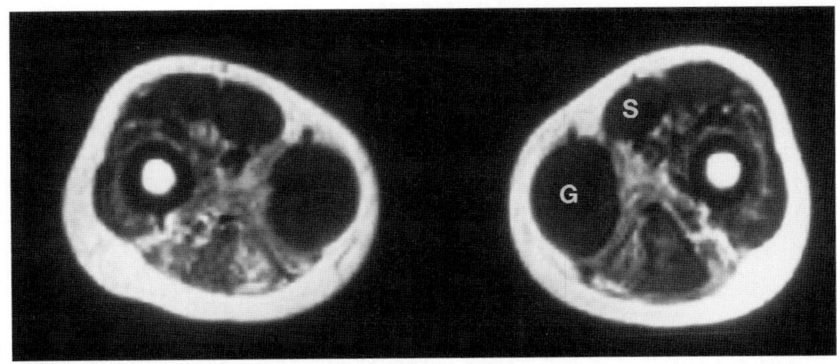

FIG. 1. E: Characteristic distribution of muscle atrophy: Duchenne's muscular dystrophy. Axial T1-weighted image (TR/TE 500/30 msec) of thighs in a 6-year-old boy shows highly symmetrical atrophy with peculiar sparing and hypertrophy of several muscles, in particular the sartorius (S) and gracilis (G).

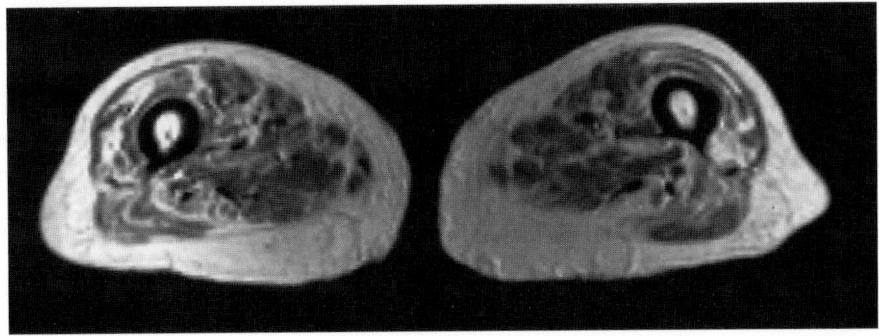

FIG. 1. F: Chronic neuropathic muscle changes: spinal muscular atrophy. In a patient with many years of proximal weakness, T1-weighted TR/TE (500/30 msec) axial image shows classical diffuse pattern of muscular atrophy and fatty change.

CASE 2

History: A 38-year-old alcoholic was found lying collapsed in his apartment. The serum concentration of creatine kinase was 140,000 IU/ml on admission when he complained of severe groin pain. One week later, at the time of imaging, he was noted to be paretic in the right lower extremity.

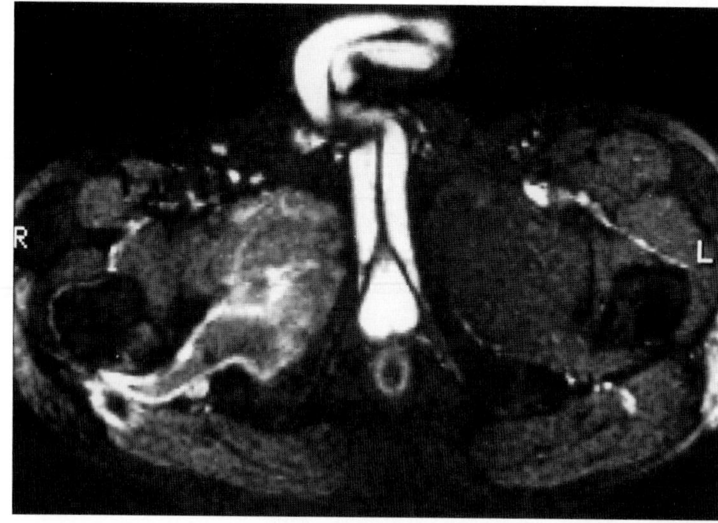

A

FIG. 2.

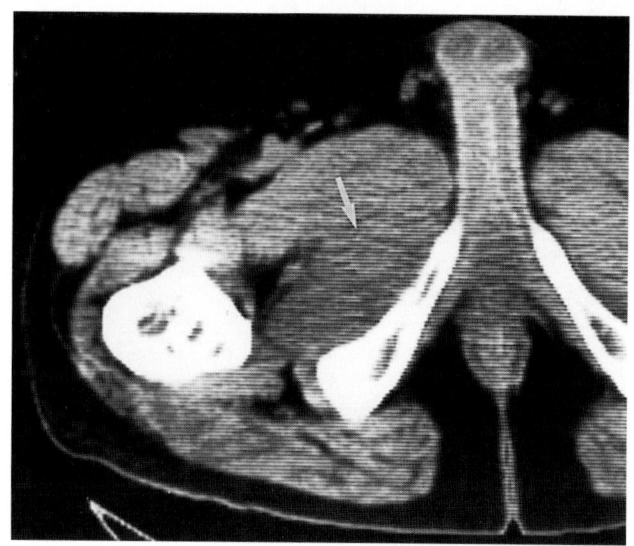

B

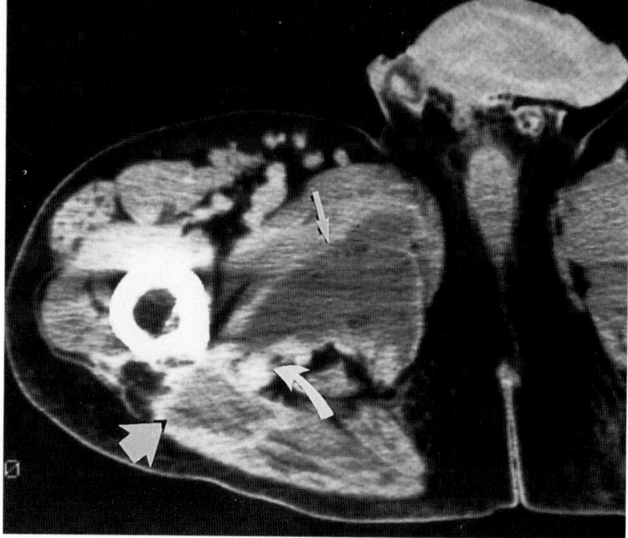

C

Findings: In Figs. 2 A–C the axial images of the proximal thighs demonstrate the relative abilities of MR and computed tomography (CT) to show muscle edema. In Fig. 2A, a short tau inversion recovery (STIR, TR/TE/TI 1500/30/150) image shows markedly increased signal intensity of multiple muscles, indicating edema-like change. Note also hyperintense and swollen sciatic nerve. In Fig. 2B depiction of normal and affected muscles with CT is hindered by inherently low sensitivity, coupled with beam hardening artifacts (arrows). In Fig. 2C note that CT is able to detect the same findings as MR only after intravenous contrast is given. Sciatic nerve (curved arrow).

Diagnosis: Rhabdomyolysis.

Discussion: Rhabdomyolysis is a clinical syndrome characterized by massive lysis of skeletal muscle fibers in which constant clinical features include elevated serum levels of creatine kinase and other muscle proteins and solutes (1). Life-threatening consequences include hyperkalemia and acute renal failure due to myoglobinuria.

Treatment of the acute event is largely supportive with hydration and temporary renal dialysis, as indicated.

A search for an underlying cause is critical for a variety of reasons, not the least of which is to avert subsequent episodes. The number of potential causes is prodigious, however. Fortunately, in many cases the etiology is obvious, such as following an epileptic seizure or when a substantial volume of the body is crushed, burned, or held in a fixed position for long periods of time, such as in an overdosed heroin addict or alcoholic. In the latter event, as in the presented case, the relative contribution of cellular toxicity and applied pressure in causing the necrosis is difficult to ascertain. It is interesting that imaging studies, such as shown here, frequently reveal that the site of necrosis is not necessarily that which would be predicted based solely on which muscle is pressed upon the most severely. The reason for this distribution is unknown.

The ability to detect necrosis-related edema introduces two questions of importance to the radiologist:

1. Is there a pattern of distribution that can help diagnose the cause of the muscle necrosis?
2. How does imaging detection of the necrosis aid in the clinical management of rhabdomyolysis and related conditions?

With regard to the first question, it is clear from the literature that the pattern of muscle abnormalities is of limited helpfulness in providing a diagnosis for the cause of muscle damage since most myopathies are symmetrical in distribution (2,3). Also, the signal intensity characteristics are poorly predictive of predisposing muscle disease since edema-like changes may be expected in neuropathies, inflammatory myopathy (e.g., polymyositis, dermatomyositis), and inherited disorders of muscle energy metabolism, and to a lesser extent, dystrophies (4). In all these diseases the edematous change may be associated with progression to fatty infiltration, so that a constellation of edema and fat is not specific. However, the development of fatty infiltration is a serious sequela of muscle necrosis and its presence strengthens the probability that a significant neuropathic or myopathic process is present. However, fatty degeneration is by no means specific for a myopathy and many myopathies may return to a normal appearance after healing (Fig. 2D).

A specific characteristic that is intriguing has been identified through imaging studies of myopathies. Selective sparing of the sartorius and gracilis muscles, noted first in Duchenne's muscular dystrophy but then also seen in congenital myopathies and inflammatory myopathies, suggests that there is some feature of these muscles that makes them particularly resistant to muscle damage (Fig. 2E). While intriguing and a potential clue to the nature of the disease process, it is clearly not specific. Alternatively, selective involvement of the same muscles has been reported in mitochondrial myopathies and centronuclear myopathy.

Although MR alone is unable to accurately diagnose the cause of muscle necrosis in most cases, data continue to accumulate that indicate that MR is very useful and cost-effective by its simple detection and localization of abnormalities to be biopsied (5). In general, it is most desirable to biopsy a muscle that is involved in an ''active'' phase of necrosis, denoted by edema-like change (Fig. 2F). Fatty-replaced and normal muscles are to be avoided. An important exception to this rule is the inherited myopathy, myophosphorylase deficiency (McArdle's disease). In this disease, necrotic muscle must be avoided since it may produce a fetal form of phosphorylase and thereby yield a false-negative test result when muscle fibers are assayed. Since muscle necrosis may show abnormality on MR for some weeks after all other clinical evidence of injury has resolved, MR can help avoid this pitfall in cases in which McArdle's disease is suspected clinically (6). It should be understood, however, that in most cases, complete recovery from muscle necrosis is the rule (7). Progression to fatty alteration, particularly in multiple regions, should alert the radiologist to a neuromuscular disease.

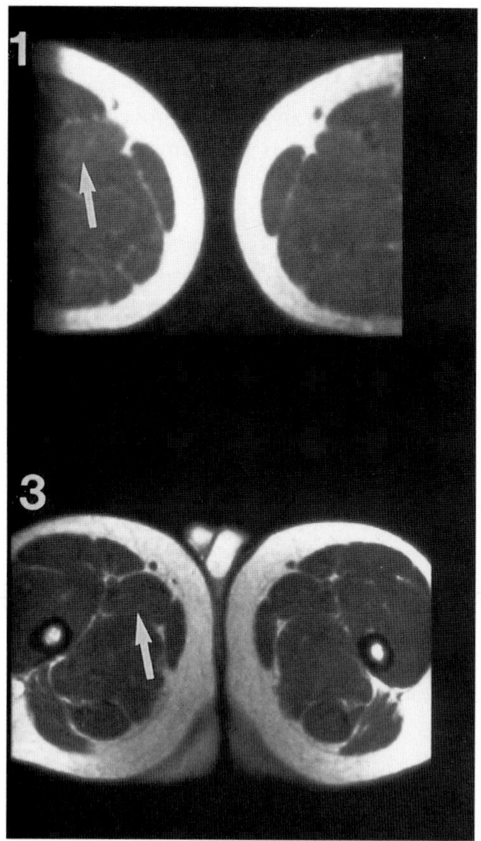

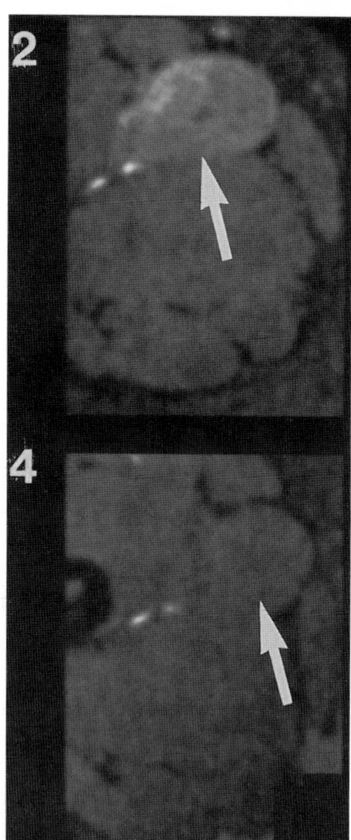

D

FIG. 2. D: Muscle necrosis in myophosphorylase deficiency. One day after sustaining a "groin pull" during sexual intercourse, the patient underwent MR evaluation. A spin density MR sequence of the thighs shows subtle increase in signal intensity in the right adductor longus (TR/TE 1,500/30 msec, *arrow,* 1). Using the highly sensitive, fat-suppression STIR sequence, the edematous abnormality is more conspicuous (TR/TE/TI 1,500/30/100, *arrow,* 2) Later, a moderately T2-weighted sequence showed resolution of the lesion (TR/TE 2,000/60, *arrow,* 3), an impression supported by a normal appearance using STIR (TR/TE/TI 1,500/30/100, *arrow,* 4). Edematous areas such as that illustrated should be avoided during biopsy when McArdle's disease is suspected because necrotic muscle produces a small amount of fetal phosphorylase, thus providing a cause for a false-negative histopathologic diagnosis. In these patients, relatively normal-appearing muscle is favored as a site for biopsy.

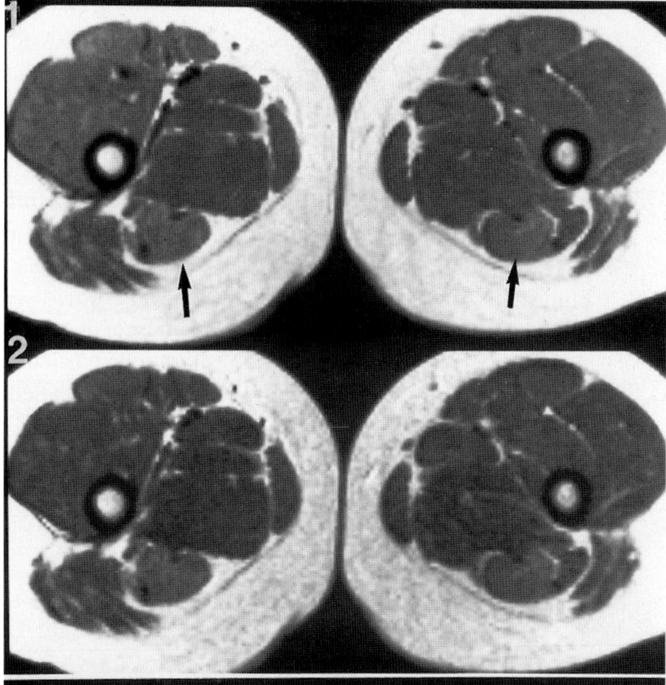

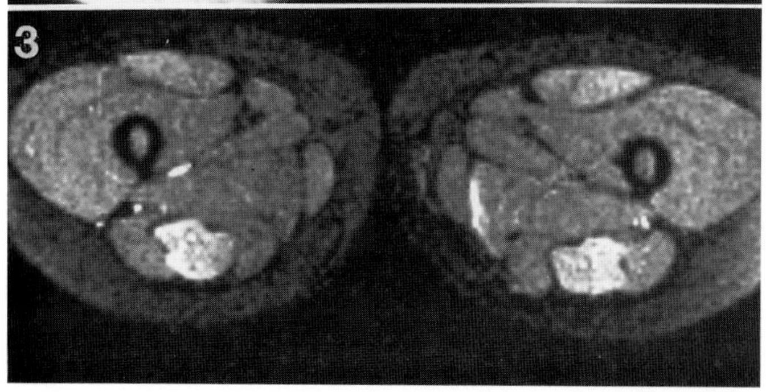

E

FIG. 2. E: Necrosis-related edema in phosphofructokinase deficiency. This patient presented for MR with a serum creatine kinase level of 45,000 IU/ml and mild hamstring pain sustained while dancing 3 days earlier. Symmetrically increased signal intensity is subtle in the semitendinosis muscles, on spin density (TR/TE 2,000/30 msec, 1) and T2-weighted images (TR/TE 2,000/60 msec, 2) but is obvious using STIR (TR/TE/TI 1,500/30 msec/100, 3).

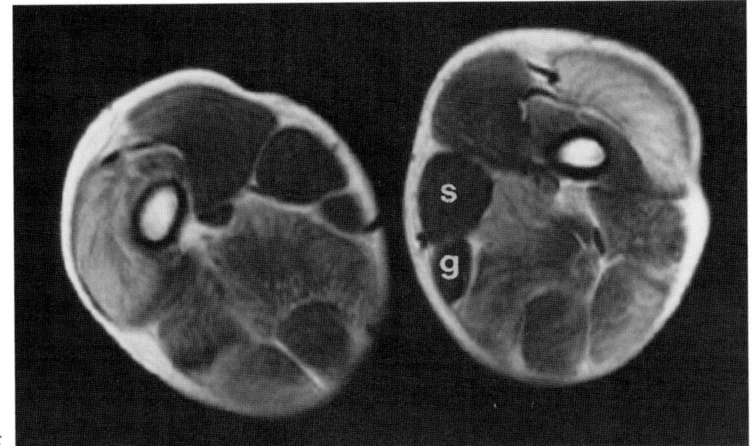

F

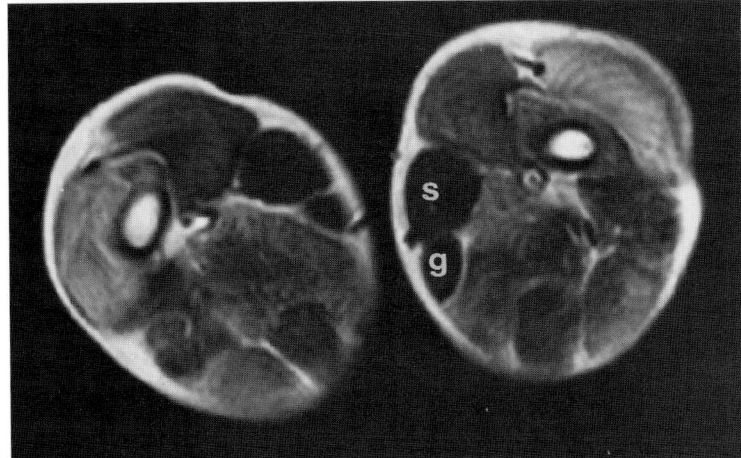

G

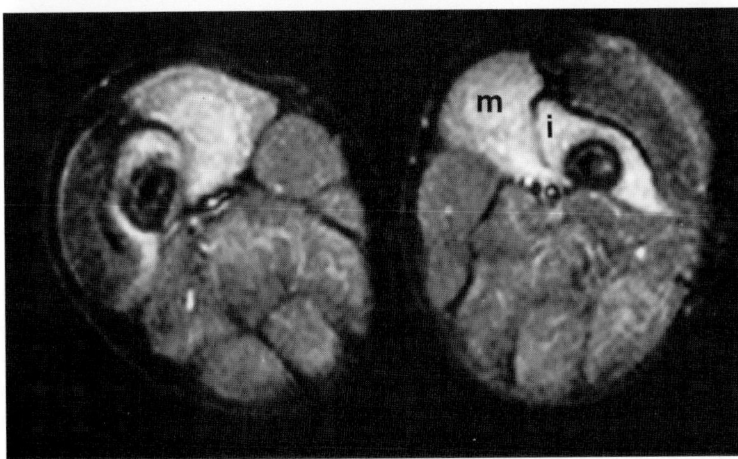

H

FIG. 2. F–H: Necrosis-related edema in Duchenne's muscular dystrophy. This child presented with myalgia and creatine kinase elevation greater than that which usually occurs in this dystrophy. As is typical in dystrophies, T1-weighted (TR/TE 500/30 msec, **F**) and T2W (TR/TE 2,000/60 msec, **G**) images appear nearly identical, revealing extensively fatty-replaced muscles with characteristic sparing of the sartorius (s) and gracilis (g), among others. The STIR sequence (TR/TE/TE 1,500/30/150, **H**), shows more edema-like change than is usual in dystrophies, particularly in the vastus medialis (m) and vastus intermedius (I), indicating preferred muscles to biopsy to assess for additional pathology. The greater conspicuity of muscle edema on STIR compared with other sequences is typical.

CASE 3

History: A 48-year-old patient with recurrent low back pain following previous L3-4 and L4-5 lumbar diskectomies. What is the significance of muscle changes?

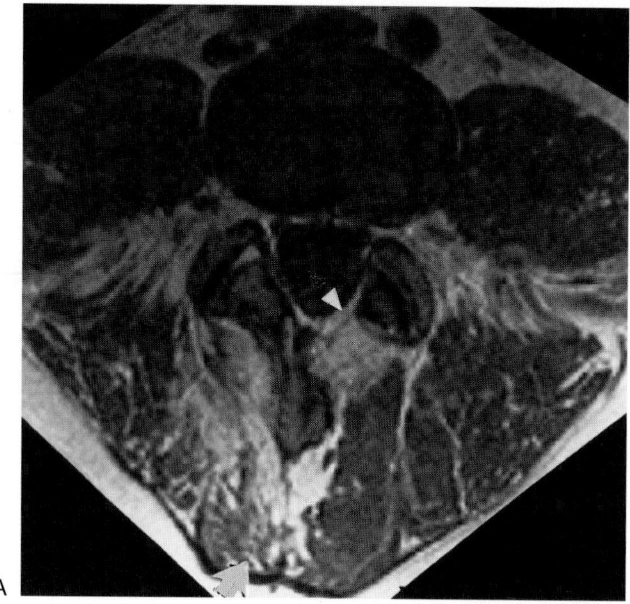

FIG. 3.

Findings: Figure 3A is a T1-weighted axial image (TR/TE 600/15) of lumbar spine at L4-5 level. Note left laminectomy site (arrow). Note fatty infiltration of the right-sided erector spinae muscles, particularly the multifidus (arrowhead).

Diagnosis: Postsurgical paraspinal atrophy.

Discussion: The prevalence of atrophy of the erector spinae muscles in patients with low back pain is low, accounting for less than 5% of all patients with low back pain whose lumbar regions are scanned with MR (unpublished observations). Following lumbar surgery, however, substantial atrophy is present in 30% of patients who present with persistent or recurrent postoperative back pain (the failed back syndrome) (1). This atrophy is generally caudal to the operated level. This case is unusual in that atrophy is on the side opposite a laminectomy. This is because the causative ipsilateral laminectomy was at a level above and cannot be seen on the image shown.

In many patients who have postoperative back pain and a recurrent disk herniation is suspected, postgadolinium infusion images, like other edema-sensitive sequences, suggest unilateral edema-like change that may precede the end-stage fatty alteration (Fig. 3B,C). Whether any of these changes contributes to the failed back syndrome is unproven but the coincidence of atrophy and pain lead to studies of both the structure and function of the involved muscles (2,3).

One study revealed selective atrophy of type 2 muscle fibers, a finding that occurs in the setting of denervation. This supports the notion that severance of motor nerves to the erector spinae causes the muscle atrophy in at least some cases (2).

That a loss of function of the erector spinae may be an important effector of the rehabilitative potential of the patient was suggested by a study in which muscle recruitment, as judged by MR, was reduced in patients with low back pain compared with normals. The impairment was particularly severe in patients with back pain who had had lumbar surgery (3). The exercise used in that study, Roman chair trunk extension, was shown by MR to particularly stress the multifidus, iliocostalis, and longissimus, i.e., the same muscles that are frequently atrophic in the failed back syndrome. Since Roman chair exercise is one of the few exercises that is successful in returning patients with low back pain to the work force, it was concluded that preservation of the paraspinal muscles during surgery might facilitate rehabilitation.

Although far less common than lumbar surgery as a cause of erector spinae muscle atrophy, neuromuscular disease may go undiagnosed until first suggested by a radiologist diligently inspecting an MR or CT scan of the soft tissues of the back. Unfortunately, no specific feature of the pattern of muscle distribution of signal intensity changes has been reported to eliminate the need for a muscle biopsy to provide a definitive diagnosis. Interestingly, however, one large CT study did report a high frequency of muscle abnormalities in familial disorders of muscle energy metabolism (4).

In that report, fatty deterioration of the psoas and erector spinae muscles was found to be very frequent in adult acid maltase deficiency (16/18 patients) (Fig. 3D). Erector spinae atrophy was less frequent in the similar disease, myophosphorylase deficiency (8/16 patients) (Fig. 3E). However, in no patient with the latter disease was there atrophy of the psoas. The authors concluded with the optimistic proposition that radiologists may be able to suggest the type of myopathy present by the distribution of deteriorated muscles seen in patients whose lumbar spine MR or CT scans disclosed selective atrophy (4).

Importantly, this conclusion does not consider that other neuromuscular diseases may also preferentially involve the psoas and erector spinae, including muscular dystrophies, myositis, and mitochondrial myopathies. In the majority of these diseases, the muscles are involved bilaterally and generally symmetrically. A notable exception is the highly asymmetrical atrophic appearance of many neuropathic processes, such as postpoliomyelitis syndrome (Fig. 3F). As is the case in other parts of the body, a focal atrophy restricted to a single muscle or group of muscles should elicit a directed search for focal nerve lesions (Fig. 3G).

Hence, the occurrence of a selective distribution of paraspinal muscle deterioration appears to be nonspecific in establishing diagnoses. However, it remains noteworthy that focal muscle deterioration in the paraspinal region occurs in glycogenoses, as well as in other myopathies, since such findings offer radiologists the opportunity to suggest the presence of a significant disease in patients in whom the spine is being evaluated to search for a cause of weakness.

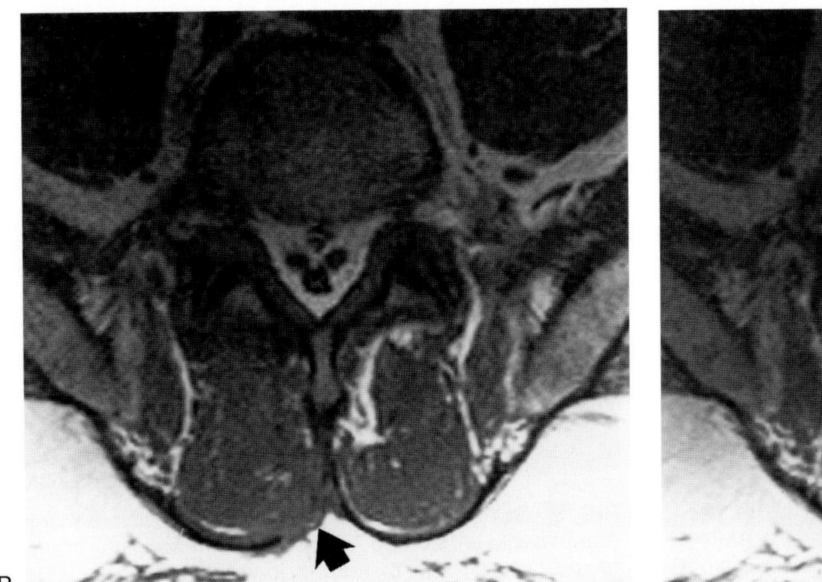

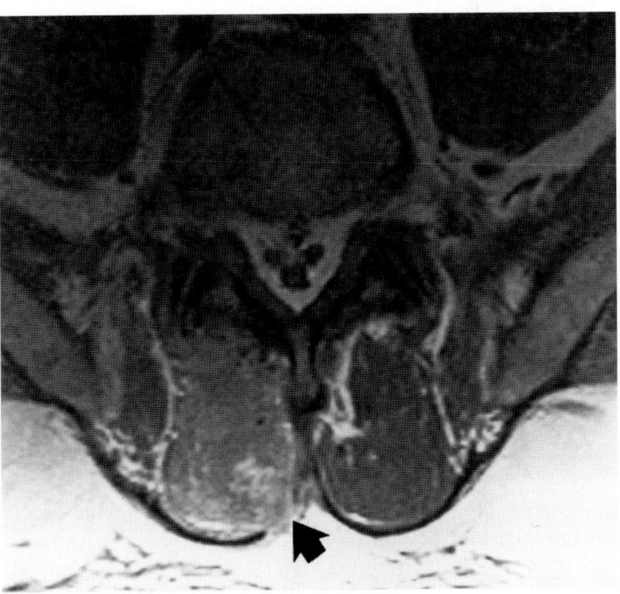

FIG. 3. B,C: Early changes of the erector spinae following lumbar surgery. **B:** T1-weighted axial image of lumbar spine at L5-S1 level (TR/TE 600/15 msec). Note that edema-like change of right multifidus is isointense with normal muscle *(arrow)*. **C:** Following intravenous gadolinium infusion, enhancement occurs in the edematous muscle *(arrow)*. Whether this muscle is destined to become replaced by fat or will recover completely is not known at the time of the scan.

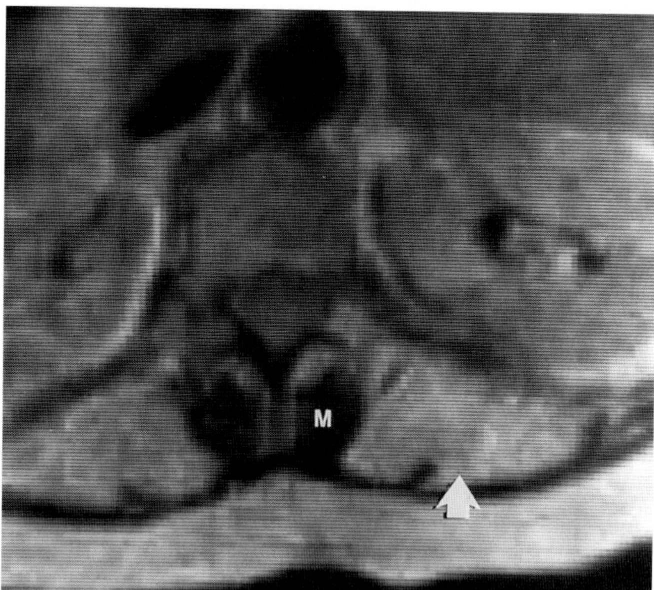

FIG. 3. D: Trunk muscle atrophy in the unoperated back: a hint of myopathy. T1-weighted (TR/TE 500/30) image of a patient with acid maltase deficiency shows marked deterioration and fatty replacement of the erector spinae (*arrow*). The multifidus (m) is spared. The psoas muscles were atrophic on other images (not shown). Selective erector spinae muscle atrophy is a nonspecific finding, however, and may be seen in other myopathies.

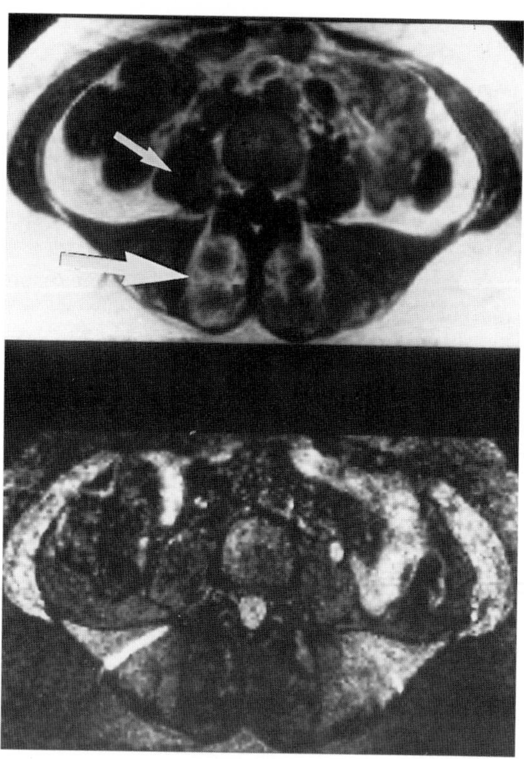

FIG. 3. E: Trunk muscle atrophy in myophosphorylase deficiency. T1-weighted image (TR/TE 500/30, *top*) shows selective deterioration and fatty replacement of the multifidus (*arrow*). Note normality of psoas *(small arrow).* STIR image (TR/TE/TI 1500/30/100, *bottom*) at the same level shows only fatty change, without coexistent edema, in the atrophied muscles.

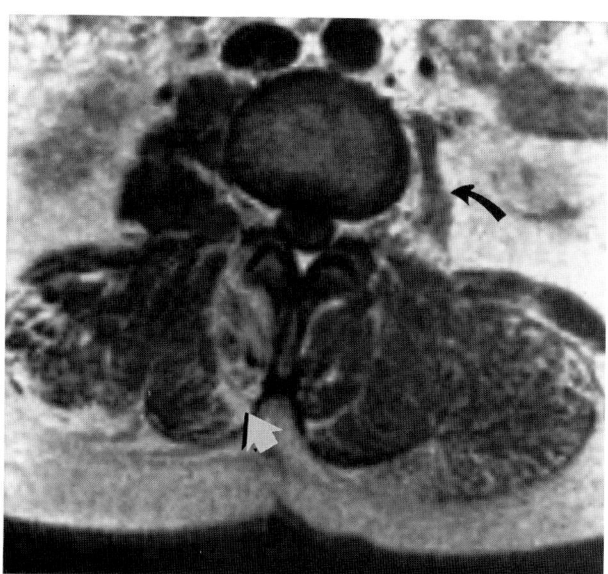

FIG. 3. F: Asymmetrical trunk muscle atrophy in the unoperated back: a hint of neuropathy. T1W image of a patient post–poliomyelitis syndrome (TR/TE 500/30 msec). Note markedly asymmetrical atrophy and sparing of the various muscles, notably the multifidus *(arrow)* and psoas *(curved arrow).* This appearance is highly characteristic of the randomly distributed insult to the spinal cord in the distant past.

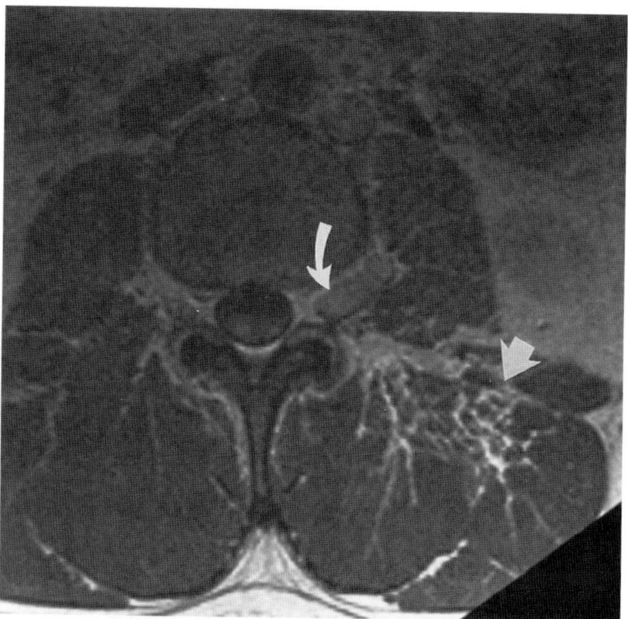

FIG. 3. G: Asymmetrical trunk muscle atrophy in the unoperated back: a "red flag." T1-weighted image shows asymmetric fatty change of the left longissimus (*arrow,* TR/TE 500/30 msec). Such an asymmetric finding should alert one to examine all nerves carefully to assess for possible masses and/or compressive lesions. In this patient with neurofibromatosis, small neuromas were seen at multiple levels, including the one shown here *(curved arrow).*

CASE 4

History: A 53-year-old man presents with several months' history of muscle weakness. Can you offer a diagnosis?

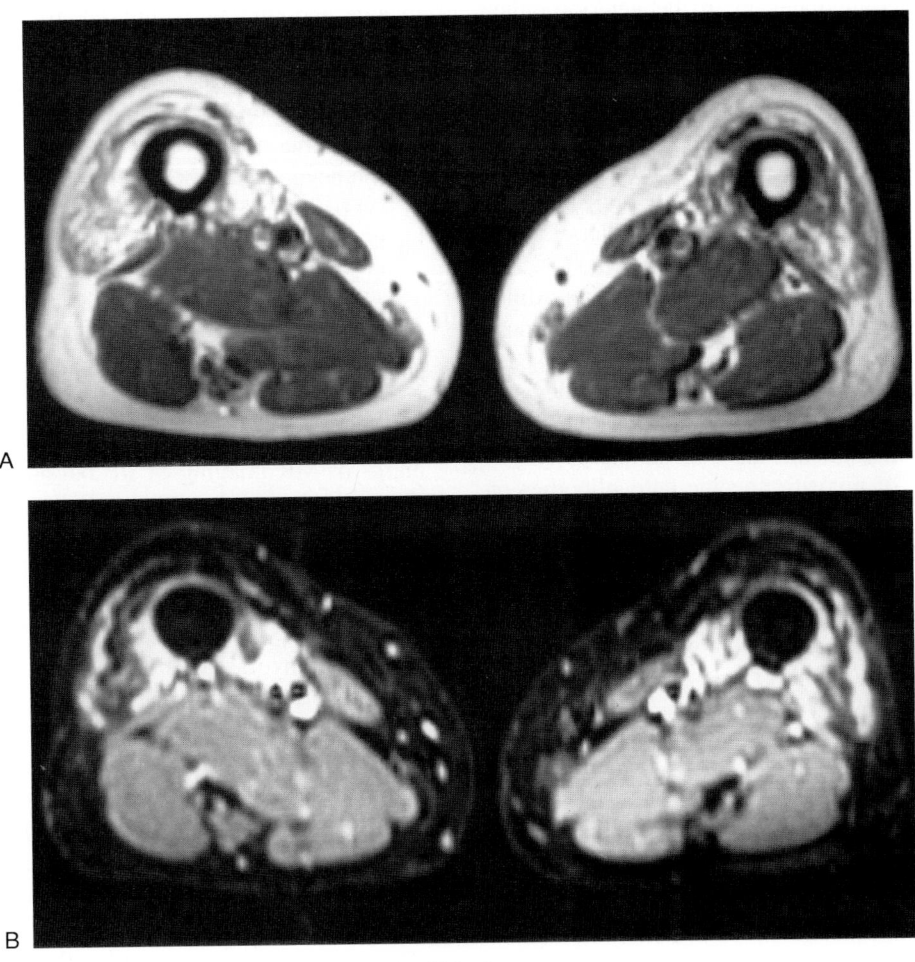

FIG. 4.

Findings: Figure 4A,B: (A) Axial (TR/TE 700/20 msec); (B) axial (TR/TE/TI 2200/35/160). In Fig. 4A, on the T1-weighted sequence there is marked atrophy of the quadriceps muscle bilaterally with evidence of extensive fatty atrophy. In Fig. 4B, the STIR sequence, the areas of high signal intensity are identified and presumably reflect areas of active inflammation interspersed between the areas of atrophy. There is striking anterior compartment predominance of the disease.

Diagnosis: Inclusion body myositis.

Discussion: Inclusion body myositis is a rare disease that occurs predominantly in men (1,2). The age of onset is variable with presentation ranging from the second to the eighth decade (1,2). The clinical findings are similar to those of polymyositis, although some patients demonstrate a more insidious onset and distal or asymmetric weakness, and neuropathic changes on electromyography may be evident (1,2). Patients do not have a rash or autoimmune features (3). The course of the disease is variable. Muscle weakness may progress or plateau. Laboratory analysis reveals minimal to mild enzyme elevation (1). In distinction to polymyositis, there is little or no association with malignancy (1). There is also no direct association with collagen vascular disease and patients with inclusion body myositis may show resistance to high-dose corticosteroid and immunosuppressive therapy (1).

Inclusion body myositis is characterized histologically by the presence of intranuclear and cytoplasmic inclusions (rimmed vacuolar inclusions) within cells derived from muscle biopsy (1,2). Electron microscopic evaluation of these inclusions has demonstrated structures resembling paramyxovirus nucleocapsid (1). Patients with inclusion body myositis demonstrate significant fatty replacement and apparent loss of muscle bulk (3). These characteristic histologic features distinguish inclusion body myositis from polymyositis. MR imaging may reveal distinctive changes in this disease (4). In distinction to polymyositis, Fraser et al. (4) found inclusion body myositis to be characterized by greater involvement in the anterior rather than the posterior compartment in the leg and a more focal pattern of increased signal was observed. A ''ghost'' appearance has been described in inclusion body myositis as a consequence of decrease in overall muscle mass with preservation of the outline of the muscles (3).

Inclusion body myositis, polymyositis, and dermatomyositis are part of a group of muscle disorders known as the idiopathic inflammatory myopathies (IIMs) (3). The condition is manifest clinically by the onset of proximal muscle weakness, which at times is associated with muscle tenderness, and in the case of dermatomyositis, a spectrum of rashes (3). IIM patients typically have elevated levels of muscle enzymes—aldolase and creatine kinase (CK)—and electromyographic analysis indicates myopathy. The presence of myositis specific autoantibodies or interstitial lung disease may support the diagnosis of polymyositis or dermatomyositis rather than another myopathy (3). Myositis-specific autoantibodies include anti Jo-1 and related antibodies directed at aminoacyl-tRNA synthetases, anti-signal recognition particle antibodies, and anti-Mi2 antibodies (3,5). These autoantibodies can define specific syndromes within polymyositis and dermatomyositis that have different prognoses and responses to medication.

Clinical features and muscle biopsy have constituted the principal means of diagnosis of the IIMs, although current methods of diagnosis have been demonstrated to have limitations (3). Electromyography, another mainstay of diagnosis, is time-consuming, may be painful, and can evoke needle-induced myositis that may complicate interpretation of subsequent muscle biopsy (3). Results of muscle biopsy in myositis may be normal in up to 10% to 15% of cases, a finding reflecting both the patchy nature of the cellular infiltrate and possible suppression of inflammation related to prior therapy (3,6,7).

MR may be successfully directed at overcoming several of the previously alluded to limitations. Several studies have reported on the value of MR including STIR for directing biopsy to sites of apparent maximal involvement (3,4,8–11). MR findings can potentially allow therapy to be instituted even in the presence of equivocal biopsy results or, alternatively, can redirect clinical evaluation in patients determined not to have findings of IIM (3). Increased signal intensity from edema within muscle has been demonstrated to correlate with disease activity (3,4,11,12). MR imaging changes indicative of edema (increased signal on T2-weighted and/or STIR images) within muscle correlate better with disease activity than do elevated CK levels (3,12,13). MR, utilizing T1-weighted and STIR imaging, may be of value in the assessment of patients with long-standing disease in differentiating between active inflammation, necrosis, muscle atrophy, and steroid-induced myopathy as the cause of muscle weakness (3). In myositis patients, histologic findings observed in areas of hyperintense signal depicted on STIR MR sequences include endomysial, perimysial, perifascicular, and perivascular inflammation (3). Changes depictable on MR may precede frank clinical deterioration and facilitate institution of earlier therapeutic intervention (3). The use of T2-weighted images does not appear to add significant information beyond that obtained with T1 and STIR sequences in IIM patients (3). This appears to be particularly applicable to patients with chronic disease in whom fatty changes could not be distinguished from edema within muscle on T2-weighted images (2). Imaging following the administration of gadopentetate dimeglumine also does not appear to add significant diagnostic information in IIM patients (3, 12–14).

Increased signal intensity represents one of the hallmarks of MR imaging of IIF but are not specific. Other causes of abnormal cellular infiltrate (e.g., bacterial myositis, lymphoma) as well as causes of edema (muscle strains), subacute or chronic denervation (see Case 1), and rhabdomyolysis (see Case 2) may demonstrate a similar appearance (3). In addition to muscle hyperintensity, edema in a myofascial distribution has also been depicted on occasion in patients with IIM. This pattern is similar to that depicted in sports-related injuries (see Case 5). The extent of fascial change is better depicted on MR than by muscle biopsy. Subcutaneous changes with abnormal reticulation of the subcutaneous soft tissues may also been seen in cases of inflammatory myositis (3). This pattern is similar to that depicted in lymphedema. In amyloidosis, one of the most prominent findings on MR may be a hypointense reticular pattern of the subcutaneous fat on all pulse sequences (see Case 5). Muscle calcification can occasionally be seen as hypointense areas on all pulse sequences in patients with IIM. This finding is most common in juvenile dermatomyositis (see Case 10). Dense fibrous or hemosiderin-laden tissues can also produce a similar pattern (see Case 12). Suspected areas of calcification should be correlated with plain radiographs to assist in differential diagnosis. Hemosiderin deposits are an unusual finding in myositis patients and their presence would suggest another diagnosis.

In summary, MR imaging findings in IIM include increased signal in and around muscle (correlating with increased water secondary to necrosis or inflammation), subcutaneous reticulation, occasional intramuscular calcium deposition, and fatty infiltration or replacement (3). T1-weighted and STIR images are often sufficient for evaluation. MR is useful for both detection and characterization of the extent of disease, for selection of sites for muscle biopsy, and for assisting in noninvasive follow-up.

CASE 5

History: A 43-year-old physician/jogger experienced the sudden development of pain during the final "sprint" stage of a run. Physical examination demonstrated a mass-like fullness in the region of the distal Achilles tendon. What is your diagnosis?

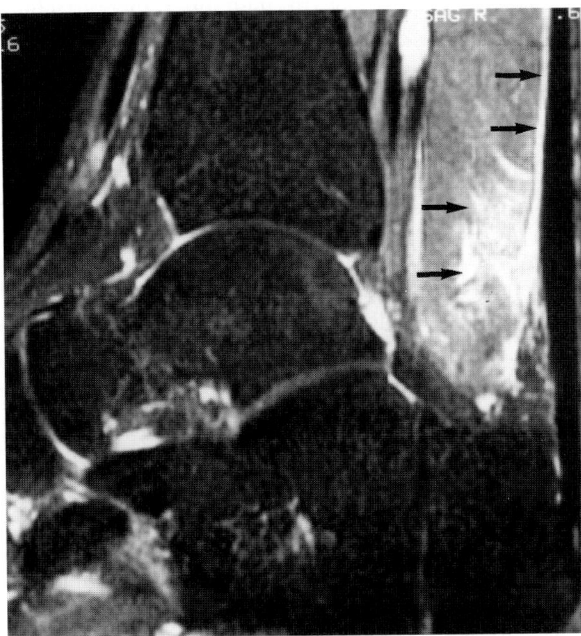
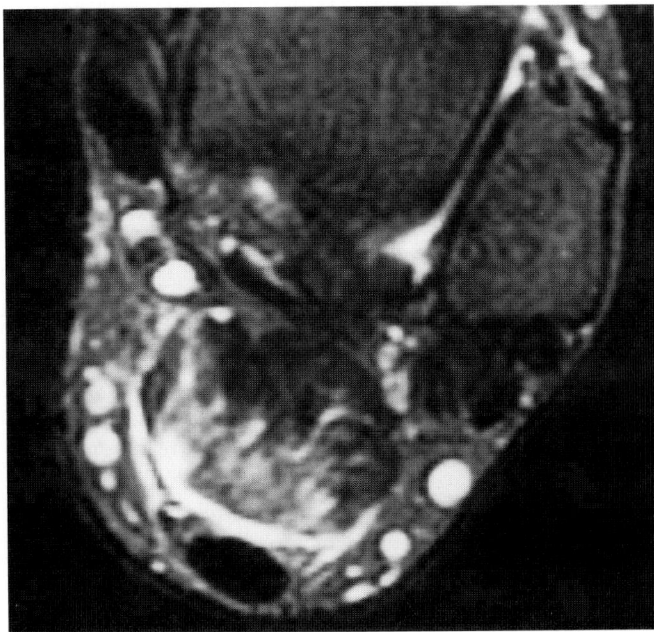

A B

FIG. 5.

Findings: In Fig. 5A, the sagittal inversion recovery fast spin echo (TR/TE/TI 3,000/32/30) image, there is an intermediate signal intensity mass (isointense with normal muscle) projecting anterior to an intact Achilles tendon and extending to the level of the calcaneus. There is near complete replacement of the normal pre-Achilles fat pad. Within the mass, there is streaky high signal intensity consistent with edema and possible interstitial hemorrhage (small arrows). There is linear high signal intensity tracking along the fascial interface between the muscle and Achilles tendon (small arrows). In Fig. 5B, the axial inversion recovery fast spin echo (TR/TE/TI 3,000/32/130) image, the streaky high signal within the muscle anterior to the Achilles tendon is again well demonstrated as is the linear high signal interface between the muscle and tendon.

Diagnosis: Grade 2 strain within an accessory soleus muscle.

Discussion: The case is unique in that MR was able not only to define and characterize the patient's injury, but also to explain the soft tissue mass that contributed to clinical confusion. An accessory soleus muscle typically extends from the deep surface of the soleus to the superior or medial aspect of the calcaneus. While often asymptomatic, affected individuals may present with pain and swelling with exercise. This is thought to relate to a closed compartment ischemia, and fasciotomy or excision of the accessory muscle may be performed. On physical examination, the muscle may be confused with a soft tissue mass within the posterior medial aspect of the ankle. In the present case, the soft tissue mass in association with the apparent acute injury raised the concern regarding both an Achilles tendon injury as well as a possible contained hematoma. MR graphically differentiated between these different possibilities and provided a precise diagnosis.

An acute strain of a muscle is defined as a painful stretch-induced injury resulting from a single applied violent force (1–3). Muscles prone to strain injury share three common characteristics (1–3). The first of these is that they commonly perform eccentric actions. The term *eccentric contraction* refers to a type of isotonic muscle contraction characterized by lengthening of the muscle during contraction in distinction to concentric contraction in which the muscle shortens during contraction. Muscles produce greater tension when they stretch than when they shorten and eccentric action is capable of producing greater force than concentric contraction. Eccentric actions of muscles are typically those that limit, restrict, or regulate motion. As an example, hamstring strains most frequently occur during intense bursts of speed when the muscle is functioning to decelerate the rapidly extending knee. The muscle is therefore contracting while being forcefully extended. Eccentric contractions are considered the primary cause of exertion-related sports injury. The second characteristic of muscles prone to strain injury is that they typically cross two joints and are subject to stretch at more than one site (e.g., hamstring group, gastrocnemius). Third, strains are most likely to occur in those muscles used in sports or in activities requiring sudden increases or decreases in speed (1).

Muscle strains are divided into three grades that are distinguished from one another by the degree of disruption that occurs. The distinction may, however, be difficult to reliably make both clinically as well as with MR (1,2). Grade 1 strains are characterized by a minor degree of muscle fiber disruption; pathologically there is edema and low-grade inflammation. Grade 2 strains represent a more significant but still incomplete disruption of the muscle (involving up to 50% of the cross-sectional area of the muscle) and are best characterized as partial tears. Grade 3 strains are complete ruptures and appear as extensive interruption of the muscle with characteristically a ''mop-end'' appearance to the torn edges. Therapy in these cases must be directed toward prompt restoration of the integrity of muscle tendon unit.

Contusions produce injury at the point of impact; strains produce lesions at the histologic musculotendinous junction where injury to elastic, noncontractile tissue appears to be responsible for symptoms. In making reference to the musculotendinous junction, it should be noted that this does not exclusively refer to the macroscopic transition. In the biceps femoris, for example, the proximal muscle and muscle/tendon junction extend nearly 60% of the total length of the muscle. As a consequence, the anatomic and functional muscle tendon junction may extend nearly the entire length of the hamstring group and injury can occur anywhere along the musculotendinous junction (4).

The appearance of a muscle strain on MR is dependent on the degree of the injury and the time from the injury that the study is obtained. The patterns of injury to muscle make their appearance 24 to 48 hours after the event and measurement of relaxation times in muscle strains has demonstrated a prolongation of both T1 and T2 of approximately two times normal.

Grade 1 strains appear similar on MR to muscle contusions and are distinguished by clinical history and mechanism of injury. The affected muscle is enlarged, has a feathery, interstitial increase in signal, and demonstrates no interruption of muscle continuity (5,6). Perifascial fluid collections are common. The MR appearance of grade 2 strains reflects the greater degree of injury. The feathery MR appearance of interstitial blood/edema is seen in the background but additionally, a focal mass-like collection or defect can be seen within the muscle as in the present case (1). The MR signal character of these lesions is dominated by the presence of water/edema. Signal increases on short TR sequences, typical of blood, may on occasion be seen and have been associated with a poorer prognosis and longer recovery.

Longitudinal follow-up of grade 1 and 2 strain injuries have demonstrated a predictable sequence of events (1,7). Initially, there is a diffuse increase in signal (T2) within multiple muscles in the same compartment but rapidly the pattern of injury becomes localized to the muscle/muscles actually injured. The signal increase is often most evident within the center of the affected muscle. The signal abnormality spreads centrifugally and becomes maximal by day 3 to 5. By 36 hours, a thin rim of increased signal (T2 weighted images) appears in the perifascial space and intermuscular septa, closely applied to the injured muscle (Fig. 5C). By 2 weeks, the abnormal signal within a strained muscle will often nearly resolve, although signal changes may persist well beyond the time of clinical recovery (Fig. 5D). Grade 3 strains reveal definite interruption in continuity and the torn edges of the muscle retract and assume a lobulated or serpiginous appearance. The gap between the torn edges may fill in with blood in sufficient quantity that a true hematoma forms.

Pomeranz and Heidt (8) have attempted to identify features of muscle strains that may have prognostic value with regard to likely convalescence interval and return to activity. Factors correlating with a poorer prognosis included complete tendinous or myotendinous junction rupture, hyperintensity or hemorrhage-like signal intensity on T1-weighted images, large cross-sectional area involvement (>50%), distal tendinous or myotendinous lesions, and localized ganglion-like peritendinous fluid collections. Superficial muscular tears without hyperintensity on T1-weighted images had the shortest recovery periods and complete transections the longest. When more than one poor prognostic feature was present, the convalescence interval usually correlated with the poorest prognostic factor.

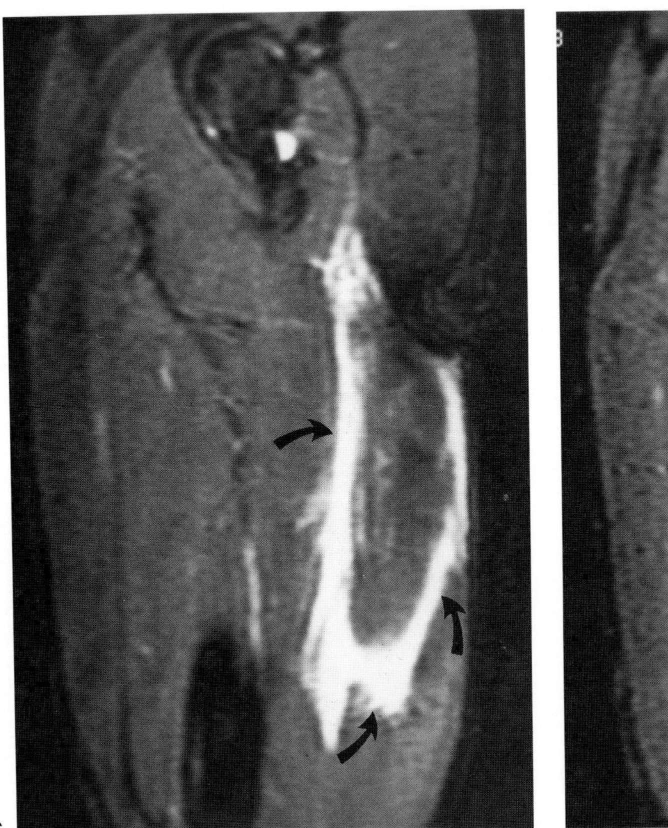
C

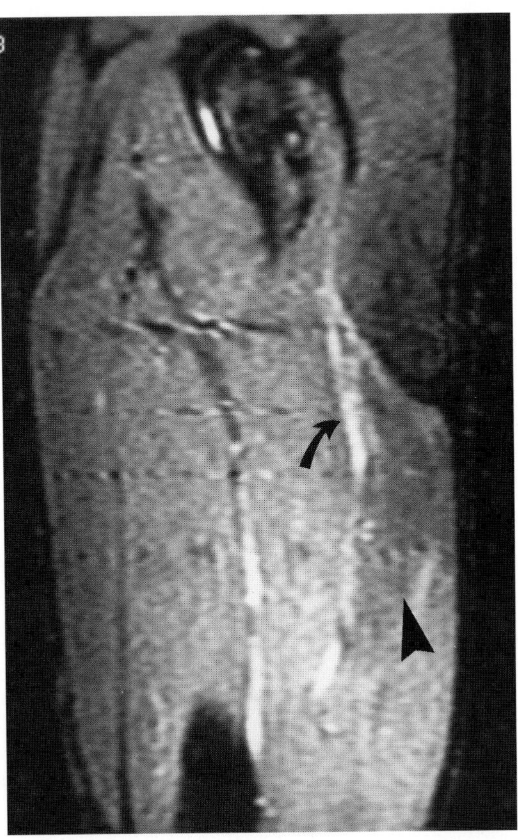
D

FIG. 5. C,D: Follow-up imaging of grade 2 hamstring strain. **C:** Sagittal STIR (TR/TE/TI 2,200/35/160) image demonstrates edema within the biceps femoris muscle as well as a striking rim of perifascial edema. Such an appearance is nonspecific and could be seen with an inflammatory myositis **D:** Sagittal STIR (TR/TE/TI 2,200/35/160) image 3 weeks later following conservative management. The ring of edema is virtually gone and the patient was asymptomatic. The biceps muscle is of slightly lower signal intensity a finding possibly reflecting diffuse hemosiderin deposition. Mink JH. Muscle injuries. In: Deutsch AL, Mink JH, Kerr R, eds. *MRI of the foot and ankle.* New York: Raven Press, 1992;281–313, with permission.

CASE 6

History: A 34-year-old HIV-positive man presented with a 3-week history of a painful leg mass. What is the most likely diagnosis? What are the typical appearances of this condition on MR examinations?

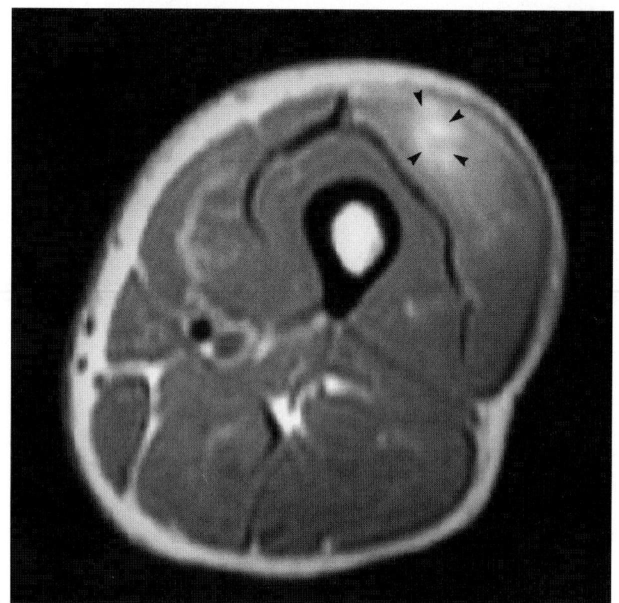

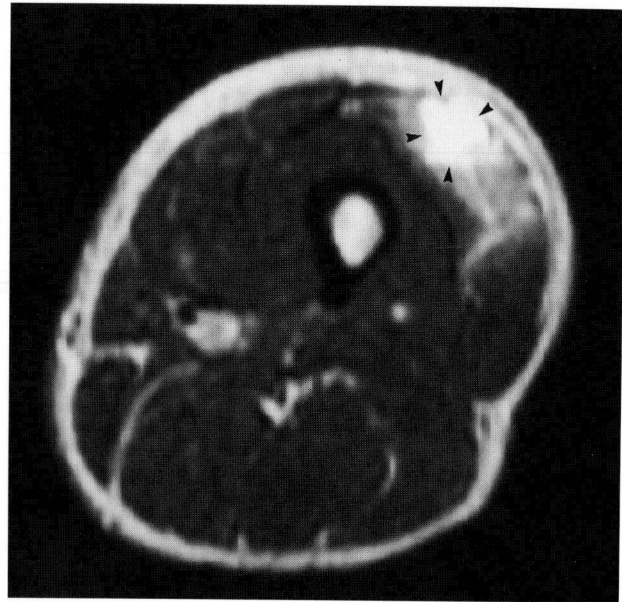

FIG. 6.

Findings: In Fig. 6A the axial proton density (PD) (TR/TE 2,000/20 msec) image demonstrates a small mass within the vastus lateralis muscle that is of increased signal intensity compared with normal muscle and suggests a slight rim (arrowheads). In Fig. 6B the axial T2-weighted image (TR/TE 2,000/80 msec) demonstrates the mass to become strikingly hyperintense (arrowheads) with surrounding slightly less intense edema.

Diagnosis: Bacterial pyomyositis associated with HIV infection.

Discussion: Bacterial and granulomatous muscle infections have been identified with increased frequency in patients with AIDS and HIV infection (1–4). The muscles of the thighs, calves, and legs are most commonly affected and multifocal disease is not uncommon (1–3). Clinical signs of inflammation may be mild in the early stages, requiring a high clinical index of suspicion (1). Low-grade fever, weakness, and leukocytosis may be the only positive findings. The skeletal muscle may be rapidly destroyed and replaced by pus (1,3). Prompt diagnosis is critical as bacterial myositis is one of the more readily treatable complications of HIV disease. While scintigraphic methods, sonography, and CT have all been utilized to diagnose pyomyositis, MR may be more sensitive and specific than these other techniques (1,3).

Pyomyositis may demonstrate a variety of patterns on MR. In some patients with bacterial pyomyositis, a rim of high signal intensity separating a central region that is isointense with surrounding normal muscle has been described (1) (Fig. 6C). Potential etiologies for this high-signal region include methemoglobin production as a result of subacute bleeding, or bacterial or macrophage sequestration of iron (1). Contrast enhancement may also be utilized to define a vascularized rim in the periphery of muscle abscesses (4). In other patients, the abscess does not have this high intensity rim but rather the entire mass has a slightly increased signal intensity compared with normal skeletal muscle (Fig. 6D,E) (3). On T2-weighted and inversion recovery fast spin echo sequences, patients with bacterial myositis demonstrate well-defined central regions of markedly augmented signal intensity surrounded by a broad zone of diffuse muscle edema.

The predominant involvement of muscle with absent or significantly lesser involvement of subcutaneous tissue has been reported as a diagnostic feature favoring the diagnosis of pyomyositis (1,2). Significant or predominant involvement of the subcutaneous tissue suggests the likelihood of cellulitis, dermal infection, lymphedema, or Kaposi sarcoma (KS). Lymphatic obstruction is common in KS and may occur in the presence of minimal skin lesions (1). Muscle edema in association with subcutaneous tissue alteration may also be seen with venous thrombosis. Finally, polymyositis, which is more frequent in the AIDS population than pyomyositis, and which may demonstrate gross abnormalities on MR images, should be readily differentiable on the basis of the clinical features, bilateral symmetry, proximal weakness, and elevated serum creatine kinase levels (1).

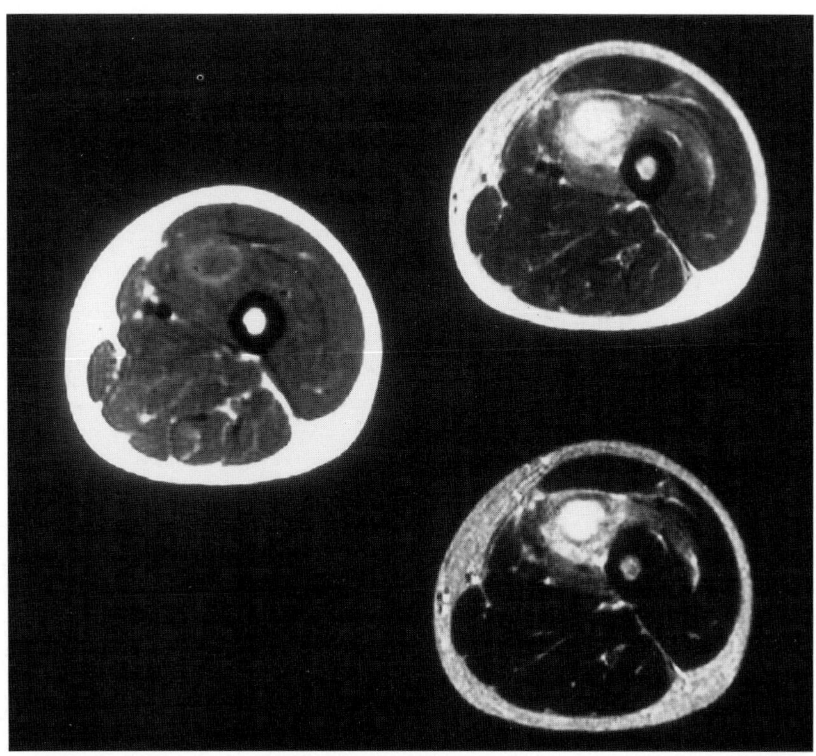

FIG. 6. C: (TR/TE 500/30 msec) Pyomyositis in an HIV-positive patient. Axial T1-weighted image of the thigh demonstrates a lesion with the vastus musculature that demonstrates a high signal intensity rim (left-sided image). On the T2-weighted spin echo (TR/TE 2,000/80 msec) (upper right) and STIR sequence (bottom right) there is strikingly augmented signal intensity to the lesion. The findings are typical of bacterial myositis. Subcutaneous fat involvement is minimal in the primary muscle problems associated with HIV disease.

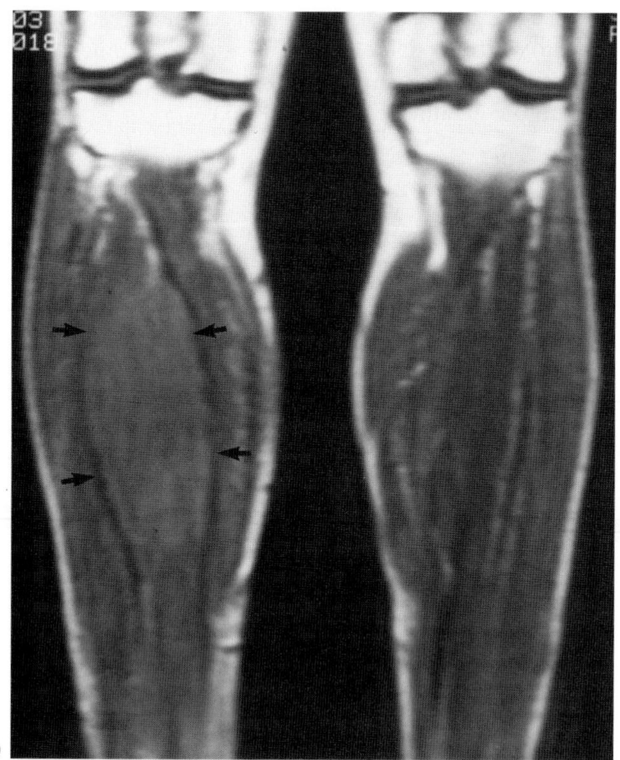

D

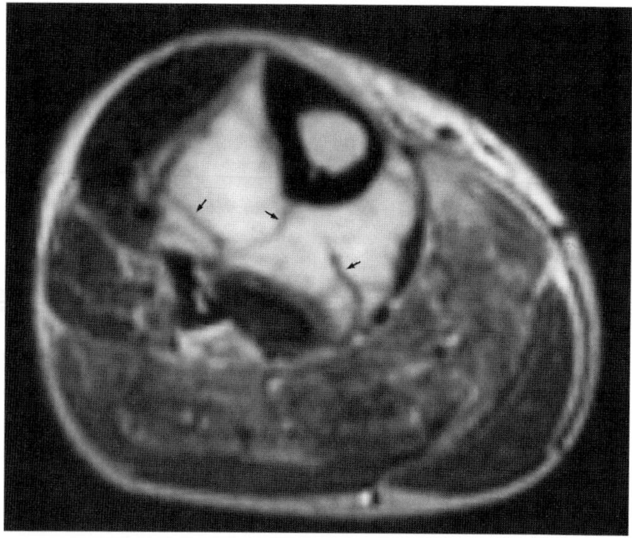

E

FIG. 6. D,E: Tuberculous pyomyositis in an HIV-positive patient. **D:** Coronal T1-weighted spin echo (TR/TE 500/20 msec) of both lower legs demonstrates a relatively well-defined mass of slightly increased signal intensity compared with normal muscle *(arrows)*. **E:** Axial T2-weighted (TR/TE 2,000/80 msec) image demonstrates areas of apparent septation *(arrows)* within the high signal intensity collection, which crosses the interosseous membrane between the tibia and fibula.

CASE 7

History: Images from the upper arm of a study subject are illustrated prior to and 5 days following exercise involving eccentric contraction. What is the significance of the signal changes identified in the brachialis muscle? What underlying mechanisms are currently believed responsible for such changes?

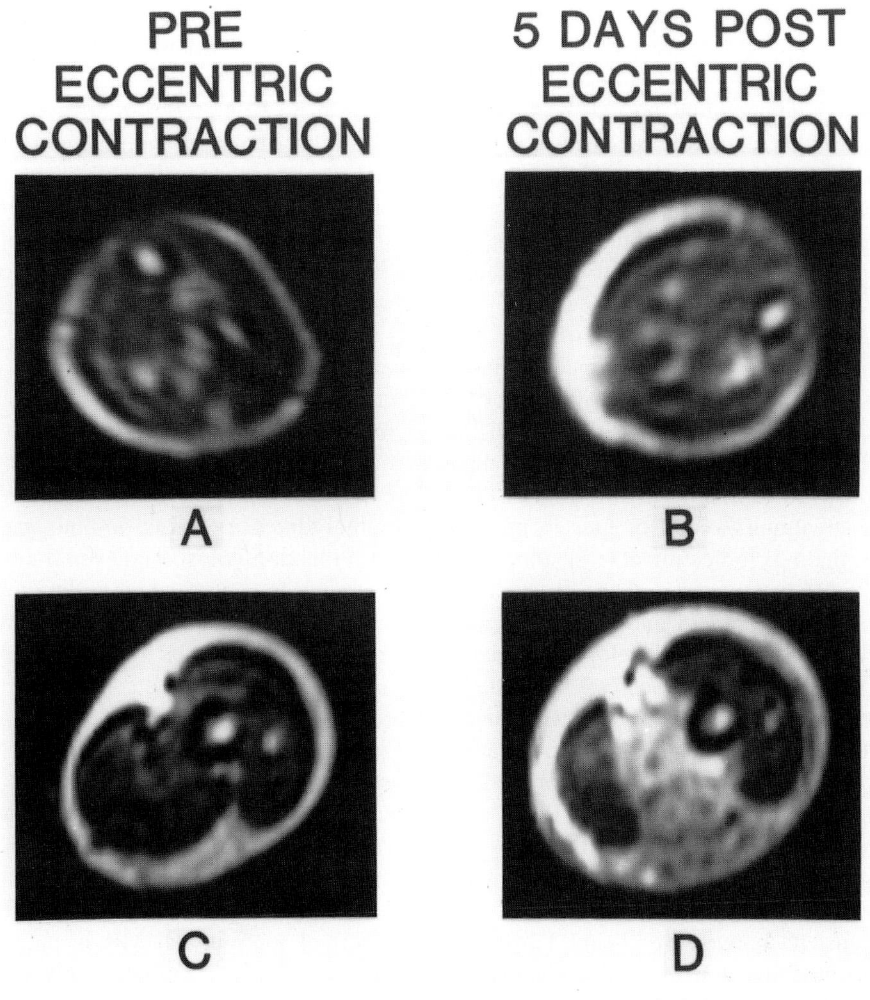

FIG. 7.

Findings: In Fig. 7A, axial T2-weighted (TR/TE 2,000/80 msec) spin echo images obtained from the middle upper arm and mid-forearm before (A,C) and on day 5 following the performance of eccentric contractions (B,D) until exhaustion. There is increased signal intensity involving the brachilalis and deep biceps musculature as well as increased signal intensity localized to the subcutaneous tissue of the forearm that is distal to the site of signal alteration within the muscle. (From Shellock FG, Fukunaga T, Mink JH, et al. Exertional muscle injury: evaluation of concentric versus eccentric actions with serial MR imaging. *Radiology* 1991;179:659–664, with permission.)

Diagnosis: Delayed-onset muscle soreness.

Discussion: Delayed-onset muscle soreness (DOMS) is defined as pain in skeletal muscle that occurs some time after exercise (1). The clinical feature that distinguishes DOMS from a muscle strain is the time of onset of pain and dysfunction. The soreness (which is usually localized to the insertion or attachment of the muscle) progressively increases in intensity over the first 24 to 48 hours following exercise, peaks at 2 to 3 days, and usually resolves by

7 days (2,3). The most predictable method of producing DOMS is to have a muscle perform eccentric contractions. Eccentric muscular actions involve the forced active lengthening or stretching of muscles and are considered the primary cause of exertion-related muscle injuries because excessive force develops in the muscle during this type of muscular action (i.e., greater tension is produced in muscle fibers when they stretch than when they shorten) (see Case 5). Although patients experience acute fatigue sooner after performing concentric rather than eccentric work, it is with the latter that they will develop DOMS (4).

Although DOMS is quite prevalent, and has been experienced by most adults on numerous occasions, rather little is known about the basic cellular mechanism involved with its production. Possible explanations for the observed muscle soreness include accumulation of metabolic waste, structural damage, and elevated muscle temperature. Muscle-based enzymes such as creatine phosphokinase (CPK) and lactic dehydrogenase (LDH) are routinely elevated in patients with DOMS, and the peak elevation of enzymes occurs from 18 to 30 hours postexercise, the same time at which DOMS is maximal (3). Histologic studies have demonstrated no gross fiber disruption, but on the ultrastructural level myofibrillar disturbances and Z-band streaming (a sign of muscle injury) have been observed, findings indicating at least some distortion of the contractile machinery in these overloaded muscles (5). The pain and soreness may in part be accounted for by increased tension that develops secondary to edema within intramuscular connective tissue (2). There is no apparent treatment for DOMS with anti-inflammatory medications, liniments, methanol, and camphor all having little value in reducing actual soreness.

The MR findings of DOMS are similar to those described for mild muscle strains. Edema within muscle bundles and perifascial fluid collections dominate the MR picture. Within 24 hours of exhaustive exercise, there is a predictable but transient increase in T1 and T2 relaxation times that can be visualized on T2-weighted and STIR sequences (6,7). These findings generally precede patient symptomatology. While initially more diffuse, within 24 to 72 hours, the signal becomes more defined, involving only those muscles that are sore. The MR signal intensity is proportional to the extent of ultrastructural myofibrillar alterations and to the volume of abnormal muscle on MR (8). The relative increase in T2 values correlates with the degree of functional impairment and pain, all of which become maximally severe by day 3 (6). T2 relaxation times remain statistically significantly increased for prolonged periods of time following the resolution of soreness. In one study, subclinical signal abnormalities were demonstrated as long as 75 days after the disappearance of symptoms (Fig. 7B). These abnormal MR findings are considered to be predominantly the result of accumulated edema, with a minor contribution from hemorrhage that is produced in response to muscle injury (2). Correlation of serum creatine kinase and MR signal alterations demonstrated that MR findings both preceded and persisted longer than other commonly used indicators of injury (3,7). Based on this MR experience, it has been suggested that clinical assessment of muscle injuries may be insufficient to demonstrate the total recovery of damaged muscles after exercise-induced muscle injury (2,7).

Exertional rhabdomyolysis may be thought of as an extreme degree of DOMS that results in release of cellular contents into the general circulation (9). Pathologically, myocyte swelling, inflammation, and hyaline degeneration are present (9). Crush injuries, burns, prolonged muscular compression, toxin exposure, hypoxia, and extremely intense exercise in hot weather can be associated with rhabdomyolysis and even frank muscle necrosis. The diagnosis is confirmed by finding myoglobin in the urine. Patients experience intense pain and weakness following exercise. Acute renal failure, tetany, compartment syndromes, and disseminated intravascular coagulation may complicate rhabdomyolysis.

In patients with rhabdomyolysis, MR demonstrates abnormal signal in affected muscle. Both T1 and T2 values are increased in affected muscles. MR can be extremely valuable in patients whose clinical syndrome suggests that decompressive fasciotomy is indicated by preoperatively localizing affected muscle groups (see Case 2). In a recent report of longitudinal MR in patients with rhabdomyolysis of diverse etiologies, the MR findings correlated precisely with the clinical and neurologic deficits of the patients. The signal changes resolved over time in all patients, suggesting that they are more likely reflective of edema and inflammation in the acute phases of rhabdomyolysis rather than permanent histologic changes (10).

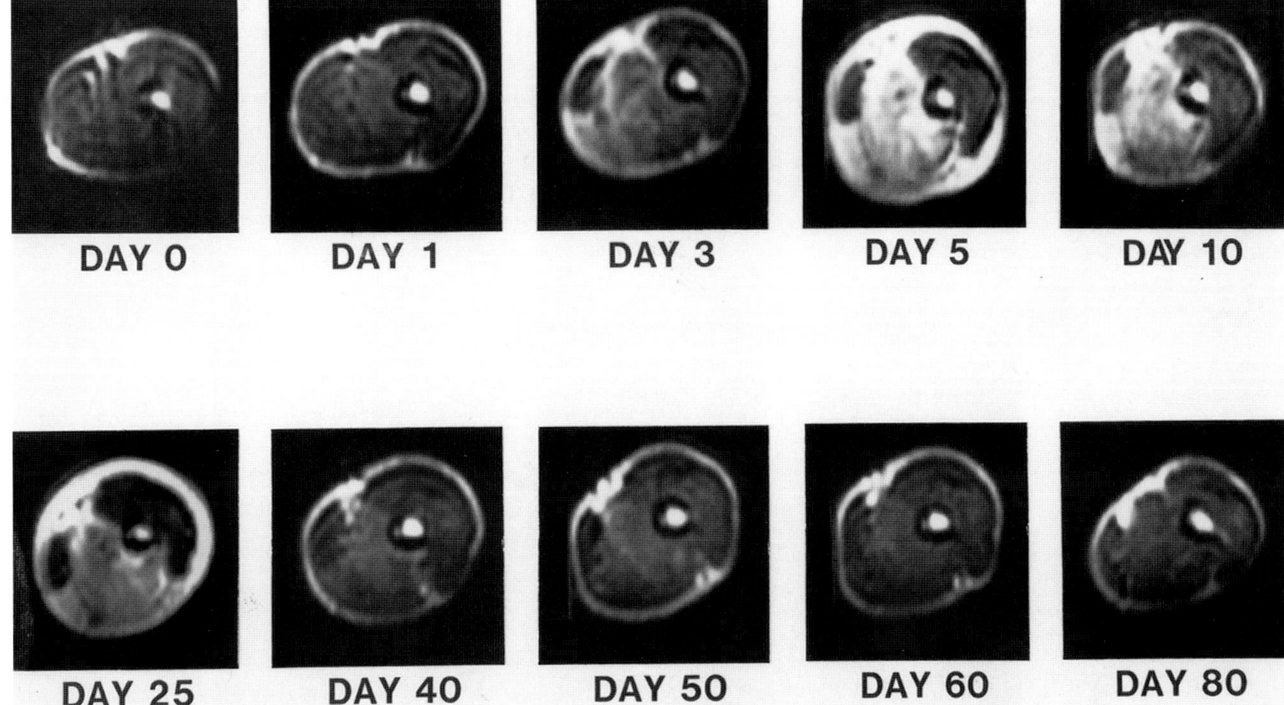

FIG. 7. B: Delayed-onset muscle soreness. Axial T2-weighted (TR/TE 2,000/80 msec) spin echo images obtained from the middle upper arm from a subject following strenuous eccentric muscle contraction. Minimal increased signal is seen within the biceps and brachialis muscles on day 1 with progressive increase by day 3. The greatest signal and morphologic abnormality is seen by day 5. Note the marked increase in overall circumference of the muscle most apparent on images from day 3 to 25. The signal progressively decreases following day 5, with a return to baseline in signal and circumference by day 80. (From Shellock FG, Fukunaga T, Mink JH, et al. Exertional muscle injury: evaluation of concentric versus eccentric actions with serial MR imaging. *Radiology* 1991;179:659–664, with permission.)

CASE 8

History: A 65-year-old woman presented with a progressively enlarging calf mass. Laboratory studies were unremarkable and a duplex study to exclude deep vein thrombosis was normal. Clinical suspicion centered on ruptured Baker's cyst. What are the principal differential considerations that should be considered for abnormalities involving muscle that remain relatively low signal intensity on T2-weighted spin echo sequences?

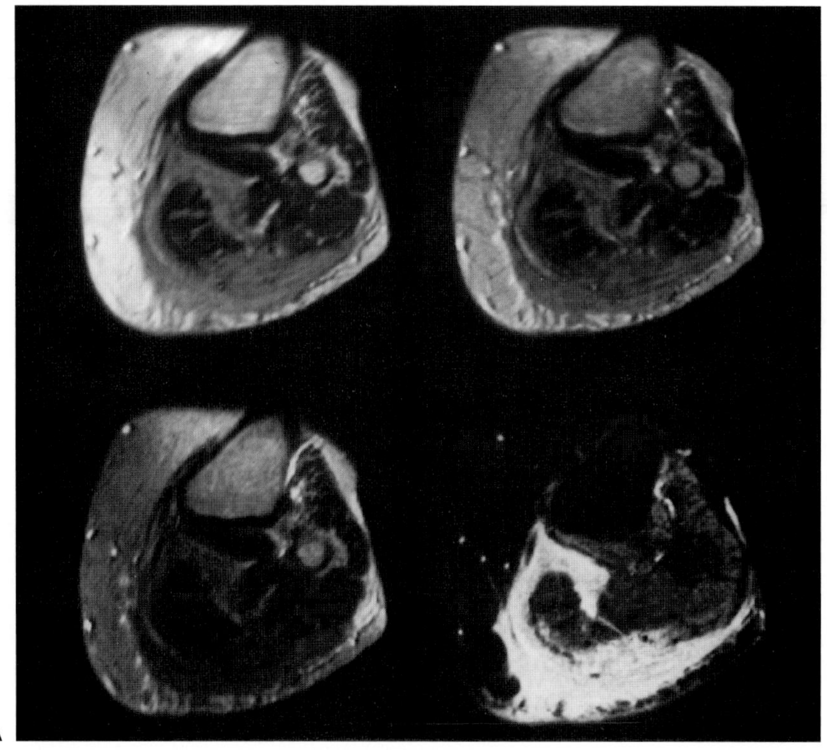

FIG. 8.

Findings: In Fig. 8A the axial PD (TR/TE 2,000/40 msec) (upper left), T2-weighted conventional spin echo (TR/TE 2,000/80 msec) (upper right), and heavily T2-weighted conventional spin echo (TR/TE 2,000/120 msec) (lower left) images of the calf demonstrate abnormal tissue that is hypointense to fat infiltrating the gastrocnemius muscle and surrounding fascial planes. Axial STIR image (lower right) demonstrates striking increased signal intensity of the infiltrating lesion contained within the medial head of the gastrocnemius muscle. (From Metzler JP, Fleckenstein JL, Vuitch F, et al. Skeletal muscle lymphoma. *Mag Reson Imag* 1992;10:491–494, with permission.)

Diagnosis: Skeletal muscle lymphoma.

Discussion: When a mass in or around skeletal muscle is hypointense to fat on conventional T2-weighted spin echo sequences, the possibility of lymphoproliferative disease should be one of the principal diagnostic considerations. As special stains are often required to differentiate soft tissue lymphoma from sarcomas when analyzing needle biopsy specimens, the importance of considering this possibility is underscored.

Muscle involvement in lymphoma may be seen in both Hodgkin's and non-Hodgkin's disease. Lymphoma can present either as a discrete mass within muscle or more commonly as diffuse muscle infiltration (2). Involvement of muscle may occur as a primary site of disease, although more commonly results from contiguous extension of a tumor mass.

On CT, the attenuation of the tumor is lower than that of normal muscle and the distinction is magnified when contrast material is utilized (2). On MR, lymphoma is most commonly isointense to hypointense in relation to fat on T2-weighted conventional spin echo sequences demonstrating a T2 relaxation time intermediate between that of normal muscle (30 msec) and that of fat (60 msec).

Other conditions that affect muscle and that may exhibit relatively short T2 relaxation times include aggressive fibromatosis, malignant fibrous histiocytomas, malignant schwannomas, scar tissue, and densely mineralized and/or hemosiderin-laden masses (1,3).

Another condition that might be considered in the differential diagnosis of conditions that do not demonstrate striking increased signal intensity on T2-weighted images is that of amyloid myopathy. This condition represents a rare manifestation of primary amyloidosis and is characterized clinically by progressive muscle weakness, stiffness, generalized enlargement of muscles, and a "woody" consistency of the limbs (4). The MR appearance of muscles infiltrated with amyloid is remarkable for the relatively minimal signal intensity changes observable. This is in contrast to the vast majority of myopathies in which the presence of focal fat and/or edema contributes to a remarkable appearance on MR images. Low signal intensity has been reported on both T1- and T2-weighted images in a patient with focal nasopharyngeal accumulation of amyloid (5) and MR reports of amyloid involvement of bone have been associated with decreased signal on T1-weighted and only minimally increased signal on conventional T2-weighted spin echo sequences (Fig. 8B,C) (6). As such, amyloidosis appears to represent another of the conditions that should be considered when abnormal muscle fails to demonstrate either markedly edematous or fatty changes on MR imaging. The paucity of MR changes in involved muscle prompts speculation that the firm and stiff extremities encountered clinically may not be as much on account of abnormal muscle but rather be reflective of amyloid deposition into subcutaneous tissue (4). This speculation is based on the strikingly recticulated and hypointense appearance of the subcutaneous fat on both T1- and T2-weighted images in patients studied on MR with amyloid myopathy (Fig. 8D) (4). Amyloidosis of the fat is supported by the fact that fat biopsy is a frequent test used to diagnose amyloidosis, as the amyloid is deposited in rings around fat cells (4). It is emphasized that the MR findings in muscle in patients with amyloid myopathy pale in contrast to the markedly abnormal physical findings. A hypointense streaky appearance to the subcutaneous fat may be the only striking visible abnormality.

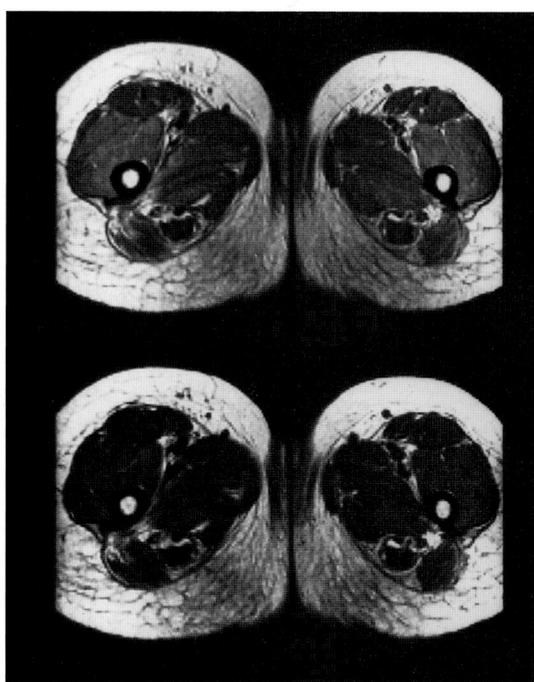
B

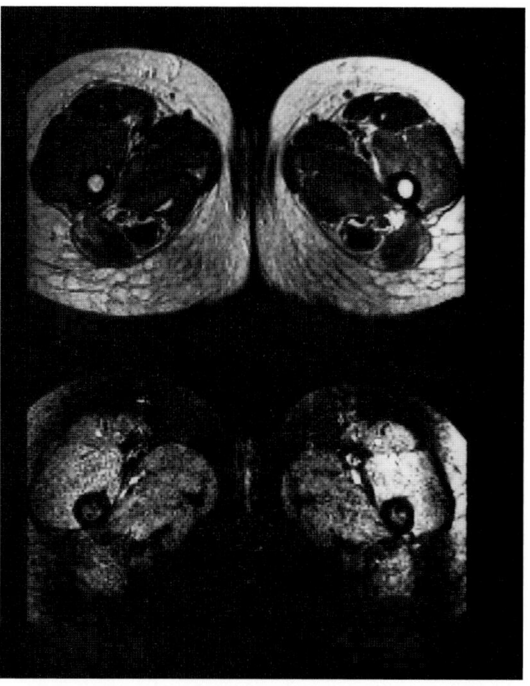
C

FIG. 8. B,C: Axial images at the mid-thigh level of a 57-year-old woman with an 8-year history of progressive muscle weakness and stiffness and biopsy-proven amyloid infiltration of muscle. **B:** On the T1-weighted spin echo image (TR/TE 500/30 msec) *(bottom)* the involved vastus muscles are essentially isointense with the surrounding noninvolved muscles. There is, however, a coarse reticular pattern of decreased signal intensity with the subcutaneous fat. There is only minimally increased signal intensity of the vastus muscles compared with the rectus femoris muscle appreciable on the PD image (TR/TE 2,000/30 msec *(upper image)*. **C:** On the T2-weighted (TR/TE 2,000/60 msec image *(upper)*, the muscles remain of only slightly increased signal intensity. On the STIR image *(lower)*, there is mildly augmented signal intensity.)

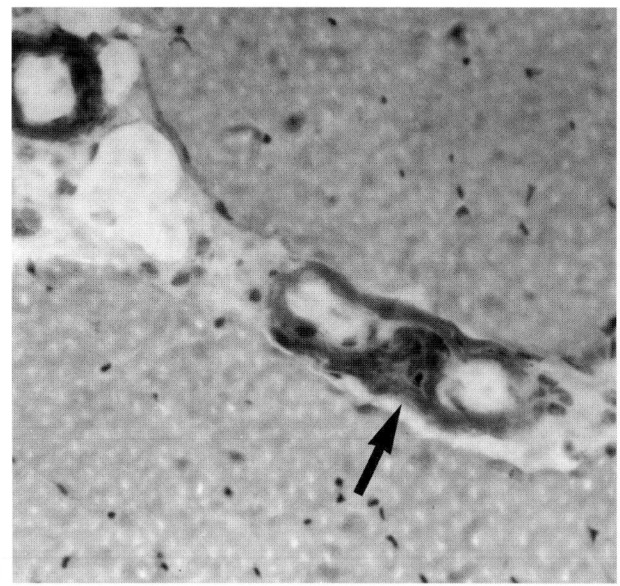

FIG. 8. D: Amyloid deposition. Amorphous deposits of congophilic proteinaceous material are present within the walls of perimysial blood vessels. These deposits demonstrated green birefringence when viewed with cross-polarizing light and were metachromatic on crystal violet stained preparations; findings characteristic of amyloid. No amyloid is visible inside myofibers. (From Metzler JP, Fleckenstein JL, White CL et al. MRI evaluation of amyloid myopathy. *Skeletal Radiol* 1992;21:463–465, with permission)

CASE 9

History: A 30-year-old man with long history of insulin-dependent diabetes mellitus presented with a 3-week history of painful swelling of the proximal calf.

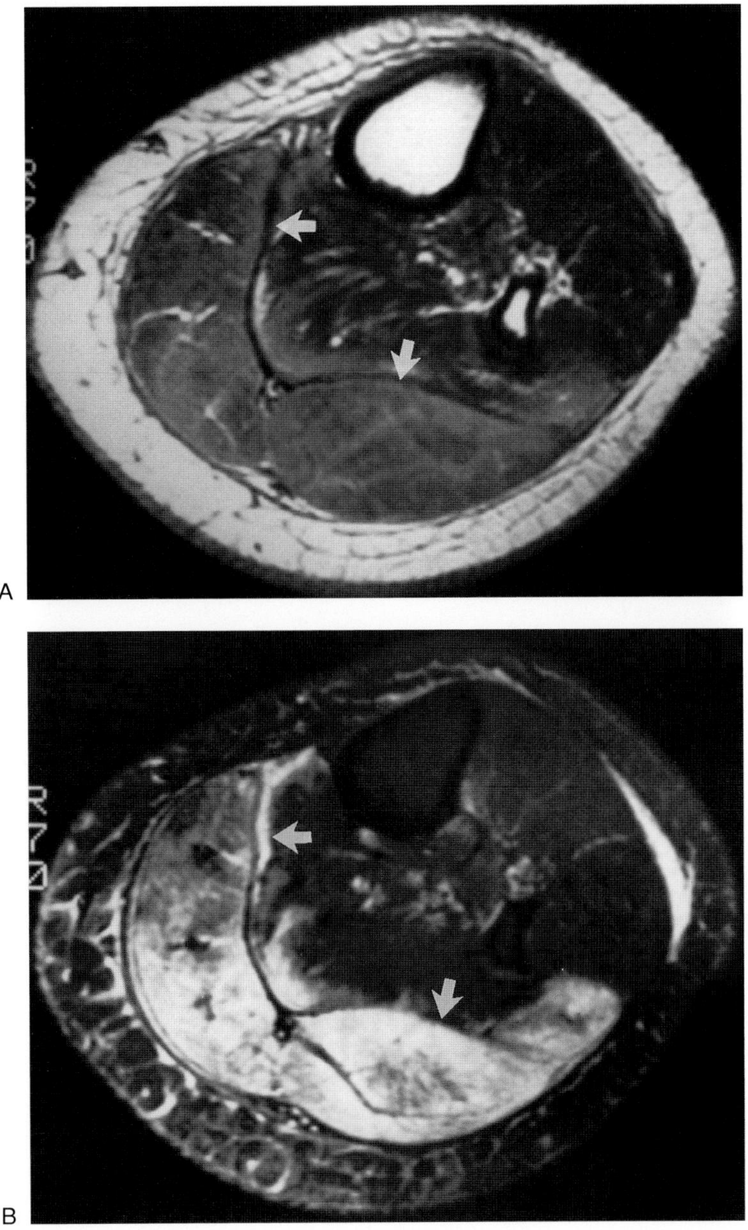

FIG. 9.

Findings: In Fig. 9A the axial T1-weighted (TR/TE 500/13 msec) image demonstrates mildly diffuse increased signal within the gastrocnemius muscle (arrows). There is a reticular appearance to the subcutaneous fat. In Fig. 9B the axial T2-weighted (TR/TE 2,500/60 msec) image demonstrates striking increased signal intensity throughout the affected muscle (arrows). The edematous appearance throughout the subcutaneous fat is well depicted. (Images courtesy of Leanne Seeger, M.D. Los Angeles, California.)

Diagnosis: Skeletal muscle infarction in diabetes mellitus.

Discussion: Skeletal muscle infarction is an uncommon but recognized complication of poorly controlled diabetes mellitus that has received little attention in the radiologic literature (1–3). It is generally seen in advanced disease after microvascular and neuropathic complications have developed and has been attributed to atherosclerosis obliterans (1,2). Patients present with excruciating pain and tender swelling of the involved muscle. The quadriceps musculature is the most commonly reported site of involvement and involvement may be bilateral (1–3). Prior involvement of the calf musculature has also been described.

With awareness of this entity, the diagnosis may be suspected on clinical grounds. Differential diagnosis includes deep venous thrombosis, myositis, abscess, ruptured Baker's cyst and neoplasm (1). MR is well suited to detection of muscle infarction, although the appearance is not specific. The affected muscle group demonstrates typical signal changes of edema with decreased or isointense signal to muscle on T1-weighted sequences and increased signal on T2-weighted and short inversion time inversion recovery sequences. Experimental work in rat models has also demonstrated similar signal changes (4).

The appearance is of feathery edema without mass and no discrete ''fluid'' collections that would be suggestive of abscess are noted. MR can exclude marrow involvement and define the extent of the process. Familiarity with this entity may allow the diagnosis to be strongly suggested in the appropriate clinical setting. Percutaneous biopsy can be performed for histologic verification, although some authorities have advised against biospy in favor of imaging procedures out of concern for possible biopsy complications (1). Follow-up studies have demonstrated resolution of the changes over time although experience in this regard is quite limited.

CASE 10

History: A 63-year-old man underwent MR evaluation for hip pain. What is the differential diagnosis for the illustrated muscle abnormality?

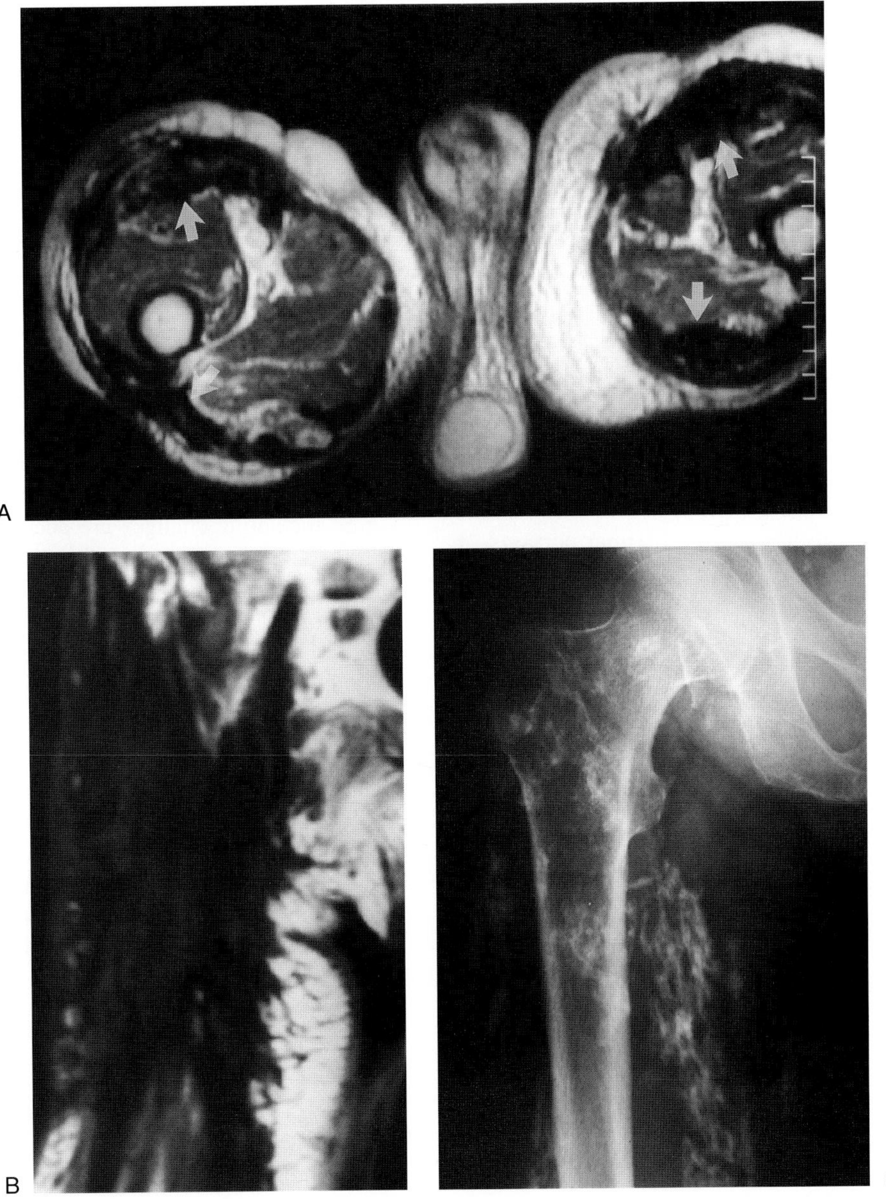

FIG. 10.

Findings: In Fig. 10A the axial PD-weighted fast spin echo (TR/TE 4,500/22 msec) image demonstrates striking relatively peripheral low signal intensity areas circumferentially involving the proximal thigh musculature bilaterally (arrows). In Fig. 10B the coronal T1-weighted (TR/TE 540/15 msec) image through the anterior thigh demonstrates the sheet-like areas of low signal intensity. In Fig. 10C the AP radiograph of the right thigh demonstrates diffuse longitudinally oriented areas of calcification.

Diagnosis: Dermatomyositis.

Discussion: Dermatomyositis represents a disorder of striated muscle characterized by nonsuppurative inflammation and degeneration. In contrast to polymyositis, in dermatomyositis there is involvement of skeletal muscle and skin. Muscle weakness eventually occurs in almost all affected individuals and is the presenting complaint in almost 50% of patients (1). Symmetric proximal muscle involvement is most common and muscle weakness can occur slowly or rapidly (1). Typical or atypical skin rashes eventually occur in 40% to 60% of patients and are an initial manifestation of the disease in 25% of patients (1). Arthralgias and arthritis are present in 20% to 50% of patients. In childhood dermatomyositis, which represents approximately 20% of all cases, the disease may be severe and the extent of soft tissue calcification is typically more extensive (1). In patients presenting over the age of 40 (particularly men) there is a significant increased incidence of malignancy (most often carcinoma) (1).

Radiographically demonstrable abnormalities include those of soft tissue and those of articulations. As previously indicated, soft tissue changes are typically more common and severe in the younger age groups. The most common soft tissue abnormality is calcification, which may be highly variable (1). Small or large calcareous intermuscular fascial plane calcification is distinctive, although it may not be as common as subcutaneous calcification. The most common sites of intermuscular calcification are the large muscles of the proximal portions of the limbs. The appearance of subcutaneous calcification may simulate those in scleroderma including linear and curvilinear calcifications with predilection for elbows, knees, and fingers (1).

Several studies have addressed the utility of MR for assessment of the distribution, extent, and activity of dermatomyositis (2–8). Differences in patterns of muscle involvement in dermatomyositis as compared with congenital myopathies and muscular dystrophies have been noted (4). Predominant involvement of the vastus lateralis and to a lesser extent the vastus intermedius with relative sparing of the rectus femoris and biceps femoris muscles has been reported in patients with dermatomyositis (4). This ability of MR to assess the distribution of disease activity allows for greater precision in selection of sites for muscle biopsy as well as for correlation of MR imaging findings. Fraser and colleagues (6) have admonished against simply equating MR signal alteration with inflammation, emphasizing that patients with dermatomyositis may demonstrate a broad spectrum of histologic findings including fiber destruction, perifascicular atrophy, and perivascular inflammation that can contribute to MR signal changes. Correlation between MR imaging studies utilizing short inversion time inversion recovery sequences and muscle biopsy has revealed a correlation between active inflammation and increased signal intensity on this pulse sequence (6). Studies in patients with juvenile dermatomyositis utilizing visual inspection of images without histologic correlation have also reported correlation between increased signal on T2-weighted images and apparent disease activity. Increased signal intensity within affected muscles has been identified at the onset of disease as has diminution to resolution of signal changes with evidence of clinical improvement. Return of signal alteration with clinical evidence of recurrence has also been reported (2,3). It has been postulated that shifts in the distribution of water owing to muscle inflammation and infarction are primarily responsible for the imaging abnormalities.

Quantitative analysis of T1 and T2 values has also been explored in the assessment of dermatomyositis and degree of inflammation (4). Park and colleagues (4) found that it was readily possible to distinguish between inflammation in muscle (with T1 values in the approximate range of 1500 to 2000 msec) from normal muscle (1300 msec) and fat (350 msec). Magnetic resonance spectroscopy (MRS) has also been utilized in the assessment of disease activity in dermatomyositis (4,6). MRS allows in vivo monitoring of relative concentrations of high energy phosphate metabolites, such as phosphocreatine, ATP, ADP, and AMP. Quantitative analysis of 31-P spectra follows the bioenergetics of exercise response by using the fact that the concentration of phosphocreatine is relatively high in resting muscle but decreases rapidly with exercise. These metabolic changes are accompanied by a complementary increase in inorganic phosphate and a reduction in pH. Within minutes, all of these parameters return to their normal resting values. Park and colleagues (4), in evaluating the 31-P spectra of the quadriceps muscles in patients with dermatomyositis, found that concentrations of adenosine triphosphate and phosphocreatine (PCr) in diseased muscles were 30% below normal values, and that the inorganic phosphate/PCr ratios were increased in the patients muscles at rest and throughout exercise. Although there is a lack of specificity to this parameter, the Pi/PCr can be a useful indicator of the clinical status of patients with a given disease (4). In the study of Park et al., Pi/PCr ratios were more elevated in the severely debilitated subjects and the increase in the ratio was due to lower levels of PCr in the patients muscles and not to the Pi concentrations, which were the same in diseased and normal muscle. In muscular disorders, the levels of PCr are always lower than normal (4). ATP concentrations, in contrast, are less predictable. As previously mentioned, both ATP and PCr values were significantly decreased in the dermatomyositis patients studied by Park and colleagues. The absolute decrease in PCr was greater and more readily detectable than that of ATP, suggesting that PCr may be a better indicator of the clinical status of patients with dermatomyositis (4).

While much work remains to be done, it appears evident that MR demonstrates great promise in the evaluation of inflammatory myositis. Signal intensity changes in muscle appear to correlate with degrees of inflammation, changes in signal patterns appear to correlate with clinical disease activity, and sites of biopsy can be ascertained with greater certainty utilizing MR.

CASE 11

History: A 50-year-old woman presented with pain and fullness in the region of the posterior left shoulder of several weeks' duration. What is the most likely diagnosis?

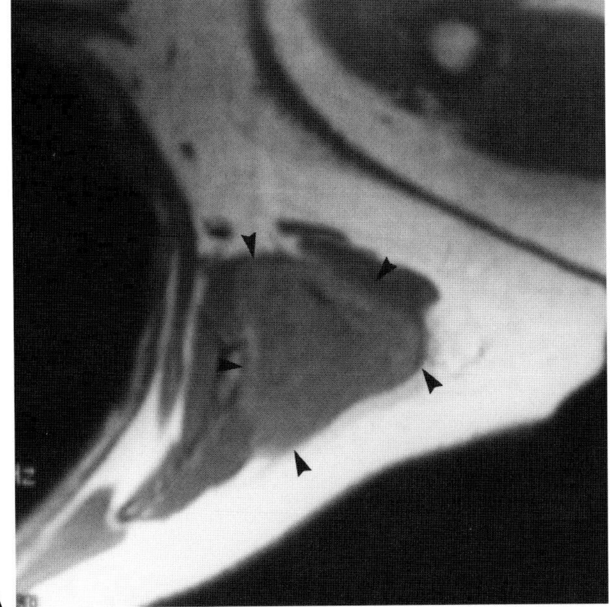
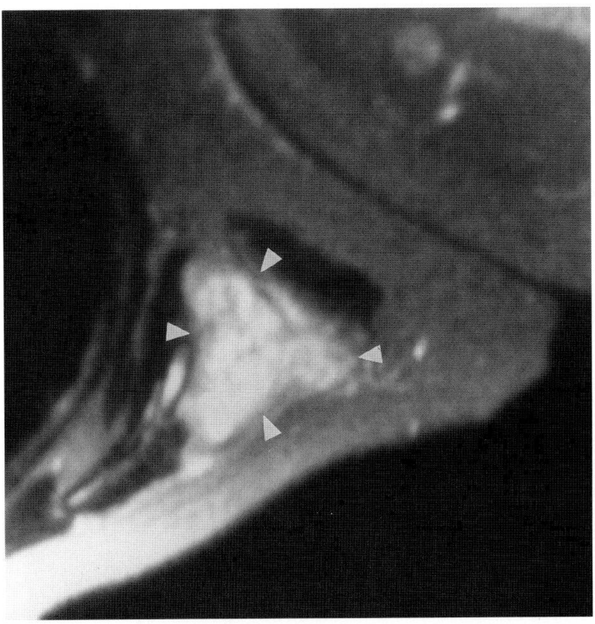

FIG. 11.

Findings: In Fig. 11A the axial T1-weighted (TR/TE 800/20 msec) image of the left scapula region demonstrates an ill-defined area of slightly increased signal intensity compared with normal muscle (arrowheads). In Fig. 11B the axial T2-weighted fat-suppressed fast spin echo (TR/TE 4,000/96 ef) image better demonstrates the intramuscular process, which becomes of relatively uniform increased signal intensity (arrowheads).

Diagnosis: Myositis ossificans.

Discussion: Myositis ossificans represents a benign, self-limited condition characterized by an ossifying soft tissue mass usually occurring within muscle (1–4). The term *myositis* is a misnomer as no primary inflammation of skeletal muscle is seen with the process (4). A history of trauma is only inconsistently obtained. On conventional radiographs, myositis ossificans may become apparent within 2 to 6 weeks following the onset of symptoms and is initially characterized by faint areas of calcification (1,2). The more classic appearance associated with this entity (e.g., well-defined ossific area with dense peripheral rim oriented parallel to the long axis of an adjacent tubular bone) is usually apparent by 6 to 8 weeks (1,2,4). The ossific shell or rim is best demonstrated on plain radiographs and computed tomography (Fig. 11C).

The MR appearance of myositis ossificans has been reported in several studies, the most extensive of which was reported by Kransdorf and colleagues (4). Myositis ossificans is most often seen as a relatively well defined, inhomogeneous soft tissue mass. Diffuse surrounding edema can be quite prominent in lesions imaged within 8 weeks of the onset of symptoms and the edema may last for several months (4). Surrounding edema is not observed in mature lesions (4). Areas of mineralization are seen as curvilinear and irregular regions of decreased signal within and along the periphery of the mass (4). These findings are typically subtle and much less apparent than on CT examinations.

Early lesions of myositis ossificans are characterized histologically by a nonossified core of proliferating fibroblasts and myofibroblasts with only a minor component of osteoid and mature lamellar bone at the periphery (Fig. 11D,E) (4). Intermediate lesions are characterized by either minor or no proliferating fibroblastic core, and consist almost entirely of osteoid rimmed by active osteoblasts surrounded by a shell of mature lamellar bone (4). On MR, early and intermediate lesions demonstrate high signal intensity on T2-weighted spin echo images (4). The margins of early lesions may be difficult to differentiate from the surrounding edema in some cases. T1-weighted images may appear grossly normal or demonstrate the presence of a mass by distortion of fascial planes (4). Marked enhancement may be seen following the administration of gadopentetate dimeglumine. Marrow edema may be depicted in bones contiguous to the lesions (3,4).

Late lesions consist of mature lamellar bone. On MR they appear as well-defined inhomogeneous masses with signal intensity approximating that of fat on both T1- and T2-weighted sequences. There is no associated surrounding edema (4). A rim of decreased signal intensity may be depicted around the lesion and similar areas of decreased signal may be seen within the lesion and are believed to represent bone trabeculae (4). Areas of hemosiderin deposition from previous hemorrhage and fibrosis may also contribute to the areas of low signal intensity seen within mature lesions (4).

The nonspecificity of MR in regard particularly to imaging early and intermediate-age lesions (which may display imaging characteristics commonly associated with malignant neoplasms) has been emphasized by Kransdorf and colleagues (4). Such lesions must be approached with a high index of suspicion to suggest the correct diagnosis if MR is the primary imaging examination utilized. Both conventional radiographs and particularly CT can be of considerable value as a consequence of their greater sensitivity to depicting the characteristic mineralization pattern associated with myositis ossificans.

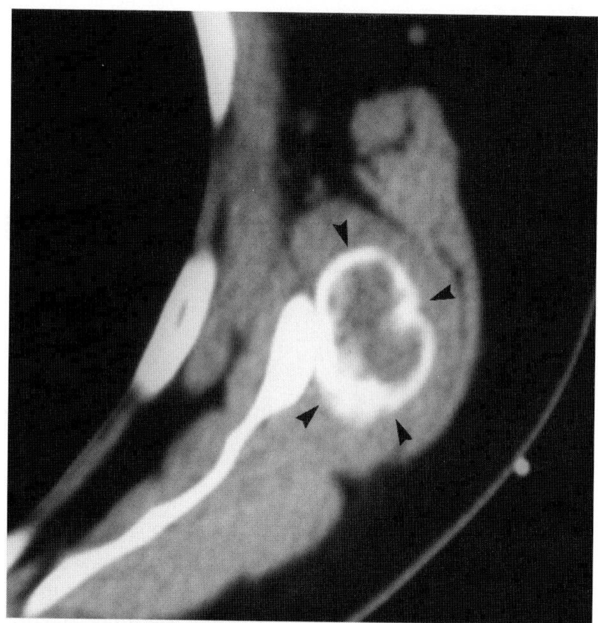

FIG. 11. C: Myositis ossificans. Axial section from a CT examination performed 10 days after the MR study and which was obtained for the purposes of imaging directed biopsy. A well-defined ossific rim is present *(arrowheads).* Given the characteristic appearance on CT no biopsy was performed and conservative management elected.

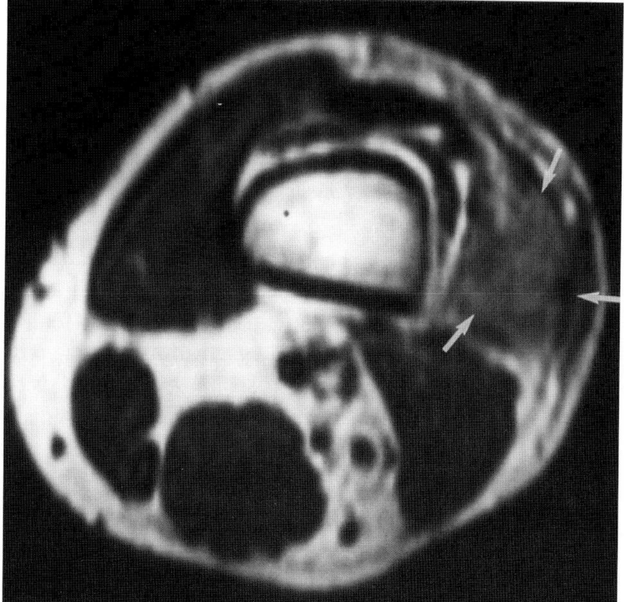

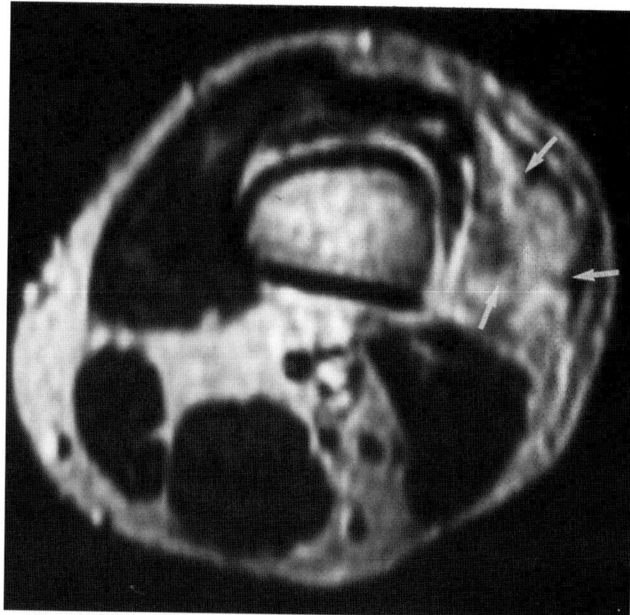

FIG. 11. D,E: Myositis ossificans in a different patient. A 36-year-old man presented with a palpable abnormality in the medial thigh. No history of trauma was obtained. **D:** Axial proton density (TR/TE 2,000/30 msec) sequence demonstrates poorly defined increased signal intensity throughout the vastus medialis muscle *(arrows).* No discrete mass is noted. **E:** Axial T2-weighted (TR/TE 2,000/80 msec) sequence demonstrates the lesion to become diffusely hyperintense *(arrows).* The case illustrates the nonspecific appearance of early myositis ossificans.

CASE 12

History: A 29-year-old African-American man presented with bilateral thigh pain. What is your diagnosis? What is the role of MR in the evaluation of this condition?

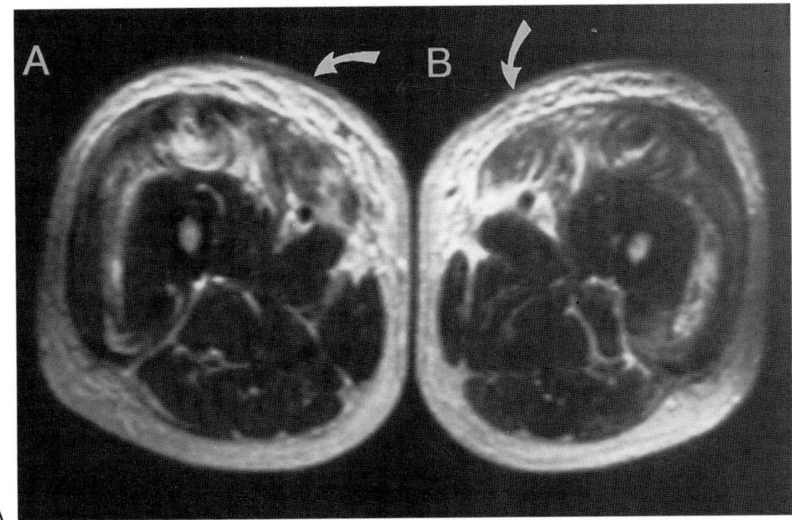

FIG. 12.

Findings: In Fig. 12A an axial T2-weighted SE (TR/TE 2,100/100 msec) image demonstrates markedly increased signal intensity with the vastus medialis and lateralis musculature bilaterally. T1-weighted sequences (not shown) demonstrated slightly increased signal corresponding to these areas. (From Feldman F, Zwass A, Staron RB et al. MRI of soft tissue abnormalities: a primary cause of sickle cell crisis. *Skeletal Radiol* 1993;22:501–506, with permission).

Diagnosis: Acute myositis reflecting a soft tissue manifestation of sickle cell crisis.

Discussion: Accurate assessment of the soft tissues in patients presenting with possible sickle cell crisis is critical to the determination as to whether conservative management can be employed or more aggressive methods (including fasciotomy) might be necessitated. The clinical documentation of sickle cell crisis remains a challenge. Clinical (e.g., swelling, warmth, tenderness) and laboratory findings (elevated temperature, leukocyte count, sedimentation rate, serum phosphokinase, and decreased circulating dense red blood cells) are nonspecific and inconstantly present (1–3). Patients commonly present for emergency treatment with repetitive attacks, and frequent usage of narcotics may further complicate the clinical picture.

MR has been reported as useful in assessment of the soft tissues in patients presenting with possible acute sickle cell crisis (1,4). A variety of patterns of soft tissue involvement with or without concomitant bone changes have been described (1). The principal criteria for acute insult as a manifestation of sickle cell crisis has consisted of augmented signal intensity on T2-weighted spin echo sequences, either focal or diffuse, as compared with the appearance on T1-weighted images. Changes within soft tissue can be seen adjacent to coexistent marrow changes, in the same extremity but isolated from involved marrow (Fig. 12B–D), or remote from involved marrow (e.g., in a different extremity) (Fig. 12E). A pattern of soft tissue changes alone without marrow involvement can also be identified on MR. The lack of signal increase on T2-weighted images can be utilized to exclude acute crisis (Fig. 12F). MR has established that symptoms in patients with sickle cell disease, while previously attributed to arthritis, osteomyelitis, or osteonecrosis, may be predominantly or solely related to soft tissue pathology. Negative MR studies may expedite conservative treatment and/or discharge of patients with a history or repeated emergency room visits and no objective laboratory findings and/or instances of suspected drug abuse (1).

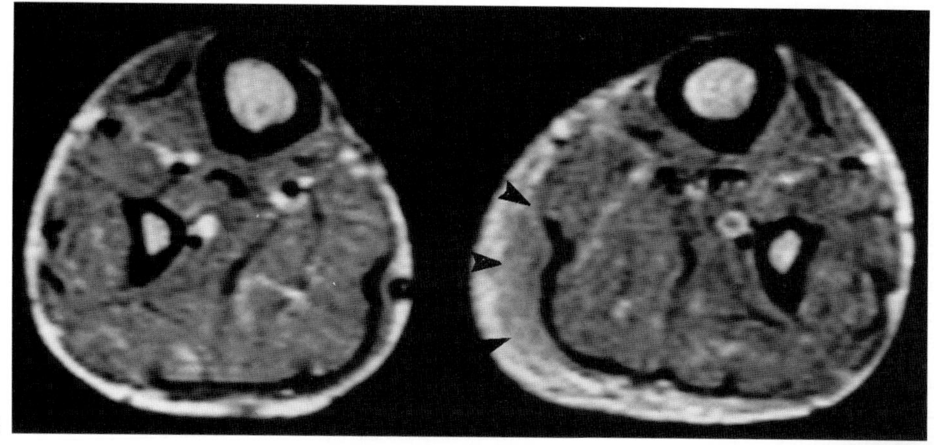

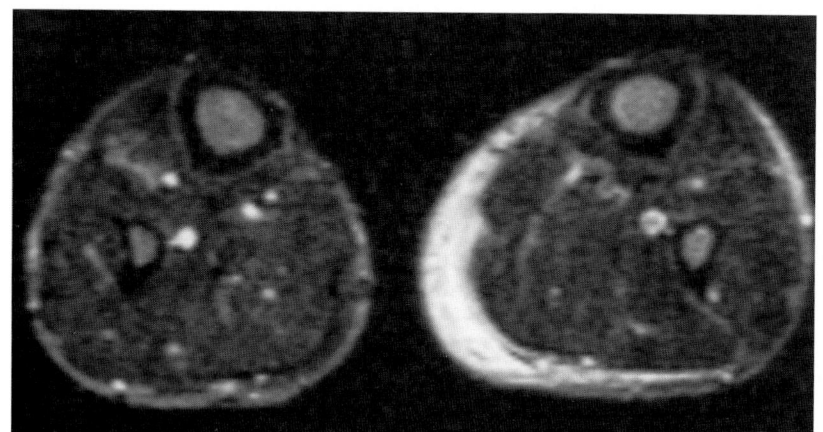

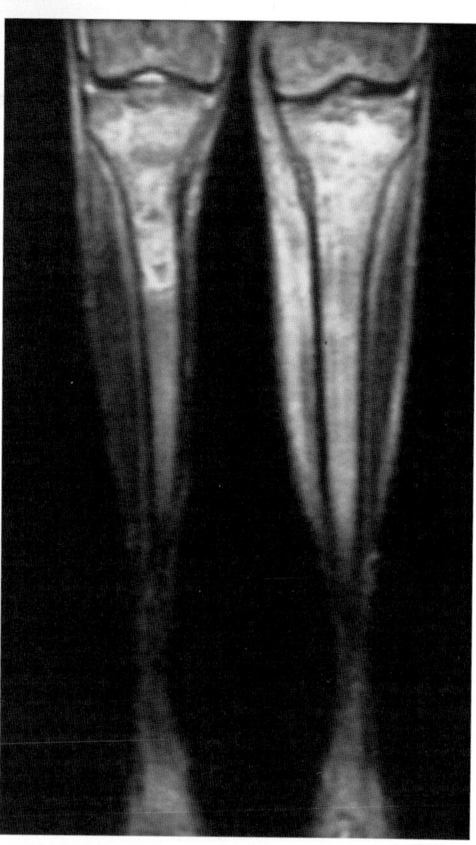

FIG. 12. B–D: A 28-year-old African-American man presented with painful legs and swollen left calf. **B:** Axial T1-weighted (TR/TE 866/20 msec) demonstrates locally increased girth involving the posteromedial subcutaneous tissues and fat on the left *(arrowheads).* This area demonstrates moderately increased signal intensity that has been variably attributed to proteinaceous edema, blood, and cellular infiltrates associated with inflammation and/or ischemia. The marrow at this level appears normal. **C:** Axial T2-weighted (TR/TE 2,233/100 msec) MR. The previously defined area demonstrates further augmentation in signal intensity. **D:** Coronal T2-weighted (TR/TE 2016/100 msec) image demonstrates asymmetric soft tissue swelling with increased signal intensity on the left, which extends from the knee to the calf. Increased signal in the proximal tibias (not included in B and C) are compatible with marrow edema ischemia and or infarction. Right-sided soft tissues are unaffected. (From Feldman F, Zwass A, Staron RB, et al. MRI of soft tissue abnormalities: a primary cause of sickle cell crisis. *Skeletal Radiol* 1993;22:501–506, with permission.)

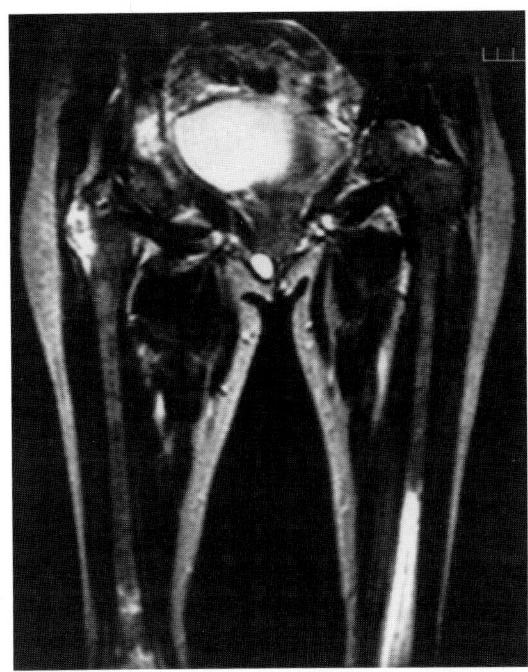

E

FIG. 12. E: A 39-year-old African-American woman with pain in the right hip and thighs. Coronal T2-weighted image demonstrates diffuse increased signal intensity involving the distal left femur consistent with an acute infaction. High signal intensity in the right trochanteric bursa, intermuscular septa, and fascia suggest either primary bursitis or sympathetic effusion due to neighboring fascitis. Bilateral abnormalities were compatible with clinical symptoms.

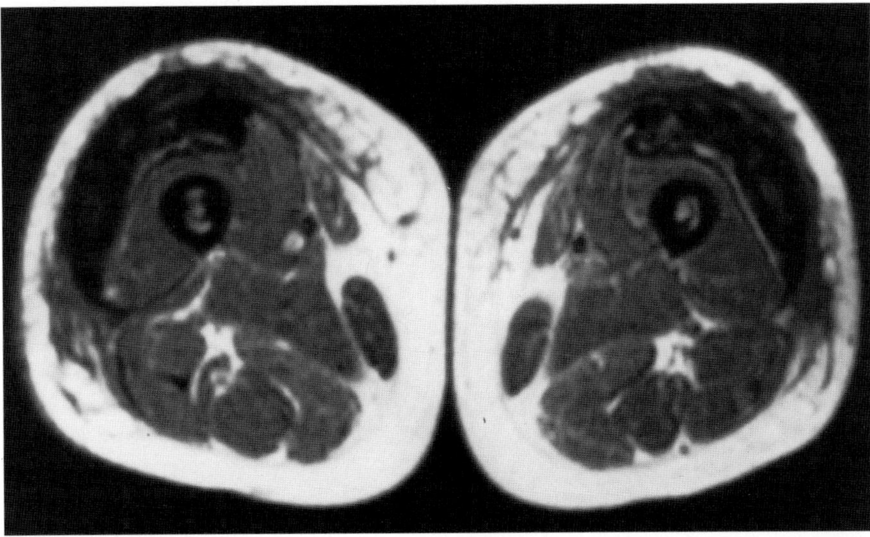

F

FIG. 12. F: A 41-year-old African-American woman complained of severe bilateral thigh pain. Axial T1-weighted spin echo (TR/TE 533/30 msec) demonstrates large sausage-shaped areas of decreased signal intensity in the anterolateral aspect of both thighs as well as low signal hematopoietic marrow. No change in signal intensity was demonstrated on T2-weighted images (not shown). Note similarity in appearance to case of dermatomyositis (see Case 10). The signal changes were attributed to prior infarction and fibrosis. The patient was discharged on the basis of the MR. (From Feldman F, Zwass A, Staron RB et al. MRI of soft tissue abnormalities: a primary cause of sickle cell crisis. *Skeletal Radiol* 1993;22:501–506, with permission).

CASE 13

History: A 28-year-old runner complained of intermittent claudication. What is your diagnosis?

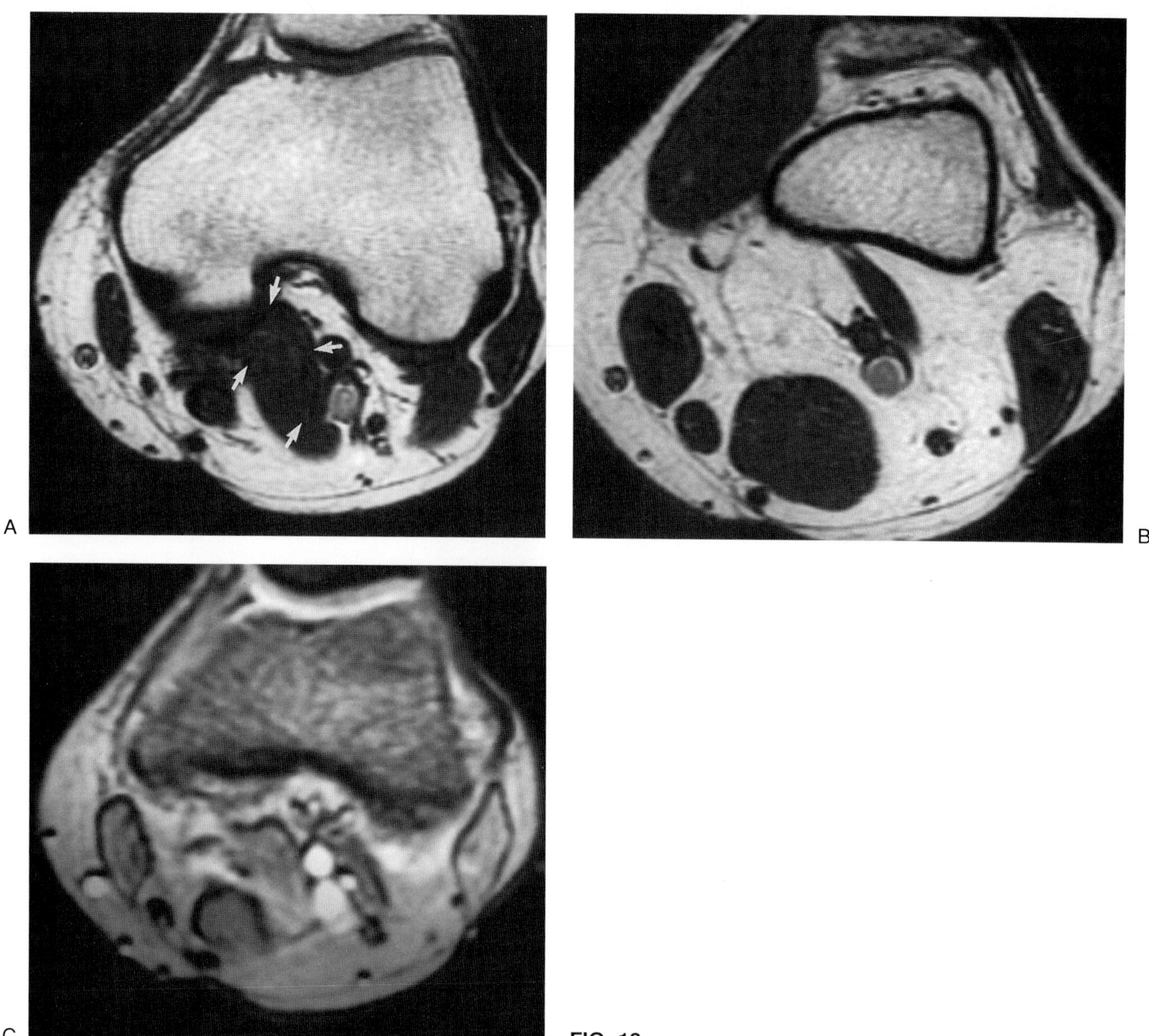

FIG. 13.

Findings: In Fig. 13A the axial T1-weighted (TR/TE 600/20 msec) image through the distal femoral condyles demonstrates a prominent medial head of the gastrocnemius muscle (arrows). In Fig. 13B the axial T1-weighted (TR/TE 600/20 msec) image is obtained more proximally than in Fig. A through the distal femur. A band-like slip of muscle is seen attaching to the femur and extending medial to the popliteal vessels. In Fig. 13C the axial gradient echo image is obtained at a level between Figs. A and B. The popliteal vessels demonstrate high signal intensity. Both the medial head of the gastrocnemius and the band-like slip are identified and form a sling around the popliteal vessels.

Diagnosis: Anomalous laterally inserting slip of the medial head of the gastrocnemius muscle.

Discussion: MR is excellently suited for detection and precise characterization of muscle anomalies that may simulate soft tissues masses (e.g., accessory soleus muscle) or contribute to symptomatic conditions (e.g., neurocompression or vascular compression). One such entity that can be evaluated by MR is that of popliteal artery entrapment produced by compression of the popliteal artery by the medial head of the gastrocnemius muscle (1,2). This condition is generally manifest as intermittent claudication in a young, otherwise normal individual. The compression may be due to either an abnormal position of the artery medial to the medial head of the gastrocnemius muscle or to compression of a normally situated popliteal artery by an anomalous laterally inserting slip related to the medial head of the gastrocnemius muscle, producing a ''sling'' around the artery (1,2).

In the differential diagnosis of claudication in young physically active patients is adventitial cystic disease of the popliteal artery. Adventitial cystic disease is characterized by a thickened cystic region of the arterial adventitia filled with mucoproteins and mucopolysaccharides (3–5). This condition is usually seen in men under the age of 30 and is usually unilateral, in contrast to popliteal artery entrapment, which may be bilateral in up to 25% of cases. In adventitial cystic disease, peripheral pulses are usually lost in flexion, whereas popliteal artery entrapment often causes loss of peripheral pulses with knee hyperextension. MR can be of value in differentiating these two conditions by demonstrating the anomalous muscle in popliteal artery entrapment and the cystic collection with the wall of affected popliteal artery in cystic adventitial degeneration (Fig. 13D–F).

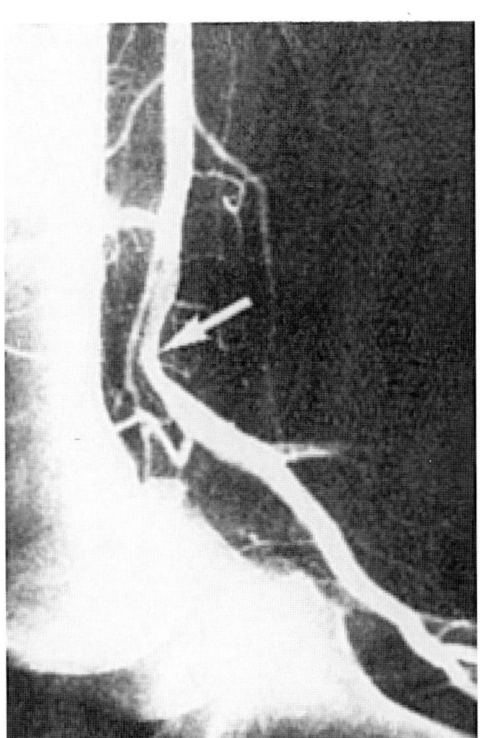
D

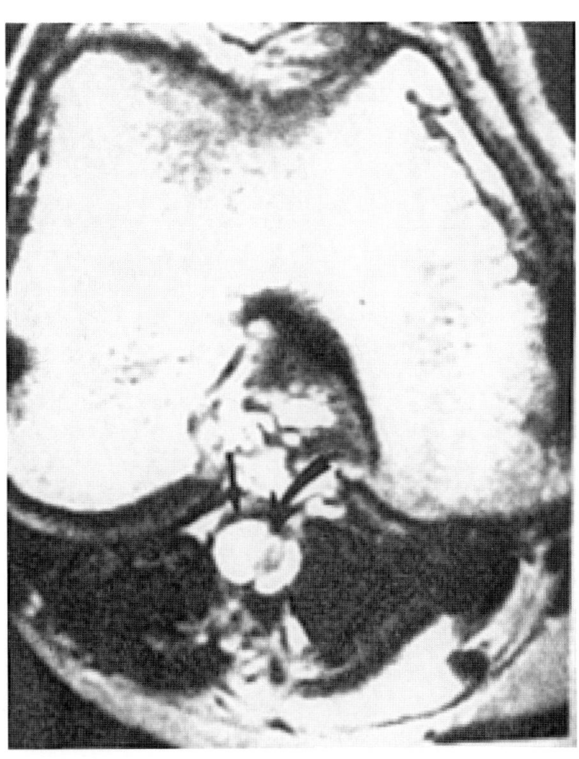
E

FIG. 13. D–F: Cystic adventitial disease of the popliteal artery. **D:** Conventional angiogram discloses focal narrowing of the high popliteal artery *(arrow)*. **E:** Axial T2-weighted MR discloses high signal intensity intramural cystic collections *(straight arrows)* and associated popliteal luminal narrowing *(curved arrow)*.

F 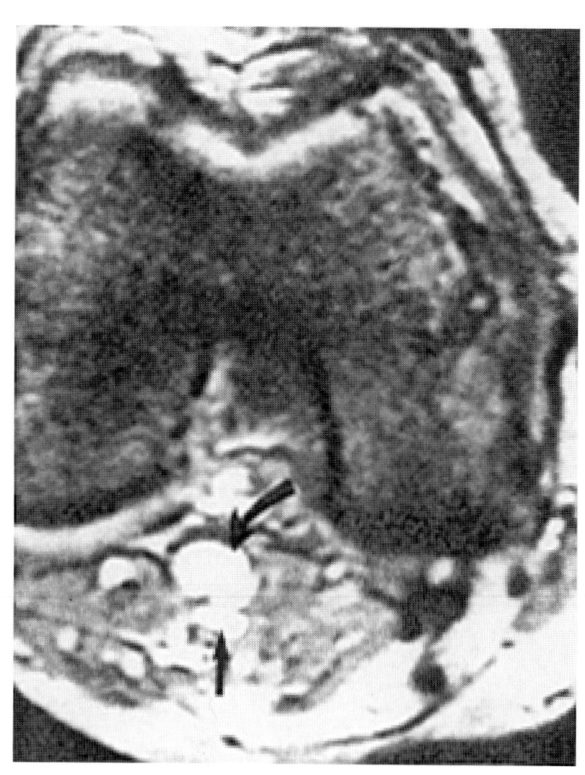

FIG. 13 F: Axial gradient-echo image again demonstrates cystic adventitial disease, with high signal intensity signifying flow in the narrowed popliteal artery *(curved arrow)* and vein *(straight arrow).* (Courtesy of Mark S. Schechter, M.D. From Reicher MA. Knee joint disorders. In: Mink JH, Reicher MA, Crues JV, Deutsch AL, eds. *MRI of the knee, 2nd edition.* New York: Raven Press, 1933, with permission.)

CASE 14

History: A 68-year-old woman presented with a 2-month history of hip and proximal thigh pain. Plain radiographs obtained at her orthopedic surgeon's office reportedly revealed poorly defined calcification or ossification and apparent cortical irregularity. The patient was referred for MR imaging to evaluate for associated soft tissue mass and to assess for possible neoplasm.

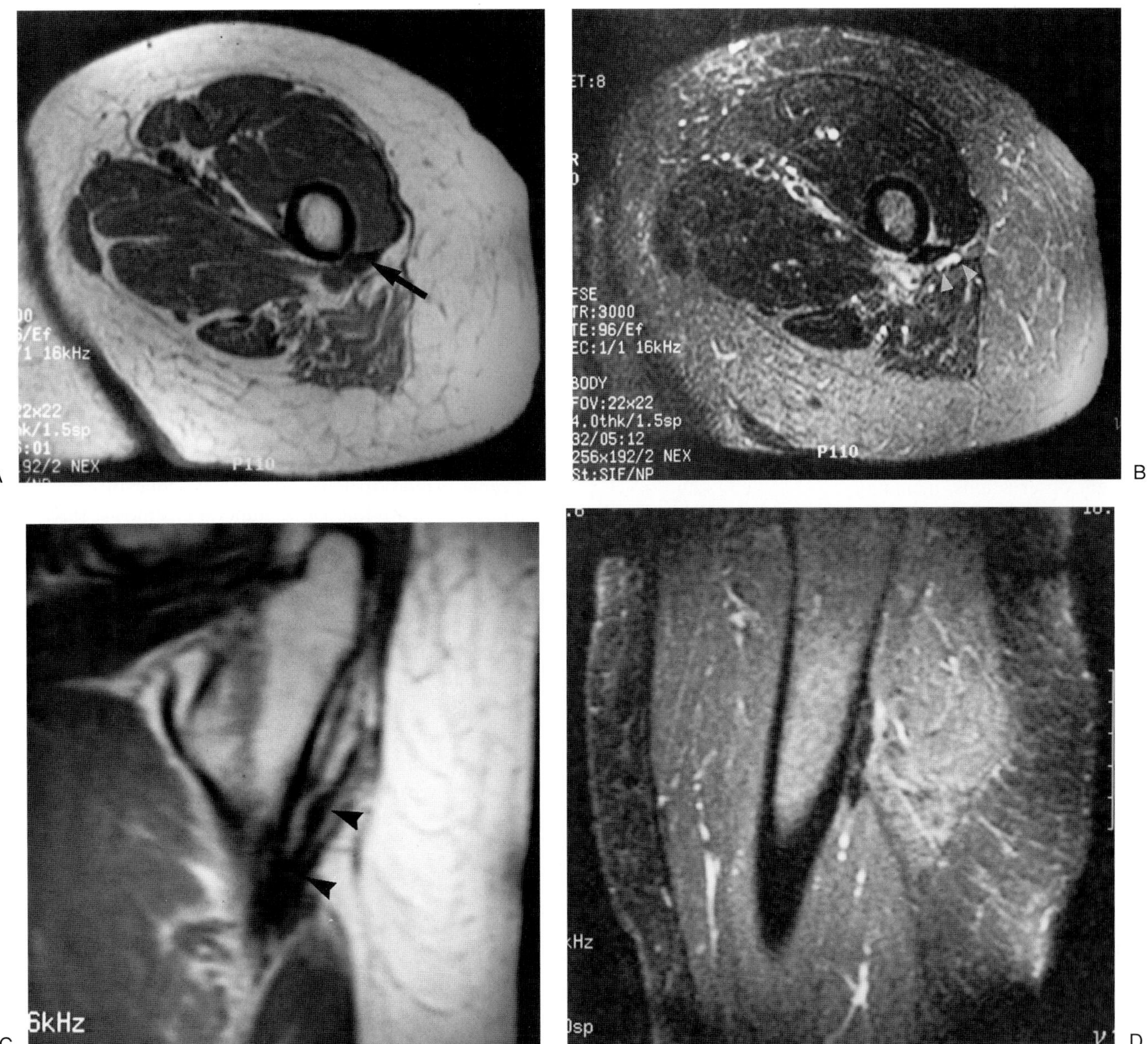

FIG. 14.

Findings: In Fig. 14A an axial T1-weighted fast spin echo (TR/TE 600/16 eff) demonstrates a small focus of low signal intensity immediately posterior to the femur (arrow). In Fig. 14B an axial T2-weighted fast spin echo (TR/TE 3,000/96 msec) demonstrates a small focus of increased signal intensity in the immediately surrounding soft tissue (small arrowheads). In Fig. 14C a coronal T1-weighted fast spin echo (TR/TE 600/17 eff) demonstrates the gluteus maximus tendon as it courses to attach to the femur (arrowheads). In Fig. 14D a sagittal fast inversion recovery (TR/TE/TI 2,800/36 msec/130) sequence well demonstrates the longitudinal extent of the low signal intensity focus.

Diagnosis: Hydroxyapatite deposition (calcific tendinitis) involving the gluteus maximus tendon insertion with underlying cortical erosion.

Discussion: The MR examination in this case was prompted by the clinical concern with regard to possible neoplasm. Familiarity with the more unusual sites of involvement, less common findings (e.g., cortical bone erosion), and expected MR findings in this otherwise common condition can help in establishing the correct diagnosis and avoidance of further diagnostic evaluation including biopsy.

Periarticular crystal deposition in tendons and soft tissues is a common condition in both men and women and is most frequently seen between the ages of 40 and 70 years (1–4). Acute symptoms may include pain and tenderness and these symptoms may be marked in severity (3,4). Symptoms may spontaneously resolve and recur over time. Alternatively, calcific deposits may be demonstrated radiographically in entirely asymptomatic individuals. The pathogenesis of hydroxyapatite remains incompletely understood (3,4). While it has previously been assumed that calcium was only deposited in necrotic tissue, areas of avascularity, and/or hypoxic regions, more recent observations, including familial cases, polyarticular distribution, and increased frequency of certain histocompatibility antigens in affected individuals, have suggested that systemic factors may be important in contributing to connective tissue transformation prior to crystal deposition (3,4).

The most common site of involvement with calcific tendinitis is the shoulder (1–4). Less common sites of involvement include the wrist, elbow, deltoid insertion, knee, and neck (1,3,4). Calcific deposits around the hip and pelvis are typically seen in relation to the gluteal insertions, particularly around the greater trochanter and in the surrounding bursae. Involvement of adductor magnus insertion has also been previously recognized (1,5).

While not generally considered as a manifestation of calcific tendonitis, bone destruction has been previously recognized pathologically and surgically (6). Radiographic evidence of cortical bone erosion, as illustrated in the present case, may raise the clinical concern for malignancy and lead to biopsy if not recognized as an associated manifestation of periarticular crystal deposition. Bone resorption in these cases may occur as a result of increased local vascularity and active inflammation at the tendon insertion, or alternatively may be due to the mechanical effects from the pull of the associated muscle.

On MR, the actual crystal deposition may not be directly appreciated or seen as a localized focus of low signal intensity on all pulse sequences. Inflammatory changes in the surrounding soft tissues and muscles may be the primary manifestation of the condition and appear as ''edema''-like changes with infiltrative-type high signal intensity on T2-weighted spin echo and STIR sequences (1,3,7,8). No ''mass'' is identified, which should allow differentiation from neoplasm. Cortical bone erosive changes are less demonstrable than on CT (Fig. 14E), but marrow signal intensity alterations may be evident secondary to intraosseous edema (1,3,8).

Calcific tendinitis is a common condition that should be differentiated from other conditions associated with periarticular calcification including hyperparathyroidism and renal osteodystrophy, collagen vascular disease, hypoparathyroidism, idiopathic tumoral calcinosis, sarcoidosis, and articular disorders such as gout and calcium pyrophosphate deposition disease (3). Calcific tendinitis is also to be distinguished from causes of soft tissue ossification, principally myositis ossificans and neurologic injury. This distinction can be established radiographically by the presence of trabeculae within the ossified tissue and on MR by the presence of high signal intensity reflective of fatty marrow in contrast to the low signal intensity associated with calcification (3). The appearance of myositis ossificans on MR, however, will directly depend on the age and level of maturity of the process (see Case 11).

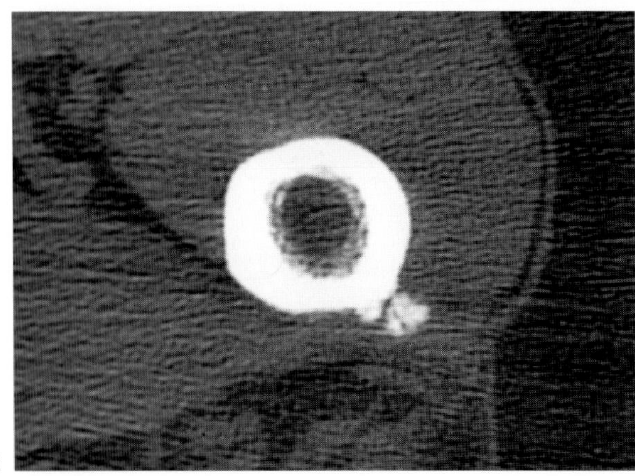

FIG. 14. E. An axial CT demonstrates a small focus of calcification immediately adjacent to a focal area of erosion within the adjacent femur.

REFERENCES

Introduction

1. Fleckenstien JL, Weatherall PT, Bertocci LA, et al. Locomotor system assessment by muscle magnetic resonance imaging. *Magn Reson Q* 1991;7(2):79–103.
2. Fleckenstein JL, Archer B, Barker B, et al. Fast short tau inversion recovery imaging. *Radiology* 1991;179:499–504.
3. Dwyer RJ, Frank JA, Sank VJ, et al Short T1 inversion recovery pulse sequence; analysis and initial experience in cancer imaging. *Radiology* 1988;168:827–36.
4. Shellock FG, Fukunaga T, Day K, et al. Serial MRI and Cybex testing evaluations of exertional muscle injury: concentric vs eccentric actions. *Med Sci Sports Med* 1991;23:110.
5. Fleckenstein JL, Canby RC, Parkey RW. Acute effects of exercise on MR imaging of skeletal muscle in normal volunteers. *AJR* 1988; 151:231–237.
6. Fisher MJ, Meyer RA, Adams GR, et al. Direct relationship between proton T2 and exercise intensity in skeletal muscle MR images. *Invest Radiol* 1990;25:480–95.
7. Fleckenstein JL, Haler RG, Lewis SF, et al. Absence of exercise induced MRI enhancement of skeletal muscle in McArdle's disease. *J Appl Physiol,* in press.
8. Shellock FG, Fukunaga T, Mink JH, et al. Accute effects of exercise on skeletal muscle: concentric vs eccentric actions. *AJR* 1991;156: 765–768.
9. Fleckenstein JL, Bertocci LA, Nunnally RL. Exercise inhanced MR imaging of variations in forearm muscle anatomy and use: importance of MR spectroscopy. *AJR* 1989;153:693–698.
10. Sjogaard G, Adams RP, Saltin B. Water and ion shifts in skeletal muscle of humans with intense dynamic knee extension. *Am J Physiol* 1985;248:190–196.
11. LeRumer E, DeCertaines J, Toulouse P, et al. Water phases in rat striated muscles as determined by T2 proton NMR relaxation times. *Magn Reson Imaging* 1987;5:267–272.
12. Lundvall J, Mellander S, Westling H. Fluid transfer between blood and tissues during exercise. *Acta Physiol Scand* 1972;85:258–269.
13. Horber FF, Hoopeler H, Schneidegger JR, et al. Impact of physical training on the ultrastructure of midthigh muscle in normal subjects and in patients treated with glucocorticoids. *J Clin Invest* 1987;79: 1181–1191.
14. Murphy WA, Totty WG, Carroll JE. MRI of normal and pathologic skeletal muscle. *AJR* 1986;146:565–574.
15. Lamminen AD. Magnetic resonance imaging of primary skeletal muscle disease: patterns of distribution and severity of involvement. *Br J Radiol* 1990;63:946–950.
16. Schwartz MS, Swash M, Ingram DA, et al. Patterns of selective involvement of thigh muscles in neuromuscular disease. *Muscle Nerve* 1988;11:1240–1245.
17. Schreiber A, Smith Wl, Zellweger H, et al. Magnetic resonance imaging of children with Duchenne muscular dystrophy. *Pediatr Radiol* 1987;17:495–497.
18. Kaufman LD, Gruber BL, Gerstman DP, et al. Preliminary observations on the role of magnetic resonance imaging for polymyositis and dermatomyositis. *Ann Rheum Dis* 1987;46:569–572.
19. Kakulas BA, Adams RD. *Disease of muscle: pathological foundations of clinical myology,* 4th ed. Philadelphia: Harper & Row, 1985.
20. Fleckenstein JL, Peshock RM, Lewis SF, et al. Magnetic resonance imaging of muscle injury and atrophy in flucolytic myopathies. *Muscle Nerve* 1989;12:849–855.
21. Shabas D, Gerard G, Rossi D. Magnetic resonance imaging examination of denervated muscle. *Comput Radiol* 1987;11:9.

Case 1

1. Fleckenstein JL, Watumull D, Conner K, et al. Denervated human skeletal muscle: MRI evaluation. *Radiology* 1993;187:213–218.
2. West GA, Haynor DR, Goodkin R, Tsuruda JS, Bronstein AD, Kraft G, Winter T, Kliot M. Magnetic resonance imaging signal changes in denervated muscles after peripheral nerve injury. *Neurosurgery* 1994;35(6):1077–1086.
3. Chason DP, Fleckenstein JL, Burns DK, Rojas G. Diabetic muscle infaction: radiologic evaluation. *Skel Radiol* 1995; in press.
4. De Visser. Muscular dystrophies. In: Fleckenstein JL, Reimers C, Crues JV, eds. *Muscle radiology.* New York: Springer-Verlag, in press.
5. Lamminen AE, Pihko H. Congenital myopathies. In: Fleckenstein JL, Reimers C, Crues JV, eds. *Muscle radiology.* New York: Springer-Verlag, in press.
6. Fleckenstein JL. Inherited defects of muscle energy metabolism. In: Fleckenstein JL, Reimers C, Crues JV, eds. *Muscle radiology.* New York: Springer-Verlag, in progress.
7. Fisher AQ. Spinal muscular atrophies. In: Fleckenstein JL, Reimers C, Crues JV, eds. *Muscle radiology.* New York: Springer-Verlag, in press.
8. Suput D, Zupan A, Sepe A, Demsar F. Discrimination between neuropathy and myopathy by use of magnetic resonance imaging. *Acta Neurol Scand* 1993;87:118–123.
9. Schedel H, Reimers CD, Vogl T, Witt TN. Muscle edema in MR imaging of neuromuscular diseases. *Acta Radiol* 1995;36:228–232.
10. Schalke, et al. Peripheral neuropathies. In: Fleckenstein JL, Reimers C, Crues JV, eds. *Muscle radiology.* New York: Springer-Verlag, in press.

Case 2

1. Penn AS, Myoglobinuria. In: Engel AG, Franzini-Armstrong, C, eds. *Myology,* 2nd ed. New York: McGraw-Hill, 1994;807–821.
2. Fleckenstein JL, Weatherall PT, Bertocci LA, Ezaki M, Haller RG, Greenlee R, Bryan WW, Peshock RM. Locomotor system assessment by muscle magnetic resonance imaging. *Magn Reson Q* 1991; 7:79–103.
3. De Visser. Muscular dystrophies. In: Fleckenstein JL, Reimers C, Crues JV, eds. *Muscle radiology.* New York: Springer-Verlag, in press.
4. Schedel H, Reimers CD, Vogl T, Witt TN. Muscle edema in MR imaging of neuromuscular diseases. *Acta Radiol* 1995;36:228–232.
5. Schweitzer ME, Fort J. Cost-effectiveness of MR imaging in evaluating polymyositis. *AJR* 1995;165:1469–1471.
6. Fleckenstein JL. Inherited defects of muscle energy metabolism. In: Fleckenstein JL, Reimers C, Crues JV, eds. *Muscle radiology.* New York: Springer-Verlag, in progress.
7. Shintani S, Shiigai T. Repeat MRI in acute rhabdomyolysis: correlation with clinicopathological findings. *J Comput Assist Tomogr* 1993;17(5):786–791.

Case 3

1. Laasonen EM. Atrophy of sacrospinal muscle groups in patients with chronic, diffusely radiating lumbar back pain. *Neuroradiology* 1984;26:9–13.
2. Flicker PL, Fleckenstein JL, Ferry K, Payne J, Ward C, Mayer T, Parkey RW, Peshock RM. Lumbar muscle usage in chronic low back pain: magnetic resonance image evaluation. *Spine* 1993;18: 582–586.
3. Sihvonen T, Herno A, Paljärvi L, Airaksinen O, Partanen J, Tapaninaho A. Local denervation atrophy of paraspinal muscles in postoperative failed back syndrome. *Spine* 1993;18:575–581.
4. Cinnamon J, Slonin AE, Black KS, Gorey MT, Scuderi DM, Hyman RA. Evaluation of the lumbar spine in patients with glycogen storage disease: CT demonstration of patterns of paraspinal muscle atrophy. *Am J Neuroradiol* 1991;12:1099–1103.

Case 4

1. Resnick D. Dermatomyositis and polymyositis. In: Resnick D, ed. *Diagnosis of bone and joint disorders,* 3rd ed. Philadelphia WB Saunders, 1995;1226–1231.
2. Sayers ME, Chou SM, Calabrese LH. Inclusion body myositis: analysis of 32 cases. *J Rheumatol* 1992;19:1385.
3. Adams EM, Chow CK, Premkumar A, et al. The idiopathic inflammatory myopathies: spectrum of MR imaging findings. *Radiographics* 1995;15:563–574.

4. Fraser DD, Frank JA, Dalakas M, et al. Magnetic resonance imaging in idiopathic inflammatory myopathies. *J Rheumatol* 1991;18:1693.
5. Love LA, Leff RL, Fraser DD, et al. A new approach to the classification of idiopathic inflammatory myopathy: myositis specific autoantibodies define useful homogeneous patient groups. *Medicine* 1991;70:360–374.
6. Bohan A, Peter JB. Polymyositis and dermatomyositis (Part I). *N Engl J Med* 1975;292:344–347.
7. Bohan A, Peter JB. Polymyositis and dermatomyositis (Part II). *N Engl J Med* 1975;292:403–407.
8. Kaufman LD, Gruber BL, Gerstman DP, et al. Preliminary observations on the role of magnetic resonance imaging for polymyositis and dermatomyositis. *Ann Rheum Dis* 1987;46:569–572.
9. Pitt AM, Fleckenstein JL, Greenlee RG Jr, et al. MRI guided biopsy in inflammatory myopathy: initial results. *Magn Reson Imaging* 1993;11:1093–1099.
10. Dunn CL, James WD. The role of magnetic resonance imaging in the diagnostic evaluation of dermatomyositis. *Arch Dermatol* 1993; 129:1104–1106.
11. Beese MS, Winkler G, Nicolas V, et al. The diagnosis of inflammatory muscular and vascular diseases using MRI with STIR sequences. *ROFO* 1993;158:542–549.
12. Reimers CD, Schedel H, Fleckenstein JL, et al. Magnetic resonance imaging of skeletal muscles in idiopathic inflammatory myopathies of adults. *J Neurol* 1994;241:306–314.
13. Hernandez RJ, Sullivan DB, Chenevert TL, et al. MR imaging in children with dermatomyositis: musculoskeletal findings and correlation with clinical and laboratory findings. *AJR* 1993;161:359–366.
14. Stiglbauer R, Graaninger W, Prayer L, et al. Polymyositis: MRI appearance at 1.5T and correlation to clinical findings. *Clin Radiol* 1993;48:244–248.

Case 5

1. Mink JH. Muscle injuries. In: Deutsch AL, Mink JH, Kerr R, eds. *MRI of the foot and ankle.* New York: Raven Press, 1992;281–313.
2. O'Donoghue DH. *Treatment of injuries to athletes,* 4th ed. Philadelphia: WB Saunders, 1984.
3. Zarins B, Ciullo JV. Acute muscle and tendon injuries in athletes. *Clin Sports Med* 1983;2(1):167–182.
4. Garrett WE, Rich FR, Nikolaou PK, et al. Computed tomography of hamstring muscle strains. *Med Sci Sports Exerc* 1989;21(5):506–514.
5. Ehman RL, Berquist TH. Magnetic resonance imaging of musculoskeletal trauma. *Radiol Clin North Am* 1986;24(2):291–319.
6. DeSmet AA, Heiner JP, et al. Magnetic resonance imaging of muscle tears. *Skel Radiol* 1990;19:283–286.
7. Fleckenstein JL, Weatherall PT, Parkey RW, et al. Sports-related muscle injuries: evaluation with MR imaging. *Radiology* 1989;172: 793–798.
8. Pomeranz SJ, Heidt RS. MR imaging in the prognostication of hamstring injury. *Radiology* 1993;189:897–900.

Case 6

1. Fleckenstein JL, Burns DK, Murphy FK, et al. Differential diagnosis of bacterial myositis in AIDS: evaluation with MR imaging. *Radiology* 1991;179:653–658.
2. Yuh WTC, Schreiber AE, Montgomery WJ, et al. Magnetic resonance imaging of pyomyositis. *Skel Radiol* 1988;17:190–193.
3. Mink JH. Muscle injuries. In: Deutsch AL, Mink JH, Kerr R, eds. *MRI of the foot and ankle.* New York: Raven Press, 1992.
4. Paajanen H, Grodd W, Revel D. Gadolinium DTPA enhanced MR imaging of intramuscular abscess. *Magn Reson Imaging* 1987;5: 109–115.

Case 7

1. Mink JH. Muscle injuries. In: Deutsch AL, Mink JH, Kerr R, eds. *MRI of the foot and ankle.* New York: Raven Press, 1992;281–313.
2. Shellock FG, Fukunaga T, Mink JH, et al. Exertional muscle injury: evaluation of concentric versus eccentric actions with serial MR imaging. *Radiology* 1991;179:659–664.
3. Tiidus PM, Ianuzzo CD. Effects of intensity and duration of muscular exercise on delayed soreness and serum enzyme activities. *Med Sci Sports Exerc* 1983;15:451–465.
4. Schwane JA, Johnson SR, Vandenakker CB, et al. Delayed-onset muscular soreness and plasma CPK and LDH activities after downhill running. *Med Sci Sports Exerc* 1983;15:51–56.
5. Armstrong RB. Mechanisms of exercise induced delayed onset muscular soreness; a brief review. *Med Sci Sports Exerc* 1984;16(6): 529–538.
6. Shellock FG, Fukunaga T, Day K, et al. Serial MRI and cybex testing evaluations of exertional muscle injury: concentric vs eccentric actions. *Med Sci Sports Exerc* 1991;23:S110.
7. Fleckenstein JL, Weatherall PT, Parkey RW, et al. Sports-related muscle injuries: evaluation with MR imaging. *Radiology* 1989;172: 793–798.
8. Giddings CJ, Nurenberg P, Stray-Gunderson J, et al. Muscle injury assessed with magnetic resonance imaging and electron microscopy following downhill running in humans. *Med Sci Sports Exerc* 1991; 23(suppl 4).
9. Zagoria RJ, Karstaedt N, Koubex TD. MR imaging of rhabdomyolysis. *J Comput Assist Tomogr* 1986;10:286–270.
10. Shintani S, Shiigai T. Repeat MRI in acute rhabdomyolysis: correlation with clinicopathological findings. *J Comput Assist Tomogr* 1993;17(5):786–791.

Case 8

1. Metzler JP, Fleckenstein JL, Vuitch F, et al. Skeletal muscle lymphoma. *Magn Reson Imaging* 1992;10:491–494.
2. Malloy PC, Fishman EK, Magid D. Lymphoma of bone, muscle, and skin: CT findings. *AJR* 1992;159:805–809.
3. Sundaram M, McLeod RA. MR imaging of tumor and tumorlike lesions of bone and soft tissue. *AJR* 1990;155:817–824.
4. Metzler JP, Fleckenstein JL, White CL, et al. MRI evaluation of amyloid myopathy. *Skel Radiol* 1992;21:463–465.
5. Gean-Marton AD, Krsch CFE, Vezina LG, et al. Focal amyloidosis of the head and neck: evaluation with CT and MR imaging. *Radiology* 1991;181:521.
6. Oliff JFC, Hardy JR, Williams MP, et al. Case report: magnetic resonance imaging of spinal amyloid. *Clin Radiol* 1989;40: 632.

Case 9

1. Nunez-Hoyo M, Gardner CL, Motta AO, et al. Skeletal muscle infarction in diabetes: MR findings. *J Comput Assist Tomogr* 1993; 17(6):986–988.
2. Banker BQ, Chester CQ. Infarction of thigh muscle in the diabetic patient. *Neurology* 1973;23:667–77.
3. Reich S, Wiener SN, Chester S, et al. Clinical and radiologic features of spontaneous muscle infarction in the diabetic. *Clin Nucl Med* 1985;10:876–879.
4. Herfkens RJ, Seivers R, Kaufman L, et al. Nuclear magnetic resonance imaging of the infarcted muscle: a rat model. *Radiology* 1983; 147:761–764.

Case 10

1. Resnick D. Dermatomyositis and polymyositis. In: Resnick D, ed. *Diagnosis of bone and joint disease,* 3rd ed. Philadelphia: WB Saunders, 1995;1218–1231.
2. Keim DR, Hernandez RJ, Sullivan DB. Serial magnetic resonance imaging in juvenile dermatomyositis. *Arthritis* Rheum 1991;34: 1580.
3. Hernandez RJ, Keim DR, Sullivan, et al. Magnetic resonance imaging appearance of the muscles in childhood dermatomyositis. *J Pediatr* 1990;117:546.
4. Park JH, Vansant JP, Kumar NG, et al. Dermatomyositis: correlative

MR imaging and P-31 MR spectroscopy for quantitative characterization of inflammatory disease. *Radiology* 1990;177:473.
5. Fraser DD, Frank JA, Dalakas M, et al. Magnetic resonance imaging in the idiopathic inflammatory myopathies. *J Rheumatol* 1991;18:1693.
6. Fraser DD, Frank JA, Dalakas MC. Inflammatory myopathies: MR imaging and spectroscopy. *Radiology* 1991;179:341–344.
7. Park JH, Gibbs SJ, Price RR, et al. Inflammatory myopathies: MR imaging and spectroscopy reply. *Radiology* 1991;179:341–344.
8. Hernandez RJ, Keim DR, Chenevert TL. Fat-suppressed MR imaging of myositis. *Radiology* 1992;182:217–219.

Case 11

1. Resnick D. Traumatic, iatrogenic and neurogenic diseases. In: Resnick D, ed. *Diagnosis of bone and joint disorders.* Philadelphia: WB Saunders, 1995;3248–3249.
2. Goldman AB. Myositis ossificans circumscripta: a benign lesion with a malignant differential diagnosis. *AJR* 1976;126:32–40.
3. Hanna SL, Magill HL, Brooks MT, et al. Case of the day. pediatrics. Myositis ossificans circumscripta. *Radiographics* 1990;10:945–949.
4. Kransdorf M, Meis J, Jelinek JS. Myositis ossificans: MR appearance with radiologic-pathologic correlation. *AJR* 1991;157:1243–1249.

Case 12

1. Feldman F, Zwass A, Staron RB, et al. MRI of soft tissue abnormalities: a primary cause of sickle cell crisis. *Skel Radiol* 1993;22:501–506.
2. Ballas SK. Treatment of pain in adults with sickle cell disease. *Am J Hematol* 1990;34:49.
3. Fabry ME, Benjamin L, Lawrence C, et al. An objective sign in painful crisis in sickle cell anemia: the concomitant reduction of high density red cells. *Blood* 1984;64:559.
4. Dorwart BB, Gabuzda TG. Symmetric myositis and fasciitis: a complication of sickle cell anemia during vasoocclusion. *J Rheumatol* 1985;12:590.

Case 13

1. Inada K, Hirose M, Iwashima Y, et al. Popliteal artery entrapment syndrome: a case report. *Br J Surg* 1978;65:613.
2. Insua JA, Young JR, Humpries AW. Popliteal artery entrapment syndrome. *Arch Surg* 1970;101:771.
3. Bergan JJ. Adventitial cystic disease of the popliteal artery. In: Rutherford RB, ed. *Vascular surgery.* Philadelphia: WB Saunders, 1977;569–576.
4. Jasinski RW, Masselink BA, Partridge RW, et al. Adventitial cystic disease of the popliteal artery. *Radiology* 1987;163:153–155.
5. Reicher MA. Knee joint disorders. In: Mink JH, Reichter MA, Crues JV, Deutsch AL, eds. *MRI of the knee,* 2nd ed. New York: Raven Press, 1993;392–395.

Case 14

1. Hayes CW, Rosenthal DI, Plata MJ, et al. Calcific tendinitis in unusual sites associated with cortical bone erosion. *AJR* 1987;149:967–970.
2. Seeger LL, Butler DL, Eckardt JJ. Tumoral calcinosis like lesion of the proximal linea aspera. *Skel Radiol* 1992;19:579–583.
3. Resnick D. Calcium hydroxyapatite crystal deposition disease. In: Resnick D, ed. *Diagnosis of bone and joint disorders,* 3rd ed. Philadelphia: WB Saunders, 1995;1632–1638.
4. Faure G, Daculsi G. Calcified tendinitis: a review. *Ann Rheum Dis* 1983;42(suppl):49–53.
5. Berney J. Calcifying peritendinitis of the gluteus maximus tendons. *Radiology* 1972;102:517–518.
6. Moseley HF. *Shoulder lesions,* 3rd ed. Edinburgh: Livingston, 1989; Chapter 6.
7. Ramon FA, Degryse HR, DeScheppeer AM, et al. Calcific tendinitis of the vastus lateralis muscle. A report of three cases. *Skel Radiol* 1991;20:21.
8. Hayes CW, Conway WF. Calcium hydroxyapatite deposition disease. *Radiographics* 1990;10:1031.

CHAPTER 9

Marrow

Bruce A. Porter and Andrew L. Deutsch

MR imaging of bone marrow represents one of the most powerful and important applications of the technique. While MR can demonstrate abnormalities involving the periosteum and cortex of bone, overwhelmingly it is the cancellous bone and marrow manifestations of a wide spectrum of disorders that are detected utilizing this technique. These uses (including trauma detection, infection evaluation, and tumor detection) have become well-established applications for MR imaging and are extensively illustrated throughout the text. This chapter highlights applications of MR to primary malignant disorders of marrow including lymphoma, leukemia, myeloma, and metastatic disease. A brief review of general considerations of marrow imaging is offered here as a complement to the treatment of this important subject both in this chapter and throughout the text.

The optimal use of MR for assessment of marrow disorders requires an understanding of the normal anatomy and physiology of bone marrow and its expected appearance on MR utilizing different pulse sequences (1–6). From a functional perspective, bone marrow is often divided into "red marrow" (hematopoietically active in the production of red cells, white cells, and platelets) and "yellow marrow" (hematopoietically inactive and composed predominantly of fat cells) (1,2). Both types of marrow are, however, commonly encountered at any particular anatomic site. The chemical composition of red marrow consists of approximately 40% water, 40% fat, and 20% protein; that of yellow marrow approximates 15% water, 80% fat, and 5% protein (2). The precise admixture of red and yellow marrow is dependent on multiple factors including the age of the person, anatomic site, and hematopoietic demand of the body.

At birth, the marrow is composed overwhelmingly of hematopoietic elements. Under normal circumstances of growth and development, a well-documented and predictable transition or "conversion" of red to yellow marrow takes place. This begins in the immediate postnatal period and progresses from appendicular to axial sites and in the tubular bones from diaphyseal to metaphyseal locations (1–3). The most dramatic changes are seen in the tubular bones within the first two decades of life. The transition process begins in the bones of the hands and feet, progresses to the distal and then proximal long tubular bones, and finally occurs in the flat bones and vertebral bodies until the adult pattern of proximal humeral, femoral, and axial hematopoietic marrow is established (generally by age 25 years) (1–3). At all ages, the apophyses and epiphyses contain predominantly yellow marrow.

The MR appearance of the marrow is influenced not only by its composition (fat, water, minerals) but also by the pulse sequences utilized to image it (7–12). The short T1 and long T2 relaxation time of adipose tissue are the factors that are primarily responsible for its appearance on spin echo sequences. The contribution of fat to the MR signal pattern of red marrow is also significant but is averaged with the longer T1 and T2 relaxation times of protein and water to account for the final MR appearance. T1-weighted spin echo sequences depict yellow marrow as high signal intensity and red marrow as more variable intermediate signal intensity (ranging from less than to greater than that of muscle). On T2-weighted spin echo sequences, the discrimination between red and yellow marrow is less related to the relatively smaller differences in T2 relaxation times. Fast spin echo techniques demonstrate a number of unique tissue contrast properties (7,8,12). Among these is the phenomenon that fat remains of high signal intensity even with prolonged effective T2. The reason for this remains debated, but may involve an effect of multiline imaging on J-coupling (7). As a consequence of this effect, cellular infiltration within the fatty marrow may be less conspicuous or entirely obscured on T2-weighted fast spin echo studies unless some form of fat suppression is employed (7,8,12).

The most widely used form of fat suppression is frequency selective presaturation (7). With this method, however, local field inhomogeneities that arise in areas of significantly varying morphology (e.g., craniovertebral junction, ankle) or near collections of gas or metal often result in uneven fat suppression that can lead to areas of apparent increased signal intensity on T2-weighted fast spin echo images that simulate areas of marrow infiltra-

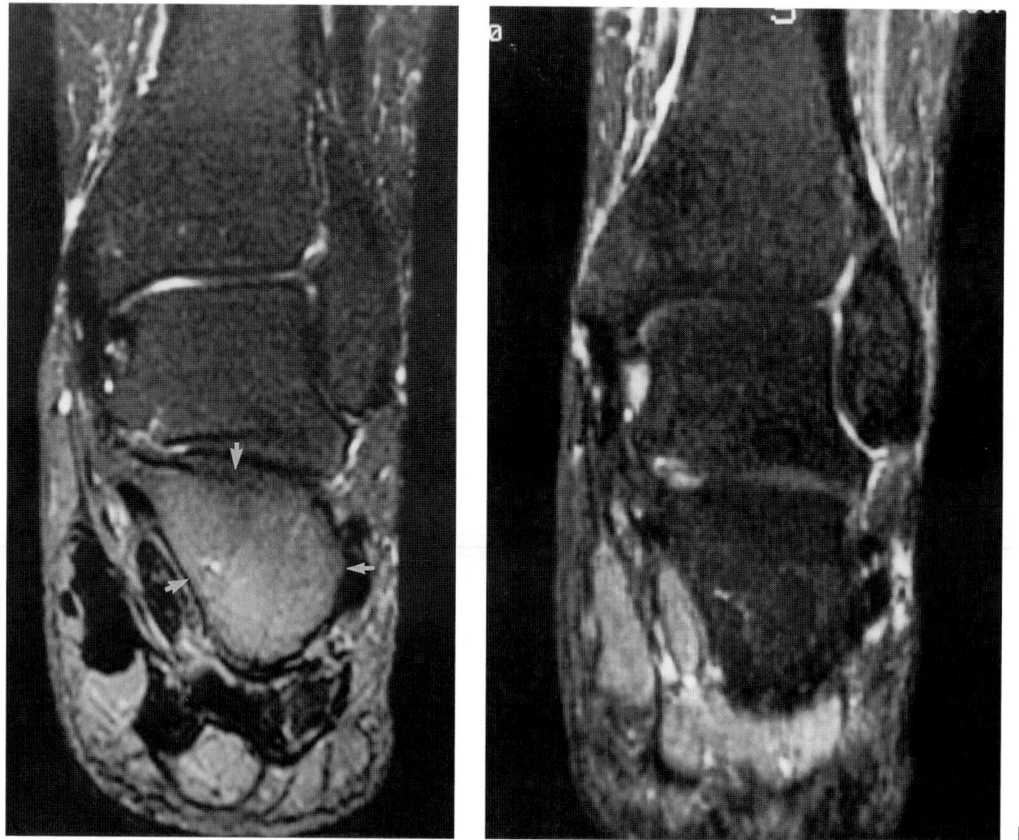

FIG. 1. A,B: Artifactual increased signal related to incomplete fat suppression. **A:** Coronal fat-suppressed fast spin echo (TR/TE 4,500/105 eff) image of a 42-year-old woman with previous documented transient marrow edema of the hip. The patient complained of ankle pain at this time. There is apparent high signal intensity throughout the calcaneus; a finding suggesting marrow edema. **B:** Coronal conventional STIR sequence (TR/TE/TI 3,500/30/160 msec at 1.5 T) obtained following Fig. A demonstrates a normal appearance to the calcaneus confirming the suspicion that the findings in Fig. A were artifactually related to incomplete fat suppression.

tion (7,9) (Fig. A and B). Field inhomogeneity can lead not only to failure of fat suppression but to inadvertent saturation of the water peak, which can paradoxically obscure areas of true edema. These implications are particularly important when fat saturation is used in conjunction with administration of contrast to improve conspicuity on T1-weighted images. Phase-contrast methods (e.g., three-point Dixon) exploit differences between fat and water by selecting different values for TE; in-phase and opposed-phase MR data are acquired and can be utilized to generate fat or water selective images. Phase-contrast techniques, alone or in combination with presaturation, may provide more homogeneous fat suppression but require additional imaging time and are more vulnerable to patient motion (7). Both methods can be utilized with contrast-enhanced studies.

Short inversion time inversion recovery (STIR) is a widely utilized technique for fat suppression and marrow imaging (10). With this technique, fat is suppressed on the basis of its rapid T1 recovery rather than on the basis of chemical shift phenomenon (7,10). It is less affected by field inhomogeneities. Multiple studies have demonstrated the high sensitivity of STIR for depiction of marrow infiltration (7,10). The relates not only to its powerful fat suppression but because both T1 and T2 relaxation are additive with this sequence (i.e., processes characterized by prolonged T1 and T2 recovery are depicted to advantage). The high sensitivity of STIR for depiction of marrow processes is offset somewhat by markedly longer imaging times, which result in limited coverage due to reductions in slice number. Additionally STIR is incompatable with contrast studies utilizing gadopentetate dimeglumine, which acts by shortening T1. Recently introduced "fast" STIR techniques (e.g., inversion recovery fast spin echo) appear to overcome some of these limitations, providing apparently equivalent sensitivity while substantially decreasing acquisition time (allowing for increased coverage and/or resolution) (7,12).

Gradient echo sequences are most commonly utilized in musculoskeletal imaging as a method for rapidly acquiring images in which structures with long T2 relaxation appear bright (7,11). As these sequences are in actuality T2* weighted, they are sensitive to field inhomogeneity. With regard to marrow imaging, differences in magnetic susceptibility of cancellous bone and marrow result in local field inhomogeneities developing at the boundaries between trabeculae and marrow, warping the magnetic lines of force and causing protons to dephase and lose signal (7). This is the basis for the intrinsic insensitivity of gradient echo imaging to marrow edema and infiltration (7,11). For this reason, STIR and spin echo sequences are preferred to gradient echo sequences for imaging marrow disorders. Exceptions to this rule are marrow-based processes that result in the loss of trabeculae. This acts to lessen the susceptibility artifact and actually lengthen T2*, which may render focally destructive processes, such as metastases, more conspicuous (7).

CASE 1

History: A pre–bone marrow transplant marrow MR was requested for a 61-year-old man with refractory non-Hodgkin's lymphoma. Left posterior iliac crest biopsy was negative for marrow tumor. Was the marrow sample adequate? What is the role of MR in tumor staging in lymphoma?

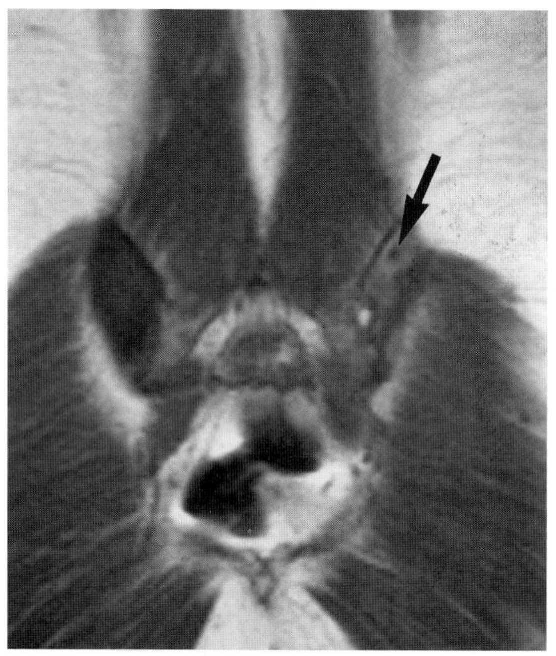

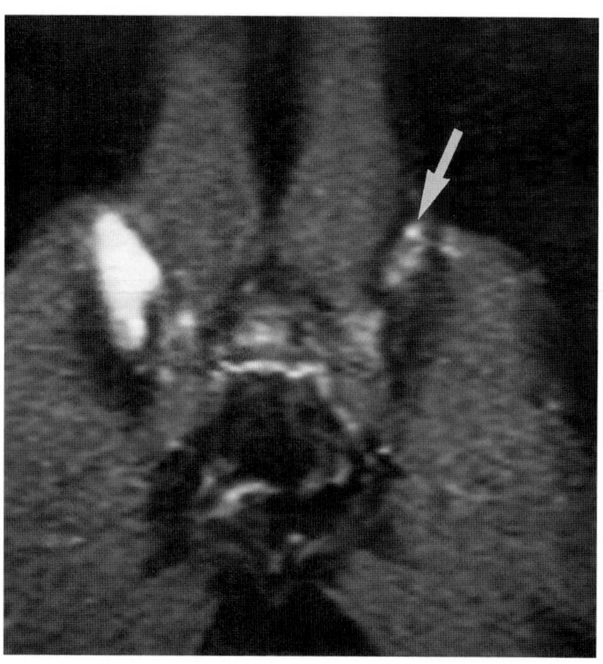

FIG. 1.

Findings: In Fig. 1A the coronal T1-weighted (TR/TE 450/10 msec), spin echo image of the posterior pelvis done at 0.5 T shows the biopsy site (arrow) on the left, in an area of normal-appearing marrow. The right posterior iliac crest, however, has markedly abnormal confluent low signal intensity, indicating either sclerotic bone or tumor. In Fig. 1B the corresponding coronal short TI inversion-recovery (STIR) image (TR/TE/TI 2,000/30/110 msec at 0.5 T) reveals focal high signal at the left biopsy site (arrow), with subtle surrounding high signal edema indicating that the biopsy is recent. The nonsampled right posterior iliac crest is hyperintense on STIR, indicating dense tumor infiltration. Sclerotic bone or marrow fibrosis would be dark on STIR.

Diagnosis: Inadequate blind marrow sampling. MR-directed evaluation significantly "upstaged" the patient's disease.

Discussion: Detection of marrow involvement is a critical determinant in staging and therapeutic decision making in patients with lymphoma (1–6), since its presence denotes stage IV disease. Clinical marrow assessment is conventionally done with marrow aspirates or biopsies from the posterior iliac crest or sternum; yet it is well known that these are subject to significant sampling error (2), and a negative marrow biopsy clearly does not exclude marrow tumor. Bone scans and plain films are often falsely negative even in patients with known lymphomatous marrow infiltration.

As new and diverse therapeutic options such as bone marrow transplant or monoclonal antibody therapy (4) become available for lymphoma patients, the ability to more accurately assess the marrow tumor burden becomes of increasing importance. MR has been shown to be complementary to marrow biopsy for improving detection of marrow tumor and for assessing whether the marrow sample accurately represents the tumor within the marrow space (2,3) (Fig. 1C–E).

Since the predominant areas of lymphomatous involvement are in the red marrow areas of the axial skeleton, the MR staging examination covers the chest, abdomen, and proximal femurs, thus showing most of the red marrow space. The highest yield for MR in lymphoma is for those patients with Hodgkin's or intermediate- to high-grade non-Hodgkin's lymphomas (2,3). These higher-graded tumors are more likely to focally involve the marrow, and hence are more often missed by random marrow biopsies. MR can significantly impact clinical management and determination of appropriate therapy in patients with lymphoma (Fig. 1F–L). Low-grade non-Hodgkin's lymphomas, on the other hand, more commonly diffusely involve the marrow, often with microscopic infiltration below the detection threshold of MR (2). This microscopic disease is best diagnosed by marrow biopsy. Using the combined T1 and STIR method, even with marrow-oriented protocols, MR is equivalent to CT for detection of lymphadenopathy and superior for focal marrow lesions (5) Hence, marrow biopsy and MR appear to be the optimal combination for staging lymphoma, and are complementary examinations.

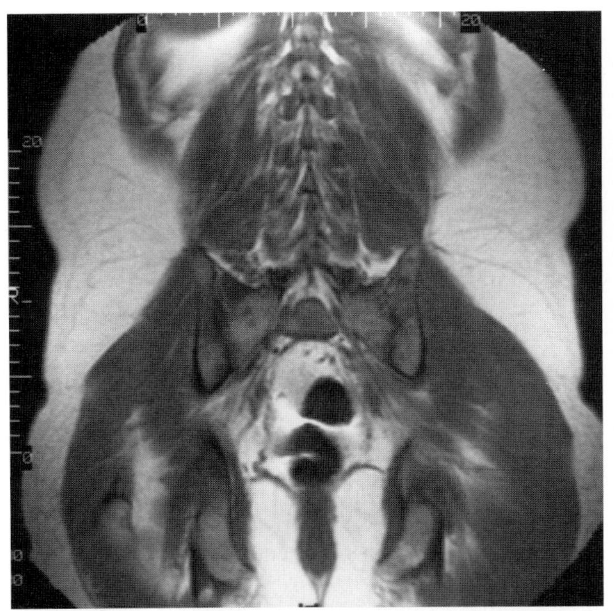
C

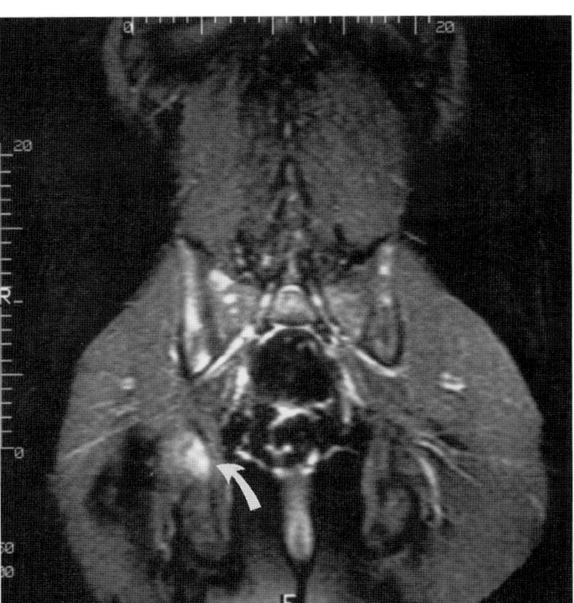
D

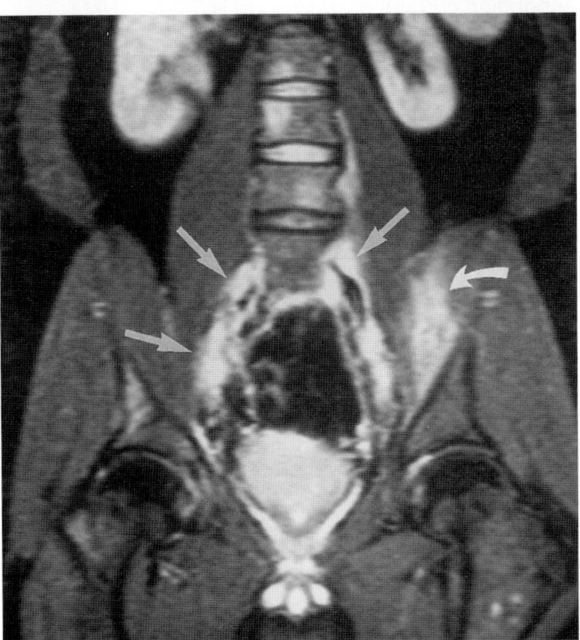
E

FIG. 1. C: Coronal T1-weighted image 2 cm anterior to Fig. 1A,B reveals additional subtle focal areas of abnormally low signal. **D:** High signal intensity on the corresponding coronal STIR image confirms the hypercellularity of these lesions. Although some are only 3 to 4 mm in diameter, they are significantly more conspicuous on STIR than on T1 spin echo image. Tumor involvement of the right ischium *(arrow)* is evident because of high lesion contrast. The corresponding area on T1-weighted imaging (Fig. 1C) could, and probably should, be interpreted as partial volume averaging of cortex or muscle. **E:** Coronal STIR image of the pelvis and lower lumbar spine reveals multiple focal areas of tumor in the lumbar vertebral bodies and proximal left femur. There is also viable, bilateral external iliac adenopathy, also typically very intense on STIR *(arrows).* Edema of the left iliac muscle *(curved arrow)* is commonly seen with processes that interfere with the innervation or blood supply of muscle groups as well as surrounding tumor (peritumoral edema). The active marrow tumor in this case was well seen, despite the small size of many of the lesions, due to the fat suppression and additive T1 and T2 contrast provided by the short TI inversion recovery technique (1).

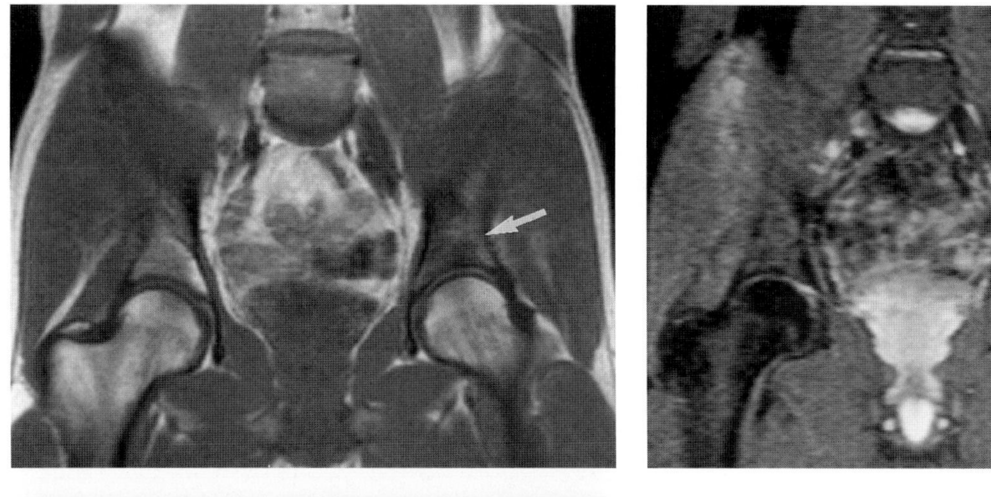

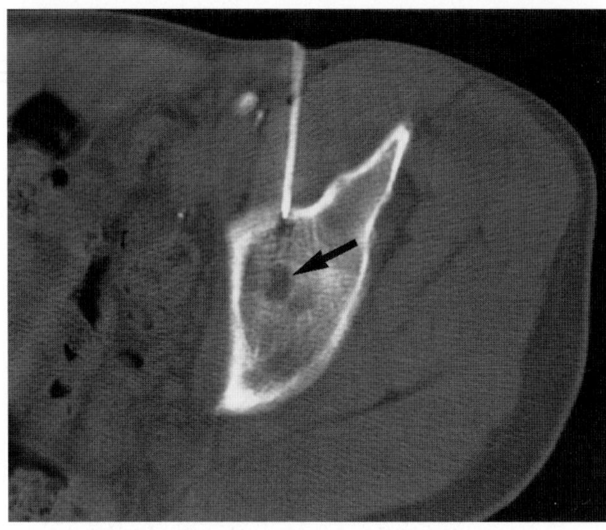

FIG. 1. F–H: A young man with nodular sclerosing Hodgkin's underwent MR imaging following radiation therapy to the chest for what was thought to be stage II Hodgkin's lymphoma confined to the mediastinum. **F:** On the MR examination performed 6 months post-therapy, abnormal signal is demonstrated in the left supra-acetabular marrow on a coronal T1 spin echo (TR/TE 600/12 msec) image of the pelvis. **G:** High marrow signal on a corresponding coronal STIR (TR/TE/TI 1,800/30/110 msec at 0.5 T) further raises the suspicion of active tumor in the left ilium. The STIR abnormality indicates a high probability of viable tumor, although trabecular injury could have a similar appearance. No other marrow lesions were identified elsewhere. **H:** Because of the substantial potential impact of this lesion on the patient's staging and therapy if it were malignant, a CT-guided biopsy was performed. Subtle lytic changes are noted on the CT in trabecular bone when bone algorithms and specific attention is focused on this region *(arrow).* Commonly, no alteration of the trabecular or bony architecture is seen with marrow lymphoma unless there is gross destructive tumor or osteoblastic stimulation ("ivory vertebra"). The biopsy confirmed viable marrow lymphoma, advancing the patient to stage IV and indicating that chemotherapy was required in addition to radiation.

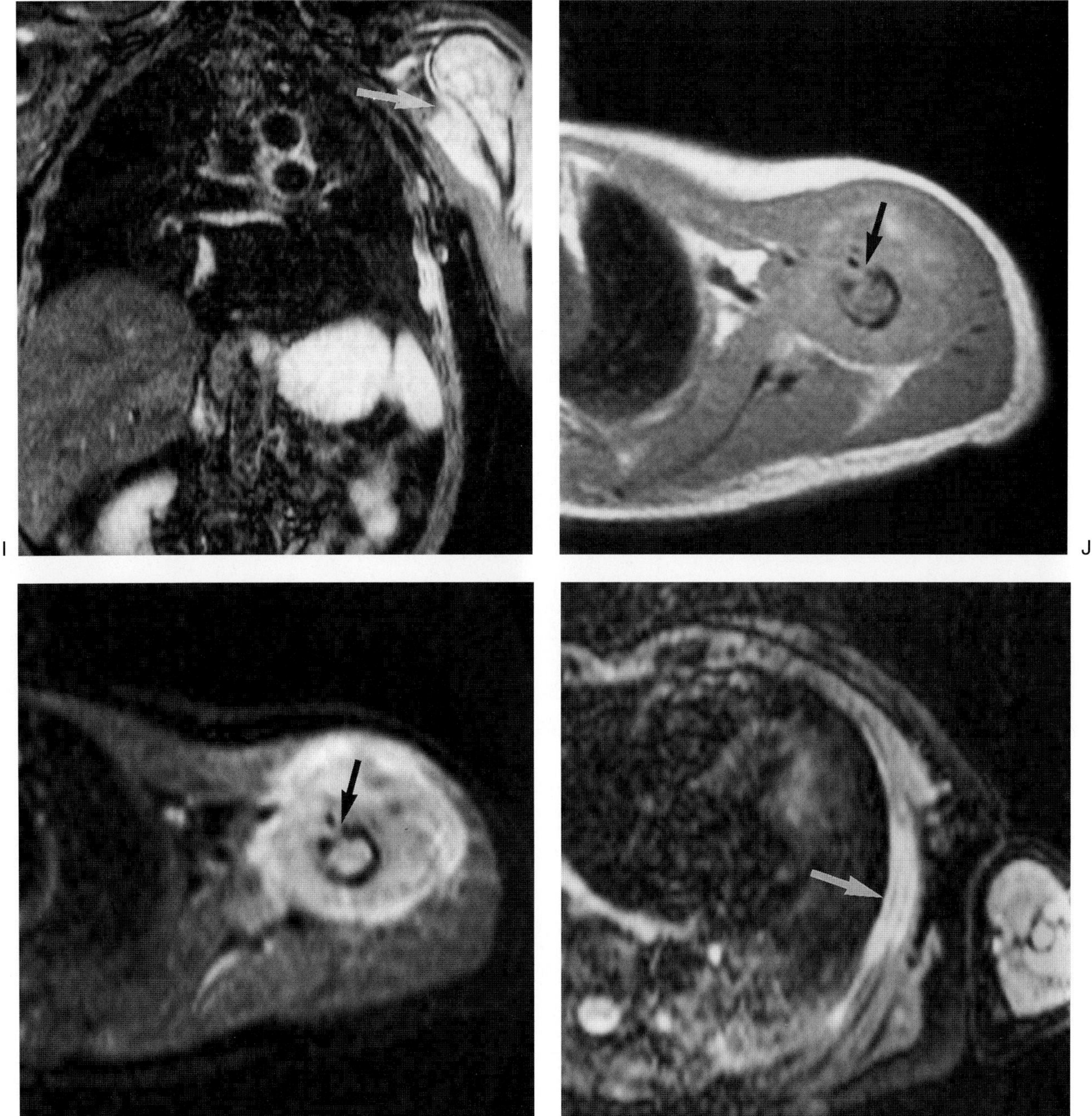

FIG. 1. I–L: A 68-year-old man with left shoulder pain. Rotator cuff surgery revealed an intermediate-grade non-Hodgkin's lymphoma. The borderline cardiac status was of concern for chemotherapy (vs. local radiation alone). MR was requested to assess the local extent of tumor and for additional lesions. **I:** Coronal STIR image of the chest (TR/TE/TI 1,950/30/110 msec, 45 cm field-of-view at 0.5 T) reveals expansile tumor of the left humeral head, with penetration into the muscle *(arrow)*. **J:** The abnormal marrow in the humeral head also noted on (axial T1 spin echo (TR/TE 500/12 msec, 256 × 256)) image *(arrow)*. **K:** The matched axial STIR image (TR/TE/TI 3000/30/110 msec, 128 × 256) clearly defines the tumor extent both in the marrow and the adjacent soft tissues, as well as disruption of the cortex *(arrow)*. The contrast is best with the inversion recovery technique, although the spatial resolution of the 128 × 256 images is lower. Some peritumoral edema is commonly seen with STIR; this may result in overestimation of tumor extent by observers with limited experience. In this case the sharp margins most likely represent the actual edges of the tumor. No adenopathy was noted. **L:** An axial STIR image more caudally reveals marrow hyperintensity, expansion, and periosteal reaction in a left mid-lateral rib *(arrow)*, also seen on Fig. I. This is a common appearance of marrow lymphoma, with no history of rib trauma. The patient received chemotherapy, with resolution of both the humeral and rib lesions.

CASE 2

History: A 36-year-old man with Hodgkin's lymphoma was imaged at the time of discharge following bone marrow transplant. Follow-up imaging was performed 9 months later. What is your diagnosis?

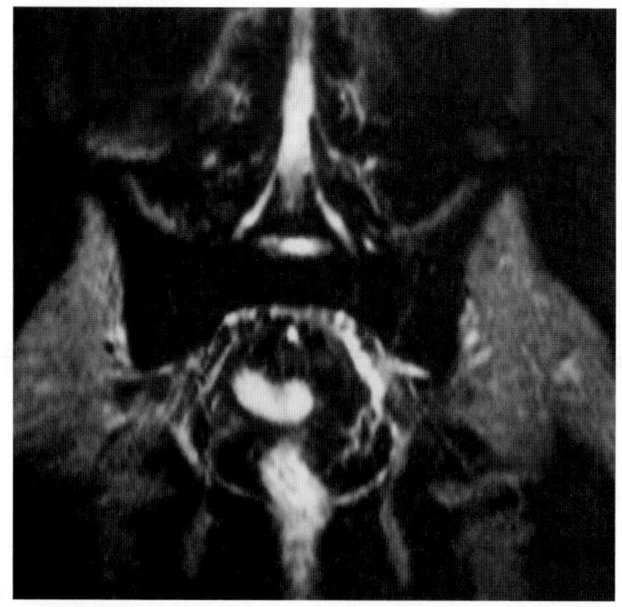

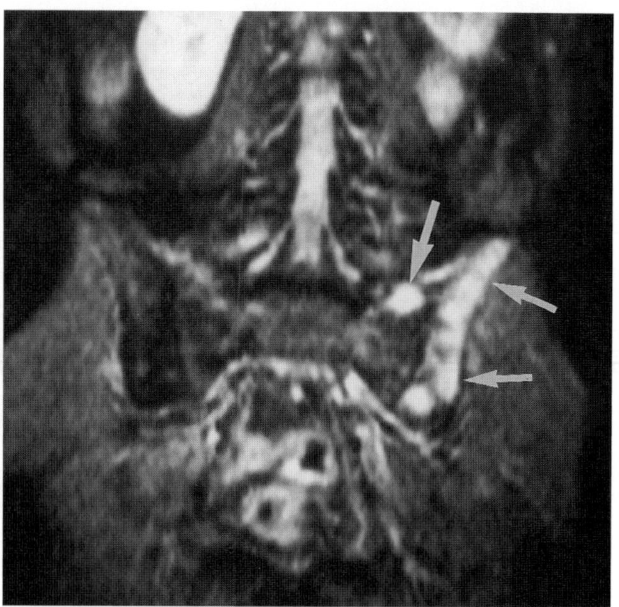

FIG. 2.

Findings: In Fig. 2A the coronal STIR image of the pelvis (TR/TE/TI 1,600/40/100 msec) shows extremely low signal in the fatty hypocellular marrow after marrow transplant. In Fig. 2B the follow-up examination at 9 months demonstrates the marrow of the sacrum to be of intermediate signal intensity. In the left sacrum and left posterior iliac crest, there are more focal areas of high signal intensity (arrows).

Diagnosis: Marrow repopulation and tumor relapse 9 months following therapy.

Discussion: This case illustrates the striking findings in hypoplastic or aplastic marrow that may be seen after high-dose radiation and chemotherapy (Fig. 2A). The resulting marrow hypocellularity is reflected on STIR by virtual absence of signal in the marrow, reflecting its fatty state. Follow-up imaging at 9 months (Fig. 2B) portrays both the expected marrow repopulation as well as the unfortunate marrow relapse of the tumor. The marrow of the central sacrum has repopulated and is intermediate gray on STIR; however, the relapse of Hodgkin's lymphoma, here noted in the left sacrum and left posterior iliac crest, is much more focal and intense (arrows). Multiple similar lesions were seen in the marrow elsewhere. Similar findings are seen in early relapse of leukemia, which can often be similarly detected by MR before clinical data or marrow biopsies are positive.

Advantages of MR over other methods including computed tomography (CT) and bone scan are not only its superiority for detecting marrow tumor infiltration, as in the above case with marrow relapse, but also its sensitive detection of epidural or soft tissue involvement. This is not uncommon with lymphoma (1) (Fig. 2C–G). MR imaging yields much greater tissue contrast than CT.

Although not commonly appreciated, MR is at least equivalent to CT for detecting lymphadenopathy, even when marrow-oriented imaging protocols are used (2). As a result, in appropriate patients, improved lymphoma staging can be provided with MR, particularly with the STIR sequence. The modest additional cost of MR can be recouped by improved staging, appropriateness of therapy, accurate guidance of radiation therapy to critical marrow, soft tissue, or epidural lesions, and better monitoring of response to treatment.

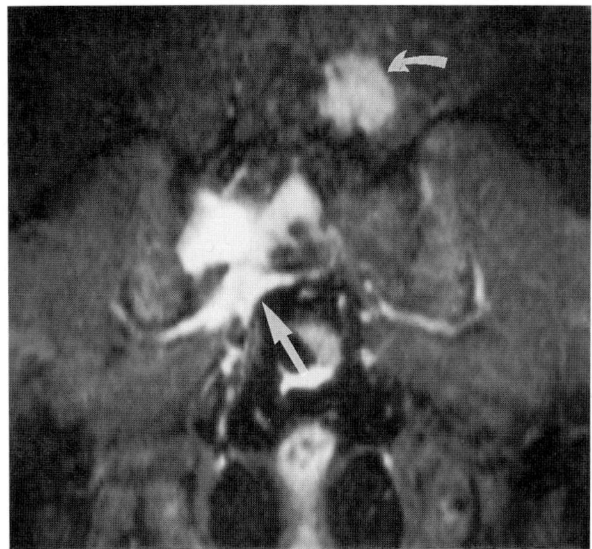
C

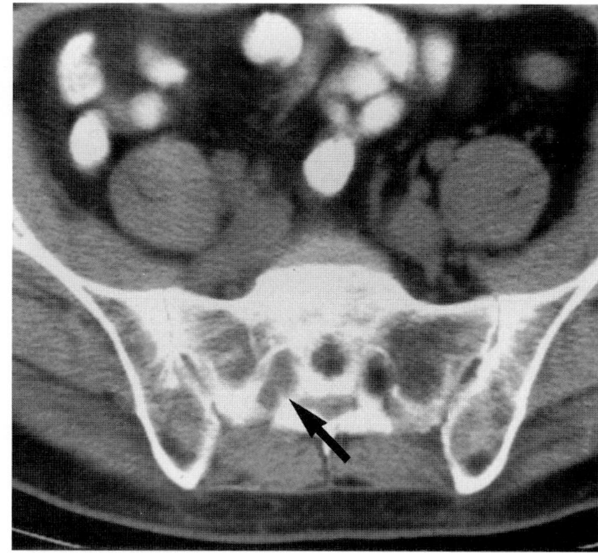
D

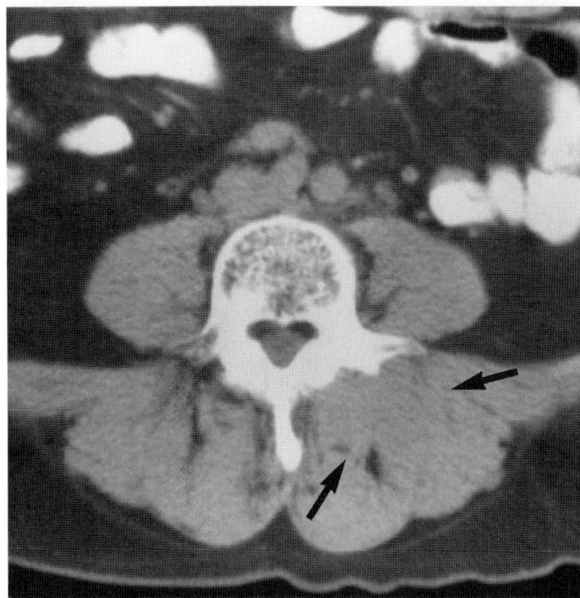
E

FIG. 2. C–G: A 47-year-old man with intermediate-grade non-Hodgkin's lymphoma with mild, nonradicular low back and abdominal pain, as may be associated with his known retroperitoneal adenopathy. **C:** Coronal STIR pelvic image (TR/TE/TI 1,600/40/100 msec at 0.15 T) revealed marked signal abnormality of the right sacral ala, and a soft tissue mass *(arrow)* extending into the presacral plexus and right sacrosciatic notch. The remaining marrow signal is normal intermediate to gray on STIR. An additional 3-cm hyperintense nodule was also detected in the left paraspinous soft tissue just above the level of the posterior iliac crest *(curved arrow)*. There was minimal retroperitoneal tumor. **D:** These findings are subtle on CT, and were not diagnosed prospectively. This is an axial CT through the mid-right sacrum; it reveals only slight asymmetrical density in the right sacral ala vs. the left. In retrospect, soft tissue in the right S1 neural foramen can also be seen *(arrow)* in addition to low-volume adenopathy posterior to the right common iliac artery and vein. **E:** CT at a level 1 cm above the left posterior-superior iliac crest (through the midportion of the hyperintense area seen on Fig. 2C). The only indication of the paraspinous mass seen on MR is subtle asymmetry of the fascial planes of the left paraspinous muscles *(arrows)*.

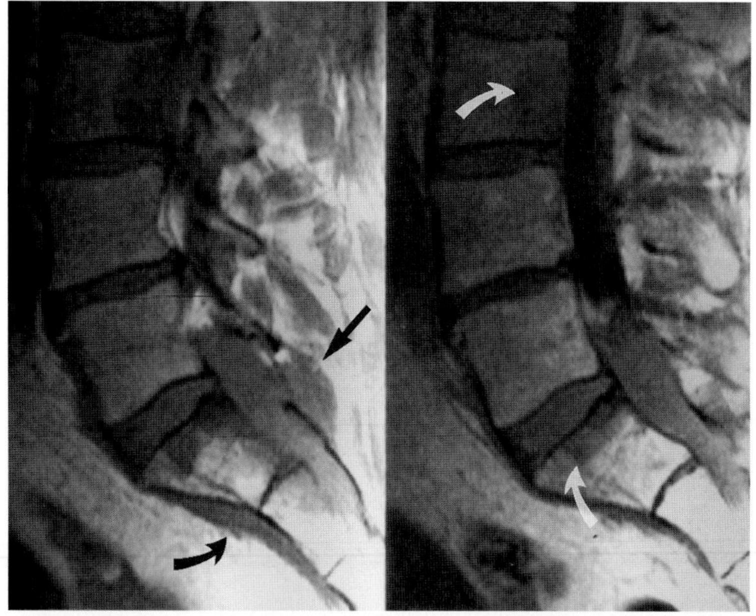
F

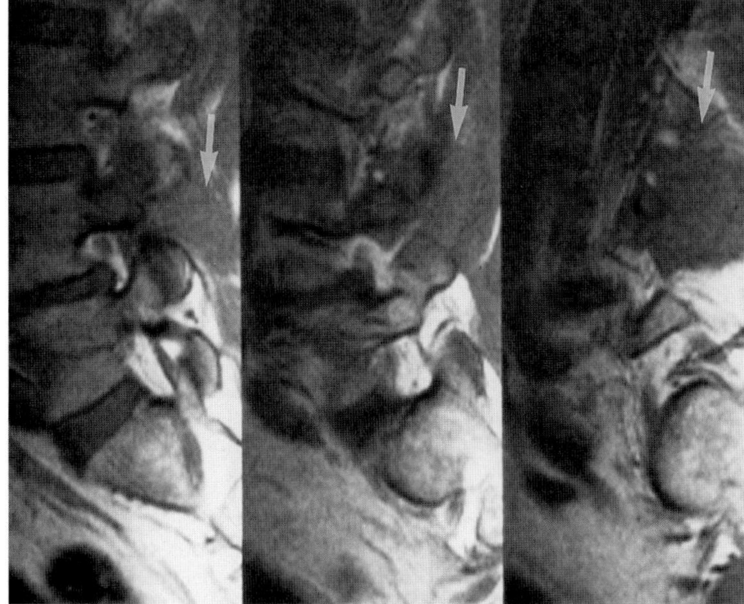
G

FIG. 2. F: Sagittal T1-weighted spin echo images (TR/TE 650/20 msec) reveal tumor obliteration of the caudal spinal canal from mid-L5 caudally, with extension into the soft tissues posteriorly *(arrow)* as well as into the presacral plexus *(curved arrow).* Marrow lymphoma is also noted in S1 and L3 *(white arrows).* **G:** A series of left parasagittal T1-weighted image confirms the paraspinous soft tissue mass *(arrows),* which also involves the left transverse process of L4.

CASE 3

History: A 46-year-old premenopausal woman in good health 2 years after mastectomy for stage II breast carcinoma. She was active in sports, with no specific complaints, but anemia and an increased reticulocyte count were noted on clinical evaluation. Despite a reportedly unimpressive bone scan and absence of bone pain, MR of the marrow was recommended to assess for marrow tumor. What is your diagnosis? What is the role of MR in the assessment of possible metastatic disease to the skeleton?

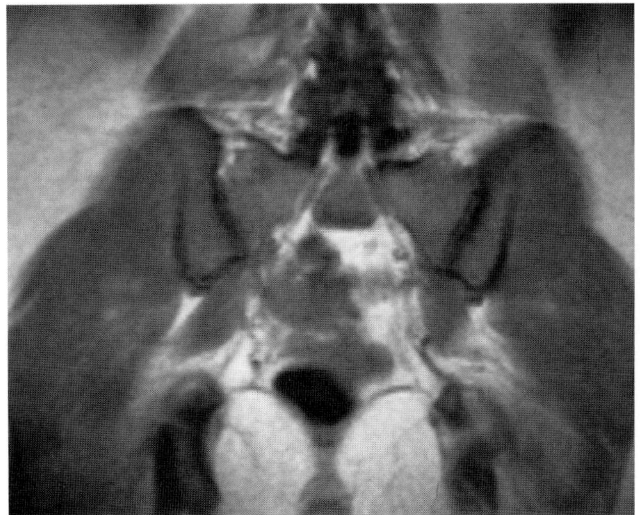

A

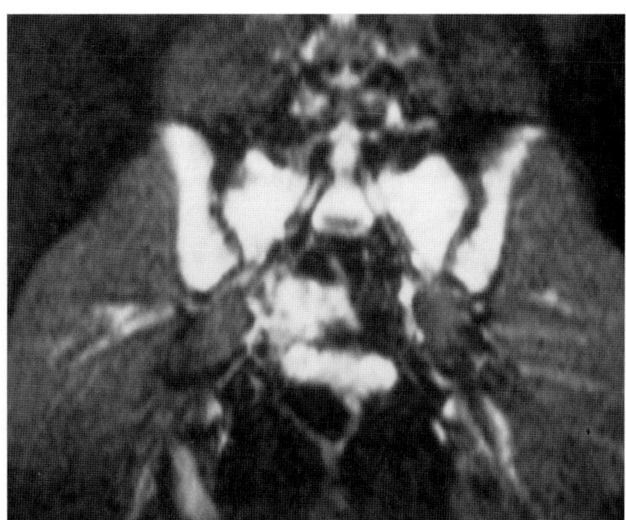

B

FIG. 3.

Findings: In Fig. 3A the coronal pelvis MR with a T1-weighted (TR/TE 700/22 msec) spin echo shows the marrow signal intensity is generally decreased and similar to that of muscle. In Fig. 3B the corresponding STIR image (TR/TE/TI 1,500/40/100 msec) at 0.15 T reveals markedly hyperintense confluent marrow signal abnormality.

Diagnosis: Diffuse marrow infiltration consistent with metastatic disease.

Discussion: On initial cursory observation, the marrow appears unremarkable in Fig. 3A. On a T1-weighted sequence, however, the marrow should have signal intensity closer to fat than to muscle, and hence this marrow is diffusely abnormal. This finding is more difficult to perceive in the pelvis than in the lumbar spine, where comparison can be made with the signal intensity of intervertebral disks. The marrow in this case is isointense to muscle. On the basis of the STIR findings, virtually complete marrow tumor infiltration was diagnosed, and this was confirmed by marrow biopsy.

The referring oncologist found the MR diagnosis surprising, since the only notable findings on the planar technetium 99-m bone scan done 3 weeks earlier (Fig. 3C) were two small and indeterminate areas of increased uptake in left anterior and lateral ribs (arrows). The nuclear medicine study was not a "superscan," since normal renal activity was demonstrable on these images (curved arrow) as well as on the posterior views of the lower thoracic and lumbosacral spine, which also showed no abnormal uptake despite the extensive marrow tumor.

Although the radionuclide bone scan has long been considered the clinical gold standard for detection of metastatic disease, particularly from breast and prostate carcinoma, in some situations it is being replaced by MR imaging (1). The discordance between the marrow biopsy and MR findings and bone scan in this case would not be surprising for a patient with myeloma, leukemia, or possibly lymphoma, but it is not in agreement with the conventional concept of very high sensitivity of bone scans to metastatic breast carcinoma. However, experience indicates that both for diffuse as well as some focal tumors the bone scan may be falsely negative, even with significant volumes of tumor (2). An intrinsic advantage of MR is that it images the marrow as well as the tumor directly (3,4), rather than being a secondary imaging method dependent on increased osteoblastic activity with uptake of technetium-labeled bone-seeking agents. MR has better spatial resolution as well, and can detect soft tissue and nodal abnormalities at the same time (5). Because of its lower cost and general availability, however, the bone scan is still the preferred examination for general screening for bone metastatic disease.

Although metastatic breast cancer is the most common metastatic bone malignancy, it is well known that the diagnostic yield from bone scans in patients with stage I or II breast cancer is low (only 2% to 6%) (6). However, the percentage of patients with bone metastases rises rapidly with stage, reaching almost 50% in patients at stage IV disease. Bone metastases are a major cause of morbidity in cancer patients, and early detection can improve clinical management and morbidity (4).

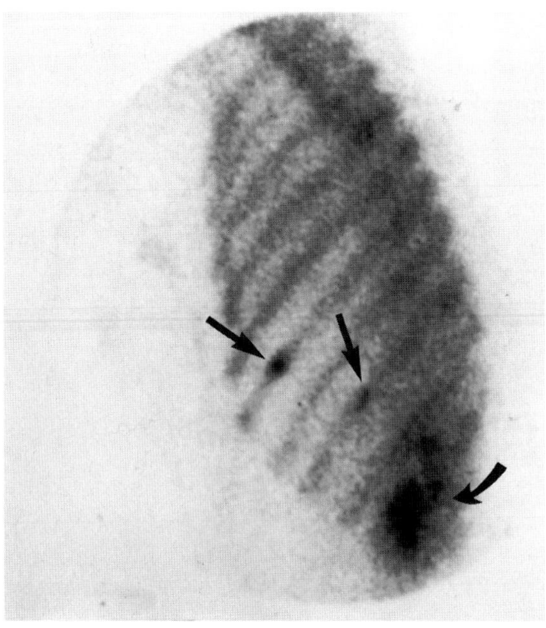

FIG. 3. C: Oblique view from technetium 99-m bone scan. Two small focal areas of increased uptake are seen in left anterior and lateral ribs *(arrows)*. Renal activity *(curved arrow)* is identified indicating that the examination is not a "superscan."

CASE 4

History: A 39-year-old woman with a biopsy-proven sternal metastasis from breast carcinoma underwent MR in addition to evaluation with radionuclide bone scan. What is your diagnosis?

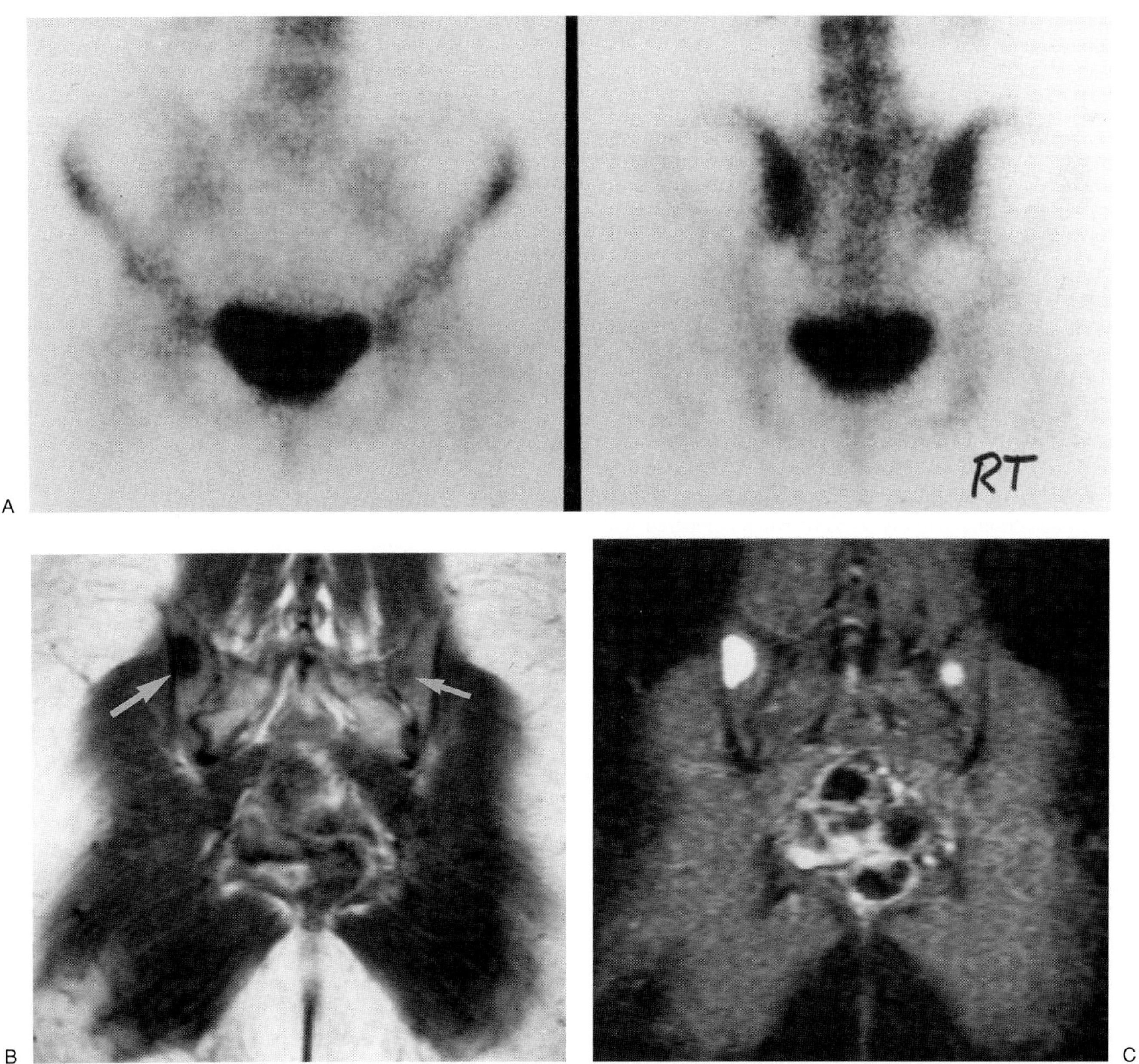

FIG. 4.

Findings: Figure 4A presents anterior and posterior views (left and right respectively) of the pelvis on a planar technetium 99-mTc bone scan. These views were interpreted as normal, and no areas of abnormal activity were defined outside the lesion detected in the sternum. In Fig. 4B, a posterior pelvic, coronal T1-weighted spin echo image (TR/TE 400/12 msec) shows a sharply marginated 2 × 1.5 cm focal lesion (arrow) in the right posterior iliac crest. A more subtle additional lesion is seen in the contralateral left posterior iliac crest (small arrow). In Figure 4C the corresponding STIR image (TR/TE/TI/T 2,000/30/110 msec at 0.5 T) shows that both lesions are markedly intense.

Diagnosis: Focal metastatic disease demonstrated on MR but not on radionuclide bone scan.

Discussion: Focal marrow metastatic lesions may also be missed on bone scan (see Case 3). The clinician requested biopsy confirmation, and the right posterior iliac crest lesion yielded viable adenocarcinoma compatible with breast origin. A coronal STIR image more anteriorly (Fig. 4D) revealed an additional lesion in the proximal right femur of just over 1 cm diameter (arrow). This lesion was not detected on the bone scan either, despite its size and location (compare with Fig. 4A). On this image notice the ability of STIR to define the subtle difference between the inactive yellow marrow of the capitofemoral epiphyses and greater trochanters vs. the hematopoietically active red marrow of the femoral shafts, supra-acetabular region, and proximal femurs. The marrow elsewhere is normal on STIR. Compare the signal intensity on this image with the diffuse marrow tumor in Fig. 3B.

Additional STIR images in the chest (Fig. 4E) revealed metastases to the right pedicle of approximately T5 as well as to the spine of the left scapula (arrows). The bone scan in these regions was normal as well despite local symptoms. These areas were tender to palpation, as was the 5- to 6-mm lesion (arrow) in the right lateral margin of C3 (Fig. 4F). Notice the much higher contrast and lesion detection capability of STIR when compared with the corresponding T1-weighted spin echo image (Fig. 4G). These lesions were painful, tender to palpation, and progressed; the patient failed therapy and subsequently died of metastatic stage IV breast cancer.

The combination of T1 spin echo and STIR is necessary, despite the frequent advantage of the latter for marrow tumor, in that some metastases (notably prostate and some breast) may be densely osteoblastic bone, with low lesion tumor cellularity. Hence, these predominantly bony lesions are more readily detected by T1 spin echo images because of their high contrast with surrounding fat. On STIR they may be dark or have a halo of increased T2 signal, perhaps representing a rim of active osteoblasts or local lytic change at the interface of active tumor and bone (1,2). The halo sign is more commonly seen with prostate cancer than with other metastatic processes, but it can also be present with blastic breast metastases.

Solitary bone lesions, whether on bone scan or MR, in patients with malignancies are always problematic. Patients with malignancies may also have benign incidentally detected bone lesions, as can the general population. Hence, correlation with plain radiographs, clinical features, and other methods is required with all solitary lesions, because of their profound effect on staging and therapy. This is illustrated in Fig. 4H, in which the isolated abnormal STIR signal in the left proximal humerus (arrow) in this patient with breast cancer on a coronal STIR image was found on x-ray examination to have typical features of an enchondroma. Since no other marrow abnormalities were noted elsewhere, the patient was not misdiagnosed to stage IV on the basis of an isolated MR signal abnormality. Edema of the left breast and a malignant left axillary node are also seen (small arrows). Another breast cancer patient was imaged with STIR (Fig. 4I), which detected a solitary marrow abnormality in the left inferior sacral ala (arrow). On biopsy this was a stress reaction, and not metastatic disease. Isolated lesions in the sternum in patients with breast cancer, however, are more likely to be metastatic than solitary lesions elsewhere, particularly ribs. As many as 76% of solitary sternal lesions are metastatic in patients with breast cancer (2).

Contrast materials may help improve the diagnostic specificity of the MR examination and may be of particular value in the assessment of solitary lesions (Fig. 4J–L). A 49-year-old woman was imaged for restaging prior to bone marrow transplant for breast carcinoma. A solitary lesion was detected in the lower left sternum on MR (Fig. 4J, arrow) and bone scan (Fig. 4K). Sequential MR imaging was also done using a dedicated breast imaging coil with precontrast STIR, T1 spin echo, postcontrast serial T1 spin echo images and image subtraction (3) (Fig. 4L). The sternal lesion depicted on the coronal acquisition is confirmed on the top (STIR) image (arrow), and it is of low signal intensity on T1 spin echo (second image). With contrast administration the lesion becomes isointense with surrounding marrow (third image). This illustrates the importance of STIR and T1-weighted noncontrast images prior to any contrast-enhanced evaluation of areas of concern for marrow metastatic disease (4,5). Subtraction of image #2 from image #3 produces a very high contrast depiction of the area of enhancement (bottom image) and subtracts normal surrounding tissues. Notice that the postoperative, STIR hyperintense left breast seroma (curved arrow) is subtracted from the bottom image as are other normal tissues. The sternal lesion, however, is conspicuous. Greater than 100% contrast enhancement was measured in the first 2.5 minutes after contrast injection with this lesion; this is similar to the enhancement degree and timing that we have seen for other breast cancers (3). Although highly compatible with a malignant process, this pattern is not pathognomonic, since benign hyperperfused lesions may appear similar (6). Again, because of the marked impact it would have on therapy, biopsy of this region was elected, which confirmed metastatic breast carcinoma. No other lesions were seen elsewhere on MR or bone scan. Changes in signal intensity or perfusion can be used to monitor therapy, and again, since they represent primary observations, they should prove more reliable than bone scan changes, which may appear worse (flare phenomenon) with successful treatment (7).

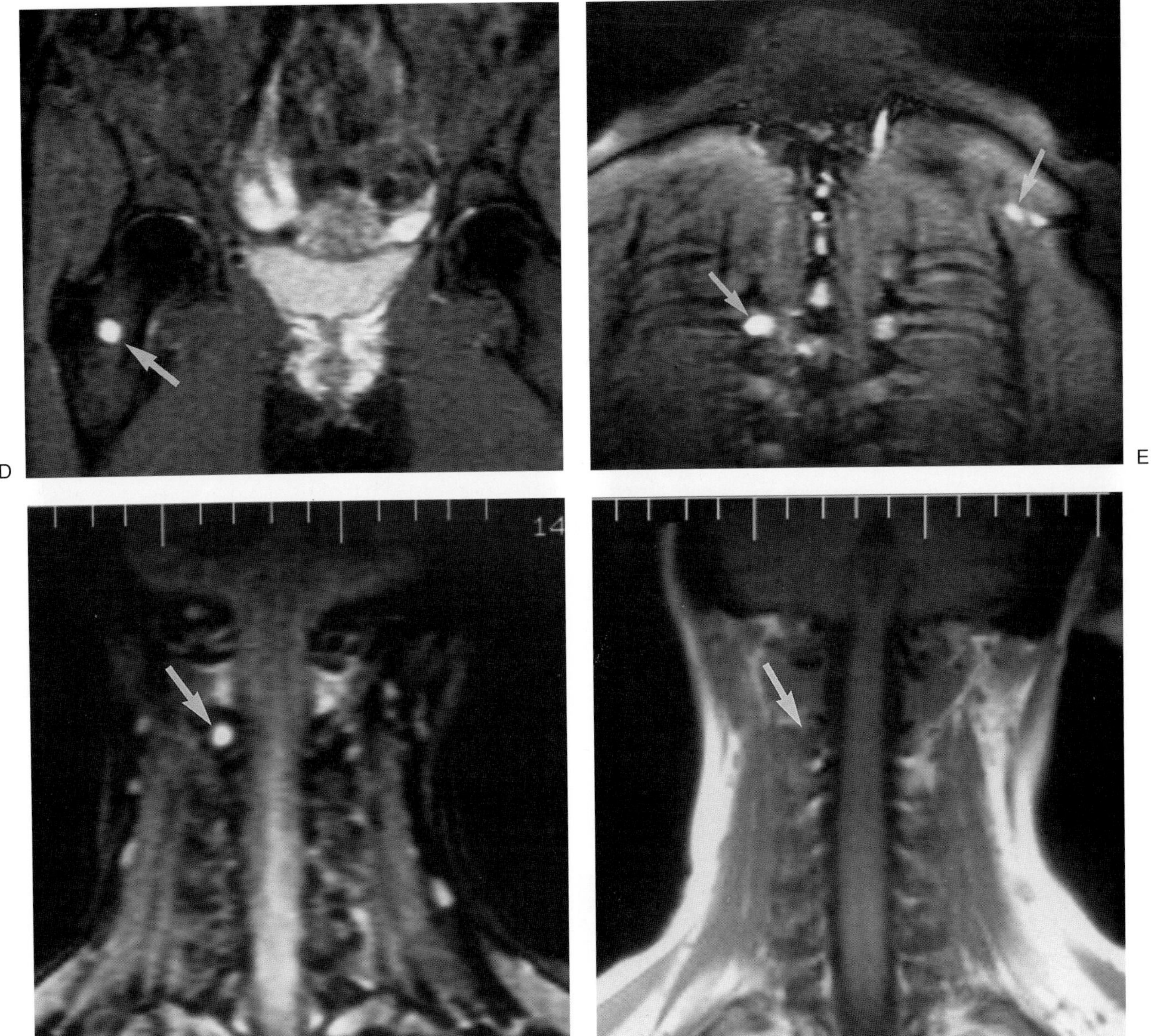

FIG. 4. D: Coronal STIR (TR 2000/TE 30/TI 110 at 0.5 T) obtained slightly more anteriorly than Fig. 4C. There is an additional approximately 1 cm focus of abnormal signal in the proximal right femur *(arrow)*. The area was not detected on the radionuclide bone scan (compare with Fig. 4A). **E:** Coronal STIR (TR 2000/TE 30/TI 110 at 0.5 T) of the chest in the same patient revealed additional metastatic lesions that were not demonstrable on the bone scan *(arrows)*. **F,G:** (F) Coronal STIR (TR 2000/TE 30/TI 110 at 0.5 T) demonstrates another metastatic lesion in this patient that was not evident on scintigraphic examination *(arrow)*. Notice the increased conspicuity of this lesion compared with its appearance on a T1-weighted spin echo image (G) where it is seen as a subtle focus of decreased signal *(arrow)*.

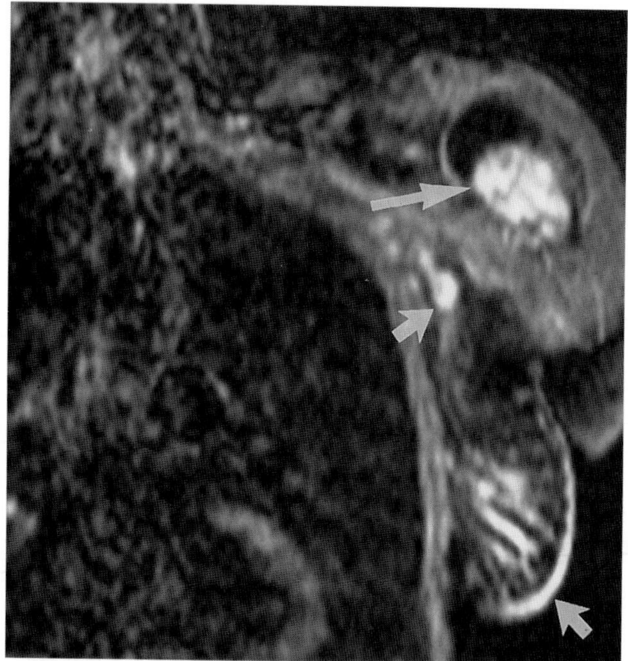

H

FIG. 4. H: Coronal STIR (TR/TE/TI 2,000/30/110 msec at 0.5 T). An area of increased signal intensity is identified in the proximal humerus of this patient undergoing evaluation for breast cancer *(arrow)*. It is critical to correlate lesions, particularly solitary lesions, with conventional radiographs in order to increase specificity. In this case the finding related to a benign enchondroma and not a metastatic lesion. Also note edema of the left breast and axillary lymph node *(short arrows)*.

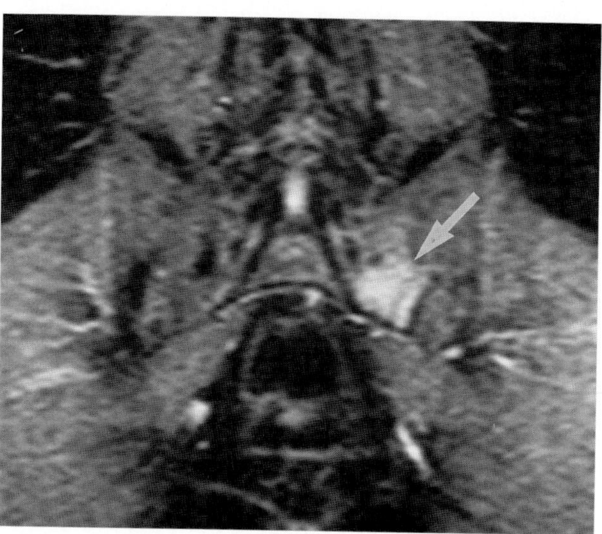

I

FIG. 4. I: Coronal STIR (TR/TE/TI 2,000/30/110 msec at 0.5 T) of another patient with breast cancer undergoing marrow screening. A solitary abnormality was identified in the pelvis *(arrow)*. This has a characteristic appearance and location for stress fracture (see Case 4-14).

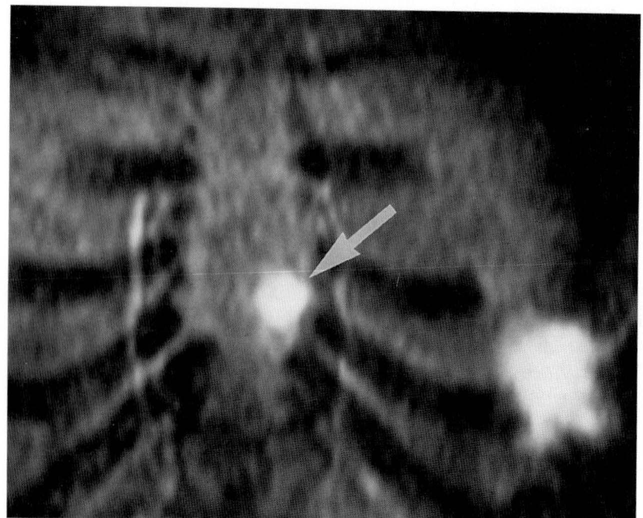

J

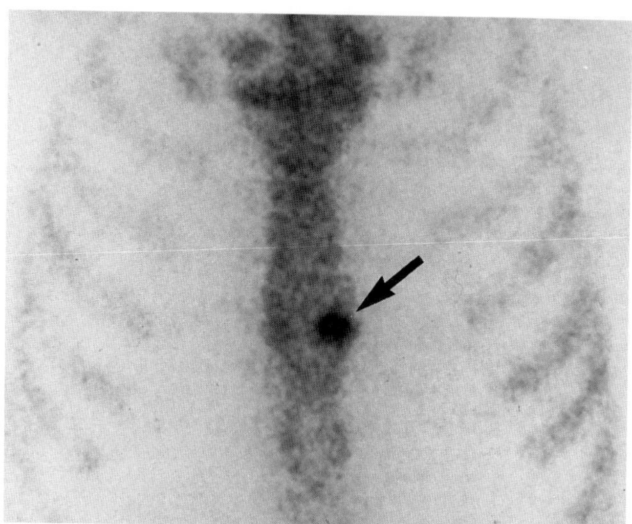

K

FIG. 4. J–K: J: Coronal STIR (TR/TE/TI 2,000/30/110 msec at 0.5 T). A solitary lesion *(arrow)* was detected in this 49-year-old woman being restaged prior to bone marrow transplantation for breast cancer. Isolated lesions in the sternum in patients with breast cancer are more likely malignant than benign. K: Technetium 99m bone scan demonstrates increased activity to the lesion *(arrow)*.

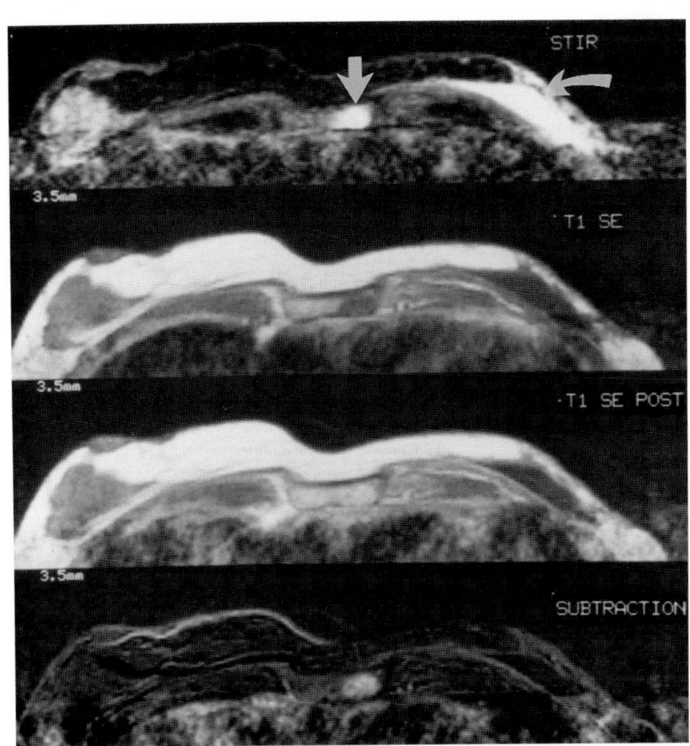

FIG. 4. L: Composite image. The top image is an axial STIR (coronal STIR—(TR/TE/TI 2,000/30/110 msec at 0.5T) sequence that again confirms the sternal lesion. The lesion is of low signal on the T1-weighted sequence (second from the top). Following the administration of contrast, the lesion becomes isointense to marrow on a T1-weighted image (third from top). A subtraction image (bottom) demonstrates high conspicuity of the lesion.

CASE 5

History: A 77-year-old man complained of pain throughout his spine, pelvis, and proximal lower extremities. X-rays of these regions showed severe osteopenia and multiple compression fractures of his thoracic and lumbar vertebrae. A serum protein electrophoresis showed a monoclonal spike and he was referred for an MR examination. What is the differential diagnosis based on the MR examination?

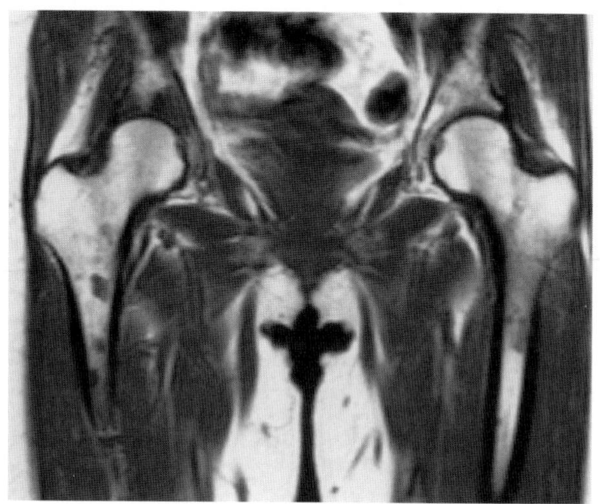
A

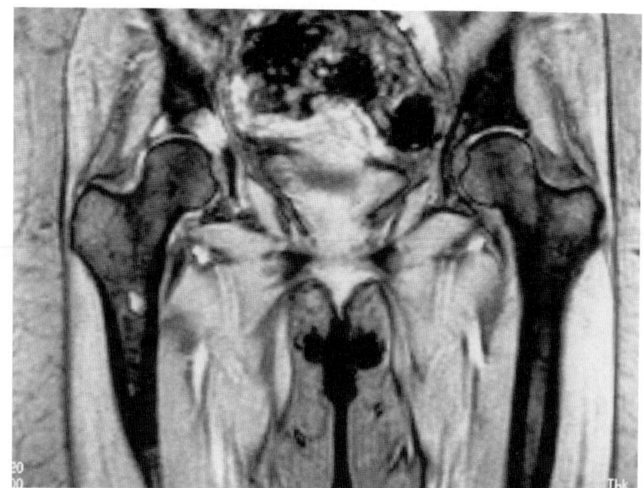
B

FIG. 5.

Findings: In Fig. 5A a T1-weighted (TR/TE 600/20 msec) coronal image of the proximal femora shows multiple, small round foci of low signal. There is a larger lesion in the right acetabular roof. In Fig. 5B a T2*-weighted (TR/TE/° 400/10/20) GE image shows high signal foci corresponding to the larger lesions seen on the T1-weighted images.

Diagnosis: Multiple myeloma.

Discussion: Primary tumors of bone that arise from the reticuloendothelial system include multiple myeloma, some forms of leukemia, and primary osseous lymphoma. Ewing sarcoma, which has traditionally been included in this family of lesions, is now thought to be of neurogenic origin (1). The imaging appearance of Ewing sarcoma, however, shares many features with the neoplasms that truly arise from the reticuloendothelial components of the bone marrow (Fig. 5C,D). All these lesions manifest a tendency to be infiltrative rather than osteolytic and therefore produce minimal radiographic changes while confined to the medullary space. Radiography is insensitive prior to the development of cortical destruction and soft tissue extension. MR is the imaging modality of choice for assessing marrow involvement by this group of malignant neoplasms because of its high sensitivity for replacement of the normal marrow by abnormal cellular infiltration.

Multiple myeloma is a bone marrow malignancy caused by focal, disseminated, or diffuse infiltration of the bone marrow by a monoclonal proliferation of malignant plasma cells (2,3). This is a disorder of the older patient, with a peak incidence in the sixth and seventh decades of life (2). The most common symptom of the disorder is nonspecific bone pain (4). Pathologic fractures of the weakened bone, particularly the vertebrae, are another common reason for the patient to seek medical attention. Less common clinical manifestations of multiple myeloma include renal insufficiency, weakness, soft tissue masses, neurologic symptoms, and frequent infections (5). Laboratory findings include anemia, hypergammaglobulinemia, hypercalcemia, and an abnormal serum and/or urine electrophoresis in most cases (5). Long-term prognosis correlates with the overall tumor cell mass and renal function (2).

Multiple myeloma can result in either multiple nodular deposits of malignant cells that replace the normal hematopoietic marrow or in diffuse infiltration, in which the myeloma cells are admixed with the normal marrow elements (2). In the nodular form, malignant plasma cell infiltration may be limited to a solitary focus, known as a plasmacytoma, early in the course of the disease. Plasmacytomas are solitary, osteolytic lesions of the axial skeleton that may show extensive bone expansion, cortical destruction, and soft tissue extension (Fig. 5E,F). Patients with solitary plasmacytoma tend to develop multiple myeloma within 2 to 3 years, although cases with long-term survival without development of myeloma have been reported (5). Usually, multiple nodules of malignant plasma cells are widely disseminated in the bone marrow at the time of diagnosis. These focal nodules replace the marrow and destroy the trabeculae, and large foci can be recognized on radiographs as areas of osteolysis and bone destruction. Focal lesions are easier to identify if the background normal bone is not demineralized due to osteoporosis. The diffusely infiltrating form is more difficult to recognize, regardless of the imaging modality employed, because the tumor cells are interspersed among the normal marrow elements. Conventional radiographs are very insensitive for multiple myeloma in its diffusely infiltrating form, particularly in cases without cortical involvement. In this situation, x-rays are frequently normal or manifest only generalized osteopenia. Scintigraphy is also insensitive for myeloma because the malignant infiltration fails to incite a reactive osteoblastic response in most cases. CT shows generalized diminution of trabecular bone and increased attenuation of the bone marrow, but the findings are often subtle, particularly in patients with diffuse infiltrative disease. Focal osteolytic lesions are well evaluated with CT but the disease extent is easily underestimated with this modality.

MR has been shown to be more sensitive than conventional radiography and CT for detection of multiple myeloma (2,3). Despite its relatively high sensitivity, a significant proportion of patients with multiple myeloma cannot be identified with routine MR imaging. With standard SE imaging, 6% to 34% of patients with myeloma will have a normal MR examination, even in the presence of widespread marrow involvement and compression fractures (2,3,6). T1-weighted images are relatively insensitive for multiple myeloma, showing focal lesions in 25% to 31% of patients and a variegated, inhomogeneous or diffuse decrease in marrow signal in 38% to 53% of patients (2,3,6). On T1-weighted images, the most common appearance of multiple myeloma is a homogeneous or inhomogeneous decrease in marrow signal caused by interstitial infiltration by the malignant plasma cells. In these patients, the low signal marrow approximates the signal intensity of muscle (2). On T2- and T2*-weighted images, focal nodules of high signal intensity can be seen in over 50% of patients with multiple myeloma (2,3). In the remaining patients, the T2- and T2*-weighted images show persistent low or intermediate signal marrow, which is difficult to recognize as abnormal. The nodules of myeloma are of homogeneous high signal on STIR images and are more conspicuous than on T2-weighted SE images (6). Multiple myeloma enhances homogeneously following intravenous gadolinium administration but contrast enhancement does not increase the sensitivity of MR for detection of myeloma deposits (6). There are limited data about the accuracy of MR in determining the overall tumor burden in myeloma, a factor critical to accurate staging in this disease. There appears to be only a weak correlation between the MR estimates of tumor volume and percentage of marrow plasmacytosis (2,3). Despite these limitations, MR is the most useful imaging modality for assessment of multiple myeloma because of its ability to detect marrow abnormalities in a lesion that is difficult to evaluate with alternative imaging methods.

The major differential diagnosis in patients with disseminated or diffuse multiple myeloma is metastatic disease to the skeleton. Like myeloma, metastases may be nodular or diffusely infiltrative, although the former appearance is more common. Solitary metastases are nonspecific in appearance and difficult to differentiate from other tumors or tumor-like processes of bone. Disseminated metastases, with multiple discrete lesions in the axial skeleton, are the most common manifestation of this common disorder on MR imaging (7) (Fig. 5G,H). Like myeloma, metastases are low signal on T1-weighted images and enhance with gadolinium. Most metastases are bright on T2-weighted imaging but densely sclerotic lesions may be isointense or hypointense. Differentiation from multiple myeloma is difficult on the basis of the MR images and correlation with clinical and laboratory findings is necessary. In many cases, a percutaneous or open biopsy of the lesion is necessary to establish a specific histologic diagnosis.

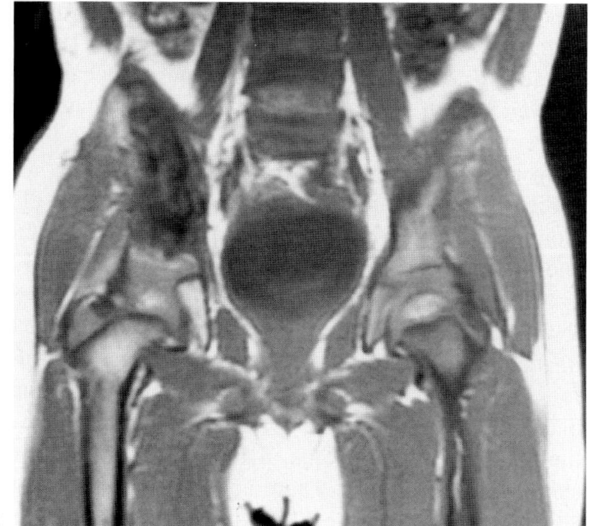

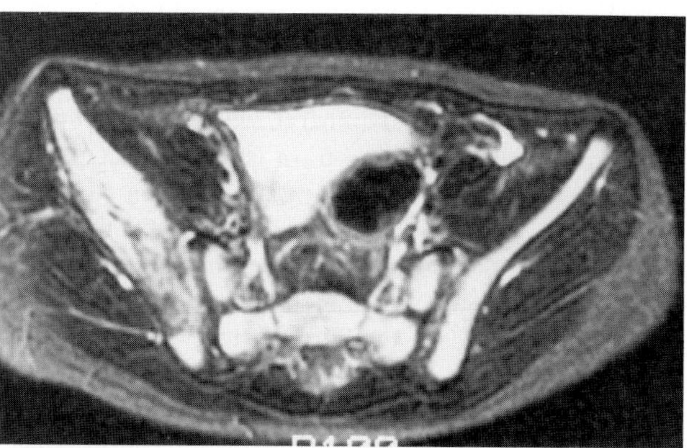

C D

FIG. 5. C,D: A 2-year-old boy with Ewing sarcoma of the right ilium was imaged following a full course of chemotherapy. **C:** The T1-weighted (TR/TE 600/12 msec) SE image shows a low signal mass within the right iliac bone. The cortex is thickened due to extensive periosteal proliferation and bone sclerosis due to partial healing of the bone marrow. **D:** A fat-suppressed T2-weighted (TR/TE 4,000/102 msec) FSE axial image of the pelvis shows the bone expansion and inhomogeneity commonly seen with Ewing sarcoma. This neoplasm is better marginated than is typically seen with untreated Ewing sarcoma.

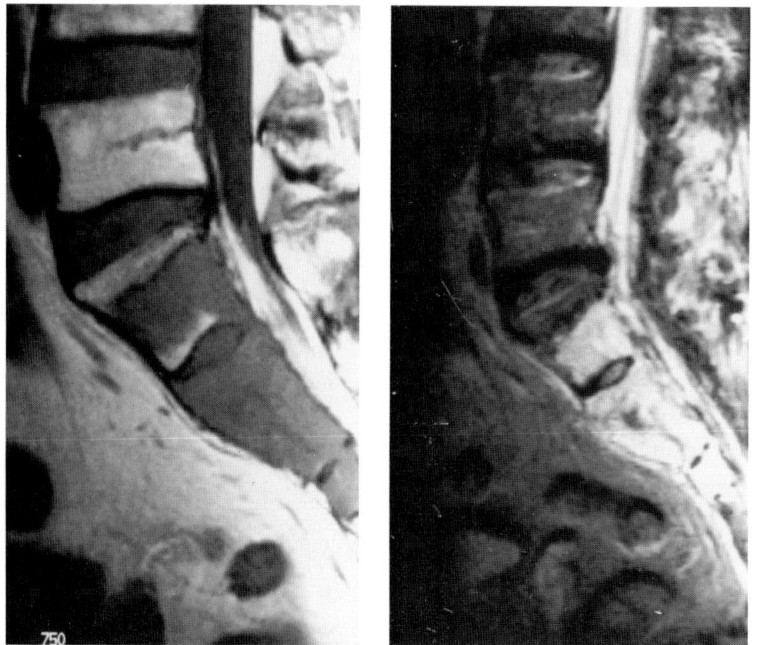

E F

FIG. 5. E,F: This 59-year-old man complained of pain in his lower back. Radiographs showed an osteolytic lesion of the sacrum. No other lesions were identified on his x-rays or a bone scan. Biopsy of the lesion showed malignant plasmacytoma. **E:** A T1-weighted (TR/TE 750/15 msec) sagittal image of the lumbosacral junction shows a focal, low signal mass within the sacrum. **F:** The T2-weighted (TR/TE 2,500/90 msec) image of the sacrum shows high signal throughout the lesion.

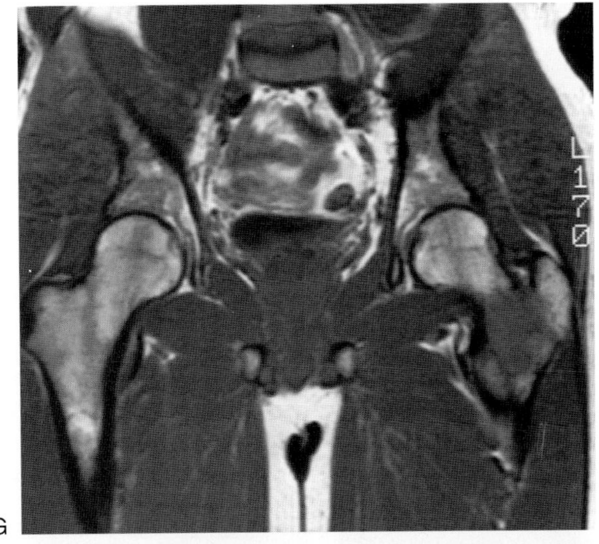

G

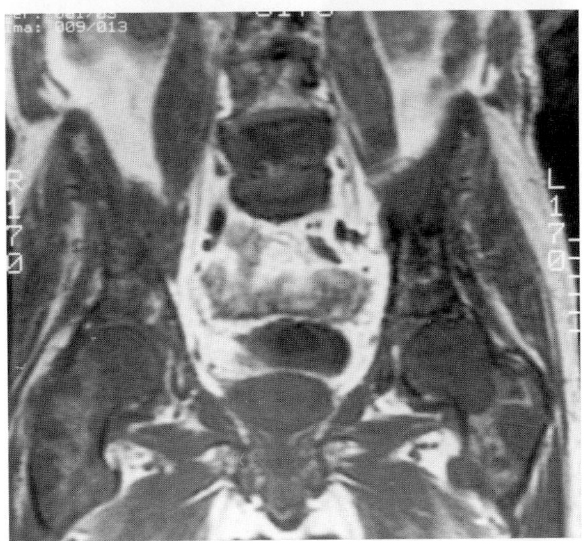

H

FIG. 5. G,H: Images from two different patients with metastatic disease are illustrated. **G:** A T1-weighted (TR/TE 650/10 msec) SE image of the pelvis shows a solitary low signal metastasis from carcinoma of the thyroid involving the intertrochanteric region of the left femur. **H:** A T1-weighted (TR/TE 600/30 msec) SE image of the pelvis from a patient with extensive metastatic disease from malignant melanoma shows numerous low signal lesions in the skeleton. Note the large lesions in both femoral heads and the involvement of the lower lumbar vertebrae. On the basis of the MR findings alone, multiple myeloma would also need to be strongly considered to account for the marrow infiltration.

CASE 6

History: A 65-year-old woman presented with a 1-year history of progressively increasing leg weakness and polyneuropathy. Recent plain films and MR examination revealed multiple osteosclerotic lesions in the spine and pelvis. The patient is mildly hypothyroid, and has noticed increased thickness of the skin, particularly on the lower extremities. What is your diagnosis?

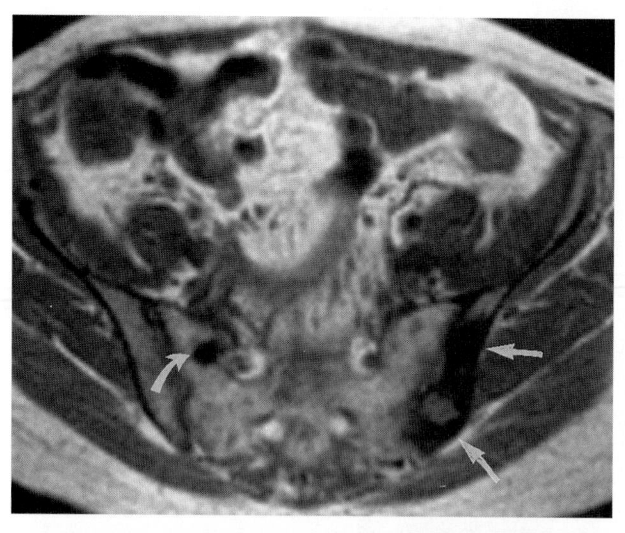

A

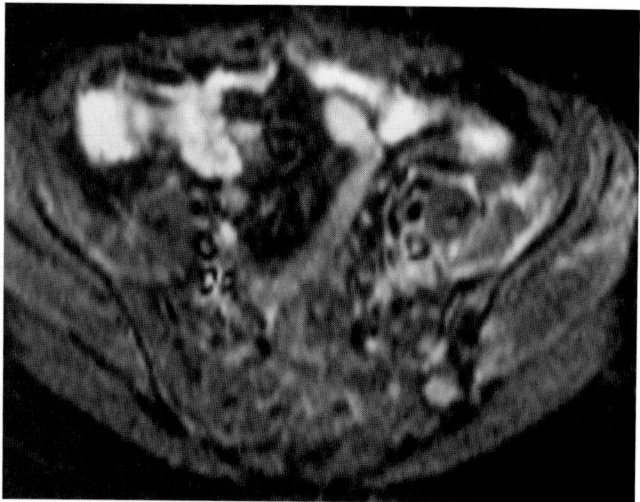

B

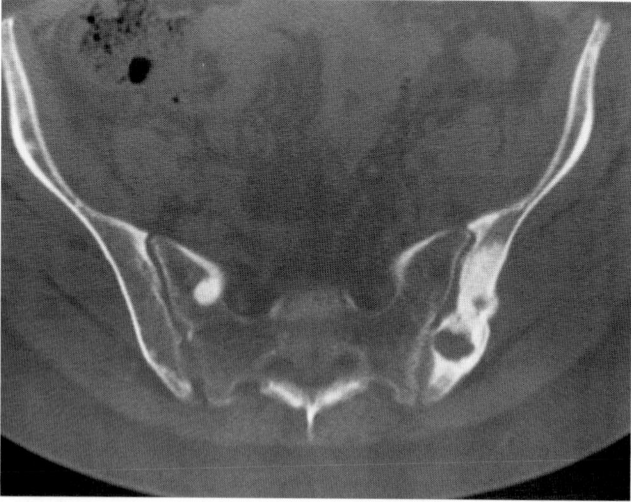

C

FIG. 6.

Findings: In Fig. 6A the axial T1-weighted (TR/TE 600/12 msec) spin echo image of the pelvis shows an area of markedly decreased signal intensity in the left posterior iliac crest, with a central area of intermediate signal intensity more posteriorly (arrow). Another focal sclerotic lesion is present in the right sacral ala (curved arrow). In Fig. 6B corresponding STIR image reveals complete absence of signal intensity in the areas of sclerotic bone, with low to intermediate STIR signal intensity in the cavity defined on the T1-weighted image. The sclerotic focus in the right sacrum is not detectable, and therefore of very low cellularity. In Fig. 6C the CT image with bone algorithms reveals the high density of the sclerotic bone as well as the central lytic cavity in the left posterior iliac crest and the sclerotic focus in the right sacrum.

Diagnosis: Multiple myeloma/POEMS syndrome.

Discussion: At this time, diagnostic considerations included metastatic breast cancer, lymphoma, or an unusual form of sclerotic myeloma. Review of clinical data revealed that overall serum proteins were normal, with only a small increase in M protein, which was considered nonspecific. As a result, biopsy of the left posterior iliac crest lesion was requested and done with CT guidance (Fig. 6D). The needle was directed into the more cellular-appearing cavity within the central portion of the lesion. The bone was extremely hard and difficult to penetrate, and initial CT biopsy produced nondiagnostic material. Subsequent additional attempts yielded a small volume of myeloma cells.

The sclerotic form of multiple myeloma is a rare presentation of an otherwise common primary marrow malignancy in elderly patients (1% to 3% reported cases) (1). This form of myeloma has been termed the POEMS syndrome because it is associated with progressive sensorimotor polyneuropathy (P), osteosclerotic (O) bone lesions, endocrine (E) abnormalities such as diabetes mellitus, adrenal insufficiency, and hypothyroidism, an increase in serum M protein (M), and skin (S) thickening with increased pigmentation (2).

MR is more sensitive than conventional x-ray studies for detection of myeloma in the marrow space (3,4). Myeloma, a malignant plasma-cell dyscrasia, has multiple MR manifestations and presentations. It may present with an isolated marrow-based lesion (plasmacytoma) and clinical manifestations of cord compression a diffuse form that may simulate leukemia on MR and a patchy or heterogeneous pattern that on T1- and T2-weighted imaging, can be difficult to differentiate from normal marrow variations in elderly patients. Commonly, also, myeloma lesions are very discrete (''punched out'') on MR, as on plain films and CT (1,3).

MR is capable of detecting the various patterns of myeloma and allows monitoring of therapy (5,6). Because of the often patchy and heterogeneous nature of marrow myeloma, however, as well as the variable secretion of abnormal paraproteins, there is often a poor correlation between the appearance of myeloma on MR imaging and its clinical laboratory findings (3). Contrast material is of no advantage for detection of additional lesions (5); however, changes in contrast enhancement do appear to correlate well with response to therapy (5,6). MR may be particularly helpful in determining response to therapy in minimally symptomatic patients with nonsecretory myeloma (7).

MR marrow findings may be helpful in estimating prognosis. The outlook for patients with diffuse involvement is generally poor, whereas lower volume or patchy, lesser quantities of marrow tumor may have better response to therapy. With response to therapy, decreased signal intensity on T2 spin echo or STIR correlates with decrease in tumor cellularity and water content. This method allows one to qualitatively determine the changes in marrow tumor burden and its possible viability. As tumors desiccate in response to therapy their T2 or STIR signal generally diminishes; this is particularly helpful with nonsecretory tumors, which are difficult to monitor with clinical laboratory data (7).

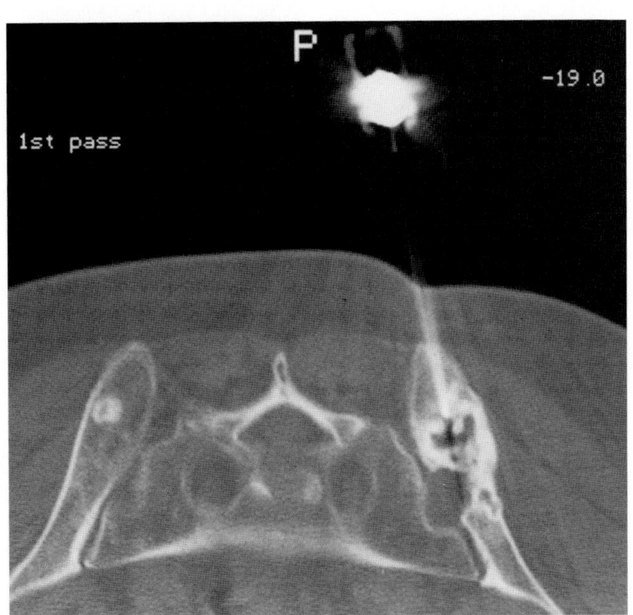

FIG. 6. D: Axial CT scan obtained for purposes of directed biopsy reveals the biopsy needle to be within the lesion.

CASE 7

History: Images were obtained both prior to and following therapy for multiple myeloma. What is the significance of the findings in Fig 7B?

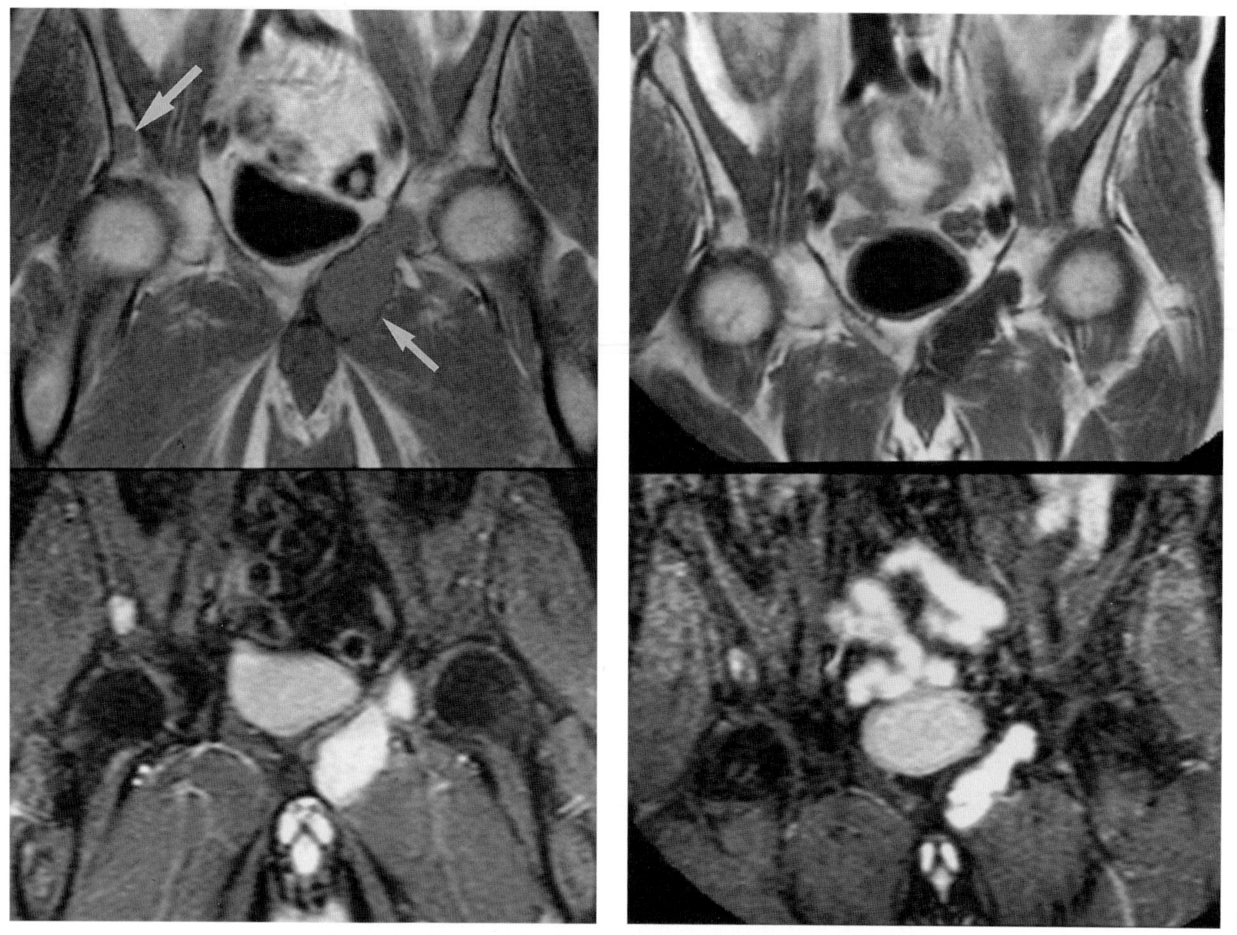

FIG. 7.

Findings: Figure 7A is a coronal T1 spin echo (TR/TE 500/12 msec) (top) with the corresponding STIR (TR/TE/TI 2,000/30/110 msec at 0.5 T) below. Figure 7B is the corresponding post-therapy image. There has been a marked decrease in signal intensity following therapy on the T1-weighted image. The lesion appears sharply circumscribed but is smaller in size. The lesion appears of uniform increased signal on the STIR sequence.

Diagnosis: Effective response to therapy with residual consistent with necrotic fluid.

Discussion: Response to therapy can also be documented by changes of signal of the myelomatous tissue over time, although this common clinical observation has not been prospectively documented (1,2). Necrosis of larger focal and expansile lesions with effective therapy may leave a fluidlike cavity that may persist unchanged for years as illustrated in this case. The STIR sequence is very sensitive to the hypercellular lesions of myeloma, either focal or diffuse, as illustrated in Figure 7C. This 57-year-old man had an expansile tender lesion of a right anterolateral rib noted as an ovoid area of very high STIR signal intensity (TR 2000/TE 30/TI 110 at 0.5 T) overlying the liver (arrow). An additional subcentimeter lesion was sensitively detected in a left anterior rib (curved arrow), and later confirmed as a "punched out" lytic lesion on CT. The expansile lesion of the rib was also seen on CT (Fig. 7D, arrow). The chief advantage of STIR in our clinical experience is its great contrast, especially for small focal or diffuse low-level tumor infiltration. In this case the T1-weighted images (TR 500/TE

12) reveal a mildly variegated, heterogeneous pattern of low signal intensity in the iliac, supra-acetabular, and proximal femoral marrow (Fig. 7E), which could be within the range of normal variance for a patient of this age. The corresponding STIR images (Fig. 7F), however, show a level of signal intensity in these areas substantially higher than is seen in hematopoietic marrow (compare with Fig. 4D). This reflects the increased water content and cellularity of the myelomatous lesion, increasing the diagnostic confidence.

Another important aspect of MR in comparison to other modalities is its ability to define spinal canal compromise (Fig. 7G). This patient has pathologic compression fractures of L4 and L1. Sagittal T1-weighted spin echo images of the lumbar spine demonstrate marked compression and complete replacement of the L1 and L4 vertebral bodies. At the L1 level, there is extension into the pedicle (arrow). Tumor involvement in the left pedicle of L1 (arrow) is much more common with malignant than benign processes. However, especially since many myeloma patients are elderly, pathologic compression fractures, possibly on an osteoporotic basis, may be difficult to differentiate in the absence of either complete tumor infiltration or (as seen here) involvement of the marrow of the pedicles and lamina. The multiplanar imaging capability and high soft tissue contrast of MR allows better clinical management and earlier treatment of myeloma patients, particularly those with new back pain and concern for spinal canal compromise (3). Early identification of cord-threatening or compressing lesions is important, since early treatment substantially decreases the incidence of paraplegia (3).

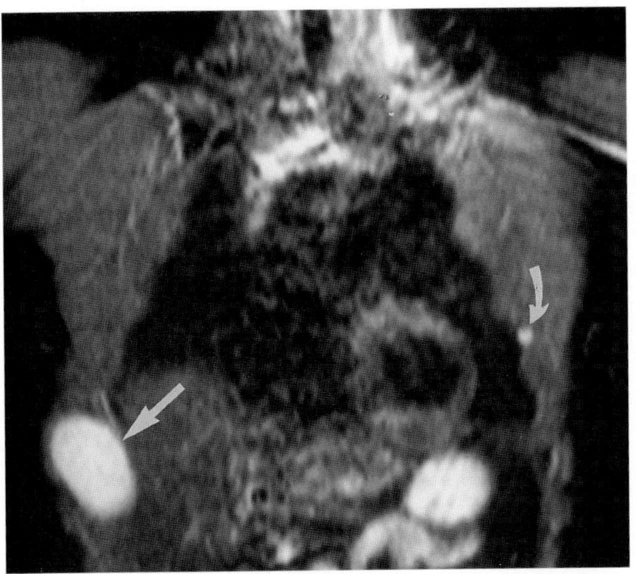
C

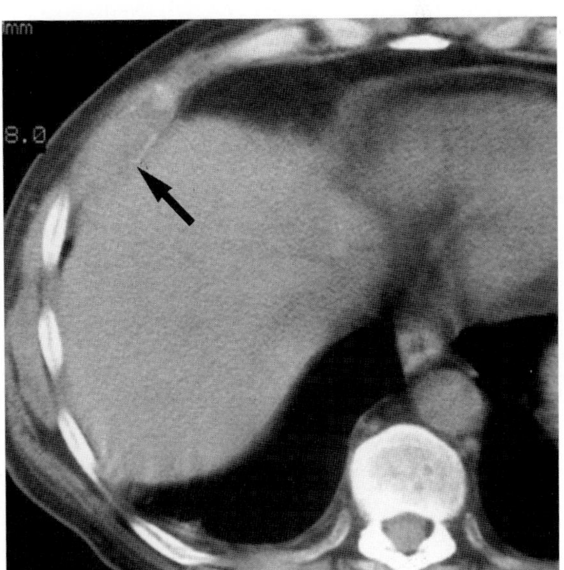

D

FIG. 7. C–F: A 57-year-old man with expansile tender lesion of the right anterolateral rib. On a coronal STIR (TR/TE/TI 2,000/30/110 msec at 0.5 T) image, the lesion demonstrates marked increased signal *(long arrow)*. An additional small lesion is detected within a left-sided rib *(curved arrow)*. **D:** Axial CT scan also well demonstrates the expansile lesion *(arrow)*.

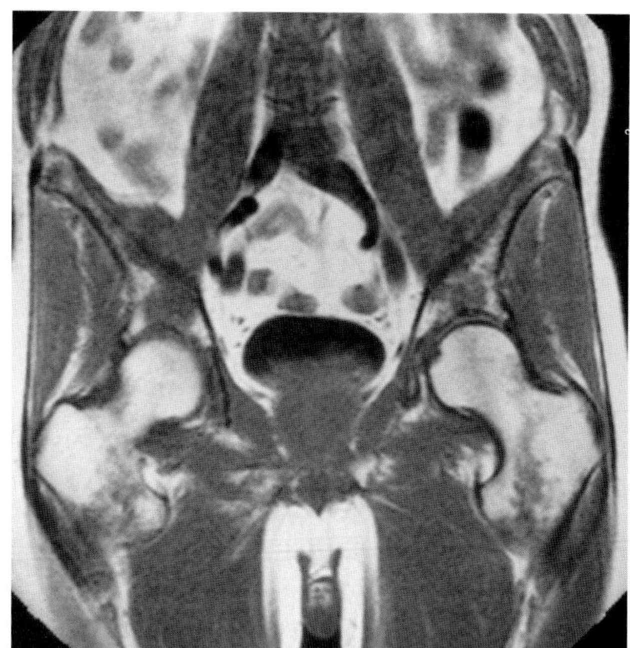

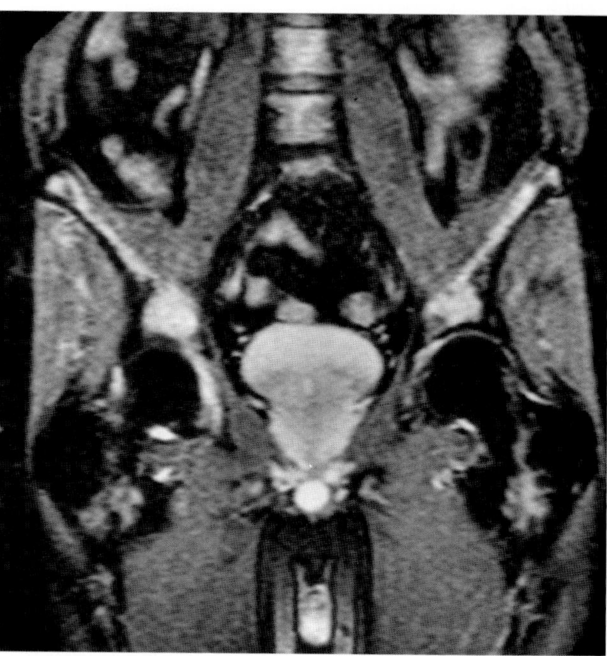

FIG. 7. E: Coronal T1-weighted spin echo image of this patient demonstrates a mildly heterogeneous marrow pattern throughout the pelvis and proximal femora that is difficult to assess and could potentially be within normal limits. **F:** The corresponding STIR image (TR/TE/TI 2,000/30/110 msec at 0.5 T) demonstrates high signal intensity corresponding to the areas depicted in Fig. E. This degree of intensity is significantly greater than one would expect with simple hematopoietic marrow (compare with Fig. 4D) and suggests tumor.

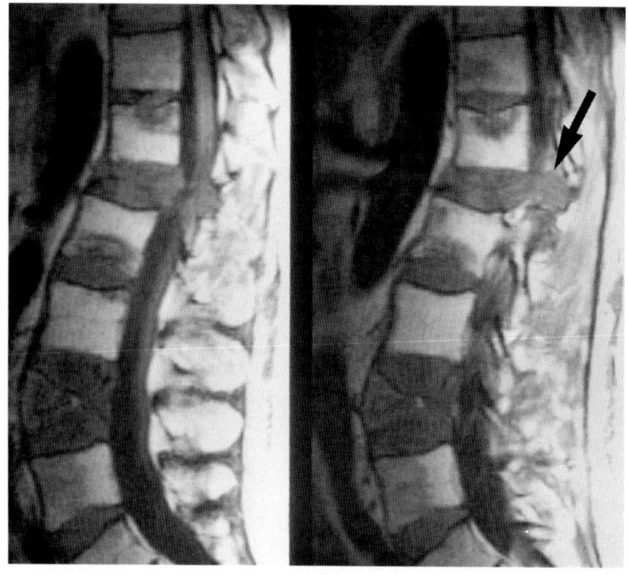

FIG. 7. G: Sagittal T1-weighted images through the lumbar spine of a patient with myeloma. There are pathologic compression fractures of L1 and L4 with complete marrow replacement and extension into the pedicle at the L1 level *(arrow)*. This latter finding is strongly suggestive of neoplasm and is a helpful differential feature when dealing with the question of possible pathologic compression fractures of the spine.

REFERENCES

Introduction

1. Resnick D. Hemoglobinopathies and other anemias. In: Resnick D, ed. *Diagnosis of bone and joint disorders,* 3rd ed. Philadelphia: WB Saunders, 1995;2120–2125.
2. Vogler JB III, Murphy WA. Bone marrow imaging. *Radiology* 1988;168:679.
3. Kricum ME. Red-yellow marrow conversion: its effects on the location of some solitary bone lesions. *Skel Radiol* 1985;14:10.
4. Dawson KL, Moore SG, Rowland JM. Age related marrow changes in the pelvis: MR and anatomic findings. *Radiology* 1992;47:183.
5. Moore SG, Dawson KL. Red and yellow marrow in the femur: age related changes in appearance at MR imaging. *Radiology* 1990;175:219.
6. Ricci C, Cova M, Kang YS, et al. Normal age-related patterns of cellular and fatty bone marrow distribution in the axial skeleton: MR imaging study. *Radiology* 1990;177:83.
7. Peterfy CG, Linares R, Steinbach LS. Recent advances in magnetic resonance imaging of the musculoskeletal system. *Radiol Clin North Am* 1994;32(2):291–309.
8. Constable RT, Anderson AW, Zhong J, et al. Factors influencing contrast in fast spin-echo MR imaging. *Magn Reson Imaging* 1992; 10:497–511.
9. Anzai Y, Lufkin RB, Jabour BA, et al. Fat-suppression failure artifacts simulating pathology on frequency-selective fat-suppression MR images of the head and neck. *AJNR* 1992;13:879–884.
10. Fleckenstein JL, Archer BT, Barker BA, et al. Fast short-tau inversion-recovery MR imaging. *Radiology* 1991;197:499–504.
11. Sebag GH, Moore SG. Effect of trabecular bone on the appearance of marrow in gradient-echo imaging of the appendicular skeleton. *Radiology* 1990;174:855.
12. Mehta RC, Marks MP, Hinks RS. MR evaluation of vertebral metastases: T1-weighted, short-inversion time inversion recovery, fast spin echo, and inversion-recovery fast spin echo sequences. *AJNR* 1995;16:281.

Case 1

1. Bydder GM, Young IR. MR imaging: clinical use of inversion recovery sequence. *J Comput Assist Tomogr* 1985;9:659–675.
2. Shields AF, Porter BA, Churchley S, et al. The detection of bone marrow involvement by lymphoma using magnetic resonance imaging. *J Clin Oncol* 1987;5:225–230.
3. Hoane BR, Shields AF, Porter BA, et al. Detection of lymphomatous bone marrow involvement with magnetic resonance imaging. *Blood* 1991;78(3):728–738.
4. Press OW, Eary JF, Appelbaum FR, et al. Radiolabeled antibody therapy of B cell lymphomas with autologous bone marrow support. *N Engl J Med* 1993;329(17):1219–1224.
5. Hoane BR, Shields AF, Porter BA, Borrow JW. Comparison of initial lymphoma staging using computed tomography (CT) and magnetic resonance (MR) imaging. *Am J Hematol* 1994;47:100–105.
6. Malloy PC, Fishman EK. Magid D. Lymphoma of bone, muscle, and skin: CT findings. *AJR* 1992;159:805–809.

Case 2

1. Malloy PC, Fishman EK, Magid D. Lymphoma of bone, muscle, and skin: CT findings. *AJR* 1992;159:805–809.
2. Hoane BR, Shields AF, Porter BA, Borrow JW. Comparison of initial lymphoma staging using computed tomography (CT) and magnetic resonance (MR) imaging. *Am J Hematol* 1994;47:100–105.

Case 3

1. Jahre C, Sze G. Magnetic resonance imaging of spinal metastases. *Top Magn Reson Imaging* 1988;1(1):63–70.
2. Hortobagyi GN. Bone metastases in breast cancer patients. *Semin Oncol* 1991;18(4)(suppl 5):11–15.
3. Colman LK, Porter BA, Redmond J, Olson DO, Stimac GK, Dunning DM, Friedl KE. Early diagnosis of spinal metastases by CT and MR studies. *J Comput Assist Tomogr* 1988;12(3):423–426.
4. Olson DO, Shields AF, Scheurich CD, Porter BA, Moss AA. Magnetic resonance imaging of bone marrow in patients with leukemia, aplastic anemia and lymphoma. *Invest Radiol* 1986;21(7):540–546.
5. Hoane BR, Shields AF, Porter BA, Borrow JW. Comparison of initial lymphoma staging using computed tomography (CT) and magnetic resonance (MR) imaging. *Am J Hematol* 1994;47:100–105.
6. Brown ML. Bone scintigraphic scanning in malignancies. In: Taveras JM, Ferruci JT, eds. *Radiology diagnosis—imaging-intervention.* Philadelphia: JB Lippincott, 1994.

Case 4

1. Schweitzer ME, Levine C, Mitchell DG, et al. Bull's eyes and halos: useful MR discriminators of osseous metastases. *Radiology* 1993;188:249–252.
2. Kwai AH, Stomper PC, Kaplan WD. Clinical significance of isolated scintigraphic sternal lesions in patients with breast cancer. *J Nucl Med* 1988;29:324.
3. Porter BA, Taylor V, Smith JP, Tsao V. Contrast-enhanced MR mammography. *Acad Radiol* 1994;1:S36–S50.
4. Stimac GK, Porter BA, Olson DO, Gerlach R, Genton M. Gd-DTPA enhanced MR of spinal neoplasms: preliminary investigation and comparison with unenhanced spin-echo and STIR. *AJNR* 1988; 9:839–846.
5. Mehta RC, Marks MP, Hinks RS, et al. MR evaluation of vertebral metastases: T1-weighted, short-inversion-time inversion recovery, fast spin-echo, and inversion-recovery fast spin-echo sequences. *AJNR* 1995;16:281–288.
6. Verstraete KL, De Deene Y, Roels H, et al. Benign and malignant musculoskeletal lesions: dynamic contrast-enhanced MR imaging—parametric "first-pass" images depict tissue vascularization and perfusion. *Radiology* 1994;192:835–843.
7. Janicek MJ, Hayes DF, Kaplan WD. Healing flare in skeletal metastases from breast cancer. *Radiology* 1994;192:201–204.

Case 5

1. Frouge C, Vanel D, Coffre C, Couanet D, Contesso G, Sarrazin D. The role of magnetic resonance imaging in the evaluation of Ewing sarcoma. *Skel Radiol* 1988;17:387–392.
2. Libshitz HI, Malthouse SR, Cunningham D, MacVicar AD, Husband JE. Multiple myeloma: appearance at MR imaging. *Radiology* 1992;182:833–837.
3. Moulopoulos LA, Varma DGK, Dimopoulos MA, Leeds NE, Kim EE, Johnston DA, Alexanian R, Libshitz HI. Multiple myeloma: spinal MR imaging in patients with untreated newly diagnosed disease. *Radiology* 1992;185:833–840.
4. Rahmouni A, Divine M, Mathieu D, Golli M, Haioun T, Anglade MC, Reyes F, Vasile N. MR appearance of multiple myeloma of the spine before and after treatment. *AJR* 1993;160:1053–1057.
5. Meszaros WT. The many facets of multiple myeloma. *Semin Roentgenol* 1974;9:219–228.
6. Rahmouni A, Divine M, Mathieu D, Golli M, Dao TH, Jazaerli N, Anglade MC, Reyes F, Vasile N. Detection of multiple myeloma involving the spine: efficacy of fat-suppression and contrast-enhanced MR imaging. *AJR* 1993;160:1049–1052.
7. Algra PR, Bloem JL, Tissing H, Falke THM, Arndt JW, Verboom LJ. Detection of vertebral metastases: comparison between MR imaging and bone scintigraphy. *Radiographics* 1991;11:219–232.

Case 6

1. Goldman AB. Multiple myeloma. In: Taveras JM, Ferrucci JT, eds. *Radiology—diagnosis, imaging intervention.* Philadelphia: JB Lippincott, 1994.
2. Resnick D, Greenway CD, Bardwick P, et al. Plasma cell dyscrasia with polyneuropathy, organomegaly, endocrinopathy, and skin changes: the POEMS syndrome. *Radiology* 1981;140:17–22.

3. Libshitz HL, Malthouse SR, Cunningham D, et al. Multiple myeloma: appearance at MR imaging. *Radiology* 1992;182:833–837.
4. Reibel T, Knop J, Winkler K, et al. Vergleich Rontgenologische und Nucklearmedizinische Untersuchungen beim Osteosarkom zur Beurteilung der Effektivitat einer Praoperativen Chemotherapie. *Forschr Rontgenstr* 1986;145:365–372.
5. Rahmouni A, Divine M, Mathieu D, et al. Detection of multiple myeloma involving the spine: efficacy of fat-suppression and contrast-enhanced MR imaging. *AJR* 1993;160:1049–1052.
6. Rahmouni A, Divine M, Mathieu D, et al. MR appearance of multiple myeloma of the spine before and after treatment. *AJR* 1993; 160:1052–1057.
7. Moulopoulos LA, Dimopoulos MA, Alexanian R, et al. Multiple myeloma: MR patterns of response to treatment. *Radiology* 1994; 193:441.

Case 7

1. Rahmouni A, Divine M, Mathieu D, et al. MR appearance of multiple myeloma of the spine before and after treatment. *AJR* 1993; 160:1052–1057.
2. Moulopoulos LA, Dimopoulos MA, Alexanian R, et al. Multiple myeloma: MR patterns of response to treatment. *Radiology* 1994; 193:441.
3. Colman LK, Porter BA, Redmond J, Olson DO, Stimac GK, Dunning DM, Friedl KE. Early diagnosis of spinal metastases by CT and MR studies. *J Comput Assist Tomogr* 1988;12(3):423–426.

CHAPTER 10

Soft Tissue Tumors

Mark J. Kransdorf and Mark D. Murphey

The imaging evaluation of soft tissue tumors has undergone dramatic evolution with the advent of magnetic resonance (MR) imaging. Despite this, the objectives of initial radiologic evaluation remain unchanged: (a) detecting the suspected lesion; (b) establishing a diagnosis or, more frequently, formulating an appropriate differential diagnosis; and (c) radiologic staging of a lesion (1). This chapter is not intended as a summary of the radiologic manifestations of soft tissue tumors and tumorlike lesions, but as a teaching approach to some common lesions likely to be encountered in everyday practice as well as some differentiating features of some closely related entities.

It must be emphasized that the radiologic evaluation of a suspected soft tissue mass must begin with the radiograph. Radiographs may be diagnostic of a palpable lesion caused by an underlying skeletal deformity (such as exuberant callus related to prior trauma) or exostosis, which may masquerade as a soft tissue mass. Radiographs may also reveal soft tissue calcifications, which can be suggestive, and at times very characteristic, of a specific diagnosis. For example, they may reveal the phleboliths within a hemangioma, the juxtaarticular osteocartilaginous masses of synovial osteochondromatosis, or the peripherally more mature ossification of myositis ossificans. In addition, radiographs are the best initial method of assessing coexistent osseous involvement, such as remodeling, periosteal reaction, or overt destruction (2). Unlike their intraosseous counterpart, however, the biologic activity of a soft tissue mass cannot be reliably assessed by its growth rate. A slow-growing soft tissue mass that may remodel adjacent bone (causing a scalloped area with well-defined sclerotic margins), may still be highly malignant on histologic examination (3).

MR has emerged as the preferred modality for evaluating soft tissue lesions, and should be obtained following plain film evaluation. MR provides superior soft tissue contrast, allows multiplanar image acquisition, obviates the need for iodinated contrast agents or for ionizing radiation, and is devoid of streak artifact commonly encountered with computed tomography (CT). Critical assessment by multiple investigators has demonstrated the superiority of MR over CT in delineating the extent of a musculoskeletal lesion and in defining its relationship to adjacent neurovascular structures. An important limitation of MR, however, is its relative inability to detect soft tissue calcification, again underscoring the necessity of correlating radiographs.

Despite the superiority of MR in identifying, delineating, and staging soft tissue tumors, it remains limited in its ability to precisely characterize soft tissue masses, with most lesions demonstrating prolonged T1 and T2 relaxation times (4). The vast majority of lesions remain nonspecific, with a correct histologic diagnosis reached on the basis of imaging studies in only approximately one-quarter to one-third of cases (5–7). There are instances, however, in which a specific diagnosis may be made or strongly suspected. These are listed in Table A.

While there is general agreement on the diagnostic value of MR in many cases, the issue of whether MR can reliably predict benign from malignant is much less clear. One study has suggested that MR can differentiate benign from malignant masses in greater than 90% of cases based on the morphology of the lesion (6). Criteria used for benign lesions included smooth, well-defined margins, small size, and homogeneous signal intensity, especially on T2-weighted images. Other studies, however, note that malignant lesions may appear as smoothly marginated, homogeneous masses and MR cannot reliably distinguish benign from malignant processes (5,7,8). This discrepancy likely reflects differences within the studied populations.

When the MR images of a lesion are not sufficiently characteristic to suggest a specific diagnosis, we tend to be conservative in our approach. Malignancies, by virtue of their very nature and potential for autonomous growth, are generally larger and more likely to outgrow their vas-

TABLE A. *Specific diagnoses that may be made or suspected on the basis of MR imaging*

Vascular lesions
Hemangioma
Hemangiomatosis (angiomatosis)
Arteriovenous hemangioma (arteriovenous malformation)
Lymphangioma
Lymphangiomatosis
Bone and cartilage forming lesions
Myositis ossificans
Panniculitis ossificans
Fibrous lesions
Elastofibroma
Musculoaponeurotic fibromatosis
Superficial fibromatosis
Lipomatous lesions
Lipoma
Lipomatosis
Intramuscular lipoma
Neural fibrolipoma
Lipoblastoma
Lipoblastomatosis
Liposarcoma
Tumorlike lesions
Ganglion
Popliteal (synovial) cyst
Hematoma (subacute)
Peripheral nerve lesions
Neurofibroma (+/−)
Schwannoma (+/−)
Synovial lesions
Pigmented villonodular synovitis
Giant cell tumor of tendon sheath
Synovial sarcoma (+/−)

cular supply (with subsequent infarction and necrosis and inhomogeneous signal intensity on T2-weighted SE MR image). Consequently, the larger a mass, the greater heterogeneity, the greater is our concern. Only 5% of benign soft tissue tumors measure 5 cm or more (9,10). In addition, most malignancies are deep lesions, whereas only about 1% of all benign soft tissue tumors are deep (9,10). When sarcomas are superficial, they generally have a less aggressive biologic behavior than do deep lesions (11). As a rule, most malignancies grow as space-occupying intramuscular lesions, enlarging in a centripetal fashion (11), pushing rather than infiltrating adjacent structures (although clearly there are exceptions to this general rule). As they enlarge, a pseudocapsule of fibrous connective tissue is formed around them by compression and layering of normal tissue, associated inflammatory reaction, and vascularization (11). As a rule, they respect fascial borders and remain within anatomic compartments, until late in their course (11).

Increased signal intensity on T2-weighted SE MR images in the skeletal muscle surrounding a musculoskeletal mass has also been suggested as a reliable indicator of malignancy (12,13). These results are based on studies in which both bone and soft tissue lesions were evaluated. Although this increased signal intensity may be seen with malignancy, in our experience when one is evaluating only soft tissue masses, this finding is quite nonspecific. We have found it commonly present in inflammatory processes, abscesses and myositis ossificans, and it may also be present following local trauma, biopsy, radiation therapy, and internal hemorrhage.

In the final analysis, unfortunately, there are no absolutes. We feel quite strongly that *benign and malignant lesions cannot be reliably distinguished on the basis of their MR appearance,* and one is ill-advised to suggest a nonspecific lesion's potential based solely on its MR characteristics.

CASE 1

History: A 22-year-old man presented with a slowly growing, painless mass in the popliteal fossa. There is no history of trauma.

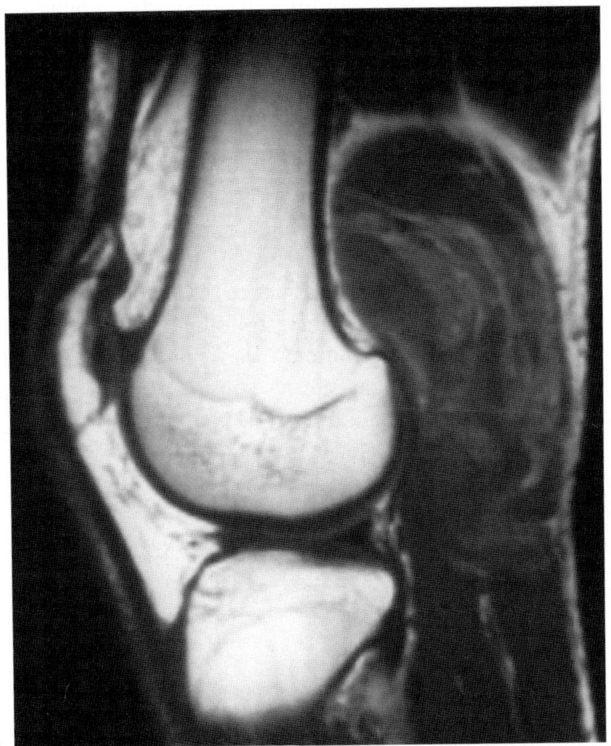

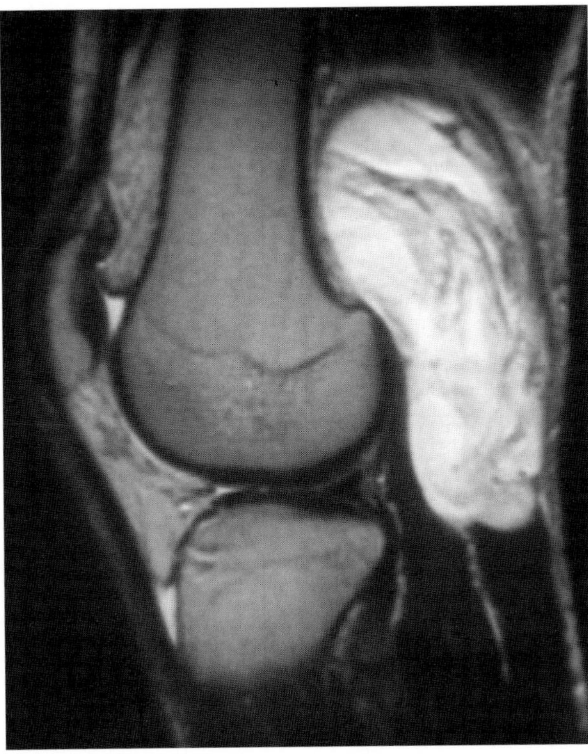

FIG. 1.

Findings: Sagittal T1-weighted (TR/TE 650/20 msec) (Fig. 1A) and T2-weighted (TR/TE 1,800/80 msec) (Fig. 1B) SE images show a soft tissue mass in the popliteal fossa within which are linear and amorphous areas, with signal intensity similar to that of fat. The remainder of the mass has a nonspecific appearance. Radiographs of the knee (not shown) demonstrate a nonspecific mass. There is no evidence of bone involvement or soft tissue calcification.

Diagnosis: Myxoid liposarcoma.

Discussion: The identification of fat within the mass is an important radiologic finding. The lesion clearly does not meet the criteria for an intramuscular lipoma. Its shape, growth pattern, and pattern of fat distribution are also not compatible with a hemangioma. The diagnosis of exclusion is a liposarcoma.

Liposarcoma is a malignant tumor of mesenchymal origin. After malignant fibrous histiocytoma, liposarcoma is the second most common soft tissue sarcoma encountered in adults, accounting for 16% to 18% of all malignant soft tissue tumors (1,2). Most patients with liposarcoma present in the fifth and sixth decades (an age distribution that parallels lipoma). Liposarcoma is exceedingly rare in infants and children, and almost all lipomatous tumors in the pediatric population are lipoblastoma (3). Liposarcoma is usually located in the extremity, particularly the thigh, and retroperitoneum. In a series of 1,755 soft tissue liposarcomas, 50% occurred in the lower extremities, 33% occurred in the trunk (including the retroperitoneum), 12% in the upper extremities, and 4% in the head and neck. Liposarcoma is rare in the distal aspect of the extremities, with only 21 (1.2%) occurring in the hand and wrist and 31 (1.8%) occurring in the foot and ankle. Most patients present with a painless mass, but pain and tenderness are reported in 10% to 15% of cases (4). On average, patients with lesions arising in the extremities present 5 to 10 years earlier than their counterparts with retroperitoneal tumors, presumably since the former are more readily detected (1).

The radiologic appearance of liposarcoma varies according to the histologic type and degree of differentiation of the tumor. The four histologic categories of liposarcoma are (a) well differentiated, (b) myxoid, (c) round cell, and (d) pleomorphic. In many instances, well-differentiated liposarcoma closely mimics lipoma both grossly and microscopically. The myxoid subtype is a low- to intermediate-grade lesion, while the round cell and pleomorphic subtypes are considered to be high-grade sarcomas.

Routine radiographs of patients with liposarcoma may identify a soft tissue mass, but are rarely specific. Calcification is uncommon, but has been reported in up to 10% of cases (4). Rarely, well-defined ossification may be seen. The MR appearance of liposarcoma reflects its degree of differentiation. The more differentiated the tumor, the more the signal intensity of the tumor approaches that of fat and, therefore, the more "fat-like" it appears. On MR, a well-differentiated (lipoma-like) liposarcoma will image as a predominantly fatty mass with irregularly thickened linear or nodular septa, which demonstrate a nonspecific, decreased signal on T1- and an increased signal on T2-weighted spin echo (SE) images (Fig. 1C,D) (5,6). The myxoid, pleomorphic, and round cell liposarcomas generally do not contain substantial amounts of fat, and only approximately 50% will demonstrate fat radiologically (5–8). When fat is present, it is usually in a lacy, amorphous, clump-like, or linear pattern (Fig. 1A,B) (6–10). Myxoid liposarcoma is typically more homogeneous than the high-grade (round cell and pleomorphic) liposarcomas. Myxoid liposarcoma may appear deceptively innocent at MR imaging and an appearance similar to that of a cyst has been reported in as many as 20% of cases (Fig. 1E,F) (6). The pleomorphic and round cell types are more heterogeneous (Fig. 1G–I).

In general, a lipoma is easily differentiated from a liposarcoma, with the exception of the well-differentiated liposarcoma. A lipoma is well characterized on MR, with the lesion having an appearance identical to that of subcutaneous fat on all pulse sequences (Fig. 1J) (11,12), without discernible enhancement following the administration of intravenous gadolinium (13). When a lipoma is encapsulated, a surrounding fibrous capsule of low signal intensity may be delineated. Uncommonly, there may be predominant fibrous septa within the lesion (Fig. 1K).

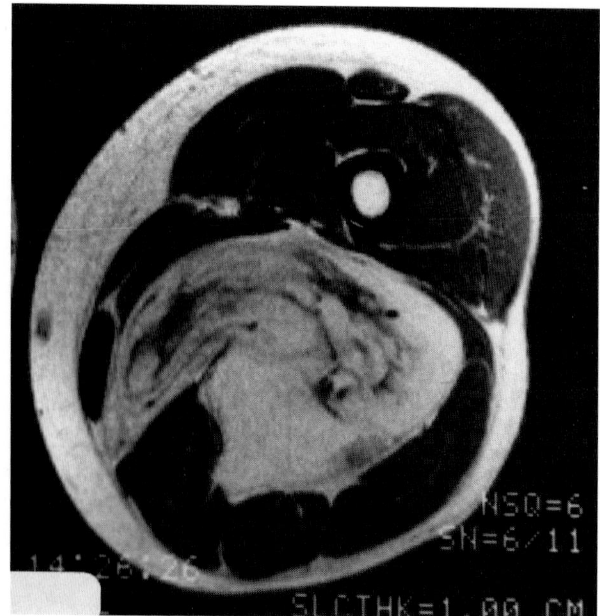

C

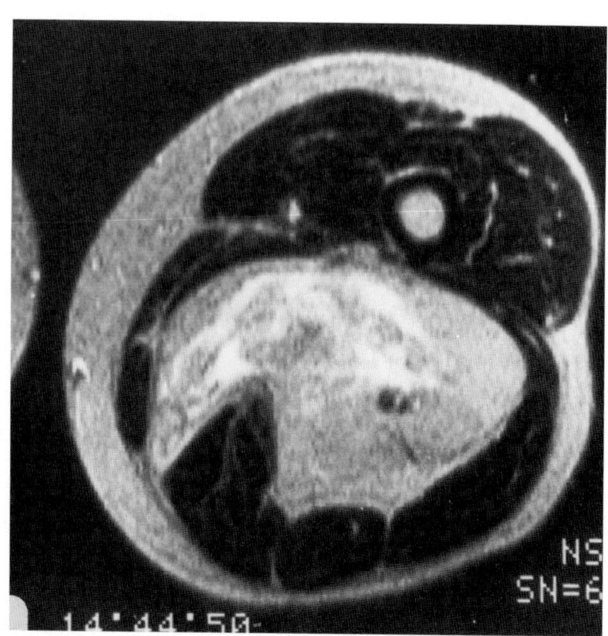

D

FIG. 1. C,D: Well-differentiated liposarcoma/atypical lipoma in the thigh of a 53-year-old man. Axial T1-weighted (TR/TE 500/30 msec) **(C)** and T2-weighted (TR/TE 2,500/90 msec) **(D)** SE images of the left thigh shows a predominantly fatty mass, containing irregularly thickened linear or nodular septa, which demonstrate a nonspecific MR appearance.

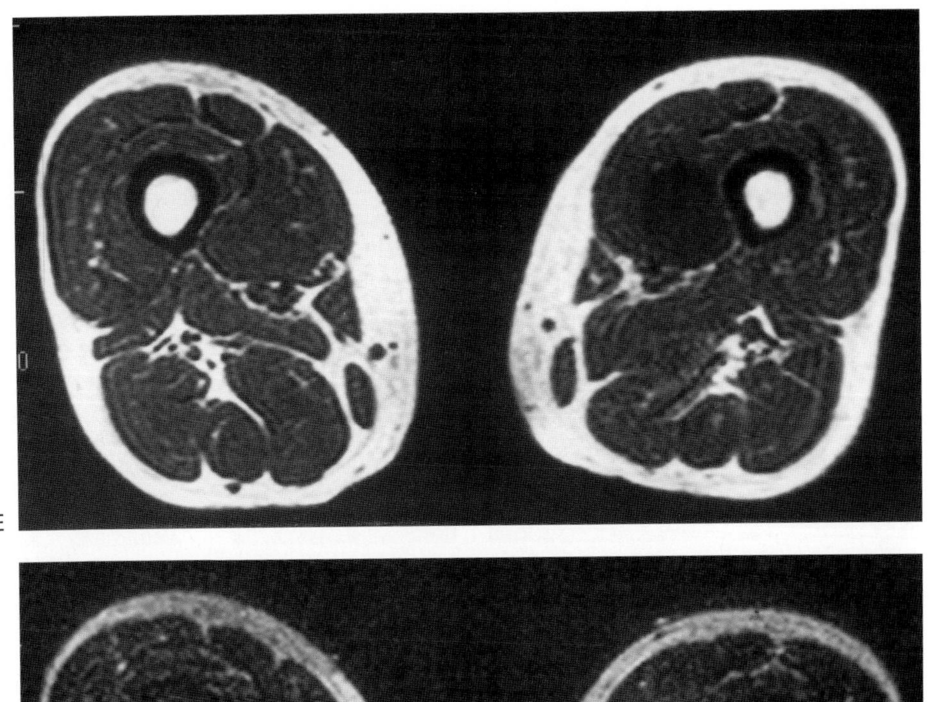

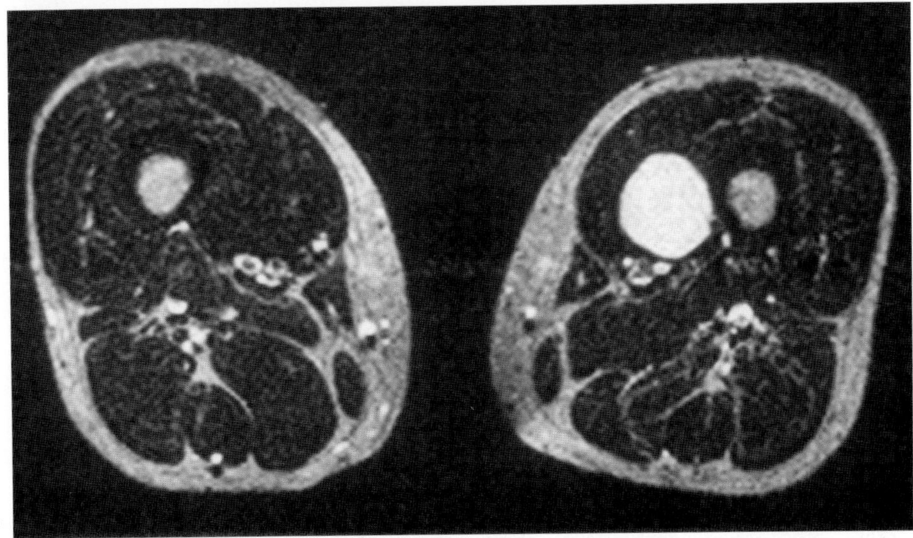

FIG. 1. E,F: Cyst-like myxoid liposarcoma in the thigh of a 56-year-old woman. Axial T1-weighted (TR/TE 650/20 msec) **(E)** and T2-weighted (TR/TE 2,000/80 msec) **(F)** SE images show a rounded, well-defined lesion in the medial aspect of the left thigh. The lesion has imaging characteristics identical to those of a cyst.

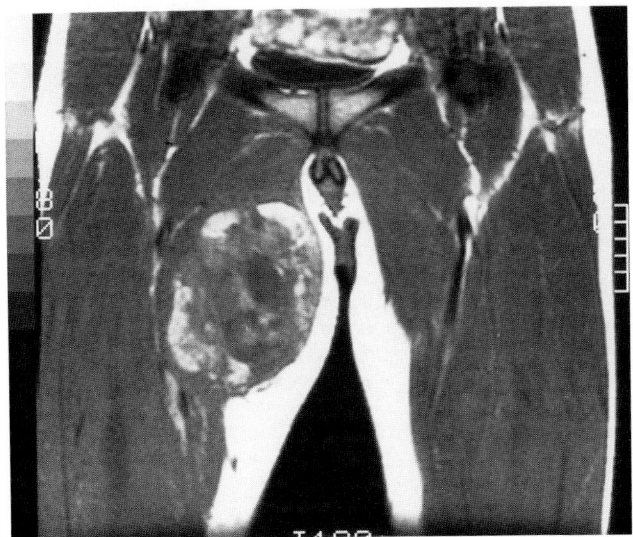

FIG. 1. G–I: Pleomorphic liposarcoma in the thigh of an 32-year-old man. Coronal T1-weighted (TR/TE 600/20 msec) **(G)** and T2-weighted (TR/TE 2,100/80 msec) **(H)** SE images shows a large mass in the adductor compartment of the right thigh.

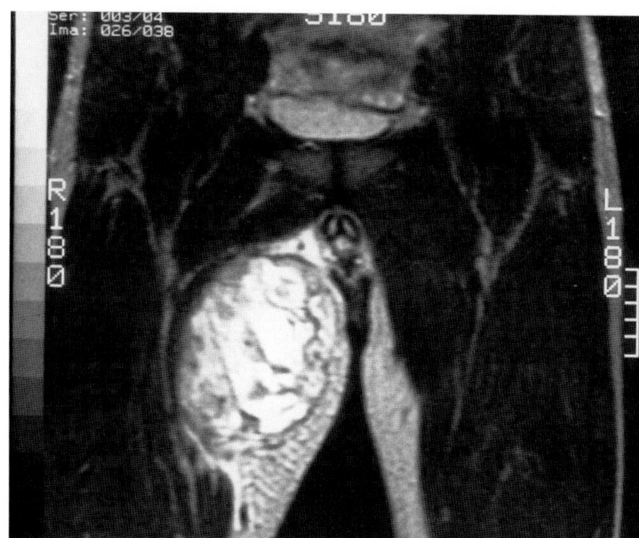

H

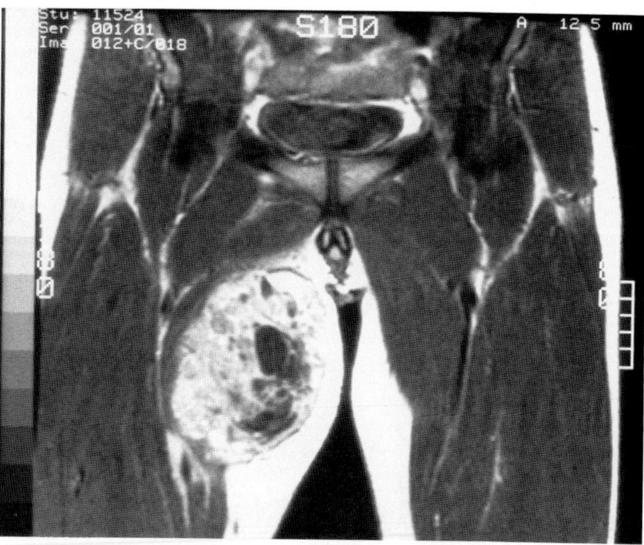

I

FIG. 1. G–I (*continued.*) Scattered areas within the mass show increased signal intensity in (G) and intermediate signal in (H), representing areas of mature fat within the mass. Coronal T1-weighted (TR/TE 600/16 msec) (**I**) SE image following gadopentetate dimeglumine shows marked inhomogeneous contrast enhancement. Nonenhancing areas reflect underlying necrosis.

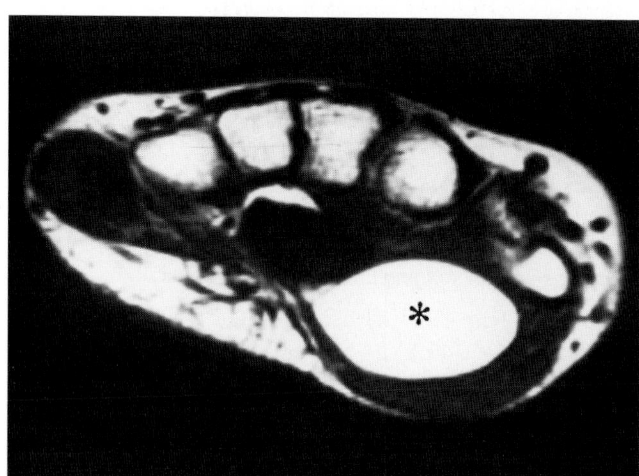

J

FIG. 1. J: Intramuscular lipoma in the thenar eminence *(asterisk)* of the hand in a 31-year-old woman. Axial T1-weighted (TR/TE 600/20 msec) SE image shows a well-defined, homogeneous mass with signal intensity identical to fat. The signal intensity was identical to that of subcutaneous fat on all pulse sequences.

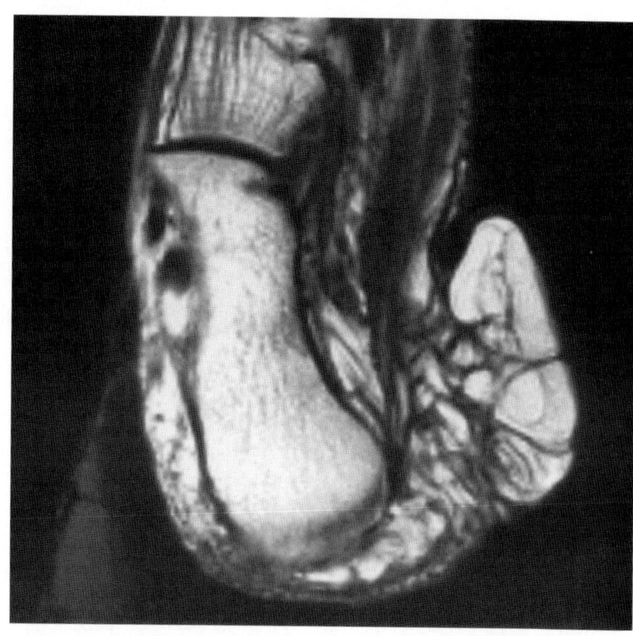

K

FIG. 1. K: Recurrent superficial lipoma in the subcutaneous tissue of the heal in a 69-year-old woman. Axial T1-weighted (TR/TE 600/15 msec) SE image shows the mass to have a lobulated contour with a signal intensity identical to that of subcutaneous fat with multiple linear septations of decreased signal intensity, which corresponded to fibrovascular connective tissue. T2-weighted image (not shown) showed an identical appearance. Lesions such as this are often referred to as a fibrolipoma.

CASE 2

History: A 22-year-old man presented with a painless mass in the popliteal fossa, which has been present for 2 years. The mass has increased significantly in size over the preceding 5 months.

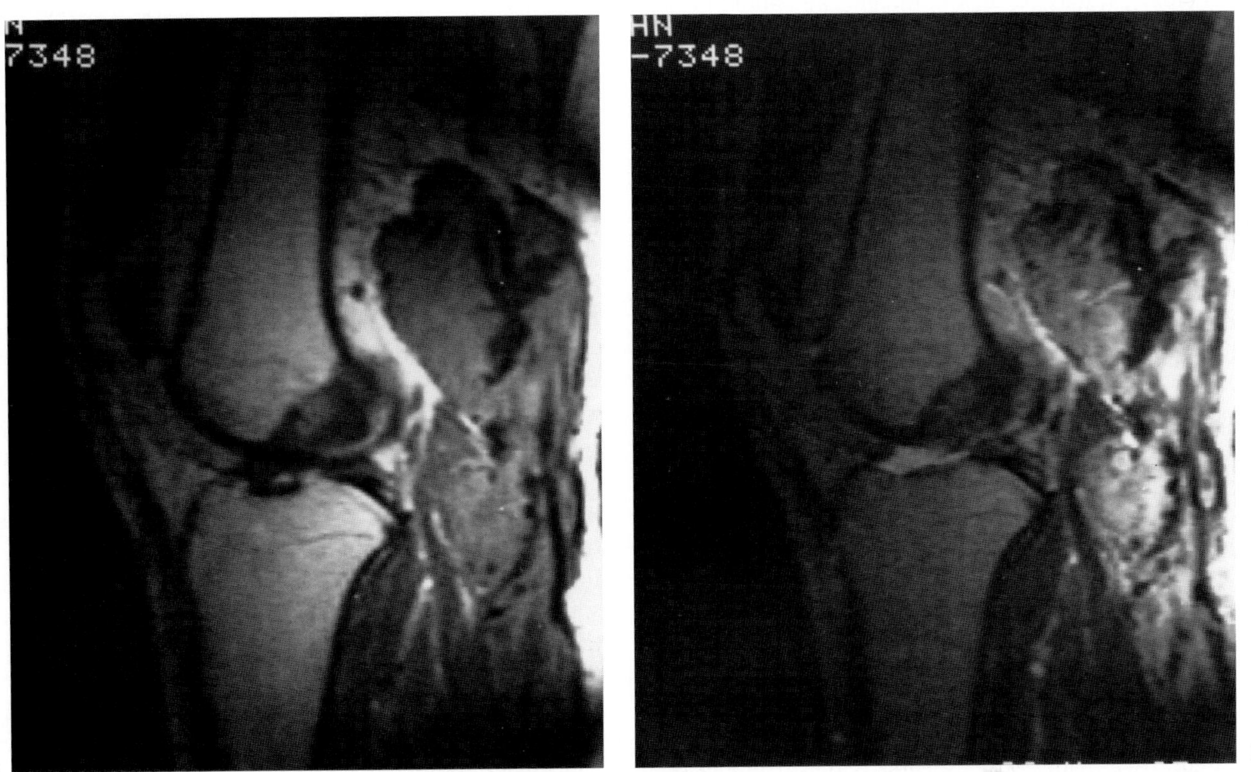

FIG. 2.

Findings: Sagittal T1-weighted (TR/TE 700/34 msec) (Fig. 2A) and T2-weighted (TR/TE 2,000/80 msec) (Fig. 2B) SE image shows well-defined soft tissue mass in the popliteal fossa with signal intensity slightly greater than that of skeletal muscle on T1-weighted images and similar to that of fat on T2-weighted images. Within the mass are corresponding areas of markedly decreased/absent signal on both pulse sequences. Radiographs of the knee (not shown) demonstrate a nonspecific mass.

Diagnosis: Musculoaponeurotic fibromatosis (extra-abdominal desmoid).

Discussion: What could account for the areas of decreased signal on MR images? Radiographs show no calcification. The corresponding areas of decreased signal within the mass suggest hypocellular, densely collagenous tissue. Hemosiderin-laden tissue could show decreased signal on all pulse sequences, however, this would be expected to have a greater effect on T2-weighted images. The decreased signal intensity of the lesion on T2-weighted MR images is also somewhat unusual in that it approximates that of skeletal muscle, suggesting it is fibrous (collagenous) in nature.

Fibromatosis refers to a family of soft tissue lesions that are characterized by a proliferation of benign fibrous tissue, which are composed of uniform, elongated, fusiform, or spindle-shaped cells surrounded and separated by abundant collagen (1). Their biologic behavior is intermediate between that of benign fibrous lesions and that of fibrosarcoma, although they never metastasize (1).

The fibromatoses are classified on the basis of their anatomic location as either superficial or deep. The superficial group includes palmar fibromatosis (Dupuytren's contracture), plantar fibromatosis (Ledderhose's disease), penile fibromatosis (Peyronie's disease), and knuckle pads. These are typically diagnosed clinically, and are not routinely imaged. The deep, or musculoaponeurotic fibromatoses, include extraabdominal fibromatosis (aggressive fibromatosis), abdominal fibromatosis, and intraabdominal fibromatosis. They were initially described in 1832 in the abdominal wall. The descriptive term *desmoid tumor* was coined in 1838 to emphasize the band-like or tendon-like character of the lesions (from the Greek *desmos* meaning band or tendon) (1–5).

The deep, or musculoaponeurotic fibromatosis typically present in young adults between puberty and 40 years of age, with a peak incidence between 25 and 35 years. Uncommonly, it may be seen in infants and children (4). Reports in the literature indicate that men and women are almost equally affected or there is a slight female predominance (1,2). The fibromatoses may occur anywhere. These tumors are usually solitary, however, synchronous multicentric lesions have been reported with a prevalence of 10% to 15% in two large series of 192 and 110 patients (2,6). Synchronous lesions are confined to the same extremity in 75% to 100% of cases (2,6), and a second soft tissue mass in the extremity of a patient with a previously confirmed desmoid tumor should be regarded as a second desmoid tumor until proven otherwise (7).

Local recurrence is common, occurring in as many as 77% of patients. Recurrence is usually within 18 months of the original surgery, although it may not present for several years following the initial surgery (6,8). Fatalities secondary to direct invasion of the chest wall or neck, and infiltration of vital organs have been reported (6–9).

Prognosis is related to the age of the patient, with younger individuals (those less than 20 to 30 years of age) having a longer duration of tumor activity and a higher recurrence rate (6,9). Similarly, younger patients with recurrence require more aggressive treatment to affect a cure (6). Wide local excision is the treatment of choice, although adjuvant radiation has been used. In extremity lesions, amputation may be required for local control.

Fibromatosis of the abdominal wall (abdominal desmoid) is distinguished from other musculoaponeurotic fibromatoses because of its distinct predilection to develop in women of child-bearing age (10). Typically, the lesion develops following, or less often, during pregnancy. Approximately 87% of cases occur in women and 95% in those with at least one child (1).

Intraabdominal fibromatosis (intraabdominal desmoid) refers to those lesions occurring in the pelvis, mesentery, and retroperitoneum. Of these, the latter two are associated with Gardner's syndrome in approximately 15% of patients (11,12). Patients with Gardner syndrome may demonstrate both intraabdominal fibromatosis as well as coincident extraabdominal lesions (13). Desmoid tumors associated with Gardner syndrome cannot be differentiated from other lesions of fibromatosis, although the former are smaller and more likely to be multiple than isolated lesions and occur in younger patients (14).

On MR, the deep musculoaponeurotic fibromatoses were initially described as demonstrating decreased signal intensity on all pulse sequences, reflecting the fibrous nature of the tumor (Fig. 2C,D) (15,16). Sundaram et al. (17) described three cases of "aggressive fibromatosis," two demonstrating a decreased signal on T2-weighted pulse sequences and one showing a paradoxical increased signal. Two of these three tumors were hypocellular and had abundant collagen. The lesion that showed high signal on T2-weighted MR images had marked cellularity as well as abundant collagen (Fig. 2E,F). Sundaram et al. concluded that the combination of marked hypocellularity and abundant collagen produces decreased signal on T2-weighted pulse sequences and that the decreased cellularity is of prime importance.

Subsequent reports have shown great variability of the MR imaging characteristics of these lesions (18–21). More typically, the lesion has an inhomogeneous signal intensity, approximating that of fat on T2-weighted images, and similar to or slightly greater than that of skeletal muscle on T1-weighted SE images. Considerable variation may be noted. The inhomogeneous signal likely reflects varying proportions and distribution of collagen, spindle cells, and mucopolysaccharide within the tumor (Fig. 2A,B). Corresponding areas of decreased signal intensity have been noted on all pulse sequences, likely reflecting areas of dense collagen within the lesions. Fibromatosis will demonstrate moderate to marked enhancement following the administration of intravenous gadolinium–diethylenetriamine pentaacetic acid (Gd-DTPA), with enhancement corresponding to the cellular portions of the lesion (Fig. 2G,H) (19). Tumor margins vary greatly, although they are usually well-defined at initial presentation (18,21). Bone involvement is less common but has been reported in up to 37% of cases (22).

It must be emphasized that decreased signal intensity on all pulse sequences only reflects the gross morphology of a lesion. Decreased signal may be seen in lesions that are densely mineralized, as well as in those that are hemosiderin-laden, such as pigmented villonodular synovitis (PVNS), or in those that are relatively acellular with large amounts of collagen (fibromatosis) (17,23). It is not characteristic for a specific histology and has also been reported in malignant fibrous histiocytoma as well as in other malignancies (Fig. 2I,J) (23,24).

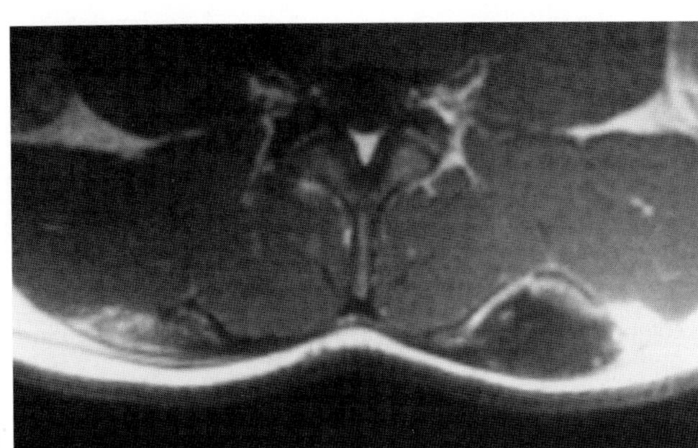

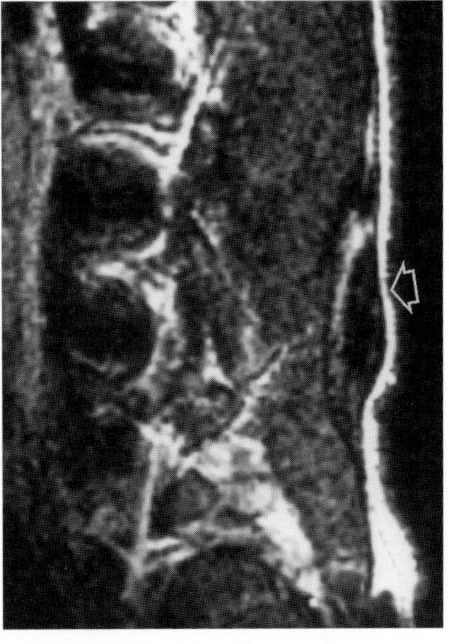

C D

FIG. 2. C,D: Musculoaponeurotic fibromatosis in the paraspinal region in a 20-year-old man. Axial T1-weighted (TR/TE 800/20 msec) **(C)** and sagittal T2-weighted (TR/TE 2,500/80 msec) **(D)** *(arrow)* SE image shows markedly decreased signal intensity within a relatively well-defined mass, compatible with a densely collagenous, hypocellular mass.

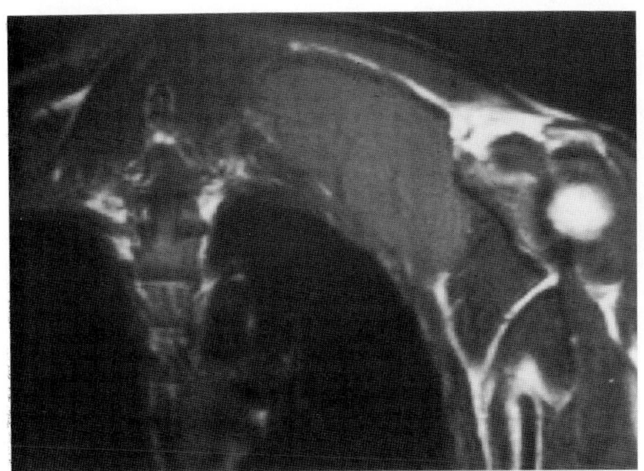

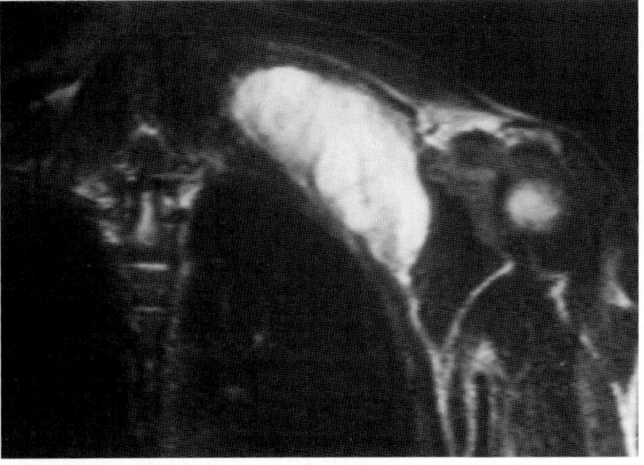

E F

FIG. 2. E,F: Musculoaponeurotic fibromatosis in the axilla and supraclavicular region of a 24-year-old man. The was initially noted following trauma, and has doubled in size over the preceding year. Coronal T1-weighted (TR/TE 700/33 msec) **(E)** and T2-weighted (TR/TE 2,000/100 msec) **(F)** SE images of the shoulder show a nonspecific well-defined, mildly inhomogeneous mass. The signal intensity of the mass is greater than that of skeletal muscle on T1-weighted images and much greater than that of fat on corresponding T2-weighted images.

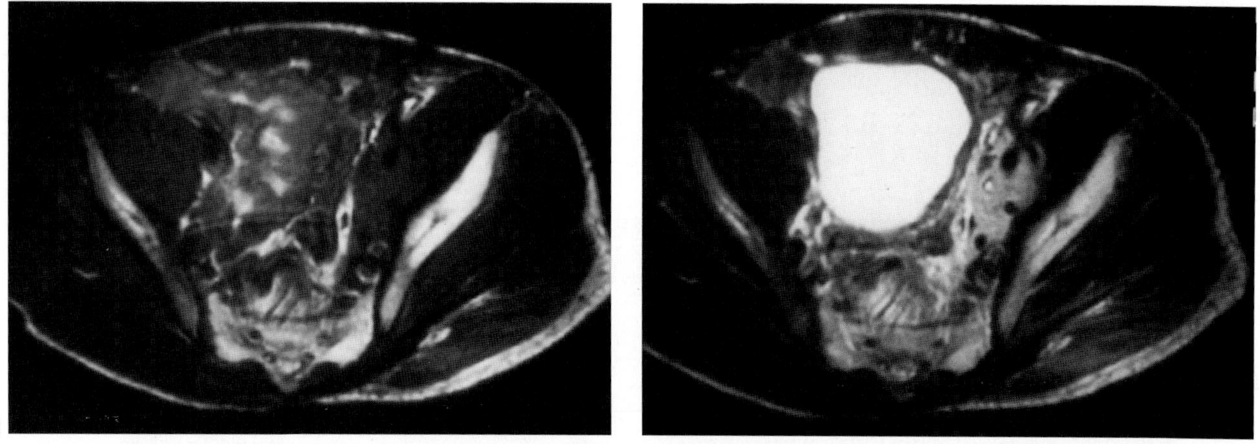

FIG. 2. G,H: Recurrent retroperitoneal fibromatosis in a 23-year-old man. Axial T1-weighted (TR/TE 450/20 msec) SE images of the pelvis preceding **(G)** and following **(H)** gadopentetate dimeglumine administration shows marked enhancement of the mass.

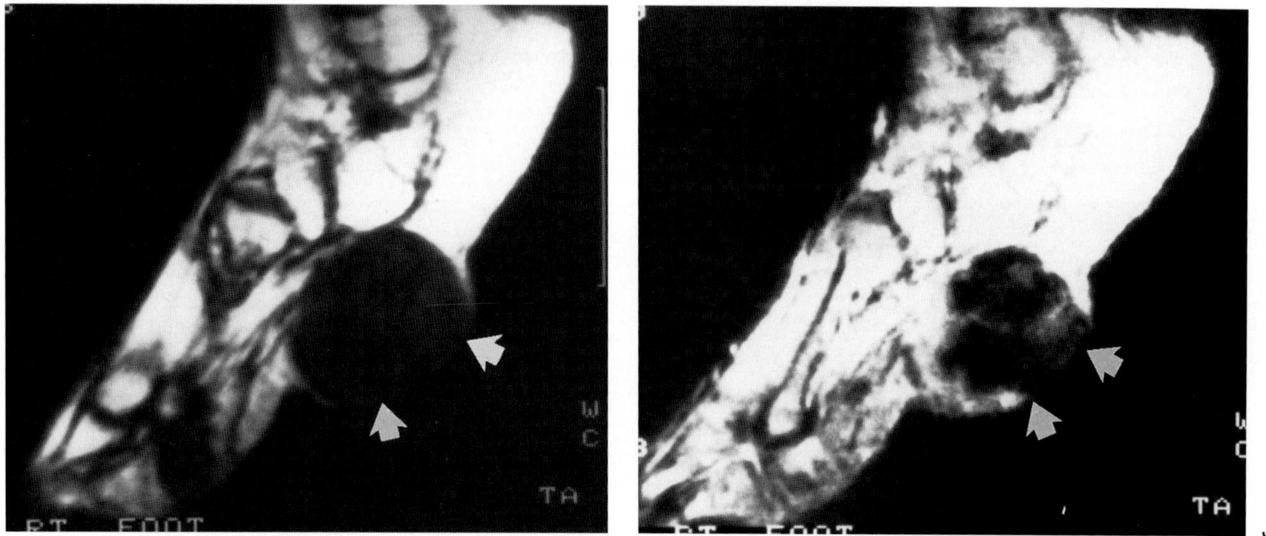

FIG. 2. I,J: Rhabdomyosarcoma of the foot in a 15-year-old girl. Sagittal T1-weighted (TR/TE 600/20 msec) **(I)** and T2-weighted (TR/TE 2,000/90 msec) **(J)** SE images show a relatively well-defined mass with markedly decreased signal intensity on all pulse sequences. (From Kransdorf MJ. Magnetic resonance imaging of musculoskeletal tumors. *ACR Categorical Course on Imaging of Cancers,* 1992:47–54, with permission.)

CASE 3

History: The patient is a 33-year-old man who has a mass in the thigh that has been rapidly growing over the previous 3 weeks. There is no significant pain, although the mass is tender to deep palpation.

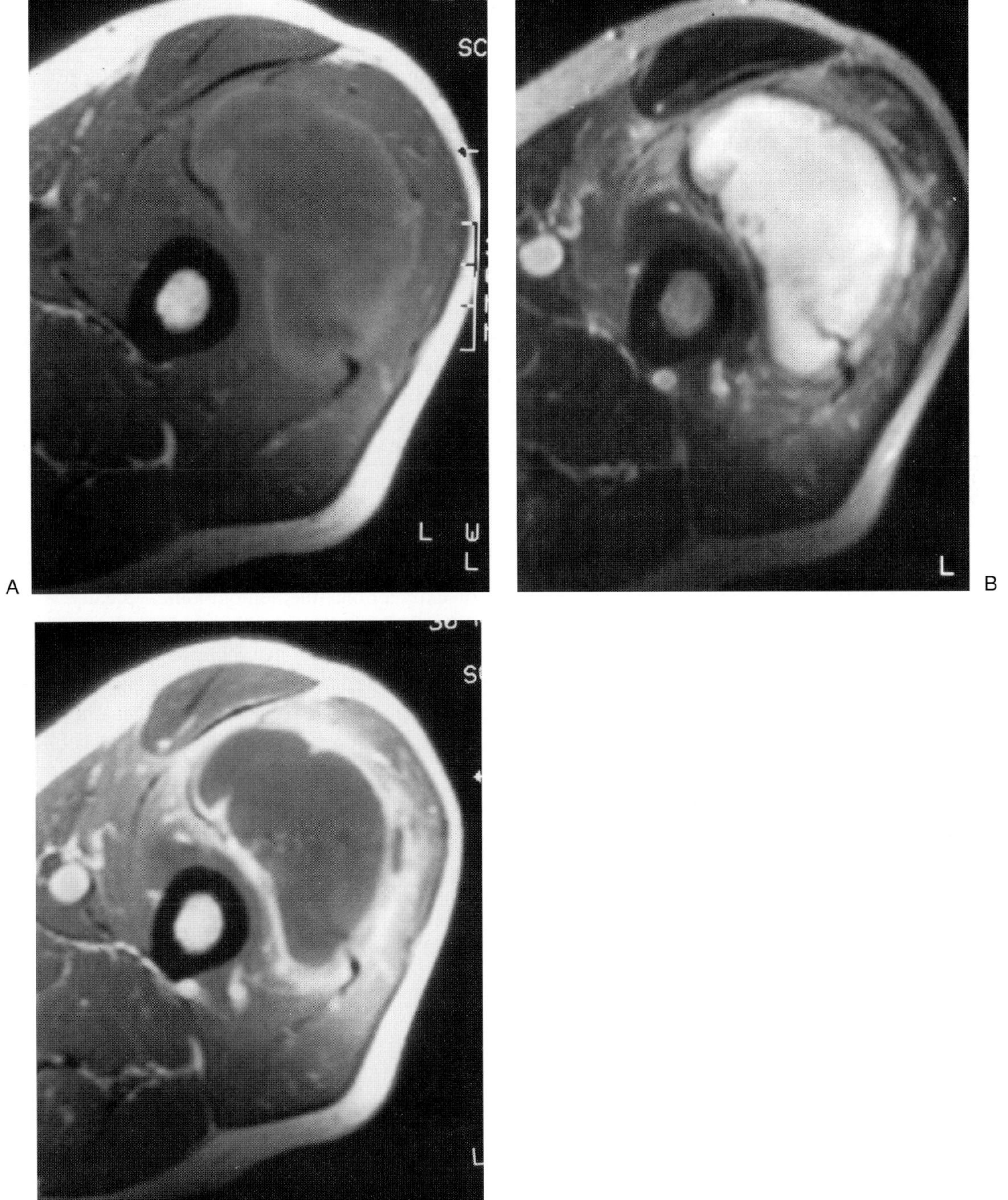

FIG. 3.

Findings: Axial T1-weighted (TR/TE 522/25 msec) (Fig. 3A) and T2-weighted (TR/TE 2,303/90 msec) (Fig. 3B) images show a well-defined, relatively homogeneous mass in the anterior compartment of the left thigh. The mass shows a signal intensity very slightly less than that of the surrounding muscle on T1-weighted images, and greater than that of fat and muscle on corresponding T2-weighted images. There is some inhomogeneity, as well as diffuse increased signal intensity in the entire anterior compartment, suggesting diffuse muscle edema. Axial T1-weighted (TR/TE 522/25 msec) (Fig. 3C) image, following intravenous gadopentetate dimeglumine administration, shows a rind of uniform peripheral enhancement.

Diagnosis: *Staphylococcus aureus* abscess.

Discussion: MR imaging suggests a cystic mass. The edema pattern suggests an inflammatory process. The two combined make an abscess the prime differential process. The history also suggests this, due to the marked increase in size over a short time interval.

There is a wide spectrum of soft tissue inflammatory disease that ranges from a diffuse cellulitis to a discrete abscess. In general, an abscess is defined as a localized pus-filled cavity, whereas the term *cellulitis* is applied to a diffuse inflammation (1). Cellulitis is sometimes used to specifically denote hemolytic group A streptococcus infection, in that this infection behaves somewhat differently from other pyogens, spreading between cells and tissue planes (1). Others have used this term to denote inflammation of loose subcutaneous tissue; however, common usage would give a broader definition of any poorly defined soft tissue pyogenic process.

A soft tissue abscess is characterized on MR imaging as a well-demarcated collection of increased signal intensity on T2-weighted pulse sequences and decreased signal intensity on T1-weighted images (2). The internal portion of the lesion is usually relatively homogeneous; however, homogeneity and signal intensity will vary depending on the amount of internal proteinaceous debris, necrosis, foreign matter, gas, etc. (Fig. 3D–F). This is surrounded by a rind of variable signal intensity, which will show marked enhancement following intravenous gadolinium administration. This pattern of enhancement is not surprising, in that the wall of an abscess cavity may contain highly vascularized connective tissue (Fig. 3C).

A discrete abscess can be differentiated from cellulitis in that the latter will show poorly defined increased signal intensity on T2-weighted images, with indistinct margins (an edema pattern), but no focal fluid collection. This edema pattern is frequently seen in association with an abscess, as noted in the current cases.

Increased MR signal intensity on T2-weighted images in skeletal muscle or subcutaneous tissue adjacent to a musculoskeletal lesion is an important radiologic finding. It has also been suggested as a reliable indicator of malignancy (3,4). These results are based on studies in which both bone and soft tissue lesions were evaluated. Although this increased signal intensity may be seen with malignancy, in our experience, when one is evaluating only soft tissue masses, this finding is quite nonspecific. We have found it commonly present in inflammatory processes, abscesses, and myositis ossificans, and it may also be present following local trauma, biopsy, radiation therapy, and internal hemorrhage. In our experience, it is quite unusual in an uncomplicated malignancy.

An abscess may be successfully differentiated from a hematoma in the vast majority of cases. The MR appearance of a hematoma is variable and will be a function of the age of the lesion. Most subacute hematomas (weeks to months old) will demonstrate increased signal on both T1- and T2-weighted SE MR images, due to the presence of extracellular methemoglobin, typically with a rim of decreased signal intensity due to the accumulation of hemosiderin-laden macrophages (Fig. 3G,H) (5). Care must be taken to differentiate malignant soft tissue tumor with hemorrhage from a hematoma. The presence of a tumor nodule of rim of tumor may be helpful in distinguishing a simple hematoma from a hemorrhagic neoplasm (6).

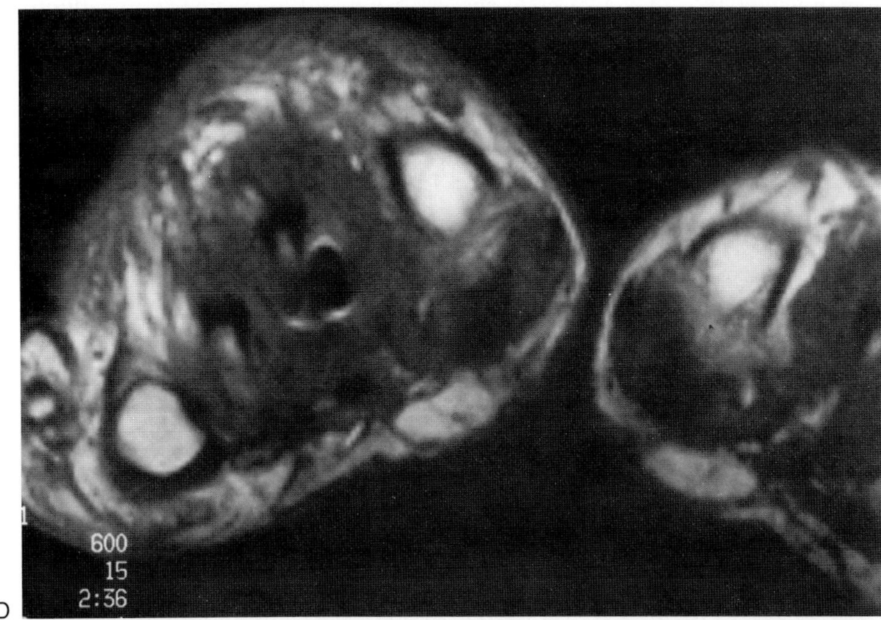

D

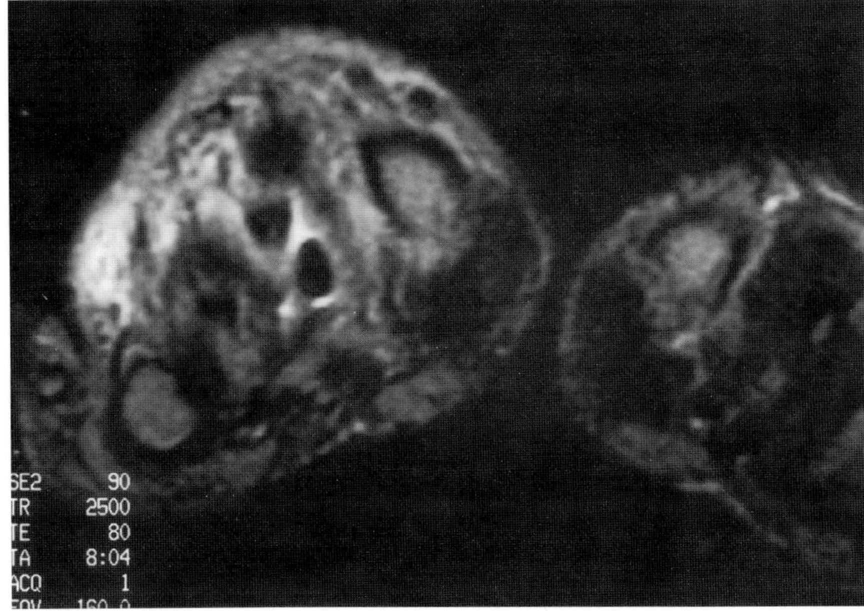

E

FIG. 3. D–F: Cellulitis and abscess associated with a foreign body in the foot of a 27-year-old woman. Coronal T1-weighted (TR/TE 600/15 msec) **(D)** and T2-weighted (TR/TE 2,500/80 msec) **(E)** SE images show a poorly delineated area of abnormal signal intensity in the right forefoot. The signal intensity is nonspecific, but the pattern of abnormality is highly suggestive of an inflammatory process. Note parenthetical artifact from foreign body.

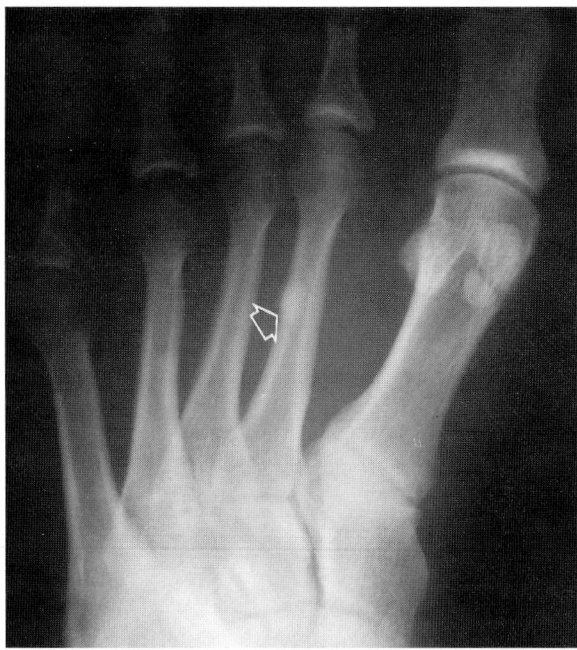

FIG. 3. **C–F** (*continued.*) **F:** Corresponding radiograph shows forefoot foreign body, which was a small stone.

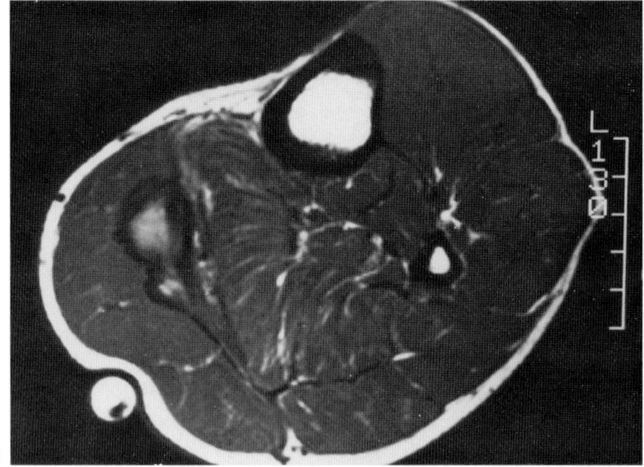

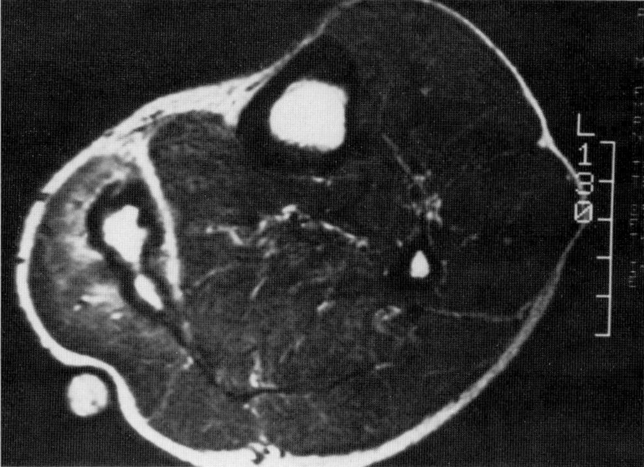

FIG. 3. **G,H:** Subacute hematoma in the calf of a 53-year-old man. Axial T1-weighted (TR/TE 650/20 msec) **(G)** and T2-weighted (TR/TE 2,000/80 msec) **(H)** SE images demonstrate an irregularly shaped mass with increased signal intensity on both pulse sequences compatible with subacute blood. The rim of decreased signal intensity surrounding the hematoma represents hemosiderin-laden areas. Note increased signal intensity surrounding the lesion on the T2-weighted image **(H),** compatible with edema. No biopsy was done. Repeat imaging demonstrated resolution of the mass and edema with time, with a small residual area of decreased signal intensity on all pulse sequences compatible with residual hemosiderin-laden tissue.

CASE 4

History: A 27-year-old woman presented with a mass in the lateral aspect of the thigh that had been slowly growing. What is your diagnosis?

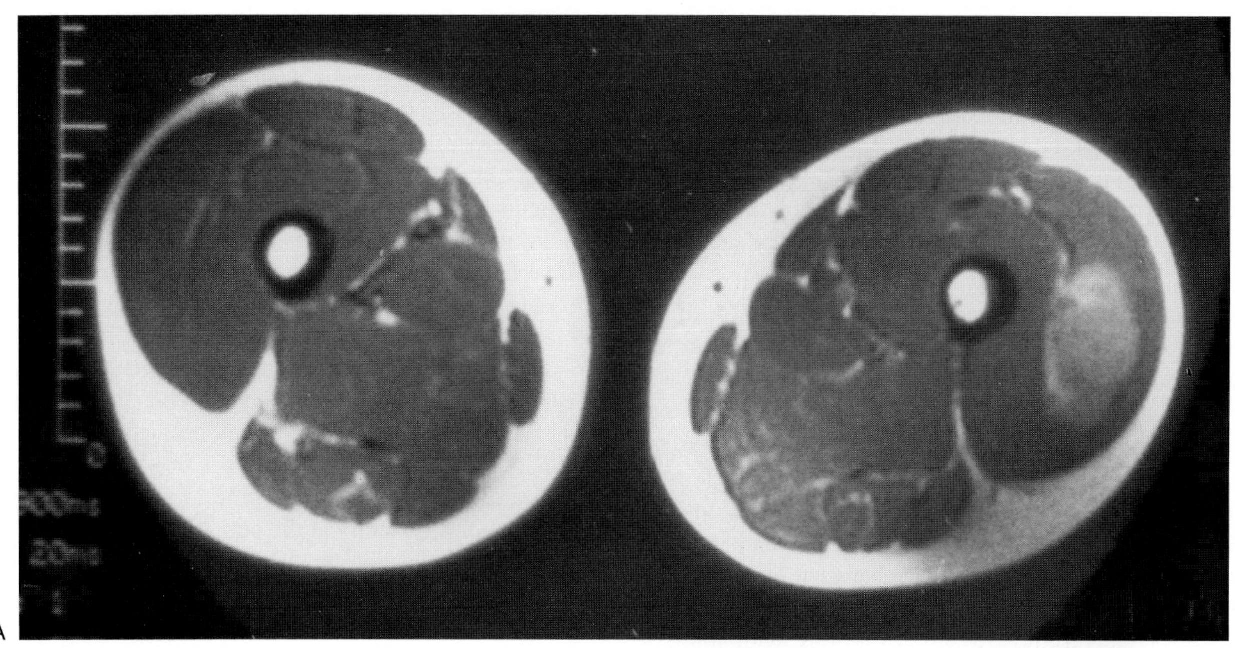

A

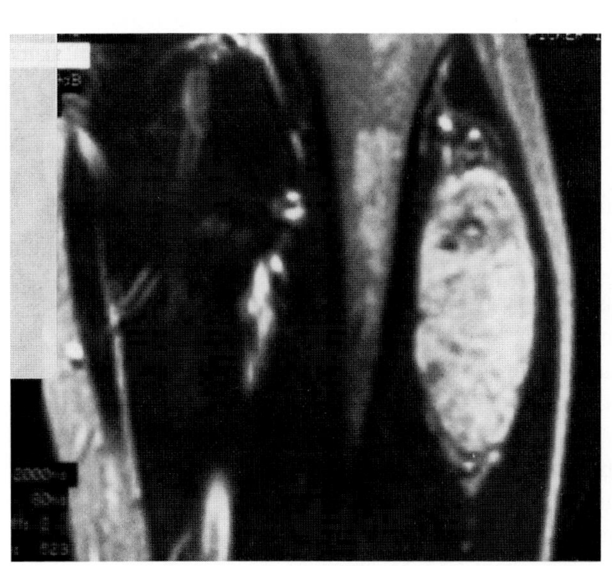

B

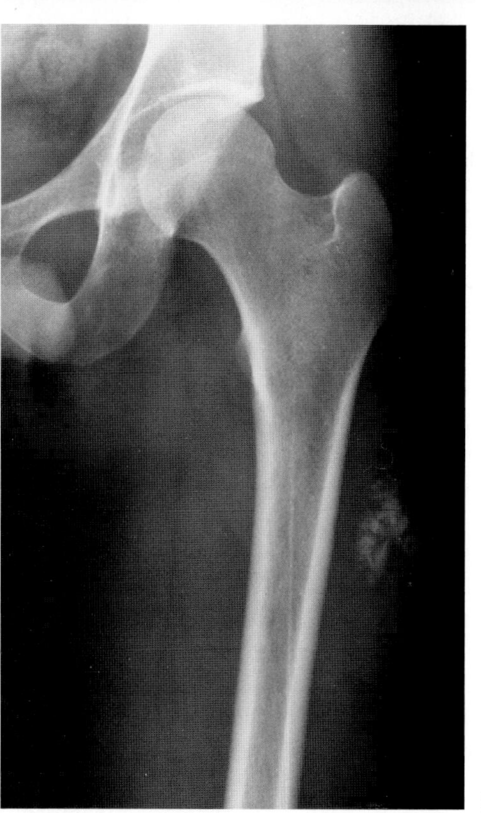

C

FIG. 4.

Findings: Axial T1-weighted (TR/TE 500/20 msec) (Fig. 4A) and coronal T2-weighted (TR/TE 1,800/80 msec) (Fig. 4B) SE images show a well-defined, somewhat inhomogeneous, soft tissue mass in the anterior compartment of the thigh. The signal intensity of the mass is nonspecific. Corresponding anteroposterior radiograph of the thigh (Fig. 4C) shows extensive calcification within the mass.

Diagnosis: Synovial sarcoma.

Discussion: In this case, the diagnosis is strongly suggested when all information is considered, despite a nonspecific MR appearance. The patient's age, the lesion's location, and the radiographic mineralization are all quite typical of synovial sarcoma. This case also emphasizes the need to evaluate all available information when formulating a differential diagnosis, as well as the indispensable importance of radiographs in the evaluation of all musculoskeletal lesions (and one is ill-advised to interpret MR examinations in their absence).

Synovial sarcoma is a well-recognized soft tissue malignancy that typically arises in young adults. First reported in 1893, it is a relatively common primary soft tissue sarcoma, accounting for approximately 5% to 10% of all malignant mesenchymal neoplasms, and it is the most common sarcoma in patients between 16 to 25 years of age (1). It has a striking predilection for the extremities, which accounts for 80% to 95% of lesions, with approximately two-thirds in the lower limbs (1–4). It is the most common malignancy of the foot and ankle in patients between the ages of 6 and 45 years, accounting for almost one-third of sarcomas in this location for young adults (1).

Although the tumor resembles synovial tissue at light microscopy, its origin is likely from undifferentiated mesenchymal tissue. Although the lesion is commonly around joints, intraarticular involvement is uncommon, and lesions limited to the joint (true intraarticular lesions) are quite unusual. Rare sites of involvement include the neck, pharynx, larynx, precoccygeal and paravertebral regions, thoracic and abdominal wall, and heart (2,5,6).

The patient generally presents with a palpable soft tissue mass, which may be quite slow growing, and may clinically simulate a benign process (7). Pain is often present. Additional complaints include sensory and/or motor dysfunction distal to the lesion. The duration of symptoms is quite variable and may be present for days to weeks, or as long as 20 years prior to initial diagnosis (3).

Metastases or local recurrence are seen in approximately 80% of patients (8). Metastases are present at the time of initial diagnosis in about one-quarter of patients (8), but have been reported as long as 35 years following initial diagnosis (9). Metastases are predominantly pulmonary, occurring in approximately 59% to 94% cases of those with metastases (4,8), although metastases may be to multiple sites. After the lungs, metastases to lymph nodes (4–18%) and bone (8–11%) are most common (4,8,10,11). Local recurrence is frequent, seen in about 20% to 26% (4,11) of patients, occurring in the excision scar or the amputation stump, often within 2 years of initial presentation. The prognosis remains guarded, although the biologic activity of the tumor is variable. Lesions that demonstrate extensive calcification have been reported to have a more favorable prognosis (12), as do younger patients, those with tumors smaller than 5 cm, and those lesions located in the extremities (4). The size of the tumor is the most important variable in determining prognosis (4).

In most patients, lesions are relatively deep; however, superficial subcutaneous lesions may be seen (13). Monophasic synovial sarcomas behave more aggressively and metastasize earlier than biphasic tumors (14,15).

Routine radiographs may be interpreted as normal in approximately half of patients (4). When a mass is identified, it is most commonly a well-defined round or lobulated soft tissue mass (2,16). As many as one-third demonstrate some internal calcification (less commonly ossification), typically at the periphery of the tumor (2,3). Coexistent adjacent bony involvement, manifested by periosteal reaction, bony remodeling (due to pressure from the adjacent tumor), or frank bony invasion, is seen in 11% to 20% of cases (2–4,16).

On MR imaging the lesion is usually a nonspecific inhomogeneous mass, with signal intensity approximately equal to that of skeletal muscle on T1-weighted and brighter than that of subcutaneous fat on T2-weighted SE MR images (13,17). Lesions are frequently inhomogeneous on T2-weighted images, likely secondary to fibrous, hemorrhagic, cystic, and necrotic areas within the tumor (Fig. 4D–F) (20). The lesion may demonstrate a multilocular configuration with internal septation, and changes compatible with previous hemorrhage, to include fluid-fluid levels, have been reported in 10% to 44% of lesions on MR imaging (13,17,20). These changes are likely due to the pronounced vascularity of most lesions. Fluid-fluid levels are a nonspecific finding and have been reported in other soft tissue lesions including hemangioma and myositis ossificans (18). Margins are usually well-defined or relatively well-defined (19,20), and small lesions may look deceptively innocent (Fig. 4G,H). Less commonly, margins may be poorly defined or infiltrating (6). Homogeneous well-defined lesions with signal intensity similar to that of skeletal muscle on both T1- and T2-weighted images have been reported (17). The soft tissue calcifications frequently seen on radiographs may not be detected on MR (13,17), although larger calcifications may be identified as areas of decreased signal intensity on all pulse sequences (Fig. 4I–K).

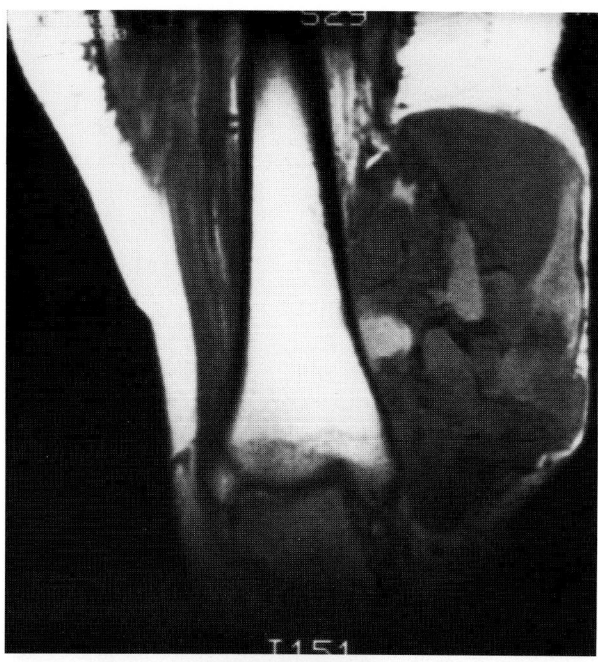

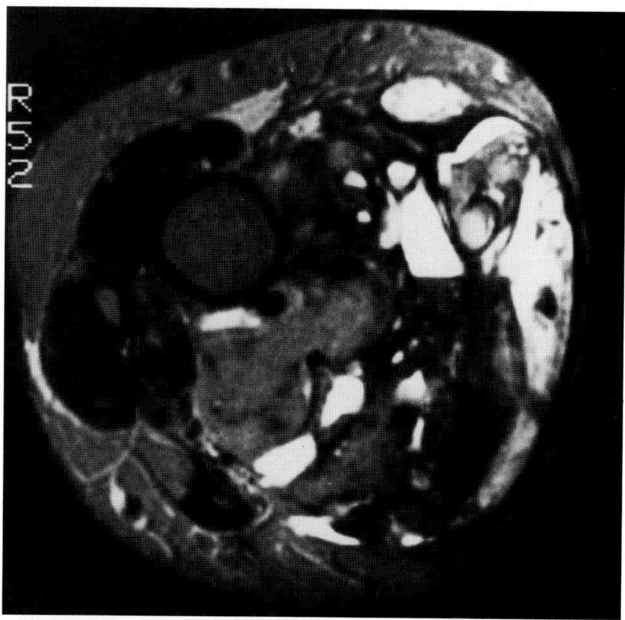

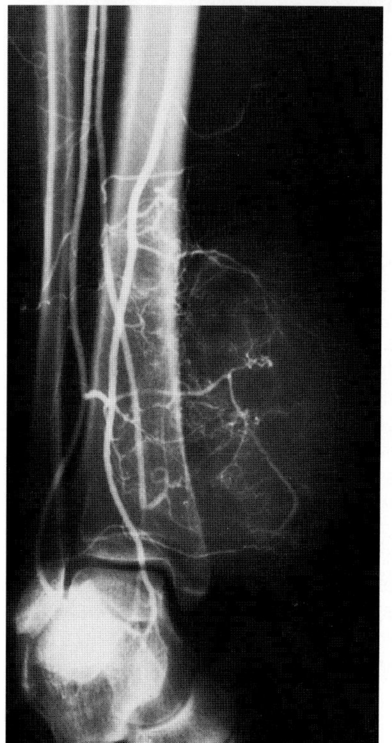

FIG. 4. D–F: Synovial sarcoma in the ankle of a 37-year-old man. Coronal T1-weighted (TR/TE 600/20 msec) **(D)** and axial T2-weighted (TR/TE 2,000/80 msec) **(E)** SE images show a well-defined soft tissue mass adjacent to the ankle. There is evidence of hemorrhage within the tumor. Arteriogram **(F)** shows the lesion to be markedly hypervascular.

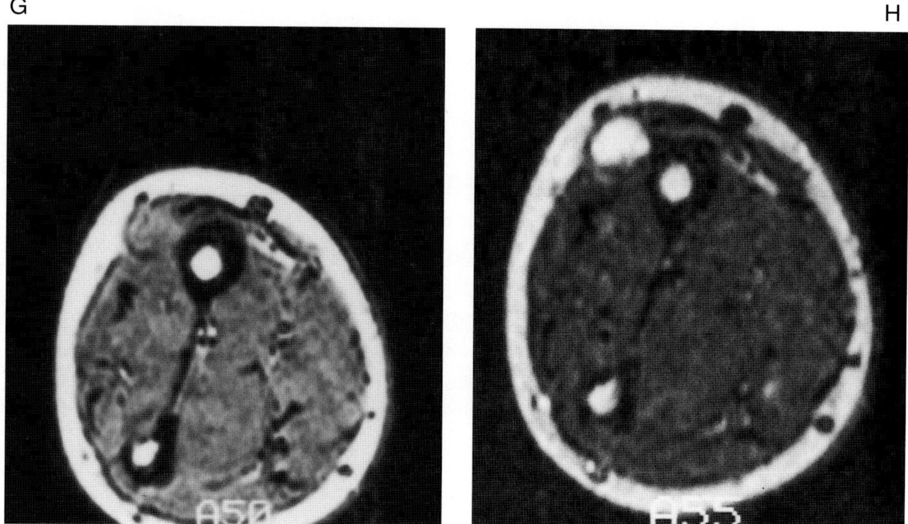

FIG. 4. G,H: Synovial sarcoma in the forearm of a 9-year-old boy, presenting with a slowly growing nodular mass. Axial T1-weighted (TR/TE 500/16 msec) **(G)** and T2-weighted (TR/TE 2,500/70 msec) **(H)** SE images show a small, innocent appearing, nonspecific nodular mass.

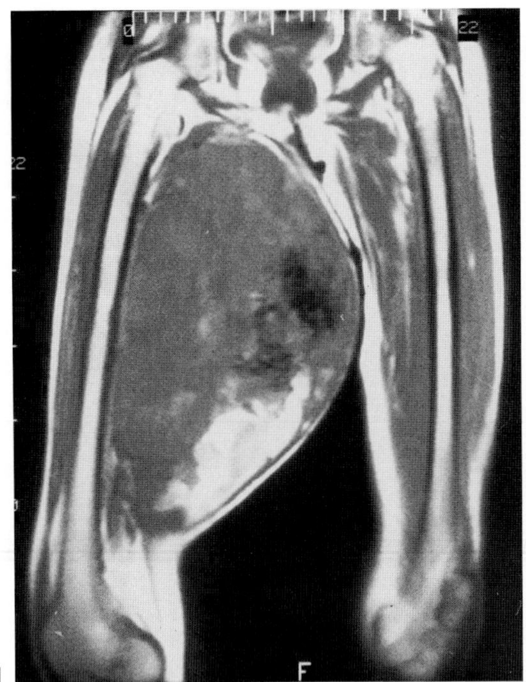

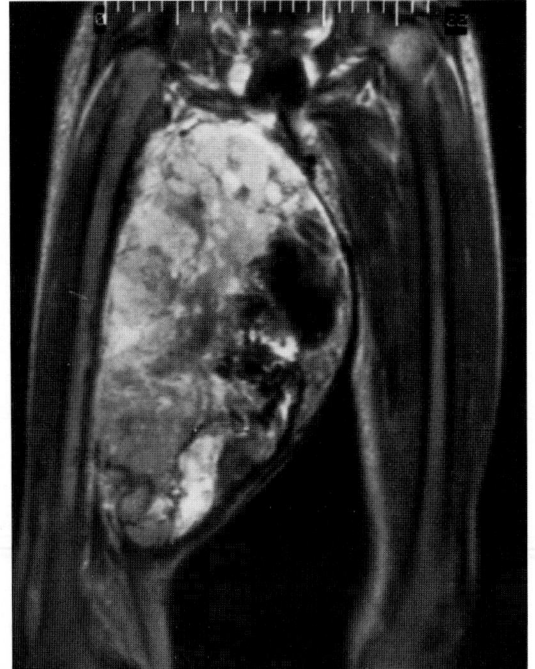

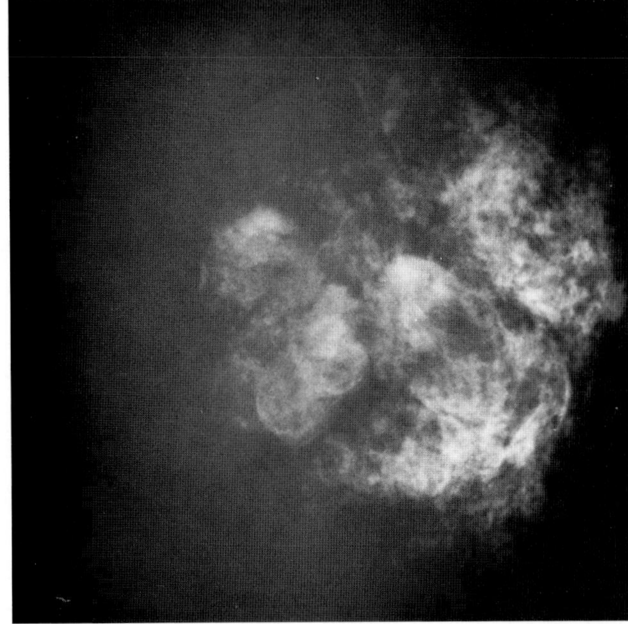

FIG. 4. I–K: Synovial sarcoma in the thigh of a 52-year-old woman, presenting with a mass that had been growing for approximately 3 years, but had recently increased dramatically in size. Coronal T1-weighted (TR/TE 600/20 msec) **(I)** and T2-weighted (TR/TE 2,050/80 msec) **(J)** SE images show a large complex soft tissue mass in the medial aspect of the thigh. The lesion is markedly inhomogeneous in T2-weighted images and there is evidence of subacute hemorrhage. The lesion is densely mineralized along its medial aspect. Corresponding anteroposterior radiograph **(K)** shows dense mineralization in the mass.

CASE 5

History: A 58-year-old man presented with a slowly enlarging painless mass in the right thigh without other contributory history. What is your diagnosis?

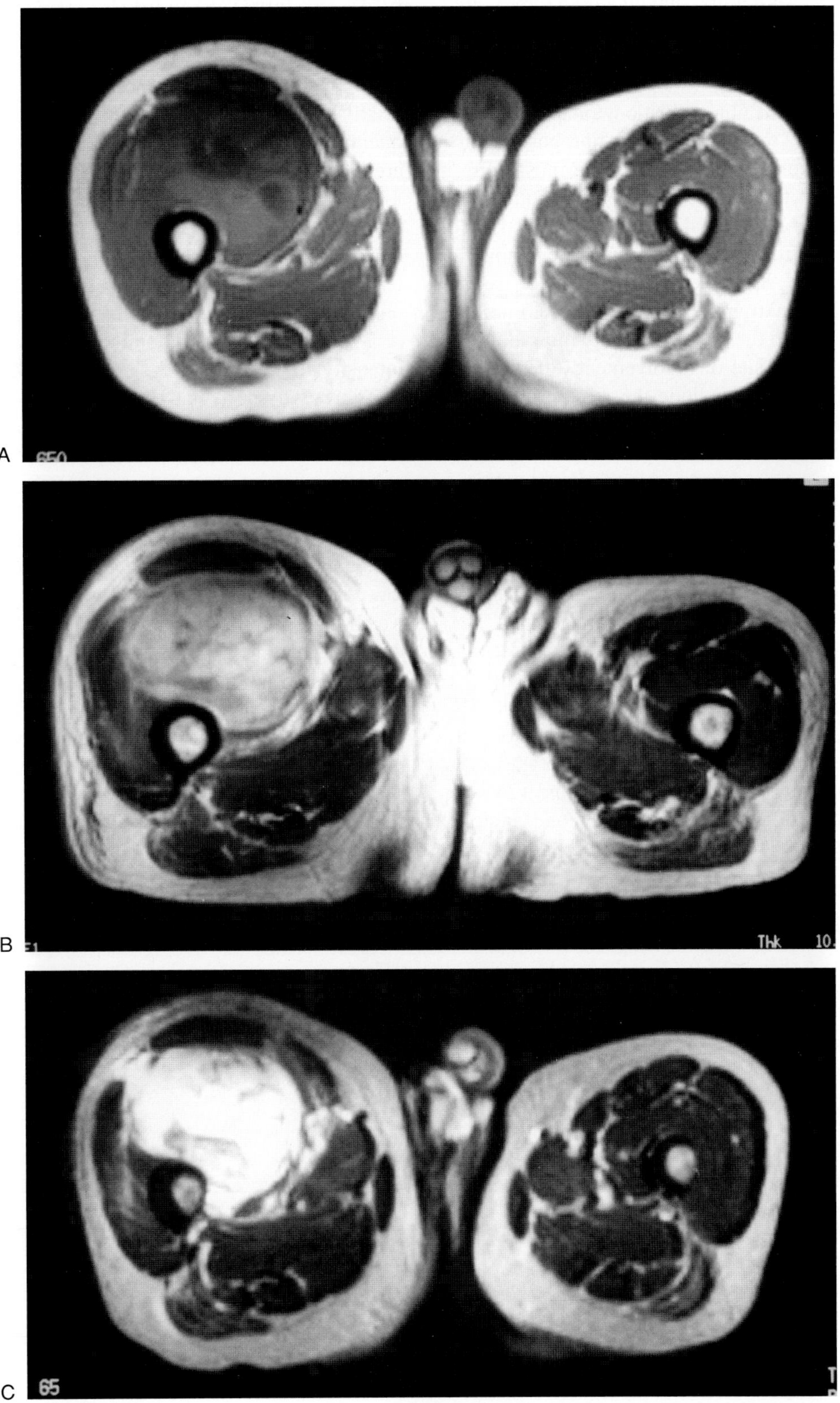

FIG. 5.

Findings: Axial T1-weighted (TR/TE 615/15 msec) before (Fig. 5A) and after intravenous gadolinium (Fig. 5B) and T2-weighted (TR/TE 2,000/80 msec) (Fig. 5C) images reveal a large heterogeneous soft tissue mass in the anterior compartment of the thigh. Overall the mass is similar to muscle on the T1-weighted image and higher signal than fat on T2-weighting, and enhances after contrast administration.

Diagnosis: Malignant fibrous histiocytoma (storiform/pleomorphic type).

Discussion: Malignant fibrous histiocytoma (MFH) is a pleomorphic sarcoma that contains both fibroblastic and histiocytic-like components (1–4). It is the most commonly occurring soft tissue sarcoma seen in adults, accounting for 20% to 30% of all malignant soft tissue tumors. MFH is most frequent in the fifth decade although the age range is 10 to 90 years and two-thirds of cases are seen in men. The extremities are the most common location (70–75%), with the lower extremity musculature the single most frequent site of involvement (50% of all cases). Additional locations include the retroperitoneum (15%), head and neck (5%), and other unusual sites (10–15%) such as the thorax and intraabdominal and pelvic structures (5–8).

Clinical presentation is usually a painless slowly enlarging soft tissue mass. Occasionally rapid growth is encountered and acute hemorrhage can obscure the underlying soft tissue mass, both clinically and radiologically. Thus a history of ''spontaneous'' hemorrhage into the deep soft tissues should be viewed with great suspicion of harboring a neoplasm such as MFH (Fig. 5D,E) (9).

MFH was described in 1964 by O'Brien and Stout (3) and the cell of origin is believed to be either the histiocyte or a primitive mesenchymal cell. Histologic subtypes include storiform-pleomorphic (50–60%), myxoid (25%), giant cell (5–10%), inflammatory (5–10%), and angiomatoid (less than 5%). The angiomatoid MFH occurs in younger adults and children often involving the subcutaneous tissues (Fig. 5F) (10).

Radiographs of MFH usually reveal a nonspecific soft tissue mass. Calcification may be apparent in 5% to 20% of cases (11,12). Involvement of adjacent bone with either erosion (Fig. 5G–I) or direct invasion is not uncommon in MFH. This propensity of MFH for osseous involvement is unusual for other soft tissue sarcomas except synovial sarcoma (13).

CT and MR imaging generally show a large lobulated relatively well-defined soft tissue mass. There is frequently a heterogeneous appearance corresponding to areas of hemorrhage, necrosis, or myxomatous tissue (13–17). CT can be helpful to identify calcification and involvement of bone. MR imaging most frequently demonstrates nonspecific features with intermediate signal intensity on T1-weighted images and high signal intensity on T2-weighting. Occasionally T2-weighted MR images of MFH will show predominantly low to intermediate intensity, likely reflecting a higher collagen content of the fibrous tissue (Fig. 5J,K). Fluid-fluid levels may be seen as a result of hemorrhage, particularly in the angiomatoid MFH (Fig. 5F). Solid components of MFH will enhance after intravenous contrast (CT or MR imaging) (Fig. 1B).

Myxoid MFH characteristically demonstrates higher extracellular fluid content on imaging. A cyst-like appearance with low attenuation on CT and low signal intensity on T1-weighed MR image and very high signal intensity on T2-weighting may be apparent (Fig. 5G,I) (18). However, after intravenous contrast nodular enhancement (often peripheral) is usually seen in myxoid MFH (Fig. 5H) as opposed to rim enhancement in a myxoma or other benign cystic mass (ganglion, abscess, synovial cyst, or liquefied hematoma) (Fig. 5L–N). Differentiation from myxoid liposarcoma is dependent on identification of adipose elements, a feature not apparent in MFH (19,20).

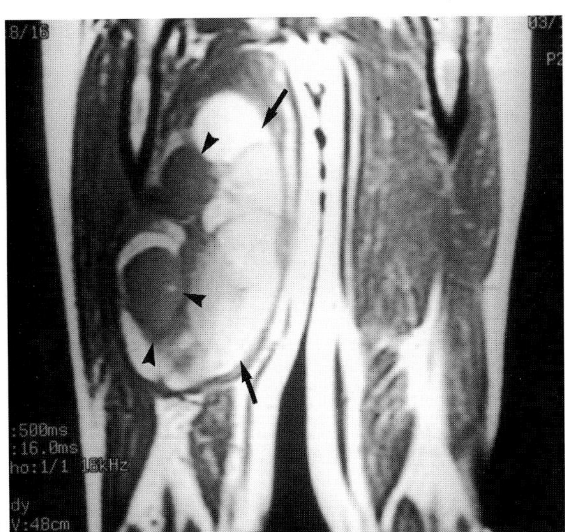

D

FIG. 5. D,E: Storiform/pleomorphic MFH presenting as "spontaneous" hemorrhage in a 45-year-old man with rapidly enlarging thigh mass. Coronal T1-weighted (TR/TE 500/16 msec) **(D)** and axial T2-weighted (TR/TE 2,500/80 msec) **(E)** images show large areas of hemorrhage (high intensity on T1 and T2 weighting with fluid-fluids levels-arrows). Area of neoplasm *(arrowheads)* is isointense to muscle on T1-weighted MR image (D).

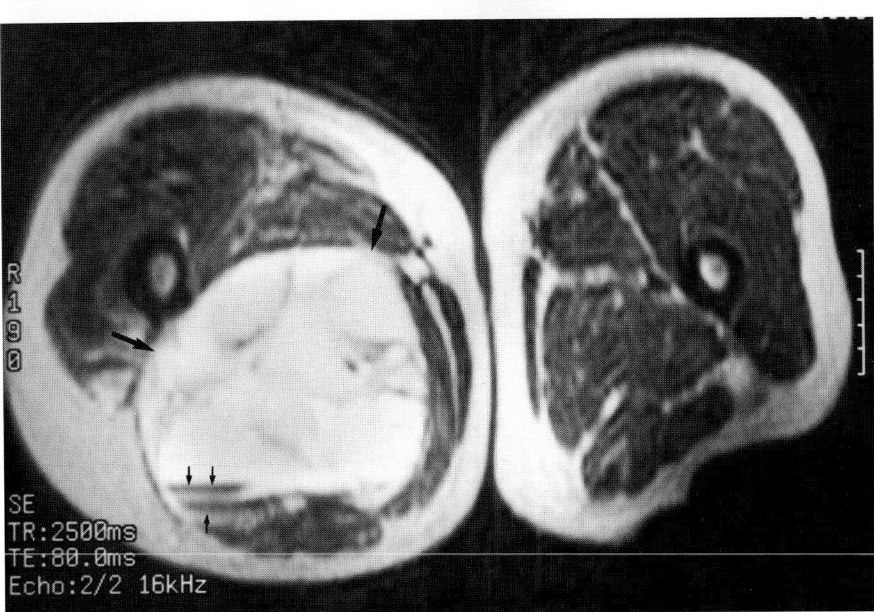

E

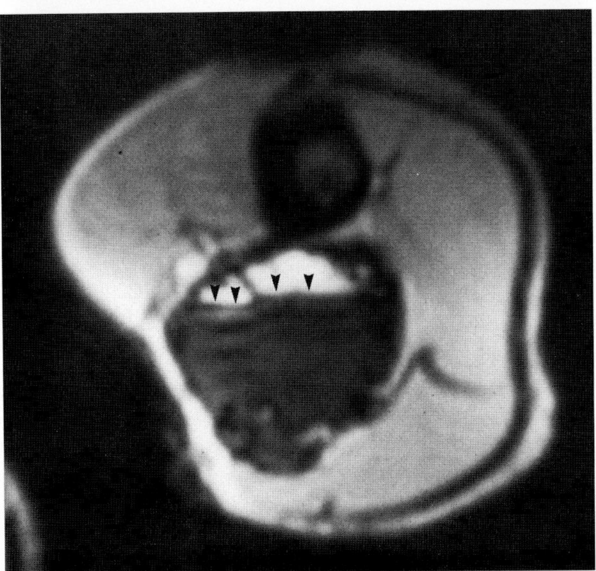

F

FIG. 5. F: Angiomatoid MFH in a 10-year-old boy with a lump medial to the upper arm. Axial gradient echo image (TR/TE/°7 600/30/25°7 flip angle) shows multiple fluid-fluid levels *(arrowheads)* in a hemorrhagic cystic mass.

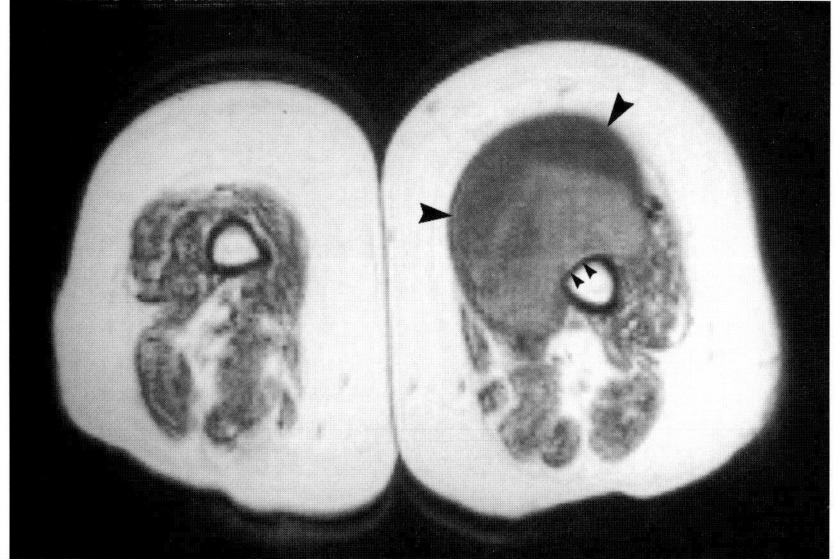
G

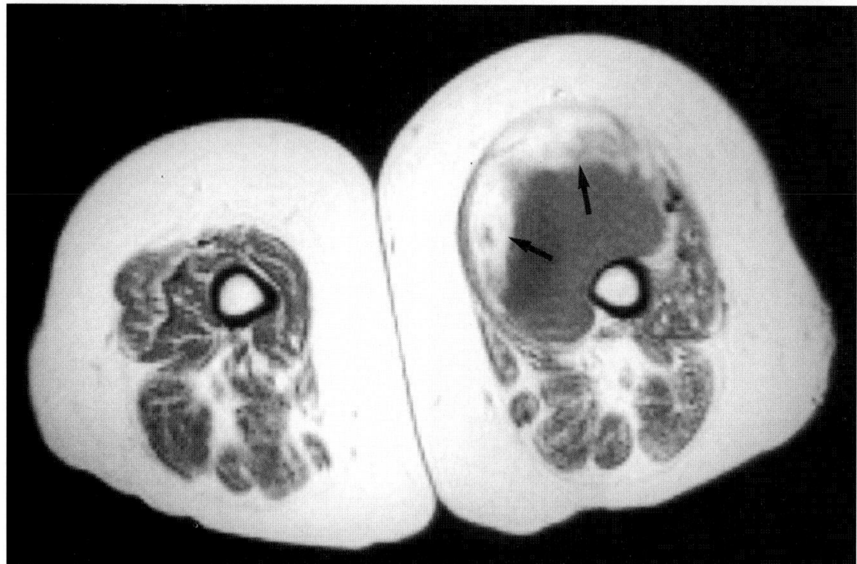
H

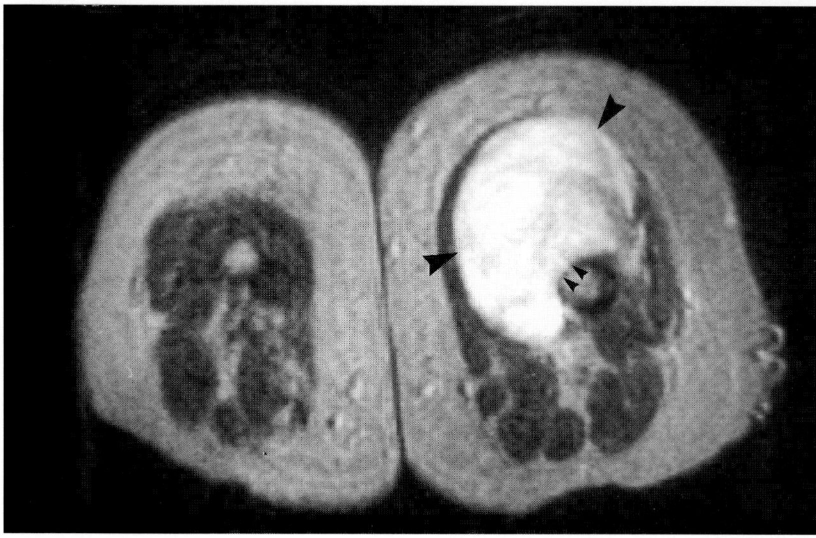
I

FIG. 5. G–I: Myxoid MFH of the thigh with cortical erosion in a 65-year-old woman with a 3-month history of enlarging thigh mass. Axial T1-weighted (TR/TE 500/17 msec) images before **(G)** and after intravenous gadolinium **(H)** and T2-weighted (TR/TE 2,100/90 msec) **(I)** image show a cystic anterior thigh mass *(large arrowheads)* with extrinsic erosion of adjacent femur *(small arrowheads).* Nodular peripheral enhancement after contrast is seen in the more solid peripheral portion of the mass *(arrows)* (H).

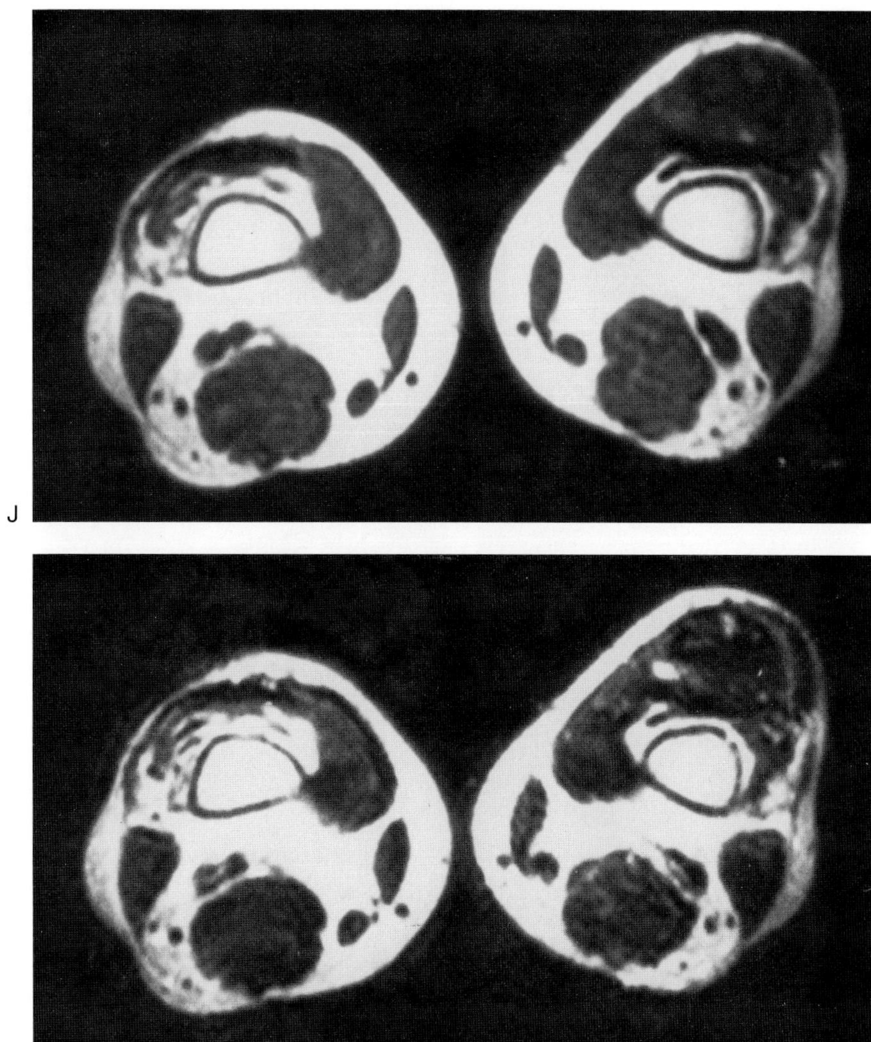

FIG. 5. J,K: MFH in a 62-year-old man with an anterior distal thigh mass. Axial T1-weighted (TR/TE 600/20 msec) **(J)** and T2-weighted (TR/TE 2,000/80 msec) **(K)** images show a large predominantly low intensity soft tissue mass anterior to the distal femur.

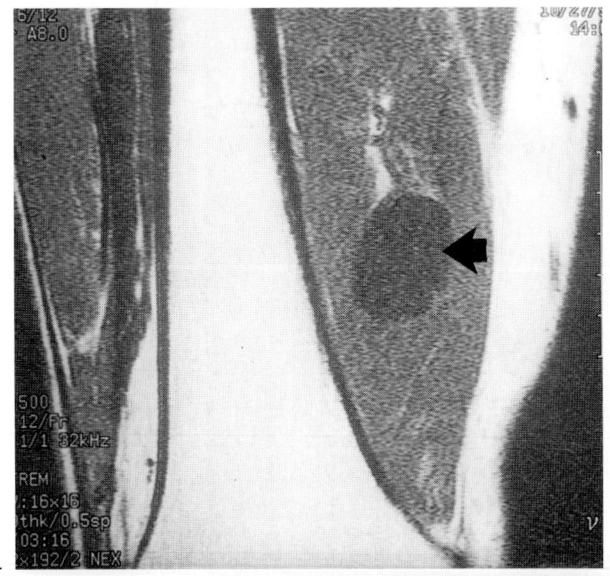

L

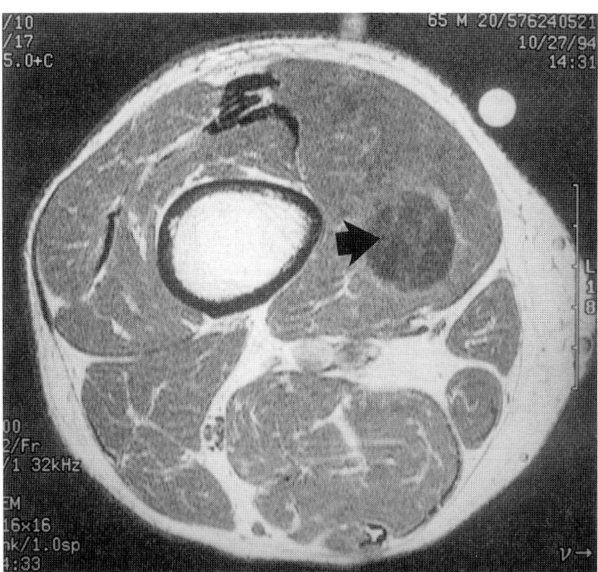

M

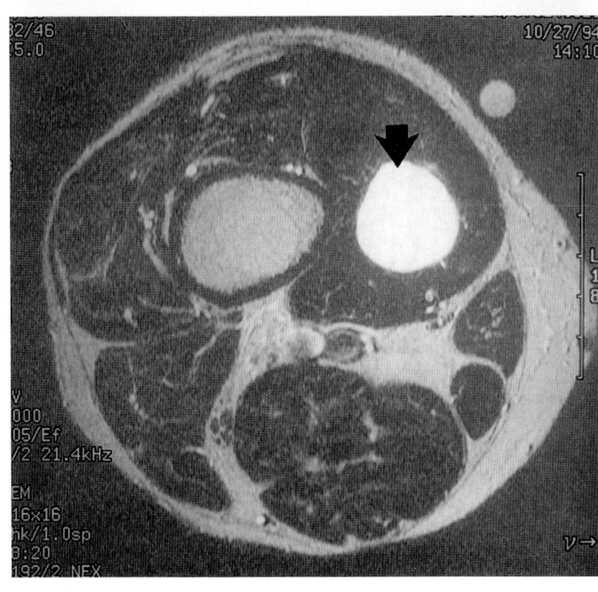

N

FIG. 5. L–N: Myxoma of the thigh in a 65-year-old man with a vague mass. Coronal T1-weighted (TR/TE 500/12 msec) **(L),** axial T1-weighted after intravenous gadolinium (TR/TE 700/12 msec) **(M)** and axial T2-weighted (TR/TE 5,000/102 msec) **(N)** images show a cystic mass with mild rim enhancement after contrast administration *(arrowheads).*

CASE 6

History: A 27-year-old woman with a 4-year history of progressive knee pain and decreasing range of motion.

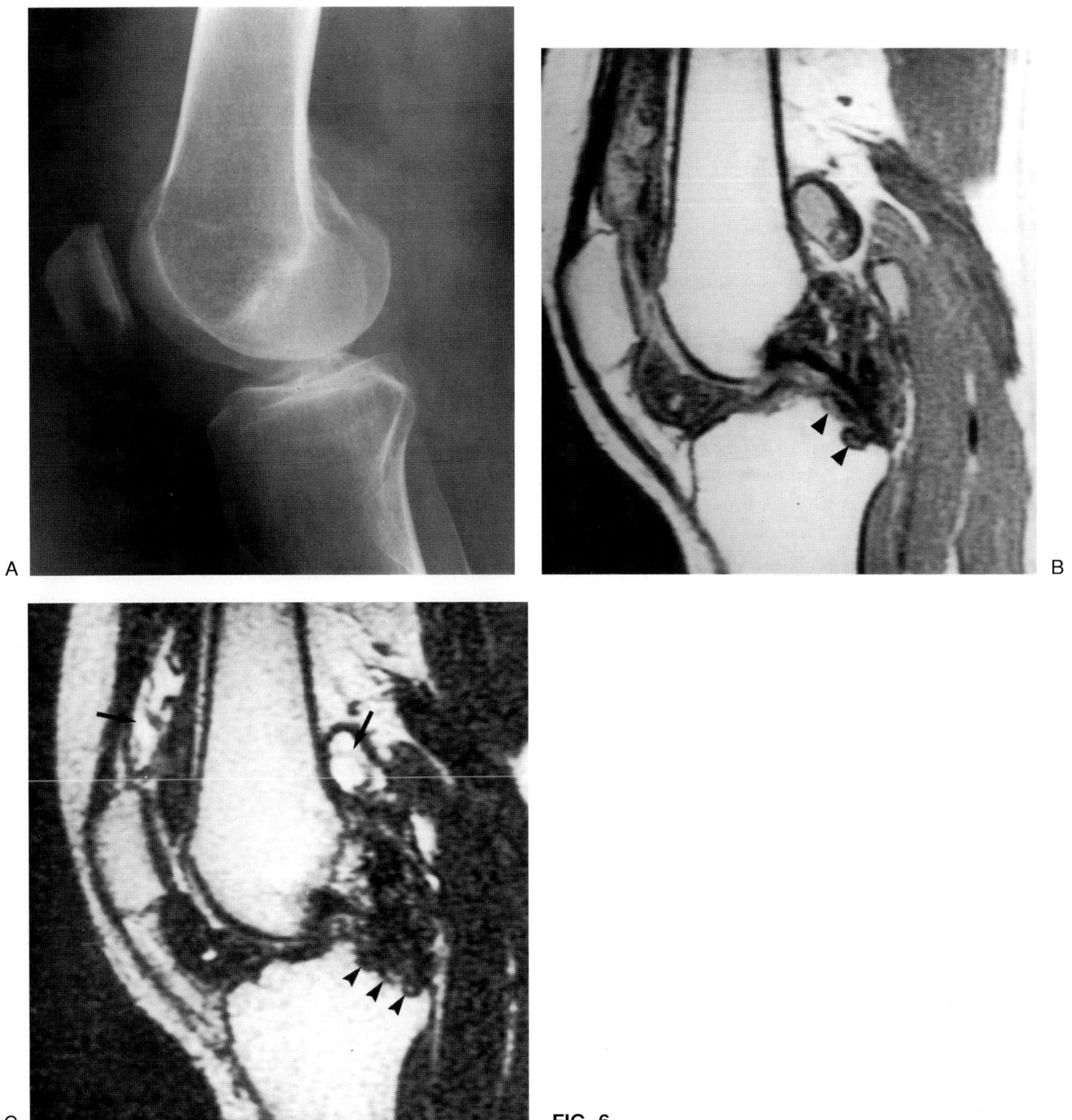

FIG. 6.

Findings: Lateral knee radiograph (Fig. 6A) shows fullness in the region of the suprapatellar bursa. Sagittal proton density (TR/TE 2,000/30 msec) (Fig. 6B) and T2-weighted (TR/TE 2,000/90 msec) (Fig. 6C) images reveal diffuse thickening within the knee joint with tissue that is predominantly low intensity on the T2-weighted images. Additional small pockets of fluid (arrows) are seen within the joint and an intraarticular process has caused extrinsic erosion of bone in the posterior tibia (arrowheads).

Diagnosis: Pigmented villonodular synovitis.

Discussion: Pigmented villonodular synovitis (PVNS) represents a disease resulting in synovial proliferation.

Pathologically PVNS should be considered a family of disorders with the synovial proliferation showing variable degrees of villous and or nodular hypertrophy and pigment (hemosiderin) deposition (1). PVNS may be diffuse, extensively infiltrating a joint, bursa, or tendon sheath, or it may be localized, often within a tendon sheath [also referred to as giant cell tumor of tendon sheath (GCTTS)].

Adults in the third to fourth decade are typically affected and a history of trauma is present in nearly half of patients. The diffuse form of PVNS is almost exclusively a monoarticular disease, and most frequently involves the knee (1–8). Although almost any joint can be affected, the hip, elbow, and ankle are other common sites of involvement. GCTTS most frequently affects the hand (80% of cases) followed by the foot, ankle, and knee (9,10). Clinical symptoms associated with intraarticular involvement by PVNS are joint pain and swelling with decreased range of motion and are often of long duration before the diagnosis is established. The localized form may present only as a soft tissue mass.

Radiographs may be normal, reveal changes suggesting a joint effusion, or show extrinsic osseous erosions. Differential diagnosis from the radiographs alone includes synovially based processes such as trauma with hemarthrosis [often associated with internal derangement such as anterior cruciate ligament (ACL) tear when at the knee], infection, inflammatory arthritis, amyloid deposition, synovial chondromatosis, and PVNS. Extrinsic erosions are infrequent in the diffuse form of PVNS when involvement is in larger capacity joints such as the knee (25%) (1). In contradistinction PVNS of small-capacity joints such as the hip (Fig. 6D,E) almost always demonstrate extrinsic erosions (93%) (1–8). Localized PVNS reveals a noncalcified mass on radiographs with adjacent extrinsic erosion of bone present in a minority of cases (10–15%) (9,10).

Arthrography in PVNS shows multinodular filling defects and aspirated fluid is xanthochromic to brownish. While CT will demonstrate synovial thickening that may be high attenuation (due to iron deposition) MR imaging is currently the modality of choice to evaluate PVNS. The characteristic MR imaging appearance of the diffuse form of PVNS is extensive synovial thickening with prominent areas of low to intermediate signal intensity on all pulse sequences. Hypointensity is caused by iron in hemosiderin (more than 25% in the ferric, Fe^{+3}, state with five unpaired electrons) shortening the adjacent water molecules T2 relaxation time (11). These changes are most pronounced on high field strength MR units because the T2 relaxation time shortening is proportional to the square of the magnetic field. Focal areas of fluid may also be present; however, these are generally surrounded by low signal intensity hemosiderin-laden synovium. PVNS will demonstrate enhancement after intravenous gadolinium injection because of the hypervascularity of the synovial tissue.

MR imaging of GCTTS is more variable than the diffuse form of PVNS, reflecting more variation in the amount of hemosiderin deposition seen pathologically. MR imaging may be nonspecific showing only a soft tissue mass with low to intermediate signal intensity on T1-weighted MR images and intermediate to high signal intensity on T2-weighting. Jelinek and coworkers (10), however, have recently reported more distinctive findings on MR imaging in nine cases of GCTTS with low to intermediate signal intensity on all pulse sequences similar to the diffuse form of PVNS. This appearance on MR imaging of a mass involving a tendon should suggest GCTTS (Fig. 6F,G).

Treatment of PVNS is usually surgical with synovectomy. However, because of the difficulty in resecting all synovial tissue (particularly in large joints) recurrence of the diffuse form of PVNS is common (40–50%) (1). GCTTS is less frequent to recur (10–14%) because it is focal and is often surrounding by a capsule or is within a tendon sheath (1). Radiologists may become involved in treatment of recurrent PVNS by performing a radionuclide synovectomy. Yttrium-90 and dysprosium-165 attached to radiocolloid can be injected into joints with the beta irradiation resulting in a nonsurgical synovectomy. MR imaging is useful to evaluate for recurrent PVNS with identical characteristics as previously described.

An MR imaging appearance very similar to the diffuse form of PVNS can also be seen in amyloidosis usually associated with chronic renal insufficiency and in hemophiliac arthropathy (12–17). In hemophiliac arthropathy repetitive intraarticular episodes of bleeding result in synovial hemosiderin deposition (low signal on all MR pulse sequences) with hypertrophy and inflammation and progressive joint destruction (15–17). The synovial deposition of amyloid (β_2-microglobulin) shows persistent low signal intensity on all MR pulse sequences owing to its fibrillar collagen-like composition (Fig. 6H,I) (12–14). Synovial involvement by amyloid deposition and in hemophiliac arthropathy is usually polyarticular as opposed to monoarticular in PVNS.

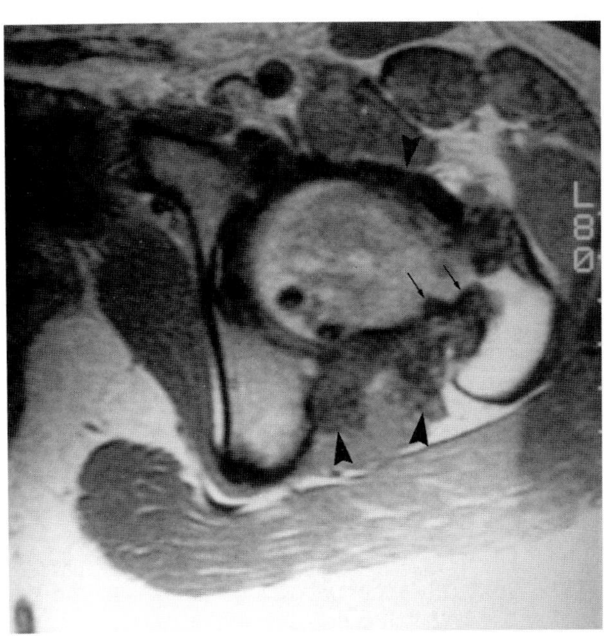
D

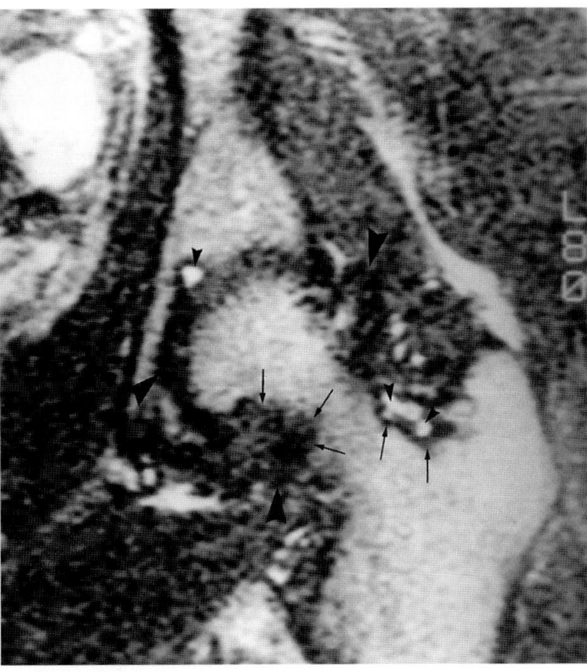
E

FIG. 6. D,E: Diffuse form of PVNS in the hip of a 27-year-old woman with chronic pain for 5 years. Axial T1-weighted (TR/TE 1,000/20 msec) **(D)** and coronal T2-weighted (TR/TE 2,000/80 msec) **(E)** images show extensive and diffuse intraarticular synovial thickening that stays predominantly low intensity on long TR image *(large arrowheads)*, small, pockets of fluid *(small arrowheads)* and extrinsic erosions of bone *(arrows)*.

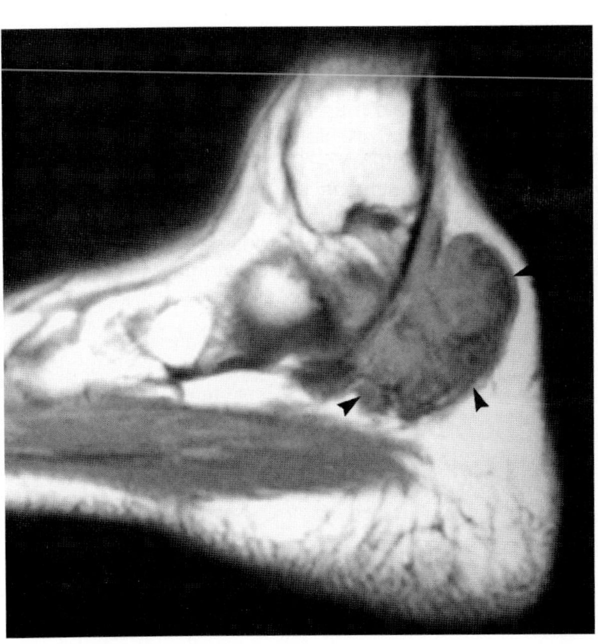
F

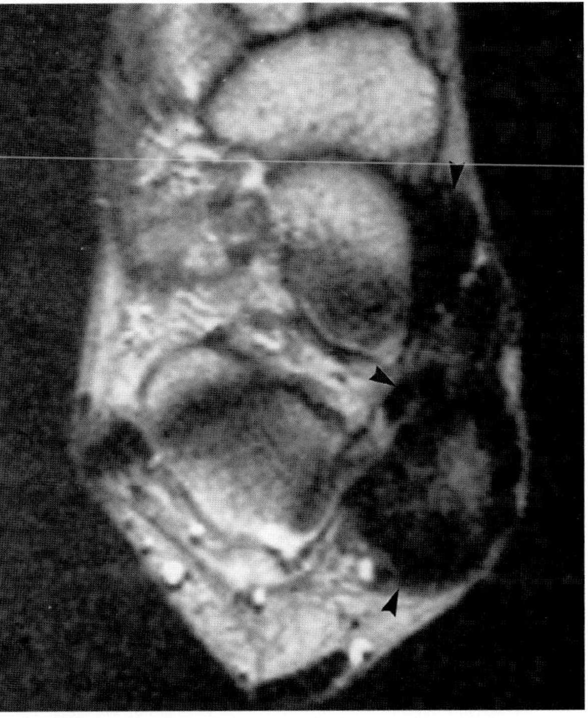
G

FIG. 6. F,G: Localized form of PVNS (GCTTS) involving the posterior tibial tendon in a 29-year-old woman. Sagittal T1-weighted (TR/TE 580/15 msec) **(F)** and axial T2-weighted (TR/TE 1,800/90 msec) **(G)** images show a lobulated low to intermediate signal intensity mass *(arrowheads)* involving the posterior tibial tendon.

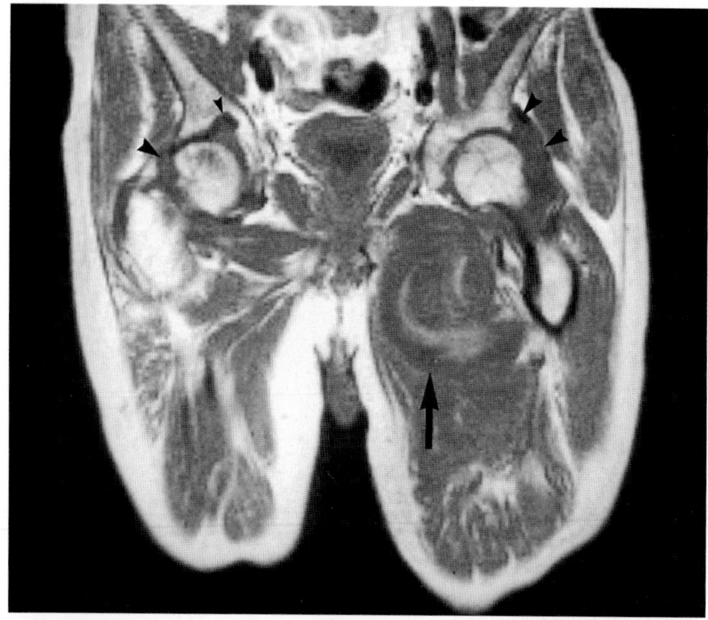

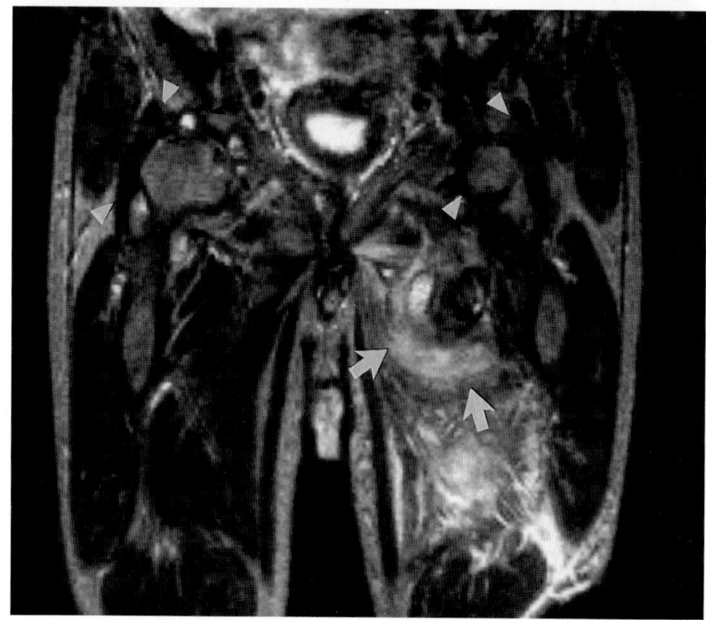

FIG. 6. H,I: Secondary synovial deposition of amyloid in a 43-year-old man on chronic hemodialysis for renal failure. Coronal T1-weighted (TR/TE 500/17 msec) **(H)** and T2-weighted (TR/TE 2,100/90 msec) **(I)** images show diffuse synovial thickening *(large arrowheads)* in both hips staying predominantly low intensity on long TR image and causing extrinsic osseous erosion *(small arrowhead)*. A left thigh hematoma *(arrows)* related to hemodialysis is also present.

CASE 7

History: A 24-year-old woman with slowly enlarging forearm mass associated with intermittent swelling.

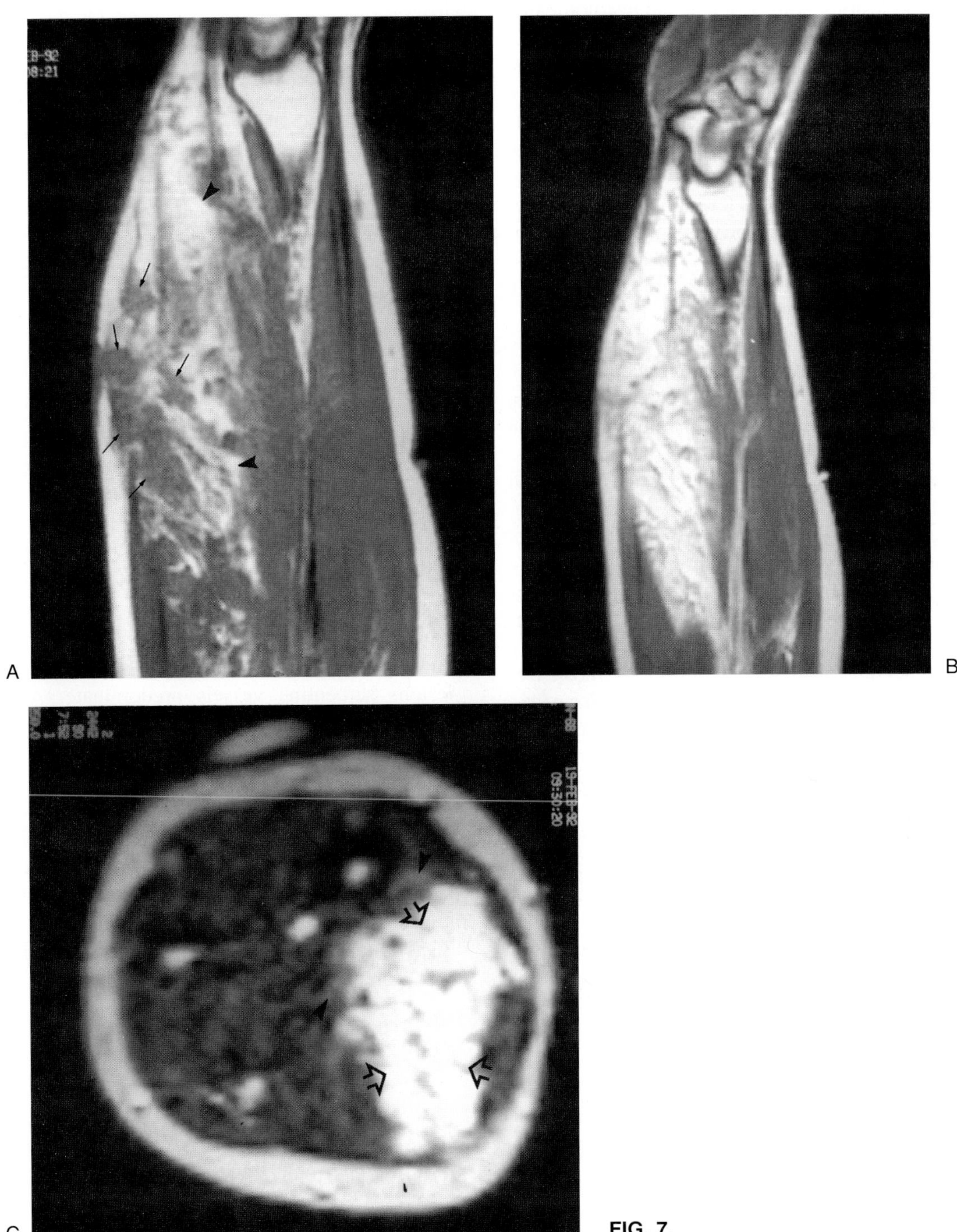

FIG. 7.

Findings: Sagittal T1-weighted (TR/TE 500/15 msec) images before (Fig. 7A) and after intravenous gadolinium (Fig. 7B) and axial T2-weighted (TR/TE 2,400/90 msec) (Fig. 7C) image reveal a large heterogeneous mass with areas isointense with fat on all pulse sequences (arrowheads). Other regions of the mass show low intensity on T1-weighted MR images becoming very high signal on T2-weighting and seen as circular components on the axial image (open arrows). These areas appear as large serpentine cavernous spaces on sagittal images (arrows). There is prominent enhancement after contrast administration.

Diagnosis: Cavernous hemangioma, intramuscular.

Discussion: Hemangiomas represent benign neoplasms composed of vessels that closely resemble normal vasculature (1). These lesions, while often mixed histologically, are classified pathologically by the predominant type of vascular component (capillary, cavernous, arteriovenous, or venous). Hemangioma is the most common tumor of infancy and childhood. It is also a common soft tissue tumor in young adults accounting for 7% of all benign soft tissue tumors (1,2). Hemangiomas in the soft tissue may be superficial or deep. Deep lesions are usually intramuscular and are more frequently evaluated radiologically than superficial hemangiomas because patients present as a diagnostic dilemma clinically, with a nonspecific soft tissue mass.

Clinical presentation of superficial lesions are usually characteristic with skin discoloration (strawberry nevus). These lesions are usually capillary-type hemangiomas, are not evaluated radiologically, and involute by age 7 in up to 90% of cases (1,2). Intramuscular hemangiomas, on the other hand, present with a painful mass that can change intermittently in size and may have overlying bluish skin discoloration. These lesions are more frequent in young women and may markedly increase in size during pregnancy.

Radiographs of hemangioma may be normal, or be nonspecific showing only a soft tissue mass. Phleboliths may be identified as characteristic calcification, most frequently, although not exclusively, associated with cavernous hemangioma (Fig. 7D,E). Soft tissue hemangioma may also extend into bone or affect osseous structures secondarily with overgrowth owing to chronic hyperemia (Fig. 7F–H). Soft tissue hemangioma may also involve the synovium (less than 1% of cases) and repetitive intraarticular bleeding can result in an appearance identical to that seen in hemophilia (3,4). Differentiation is not difficult by clinical history and radiologically hemophilia is a polyarticular disease versus monoarticular changes with synovial hemangioma.

CT and sonography can demonstrate soft tissue hemangiomas, however, MR imaging is considered the modality of choice to evaluate these lesions (5–18). MR imaging characteristics are frequently diagnostic of intramuscular hemangioma. Hemangiomas are usually heterogeneous masses on all MR pulse sequences. On T1-weighted MR images low to intermediate signal intensity is seen in much of the mass; however, high signal regions are also frequently present. These areas of high signal intensity are almost always due to fat overgrowth, in our experience, and represent a reactive phenomena pathologically. This appearance should not imply a diagnosis of angiolipoma, which is a distinct pathologic lesion typically present in the subcutaneous tissue of the forearm. The imaging appearance of true angiolipoma has not been described to the best of our knowledge. The fat is often most prominent in the periphery of the mass, as in our case.

On T2-weighted MR images the vascular components of intramuscular hemangioma usually show marked increased signal intensity while the associated adipose tissue remains isointense to subcutaneous fat (5–18). The vascular component of the mass, when looked at more carefully, often has a very characteristic appearance. These regions are composed of circular areas (vessels seen en face) and linear/serpentine channels (vessels seen longitudinally), reflecting the underlying pathology. In some cases it is possible to differentiate histologic subtypes of soft tissue hemangiomas with cavernous lesions composed of large cystic space while in arteriovenous lesions serpentine vessels predominate. Arteriovenous hemangiomas (often referred to as arteriovenous malformation) may show rapid flow in serpentine vessels with persistent low signal intensity on all pulse sequences (Fig. 7F–H) (19). Intramuscular hemangiomas demonstrate marked enhancement after intravenous gadolinium injection. Phleboliths are more easily recognized on CT or radiographs; however, they appear as circular areas of low signal intensity on all MR pulse sequences (Fig. 7D,E). Fluid-fluid levels may be present in hemangiomas as a result of hemorrhage (20).

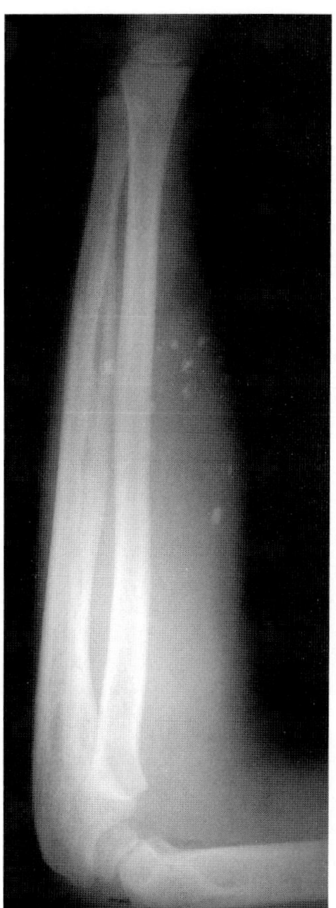

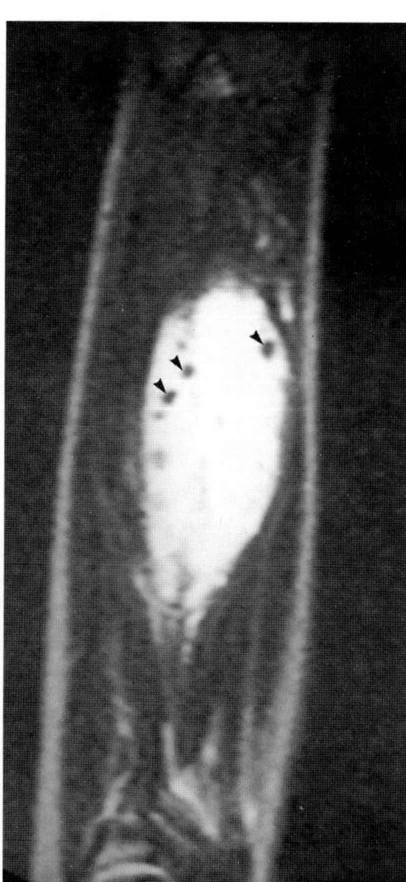

D,E

FIG. 7. D,E: Capillary hemangioma in the forearm of an 8-year-old boy with phleboliths. Multiple phleboliths are seen on the lateral radiograph **(D).** The T2-weighted (TR/TE 2,000/100 msec) sagittal image **(E)** reveals marked hyperintensity of the mass with circular low intensity foci representing the phleboliths *(arrowheads).*

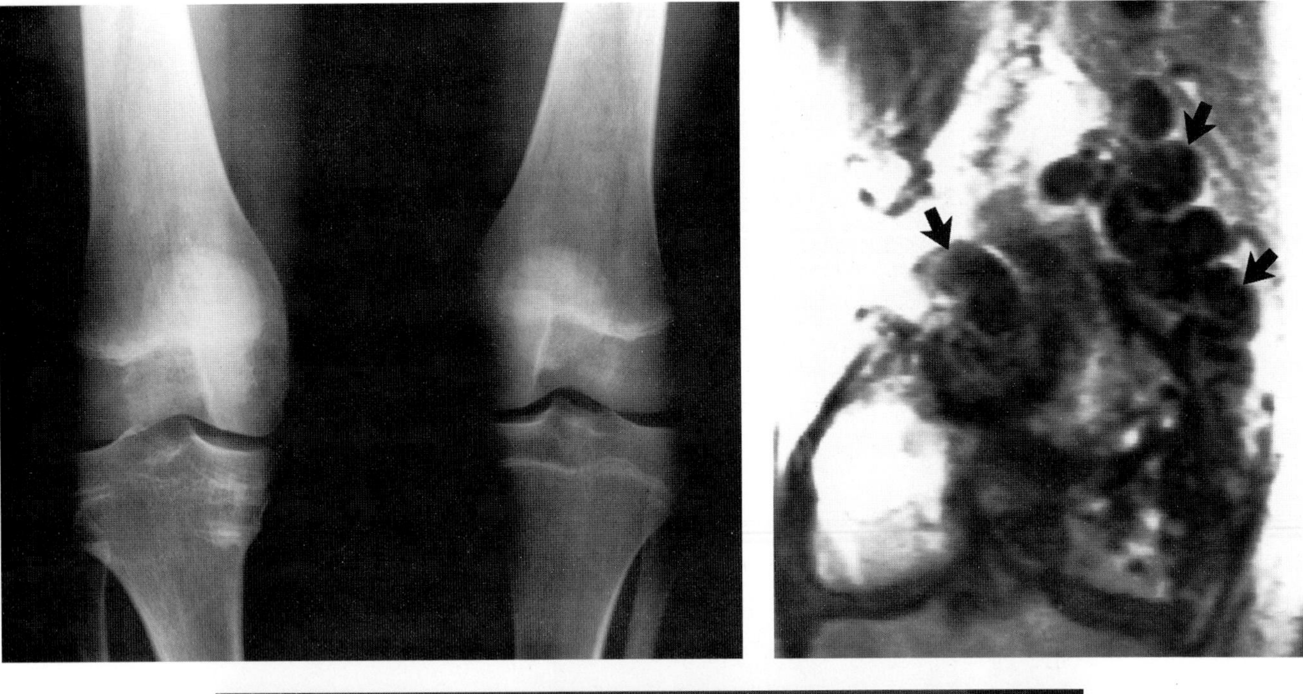

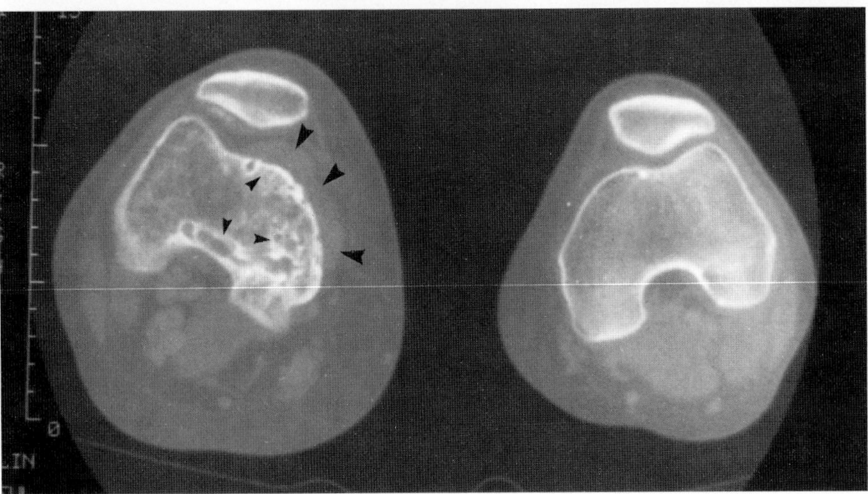

FIG. 7. F–H: Arteriovenous hemangioma of the knee with secondary osseous overgrowth and synovial involvement in a 31-year-old woman. Anteroposterior radiograph **(F)** shows overgrowth of bone. CT **(G)** shows soft tissue mass involving the suprapatellar portion of the joint *(large arrowheads)* and osseous involvement *(small arrowheads)*. Coronal T2-weighted (TR/TE 1,800/180 msec) **(H)** image reveals the large serpentine arteriovenous channels *(arrows)* with rapid flow resulting in signal void.

CASE 8

History: A 28-year-old woman with vague shoulder pain of long duration.

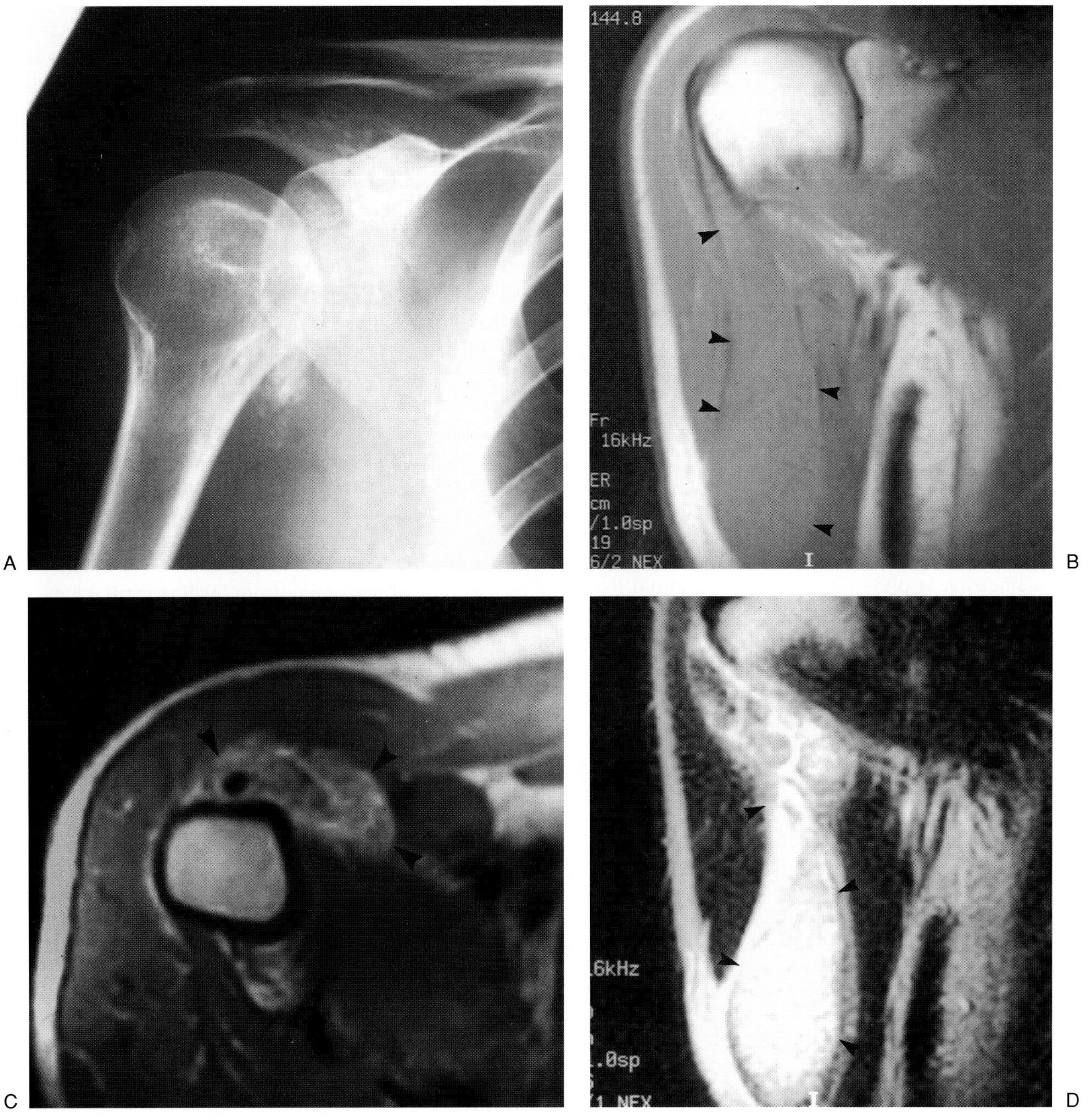

FIG. 8.

Findings: Shoulder radiograph with internal rotation (Fig. 8A) shows calcifications medial to the humeral head/neck junction. MR images include coronal T1-weighting (TR/TE 700/15 msec) (Fig. 8B), axial proton density (TR/TE 4,000/13 msec) (Fig. 8C), and coronal T2-weighting (TR/TE 2,000/90 msec) (Fig. 8D), and reveal a mass (arrowheads) extensively involving and within the bicipital tendon sheath. The mass is partially calcified in correlation with the radiograph and shows low to intermediate intensity on T1-weighted MR images and high signal intensity on T2-weighting with some surrounding fluid. No intraarticular abnormality is seen.

Diagnosis: Tenosynovial chondromatosis.

Discussion: Synovial chondromatosis represents a primary synovial disease in which cartilage formation occurs within the synovial membrane. This process is likely metaplastic or neoplastic and usually involves joints; however, synovial tissue of bursae and tendon sheaths can also be affected (tenosynovial chondromatosis) (1,2). Cartilaginous nodules are produced that can detach to become free bodies within the joint, bursae, or tendon sheath. These fragments may increase in size (nourished from synovial fluid) or be resorbed and may undergo endochondral ossification. Synovial chondromatosis is a benign process, although pathologically the cartilaginous tissue may appear aggressive leading to a misinterpretation of chondrosarcoma. True malignant degeneration of synovial chondromatosis to chondrosarcoma is rare.

Clinical presentation is usually in the third to fifth decade and men are affected twice as frequently as women. Symptoms are often of long duration before diagnosis, with pain, joint locking, and reduced range of motion. The disease is typically monoarticular with the knee joint most commonly affected.

Radiographs and CT are optimal to detect calcification within the cartilage seen in 75% to 90% of cases (3–11). The calcification is often diagnostic of chondroid tissue with a ring-like appearance. The calcification may also be less characteristic and punctate or have a target appearance (Fig. 8E–G). Noncalcified areas of cartilage metaplasia on CT show soft tissue thickening of similar or lower attenuation to muscle within the joint, bursae, or tendon sheath. Differential diagnosis of calcified masses in and about joints includes tumoral calcinosis, hyperparathyroidism, collagen vascular disease [scleroderma, dermatomyositis, and systemic lupus erythematosus (SLE)], neuropathic joint, synovial sarcoma, and synovial chondromatosis.

The MR imaging appearance of synovial chondromatosis is variable, depending on the amount and degree of calcification and ossification that is present (4,5,8). Areas of hyaline cartilage (composed of 75% to 80% water) metaplasia without mineralization show intermediate intensity on T1-weighted MR images and become hyperintense on T2-weighting. Small mineralized cartilage nodules may be very difficult to detect, on MR images, as in this case, because of their size. It is important to correlate the radiographs and/or CT in these cases with the MR images in order to arrive at the appropriate diagnosis. Larger mineralized fragments show low intensity on all pulse sequences when diffusely mineralized or a low intensity periphery and high intensity central areas on long TR images with characteristic ring-like chondroid calcification. Ossification of fragments results in MR images showing an outer cortex (low intensity all pulse sequences) and inner marrow element (isointense to fat). Mixture of chondroid calcification and ossification peripherally can cause a target appearance on MR images (Fig. 8E–G). In general the mineralized areas are central with synovial thickening from hyaline cartilage metaplasia peripherally. This creates an MR imaging appearance opposite that of PVNS on long TR images with low signal centrally (calcified fragments) and high intensity peripherally (hyaline cartilage metaplasia). Joint fluid and extrinsic erosion of bone may be identified.

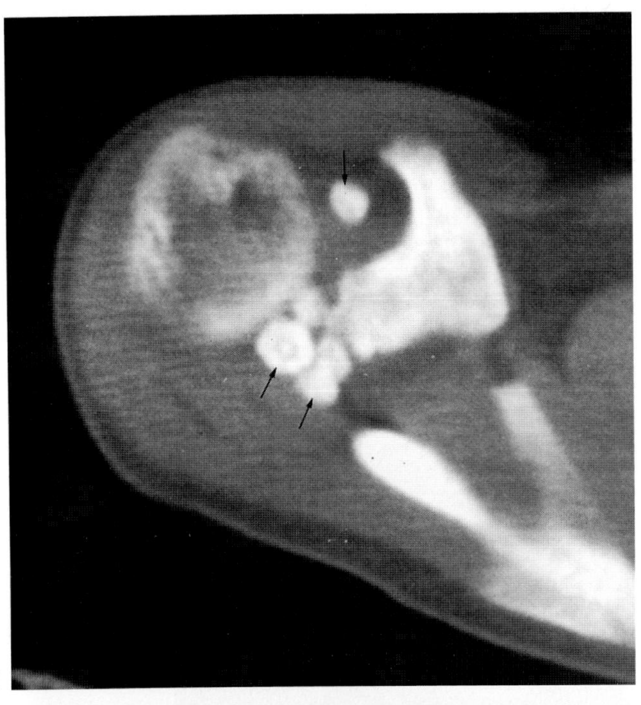

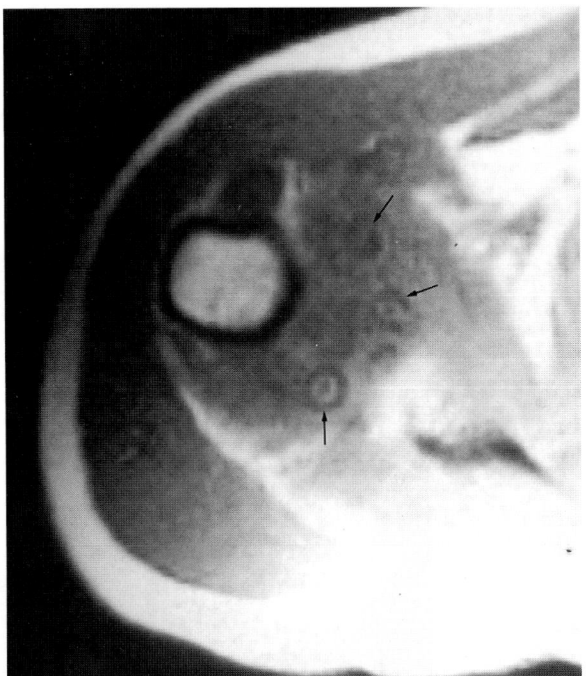

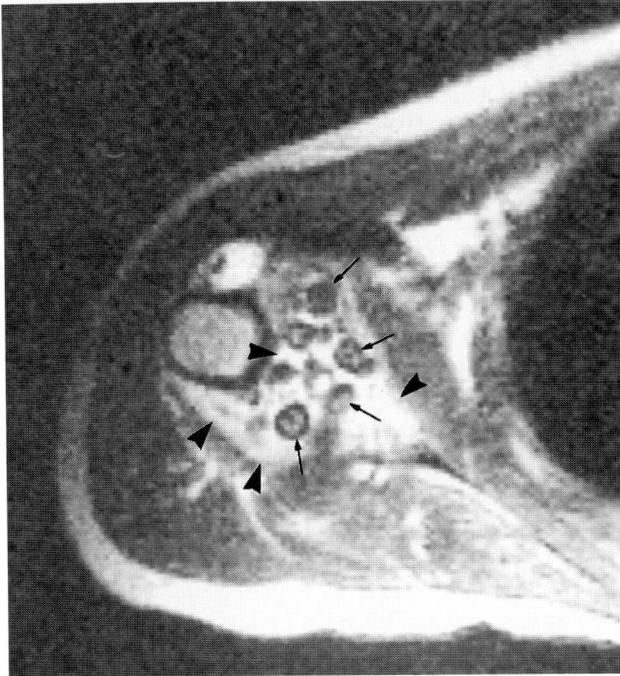

FIG. 8. E–G: Synovial chondromatosis in a 19-year-old woman with pain and limited range of motion. CT **(E)** and axial images with T1-weighting (TR/TE 500/17 msec) **(F)** and T2-weighting (TR/TE 2,100/90 msec) **(G)** show multiple mineralized intraarticular fragments having both cartilage and osteoid features causing a target appearance *(arrows)*. Central dot of chondroid calcification, surrounded by a ring of yellow marrow with peripheral cortex is seen. Nonmineralized cartilage metaplasia of synovium is seen as high signal intensity areas on T2-weighted **(G)** MR image *(arrowheads)* peripherally with mineralized fragments centrally.

CASE 9

History: A 70-year-old woman presented with a painless mass in the buttocks. A malignant fibrous histiocytoma had been resected approximately 2 years earlier, and she has received radiation therapy and chemotherapy.

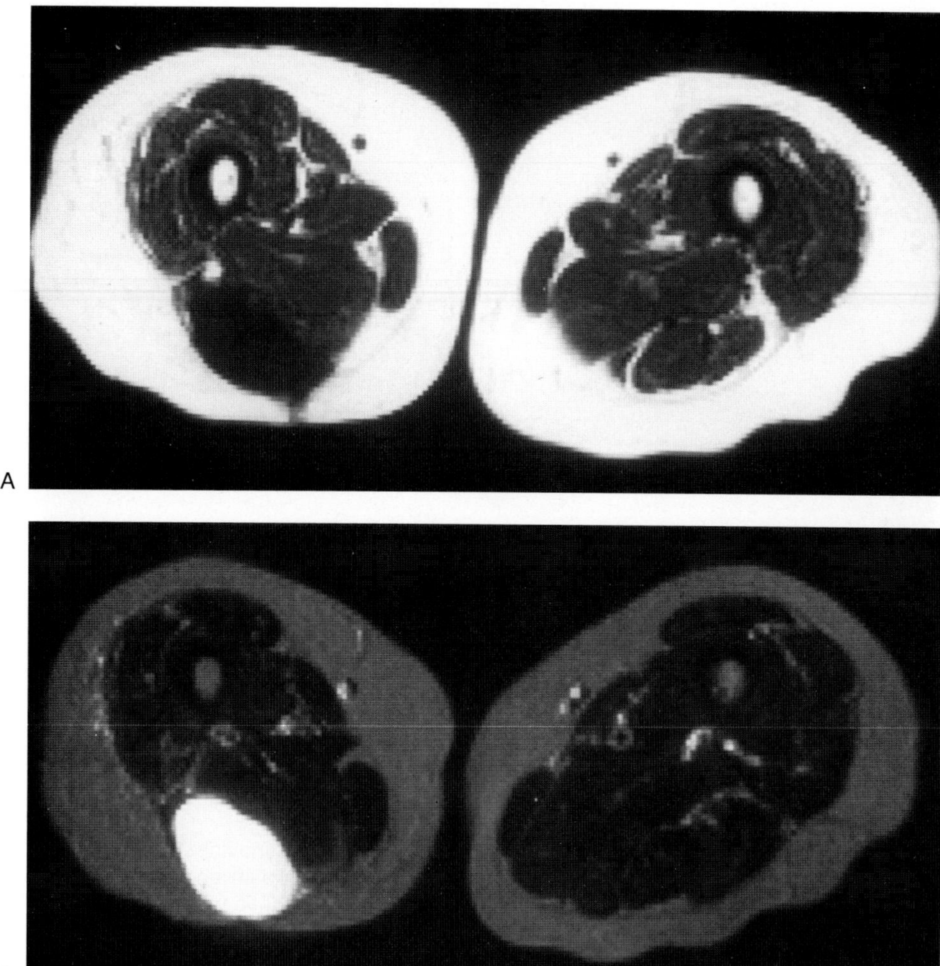

FIG. 9.

Findings: Axial T1-weighted (TR/TE 500/17 msec) (Fig. 9A) and T2-weighted (TR/TE 2,100/90 msec) (Fig. 9B) SE images show a well-defined, homogeneous, soft tissue mass in the posterior aspect of the right thigh. The lesion has a signal intensity similar to that of fluid, being greater than that of fat on T2-weighted images and less than that of skeletal muscle on T1-weighted images. Radiographs of the pelvis (not shown) were unremarkable, with no evidence of bone involvement or soft tissue calcification.

Diagnosis: Postoperative fluid collection (seroma).

Discussion: As many as half of patients with soft tissue sarcomas will have local recurrence and routine follow-up is essential. MR is an ideal modality for this evaluation and the identification of a discrete mass is always of concern.

Recurrent tumor is characterized by the presence of a discrete nodule or mass with prolonged T1 and T2 relaxation times (Fig. 9C,D), whereas fluid collections without nodule or areas of low or intermediate signal intensity suggest postsurgical change (Fig. 9A,B) (1,2). When a mass with high signal intensity on T2-weighted images is found, tumor must be differentiated from postoperative hygroma. In most cases (as this example), this differentiation is straightforward with fluid characterized by a homogeneous, well-defined mass with prolonged T1- and T2-

weighted SE images. Tumor margins are usually irregular and signal intensity is inhomogeneous.

There are some important caveats. To effectively evaluate a patient for recurrence, one must know the original tumor histology. For example, myxoid tumors may mimic cysts on MR imaging (see Case 1), and concern for recurrence would be considerably greater if the patient's original tumor were a myxoid liposarcoma. When there is question as to whether an area of high signal intensity on T2-weighted images represents fluid, ultrasound examination is an ideal method for further evaluation. It is easy, inexpensive, and highly accurate. Alternatively, gadolinium-enhanced MRI may be used. Gadolinium-enhanced imaging is not without a price, however, in that intravenous contrast significantly increases the length and cost of the examination.

Moreover, although the incidence of untoward reaction as a result of contrast administration is small, it is real. Severe reactions have been reported with both gadopentetate dimeglumine (Magnevist; Berlex Laboratories, Wayne, NJ) and gadoteridol (ProHance; Squibb Diagnostics, Princeton, NJ), including hypotension, laryngospasm, bronchospasm, anaphylactoid reaction, and anaphylactic shock (3–7), as have a full spectrum of less serious reactions. Recently, Jordan and Mintz (8) reported a fatal reaction to gadopentetate dimeglumine presumed due to anaphylactic reaction with associated bronchospasm. Consequently, when confirmation of a suspected fluid accumulation is required, we prefer to use ultrasound when possible, reserving gadolinium-enhanced imaging for lesions not readily evaluated by ultrasound.

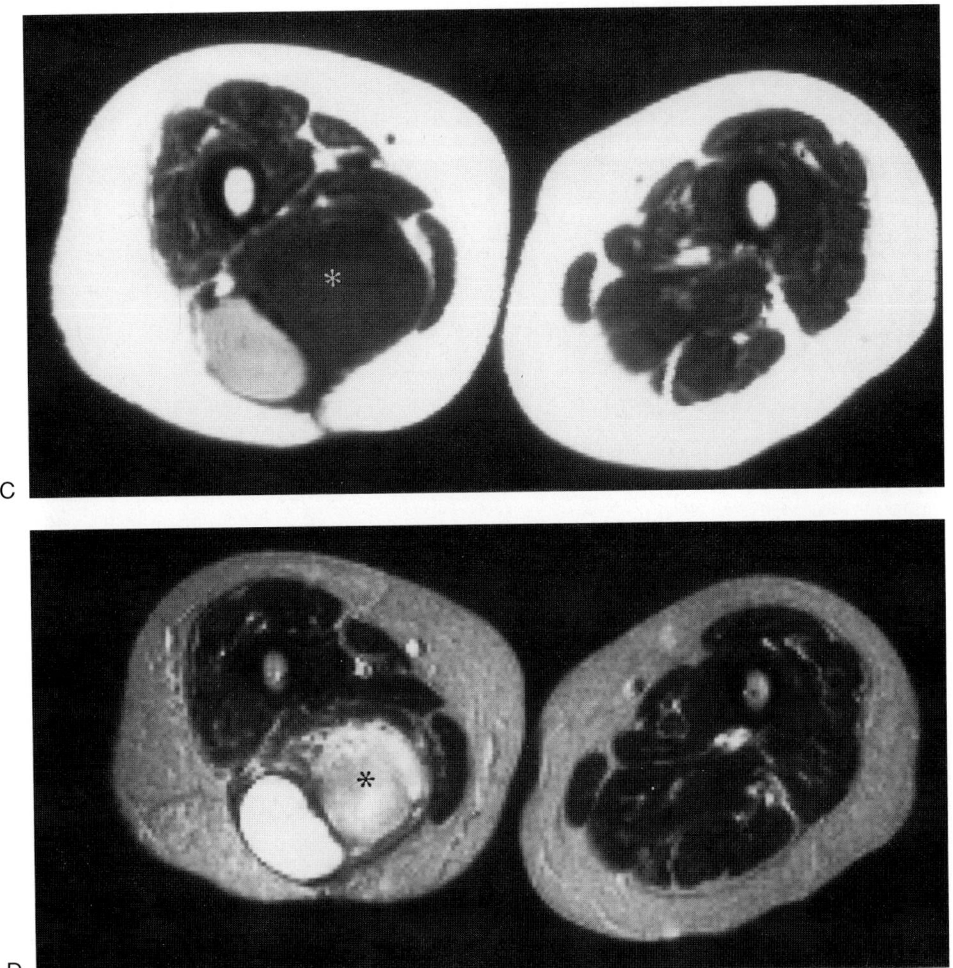

FIG. 9. C,D: Recurrent malignant fibrous histiocytoma and postoperative fluid collection, approximately 18 months following previous images. Axial T1-weighted (TR/TE 500/17 msec) **(C)** and T2-weighted (TR/TE 2,100/90 msec) **(D)** SE images of the right thigh show a discrete mass *(asterisk)* as well as the previously identified fluid collection. Note that the signal intensity of the fluid has increased significantly on T1-weighted images, likely reflecting increased protein content.

CASE 10

History: A 62-year-old woman presented with a palpable mass in the region of the right scapula. She was referred for MR with the clinical diagnosis of possible soft tissue sarcoma.

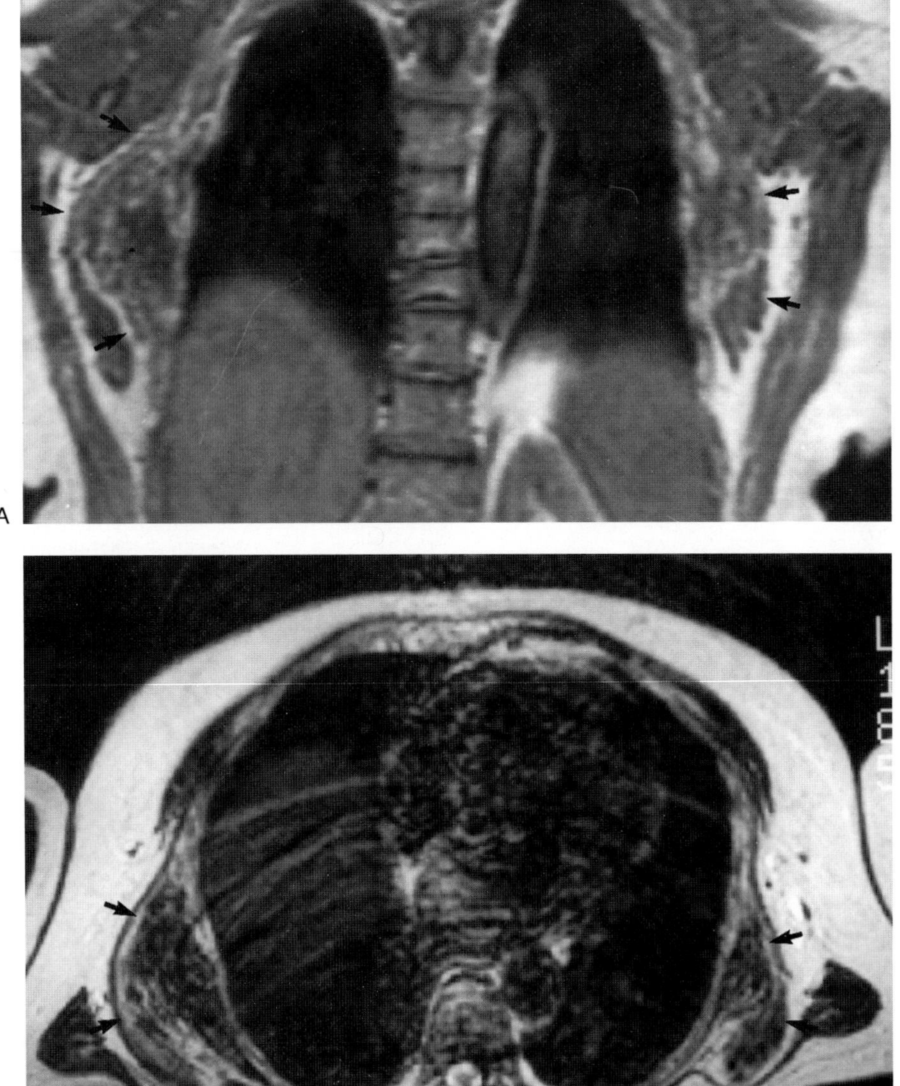

FIG. 10.

Findings: In Fig. 10A coronal T1-weighted gated acquisition (TR/TE 895/15 msec) demonstrates bilateral soft tissue masses (small arrows) projecting along the posterolateral aspect of the chest wall bilaterally. The masses are of low to intermediate signal intensity with interspersed areas of high signal intensity suggesting fat. In Fig. 10B axial T2-weighted gated acquisition (TR/TE 2,727/102 msec) demonstrates the masses along the posterolateral chest wall deep to the scapula (arrows). The masses do not significantly increase in signal intensity.

Diagnosis: Bilateral elastofibroma.

Discussion: Elastofibroma is one of a number of conditions that can be generally classified as reflecting lesions of benign fibrous proliferation (1,2). The tumors may be unilateral or bilateral, occur more often in women (8:1 female predominance), and are most prevalent in patients with a history of hard manual labor and in weightlifters. It has been suggested that the tumor arises as a reactive fibrous lesion that occurs as a result of repetitive friction between the scapula and chest wall (1). The tumor typically occurs at the inferomedial scapular border, beneath the rhomboid major and latissimus dorsi muscles. It extends into the serratus posterior muscle and even into the ribs and chest wall. It almost always involves the subscapularis muscle, which can be displaced or infiltrated by the tumor. Less common locations reported for elastofibroma include beneath the olecranon, the deltoid axilla, greater trochanter, ischial tuberosity, brachial plexus, and foot (1).

Histologically, the tumors contain dense fibrous tissue and moderate elastinophilic fibers (1,2). On MR, the lesions are typically of intermediate to low signal intensity but may contain areas of high signal intensity consistent with fat. This feature of entrapped fat within a predominately fibrous lesion is reported to be characteristic of elastofibroma (2). This MR feature, coupled with the characteristic location, should allow for a strongly suggestive diagnosis.

REFERENCES

Introduction

1. Hudson TM. *Radiologic-pathologic correlation of musculoskeletal lesions.* Baltimore: Williams & Wilkins, 1987.
2. Madewell JE, Moser RP. Radiologic evaluation of soft tissue tumors. In: Enzinger FM, Weiss SW, eds. *Soft tissue tumors,* 2nd ed. St. Louis: CV Mosby, 1988;43–82.
3. Enzinger FM, Weiss SW. *Soft tissue tumors,* 2nd ed. St. Louis: CV Mosby, 1988.
4. Sundaram M, McLeod RA. MR imaging of tumor and tumorlike lesions of bone and soft tissue. *AJR* 1990;155:817–824.
5. Kransdorf MJ, Jelinek JS, Moser, et al. Soft-tissue masses: diagnosis using MR imaging. *AJR* 1989;153:541–547.
6. Berquist TH, Ehman RL, King BF, Hodgman CG, Ilstrup DM. Value of MR imaging in differentiating benign from malignant soft tissue masses: study of 95 lesions. *AJR* 1990;155:1251–1255.
7. Crim JR, Seeger LL, Yao L, Chandnani C, Eckardt JJ. Diagnosis of soft-tissue masses with MR imaging: can benign masses be differentiated from malignant ones? *Radiology* 1992;185:581–586.
8. Sundaram M, McGuire MH, Herbold DR. Magnetic resonance imaging of soft tissue masses: an evaluation of fifty-three histologically proven tumors. *Magn Reson Imaging* 1988;6:237–248.
9. Myhre-Jensen O. A consecutive 7-year series of 1331 benign soft tissue tumors. *Acta Orthop Scand* 1981;52:287–293.
10. Rydholm A. Management of patients with soft-tissue tumors. Strategy developed at a regional oncology center. *Acta Orthop* 1983; 54(suppl 203):1–77.
11. Peabody TD, Simon MA. Principles of staging of soft-tissue sarcomas. *Clin Orthop* 1993;289:19–31.
12. Beltran J, Simon DC, Katz W, Weis LD. Increased MR signal intensity in skeletal muscle adjacent to malignant tumors: pathologic correlation and clinical relevance. *Radiology* 1987;162:251–255.
13. Hanna SL, Fletcher BD, Parham DM, Bugg MF. Muscle edema in musculoskeletal tumors: MR imaging characteristics and clinical significance. *J Magn Reson Imaging* 1991;1:441–449.

Case 1

1. Enzinger FM, Weiss SW. *Soft tissue tumors,* 2nd ed. St. Louis: CV Mosby, 1988.
2. Kransdorf MJ. Malignant soft-tissue tumors in a large referral population: distribution of diagnoses by age, sex and location. *AJR* 1995; 164:129–134.
3. Chung EB, Enzinger FM. Benign lipoblastomatosis: an analysis of 35 cases. *Cancer* 1973;32:482–492.
4. Reszel PA, Soule EH, Coventry MB. Liposarcoma of the extremities and limb girdles. A study of two hundred twenty-two cases. *J Bone Joint Surg [Am]* 1966;48-A:229–244.
5. Bush CH, Spanier SS, Gillespy T. Imaging of atypical lipomas of the extremities: report of three cases. *Skel Radiol* 1988;17:472–475.
6. Jelinek JS, Kransdorf MJ, Shmookler BM, Aboulafia A, Malawer M. Liposarcoma of the extremities: MR and CT findings of the histologic subtypes. *Radiology* 1993;186:455–459.
7. London J, Kim EE, Wallace S, Shirkhoda A, Coan J, Evans H. MR imaging of liposarcomas: correlation of MR features and histology. *J Comput Assist Tomogr* 1989;15:832–835.
8. Petasnick JP, Turner DA, Charters JR, Gitelis S, Zacharias CE. Soft-tissue masses of the locomotor system: comparison of MR imaging with CT. *Radiology* 1986;160:125–133.
9. Sundaram M, Baran G, Merenda G, McDonald DJ. Myxoid liposarcoma: magnetic resonance imaging appearances with clinical and histological correlation. *Skel Radiol* 1990;19:359–362.
10. Sundaram M, McLeod RA. MR imaging of tumor and tumorlike lesions of bone and soft tissue. *AJR* 1990;155:817–824.
11. Dooms GC, Hricak H, Sollitto RA, Higgins CB. Lipomatous tumors and tumors with fatty component: MR imaging potential and comparison of MR and CT results. *Radiology* 1985;157:479–483.
12. Sundaram M, McGuire MH, Herbold DR. Magnetic resonance imaging of soft tissue masses: an evaluation of fifty-three histologically proven tumors. *Magn Reson Imaging* 1988;6:237–248.
13. Erlemann R, Reiser MF, Peters PE, et al. Musculoskeletal neoplasms: static and dynamic Gd-DTPA-enhanced MR imaging. *Radiology* 1989;17:767–773.

Case 2

1. Enzinger FM, Weiss SW. *Soft tissue tumors,* 2nd ed. St. Louis: CV Mosby, 1988.
2. Disler DG, Alexander AA, Mankin HJ, O'Connell JX, Rosenberg AE, Rosenthal DI. Multicentric fibromatosis with metaphyseal dysplasia. *Radiology* 1993;187:489–492.
3. Mueller J. *Uber den feineren Bau der Krankhaften Geschwukste.* Berlin: Breicht, 1836;107–113.
4. Taylor LJ. Musculoaponeurotic fibromatosis. A report of 28 cases and review of the literature. *Clin Orthop* 1987;224:294–302.
5. Macfarlane J. *Clinical reports of the surgical practice of the Glasgow Royal Infirmary.* Glasgow, Scotland: Robertson, 1832;63–66.
6. Rock MG, Pritchard DJ, Reiman HM, Soule EH, Brewster RC. Extra-abdominal desmoid tumors. *J Bone Joint Surg* 1984;66A: 1369–1374.
7. Sundaram M, Duffrin H, McGuire MH, Vas W. Synchronous multicentric desmoid tumors (aggressive fibromatosis) of the extremities. *Skel Radiol* 1988;17:16–19.
8. Griffiths HJ, Robinson K, Bonfiglio TA. Aggressive fibromatosis. *Skel Radiol* 1983;9:179–182.
9. Enzinger FM, Shiraki M. Musculo-aponeurotic fibromatosis of the shoulder girdle (extra-abdominal desmoid). *Cancer* 1967;20:1131–1140.
10. Shiu MH, Flancbaum L, Hajdu SI, Fortner JG. Malignant soft-tissue tumors of the anterior abdominal wall. *Arch Surg* 1980;115:152–155.
11. Burke AP, Sobin LH, Shekitka KM. Mesenteric fibromatosis. *Arch Pathol* 1990;114:832–835.
12. McAdam WAF, Goligher JC. The occurrence of desmoid tumors in patients with familial polyposis coli. *Br J Surg* 1970;57:618–631.
13. Bessler W, Egloff B, Sulser H. Case report 253. Gardner syndrome with aggressive fibromatosis. *Skel Radiol* 1984;11:56–59.

14. Kawashima A, Goldman SM, Fishman EK, Kuhlman JE, Onitsuka H, Fukuya T, Masuda K. CT of intraabdominal desmoid tumors: is the tumor different in patients with Gardner's disease. *AJR* 1994; 162:339–342.
15. Aisen AM, Martel W, Braunstein EM, McMillin KI, Phillips WA, Kling TF. MRI and CT evaluation of primary bone and soft-tissue tumors. *AJR* 1986;146:749–756.
16. Wetzel LH, Levine E, Murphey MD. A comparison of MR imaging and CT in the evaluation of musculoskeletal masses. *RadioGraphics* 1987;7:851–874.
17. Sundaram M, McGuire MH, Schajowicz F. Soft tissue masses: histologic basis for decreased signal (short T2) on T2-weighted MR images. *AJR* 1987;148:1247–1250.
18. Kransdorf MJ, Jelinek JS, Moser RP, et al. MR appearance of fibromatosis: a report of 14 cases and review of the literature. *Skel Radiol* 1990;19:495–499.
19. Hawnaur JM, Jenkins JPR, Isherwood I. Magnetic resonance imaging of musculoaponeurotic fibromatosis. *Skel Radiol* 1990;19: 509–514.
20. Feld R, Burk L, McCue P, Mitchell DG, Lackman R, Rifkin MD. MRI of aggressive fibromatosis: frequent appearance of high signal intensity on T2-weighted images. *Magn Reson Imaging* 1990;8: 583–588.
21. Quinn SF, Erickson SJ, Dee PM, et al. MR imaging in fibromatosis: results in 26 patients with pathologic correlation. *AJR* 1991;156: 539–542.
22. Hartman TE, Berquist TH, Fetsch JF. MR imaging of extraabdominal desmoids: differentiation from other neoplasms. *AJR* 1992;158: 581–585.
23. Sundaram M, McLeod RA. MR imaging of tumor and tumorlike lesions of bone and soft-tissue. *AJR* 1990;155:817–824.
24. Kransdorf MJ. Magnetic resonance imaging of musculoskeletal tumors. *ACR Categorical Course on Imaging of Cancers* 1992;47–54.

Case 3

1. Robbins SL. *Pathology of disease.* Philadelphia: WB Saunders, 1974:55–105.
2. Beltran J, Noto AM, McGhee RB, Freedy RM, McCalla MS. Infections of the musculoskeletal system: high-field-strength MR imaging. *Radiology* 1987;164:449–454.
3. Beltran J, Simon DC, Katz W, Weis LD. Increased MR signal intensity in skeletal muscle adjacent to malignant tumors: pathologic correlation and clinical relevance. *Radiology* 1987;162:251–255.
4. Hanna SL, Fletcher BD, Parham DM, Bugg MF. Muscle edema in musculoskeletal tumors: MR imaging characteristics and clinical significance. *J Magn Reson Imaging* 1991;1:441–449.
5. Rubin JI, Gomori JM, Grossman RI, Gefter WB, Kressel HY. High-field MR imaging of extracranial hematomas. *AJR* 1987;148:813–817.
6. Sundaram M, McLeod RA. MR imaging of tumor and tumorlike lesions of bone and soft tissue. *AJR* 1990;155:817–824.

Case 4

1. Kransdorf MJ. Malignant soft-tissue tumors in a large referral population: distribution of diagnoses by age, sex and location. *AJR* 1995; 164:129–134.
2. Enzinger FM, Weiss SW. *Soft tissue tumors,* 2nd ed. St. Louis: CV Mosby, 1988;659.
3. Cadman NL, Soule EH, Kelley PJ. Synovial sarcoma: an analysis of 134 tumors. *Cancer* 1965;18:613–627.
4. Wright PH, Sim FH, Soule EH, Taylor WF. Synovial sarcoma. *J Bone Joint Surg* 1982;64A:112–122.
5. Roth JA, Enzinger FM, Tannenbaum MT. Synovial sarcoma of the neck: a follow up study of 24 cases. *Cancer* 1975;35:1243–1253.
6. Tahir T, Sanjiv G. Synovial sarcoma of the right ventricle. *Am Heart J* 1991;121:933–938.
7. Bogumill GP, Bruna PD, Barrick EF. Malignant lesions masquerading as a popliteal cysts. *J Bone Joint Surg* 1981;63-A:474–477.
8. Ryan JR, Baker LH, Benjamin RS. The natural history of metastatic synovial sarcoma. The experience of the Southwest Oncology Group. *Clin Orthop* 1982;164:257–260.
9. Sutro J. Synovial sarcoma of the soft parts in the first toe: recurrence after thirty-five year interval. *Bull Hosp Joint Dis* 1976;37:105–109.
10. Mazeron JJ, Suit HD. Lymph nodes as sites of metastases from sarcomas of soft tissue. *Cancer* 1987;60:1800.
11. Pack GT, Ariel IM. Treatment of cancer and allied diseases. In: Pack GT, Ariel IM, eds. *Tumors of the soft somatic tissues and bone,* vol 8. New York: Harper and Row, 1965;8–39.
12. Varela-Duram J, Enzinger FM. Calcifying synovial sarcoma. *Cancer* 1982;50:345–352.
13. Mahajan H, Lorigan JG, Shirkhoda A. Synovial sarcoma: MR imaging. *Magn Reson Imaging* 1989;7:211–216.
14. Evans HL. Synovial sarcoma. A study of 23 biphasic and 17 probable monophasic examples. *Pathol Ann* 1980;15:309–331.
15. Hajdu SI, Shiu MH, Fortner JG. Tenosynovial sarcoma: a clinicopathological study of 136 cases. *Cancer* 1977;39:1201–1217.
16. Horowitz AL, Resnick D, Watson RC. The roentgen features of synovial sarcoma. *Clin Radiol* 1973;24:481–484.
17. Morton MJ, Berquist TH, McLeod RA, Unni KK, Sim FH. MR imaging of synovial sarcoma. *AJR* 1990;156:337–340.
18. Tsai JC, Dalinka MK, Fallon MD, Zlatkin MB, Kressel HY. Fluid-fluid level: a nonspecific finding in tumors of bone and soft tissue. *Radiology* 1990;175:779–782.
19. DeCoster TA, Kamps BS, Craven JP. Magnetic resonance imaging of a foot synovial sarcoma. *Orthopedics* 1991;14:169–171.
20. Jones BC, Sundaram M, Kransdorf MJ. Synovial sarcoma: MR imaging findings in 34 patients. *AJR* 1993;161:827–830.

Case 5

1. Enzinger FM, Weiss SW. Malignant fibrohistiocytic tumors. In: Enzinger FM, Weiss SW, eds. *Soft tissue tumors,* 3rd ed. St. Louis: CV Mosby, 1988;325–349.
2. Enzinger FM. Malignant fibrous histiocytoma 20 years after Stout. *Am J Surg Pathol* 1986;10:43–53.
3. O'Brien JE, Stout AP. Malignant fibrous xanthomas. *Cancer* 1964; 11:1445–1455.
4. Kearney MM, Soule EH, Ivins JC. Malignant fibrous histiocytoma: a retrospective study of 167 cases. *Cancer* 1980;45:167–178.
5. Barnes L, Kanbour A. Malignant fibrous histiocytoma of the head and neck. *Arch Otolaryngol Head Neck Surg* 1988;114:1149–1156.
6. Goldman SM, Hartman DS, Weiss SW. The varied radiographic manifestations of retroperitoneal malignant fibrous histiocytoma revealed through 27 cases. *J Urol* 1986;135:33–38.
7. Bruneton JN, Drouillard J, Rogopoulos A, et al. Extraretroperitoneal abdominal malignant fibrous histiocytoma. *Gastrointest Radiol* 1988;13:299–305.
8. Vera-Donoso CD, Llopis B, Froufe M, Boronat F, Oliver F, Jimenez-Cruz JF. Retroperitoneal malignant fibrous histiocytoma. *Eur Urol* 1988;15:302–305.
9. Panicek DM, Casper ES, Brennan MF, Hajdu SI, Heelan RT. Hemorrhage simulating tumor growth in malignant fibrous histiocytoma at MR imaging. *Radiology* 1991;181:398–400.
10. Enzinger FM. Angiomatoid malignant fibrous histiocytoma: a distinct fibrohistiocytic tumor of children and young adults simulating a vascular neoplasm. *Cancer* 1979;44:2147–2157.
11. Dorfman HD, Bhagavan BS. Malignant fibrous histiocytoma of soft tissue with metaplastic bone and cartilage formation: a new radiologic sign. *Skel Radiol* 1982;8:145–150.
12. Bhagavan BS, Dorfman HD. The significance of bone and cartilage formation in malignant fibrous histiocytoma of soft tissue. *Cancer* 1982;49:480–488.
13. Murphey MD, Gross TM, Rosenthal HG. Musculoskeletal malignant fibrous histiocytoma: radiologic-pathologic correlation. *RadioGraphics* 1994;14:807–826.
14. Feldman F, Norman D. Intra-and extraosseous malignant histiocytoma (malignant fibrous xanthoma). *Radiology* 1972;104:497–508.
15. Fisher HJ, Lois JF, Gomes AS, Mirra JM, Deutsch LS. Radiology and pathology of malignant fibrous histiocytomas of the soft tissues: a report of ten cases. *Skel Radiol* 1985;13:202–206.
16. Mahajan H, Kim EE, Wallace S, Abello R, Benjamin R, Evans

HL. Magnetic resonance imaging of malignant fibrous histiocytoma. *Magn Reson Imaging* 1989;7:283–288.
17. Aisen AM, Martel W, Braunstein EM, McMillin KI, Phillips WA, King TF. MRI and CT evaluation of primary bone and soft-tissue tumors. *AJR* 1986;146:749–756.
18. Peterson KK, Renfrew DL, Feddersen RM, Buckwalter JA, El-Khoury GY. Magnetic resonance imaging of myxoid containing tumors. *Skel Radiol* 1991;20:245–250.
19. Kransdorf MJ, Moser RP Jr, Meis JM, Meyer CA. Fat-containing soft-tissue masses of the extremities. *RadioGraphics* 1991;11:81–106.
20. Jelinek JS, Kransdorf MJ, Schmookler BM, Aboulafia AJ, Malawar MM. Liposarcoma of the extremities: MR and CT findings in the histologic subtypes. *Radiology* 1993;186:455–459.

Case 6

1. Flandry F, Hughston JC. Current concepts review pigmented villonodular synovitis. *J Bone Joint Surg* 1987;69-A:942–949.
2. Spritzer CE, Dalinka MK, Kressel HY. Magnetic resonance imaging of pigmented villonodular synovitis: a report of two cases. *Skel Radiol* 1987;16:316–319.
3. Jelinek JS, Kransdorf MJ, Utz JA, et al. Imaging of pigmented villonodular synovitis with emphasis on MR imaging. *AJR* 1989; 152:337–342.
4. Kottal RA, Vogler JB, Matamoros A, Alexander AH, Cookson JL. Pigmented villonodular synovitis: a report of MR imaging in two cases. *Radiology* 1987;163:551–553.
5. Butt YP, Hardy G, Ostlere SJ. Pigmented villonodular synovitis of the knee: computed tomographic appearances. *Skel Radiol* 1990; 19:191–196.
6. Jelinek JS, Kransdorf MJ, Utz JA, et al. Imaging of pigmented villonodular synovitis with emphasis on MR imaging. *AJR* 1989; 152:337–342.
7. Balsara AN, Stainken BF, Martinez AJ. MR image of localized giant cell tumor of the tendon sheath involving the knee. *J Comput Assist Tomogr* 1989;13:159–162.
8. Goldman AB, DiCarlo EF. Pigmented villonodular synovitis: diagnosis and differential diagnosis. *Radiol Clin North Am* 1988;26: 1327–1347.
9. Karasick D, Karasick S. Giant cell tumor of tendon sheath: spectrum of radiologic findings. *Skel Radiol* 1992;21:219–224.
10. Jelinek JS, Kransdorf MJ, Shmookler BM, Aboulafia AA, Malawer MM. Giant cell tumor of the tendon sheath: MR findings in nine cases. *AJR* 1994;162:919–922.
11. Gomori JM, Grossman RJ, Goldberg HI, et al. Intracranial hematomas: imaging by high field magnetic resonance. *Radiology* 1985; 157:87–93.
12. Camacho CR, et al. Radiological findings of amyloid arthropathy in long-term hemodialysis. *Eur Radiol* 1992;305:2–4.
13. Cobby MJ, Adler RS, Swartz R, Martel N. Dialysis related amyloid arthropathy: MR findings in four patients. *AJR* 1991;157:1023–1027.
14. Bardin T, Kuntz D, Zingroff J, Voison MC, Zelmar A, Lansama J. Synovial amyloidosis in patients undergoing long-term hemodialysis. *Arthritis Rheum* 1985;28:1052–1058.
15. Hermann G, Gilbert MS, Abdelwahab IF. Hemophilia: evaluation of musculoskeletal involvement with CT, sonography, and MR imaging. *AJR* 1992;158:119–123.
16. Kulkarni MV, Drolshagen LF, Kaye JJ, et al. MR imaging of hemophiliac arthropathy. *J Comput Assist Tomogr* 1986;10:445–449.
17. Yulish BS, Lieberman JM, Strandjord SE, Bryan PJ, Mulopulos GP, Modic MT. Hemophiliac arthropathy: assessment with MR imaging. *Radiology* 1987;164:759–762.

Case 7

1. Enzinger FM, Weiss SW. Benign tumors and tumor like lesions of blood vessels. In: *Soft tissue tumors,* 3rd ed. St. Louis: CV Mosby, 1995;579–626.
2. Resnick D, Kyriakos M, Greenway GD. Tumors and tumor-like lesions of bone: imaging and pathology of specific lesions. In: *Diagnosis of bone and joint disorders,* 3rd ed. Philadelphia: WB Saunders, 1995;3821–3840.
3. Resnick D, Oliphant M. Hemophilia-like arthropathy of the knee associated with cutaneous and synovial hemangiomas. *Radiology* 1975;114:323–326.
4. Devaney K, Vinh TN, Sweet DE. Synovial hemangioma: report of 20 cases with differential diagnostic considerations. *Hum Pathol* 1993;24:737–745.
5. Allen PW, Enzinger FM. Hemangioma of skeletal muscle: an analysis of 89 cases. *Cancer* 1972;29:8–22.
6. Levine E, Wetzel LH, Neff JR. MR imaging and CT of extrahepatic cavernous hemangiomas. *AJR* 1986;147:1299–1304.
7. Shallow TA, Eger SA, Wagner FB Jr. Primary hemangiomatous tumors of skeletal muscle. *Ann Surg* 1944;119:700–704.
8. Madewell JE, Sweet DE. Tumors and tumor-like lesions in or about joints. In: *Diagnosis of bone and joint disorders,* 3rd ed. Philadelphia: WB Saunders, 1995;3950–3956.
9. Rauch RF, Silverman PM, Kurobkin, et al. Computed tomography of benign angiomatous lesions of the extremities. *J Comput Assist Tomogr* 1984;8:1143–1146.
10. Hill JH, Mafee MF, Chow JM, Applebaum EL. Dynamic computerized tomography in the assessment of hemangioma. *Am J Otolaryngol* 1985;6:23–28.
11. Greenspan A, McGahan JP, Vogelsang P, Szabo RM. Imaging strategies in the evaluation of soft tissue hemangiomas of the extremities: correlation of the findings of plain radiography, angiography, CT, MRI, and ultrasonography in 12 histologically proven cases. *Skel Radiol* 1992;21:11–18.
12. Hawnaur JM, Whitehouse RW, Jenkins JPR, Isherwood I. Musculoskeletal hemangiomas: comparison of MRI with CT. *Skel Radiol* 1990;19:251–258.
13. Derchi LE, Balconi G, De Flaviis L, Oliva A, Rosso F. Sonographic appearances of hemangiomas of skeletal muscle. *J Ultrasound Med* 1989;8:263–267.
14. Buetow PC, Kransdorf MJ, Moser RP, Jelinek JS, Berrey BH. Radiologic appearance of intramuscular hemangioma with emphasis on MR imaging. *AJR* 1990;154:563–567.
15. Yuh WTC, Kathol MH, Sein MA, Ehara S, Chiu L. Hemangiomas of skeletal muscle: MR findings in five patients. *AJR* 1987;149: 765–768.
16. Kaplan PA, Williams SM. Mucocutaneous and peripheral soft tissue hemangiomas: MR imaging. *Radiology* 1987;163:163–166.
17. Cohen EK, Kressel HY, Perosio T, et al. MR imaging of soft-tissue hemangiomas: correlation with pathologic findings. *AJR* 1988;150: 1079–1081.
18. Nelson MC, Stull MA, Teitelbaum GP, et al. Magnetic resonance imaging of peripheral soft-tissue hemangiomas. *Skel Radiol* 1990; 19:447–482.
19. Bradley WG, Waluch V, Lai KS, Fernandez EJ, Spalter C. The appearance of rapidly flowing blood on magnetic resonance images. *AJR* 1984;143:1167–1174.
20. Tsai JC, Dalinka MK, Fallon MD, Zlatkin MB, Kressel HY. Fluid-fluid level: a nonspecific finding in tumors of bone and soft tissue. *Radiology* 1990;175:779–782.

Case 8

1. Jaffe HL. Snovial chondromatosis and other articular tumors. In: *Tumors and tumorous conditions of the bones and joints.* Philadelphia: Lea & Febiger, 1958;566–567.
2. Milgram JW. Synovial osteochondromatosis. A histopathological study of thirty cases. *J Bone Joint Surg* 1977;59:792–801.
3. Karlin CA, De Smet AA, Neff J, Lin F, Horton W, Wertzberger JJ. The variable manifestations of extraarticular synovial chondromatosis. *AJR* 1981;137:731–735.
4. Burnstein MI, Fisher DR, Yandow DR, Hafez GR, De Smet AA. Case report 502. *Skel Radiol* 1988;17:458–461.
5. Herzog S, Mafee M. Synovial chondromatosis of the TMJ: MR and CT findings. *AJNR* 1990;11:742–745.
6. Liu S-K, Moroff S. Case report 733. *Skel Radiol* 1993;22:50–54.
7. Baird RA, Schobert WE, Pais MJ, et al. Radiographic identification of loose bodies in the traumatized hip joint. *Radiology* 1982;145: 661–665.
8. Kransdorf MJ, Meis JM. From the archives of the AFIP: extraskele-

tal osseous and cartilaginous tumors of the extremities. *Radio-Graphics* 1993;13:855–886.
9. Szypryt P, Twining P, Preston BJ, Howell CJ. Synovial chondromatosis of the hip joint presenting as a pathological fracture. *Br J Radiol* 1986;59:399–401.
10. Norman A, Steiner GC. Bone erosion in synovial chondromatosis. *Radiology* 1986;161:749–752.
11. Villacin AB, Brigham LN, Bullough PG. Primary and secondary synovial chondrometaplasia. Histopathologic and clinicoradiologic differences. *Hum Pathol* 1979;10:439–451.

Case 9

1. Vanel D, Shapeero LG, De Baere T, et al. MR imaging in the follow-up of malignant and aggressive soft-tissue tumors: results of 511 examinations. *Radiology* 1994;190:263–268.
2. Choi H, Varma DGK, Fornage BD, Kim EE, Johnston DA. Soft-tissue sarcoma: MR imaging vs sonography for the detection of local recurrence after surgery. *AJR* 1991;157:353–358.
3. Takebayashi S, Sugiyama M, Nagase M, Matsubara S. Severe adverse reaction to IV gadopentetate dimeglumine. *AJR* 1993;160:659.
4. Tardy B, Guy C, Barral G, Page Y, Ollagnier M, Bertrand C. Anaphylactic shock induced by intravenous gadopentetate dimeglumine. *Lancet* 1992;339:494.
5. Tisher S, Hoffman JC. Anaphylactoid reaction to IV gadopentetate dimeglumine. *AJNR* 1990;174:17–23.
6. Omohundro JE, Elderbrook MK, Ringer TV. Laryngospasm after administration of gadopentetate dimeglumine. *J Magn Reson Imaging* 1992;2:729–730.
7. Shellock FG, Hahn HP, Mink JH, Itskovich E. Adverse reaction to intravenous gadoteridol. *Radiology* 1993;189:151–152.
8. Jordan RM, Mintz RD. Fatal reaction to gadopentetate dimeglumine. *AJR* 1995;164:743–744.

Case 10

1. Pechman D, Kenan S, Abdelwahab IF, et al. Case report 839. *Skel Radiol* 1994;23:459–461.
2. Kransdorf MJ, Meis JM, Montgomery E. Elastofibroma: MR and CT appearance with radiologic-pathologic correlation. *AJR* 1992;159:575.

CHAPTER 11

Bone Tumors

Johan L. Bloem, Henk-Jan van der Woude, and Pancras C. W. Hogendoorn

MR imaging has gained a prominent position in diagnosis and management of patients with bone tumors. It directly exhibits the lesion in relationship to surrounding normal structures with exquisite anatomic detail. This has had an unparalleled impact on detection, diagnosis, treatment planning and treatment monitoring. MR imaging has, for instance, facilitated the development of new therapy strategies. Treatment of bone sarcoma has been improved considerably by increasingly effective chemotherapy and by modern operative techniques, often allowing reconstructive and limb salvage procedures instead of amputation or disarticulation (1–5). Only after meticulous preoperative staging is it possible to execute limb-salvage surgery resulting in local control of the tumor and good residual function.

While acknowledging the prominence of MR imaging, one should not forget the importance of other imaging techniques, in particular conventional radiography. The relationship of MR imaging with other imaging techniques can be summarized in the following five critical moments that occur during diagnosis and treatment of patients with primary or metastatic bone tumors.

1. Detection and characterization of primary bone tumor: High-quality radiographs are as a rule sufficient for detection of tumor. Computed tomography (CT) (vertebral column, pelvis), ultrasound (soft tissue), MR imaging (bone marrow and soft tissue), or bone scintigraphy are sometimes needed to detect tumors when radiographs are negative or equivocal. A radiologic diagnosis is initially made on plain radiographs. Although MR imaging, as will be discussed, can be used to further narrow the differential, one should never attempt to make a specific diagnosis, or even execute a MR study without conventional radiographs. A trocar or open biopsy is (with some exceptions such as fibrous cortical defect or nonossifying fibroma), necessary to make a histologic diagnosis. Histologic and radiologic findings (radiographs and MR images) combined are used to make a final diagnosis. Imaging studies are performed prior to biopsy and can thus guide the biopsy needle. Furthermore, biopsy induces reactive changes (edema, hemorrhage) that interfere with accuracy of staging.

2. Local staging of bone sarcoma: Reconstructive, limb salvage procedures can only be performed when the surgeon is informed in detail about the intra- and extraosseous extension of tumor growth (6,7). MR imaging can be used to stage benign tumors. This is, however, of limited clinical importance since these well-contained benign tumors are usually accurately staged by radiographs and CT. The major indication for MR imaging is staging bone sarcoma. MR imaging is superior in displaying bone marrow involvement (7). MR has an almost perfect correlation (r = .99) with pathologic/morphologic examination for bone marrow involvement. CT has a less substantial correlation (r = .93), while technetium (Tc)-99m-MDP scintigraphy has a weak correlation (r = .69). Two potential problems that can be dealt with effectively are bone marrow edema beyond the true tumor margin (dual echo techniques, morphology, or dynamic gadolinium-enhanced MRI) and skip metastases (large field of view).

Invasion of cortex by tumor is best shown on T2-weighted images as a disruption of the cortical line and replacement of cortex by high signal intensity of tumor. The sensitivity and specificity of MRI (respectively 92% and 99%) are not significantly higher than the sensitivity and specificity (respectively 91% and 98%) of CT (7).

MRI is significantly superior (sensitivity 97%, specificity 99%) to CT in identifying muscle compartments containing tumor (7).

CT (sensitivity 36%, specificity 94%) and MRI (sensitivity 92%, specificity 98%) provide more information than angiography (sensitivity 75%, specificity 71%) because large vessels are, especially with MRI, well visualized in relation to the tumor (7).

CT (sensitivity 94%, specificity 90%) and MRI (sensitivity 95%, specificity 98%) are both able to demonstrate joint involvement with high accuracy (7). Joint involvement is sometimes more accurately demonstrated on MR images than on CT, because the articular surfaces may be parallel to the transverse CT plane. Cartilage is an effective barrier that is not easily permeated by tumor.

However, osteosarcoma and giant cell tumor frequently cross the cartilage. When assessing possible joint involvement, the number of false-positive readings is higher than the number of false-negative readings (*unpublished data*). When in doubt, the joint usually is not affected. Joint effusion with or without hemorrhage, often but not always a secondary sign of a contaminated joint, is easily identified on CT and MRI.

3. Monitoring of preoperative (neo-adjuvant) chemotherapy: The holy grail in monitoring chemotherapy is the ability to select patients for specific forms of chemotherapy. The potential benefits are reduction of therapy-related morbidity and mortality, optimizing the moment of surgical intervention, assisting the pathologist in localizing possible viable tumor in the resected specimen for histologic grading of response to chemotherapy, and comparing and optimizing various schedules of neoadjuvant (presurgical) and even adjuvant (postsurgical) chemotherapy. The ultimate goal, obviously, is to contribute to increasing disease-free survival.

Although some progress has been made (8), it is currently not proven that we can accurately predict prior to start of chemotherapy, or shortly after the start of chemotherapy, if an individual patient will benefit from a certain schedule of chemotherapy. The current status of MR imaging in this respect will be illustrated in several cases.

4. Detection of osseous metastases: Currently, Tc-99m-MDP is the method of choice for screening for osseous metastases, because of its capability to image the entire skeleton. However, MR imaging is superior to planar bone scintigraphy in the detection of metastases in the vertebral column (9). If Tc-99m-MDP scintigraphy is negative in a patient with a high clinical suspicion, we like to perform MR imaging. Also in major treatment decisions, MR imaging is used instead of, or in combination with, bone scintigraphy.

5. Detection of recurrent disease: Clinical examination is not sufficient to detect early recurrent disease, and reliable tumor markers are not available for bone sarcoma. Imaging studies will be needed to address the problem of tumor recurrence. The application of imaging studies depends on the histologic grade of the primary tumor, its location, type of margins achieved at surgery and therapeutic options when recurrence is detected.

Without attempting to be complete, cases illustrating possibilities, limitations, and pitfalls of MR imaging in patients with primary and secondary bone tumors will be presented. Emphasis is on MR-related issues, and not on disease specific information, as this can be found elsewhere (10–12).

CASE 1

History: This 10-year-old boy presented with pain in the left pelvis. The radiologic diagnosis of Ewing's sarcoma (Fig. 1A) was histologically confirmed. After local staging with MR imaging the tumor was treated with neo-adjuvant chemotherapy and resection. Several months later, the patient complained of pain in the right hip region. Can you diagnose or exclude osseous metastatic disease in the structures visualized on this T1-weighted (TR/TE 600/20 msec) coronal SE image (Fig. 1B)?

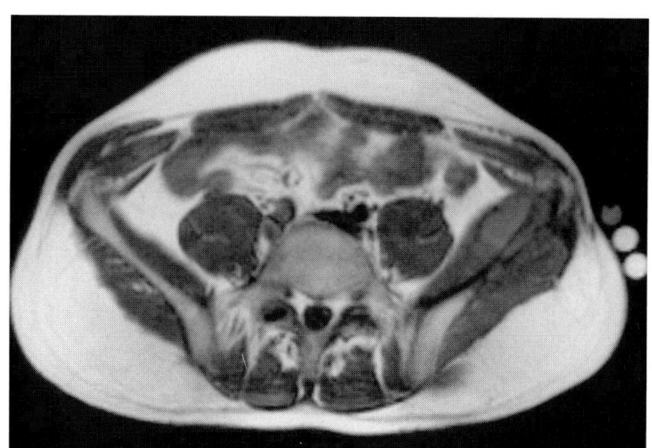

A

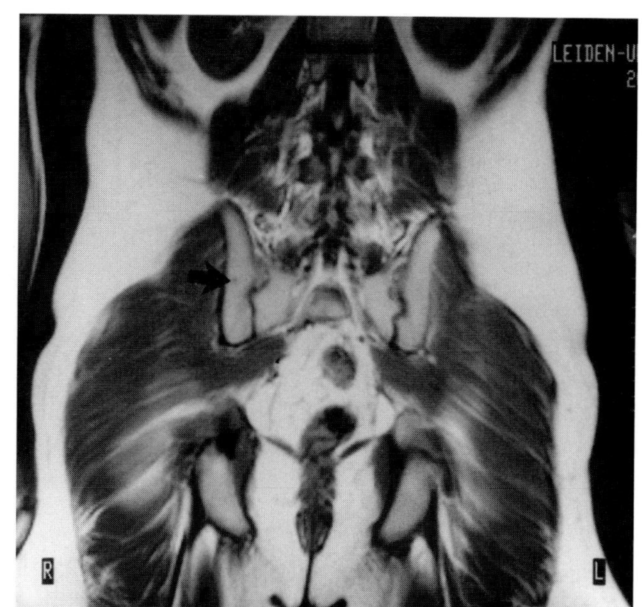

B

FIG. 1.

Findings: In Fig. 1A the axial T1-weighted (TR/TE 600/200 msec) sequence shows intermediate signal intensity in the left iliac bone and surrounding tissues consistent with Ewing's sarcoma. In Fig. 1B the coronal T1-weighted sequence made 8 months after Fig. 1A, shows normal high signal intensity in left and right tuber ischia. A small low signal intensity area (arrow) is seen in the right iliac bone.

Diagnosis: Metastases of Ewing's sarcoma in right tuber ischia, and right iliac bone. (Original tumor in left iliac bone.)

Discussion: T1-weighted SE sequences have been proven to be very sensitive in visualizing focal tumor, especially in yellow bone marrow in the adult population (1,2). This case is one of the few examples demonstrating that the T1-weighted SE sequences is not 100% sensitive. The T1-weighted and T2-weighted (not shown) sequences only showed the original tumor in the left iliac bone and a small lesion in the right iliac bone suspected to be a metastatic deposit. A TC-99m bone scintigram (Fig. 1C) was performed on the same day as the MR examination (Fig. 1B). Increased tracer uptake in the right ischium is consistent with metastatic disease. The tumor in the left ilium also shows increased uptake. Three months later the MR examination was repeated (Fig. 1D). On this image the histologically proven metastatic deposits in the right tuber ischia and iliac crest are easily appreciated. Probably the amount of tumor in the tuber ischia relative to the amount of normal bone marrow was too small at the time of the first MR examination. A negative MR scan and a positive bone scintigram is, however, an exception, since the sensitivity of MR is higher than that of bone scintigram (1–5). This case demonstrates that additional pulse sequences, discussed elsewhere in this chapter, are often needed to increase sensitivity.

Additional pulse sequences are also useful to define tumor in red bone marrow in children or adolescents, since in this age group normal red bone marrow has a fairly low signal intensity (6–10). In a 24-year-old man, the Ewing's sarcoma in the right liac bone is easily appreciated on this T1-weighted (600/20) sequence (Fig. 1E) because of its volume and signal void. The histologically proven diagnosis of diffuse marrow disease, however, is challenging on this sequence, since the signal intensity of normal red marrow and diffuse Ewing's is very similar. Ewing's sarcoma was present in all visualized marrow components including the left iliac wing, both proximal femurs, and the lumbar spine. Despite the poor contrast between red marrow and tumor, the MR images show diffuse metastases better than Tc-bone scintigram (Fig. 1F). Radiographs (not shown) only showed the primary Ewing's sarcoma in the right iliac bone.

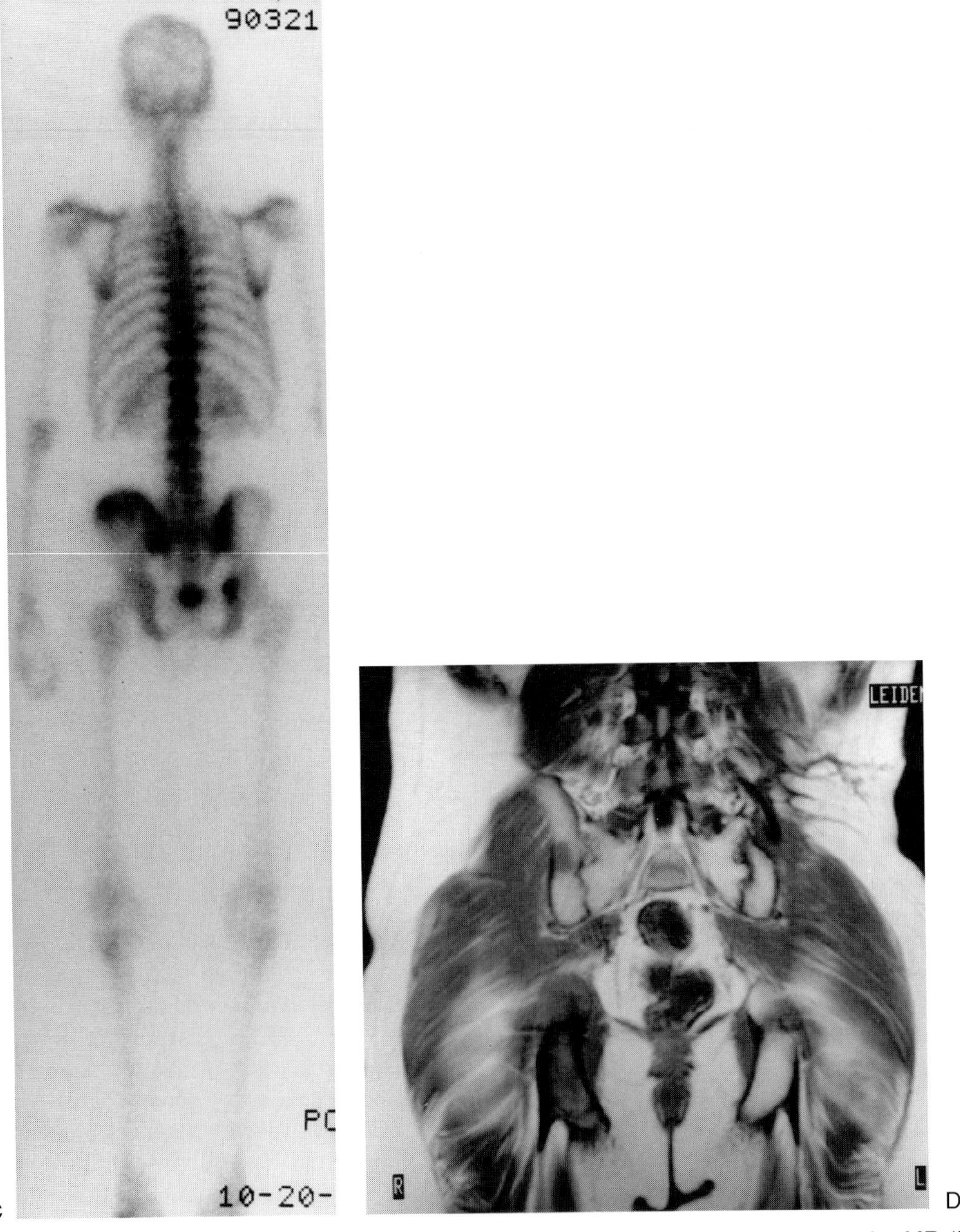

FIG. 1. C: On this posterior view of the Tc-bone scintigram, made at the same day as the MR (fig. 1B) increased tracer uptake consistent with tumor is seen not only in the left iliac wing, but also in the right tuber ischia. **D:** The low signal intensity in the tuber ischia representing metastatic disease is 3 months later also visualized on this coronal T1-weighted (TR/TE 600/20 msec) sequence. The metastasis in the right iliac bone has increased in size.

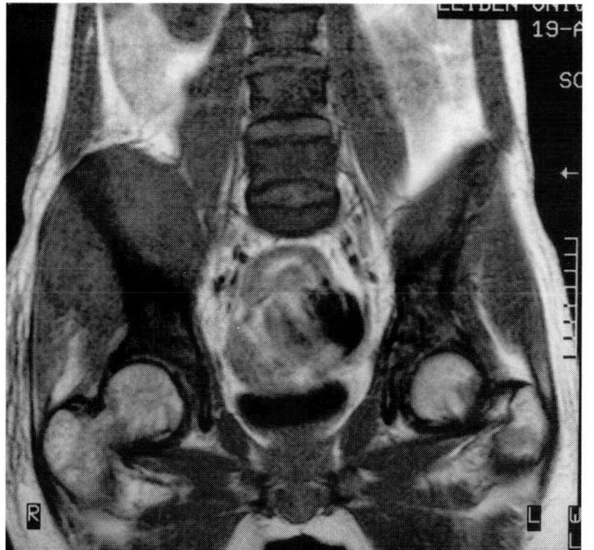

E

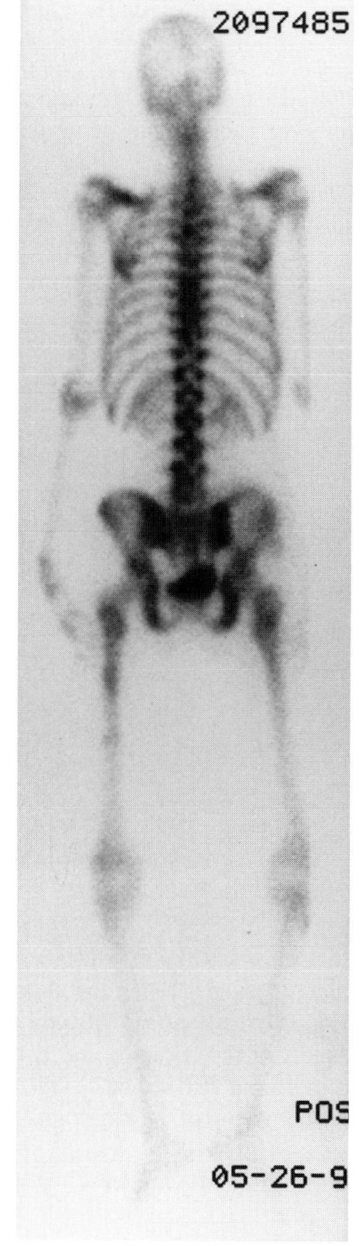

F

FIG. 1. E: A low signal intensity mass, representing Ewing's sarcoma is easily appreciated in the right iliac wing. The low signal intensity in the spine, left iliac wing and proximal femurs is secondary to diffuse marrow involvement, and to presence of normal red bone marrow. **F:** Posterior view of Tc bone scintigram shows multiple areas of abnormal tracer uptake, but the abnormalities do not seem to be so extensive as shown on MR.

CASE 2

History: This 34-year-old patient with extranodal osseous non-Hodgkin's lymphoma in the pelvis complained of pain in the left shoulder. The radiograph (not shown) was normal. MR imaging was performed to rule out or diagnose non-Hodgkin's lymphoma in the shoulder region. What is your diagnosis on this axial T2-weighted (5000/150) TSE sequence (Fig. 2A) obtained on a 0.5 T system? If any, what additional pulse sequences would you obtain?

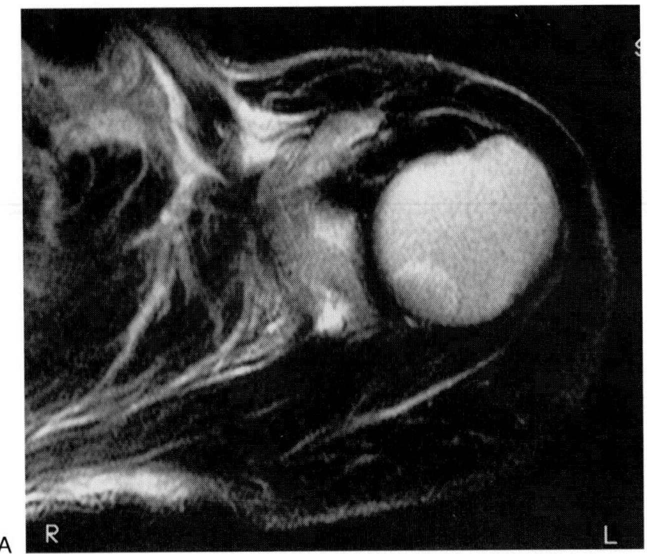

FIG. 2.

Findings: In Fig. 2A the contrast between the tumor in the posterior aspect of the humerus and normal bone marrow is poor on this axial TSE sequence (TR/TE 5,000/150 msec). Only visualization of the tumor margin allows a correct diagnosis.

Diagnosis: Non-Hodgkin's lymphoma within the humeral head.

Discussion: The lesion is not well visualized on the TSE sequence because of insufficient contrast between lesion and normal bone marrow. This is similar to the problem at CT (Fig. 2B). On CT the lesion is also poorly visualized. The tumor scintigram (^{67}Ga-citrate) was negative for the left humerus (Fig. 2C).

There are various methods to improve MR contrast. Other pulse sequences such as T1-weighted SE (Fig. 2D) or STIR can be obtained (1–3). One can also address the problem of the TSE sequence. In the case presented, both the tumor and normal yellow bone marrow have a high signal intensity. The lesion has a high signal intensity because of a prolonged T2 relaxation time. Yellow bone marrow has a high signal intensity because fat has a high signal intensity on TSE secondary to decoupling of J-modulation effects (4,5). A solution would thus be to reduce the signal intensity of normal bone marrow by a STIR-TSE sequence or by fat selective presaturation (Fig. 2E) (1–3,6,7). On the TSE sequence with fat selective presaturation the low signal intensity of normal marrow contrasts maximally with the high signal intensity of tumor. Note that normal red marrow of the scapula has a somewhat higher signal intensity than that of the fully suppressed fat of yellow bone marrow in the humerus.

The use of TSE sequences in tumor imaging is controversial. The disadvantage of TSE is the high signal intensity of fat. As illustrated in this case this disadvantage is easily solved. In TSE sequences, fat can be suppressed either by STIR or by fat selective presaturation (Fig. 2F,G). Furthermore, the high signal intensity of fat is rarely a problem if TSE sequences are used in combination with T1-weighted SE sequences. The advantage of TSE relative to conventional SE is demonstrated in a patient with malignant fibrous histiocytoma (MFH) in the spine. The TSE sequence (Fig. 2H) displays sufficient contrast compared to the conventional SE (TR/TE 2,000/100 msec) sequence (Fig. 2I). Advantages of the TSE sequence are substantially reduced acquisition time, fewer artifacts, and good signal-to-noise ratio (SNR) and contrast-to-noise ratio (CNR). Because of saturation effects tumor may have a relatively intermediate signal intensity when turbo factor is high and TE is short. Therefore, we usually use a turbo factor of 8 with a TE of at least 80 ms. If fat is suppressed, TE can be shortened. When TE is chosen too short, for instance 20 ms or less, blurring degrades the images (4,5).

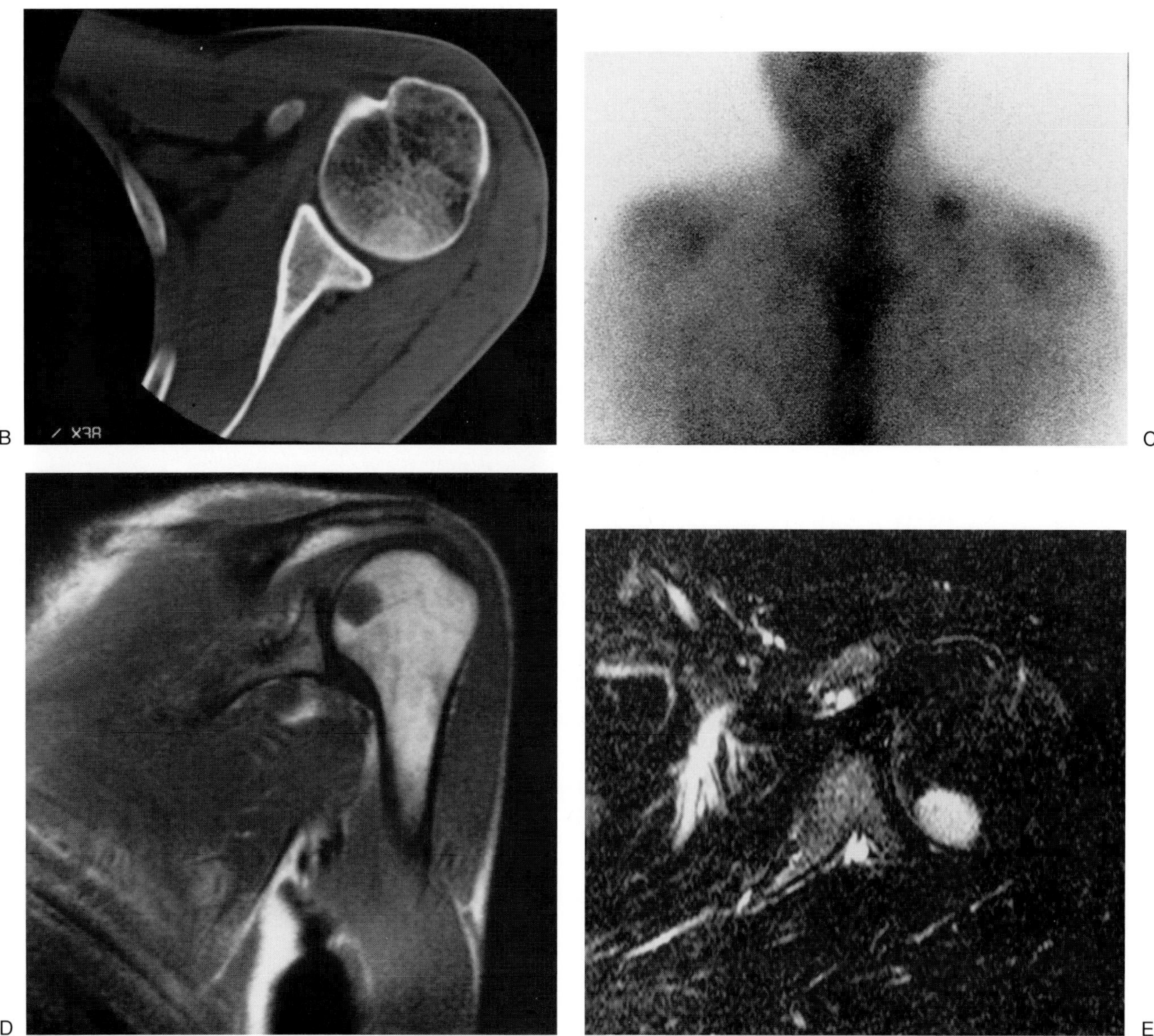

FIG. 2. B: The tumor is seen because of a minor increase in density posteriorly in the humerus. **C:** Tumor scintigram (180 MBq of ^{67}Ga-citrate) only shows increased tracer uptake in a supraclavicular lymph node; the humerus is normal. **D:** Contrast between the low signal intensity of tumor and high signal intensity of normal yellow bone marrow is excellent on this coronal T1-weighted (TR/TE 600/20 msec) sequence. **E:** On this fat selective presaturation TSE image (TR/TE 5000/150 msec) obtained on a 0.5 T system, contrast between the high signal intensity of tumor, and the low signal intensity of marrow is excellent.

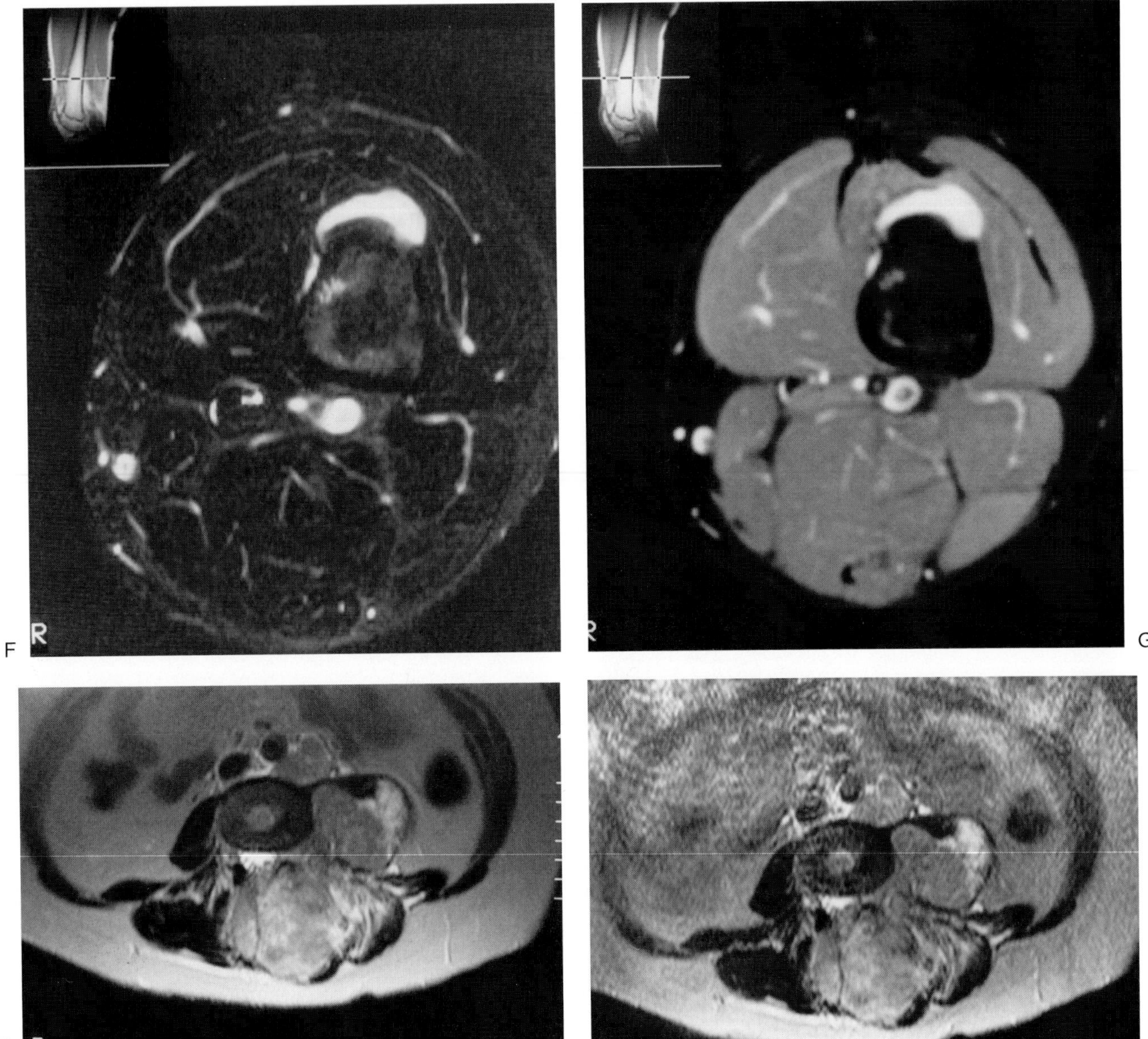

FIG. 2. F,G: Axial TSE sequence of a mature osteochondroma with bursa made on a 0.5 T system with fat selective presaturation **(F),** and STIR **(G). H:** Axial TSE sequence (TR/TE 5,000/150 msec) obtained on a 0.5 T system clearly visualizes tumor (MFH) extent. **I:** Axial T2-weighted (TR/TE 2,000/120 msec) SE sequence obtained several minutes after Fig. 2H shows similar contrast as the TSE sequence, but acquisition time was three times as long and artifacts are much more prominent.

CASE 3

History: This 39-year-old woman was sent to MR for evaluation of metastatic disease secondary to breast carcinoma. Can you identify unaffected vertebral bodies on this TSE sequence of the lumbar segment obtained with fat selective presaturation at 0.5 T (Fig. 3A)?

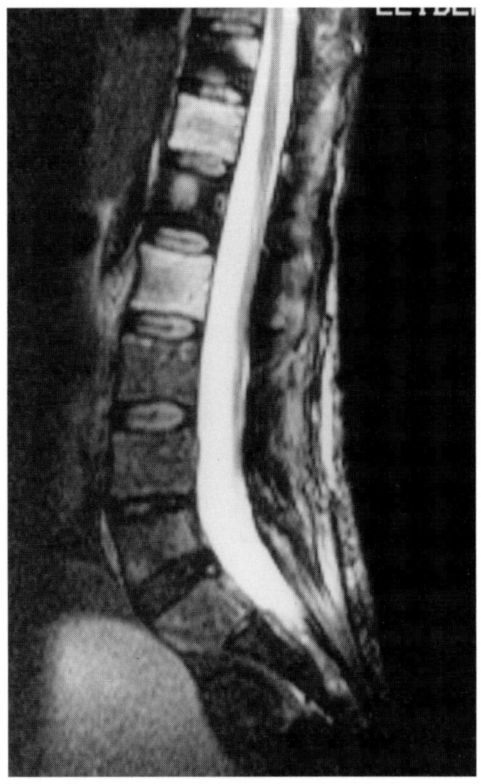

A

FIG. 3.

Findings: In Fig. 3A, sagittal TSE (TR/TE 4,400/150 msec) with fat selective presaturation, increased signal intensity consistent with metastases are seen in all vertebral bodies visualized with the exception of L3, L4, and S2–3.

Diagnosis: Metastases of breast carcinoma in all levels from D11 to S2.

Discussion: MR imaging is very effective in detecting metastatic disease (1–4). In the case demonstrated we can identify definite metastases (D11, D12, L1, L2, L5, S1). However, we also have equivocal findings (L3–4), which we would call suspected disease, and normal findings (S2, S3), which we would call negative. The accuracy further increases when we use a second sequence, in this case a T1-weighted SE sequence (Fig. 3B). Signal intensity of L3 and L4 is also abnormal and consistent with metastases. Signal intensity of S2 and S3 is normal. Radiographs, technetium (Tc) bone scintigram, and follow-up examinations confirmed metastatic disease at all levels with the exception of S2 and S3. Our routine protocol for metastatic disease consists of T1-weighted SE and fat-suppressed TSE. This combination is more accurate than the combination of T1-weighted SE and GE, which in its turn is more sensitive than Tc bone scintigraphy (5). Fat-selective presaturation can be used not only at high field, but also at mid-field.

Good alternatives to our routine approach are available (6–9). Although optimization of sequences for detection of metastatic disease depends on many machine-related parameters such as field strength, homogeneity, soft ware coils, etc., some rules of thumb can be used. A STIR-TSE sequence, as demonstrated in the same patient (Fig. 3C), is also very accurate. Although field homogeneity is important in STIR imaging, it is not as critical as in fat-selective presaturation (9). STIR sequences show fewer artifacts than fat-selective suppressed images in the presence of metallic hardware left behind at reconstructive surgery. Long acquisition times (although this is not a major problem when STIR is combined with TSE), and/ or limited number of slices, and patient to patient variation in optimal inversion time (10) are disadvantages of STIR sequences. For this reason we prefer to use fat-selective presaturation sequences in children and young adults. Another disadvantage of STIR sequences may be the high sensitivity (11,12). We have anecdotal evidence that although sensitivity of STIR-TSE is somewhat higher than that of fat-suppressed TSE, the disadvantage of false positives outweighs the advantage of increased sensitivity. In Fig. 3D a STIR-TSE sequence of patient with metastatic disease secondary to breast carcinoma is presented. Bone scintigram (Fig. 3E) was interpreted as negative for metastases. Low signal intensity, indicating absence of disease, is only seen at the L5-S1 level (Fig. 3D). However, the T1-weighted SE (Fig. 3F), and fat-suppressed TSE sequence (Fig. 3G) show that only three well-defined lesions, proven to be metastatic disease during follow-up, are present.

When using STIR sequences one should be aware of the fact that enhancing lesions after gadolinium (Gd) administration may be masked. This is because the inversion pulse is not selective. On the other hand, fat-selective presaturation can be used in combination with Gd, and further increase contrast between enhancing lesions and normal tissue.

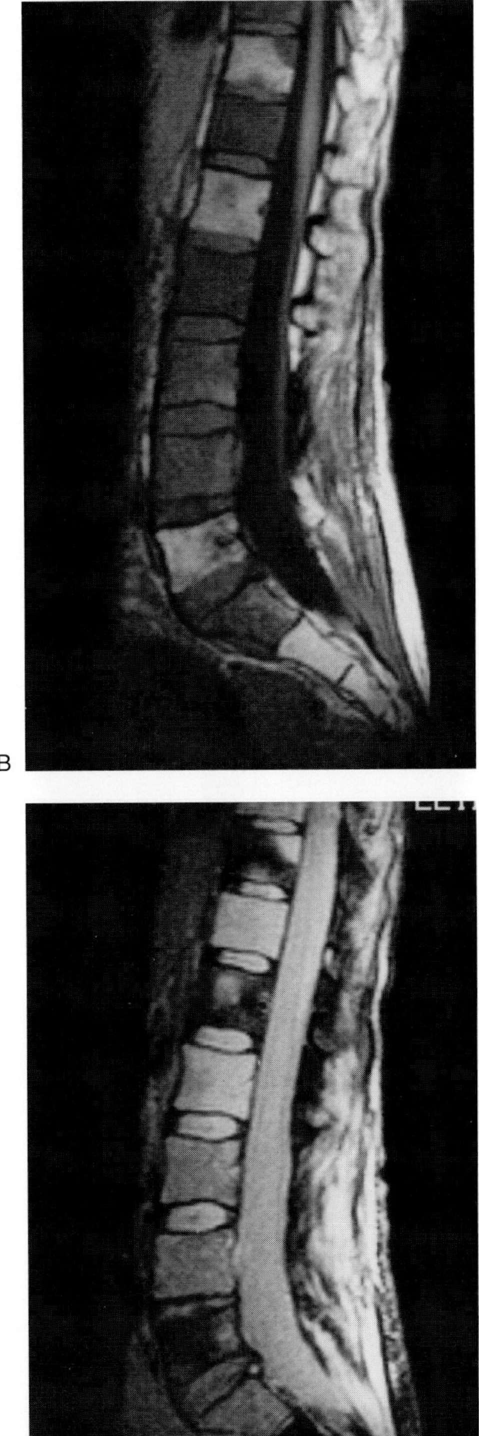

FIG. 3. B: On this T1-weighted sequence, only S2 and 3 are normal. **C:** On this STIR-TSE (TI 120, 1,400/20) all vertebral bodies with the exception of S2 and S3 are abnormal.

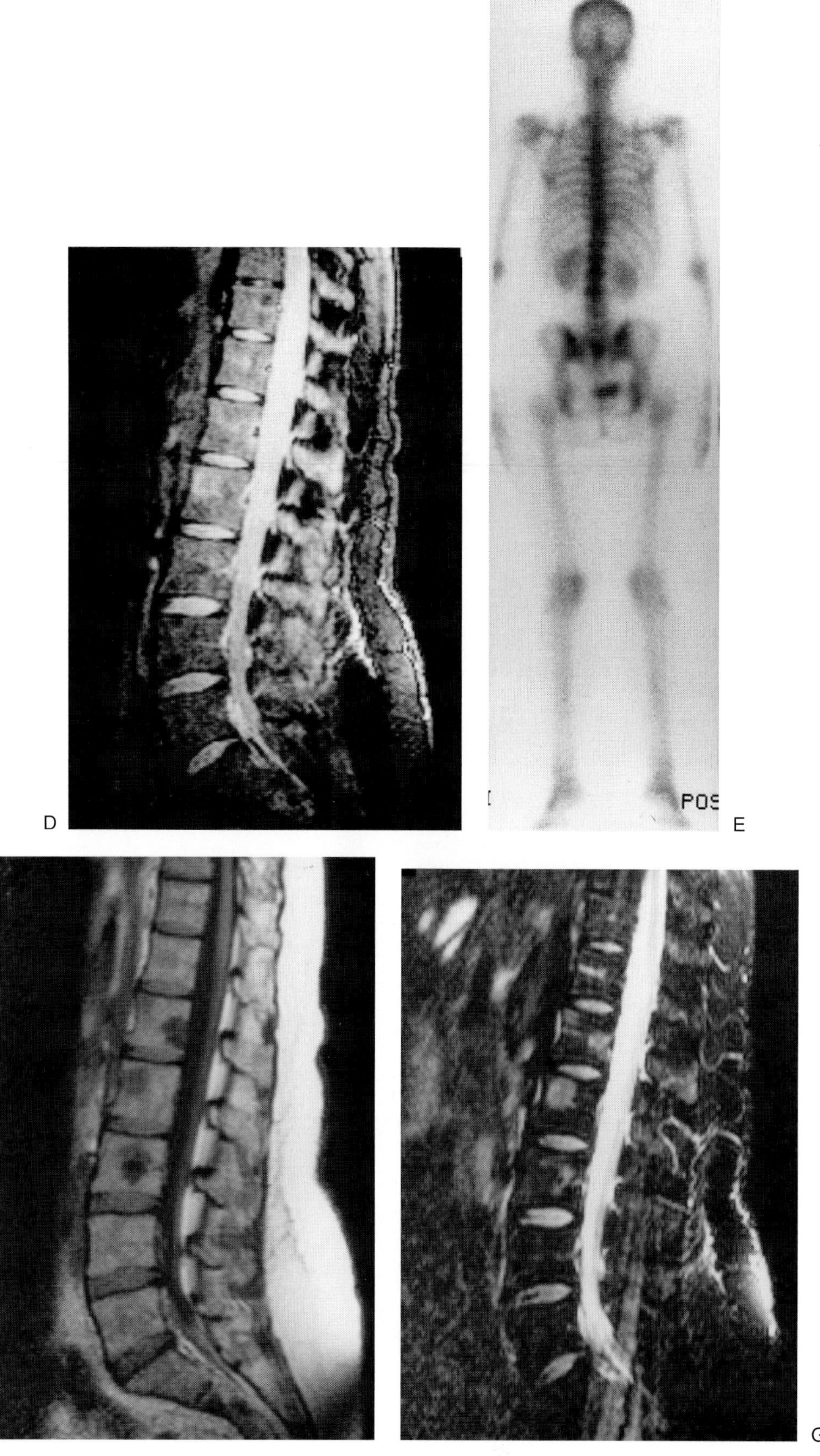

FIG. 3. D: Poor quality STIR-TSE sequence suggests the presence of metastatic disease in D10–12, L1–3. **E:** Bone scintigram was thought to be negative for metastatic disease. Posterior view only is shown. **F:** T1-weighted sequence shows metastases in L1, 2, 3. **G:** Although the lesion in L3 is not as well visualized on this TSE with fat selective presaturation, all three lesions can be identified.

CASE 4

History: A 22-year-old man is referred to our hospital because of a large painful mass in the foot. Pain following trauma had been persistent for 2 years. The initial radiograph is available and was at the time of the traumatic event reported to be normal. What would be your MR diagnosis? Is your MR diagnosis supported by the 2-year-old radiograph?

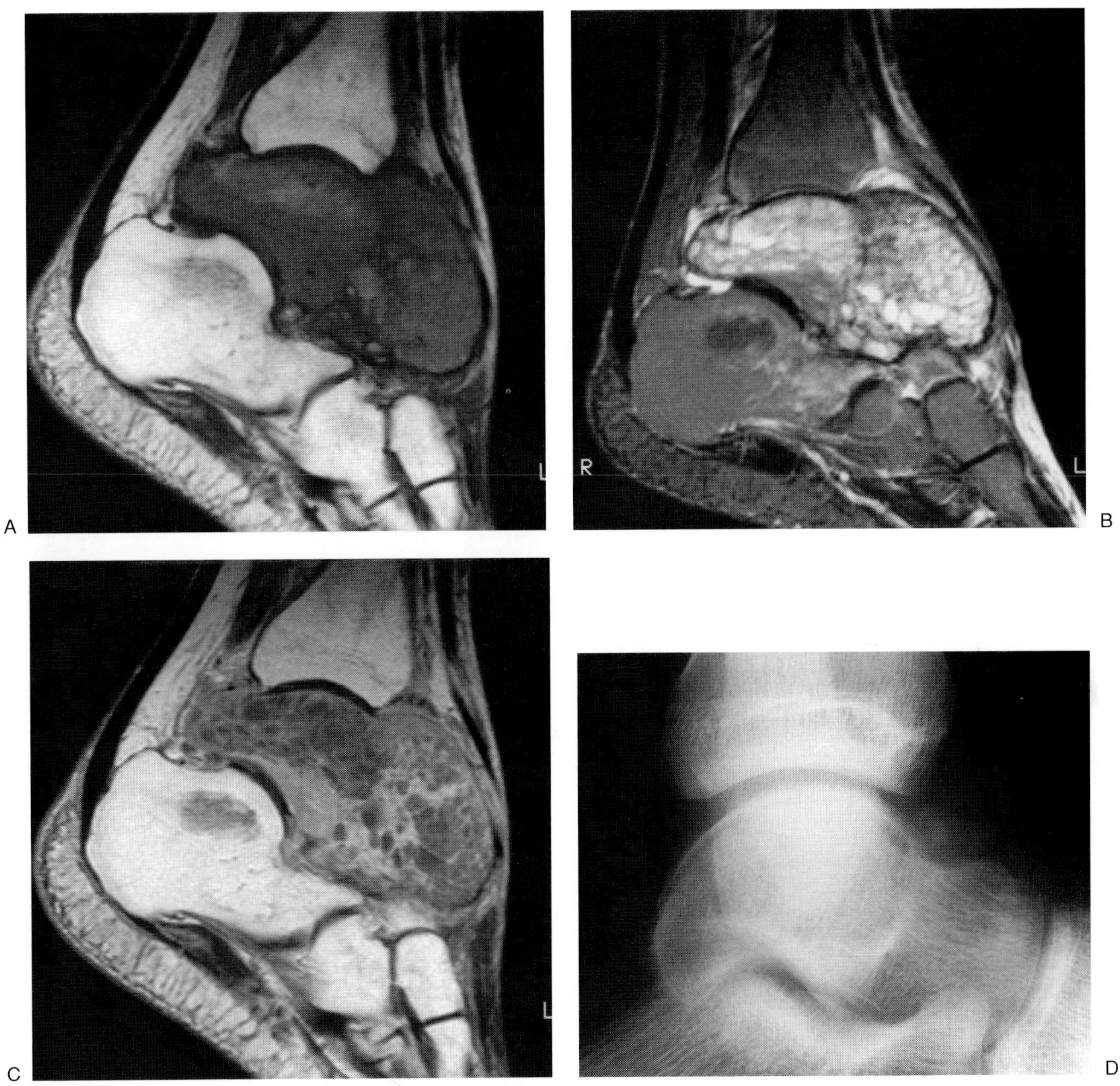

FIG. 4.

Findings: In Fig. 4A, the sagittal T1-weighted SE (TR/TE 600/20 msec) image, the mass originates from the talus and has some areas with high signal intensity. In Fig. 4B, the T2-weighted image shows a nonspecific high signal intensity. In Fig. 4C, the static Gd-enhanced sagittal image, a septal-like enhancement pattern can be appreciated. In Fig. 4D, the original radiograph is not normal, but shows a blister-like lesion on the surface of the talar neck.

Diagnosis: Osteoblastoma with secondary aneurysmal bone cyst.

Discussion: The diagnosis of osteoblastoma with secondary aneurysmal bone cyst was confirmed after curettage. The blister-like lesion that was missed on the initial radiograph (Fig. 4D) is typical for an osteoblastoma (1–3). The high signal intensity areas on the T1-weighted MR image (Fig. 4A) is consistent with hematoma. The septal-like enhancement pattern (Fig. 4C) is only seen in a limited number of lesions like well-differentiated chondrosarcoma, telangiectatic osteosarcoma and aneurysmal bone cyst. Osteoblastoma is one of the benign bone tumors that are associated with aneurysmal bone cyst (1–3).

This case demonstrates that even in complex cases MR findings may play an important role in making a diagnosis. However, the MR findings alone are not specific enough. The MR images are not specific for osteoblastoma, and hematoma and even fluid levels can be found in association with many benign and malignant tumors such as chondroblastoma, giant cell tumor, fibrous dysplasia, myositis ossificans, osteosarcoma, MFH, and synovial sarcoma (4,5).

High signal intensity on T1-weighted SE images is unusual and may assist in suggesting a specific diagnosis. It can be seen not only with hematoma, but also in fat or bone marrow (Fig. 4E) (5–9). Fluid with a high viscosity and protein concentration may also exhibit a high signal intensity. This is, for instance, occasionally seen in Ewing's sarcoma, especially after chemotherapy (Fig. 4F) (5).

A low signal intensity on T2-weighted sequences also may assist in suggesting a specific diagnosis, especially in soft tissue lesions and may represent cholesterol, mature fibrosis, foreign body, air, flow void, hemosiderin, and intracellular deoxyhemoglobin in acute hematoma (5,6). A low signal intensity in bone is usually secondary to calcification, and thus does not contribute anything to radiographic findings.

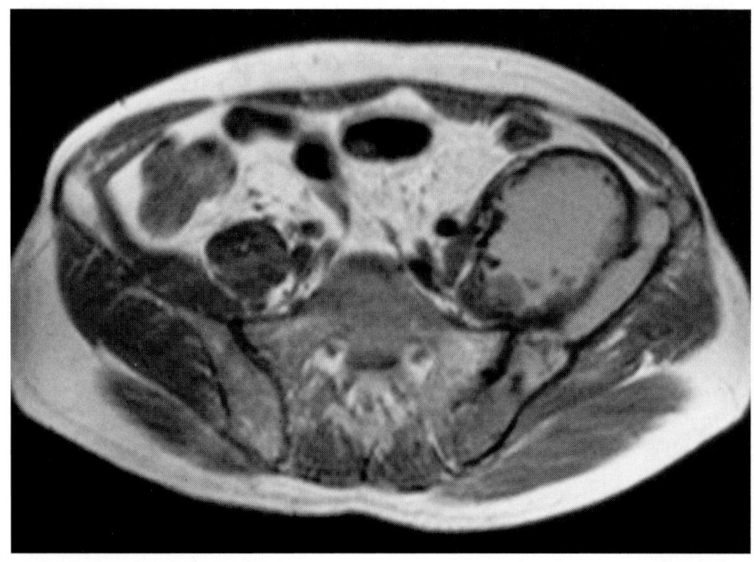

E

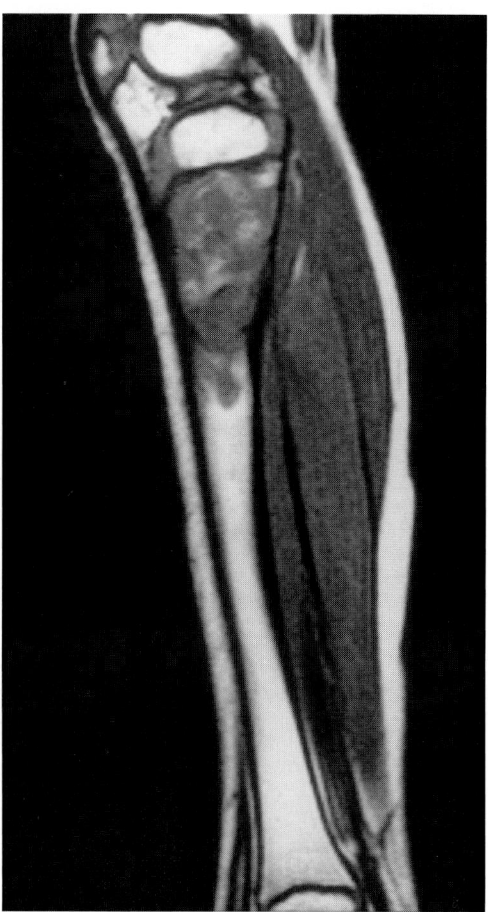

F

FIG. 4. E: Pseudotumor in a patient with hemophilia. On this axial T1-weighted image the pseudotumor in the iliac bone and iliac muscle has a high signal intensity suggesting hematoma. (From Bloem JL, Holscher HC, Taminiau AHM. Magnetic resonance imaging and computed tomography of primary malignant musculoskeletal tumors. In: Bloem JL, Sartoris DJ, eds. *MRI and CT of the musculoskeletal system.* Baltimore: Williams and Wilkins, 1992;189–217, with permission.)

FIG. 4. F: Sagittal T1-weighted (TR/TE 600/20 msec) image of a 4-year-old with Ewing's sarcoma obtained prior to chemotherapy. Note the high signal intensity areas within the low signal intensity tumor in the proximal tibia metaphysis. The high signal intensity was secondary to the presence of mucoid substance.

CASE 5

History: A 31-year-old man presented with impaired function and pain in the hip following trauma 5 months ago. MR was requested to visualize posttraumatic abnormalities that could explain the clinical symptoms. Plain film (not shown) was reported to be normal. What is your diagnosis?

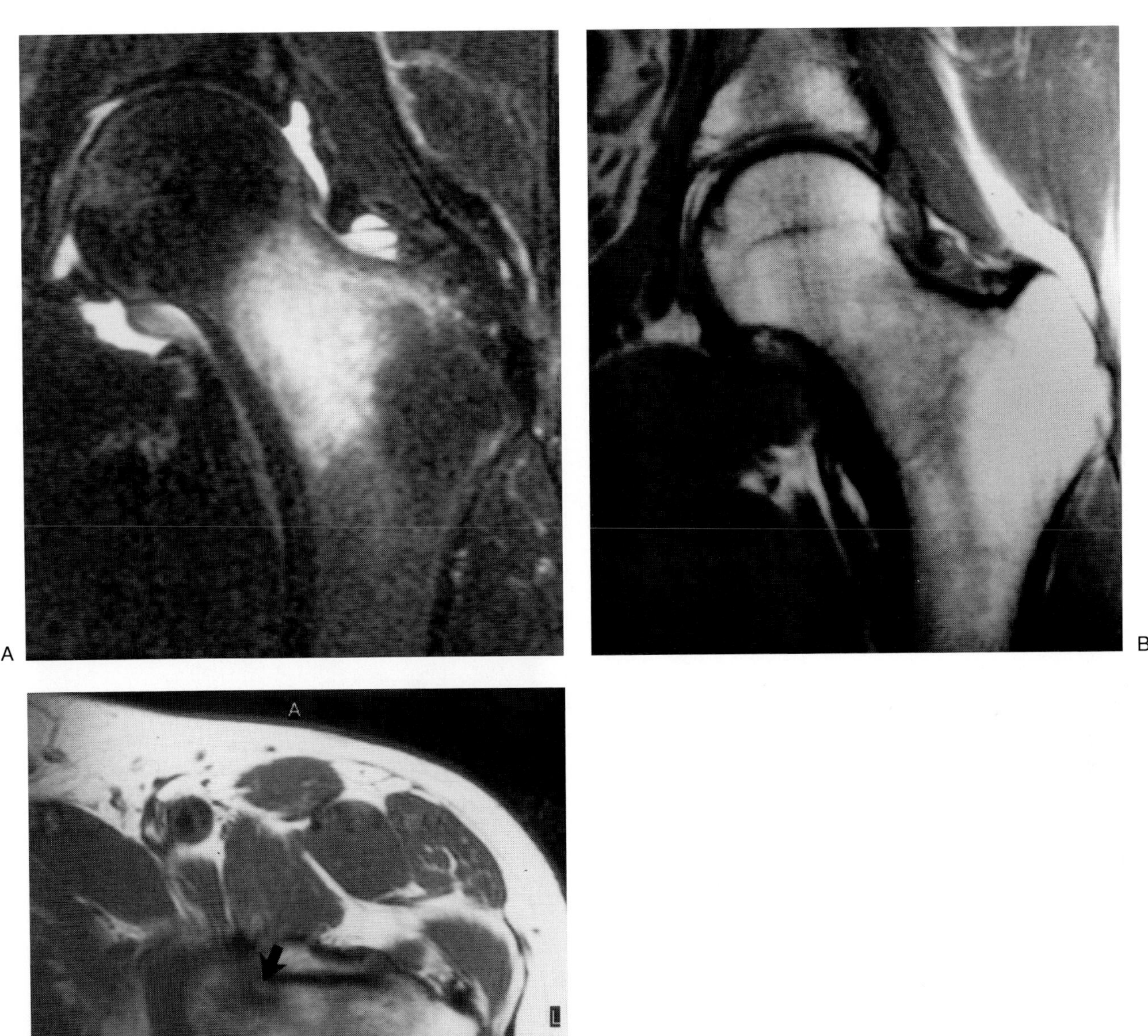

FIG. 5.

Findings: In Fig. 5A, a coronal TSE (TR/TE 3,250/100 msec) image with fat selective presaturation, a joint effusion and/or synovial thickening and high signal intensity within the femoral neck, consistent with edema, are seen. There also is some periosteal reaction, visualized as low signal intensity thickening and subtle increase of signal intensity, on the medial side. In Fig. 5B, the bone marrow abnormality is not so well visualized on the coronal T1-weighted sequence (TR/TE 600/20 msec). In Fig. 5C, an axial T1-weighted sequence (TR/TE 600/20 msec), a small focal area of high signal intensity, surrounded by a low signal intensity rim (arrow), is depicted on the ventral surface of the femoral neck.

Diagnosis: Osteoid osteoma.

Discussion: Because of the marked bone marrow edema and synovial reaction, a diagnosis of osteoid osteoma was considered. After careful screening of the images a small nidus was identified (Fig. 5C). The diagnosis was confirmed on CT (Fig. 5D) (1). The patient was subsequently treated with percutaneous thermocoagulation (Fig. 5E) (2).

Marked reactive changes are not an accurate indicator for malignancy (3,4). Reactive changes rather reflect the biologic activity of the lesion. We have encountered marked reactive changes more often in benign than in malignant lesions. Characteristically these secondary changes indicate the presence of osteoblastoma (Fig. 5F,G), osteoid osteoma, chondroblastoma, or active Langerhans cell histiocytosis (Fig. 5H).

Many lesions originate in, or are even confined to, the cortex; these include metastases (Fig. 5I), lymphoma (Fig. 5J), brown tumor in hyperparathyroidism, (osteo)-fibrous dysplasia, osteosarcoma, aneurysmal bone cyst, osteomyelitis, osteoblastoma, osteoid osteoma, and adamantinoma (4–8).

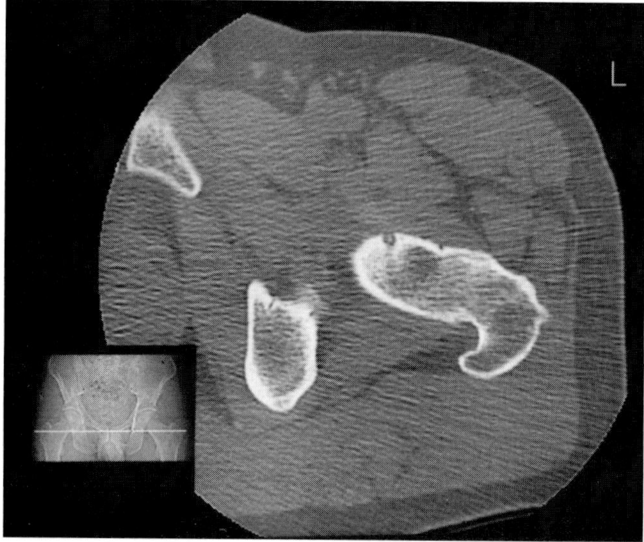

D

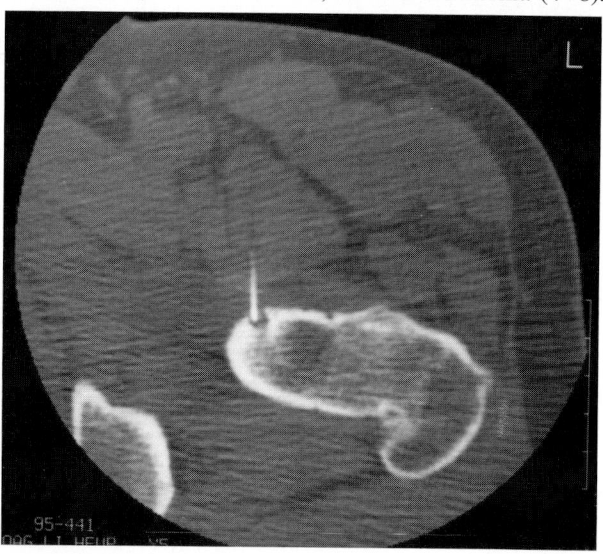

E

FIG. 5. D: A small calcified nidus consistent with osteoid osteoma is visualized on this 2 mm CT cut. **E:** During percutaneous thermo-coagulation the position of the needle is confirmed to be in the center of the lesion. (Courtesy Wim Obermann, MD.)

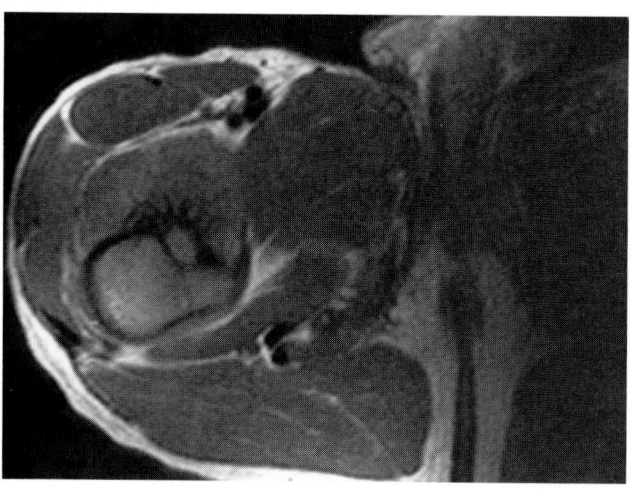

F

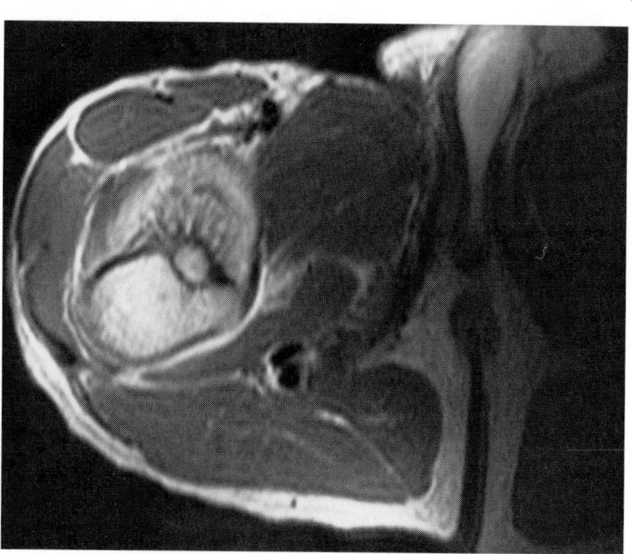

G

FIG. 5. F: In another patient on the T1-weighted image (TR/TE 600/20 msec) an intracortical lesion is appreciated. Surrounding this osteoblastoma an ill-defined low signal intensity is seen within the bone marrow. **G:** On the T2-weighted sequence (TR/TE 2,000/100 msec) an increased signal intensity surrounding the lesion is seen not only in bone marrow, but also in the surrounding soft tissue. Note the periosteal reaction. (Figs. F and G Kroon HM, Schuurmans J. Osteoblastoma: *clinical and radiologic findings in 98 new cases. Radiology* 1990;175:783–790, with permission.)

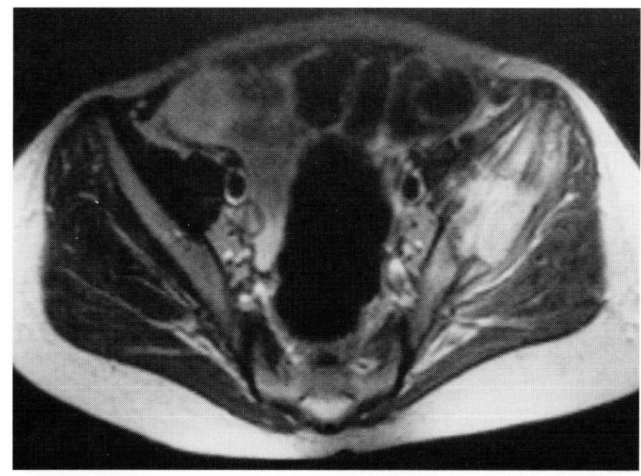

H

FIG. 5. H: Axial T2-weighted image (TR/TE 1,800/90 msec) show eosinophilic granuloma (Langerhans cell histiocytosis) in the left iliac bone and marked reactive changes in the gluteus and iliacus muscle compartments. The differential should include Ewing's sarcoma.

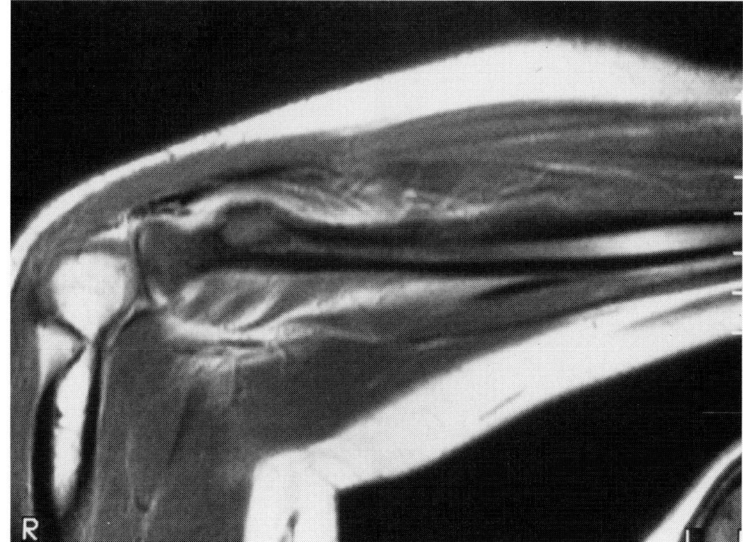

I

FIG. 5. I: This metastasis in the cortex of the radius was the presenting symptom in a 25-year-old woman with carcinoma of the breast. Marked intra- and extraosseous changes are seen on this sagittal T1-weighted image and were confirmed on T2-weighted (not shown), and Gd-enhanced images (not shown).

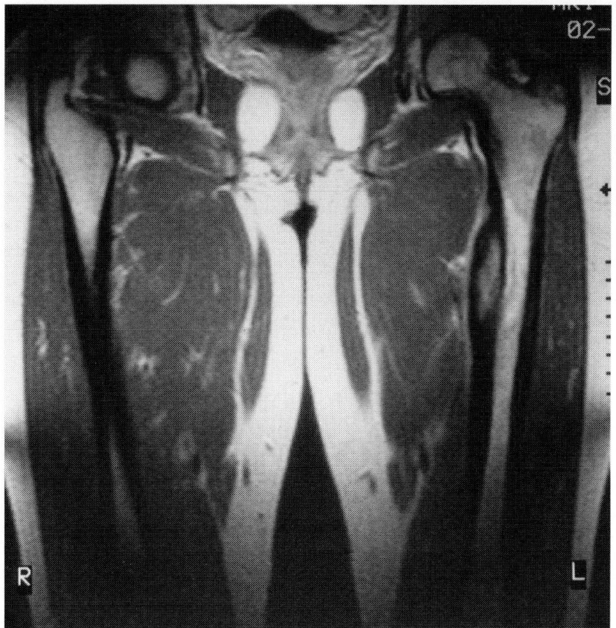

J

FIG. 5. J: Coronal T1-weighted image (TR/TE 600/20 msec) shows intracortical NH-lymphoma.

CASE 6

History: A 22-year-old man was sent to MR for local staging of what was thought radiographically to be an osteosarcoma. The sagittal T1-weighted sequence is obtained to define proximal and distal osseous margins. What would be your conclusion concerning these margins?

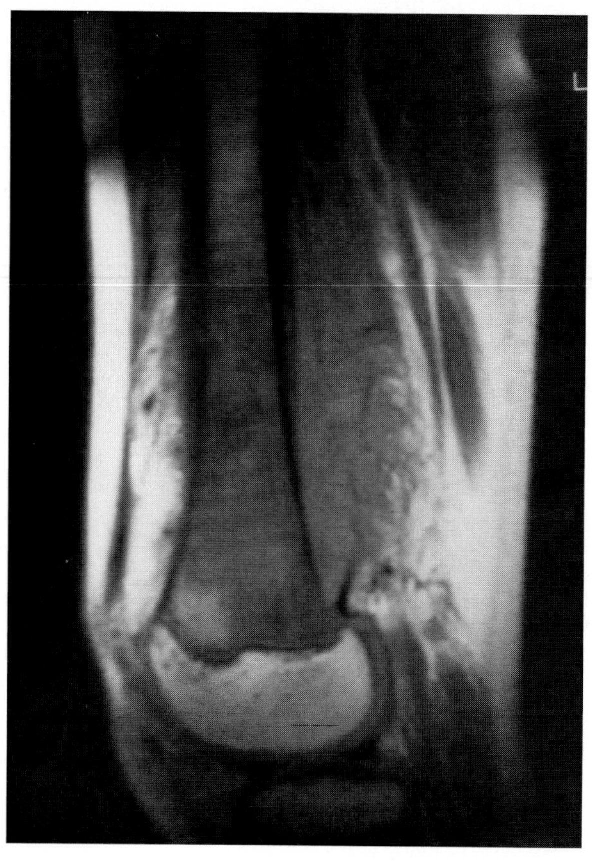

FIG. 6.

Findings: In Fig. 6A the growth plate is the distal margin, and the epiphysis is not involved. The proximal margin is visualized, but is close to the edge of the surface coil.

Diagnosis: Skip metastasis of osteosarcoma.

Discussion: There are two major pitfalls while using sagittal T1-weighted SE sequences for defining osseous tumor margins. One is lack of contrast, resulting in poor definition. Contrast can be poor if the tumor has a relatively high signal intensity such as in hematoma, some Ewing's sarcomas (Case 4), and ossification, or when normal bone marrow has a relatively low signal intensity such as occurs in red bone marrow in the young, partial volume effects with cortical bone, and reactive changes like edema and inflammatory response.

A second pitfall is too small a field of view. We try to use surface coils as much as possible to increase signal to noise ratios; the trade-off is a small field of view. In Fig. 6A the proximal margin is unacceptably close to the sensitive area of the coil. The sequence was repeated with the body coil, allowing a large field of view (Fig. 6B). The skip lesion is now easily detected. The true incidence of skip metastases of osteosarcoma and Ewing's sarcoma is not known, but a prevalence of up to 20% has been described (1,2). Also small extensions of the primary tumor may escape detection if the field of view is too small. In Ewing's sarcoma diffuse involvement of bone marrow is not an exception (see also Case 1); therefore screening of bone marrow in patients with Ewing's sarcoma is indicated either with MR imaging or bone scintigraphy. Large skip lesions, or osseous metastases of osteosarcoma, can also be detected with bone scintigraphy.

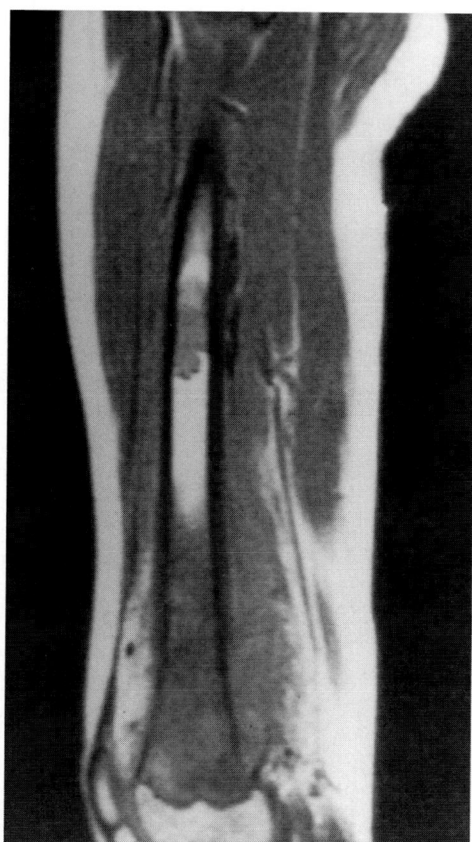

FIG. 6. B: In the same imaging session a large field of view T1-weighted sequence shows a large skip lesion proximal to the primary osteosarcoma. (Figs. A and B from Bloem JL, Holscher HC, Taminiau AHM. Magnetic resonance imaging and computed tomography of primary malignant musculoskeletal tumors. In: Bloem JL, Sartoris DJ, eds. *MRI and CT of the Musculoskeletal system.* Baltimore: Williams and Wilkins 1992, 189–217.)

CASE 7

History: An 18-year-old who recently emigrated from southeast Asia to the Netherlands, was referred to our MR unit for staging of a presumed osteosarcoma (Fig. 7A). What would be your diagnosis?

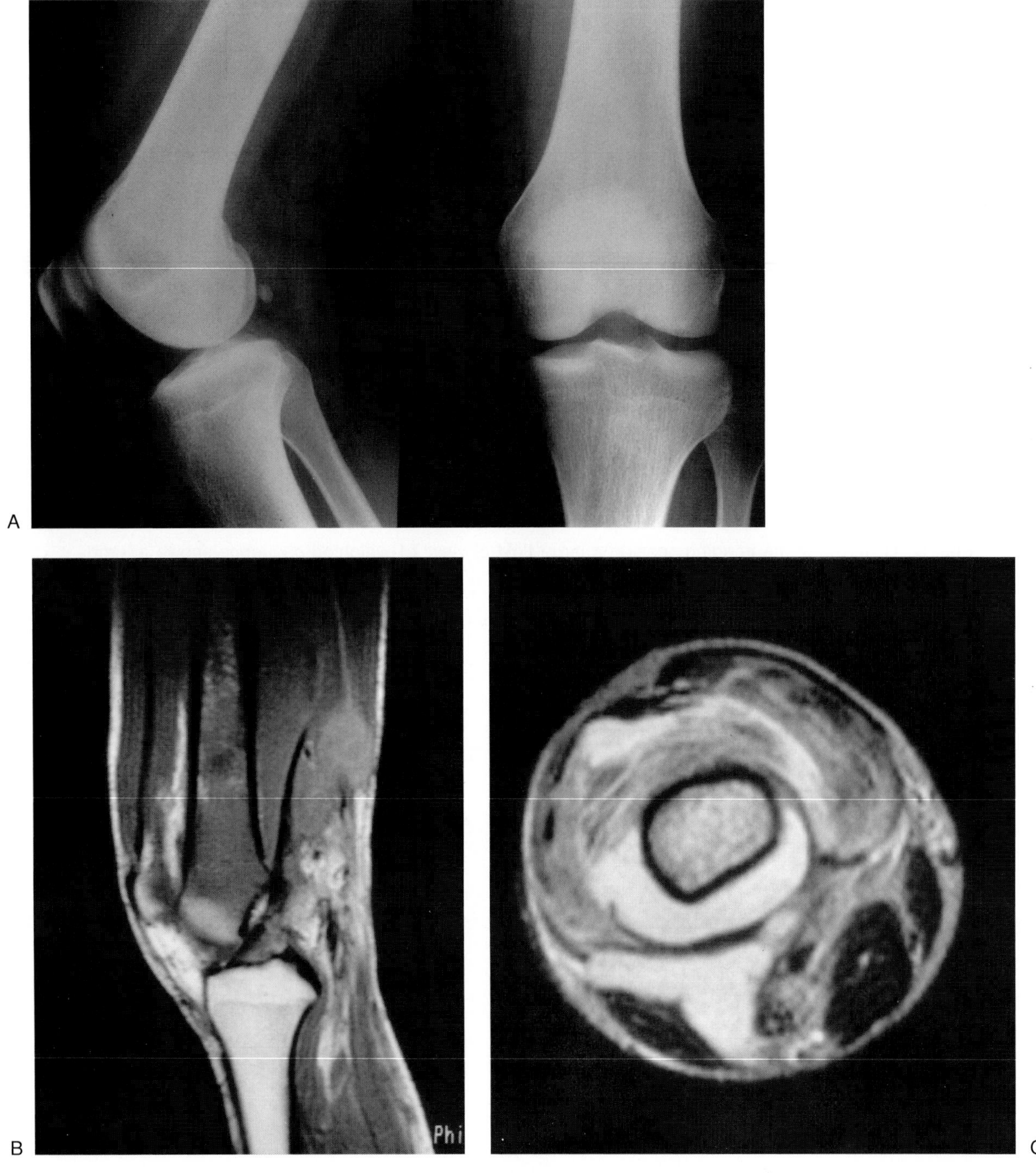

FIG. 7.

Findings: In Fig. 7A, because of suboptimal quality of the plain films, the diffuse sclerosis in the distal femur is not well seen. In Fig. 7B, the sagittal T1-weighted sequence, low signal intensity in the femur and elevated periosteum are easily appreciated. In Fig. 7C, the axial T2-weighted SE sequence, the high signal intensity of the soft tissue extension is depicted. (Figs. A–C from Bloem JL, Holscher HC, Taminiau AHM. Magnetic resonance imaging and computed tomography of primary malignant musculoskeletal tumors. In: Bloem JL, Sartoris DJ, eds. MRI and CT of the musculoskeletal system. Baltimore: Williams and Wilkins, 1993;3:439–446, with permission.)

Diagnosis: Osteomyelitis with soft tissue extension.

Discussion: The signal intensities of the lesion are not specific. However, the morphology of the soft tissue extension is not in accordance with the diagnosis osteosarcoma. The soft tissue component has concave instead of convex margins; the lesion takes only the space available, and follows fascial planes. These findings favor infection rather than neoplasm. Diagnosis of soft tissue abscesses accompanying osteomyelitis was made and confirmed at surgery.

The various phases of osteomyelitis and spondylodiscitis are usually characteristic enough to allow differentiation from neoplasm (Fig. 7D–F) (1). In some instances, however, differentiation is challenging. In particular, Ewing's sarcoma may mimic infection both from a radiographic and clinical point of view. Fortunately, a large soft tissue mass is seen with MR imaging allowing the diagnosis. However, one should be aware of the fact that sometimes Ewing's sarcoma does not present with a large soft tissue mass (Case 13).

A second problem may arise in the spine especially in children. In the adult, involvement of the disk usually allows a confident diagnosis (2). Only chordoma or giant cell tumor may extend over multiple levels (Fig. 7G). In children the disk and end plates may be completely normal since the infection initially behaves like osteomyelitis (Fig. 7H), and may thus mimic other lesions such as Langerhans cell histiocytosis. Later on the infection has the typical features of spondylodiscitis (Fig. 7I). Also in children blood cultures may remain negative for some time.

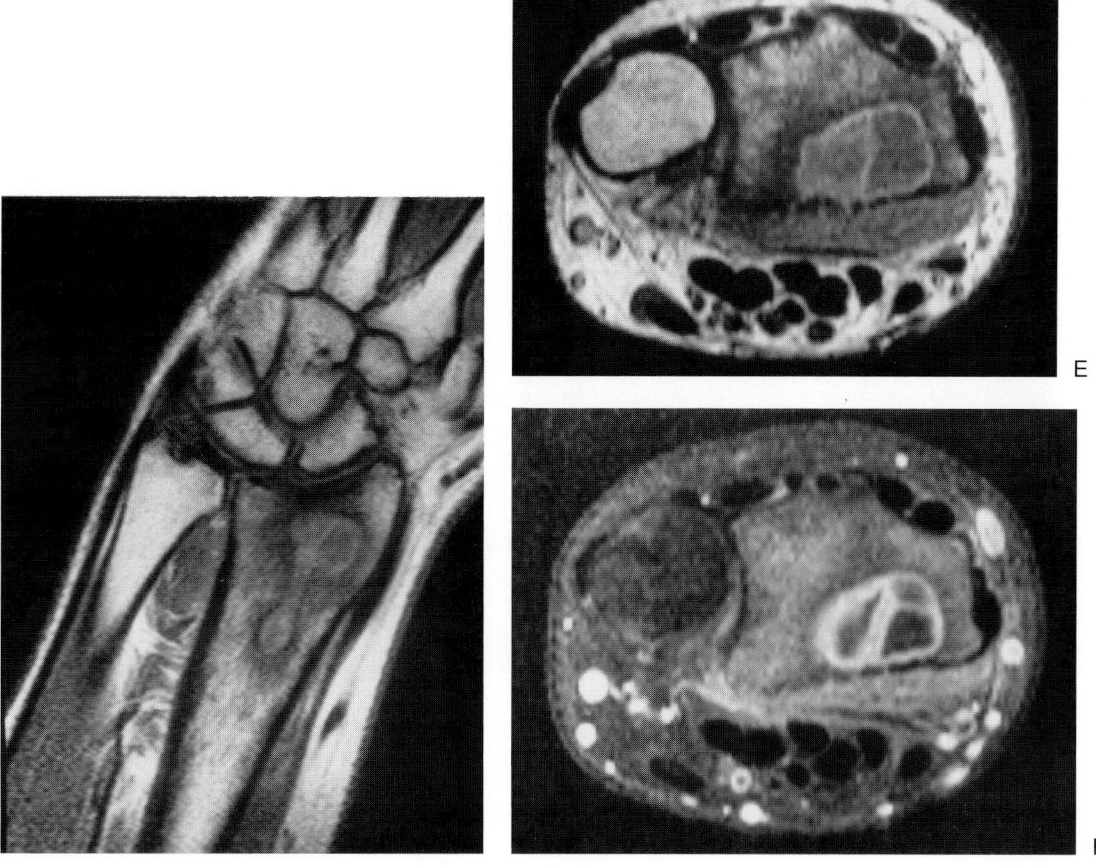

FIG. 7. D,E: Brodie's abscess in the distal radius has, on these T1-weighted coronal (D), and axial (E) images, a high signal intensity margin surrounding the low signal intensity abces. This indicates the presence of granulation tissue marginating the abscess (1). Low signal intensity edema surrounds the abscess. **F:** The granulation tissue and marked reactive changes that are in favor of the diagnosis of infection rather than neoplasm are well exhibited on this fat suppressed Gd-enhanced short (TR/TE 600/20 msec) image.

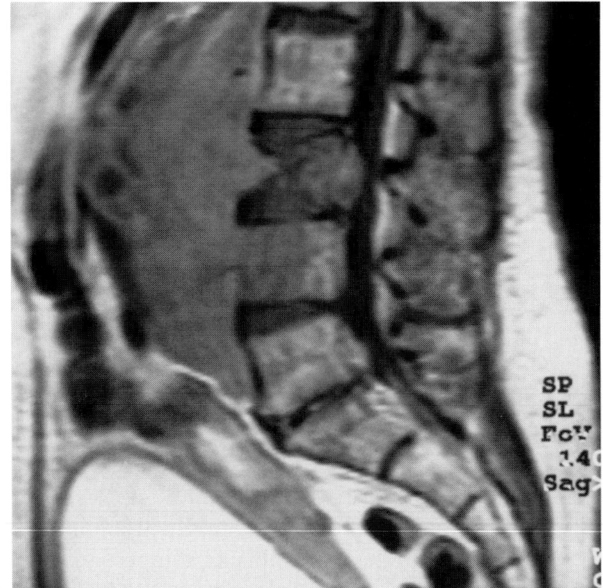

G

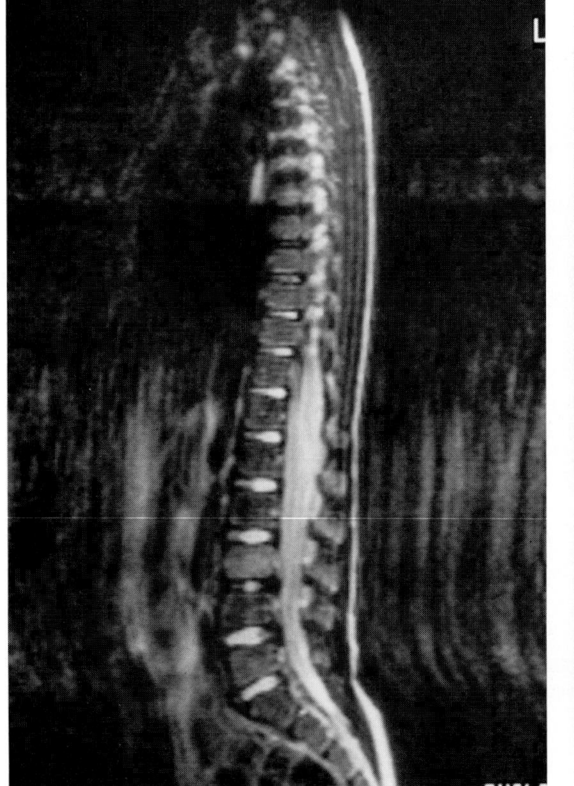

H

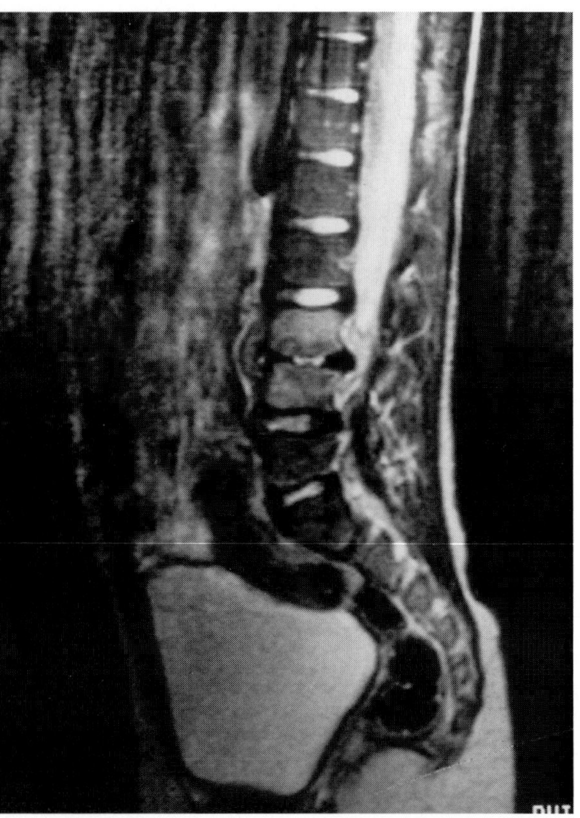

I

FIG. 7. G: Sagittal Gd-enhanced (TR/TE 600/20 msec) image of a chordoma shows involvement of multiple levels with intact disks. **H:** Despite the motion artifacts in this child with lumbar pain, a high signal intensity lesion confined to the vertebral body is seen. The L3–L4 disk has a slightly smaller sized high signal intensity disk. **I:** On the follow-up MR, obtained after biopsy, positive culture for staphylococcus aureus, and antibiotic treatment, a classic extension of adult type spondylodiscitis with involvement of the disk and two adjacent vertebral bodies is appreciated.

CASE 8

History: A 20-year-old man presents with a large mass in the area of the right hip. What is your diagnosis and how do you explain the very high signal intensity that is seen in part of the lesion? What is the significance of this component?

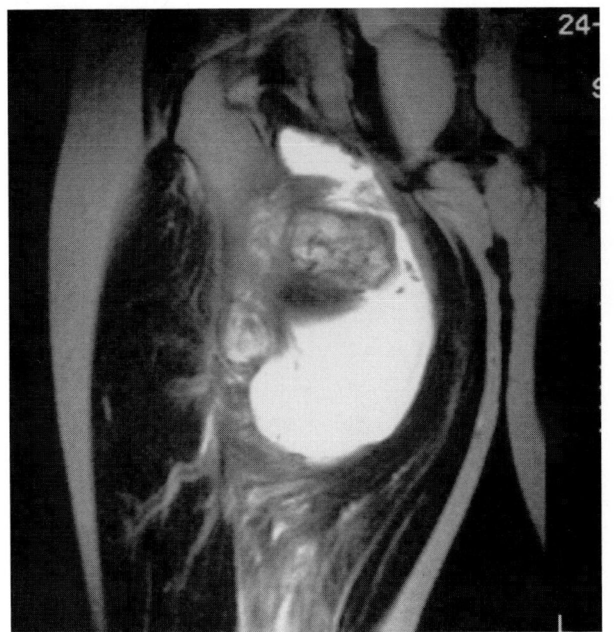

A

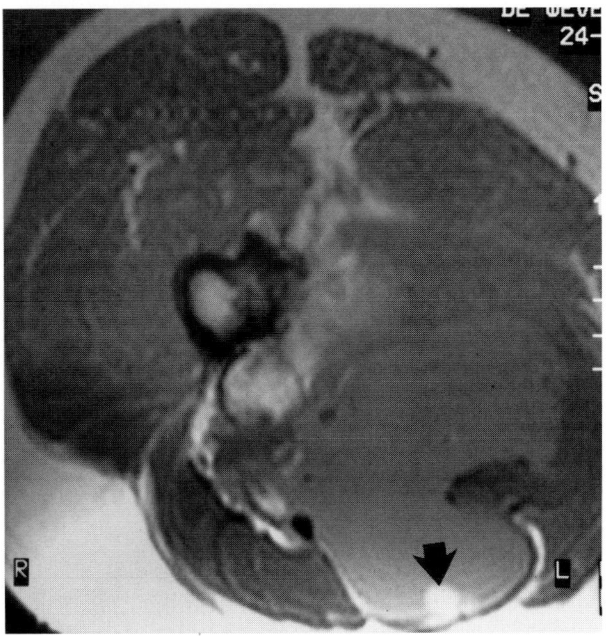

B

FIG. 8.

Findings: In Fig. 8A, the coronal T2-weighted SE sequence, an exostoses is seen on the medial side of the femur. Bone marrow is contiguous with the lesion and cortical bone flares into the lesion. A large inferiorly located homogeneous high signal intensity area is seen. In Fig. 8B, a small high signal intensity area (arrow) is seen within the large low signal intensity area on an axial T1-weighted sequence.

Diagnosis: Osteochondroma with large bursa containing cartilaginous bodies.

Discussion: The lesion has all the characteristics of an osteochondroma. Of interest is the focal high signal intensity seen on the T2-weighted sequence. This fluid-like appearance represents a bursa. Within a bursa accompanying an osteochondroma, cartilaginous particles may be present. Such a particle is seen posteriorly as a high signal intensity focus on the axial T1-weighted sequence. Enchondral ossification accounts for the high signal intensity. The clinical significance of a bursa is twofold. It may cause sampling errors when biopsied. Fluid from the bursa may be nondiagnostic, or the earlier described cartilaginous particles may be found. It is important that the pathologist knows when he/she is dealing with cartilage from the bursa, because this cartilage may mimic malignancy (1,2). The second point concerns planning surgery. To avoid spilling, the bursa either has to be removed without opening it, or the surgeon has to empty it before removing it together with the osteochondroma.

As in conventional radiography diagnosis of an osteochondroma is straightforward. The same features, i.e., continuity of osteochondroma with marrow, and flaring of cortical bone in the stalk are also seen in dysplasia epiphysealis hemimelica (also called Trevor's disease) (3). Dysplasia epiphysealis hemimelica is an intraarticular osteochondroma that is often located in the lower extremity of children, frequently causing mechanical problems.

A second mimic of classic osteochondroma is juxtacortical osteosarcoma (4,5). These lesions also contain (large) amounts of cartilage. The relationship with cortex and medulla is, however, not the same as that of osteochondroma. Also juxtacortical osteosarcoma mainly occurs on the posterior aspect of the femur.

CASE 9

History: A 33-year-old woman has been complaining of intermittent pain in the right knee. What is your diagnosis? Is the joint involved?

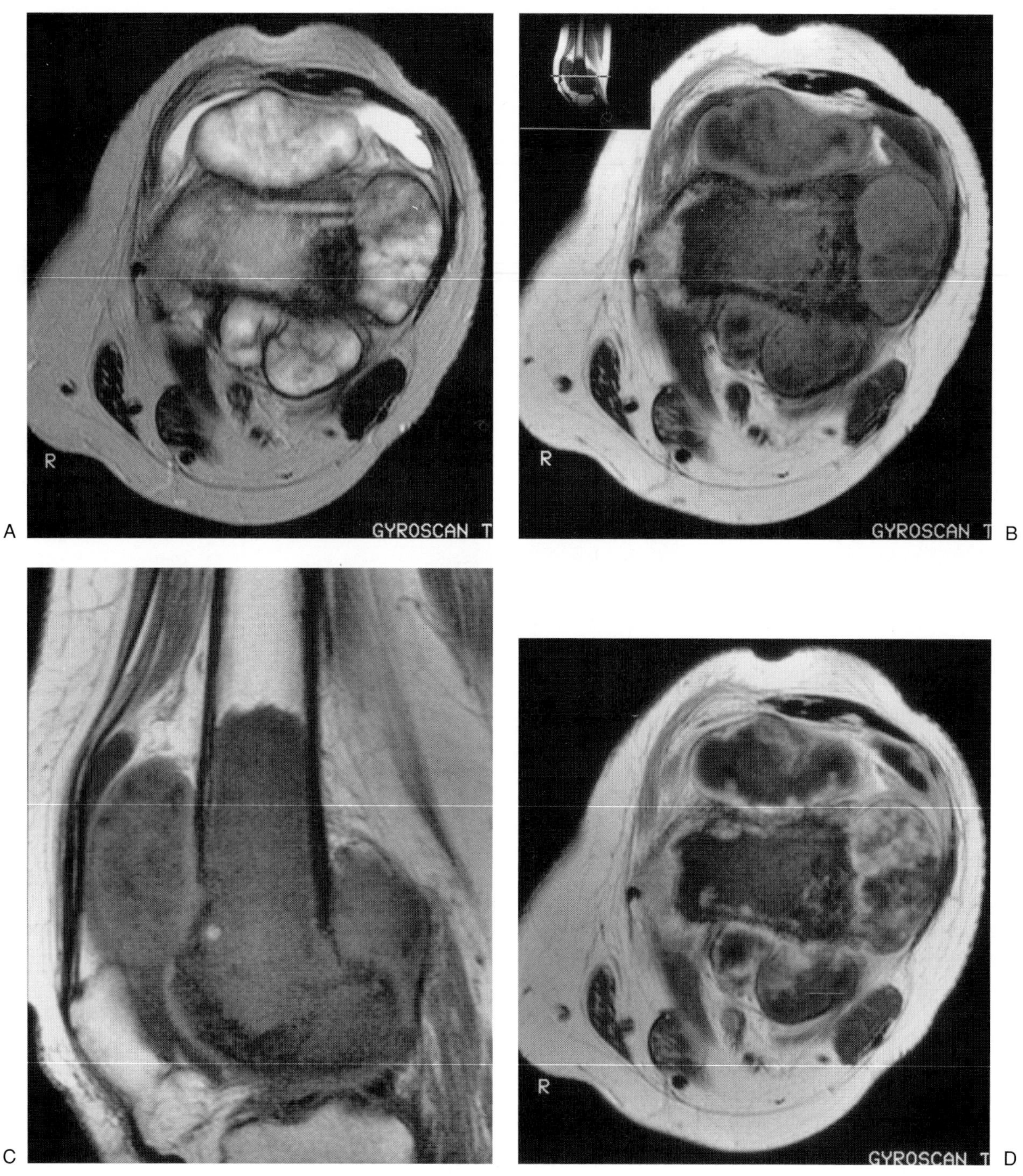

FIG. 9.

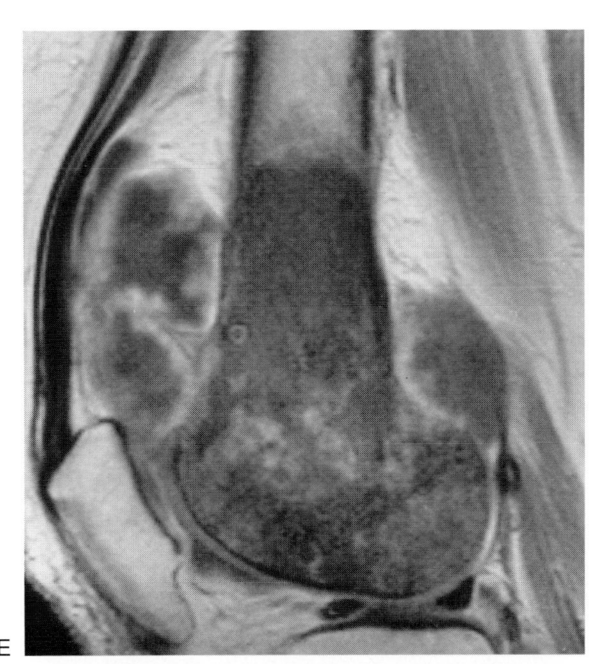
E

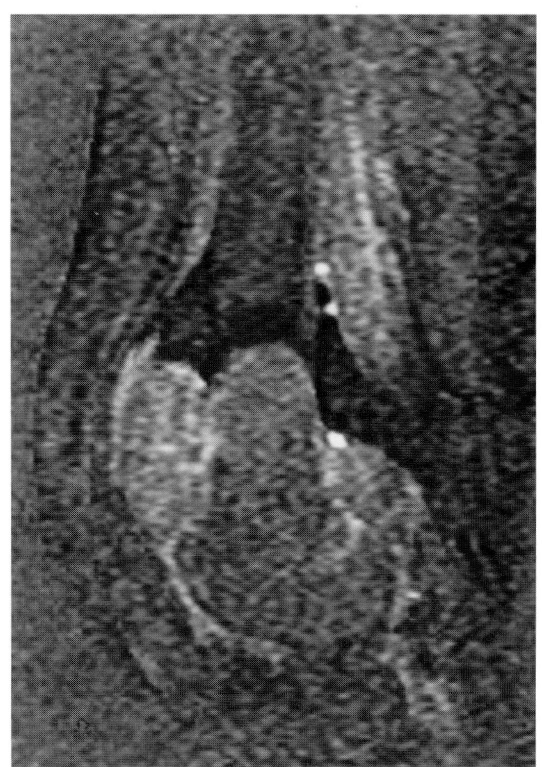
F

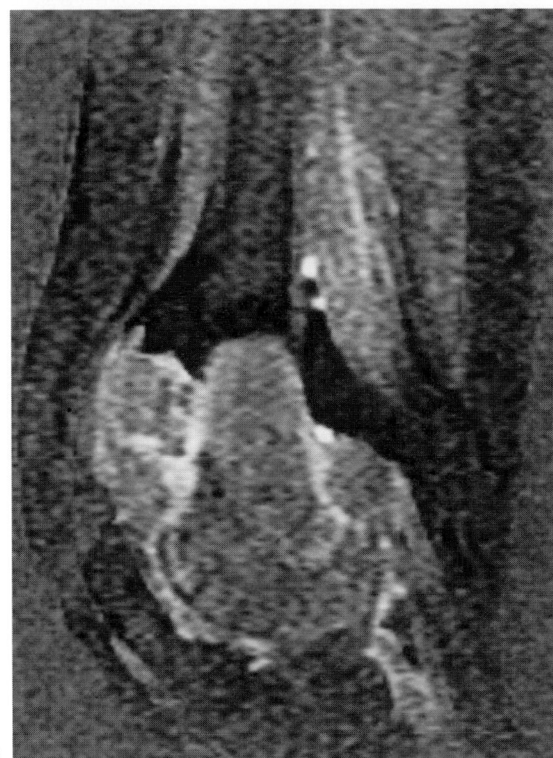
G

FIG. 9.

Findings: In Fig. 9A, a lobulated mass arising from the distal femur is seen on this axial TSE (4500/150) sequence. Joint effusion is present. The tumor extends into the prefemoral fat. In Fig. 9B,C axial (B) and sagittal (C) T1-weighted sequences (TR/TE 600/20 msec) demonstrate intraosseous extension. The small high signal intensity dot represents the biopsy site. In Fig. 9D,E, a serpentiginous and nodular enhancement pattern is noted on these static Gd-enhanced axial (D) and sagittal (E) images. In Fig. 9F a sagittal magnetization prepared (MP) GE (15/6.8/30°) sequence taken 34 seconds after Gd bolus injection shows marked enhancement of arteries posteriorly, and already some enhancement of the tumor. In Fig. 9G only after 60 seconds, substantial tumor enhancement is noted on the MPGE sequence. This enhancement slope is not consistent with high-grade malignancy.

Diagnosis: Grade II chondrosarcoma, extending into the joint.

Discussion: The extension in the prefemoral fat is extraarticular. This is often erroneously called intraarticular. In this patient, joint extension is present on the ventral, dorsal, and lateral sides (Fig. 9H).

In well-differentiated chondrosarcoma (i.e., grade I and II), cellular tumor nodules, perichondrium, and fibrovascular septation enhances (1–3). This is well seen on static images taken approximately 5 minutes after Gd injection. When this enhancement pattern is seen one should always consider well-differentiated chondrosarcoma. High-grade chondrosarcoma, and most other sarcomas, or secondary osseous neoplasm do not enhance in a nodular-septal pattern (4). Enhancement of grade 3 chondrosarcoma is more homogeneous and nonspecific on static Gd-enhanced images (Fig. 9I–K). Whether this sign can be used to differentiate benign from malignant cartilaginous tumor is still under investigation. Many enchondromas do not enhance, but others do show some enhancement. Currently a prospective study, using static and dynamic Gd-enhanced imaging techniques, is being conducted to address this problem. Unfortunately radiographic criteria are not accurate enough to differentiate enchondroma from chondrosarcoma (5). Pending the results of the prospective MR study, we currently biopsy enhancing components when dealing with cartilaginous lesions. The yield of this strategy is very high.

Septal enhancement without the nodular component can sometimes be seen in other tumors as well. The most important ones are telangiectatic osteosarcoma (Fig. 9L,M), aneurysmal bone cyst (see Case 4), giant cell tumor (Fig. 9N,O), and necrosis in high-grade sarcoma. In cartilaginous-containing lesions such as chondroblastic osteosarcoma, fibrosarcoma, juxtacortical osteosarcoma, septal enhancement representing the cartilaginous component may be observed. When the cartilage is cellular, which often is the case in these lesions, enhancement is more of the homogeneous type.

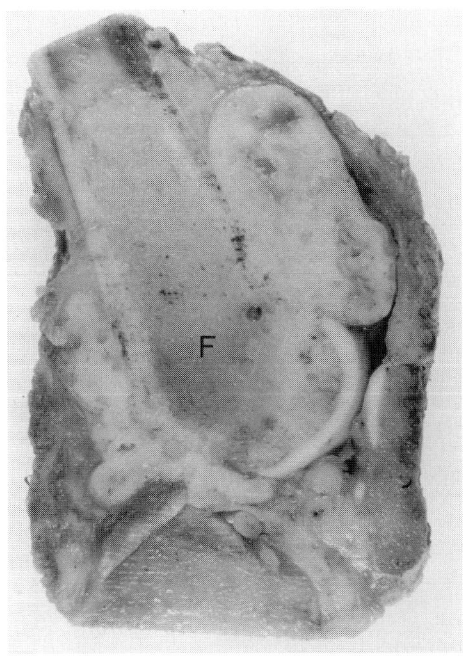

FIG. 9. H: Sagittal cut of specimen (F, femur) shows extension into the joint, both on the ventral and dorsal side. This corresponds with extension seen on the sagittal images (Fig. 9C,E–G).

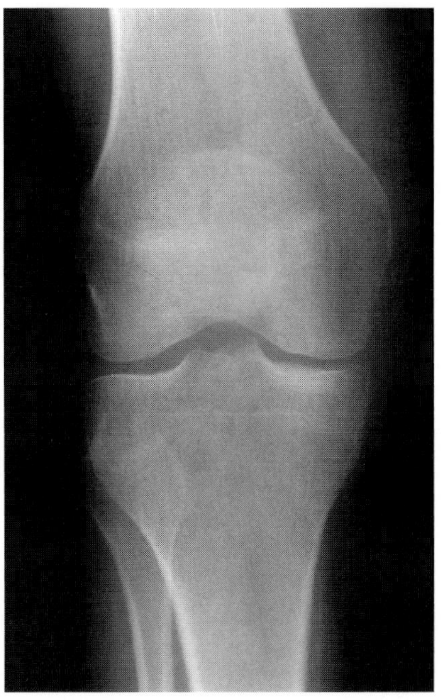

FIG. 9. I: Only subtle osteolysis is seen in the medial condyle on this radiograph of a patient with grade III chondrosarcoma.

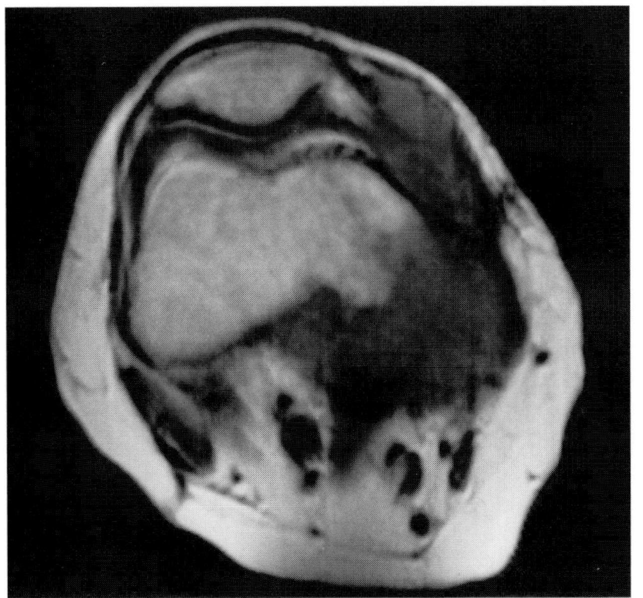

FIG. 9. J: An nonspecific low signal intensity mass is noted on this axial T1-weighted sequence (TR/TE 600/20 msec).

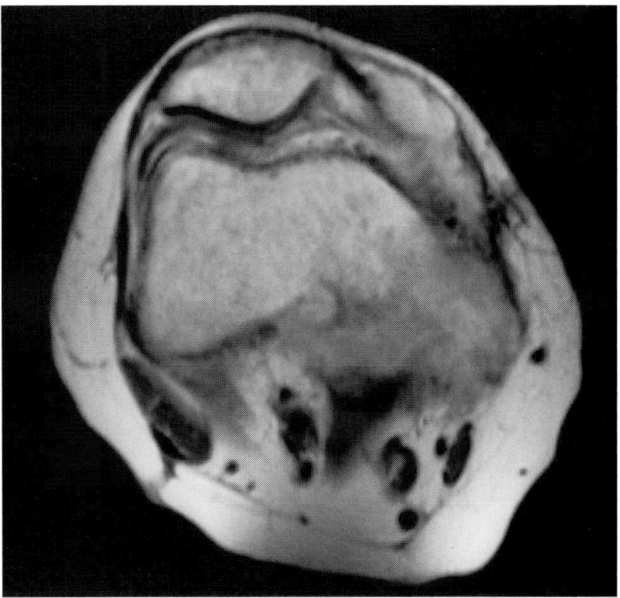

FIG. 9. K: On the static Gd-enhanced image taken 5 minutes after injection, an nonspecific homogeneous enhancement is exhibited.

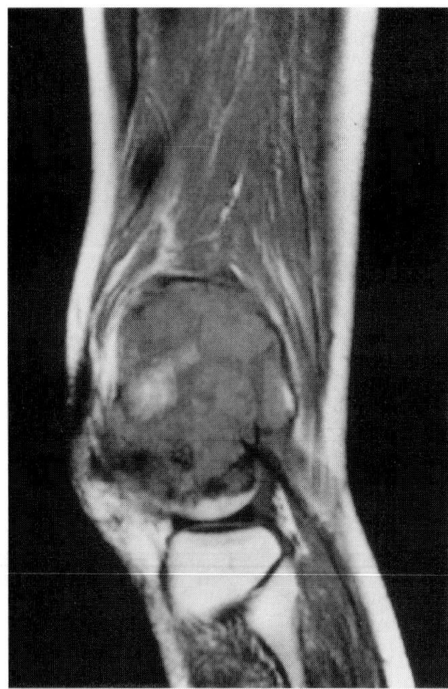

FIG. 9. L: Sagittal T1-weighted sequence (TR/TE 600/20 msec) of a telangiectatic osteosarcoma shows some high signal intensity areas indicating tumor hemorrhage.

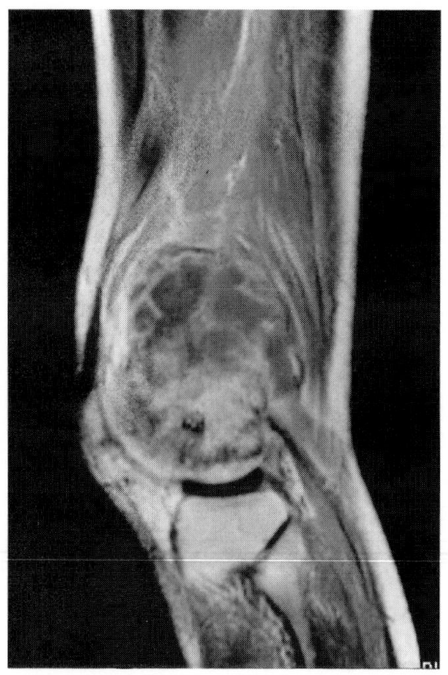

FIG. 9. M: The septal enhancement in the proximal part as visualized on this static (5 min after injection) Gd-enhanced sequence (TR/TE 600/20 msec) is similar to the enhancement pattern seen in grade I and II chondrosarcoma.

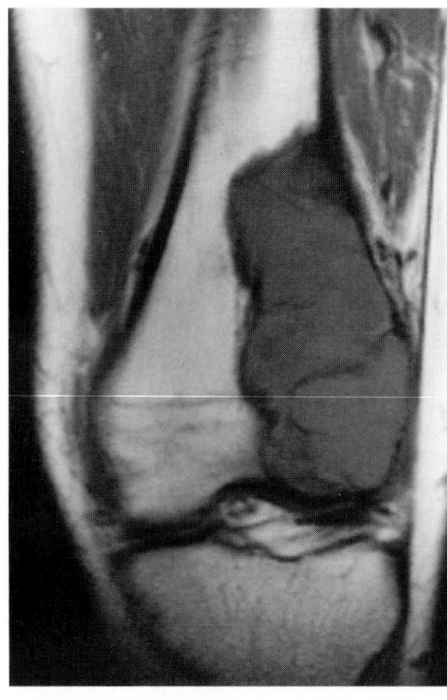

FIG. 9. N: Coronal T1-weighted sequence (TR/TE 600/20 msec) of a giant cell tumor has a nonspecific low signal intensity, but shows the extension toward the subchondral bone, which is one of the characteristics of giant cell tumor.

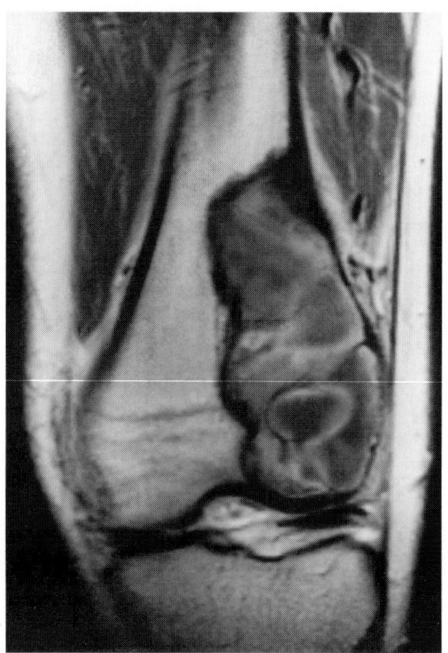

FIG. 9. O: The enhancement pattern visualized on this static (5 min after injection) Gd-enhanced coronal image is septal, and therefore mimics that of low-grade chondrosarcoma. Frequently giant cell tumor enhances in a homogeneous fashion. (Fig. 9H and J from Bloem JL, Holscher HC, Taminiau AHM. Magnetic resonance imaging and computed tomography of primary malignant musculoskeletal tumors. 189–217. In: J.L. Bloem, D.J. Sartoris, eds. *MRI and CT of the musculoskeletal system.* Baltimore: Williams and Wilkins, 1992; 189–217, with permission.) (Fig. M and N from Bloem JI, van der Woude HJ. Tumoren und Tumoranhlige lasionen des Kniegelenkes. In: Nagele M, Adam G, eds. *Moderne kniegelenkendiagnostik.*)

CASE 10

History: A 49-year-old man presents with multiple soft tissue masses in left upper and lower extremity. What is your diagnosis?

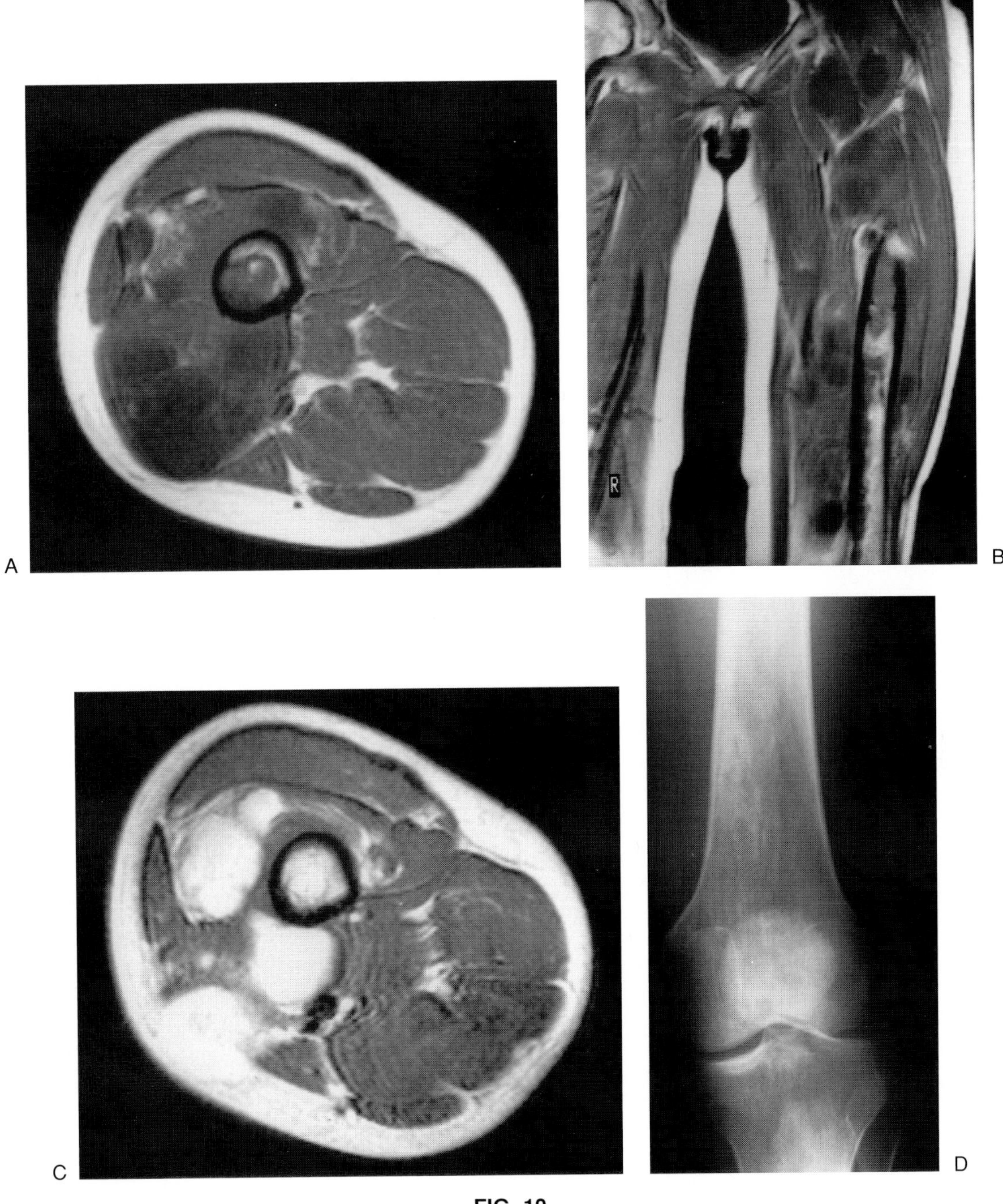

FIG. 10.

Findings: In Fig. 10A,B the intra- and extraosseous lesions seen in the lower extremity have a nonspecific low signal intensity on the axial (A) and coronal (B) T1-weighted (600/20) images. In Fig. 10C the intraosseous lesion and the multiple soft tissue lesions have a nonspecific high signal intensity on this T2-weighted sequence (TR/TE 2000/50 msec). The multiple soft tissue masses do not connect with the osseous lesion. In Fig. 10D a radiograph of the knee shows ill-defined sclerosis and coarse osseous ridges in the distal femur and proximal tibia. Similar lesions were found in the humerus. (Figs. A, C, and D from Bloem JL, van der Woude HJ. MR imaging of the musculoskeletal tumors: In: *Syllabus for the categorical course on body magnetic resonance imaging.* Reston, VA: American College of Radiology 1993;19–33, with permission.)

Diagnosis: Multifocal fibrous dysplasia with multiple myxomas.

Discussion: When multiple osseous and soft tissue tumors are present, as in this case, one should consider multifocal fibrous dysplasia with multiple myxomas. This case demonstrates that although signal intensities are not specific, a specific diagnosis of fibrous dysplasia with myxoma, can still be made (1,2).

One should never forget to apply conventional radiologic and clinical knowledge on MR images. Multiplicity of intra- and extraosseous lesions is in this case the tipoff. Not only metastatic disease and myeloma are multiple. Examples of osseous lesions that can be multifocal are giant cell tumor, fibrous dysplasia, enchondromatosis, and Langerhans cell histiocytosis (3–5).

CASE 11

History: A 26-year old man presented with acute severe pain in the right flank. MR imaging was performed, because a mass was seen on conventional radiographs. What is your differential diagnosis based on the native spin echo images? What kind of additional information could be obtained by performing contrast-enhanced pulse sequences?

A
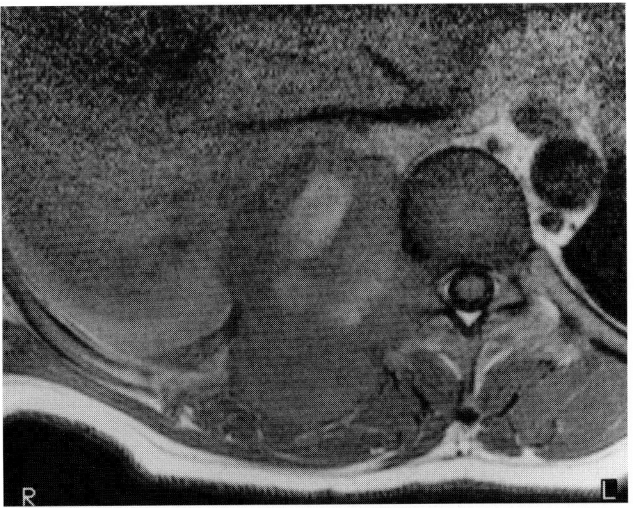

B
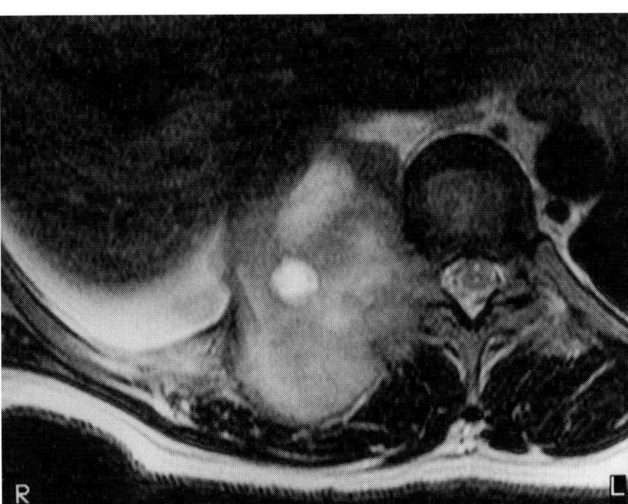

C

FIG. 11.

Findings: In Fig. 11A there is a mass extending from the 10th rib with central areas of intermediate and high signal intensity on the axial T1-weighted SE (TR/TE 600/20 msec) image. In Fig. 11B, the same lesion on axial T2-weighted TSE (TR/TE 5,000/150 msec) images, note the presence of (hemorrhagic) effusion in the pleural space. In Fig. 11C axial dynamic turbo MP GE contrast-enhanced images (15/6.8, prepulse delay time 741, flip angle 30°) reveal rapidly progressive enhancement of the tumor periphery *(1)* with early washout of contrast medium and lack of enhancement in the center of the lesion *(3); (2)* is muscle.

Diagnosis: Giant cell tumor of bone.

Discussion: The specificity of MR imaging in characterization of musculoskeletal lesions is limited (1). The high signal intensity on the T1-weighted native SE images only suggests presence of a central hemorrhagic component of the lesion (Fig. 11A,B).

Based on tumor site, gender, and age, the differential diagnosis of a mass from a rib includes both benign (hemangioma, fibrous dysplasia, eosinophilic granuloma, brown tumor) and malignant tumors (chondrosarcoma, Ewing's sarcoma, osteosarcoma, hemangiopericytoma) (2,3).

There are contradictory reports on the use of gadolinium–diethylenetriamine pentaacetic acid (Gd-DTPA)-enhanced sequences in characterizing musculoskeletal masses (4–6). The dynamic enhancement pattern, using a fast turbo gradient echo sequence, favors a high-grade malignant bone tumor or an active highly vascularized and perfused benign lesion, such as giant cell tumor or osteoblastoma. The lesion is characterized by early progressive peripheral enhancement, immediately after arrival of the bolus in the aorta (Fig. 11C). Moreover, early washout is noted, as depicted in the time-signal intensity diagram, representing an early equilibrium between the concentration of Gd-DTPA in the circulation and the tumoral interstitium.

The histologic diagnosis of this tumor was giant cell tumor, despite the rather uncommon tumor location (<1% of the tumors of the rib, Netherlands Committee of Bone Tumors) (3). The pattern of early wash-in and washout appears to be a common finding in patients with giant cell tumor. The dynamic images, indicating the most viable components of the lesion, assist in targeting the optimal site for a representative biopsy, i.e., the periphery in this particular case (7). In addition, the high rate of perfusion allows the performance of tumor embolization preceding resection of the rib in order to decrease the risk of massive spread in the thoracic space during surgery.

A similar enhancement pattern was found during follow-up of this patient, 6 months after surgery (Fig. 11D–G). This favored the diagnosis of recurrence rather than postoperative, reactive changes. Histologic evidence of recurrence was obtained. In selected cases, dynamic contrast-enhanced sequences can assist in identifying recurrent tumor, if findings of a mass and high signal intensity on T2-weighted SE images are equivocal (8).

In conclusion, fast dynamic contrast-enhanced sequences can be used in selected cases to restrict your differential diagnosis.

D

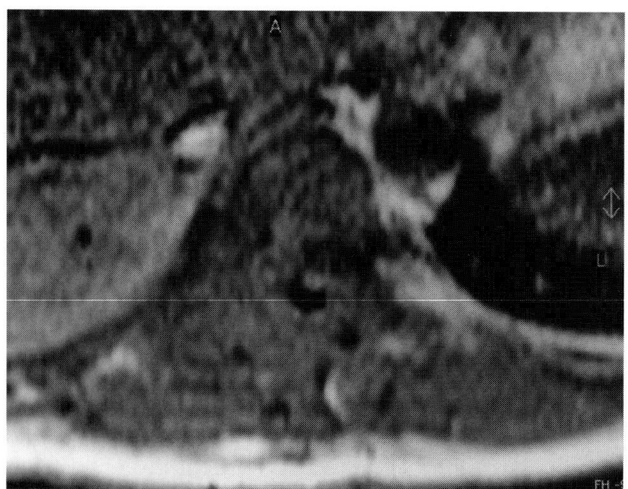

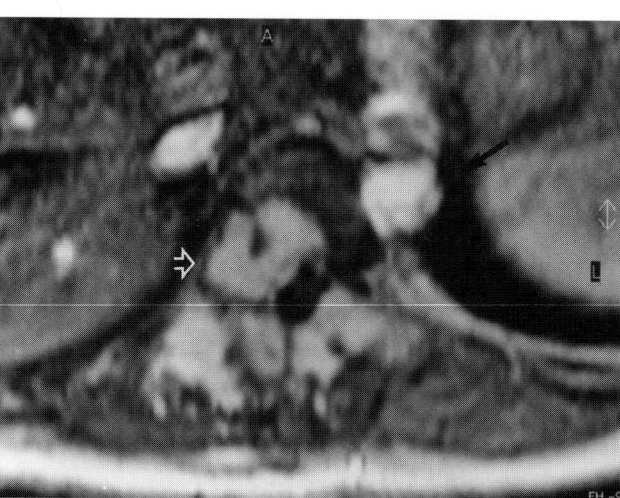

E

FIG. 11. D,E: Axial dynamic contrast-enhanced MP GE (15/6.8, 30°) images, acquired 6 months after resection of the primary lesion reveal a mass. Images were obtained before **(D)** and at 17 sec after **(E)** bolus injection. The mass showed very rapid and complete enhancement *(open arrow),* synchronously with arrival of the bolus in the aorta.

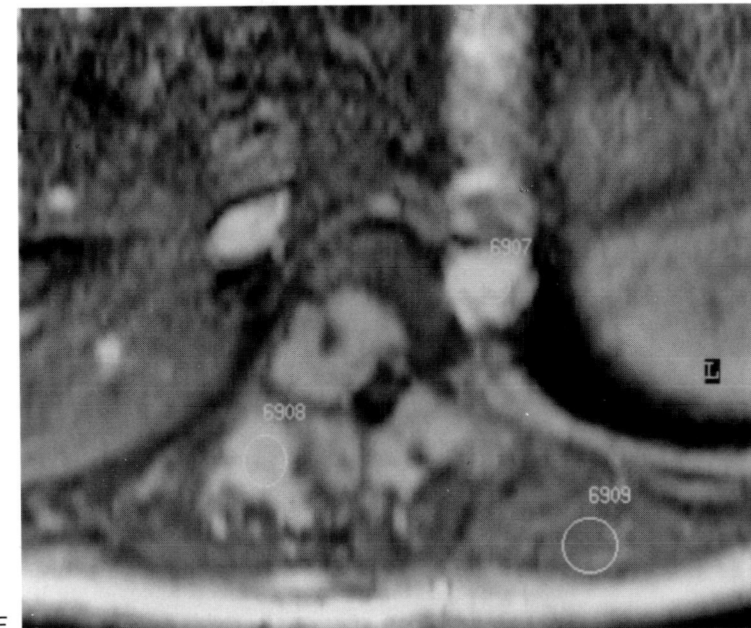

F

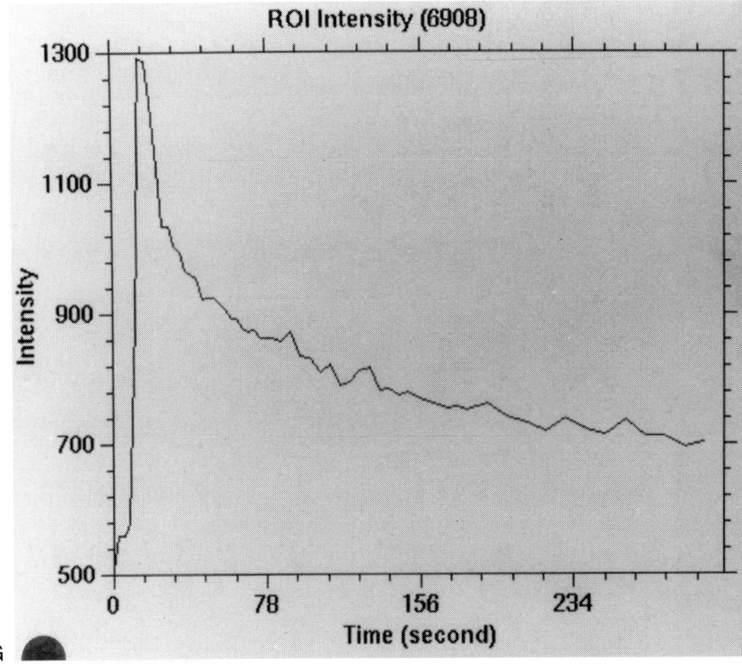

G

FIG. 11. F,G: The time-signal intensity curve of this recurrent mass (ROI 6908) is similar to that of the primary lesion. Recurrence of giant cell tumor was histologically confirmed.

CASE 12

History: A low-grade pelvic chondrosarcoma was resected 10 years ago in this 62-year-old man. A (sub)pleural mass was noticed during routinely performed follow-up. Can you characterize the mass with MR imaging in view of the clinical information?

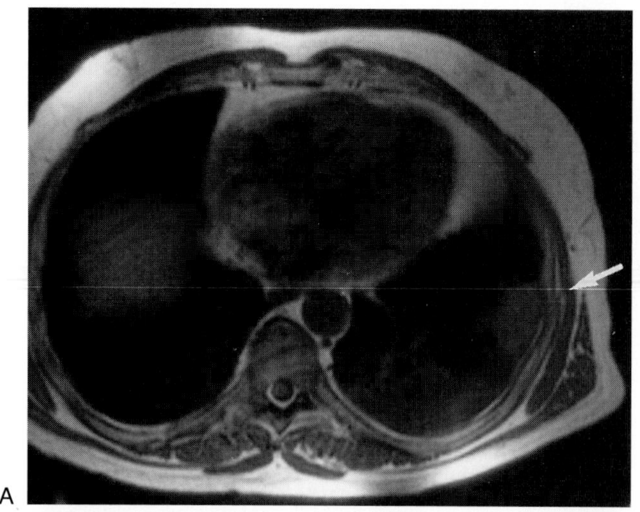

A

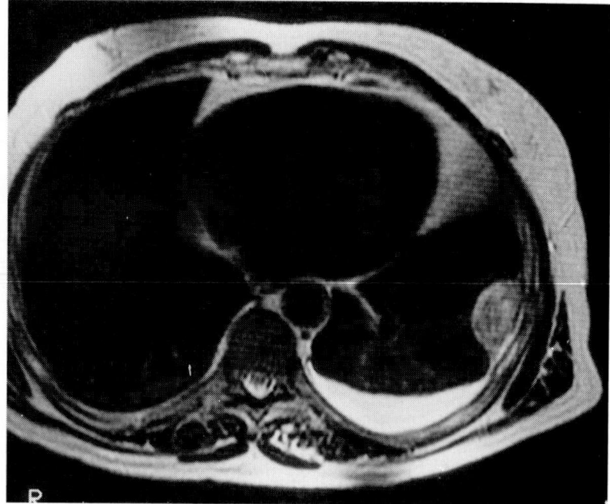

B

C

FIG. 12.

Findings: In Fig. 12A the axial T1-weighted SE (TR/TE 600/20 msec) image shows a well-defined (sub)pleural mass with intermediate signal intensity (arrow). In Fig. 12B, the axial T2-weighted TSE (TR/TE 4500/150 msec) image, the lesion has a heterogeneous appearance. Note there is also pleural effusion. In Fig. 12C, a dynamic contrast-enhanced (turbo) MP GE image (15/6.8/741, 30°) with associated time-signal intensity curves of the entire lesion *(1)* the periphery of the lesion *(2)*, and normal muscle *(3)*, a solid part of the lesion shows rapidly progressive enhancement with a time-intensity curve that paralleled the curve of the aorta.

Diagnosis: Metastasis of chondrosarcoma (high-grade).

Discussion: A low-grade (grade 1) chondrosarcoma is a relatively nonaggressive tumor that, if resected properly, rarely metastasizes (1). Like its benign counterparts (cartilaginous exostosis, enchondroma), low-grade chondrosarcoma is a poorly vascularized lesion, although the vascular response surrounding a chondrosarcoma is greater than that around a mature benign lesion (2). The enhancement pattern of the mass (Fig. 12C) excludes a poorly vascularized lesion (3). The differential diagnosis should therefore include metastasis with a high-grade, probably dedifferentiated, chondrosarcoma component. Alternatively, a biologically active new primary tumor could be considered. A histologic tissue sample of the periphery of the mass revealed high-grade chondrosarcoma, with high mitotic rate.

Lack of biologic activity of a musculoskeletal lesion can be suggested on the basis of gradually increasing signal intensity, compared with the first pass of the bolus of contrast-agent through an artery. On the other hand, the presence of a rapidly enhancing component is no guarantee for high-grade malignancy, as there are various benign lesions (e.g., osteoblastoma, eosinophilic granuloma) that may mimic sarcoma on the basis of their dynamic enhancement pattern (3–5). Enhancement patterns are no substitute for conventional radiographs.

Preferably, dynamic imaging is performed in a plane encompassing the greatest tumor cross section. Even better is to perform multislice imaging, maintaining, however, a high temporal resolution, for instance three seconds per slice. At this moment it is not known if dynamic MR imaging can contribute in differentiating between low-grade chondrosarcoma and enchondroma. Static T1-weighted contrast-enhanced images with or without fat saturation may assist in suggesting the diagnosis of chondrosarcoma, based on the serpentiginous enhancement pattern of the fibrovascular septa (6) (see also Case 9).

CASE 13

History: A 21-year-old woman was treated with neoadjuvant chemotherapy for a Ewing's sarcoma of the proximal femur prior to surgery. Can you predict the response to chemotherapy based on the native and static contrast-enhanced MR sequences?

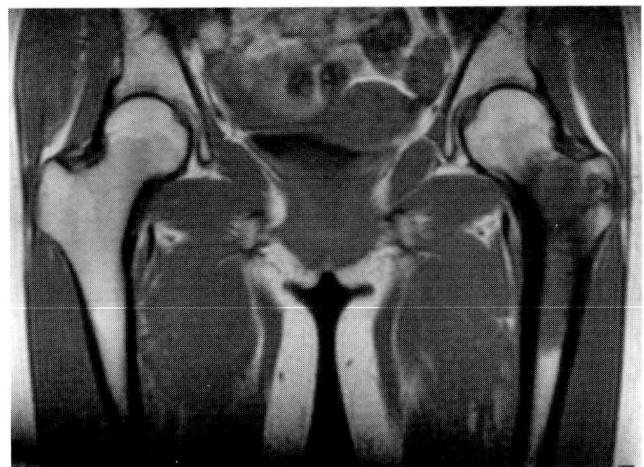
A

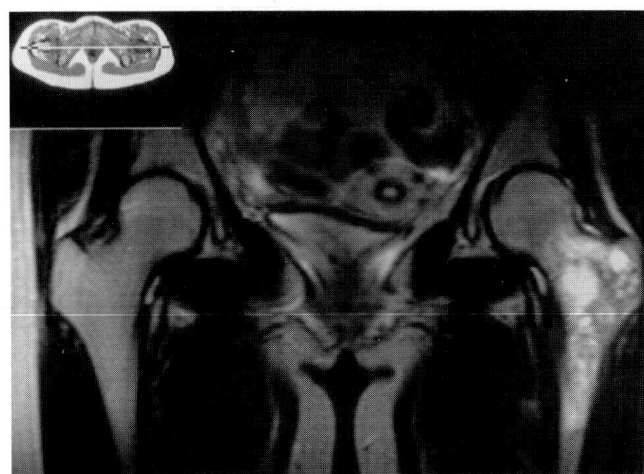
B

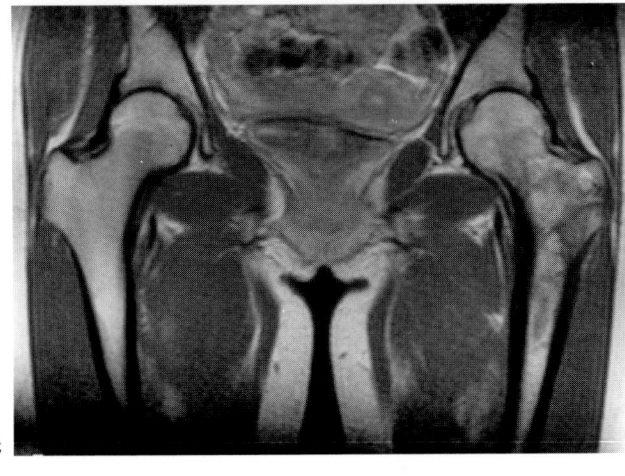
C

FIG. 13.

Findings: In Fig. 13A the coronal native T1-weighted SE (TR/TE 600/20 msec) image shows a relatively low-volume lesion with intermediate signal intensity in the meta-diaphyseal part of the proximal femur. In Fig. 13B the coronal T2-weighted TSE image at the same level reveals nonspecific areas of high signal intensity. In Fig. 13C, the coronal T1-weighted SE image after administration of 0.1 mmol/kg gadopentetate dimeglumine, there is inhomogeneous enhancement of the lesion. There appears to be no soft tissue extension of the tumor.

Diagnosis: Good response, but presence of residual viable tumor.

Discussion: There are several targets to monitor the effect of chemotherapy with MR imaging in patients with high-grade bone sarcoma (1). Morphologic changes secondary to chemotherapy can be evaluated, including changes in tumor volume, signal intensity on T2-weighted images, and changes in edema surrounding the tumor (2–5). Increase in tumor volume, measured on the combination of longitudinal T1-weighted and transverse T2-weighted images, is a reliable predictor of a poor histologic response (3,5). Occasionally, hemorrhagic necrosis, seen as high signal intensity on T1-weighted images, results in increased tumor volume. Unaltered or decreased tumor volume cannot be used to assess the final response in osteosarcoma. Moreover, low-volume intraosseous tumors, like in the presented case, usually do not change substantially in volume, irrespective of the effect of therapy. The situation in patients with Ewing's sarcoma is somewhat different from that in patients with osteosarcoma. Ewing's sarcoma is usually, but not always, accompanied by a soft tissue mass. In cases of Ewing's sarcoma with a large associated soft tissue mass, substantial decrease in tumor volume (>75%) is an indicator of good response, particularly when the soft tissue mass has disappeared, or reduced to a cuff with a low signal intensity rim (6). However, small remnant foci of viable tumor cannot be excluded using these morphologic estimations (6,7). High signal intensity on T2-weighted images (Fig. 13B) after chemotherapy may reflect both tumoral necrosis, compatible with a favorable response, and cellular viable tissue (6–8). In selected cases only, homogeneous high water-like signal intensity in the marrow compartment represents complete mucomyxoid degeneration and thus good response (6). MR signal changes of the soft tissue component on T2-weighted images are not predictive for the ultimate response to chemotherapy in patients with Ewing's sarcoma (8).

Another target for monitoring chemotherapy is tumor vascularization, which can be studied with angiography, (three-phase) bone scintigraphy, color Doppler ultrasound (US), and contrast-enhanced MR (1,9–11). A limitation of static T1-weighted contrast-enhanced images is the overlap of enhancement of (remnant) viable tumor, reactive edema, and chemotherapy-induced vascularized granulation tissue (11,12). This warrants the use of a dynamic sequence that allows distinction between viable and nonviable neoplastic tissue (9,13–15). Fast dynamic subtraction images (Fig. 13D,E) of this case focus on the presence of small areas of early enhancement that correspond to small clusters of residual, morphologically viable tumor in the macrosection of the resected tumor (Fig. 13F). Subtracting the enhanced images from the selected native image(s) facilitates identification of small tumoral nests, by suppressing substrates with high signal intensity on T1-weighted images, like yellow bone marrow (16). Small foci of minimal residual disease stresses the need for surgical removal of the affected bone, despite overall excellent response.

The presence of areas of residual viable tumor should be assumed when (small) areas show rapidly progressive enhancement, starting within 6 seconds after arrival of the contrast-medium in a (feeding) artery (15). Fibrovascular tissue and edema will show gradual enhancement, starting after the progressive enhancement of solid viable clusters. Presence of scattered, isolated viable-appearing cells cannot be excluded on the basis of dynamic contrast-enhanced images (15).

In conclusion, for the assessment of response to chemotherapy in small intraosseous bone sarcomas, fast dynamic (subtraction) imaging is the modality of first choice.

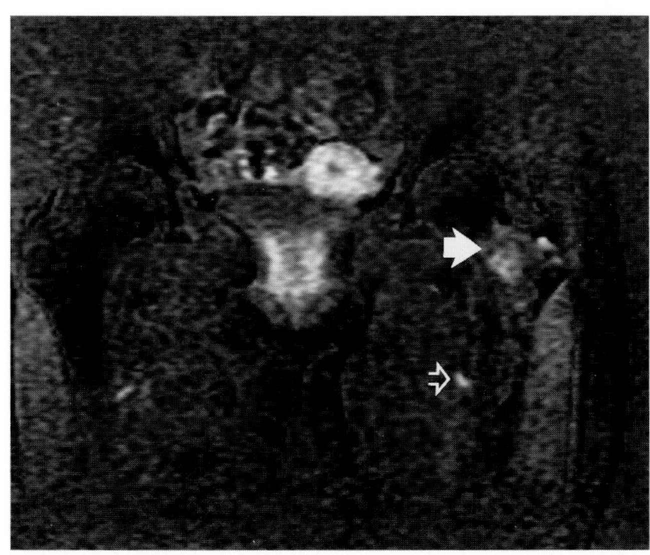
D

FIG. 13. D: Coronal dynamic contrast-enhanced GE image (15/6.8/741/30°, temporal resolution 1 sec) obtained 17 sec after bolus injection and 3 sec after start of arterial enhancement *(arrows)* reveals early enhancement of small areas in the proximal part of the tumor.

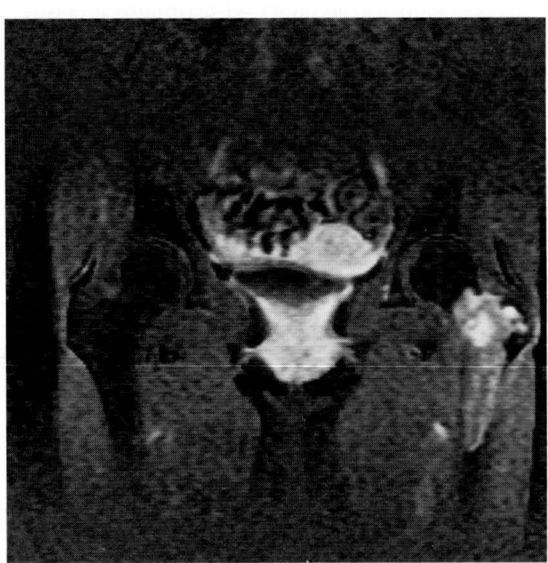

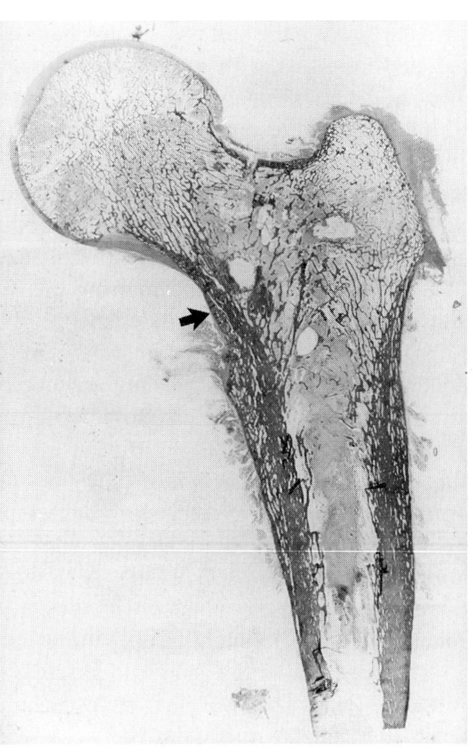

E F

FIG. 13. E: Dynamic image obtained at the end of the GE sequence (5 min after injection). Besides enhancement of the small solid areas, there is heterogeneous enhancement of the entire lesion, corresponding with vascularized granulation tissue. **F:** Macrosection of the surgical specimen cut in the plane of dynamic imaging shows foci of residual tumor at the corresponding sites of early enhancement *(arrow).* (Figs. D–F from van der Woude HJ, Bloem JL, Verstraete KL, Taminiau AHM, Nooy MA, Hogendoorn PCW. Osteosarcoma and Ewing's sarcoma after neoadjuvant chemo-therapy: value of dynamic MR imaging in detecting viable tumor before surgery. *AJR* 1995;165: 593–598).

CASE 14

History: This 15-year-old girl with an osteosarcoma of the proximal tibia was treated with three cycles of neoadjuvant chemotherapy. Can you differentiate the viable tumor components from (chemotherapy-induced) necrosis on the early and late dynamic contrast-enhanced images following completion of chemotherapy?

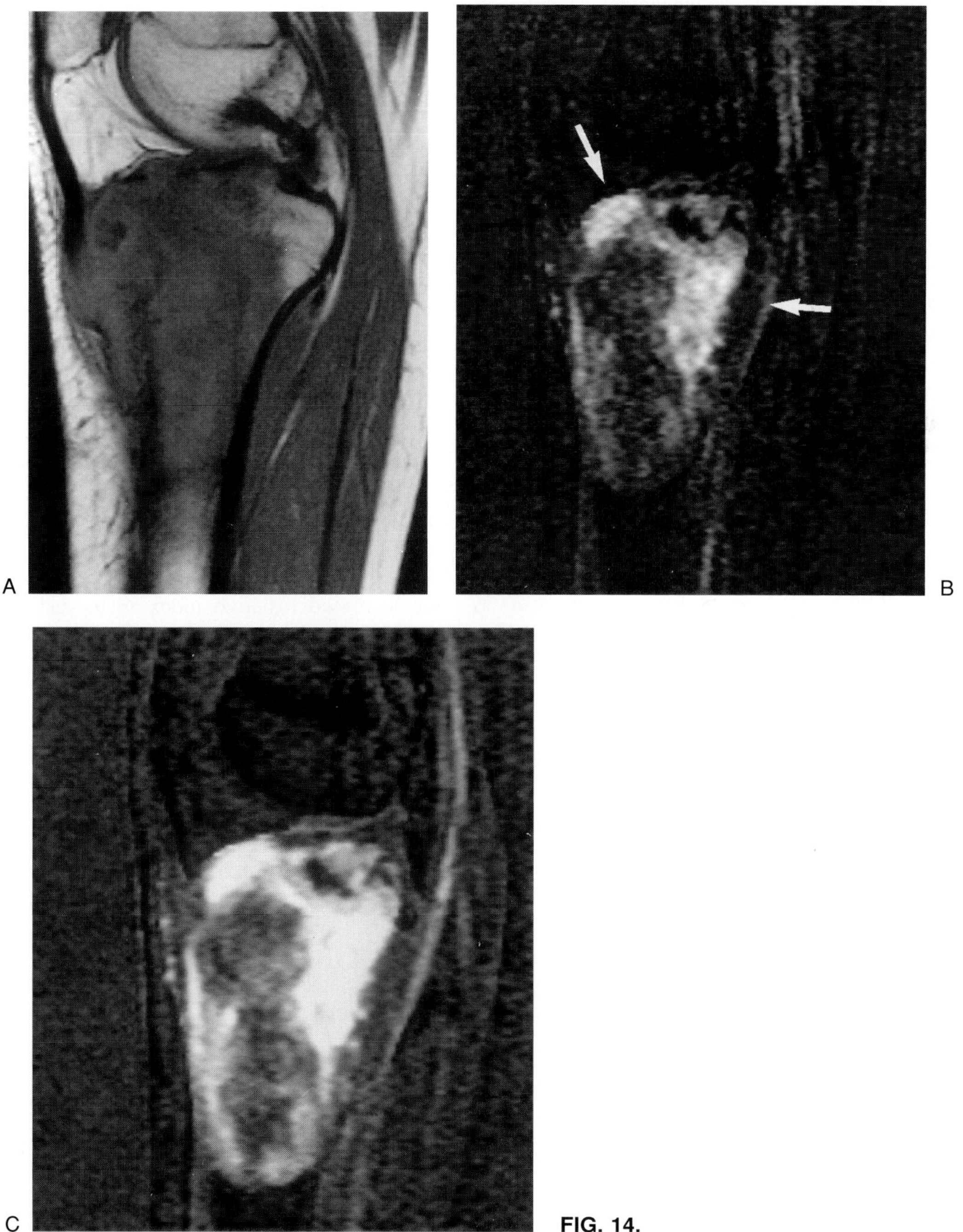

FIG. 14.

Findings: In Fig. 14A, the sagittal T1-weighted SE (TR/TE 600/20 msec) image of a tumor with predominantly intermediate signal intensity in the proximal tibia, there is a soft tissue mass associated with the intramedullary part of the tumor, abutting the infrapatellar fat of Hoffa. In Fig. 14B, the dynamic contrast-enhanced MP GE (15/6.8/741, 30°) image obtained 17 seconds after contrast injection, there is early enhancement of solid peripheral parts of the soft tissue mass (arrows), whereas the central part of the tumor lacks enhancement in the early phase of the dynamic sequence. In Fig. 14C a dynamic GE image taken at the end (5 min) of the sequence depicts larger areas of enhancement.

Diagnosis: Remnants of viable malignant fibrous histiocytoma-like osteosarcoma are present in the periphery of the tumor (Fig. 14D).

Discussion: The combined use of a surface coil and subtracted images of a fast dynamic MP GE sequence (temporal resolution, 1–3 sec) with relatively large field of view offers great potential to identify residual viable tumor foci of at least 3 mm^2 (1–3). Location of the areas with the highest cellularity of this malignant fibrous histiocytoma-like osteosarcoma on the early dynamic images (Fig. 14B) correspond to those found in the macrosection of the surgical specimen (Fig. 14D). The central part of the tumor, composed of chemotherapy-induced necrosis or tumoral necrosis, based on tumor weight-adjusted decline of flow and surrounded by a zone of granulation tissue, is distinguished by gradual enhancement or absence of enhancement, depending on the amount of vascularity. Moreover, start of enhancement of nonviable tumor, or reactive tissue will occur later than enhancement of viable tumor (later and earlier than 6 seconds after arterial enhancement, respectively). Presumably the morphologically and functionally abnormal vascular network associated with viable tumor, contributes to the preferential blood flow, reflected by the early progressive signal intensity increase of viable tumor in the time-intensity diagrams (2,4) (Fig. 14E).

The theory of a low-resistance vascular network is supported by the flow pattern in the popliteal artery feeding the tumor-bearing limb, compared with the contralateral normal artery (Fig. 14F). Such a flow (velocity) measurement can be easily performed, using a 2D cine phase contrast sequence with a large field of view. In this particular case, a normal triphasic pattern with a reverse flow component in diastole is encountered in the normal popliteal artery. A higher systolic flow with absence of a reverse flow component in diastole is found in the popliteal artery proximally to the tumor at the same level as the contralateral artery. In selected cases with a soft tissue tumor component, color Doppler flow imaging can also be used as a noninvasive tool to assess the potential viability of the tumor, based on the presence or absence of characteristic features accompanying highly vascularized viable tumor; these include high-frequency Doppler shifts (at the periphery) of the soft tissue mass (Fig. 14G,H) and decreased resistive index in the artery feeding the tumor bearing limb compared to the opposite artery (4,5).

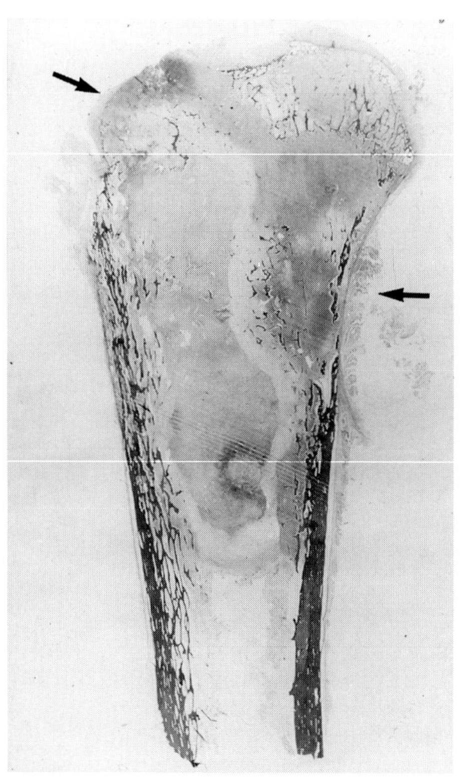

FIG. 14. D: Macrosection of the resected specimen shows that the earliest enhancing areas (Fig. 14B) correspond to the most cellular remnant viable tumor parts at the periphery *(arrows).* The tumor center is composed of acellular necrosis surrounded by a zone of granulation tissue.

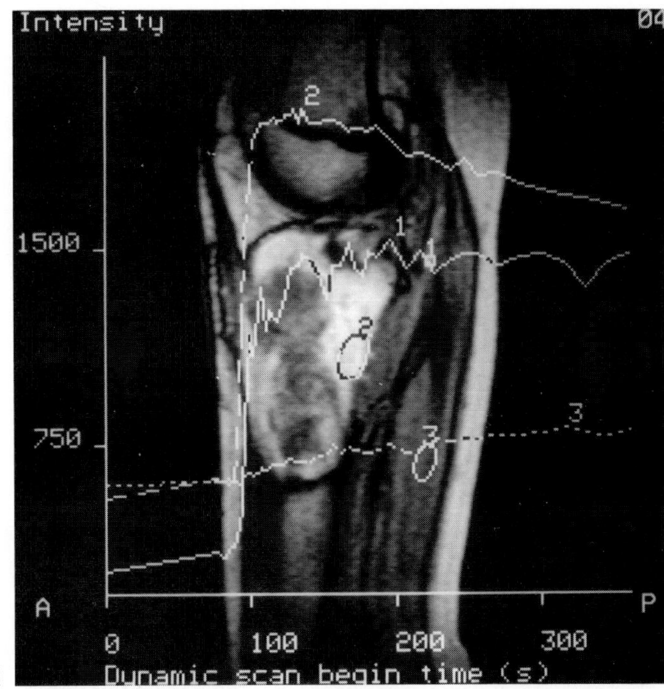

FIG. 14. E: Time-intensity curves of the peripheral early enhancing parts show rapidly progressive increase of signal intensity, reaching an early plateau-phase and even decline (washout) shortly after the point of maximum intensity. *1,* artery; *2,* tumor in periphery; *3,* muscle. (Figs. D and E from van der Woude HJ, Bloem JL, Verstraete KL, Taminiau AHM, Nooy MA, Hogendoorn PCW. Osteosarcoma and Ewing's sarcoma after neoadjuvant chemo-therapy: value of dynamic MR imaging in detecting viable tumor before surgery. *AJR* 1995;165:593–598; with permission.)

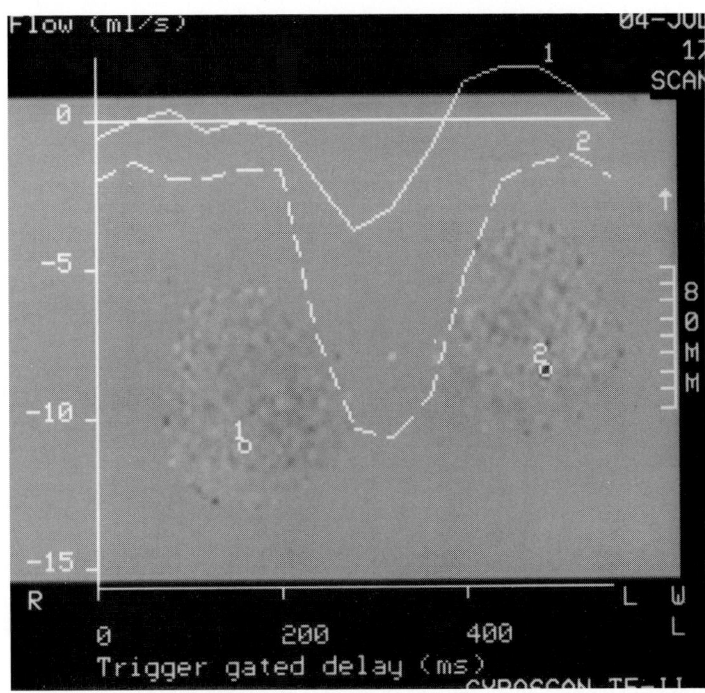

FIG. 14. F: Flow pattern obtained in the popliteal artery feeding the tumor-bearing limb *(2),* compared to the opposite normal limb *(1),* using a 2D phase contrast sequence. There is a usual triphasic pattern in the normal popliteal artery. The artery feeding the tumor-bearing limb lacks a reverse flow component in diastole, consistent with a low-resistant peripheral tumoral vascular network. Moreover, there is higher forward flow in systole.

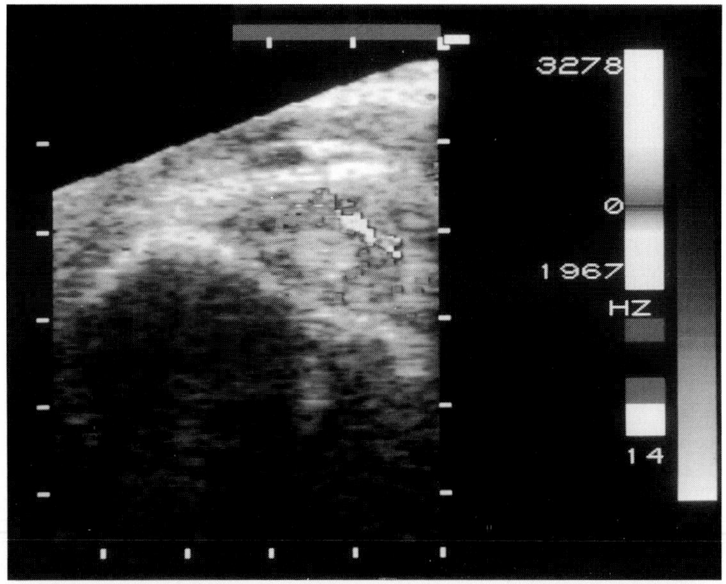

FIG. 14. G,H: Transverse color Doppler flow images of the soft tissue mass depict peripheral flow **(G)** with high frequency Doppler shifts **(H)**, consistent with viable tumor.

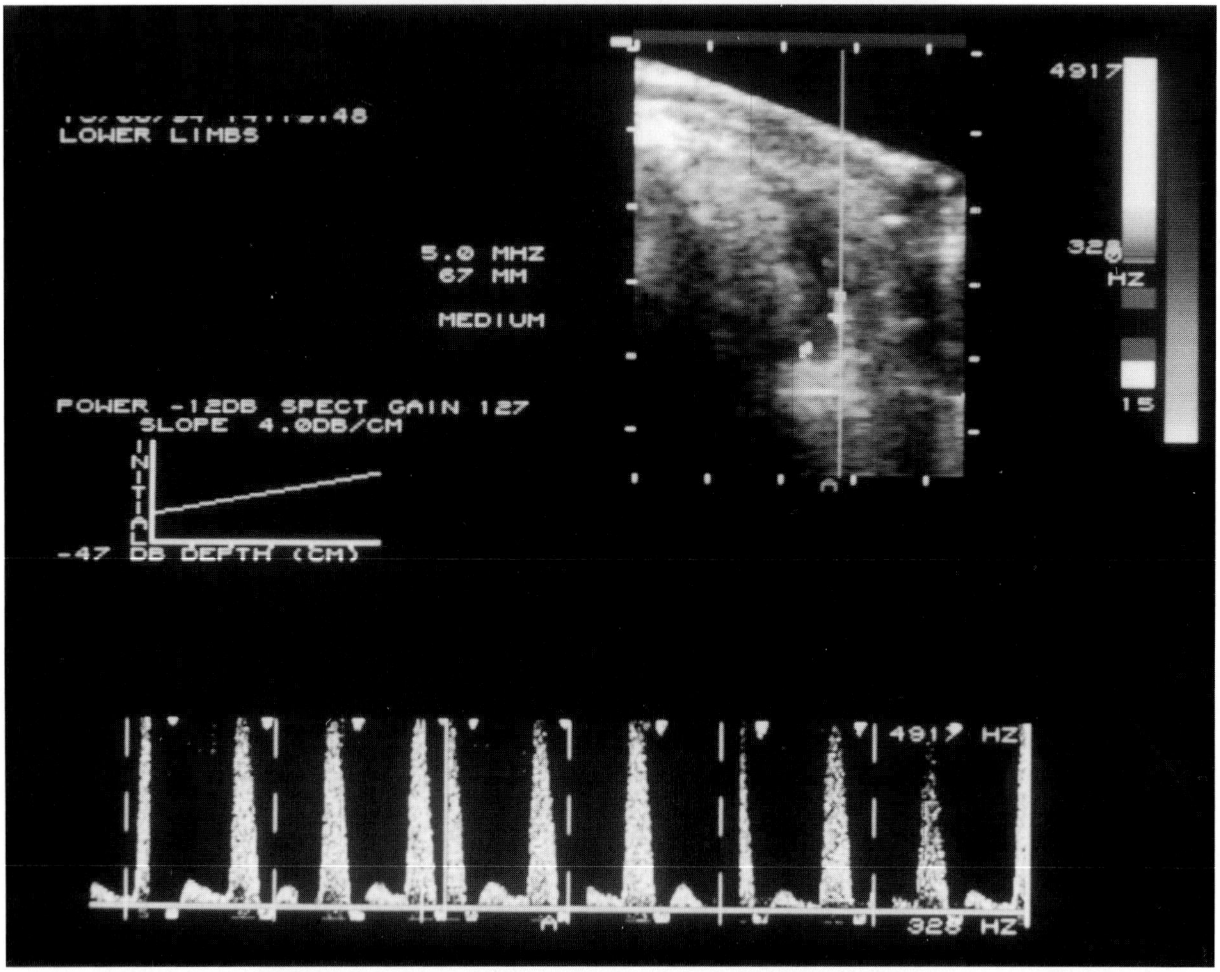

CASE 15

History: This 16-year-old girl was treated with chemotherapy for a chondroblastic osteosarcoma of the proximal tibia. Are there signs for the presence of a chondroblastic phenotype on the pre- and postchemotherapy MR images? How would you classify the response to chemotherapy?

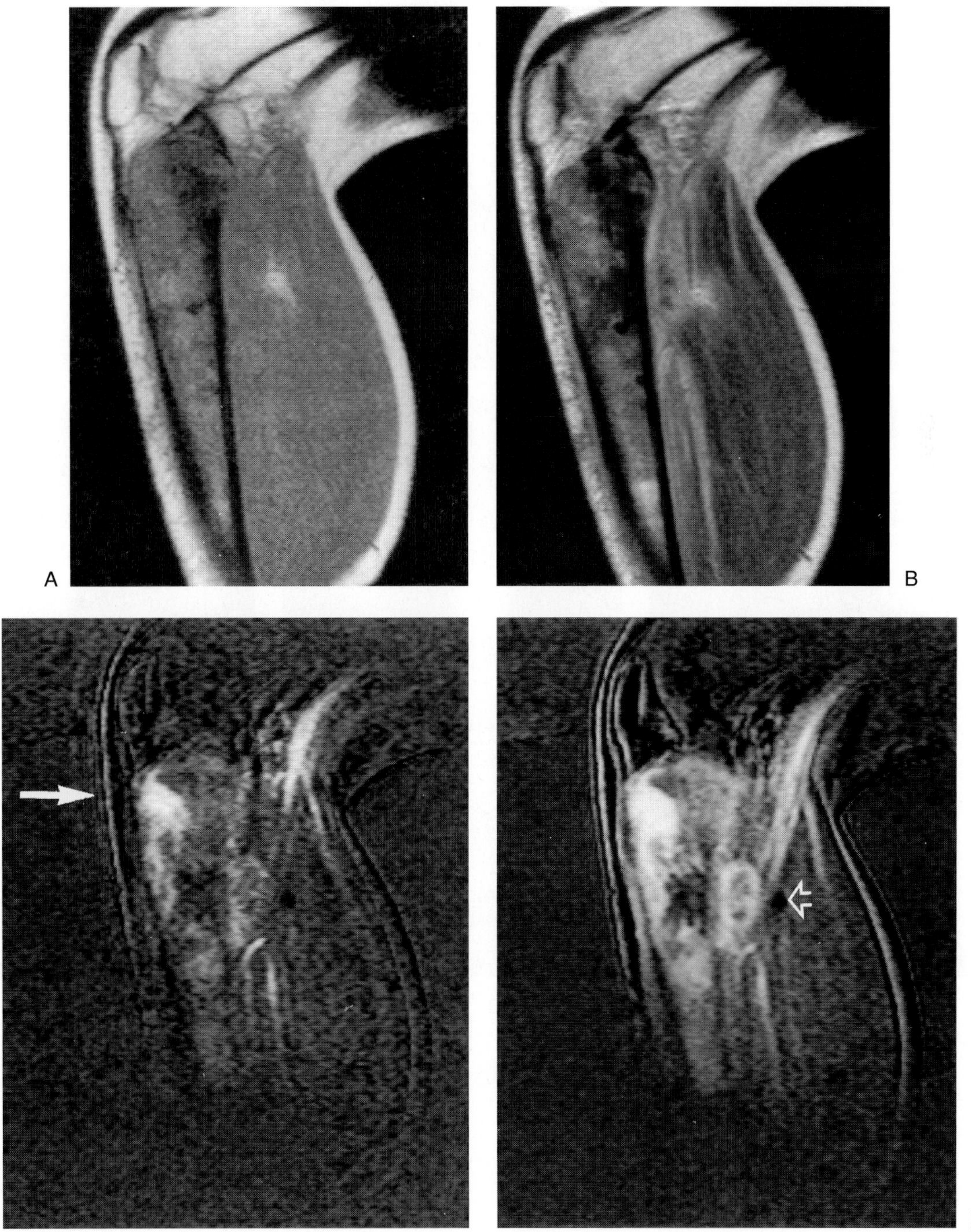

FIG. 15.

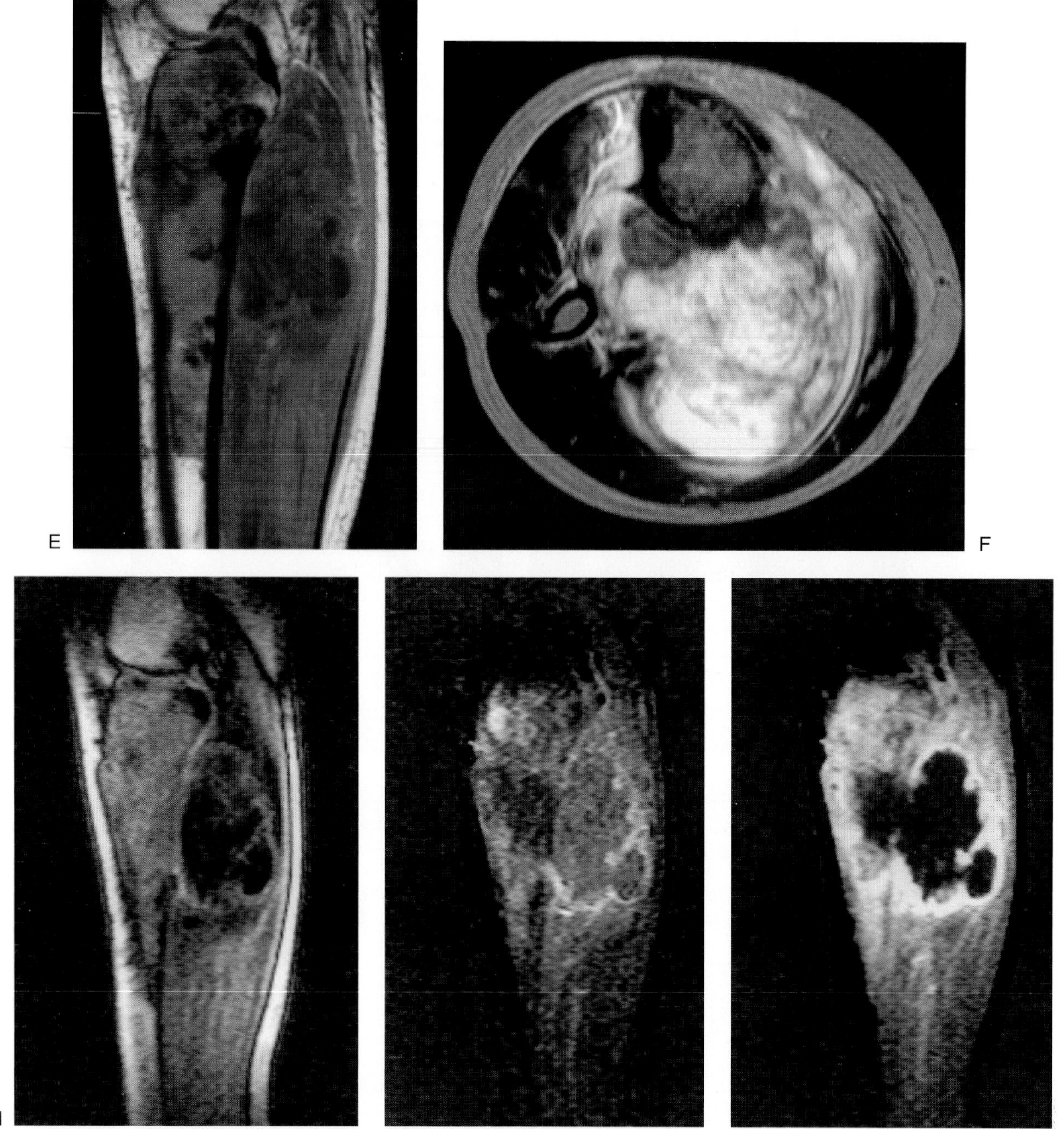

FIG. 15.

Findings: In Fig. 15A, the prechemotherapy sagittal T1-weighted SE (TR/TE 600/20 msec) image reveals a large mass arising from the proximal tibia. In Fig. 15B, the corresponding static T1-weighted image after the administration of gadopentetate dimeglumine, there is heterogeneous enhancement. In Fig. 15C,D, there is early and progressive enhancement of a solid area in the proximal part of the tumor (arrow) on these fast dynamic MP GE (15/6.8/741, 30°) images, taken at 17 (C) and 50 seconds (D) after bolus injection of 0.1 mmol/kg gadopentetate dimeglumine. Note the presence of contrast medium in the (feeding) popliteal artery. A relatively limited soft tissue mass is appreciated posteriorly, with a solid enhancing peripheral rim (open arrow). In Fig. 15E, the sagittal T1-weighted SE (TR/TE 600/20 msec) image after chemotherapy, there is an increased tumor volume. Note the relation of the cruciate ligament to the extension of the tumor. In Fig. 15F, the axial T2-weighted turbo SE (TR/TE 4,500/150 msec) image after chemotherapy shows a large posterior soft tissue mass, predominantly of high or very high signal intensity. In Fig. 15G–I, the sagittal dynamic MP GE images after chemotherapy, acquired before (G), at 17 (H) and 50 seconds (I) after bolus injection, display persistent early enhancement of a small area superiorly (H), and a lobulated soft tissue mass posteriorly that has increased in size and only enhances at the periphery (I).

Diagnosis: The nonenhancing, lobulated mass with high signal intensity is consistent with presence of cartilage. Histologic response was poor.

Discussion: Occasionally, a major part of an osteosarcoma may show differentiation into cartilaginous phenotype (1). According to Mirra (2), an osteosarcoma is referred to as chondroblastic osteosarcoma if more than 90% of the tumor is cartilaginous. These prominent cartilage nodules, however, do not behave in the relatively nonaggressive fashion of the classic chondrosarcoma (1–3). The response to chemotherapy of these chondroblastic osteosarcomas is in general heterogeneous. The high-grade osteoblastic part, that usually shows early and progressive enhancement, may reveal a (partial) response (Fig. 15C,D,H).

In the present case, there is a persistent, although smaller, solid enhancing area after chemotherapy in the proximal intramedullary part of the tumor, corresponding to osteoblastic osteosarcoma in the macrosection of the surgical specimen (Fig. 15C,D,H,J). The chondroblastic portion is not sensitive to chemotherapy. This is reflected by the increase of the posterior mass during chemotherapy (Fig. 15A,E). Characteristically, the central hypocellular chondromyxoid part (high signal intensity on T2 weighted images, Fig. 15F) lacks enhancement, whereas the peripheral proliferation zone, composed of cellular cartilage, shows early enhancement (4). Another type of tumor matrix with absence of enhancement is depicted in the central intramedullary part (Fig. 15I,J). Abundant (calcified) osteoid should be considered when lack of enhancement is noted, in accordance with low signal intensity on T1- and T2-weighted image (4). Alternatively, central lack of enhancement can be secondary to hemorrhage (high signal intensity on T1-weighted images), liquefaction or mucomyxoid degeneration (high signal intensity on T2-weighted images) or occasionally a focus of osteomyelitis (4,5).

The optimum determination of morphologic substrates based on MR features is achieved by combining T1 SE, T2 (TSE), and fast GE dynamic contrast-enhanced sequences, using a surface coil with a relatively large field of view.

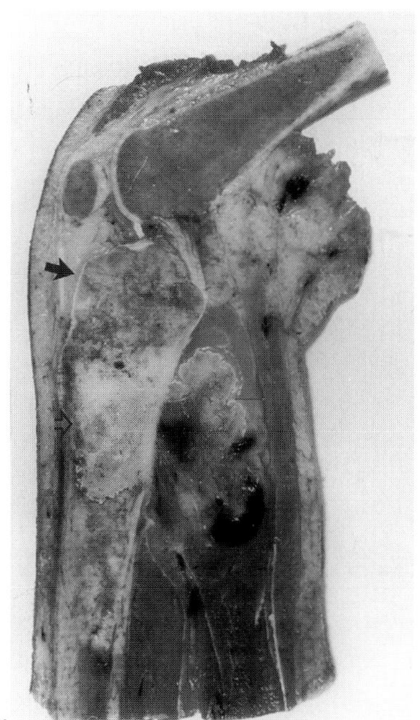
J

FIG. 15. J: The macrosection of the resected specimen cut in the plane of dynamic imaging shows that the proximal early enhancing focus corresponds to an area of viable osteoblastic osteosarcoma *(arrow)*. The posterior soft tissue mass is composed of low cellular and myxoid (nonenhancing) cartilage, margined by a cellular (enhancing) cartilaginous proliferation zone. The central intramedullary part consists of (nonenhancing) osseous matrix without viable tumor cells *(open arrow)*. The histologic response was poor.

CASE 16

History: A 56-year-old man presented with pain in the lower leg. Conventional radiographs were made (not shown) and MR imaging was performed. What would be your MR diagnosis on T1- and T2-weighted images? Can you restrict your differential diagnosis after the contrast-enhanced sequences?

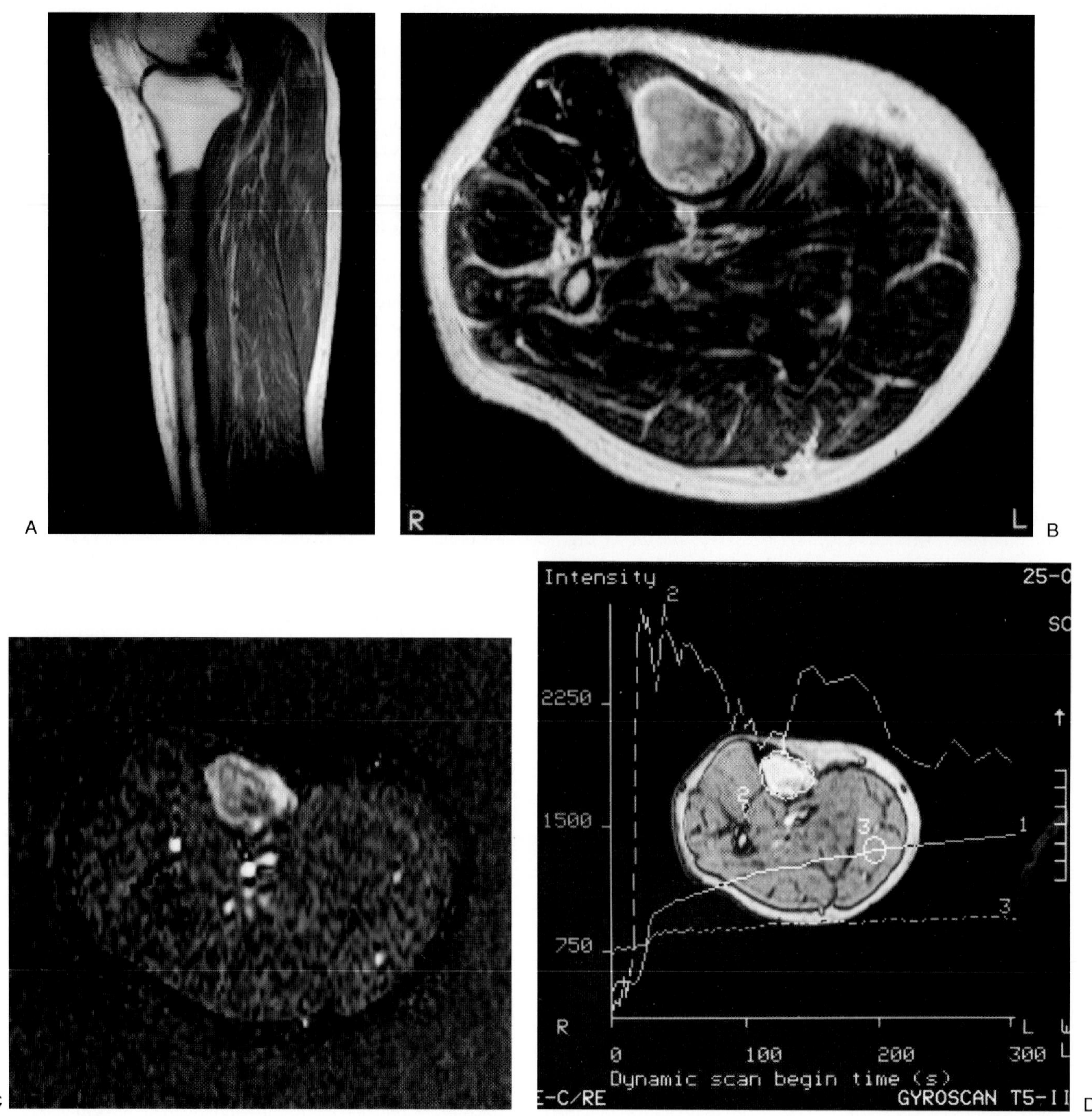

FIG. 16.

E 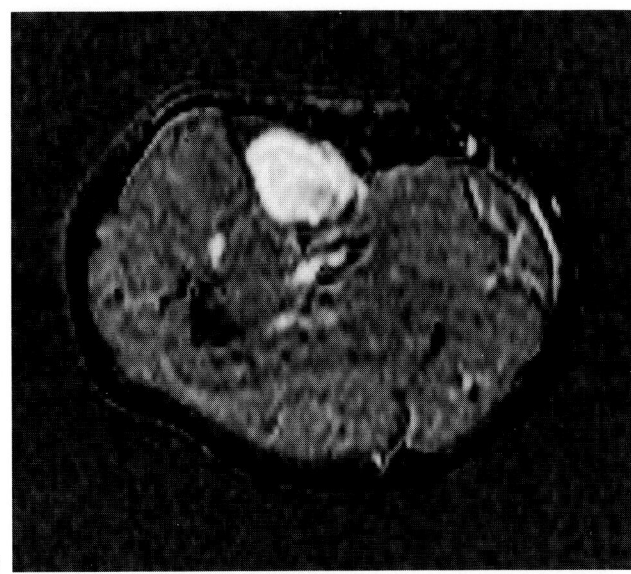

FIG. 16.

Findings: In Fig. 16A the sagittal T1-weighted SE (TR/TE 600/20 msec) image depicts an intraosseous lesion in the diaphyseal part of the tibia with intermediate signal intensity. There is endosteal invasion of the cortical bone. In Fig. 16B, the axial T2-weighted turbo SE (TR/TE 5,800/150 msec) images, the lesion has intermediate signal intensity. In Fig. 16C, after bolus injection of Gd-DTPA, there is initially peripheral enhancement, starting at 6 seconds after arrival of the bolus in the artery on this axial dynamic MP GE (15/6.8/741, 30°) image. In Fig. 16D the signal intensity of the lesion *(1)* continues to increase instead of reaching an early plateau phase relative to the arterial curve *(2)*; *(3)* is muscle. In Fig. 16E there is homogeneous enhancement of the lesion at the end of the dynamic sequence (5 min after bolus injection).

Diagnosis: Low-grade malignant fibrous histiocytoma of bone.

Discussion: The signal intensity characteristics of the lesion are not specific (Fig. 16A,B). The endosteal scalloping suggests an active rather than a biologically inert lesion. With the patient's age and the localization of the lesion in mind, one should consider in the first place a solitary metastasis, lymphoma, or myeloma localization (1). Fibrous dysplasia presenting symptomatically at this age is not likely, regarding the cortical attack as well. The time-signal intensity curve associated with the dynamic contrast-enhanced sequence displays a short track of rapidly increasing signal intensity, relative to the arterial curve. This, however, does not reach an early plateau phase (Fig. 16E). This reduces the possibility of a high-grade primary bone sarcoma or highly vascularized metastasis, which usually shows early, and rapidly progressive enhancement, with early washout of contrast agent.

The lesion proved to be low-grade malignant fibrous histiocytoma (MFH), dermatofibrosarcoma of bone. It is mostly seen in the long bones, sometimes as a complication of Paget's disease or occasionally arising in or around old bone infarcts. Usually, the high-grade MFH pursues an aggressive course (2). Presumably, the less aggressive character of this tumor, expressed histologically by a low mitotic rate and subtle nuclear pleomorphism, is represented by a more gradual and delayed enhancement pattern, related to the arterial first pass of contrast agent. Since certain active benign lesions may exhibit enhancement patterns resembling malignant bone tumors, dynamic enhancement features alone cannot be used to differentiate between benign and malignant masses (3,4). These should be analyzed in addition to known radiographic features and clinical data.

Although metastasis is not typical, wide local excision must be performed in patients with low-grade MFH to avoid a more aggressive clinical course in case of transformation. The macrosection of the surgical specimen (Fig. 16F) and the specimen radiograph (Fig. 16G) reflect the endosteal scalloping, as can be seen on the sagittal T1-weighted SE image (Fig. 16B).

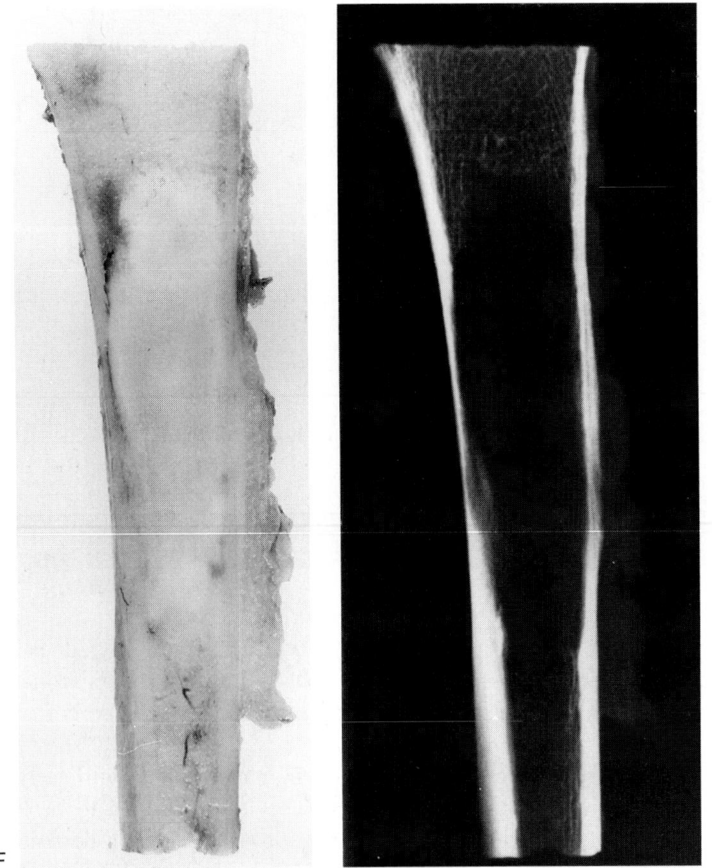

FIG. 16. F: Corresponding macrosection of the surgical specimen. **G:** The specimen radiograph also shows the lytic intraosseous lesion with ill-defined margins. There is partial destruction of the cortical bone.

CASE 17

History: In this 12-year-old boy with a common type high-grade central osteosarcoma of the proximal tibia, MR imaging was performed before and after three cycles of neoadjuvant chemotherapy. Can you exclude involvement of the epiphysis before and after chemotherapy? Is there a good or poor response?

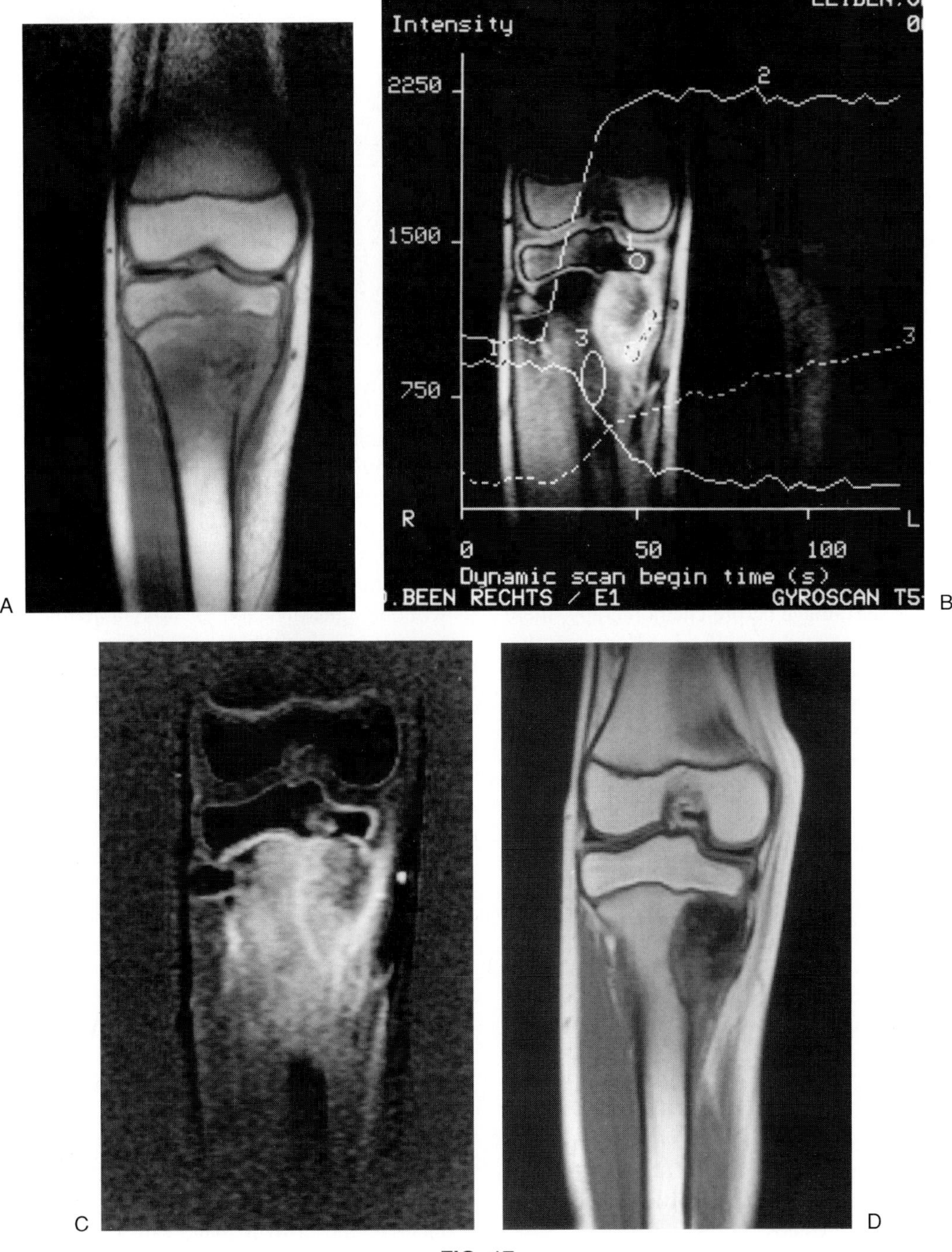

FIG. 17.

E 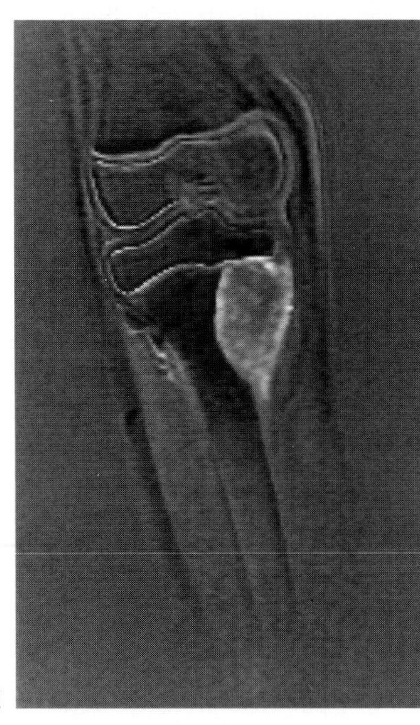

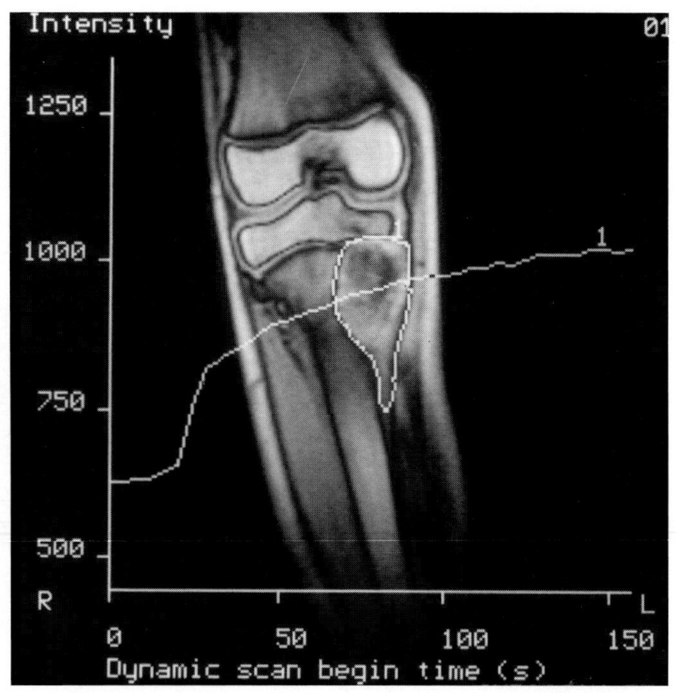

F

FIG. 17.

Findings: In Fig. 17A, the coronal T1-weighted SE (TR/TE 600/20 msec) image before chemotherapy, there is a low signal intensity lesion abutting the physeal plate. There is an ill-defined area of homogeneous intermediate signal intensity adjacent (lateral) to the tumor, and a smaller similar area in the epiphysis. In Fig. 17B, the dynamic contrast-enhanced MP GE (15/6.8/741, 30°, temporal resolution 3 sec) image before chemotherapy, there is normal bone marrow of the epiphysis *(1)* progressive enhancement of the lesion *(2),* and delayed, gradual enhancement of the bordering areas *(3),* consistent with edema. In Fig. 17C the late dynamic subtraction image (5 min after bolus injection) illustrates the reduced potential of contrast-enhanced MR to differentiate between viable tumor and reactive changes in the late phase of a dynamic sequence. In Fig. 17D, the coronal T1-weighted SE (TR/TE 600/20 msec) image after chemotherapy, the tumor volume did not change, but is better defined. There is obviously normal bone marrow adjacent to the tumor and in the epiphysis. In Fig. 17E there is only slight peripheral enhancement on this dynamic subtraction image after chemotherapy. In Fig. 17F, the nonsubtracted dynamic GE image made after chemotherapy, the time-signal intensity curve of the tumor has changed to a gradual enhancement pattern, with reduced maximum signal intensity and absence of an early plateau phase relative to the prechemotherapy curve (Fig. 17B).

Diagnosis: The epiphysis is not involved in this good-responding osteoblastic osteosarcoma.

Discussion: Marrow edema is generally noted as an ill-defined area of intermediate signal intensity on T1- and high signal intensity on T2-weighted, fat saturation, and STIR sequences (1–4). Probably due to the expanded interstitial space, there is gradual and homogeneous enhancement of the area referred to as the reactive zone. Dynamic images thus assist in distinguishing viable, progressively enhancing tumor from edema (5,6). Although the presence of scattered microscopic tumor cells within the areas of edema cannot be excluded in high-grade tumors, there are reports describing normal bone marrow without signs of viable or nonviable tumor or inflammatory reaction in the regions of previously presumed edema (3). In the present case, the tumor cross-area dimensions on the MR images after chemotherapy exactly corresponded to the dimensions of the resected specimen (7) (Fig. 17G). Normal bone marrow was found adjacent to the tumor in the metaphysis and epiphysis.

Disappearance of marrow edema secondary to chemotherapy is not a reliable indicator of a good response (3). Similarly, decrease of soft tissue edema, noticed as ill-defined areas of high signal intensity on T2-weighted images with a feathery aspect, is found in both good and poor respondents. There is, however, evidence that increase in edema is associated with poor response (8).

In the present case, good response of the tumor is obvious, on the basis of the altered enhancement pattern after chemotherapy. This was histopathologically confirmed. Involvement in the epiphysis is not an uncommon finding in patients with osteosarcoma, which contradicts the concept that the physis acts as a ''barrier'' to tumor spread (9,10). Occasionally, it can be difficult to exclude involvement of the growth plate with dynamic imaging, since flow through the physeal vessels contributes to early enhancement as well (Fig. 17H–J). However, longitudinal T1-weighted images will usually be sufficient to determine the physeal or epiphyseal extent of the tumor.

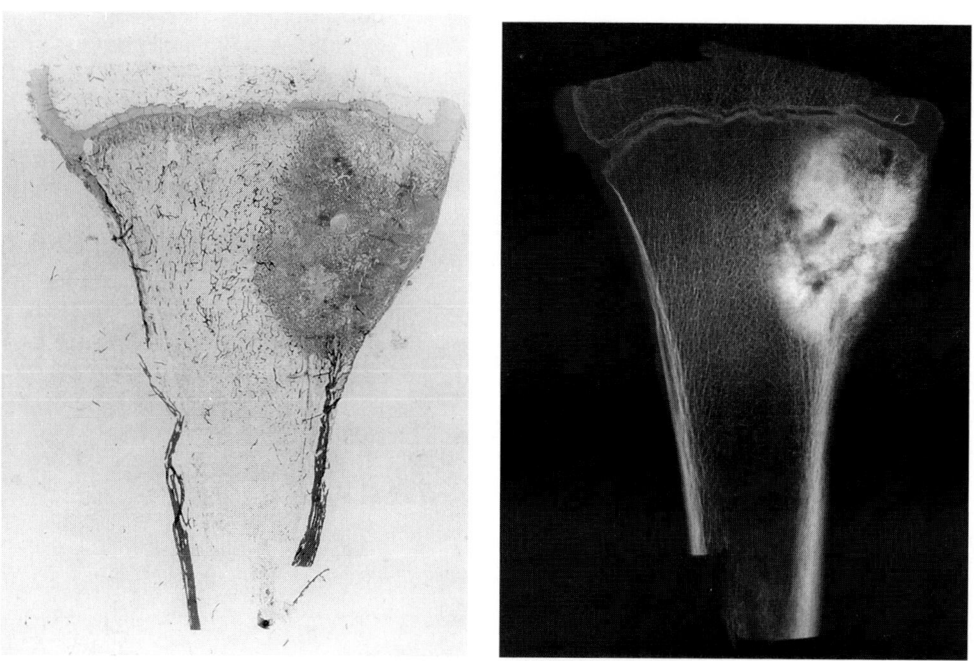

FIG. 17. G: Macrosection of the resected specimen cut in the plane of dynamic imaging; the dimensions of the intraosseous tumor closely correlate with the dimensions of the lesion on the (dynamic) images. **H:** Specimen radiograph also nicely demonstrates the relation of the residual tumor to the physeal plate.

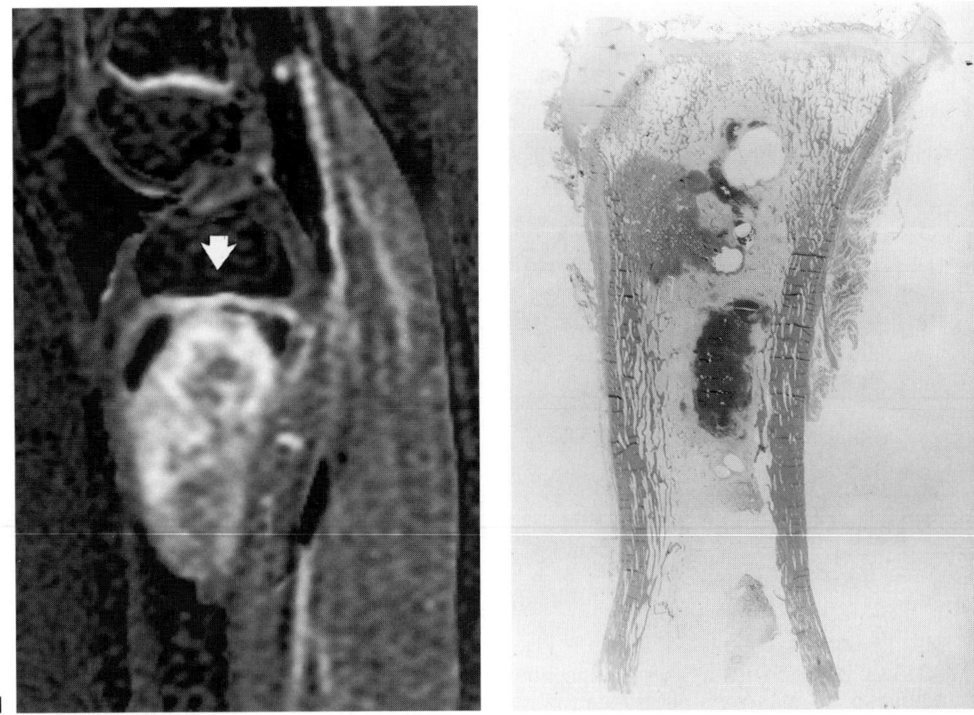

FIG. 17. **I:** Sagittal dynamic contrast-enhanced MP GE (15/6.8/741, 30°) image of an 11-year-old boy with an osteosarcoma of the proximal tibia. There is enhancement of physeal vessels *(arrow)* synchronously with the tumor, such that physeal involvement by the tumor cannot be excluded. **J:** Macrosection of the resected specimen corresponding to Fig. 17I, demonstrating the tumor abutting, but not breaking through the physeal plate. (Figs. H and J from van der Woude HJ, Bloem JL, Verstraete KL, Taminiau AHM, Nooy MA, Hogendoorn PCW. Osteosarcoma and Ewing's sarcoma after neoadjuvant chemo-therapy: value of dynamic MR imaging in detecting viable tumor before surgery. *AJR* 1995;165:593–598, with permission).

CASE 18

History: A 27-year-old male patient was admitted with persistent headache. What would be your diagnosis on the native and contrast-enhanced MR images?

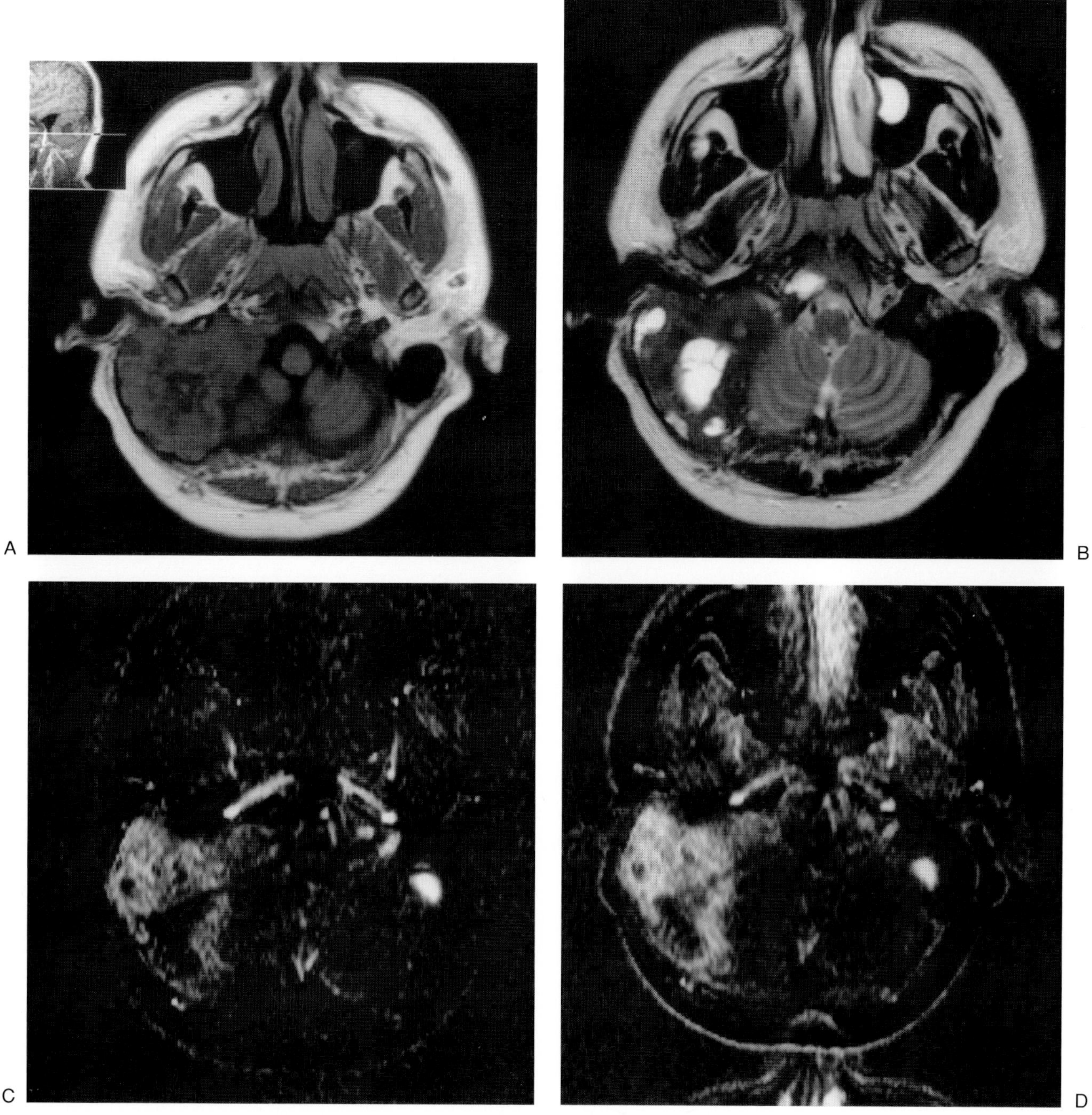

FIG. 18.

Findings: In Fig. 18A the axial T1-weighted SE (TR/TE 600/20 msec) image reveals a more or less lobulated mass of the temporal-occiptal area with intermediate signal intensity. In Fig. 18B, the T2-weighted TSE (TR/TE 3,200/120 msec) image in the same plane as A, there are well-defined areas with very high, water-like signal intensity and areas of intermediate signal intensity. In Fig. 18C,D, the dynamic GE (15/6.8/741, 30°) images obtained at 17 seconds (C) and 3 minutes (D) after bolus injection of 0.1 mmol/kg gadopentetate dimeglumine, there is a sudden increase in signal intensity of the lesion, which enhances synchronously with the neighboring arteries. The areas corresponding with the high signal intensity on T2 do not enhance. There are no significant changes in the appearance of the tumor at 3 minutes after bolus injection.

Diagnosis: Active fibrous dysplasia.

Discussion: Fibrous dysplasia is a developmental condition in which areas of the skeleton fail to mature normally (1). Clinically, the disease is divided in monostotic and polyostotic (multiple lesions) types, the basic pathology of the two being the same. On conventional radiographs, characteristically the lesions have a ''ground glass'' appearance, reflecting the homogeneous distribution of immature trabeculae that replace either cortical or trabecular bone. Although fibrous dysplasia tends to regress spontaneously, infrequently sarcomatous transformation may occur (1–3). Fibrous dysplasia is not infrequently seen in the skull (9%, Netherlands Committee of Bone Tumors) (2). Fibrous dyspasia may be accompanied by neurologic symptoms due to involvement of the inner table of the skull. Other tumor-like lesions that should be considered in the skull include hemangioma, ossified hematoma, and eosinophilic granuloma. Moreover, fibrous dysplasia of the skull may mimic morbus Paget or meningioma *en plaque* (3).

The homogeneous high signal intensity areas, seen on T2-weighted (turbo) SE images (Fig. 18B) and nonenhancing on contrast-enhanced sequences (Fig. 18C,D), are a rather common finding in fibrous dysplasia, representing a substantial cystic component.

Presumably, the intense osteoblastic activity and the abundant blood supply contribute to the progressive dynamic enhancement after bolus injection of gadopentetate dimeglumine that can be encountered in some patients with fibrous dysplasia, as in the present case.

CASE 19

History: A 25-year-old man complained of pain in the right buttock area. MR imaging was performed. Is this a benign or a malignant lesion? What are your diagnostic considerations on the native and contrast-enhanced MR images?

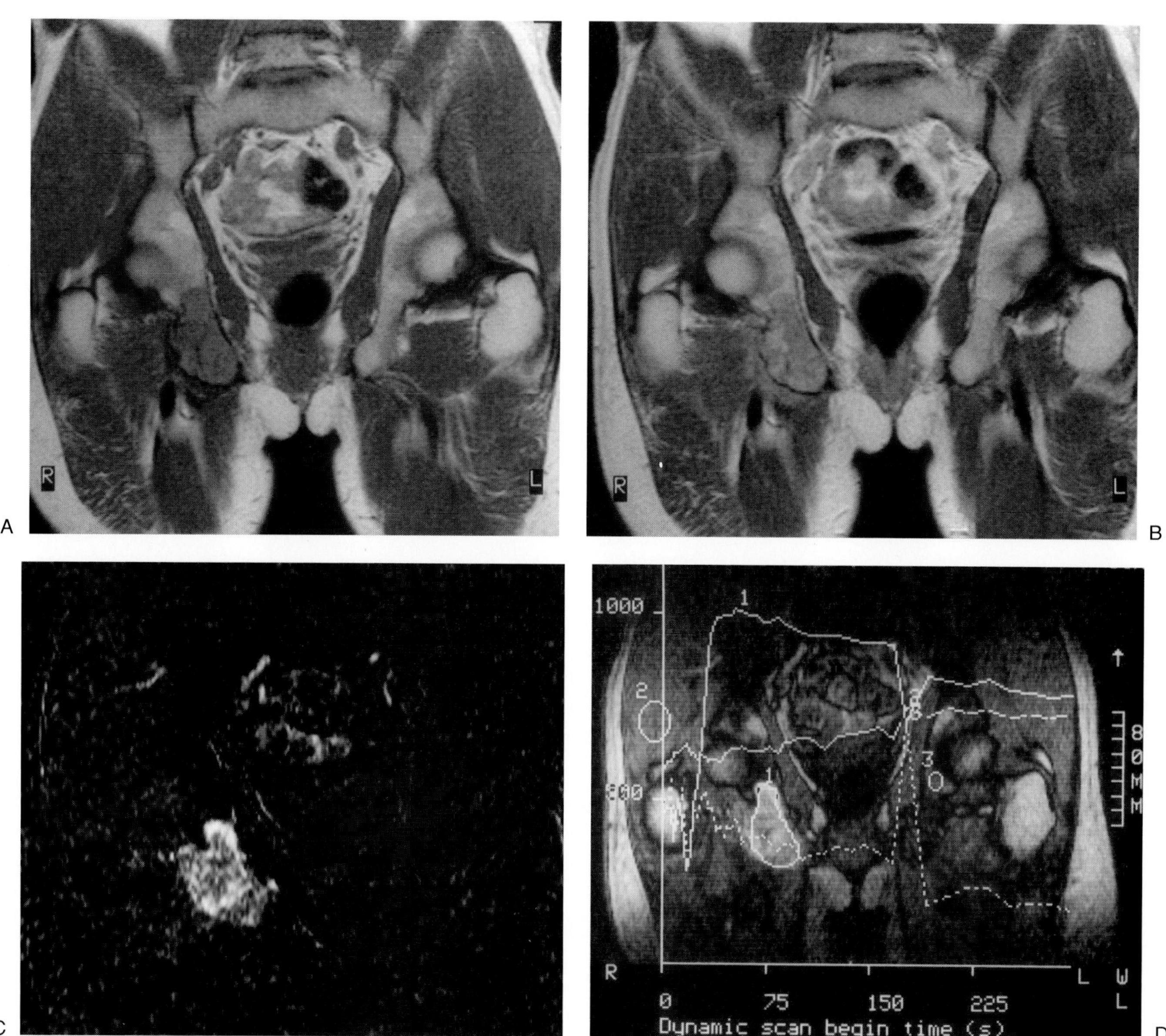

FIG. 19.

Findings: In Fig. 19A, the coronal T1-weighted SE (TR/TE 600/20 msec) image, a somewhat lobulated lesion is appreciated in the ischial bone. The lesion appears to be intraosseous and has intermediate signal intensity. In Fig. 19B, the static contrast-enhanced SE image, the lesion shows rather homogeneous enhancement. In Fig. 19C,D early enhancement of the tumor is seen on this subtracted dynamic MP GE image (15/6.8, 30°), obtained at 17 seconds after bolus injection of Gd-DTPA, and reflected by the rapidly progressive increase of SI in the corresponding time-intensity curve (D).

Diagnosis: Chondroblastoma of the ischial bone.

Discussion: Chondroblastoma is a rare, usually benign lesion of childhood and adolescence, predominantly arising in the secondary ossification center of major long bones, particularly in the head of the humerus (1–4). The ischial bone is an extremely rare location for chondroblastoma.

On conventional radiographs, chondroblastoma usually presents as a lytic lesion with a thin sclerotic margin, typically involving the epiphysis or neighboring metaphysis of a long bone. Metaphyseal extension of a lesion results in bulging expansion of the overlying cortex. Uncommonly, chondroblastoma presents as a large size lesion. These lesions can be mistaken for a malignant process, such as osteosarcoma or lymphoma. If only small amounts of chondroid are present, chondroblastoma may radiographically resemble giant cell tumor, although sclerotic margins are very unusual in giant cell tumors. Also, the enhancement patterns on dynamic MR images after administration of gadopentetate dimeglumine can be the same in the two tumors. In the present case, steep progression of the time-intensity curve was followed by early washout. This is also a frequent finding in giant cell tumor (see Case 11).

Other important radiographic differential diagnoses include enchondroma or central chondrosarcoma. Rapidly progressive enhancement is only seen in grade III chondrosarcoma, and not in enchondroma, or in grade I or II chondrosarcoma. Monostotic fibrous dysplasia appearing in the flat bones of the pelvic girdle cannot be separated radiographically from a chondroblastoma (1). Although not present in this case, chondroblastoma is, as opposed to fibrous dysplasia, frequently accompanied by edema (5,6).

It seems possible to distinguish chondroblastoma from the above-mentioned tumors by MR imaging, as intra- and extramedullary edema, frequently present adjacent to chondroblastoma, can be demonstrated (5,6).

CASE 20

History: This 40 year old man had a routine MR examination 3 years after resection of a large peripheral chondrosarcoma grade II. Is there recurrent grade II chondrosarcoma?

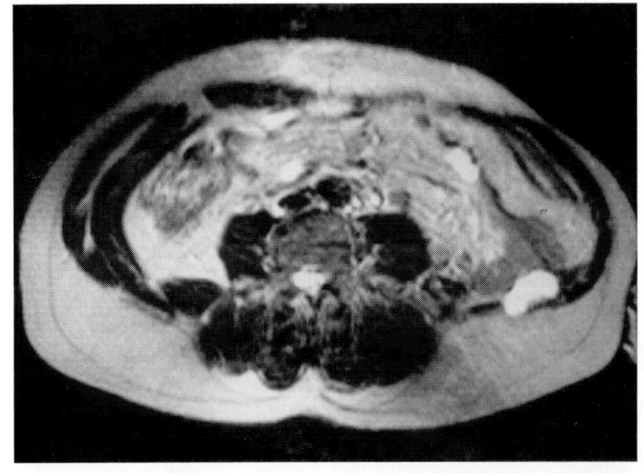

A

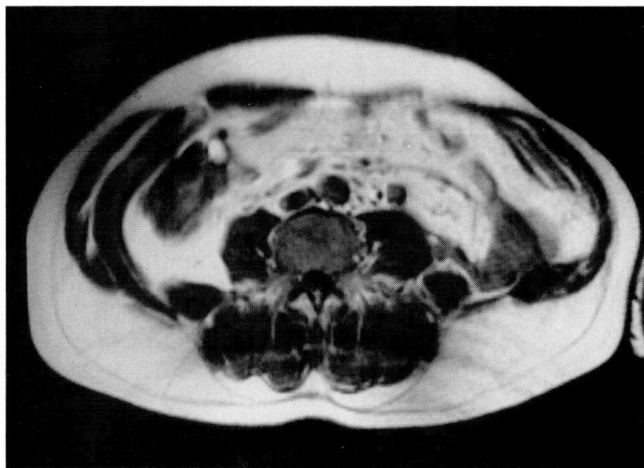

B

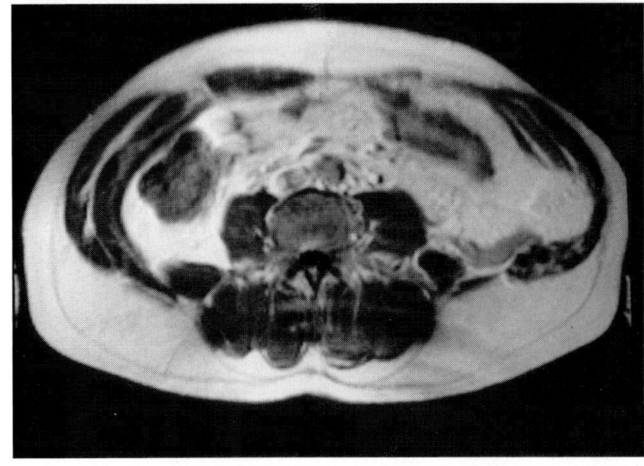

C

FIG. 20.

Findings: In Fig. 20A the axial T2-weighted SE image (TR/TE 2000/100 msec) shows a high signal intensity mass at the left iliac crest. In Fig. 20B,C the Axial T1 (B) and Gd-enhanced (C) MR images show only minor serpentiginous enhancement.

Diagnosis: Recurrent grade II chondrosarcoma.

Discussion: Three years ago a huge peripheral chondrosarcoma was resected (Fig. 20D–F). The enhancement pattern suggests the presence of a grade I or II chondrosarcoma (see Case 9). The high signalintensity mass seen on the follow-up scan allows diagnosis of recurrent tumor because of its space-occupying morphology and its high signal intensity (1,2). Differentiation between grade II and grade III chondrosarcoma is not possible on these native images. The enhancement pattern exhibited in Fig. 20C is that of grade II or I chondrosarcoma; grade III can be excluded. MR imaging is also very effective in detecting and characterizing lesions when extensive susceptibility artifacts are present (Fig. 20G).

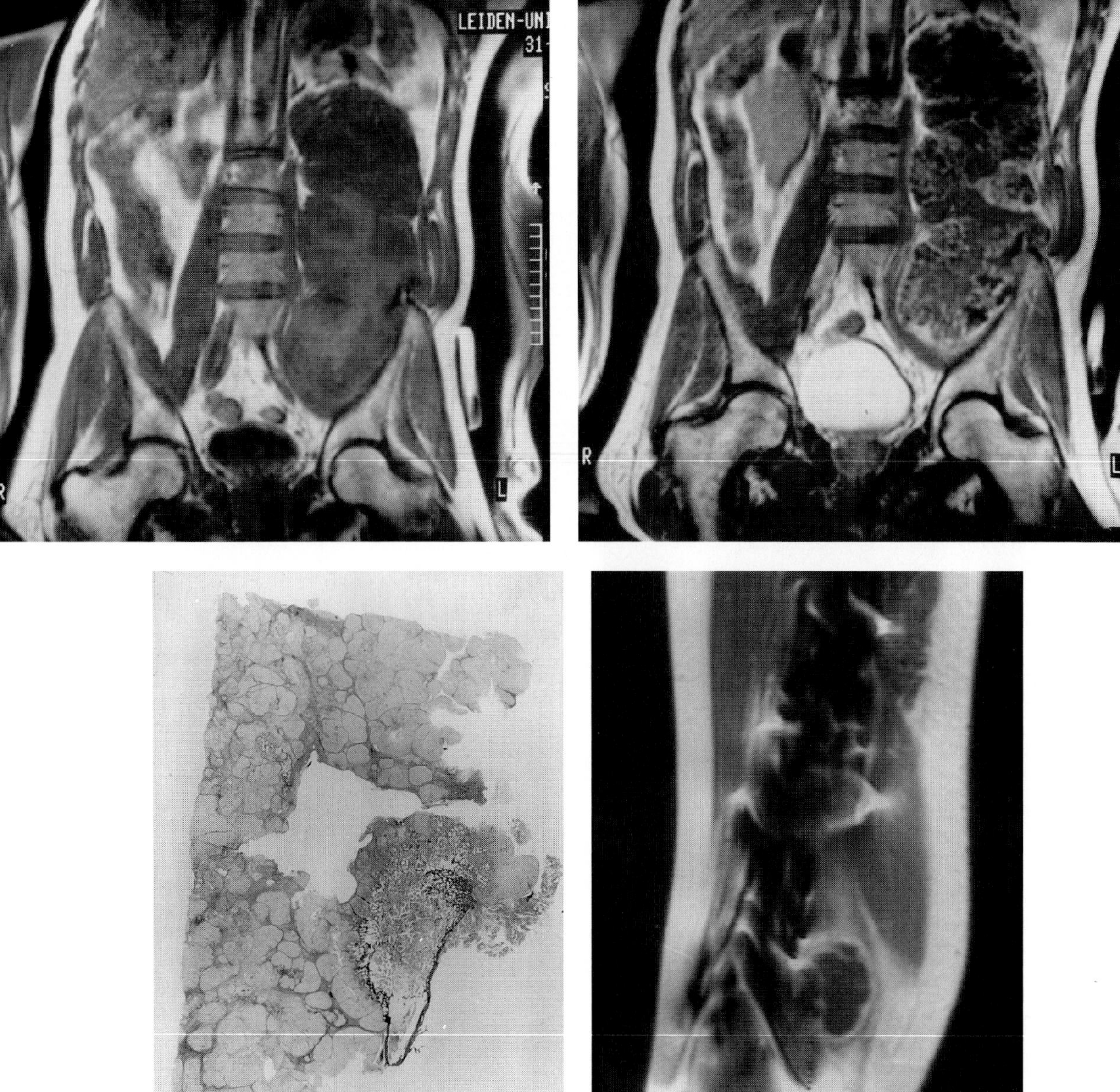

FIG. 20. D: Huge inhomogeneous tumor is exhibited, on this coronal T1-weighted sequence, filling the left side of the abdomen. **E:** Serpentiginous enhancement is appreciated on this static (5 min after injection) Gd-enhanced image consistent with chondrosarcoma grade II. **F:** Coronally sliced specimen shows the typical lobulated architecture of cartilage. The lesion arises from the iliac bone (chondrosarcoma secondary to an osteochondroma). The centrally located mucoid degeneration does not enhance and emptied itself during cutting. **G:** Sagittal Gd-enhanced image of a patient with enchondromatosis, after resection of grade I chondrosarcoma. Image quality is degraded due to susceptibility artifacts secondary to hardware left behind after reconstructive surgery. Rim enhancement is seen in the posteriorly located tumor, allowing a diagnosis of recurrent, well-differentiated chondrosarcoma. (Figs. D and E reprinted with permission from Bloem JL, Holscher HC, Taminieu AHM. Magnetic resonance imaging and computed tomography of primary malignant musculoskeletal tumors: In J. L. Bloem, D. J. Sartoris eds. *MRI and CT of the musculoskeletal system.* Baltimore: Williams and Wilkins 1992, 189–217.) Fig. G was reprinted with permission from: Bloem JL, van der Woude HJ. MRI of primary and secondary tumors of bone. In: Bloem JL, ed. Musculoskeletal imaging an update. New York: Springer. *Categorical course ECR '95,* 1995.

CASE 21

History: MR imaging was performed in this 14-year-old child to locally stage the tumor. What is the most likely histologic diagnosis, and is the physeal plate involved or not?

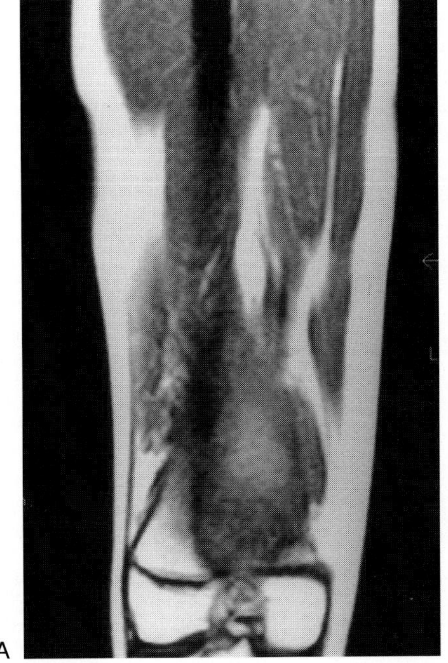

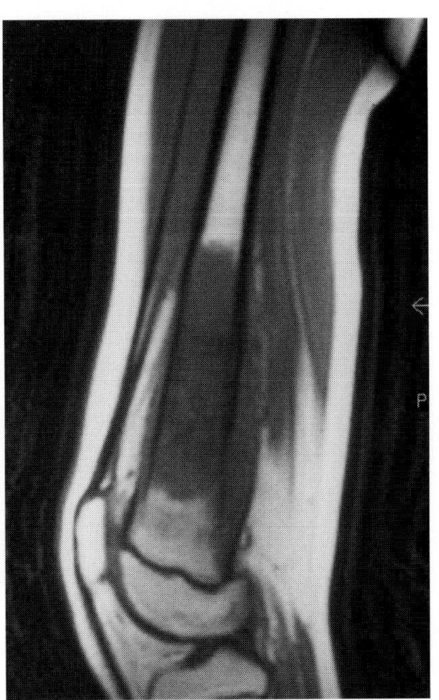

FIG. 21.

Findings: In Fig. 21A,B, the coronal (A) and sagittal (B) T1-weighted images (TR/TE 600/20 msec), the intraosseous extension of the intermediate signal intensity of tumor is well visualized.

Diagnosis: Osteosarcoma abutting, but not invading, the physeal plate.

Discussion: Osteosarcoma frequently extends into the epiphysis (1–3). In this patient, the tumor abuts the physeal plate on the posterior side. The epiphysis is not involved. Although the level of confidence is usually not high in these situations, the intermediate signal intensity of osteosarcoma does not extend into the low signal intensity of the physeal plate. The microsection of the resected specimen confirms the MR findings (Fig. 21C).

In contrast to osteosarcoma, Ewing's sarcoma rarely crosses the physeal plate (2,3). Therefore, with a similar image (Fig. 21D,E) of a Ewing's sarcoma the level of confidence is much higher. Knowledge of biologic behavior of the histologic types of sarcoma does improve accuracy of staging.

In equivocal cases, there is a tendency to overstage the tumor. This is especially a point of concern in the joint. When sarcoma extends into the joint overstaging occurs in particular along capsule and ligamentous insertions (Fig. 21D–J). Staging in relation to other structures like bone marrow, muscle and neurovascular bundle remains very accurate (1–4) (Fig. 21K,L).

ACKNOWLEDGMENTS

We thank the division of nuclear medicine, Herman Kroon MD, Maartje Geirnaerdt MD, and Anthonie Taminiau MD for their contributions and support.

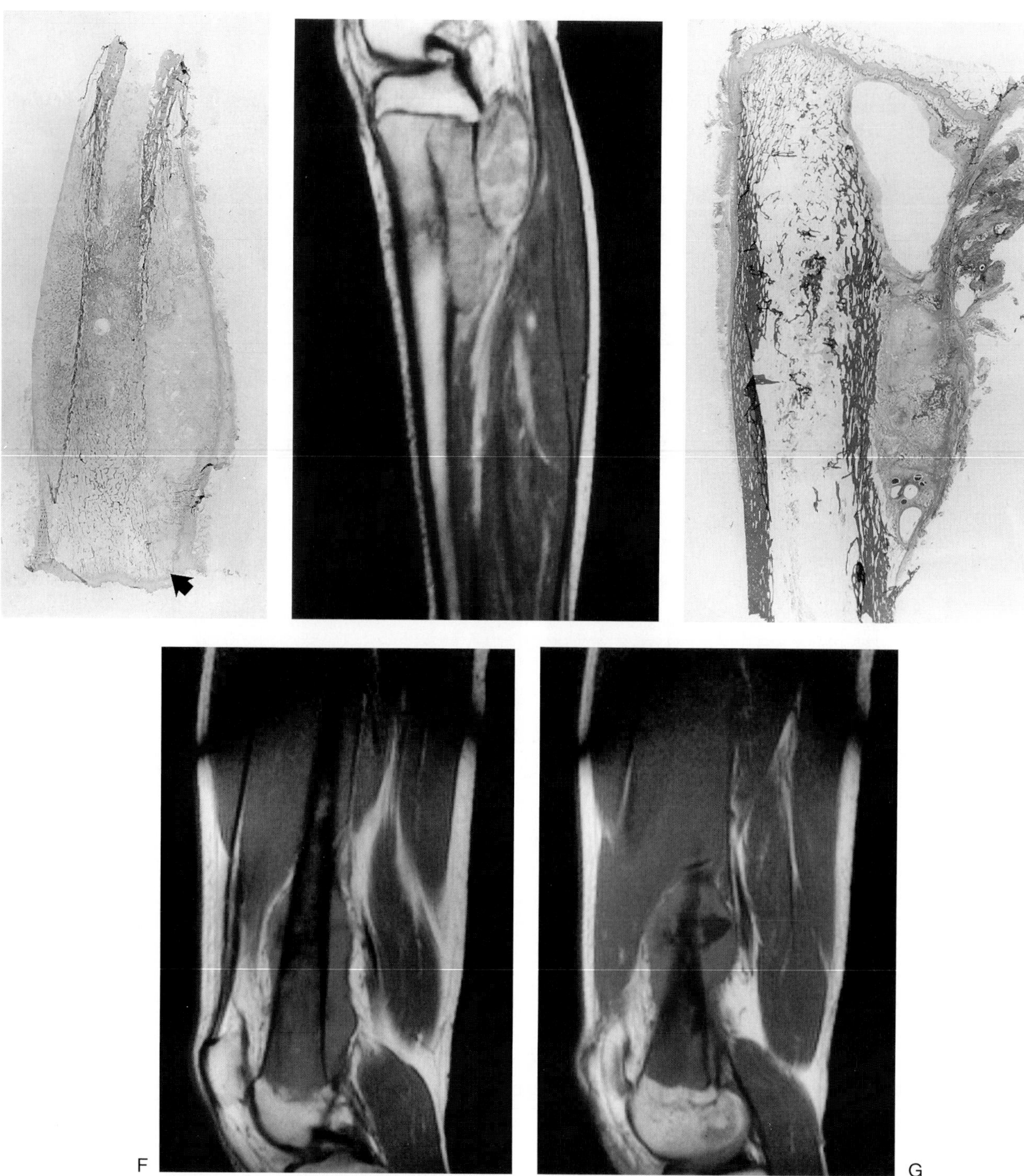

FIG. 21. **C:** Sagittal microsection shows tumor abutting the posterior part of the physis *(arrow)*. **D:** Sagittal Gd-enhanced image of a 15-year-old patient with Ewing's sarcoma of the tibia. The enhancing tumor abuts the growth plate but does not cross it. Posteriorly the tumor extends into the soft tissue and into the joint, using the posterior cruciate ligament as a scaffold. **E:** Sagittal microsection exhibits viable tumor extending toward, but not crossing, the physeal plate. Viable tumor marginates the empty space, which was, also on T1-weighted and dynamic Gd-enhanced images (not shown), diagnosed as mucoid degeneration. **F,G:** Two adjacent sagittal T1-weighted (TR/TE 600/20 msec) images show low signal intensity osteosarcoma extending into the area of the capsular insertion posteriorly.

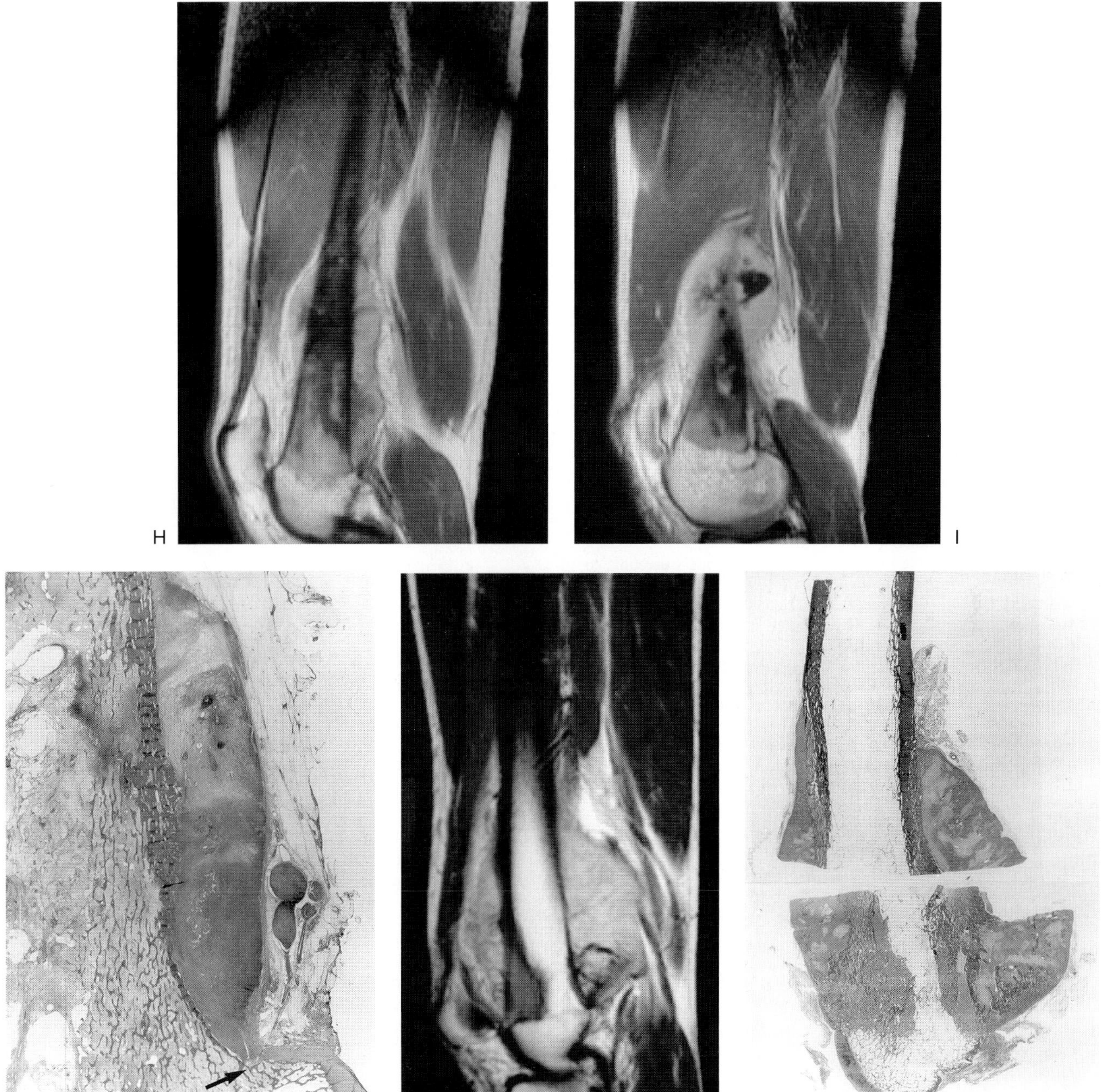

FIG. 21. H,I: On these static Gd-enhanced corresponding images a positive diagnosis of joint invasion cannot be made. The enhancing tumor does not seem to extend beyond the capsular-cartilaginous junction. **J:** Sagittal microsection shows the soft tissue extension of viable tumor. The tumor does not extend beyond the capsule, or along the femoral cartilage *(arrow)*. Also note the presence of tumor in two vessels posteriorly. This angio-invasion in small or intermediate sized vessels can usually not be separated from soft tissue extension on MR images. **K:** Gd-enhanced sagittal image of a juxtacortical osteosarcoma shows tumor on the outer surface of cortical bone posteriorly. Anteriorly the tumor, extending along the femur laterally, does invade cortical bone and marrow space. **L:** Sagittal microsection confirms the MR findings. Inferiorly, tumor also invades the posterior aspect of bone marrow, this was also diagnosed on the complete set of sagittal and axial images (not shown).

REFERENCES

Introduction

1. The National Institutes of Health, Bethesda. Limb-sparing treatment of adult soft-tissue sarcomas and osteosarcomas, concensus conference. *JAMA* 1985;254:1791–1794.
2. Simon MA, Aschliman MA, Thomas N, Mankin HJ. Limb-salvage treatment versus amputation for osteosarcoma of the distal end of the femur. *J Bone Joint Surg* 1986;68-A:1331–1337.
3. Simon MA. Limb salvage for osteosarcoma. *J Bone Joint Surg* 1988;70-A:307–310.
4. Gherlinzoni F, Picci P, Bacci G, Campanacci D. Limb sparing versus amputation in osteosarcoma. *Ann Oncol* 1992;3:23–27.
5. Bramwell VH, Burgers M, Sneath R, et al. A comparison of two short intensive adjuvant chemotherapy regimens in operable osteosarcoma of limbs in children and young adults: the first study of the European osteosarcoma intergroup. *J Clin Oncol* 1992;10: 1579–1591.
6. Enneking WF, Spanier SS, Goodman MA. The surgical staging of musculoskeletal sarcoma. *J Bone Joint Surg* 1980;62-A:1027–1030.
7. Bloem JL, Taminiau AHM, Eulderink F, Hermans J, Pauwels EKJ. Radiologic staging of primary bone sarcoma: MR imaging, scintigraphy, angiography, and CT correlated with pathologic examination. *Radiology* 1988;169:805–810.
8. Fletcher BD. Response of osteosarcoma and Ewing's sarcoma to chemotherapy: imaging evaluation. *AJR* 1991;157:825–833.
9. Algra, PR, Bloem, JL, Tissing H, et al. Detection of vertebral metastases: comparison between MR imaging and bone scintigraphy. *Radiographics* 1991;11(2):219–232.
10. Mirra JM. *Bone tumors: clinical, radiologic and pathologic correlations.* Philadelphia: Lea & Febiger, 1989.
11. Mulder JD, Schutte HE, Kroon HM, Taconis WK. *Radiologic atlas of bone tumors.* Amsterdam: Elsevier, 1993.
12. Huvos AG. *Bone tumors. Diagnosis, treatment and prognosis.* Philadelphia: WB Saunders, 1991.

Case 1

1. Gillespy T III, Manfrini M, Ruggieri P, Spanier SS, Pettersson H, Springfield DS. Staging of intraosseous extent of osteosarcoma: correlation of preoperative CT and MR imaging with pathologic macroslides. *Radiology* 1988;167:765–767.
2. Bloem, JL, Taminiau, AHM, Eulderink F, et al. Radiologic staging of primary bone sarcoma: MR imaging, scintigraphy, angiography, and CT correlated with pathologic examination. *Radiology* 1988; 169:805–810.
3. Algra, PR, Bloem, JL, Tissing H, et al. Detection of vertebral metastases: comparison between MR imaging and bone scintigraphy. *Radiographics* 1991;11(2):219–232.
4. Avra-Hami E, Tadmor R, Dally O, et al. Early MR demonstration of spinal metastases in patients with normal radiographs and CT and radionuclide bone scan. *J Comput Assist Tomogr* 1989;13:598–602.
5. Kattapuram SV, Khurana JS, Scott JA, El-Khoury GY. Negative scintigraphy with positive magnetic resonance in bone metastases. *Skel Radiol* 1990;19:113–116.
6. Kricun ME. Red and yellow marrow conversion: its effect on the location of some solitary bone lesions. *Skel Radiol* 1985;14:10–19.
7. Moore SG, Dawson KL. Red and yellow marrow in the femur: age related changes in the appearance of MRI. *Radiology* 1990;175: 219–223.
8. Moore SG, Bisset GS, Siegel MJ, et al. Pediatric musculoskeletal MRI. *Radiology* 1991;179:345–360.
9. Ricci C, Cova M, Kang YS, et al. Normal age related patterns of cellular and fatty bone marrow distribution in the axial skeleton: MRI study. *Radiology* 1990;177:83–88.
10. Richards MA, Webb JAW, Jewell SE, et al. In vivo measurement of spin-lattice relaxation time (T1) of bone marrow in healthy volunteers: the effects of age and sex. *Br J Radiol* 1988;61:30–33.

Case 2

1. Golfieri R, Baddeley H, Pringle JS, Souhami R. The role of the STIR sequence in magnetic resonance imaging examination of bone tumors. *Br J Radiol* 1990;63:251–256.
2. Shuman WP, Patten RM, Baron RL, Liddell RM, Conrad EU, Richardson ML. Comparison of STIR and spin-echo MR imaging at 1.5 T in 45 suspected extremity tumors: lesion conspicuity and extent. *Radiology* 1991;179:247–252.
3. Weinberger E, Shaw DWW, White KS, et al. Nontraumatic pediatric musculoskeletal MR imaging: comparison of conventional and fast spin-echo short inversion time recovery technique. *Radiology* 1995;194:721–726.
4. Listerud J, Einstein S, Outwater E, Kressel HY. First principles of fast spin echo. *Magn Reson Q* 1992;8:199–244.
5. Constable RT, Zhong J, Gore JC. Factors influencing contrast in fast spin-echo MR imaging. *Magn Reson Imaging* 1992;10:497–511.
6. Mirowitz SA, Apicella P, Reinus WR, Hammerman AM. MR imaging of bone marrow lesions: relative conspicuousness on T1-weighted, fat-suppressed T2-weighted and STIR images. *AJR* 1994; 162:215–221.
7. Kapelov SR, Teresi LM, Bradley WG, et al. Bone contusions of the knee: increased lesion detection with fast spin-echo MTR imaging with spectroscopic fat saturation. *Radiology* 1993;189:901–904.

Case 3

1. Frank JA, Ling A, Patronas NJ, et al. Detection of malignant bone tumors: MR imaging vs scintigraphy. *AJR* 1990;155:1043–1048.
2. Algra, PR, Bloem, JL, Tissing H, et al. Detection of vertebral metastases: comparison between MR imaging and bone scintigraphy. *Radiographics* 1991;11(2):219–232.
3. Avra-Hami E, Tadmor R, Dally O, et al. Early MR demonstration of spinal metastases in patients with normal radiographs and CT and radionuclide bone scan. *J Comput Assist Tomogr* 1989;13:598–602.
4. Kattapuram SV, Khurana JS, Scott JA, El-Khoury GY. Negative scintigraphy with positive magnetic resonance in bone metastases. *Skel Radiol* 1990;19:113–116.
5. Algra PR, Bloem JL. Magnetic resonance imaging of metastatic disease and multiple myeloma. In: Bloem JL, Sartoris DJ, eds. *MRI and CT of the musculoskeletal system.* Baltimore: Williams & Wilkins.
6. Graif M, Pennock JM, Pringle J, et al. Magnetic resonance imaging: comparison of four pulse sequences in assessing primary bone tumors. *Skel Radiol* 1989;18:439–444.
7. Baker LL, Goodman SB, Perkash I, Lane B, Entzman DR. Benign versus pathologic compression fractures of vertebral bodies: assessment with conventional spin-echo, chemical shift, and STIR MR imaging. *Radiology* 1990;174:495–502.
8. Mirowitz SA. Fast scanning and fat-suppression MR imaging of musculoskeletal disorders. *AJR* 1993;161:1147–1157.
9. Weinberger E, Shaw DWW, White KS, et al. Nontraumatic pediatric musculoskeletal MR imaging: comparison of conventional and fast spin-echo short inversion time recovery technique. *Radiology* 1995;194:721–726.
10. Shuman WP, Lambert DT, Patten RM, Baron RL, Taziola PK. Improved fat suppression in STIR MR imaging: selective inversion time through spectral display. *Radiology* 1991;178:885–887.
11. Golfieri R, Baddeley H, Pringle JS, Souhami R. The role of the STIR sequence in magnetic resonance imaging examination of bone tumors. *Br J Radiol* 1990;63:251–256.
12. Shuman WP, Patten RM, Baron RL, Liddell RM, Conrad EU, Richardson ML. Comparison of STIR and spin-echo MR imaging at 1.5 T in 45 suspected extremity tumors: lesion conspicuity and extent. *Radiology* 1991;179:247–252.

Case 4

1. Mulder JD, Schütte HE, Kroon HM, Taconis WK, eds. *Radiologic atlas of bone tumors.* Amsterdam: Elsevier, 1993;399–418.

2. Kroon HM, Schuurmans J. Osteoblastoma: clinical and radiologic findings in 98 new cases. *Radiology* 1990;175:783–790.
3. Bloem, JL, Kroon HM. Osseous lesions. *Radiol Clin North Am* 1993;31:261–278.
4. Tsai JC, Dalinka MK, Fallon MD, Zlatkin MB, Kressel HY. Fluid-fluid level: a non specific finding in tumors of bone and soft tissue. *Radiology* 1990;175:779–782.
5. Bloem JL, Holscher HC, Taminiau AHM. Magnetic resonance imaging and computed tomography of primary malignant musculoskeletal tumors. In: Bloem JL, Sartoris DJ, eds. *MRI and CT of the musculoskeletal system.* Baltimore: Williams & Wilkins, 1992;189–217.
6. Bloem JL, van der Woude HJ. MR imaging of the musculoskeletal tumors. *ACR Categorial Course on Body Magnetic Resonance Imaging* 1993;19–33.
7. Jelinek JS, Kransdorf MJ, Shmookler BM, Aboulafia AJ, Malawer MM. Liposarcoma of the extremities: MR and CT findings in the histologic subtypes. *Radiology* 1993;186:455–459.
8. Laredo JD, Assouline E, Gelbert F, Wybier M, Merland JJ, Tubiana JM. Vertebral hemangiomas: fat content as a sign of aggressiveness. *Radiology* 1990;177:467–472.
9. Kransdorf MJ, Meis JM, Jelinek JS. Myositis ossificans: MR appearance with radiologic-pathologic correlation. *AJR* 1991;157: 1243–1248.

Case 5

1. Assoun J, Richardi G, Railhac JJ, et al. Osteoid osteoma: MR imaging versus CT. *Radiology* 1994;191:217–223.
2. De Berg Y, Pattynama P, Obermann WR, et al. Percutaneous thermo-coagulation of osteoid osteoma, a minimally invasive procedure. *Lancet,* 1995;346:350–351.
3. Kroon HM, Bloem JL, Holscher HC, Van der Woude HJ, Reynierse M, Taminiau AHM. MR imaging of edema accompanying benign and malignant bone tumors. *Skel Radiol* 1994;23:261–269.
4. Bloem JL, Holscher HC, Taminiau AHM. Magnetic resonance imaging and computed tomography of primary malignant musculoskeletal tumors. In: Bloem JL, Sartoris DJ, eds. *MRI and CT of the musculoskeletal system.* Baltimore: Williams & Wilkins, 1992;189–217.
5. Coerkamp EG, Kroon HM. Cortical bone metastases. Radiology 1988;169:525–528.
6. Bloem JL, van der Woude HJ. MR imaging of the musculoskeletal tumors. *ACR Categorical Course on Body Magnetic Resonance Imaging* 1993;19–33.
7. Bloem JL, Kroon HM. Osseous lesions. *Radiol Clin North Am* 1993;31:261–278.
8. Klein MH, Skankman S. Osteoid osteoma: radiologic and pathologic correlation. *Skel Radiol* 1992;21:23–31.

Case 6

1. Enneking WF, Kagan A. Skip metastases in osteosarcoma. *Cancer* 1975;36:2195–2205.
2. Wetzel LH, Schweiger GD, Levine E. MR imaging of transarticular skip metastases from distal femoral osteosarcoma. *J Comput Assist Tomogr* 1990;14:315–317.

Case 7

1. Beltran J. Magnetic resonance imaging and computed tomography of musculoskeletal infection. In: Bloem JL, Sartoris DJ, eds. *MRI and CT of the musculoskeletal system.* Baltimore: Williams & Wilkins, 1992;153–171.
2. Haddad MC, Sharif HS, Aideyan OA, Clark DC, Al Shahed MS, Qureshi ZG, Aabed MY, Sammak B, Baydoen TM, Ross JS, Bloem JL. Infection versus neoplasm in the spine: differentiation by MRI and diagnostic pitfalls. *Eur Radiol* 1993;3:439–446.

Case 8

1. Moser RP, Gilkey FW, Kransdorf MJ. Osteochondroma. In: Moser RP, ed. *Cartilaginous tumors of the skeleton. AFIP fascicle II.* Philadelphia: Hanley and Belfus, Inc. 1990;35–73.
2. El-Khoury GY, Bassett GS. Symptomatic bursa formation with osteochondromas. *AJR* 1979;133:895–899.
3. Bloem JL, Holscher HC, Taminiau AHM. Magnetic resonance imaging and computed tomography of primary malignant musculoskeletal tumors. In: Bloem JL, Sartoris DJ, eds. *MRI and CT of the musculoskeletal system.* Baltimore: Williams & Wilkins, 1992;189–217.
4. Mulder JD, Schütte HE, Kroon HM, Taconis WK, eds. *Radiologic atlas of bone tumors.* Amsterdam: Elsevier, 1993;399–418.
5. Bloem JL, Kroon HM. Osseous lesions. *Radiol Clin North Am* 1993;31:261–278.

Case 9

1. Geirnaerdt MJA, Bloem JL, Eulderink F, Hogendoorn PCW, Taminiau AHM. Cartilaginous tumors: correlation of gadolinium-enhanced MR imaging and histopathologic findings. *Radiology* 1993;186:813–817.
2. Crim JR, Seeger LL. Devil's advocate. Diagnosis of low grade chondrosarcoma. *Radiology* 1993;189:503.
3. Geirnaerdt MJA, Bloem JL, Eulderink F, Hogendoorn PCW, Taminiau AHM. Devil's advocate reply; diagnosis of low-grade chondrosarcoma. *Radiology* 1993;189:504.
4. Bloem JL, Holscher HC, Taminiau AHM. Magnetic resonance imaging and computed tomography of primary malignant musculoskeletal tumors. In: Bloem JL, Sartoris DJ, eds. *MRI and CT of the musculoskeletal system.* Baltimore: Williams & Wilkins 1992;189–217.
5. Geirnaerdt MJA, Bloem JL, Hermans J, et al. How useful are radiographs in differentiating enchondroma from grade 1 chondrosarcoma? Submitted.

Case 10

1. Mazabraud A, Semat P, Roze R. A propos de l'association de fibromyxomes de tissus mous a la dysplasie fibreuse des os. *Presse Med* 1967;75:2223–2228.
2. Sundaram M, McDounald DJ, Merenda G. Intramuscular myxoma: a rare but important association with fibrous dysplasia. *AJR* 1989; 153:107–108.
3. Hudson TM, Stiles RG, Monson DK. Fibrous lesions of bone. *Radiol Clin North Am* 1993;31:279–297.
4. Sim FH, Dahlin DC, Beabout JW. Multicentric giant-cell tumor of bone. *J Bone Joint Surg* [*Am*] 1977;59:1052–1060.
5. Manaster BJ, Doyle AJ. Giant cell tumors of bone. *Radiol Clin North Am* 1993;31:299–323.

Case 11

1. Bloem JL, Holscher HC, Taminiau AHM. Magnetic resonance imaging and computed tomography of primary musculoskeletal tumors. In: Bloem JL, Sartoris D, eds. *MR imaging and CT of the musculoskeletal system.* Baltimore: Williams & Wilkins, 1992
2. Mirra JM. *Bone tumors: clinical, radiologic and pathologic correlations.* Philadelphia: Lea & Febiger, 1989.
3. Mulder JD, Schütte HE, Kroon HM, Taconis WK. *Radiologic atlas of bone tumors.* Amsterdam: Elsevier, 1993.
4. Verstraete KL, De Deene Y, Roels H, Dierick A, Uyttendaele D, Kunnen M. Benign and malignant musculoskeletal lesions: dynamic contrast-enhanced MR imaging—parametric "first-pass" images depict tissue vascularization and perfusion. *Radiology* 1994;192: 835–843.
5. Mirowitz SA, Totty WG, Lee JKL. Characterization of musculoskeletal masses using dynamic Gd-DTPA enhanced spin-echo MRI. *J Comput Assist Tomogr* 1992;16:120–125.
6. Erlemann R, Reiser MF, Peters PE, et al. Musculoskeletal neo-

plasms: static and dynamic Gd-DTPA-enhanced MR imaging. *Radiology* 1989;171:767–773.

7. Verstraete KL, Vanzieleghem B, De Deene Y, et al. Static, dynamic and first-pass MR imaging of musculoskeletal lesions by use of gadodiamide injection. *Acta Radiol* 1995;36:27–36.
8. Vanel D, Lacombe MJ, Couanet D, Kalifa C, Spielmann M, Genin J. Musculoskeletal tumors: follow-up with MR imaging after treatment with surgery and radiation therapy. *Radiology* 1987;164:234–245.

Case 12

1. Huvos AG. *Bone tumors. Diagnosis, treatment and prognosis.* Philadelphia: WB Saunders, 1991.
2. Enneking WF. Malignant skeletal neoplasms. In: Enneking WF, ed. *Clinical musculoskeletal pathology.* Gainesville: University Presses of Florida, 1990.
3. Verstraete KL, De Deene Y, Roels H, Dierick A, Uyttendaele D, Kunnen M. Benign and malignant musculoskeletal lesions: dynamic contrast-enhanced MR imaging—parametric "first-pass" images depict tissue vascularization and perfusion. *Radiology* 1994;192:835–843.
4. Verstraete KL, Dierick A, De Deene Y, et al. First-pass images of musculoskeletal lesions: a new and useful diagnostic application of dynamic contrast-enhanced MRI. *Magn Reson Imaging* 1994;12:687–702.
5. Verstraete KL, Vanzieleghem B, De Deene Y, et al. Static, dynamic and first-pass MR imaging of musculoskeletal lesions by use of gadodiamide injection. *Acta Radiol* 1995;36:27–36.
6. Geirnaerdt MJA, Bloem JL, Eulderink F, Hogendoorn PCW, Taminiau AHM. Cartilaginous tumors: correlation of gadolinium-enhanced MR imaging and histopathologic findings. *Radiology* 1993;186:813–817.

Case 13

1. Fletcher BD. Response of osteosarcoma and Ewing's sarcoma to chemotherapy: imaging evaluation. *AJR* 1991;157:825–833.
2. Bonnerot V, Charpentier A, Frouin F, Kalifa C, Vanel D, Di Paola R. Factor analysis of dynamic magnetic resonance imaging in predicting the response of osteosarcoma to chemotherapy. *Invest Radiol* 1992;27:847–855.
3. Holscher HC, Bloem JL, Vanel D, et al. Osteosarcoma: chemotherapy-induced changes at MR imaging. *Radiology* 1992;182:839–844.
4. Golfieri R, Baddeley H, Pringle JS, Leung AWL, Greco A, Souhami R. MRI in primary bone tumors: therapeutic implications. *Eur J Radiol* 1991;12:201–207.
5. Holscher HC, Bloem JL, Nooy MA, Taminiau AHM, Eulderink F, Hermans J. The value of MR imaging in monitoring the effect of chemotherapy on bone sarcomas. *AJR* 1990;154:763–769.
6. Van der Woude HJ, Bloem JL, Holscher HC, et al. Monitoring the effect of chemotherapy in Ewing sarcoma of bone with MR imaging. *Skel Radiol* 1994;23:493–500.
7. MacVicar AD, Olliff JFC, Pringle J, Ross Pinkerton C, Husband JES. Ewing's sarcoma: MR imaging of chemotherapy-induced changes with histologic correlation. *Radiology* 1992;184:859–864.
8. Lemmi MA, Fletcher BD, Marina NM, et al. Use of MR imaging to assess results of chemotherapy for Ewing's sarcoma. *AJR* 1990;155:343–346.
9. Carrasco CH, Charnsangavej C, Raymond AK, et al. Osteosarcoma: angiographic assessment of response to preoperative chemotherapy. *Radiology* 1989;170:839–842.
10. Bloem JL, Van der Woude HJ. MR imaging of the musculoskeletal tumors. In: *Syllabus for the categorial course on body magnetic resonance imaging.* Reston, VA: American College of Radiology, 1993;19–33.
11. Van der Woude HJ, Bloem JL, Schipper J, et al. Changes in tumor perfusion induced by chemotherapy in bone sarcomas: color Doppler flow imaging compared with contrast-enhanced MR imaging and three-phase skeletal scintigraphy. *Radiology* 1994;191:421–431.
12. Seeger LL, Widoff BE, Bassett LW, Rosen G, Eckardt JJ. Preoperative evaluation of osteosarcoma: value of gadopentetate dimeglumine-enhanced MR imaging. *AJR* 1991;157:347–351.
13. Erlemann R, Sciuk J, Bosse A, et al. Response of osteosarcoma and Ewing sarcoma to preoperative chemotherapy: assessment with dynamic and static MR imaging and skeletal scintigraphy. *Radiology* 1990;175:791–796.
14. Verstraete KL, Vanzieleghem B, De Deene Y, et al. Static, dynamic and first-pass MR imaging of musculoskeletal lesions by use of gadodiamide injection. *Acta Radiol* 1995;36:27–36.
15. Van der Woude HJ, Bloem JL, Verstraete KL, Taminiau AHM, Nooy MA, Hogendoorn PCW. Osteosarcoma and Ewing's sarcoma after neoadjuvant chemotherapy: value of dynamic MR imaging in detecting viable tumor before surgery. *AJR* 1995;165;593–598.
16. Hanna SL, Langston JW, Gronemeyer SA, Fletcher BD. Subtraction technique for contrast-enhanced MR images of musculoskeletal tumors. *Magn Reson Imaging* 1990;8:213–215.

Case 14

1. de Baere T, Vanel D, Shapeero LG, Charpentier A, Terrier P, di Paola M. Osteosarcoma after chemotherapy: evaluation with contrast material-enhanced subtraction MR imaging. *Radiology* 1992;185:587–592.
2. Van der Woude HJ, Bloem JL, Verstraete KL, Taminiau AHM, Nooy MA, Hogendoorn PCW. Osteosarcoma and Ewing's sarcoma after neoadjuvant chemotherapy: value of dynamic MR imaging in detecting viable tumor before surgery. *AJR* 1995;165;593–598.
3. Hanna SL, Langston JW, Gronemeyer SA, Fletcher BD. Subtraction technique for contrast-enhanced MR images of musculoskeletal tumors. *Magn Reson Imaging* 1990;8:213–215.
4. Van der Woude HJ, Bloem JL, Schipper J, et al. Changes in tumor perfusion induced by chemotherapy in bone sarcomas: color Doppler flow imaging compared with contrast-enhanced MR imaging and three-phase skeletal scintigraphy. *Radiology* 1994;191:421–431.
5. Van der Woude HJ, Bloem JL, Van Oostayen et al. Treatment of high-grade bone sarcomas with neoadjuvant chemotherapy: the utility of sequential color Doppler sonography in predicting final histologic response. *AJR* 1995;165:593–598.

Case 15

1. Enneking WF. Malignant skeletal neoplasms. In: Enneking WF, ed. *Clinical musculoskeletal pathology.* Gainesville: University Presses of Florida, 1990.
2. Mirra JM. *Bone tumors: clinical, radiologic and pathologic correlations.* Philadelphia: Lea & Febiger, 1989.
3. Huvos AG. *Bone tumors. Diagnosis, treatment and prognosis.* Philadelphia: WB Saunders, 1991.
4. Van der Woude HJ, Bloem JL, Verstraete KL, Taminiau AHM, Nooy MA, Hogendoorn PCW. Osteosarcoma and Ewing's sarcoma after neoadjuvant chemotherapy: value of dynamic MR imaging in detecting viable tumor before surgery. *AJR* 1995;165:593–598
5. Bloem JL, Reiser MF, Vanel D. Magnetic resonance contrast agents in the evaluation of the musculoskeletal system. *Magn Reson Q* 1990;6:136–163.

Case 16

1. Mulder JD, Schütte HE, Kroon HM, Taconis WK. *Radiologic atlas of bone tumors.* Amsterdam: Elsevier, 1993.
2. Enneking WF. Malignant skeletal neoplasms. In: Enneking WF, ed. *Clinical musculoskeletal pathology.* Gainesville: University Presses of Florida, 1990.
3. Erlemann R, Reiser MF, Peters PE, et al. Musculoskeletal neoplasms: static and dynamic Gd-DTPA-enhanced MR imaging. *Radiology* 1989;171:767–773.
4. Verstraete KL, De Deene Y, Roels H, Dierick A, Uyttendaele D, Kunnen M. Benign and malignant musculoskeletal lesions: dynamic contrast-enhanced MR imaging—parametric "first-pass" images depict tissue vascularization and perfusion. *Radiology* 1994;192:835–843.

Case 17

1. Bloem JL, Holscher HC, Taminiau AHM. Magnetic resonance imaging and computed tomography of primary musculoskeletal tumors. In: Bloem JL, Sartoris D, eds. *MR imaging and CT of the muskuloskeletal system.* Baltimore: Williams & Wilkins, 1992.
2. Pan G, Raymond AK, Carrasco CH, et al. Osteosarcoma: MR imaging after preoperative chemotherapy. *Radiology* 1990;174:517–526.
3. Kroon HM, Bloem JL, Holscher HC, Van der Woude HJ, Reynierse M, Taminiau AHM. MR imaging of edema accompanying benign and malignant bone tumors. *Skel Radiol* 1994;23:261–269.
4. Moore SG, Bisset GS, Siegel MJ, Donaldson JS. Pediatric musculoskeletal MR imaging. *Radiology* 1991;179:345–360.
5. Bloem JL, Reiser MF, Vanel D. Magnetic resonance contrast agents in the evaluation of the musculoskeletal system. *Magn Reson Q* 1990;6:136–163.
6. Van der Woude HJ, Bloem JL, Verstraete KL, Taminiau AHM, Nooy MA, Hogendoorn PCW. Osteosarcoma and Ewing's sarcoma after neoadjuvant chemotherapy: value of dynamic MR imaging in detecting viable tumor before surgery. *AJR* 1995;165:593–598.
7. Gillespy T, Manfrini M, Ruggieri P, Spanier SS, Pettersson H, Springfield DS. Staging of intraosseous extent of osteosarcoma: correlation of preoperative CT and MR imaging with pathologic macroslides. *Radiology* 1988;167:765–767.
8. Holscher HC, Bloem JL, Vanel D, et al. Osteosarcoma: chemotherapy-induced changes at MR imaging. *Radiology* 1992;182:839–844.
9. Norton KI, Hermann G, Abdelwahab IF, Klein MJ, Granowetter LF, Rabinowitz JG. Epiphyseal involvement in osteosarcoma. *Radiology* 1991;180:813–816.
10. Panuel M, Gentet JC, Scheiner C et al. Physeal and epiphyseal extent of primary malignant bone tumors in childhood. *Skel Radiol* 1993;23:421–424.

Case 18

1. Enneking WF. Skeletal diseases. In: Enneking WF, ed. *Clinical musculoskeletal pathology.* Gainesville: University Presses of Florida, 1990.
2. Mulder JD, Schütte HE, Kroon HM, Taconis WK. *Radiologic atlas of bone tumors.* Amsterdam: Elsevier, 1993.
3. Mirra JM. *Bone tumors: clinical, radiologic and pathologic correlations.* Philadelphia: Lea & Febiger, 1989.

Case 19

1. Huvos AG. *Bone tumors. Diagnosis, treatment and prognosis.* Philadelphia: WB Saunders, 1991.
2. Bloem JL, Mulder JD. Chondroblastoma, a clinical and radiological review of 104 cases. *Skel Radiol* 1985;14:1–9.
3. Mirra JM. *Bone tumors: clinical, radiologic and pathologic correlations.* Philadelphia: Lea & Febiger, 1989.
4. Mulder JD, Schütte HE, Kroon HM, Taconis WK. *Radiologic atlas of bone tumors.* Amsterdam: Elsevier, 1993.
5. Kroon HM, Bloem JL, Holscher HC, Van der Woude HJ, Reijnierse M, Taminiau AHM. MR imaging of edema accompanying benign and malignant bone tumors. *Skel Radiol* 1994;23:261–269.
6. Brower AC, Moser RP, Kransdorf MJ. The frequency and diagnostic significance of periostitis in chondroblastoma. *AJR* 1990;154:309–314.

Case 20

1. Bloem JL, Holscher HC, Taminiau AHM. Magnetic resonance imaging and computed tomography of primary malignant musculoskeletal tumors. In: Bloem JL, Sartoris DJ, eds. *MRI and CT of the musculoskeletal system.* Baltimore: Williams & Wilkins, 1992;189–217.
2. Vanel D, Lacombe, MJ, Couanet D, Kalifa C, Spielmann M, Genin J. Musculoskeletal tumors: follow-up with MR imaging after treatment with surgery and radiation therapy. *Radiology* 1987;164:243–245.

Case 21

1. Gillespy T. III, Manfrini M, Ruggieri P, Spanier SS, Pettersson H, Springfield DS. Staging of intraosseous extent of osteosarcoma: correlation of preoperative CT and MR imaging with pathologic macroslides. *Radiology* 1988;167:765–767.
2. Bloem JL, Taminiau AHM, Eulderink F, et al. Radiologic staging of primary bone sarcoma: MR imaging, scintigraphy, angiography, and CT correlated with pathologic examination. *Radiology* 1988;169:805–810.
3. Panuel M, Gentet JC, Scheiner C, et al. Physeal and epiphyseal extent of primary malignant bone tumors in childhood: correlation of preoperative MRI and pathologic examination. *Pediatr Radiol* 1993;23:421–424.

Subject Index

Page numbers followed by f refer to illustrations.

A

Abscess
  of foot, 624, 625f
  imaging characteristics, 624
  Staphylococcus auureus, 624–625
  of thigh, 623f
Accessory navicular
  navicular in, 502f
  pain basis in, 502–503
  PTT in, 502
  synchondrosis and, 502f, 503f
Accessory solues muscle
  Achilles tendon and, 555f
  acute strain, 555
Acetabulum, *see also* Periacetabular
  anatomy, 233
  biopsy of, 241f
  CT imaging of, 209
  in DDH, 220f, 222f, 223f
  erosion, 234,236f
  fossa, 233f
  fractures, 208–209
  Gd imaging of, 232–233
  hyaline cartilage of, 225f
  hypoplastic, 222f
  labra imaging, 233f
  labral tears
    anterior, 225f
    clinical presentation, 217, 224
    differential diagnosis, 224–225
    Gd imaging, 224f, 225f
    management, 217–218
    supralateral, 217f, 219f
  labrum enlargement, 219f
  in LCPD, 248-249
  in osteochondromatosis, 238f
  in Paget's disease, 242f
  in septic arthritis, 234f
Achilles tendon
  accessory soleus muscle and, 555f
  calcaneal insertion, 480f, 555f
  disruption, 480–482
  in failed repair, 504–505
  partial tear, 481f
  repair with implant, 505, 507f
  rupture, 480, 482f
  rupture healed, 505, 506f
  rupture treatment, 505
  spin-echo image of, 504f
  STIR image of, 506f
  tear, 480–481
  tendon healing process, 505, 506f, 507f
Acquired immune deficiency syndrome (AIDS)
  bacterial pyomyositis in, 558–560
  KS in, 558–559
Acromioclaviclar joint
  arthritis, 23–26
  articular disk degeneration, 23f, 26f
  articulation, 24
  glenohumeral joint and, 24
  inflammation, 23f, 24
  soft tissue changes in, 24, 25f
  supraspinatus tendon and, 25f
  synovial cysts, 24, 26f
  trauma, 23, 24
Acromion, in failed rotator cuff repair, 46f, 47, 48f, 49f
Acromion process
  anatomic variations in, 9–13
  clavicle relationship, 9
  hooked, 9f, 10, 11f, 12f
  in impingement syndrome, 9–13, 30
  os acromiale, 10, 13f
  osseous changes in, 30
  osteophytes, 10, 11f
  spurs, 9–10, 11f, 12f, 30f, 41f
  in supraspinatus tendon tear, 19f, 20f, 28f, 34f, 35f
Adductor thigh muscle tear, 212, 213f
Adventitial cystic disease, popliteal artery narrowing in, 577, 578f
Amyloid myopathy
  imaging characteristics, 565, 566f
  of vastus muscle, 565f
Amyloidosis, PVNS differentiation, 638, 640f
Anconeus muscle
  anatomy, 78
  atrophy, 127f
  cubital tunnel and, 125f
Aneursymal bone cyst (ABC)
  diagnosis, 478–479
  of distal tibia, 478f
  solid, 478–479
  treatment, 479
Ankle, *see also* Foot
  in achilles tendon disruption, 480–482
  in accessory navicular, 502–503
  in Achilles tendon repair, 504–507
  anatomy, 465f, 475–477
  in anteriolateral ankle impingement syndrome, 487–489
  in ATAF tear, 474f, 475, 488
  eversion injuries, 526
  FHL tendons, 462f
  ganglion cyst, 499, 501f
  Gd imaging, 487
  idiopathic synovial chondromatosis of, 466–467
  imaging techniques, 461–462
  innervation, 521–522
  inversion injuries, 475, 488
  in Kohler's disease, 524–525
  MCL of, 468, 470f
  in OCD, 513–517
  in peroneal tenosynovitis, 490–491
  in peroneus brevis tendon tear, 490f
  schwannoma, 521f
  sprains, 475, 487, 488, 513, 526
  synovial sarcoma, 629f
  synovitis, 487–488
  in talus fracture, 474f
  in tarsal coalition, 483–486
  in tarsal tunnel syndrome, 521–523
  in tarsal sinus syndrome, 487–488, 498–499
  in tibial bone cyst, 478–479
  in tibialis anterior muscle laceration, 470, 473f
  in tibialis anterior tendon tear, 471f, 472
  in tibiospring ligament disruption, 470f
  in tibiospring ligament, normal, 468f
Annular ligament anatomy, 78
Anterior colliculus, MCL and, 468
Anterior cruciate ligament (ACL)
  avulsion, 352, 354f, 355f, 444
  associated ganglion cyst, 416
  Gd imaging of, 419, 420f
  graft impingement, 418
  intact, 353f
  insertion fracture, 449f
  reconstruction evaluation, 418–420
  roof impingement, 418–419, 420f
  unstable, 373f
Anterior cruciate ligament tear
  acute, 351f, 352, 353f, 447f
  angular, 353f, 375f
  autografts, 403f
  bony lesions in, 443
  buckled PCL and, 384, 397f
  chronic, 351, 354f, 397f
  diagnosis, 384, 443
  edema in, 447f
  empty lateral wall and, 354f
  in failed reconstruction, 400f
  femoral attachment, 352
  femur fragment and, 354f
  focal mass-like, 352
  grafts, 400–403
  interstitial, 352
  LFC injury and, 442–443, 445f
  mechanisms, 351–352
  MRI diagnostic accuracy for, 353, 396
  partial, 375f, 396
  patellar dislocation and, 390
  PCL buckling and, 384–386, 395
  reattachment to PCA, 351f
  reconstruction, 389, 400–403
  secondary signs in, 384–387
  Segond fracture and, 444, 449f
  tibeal plateau fracture and, 384, 443

Anterior talofibular ligament (ATAF)
acute tear, 474f, 475, 477f
anatomy, 475, 476f, 526, 528f
disruption, 526f
in impingement syndrome, 487–489
injury, 527
normal, 528f
partial tear, 488, 489f
in syndesmotic complex injury, 526–529
tibial origin, 528f
Anterolateral ankle impingement syndrome
schematic of, 489f
tarsal sinus syndrome and, 487–488, 489f
Arachnoiditis
adhesive, 325, 326f
intradural mass of, 327f
myelography of, 325, 326f, 327f
presentation patterns, 325
Arthritis, *see also* Osteoarthritis
of acromioclavicular joint, 23–26
biceps tendon subluxation in, 44f
of gout, 509
of hip, 215, 216f, 234-239
image assessment, 215, 2116f
juvenile chronic, 236
septic, 173, 215, 216f, 234–237
Arthritis, rheumatoid
cystic bone erosion in, 167
diagnosis, 167
of elbow, 100–102
of hip, 236, 239f
juvenile, 236, 237f
pathology, 166–167
synovitis, 101–102, 166
of wrist, 166–167
Arthroscopic decompression
imaging of, 47–48
in impingement syndrome, 28, 30
subdeltoid space in, 48f
Arthroscopy
of knee vs. MRI, 347
of meniscal tears, 378–379
of PCL tears, 396
Articular cartilage
in chondral fractures, 448f
in DDP, 404f
histology, 382f
imaging characteristics, 414
of knee, 380, 382f, 383f, 448f
lesion classification, 381
matrix injury, 414
normal, 415f
in osteoarthritis, 415
posttraumatic defects, 414–415
spin-echo imaging, 414–415
structure, 380, 382f
Avascular necrosis (AVN)
classification, 256
diagnosis, 158, 256–257, 262–263
of femoral head, 256–258, 262–264
fracture nonunion and, 158
of hip, 231, 256–258
histopathology, 263
occult, 256
prognosis in, 256–257
of scaphoid, 158, 159
treatment, 159
Avulsion
of ACL, 352, 354f, 355f, 444
ACL-associated fractures and, 444
of FCL, 405
fragments, 444
of intercondylar eminence, 355f
of PCL, 396, 398f, 399

B

Bankard lesion
glenohumeral instability and, 58, 59f
osseous, 57f
partial, 31, 32f
reversibility of, 54–55, 62f
Biceps brachii, anatomy, 78, 81
Biceps femoris muscle, edema, 557f
Biceps femoris tendon
FCL and, 405, 406f–408f
lateral knee support and, 405, 406f, 407f
Biceps tendon
anatomy, 81
degeneration, 35f, 36
in degenerative arthritis, 44f
distal, 81–85
head dislocation, 41f, 42, 43f
impingement syndrome, 85f
injury mechanisms, 81
intracapsular tear, 45f
long head instability, 42
normal, 53f, 82f
old partial tear, 84f
partial tear, 83f
pathology, 42
proximal, 84f
radial nerve and, 82
reattachment, 82
in SLAP lesion, 64
subacute rupture, 80–82
subscapularis tendon and, 42, 43f
tendinosis, 81, 84f
tenodesis, 49f
tear, 35f, 36
Biceps tendon sheath
fluid, 41f
loose bodies in, 44f
Biceptal grove, in biceps tendon tear, 41f, 42, 45f
Bicipital radial bursa
anatomy, 81
enlargement, 81–82
Bone contusion
of knee, 90
STIR imaging of, 90
Bone cyst, aneurysmal, 667f, 668
Bone marrow
artifacts, 259f
in AVN, 263, 264f
biopsy, 588
in chemotherapy, 704
clinical assessment, 588
composition, 585
contrast-enhanced imaging, 158–159
in diabetes, 519
edema, 91, 198, 228f, 263, 264, 498f, 519, 669f, 704,
Ewing's sarcoma metastasis to, 609f
in fibrous dysplasia, 251, 252f
in Gaucher's disease, 266–267, 365, 367
hematopoietic expansion of, 265–267
of iliac crest, 588f
imaging techniques, 2, 443, 446f
of ischium, 589f
loss in femur, 230f
meta-epiphyseal edema, 443, 447f
normal hip, 200f
normal humeral, 591f
in normal radius, 99f
in osteomyelitis, 259f
in Paget's disaese, 242f, 243, 244f
pre-transplant staging, 588
"red", 585, 672
in sickle cell disease, 265–267, 573–574, 575f
in stress response, 91
subcortical injury, 444, 448f
in TOH, 198–201
trabecular injury, 444, 448f
transient edema, 198, 199
in triceps tendon avulsion, 91f
"yellow", 585, 660
Bone marrow tumors
benign endochondroma, 600f
biopsy of, 588, 590, 603, 606, 607f
in breast cancer metastases, 596, 597–598
of caudal spinal canal, 594f
CT vs. MR assessment, 593, 593f, 608–609
edema, 591f
Ewing's sarcoma as, 602, 604f
of femur, 599f, 602f, 605f, 610f
Hodgkin's lymphoma as, 590f
of humerus, 591f, 598, 600f
of iliac crest, 588, 593f, 597f, 606–607
of ilium, 590f, 604f
of ischium, 589f
of lumbar spine, 589f, 594f, 605f, 610f
lymphoma as, 588, 590f
metastatic, 595–596
multiple myeloma as, 602–603, 604f, 605f, 606–607
non-Hodgkin's lymphoma as, 591f, 593f, 660f
origin sites, 602
of pelvis, 589f, 592f, 593f, 597–598, 600f, 604f, 605f, 610f
post-treatment, 590f, 592f, 608
of rib, 591f, 595f, 596f, 609f
of sacrosciatic notch, 593f
of sacrum, 604f, 606f
of scapula, 598, 599f
of spinal canal, 609f
staging, 588, 596
of sternum, 600f, 601f
STIR vs spin-echo imaging, 598, 599f
Tc 99m scanning, 595, 596f, 597f, 600f
in treatment relapse, 592
Bone tumors
abscess differentiation, 675–676
breast carcinoma metastases as, 663–666, 671f
chemotherapy monitoring, 690–691, 693–694, 697–699, 703–705
chondroblastoma as, 709–710
chondrosarcoma as, 678–681
chondrosarcoma metastasis and, 688–689
chondrosarcoma, recurrent, 711–712
CT imaging merits, 655–656, 669
defining margins of, 672
Doppler imaging, 694, 696f
dynamic subtraction imaging, 691, 692f
dynamic MPGE imaging of, 679, 680f, 685f, 686f, 688f, 693f, 697f, 698, 700f, 709f
Ewing's sarcoma metastases as, 657–659, 668f
Ewing's sarcoma, post-treatment, 690–692
fat suppression imaging, 660, 663, 664f, 669
in femur, 659f, 679, 680f, 681f, 690–692
Ga^{67} scintigram of, 660, 661f
Gd imaging, 678f, 679, 680f, 681f, 686, 700, 707f, 709f
hemorrhage, 682f
of ilium, 657f, 658f
imaging techniques, for, 655–656
intraosseous lesions in, 683–684, 700, 705f, 709f
of ischium, 657f, 658f, 709–710

limb-salvage and, 655
metastases detection in, 656, 663–666
MR imaging merits for, 655–656
multiple osseous, 683–684
non-Hodgkin's lymphoma as, 591, 593f, 660f, 664, 671f
osteoblastoma as, 667–668
osteochondroma as, 662, 677
osteoid osteoma as, 669–670
osteosarcoma as, 672–673, 679f, 682f, 690, 693–696, 697–699, 703–706, 713–715
preoperative monitoring, 656
pseudotumers as, 668
radiography of, 655, 689
reactive changes in, 669
recurrence detection, 656
septal enhancement, 679, 682
skip metastases, 672–673
of spine, 663–666
staging of, 713, 714f
STIR sequences of, 663–664, 665f, 666f
of talus, 667f
Tc-99m imaging, 657, 658f, 659f, 665f
of tibia, 668f, 681f, 700–702, 703–706
TSE sequencing, 660, 662f
vascular network of, 694, 695f, 696f
Brachialis muscle
anatomy, 78
in elbow dislocation, 139
Breast cancer metastasis
to bone marrow, 596, 597–598, 599f, 600f, 6601f, 663–666
to radius, 671f
to spine, 663–666
STIR sequences of. 663–664, 665f, 666f
Tc-99m scans vs. MRI of, 597–598
Brodie's abscess, in radius, 675f
Buford complex, sublabral hole and, 52, 53f
Bursitis
of biceps tendon, 85f
chronic, 97
olecranon, 87, 96–97
peribursal fat signal in, 40f
septic, 96–97
in sickle cell disease, 575f
subacromial-subdeltoid, 6f
superficial, 96

C

Caffey's disease
clinical presentation, 68–69
pathology, 69
scapula in, 68f
Calcaneofibular (CF) ligament
anatomy, 475, 476f
function, 475
tears, 475
Calcaneus
normal, 500f, 555f
stress fractures, 494–495
in tarsal coalition, 485f
Calcific tendinitis
common sites for, 580
hydroxyapatite deposition in, 579–580
imaging techniques, 39
pathology in, 39
of rotator cuff, 38–40
Calcium pyrophosphate dihydrate (CPPD) deposition disease, osteonacrosis differentiation, 441
Capitellum
contusion of, 142
osteochondral defect in, 110f
osteochondritis dissecans of, 113–114
pseudodeficit of, 116–118
radiocapitellar articulation, 116f
Carpal tunnel syndrome
carpal instability and, 175–177
clinical presentation, 154
criteria for, 55
flexor tendons in, 154–155, 156f
management, 155
median nerve and, 152, 154–155, 156f
release image, 156f
synovitis and, 156f
Carpus, in TFC tear, 165f
Cartilaginous coalitions, in ankle, 484, 485f, 486f
Cat-scratch disease
clinical findings in, 119–120
elbow manifestations, 119–121
lymph nodes in, 119f, 120, 121f
Cauda equina
adhesive arachnoiditis, 325–327
nerve root clumping, 325
Cellulitis
defined, 624
of foot, 625f
Cerebral spinal fluid (CSF)
cytology, 329
in leptomininageal tumor, 328, 329
MR signal of, 325
Chemotherapy monitoring
bone marrow edema in, 704
of chondroblastic osteosarcoma, 697–699
Doppler flow imaging in, 694, 696f
of Ewing's sarcoma, 690–691
macrosection evaluation, 695f, 705f
of necrosis, 693–694, 695f
of osteosarcoma, 693-695, 703–705
vascular assessment in, 694
Chondral defects, diagnosis, 245, 247f, 415
Chondroblastoma
characteristics of, 709–710
differential diagnosis, 710
intraosseous, 709f
of ischium, 709–710
MPGE image of, 709f
Chondromatosis
of ankle, 466–467
characterization, 103
of elbow, 103–105
of hip, 240f
histopathology in, 467
idiopathic synovial, 104, 466–467
MRI findings in, 467
primary, 104
synovial, 103
Chondrosarcoma
cartilage architecture in, 712f
differential diagnosis, 688–689
endochondromatosis and, 712f
endochondroma differentiation, 679
of femur, 678–681
of iliac crest, 711f
metastasis of, 688–689
MPGE imaging, 688f
prefemoral fat extension, 678f
recurrent, grade II, 711-712
Clavicle
acromion process relationship, 9
displacement, 46f
osteolysis, 24
Common extensor tendon,
calcification of, 130f
complete tear, 138f
in elbow dislocation, 129
normal, 137f
rupture, 129, 136
small tear of, 137f
in tennis elbow, 136
Common flexor tendon
in elbow dislocation, 129, 141f
muscle tear, 133f
tendonosis, 132
Condylopatellar sulcus, in LFC fractures, 442f, 443
Coracoacromial ligament, in impingement syndrome, 11f, 12f
Coracohumeral ligament, normal, 53f
Coronoid process
in elbow dislocation, 139, 141f
fracture, 139, 141f
Coxa vara
developmental, 254–255
femoral neck angle in, 254
Cubital tunnel, nerve entrapment, 125–127

D

Delayed-onset muscle soreness (DOMS)
of brachialis, 561–563
clinical features, 561–562
MRI findings in, 561f, 562, 563f
de Quervain's stenosing tenosynovitis
clinical presentation, 171–172
etiology, 171–172
synovitis and, 172
Dermatomyositis
inflammation in, 569f, 570
MRS assessment, 570
polymyositis differentiation, 569–570
Developmental dysplasia of hip, (DDH)
coxa valga in, 222f
diagnosis, 220–221
etiology, 220
in failed treatment, 221f, 222f, 223f
femoral head subluxation in, 220f
reduction impediments, 221
Diabetes mellitus, gastrocnemius infarction in, 567–568
Diabetic amyotrophy
denervation effects in, 542–543
edema in, 543, 544f
STIR imaging of, 542f, 543
Dialysis-related arthropathy, (DRA), characterization, 311
Dialysis-related spondylarthropathy (DRSA)
clinical presentation, 310–311
DRA and, 311
at LS level, 310–311
pathogenesis, 311
Dislocation, of elbow, 129, 139–142
Doppler color flow imaging, of tumors, 694, 696f
Dorsal defect of patella (DDP)
description of, 404
GRASS imaging of, 404f
subchondral bone plate in, 404f
Duchenne's muscular dystrophy
muscle alterations in, 545, 547, 549f
vastus medialis in, 549f

E

Edema
in ACL tears, 443
of bone marrow, 91, 198, 228f, 263, 443
in denervated muscle, 542, 544f
in diabetic muscle, 543, 544f
of hip muscle, 210f
in LFCfracture, 442f
in muscle pathology, 210f, 539, 544f, 546f, 547f, 548f

Edema (*contd.*)
in osteoid osteoma, 203–204
peritumeral, 203
STIR imaging of, 203, 539
Elastofibroma
characterization, 650–651
histopathology, 651
in scapula region, 650f
Elbow, 77–142
in anabolic steroidabuse, 81, 86–87
anatomy, 78–79, 125
bone marrow edema, 91
bursitis, 85f, 96–97
capitellum pseudodefect of, 116–118
in cat-scratch disease, 119–121
clear cell sarcoma, 98f
common extensor tendon, 129, 130f,136f
common flexor tendon, 129,132
cubital tunnel retinaculum, 125f, 126f
dislocation, 129,139–142
in distal biceps tendon rupture, 80–85
fat pad displacement, 103, 121f, 124f
fracture, 122–124
imaging pitfalls, 116–118
imaging procedures, 77–78
instability, 140, 142f
locking/catching of, 109–111
ligaments, 78
loose bodies, 109–111, 127f
MCL of, 93f, 125f, 128–131, 140
medial epicondyle fracture, 122–123
muscles of, 768–769
in muscle strain injury, 133f
olecranon contusion, 90–91
os supratrochleare dorsale of, 106–108
osteochondral loose bodies in, 109–112
osteochondritis dissecans of, 113–114, 117f
in radial head contusion, 134
in radial neck fracture, 94
rheumatoid arthritis of, 100–102
sarcoma suspected in, 121f
soft tissue swelling, 119–121
STIR imaging, 84f, 87, 89f, 90, 91f, 92f, 93f, 94, 95f, 130f, 133f, 136f
supracondylar fractures, 123, 124f
synovial chondromatosis of, 103–105
synovial thickening, 100f
in tennis elbow, 136–138
triceps tendon rupture, 86–89
ulner neuropathy and, 122, 123f, 125–127
Electromyography (EMG), of neuromuscular disorders, 542
Ependymoma
diagnosis, 303
of conus medullaris, 302–303
Epidural hematoma
etiology, 309
of lumbar spine, 308–309
spontaneous, 309
Epidural lipomatosis
contributing factors in, 315–316
of lumbar spine, 315–316
Erector spinae muscles
in acid maltase deficiency, 552
postsurgical atrophy, 550–551
Ewing's sarcoma
chemotherapy monitoring, 690–691
of femur, 659f, 690–692
ileum marrow in, 602, 604f
iliac metastasis, 657f, 658f
macrosection of tumor, 692f
in marrow, 659f
osteosarcoma differentiation, 713,714f
pelvic marrow in, 604f
post treatment, 690–692
Tc-99m imaging of, 657,658f, 659f
of tibia, 668f, 714f
tuber ischia metastasis, 657f, 658f
Excessive lateral pressure syndrome (ELPS)
patellar malalignment and, 432, 435f
spoiled GRASS images of, 435f
Exit foraminal obliteration, post fusion, 334, 335f
Extensor carpi radialis longus tendon, fluid, 183f

F

Failed back surgery syndrome (FBSS), causes of, 293
Fat suppression imaging
of elbow, 78, 94
of fractures, 94, 95f
of marrow, 2, 158–159, 443, 446f, 586–587, 660, 663
of muscle edema, 539
of osteochondroma, 660, 662f
of shoulder, 2
STIR imaging in, 586, 660
Femoral condyle, *see also* Lateral femoral condyle
contusional injury, 390–391
edema, 391f
in gastrocnemious slip, 576f
in lateral patellar dislocation, 390–391
osteochondritis dissecans of, 421–422
spontaneous ostenecrosis of, 440
Femoral trochlear grove, in patellar subluxation, 431, 433
Femoral tunnel
abnormal position of, 400f
in ACL reconstruction, 400–403
in failed ACL reconstruction, 400
isometric position of, 400
MCL attachment sites, 402f
normal position, 400, 402f
Femur
ACL detachment, 354f
anterior lateral, 449f
atrophic destruction, 238f
AVN of, 256–258, 262–264
bone marrow edema, 198, 228f, 263, 264f
bone marrow expansion, 265–267
chondrosarcoma of, 678–681
in coxa vara, 254
CT imaging, 208, 579, 581f
diffuse sclerosis in, 674f
epiphysis, 248–250, 254, 257f
Ewing's sarcoma of, 659f, 690–692
fiberous dysplasia, 251–253
gastrocnemious muscle and, 576f
in Gaucher's disease, 365
gluteus maximus insertion, 579f
infection, 235–236
intracortical osteoid osteoma of, 202–203
in juvenile rheumatoid arthritis, 237f
in LCPD, 248–250
marrow replacement in, 251–252
marrow tumors, 599f, 602f, 605f
meta-epiphyseal lesions, 444, 449f
neck fatigue fracture, 226–228
neck occult fracture, 208, 209f
neck sclerosis, 228f
osteochondroma of, 677f
osteoid osteoma of, 669f
osteomyelitis, 259–261, 674–675
osteonecrosis, 159, 248–250
in patellar dislocation, 390–391
PCL insertion point, 386
proximal focal deficiency, 205–206
proximal occult fracture, 207
radiography of, 219f, 228f, 251f
in SCFE, 230–231
in septic arthritis of hip, 235–236, 237f
in sickle cell disease, 265–267, 575f,
STIR sequences of, 228f
subluxation in DDH, 220–223
in TOH, 198–201
Fibromatosis
abdominal, 620
classification, 619
Gd imaging of, 620, 622f
musculoaponeurotic, 619, 620, 621f
of pelvis, 622f
of popliteal fossa, 619f
prognosis in, 620
recurrence, 620, 622f
in shoulder area, 621f
in supraclavicular region, 621f
Fibrous dysplasia
bowing deformities in, 252, 253f
characterization, 252, 708
of femur, 251–253
Gd imaging of, 707f, 708
sites, 252
of skull, 707f
Fibula
in FCL injury, 406, 408f
tunnel, 487f
Fibular collateral ligament (FCL)
in ACL tear, 387f
anatomy, 405
avulsion, 405
biceps tendon and, 405, 406f, 407f, 408f
conjoint tendon and, 406f
fibular head attachment, 407f, 408f
PCL imaging and, 385, 387f
rupture, 405
Fibular head, FCL attachment, 407f
Finger, *see also* Hand
deep flexor tendon re-tear, 180–181
distal flexor tendon, 180f
FDP of, 81
FDS of, 81
glomus tumor of, 188–190
MCP joint, 180–181
tendon alignment, 180–181
toe tendon similarities, 511
Flexor digital tendon, deep (FDP)
anatomy, 181
injured, 181f
normal, 181f,182f
Flexor digital tendon, superficial (FDS)
anatomy, 181
injury, 181f
normal, 181f, 182f
Flexor digitorum tendon, synovitis of, 166f
Flexor hallucis longus (FHL) tendon, gradient-echo image of, 362f
Flexor retinaculum tendon, in carpal tunnel syndrome, 155,156f
Foraminal encroachment, in spondylolysis, 320, 321f
Foraminal extrusions, *see also* Exit foramen stenosis, nerve root compression in, 331
Foot, *see also* Toe
abscess, 625f
in calcaneus stress fracture, 494–495
imaging techniques, 461–462
in metatarsal osteomyelitis, 518–520
in metatarsal stress fracture, 497f
in MTP joint gout, 508–510
in phalanx lesion, 508–510
in planter fasciitis, 492–493

PTT disruption in, 463
rhabdomyosarcoma of, 622f
synchronus interdigital neuroma of, 532–533
in tarsal coalition, 483–486
in tarsal tunnel syndrome, 521–523
toe of, 508–510, 511–512, 530–531
Fractures, *see also* Insufficiency fractures, Stress fractures
of acetabulum, 208–209
avulsion and, 444, 449f
in avulsion of iliac spine, 229f
chondral, 444, 448f
complications of, 122, 123f
of coronoid process, 139, 141f
of elbow, 122–123
fatigue type, 226–228
of femur, 207–209, 226–228
of hamate, 169–170
of LCF, 442
micro, 444, 447f
of pelvis, 240
prognosis in, 443
of radial head, 139
radiographic vs. MR imaging, 226–227
in sacrum insufficiency, 239–241
Salter-Harris type, 124f
of scaphoid, 157–158
Segong, 444, 449f
STIR sequences of, 227, 228f
subchondral, 444, 448f
in supra-acetabular insufficiency, 240, 241f
terminology of, 444
trabecular, 444
treatment, 122,124

G

Gadolinium (Gd) imaging
of elbow synovitis, 100, 102f
of foot, 519, 520f
Gd concentration effects on, 232
of glomus tumors, 189f
intra-articular injection, 245–246
of loose bodies misdiagnosis, 245–246
of lumbar spine, 273, 274f
of rheumatoid arthritis, 167, 168f
of soft tissue, 519
of spinal free disk fragment, 292f, 304f
of spinal nerve roots, 284–285
of spine, 284f, 286f, 291f, 292f, 328f
technique pitfalls, 232–233, 245–246
of synovial envelope, 419, 420f
untoward reactions to, 649
of wrist, 171f
Gallium^{67} (Ga^{67}) scintigram, of non-Hodgkin's lymphoma, 660, 661f
Gamekeeper's thumb (Skier's thumb)
first MCP joint in, 150–151
UCL in, 150–151
Ganglion, periacetabular, 217f, 218, 219f
Ganglion cysts
ACL associated, 416f
attachment sites, 417f
benign, 217
diagnosis, 438
favored sites for, 438
intraosseous, 193f
pathogenesis, 416
PCL associated, 416, 417f
periosteal, 438–439
spin-echo imaging, 438f, 501f
of tarsal sinus, 499, 501f
of wrist, 184, 193
Gastrocnemius muscle
in diabetes mellitus, 567–568
edema, 567–568
lymphoma in, 564f
medial head slip, 576f
popliteal artery entrapment, 577
Gaucher's disease
clinical forms, 365–366
femur marrow in, 365f
hematopoietic marrow expansion in, 266–267, 366
hip in, 267f
marrow lipids in, 266–267, 365–366
osteomyelitis and, 366
Giant cell tumor (GCT)
of hand tendon, 183–185
MRI variability, 638, 686
postsurgery, 686f, 687f
PVNS and, 183–185, 638, 639f
recurrence, 686, 687f
of rib, 685–687
septal enhancement, 679, 682f
of tibial tendon, 639
Glenohumeral joint
acromioclavicular joint and, 24
in biceps dislocation, 43f
capsular anatomy, 51
evaluation, 1, 51
impingement syndrome, 30
instability, 6, 30, 58, 62f, 65, 66f
in rotator cuff repair, 449f
in supraspinatus cuff tears, 21f, 27f, 29f, 36,41f
transient subluxation, 58
Glenohumeral ligament
anterior, 54f, 55, 57f, 58f
Bankart lesion and, 54–55, 57f, 58, 59f
capsular tear, 56f
deformity, 57f
glenoid labrum and, 51–52, 53f
Hill-Sachs lesion, 57f
inferior, 55f, 56f, 57f, 61f
instability, 50–51, 54–57, 58–60
middle, 53f
shoulder dislocation and, 54
soft tissue pathology, 58
superior, 50f
Glenoid labrum
aging and, 59
anterior instability, 61
detachment, 58f, 59, 60f
evaluation, 1, 2, 65
in glenohumeral instability, 50–51, 58
Hill-Sachs lesion and, 60
instability, 50–51, 55f
normal, 51, 52f, 53f
posterior instability , 61, 62f, 63f
SLAP lesions and, 61, 64
sublaberal holes, 52, 53f
subluxation, 54, 58–60, 61
superior detachment, 64, 65, 67f
tears, 1, 54f, 55, 61f, 62f
Glomus tumors
of finger, 188–190
normal glomas bodies and, 188, 189
Gluteus minimus muscle, normal, 217f
Gluteus maximus tendon
calcific tendinitis, 579–581
femur attachment, 579f
spin-echo image, 579f
Gout
acute tophaceous, 508–510
arthritis of, 509
lower extremity expression, 505
tophaceous deposits in, 509
Gradient-echo imaging
of elbow,77–78
of shoulder, 1, 2
versus fast spin-echo imaging, 2, 77–78
of wrist, 162
Gradient-recalled acquisition in steady state (GRASS) sequences
of DDP, 404f
of ELPS, 435f
of patellofemoral joint, 431, 434f, 435f
spoiled, 431, 434f, 435f
of vascular pool, 293, 295f
Gradient recalled (GRE) sequences, of PCL, 395

H

Hamate, hook fracture, 169–170
Hamstring sprain, 557f
Hand
flexor tendons of, 184f
foreign bodies in, 184
GCT of, 183–184
hemangioma of, 184, 185f
lipomas, 184
in palmer injury, 181f
PVNS of, 183–185
volar flexor tendons of, 184f
Heal pain
incalcaneal stress fracture, 494–495
in planter fasciitis, 492–493
Hemangioma
arteriovenous, 642, 644f
capillary, 643f
cavernous, 641–642
phleboliths differentiation, 642,643
radiography of, 642, 643f, 644f
Hematoma
acute, 212, 213f
evaluation of, 212, 626f
spontaneous epidural, 308–309
subacute, in calf, 626f
TCL tear related, 370f
of thigh, 211, 213f
Hemorrhagic bone lesions, posttraumatic, 443
Hill-Sachs lesion
glenohumeral instability and, 55, 57f, 59
imaging procedures, 59, 60f
reversal, 63f
Hip, 197–267, *see also* Femur, Pelvis
acetabular dysplasia, 218, 219f
acetabular labral tear, 217–218, 219f, 224, 225
apophyseal fatigue fractures, 227, 229f
in AVN, 231, 256–258, 262–264
chondral defects, 245–247
in coxa vara, 254–255
DDH of, 220
fatigue fracture, 226–228
fractures, 210f
in Gaucher's disease, 266–267
Gd imaging of, 224f, 232–233
hematoma, 211
ilium and, 219f, 227, 229f, 242–244
imaging techniques, 197, 245
joint effusion, 230
joint fluid imaging, 232–233
labral tear, 217–219
in LCPD disease, 248–250
loose bodies, misdiagnosed, 245–246, 247f
normal, 207f, 208f
osteoarthritis, 238f
in Paget's disease, 242–244
periacetabular ganglion of, 217–219
posterosuperior labrum tears, 224–225
in sacrum insufficiency fractures, 239–241

Hip (*contd.*)
SCFE of, 230–231
in septic arthritis, 215, 216f, 234–237
in sickle cell disease, 265–267
spin-echo imaging of, 200f, 201f, 219f, 245, 247f, 261f, 262f, 264f
STIR imaging of, 208
subluxation, 220
TOH and, 198–201
zona orbicularis, 233f
Humeral head
compression fracture, 63f
deformity, 34f, 46f, 47
marrow tumor, 591f
medullary lesion, 63f
migration, 32f, 36, 41f
subcortical cyst, 38f, 39
in supraspinatus tendon tear, 32f, 34, 36, 37f
Humerus
bone marrow, normal, 591f
bone marrow tumor, 591f, 598, 600f

I
Idiopathic inflammatory myopathies, (IIMs)
clinical presentation, 554
myositis of, 554
Ileofemoral ligament, 233f
Iliopsoas muscle, in septic arthritis, 215, 216f, 234f
Iliopsoas tendon
in DDH, 223f
snapping, 224–225
Iliotibial band, in lateral knee support, 405, 407f
Ilium
apophyseal fatigue fractures, 227
avulsion fracture, 229f
chondrosarcoma recurrence in, 711–712
cortical irregularity, 219f
crest, 588, 593f, 597f, 606–607, 711f
Ewing's sarcoma metastasis to, 657f, 658f, 659f
marrow tumors, 588, 590f, 593f, 604f, 606–607
in Paget's disease, 242–244
pseudotumor of, 668f
Imaging techniques,
angle dependent signals in, 530–531
for ankle, 461–462
for bone marrow, 2,158–159, 443, 446f, 585–587
forchondral defects, 380–383
CT- arthrography as, 58
for denervated muscle, 542–543
3D gradient-echo, 462f
Doppler color flow, 694, 696f
for elbow, 77–78
exercise-enhancement MR, 539
fast spin-echo sequences in, 2, 77–78, 273
fat suppression in, 2, 78, 94, 95f, 539, 585–586
for foot, 461–462, 518–519
for fractures, 226–227
for glenohumeral joint, 51
gradient-echo sequences in, 2, 77–78, 149, 462f
for hip, 197, 245
for IIMs, 554
kinematic, 431–437
for knee, 347, 431–437
for lumbar spine, 273
"magic angle" phenomenon in, 530–531
pediatric, 356
phased-array coils in, 273
postoperative, 47–48
receiver coil in, 149
for shoulder, 1–2, 18
for soft tissue tumors, 613–614
STIR sequences as, 77,78, 197, 539, 586
surface coils in, 1, 197, 245, 672
ultrasound as, 150–151
of wrist, 149
Impingement syndrome
acromion slope in, 9–11, 30
coracoacromial ligament and, 11f
glenohumeral instability in, 20
management, 28, 30
rotator cuff tendinitis and, 6
second stage, 30–31
surgical procedue for, 9,30
Inclusion body myositis
clinical findings in, 553, 554
histopathology, 553–554
quadriceps atrophy in, 553–554
Infection, *see also* Osteomyelitis
septic arthritis as, 174
of wrist soft tissue, 173–174
Infraspinatus tendons
of normal shoulder, 3, 4f, 5f
rotator cuff tears and, 35–36
Insufficiency fractures
etiology, 239–240
of pelvis, 240, 241f
of sacrum, 239–241
supra-acetabular, 240, 241f
Interdigital neuroma (Morton neuroma)
planter digital nerve in, 532, 533f
synchronus, 532–533
Intertrochantric ridge, fracture, 210f
Intervertebral disc
calcification, 322–323
hyperintense signal of, 322–323
Ischial apophysis, fatigue fracture, 227, 229f
Ischiofemoral ligament, 233f
Ischium
chondroblastoma of, 709–710
Ewing's sarcoma metastasis in, 657f, 658f
marrow tumor, 589f

J
"Jumper's knee", patellar tendinitis as, 429, 430f

K
Kaposi sarcoma (KS), connective tissue in, 558–559
Kienbock's disease, *see also* Avascular necrosis
etiology, 192
of lunate, 191–193
presentation, 191–192
Kinematic MRI
of knee flexion, 431–437
of patellofemoral joint , 431–432
Kissing lesion, of lateral tibeal plateau, 442f, 443
Knee, 347–453
in ACL rear
acute, 351, 352, 388–389
buckled PCL and, 397f
chronic, 351, 397f
interstitial, 352
lateral miniscus in, 388–389
reconstruction evaluation, 418
tibia shift in, 384f, 397f
arthroscopy vs. MRI evaluation, 347, 378–379
articular cartilage, 380, 382f, 383f, 414
cartilage abnormality diagnosis, 380
cartilage lesion classification, 381
cartilage structure, 380, 382f
chondral degeneration, 380
chondral fracture, 449f
in DDP, 404
dislocation, 406
in failed patellar tendon repair, 428–429
in FCL avulsion, 405–408
ganglion cysts, 416–417, 438–439
in Gaucher's disease, 365–366
graft isometry, 400
in hyperextension, 444
imaging procedures, 348–350
instability, 351
intercondyler eminence avulsion, 355f
intercondyler notch, 353f
intermittent locking, 461
intraarticular injury incidence, 395
kinematic MRI of, 431–437
lateral compartment disruption, 405–406
in LFC fracture, 442–449
lateral capsular ligament, 449f
in lateral patellar dislocation, 390–391
ligament of Wrisberg, 453f
meniscal repair, 423–427
meniscus, lateral pseudotear, 450
meniscus, medial
ossicle of, 376–377
pseudotear, 392–393
retear , 409
tear, 361f, 362f, 363f
MRI limiations, 348–349
MRI truncation artifacts, 392–393
osteochondritis dissecans of, 421–422
patella chondral defect, 380, 416f
in patellar subluxation, 431–432, 434f–437f
in PCL avulsion, 396, 398f, 399f
in PCL buckling, 384–385
in PCL tear, 394–395, 398f, 399f
PVNS of, 637–638
Salter-Harris fracture of, 356–360
spin-echo imaging, 348
stable, 351, 373f
subchondral collapse, 449f
in tibeal plateau fracture, 384f, 386f, 442f
tibeal osteonecrosis, 440–441
in TCL sprain, 367–368, 369f, 370f
in TCL tear, 371–375
unstable fragment in, 421f
Kohler's disease
clinical manifestations, 524–525
navicular in, 524f
osteochondrosis and, 524
osteonecrosis and, 524

L
Lateral collateral ligament (LCL)
in ACL reconstruction, 402f
anatomy, 78, 475–476
disruption effects, 78
PTAF and, 475, 476f
reconstruction, 487f
tarsal sinus syndrome and, 498–499
Lateral epicondylitis (tennis elbow)
in common extensor tendon rupture, 136, 137
in common extensor tendon tear, 138f
extensor carpi radialis brevis muscle in, 137f
management, 136
pathology, 136
STIR imaging of, 137f
Lateral femoral condyle (LFC)
in ACL tear, 442–443, 447f
edema, 442f, 443, 446f, 447f
impaction fracture, 442–443

lateral tibeal plateau impact, 443, 445f, 446f
marrow imaging, 443, 446f, 447f
in patellar dislocation, 446f
posttraumatic hemorrhagic lesions, 443
subcortical lesions, 444, 447f
tibial lateral plateau and, 442–443
Lateral suprapatellar bursa, in FCL avulsion, 405f
Legg-Caloe-Perthes disease (LCPD)
diagnosis, 248–249
of femur epiphysis, 248–250
lateral head cartilage in, 249f, 250f
Leptomeningeal membranes
carcinomatosis, 329, 330f
CSF analysis and, 328–329
melanoma spread to, 328–329, 330f
Ligament grafts
materials for, 401
MRI assessment of, 401, 403f
temporal changes in, 401, 403f
Ligament of Humphry
in acute PCL tear, 398f
double PCL and, 395, 397f
Ligamentous sprains, classification, 475
Lipomas
fibrolipoma as, 618f
of hand, 184
intramuscular, 618f
liposarcoma differentiation, 616
recurrent superficial, 618f
in thigh, 616f
Liposarcoma
cyst-like, 617f
Gd imaging, 618f
MRIcharacterization, 615–616
myxoid, 615f, 617f
pleomorphic, 618f
of popliteal fossa, 615f
of thigh, 616f, 618f
Long flexor tendon
angle dependent signal of, 530–531
laceration, 511–512
normal, 530–531
Loose bodies, *see also* Osteochondral loose bodies, in hip joint misdiagnosis, 245–246, 247f
Lumbar spine, *see* Spine, lumbar
Lunate
alignment, 177f, 179f
AVN of, 191–193
chondromalacia of, 192–193
in scapholunate dissociation, 176, 177f
silicone prosthesis, 186–187
Lunotriquetral ligament
tear, 178–179
VISI and, 178–179
wrist stabilization, 176, 177
Lymphadenopathy, MR versus CT imaging of, 592
Lymphoma, *see also* non-Hodgkin's lymphoma
bone marrow in, 588, 590f, 591f, 593f
differential diagnosis, 564–566
gastrocnemius muscle in, 564

M

Magnetic resonance spectroscopy (MRS)
dermatomyositis assessment, 570
metabolic activity assessment, 570
Magnetization prepared (MP) GE images
of chondrosarcoma, 680f, 697f, 698
of MFH of tibia, 700f, 701f
of osteosarcoma, 693f, 703f, 704, 705f
tumorimage enhancement, 679, 680f, 685
Malignant fibrous histocytoma (MFH)
angiomatoid, 633f
of arm, 633f
calcification in, 631
characteristics, 631, 701
cystic, 634f
differential diagnosis, 632, 648–649
dynamic MPGE imaging, 700f, 701f
fluid in, 632, 648–649
Gd imaging, 700f
hemorrhage in, 631, 633f
imaging techniques, 631–632
intraosseous, 700
low-grade, 700–702
macrosection of, 702f
myxoid, 632, 634f
radiograph of, 702f
recurrent, 648, 649f
of spine, 662f
of thigh, 631f–635f
TSE sequencing of, 660, 662f
McArdle's disease, muscle necrosis in, 547, 548f
Medial collateral ligament (MCL)
anatomy, 128, 468–469
avulsion, 144f
CT imaging, 130
detachment, 128, 130f, 131f
in elbow dislocation, 140
flexor digitorum superficialis muscle and, 129
MRI of, 469f
normal, 129f
ossification, 131f
reconstruction, 128
rupture, 93f, 128–131
schematic of, 470f
STIR imaging, 130f
ulner nerve and, 125
Medial collateral ligament (MCL) complex
in ACL reconstruction, 402f
injury grading, 368–369
injury management, 369
lateral bone contusions and, 369
MRI assessment of, 369
TCL and, 367
Medial epicondylitis (golfer's elbow), of common flexor tendon of elbow, 132
Medial femoral condyle
fracture, 443, 446f
in LFC fractures, 443, 446f
Medial malleolus
gradient-echo imaging, 465f
MCL and, 468, 470f
Medial navicular bone
gradient-echo image, 465f
normal, 465f
Median nerve
carpal tunnel syndrome and, 152, 154–155, 156f
entrapment, 126
fibrolipomatous infiltration, 152–153
Melanoma, lipomeningeal spread in, 328–329
Meniscofemoral femoral ligament (MFL)
anatomy, 395
MRI of 395
Meniscus, of knee
apex of, 410f
arthrogram of, 412f, 413f
arthroscopy of, 378–379, 412f
bucket-handle tear repair, 423
2 DFT imaging, 349
3DFT imaging, 349
displaced fragment and, 429f
of dog, 424, 426f
Gd imaging, 413f
imaging techniques, 348–349
intraarticular gas and, 392
intrameniscal signal conversion, 409, 410f
lateral
in ACL tear, 388–389
anterior horn of, 364f
normal anatomy, 450, 451f
radial tear, 348–350, 361, 362, 364f
posterior horn, 389, 450–451
posterior peripheral tear, 388–389
pseudotear of, 450f, 452f, 453f
surgical tear, 379
in tibial subluxation, 388
uncovered, 384f, 385
medial
in ACL tear, 389
bucket-handle tear, 363f
flap tear, 361
ossicle, 376–377
possible tear, 378f, 379
posterior horn, 389, 412f
posterior horn retear, 409f
pseudotear, 392–393
radiography, 377
surgical tear, 379
MRI interpretation, 450–451
Meniscus of knee, repair
candidates for, 423
complications of, 423
by conservative management, 425f
experimental, in dog, 424, 426f
failed, 424f
histopathology in, 426f, 427f
methods, 423
recent, 423f
re-tears of, 423–424
success rates, 423–424
suture images in, 424f
Meniscus of knee, tears
classification, 362, 378, 379
false positives, 378–379
full thickness, 423
horizontal cleavage, 364
incidence of, 388–389
longitudinal, 361, 362f
oblique, 361
parrot-beak , 379
peripheral, 388f, 425f
post menisectomy assessment, 409–413
radial, 349, 361, 362, 363f, 364f
retear grading system, 409
simulated, 392–393
of tibial attachment, 350
Metacarpal-phalangear (MCP) joint
flexor tendon and, 186f
infection, 173–174
UCL disruption and, 150–151
Metastatic carcinoma
of breast, 576, 597–598, 599f, 660f, 661f, 663–666, 671f
to femur, 599f
to humerus, 600f
to radius, 671
to spine, 663–666
to sternum, 601f
Metatarsal
head osteomyolitis, 518–520
spin-echo imaging, 518f, 519
soft tissue of, 519f, 520f
stress fracture of, 497f
Metatarsophalangeal (MTP) joint, in gout, 508f, 509, 510f

Multiple myeloma
clinical manifestations, 602–603, 607
CT versus MR imaging of, 603
differential diagnosis, 603, 605f, 606
femoral marrow metastasis, 602–603, 604f
pelvic metastasis, 605f
plasmacytomas and, 602–603, 604f
posttherapy image, 608f
sclerotic form, 606–607
Muscle, *see also* Specific muscle
in accessory soleus strain, 555-557
in acute myositis, 573
in AIDS, 558–560
amyloid myopathy of, 565, 566f
anomalies, 576–578
atrophy, 542–545, 550–552
in bacterial pyomyositis, 558–560
biopsy indications, 547
denervation effects, 542–543
dermatomyositis, 569–570
in diabetes mellitus, 567–568
in diabetic amytrophy, 542–545
disorders of, 540
in DOMS, 561–563
in Duchenne's dystrophy, 545f, 547, 549f
edema, 210f, 539, 544f, 548f, 549f, 551f, 567–568
EMG of, 542
enzyme alterations in, 554
exertion injury, 561–563
fat imaging, 539
fatty atrophy of, 544f, 550f, 551f
focal atrophy, 551, 552f
Gd imaging, 551f
IIMs of, 554
imaging procedures, 539–541, 554
in inclusion body myositis, 553
inflammation, 569–570
in McArdle's disease, 547, 548f
in musculoskeletal disorders, 540
in myophosphorylase deficiency, 548f
in myositis ossificans, 571–572
necrosis, 547, 548f, 549f
in neuromuscular dysfunction, 540–541
in peripheral neuropathy, 543, 544f
in postpoliomyelitis syndrome, 545f, 551, 552f
in postsurgical paraspinal atrophy, 550–551
in rhabdomyolysis, 546–549, 562
in sickle cell disease, 573–575
in skeletal muscle lymphoma, 564
in SMA, 543, 545f
STIR imaging of, 132, 133f, 539f, 542f, 543, 546f, 548
of thigh, 569f
water content, 539, 540
Muscle strain
of accessory soleus , 555–557
acute, 555–556
contusions and, 556
eccentric contraction and, 555
follow-up imaging, 556, 557f
grading of, 556
of hamstring, 557
muscles prone to, 555–556
prognostic factors in, 556
Musculoskeletal disorders, tabulation of, 540
Myxomas
MFH differentiation, 632, 636f
multiple, 683–684
of thigh, 636f
Myophosphorylase deficiency, muscle necrosis in, 548
Myositis ossificans
CT evaluation, 571, 572f
early lesion in, 571, 572f
histopathology, 571
in scapula region, 571f
in thigh, 572f

N

Navicular, *see also* Accessory navicular, Scaphoid bones
in Kohler'sdisease, 524f
spin-echo imaging, 502f, 503f
Nerve entrapment
diagnosis, 126
in elbow, 125–127
foraminal, 320, 321
in interdigital neuroma, 532–533
in spondylolysis, 319–321
Nerve root clumping
in arachnoiditis, 325–327
of cauda equina, 325
Nerve root compression
in foraminal extrusion, 331–332
in lumbarependymoma, 302–303
in LS disk protrusion, 280–281
Nerve root enhancement (radiculitis)
differential diagnosis, 334
in intervertebral disk herniation, 284–285
in neurogenic claudiction, 286f
postsurgical, 333–335
prognosis in, 334
retrospective study, 334
Nerve root impingement
in disk herniation, 284–285
in disk protrusion, 280–281
in ependymoma of cones medullaris, 302–303
Neurogenic claudication
Gd image of, 286f
intrathecal nerve root enhancement in, 285, 286f
Neuroma
imaging procedures, 533
interdigital, 532–533
Neuromuscular disorders, tabulation of, 540
Neuropathy, *see* Peripheral neuropathy
Non-Hodgkin's lymphoma
CT misdiagnosed, 593f
of humeral head, 591f, 660f, 661f
intracortical, 671f
normal marrow comparison, 591f
of pelvis, 593f
STIR imaging of, 591f, 593f
Noyes classification, of articular cartilage lesions, 381, 383f

O

Olecranon
bone contusion, 90–91
bursa removal, 88f
bursites, 87, 96–97
fossa, 106f, 107f, 108f
osteomyelitis of, 92f
spurs, 108f
stress reaction, 92f
in triceps tendon rupture, 86f, 88f
Os acromiale, in impingement syndrome, 10, 13f
Ossification
clinical presentation, 376–377
of knee meniscus, 376–377
pathogenesis, 377
radiography of, 376f, 377
Os supratrochleare dorsale
loose bodies in, 106, 107f, 108f
olecranon fossa in, 106f–108f
pathology, 106
symptomatic, 106–107
Osteoarthritis, articulate cartilage in, 415
Osteoblastoma
diagnosis, 668
Gd image of, 667f
MRI non-specificity for, 668
of talus, 667f
Osteochondral fracture, of LFC, 442
Osteochondral loose bodies
arthroscopic removal, 109–110
in elbow, 109–111, 127f
etiology, 109
in OCD, 513–514, 516f, 517f
synovitis and, 111f
synovium attachment, 109f
trochler notch site, 112f
in ulner trochlear notch, 111f
Osteochondritis dissecans (OCD)
acute injury in, 516f
articular cartilage in, 513–514, 515, 517f
assessment procedures, 422, 515
of capitellum, 113–114, 117f
capitellum pseudodefect and, 116, 117f
chrondal fragments in, 513, 516f
chronic lateral impaction and, 113
classification of, 422, 514, 516f
etiology, 421–422, 513–514
of femoral condyle, 421
Gd imaging, 517f
management, 514
osteochondral fragments in, 513–515, 516f, 517f
Panner's disease differentiation, 113–114,115f
schematic of stages of, 514, 516f
STIR sequencing of, 516f, 517f
subchondral bone in, 513f, 517f
talar dome in, 513–517
talar dome articular cartilage in, 515
unstable fragment and, 513f, 516f, 517f
Osteochondroma
bursain, 662f, 677f
fat suppression imaging, 660, 662f
of femur, 677f
Osteochondrosis, Kohler's disease and, 524
Osteoid osteoma
clinical presentation, 203
cortical lesions in, 203
CT imaging of, 203, 204f, 669, 670f
edema in, 203–204
of femoral neck, 669f
of proximal femur, 202–204
of spine, 204f
Osteomyelitis
acute phase of, 260
bone marrow in, 519, 520f
chronic, 261f
coccidiomycosis in, 261f
differential diagnosis, 261, 675–676
of femur, 259–261, 674–675
Gaucher's disease and, 366
imaging procedures, 174, 260, 518–519, 520f
intracortical abscess in, 259f
of metatarsal head, 518–520
of olecranon, 92f
sacrum in, 261f
septic arthritis and, 173, 234–236
of wrist, 173–174
Osteonecrosis,
diagnosis, 440–441
of femoral epiphysis, 248
of femoral head, 159, 248

idiopathic, 248–250
Kohler's disease and, 524
pathogenesis, 440
of scaphoid, 161f
spontaneous, 440
of tibial plateau, 440–441
Osteopenia, of ankle, 494f
Osteoporosis
regional migratory, 198, 199, 201f
transient, of hip, 198–201
Osteosarcoma
cartilagenous proliferation in, 698, 699f
chemotherapy induced necrosis in, 693–694, 695f
chemotherapy monitoring of, 697–699, 703–705
chondroblastic, 697–699
dynamic MPGE imaging, 697, 698, 703f, 704, 705f
Ewing's sarcoma differentiation, 713, 714f
Gd imaging of, 714f, 715f
intraosseous, 705f
juxtacortical, 715f
macrosection of, 695f, 698, 699f, 705f, 706f
marrow edema in, 703f
MFS-like, 693
microsection of, 714f
misdiagnosed, 674–675
monitoring 690
osteoblastic, 703–704
physeal plate abutting, 706f, 713f, 714f, 715f
skip metastasis of, 672–673
staging of, 713
telangiectatic, 682f
of tibia, 693–696, 697–699, 703–706
time-intensity diagrams of, 694, 695f
vascular network of, 694

P

Paget's disease
bone marrow in, 242f, 243, 244f
characterization of, 242–243
ilium in, 242–243
neoplasm in, 243
radiographic appearance, 242f, 243
sarcoma of, 243, 244f
Panner's disease, as osteochondrosis of capitellum, 113–114, 115f
Pars interarticularis, in spondylolysis, 320, 321f
Patella, *see also* Knee
ACL disruption and, 390
articular cartilage lesion, 383f
chondral defect, 380
in DDP, 404
ELPS and, 432
femoral condyle and, 390–391
femoral trochlear groove and, 431, 433
injury combinations, 390–391
in knee flexion, 431
lateral dislocation, 390–391
lateral to medial subluxation, 433
lateral subluxationof, 431, 432
medial facet injury, 415f
medial subluxation of, 432–433, 436f, 437f
patellofemoral joint and, 431
SPGR imaging, 436f, 437f
STIR sequences of, 391f
tracking, 431–433
tracking bracing, 437f
Patellar cartilage
FLASH imaging, 380
histology, 382f
lesion, 383f
Patellar tendon
degeneration, 429, 430f
disruption, 428–429
evaluation, 429
failed repair, 428–429
in "jumper's knee" 429, 430f
tendinitis, 429
Patellofemoral joint
GRASS imaging, 431, 434f
in knee flexion, 431
kinematic study, 431, 434f
patellar alignment and, 431
radiography of, 431f
Pediatric imaging, MRI advantages, 356
Pelvis
acetabular roof, 208, 209f
fracture, 208–209
insufficiency fractures of, 240, 241f
in LCPD, 249f
marrow tumors, 589f, 593f, 597–598, 600f, 604f, 605f
normal, 200f, 201f, 207f
in Paget's disease, 242f
radiography of, 226f
sarcoma, 213f
spin-echo image of, 207f
STIR sequences of, 208, 241f
supraacetabular region, 209f, 240, 241f
Percutaneous diskectomy
candidates for, 314
methods, 313
predicting outcome in, 312–314
pre, postoperative imaging, 312f
success hypothesis, 313–314
Periacetabular ganglion
etiology, 217
of hip, 217–219
Periosteal ganglion
chondroma differentiation, 438–439
tibial site, 438–439
treatment, 439
Peripheral neuropathy, neuromuscular changes in, 543, 544f
Peroneus brevis tendon
normal, 491f
split tear, 490
tenosynovitis, 490–491
Phleboliths, hemangioma differentiation, 642, 643
Phosphofructokinase deficiency, muscle necrosis in, 548f
Pigmented villonodular synovitis (PVNS)
amyloidosis differentiation, 638, 640f
arthrography of, 638
clinical presentation, 638
fibromatosis and, 620
GCT and, 183–185, 638, 639f
of hand tendon, 183–185
of knee, 637f
septic arthritis differentiation, 236
of tibial tendon, 639f
treatment, 638
Planter calcaneonavicular ligment, anatomy, 468f, 469, 470f
Planter digital nerve
anatomy, 521–522, 533f
interdigital neuroma, 532–533
in tarsal tunnel syndrome, 521–522
Planter fasciitis
differential diagnosis, 492
heal pain of, 492
imaging procedures for, 493
pathogenesis, 492–493
Popliteal artery
cystic adventitial disease of, 577f
entrapment, 576, 577
Popliteal fossa, fibromatosis, 619
Popliteus bursa, in meniscus pseudotear, 452f
Popliteus muscle, capsular rupture, 387f
Popliteus tendon
lateral knee support, 405
in lateral meniscus pseudo tear, 452f
rupture, 406, 408f
Posterior cruciate ligament (PCL)
ACL tear and, 384–386, 387f, 395
ACL tear reattachment to, 351
acute tear, 394–395, 398f
anatomy, 394–395
angle defined, 385, 386f
arthroscopy, 396
associated ganglion cyst, 416, 417f
avulsion, 395, 398f, 399f
bucket handle tear, 395, 396f
buckled, 384f
curvature value determination, 384–385, 386f
double, 395, 396f, 397f
edema, 344f
hemorrhage, 394f
injury mechanisms, 395
ligament of Humphry and, 395, 397f, 398f
MRI diagnosis accuracy, 396
osteosarcoma and, 714f
rupture, 395
seahorse configuration, 395, 397f
spin-echo imaging, 395
Posterior malleolar fracture, 526f
Posterior talofibular ligament (PTAF)
anatomy, 475–476,
attachment site, 474f
Posterior tibial tendon (PTT)
in accessory navicular, 502
anatomy, 463, 465f, 470f, 522f
3D-gradient echo imaging of, 462f, 463f, 464, 465f
distal disruption, 463
function, 463
MRI accuracy, 464
normal, 465f
PVNS of, 639f
rupture, 464
tear, 464, 465f
tear classification, 464
Posterior tibofibular ligament (PTIF)
anatomy, 526, 529f
injury, 527
normal, 529f
in syndesmotic complex injury, 526–529
Postpoliomyelitis syndrome
muscle alterations in, 545f, 551, 552f
trunk muscle atrophy in, 552f
Postsurgical paraspinal atrophy
of erector spinae muscles, 550f, 551f
fatty infiltration in, 550f, 551f
rehabilition, 550
Pronator teres muscle, strain injury, 133f
Proximal focal femoral deficiency (PFFD)
classification, 205
coxa vara differentiation, 206
presentation, 205–206
treatment, 205–206
varus deformity in, 206f
Pyomyositis
in AIDS, 558–560
etiology, 558
muscle imaging patterns in, 558, 559f, 560f
tubercular, 560f

Q

Quadriceps muscle
  hematoma compression of, 211
  inclusion body myositis, 553–554

R

Radial articular cartilage, TFC attachment, 164
Radial collateral ligament, anatomy, 78
Radial nerve
  in biceps tendon rupture, 82
  entrapment, 126
  triceps muscle and, 86
Radiculitis, *see* Nerve root entrapment
Radiohumeral meniscus, normal varient of, 117, 118f
Radioscaphocapitate ligament, in scapholunate dissociation,175, 176
Radioulner joint
  annular ligament disruption and, 78
  fluid collection in, 162f, 164f, 165f, 168f, 170f
  in TFC tear, 162f, 164f, 165f, 170f
Radioulner ligament
  anatomy, 163
  dorsal, 163, 164f
  volar, 163, 164f
Radiounotriquetral ligament
  in scapholunate dissociation, 175
  tear, 178–179
  VISI and, 178–179
Radius
  Brodie's abscess of, 675f
  in elbow dislocation, 149–140, 142f
  head fractures, 139–140, 142f
  metastatic carcinoma, 671f
  normal radiolucency of, 99f
  occult fracture, 170f
  subluxation, 142f
Rhabdomyolysis
  clinical syndrome, 546–547
  DOMS and, 562
  edema in, 546f
  exertional, 562
  of thigh, 546
Rhabdomyosarcoma, of foot, 622f
Rheumatoid arthritis, *see* Arthritis
Rotator cuff, calcific tendinitis of, 38–40
Rotator cuff tear
  acromioclavicular joint in, 24, 25f
  acromiohumeral distance in, 36
  in aging, 3,15, 34
  anatomy, 3
  biceps tendon and, 35–36, 42
  classification, 15
  evaluation, 1–2, 4, 15–16, 18–19, 21–22, 33–34
  failed repair, 46–50
  full-thickness 14–17, 21–22, 30, 32–34, 65, 66f
  glenohumeral joint in, 27f, 30, 31, 33f, 36, 49f
  humeral head and, 36
  impingement, 6
  infraspinatus tendon and, 35f
  muscle atrophy in, 34, 35
  normal, 3, 4f, 5f, 7
  partial, 1, 16f, 30, 31f, 64
  pathology in, 6–7, 18
  peribursal fat in, 33f
  progression of, 30, 36
  repair outcome for, 47–48
  small, 17f
  subacromial-subdeltoid bursa in, 27f, 28f
  subcoracoid bursa in, 33
  supraspinatus muscle in, 31f, 32f, 33–34
  supraspinitus tendon tear and, 14–17, 18, 21–22, 27–29, 33
  synovial surface in, 28f
  tendinitis, 6, 7f, 8f
  tendon degeneration in, 15
  treatment, 16, 21–22, 28, 29f, 32f

S

Sacroiliac joints
  in ankylosing spondylitis, 214
  CT vs. MR assessment, 214–215
  septic arthritis, 215, 216f
Sacroiliitis
  articular cartilage in, 215
  Gd imaging of, 215
  subchondralbone in, 215
Sacrum
  Gd image of, 239f, 241f
  insufficiency fractures of, 239–241
  osteolytic lesions of, 261
  STIR images of, 239f, 241f
Salter-Harris fracture
  edema in, 358f
  epiphysis fracture and, 358f
  growth center in, 359f
  of growth plate, 356–357, 359f, 360f
  growth recovery lines in, 357, 360f
  reclassification, 357
  type II, 356f, 357
  type III, 358f, 359f
  type IV, 356f, 357, 360f
  of ulna, 359f
Sarcoma, *see also* Ewing's sarcoma
  adenopathy differentiation, 120,121f
  clear cell, 98f
  of foot, 622f
  hemorrhage in, 213f
  MFH as, 631
  misdiagnosis of, 98f
  MR staging of, 655
  of Paget's disease, 243, 244f
  in pelvis, 213f
  synovial, 627–628, 630f
Sartorius muscle, hematoma, 211
Scaphoid (navicular)
  AVN of, 158
  cyst, 160f, 161f
  fracture, 157–159
  fracture nonunion, 157f
  healed fracture, 159f, 160f
  lunate position and, 176
  normal, 170f
  osteonecrosis, 161f
  palmarflexion tendency of, 176, 177f, 179f
  sclerosis, 157f
  waist fracture, 157, 158, 159f, 161f
Scapholunate, DISI and, 175–177
Scapholunate ligament
  carpal instability and, 175, 176, 177f
  tear, 175, 176, 177f
Scapula
  in infantile cortical hyperostosis, 68–69
  marrow tumor, 598, 599f
Schwannoma
  of ankle, 521f
  cystic degeneration in, 339f
  Gd images of, 336f, 339f
  imaging characteristics, 337–339
  intradural, 336–339
  of knee, 339f, 340f
  of neuroforamen, 338f
  spinal, 336–339
  spin-echo imaging, 359f, 340f
  of thecal sac, 336f
Sciatica, disk prolapse and, 280
Sciatic nerve, swollen, 546f
Segond fracture, ACL tears and, 444, 449f
Septic arthritis
  acetabulum in, 234f
  complications of, 235
  effusion in, 234f, 236, 237f
  femur infection in, 235-236, 237f, 238f
  of hip, 215, 216f, 234–237
  iliopsoas muscle in, 215, 216f, 234f
  imaging procedures, 215, 235
  in infants, 235–236
  osteomyelitis and, 173, 234–236
  presentation, 234
  PVNS differentiation, 236
  synovial infection in, 234, 235
Septic sacroiliitis, of hip, 237f
Short T1 inversion recovery (STIR) sequences
  in elbow imaging, 77, 84f, 87, 89f, 90, 91f, 92f, 95f, 115f, 130f, 133f, 136f
  in fat suppression, 586
  of femur fracture, 227, 228f
  of foot, 519
  of marrow, 586, 588
  of muscle pathology, 132, 133f, 539, 542f, 543
  of pelvis, 208
  of stress fractures, 495, 496
  of wrist, 158
Shoulder, 1–69, *see also* Rotator cuff tear
  acromioclavicular joint arthritis, 23–26
  acute injury, 61–63
  in aging, 59
  in biceps tendon tear, 41–45
  in Caffey's disease, 698–699
  capsulolabral lesions, 1, 2
  dislocation, 54–57, 61
  in failed rotator cuff repair, 46–49
  focal cuff degeneration, 3
  in glenohumeral instability, 50–53, 54–57, 58–60
  imaging techniques, 1–2
  impingement syndromes, 6, 9
  infraspinatus tendon of, 3
  instability, 1
  limited motion in, 3
  musculotendinous cuff of, 3
  in rotator cuff impingement, 30
  rotator cuff tear diagnosis, 1,14–17, 21–22, 27–29, 33–34
  in rotator cuff tendinitis, 38–40
  SLAP lesion, 61–62, 64–67
  in subscapularis tear, 41–45
  in supraspinatus cuff tendinitis, 6–8, 9–13
  supraspinatus tendon of, 3f
  supraspinatus tendon tear and, 18–20, 27–29, 35–38, 41–45
  synovial chondromatosis, 645–646
  weakness, 14, 27–28, 33, 35
Sickle cell disease
  AVN, 265–267
  bursitis, 575f
  femoral head in, 265–267
  femur infarction in, 575
  hematopoietic marrow expansion in, 265–267
  marrow alterations in, 265–267, 573, 574f, 575f
  MRI assessment of, 573
  muscle inflammation in, 574f
  myositis of, 573–575
Slipped capital femoral epiphysis (SCFE)
  complications of, 231

etiology, 231
presentation, 230–231
Soft tissue tumors, *see also* Tumors
abscess differentiation, 675
cavernous hemangioma as, 641–642
chondroma as, 676f
cystic mass differentiation, 632, 636f
elastofibroma as, 650–651
fibromatosis as, 619–622
of forearm, 641–643
Gd imaging of, 620, 622f, 631f, 634f, 649
hemangioma as, 642–644
of hip, 639f
hygroma differentiation, 648–649
imaging techniques, 613–614
inflammation differentiation, 623–626
of knee, 637f, 644f
lipoma as, 617f, 618f
liposarcoma as, 616–618
MFH as, 631–636
MRI dignostic specificityfor, 613, 614
myxoid liposarcoma as, 615f, 617f
myxoma as, 632, 636f, 683–684
postoperative fluid differentiation, 648–649
PVNS as, 620, 637–639
radiography of, 613
recurrence evaluation, 648–649
recurrent fibrous histocytoma as, 648, 649f
sarcoma as, 648
in scapula region, 650f
of shoulder, 645f
soft tissue lesions and, 683–684
synovial chondromatosis as, 645–647
synovial sarcoma as, 627–628, 629f, 630f
of thigh, 616f, 618f, 627–628, 630f, 631f, 632f, 633f, 634f, 635f
SPGR sequences, of patella subluxation, 436f, 437f
Spinal fixation
evaluation, 317f, 318
methods, 317
pedicle screws in, 317f, 318f
Spinal muscular atrophies (SMA), muscle changes in, 543, 545f
Spine
annular imaging, 275, 278f
avascular disk, 294f
blood-nerve barrier, 286f, 334
breast carcinoma metastasis to, 663–666
cauda equina, 325–326
disk
extrusion, 276, 290f
herniation, 276, 294, 295f
hyperintense, 322–324
sequestration, 276
disorder terminology, 273, 275–276
in epidural fibrosis, 287, 289, 291f, 293
in FBSS, 293
foramen orientation, 320, 332f
Gd imaging, 284, 286, 291f, 292f,
GRASS sequences of, 293, 295f
herniated disk fragment enhancement, 305–307
intervertebral disk, 322–323
intrathecal nerve roots, 284f
low back pain and, 280, 281, 282
MRI limitations, 288
myelogram studies, 280, 325, 326f, 327f
nerve root compression, 280
nerve root enhancement, 284–285
osteoid osteoma of, 203, 204f
paracentral protrusion, 275
in percutaneous diskectomy, 312–314
post laminectomy, 325f
postoperative disk contour, 290
postoperative disk fragment, 292–295, 304–305
postoperative imaging, 287f, 288–289, 290f, 291f
prolapse, 280
soft tissue masses, 288–289, 291f, 294f
spin-echo imaging, 294, 304f, 306f, 308f, 321f, 322f, 323, 333f
stenosis, 285, 286
surgical infection, 300
thecal sac, 291f, 309, 329, 330f
Spine, lumbar (L)
annular fissure, 275, 278f
annular tear, 278f
arachnoiditis, 325–327
avascular herniated disk, 305, 307f
back pain and, 280, 282–283
chondroma, 676f
disk
bulge, 275–276, 278f
bulge incidence, 280–281
degeneration, 275, 277f, 281
disorder terminology, 273, 275–276
extrusion, 276, 277f, 279f, 282–283, 302–303, 332f
extrusion, conservative management, 283
extrusion regression, 283
herniation, 307f
protrusion, 274f, 275, 277f, 280
space narrowing, 274f, 278f
epidural hematoma, 308–309
epidural lipomatosis, 315–316
in failed diskectomy, 331–332
fixation, 317–318
foraminal extrusion, 331–332
Gd imaging, 274f, 304f, 305f, 328f, 332f
leg pain and, 282–283
lobulated extrusion, 274
nerve root impingement, 302–303
paracentral protrusion, 274f, 275
in postoperative diskectomy, 287f
in postoperative paraspinal atrophy, 550–551
postoperative protocol, 293
in pre/postoperative diskectomy, 331–332
schwannoma, 336–339
sequestration, 279f
spondylolysis, 319–321
stenosis, 286f
synovial cyst, 296–297, 298f
thecal sac, 298f, 315f, 332f, 336f
vascularized free disk fragment, 304
Spine, lumbosacral (LS)
disk
extrusion, 287
herniation, 284, 312–314, 334
protrusion, 299f, 334, 335f
vascularizedherniated, 305, 306f
diskectomy, postoperative, 299f, 300, 324f, 328f
DRSA of, 310–311
epidural abscess, 299f, 300, 301f
Gd imaging of, 324f, 333f
infective spondylitis of, 299–301
marrow tumor, 589f, 594f, 605f, 610f
nerve entrapment, 319–321
nerve roots, 284
postsurgical syndrome, 333–335
rediculitis, 333–335
spondylolisthesis, 328f
tumors, 328–329
Spin-echo sequencing
hyperintense signal in, 322–323
versus gradient echo imaging, 2, 77–78
Spondylitis
ankylosing, 214
degenerative, 311
diagnosis, 300
DRSA and, 311
Gd imaging, 301f
infective, 299–301
MR versus CT imaging, 300
of sacroiliac joint, 214
Spondylolisthesis
characterization, 297
degenerative, 297, 328f
SAP in, 297
schematic of, 321f
spondylolysis and, 319
Spondylolysis
characterization, 319–321
foraminal encroachment in, 320, 321
LS nerve entrapment in, 319–321
schematic of, 321
spinal canal in, 320
spondylolisthesis and, 319
Stener lesions, in UCL ruptures, 150,151f
Subacromial space, calcific deposits, 38f
Subacromial-subdeltoid bursa
fluid, 8f
perforation, 22
in rotator cuff tears, 15, 17f, 22, 27f, 29f
in supraspinatus tears, 27f, 29f
Spondylodiscitis, neoplasm differentiation, 675, 676f
Spring ligament, *see* Plantar calcaneonavicular ligament
Stress fractures
amorphous type, 495–496
of calcaneus, 494–495
of distal tibia, 496f
etiology, 495
healing image of, 497f
of second metatarsal, 497f
spin-echo image of, 497f
STIR imaging of, 495, 496
streesresponses and, 495, 496
types, 495
Subchondral bone
articular cartilage defect and, 448f
erosion, 214f
fracture, 444, 448f
Subcortical bone lesions
histopathology in, 444
of LFC, 444, 447f
Subdeltoid bursa
fluid, 14f, 15f
inflammation, 38f, 39
in supraspinatustendon tears, 18f
Subdeltoid peribursal fat
obliteration, 38f
in rotator cuff tears, 16, 22f
in supraspinatus tendinitis, 6f, 9f
Subscapular bursa, normal, 50f
Subscapularis tendon
biceps tendon dislocation and, 42, 43f
partial thickness tear, 41–45
tear, 35f, 36
Subtalar ligamentcomplex
acute sprains, 499
anatomy, 499
normal image, 499, 500f
tarsalsinus syndrome and, 498–499
Superior articular facet (SAP), in degenerative spondylolthesis, 297

Superior labrum-biceps anchor tear (SLAP lesion)
  biceps anchor in 64
  classification, 64–65
  diagnosis, 65
  gangion cysts in, 65, 67f
  injury mechanisms, 64
  labrum tears and, 61–62, 64–67
Supra-acetabular, insufficiency fractures, 240–241
Supraspinatus cuff tendinitis
  acromion process in, 9f, 11f, 12f
  rotator cuff impingement in, 6–7
Supraspinatus muscle
  atrophy, 28f, 34, 35f
  in full-thickness tears, 25f, 27f, 28f, 33f
  in partial rotator cuff tears, 31f
  in rotator cuff tears, 34
  supraspinatus tendon junction, 33–34
Supraspinatus tendon
  acromioclavicular joint and, 25f
  acronion process and, 9, 12f, 19f
  in aging, 15
  anatomy, 3
  atrophy, 21, 37f
  bursa, 18f, 19, 20f, 22f, 30f, 33, 40f
  critical zone of, 3–4
  degeneration, 5f, 15
  detachment, 14–17, 20–21
  displacement, 46
  full-thickness tears, 14f, 15,16, 18–19, 20–21, 27–29, 33–34, 41–45
  glenohumeral joint and, 21f, 27f, 36
  massive tear, 35–37
  medium-thickness tear, 21–22
  morphology, 18–19, 33
  normal, 3f, 4f
  partial tear, 18–20
  pathology, 7, 12f, 15
  peribursal fat and, 22f, 31f, 33f
  in rotator cuff impingement, 30
  rotator cuff tears and, 3–4, 14–17, 21–22, 21–22, 64
  scaring, 46f
  supraspinatus muscle junction, 33–34
  synovial surface, 19f, 30, 31f, 32f
Synchondrosis, in accessory navicular, 502f, 503f
Syndesmotic ligament complex injury
  assessment, 527
  ATIF in, 526–529
  chronic, 529f
  CT imaging of, 529f
  PTIF in, 526–529
Synovial chondromatosis
  benign cartilage deposition in, 646
  clinical presentation, 646
  mineralization in, 646, 647f
  radiography of, 646, 647f
  of shoulder, 645–646
Synovial cysts
  CT image of, 298f
  degenerative spondylolisthesis and, 296–297
  image of, 296f, 297
  intraosseous, 102f
  of lumbar facet joint, 296–297, 298f
  of olecranon, 102f
  in rheumoatoid synovitis, 100–101
  synovial condroma differentiation, 297
Synovial envelope, Gd enhancement, 419, 420f
Synovial sarcoma
  of ankle, 629f
  calcification, 628, 630f
  characterization, 628
  of forearm, 629f
  hemorrhage in, 629f
  metastases, 628
  MRI nonspecificity for, 628
  radiography, 628, 630f
  of thigh, 629f, 630f
Synovitis
  of ankle, post-reconstruction, 487–488, 489f
  de Quervain's disease and, 172
  diagnosis, 100, 167
  of elbow joint, 100–102, 111f
  in failed rotator cuff repair, 47
  Gd imaging of, 100
  loose body and, 111f
  PVNS and, 637–638
  rheumatoid, 100–102
  silicone induced, 186–187
  of wrist, 186–187
Synovium
  chondromatosis of, 103–105, 246f, 466–467
  CT imaging of, 104
  of elbow, 100–101, 103–105, 109, 117
  filling defects, 245, 246f, 247f
  ganglia attachment, 217
  hypertrophy, 171
  infection, 234
  loose body attachment, 109f
  osteochondromatosis of, 104, 105f, 238f
  pannus, 167, 168f
  pathology, 103–104
  plica of, 117
  in septic arthritis, 234, 235
  thickened, 100f, 101f

T

Talar dome articular cartilage, assessment in OCD, 515
Talocalcaneal interosseous ligament, normal, 500f
Talofibular anterior ligament, 487f
Talus
  bone marrow edema, 498f, 515
  dome OCD, 515–517
  osteoblastoma, 667f, 668
  posterior tubercle fracture, 474f, 475
  in tarsal coalition, 483f, 484, 485f
Tarsal coalition
  cartilaginous, 484, 485f
  clinical presentation, 483, 484
  CT imaging of, 484, 485f, 486f
  fiberous type, 483–486
  talus in, 483f
Tarsal sinus
  anatomy, 499
  ganglion cysts, 499, 501
Tarsal sinus syndrome
  in ankle, post-reconstruction, 487–488
  characterization, 499
  fluid collections in, 499, 501f
  LCL injuries and, 499
  in subtalar ligament injury, 498–499
  tarsal sinus in, 498f
  treatment, 499
Tarsal tunnel syndrome
  anatomy, 521, 522f
  diagnosis, 522
  posterior tibial nerve in, 522
  recurrent, 523f
  schematic of, 522f
  schwannoma and, 521f
  treatment, 522
Technetium (Tc) 99m bone scan
  of Ewing's sarcoma metastasis, 657, 658f, 659f
  of pelvis, 597f
  of rib marrow tumor, 595, 598f
  of sternum, 600f
Tendinitis
  of gluteus maximus tendon, 579–581
  of patellar tendon, 429, 430f
  pathology of, 6–7
  of rotator cuff, 6, 7f, 8f
Tendinosis, Achilles tendon tear and, 480, 481f
Tennis elbow
  common extensor tendon rupture in, 136–138
  extensor carpiradialis brevis tendon degeneration in, 136, 137f
  LCl complex in, 138f
Tenosynovitis
  diagnosis, 491
  peroneal, 490–491
  of wrist, 171–172
Three-dimensional Fourier transform (3DFT) imaging, in meniscal evaluation, 349
Tibia
  in ACL avulsion, 352
  in ACL reconstruction, 400, 403f
  in ACL tear, 384, 442–443, 445f
  articular surface, 409f
  bone cyst, 478–479
  bone tumors of, 668f, 681f
  in buckled PCL, 384, 387f
  CT scan of, 445
  Ewing's sarcoma of, 713, 714f
  fracture, 443, 447f
  ischemic necrosis, 441
  kissing lesions of, 442, 443
  lateral plateau, 443
  in LFC fractures, 443
  medial plateau, 443
  metaphyseal fracture resolution, 443, 447f
  MFH of, 700–702
  osteonecrosis of, 440–441
  osteosarcoma , 693–696, 697–699
  in PCL avulsion, 399f
  plateau fracture, 384f, 386f, 442f
  posterolateral plateau fracture, 445f
  shift, 397f
  shift measurements, 384f, 387f
  STIR sequences of, 446f
  subluxation, 384–385, 445f
  tunnel, 400, 403f, 420f
Tibial colateral ligament (TCL)
  ACL and, 373f, 375f
  acute tear assessment, 371–372
  anatomy, 367–368
  complete tear, 371f
  diagnosis, 371–372
  discontinuity, 370f
  function, 368
  hematoma, 370f
  injury grading, 368–369
  knee joint capsule and, 368
  MCL complex and, 367
  partial tear, 371f– 375f
  partial tear follow-up, 373f, 374f
  repair re-rupture, 364f
  spin-echo image, 370f
  sprain, 367, 369f
Tibialis anterior muscle
  edema, 473f
  hemorrhage, 473f
  laceration, 472, 473f
Tibialis anterior tendon
  laceration, 472, 473f
  spontaneous tear, 471f, 472
Tibialis posterior muscle, anatomy, 465f

Tibiospring ligament
  anatomy, 469, 470f
  3D gradient-echo imaging of, 469, 470f
  disruption, 470f
  edema, 470f
  normal, 468f
Tibiotaler ligaments
  anatomy, 469
  anterior, 469, 470f
  posterior, 469, 470f
Toe
  in acute tophaceous gout, 508–510
  in long flexor tendon laceration, 511–512
  long flexor tendon signal, 530–531
  normal, 530–531
  postoperative evaluation, 511, 512f
Transient osteoporosis of hip (TOH)
  clinical presentation, 198–199
  marrow edema in, 198
  osteonecrosis differentiation, 199
  pathology, 198
  pregnancy and, 198
Triangular fibrocartilage(TFC)
  anatomy, 162
  avulsion, 170f
  complex tears, 162, 163, 164f
  degenerative tear of, 162–165
  GRE imaging of, 162
  lesion classification, 163
  normal, 164f
  radial articular cartilage and, 164
  ulnar variance and, 163–164
Triceps muscle, anatomy, 86
Triceps tendon
  anatomy, 86
  avulsion, 89f, 91f
  STIR sequences of, 91f
Triceps tendon rupture
  acute, 86–87, 88f
  assessment, 87
  mechanisms, 86–87
  normal, 89f
  olecranon in, 86f, 88f
  ossification, 89f
  radial head fractures and, 87
  retraction in, 88f
  tendinosis, 87, 89f
Trochlear cartilage, normal, 124f
Tumors, *see also* Bone tumors, Bone marrow tumors, Soft tissue tumors
  melanoma of leptomeningeal membranes, 328–329
  neurofibromas as, 336–337
  neuroma as, 532–533
  subarachnoid, 329
  vascularization monitoring in, 691, 692f
Two-dimensional Fourier transform (2DFT) imaging, in meniscal evaluation, 349

U
Ulna
  coronoid process fracture, 139, 141f
  edema in, 359f
  lunate abutment syndrome, 162–163
  Salter-Harris fracture of, 309f
  trochlear notch loose body, 111f
  variance assessment, 163–164
Ulnar collateral ligament (UCL)
  in elbow dislocation, 140, 142f
  in Gamekeeper's thumb, 150–151
  lateral, 136, 138f, 140, 142f
  normal, 130f, 131f
  ossification, 123f
  rupture, 142f
  Stener lesions of, 150
  tears, 138f, 140, 150–151
  ultrasound imaging of, 150
Ulner nerve
  common flexor tendinitis and, 133, 134f
  in cubital tunnel narrowing, 125f, 126
  entrapment, 125–126
  neuropathy, 122, 123f, 125–127, 133, 134f
  normal, 169f
Ultrasound imaging
  MRI comparison, 150–151
  of Stener lesions, 150–151

V
Vastus medialis obliqus (VMO), patellar subluxation and, 432
Vastus muscle
  amyloid infiltration of, 565f
  bacterial pyomyositis of, 558, 559f
Volar flexor tendons, of hand, 184f
Volar intercalated segment instability (VISI)
  lunotriquetral ligament tear and, 178–179
  radiocarpal ligament in, 179

W
Wisberg's ligament, 453f
Wrist, 149–193, *see also* Hand
  carpal tunnel syndrome and, 152, 154–155
  DISI of, 175–177
  fractures, 157–158, 169–170
  ganglion cysts, 184, 193
  imaging procedures, 149
  infections, 173–174
  lunate avascularnecrosing, 191–193
  osteomyelitis, 173–174
  rheumatoid arthritis of, 166–168
  scapholunate disassociation, 175–177
  silicone implants in, 186–187
  silicone synovitis, 186–187
  in skier's thumb, 150–151
  soft tissue infection, 173–174
  STIR imaging, 158
  synovitis, 171–172
  tendon sheath tumor, 183–184
  tenosynovitis, 171–172
  in TFC tear, 162–165, 166
  UCL tears and, 150–151
  in ulnar-lunate abutment syndrome, 162
  in VISI, 178–179

ISBN 0-397-51672-X

9 780397 516728